STANDARD METHODS

OF

CHEMICAL ANALYSIS

*A Manual of Analytical Methods and General
Reference for the Analytical Chemist
and for the Advanced Student*

EDITED BY

WILFRED W. SCOTT, Sc.D.

*Professor of Chemistry, University of Southern California. Author of "Qualitative
Chemical Analysis," "Technical Methods of Metallurgical Analysis,"
"Inorganic Quantitative Chemical Analysis."*

IN COLLABORATION WITH EMINENT SPECIALISTS

In Two Volumes, Illustrated

VOLUME ONE—THE ELEMENTS

FOURTH EDITION REVISED
FOURTH PRINTING

Total Issue, Sixteen Thousand

NEW YORK

D. VAN NOSTRAND COMPANY, Inc.

250 FOURTH AVENUE

Copyright, 1917, 1922, 1925, by
D. Van Nostrand Company

THIS BOOK IS AFFECTIONATELY DEDICATED

TO MY FATHER,

Rev. Thomas Jefferson Scott, D.D

PREFACE TO THE FOURTH EDITION

In this fourth edition the writer and his colleagues have made changes throughout the work in presenting standard methods that have been adopted as such by the American Chemical Society, by the American Society for Testing Materials, and other organizations conducting method investigations. The large amount of work done in this line since the publication of the last edition of Standard Methods has necessitated addition of considerable matter and the deletion of obsolete methods. To accomplish this work we have increased our staff from thirty-seven to fifty-two members.

The following brief survey takes up the more important additions and changes of the fourth edition—Dr. H. V. Churchill, Chief Chemist of the Aluminum Company of America, has contributed towards the chapter on Aluminum and has revised this section; Dr. W. R. Schoeller and A. R. Powell, authors of The Analysis of Minerals and Ores of the Rarer Elements, have rewritten the chapter on Cerium and the Rare Earths and have added a chapter on Indium, Thallium and Other Rare Elements; the writer has contributed a new method for the determination of Fluorine in the chapter on the subject and has added to the chapter on Iron, including the diphenylamine internal indicator method for dichromate titrations; the section on steel analysis has been largely transferred to a special chapter in Vol. 2; additional methods have been added to the chapter on Lead and the A.S.T.M. method for the analysis of pig lead has been included; in the chapter on Manganese, in addition to approved A.S.T.M. methods, we have included the bismuthate method as modified by R. W. Coltman and T. R. Cunningham, which makes this method applicable to large amounts of manganese as well as the minute amounts for which it was originally recommended; the thiocyanate method for mercury as recommended by A. H. Low has been included in the chapter on Mercury. Dr. Paul D. Mercia, Director of Research, and Mr. T. Fudge, Chief Chemist, The International Nickel Company, have contributed to the chapter on Nickel. Mr. L. E. Barton has contributed his paper on the subject of the determination of nitrogen in steel to the chapter on Nitrogen and has revised the chapter on Titanium. The writer has added to the chapter on Phosphorus and has rewritten a portion of this section. The chapters on Platinum, Palladium and the Rarer Elements of the Platinum Group have been carefully revised by Mr. R. E. Hickman and contributions made to this section by Edward Wichers, U. S. Bureau of Standards. The chapter on Radium, by Professor L. D. Roberts, has been placed in Volume 1. We are grateful to Dr. Victor Lenher, Head of the Analytical Department, University of Wisconsin, for the complete revision of the chapter on Selenium and Tellurium. Some attention has been given to the chapter on silver, a portion of this being rewritten and the U. S. Mints method being added, together with tables for rapid calculating of results, for which we are indebted to Mr. A. E. Moynahan, Chief Assayer, Denver Mint. A new chapter on the Fire Assay of Gold and Silver, by Professor Irving A. Palmer, follows the chapter on Silver. Mr. C. A. Newhall, Consulting Chemist, Seattle, Washington, has contributed several methods of the determination of Sulphur. The chapters on Thorium and Zirconium have been completely rewritten by Dr. H. M. P. Brinton, Head of the Division of Analytical Chemistry, University of Minnesota. All the chapters have been reviewed by their respective contributors and revised where necessary.

In Volume 2 some of the subject matter has been rearranged on account of the growth of certain sections to avoid repaging of the entire volume. The volume opens with the chapter on Sampling by Mr. J. B. Barnitt, Chief of Development Division, Aluminum Company of America, replacing the chapter on Acidimetry and Alkalimetry which has been placed in the latter portion of the volume. The writer has contributed a few pages on methods of decomposition of material for analysis. Mr. W. B. Price, author of Technical Analysis of Brass and Non-Ferrous Alloys, has completely rewritten the chapter on Alloys, adding a number of methods, approved as Standards by the American Society of Testing Materials. We are indebted to Dr. A. Campbell, President of the Globe Soap Company, for the latest Standard method in the analysis of Soap and Soap Products; the chapter has been placed later in the volume owing to expansion. Mr. R. K. Meade has completely rewritten his chapter on Cement and has included the latest approved methods recommended by the American Engineering Standards Committee. The chapter on Paint and Paint Pigments has been revised by the authors— Mr. H. A. Gardner, Director of the Institute of Paint and Varnish Research, and Dr. J. A. Schaeffer, Vice-President and Chief Chemist, Eagle-Picher Lead Company. Professor A. H. Gill, Massachusetts Institute of Technology, has thoroughly revised his chapters on Fixed Oils, Fats and Waxes and Gas Analysis. Dr. G. A. Burrell and Mr. J. M. Morehead have contributed to the chapter on Gas Analysis. A number of new methods have been included in these sections. Mr. S. Collier, Chemist in Charge of the Rubber Laboratory, Waukegan Plant, Johns-Manville, Inc., and Chairman of the Committee on Revision of method for rubber analysis, has given us the latest American Standard Methods on this subject; Mr. L. E. Salas has presented the latest English procedures. The writer has added a chapter on the Analysis of Iron and Steel and a chapter on Slag Analysis to the latter portion of the work. The methods adopted as Standards by the A.S.T.M. for analysis of Steel have been followed in the subject on Steel. The methods for Slag Analysis are largely those of the Anaconda Copper Company. The chapters on Water Analysis have been revised and portions rewritten by their authors Mr. D. K. French and Dr. F. E. Hale. Mr. H. Abraham has brought his chapter on Bituminous Substances to date. A new chapter on Standard Apparatus, by Mr. R. M. Meiklejohn, Chief Chemist, General Chemical Company, is included in Volume 2. Drs. G. L. Kelley and J. S. Coye have added to their chapter on Electrometric Titrations. A number of procedures, which could not well be classed under chapters mentioned, have been included by the writer in the Miscellaneous Chapter. The growth in the importance of Metallography in the analytical laboratory has induced us to include this subject in our work. The chapter is contributed by Mr. Joseph Winlock, Research Chemist, Edward G. Budd Mfg. Co., and includes the latest developments in the subject. A chapter on Determination of Poisons has been added by Dr. Arthur R. Maas, Consulting Chemist, Los Angeles, California. Prof. R.

C. Beckstrom, head of the Department of Petroleum Engineering, Colorado School of Mines, has contributed a section dealing with certain chemical and physical tests of oil that are not included in Prof. Gill's chapter on oil analysis.

The excellent cooperation of the contributors to Standard Methods of Chemical Analysis has made this work possible. We have been ably supported by a staff of co-workers, who are recognized specialists in the subjects that they have presented. The names of a number of the staff have become familiar to us through their contributions to chemical literature. Some have published works that are considered standards in their special lines, and have received a wide circulation. It gives us considerable pleasure to offer to our chemical fraternity the reliable information of these men. We are also indebted to those who have called to our attention certain errors appearing in the third edition and those who have offered suggestions and criticisms of previous editions. We are grateful for all such suggestions and criticisms and will welcome these on the present edition. We realize that absolute perfection can not be expected, but in the fourth edition of Standard Methods of Chemical Analysis, we have made an effort to present what is considered thoroughly up to date and reliable.

WILFRED W. SCOTT.

UNIVERSITY OF SOUTHERN CALIFORNIA,
LOS ANGELES, CALIFORNIA,
August, 1925.

PREFACE TO FIRST EDITION

This book is a compilation of carefully selected methods of technical analysis that have proven of practical value to the professional chemist. The subjects have been presented with sufficient detail to enable one with an elementary knowledge of analytical processes to follow the directions; on the other hand, lengthy exposition, theoretical dissertation and experimental data are purposely avoided, in order to include a large amount of information in a compact, accessible form. References to original papers are given when deemed advisable.

For methodical arrangement the material is grouped under three major divisions—Part I. Quantitative determination of the elements. Part II. Special subjects. Part III. Tables of information.

In the first division the elements are generally taken up in their alphabetical order, each chapter being fairly complete in itself, cross-references being given to certain details included elsewhere to avoid repetition. For example, the complete directions for separation of the halogens are given in the chapter on chlorine, and references to these details are given in the chapters dealing with the other members of this group. Occasionally it has been deemed advisable to place several related elements together in the same chapter.

Each chapter on the elements is generally arranged according to the following outline:

Physical Properties. Atomic weight; specific gravity; melting-point; boiling-point; oxides.

Detection. Characteristic reactions leading to the recognition of the element.

Estimation. The subject is introduced with such information as is useful to the analyst.

Preparation and Solution of the Samples. Here directions are given for the preparation and decomposition of characteristic materials in which the element occurs.   Recommendations to the best procedures are included to assist the analyst in his choice.

Separations.   This section is devoted to procedures for the removal of substances, commonly occurring with the element, that may interfere with its estimation.   In the absence of such substances, or in case methods are to be followed by which a direct estimation of the element may be made in the presence of these substances, this section on separations may be omitted in the course of analysis.   Here the discretion of the chemist is necessary, and some knowledge of the substance examined essential.

Methods.   The procedures are grouped under gravimetric and volumetric methods.   Several processes are generally given to afford the opportunity of selection for particular cases and for economical reasons where special reagents may not be available.

In many of the chapters methods for determining traces of the element are given, and the subjects are concluded by typical examples of complete analysis of substances containing the elements.

The titles to the procedures generally give a clue to the processes. Names of originators are occasionally retained where common usage makes the methods generally known by these.

Although the combined acid radicals are taken up with the elements to which they may be assigned, a chapter is devoted to the more important of the acids in their free state, and is placed with the other special subjects in the second division of the book. Here are found chapters on water, paint, oil, alloys, coal, cement, gas, and such subjects as are best classed in sections apart from simple substances dealt with in the first portion of the work.

The last portion of the book is devoted to tables of the more important arithmetical operations.   These are designed to assist the analyst to greater accuracy of calculations, as well as to relieve him of needless expenditure of time and energy.

The material herein included has been carefully selected, an effort having been made to obtain the more trustworthy methods that will meet the general needs of technical chemists.

A list of the majority of publications consulted is given in alphabetical order in the appendix of this volume.   Reference to these authorities will be found throughout the book.

W. W. SCOTT.

NEW YORK CITY,
January, 1917.

CONTRIBUTORS TO VOLUME ONE

Frank G. Breyer, A.B., A.M. Chief of Research Division, New Jersey Zinc Company. Author of a number of technical papers and numerous patents dealing with zinc, lithopone and allied industry. Contribution—Chapter on Zinc.

Paul H. M. P. Brinton, B.S., M.S., Ph.D. Professor and Head of Division of Analytical Chemistry, University of Minnesota. Author of Analytical and Physical Inorganic Chemistry, Rare Elements and a number of technical papers. Contribution—Chapters on Thorium, Zirconium.

H. V. Churchill, A.M. Chief Chemist, Aluminum Company of America, Pittsburgh, Pa. Contribution—Chapter on Aluminum.

B. S. Clark, B.S. Research Chemist, American Can Company. Contribution —Chapter on Tin.

Wallace G. Derby, A.M. Chemist and Assistant Superintendent, Wells and Richardson Company, Inc., Burlington, Vermont. Contribution— Chapters on Copper, Gold and Silver.

T. Fudge, B.S. Chief Chemist, The International Nickel Company. Contribution—Chapter on Nickel.

Harry D. Greenwood, B.S. Chief Chemist, U. S. Metals Refining Company, Carteret, N. J. Author of technical papers. Contribution—Chapter on Selenium and Tellurium.

R. E. Hickman, B.S. Chief Chemist, J. Bishop and Company Platinum Works. Contribution—Chapters on Platinum; Palladium; Rarer Elements of the Allied Platinum Metals.

William B. Hicks, A.B., A.M., Ph.D. Chief of Analytical Department, Solvay Process Company, Syracuse, N. Y. Author of a number of technical papers on mineralogy; potassium; muds; action of salt solution on clay filters, etc. Contribution—Chapter on Sodium, Potassium and Other Alkalies.

James A. Holladay. Chief Chemist, Electro Metallurgical Company. Author of technical papers. Advisory Editor. Contribution—Chapters on Molybdenum and Tungsten.

L. S. Holstein, A.M. New Jersey Zinc Company, New York. Contribution —Chapter on Zinc.

Victor Lenher, Ph.D. Professor of Chemistry, University of Wisconsin. Former chairman of Committee on National Research Council on the uses of selenium and tellurium. Author of Inorganic Chemistry; Rare Elements; Selenium; Tellurium; Rare Earths; Geochemistry, etc. Contribution—Chapter on Selenium and Tellurium.

Albert H. Low, A.B., Sc.D. Professor of Chemistry, Colorado School of Mines. Author of Technical Methods of Ore Analysis and papers on metallurgical chemistry, analytical methods, poisonous minerals in water, azo colors in butter, etc. Contribution—Chapter on Uranium, and a number of methods in text.

Paul D. Mercia, Ph.D.   Director of Research, The International Nickel Company.   Author of Metallography and Physical Chemistry of Ferrous and Non-Ferrous Alloys and Metals.   Contribution—Chapter on Nickel.

John J. Mulligan, B.S.   Superintendent, U. S. S. Lead Refinery, Inc., East Chicago, Ind.   Contribution—Chapter on Bismuth.

Charles A. Newhall, B.Sc. Chemical Engineer, Seattle, Washington. Author— Growth of the Cement Industry; Strength of Highway Surfacing, etc. Advisory Editor.   Contribution—Chapter on Sulphur.

R. S. Owens, A.B.   Consulting Chemist, New York City.   Contribution —Chapter on Cerium.

Irving A. Palmer, B.S., M.S.   Professor of Metallurgy, Colorado School of Mines.   Author of a number of papers on chemistry and metallurgy in various technical periodicals.   Contribution—Chapter on Fire Assay of Gold and Silver.

A. R. Powell, A.B.   Metallurgical Chemist, London, England.   Author of a number of technical papers, and " Analysis of Minerals and Ores of the Rarer Elements."   Contribution—Chapter on Cerium and Rare Earths.

Lewis D. Roberts, A.B., S.M.   Associate Professor of Chemistry, Colorado School of Mines.   Contribution—Chapter on Radium.

W. L. Savell, Ph.D.   Director and President, Savell, Sayer and Company, Inc. Author—Preparation of Metallic Cobalt from Oxides; Electroplating with Cobalt, etc.   Contribution—Chapter on Cobalt and Nickel.

W. R. Schoeller, Ph.D.   Metallurgical Chemist, Daniell C. Griffeth and Company, London, England.   Author of a number of scientific papers, The Analysis of Minerals and Ores of the Rarer Elements.   Contribution— Chapters on Cerium and the Rare Earths, Thorium, Indium and Other Rare Elements.

Wilfred W. Scott, A.B., A.M., Sc.D.   Professor of Analytical Chemistry, University of Southern California.   Editor in Chief.   Author of a number of technical papers on chemical research and chemical methods of analysis (see also title page).   Contribution—A number of chapters throughout the book.

S. Skowronski, B.S.   Research Chemist, Raritan Copper Works, Perth Amboy, N. J.   Advisory Editor.   Contribution—Analytic Methods for Gold and Silver.

Albert M. Smoot.   Technical Director, Ledoux and Company, New York City. Author of a number of technical papers on analytical methods.   Advisory Editor.   Contribution—Chapters on Molybdenum and Tungsten.

L. A. Wilson, M.S.   Chief of Testing Dept., New Jersey Zinc Company, Palmerton, Pa.   Contribution—Chapter on Zinc; also chapter on Cadmium.

Edward Wichers, Ph.D.   Chemist, U. S. Bureau of Standards, Washington, D. C.   Advisory Editor.   Contributions—Chapters on Platinum and the Rarer Elements of the Allied Platinum Metals.

CONTENTS

VOL. I. THE ELEMENTS

ALUMINUM

ANTIMONY

CONTENTS

ARSENIC

BARIUM

BERYLLIUM (GLUCINUM)

BISMUTH

BORON

Detection—flame test, borax bead and turmeric tests, 81. Estimation, 81. Preparation and solution of the sample—boric acid in silicates and enamels; boronatrocalcite, borocalcite, boracite, calcium borate; borax and boric acid; boric acid in mineral water; in carbonates; in foods—milk, butter, meat, etc., 82–83. Gravimetric methods—distillation of methyl borate and fixation with lime, 84–85. Volumetric methods—titration of boric acid in presence of mannitol or glycerole in evaluation of borax or boric acid. Chapin's method. Robin's test for traces, 86–91.

BROMINE

Detection—by silver nitrate, by absorption in carbon tetrachloride or disulphide, by magenta test, bromates, 92. Estimation, 93. Preparation and solution of the sample—bromides, bromine in organic matter, 93. Separation of bromine from the heavy metals, from silver, cyanides, chlorine and iodine, 93–94. Gravimetric methods—precipitation as silver bromides, (1) hydrobromic acid and bromides of the alkaline earths and alkalies; (2) treatment in presence of heavy metals, 94. Volumetric methods—determination of free bromine with potassium iodide; soluble bromides by chlorine method, Volhard's method; traces of bromine, 94–96. Arsenous acid method for bromates, 96. Analysis of crude potassium bromide and commercial bromine, 96–97. Determination in mineral water; separation from iodine, 97.

CADMIUM

Detection—as sulphide, tube test, blowpipe tests, 99. Estimation, 99. Preparation and solution of the sample, 100. Separations from silica, arsenic, antimony, tin, bismuth, copper, mercury, zinc, 100–101. Gravimetric methods—Procedure for ores; determination as sulphate, electrolytic method, 101–103. Volumetric method, 103.

CALCIUM

Detection—as oxalate, flame test, spectrum of calcium, 104. Estimation and occurrence, 104. Preparation and solution of the sample—limestone, dolomite, magnesite, cement, lime, gypsum, Plaster of Paris, sulphates, silicates, chlorides, nitrates and other water soluble salts, sulphides, pyrites ore, 105. Separation of calcium from—silica, iron, alumina, copper, nickel, cobalt, manganese, zinc, barium, strontium and the alkalies, treatment in presence of phosphates of iron and aluminum, decomposition of material, 105, 106. Gravimetric methods—precipitation as calcium oxalate; other methods, 107–108. Volumetric method by titration of the oxalate with standard potassium permanganate solution, 108. Precipitation from acetic acid solution, Builar's rapid iodine method, 108a. Available lime, 108. Methods for analysis of gypsum and gypsum products, 108b–108d.

CARBON

Detection—element, carbon dioxide in carbonates and in gas, free carbonic acid in water, distinction between carbonates and bicarbonates, carbon monoxide, 109. Estimation, 110. Preparation of the sample—iron, steel and alloys; organic matter; carbonates and bicarbonates, 110. Separation of carbon from other substances; separation from iron by the cupric potassium chloride method, 110–111. Gravimetric method—combustion furnaces, types of absorption apparatus, general procedure for determining carbon by combustion, 111–115. Graphitic carbon in iron and steel, 115–116. Combined carbon, 116–117. Determination of carbon in organic substances—organic

2

FLUORINE

GOLD

INDIUM, SCANDIUM, THALLIUM

IODINE

IRON

LEAD

MAGNESIUM

MANGANESE

PHOSPHORUS

PLATINUM

RADIUM

SELENIUM AND TELLURIUM

SILICON

SILVER

THE FIRE ASSAY FOR GOLD AND SILVER

TIN

TITANIUM

TUNGSTEN, TANTALUM AND COLUMBIUM

CONTENTS

URANIUM

VANADIUM

ZINC

CONTENTS

LIST OF ILLUSTRATIONS

VOLUME I

PART I

TECHNICAL METHODS FOR THE DETECTION AND DETERMINATION OF THE MORE IMPORTANT ELEMENTS

ALUMINUM [1]

Al, *at.wt.* 26.96; *sp.gr.* 2.583; *m.p.* 658.7°; *b.p.* 2200°; *oxide* Al_2O_3.

DETECTION

General Procedure. The sample is prepared by one of the procedures outlined under "Preparation and Solution of the Sample." Silica is removed by taking the solution to dryness, boiling the residue with hydrochloric acid and filtering. The members of the hydrogen sulphide group are removed as usual with H_2S, the filtrate boiled to expel the excess of H_2S, iron oxidized with nitric acid, and aluminum, iron and chromium precipitated as hydroxides by addition of ammonium hydroxide in presence of ammonium chloride. On treating the precipitate with sodium peroxide, aluminum and chromium hydroxides dissolve, whereas ferric hydroxide remains insoluble. Aluminum hydroxide is precipitated by acidifying the alkaline solution with hydrochloric or nitric acid, and neutralizing with ammonia; chromium remains in solution.

Cobalt Nitrate Test. The white gelatinous precipitate of aluminum hydroxide may be confirmed by adding a drop of cobalt nitrate solution and burning the filter. The residue will be colored blue by the resulting aluminum cobalt compound.

Sodium Thiosulphate Test. $Na_2S_2O_3$, added to a neutral or slightly acid solution, containing aluminum, precipitates aluminum hydroxide, upon boiling the solution. Sodium sulphite, or ammonium chloride added in large excess, will also cause this precipitation.

Alizarin S Test. The reagent used is a 0.1% filtered solution of commercial alizarin S, the sodium salt of alizarin monosulphonic acid (yellow with acids, purple with alkalies).

To 5 cc. of the neutral or acid solution under examination is added 1 cc. of the reagent, and then ammonia until the solution is alkaline, as shown by the purple color. The solution is boiled for a few moments, allowed to cool, and then acidified with dilute acetic acid, when red coloration or precipitate remaining is conclusive evidence of the presence of aluminum. The red calcium, strontium, barium, zinc and magnesium salts, and salts of other metals later than Group II are readily soluble in cold dilute acetic acid, and do not interfere with the coloration.

Phosphates or chromium do not interfere and comparatively large amounts of iron may be present (0.003 milligram Al in presence of 1 milligram ferric iron, 10 milligrams chromium salt). In presence of greater quantities of iron citric acid is added to keep this in solution. One part of aluminum may be detected in 10 million parts of water.

[1] Also spelled Aluminium.

By W. W. Scott and H. V. Churchill.

ESTIMATION

The determination of aluminum, in terms of alumina, Al_2O_3, is required in the evaluation of aluminum ores, bauxite, $Al_2O(OH)_4$; diaspore, $AlO(OH)$; alunite, $K_2O.3Al_2O_3.4SO_3.6H_2O$, etc. It is determined in the analysis of feldspar, halloysite, clays, granite, gneiss, porphyry, mica schist, slate, obsidian or pumice stone, cryolite, limestone, and in the complete analysis of a large number of mineral substances. The estimation of alumina is required in the analysis of cements, plaster, ceramic materials, aluminum salts, and is especially important in the control of processes in the manufacture of aluminum products. As a metal it is determined in commercial aluminum, and its alloys.

Preparation and Solution of the Sample

In dissolving substances containing aluminum it will be recalled that alumina, although ordinarily soluble in acids, is very difficult to dissolve when it is highly heated. It may be best dissolved, in this case, by fusion with sodium carbonate or with acid potassium sulphate, followed by an acid extraction. The metal is scarcely acted upon by nitric acid, but is readily soluble in the halogen acids and in hot concentrated sulphuric acids.

General Procedure for Ores. One gram of the finely powdered ore, taken from a representative sample, is placed in a platinum dish, 5 cc. of concentrated sulphuric acid are added, followed by about 20 cc. of strong hydrofluoric acid. The mixture is evaporated over a steam bath as far as possible and then taken to SO_3 fumes on the hot plate (*Hood*). Upon cooling, a little dilute hydrochloric acid is added and the mixture warmed. The solution is diluted with distilled water and filtered if any residue remains.

The insoluble residue remaining on the filter may be brought into solution by fusing the ignited residue with sodium carbonate or acid potassium sulphate. If barium is present sodium carbonate fusion is made and the melt extracted with water to remove the sodium sulphate. The residual carbonates may now be dissolved with hydrochloric acid.

SULPHIDE ORES should be oxidized with nitric acid and bromine according to the general procedure for decomposing pyrites in the determination of sulphur.

The solution of the sample having been effected, aluminum is separated from elements that interfere in its estimation. Directions for the removal of these substances are given under "Separations." The element is now in solution in such form that it may be determined gravimetrically or volumetrically.

Fusion Method. Sodium Carbonate. The air-dried material, ground to a fine powder, is placed in a glass-stoppered bottle. If the determination is to be made on the dry basis, moisture is driven out by placing the material in the hot air or steam oven for an hour (100 to 110° C.). One gram sample, placed in a large platinum crucible, is mixed with 4 to 5 grams of anhydrous sodium carbonate and the material heated to fusion, the heating being continued until the molten mass appears clear. The liquid mass may be poured on a large platinum crucible lid, or if preferred, allowed to cool in the crucible, a platinum prod being held in the fusion until it solidifies. By gently heating the crucible over a flame the fusion loosens from the sides and may be lifted out on the prod. In either case the cooled mass is dissolved by placing it, together with the crucible in which the fusion was made in a casserole, and treating with hydrychloric acid, the casserole being covered with a clock glass during the reaction.

Silica is removed by evaporating the solution to dryness on the water or steam bath and drying in the oven at 110° C. for an hour or more. The residue is extracted with hot dilute hydrochloric acid and silica filtered off.

If the solution is cloudy upon treatment of the fusion with acid, it indicates either the presence of barium sulphate or incomplete decomposition of the sample. In the latter case the residue is gritty and the fusion of this material should be repeated.

Solution of Metallic Aluminum and its Alloys. The metal may be dissolved in dilute hydrochloric acid, 1 : 1, or in a solution of sodium hydroxide or potassium hydroxide.

Alloys of aluminum are best brought into solution with a mixture of hydrochloric and nitric acids.

SEPARATIONS

General Considerations. In the usual course of analysis, aluminum is in solution as a sulphate or as a chloride, silica having been removed by dehydration, as described under "Preparation and Solution of the Sample." The following interfering elements may be present in the solution: iron, manganese, arsenic, antimony, titanium, phosphoric acid, and more rarely chromium and zirconium. In alloys of aluminum other elements may be added to this list. The separation more commonly required is from iron, aluminum and iron being precipitated together as the hydroxides. In usual practice the two are weighed together as Fe_2O_3 and Al_2O_3, after ignition to this form, and iron then determined, either on a separate portion of the sample, or by solution of the precipitate by fusion with sodium carbonate or potassium bisulphate and subsequent extraction with hydrochloric acid.

Removal of Silica. This compound has already been considered under 'Preparation and Solution of the Sample," SiO_2 being removed by taking the solution to dryness, dehydrating the oxide by additional heating in the oven, followed by extraction of the soluble constituents with dilute hydrochloric acid and filtration. Under the first procedure for solution of the ore by sulphuric and hydrofluoric acids silica is expelled as gaseous SiF_4.

Separation from Iron. 1. Aluminum hydroxide is precipitated by the addition of a salt of a weak acid to the neutral or slightly acid aluminum solution; iron remains in solution. Details of the procedure for precipitation of aluminum hydroxide by means of sodium thiosulphate are given under " Gravimetric Methods for Determination of Aluminum," page 9.

2. Aluminum chloride is precipitated from a concentrated solution of hydrochloric acid and ether saturated with HCl gas. Details of the procedure are given under the gravimetric methods for aluminum, page 10.

NOTE. The following additional procedures for separation of iron and alumina have been suggested:

(a) Precipitation of iron as FeS in presence of organic acids, citric, tartaric, salicylic, etc., aluminum remaining in solution.

(b) Precipitating iron by adding sodium peroxide to a cold neutral solution of the elements until the precipitate first formed dissolves, then decomposing the sodium ferrate by boiling, $Fe(OH)_3$ precipitates, Al remaining in solution. (Glaser, J. S. C. I., 1897, 936.)

(c) The neutral solution of the elements is boiled with freshly precipitated MnO_2, which causes the precipitation of iron as $Fe(OH)_3$, while aluminum remains in solution, (chromium also passes into the filtrate).

(d) Precipitation of iron from acid solutions by means of amino-nitrosophenyl-hydroxylamine, (cupferron), aluminum remaining in solution. (O. Baudisch, Chem. Ztg., **33**, 1298, 1905. *Ibid.*, **35**, 913, 1911; O. Baudisch and V. L. King, J. I. E. C., 3, 627, 1911).

(e) Precipitation of aluminum (together with phosphoric acid, if present), by phenylhydrazine, added to the reduced, weakly acid or neutral solutions. Iron, cobalt, nickel, calcium, and magnesium remain in solution. (Hess and Campbell. C. N., lxxxi, 158. Engles, J. S. C. I., 1898, 796.)

(f) Electrolytic separation of iron by amalgamation with mercury cathode and determining aluminum in the solution. (Kretzschmar, J. S. C. I., 1890, 1064; Kolin and Woodgate, J. S. C. I., 1889, 260.)

Phosphoric Acid. In presence of phosphoric acid, the phosphates of iron and alumina together with the phosphates of the other elements of the group and those of the alkaline earths will be precipitated upon making the solution alkaline with ammonia. Should iron and alumina be the only elements of these two groups present in the solution, they may be precipitated together as phosphates, iron determined by titration and calculated to the phosphate salt, and alumina obtained by difference. Occasionally, however, it is necessary to remove phosphoric acid.

Removal of Phosphoric Acid. The material is fused with about six times its weight of a mixture of 4 parts Na_2CO_3 and 1 part SiO_2 (silex), and the melt extracted with water containing ammonium carbonate. Iron and aluminum remain on the filter, upon filtration, while sodium phosphate passes into solution. Both the precipitate and filtrate contain silica. The precipitate of iron and alumina is dissolved in hydrochloric acid and taken to dryness, the residue dehydrated as usual, then treated with dilute hydrochloric acid and silica filtered off. The solution contains iron and aluminum in form of chlorides.

Separation of Aluminum from Chromium. The solution is made strongly alkaline with sodium or potassium hydroxides and chromium oxidized by passing in chlorine gas or by adding bromine. The solution is now acidified with nitric acid and aluminum hydroxide precipitated by addition of ammonium hydroxide, chromium remaining in solution as a chromate.

Separation of Aluminum from Manganese, Cobalt, Nickel, Zinc, the Alkaline Earths, and Alkalies. Iron and aluminum are precipitated as basic acetates, the other elements passing into solution. Details of the procedure are given under the basic acetate method on page 298.

In absence of phosphates, these elements do not interfere in the determination of aluminum by precipitation as the hydroxide.

Separation of Aluminum from Titanium. Details of the procedure are given under "Titanium."

Separation of Aluminum from Uranium. Aluminum is precipitated as $Al(OH)_3$ in presence of a large amount of ammonium salts by addition of a large excess of ammonium carbonate and ammonium sulphide, while uranium remains in solution as the complex compound $UO_3(CO_3)_3(NH_4)_4$.

Separation from Glucinum. Aluminum is soluble in the fixed alkalies and remains in solution on boiling; glucinum also dissolves, but is precipitated on boiling. Glucinum is soluble in an excess of ammonium carbonate, aluminum is not.

For additional separations see chapter on element in question.

GRAVIMETRIC METHODS FOR THE DETERMINATION OF ALUMINUM

There are two general procedures for the gravimetric determination of aluminum. *A.* Direct determination, when it is possible to precipitate the hydroxide or phosphate of the element, free from impurities. *B.* Indirect determination when the element is precipitated and weighed along with iron, the latter then determined by titration and aluminum estimated by difference.

Determination by Hydrolysis of an Aluminum Salt with Ammonium Hydroxide

Principle. The method depends upon the hydrolysis of a soluble salt of aluminum by neutralizing the free and combined acid with ammonia. This hydrolysis takes place in presence of ammonium chloride, which prevents the precipitation of magnesium hydroxide by NH_4OH, the common ion, NH_4^+, repressing the ionization of the base, NH_4OH. (See Notes.) The direct determination of aluminum by this procedure excludes the presence of elements undergoing hydrolysis with similar conditions. Iron, chromium, titanium, zirconium, thallium, cerium interfere. In their presence a separation must be made.

Reaction. $AlCl_3 + 3NH_4OH = Al(OH)_3 + 3NH_4Cl.$

If phosphoric acid is present in the solution aluminum will be precipitated as the phosphate, $AlPO_4$.

Procedure. To the solution, containing aluminum, free from phosphoric acid and the elements precipitated by ammonium hydroxide, are added 10 cc. of ammonium chloride (10%) and 5 cc. of concentrated nitric acid. The solution is diluted to about 150 cc. and heated to boiling. Upon cooling slightly, carbonate-free ammonium hydroxide is added slowly from a burette until a slight permanent precipitate forms, and then drop by drop until the solution reacts alkaline to litmus paper and the odor of ammonia is faintly perceptible. The precipitate is allowed to settle on the water bath for a few minutes, then filtered hot and washed first several times by decantation and finally on the filter with a hot solution of ammonium nitrate. (Twenty cc. strong nitric acid diluted and neutralized with ammonium hydroxide and made to 1000 cc.)

The precipitate is purified, if other members of the ammonium sulphide group and following groups are present, as the gelatinous precipitate is apt to occlude some of these. This is accomplished by dissolving the precipitate in a small amount of hot, dilute hydrochloric acid, 1 : 1, the solution being caught in the beaker in which the first precipitation was made. The precipitation of the hydroxide is repeated exactly as is stated above. The precipitate, washed free of chlorides ($AgNO_3$ test), is drained of water and placed together with the filter paper in a platinum crucible.

The ignition of the precipitate is conducted slowly at first until the paper is thoroughly charred, the heat is now increased to the full power of the Meker blast, the crucible being covered to prevent mechanical loss. Blasting for thirty minutes is generally sufficient to dehydrate the oxide, Al_2O_3. It is advisable, however, to repeat the heating until the weight becomes constant. The residue is weighed as Al_2O_3.

$$Al_2O_3 \times 0.5303 = Al.$$

NOTES. Ammonia should be free from carbonates. Upon long standing with frequent exposure to air the ammonia takes up CO_2, forming carbonate of ammonia. Freshly distilled ammonia will be pure, the carbonate being precipitated by addition of lime in the distilling flask. Ammonia is best kept in a ceresine or paraffine bottle. It will then remain free from silica, which it invariably contains when confined in glass bottles.

Long heating of the mixture containing the aluminum precipitate is objectionable.

1. The solution is apt to become acid owing to the decomposition of ammonium salts and the volatilization of ammonia.

2. The precipitate will become slimy and will be difficult to wash and filter. It is preferable to redissolve and again precipitate if this condition occurs.

3. The CO_2 of the air is apt to be absorbed by the solution, causing the precipitation of calcium carbonate, etc., should the solution be exposed for any length of time.

4. Silica from the beaker will contaminate the precipitate.

Hence it is advisable to filter as soon as possible after making the precipitation of $Al(OH)_3$.

S. and S. No. 589, black band filter paper filters well and may be used to advantage with precipitates of the nature of aluminum hydroxide. B. and A. No. B. filter is also good.

Washing the precipitate with ammonium nitrate prevents the aluminum from passing through the filter and keeps it from packing. It favors the formation of the insoluble hydrogel form of the hydrate while preventing the formation of the soluble hydrosol. Ammonium chloride may be used in place of nitrate.[1]

Aluminum hydroxide is soluble in acids and alkalies. The ignited oxide, Al_2O_3, is insoluble in acetic acid but is soluble in mineral acids and the fixed alkalies. It is rendered very difficultly soluble in acids by strong ignition, generally requiring fusion with sodium carbonate or potassium bisulphate with subsequent acid treatment to effect solution.

Al_2O_3, *m.w.*, 102.2; *sp.gr.*, 3.73 to 3.99; *m.p.*, 2020° C.

A yellow or reddish precipitate indicates the presence of iron, an element frequently present with aluminum. Should this be the case, iron must be determined, either in a separate portion of the sample, or in the residue obtained by the procedure outlined. The amount of Fe_2O_3 is subtracted from the total residue, and Al_2O_3 obtained by difference.

If *phosphoric acid* is present the phosphate of alumina will precipitate together with the phosphates of elements insoluble in alkaline solutions. Should phosphoric acid be present either its removal is essential, or the phosphate method for alumina should be followed.

Fluorides hinder the precipitation of aluminum.[2] Evaporation to dryness and heating the residue to redness will transform fluorides to oxides and overcome this difficulty.

Sulphates tends to hold up aluminum from precipitation and a certain amount of sulphuric acid is occluded by the aluminum hydroxide precipitate. Magnesium is more apt to precipitate with alumina in presence of sulphates.[3] Ammonium chloride greatly lessens this difficulty.

Traces of alumina may be recovered from the filtrate by evaporation to dryness, ignition and resolution with HCl. The $Al(OH)_3$ is now precipitated with NH_4OH.

Since alumina absorbs moisture from the air, the crucible containing this compound should be kept covered in a desiccator until weighed.

Ammonium hydroxide, in presence of sufficient NH_4Cl, will not precipitate $Mg(OH)_2$, since the addition of NH_4Cl increases the ammonium ions in the solution and, by the common ion effect, represses the hydroxyl ions of the base, NH_4OH, so that there are insufficient hydroxyl ions for the solubility product of $Mg(OH)_2$ to be exceeded; therefore magnesium remains in solution. A discussion of the theory of solubility product and law of mass action may be found in the author's work on Qualitative Chemical Analysis, published by D. Van Nostrand Co. Reference is also made to Vol. I of The Elements of Qualitative Chemical Analysis, by Julius Stieglitz, publ. by the Century Co.

[1] W. Blum, Jour. Am. Chem. Soc., 38, 7, 1282, 1916. C. F. Sidener and Earl Pettijohn, Jour. Ind. Eng. Chem., 8, 8, 714, Aug., 1916.

[2] E. P. Veitch, Jour. Am. Chem. Soc., 22, 246, 1900. W. R. Bloor, ibid., 29, 1603, 1907. L. P. Curtman and H. Dubin, ibid., 34, 1485, 1912.

Determination of Aluminum by Hydrolysis, Neutralizing the Mineral Acid by Addition of a Salt of a Weak Acid. Sodium Thiosulphate Method

If a salt of a weak acid and strong base is added to a neutral or slightly acid solution of an aluminum salt containing a mineral acid, transposition takes place and aluminum is hydrolyzed.

Reaction. $2AlCl_3 + 3Na_2S_2O_3 + 3H_2O = 2Al(OH)_3 + 6NaCl + 3SO_2 + 3S$.

Procedure. If the solution is acid, dilute ammonia is added until a precipitate forms that dissolves with difficulty, but not enough ammonium hydroxide to cause a permanent precipitation. The solution is diluted so that it contains about 0.1 g. Al per 200 cc., then an excess of sodium thiosulphate is added, and the solution is boiled free of SO_2. $Al(OH)_3$ precipitates along with free sulphur. If iron is absent it is advisable to add a few drops of ammonium hydroxide until the solution has a slight odor of ammonia. The mixture again boiled is filtered and the residue of $Al(OH)_3$ and sulphur washed with hot water containing ammonium chloride or nitrate. The precipitate is dried, separated from the filter, the latter ignited and the ash added to the main precipitate. Alumina is now determined by blasting to constant weight, the residue being weighed as Al_2O_3.

NOTES. The above method may be employed for separation of aluminum from iron, the addition of ammonia, following the neutralization of the mineral acid by thiosulphate being omitted. The precipitation of $Al(OH)_3$ by this procedure gives a more dense and better filtering precipitate than does ammonia alone.

NOTE. G. Wynkoop suggests the use of sodium nitrite as the salt of a weak acid for neutralizing the mineral acid. (J. Am. Ch. Soc., **19**, 434 (1897).

I. Ivanov recommends neutralizing the aluminum solution with sodium thiosulphate then diluting to 100 cc. and adding potassium iodide, followed by a 3% solution of KIO_3 and additional KI (10% sol.) until precipitation is complete. The excess of iodine is expelled by boiling. The $Al(OH)_3$ is filtered and washed with NH_4NO_3 sol. (2% sol.) and then ignited to Al_2O_3.

Precipitation of Aluminum as a Phosphate

Principle. This procedure, developed by Carnot, is of special value in determination of aluminum in iron and steel. It is founded on the reaction that aluminum is precipitated as the neutral phosphate, from a boiling solution faintly acid with acetic acid. Iron, reduced to the ferrous condition by addition of sodium hyposulphite, does not interfere.

Procedure. A sample of 10 grams of iron or steel, in a platinum dish, covered with a piece of platinum foil, is dissolved by addition of hydrochloric acid. The solution is diluted to about 100 cc. and filtered into a flask, the residue of carbon, silica, etc., is washed thoroughly and the filtrate is neutralized by addition of ammonium hydroxide and ammonium carbonate; no permanent precipitate should form. A little sodium hyposulphite is added, and when the liquid, at first violet, becomes colorless, 2 or 3 cc. of a saturated solu-

E. Schum (Chem. Zeit., 1909, XXXIII, 877) recommends neutralizing the solution of aluminum with NH_4OH, just avoiding a precipitation of $Al(OH)_3$. After diluting to 250 cc. the aluminum is precipitated by adding 20 cc. of a 6% sol. NH_4NO_2. The oxides of nitrogen are expelled by boiling and the $Al(OH)_3$ settled on hot water bath 20 min. The sol. is decanted and the precipitate washed with neutral (2% sol.) ammonium nitrite and ignited as usual. The method is considered very accurate.

tion of sodium phosphate and 5 or 6 grams of sodium acetate, dissolved in a little water, are added. The solution is boiled until free of sulphurous acid odor (about three-quarters of an hour). The solution is filtered off from the precipitated aluminum phosphate (mixed with a little silica and ferric phosphate) and washed with boiling water. The precipitate on the filter is treated with hot dilute hydrochloric acid the filtrate caught in a platinum dish, and then evaporated to dryness and heated at 110° C. for an hour to dehydrate silica. The residue is taken up with dilute hydrochloric acid and the solution filtered free of silica. Upon dilution to about 100 cc. with cold water, the solution is neutralized as before, a little hyposulphite is added to the cold solution and then a mixture of 2 grams of sodium hyposulphite and 2 grams of sodium acetate. The material is boiled for half an hour or more, filtered and the aluminum phosphate residue washed with hot water, then dried, ignited and weighed as aluminum phosphate. The residue contains 22.19% Al.

$$AlPO_4 \times 0.2219 = Al. \quad AlPO_4 \times 0.4185 = Al_2O_3.$$

NOTE. Properties of $AlPO_4$, $m.w.$, 122.14; $sp.gr.$, 2.59; infusible, insoluble, in H_2O and in $HC_2H_3O_2$, soluble in mineral acids and in alkalies; white, amorphous salt.

Precipitation of Aluminum as Aluminum Chloride [1]

Principle. Gooch and Havens found that aluminum chloride is practically insoluble in a mixture of concentrated hydrochloric acid and ether saturated with HCl gas, 5 parts of $AlCl_3.6H_2O$ equivalent to 1 part of Al_2O_3 dissolving in 125,000 parts of the mixture. The method serves for a separation of aluminum from iron, berillium, zinc, copper, mercury and bismuth, the chlorides of these elements being soluble under the above conditions. Barium, however, is precipitated as a chloride with aluminum, if it is present in the solution.

Procedure. To the concentrated aqueous solution of aluminum is added a convenient volume of strong hydrochloric acid (15 to 25 cc.) and an equal volume of ether. The mixture is best placed in a large platinum crucible, which is kept cool in running water. HCl gas is passed into the solution to saturation. The precipitated chloride of aluminum is filtered upon asbestos in a weighed Gooch crucible and then washed with a mixture of ether and water 1 : 1, saturated with HCl gas. The precipitate is dried for half an hour at 150° C., then covered with a layer of C.P. mercuric oxide (1 gram) and heated at first, gently over a low flame (hood) and then blasted to constant weight. The residue is weighed as Al_2O_3.

NOTES. HCl gas is generated by dropping strong sulphuric acid into concentrated hydrochloric acid according to the procedure described under the determination of arsenic by volatilization as arsenious chloride. The gas may be produced in a Kipp generator by the action of concentrated sulphuric acid on ammonium chloride.

The filtrate from aluminum contains iron, berillium, copper, zinc, etc., if these are present in the original solution. If much iron is present it is necessary to increase the amount of ether to prevent precipitation of the ferric salt.

[1] F. A. Gooch and F. S. Havens, Am. Jour. Sci. (4), **11**, 416. F. A. Gooch " Methods in Chemical Analysis."

VOLUMETRIC METHODS FOR THE DETERMINATION OF ALUMINUM

Volumetric Determination of Combined Alumina in Aluminum Sulphate and Aluminum Salts

Introduction. Aluminum salts dissociate in hot solutions and react acid to phenolphthalein indicator; the acid readily combines with fixed alkalies, forming the neutral alkali salt. The end point of the reaction is indicated by the pink color produced upon phenolphthalein by the excess of alkali. From the amount of caustic required the percentage of combined Al_2O_3 may be calculated. The following reaction takes place:

$$Al_2(SO_4)_3 + 6NaOH = 2Al(OH)_3 + 3Na_2SO_4.$$

Procedure. The factor weight,[1] 3.4067 grams, is dissolved in a 4-in. casserole with 100 cc. of distilled water, 1 cc. of phenolphthalein indicator added, and the sample titrated boiling hot [2] with N/2 NaOH, added from a chamber burette, graduated from 50 to 100 cc. in tenths of a cc.[1] The solution is kept boiling during the titration and is constantly stirred. Towards the end of the reaction the alkali is added cautiously drop by drop until a permanent pink color is obtained.

Cc. of NaOH required divided by 4 = per cent combined Al_2O_3.[3]
Combined Al_2O_3 + free Al_2O_3 = total Al_2O_3.

NOTES. If iron is present a correction must be made for it after determining the ferrous and ferric forms as given below.

The amount of phenolphthalein indicator used should be the same in each determination. An excess of indicator causes low results. It has been noted in case of alums where iron does not interfere that best results are obtained with three or four drops of phenolphthalein solution. Iron tends to mask the end point, hence a larger amount of indicator is necessary if this is present.

Correction for Iron if Present. Since iron salts will also dissociate and titrate with aluminum salts, by this method a correction has to be made for iron if present. Total Al_2O_3 in presence of iron =

combined $Al_2O_3 - (FeO \times .47 + Fe_2O_3 \times .64) +$ basic $Al_2O_3 +$ an additive factor.

The additive factor is obtained by subtracting

(Combined $Al_2O_3 +$ basic $Al_2O_3) - (FeO \times .47 + Fe_2O_3 \times .64)$ volumetric,

from total Al_2O_3 obtained by gravimetric analysis of an average sample.

---

[1] Large samples must be taken for salts containing less than 13 per cent Al_2O_3 if the chamber burette is to be used. E.g., potash alum twice this amount is advisable.

[2] Otto Schmatolla, Berichte, xxxviii, No. 4. C. N., 91–2375–236 (1905).

[3] If free acid is present (see next method), the equivalent volume in terms of $\frac{1}{2}$ N acid must be deducted from the total titration for combined alumina before dividing by 4.

Ferrous Iron, Ferric Iron, and Total Iron A five-gram sample is dissolved in water and the iron oxidized with a few drops of strong potassium permanganate solution; the solution should be pink; the excess of permanganate is destroyed by a drop or so of normal oxalic acid solution and the total iron determined by stannous chloride solution method for iron. On a separate sample ferric iron is determined. Ten grams of the sample are dissolved in an Erlenmeyer flask by boiling with hydrochloric acid, 2 : 1, in an atmosphere of CO_2 to prevent oxidation, and the iron titrated with standard stannous chloride. The difference between total iron as Fe_2O_3 and ferric oxide = ferrous iron in terms of Fe_2O_3. This multiplied by .9 = FeO.

Combined Sulphuric Acid

Provided no free acid is present, the per cent combined sulphuric acid in aluminum sulphate is obtained by multiplying the cc. caustic titration for total alumina by 0.72.

In case free acid is present, the per cent free acid deducted from total acid found by titration gives combined acid.

Sulphuric acid combined with the fixed alkalies is not titrated.

Determination of Free Alumina or Free Acid by the Potassium Fluoride Method

Introduction. The method suggested by T. J. I. Craig (J. S. C. I., 1911, **30**, 185), has been modified by the author,[1] after a personal investigation of the details involved. In this modified form it has been used successfully as a rapid works method. Frequent gravimetric checks on a large number of determinations have shown it to be accurate.

The procedure is based upon the fact that an excess of neutral potassium fluoride decomposes aluminum salts, forming two stable compounds, which react neutral to phenolphthalein, while the free acid remains unaltered, the following reaction taking place:

$$Al_2(SO_4)_3 + 12KF + xH_2SO_4 = 2AlF_3 3KF + 3K_2SO_4 + xH_2SO_4.$$

The precipitate $AlF_3 3KF$ is insoluble in an excess of the potassium fluoride reagent and is not appreciably attacked by acids or alkalies. Although theoretically about 7 parts by weight of potassium fluoride is sufficient to combine with 1 part of aluminum sulphate, in practice it is advisable to use twice this amount.

Reagents Required. *Half Normal* solutions of sulphuric acid and potassium hydroxide.

Phenolphthalein indicator, 0.1% alcoholic solution.

Potassium fluoride solution; made by dissolving 1000 grams of potassium fluoride in about 1200 cc. of hot, CO_2-free water, then neutralizing the solution with hydrofluoric acid or potassium hydroxide as the reagent may require, using 5 cc. of phenolphthalein as indicator. Dilute sulphuric acid may be used in place of hydrofluoric acid in the final acid adjustment to get a neutral product. One cc. of the solution in 10 cc. of CO_2-free water should appear a faint pink. The concentrated mix is filtered if necessary and then diluted to 2000 cc. with CO_2-free water. The gravity will now be approximately 1.32 or about 35° Bé. One cc. contains 0.5 g. potassium fluoride.

[1] W. W. Scott, Jour. Ind. Eng. Chem., **7**, 1059, 1915.

Method of Procedure

Solids. 3.4067 g. of the finely ground sample, or an equivalent amount in solution (100 cc. of sample containing 34.067 g. per liter), are taken for analysis The powder is dissolved by boiling with 100 cc. of distilled water in a 4-in. casserole with clock glass cover. To the hot solution 10 cc. of $N/2$ H_2SO_4 are added. and after cooling to room temperature, 20° C., 18 to 20 cc. of the potassium fluoride reagent are added and 0.5 cc. of phenolphthalein. The solution is now titrated with $N/2$ KOH, added drop by drop until a delicate pink color, persisting for one minute, is obtained. This titration shows whether the product is basic or acid.

Basic Alumina. This is indicated when the alkali back-titration is less than the amount of acid added. Free $Al_2O_3 = $ (cc. H_2SO_4 − cc. KOH) ÷ 4.

Free Acid. In case the back-titration of the alkali is greater than the cc. of acid added, free acid is present. Free acid = (cc. KOH − cc. H_2SO_4) × 0.72.

Liquors. In works control it is necessary to test the concentrated liquors to ascertain whether these are basic or acidic. The Bé. or sp.gr. of the solution having been taken, 5 cc. is diluted to 100 cc. with distilled, CO_2-free water. If H_2S is present, it is expelled by boiling the solution, which should be acid, 10 cc. of $N/2$ H_2SO_4 is added, the solution cooled, and KF and phenolphthalein added and the titration made as in case of solids.

If basic (cc. H_2SO_4 − cc. KOH) × (.0245 × .3473 × 100) ÷ (5 × sp.gr.) = Al_2O_3.

If acid (cc. KOH − cc. H_2SO_4 × 2.45) ÷ wt. of sample = per cent free acid (H_2SO_4).

If neutral, the back titration of the alkali is the same as the cc. acid added.

NOTES. CO_2-free water must always be used when phenolphthalein indicator is necessary. This may be obtained by boiling distilled water for several minutes to expel CO_2. This reagent is very sensitive to carbonic acid.

If the sample does not dissolve clear, a prolonged digestion with previous addition of the required amount of standard acid, 10 cc., is advisable. This is best accomplished in an Earlnmeyer flask with a return condenser.

Darkening of the solution during the back titration with the alkali, indicates that an insufficient amount of fluoride has been added. If this is the case it will be necessary to make a fresh determination.

The fluoride method has the following advantages. Determinations may be made by gas or electric light. The end point is easily detected. No neutral standard is necessary as in case of the tint method.

Ammonium salts, if present, must be expelled by boiling the sample with an excess of standard KOH and this excess determined.

3.4067 = 2.45225 × .3473 × 4 (i.e. gms. H_2SO_4 per 100 cc. $N/2$ acid multiplied by 4 times factor to equivalent Al_2O_3). Derived directly from mol. wt. of Al_2O_3 = (.1022 × 100 × 4) ÷ (6 × 2). 0.72 = 2.8792 ÷ 4 (i.e. factor Al_2O_3 to H_2SO_4 + 4).

The main details of the above volumetric procedures were worked out at the Laurel Hill Laboratory, General Chemical Company, and are published by courtesy of this company.

Colorimetric Estimation of Minute Amounts of Aluminum with Alizarin S. —Atack's Method[1]

Reagent. See page 3.

Procedure. The original solution (5 to 20 cc.) is acidified with hydrochloric or sulphuric acid. Ten cc. of glycerin and 5 cc. of a 0.1% solution of alizarin S are added, the solution made up to about 40 cc. with water (in presence of much iron or chromium citric acid is added to form the double citrates) and then rendered slightly ammoniacal. After standing for five minutes, the cold solution is acidified with dilute acetic acid, the alizarin S acting as indicator (red coloration) until no further change in the coloration occurs. The liquid is then made up to 50 cc. and compared with a standard. Suitable amounts of aluminum for estimation are 0.005 to 0.05 milligrams, the solution under examination being suitably diluted if necessary.

BAUXITE ANALYSIS [2]

Characteristic bauxites	H_2O	SiO_2	Fe_2O_3	Al_2O_3	TiO_2
Arkansas		6.4%	1.43%	87.3%	3.99%
Georgia	36%	9–15	1–14	42–62	1.8–2.3
Tennessee	27.6	18.4	4.1	49.9	

Sampling. The bauxite received in cars is sampled during the unloading according to the standard procedure for ores. If the sample is a composite aliquot parts of the total weights are taken and mixed, e.g., suppose three cars contained respectively 23,000, 32,500, and 26,340 pounds, then the aliquots would be 23, 32.5 and 26.34 pounds, which mixed, would make a representative sample of the shipment. The ore is broken down, quartered, ground down and again quartered. The moisture is determined on 1000 grams, dried in the oven at 100° C. for one hour, the sample being spread out on a sheet of manilla paper. The dried sample is placed in a large bottle for analysis.

Extraction of Ores of Aluminum for Their Commercial Valuation. Twenty-five grams of bauxite, alunite or clay, placed in an 600 cc. Kjeldahl flask with reflux condenser are digested for one hour with 60 cc. 10 N (48%) sulphuric acid. Time being taken when the mix begins to boil. The flame is so regulated that the water drips back into the flask from the condenser at the rate of 12 to 15 drops per minute. When the digestion is completed 150 cc. of hot water are introduced through the condenser. The mixture is now filtered using a Buchner funnel and suction. The residue is washed with six 100 cc. portions of water and the filtrate made up to 1000 cc. A 200 cc. aliquot is made up to 1000 cc. and 100 cc. of this taken for determination of alumina as stated on next page.

[1] F. W. Atack, Jour. Soc. Chem. Ind.; **34**, 936 (1915); C. A. **9**; 23; 3186 (1915).

[2] Bauxite and kaolin are two sources of alumina of commercial importance. The alumina of certain grades of clay, is made available for acid extraction by roasting at a definite temperature. 550°–700° C.

Insoluble Residue. The residue on the filter paper is ignited in a platinum dish over a low flame until the paper chars, and then over a good Meker blast for 15 to 20 minutes.

Weight of the residue $\times$ 4 = per cent insoluble residue.

Soluble Alumina. 100 cc. of the above solution (0.5 g.) is diluted with an equal volume of water, 10 cc. of hydrochloric and 2 cc. of nitric acids added and the solution boiled. Iron and alumina are now precipitated and determined in the usual way.

Soluble Iron. 200 cc. of the solution (1.0 g.), is oxidized by adding a few crystals of potassium chlorate and the solution taken to dryness. The residue is taken up with 10 to 15 cc. of concentrated hydrochloric acid and again evaporated to dryness to expel chlorine. Then taken up with 25 cc. hydrochloric acid and the iron determined by titration. The stannous chloride method is used for samples containing less than 5% iron and the dichromate method for ores containing over 5%.

Determination of Total Silica, Titanium Oxide, Ferric Oxide and Alumina

The method by the Aluminum Company of America is to digest 1 gram of the dried bauxite in 90 cc. of an acid mixture containing 12 parts of dilute sulphuric acid, 1 : 3, together with 6 parts of strong hydrochloric acid and 2 parts of nitric by volume, to this are added 10 cc. of concentrated sulphuric acid. The mixture is heated until sulphuric acid fumes are evolved, then diluted with water and filtered.

Silica. The residue is ignited and the ash fused with potassium bisulphate. The cooled fusion is taken up with 5 cc. sulphuric acid and 20 cc. of water and digested until only a white residue remains. This filtered off, washed and ignited = SiO_2.

Titanium Oxide. This is best determined colorometrically on a 0.1 gram sample according to the procedure outlined in the chapter on Titanium.

Iron and Alumina. These are determined by the usual procedure;—oxidation with potassium chlorate, precipitation with ammonium hydroxide and ignition. Iron may be determined in a separate sample (100 cc. = 0.5 g.) by titration. Al_2O_3 = difference between weighed oxides and Fe_2O_3, after subtracting TiO_2 if present.

VOLUMETRIC DETERMINATION OF THE AVAILABLE ALUMINA IN BAUXITE AND CLAYS[1]

This method is based on the solution of the sample in a known amount of sulphuric acid, and the titration of the excess acid with standard sodium hydroxide solution, and the alumina estimated from the amount of acid used to combine with it. It has been proven with long practice that this method is accurate within 0.5%, and is very satisfactory as a works control method. It is much more rapid than the regular gravimetric procedure.

Take a five gram sample that has been ground to pass through a 60-mesh sieve and put it into a 300 cc. Kjedahl flask with about 28 to 30 grams of 40° Bé sulphuric acid, which is weighed from a weighing burette. (This acid may be measured with a burette and the weight calculated if so desired with a reasonable chance of error introduced.) Digest this at a gentle boil for one hour, giving the flask an occasional shaking. Care must be taken that good condensation is effected, otherwise a loss in acid will give high results. After the digestion wash down the condenser and filter the insoluble residue off on a Bueckner funnel with suction, washing the residue acid free. Make the filtrate up to 2000 cc. in a volumetric flask. To a 200-cc. aliquot add 20 cc. KF reagent, and 1 cc. phenolphthalein indicator and titrate to a permanent pink with standard N/10 NaOH.

Reagents: The 40° acid must be made up very accurately and standardized against standard caustic.

1000 grams of potassium fluoride are dissolved in 1200 cc. of hot CO_2 free water, and then neutralized with HF or KOH as may be necessary, using phenolphthalein indicator. Filter this solution and dilute to 2000 cc. This solution should be kept in a wax-lined bottle.

Take 20 cc. of KF reagent and add 5 cc. of N/2 NaOH and one cc. of phenolphthalein indicator, and titrate with N/2 H_2SO_4. Apply this correction to the titration of the sample.

Calculations:

$$\frac{(\text{cc. N/10 } H_2SO_4 - \text{cc. N/10 NaOH}) \times .3473 \times 100}{\text{weight of sample}} = \% \ Al_2O_3 + Fe_2O_3$$

$\% \ Fe_2O_3 \times .64 = Fe_2O_3$ equivalent to Al_2O_3 which can be subtracted from the total oxides above, the result being the available alumina.

Determination of Aluminum in Bronze. A sample weighing 0.5 to 1.0 g. is dissolved in 10 cc. of HNO_3 (2 : 1) and heated to expel nitrogen oxides. 50 cc. of water are added and the precipitate allowed to settle (oxide of Sn, P_2O_5 and some Cu) and then filtered. The filtrate is treated with 5 cc. H_2SO_4 and evaporated to fumes, then taken up with about 50 cc. of water and saturated with H_2S, and the precipitate filtered off. The filtrate containing the iron, aluminum, zinc, etc. is boiled to expel H_2S and oxidized by boiling with 5 cc. HNO_3. Iron and aluminum are now precipitated as hydroxides by addition of NH_4OH and filtered off and washing as usual. Iron and aluminum are separated by dissolving the hydroxides in a little HCl and neutralizing the free acid with Na_2CO_3 solution. Any permanent precipitate is dissolved by a few drops of HCl. For each 0.1–0.2 g. of the metals present the solution is diluted to about 250 cc. and an excess of $Na_2S_2O_3$ is added. Aluminum hydroxide is precipitated, iron remains in solution in the ferrous form. $2AlCl_3 + 3Na_2S_2O_3 + 3H_2O = 2Al(OH)_2 + 3SO_2 + 3S + 6NaCl$. The $Al(OH)_3$ is filtered off and washed with hot water, then dried and the hydroxide and filter ignited separately, the ash of the paper added to the alumina and the ignition continued until a constant weight is obtained. An electric furnace of a Meker burner may be used to expel the combined water. Weigh as Al_2O_3.

$$Al_2O_3 \times 0.5303 = Al.$$

[1] By Harold E. Martin.

DETERMINATION OF ALUMINUM IN IRON AND STEEL [1]

The method is especially adapted for determination of aluminum in iron and steel, but may be extended to iron ores and materials high in iron.

Procedure. Solution. Ten grams of iron or steel are dissolved by adding about 50 cc. of hot hydrochloric acid, 1 : 1, preferably in a platinum dish, covered with a platinum foil.

Precipitation. When the solution of iron is complete, it is diluted to about 100 cc. and filtered free of carbon, silica, etc. Two grams of sodium phosphate are added and the solution neutralized with ammonium hydroxide or carbonate, then cleared by hydrochloric acid with about 1 cc. excess. Twenty cc. of acetic acid are now added and the solution diluted to 300 to 400 cc. with hot water and, on boiling, 10 grams of sodium thiosulphate added. The solution is boiled free of sulphurous acid, (no odor of SO_2) about 20 to 30 minutes being necessary. The phosphate is filtered off and washed with hot water. It is again dissolved in a little hydrochloric acid and aluminum reprecipitated by neutralizing with ammonium hydroxide and adding about 1 gram of sodium phosphate together with 10 grams of sodium thiosulphate, following the above procedure. The precipitate will now be free of iron.

Ignition and Calculation. The precipitate and filter are ignited wet, first over a low flame, then gradually increasing the heat to full blast of a Meker burner. The residue contains 22.19% Al or 41.85% of Al_2O_3.

Factor $AlPO_4$ to Al = .2219.

Factor $AlPO_4$ to Al_2O_3 = .4185.

NOTES. Interfering substances. Copper may be removed by H_2S. Other members of this group will also be eliminated.

Manganese and nickel are eliminated together with small amounts of iron at the second precipitation.

Titanium may be estimation colorimetrically or separated from alumina.

Vanadium, if present, may be separated according to directions given in the chapter on Vanadium.

Chromium is eliminated by fusion of the mixed phosphates with Na_2CO_3, extraction with water, and precipitation of aluminum phosphate by adding ammonium acetate and sodium phosphate. Chromium remains in solution.

Other Methods. L. Belasio adds crystalline tartaric acid to hold up the precipitation of other metals that commonly interfere. I. Ivanov neutralizes the slightly acid solution of aluminum with $Na_2S_2O_3$, then dilutes to 100 cc. adds KI in excess, then a 3% sol. of KIO_3 with additional KI until precipitation is complete. The excess of iodine is expelled by boiling and the $Al(OH)_3$ washed with neutral 3% NH_4NO_3 sol. and ignited as usual, to Al_2O_3. E. Schum (Chem. Zeit., **32**, 877, 1909) neutralizes the solution with NH_4OH just avoiding precipitation of $Al(OH)_3$. After dilution to 250 cc. 20 cc. 6% ammonium nitrite sol. are added, the solution boiled to expel NO_2, the $Al(OH)_3$ settled 20 min. and filtered and washed with ammonium nitrite solution and the precipitate ignited as usual to Al_2O_3.

[1] Arnold and Ibbotson, "Steel Works Materials." Stillman, "Engineering Chemistry." "A Rapid Method for the Determination of Aluminum in Iron and Steel." C. N., **61**, 313. "On the Determination of Minute Quantities of Al in Iron and Steel," J. E. Stead, J. S. C. I., 1889, 956.

ANALYSIS OF ALUMINUM AND ALUMINUM ALLOYS. METHODS OF THE ALUMINUM COMPANY OF AMERICA[1]

Determination of Silicon

Fusion Method

In some cases a portion of the silicon in aluminum and its alloys is not oxidized when the metal is dissolved in acids. This is more frequently observed when the silicon is higher than regularly found as an impurity. It is made evident by a brown or dark gray film coating the liquid and container after dissolving the metal in acid mixture, and has been referred to as graphitic silicon. In such a case a fusion of the residue is needed to give a complete oxidation. A fusion is useful also for cleaning up residues contaminated with insoluble material other than SiO_2.

Dissolve 1 gram of well-mixed drillings in 35 cc. of acid mixture No. 1 (450 cc. water, 150 cc. conc. H_2SO_4, 300 cc. conc. HCl, 100 cc. conc. HNO_3), using a 250-cc. flat bottom casserole or Pyrex beaker and cover glass. When the drillings are completely dissolved, place on a moderately heated plate and boil the solution to dryness, then increase the heat till the H_2SO_4 distills as shown by flowing on the cover. Continue till a few drops have fallen from the center of the glass. This insures the complete removal of HCl and HNO_3 and the dehydration of SiO_2. These effects can also be secured by heating at the moderate temperature for an hour after boiling dry. Cool, add 10 cc. 1 : 3 H_2SO_4 and about 100 cc. water, boil to complete solution of the sulphate, filter, wash and ignite in a platinum crucible.

Fuse the residue with 8 to 10 times its weight of Na_2CO_3 and treat the fused mass in a casserole or beaker with 20 cc. 1 : 3 H_2SO_4, washing out the crucible. Evaporate the solution until it fumes strongly to separate the SiO_2. Cool, dilute to about 100 cc., boil till the salts are dissolved, filter, wash well, ignite in a platinum crucible and weigh. Treat the ash with HF and a few drops of H_2SO_4, evaporate till dry, ignite and weigh.

From the difference between the two weights deduct a blank obtained from the reagents and filter. The rest of the loss in weight represents SiO_2 corresponding to the Si in the metal.

$$SiO_2 \times .4676 = Si.$$

Solution Method

When only a small amount of silicon is present such as the few tenths per cent commonly found in commercial aluminum, it is practically all oxidized by the acid treatment, and a direct weight of the ignited insoluble residue considered as SiO_2 gives a measure of the original Si sufficiently accurate for general use.

Proceed as in the fusion method until the first insoluble residue has been filtered and washed. Ignite it in a crucible, allow to cool, brush the residue on the balance pan and weigh directly. Deduct a determined blank. Consider the remainder SiO_2 and calculate to Si as above.

[1] By courtesy of the Aluminum Company of America, E. Blough, Technical Director, H. V. Churchill, Chief Chemist.

Determination of Iron

Permanganate Method

Cool the filtrate from the insoluble residue obtained in the determination of silicon. It should contain 3 to 5% H_2SO_4. Add $KMnO_4$ solution till a pink color persists. Reduce the iron present by passing the solution through a Jones reductor. See chapter on Iron. Titrate immediately with a solution of $KMnO_4$ of such strength that 1 cc.=.001 gram Fe. A blank, determined by carrying out the procedure without a sample, is deducted from each titration.

Cautions. (1) The $KMnO_4$ solution should not be permitted to come in contact with organic substances. It should be kept in a glass-stoppered bottle in a dark closet. Glass-stoppered burettes should be used.

(2) Always have the iron solution cool before reduction.

(3) There should be a solid column of solution or water passing during use. The zinc should be kept covered with water when not in use.

(4) When the sample contains copper, the reductor will be kept cleaner if the copper is precipitated with a little granulated zinc, and the solution decanted from the precipitated copper before passing through the reductor.

Determination of Copper

Iodide Phosphate Method: Cu up to 1%

Dissolve 2 grams of well-mixed drillings in a tall 300-cc. beaker with 40 cc. of 25% caustic soda solution. When the aluminum is completely dissolved, dilute to 200 cc. with hot water that has been boiled. Allow the residue to settle, filter and wash thoroughly with hot water.

Using as small amount as possible of hot 1:1 nitric acid, completely dissolve the residue from filter paper into a 250-cc. wide mouth Erlenmeyer flask containing 0.1 cc. of concentrated sulphuric acid. If much manganese or iron is present, more concentrated sulphuric acid (about 0.5 cc.) should be used to prevent the formation of MnO_2 or Fe_2O_3 which will obscure the end point. Place the flask on a hot plate and evaporate to complete dryness. Cool. Add 25 cc. water and 3 cc. 99.5% acetic acid. Heat to complete solution.

Remove flask from hot plate, at the same time adding 5 cc. of a saturated solution of Na_2HPO_4, shake thoroughly and cool in any convenient manner.

The ferric sulphate which later would react with potassium iodide thus:

$$Fe_2(SO_4)_3 + 2KI = 2FeSO_4 + K_2SO_4 + I_2,$$

will be precipitated and rendered inactive by the reaction,

$$Fe_2(SO_4)_3 + 2Na_2HPO_4 = 2FePO_4 + 2NaHSO_4 + Na_2SO_4.$$

To the cooled solution add 5 cc. potassium iodide solution (30 grams of potassium iodide to 100 cc. water) and shake thoroughly to mix solutions.

$$2Cu(C_2H_3O_2)_2 + 4KI = Cu_2I_2 + 4KC_2H_3O_2 + I_2.$$

Titrate the free iodine with a solution of sodium thiosulphate of 0.001 copper value which has been standardized by the copper sulphate method. A few drops of fresh starch solution, which is used as an indicator, are added near the end of the titration. This solution is prepared by dissolving 1 gram of soluble starch in 150 cc. of water.

Copper Percentage= No. of cc. Thiosulphate×Copper Value per cc.

×100÷Weight of Sample (2 grams).

NOTE. In case the residue has not been carefully washed or when the iron content is high, an additional amount of Na_2HPO_4 may be required, and must be added until all iron and aluminum salts have been precipitated.

Determination of Manganese
Sodium Arsenite Method: Mn up to 2%

Weigh out accurately a 0.2-gram sample and place in a 200-cc. Erlenmeyer flask. Add 15 cc. of acid mixture No. 2 (400 cc. conc. H_2SO_4—400 cc. conc. HNO_3—200 cc. H_2O). Place upon a hot plate and boil until solution is complete. Add 5 cc. more of acid mixture No. 2 and boil for 2 or 3 minutes to drive off nitrous fumes. Remove from the source of heat, cool, add 10 cc. of $AgNO_3$ solution (3 grams per liter) and 90 cc. of warm water containing 1 gram of ammonium persulphate Heat gently until the permanganic acid is fully developed as indicated by a ring of bubbles forming around the surface. Cool to room temperature and titrate with standard sodium arsenite solution, the approximate value of 1 cc. being equal to 0.0002 gram or 0.10% Mn.

If desired, the silver may be precipitated before titration by adding 5 cc. of NaCl solution (2 grams to 1 liter). The titration in standardizing must be done the same way as in working the samples.

Preparation of Standard Arsenite Solution

For standardizing the sodium arsenite solution for the determination of manganese a standard alloy of known manganese content must be used.

Procedure. (1) Stock Solution—stock solution is prepared by dissolving 10 grams As_2O_3 and 30 grams Na_2CO_3 in a small amount of hot water, filtering and making up to 1000 cc.

(2) Standard Solution—the standard solution is prepared by taking required amount of stock solution (about 65 cc.) and diluting to 1000 cc. After 24 hours, this solution is carefully standardized by using it to titrate the manganese in the standard sample. Three determinations should be made using exact procedure as given above.

NOTE. (1) By adding the acid mixture in two portions it is made more certain that the proper amount is present when peroxidizing the manganese.

(2) When dissolving the sample it should not be allowed to go to dryness, as some manganese oxide might be made insoluble.

Determination of Zinc—0.50% or Less
ZnO Method

Place 2 grams of sample in a 400-cc. beaker or casserole with 10 grams NaOH and 50 cc. water. When violent reaction is over, wash down and heat to near boiling till no more action can be seen. Add boiling water to make 200 cc. and bring to boil. Let settle, then filter. The filtrate is allowed to stand for several hours or is boiled a few minutes with about .5 gram Na_2O_2 to insure removal of manganese. If any precipitate forms, it is filtered out. 5 grams NaOH is added to the filtrate which is then heated to boiling and treated with H_2S for about 3 minutes. The solution is boiled for 10 minutes and filtered through a dry double filter, keeping the filter full. Do not wash. The zinc sulphide is then dissolved off the filter with warm dilute hydrochloric acid (1 : 10) and the filter washed three times with hot water. The filtrate is boiled free from hydrogen sulphide, a few drops of bromine water are added to oxidize any iron present and the solution is made slightly alkaline with ammonia, boiled and filtered. Hydrogen sulphide is then passed into the filtrate until saturated. The solution is then boiled for ten minutes and filtered through a small filter without washing. The filter is burned at low temperature in a porcelain crucible. Dissolve the residue using HCl and transfer to a beaker. Make just ammoniacal to methyl red, filter and wash. Add 10 drops of ammonia to the filtrate, heat solution to near boiling, saturate with H_2S, filter, burn the paper and precipitate in a weighed porcelain crucible. Calculate the weight of zinc oxide to zinc.

$$ZnO \times 0.8034 = Zn.$$

Determination of Calcium and Magnesium

Calcium: Permanganate Method

Place one gram of the alloy together with 5 grams of solid sodium or potassium hydrate in a porcelain dish and add just enough water to cover; when violent action has ceased, dilute to 200–250 cc. with hot water and boil until solution is complete. Filter while hot and wash the residue until free from alkali. The aluminum and most of the zinc dissolve, while the other metals remain in the metallic state.

Using a few drops of nitric acid and 40 cc. of hot 1 : 1 hydrochloric acid completely dissolve the residue obtained from the sodium hydrate solution into a 250-cc. beaker, or flask. Neutralize with ammonia and add 5 cc. excess. Precipitate with hydrogen sulphide. Filter and wash with hot water. Discard the precipitate.

NOTE. The hydrogen sulphide metals which are here discarded can be used for the determination of copper by the iodide phosphate method.

Boil off the hydrogen sulphide from the filtrate, adding a little bromine water toward the latter part of the boiling to oxidize the sulphur. Boil off the excess of bromine. Add methyl red indicator and then ammonia drop by drop until the solution is just ammoniacal; then add 2 drops excess. Add 10 cc. of saturated solution of ammonium oxalate, allow the solution to digest for 5 minutes, remove from the hot plate and filter. Wash eight times with hot water.

More or less magnesium will be precipitated with the calcium oxalate depending on the ratio in which these metals are present. If very accurate results are desired, it will be necessary to reprecipitate the calcium oxalate.

Return the paper containing the thoroughly washed calcium oxalate to the original beaker. Add 150 cc. of boiling water and 10 cc. (1 : 1) sulphuric acid and titrate with standard potassium permanganate.

Potassium permanganate in the presence of sulphuric acid reacts with calcium oxalate thus:

$$5CaC_2O_4 + 2KMnO_4 + 8H_2SO_4 = 5CaSO_4 + K_2SO_4 + 2MnSO_4 + 10CO_2 + 8H_2O.$$

The iron value $\times 0.3588 =$ the calcium value.

Magnesium: $Mg_2P_2O_7$ Method

The filtrate from the calcium oxalate which has been received into a 250-cc. beaker is made just acid with hydrochloric acid. Add 30 cc. of a saturated solution of microcosmic salt. Add ammonia drop by drop, stirring vigorously to make the precipitate crystalline. Then add an excess of ammonia equal to ten per cent by volume, and let stand for at least three hours. Filter, wash with cold water containing 5% of strong ammonia, and 5% of ammonium nitrate. Ignite until completely white, weigh as $Mg_2P_2O_7$.

$$Mg_2P_2O_7 \times 0.2185 = \text{grams Magnesium.}$$

For very accurate results the $Mg_2P_2O_7$ precipitate should be dissolved and reprecipitated, and the final precipitate should be examined for manganese or other impurities and corrected accordingly.

ANALYSIS OF SILICON ALLOYS

Determination of Silicon

Place convenient weight (1.0 gram up to 10% Si, 0.5 gram above 10% Si) of well-mixed drillings in a 250-cc. casserole and cover with a watch glass. Add 25 cc. of acid mixture No. 1 (1200 cc. 25% sulphuric acid, 600 cc. conc. hydrochloric acid, 200 cc. conc. nitric acid) from an acid burette, keeping the casserole covered and adding the acid very slowly at first until violent action has subsided. Digest on the hot plate and evaporate to dryness, taking care to prevent spattering. Cool, moisten with 5 cc. of 1 : 1 sulphuric acid, add 100 cc. water, and boil to solution of soluble salts. Filter, using a double quantitative filter paper, and filter pulp. Wash the residue six times in hot water to insure the removal of all iron salts.

Ignite the residue in a large platinum crucible. Cool. Add 5–8 grams of a mixture of 1 : 1 anhydrous sodium carbonate and sodium bicarbonate. Stir thoroughly and cover with a layer of the fusion mixture. Fuse the mixture over a Meker flame, keeping the crucible covered, until reaction is complete. This will require about 15 minutes. Cool. Place the crucible in a casserole and add 60 cc. of 1 : 3 sulphuric acid. When the fusion cake has dissolved, remove the crucible. Evaporate to heavy fumes. Cool. Take up with 100 cc. water. Heat to complete solution of soluble salts.

Filter through a quantitative paper, wash thoroughly, dry and ignite in a small platinum crucible. Weigh, moisten the residue with water, treat with a drop of H_2SO_4, and then with an excess of HF. Take down to dryness, ignite, cool, weigh again. The loss in weight is SiO_2, which multiplied by $0.4676 =$ Silicon.

ANALYSIS OF BAUXITE (R. Z. Method)

Methods used by the Aluminum Company of America, Technical Direction Bureau, Standard Methods of Analysis

Determination of Moisture

If the bauxite has not been previously dried, the moisture is determined at 105° C.

Determination of Loss on Ignition

Place one gram of dried sample in a weighed crucible and cover with platinum lid. Heat slowly to a red heat for one hour. Remove cover and heat for one hour at the hottest temperature obtainable on a blast lamp. Cool and weigh. The loss in weight represents combined water and organic matter.

Determination of SiO_2

Place in a 250-cc. casserole one gram of dried sample and moisten with 5 cc. of water. Add 90 cc. of acid mixture (900 cc. water, 300 cc. H_2SO_4, 600 cc. HCl and 200 cc. HNO_3) and 10 cc. concentrated H_2SO_4. Cover with watch glass and boil until H_2SO_4 fumes are given off copiously. Cool, and dilute carefully to 150 cc. volume, washing cover glass and sides of dish. Replace cover glass and boil the solution carefully until the solution has a clear appearance. Filter the solution through an ashless filter into a 250-cc. flask and wash the residue with hot water until free from acid. Ignite the filter in a platinum crucible, cool in desiccator and weigh. Moisten the residue in crucible with two or three drops of water, 3 drops conc. H_2SO_4, and 2–3 cc. HF. Evaporate slowly to dryness to avoid spattering, ignite, cool and weigh. The difference between the two weights obtained gives the silica content of sample.

Determination of Fe_2O_3

If any appreciable residue remains in the crucible, it is brought into solution by fusing it with a little potassium bisulphate and dissolving the resultant fusion in very dilute sulphuric acid. This solution is then added to the original filtrate, the whole cooled to room temperature and made up to a volume of 250 cc.

25 cc. of this solution is then transferred with a pipette to a 100-cc. Nessler comparison tube for use in the determination of titanium.

The remainder of the solution is transferred to a 400-cc. beaker and evaporated to approximately 100 cc., 20 cc. of conc. HCl are then added and the solution brought to boiling. The iron is reduced by adding drop by drop to the boiling hot solution a concentrated solution of stannous chloride until the solution is colorless. The end point is sharp and not more than two drops of stannous chloride should be added in excess. Cool the solution to the room temperature and add while stirring 5 cc. of a saturated solution of mercuric chloride. Stir solution for about thirty seconds. The iron must be titrated either with bichromate solution, using potassium ferricyanide as an external indicator, or with permanganate solution according to the Reinhart-Zimmerman method which is as follows:

Place 50 cc. of a preventative solution in a wide neck liter flask containing 350 cc. water. Pour the reduced iron solution into the flask and while whirling the flask, titrate with permanganate solution to the appearance of a pink tinge. The end point is sharp, but only lasts for a moment.

The preventative solution contains:

> 1750 cc. water,
> 160 grams manganous sulfate,
> 330 cc. 85% phosphoric acid,
> 320 cc. H_2SO_4 sp.gr. 1.84.

Determination of TiO_2

To the 25-cc. solution in the Nessler comparison tube add 20 cc. 25% H_2SO_4 and 3 cc. 3% hydrogen peroxide, fill to the mark with water and mix. In a similar tube place 20 cc. 25% H_2SO_4, 3 cc. hydrogen peroxide and water to slightly below the mark. To this tube add standard titanic acid solution from a burette, mixing and comparing after each addition, and continue till its color matches that of the tube containing sample. From the amount of the standard used the titanium content of the sample may be calculated, remembering that the sample taken for the titanium determination represents one-tenth of the entire sample.

If a colorimeter is at hand, it may be used in place of the second Nessler tube for comparison.

Determination of Al_2O_3

The alumina is obtained by the difference between the sum of the silica, titanium oxide, iron oxide and loss on ignition calculated as percentages and 100%.

When it is required to determine whether a material is bauxite or not, some modifications are necessary. If the unvolatilized residue from silica determination amounts to more than a few milligrams, it should be fused with sodium carbonate and examined for additional silica. Absence of any considerable amount of phosphate in the sample should be established. Also the alumina should be determined in a portion of the filtrate from silica by precipitation with ammonia, deducting from the ignited oxides the iron and titanium oxides which have been directly determined.

ANALYSIS OF ALUMINUM HYDRATE

Determination of Loss on Ignition

A one-gram sample is weighed into a platinum crucible and placed over a low flame of the bunsen burner. The temperature is gradually raised until the crucible is a bright red, when it is transferred to a muffle furnace at 1100° C. for one hour. Place in desiccator, cool and weigh. The loss in weight equals the per cent loss on ignition. This value of loss on ignition is the total of free and combined water.

NOTE. A great deal of care is necessary when starting to heat the hydrate. The temperature should be slowly increased because if the hydrate is heated too rapidly, the high rate at which the moisture is driven off will force the particles of hydrate out of the crucible.

Determination of SiO_2

A ten-gram sample is weighed into a casserole. Enough H_2O is added to make a thin paste, then 25 cc. conc. H_2SO_4 are quickly added and casserole covered. Casserole is placed on the hot plate and taken down to considerable fuming. Allow contents to cool and dilute with 150 cc. H_2O, boil until clear. Filter and wash with hot water. Place the filter paper in a platinum crucible and ignite. Cool and weigh. Add a few drops of conc. H_2SO_4 and 1 cc. of HF. Heat on hot plate until SO_3 fumes have disappeared. Then ignite for 15 minutes; cool and weigh. Difference in weights $\times 10 =$ per cent SiO_2.

NOTE. Considerable attention is required in placing the hydrate in a solution of H_2SO_4. When the mixture of hydrate and H_2SO_4 is heated slightly, violent action takes place which is caused by the reaction of hydrate of alumina and H_2SO_4 to form aluminum sulphate. After this reaction, continue to heat for one minute and cool. On taking up with water it seems to go in solution quite readily.

Determination of Fe_2O_3

To the filtrate from the silica determination add 5 cc. of conc. HCl; cool to room temperature. Make up a solution of 0.5 gram sample of cupferron in 25 cc. H_2O. Filter and add filtrate slowly to the acidified filtrate from the silica determination, stirring well. Set aside to complete precipitation for 30 minutes. Filter precipitate with gentle suction and be sure the filtrate is crystal clear. Wash 1 : 1 HCl solution; ignite, precipitate and weigh.

Per cent $Fe_2O_3 = 10 \times$ weight of ppt.

NOTE. It is important that the solution be thoroughly cooled before adding the cupferron. Also in making the cupferron solution cool water is necessary. After adding the cupferron, place the solution in a cool place such as a vessel containing circulating tap water for at least 30 minutes to complete the precipitation.

Determination of Na₂O

Weigh a 10-gram sample into a platinum dish and add 30 cc. conc. HCl. Take to dryness on a steam bath or in a hot oven 110° C. Remove to muffle furnace at 500° C. to burn off free HCl. Cool and add 50 cc. hot water; boil ten minutes and add 25 cc. saturated ammonium carbonate solution; allow to stand for one hour or longer on hot plate but not boiling; filter and wash residue with hot water. Transfer filtrate to platinum dish, cover with watch glass and evaporate to dryness. Place in muffle at 500° C. to drive off NH_3 salts. Remove, cool, and add 5 cc. hot water. Filter contents of dish, catching filtrate in a 250-cc. beaker, wash to a volume of 25 cc. of filtrate; allow to cool and add a crystal of potassium chromate and titrate with N/20 silver nitrate.

Cc. N/20 $AgNO_3$ ×value of $AgNO_3$ solution in grams Na_2O per cc.

×10= per cent Na_2O in hydrate.

NOTE. When taking the hydrate sample down to dryness with HCl, caution has to be taken so as not to heat it at too high a temperature. The use of a muffle furnace at 500° C. will prevent this. The same precaution is necessary when driving off the NH_3 salts. Too high a temperature will volatilize some of the NaCl. Always have a solution cool before titrating with N/20 silver nitrate.

Determination of Insoluble

Weigh a 10-gram sample into a casserole and add 50 cc. H_2SO_4 50° Be. and boil until solution is clear. Be careful not to boil down until fumes come off. Filter and place filter paper in platinum crucible and ignite for one hour in muffle furnace at 1100° C. Cool and weigh. This weight less weight of crucible×10= per cent of insoluble.

ANALYSIS OF CALCINED ORE

Bisulphate Fusion Methods

Preparation of Sample

The portion of ore collected is well mixed and cut down to the amount which it is wished to reserve for the determination of impurities, ordinarily about 4 oz. A separate part should be kept for mesh test, and may be used also for loss on ignition. The 4-oz. lot is quickly sifted through a 60-mesh sieve, the part retained ground till it passes the sieve and the whole thoroughly remixed. In the above grinding it is well to cover the coarse particles with a little of the sifted ore to prevent flying out and also to rinse the mortar with some of this fine part. No further grinding is to be done on the sample as there is danger of disproportionate contamination.

Determination of SiO_2

" A " (General Laboratory). Twenty-two grams of bisulfate are considered ample to make the fusion by this procedure. Weigh two-gram sample into 50 cc. platinum crucible, add about 17 grams of bisulfate and carefully fuse until violent action is over. Cool the fusion, running it up the sides of the crucible. Add the remaining 5 grams of bisulfate to be used and continue fusion until decomposition of the ore is complete.

Using procedure " A " continue from this point by cooling the crucible with air blast. Place the fused cake, which is easily removed, in a 250-cc. casserole and add 25 cc. concentrated sulfuric acid. Heat gently until the fused cake is completely broken up and copious fumes are evolved. Cool and take up with about 200 cc. of hot water. Heat gently to complete solution of soluble salts. Filter off the SiO_2, using approved quantitative filter paper, and wash with a small amount of hot water. Set aside the strong filtrate and wash the SiO_2 thoroughly until washings give no test for sulfate with barium chloride.

Ignite the filter paper in a platinum crucible at approximately 950° C. for one hour. Cool and weigh accurately on an approved balance. Treat with HF, a few drops of H_2SO_4, and evaporate to dryness. Ignite and weigh. Loss in weight is SiO_2. Corrections must be made for SiO_2 derived from reagents or apparatus.

$$\text{Per cent } SiO_2 = \frac{\text{Loss in Weight} \times 100}{\text{Weight of Sample (2 grams)}}.$$

Determinations of Fe_2O_3 (Cupferron Method)

Receive the filtrate from SiO_2 determination in a 600-cc. beaker and bring to a volume of 250 cc. to 300 cc. The solution must have sufficient volume to prevent any crystallization of salts upon cooling. Add 10 cc. concentrated HCl. To the cold solution add drop-wise—at the same time stirring vigorously —30 cc. of a 1% water solution of "Cupferron." The cloudy brown solution is filtered. Wash with 20 per cent cold HCl until the paper is free from soluble salts, then wash with cold water. Ignite the paper in platinum crucible, cool and weigh. All results must be corrected by running blank.

$$\text{Per cent } Fe_2O_3 = \frac{\text{Weight of Residue} \times 100}{\text{Weight of Sample}}.$$

Determination of Na_2O

Weigh 5 grams from sample which has been pulverized in an agate mortar into a 250-cc. casserole. Add 20 cc. of concentrated HCl and evaporate on hot plate to dryness. Care should be taken at this point not to heat the residue too high so that a minimum amount of cake will be formed. Take up with hot water and digest until all soluble salts are in solution. This operation should be assisted by thoroughly breaking up any cake that has formed. Add 50 cc. of saturated $(NH_4)_2CO_3$. Let stand for half an hour and filter with suction. Transfer the filtrate to casserole, evaporate to dryness and heat until all white fumes of $(NH_4)_2CO_3$ are driven off (500° C. to 600° C.). Cool, take up with hot water, add 5 cc. of saturated solution of $(NH_4)_2CO_3$, filter and wash into a platinum dish of 125–150 cc. capacity. Evaporate to dryness and ignite at 500° C. until all ammonium salts are driven off. Cool. Dissolve in a little water, filter and wash into a 200-cc. Erlenmeyer flask. Bring volume of solution to 150 cc. Add 6 to 8 drops of saturated solution of K_2CrO_4 and titrate with standard silver nitrate solution.

Determination of Loss on Ignition

Prepare a 10 or 15 cc. platinum crucible provided with a close-fitting cover by igniting, cooling in desiccator and weighing. Weigh into it one gram of ore. Place in an electric furnace heated to 1050–1100° C. or in a flame giving that temperature for 45 minutes. Transfer to a sulphuric acid desiccator and as soon as cool weigh quickly. The loss in weight $\times 100 =$ per cent loss on ignition.

Estimating Metallic Aluminum in Aluminum Dust[1]

The value of aluminum dust for technical purposes depends chiefly upon the amount of metallic aluminum contained. A rapid method of estimating this constituent is therefore highly desirable. A modification of Wahl's method for the determination of metallic zinc by the reduction of ferric sulphate in cold neutral solution has been found to be rapid and simple. It does not require special skill in manipulation and gives results that are of value for comparative purposes at least.

The reduction of ferric sulphate by metallic aluminum does not take place in cold neutral aqueous solution. If, however, sufficient free sulphuric acid be added and the liquid be heated, ferrous sulphate is formed proportionally to the amount of metallic aluminum present.

The results obtained with samples of known composition were usually about 5% lower than those indicated by the supposed reaction

$$3Fe_2(SO_4)_3 + 2\,Al = Al_2(SO_4)_3 + 6FeSO_4, \tag{1}$$

so that for accurate results it is necessary to standardize on a sample of pure metallic aluminum or on a sample of which the aluminum contents have been exactly determined by some other method. From equation (1) above, it follows that 1 part Al = 6.182 parts Fe reduced from ferric to ferrous iron, or 1 part Fe = 0.16177 parts Al, assuming Al = 27.1 and Fe = 55.84.

The test as generally made is as follows: 100 mg. of the aluminum dust is accurately weighed and placed in a conical flask provided with stopper and Bunsen valve to reduce risk of oxidation, together with 50 cc. of a standard solution of acid ferric sulphate[2] and the mixture heated gradually until the liquid is boiling gently. If the aluminum dissolves completely with moderate heat, it is not essential to boil. The flask is then cooled to room temperature by placing under a running tap, and 50 cc. of cold distilled water is added. The liquid is then titrated with standard permanganate.

A convenient standard solution contains 3.35 grams $KMnO_4$ per liter of this solution; 1 cc. = 1% Al, approximately.

With samples of ordinary aluminum dust the reaction takes place without visible evolution of hydrogen. When the powder contains coarse flakes or granules, some effervescence can be observed. In such cases it would seem that the titration must indicate a result too low, as part of the Al has been dissolved according to the reaction

$$3H_2SO_4 + 2Al = Al_2(SO_4)_3 + 6H, \tag{2}$$

without reducing its equivalent of ferric sulphate.

Another source of error, which acts in the opposite direction, is the presence of iron, zinc or other metal capable of reacting with ferric sulphate. For accurate work, therefore, these metals must be determined and their effect allowed for. When, however, comparative results only are required, as in checking the quality of factory products from day to day, the rapid test described is sufficient. For some purposes also metallic iron, zinc, etc., would be as effective as their equivalent of aluminum, in which case no correction ought to be applied. The reactions in the case of iron and zinc are presumably:

$$Fe + Fe_2(SO_4)_3 = 3FeSO_4, \tag{3}$$
$$Zn + Fe_2(SO_4)_3 = 2FeSO_4 + ZnSO_4. \tag{4}$$

Assuming Al = 27.1; Fe = 55.84; Zn = 65.37, 1 part metallic Fe has the same reducing power as 0.4853 part metallic Al, and 1 part metallic Zn has the same reducing power as 0.2763 part metallic Al.

In making the ordinary tests for controlling plant operations, a sample of the purest obtainable aluminum dust is accepted as a standard. Tests are then made concurrently with the standard dust and with the sample to be examined, the result of the latter being expressed as a percentage of the standard.

[1] By J. E. Clennell, Eng. and Mining Journal, May 6, 1916.
[2] To 100 grams of pure ferric sulphate distilled water is added, then 250 cc. of concentrated sulphuric acid. This mixture is heated to boiling until the ferric sulphate has completely dissolved, then cooled and finally made up to a liter with distilled water.

The revised atomic weights of Al and Zn changes slightly the ratios above.—Editor.

ANTIMONY

Sb$_2$ *at.wt.* 121.77; *sp.gr.* 6.62[1]; *m.p.* 630°C[2]; *b.p.* 1440°C[1]; *oxides,* Sb$_2$O$_3$, Sb$_2$O$_4$, Sb$_2$O$_5$.

DETECTION

Hydrogen Sulphide precipitates the orange-colored sulphide of antimony from fairly strong hydrochloric acid solutions (1 : 4) in which several members of the group remain dissolved. Arsenic is also precipitated. The latter may be removed by boiling the solution containing the trichloride, AsCl$_3$ being volatile.

If antimony is already present as a sulphide, together with other elements of the hydrogen sulphide group, it may be dissolved out by treating the precipitate with sodium hydroxide, potassium hydroxide, sodium sulphide, ammonium polysulphide in solution. Antimony sulphide is reprecipitated upon acidifying the filtrate. Arsenic and tin will also be precipitated with antimony if they are present in the original precipitate. Should a separation be necessary, the precipitate is dissolved with hot concentrated hydrochloric acid, with the addition of crystals of potassium chlorate, from time to time, until the sulphides dissolve. The solution is placed in a Marsh apparatus, pure zinc added and the evolved gases passed into a neutral solution of silver nitrate. The black precipitate of silver antimonide and metallic silver are filtered off, washed free of arsenous acid, and the antimonide dissolved in strong hydrochloric acid (silver remains insoluble). The orange-colored antimony sulphide may now be precipitated by diluting the solution with water and passing in H$_2$S gas to saturation.

Minerals which contain antimony, when heated alone or with 3 to 4 parts of fusion mixture (K$_2$CO$_3$ and Na$_2$CO$_3$), on charcoal, yield dense white fumes, a portion of the oxide remaining as a white incrustation on the charcoal. A drop of ammonium sulphide placed upon this sublimate gives a deep orange stain.

Hydrolysis. Most of the inorganic antimony salts are decomposed by water, forming insoluble basic salts, which in turn break down to the oxide of antimony and free acid. An excess of tartaric acid prevents this precipitation.

Traces of Antimony. Nascent hydrogen liberated by the action of zinc and hydrochloric or sulphuric acid reacts upon antimony compounds with the formation of stibine. This gas produces a black stain on mercuric chloride or silver nitrate paper. Details of the procedure are given under the quantitative method for determining minute amounts of antimony.

Distinction between Antimonous and Antimonic Salts.

Chromates form with antimonous salts green chromic salts and antimonic salts.

Potassium Iodide reduces antimonic salts, free iodine being liberated.

[1] Van Nostrand's Chem. Annual, Olsen, 3d Ed.
[2] Cir. 35, U. S. Bureau of Standards.

Contributed by Wilfred W. Scott.

ESTIMATION

The determination of antimony is required in the evaluation of antimony ores—stibnite, Sb_2S_3; valentinite, Sb_2O_3, etc. It is generally required in the complete analysis of minerals of nickel, lead, copper, silver, in which antimony generally occurs as a sulphide. The determination is required in the analysis of Britannia metal, bearing and antifriction metals, type metal and hard lead; in the analysis of certain mordants, antimony salts, vulcanized rubber, etc. It is looked for as an undesirable impurity in certain food products.

Preparation and Solution of the Sample

In dissolving the substance containing antimony it must be remembered that metallic antimony is practically insoluble in cold dilute hydrochloric, nitric or sulphuric acid and the oxides, Sb_2O_3 or Sb_2O_5, are precipitated in strong nitric acid. The element, however, is readily soluble in hydrochloric acid containing an oxidizing agent, such as nitric acid, potassium chlorate, chlorine, bromine, etc. The oxides of antimony are soluble in hydrochloric acid and the caustic alkalies.

Solution of Sulphide Ores, Low-grade Oxides, etc.[1]

0.5 to 1 gram of the finely ground ore, placed in a Kjeldahl flask, is mixed with 5 to 7 grams of granular or powdered potassium sulphate, and 10 cc. of strong sulphuric acid. About 0.5 gram of tartaric acid, or a piece of filter paper, is added to reduce arsenic and antimony and the mixture heated, gradually at first, and then with the full Bunsen flame. The heating is continued until the carbon is completely oxidized and most of the free acid driven off, leaving a clean fusion but not to complete expulsion of H_2SO_4. The melt is now cooled over the bottom and sides of the flask by gently rotating during the cooling.

About 50 cc. of dilute hydrochloric acid (1 : 1) are added and the melt dissolved by warming gently. The contents of the Kjeldahl flask are transferred to an Erlenmeyer flask, the Kjeldahl being rinsed out with 25 cc. of strong hydrochloric acid. Arsenic sulphide may now be precipitated with H_2S from the strongly acid solution, whereas antimony, etc., remain in solution. The sulphide is filtered off through a double filter, that has been moistened with hydrochloric acid (2 : 1), a platinum cone supporting the filter to prevent its breaking. The flask is rinsed out with hydrochloric acid (2 : 1). The precipitate is washed at least six times with the acid. Antimony passes into the filtrate together with other elements of the ore.

The filtrate is diluted with double its volume of warm water and then is saturated with hydrogen sulphide. Antimony sulphide, together with other elements of the Hydrogen Sulphide Group, will precipitate. These are washed with hydrogen sulphide water. Antimony sulphide may now be dissolved by addition of sodium sulphide and caustic solution (separation from Cu, Pb, Cd, Bi, etc.) (5 to 10 cc. of a mix of 60 grams Na_2S with 40 grams of NaOH diluted to 1000 cc.).

5

[1] Method of A. H. Low modified.

The solution containing the antimony is treated with about 2 grams of potassium sulphate and 10 cc. of strong sulphuric acid and heated as before, to destroy liberated sulphur and expel most of the free acid. The melt is dissolved in hydrochloric acid, and the antimony titrated according to one of the volumetric procedures given under "Volumetric Methods."

NOTE. An insoluble residue remaining from the acid extraction of the first melt may be dissolved by fusion with sodium hydroxide and extraction of the melt with hot water. If a precipitate forms when this alkaline solution is acidified with hydrochloric acid, the presence of barium sulphate is indicated.

Decomposition of the Ores by Fusion with Sodium Hydroxide.
Oxides. 0.5 to 1 gram of the powdered ore is mixed with about 10 grams of sodium hydroxide and placed in a thin-walled iron crucible of 60 cc. capacity. It is advisable to fuse a portion of the alkali hydroxide in the crucible with a pinch of potassium nitrate and then add the ore mixed with the remainder of the sodium hydroxide. The covered crucible is heated until the fusion becomes homogeneous. The melt is poured out on a large nickel crucible cover or shallow dish. On cooling, the cake is detached and placed in a casserole containing water, any adhering cake on the cover, or melt remaining in the iron crucible, being dissolved with dilute hydrochloric acid and added to the sample in the casserole. About 30 to 40 cc. of strong hydrochloric acid are now added and the mixture heated (casserole covered) until the melt has dissolved. Two to 3 grams of tartaric acid having been added to keep antimony dissolved, the solution is diluted to about 300 cc., and antimony is then precipitated as the sulphide with hydrogen sulphide. The treatment of the precipitate at this stage has been given in the "Solution of Sulphide Ores."

Sulphides. Howard and Harrison [1] recommend the following procedure for fusion of sulphide ores with caustic: 0.5 gram of the powdered ore is fused with a mixture of 8 grams of sodium carbonate and sodium peroxide, 1 : 1, in a nickel crucible. The cooled melt is dissolved with sufficient hydrochloric acid to neutralize the alkali and about 15 cc. of strong acid added in excess. The solution is diluted to 250 cc., antimony being kept in solution by addition of potassium chlorate. An aliquot portion of the solution is taken, antimony reduced by metabisulphite and titrated with iodine.

Treatment of Speisses, Slags, Mattes, etc.[2] 0.5 to 2 grams of the sample is treated with 10 to 15 cc. of strong nitric acid and the mixture taken to dryness. Fifteen cc. of strong hydrochloric acid are added and the sample transferred to a 350-cc. flask, additional hydrochloric acid being used to wash out the beaker. Arsenic is precipitated from the strong acid solution as the sulphide, and antimony determined in the filtrate.

Solution of Alloys. Alloys are generally decomposed by treatment with mixtures of hydrochloric acid together with an oxidizing agent—nitric acid, potassium chlorate, bromine, etc. The subject is taken up in detail in the chapter on alloys.

The alloy drillings are treated with strong hydrochloric acid, a little bromine added, and the mixture heated until the alloy dissolves, additional bromine being added from time to time if necessary. The excess bromine is removed by heating gently to boiling. The higher oxides are reduced by addition of

---

[1] Phar. Jour., 1909, **83**, 147.
[2] H. E. Hooper's method.

sodium metabisulphite and the sulphides precipitated, as usual, with hydrogen sulphide. Arsenic may now be volatilized by boiling, and antimony titrated with iodine or potassium bromate.

Alloys of Antimony, Lead and Tin. 0.5 to 1 gram of the finely divided alloy is warmed with 100 cc. of strong hydrochloric acid until the action subsides. Solid iodine is now added, in small quantities at a time, until the alloy completely dissolves. The excess of iodine is now removed by boiling and the small amount of free iodine remaining neutralized with a few drops of a weak solution of sodium thiosulphate. Although tin is oxidized to the higher state, antimony is not oxidized by iodine in acid solution beyond the trivalent form. The solution may now be titrated with standard iodine in presence of an excess of sodium bicarbonate according to the procedure given under the volumetric methods.

Hard Lead. The method of solution and titration are given under "Potassium Bromate Method for Determining Antimony."

Antimony in Rubber Goods.[1] Three grams of the finely rasped rubber are treated in a Kjeldahl flask with 40 to 45 cc. of strong sulphuric acid. A small quantity of mercury or mercury salt is added, together with a small piece of paraffine wax. The mixture is heated until the rubber is dissolved and the black liquid begins to clear. Two to 4 grams of potassium sulphate are then added and the heating continued until a colorless or pale yellow liquid is obtained. After cooling, 1 to 2 grams of potassium metabisulphite are added and an excess of tartaric acid. The liquid is diluted sufficiently to prevent the charring of the tartaric acid and boiled until the odor of sulphurous acid has disappeared. A few cc. of dilute hydrochloric acid are added, the liquid diluted to 200 cc., filtered through a dry filter, and 195 cc. titrated either with iodine or with potassium bromate (the latter in acid solution), as described under the volumetric procedures.

[1] W. Schmitz, Chem. Zentralbl., 1911, ii, 1710. Analyst, 1912, p. 64.

SEPARATIONS

Separation of Antimony (together with Members of the Hydrogen Sulphide Group), from Iron, Chromium, Aluminum, Cobalt, Nickel, Manganese, Zinc, the Alkaline Earths, and Alkalies. The acid solution of the elements is saturated with hydrogen sulphide, the elements of the Hydrogen Sulphide Group are precipitated as sulphides, the other elements remaining in solution. Antimony sulphide may be precipitated from an hydrochloric acid solution containing 15 cc. of strong acid per 100 cc. of solution; lead and cadmium are incompletely precipitated.

Separation of Antimony (together with Arsenic and Tin), from Mercury, Copper, Bismuth, Cadmium and Lead. The sulphides of antimony, arsenic, and tin are soluble in a mixture of sodium hydroxide and sodium sulphide, the soluble sulpho salts being formed, mercury, copper, bismuth, cadmium, and lead remaining as insoluble sulphides. The following procedure may be used for alloys free from members of other groups. The acid solution is treated with 3 to 5 grams of tartaric acid and diluted slightly (more tartaric acid being added if the solution becomes turbid), then poured into 300 cc. of a mixture of sodium sulphide and sodium hydroxide (150 cc. of the mix described under "Solution of Sulphide Ores" diluted to 300 cc.). The mixture is warmed and the insoluble sulphides allowed to settle out. The solution is filtered free of the precipitate and the latter washed. The filtrate is acidified with hydrochloric or sulphuric acid and saturated with hydrogen sulphide. The sulphides of arsenic, antimony and tin are now filtered off and treated as described later.

Separation of Arsenic, Antimony, and Tin. The sulphides may be dissolved in concentrated hydrochloric acid by addition of potassium chlorate to oxidize the sulphur to sulphuric acid. This oxidation may be effected in the alkaline solution of the sulpho salts by addition of 30% hydrogen peroxide in small portions until the yellow solution is completely decolorized and then 1 to 2 cc. in excess, the solution then boiled to completely oxidize the sulphides to sulphates and to remove the excess of peroxide. The solution is then acidified, the precipitation of the sulphides and the subsequent filtration and resolution being avoided.

Removal of Arsenic. This may be accomplished by volatilizing arsenic as arsenic trichloride in a strong hydrochloric solution by boiling. If arsenic is to be determined the procedure given under the chapter on arsenic is followed, the arsenic being distilled in a current of hydrochloric acid gas. If arsenic is not desired it may be expelled by reducing the solution with sodium metabisulphite or potassium iodide and boiling. Antimony and tin remain in the concentrated acid solution.

The separation of arsenic from antimony and tin may be effected by removal of the former in a strong hydrochloric acid solution as described under the section "Preparation and Solution of the Sample," arsenic being precipitated by hydrogen sulphide, whereas antimony and tin remain in solution.

Separation of Antimony from Tin. Upon the removal of arsenic, antimony may be determined directly in the presence of tin by one of the volumetric methods given later. If a gravimetric separation is desired, it may be made according to a modification of Clark's method, [1] which depends upon the fact that antimony is completely precipitated from a solution containing oxalic acid, by hydrogen sulphide, whereas tin is not. The tin must be in the stannic form, otherwise the insoluble crystalline stannous oxalate will form.

If the mixture is acid, it is neutralized with caustic and twenty times the weight of the Sn and Sb present added in excess, e.g., 2 grams potassium hydroxide in excess for every 0.1 gram of tin and antimony present in the solution. About ten times as much of tartaric acid is now added as the maximum weight of the two metals, followed by 30% hydrogen peroxide to oxidize the tin. The excess of peroxide is removed by boiling. To the slightly cooled solution a hot solution of pure oxalic acid is added, 5 grams of oxalic acid for each 0.1 gram of the mixed elements. CO_2+O_2 are evolved. The solution is boiled for about ten minutes and the volume made up to about 100 cc. Hydrogen sulphide is rapidly passed into the boiling solution until a change from a white turbidity to an orange color takes place and antimony begins to precipitate. The passage of the gas is continued for fifteen minutes, the solution diluted with hot water to a volume of 250 cc. and hydrogen sulphide passed into the boiling solution for another fifteen minutes. The flame is now removed and the H_2S " gasing " continued for ten minutes longer. The precipitated antimony pentasulphide is filtered off in a weighed Gooch crucible. It may be determined gravimetrically as Sb_2S_3, according to the procedure given later, by washing with 1% oxalic acid and dilute acetic acid, by decantation, the solutions being hot and saturated with hydrogen sulphide. The precipitate washed into the crucible is dried in a current of CO_2 at a heat of 280 to 300° and weighed as Sb_2S_3.

Tin may be determined electrolytically in the filtrate evaporated to about 150 cc., the oxalic acid being nearly neutralized with ammonia. See Electrolytic Determination of Tin.

Antimony may be separated from tin in a hot hydrochloric acid solution by addition of pure iron. The iron and tin sulphides are dissolved in concentrated hydrochloric acid plus a few crystals of potassium chlorate. The solution should contain about 10% hydrochloric acid, more hydrochloric acid being added as the iron dissolves. Antimony is precipitated as a metal.

[1] The Original procedure may be found in Chem. News, Vol. XXI, p. 124.

GRAVIMETRIC METHODS FOR THE DETERMINATION OF ANTIMONY

The accuracy and rapidity of volumetric methods for the determination of antimony leave little to be desired in the estimation of this element, so that the more tedious gravimetric methods are less frequently used. The following procedures are given in view of possible utility in certain analyses.

Determination of Antimony as the Trisulphide, Sb_2S_3 [1]

Although hydrogen sulphide passed into a cold solution tends to precipitate Sb_2S_5, in hot strongly acid solutions, the lower sulphide, Sb_2S_3, tends to form. The higher sulphide is decomposed at 230° C. with formation of Sb_2S_3 and the volatilization of sulphur. A temperature[2] of 280 to 300° is even more favorable for this transformation. The method takes advantage of these conditions for formation of antimony trisulphide, in which form it is weighed.

Procedure. The solution of antimony, free from arsenic, is treated in an Erlenmeyer flask with strong hydrochloric acid until the solution contains about 20% of the concentrated acid. The mixture is heated to boiling and a slow current of hydrogen sulphide is passed into the hot solution until the precipitate passes from a yellow color through an orange and finally becomes a dark red to black color. The flask is agitated gently to coagulate the precipitate, which settles in a crystalline form. The solution is diluted with an equal volume of water, washing down the walls of the flask. A slight turbidity is generally seen, due to precipitation of a small amount of antimony that remains in solution in a strong acid solution. H_2S is now passed into the diluted solution until it becomes clear, thirty-five to forty minutes are usually sufficient to precipitate all of the antimony. The precipitate is transferred to a weighed Gooch crucible, washed with small portions of water containing hydrogen sulphide, and finally with pure water.

It is a common practice, at this juncture, to wash the precipitate with carbon disulphide or carbon tetrachloride to remove precipitated sulphur. Alcohol is now used, followed by ether, and the precipitate sucked dry.

The Gooch crucible is placed in a large combustion tube and heated in a current of dry, pure CO_2 at 130° C. for an hour. The temperature is now raised to 280 to 300° C. and the heating continued for two hours. The residue will consist of pure Sb_2S_3.

$$Sb_2S_3 \times 0.7142 = Sb, \text{ or } Sb_2S_3 \times 0.8568 = Sb_2O_3.$$

NOTES. Antimony may be determined by oxidation of the sulphide precipitate by means of fuming nitric acid. The mixture evaporated to dryness is ignited and the residue weighed as Sb_2O_4. The temperature of the ignition should be between 750 to 800° C. The volatile trioxide forms at a little above 950°. The procedure requires greater care than the sulphide method and possesses no advantages.

Pure carbon dioxide may be obtained from limestone placed in a Kipp generator. The gas is dried by passing it through strong sulphuric acid. It should be free from oxygen of the air. It is advisable to sweep out the air from the generator before attaching it to the combustion train. The air in the tube is swept out with carbon dioxide before heating the sample.

Property of Sb_2S_3, *m.w.*, 336.61; *sp.gr.*, 4.65; fusible and volatile; solubility, 0.000175 gram per 100 cc. H_2O; decomposed by hot H_2O; soluble in alkalies, NH_4HS, K_2S, conc. HCl.

[1] Method of Vortmann and Metzel modified.
[2] Paul, Z. anal. Chem., **31**, 540 (1892).

Electrolytic Determination of Antimony [1]

The chief condition for the success of the electrolytic deposition of antimony in metallic form is the absence of polysulphides, since these substances prevent the element from being deposited, $2Sb + 3Na_2S_2 = 2Na_3SbS_3$. The formation of polysulphides may be prevented during electrolysis by addition of potassium cyanide to the solution, $Na_2S_2 + KCN = Na_2S + KCNS$.

The results of this method, according to F. Henz, are invariably 1.5 to 2% too high of the total antimony present in the solution. The sample for analysis should contain not over 0.2 gram antimony.

Procedure. Antimony precipitated as the sulphide is washed and then dissolved off the filter by pouring pure sodium sulphide solution (sp.gr. 1.14) over the precipitate, the solution being caught in a weighed platinum dish, with unpolished inner surface. The total volume of the solution should be not over 80 cc. (if less than this, additional Na_2S solution is added to make up to 80 cc.). Sixty cc. of water followed by 2 to 3 grams of potassium cyanide (C.P.) are added and the cyanide dissolved by stirring with the rotating anode. The solution heated to 60 to 70° is electrolyzed with a current of 1 to 1.5 amperes, E.M.F. = 2 to 3 volts. Two hours are generally sufficient to deposit all the antimony. The light-gray deposit adheres firmly upon the cathode. Without breaking the current the solution is siphoned off, while fresh water is being added, until the current ceases to flow through the liquid. The cathode is washed thoroughly with water, followed by alcohol and ether and then dried at about 80°, cooled in a desiccator and weighed.

The antimony deposits may be removed by heating with a solution of alkali polysulphide or by a mixture of equal parts of saturated solution of tartaric acid and nitric acid.

[1] Method first proposed by Parrodi and Mascazzini, Z. anal. Chem., **18**, 587 (1879), modified by Luckow, Z. anal. Chem., **19**, 13 (1880), and later improved by Classen and Reiss, Berichte, **14**, 1629 (1881); **17**, 2474 (1884); **18**, 408 (1885); **27**, 2074 (1894). F. Henz, Z. anorg. Chem., **37**, 31 (1903).

VOLUMETRIC METHODS
Potassium Bromate Method for Determining Antimony [1]

Outline. This method is of special value in determining antimony in hard lead and alloys. It was first suggested by Györy and later modified by Siedler, Nissensen and Rowell.[2] The process is based upon the oxidation of antimony from the trivalent to the pentavalent form by potassium bromate, the following reaction taking place:

$$KBrO_3 + 3SbCl_3 + 6HCl = 3SbCl_5 + KBr + 3H_2O.$$

Standard Solutions.

Antimony Chloride Solution. Six grams of the C. P. pulverized metal are dissolved in 500 cc. of concentrated hydrochloric acid together with 100 cc. saturated bromine solution, more acid and bromine added if necessary to effect solution. After expelling the bromine by boiling, about 200 cc. concentrated hydrochloric acid are added and the whole made up to one liter. Fifty cc. =0.3 gram antimony.

N/10 Potassium Bromate Solution. 2.82 grams of C. P. salt are dissolved in water and made up to 1 liter. Theoretically 2.7852 grams are required, but the salt invariably contains potassium bromide as an impurity. The solution is standardized against 50 cc. of the antimony chloride solution, which has been reduced with sodium sulphite according to the standard scheme. One cc. of N/10 KBrO₃ =0.006 gram Sb.

Methyl Orange. 0.1 gram M. O. per 100 cc. of distilled water. The indicator should be free from sediment.

Saturated Bromine Solution. 500 cc. concentrated hydrochloric acid saturated with 70 cc. of bromine.

Procedure. *Solution.* One gram of the finely divided alloy is brushed into a 500-cc. beaker, 100 cc. of concentrated hydrochloric acid and 20 cc. of saturated bromine solution are added. The beaker is covered and placed on the steam bath until the metal dissolves. It may be necessary to add more bromine and acid to effect complete solution. In case the oxides of antimony and tin separate out and do not redissolve, fusion with sodium hydroxide may be necessary. Bromine is now expelled by boiling the solution down to about 40 cc.

Reduction. One hundred cc. of concentrated hydrochloric acid and 10 cc. of a fresh saturated solution of Na_2SO_3 are added and the solution boiled down to 40 cc., on a sand bath, to expel arsenic and the excess of normal sodium sulphite. Samples high in arsenic may require a repetition of the reduction.

Titration. The cover and sides of the beaker are rinsed down with 20 cc. of hydrochloric acid (sp.gr. 1.2) followed by a few cc. of hot water and the solution heated to boiling on a sand bath. The standard bromate solution is now run into the hot solution of antimony to within 2 to 3 cc. of the end-

---

[1] S. Györy, Zeit. Anal. Chem., **32**, 415 (1893). J. B. Duncan, Chem. News, **95**, 49 (1907).

[2] H. W. Rowell, Jour. Soc. Chem. Ind., XXV, 1181.

point, this having been determined in a preliminary run with methyl orange added in the beginning, 4 drops of methyl orange are added and the titration completed cautiously until the color of the indicator is destroyed. If iron or copper is present the final product will appear yellow. Since the end-reaction is slow the last portion of the reagent should be added drop by drop with constant stirring.

1 cc. N/10 KBrO₃ = 0.006 gram Sb.

NOTES. Since antimony chloride begins to volatilize at 195° C. and boils at 220° C. it is advisable not to carry the concentration too far while expelling arsenic.

Lead, copper, zinc, tin, silver, chromium, and sulphuric acid have no effect upon the determination, but large quantities of calcium, magnesium, and ammonium salts tend to make the results high. Low[1] found that copper produced high results, approximately .012% too high for every 0.1% of copper present. The author (W.W.S.) finds, however, that with the procedure given above, amounts of copper as high as 15% produced no difficulty beyond a yellow coloration of the solution. With larger amounts of copper, the end-point became difficult to detect owing to the depth of this yellow color, so that in case of brass and copper alloys, the method must be modified by a procedure for removal of the copper. Lead up to 95% caused no difficulty. Iron, in amounts such as are commonly met in alloys of lead, does not interfere.

During the course of analysis antimony may be isolated as the sulphide; this is dissolved in strong hydrochloric acid, and reduced and concentrated to expel arsenic that may be present as a contamination, and the resulting solution titrated with potassium bromate as directed above.

Sources of Error. (a) Imperfect volatilization of arsenic. (b) Incomplete expulsion of SO₂. (c) Over-titration if insufficient hydrochloric acid is present.

No loss of antimony occurs at temperatures below 120° C.

Potassium Iodide Method for Determining Antimony

Procedure. To 1 gram of fine sawings or filings in a 16-oz. Erlenmeyer flask add 60 cc. of concentrated hydrochloric acid and heat on an asbestos board or on the water bath just below boiling. When hydrogen is no longer evolved, decant the liquor and wash twice with concentrated hydrochloric acid, retaining the antimony in the flask. Now dissolve the antimony by adding 15 cc. of concentrated hydrochloric acid and solid potassium chlorate, a few crystals at a time, until the antimony is in solution, the liquid being kept hot. Expel chlorine by boiling, add 50 cc. of concentrated hydrochloric acid and again bring to boiling. Cool and add 20 cc. of 20% potassium iodide solution and 1 cc. of carbon disulphide or tetrachloride. Titrate the liberated iodine with tenth-normal sodium thiosulphate. The brown color will gradually disappear from the solution and the last traces of free iodine will be collected in carbon disulphide or carbon tetrachloride, giving a pink color. When this pink color disappears the end-point has been reached.

One cc. N/10 Na₂S₂O₃ = .006 gram of Sb.

Na₂S₂O₃ is standardized against .3 gram antimony as in case of Potassium Bromate Method, the above procedure, however, being followed. Antimony must be free from copper and arsenic.

NOTES. The following reversible reaction is of interest: "R" representing a trivalent metal with oxidation to pentavalent form.

$$R_2O_3 + 2I_2 + 2H_2O = R_2O_5 + 4HI.$$

The reaction goes to the right when an alkali is present to neutralize the free acid formed; e.g., Mohr's process for determining arsenic by titration of the lower

[1] A. H. Low, "Technical Methods of Ore Analysis."

oxide with iodine in presence of sodium bicarbonate. The reaction goes to the left in presence of strong acid; e.g., Weller's process for the determination of antimony in an acid solution.

The solution should not contain more than $\frac{1}{5}$ of its volume of hydrochloric acid (sp.gr. 1.16), since too much hydrochloric acid gives high results, owing to the action of hydrochloric acid on potassium iodide. Too little acid leads to the separation of basic iodides and chlorides of antimony. The solution is best boiled down to 20% hydrochloric acid (above strength).

Stannous chloride may be used in place of thio-sulphate in titration of iodine.

$$SbCl_5 + 2KI = SbCl_3 + 2KCl + I_2 \quad \text{and} \quad I_2 + SnCl_2 + 2HCl = SnCl_4 + 2HI.$$

Determination of Antimony by Oxidation with Iodine

The procedure originated by Mohr and modified by Clark, depends upon the reaction $Sb_2O_3 + 2I_2 + 2H_2O = Sb_2O_5 + 4HI$.

The reaction takes place when iodine is added to a solution of antimonous salt in presence of an excess of alkali bicarbonate. In an acid solution oxidation with iodine does not go beyond Sb_2O_3.

Procedure. *Solution.* The sample is brought into solution by one of the procedures given under "Preparation and Solution of the Sample." Alloys of antimony, lead, and tin are treated according to directions given for this combination.

Titration. To the hydrochloric acid solution of antimony is added tartaric acid or Rochelle salts, the excess of the acid neutralized with sodium carbonate, the solution made barely acid with hydrochloric acid and a saturated solution of sodium bicarbonate added in the proportion of 10 cc. bicarbonate solution for each 0.1 gram of Sb_2O_3. Starch is added as an indicator and the solution titrated with N/10 iodine.

1 cc. N/10 iodine = 0.006 gram Sb.

NOTE. The titration should be made immediately upon addition of the sodium salts.

Antimony in Solder Metal and Alloys with Tin and Lead [1]

Procedure. Dissolve 2 grams of the sample of alloy in concentrated hydrochloric acid. When the metal is all in solution, add crystals of iodine until the solution is thoroughly permeated. The color at this point should be a deep purple. Boil until all of the iodine fumes have been driven out. The metallic antimony which did not go into solution in the hydrochloric acid should now be all dissolved. If it is not, add more iodine until the solution is complete. When all is in solution and the color changes to a straw yellow, cool, add a few cc. of starch solution. If a blue color appears, due to an excess of iodine, run in N/10 sodium thiosulphate solution until colorless. In case there is no blue color developed, add N/10 iodine until a faint blue appears. Now add 50 cc. of a saturated solution of Rochelle salts. Make alkaline to litmus by adding 25% sodium hydrate solution. Then make slightly acid with HCl and finally alkaline with sodium bicarbonate. Cool and titrate with N/10 iodine.

NOTE. "The method gives very good results. I have checked it up when there was one-tenth of a gram known antimony present and the results were within a reasonable limit of accuracy." [1]

[1] Method communicated to author by Mr. B. S. Clark.

Other Procedures
Permanganate Method

Antimonous salts may be titrated with standard potassium permanganate. The iron value for the permanganate multiplied by 1.075 or the oxalic acid ($C_2H_2O_4 \cdot 2H_2O$) value multiplied by 0.9532, will give the antimony value.

Indirect Evolution Method

The method depends upon the evolution of H_2S from the sulphides of antimony decomposed by strong hydrochloric acid, the amount of hydrogen sulphide being the same for either Sb_2S_3 or Sb_2S_5, the following reactions taking place:

1. $Sb_2S_3 + 6HCl = 2SbCl_3 + 3H_2S$.
2. $Sb_2S_5 + 6HCl = 2SbCl_3 + S_2 + 3H_2S$.

The details of the method are practically the same as determination of sulphur by the evolution method in the analysis of iron and steel. See Chapter on Sulphur. The antimony sulphide precipitate is placed in the evolution flask, strong hydrochloric acid added with an equal volume of water and the evolved hydrogen sulphide absorbed in an ammoniacal solution of cadmium chloride. The precipitated cadmium sulphide is then titrated with iodine in an acid solution.

One cc. N/10 I = 0.001604 gram S, since 3S = 2Sb, therefore Sb = S × 2.499, hence, 1 cc. N/10 I = 0.00401 gram Sb.

Preparation of Standard Iodine Solution. An approximate tenth normal solution is made by dissolving 12.7 grams of commercial iodine, roughly weighed on a watch-glass, in 200 cc. of water containing about 25 grams of potassium iodide, solution being effected in a graduated liter flask. After making up to 1000 cc. with distilled water, the reagent is transferred to a dark-colored bottle, to protect it from light. It is advisable to make up 5 to 10 liters at a time for laboratories where the solution is in constant demand. After standing several hours, the reagent is standardized by running a portion from a burette into 100 cc. of tenth normal arsenous acid (see page 240) until a faint yellow color is perceptible. In presence of starch indicator a faint blue color is obtained.

100 divided by the cc. of iodine required gives the factor for a N/10 solution.

Example. If 98.5 cc. of iodine are required, 100 ÷ 98.5 = 1.0152 N/10 or .10152 normal.

Tenth normal arsenous acid solution contains 4.953 grams of As_2O_3, per liter, dissolved in sodium hydroxide and made up according to directions given on page 204. The oxide is seldom pure, so that allowance must be made for impurities. For example, the acid containing 99.56 per cent As_2O_3 would require 4.953 ÷ .9956 = 4.97 grams per liter of solution.

Commercial iodine may contain chlorine, bromine, cyanogen and water. It may be purified by repeated sublimation (" Analytical Chemistry," Treadwell and Hall, IV Ed., page 646, or " A Treatise on Quantitative Inorganic Analysis" (1913), by J. W. Mellor, page 288). There is no advantage in taking the theoretical amount of purified iodine, however, since the reagent changes in strength on standing.

Potassium iodide augments solution of iodine, which is sparingly soluble in water. The iodine may be standardized by titrating a definite volume with N/10 sodium thiosulphate. See page 240.

VOLUMETRIC

Determination of Antimony in Brass=Permanganate Method

Reagents. *Potassium Permanganate.* 0.3 g. of $KMnO_4$ is dissolved in water and made to 1000 cc. The reagent is standardized against 25 mg. of pure antimony that has been dissolved in 15 cc. of boiling sulphuric acid and treated as described in the procedure below, under 4 and 5.

Procedure. 1. A sample of 5 grams of brass is dissolved in a 250 cc. beaker in 25 cc. of strong nitric acid (d. 1.42), and after the action has ceased the solution is boiled to expel the oxides of nitrogen. Now 125 cc. of boiling water are added and the solution allowed to settle for an hour or more, keeping the temperature just below boiling. The tin and antimony precipitates are filtered on double 9 cm. closely woven filter papers, keeping the solution hot, and then washed with boiling water. The filtrate is discarded.

2. The papers and residue, transferred to a 350 cc. beaker, are treated with 25 cc. of strong nitric acid (d. 1.42), 5 grams of ammonium persulphate, and 15 cc. of strong sulphuric acid (d. 1.84) and boiled down to strong fumes. (The reaction may be conveniently carried out in a " copper flask " of pyrex glass.) If the solution is brown, 5 cc. of strong nitric acid are added to the cooled solution and about 1 gram additional of persulphate and the boiling to fumes repeated.

4. When the solution is colorless, it is cooled, 20 cc. of water added, together with 20 cc. strong hydrochloric acid (d. 1.20) and (cautiously) 1 gram of sodium sulphite and the SO_2 completely expelled by gentle heating for 10 minutes, or longer.

5. The solution is diluted with 200 cc. of water, and cooled under running water to 10° to 12° C., then titrated with the standard potassium permanganate solution to a decided pink color.

NOTES. Antimony is precipitated quantitatively with meta-stannic acid in alloys containing a large amount of tin.

The filter paper is destroyed by ammonium persulphate and nitric acid, while tin and antimony go into solution with the sulphuric acid. Fuming nitric and sulphuric acids may be used, in place of the persulphate and nitric acid, but are not so efficient.

The solution is kept hot to prevent solution of the meta-stannic acid.

Arsenic in the alloy necessitates a correction.

In case of alloys containing considerable amounts of tin and antimony, smaller samples should be taken and stronger potassium permanganate solution than is recommended for brass.

Should the oxides remain undissolved upon fuming with sulphuric acid, a small piece of filter paper added ($\frac{1}{8}$ inch square) will effect reduction and solution of the oxides. The solution should be heated until the carbon of the filter is destroyed and the solution clears and becomes colorless.

Determination of Small Amounts of Antimony

Details of this procedure were worked out by Mr. W. Shelton, under the direction of Mr. W. C. Ferguson, chief chemist, and Mr. E. Fitzpatrick, first assistant chemist, Nichols Copper Company. The method is accurate and is of special value in determining traces of antimony in copper and in alloys. Since arsenic may also be determined a separation by distillation is necessary if the latter is present. (See pages 32, 37.)

Description of Generator

The generator consists of three separate parts:

1. Glass cap which is placed over funnel A, to hold the disc of test paper in place.

2. $F-G$, this part of the apparatus has two small parts: F, which is a tube of glass $\frac{13}{16}''$ long, $\frac{1}{16}''$ wide, fitted into a rubber tube G, $\frac{5}{16}''$ wide which in turn is fitted into the lower part of funnel A. The part F is a very important one and care should be taken to have exactly the same size glass tubing and that distance from the top of A to top of F is $\frac{5}{16}''$.

The entire apparatus consists of parts, A, B, C, D, E, F, G.

A. The funnel for test paper.

B. Bulb for holding cotton saturated with lead acetate to absorb any H_2S gas should any be present when generator is operating. Use 0.5 gms. of cotton.

C. This part extends to E, which has two purposes: No. 1. For introducing acid, H_2O, the test, etc., without opening the apparatus. No. 2. As a safety valve, should the apparatus become clogged or stopped up the pressure will exert itself in this direction.

D. Upper part acts as condenser. The lower part is ground to fit the bottle No. 3.

E. This part is explained in C.

3. This part is the bottle which has 250 cc. capacity with a ground mouth to receive the No. 2 part of generator.

Note: All generators must be made and assembled as nearly uniform as possible to assure concordant results.

4. This figure shows the manner in which the test is placed on funnel A, Fig. 2, and how the cap, Fig. 1, fits over and holds the test paper in position.

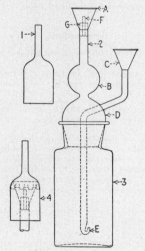

Fig. 1a. Fitzpatrick Apparatus for Determining Traces of Antimony.

Chemicals and Solutions

All chemicals and solutions must be previously tested for arsenic and antimony before using.

HCl—C.P. Conc. redistilled, As and Sb free.

HNO_3—C.P. Conc.

NH_4OH—C.P. Conc.

$Fe_2(NO_3)_3$—5 lbs. to 9 liters, about 3% solution.

$KClO_3$—use dry crystals.

$FeCl_2$—2 lbs. to 2 liters.

$ZnCl_2$—20 lbs. to 9 liters HCl (purified by dissolving 15 gms. zinc in 500 cc. of the above stock solution).

Zinc Shot—Wash in dilute HCl before using.

$SnCl_2$ solution—52.5 grs. per liter.

5% $HgCl_2$ used for test paper. 5 gr. to 100 cc. H_2O. (Cut with die into circles of $1\frac{5}{8}''$ in diameter.)

10% Pb. $(C_2H_3O_2)_2$ for cotton. 10 grs. to 100 cc. H_2O.

5% NH_4OH_2 for developer of test paper, 200 cc. NH_4OH per liter.

1% $AuCl_3$ solution.

Preparation of Test Paper. The paper used must be selected, when purchased, for evenness of thickness and texture in sheets of 24" x 40".

The above sheets are cut in half and saturated with a 5% $HgCl_2$ solution—the wet sheet is then placed on a glass plate and the surplus solution is squeezed out with a 10" rubber roller, which is rolled over the paper twice. Care must be exercised to roll the paper evenly and with good pressure using the same conditions for each sheet. The sheet is now hung over a line to dry, in a warm place away from the sunlight or any influence of hydrogen sulphide. Do not dry paper in oven. When dry the paper is cut with die into pieces of $1\frac{5}{8}''$ in diameter. Keep the discs of test paper in a dark-brown bottle and away from the light until used.

Enough test paper should be made at one time to last for about 3 months.

Each new lot made should be tested with known amounts of As and Sb and compared with standards, before using. Should they not check close it is advisable to make new set of standards from the test paper just made.

Preparation of Lead Acetate Cotton. A roll of absorbent cotton is opened and saturated with a 10% solution of lead acetate and surplus drained off, then hung on a line to dry in a warm place away from the influence of hydrogen sulphide. Do not dry in oven. When dry, place in stoppered bottle until used.

Precautions

Blank. A blank test should be run with each day's work, using all the reagents used in actual tests.

The stain obtained on test paper from blank is subtracted from the actual test.

Limits. The limits of As or Sb that can be determined by this method must be within the following figures:

As separately from .00002 grs. to .00010 grs.
Sb separately from .00002 grs. to .00015 grs.

Checks. A 10 grs. sample of standard copper known to be free from As or Sb is weighed out and known amounts of As and Sb are introduced.

Distillation. The distilling apparatus should *not* be used for any other tests when the As or Sb is known to be higher than the limits for this work.

Zinc. The zinc shot must be cleaned with dilute HCl and washed with distilled water each day to insure proper action in generator, and to expel any sulphide present which would spoil the test

Generator. The presence of nitrates, chlorates, or compounds of copper interfere with generation of arsine and stibine, so care must be exercised to have these compounds eliminated.

Large quantities of ferrous and ferric compounds interfere also in the generation of stibine to some degree. The small amount of Fe that gets into the test from the process of distillation is overcome by the addition of 2 cc. stannous chloride—at times more may be required.

Uniformity. Uniformity must be strictly adhered to throughout the test.

In the determination of antimony in presence of arsenic the removal of the latter is necessary. This is accomplished by distillation of $AsCl_3$ according to the procedure outlined on page s 32, 37.

Standard Antimony Solution and Standard Stains

Antimony Solution. A stock solution is made up by weighing out 0.553 gms. $KSbO_3C_4H_4O_6$ which is dissolved in distilled water and made up to 2000 cc. which represents 1 cc. = .0001 g. Sb.

From the above stock solution take 100 cc. and make up to 1000 cc. this solution now equals 1 cc. = .00001 g. Sb, which is used for making the standard stains and introducing into checks.

Outline of the Method

Preparation of Standard Stains. Extreme care must be taken when preparing the standard stains.

Wash the generator thoroughly with distilled water, place freshly prepared lead acetate cotton in the bulb, B, No. 2, and see that the top of part F, No. 2, is exactly $\frac{5}{16}''$ from the top of part A, No. 2.

Now introduce into bottle of generator, No. 3, the required amount of As or Sb as desired and then add 50 cc. redistilled HCl, As free, 2 cc. stannous chloride solution and make up to 220 cc. with distilled water.

The disc of mercuric chloride test paper is now placed on top of funnel A, No. 2, and the glass cap, No. 1, is forced over the paper holding it in place.

Now introduce 15 gms. metallic zinc shot and place the No. 2 section with No. 1 attached into the No. 3 or bottle of generator. The apparatus now being assembled, observe that the apparatus is fitted together tightly, because as soon as the zinc is introduced, Arsine and Stibine is generated immediately. Place the generator into the water bath to maintain constant temperature which should be about 70° F. Allow the generator to operate for 1 hour.

The glass cap, No. 1, is now removed and the test paper is developed in a No. 2 beaker with 5% NH_4OH solution for three minutes, then washed 5 times with distilled water. The test paper is now toned with a 1% $AuCl_3$ solution by allowing the test paper to remain in solution for five minutes. The test paper will now have a violet or purple stain, the intensity depending on the amount of As or Sb introduced. Wash the paper 5 times with distilled water and preserve in 50 cc. glass stoppered bottles containing about 5 cc. water. Keep bottles in dark place, because the stains darken on exposure to light.

Duplicate tests are made, finally selecting of two the one which is the most uniform.

The stains are made to represent the following amounts:

	Sb
1	.00002
2	.00004
3	.00006
4	.00008
5	.00010
6	.00012
7	.00014
8	.00016

Procedure for Refined Copper

A blank is run with all tests.

Weigh 10 gms. of the shot or drilled sample into a No. 3 beaker. Add 50 cc. Conc. HNO_3, C.P. As free, let stand covered with watch glass until the action has subsided. Now place beaker on wire gauze over Bunsen flame and heat until all the copper is dissolved.

Remove from flame, dilute to 150 cc. with distilled water (if too basic add a few drops of HCl to clear the solution). Add 2 cc. ferric nitrate solution, stir, then make ammoniacal by adding C.P. ammonium hydrate (As free). Bring to boiling. Remove from flame and filter through a 15 cm. fluted Perfection filter paper. Immediately wash the filter paper free from copper compounds with hot water, using dilute ammonia where necessary to wash out any copper salts that have crystallized.

The precipitate (which contains both As and Sb) is dissolved off the filter with hot dilute hydrochloric acid, by means of a wash bottle into a No. 4 casserole. Wash the filter three times with hot water.

Add a pinch of $KClO_3$ to the casserole, cover with watch glass, place the casserole in an asbestos cut out over Bunsen burner and boil the contents down to 10 cc., taking care that it does not roast on the sides.

Distillation. Transfer the contents of the casserole to the distilling apparatus. Add 20 cc. ferrous chloride and 20 cc. zinc chloride solution, and distill until the contents of flask begins to froth. Now add, drop at a time, 35 cc. HCl, through the dropping funnel which is connected to flask. Distil until all the HCl is out of the funnel and out of the flask.

The distillate is received in a No. 4 beaker having 40 cc. H_2O in which both the As and Sb is contained.

The above distillate is now transferred and washed from the beaker into the special designed generator. Add 2 cc. stannous chloride which insures a complete reduction of any ferric compounds present. Dilute the contents of generator to 220 cc. Place disc of $HgCl_2$ test paper on the funnel top, then put on cap to hold in place. Add 15 gms. metallic zinc or 1 No. 6 porcelain spoonful. Take care that the generating apparatus is properly closed, then place into water bath to maintain constant temperature which should be about 70° F. The apparatus is allowed to operate for one hour during which time the arsine and stibine generated shall effect the $HgCl_2$ test paper, causing a yellow or orange colored spot which varies in color and size according to the amount of As and Sb present. The paper is now removed from the apparatus and developed in a No. 2 beaker containing 5% ammonium hydrate solution for three minutes. The color of the spot now changes to a brownish black. Wash test paper five times with distilled water. Now cover the test paper with 10 cc. of 1% gold chloride solution which tones the color of the spot to a violet or purple hue that fixes it so comparison can be made with the standard stains or spots to determine the amount of arsenic or antimony in the sample.

Determination of Antimony in Tartar Emetic

Iodine in the presence of sodium bicarbonate oxidizes trivalent antimony to the pentavalent form as shown by the reaction:

$$K(SbO)C_4H_4O_6 + 6NaHCO_2 + I_2$$
$$= Na_3SbO_4 + KNaC_4H_4O_6 + 2NaI + H_2O + 6CO_2.$$

Procedure. 10 grams of tartar emetic are dissolved in water, the solution diluted to 500 cc. and 20 cc. taken for analysis. This is diluted to 100 cc. 25 cc. of 2 per cent sodium bicarbonate are added and the mixture titrated with N/10 iodine reagent.

$$1 \text{ cc. } N/10 \text{ I} = 0.016617 \text{ g. } K(SbO)C_4H_4O_6 \cdot \tfrac{1}{2}H_2O$$

ARSENIC

As_4 *at.wt.* 74.96— *cryst.* *sp.gr.* 5.73 *m.p.* 850 *b.p.* *subl.* 554°
amorp. 4.72 .... < 360°

Oxides, As_2O_3, As_2O_5.[1]

DETECTION

Hydrogen sulphide precipitates the yellow sulphide of arsenic, As_2S_3, when passed into its solution made strongly acid with hydrochloric acid. If the solution contains more than 25% hydrochloric acid, (sp.gr. 1.126) the other members of the hydrogen sulphide group do not interfere, as they are not precipitated from strong acid solutions by hydrogen sulphide. Arsenic sulphide is soluble in alkaline carbonates. (Antimony sulphide, Sb_2S_3, reddish yellow, is insoluble in alkaline carbonates.)

Volatility of the chloride, $AsCl_3$, is a means of separation and distinction of arsenic. Details of the procedure are given under "Separations." The distillate may be tested for arsenic as directed above.

Traces of arsenic may be detected by either the Gutzeit or Marsh test for arsenic. Directions for the Gutzeit test are given at the close of the volumetric procedures.

Distinction between Arsenates and Arsenites. Magnesia mixture precipitates white, $MgNH_4AsO_4$, when added to ammoniacal solutions containing arsenates, but it produces no precipitate with arsenites.

Red silver arsenate and yellow silver arsenite are precipitated from neutral solutions by ammoniacal silver nitrate. An arsenate gives a yellow precipitate with ammonium molybdate solution.

ESTIMATION

The determination of arsenic is required in the valuation of native arsenic, white arsenic, $As.O_3$; ores of arsenic—orpiment, As_2S_3; realgar, As_2S_2; pyrargyrite, As_3Sb_3; arsenopyrite, or mispickel, FeSAs; cobaltite or cobalt glance, CoSAs; smaltite, $CoAs_2$; niccolite. NiAs. The substance is estimated in copper ores, in speiss, regulus; in iron precipitates (basic arsenate). It is determined in paint pigments, Scheel's green, etc. The element is determined in shot alloy and in many metals. It is estimated in germicides, disinfectants, and insecticides—Paris green, lead arsenate, zinc arsenite. Traces are looked for in food products and in substances where its presence is not desired.

Preparation and Solution of the Sample

In dissolving arsenic compounds it will be recalled that the oxide, As_2O_3, is not readily acted upon by dilute acids—hydrochloric or sulphuric. The compound is soluble, however, in alkaline hydroxides and carbonates. Nitric

[1] Van Nostrand's Chem. Annual—Olsen—3d Ed.

Chapter contributed by Wilfred W. Scott.

acid oxidizes As_2O_3 to the higher oxide, As_2O_5, which is soluble in water. The sulphides As_2S_3 and As_2S_5 are practically insoluble in hydrochloric or sulphuric acids, but are dissolved by the fixed alkalies and alkali sulphides. All arsenites, with the exception of the alkali arsenites, require acids to effect solution.

Pyrites Ore and Arseno-pyrites. The amount of the sample may vary from 1 to 20 grams,[1] according to the arsenic content. The finely ground sample in a large casserole is oxidized by adding 10 to 50 cc. of bromine solution (75 cc. $KBr + 50$ cc. liquid $Br + 450$ cc. H_2O) covering and allowing to stand for fifteen minutes, then 20 to 50 cc. of strong nitric acid are added in three or four portions, allowing the action to subside upon each addition. The glass cover is raised by means of riders, and the sample evaporated to dryness on the steam bath; 10 to 25 cc. of hydrochloric acid are now added and the sample again taken to dryness. Again 10 to 25 cc. of hydrochloric acid are added and the sample taken to dryness. Finally 25 cc. of hydrochloric acid and 75 cc. of water are added, and the mixture digested over a low flame until all the gangue, except the silica, is dissolved. The solution is now examined for arsenic by distillation of the arsenic after reduction, the distillate being titrated with standard iodine solution according to directions given later.

Arsenous Oxide. The sample may be dissolved in caustic soda, the solution neutralized with hydrochloric acid, and the resulting sample titrated with iodine.

Fusion Method. One gram of the finely powdered mineral is fused in a nickel crucible with about 10 grams of a mixture of potassium carbonate and nitrate, 1 : 1,[1] and the melt extracted with hot water. Two hundred cc. of a saturated solution of SO_2 is added to the filtrate to reduce the arsenic, the excess of SO_2 then expelled by boiling, the solution diluted with dilute sulphuric acid, and arsenic determined in the filtrate.

Arsenic in Sulphuric Acid. *Arsenous acid* may be titrated directly with iodine in a 20- to 50-gram sample, which has been diluted to 200 to 300 cc. with water and nearly neutralized with ammonium hydroxide and then an excess of sodium acid carbonate added, followed by the iodine titration.[2]

Arsenic Acid in Sulphuric Acid. Twenty-five cc. of the acid containing about 0.1% arsenic or a larger volume in case the percentage of arsenic is less than 0.1% As_2O_3 (the sp.gr. of the acid being known) are measured out into a short-necked Kjeldahl flask. About half a gram of tartaric acid and 2 grams of fused, arsenic-free potassium bisulphate are added and the acid heated over a low flame until the liberated carbon is completely oxidized and the acid again becomes clear, e.g., a pale straw color. It is not advisable to heat to violent fuming, as a loss of arsenic is then apt to occur. The cooled acid is poured into about 300 cc. of water, the excess acid nearly neutralized with ammonia, bicarbonate of soda added in excess and the arsenous acid titrated with standard iodine. Total arsenic as As_2O_3 minus arsenous arsenic as As_2O_3 = arsenic arsenic in terms of As_2O_3. This result multiplied by 1.1616 = As_2O_5.

Arsenic in Hydrochloric Acid. The arsenic in 20 to 100 cc. sample is reduced by ferrous chloride, the arsenic distilled according to directions given later, and the distillate titrated with iodine.

---

[1] 0.1% arsenic determined on a 20-gram sample.

[2] SO_2 should be expelled by heat or by a current of air before treating with the alkali.

Arsenic in Organic Matter.[1] 0.2 to 0.5 gram of the sample finely powdered is oxidized by mixing with 10 to 15 grams of sodium carbonate and sodium peroxide, 1 : 1, in a nickel crucible, a portion of the fusion mixture being spread over the charge. After heating gently for fifteen minutes, the fusion is completed by heating to dull redness for five minutes longer. The contents of the crucible are rinsed into an Erlenmeyer flask after extraction with water, and the solution made acid with dilute sulphuric acid, 1 : 1. The mixture is boiled down to 100 cc., 1 to 2 grams of potassium iodide added and the solution further concentrated to about 40 cc. Iodine is reduced with sulphurous acid or thiosulphate, the solution diluted with hot water and saturated with hydrogen sulphide. Arsenous sulphide is filtered off, washed, dissolved in 15 to 20 cc. of half-normal sodium hydroxide and 30 cc. of hydrogen peroxide (30%) solution added, and the solution boiled. About 12 cc. of dilute sulphuric acid, 1 : 1, are added, together with 1 to 2 grams of potassium iodide, the solution concentrated to 40 cc. and free iodione reduced with thiosulphate as before. Arsenic is now titrated, with standard iodine, upon neutralization of the free acid with sodium hydroxide and sodium acid carbonate.

Lead Arsenate. Ten grams of the thoroughly mixed paste or 5 grams of the powder are dissolved by treating with 25 cc. of 10% hot sodium hydroxide solution, and diluted to 250 cc. An aliquot part, 50 cc. (= 2 grams paste and 1 gram powder) is placed in an Erlenmeyer flask and 20 cc. of dilute sulphuric acid, 1 : 1, added, and the solution diluted to 150 cc. About 3 grams of solid potassium iodide are added and the solution boiled down to about 50 cc. (but not to fumes). The liquor will be colored yellow by free iodine. Tenth normal sodium thiosulphate is added drop by drop until the free iodine is neutralized (solution loses its yellow color), it is now diluted to about 250 cc. and the free acid neutralized by ammonium hydroxide (methyl orange indicator), then made slightly acid with dilute sulphuric acid, and an excess of bicarbonate of soda added. The arsenic is titrated with standard iodine.

The arsenic may be reduced by placing the 50-cc. sample in a Kjeldahl flask, adding 25 cc. of strong sulphuric acid (1.84 sp.gr.), $\frac{1}{2}$ gram tartaric acid and 2 grams acid potassium sulphate, $KHSO_4$, and digesting over a strong flame until the organic matter is destroyed and the solution is a pale yellow color. The cooled acid is diluted and neutralized, etc., as directed above.

Water-soluble Arsenic in Insecticides. Rapid Works Test. Two grams of the paste is digested with 1000 cc. of water at 90° C. for five minutes, in a graduated 1000-cc. flask. An aliquot portion is filtered and the arsenic determined by the Gutzeit method.

Water-soluble arsenite may be titrated directly with iodine in presence of sodium bicarbonate.

Zinc Arsenite. About 5 grams of powder or 10 grams of paste are taken and dissolved in a warm solution containing 300 cc. of water and 25 cc. of strong hydrochloric acid. The cooled solution is diluted to 500 cc. and 100-cc. portions taken for analysis. The acid is partly neutralized with ammonium hydroxide and 50 cc. of a saturated solution of ammonium oxalate added (to prevent precipitation of the zinc as $ZnCO_3$), and an excess of sodium bicarbonate, $NaHCO_3$. Arsenic is now titrated with iodine as directed later.

Soluble Arsenic in Zinc Arsenite. One gram sample is rubbed into an

[1] Organic matter may be destroyed by heating the substance on addition of 10% H_2SO_4 and solid $(NH_4)_2S_2O_8$.

emulsion with several portions of water until the whole is in suspension. The cloudy liquor is diluted to 1000 cc. and a portion filtered through a $\frac{1}{4}$-in. asbestos mat on a perforated plate, the asbestos being covered with a layer of filter paper. The first 50 cc. are rejected. One hundred cc. of the clear filtrate (= 0.1 gram) is treated with 10 cc. of strong sulphuric acid, 0.05 gram, Fe_2O_3 (use ferric ammonium sulphate) and $\frac{1}{2}$ cc. of 80% stannous chloride solution and heated until colorless. Arsenic is now determined by the Gutzeit method, using the larger-sized apparatus.

Arsenic Acid, Alkali Arsenates, etc. The sample is dissolved in 20 to 25 cc. of dilute sulphuric acid, 1 : 1, in an Erlenmeyer flask, and reduced by addition of 3 to 5 grams of potassium iodide, the action being hastened by placing the mixture on a steam bath. The iodine liberated is exactly neutralized with thiosulphate and the arsenous acid titrated with iodine according to the procedure given later.

Arsenic in Steel, Iron, Pig Iron, etc. One to 50 grams of steel, etc., may be treated according to the scheme for pyrites. If a large sample is taken, it is advisable to treat it in a 500-cc. flask, connected with a second flask containing bromine, to guard against loss of arsenic by volatilization. When the sample has dissolved it is taken to dryness (the bromine in the second flask being combined with it) and treated as directed in pyrites. Arsenic chloride, $AsCl_5$, is transferred to the distilling flask with strong hydrochloric acid, and arsenic separated from the iron by volatilization of reduced chloride according to the procedure given below.

Arsenic in Copper. Arsenic is precipitated with iron by the basic acetate method, and thus freed from copper. Details of procedure are given under the determination of impurities in copper in the chapter on the subject.

NOTES. In the decomposition of the sample Low recommends the addition of a little sodium sulphide to ores containing oxides. To prevent loss of arsenic during the treatment with H_2S he uses a flask with a two-hole rubber stopper through which passes an inlet tube reaching to the bottom of the flask and an exit tube, the latter a thistle tube containing a little absorbent cotton soaked with dilute NaOH to retain any arsenic escaping from the flask.

Iron sulphate dissolves slowly, so that if much is present in the ore time must be allowed for this to dissolve.

As arsenous chloride is volatile, great care must be exercised in heating solutions containing HCl and arsenous salts as a loss will occur. B.p. 113.9°.

The ore may be brought into solution by fusion with a mixture of sodium carbonate, potassium nitrate and zinc oxide, 1 : 1 : 2. The fusion being made in a platinum dish. The potassium iodide procedure may be followed for reduction of arsenic. (See Lead Arsenate.)

SEPARATIONS

Isolation of Arsenic by Distillation as Arsenous Chloride [1]

By this method arsenic may be separated from antimony, tin, and from other heavy metals. It is of special value in the direct determination of arsenic in iron ores, copper ores, and like products and has a wide application. The procedure depends upon the volatility of arsenous chloride at temperatures lower than the other heavy metals. In a current of HCl gas, arsenous chloride begins to volatilize below 108° C., and is actively volatile at 120° C.; antimony starts to volatilize at 125° C., but is not actively volatile until a temperature of 180° has been reached. The boiling-point of arsenous chloride, $AsCl_3$, is 130.2°; antimony trichloride, $SbCl_3$, is 223.5°; and that of stannous chloride, $SnCl_2$, is over 603°; other chlorides having still higher boiling-points. Tin in its higher form, $SnCl_4$, is readily volatile, boiling-point is 114° C., so that it is necessary to have it in its divalent form to effect a separation from arsenic. When heavy metals are present in the residue remaining from the arsenic distillate, or when zinc chloride is added to raise the boiling-point, antimony may also be separated by distillation by carrying the solution to near dryness, adding concentrated HCl by means of a separatory funnel, drop by drop, during further distillation of the concentrate. Arsenic may be determined in the distillate (first portions) either gravimetrically or volumetrically.

Procedure. If arsenic is present as arsenic chloride, as prepared in the method for solution of iron ores, the sample may be transferred directly to the distillation flask by means of concentrated, arsenic-free hydrochloric acid. If a preliminary separation of other metals has been made and arsenic is present (along with antimony and tin) as a sulphide, it is oxidized by addition of concentrated HCl and sufficient potassium chlorate to cause solution and oxidation of free sulphur, and the chlorate decomposed by evaporation to dryness; or if preferred, by evaporation of the alkaline solution to dryness, oxidation with fuming nitric and re-evaporation to dryness to expel the nitric acid. The residue is taken up with hydrochloric acid and washed into the flask with strong hydrochloric acid as directed above.

Distillation. The sample, in a half-liter distilling flask (Fig. 1, "5") is made up to about 150 cc. with concentrated hydrochloric acid and about 5 grams of cuprous chloride, Cu_2Cl_2, are added. The apparatus is connected up as shown in the illustration, Fig. 1. The end of the condenser dips into 400 cc. of cold water in a large beaker (1 liter) or flask ("4"). The solution is cooled by placing it in ice-water or cold running water. The sample is saturated with dry hydrogen chloride gas generated by dropping concentrated sulphuric acid into strong hydrochloric acid ("3") and passing the gas through

[1] J. E. Stead's Method. R. C. Roark and C. C. McDonnell, Jour. Ind. Eng. Chem., VIII, **4**, 327 (1916).

sulphuric acid (sp.gr. 1.84) as shown in cut. When the point of saturation is reached the gas begins to bubble through the solution instead of being absorbed by it. When this occurs, heat is applied and the solution brought to boiling, the current of HCl gas being continued. At a temperature of 108 to 110° C. the first 100 cc. will contain practically all of the arsenic. About

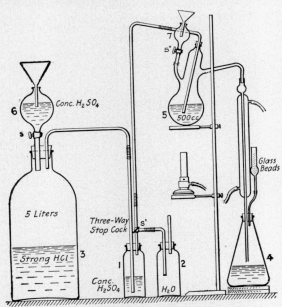

Fig. 1.—Apparatus for the Distillation of Arsenous Acid.

two-thirds of the solution is distilled off. It is advisable to add more hydrochloric acid to the residue in the flask, together with cuprous chloride, and repeat the distillation into a fresh lot of water. This may be done during the estimation of arsenic in the first distillate.

Arsenic may be determined in the distillates either gravimetrically or volumetrically. The volumetric procedures for arsenic, in this isolated form, are generally to be preferred, since they are both rapid and accurate. For amounts over 0.5% arsenic, the iodine method is recommended, for smaller amounts (arsenic in crude copper), precipitation with silver nitrate and titration of the silver salt is best. Exceedingly small amounts are best determined by the Gutzeit method, page 46.

In place of the large bottle shown, a smaller wash bottle may be used filled with concentrated hydrochloric acid, the bottle contains an inlet tube dipping to the bottom and an exit tube connected to the distillation flask containing the arsenic. The receiving flask is connected with an aspirator and air drawn through the system. HCl is swept into the distillation flask during the arsenic distillation, keeping the solution concentrated with HCl gas.

The inlet funnel is filled about half full with hydrochloric acid (sp.gr. 1.2). The outlet of the condenser tube is caused to dip just beneath the surface of 100 cc. of distilled water, containing a lump of ice. The solution in the distillation flask is heated to boiling; concentrated hydrochloric acid is introduced through the funnel drop by drop at the rate sufficient to replace the evaporation. All the arsenic usually distils over in half an hour. At this time the beaker holding the condensate is replaced by another with 100 cc.

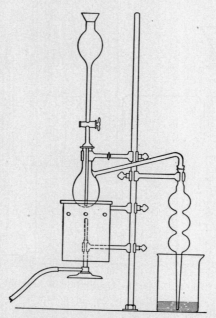

FIG. 2.—Knorr Arsenic Distillation Apparatus.

of water. And the distillation continued about 15 minutes. Test this distillate to ascertain whether any arsenic is present. The arsenous chloride thus obtained is titrated with standard iodine according to the procedure described on page 44. The free HCl is first neutralized with a fixed alkali and then made faintly acid with HCl. Sodium bicarbonate is now added in excess and the titration with iodine made according to the customary procedure.

Commercial hydrochloric acid invariably contains arsenic, so this must be purified by redistillation in presence of an oxidizing agent to oxidize the arsenic to the non-volatile arsenic pentachloride, $AsCl_5$, form, (Fig. 5) or by treatment with H_2S and filtration. A blank run should be made on the reagents used, especially when traces of arsenic are to be determined.

Cuprous Chloride. This is used to reduce arsenic. At least 2 grams CuCl should be used per each gram of iron present. In the distillation, HCl gas (generated in a flask containing strong sulphuric acid by allowing hydrochloric acid to flow in through a thistle tube, drop by drop) may be passed into the solution containing the arsenic, in place of adding strong hydrochloric acid.

Hydrazine Distillation Method. Weigh a suitable amount of sample into a 275 cc. Pyrex sulphur flask. Treat with a small amount of HNO_3 (5–10 cc.), and a pinch of potassium chlorate. Take to dryness. Add 3–5 cc. HCl and take again to dryness. Add 20 cc. 1–1 HCl and boil several minutes. Filter, if insoluble is appreciable, into another flask, and wash. Add 30 cc. HCl, $\frac{1}{2}$ gm. NaBr and $\frac{1}{2}$ gm. hydrazine sulphate, $(NH_2)_2SO_4$. Set the flask on a small electric plate, and at the same time, insert in the flask a two-hole stopper, in one hole of which is a separatory funnel, and in the other a glass tube leading to an 8-in. Allihn condenser, set vertically. The lower end of the condenser is immersed in cold water contained in a No. 3 beaker, which sits on a block of wood. Have the cock of the funnel open. Have a good stream of cold water running through the jacket of the condenser.

Distill until the volume in the flask has been reduced to 20 cc. Close the cock of the funnel, add 20 cc. HCl, remove the block of wood from underneath the beaker and hold the beaker in one hand. Holding the beaker at such a height that no liquid may be sucked back into the flask, open the funnel cock and let the acid run into the flask. Now, place the block under the beaker, and distill until liquid in flask is again reduced to 20 cc.

Remove the flask from the plate, and disconnect it from the condenser. Wash the condenser, allowing washings to run into the distillate. Remove the beaker from under the condenser. Add 8–10 drops of methyl orange (1 gm. of salt per liter of water). Make the solution alkaline with NH_4OH, then just acid with HCl. Cool. Add 10 gm. $NaHCO_3$ and 10 cc. starch solution (10 gm. soluble starch boiled in a liter of water; cooled).

Titrate with iodine solution, one cc. of which equals about .005 gm. As_2O_3. Subtract a blank determination which amounts to .4 or .5 cc.

Separation of Arsenic from Antimony and Tin by Precipitation as Sulphide in a Strong Hydrochloric Acid Solution

This procedure for isolation of arsenic depends upon the insolubility of the sulphide of arsenic in strong hydrochloric acid, whereas that of antimony dissolves. The sulphide of tin is also soluble.

Procedure. The metals present in their lower conditions of oxidation are precipitated as sulphides in presence of dilute hydrochloric acid (5% solution) to free them from subsequent groups (Fe, Al, Ca, etc.). The soluble members of the hydrogen sulphide group are now dissolved and separated from copper, lead, etc., by caustic as follows: The greater part of the washed precipitate is transferred to a small casserole, that remaining on the filter paper is dissolved off by adding to it a little hot dilute potash solution, catching the filtrate in the casserole. About 5 grams weight of solid potassium hydroxide or sodium hydroxide is added to the precipitate. Arsenic, antimony, and tin sulphides dissolve. The solution is filtered if a residue remains, and the filter washed. This preliminary treatment is omitted if alkaline earths and alkalies are the only contaminating elements present.

The casserole containing the sample is covered and placed on a steam bath. Chlorine is now conducted into the warm solution for an hour, whereby the alkali is decomposed and antimony and arsenic oxidized to their higher state. Sufficient hydrochloric acid is added to decompose the chlorate formed, and the uncovered solution evaporated to half its volume. An equal volume of hydrochloric acid is added and the evaporation repeated, to expel the last trace of chlorine. The acid solution is washed into an Erlenmeyer flask, cooled by ice to 0° C. and two volumes of cooled, concentrated, hydrochloric acid added. H_2S gas is rapidly passes into this solution for an hour and a half. The flask is now stoppered and placed in boiling water for an hour. The yellow arsenic sulphide, As_2S_5, is filtered through a weighed Gooch crucible, washed with hydrochloric acid, 2 : 1, until free from antimony, i.e., the washing upon dilution remains clear. The residue is now washed with water, followed by alcohol, and may be dried and weighed as As_2S_5, or determined volumetrically. Antimony and tin are determined in the filtrate. McCay recommends washing As_2S_5 with alcohol, CS_2 and finally alcohol.[1]

The sulphide may be dissolved in concentrated sulphuric acid by heating to sulphuric acid fumes and until the solution becomes clear. No arsenic is lost, provided the heating is not unduly prolonged. Fifteen to twenty-five minutes is generally sufficient to dissolve the sulphide and expel SO_2, etc. The acid may be neutralized with ammonia or caustic, made again barely acid and then alkaline with bicarbonate of soda, and arsenous acid titrated with iodine.[2]

[1] Le Roy W. McCay, Chem. News, **56**, 262 (1887).
[2] J. and H. S. Pattinson, Jour. Soc. Chem. Ind., 1898, p. 211.

GRAVIMETRIC METHODS FOR DETERMINATION OF ARSENIC

As in the case of antimony, the accuracy and rapidity of the volumetric methods for the determination of arsenic make these generally preferable to the more tedious gravimetric methods. The following methods, however, are of value in certain analytical procedures.

Determination of Arsenic as the Trisulphide, As_2S_3

Arsenic acid and arsenates should be reduced to the arsenous form before precipitation as the sulphide. The procedure is especially adapted to the isolation of arsenic from other elements, when this substance is present in the solution in appreciable quantities, advantage being taken of the extreme difficulty with which arsenous sulphide, As_2S_3, dissolves in hydrochloric acid solution.

Procedure. The solution containing arsenic in the arsenious form is made strongly acid with hydrochloric acid and hydrogen sulphide passed into the cold solution to complete saturation. The hydrogen sulphide pressure generator is recommended for this treatment. Figs. 3 and 4. The precipitate is filtered into a weighed Gooch crucible (previously dried at 105° C.), the compound dried at 105° C. to constant weight and weighed as As_2S_3.

Factors. $As_2S_3 \times 0.6091 =$ grams As.
$As_2S_3 \times 0.8042 =$ grams As_2O_3.
$As_2O_3 \times 1.1616 =$ grams As_2O_5.
$As_2O_5 \times 1.3134 =$ grams $H_3AsO_4 \cdot \frac{1}{2}H_2O$.
$As_2S_3 \times 1.2606 =$ grams As_2S_5.

NOTE. Arsenic may also be determined as arsenic sulphide by passing a rapid stream of H_2S into a cooled solution of arsenic acid containing at least two parts of concentrated hydrochloric acid for each part of water present in the solution.

Determination of Arsenic as Magnesium Pyroarsenate

The method worked out by Levol depends upon the precipitation of arsenic as $MgNH_4AsO_4 \cdot 6H_2O$, when magnesia mixture is added to an ammoniacal solution of the arsenate. Although 600 parts of water dissolve 1 part of the salt, it is practically insoluble in a $2\frac{1}{2}$ per cent ammonia solution, 1 part of the anhydrous salt requiring 24,558 parts of the ammonia water according to Virgili.[1] The compound loses $5\frac{1}{2}$ molecules of water at 102° C. and all of the water when strongly ignited, forming in presence of oxygen the stable magnesium pyroarsenate, $Mg_2As_2O_7$, in which form arsenic is determined.

Procedure. The solution containing the arsenic, in the form of arsenate, and having a volume not exceeding 100 cc. per 0.1 gram arsenic present, is treated with 5 cc. of concentrated hydrochloric acid, added, with constant stirring, drop by drop. Ten cc. of magnesia mixture are added (Reagent = 55 grams $MgCl_2 + 70$ grams $NH_4Cl + 650$ cc. H_2O and made up to 1000 cc. with NH_4OH, sp.gr. 0.96), for each 0.1 gram of arsenic present. Ammonia solution (sp.gr. 0.96) is added from a burette, with stirring, until the mixture is neutralized (a red color imparted to the solution in presence of phenolphthalein indicator), and

[1] Average of three results. J. F. Virgili, Z. anal. Chem., **44**, 504 (1905).

then ammonia added in excess equal to one-third the volume of the neutralized solution. The precipitate is allowed to settle at least twelve hours and is then filtered into a weighed Gooch crucible and washed with 2.5% ammonia until free from chloride. After draining as completely as possible by suction the

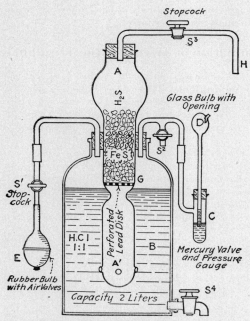

FIG. 3.—Scott's Hydrogen Sulphide Generator.

Fig. 3 shows a convenient form of a generator for obtaining hydrogen sulphide gas under pressure. The apparatus is the writer's modification of the Banks' generator sold by E and A. and is designed for large quantities of hydrogen sulphide gas. The cylinder A A' is constricted, as shown, to support perforated lead disk G, upon which rests the iron sulphide. The lower end of the chamber is closed to catch small particles of FeS that may be carried through the perforations of the disk. Small openings admit the acid to A'. The level of the acid is below the disk G, so that the acid only comes in contact with the sulphide when pressure is applied by means of the rubber bulb E, the stopcock S¹ being open and S³ closed. The mercury gauge C is adjusted to blow out at a given pressure, to prevent accident, the bulb D preventing the mercury from being blown out of the apparatus. A small opening in D allows the escape of the gas. When the apparatus is in operation, H is connected to an empty heavy-walled bottle, which in turn is attached with glass tube connection to the pressure flask in which the precipitation of the sulphide is made, the flask being closed to the outside air. By pressure on the rubber bulb E, acid is forced into the chamber A' past the disk into the sulphide in A. The entire system will now be under the pressure indicated by the gauge C. The pressure is released by opening the stopcock S² and the flask containing the precipitate then disconnected. The reservoir is designed to hold about two liters of acid, and the cylinder containing the sulphide is of sufficient capacity to hold over one pound of FeS, so that the apparatus will deliver a large quantity of hydrogen sulphide.

precipitate is dried at 100° and then heated to a dull red heat (400 to 500° C.), preferably in an electric oven, until free of ammonia. The temperature is then raised to a bright red heat (800 to 900° C.) for about ten minutes, the crucible then cooled in a desiccator and the residue weighed as $Mg_2As_2O_7$.

Factors, $Mg_2As_2O_7 \times 0.4827 = As$, or $\times 0.6373 = As_2O_3$, or $\times 0.7403 = As_2O_5$, or $\times 0.7925 = As_2S_3$.

Notes. In place of an electric furnace the Gooch crucible may be placed in a larger non-perforated crucible, the bottom of the Gooch being 2–3 mm. above the bottom of the outer crucible. The product may now be heated in presence of a current of oxygen passed through a perforation in the covering lid of the Gooch, or

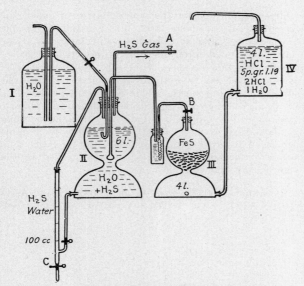

FIG. 4.—Urbasch's Hydrogen Sulphide Generator.

The apparatus designed by Urbasch (Chem. Zeit. (1910) **34**, 1040.. The Analyst (1910) **35**, 558), shown in Fig. 2, enables a constant supply of gas and its saturated aqueous solution to be obtained. The bottle IV is charged with hydrochloric acid, and iron sulphide is placed in III. The hydrogen sulphide is passed through the water in II until a saturated solution is obtained. Water is placed in I and II. If gas is required the taps A and B are opened and H$_2$S drawn from A. Hydrogen sulphide water is obtained by opening the pinch cock C of the burette, the liquid drawn off being simultaneously replaced from the vessel II. The container is made of dark-colored glass to protect the hydrogen sulphide water from light. Water may be drawn into II, when required by opening the pinch cock leading to the bottle I.

in place of the oxygen, a thin layer of powdered NH$_4$NO$_3$ may be placed on the arsenate residue and the heat gradually applied until the outer crucible attains a light red glow.

VOLUMETRIC METHODS FOR THE DETERMINATION OF ARSENIC

Oxidation of the Arsenous Acid with Standard Iodine [1]

This procedure is applicable for the determination of arsenic in acids, after reduction of arsenic to its arsenous form, for valuation of arsenic in the trioxide, for determination of arsenic isolated by distillation as arsenous chloride, for arsenic in arsenites and reduced arsenates in insecticides, etc. The method depends upon the reaction—$As_2O_3+2H_2O+2I_2=As_2O_5+4HI$. The liberated hydriodic acid is neutralized by sodium bicarbonate. The trace of excess iodine is detected by means of starch, a blue color being produced.

Procedure. If the solution is acid, it is neutralized by sodium or potassium hydroxide or carbonate (phenolphthalein indicator) then made slightly acid. If the solution is alkaline, it is made slightly acid. Two to 3 grams of sodium bicarbonate are added together with starch indicator and the solution titrated with tenth normal iodine solution, the iodine being added cautiously from a burette until a permanent blue color develops.

One cc. N/10 iodine = 0.00375 gram As, or 0.004948 gram As_2O_3.
$As_2O_3 \times 1.1616 = As_2O_5$. $As \times 1.3201 = As_2O_3$ or $\times 1.5336 = As_2O_5$.
$As_2O_3 \times 0.7575 = As$.

Potassium Iodate Method for Determining Arsenic [1]

The method is specially applicable to determining arsenic in insecticides. The reaction is represented as follows:

$$As_2O_3+KIO_3+2HCl=As_2O_5+ICl+KCl+H_2O.$$

Procedure. In determining total arsenic the sample is placed in a distilling bulb, connected to a condenser, strong hydrochloric acid added together with cuprous chloride, and arsenious chloride distilled over into an Erlenmeyer flask according to the standard procedure outlined on page 38. If arsenic is high, an aliquot portion of the distillate is taken and the titration made with standard iodate solution as stated later.

[1] Geo. S. Jamison, J. I. E. C. **10**, 290–292, 1918.

For determining arsenious oxide in Paris Green or other arsenite 0.15 to 0.4 grams of the sample may be weighed directly into a glass-stoppered bottle (500 cc.) and the titration made.

Iodate Titration. 30 cc. of hydrochloric acid sp.gr. 1.19, 20 cc. of water and 6 cc. of chloroform are added to the solid arsenite. If the arsenic is in solution, sufficient HCl should be present to have the acidity between 11 and 20 per cent HCl. (If this falls below 11% HCl hydrolysis of the iodine mono-chloride will take place. If over 20% HCl the reaction proceeds very slowly.) Potassium iodate solution is now added, rapidly at first, shaking the contents of the bottle. When the iodine that has been liberated during the first part of the titration has largely disappeared from the solution, the stopper of the bottle is inserted and the contents thoroughly shaken. The titration is now conducted cautiously, shaking thoroughly with each addition of the reagent. The titration is complete when after shaking and allowing to stand 5 minutes no color is observed in the chloroform.

Reagent. Contains 3.244 g. of KIO_3 (dried at 140° C.) per 1000 cc.— 1 cc. = 0.003 g. As_2O_3.

Volumetric Determination of Arsenic by Precipitation as Silver Arsenate

Bennett's modification of Pearce's method, combining Volhard's, depends upon precipitation of arsenic, from a solution neutralized with acetic, by addition of neutral silver nitrate solution; the silver arsenate is dissolved in nitric acid, and the silver titrated with standard thiocyanate.

Procedure. 0.5 gram, or less, of the finely powdered substance is fused with 3 to 5 grams of a mixture of sodium carbonate and potassium nitrate (1 : 1) about one-third being used on top of the charge. The cooled mass is extracted with boiling water and filtered. The filtrate, containing the alkali arsenate, is strongly acidified with acetic acid, boiled to expel the carbon dioxide, then cooled and treated with sufficient sodium hydroxide solution to give an alkaline reaction to phenolphthalein indicator. The purple red color is now discharged from the solution by addition of acetic acid. A slight excess of neutral silver nitrate is vigorously stirred in and the precipitate allowed to settle in the dark. The supernatant liquid is poured off through a filter and the precipitate washed by decantation with cold distilled water, then thrown on the filter and washed free of silver nitrate solution. The funnel is filled with water and 20 cc. of strong nitric acid added. The dissolved silver arsenate is caught in the original beaker in which the precipitation was made, the residue on the filter washed thoroughly with cold water and the filtrate and washings made up to 100 cc. The silver is now titrated by addition of standard ammonium or potassium thiocyanate, until a faint red color is evident, using ferric ammonium alum indicator, according to the procedure described for determination of silver. (See Chlorine and Silver Chapters.)

One cc. N/10 thiocyanate = 0.010788 gram Ag.

Factor. Ag × 0.2316 = As.

NOTE. The silver arsenate salt is nearly six times the weight of arsenic, so that very small amounts of arsenic may be determined by the procedure, hence it is not necessary to use over 0.5 gram of the material. For traces of arsenic the Gutzeit method, following, should be used.

DETERMINATION OF SMALL AMOUNTS OF ARSENIC
Modified Gutzeit Method

The following procedure furnishes a rapid and accurate method for determination of exceedingly small amounts of arsenic ranging from 0.001 milligram to 0.5 milligram As_2O_5. It is more sensitive and less tedious than the Marsh test. The details, given below with slight modifications, have been carefully worked out in the laboratories of the General Chemical Company [1] and have proved exceedingly valuable in estimating small amounts of arsenic in acids, bases, salts, soluble arsenic in lead arsenate and zinc arsenite and other insecticides, traces of arsenic in food products, baking powders, canned goods, etc.

The method depends upon the evolution of arsine by the action of hydrogen on arsenic compounds under the catalytic action of zinc, the reaction taking place either in alkaline or acid solutions. The evolved arsine reacts with mercuric chloride, forming a colored compound. From the length and intensity of the color stain the amount of arsenic is estimated by comparison with standard stains.

Although the acidity of the sample and the amount of zinc shot should be kept within certain limits, the results are not effected by slight variation as was formerly thought. The physical characteristics of the zinc used rather than the surface exposed to acid action appears to have an effect on the evolution of arsine. The best results are obtained with zinc having a finely crystalline structure.

Iron present in the solution tends to prevent evolution of stibine, but has no apparent effect of arsine generation.

Stannous chloride is essential to the complete evolution of arsine, hence this reagent is added to the solution in which arsenic is determined.

Antimony present in the solution in amounts less than 0.0001 gram, does not interfere with the determination of arsenic. If a greater amount of antimony is present a separation of arsenic should be made by distillation. The following modification of the method given on page 38 is recommended. In place of the generator for HCl shown in Fig. 1, air saturated with HCl, by passing it through a gas wash bottle containing concentrated hydrochloric acid, is drawn through the boiling solution containing the sample in a saturated HCl solution, reduction of arsenic to arsenious chloride having been effected with cuprous chloride as prescribed. The air sweeps the arsine into the water in the receiving flasks (Fig. 1). It is advisable to have two flasks connected in series in place of one as shown. Gentle suction is applied at the receiving end of the train. The apparatus may be made in fairly compact form.

NOTE. The above method was carefully investigated by R. M. Palmer and F. J. Seibert and found to be extremely accurate.

Special Reagents. *Standard Arsenic Solution.* One gram of resublimed arsenous acid, As_2O_3, is dissolved in 25 cc. of 20% sodium hydroxide solution (arsenic-free) and neutralized with dilute sulphuric acid. This is diluted with fresh distilled water, to which 10 cc. of 95% H_2SO_4 has been added, to a volume of 1000 cc. Ten cc. of this solution is again diluted to a liter with distilled water containing acid. Finally 100 cc. of the latter solution is diluted to a liter with distilled water containing acid. One cc. of the final solution contains 0.001 milligram As_2O_3.

Standard Stains. Two sets of stains are made, one for the small apparatus for determining amounts of As_2O_3 ranging from 0.001 to 0.02 milligram, and a second set for the larger-sized apparatus for determining 0.02 to 0.5 milligram As_2O_3. Stains made by As_2O_3 in the following amounts are convenient for the standard sets; e.g., small apparatus, 0.001, 0.002, 0.004, 0.006, 0.01, 0.15, 0.02 milligram As_2O_3. Large apparatus, 0.02, 0.05, 0.1, 0.2, 0.3, 0.4, 0.5 milligram As_2O_3.

In making the stain the requisite amount of standard reagent, As_2O_3 solution, is placed in the Gutzeit bottle with the amounts of reagents prescribed for the regular tests and the run made exactly as prescribed in the regular procedure.

Preservation of the Stains. The strips of sensitized paper with the arsenic stain are dipped in molten paraffine (free from water), and mounted on a sheet of white paper, folded back to form a cylinder. The tube is placed in a glass test-tube containing phosphorus pentoxide, which is then closed by a stopper. It is important to keep the stained strip dry, otherwise the stain soon fades, hence the paper on which the strips are mounted and the glass test-tube, etc., must be perfectly dry. It is advisable to keep the standard in a hydrometer case, while not in use, as light will gradually fade the color.

Sensitized Mercuric Chloride (or Bromide) Paper. 20×20 in. Swedish Filter Paper No. 0 is cut into four equal squares. For use in the large Gutzeit apparatus the paper is dipped into a 3.25% solution of mercuric chloride (mercuric bromide may be used in place of the chloride) or if it is to be used in the small Gutzeit apparatus it is dipped into a 0.35% mercuric chloride solution. (The weaker the solution, the longer and less intense will be the stain.) The paper should be of uniform thickness, otherwise there will be an irregularity in length of

[1] The accuracy of the method is within 10% of the truth.

Evolution of arsine by the electrolytic method, in place of the method outlined, proved to be unreliable. The evolution of arsine is effected by the slightest variation in conditions so it is extremely difficult to obtain concordant results.

Mercuric Bromide Paper.—Kemmerer and Schrenk (G. Kemmerer, H. H. Schrenk, Ind. Eng. Chem., **18**, 707, July, 1926) recommend that the paper that is to be sensitized be dried at 105° C. for 1 hour and stored in a desiccator over $CaCl_2$. The paper is cut into 4 mm. strips and saturated with a 1.5% solution of mercuric bromide in 95% ethyl alcohol. After draining the strips are dried in a desiccator for 10 minutes and used. The treated strips should not be stored for longer than 2 hrs. before use.

In the *Marsh* test arsine is passed through a glass tube constricted to capillarity. By application of heat the arsine is decomposed and metallic arsenic deposited. The tube is heated just before the capillary constriction so that arsenic deposits in the drawn out tube. Comparison is made with standards, the length of the stain being governed by the amount of arsenic in the evolved gas. Slight variations in the size of the capillary tube and rate of evolution make a notable variation in length of stain.

stain for the same amounts of arsenic. (The thicker the paper the shorter the stain. The paper is hung up and dried in the air, free from gas fumes, H₂S being particularly undesirable.) When dry, half an inch of the outer edge is trimmed off (since this is apt to contain more of the reagent), and the paper cut into strips. The paper with more concentrated reagent is cut into strips 13 cm. by 5 mm. and that with 0.5% mercuric chloride into strips 7 cm. by 4 mm. The paper is preserved in bottles with tight-fitting

FIG. 5.—Purification of Hydrochloric Acid.

stoppers. Standards should be made with each batch of paper. Paper with a white deposit of HgCl₂ should not be used.

Ferric Ammonium Alum. Eighty-four grams of the alum with 10 cc. of mixed acid is dissolved and made up to a liter. Ten cc. of this solution contains approximately 0.5 gram Fe₂O₃.

Lead Acetate. One per cent solution with sufficient acetic acid to clear the solution.

Zinc. Arsenic-free zinc shot, 3 to 6-in. mesh. The zinc is treated with C. P. hydrochloric acid, until the surface of the zinc becomes clean and dull. It is then washed, and kept, in a casserole, covered with distilled water, a clock-glass keeping out the dust.

Mixed Acid. One volume of arsenic-free H_2SO_4 is diluted with four volumes of pure water and to this are added 10 grams of NaCl per each 100 cc. of solution.

Stannous Chloride. Eighty grams of stannous chloride dissolved in 100 cc. of water containing 5 cc. arsenic-free hydrochloric acid (1.2 sp.gr.).

Arsenic-free Hydrochloric Acid. The commercial acid is treated with potassium chlorate to oxidize the arsenic to its higher form and the acid distilled. The distilling apparatus may be arranged so that a constant distillation takes place, acid from a large container dropping slowly into a retort containing potassium chlorate, fresh hydrochloric acid being supplied as rapidly as the acid distills. See Fig. 5 on page 48.

Lead Acetate Test Paper for Removal of H_2S. Large sheets of qualitative filter paper are soaked in a dilute solution of lead acetate and dried. The paper is cut into strips 7×5 cm.

Blanks should be run on all reagents used for this work. The reagents are arsenic-free if no stain is produced on mercuric chloride paper after forty-five minutes' test.

Special Apparatus. The illustration, Fig. 6 (page 52), shows the Gutzeit apparatus connected up, ready for the test. The dimensions on the left-hand side are for the small apparatus and those on the right for the large form. Rubber stoppers connect the tubes to the bottle. The apparatus consists of a wide-mouth 2-oz. or 8-oz. bottle according to whether the small or large apparatus is desired, a glass tube (see Fig. 6) containing dry lead acetate paper and moist glass wool for removal of traces of hydrogen sulphide and a small-bore tube containing the strip of mercuric chloride paper.

Preparation of the Sample

The initial treatment of the sample is of vital importance to the Gutzeit Method for determining traces of arsenic. The following procedures cover the more important materials or substances in which the chemist will be called upon to determine minute amounts of arsenic.

Traces of Arsenic in Acids. The acid placed in the Gutzeit apparatus should be equivalent to 4.2 grams of sulphuric acid or 3.1 grams of hydrochloric acid and should contain 0.05 to 0.1 gram Fe_2O_3 equivalent. If large samples are required for obtaining the test it is necessary either to expel a portion of the acid in order to obtain the above acidity or to make standard stains under similar conditions of acidity. It must be remembered that arsenous chloride is readily volatile, whereas the arsenic chloride is not, hence it is necessary to oxidize arsenic before attempting to expel acids. If nitric acid or bromine or chlorine (chlorate) be added for this purpose, it must be expelled before attempting the Gutzeit test. Nitric acid may be expelled by adding sulphuric acid and taking to SO_3 fumes. Free chlorine, bromine, or iodine will volatilize on warming the solution. Chlorine in a chlorate is expelled by taking the sample to near dryness in presence of free acid. Sulphurous acid or hydrogen sulphide, if present, should be expelled by boiling the solution, then making faintly pink with $KMnO_4$ and destroying the excess with a drop or so of oxalic acid. SO_2 is reduced by zinc and hydrogen to H_2S, which forms black HgS with mercuric chloride, hence removal of SO_2 and H_2S are necessary before running the test.

Sulphuric Acid. With amounts of arsenic exceeding 0.00005% As_2O_3, 5 to 10 grams of acid, according to its strength, are taken for analysis and diluted to 15 or 20 cc. If H_2S or SO_2 are present, expel by boiling for fifteen or twenty minutes. Prolonged fuming of strong acid should be avoided by previously diluting the acid with sufficient water. In mixed acid containing nitric acid the sample is taken to SO_3 fumes to expel nitric acid. The procedure given later for the regular determination is now followed.

For estimating very minute amounts of arsenic, 0.000005 to 0.00005% As_2O_3, it is necessary to take a 25- to 50-gram sample for analysis. The acid is treated as directed above for removal of H_2S or SO_2 or nitric acid and diluted in the Gutzeit apparatus to at least 130 cc., using the large apparatus. Upon the addition of iron and stannous chloride as directed in the procedure described on page 52 for large Gutzeit test. The stains are compared with standard stains produced by known amounts of arsenic added to 50-gram portions of arsenic-free sulphuric acid of strength equal to that of the sample. The stains are longer and less intense than those produced by less acid.

Hydrochloric Acid. Twenty cc. is taken for analysis (sp.gr. being known); the sample should contain an acid equivalent of about 3.1 grams of hydrochloric acid. Chlorine is expelled by bubbling air through the acid before taking a sample. The procedure is given for further treatment of the sample following the section on preparation of the sample.

Nitric Acid. One hundred cc. of the acid (sp.gr. being known) is evaporated with 5 cc. of concentrated sulphuric acid to SO_3 fumes, to expel nitric acid. Arsenic is determined in the residue by the standard procedure.

Iron Ores, Pyrites, Burnt Pyrites, Cinders, etc. One gram of the finely ground ore is oxidized by treating with 5 cc. of a mixture of 2 parts liquid bromine and 3 parts of carbon tetrachloride. After fifteen minutes, 10 cc. of concentrated nitric acid are added and the mixture taken to dryness. Five cc. of concentrated sulphuric acid (95%) are added and the mixture taken to SO_3 fumes to expel the nitric acid. The cooled sample is taken up with 50 cc. of water and digested until all of the iron sulphate has dissolved; it is now washed into a 100-cc. flask, made to volume, and arsenic determined in an aliquot portion in the usual way, given later. Insoluble Fe_2O_3, briquettes, etc., is best dissolved by fusion with potassium bisulphate, $KHSO_4$. The fused mass is dissolved in warm dilute hydrochloric acid, and then washed into the Gutzeit bottle.

Alumina Ores. Bauxite. One gram of bauxite is treated with one part of concentrated nitric acid and 6 parts of concentrated hydrochloric acid, and taken to dryness on the water bath. The residue is taken up with an equivalent of 4.7 grams of hydrochloric acid or 6.3 grams of sulphuric acid in a volume of 25 cc. and the mix heated until the material has dissolved. The sample is diluted to exactly 100 cc. and arsenic determined on an aliquot portion.

Phosphates, Phosphoric Acid. Arsenic, in phosphoric acid, combined or free, cannot be determined in the usual way, as P_2O_5 has a retarding effect upon the evolution of arsine, so that the results are invariably low, and small amounts of arsenic escaping detection. Arsenic, however, may be volatilized from phosphates and phosphoric acid, as arsenous chloride, $AsCl_3$, in a current of hydrogen chloride by heating to boiling. One gram or more of the phosphate is placed in a small distilling flask, connected directly to a 6-in. coil condenser dipping into the Gutzeit bottle, containing 20 to 30 cc. of cold distilled water. A second bottle connected in series may be attached for safeguarding loss (this seldom occurs). Fifty cc. of concentrated hydrochloric acid are added to the sample and 5 grams of cuprous chloride. Arsenic is distilled into the Gutzeit bottle by heating the solution to boiling and passing a current of air through strong hydrochloric acid into the distilling flask by applying suction at the receiving end of the system. All of the arsenic will be found in the first 10 or 15 cc. of the distillate. Arsenic may now be evolved after addition of iron, stannous chloride and zinc, as directed in the procedure.

Salts, Sodium Chloride, Magnesium Sulphate, etc. One-gram samples are taken and dissolved in a little water and an equivalent of 6.3 grams of sulphuric acid added. The solution of iron and stannous chloride having been added, the run is made with 5 cc. of zinc shot, placed in the Gutzeit bottle.

Baking Powder, Other than Phosphate Baking Powder. A 10-gram sample is heated with 10 cc. hydrochloric acid, 10 cc. of ferric ammonium alum and 30 cc. of distilled water, until the starch hydrolyzes. 0.5 cc. of stannous chloride is added to the hot solution and the mixture washed into the Gutzeit apparatus. The required amount of zinc is added and the arsenic determined as usual.

Phosphate Baking Powders. Ten grams of the material mixed to a paste with about 50 cc. of hydrochloric acid are transferred to a small distilling flask with a few cc. of HCl. A tube, connected to a bottle of strong hydrochloric acid, passes into the mixture in the flask through a ground glass stopper. The flask is attached to a tube, which dips into water in a Gutzeit bottle. Two grams of cuprous chloride are added, the apparatus made tight and the flask immersed in boiling hot water. By aspirating air through the system into the Gutzeit bottle, which is water cooled, arsenic distills into the bottle and may be determined by the procedure outlined.

Arsenic in Organic Matter, Canned Goods, Meat, etc. The finely chopped, well-mixed sample is placed in a large flask and enough water added to produce a fluid mass. An equal quantity of concentrated hydrochloric acid and 1 to 2 grams of potassium chlorate are added. The flask is shaken to mix the material and it is then placed on the steam bath. Upon becoming hot, nascent chlorine is evolved and vigorously attacks the organic matter. Half-gram portions of potassium chlorate are added at five-minute intervals, shaking the flask frequently. When the organic material has decomposed and the solution becomes a pale yellow color, the mass is diluted with water and filtered. Arsenic will be found in the filtrate. A white, amorphous substance generally remains on the filter, when cadaver is being examined. The filtrate is diluted to a given volume and an aliquot portion taken for analysis. This is evaporated to near dryness to expel excess of acid and decompose chlorates. An equivalent of 4.7 grams of hydrochloric acid is added (three times this amount for the large apparatus), the volume of the solution made to about 30 cc., 10 cc. of ferric ammonium alum and 0.5 cc. of stannous chloride added, and the solution poured into the Gutzeit apparatus for the test as given on the following page.

Procedure for Making the Test

For amounts of arsenic varying from 0.001 milligram to 0.02 milligram As_2O_3, the small apparatus is used. The volume of the solution should be 50 cc. It should contain an equivalent of 4.2 to 6.3 grams sulphuric acid and should have about 0.1 gram equivalent of Fe_2O_3 reduced by 0.5 cc. of stannous chloride solution. Arsine is generated by adding one 5-cc. crucible of arsenic-free zinc shot, $\frac{1}{3}$ to $\frac{1}{6}$-inch mesh. Temperature 75 to 80° F.

For amounts ranging from 0.02 to 0.5 milligram As_2O_3,[1] the large apparatus is used. The volume of the solution should be about 200 cc. and should contain an equivalent of 18.5 grams of sulphuric acid and should have 0.1 gram equivalent of Fe_2O_3, reduced by 0.5 cc. stannous chloride solution. Arsine is generated by adding one 12-cc. crucible of zinc shot ($\frac{1}{3}$ to $\frac{1}{6}$-inch mesh.) The temperature should be 105° F. The sample taken should be of such size that a stain is obtained equivalent to that given by 0.1 to 0.5 milligram As_2O_3.

Lead acetate paper is placed in the lower portion of tube *B*; the upper portion of *B* contains glass wool moistened with lead acetate solution; the tube *A* contains the test strip of mercuric chloride paper. See Fig. 6. Immediately upon adding the required amount of zinc to the solution in the bottles, the connected tubes are put in position, as shown in the illustration, and the bottle gently shaken and allowed to stand for one hour for the small apparatus, forty minutes for the large. The test paper is removed, dipped in molten paraffine and compared with the standard stains. See Plate I.

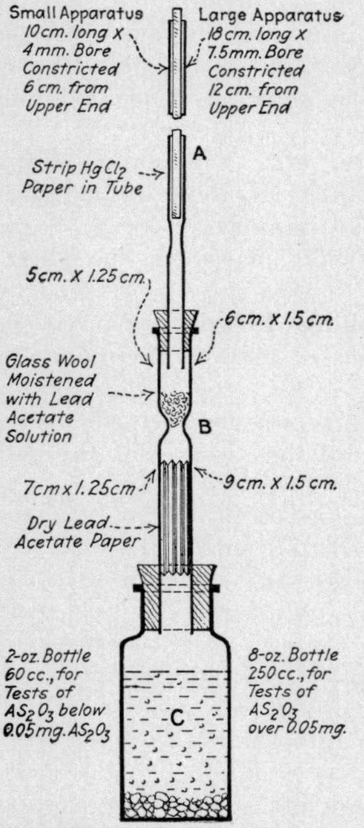

Small Apparatus
10 cm. long x
4 cm. Bore
Constricted
6 cm. from
Upper End

Large Apparatus
18 cm. long x
7.5 mm. Bore
Constricted
12 cm. from
Upper End

Strip Hg Cl₂
Paper in Tube

A

5 cm. X 1.25 cm.

6 cm. x 1.5 cm.

Glass Wool
Moistened
with Lead
Acetate
Solution

B

7 cm x 1.25 cm.

9 cm. X 1.5 cm.

Dry Lead
Acetate Paper

2-oz. Bottle
60 cc., for
Tests of
AS₂O₃ below
0.05 mg. AS₂O₃

8-oz. Bottle
250 cc., for
Tests of
AS₂O₃
over 0.05 mg.

C

Fig. 6.—Gutzeit Apparatus for Arsenic Determination.

Estimation of Per cent.

$$\frac{\text{The milligram } As_2O_3 \text{ stain} \times 100}{\text{Weight of sample taken}} = \% As_2O_3.$$

[1] It is advisable to use smaller samples when the arsenic content is over 0.3 milligram As_2O_3, as the longer stains are unreliable.

Ferrous iron prevents polarization between zinc and the acid and hence aids in the evolution of arsine.

In the analysis of baking powders, bauxite, sodium or similar salts, the distillation method is recommended. See pages **50** and **51**, " Phosphates," and " Phosphate baking powder."

Hydrochloric acid is used in place of sulphuric acid in cases where complete solution by the latter acid cannot be effected.

Standards and samples should be run under similar conditions, temperature, acidity, amount of zinc, volume of solution, etc. In place of zinc shot, zinc rods, cubes or discs may be used for generating arsine and hydrogen.

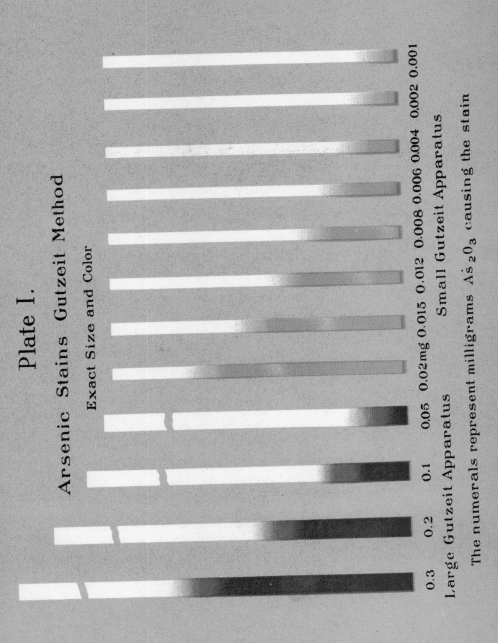

Plate I.

Arsenic Stains Gutzeit Method

Exact Size and Color

0.3 0.2 0.1 0.05 0.02mg 0.015 0.012 0.008 0.006 0.004 0.002 0.001

Large Gutzeit Apparatus   Small Gutzeit Apparatus

The numerals represent milligrams As_2O_3 causing the stain

METHOD FOR ANALYSIS OF COMMERCIAL "ARSENIC," ARSENOUS OXIDE, As_2O_3

The following constituents may be commonly present as impurities, SiO_2, Sb_2O_3, Fe_2O_3, NiO, CoO, CaO, SO_3, Cu, Pb, and Zn.

Determination of Moisture

Two 10-gram samples are dried to constant weight in the oven at 100° C. Loss in weight = moisture.

Sulphuric Acid, H_2SO_4

The samples from the moisture determination are dissolved in concentrated hydrochloric acid, heating to boiling if necessary, and the samples diluted to 300 to 400 cc. Barium chloride solution is added in slight excess to the hot solution, the precipitate, $BaSO_4$, allowed to settle and filtered and the sulphate dried and ignited as usual.

$$BaSO_4 \times 0.343 = SO_3.$$

Determination of Arsenic as As_2O_3

Duplicate 5-gram samples are dissolved in 20 grams potassium carbonate in 60 cc. of hot water, by boiling until solution is effected. The samples are made up to 1 liter and aliquots of 100 cc. (=0.5 gram) taken for analysis. The solution is made faintly acid with hydrochloric acid, testing the solution with litmus paper or by adding methyl orange directly to the solution. An excess of bicarbonate is added and the arsenic titrated with tenth-normal iodine according to the standard procedure for arsenic. One cc. N/10 I = .004948 gram As_2O_3.

Residue upon Sublimation of As_2O_3. SiO_2, Pb, Cu, Fe_2O_3, NiO, CoO, Zn

Two 5-gram samples are weighed into tared porcelain crucibles and heated gently on sand baths with the sand banked carefully around the crucible so as to heat the entire receptacle. After the greater part of the arsenous oxide has volatilized, the crucible is ignited directly in the flame to a dull red heat, until fumes are no longer given off. The residue is weighed as total non-sublimable residue.

Silica

The residues are transferred to beakers and treated with aqua regia, taken to dryness, and the silica dehydrated at 110° C. for an hour or more. The residue is taken up with hot dilute hydrochloric acid, boiled, and the silica filtered off, ignited, and weighed.

Lead and Copper

The filtrate from the silica is "gassed" with H_2S and the precipitate filtered off. The filtrate is put aside for determination of iron, etc. The precipitate is dissolved in hot dilute nitric acid, 2 to 3 cc. of concentrated sulphuric acid added, the solution taken to SO_3 fumes, the cooled concentrate diluted to 20 or 30 cc., and the lead sulphate filtered off, ignited, and weighed as $PbSO_4$.

The filtrate from the lead sulphate containing the copper is treated with aluminum powder and the copper thrown out of solution; the excess of aluminum is dissolved with a few cc. of hydrochloric acid. The filtrate should be tested for copper with H_2S and the precipitate added to the copper thrown out by the aluminum. The copper on the filter is dissolved in hot dilute nitric acid, the extract evaporated to 2 or 3 cc., the acid neutralized with ammonia and then made acid with acetic, potassium iodide added and the liberated iodine titrated with standard thiosulphate solution according to the regular scheme for copper.

Iron, Nickel, Cobalt, and Zinc

The filtrate from the H_2S Group is boiled to expel the H_2S and the iron oxidized by addition of nitric acid and boiling. The iron (and alumina) is precipitated with ammonium hydroxide and the precipitate filtered off and washed several times with hot water. If alumina is suspected (light-colored precipitate) it may be determined by the difference method—ignition of the precipitate, weighing, and finally subtracting the iron found by titration with standard stannous chloride solution. The iron is dissolved in hydrochloric acid and titrated hot with stannous chloride solution.

The filtrate from the iron is boiled and a 1% alcoholic solution of dimethylglyoxime added to precipitate the nickel. The salt is filtered on a tared Gooch, the precipitate dried at 100° C., and weighed. The weight of the salt$\times 0.2032 = Ni$.

The filtrate from the nickel is boiled until all the alcohol has been driven off and the cobalt precipitated by addition of sodium hydroxide in excess filtered, ignited, and weighed as CoO.

The filtrate is made acid with hydrochloric acid, and then alkaline with ammonium hydroxide and colorless sodium sulphide solution added to precipitate the zinc. The mixture is boiled five to ten minutes, the precipitated ZnS allowed to settle, filtered off, and washed once or twice and then dissolved in hydrochloric acid and the zinc determined by titration directly with potassium ferrocyanide, or by converting to the carbonate by addition of potassium carbonate, filtered and washed free of alkali, the precipitate dissolved in a known amount of standard acid, and the excess acid titrated with standard caustic (methyl orange indicator) according to the procedure given for zinc. $H_2SO_4 \times 0.06665 = Zn$.

Antimony and Calcium Oxides

Two 15-gram samples are treated with 300 cc. of concentrated hydrochloric acid, boiled down to 50 cc. to expel the arsenic as $AsCl_3$, an equal amount of concentrated hydrochloric acid is added, and the last traces of arsenic precipitated by H_2S passed into the hot concentrated hydrochloric acid solution. The arsenous sulphide, As_2S_3, is filtered off. Antimony is precipitated by diluting the solution with an equal volume of water, the solution having been concentrated by boiling down to about 50 cc. The Sb_2S_3 is filtered off, washed several times with hot water, dissolved by washing through the filter with concentrated hydrochloric acid, and antimony determined in the strong hydrochloric acid solution by the potassium bromate method—addition of methyl orange indicator and titration with standard potassium bromate added to the hot solution to the disappearance of the pink color of the indicator.

The filtrate from the antimony is concentrated, made slightly alkaline with ammonium hydroxide, and gased with hydrogen sulphide to remove iron, nickel, cobalt, zinc, chromium, and last traces of lead, etc. The filtrate is then concentrated and made acid with crystals of oxalic acid, boiled and methyl orange added and then ammonia drop by drop slowly until the indicator changes to an orange color. An excess of ammonium oxalate is now added and the beaker placed on the steam bath until the calcium oxalate has settled. The lime is now determined by filtering off the precipitate and washing, drying and igniting to CaO, or by titration with standard permanganate, according to the regular procedure for calcium.

Arsenic in Iron and Steel. Ten grams of sample are placed in a distillation flask and dissolved in dilute HNO_3, the solution evaporated to dryness and heated to expel oxides of nitrogen, 100 cc. of HCl and 20 grams of CuCl are added and the arsenious acid distilled and determined by the iodine method.

Arsenic in Copper. Since arsenic impairs the electrical conductivity of copper, its determination is required. One gram of the sample is placed in a distillation flask with 10 cc. $FeCl_3$ and 100 cc. HCl and 5 grams KCl and the arsenious acid distilled and determined by titration with iodine.

BARIUM

Ba, *at.wt.* 137.37; *sp.gr.* 3.78; *m.p.* 850° C.; *volatile at* 950° C.; *oxides,* BaO, BaO$_2$.

DETECTION

Barium is precipitated as the carbonate together with strontium and calcium, by addition of ammonium hydroxide and ammonium carbonate to the filtrate of the ammonium sulphide group. It is separated from strontium and calcium by precipitation as yellow barium chromate, BaCrO$_4$, from a slightly acetic acid solution.

Saturated solutions of calcium or strontium sulphates precipitate white barium sulphate, BaSO$_4$, from its chloride or nitrate or acetate solution, barium sulphate being the least soluble of the alkaline earth sulphates.

Soluble chromates precipitate yellow barium chromate from its neutral or slightly acetic acid solution, insoluble in water, moderately soluble in chromic acid, soluble in hydrochloric or nitric acid.

Fluosilicic acid, H$_2$SiF$_6$, precipitates white, crystalline barium fluosilicate, BaSiF$_6$, sparingly soluble in acetic acid, insoluble in alcohol. (The fluosilicates of calcium and strontium are soluble.)

Flame. Barium compounds color the flame yellowish green, which appears blue through green glass.

Spectrum. Three characteristic green bands (α, β, γ).

Barium sulphate is precipitated by addition of a soluble sulphate to a solution of a barium salt. The compound is extremely insoluble in water and in dilute acids (soluble in hot concentrated sulphuric acid). The sulphate is readily distinguished from lead sulphate by the fact that the latter is soluble in ammonium salts, whereas barium sulphate is practically insoluble.

ESTIMATION

The determination of barium is required in the valuation of its ores, barite, heavy spar, BaSO$_4$; witherite, BaCO$_3$; baryto calcite, BaCO$_3$CaCO$_3$. It is determined in certain white mixed paints and colored pigments, Venetian, Hamburg or Dutch whites, chrome paints, etc. In analysis of Paris green, baryta insecticides, putty, asphalt, dressings and pavement surfacings. It may be found as an adulterant in foods, wood preservatives, filler in rubber, rope, fabrics. It is determined in salts of barium. The nitrate is used in pyrotechny, in mixtures for green fire.

Chapter contributed by Wilfred W. Scott.

Preparation and Solution of the Sample

Compounds of barium, with the exception of the sulphate, $BaSO_4$, are soluble in hydrochloric and nitric acids. The sulphate is soluble in hot concentrated sulphuric acid, but is reprecipitated upon dilution of the solution. The sulphate is best fused with sodium carbonate, which transposes the compound to barium carbonate; sodium sulphate may now be leached out with water and the residue, $BaCO_3$, then dissolved in hydrochloric acid.

Solution of Ores. Sulphates. 0.5 to 1 gram of the finely divided ore is fused with 3 to 5 grams of sodium and potassium carbonate mix, 2 : 1, or sodium carbonate alone, in a platinum dish. (Prolonged fusion is not necessary.) The melt is cooled and then extracted with hot water to dissolve out the alkali sulphates. Barium carbonate, together with the other insoluble carbonates, may now be dissolved by hot dilute hydrochloric acid. From this solution barium may be precipitated by addition of sulphuric acid. If it is desired to separate barium along with strontium, calcium, and magnesium, the members of the preceding groups are removed by H_2S in acid and in ammoniacal solution, as directed under "Separations."

Sulphides. The ore is oxidized, as directed for pyrites under the subject of sulphur. After the removal of the soluble sulphates, the residue, containing silica, barium, and small amounts of insoluble oxides, is fused and dissolved according to the procedure for sulphates.

Carbonates. In absence of sulphates the material may be dissolved with hydrochloric acid, taken to dryness to dehydrate silica and after heating for an hour in the steam oven (110°±) the residue is extracted with dilute hydrochloric acid and filtered. The filtrate is examined for barium according to one of the procedures given later.

Salts Soluble in Water. Nitrates, chlorides, acetates, etc., are dissolved with water slightly acidulated with hydrochloric acid.

Material Containing Organic Matter. The substance is roasted to destroy organic matter before treatment with acids or by fusion with the alkali carbonates.

The Insoluble Residue remaining from the acid treatment of an ore may contain barium sulphate in addition to silica, etc. The filter containing this residue is burned and the ash weighed. Silica is now volatilized by addition of hydrofluoric acid with a few drops of sulphuric acid, and evaporation to dryness. If an insoluble substance still remains after taking up the remaining residue with dilute hydrochloric acid, barium sulphate is indicated. This is treated according to the method given for sulphates.

NOTE. The insoluble substance remaining is frequently ignited and weighed as barium sulphate without fusion with the carbonate.

SEPARATIONS

The Alkaline Earths

Preliminary Considerations. In the determination of barium, calcium, and strontium, the following causes may lead to loss of the elements sought:

a. Presence of Phosphates. Phosphoric acid, free or combined, has a decided influence upon the determination of the members of this group. Combined as phosphate it will cause the complete precipitation of barium, calcium, and strontium, along with iron, alumina, etc., upon making the solution alkaline for removal of the ammonium sulphide group. It is a common practice to hold up the iron+alumina by means of tartaric, citric, or other organic acids before making ammoniacal for precipitation of this group as oxalates, or again the basic acetate method is used for precipitation of iron and alumina; calcium, barium, and strontium going into solution. These procedures may be satisfactory for the analysis of phosphate rock and similar products, but do not cope with the difficulty when large amounts of phosphates are present. In samples containing free phosphoric acid, barium, calcium, and strontium, present in small amounts, may remain in solution in presence of sulphates or oxalates. Appreciable amounts of calcium, 1% or more, may escape detection by the usual method of precipitation by ammonium oxalate added to the alkaline solution, on account of this interference, so that the removal of phosphoric acid before precipitation of this group is frequently necessary. This may be accomplished by addition of potassium carbonate in sufficient excess to combine completely with the phosphoric acid and form carbonates with the bases. The material taken to dryness is fused with additional potassium carbonate in an iron crucible, and the fusion leached with hot water—sodium phosphate dissolves and the carbonates of the heavy metals remain insoluble.

b. Another source of loss is the presence of sulphates, either in the original material or by intentional or accidental addition, in the latter case due to the oxidation of hydrogen sulphide, which has been passed into the solution during the removal of elements of the hydrogen sulphide and ammonium sulphide groups, barium and strontium sulphate being precipitated along with these members. A potassium carbonate fusion will form K_2SO_4, which may be leached out with water.

c. Loss may be caused by occlusion of barium, calcium, strontium, and magnesium by the gelatinous precipitates $Fe(OH)_3$, $Al(OH)_3$, etc. A double precipitation of these compounds should be made if considerable amounts are present.

d. A large excess of ammonium salts, which accumulate during the preliminary separations, will prevent precipitation of the alkaline earths. This can be avoided by using the necessary care required for accurate work, the addition of reagents by means of burettes or according to definite measurements in graduates, etc. Careless addition of large amounts of ammonium hydroxide and hydrochloric acid should be guarded against. In case large amounts of ammonium chloride are present, time is frequently saved by a repetition of the separations. Ammonium chloride may be expelled by heating the material, taken to dryness in a large platinum dish, the ammonium salts being volatilized.

e. Carbon dioxide absorbed by ammonium hydroxide from the air will precipitate the alkaline earths with the ammonium sulphide group.

Emmissions-Spectra.

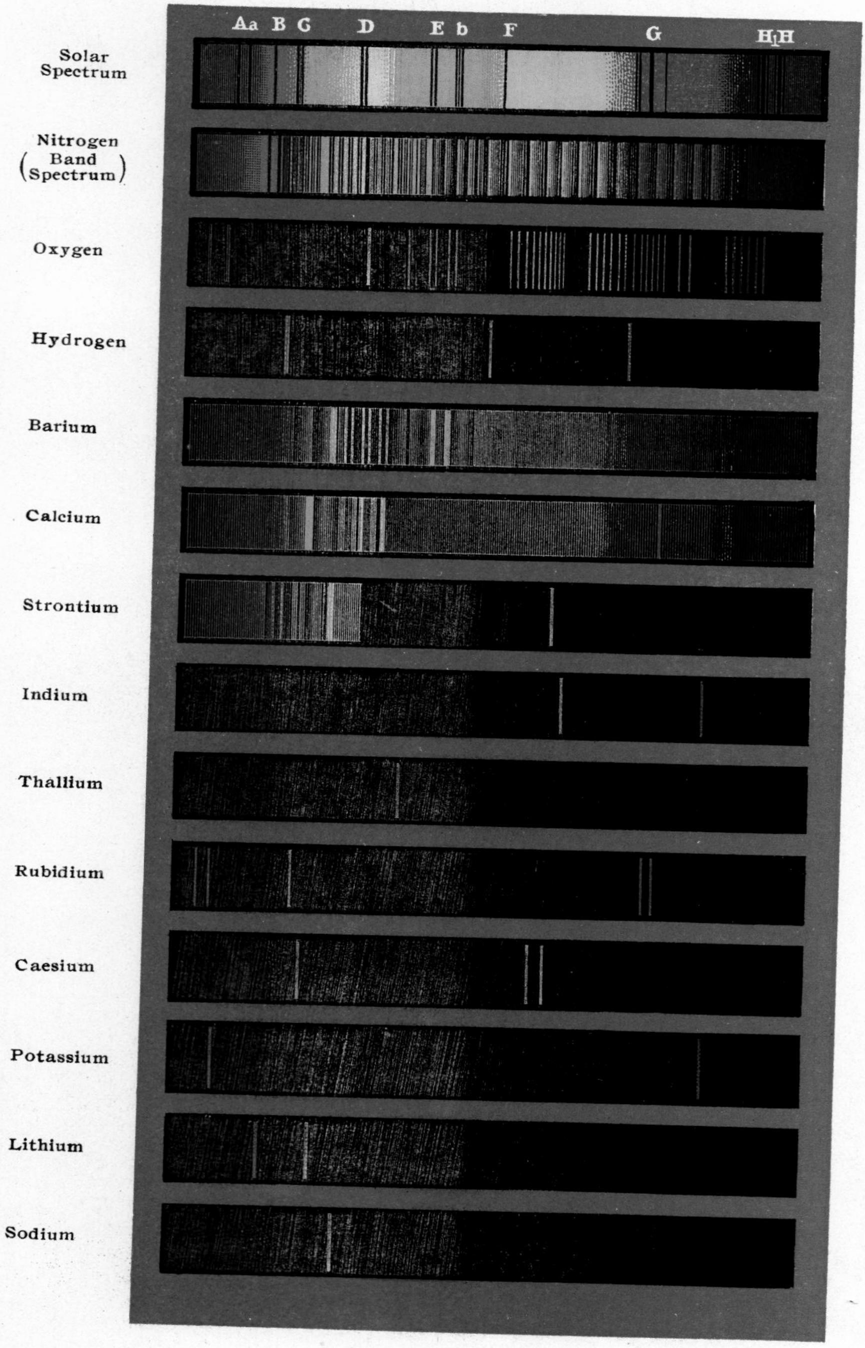

	Aa	B	G	D	E	b	F		G		H₁H

Solar Spectrum

Nitrogen Band Spectrum

Oxygen

Hydrogen

Barium

Calcium

Strontium

Indium

Thallium

Rubidium

Caesium

Potassium

Lithium

Sodium

Plate II.

Direct Precipitation on Original Sample. For the determination of barium, calcium, and strontium, it is advisable to take a fresh sample, rather than one that has been previously employed for the estimation of the hydrogen sulphide and ammonium sulphide groups, as is evident from the statements made above. The alkaline earths are isolated by being converted to the insoluble sulphates and separations effected as given later under Sulphate Method.

Preliminary Tests. Much time may be saved by making a preliminary test for barium, strontium, and calcium by means of the spectroscope and avoiding unnecessary separations.

By means of the spectroscope with the use of the ordinary Bunsen flame exceedingly minute amounts of calcium, strontium and barium may be detected per cc. The test is very much more delicate by the arc spectra method.[1] The liquid containing the substance is connected to the positive pole and an iridium needle is connected by means of an adjustable resistance of 300 to 500 ohms to the negative pole. An E.M.F. of 100 to 200 volts and 1 ampere current are required. By the arc it is possible to detect 0.002 milligram of calcium, 0.003 milligram of strontium, 0.006 milligram of barium, 0.1 milligram of magnesium per cc. In these concentrations, calcium shows one brilliant line (423 $\mu\mu$), a bright line (616 $\mu\mu$), and a faint line between them; strontium two bright lines (422 and 461 $\mu\mu$) and two fairly bright lines; barium two brilliant lines (455 and 493 $\mu\mu$), two other bright lines, and a fairly bright one; and magnesium a brilliant band composed of three lines (516.8 to 518.4 $\mu\mu$), as well as a fairly bright line further towards the violet end of the spectrum

The flame test may be of value in absence of sodium; barium giving a green flame, strontium a brilliant scarlet, and calcium an orange red.

Separation of the Alkaline Earths from Magnesium and the Alkalies. Two general procedures will cover conditions commonly met with in analytical work:

A. Oxalate Method. Applicable in presence of comparatively large portions of calcium. The acid solution containing not over 1 gram of the mixed oxides is brought to a volume of 350 cc. and for every 0.1 gram of magnesium present about 1 gram of ammonium chloride is added, unless already present. Sufficient oxalic acid is added to completely precipitate the barium, calcium, and strontium.[2] ($H_2C_2O_4 \cdot 2H_2O = 126.04$, $Ba = 137.37$, $Ca = 40.07$, $Sr = 87.63$.) The solution is slowly neutralized by addition, drop by drop, of dilute ammonium hydroxide (1 : 10), methyl orange being used as indicator. About $\frac{1}{2}$ gram of oxalic acid is now added in excess, the solution again made alkaline with ammonium hydroxide, and allowed to settle for at least two hours. The precipitate is filtered off and washed with water containing 1% ammonium oxalate, faintly alkaline with ammonia.

The precipitate contains all the calcium and practically all of the barium and strontium. If Mg is present in amounts of 10 to 15 times that of the alkaline earths a double precipitation is necessary, to remove it completely from this group. The oxalates are dissolved in hydrochloric acid and reprecipitated with ammonium oxalate in alkaline solution.

The filtrate contains magnesium and the alkalies. Traces of barium and strontium may be present. If the sample contains a comparatively large proportion of barium and strontium, the filtrate is evaporated to dryness, the ammonium salts expelled by gentle ignition of the residue, and the Ba and

[1] E. H. Riesenfeld and G. Pfützer, Ber., 1913, **46**, 3140–3144; Analyst, 1913, **38**, 584.

[2] Calcium and strontium will slowly precipitate in the oxalic acid solution. Ba oxalate will precipitate upon making the solution alkaline.

Sr recovered as sulphates according to the method described below. Magnesium is precipitated as magnesium ammonium phosphate from the filtrate.

The oxalates of barium, calcium, and strontium are ignited to oxides, in which form they may be readily converted to chlorides by dissolving in hydrochloric acid, or to nitrates by nitric acid.

B. Sulphate Method. Applicable in presence of comparatively large proportions of barium, strontium, or magnesium. The solution containing the alkaline earths, magnesium and the alkalies is evaporated to dryness and about 5 cc. concentrated sulphuric acid added, followed by 50 cc. of 95% alcohol. The sulphates [1] of barium, calcium, and strontium, are allowed to settle, and then filtered onto an S. and S. No. 589 ashless filter paper and washed with alcohol until free of magnesium sulphate. In presence of large amounts of magnesium as in case of analyses of Epsom salts and other magnesium salts it will be necessary to extract the precipitate by adding a small amount of water, then sufficient 95% alcohol to make the solution contain 50% alcohol and filter from the residue. Magnesium is determined in the filtrate.

The residue containing barium, calcium, and strontium as sulphate is fused with 10 parts of potassium carbonate or sodium acid carbonate until the fusion becomes a clear molten mass, a deep platinum crucible being used for the fusion. A platinum wire is inserted and the mass allowed to solidify. The fusion may be removed by again heating until it begins to melt around the surface next to the crucible, when it may be lifted out on the wire. The mass is extracted with hot water and filtered, Na_2SO_4 going into the solution and the carbonates of barium, strontium, and calcium remaining insoluble. The carbonates should dissolve completely in hydrochloric acid or nitric acid, otherwise the decomposition has not been complete, and a second fusion of this insoluble residue will be necessary.

Separation of the Alkaline Earths from One Another. This separation may be effected by either of the following processes:

1. Barium is separated in acetic acid solution as a chromate from strontium and calcium; strontium is separated as a nitrate [2] from calcium in ether-alcohol or amyl alcohol.

2. The three nitrates are treated with ether-alcohol in which barium and strontium nitrates are insoluble and calcium dissolves; the barium is now separated from strontium by ammonium chromate.

Procedures. 1. (a) Separation of Barium from Strontium (and from Calcium). In presence of an excess of ammonium chromate, barium is precipitated from solutions, slightly acid with acetic acid, as barium chromate (appreciably soluble in free acetic acid), whereas strontium and calcium remain in solution.

The mixed oxides or carbonates are dissolved in the least amount of dilute hydrochloric acid and the excess of acid expelled by evaporation to near dryness. The residue is taken up in about 300 cc. of water and 5–6 drops of acetic acid (sp.gr. 1.065) together with sufficient ammonium acetate (30% solution) to neutralize any free mineral acid present. The solution is heated and an excess of ammonium chromate (10% neutral sol.) [3] added (10 cc. usually sufficient).

[1] Solubility of $BaSO_4 = 0.17$ milligram, $CaSO_4 = 179$ milligram, $SrSO_4 = 11.4$ milligrams per 100 cc. sol. cold.

[2] Method of Stromayer and Rose. H. Rose, Pogg. Ann., **110**, 292, (1860).

[3] The solution is prepared by adding NH_4OH to a solution of $(NH_4)_2Cr_2O_7$ until yellow. The solution should be left acid rather than alkaline.

The precipitate of barium chromate is allowed to settle for an hour and filtered off on a small filter and washed with water containing ammonium chromate until free of soluble strontium and calcium (test—addition of NH_4OH and $(NH_4)_2CO_3$ produces no cloudiness), and then with water until practically free of ammonium chromate (e.g., only slight reddish brown color with silver nitrate solution).

To separate any occluded precipitate of strontium or calcium the filter paper is pierced and the precipitate rinsed into a beaker with warm dilute nitric acid (sp.gr. 1.20) (2 cc. usually are sufficient). The solution is diluted to about 200 cc. and boiled. About 5 cc. of ammonium acetate, or enough to neutralize the free HNO_3, are added to the hot solution and then sufficient ammonium chromate to neutralize the free acetic acid, 10 cc. usually sufficient. The washing, as above indicated, is repeated. Barium is completely precipitated and may be determined either as a chromate or a sulphate or by a volumetric procedure. Strontium and calcium are in the filtrates and may be separated as follows:

(b) Separation of Strontium from Calcium. The method depends upon the insolubility of strontium nitrate and the solubility of calcium nitrate in a mixture of ether-alcohol, 1 : 1.

Solubility of $Sr(NO_3)_2 = 1$ part $Sr(NO_3)_2$ in 60,000 parts of the mixture. Ca easily soluble.

If the solution is a filtrate from barium, 1 cc. of nitric acid is added and the solution heated and made alkaline with ammonium hydroxide followed immediately with ammonium carbonate, the carbonates of strontium (together with some $SrCrO_4$) and calcium will precipitate. The precipitate is dissolved in hydrochloric acid and reprecipitated from a hot solution with ammonium hydroxide and ammonium carbonate. The precipitate, $SrCO_3$ and $CaCO_3$, is washed once with hot water and is then dissolved in the least amount of nitric acid, washed into a small casserole, evaporated to dryness and heated for an hour at 140 to 160° C. in an oven, or at 110° C. over night. The dry mass is pulverized and mixed with 10 cc. of ether-alcohol (absolute alcohol, one part, ether-anhydrous, one part). Several extractions are thus made, the extracts being decanted off into a flask. The residue is again dried in an oven at 140 to 160° C., then pulverized and washed into the flask with the ether-alcohol mixture and digested for several hours with frequent shaking of the flask. The residue is washed onto a filter moistened with ether-alcohol mixture. Strontium nitrate, $Sr(NO_3)_2$, remains insoluble, and may be dissolved in water and determined gravimetrically as a sulphate, oxide, or carbonate or volumetrically. Calcium is in the filtrate and may be determined gravimetrically as an oxide or volumetrically.

Instead of using a mixture of ether-alcohol, amyl alcohol may be used (hood), the mixture being kept at boiling temperature to dehydrate the alcohol to prevent solution of strontium (b.p. = 130° C.).

2. Separation of Barium and Strontium from Calcium.[1] The procedure depends upon the insolubility of barium nitrate, $(BaNO_3)_2$, and strontium nitrate, $Sr(NO_3)_2$, in a mixture of anhydrous ether and absolute alcohol or anhydrous amyl alcohol, whereas $Ca(NO_3)_2$ dissolves.

The mixed oxides or carbonates are dissolved in nitric acid and taken to dryness in a beaker or Erlenmeyer flask, and heated for an hour or more in an

[1] See Fresenius, Z. anal. Chem., **29**, 413–430 (1890).

oven at 140 to 160° C. Upon cooling, the mixture is treated with ten times its weight of ether-alcohol mixture and digested, cold, in the covered beaker or corked flask for about two hours with frequent stirring. An equal volume of ether is now added and the digestion continued for several hours longer. The residue is washed by decantation with ether and alcohol mixture until calcium is removed (test—no residue on platinum foil with drop of filtrate evaporated to dryness).

Separation of Barium from Strontium. The dry mixed chlorides are dissolved in the least possible amount of water (0.2 cc., or more if necessary) the solution warmed, then cooled. More water is added if crystals appear. (The solution should be saturated.) A mixture of 4 : 1 of HCl (33%) and ether is added dropwise with stirring. Sufficient reagent is added to precipitate $BaCl_2$ and dissolve $SrCl_2$. The mixture is decanted on an asbestos filter and washed with the HCl-ether reagent. The $BaCl_2$ is dried at 150° C. and weighed. (Method of Gooch and Soderman.)

Barium and strontium may be separated by precipitation of barium as a chromate, the nitrate residue being dissolved in water and barium precipitated according to directions given under Procedure No. 1.

Amyl alcohol may be used in place of ether-alcohol by digesting the nitrates in a boiling solution (130° C.), calcium going into solution and barium and strontium remaining insoluble as nitrates.

Separation of the Alkaline Earths from Molybdenum. The substance is fused with sodium carbonate and the fusion extracted with water and filtered. Molybdenum passes into the filtrate and the alkaline earths remain in the residue.

Separation of Phosphoric Acid from the Alkaline Earths. Ammonium carbonate is added to the hydrochloric acid solution until a slight permanent turbidity is obtained, and the solution just cleared with a few drops of HCl. Ferric chloride is now added drop by drop until the solution above the yellowish white precipitate becomes brownish in color. The solution is diluted to about 400 cc. and brought to boiling and then filtered and the residue washed with water containing ammonium acetate. The filtrate contains the alkaline earths, free from phosphoric acid.

GRAVIMETRIC METHODS FOR THE DETERMINATION OF BARIUM

For reasons given under "Preliminary Considerations," it is advisable to take a special sample for the determination of barium that has not undergone treatment with hydrogen sulphide or ammonium hydroxide, since these may cause the loss of barium as stated.

Barium in Insoluble Residue. In the complete analysis of ores the residue remaining insoluble in acids is composed largely of silica, together with difficultly soluble substances, among which is barium sulphate. This residue is best fused in a platinum dish with sodium carbonate or a mixture of sodium and potassium carbonates (long fusion is not necessary). The cooled mass is digested with hot water to remove the soluble sodium compounds, silicate being included. Barium, together with the heavy metals, remains insoluble as carbonate and may be filtered off. The residue is now treated with dilute ammonia water to remove the adhering sulphates (testing the filtrate with hydrochloric acid and barium chloride solution; the washing being complete when no white precipitate of barium sulphate forms). The carbonates are washed off the filter into a 500-cc. beaker, the clinging carbonate being dissolved by pouring a few cc. of dilute, 1 : 1, hydrochloric acid on the paper placed in the funnel. This extract is added to the precipitate in the beaker and the latter covered to prevent loss by spattering. Additional hydrochloric acid is cautiously added so that the precipitate completely dissolves and the solution contains about 10 cc. of free hydrochloric acid (sp.gr. 1.2). Barium is precipitated from this solution best as a sulphate according to directions given later.

Silicates. One gram of the finely pulverized sample is treated with 10 cc. of dilute sulphuric acid, 1 : 4, and 5 cc. of strong hydrofluoric acid. The mixture, evaporated to small bulk on the steam bath, is taken to SO_3 fumes on the hot plate. Additional sulphuric acid and hydrofluoric acid are used if required. By this treatment the silica is expelled and barium, together with other insoluble sulphates, will remain upon the filter when the residue is treated with water and filtered. Lead sulphate, if present, may be removed by washing the residue with a solution of ammonium acetate. Barium sulphate may be purified by fusion with potassium carbonate as above directed or by dissolving in hot concentrated sulphuric acid, and precipitating again as $BaSO_4$ by dilution.

Ores may be decomposed by either of the above methods or a combination of the two. Sulphide ores require roasting to oxidize the sulphide to sulphate.

Barium Sulphate is decomposed by fusion with sodium and potassium carbonates. The fusion is leached with water to remove the soluble sulphate and the residue, $BaCO_3$, is dissolved in HCl. Barium is determined in this solution.

8

Determination of Barium as a Chromate

A preliminary spectroscopic test has indicated whether a separation from calcium and strontium is necessary. If these are present, barium is separated along with strontium from calcium as the nitrate in presence of alcohol-ether mixture, according to directions given under "Separations." Barium is now precipitated as the chromate, $BaCrO_4$, from a neutral or slightly acetic acid solution, strontium remaining in solution.

Precipitation of Barium Chromate. If barium is present in the form of nitrate, together with strontium, the mixed nitrates are evaporated to dryness and then taken up with water. About 10 cc. ammonium acetate (300 grams $NH_4C_2H_3O_2$ neutralized with $NH_4OH + H_2O$ to make up to 1000 cc.) added and the solution heated to boiling. Five cc. of 20% ammonium bichromate are added drop by drop with constant stirring and the precipitate allowed to settle until cold. The solution is decanted off from the precipitate through a filter and washed by decantation with dilute (0.5%) solution of ammonium acetate, until the excess chromate is removed, as indicated by the filtrate passing through uncolored. If much strontium was originally present, a double precipitation is necessary, otherwise the precipitate may be filtered directly into a Gooch crucible and dried, (120° C.) to constant weight.

Purification from Strontium. The precipitate is dissolved from the filter by running through dilute (1 : 5) warm nitric acid, poured upon the chromate, catching the solution in the beaker in which the precipitation was made; the least amount of acid necessary to accomplish this being used and the filter washed with a little warm water. Ammonium hydroxide is now added to the solution, cautiously, until a slight permanent precipitate forms and then 10 cc. of ammonium acetate solution added with constant stirring and the mixture heated to boiling. The precipitate is allowed to settle until the solution is cold and then filtered and washed by decantation as before, a Gooch crucible being used to catch the precipitate.

Ignition. The precipitate is washed with dilute alcohol once, then dried at 110° C. The Gooch containing the $BaCrO_4$ is gently heated in a larger crucible (allowing an encircling air space around the Gooch) until the color of the chromate becomes uniform.

$$BaCrO_4 \times 0.6051 = BaO. \quad BaCrO_4 \times 0.5420 = Ba.$$

NOTES. The use of sodium hydrate or acetate in place of the ammonium hydroxide and acetate is sometimes recommended, owing to the slight solubility of $BaCrO_4$ in ammonium salts, as seen by the following table, approximate figures being given:

100,000 parts of cold water dissolves	0.38 parts $BaCrO_4$
100,000 parts of hot water dissolves	4.35 parts $BaCrO_4$
100,000 of 0.5% solution of NH_4Cl dissolves	4.35 parts $BaCrO_4$
100,000 of 0.5% solution of NH_4NO_3 dissolves	2.22 parts $BaCrO_4$
100,000 of 0.75% solution of $NH_4C_2H_3O_2$ dissolves	2.00 parts $BaCrO_4$
100,000 of 1.5% solution of $NH_4C_2H_3O_2$ dissolves	4.12 parts $BaCrO_4$
100,000 of 1% acetic acid dissolves	20.73 parts $BaCrO_4$

Although the solvent action of ammonium salts is practically negligible under conditions of analysis given above, the solvent action of free acetic acid is of importance, so that it is necessary to neutralize or eliminate free mineral acids before addition of the acetate salt.

The edges of the $BaCrO_4$ precipitate upon drying may appear green, owing to the action of alcohol; upon ignition, however, the yellow chromate is obtained. The color orange yellow, when hot, fades to a light canary yellow upon cooling.

$BaCrO_4$, *mol.wt.*, 253.47; *sp.gr.*, $4.498^{15°}$; 100 cc. H_2O sol. cold will dissolve $0.00038^{18°}$ gram, hot dissolves 0.0043 gram; soluble in HCl, HNO_3, yellow rhombic plates.

Determination of Barium as Barium Carbonate. The solution free from previous groups and from calcium and strontium is made ammoniacal; after addition of ammonium chloride if not already present for the purpose of preventing precipitation of magnesium. Ammonium carbonate is now added in slight excess and the precipitated $BaCO_3$ allowed to settle on the water bath or in a warm place for an hour or preferably longer. The precipitate is filtered and washed with dilute NH_4OH, dried and ignited and weighed as $BaCO_3$. The method proposed by Fresenius, is considered by some to be more accurate than the sulphate method.

$$BaCO_3 \times 0.686 = Ba.$$

Determination of Barium by Precipitation as Sulphate, BaSO$_4$

This method depends upon the insolubility of barium sulphate in water and in very dilute hydrochloric acid or sulphuric acid, one gram of the salt requiring about 344,000 cc. of hot water to effect solution.

Reaction, $BaCl_2 + H_2SO_4 = BaSO_4 + 2HCl$.

BaSO$_4$, *mol.wt.*, 233,44; *sp.gr.*, 4.47 *and* 4.33; *m.p.*, 1580° (*amorphous decomposes*); H$_2$O *dissolves* 0.000172$_0$° *gram and* 0.0003^{34}° *per* 100 *cc.* 3% HCl *dissolves* 0.0036 *gram. Soluble in conc.* H$_2$SO$_4$. *White, rhombic and amorphous forms.*

Procedure. The slightly hydrochloric acid solution of barium chloride, prepared according to directions given, is heated to boiling (volume about 200–300 cc.) and a slight excess of dilute hot sulphuric acid added. The precipitate is settled on the water bath and the clear solution then decanted through a weighed Gooch crucible or through an ashless filter paper (S. and S. 590 quality). The precipitate is transferred to the Gooch (or paper), and washed twice with very dilute sulphuric acid solution (0.5% H$_2$SO$_4$), and finally with hot water until free of acid. The precipitate is dried and ignited, at first gently and then over a good flame to a cherry red heat, for half an hour. The residue is weighed as barium sulphate, BaSO$_4$.

$$BaSO_4 \times 0.5884 = Ba, \text{ or } \times 0.6569 = BaO, \text{ or } \times 0.8455 = BaCO_3.$$

NOTES. The determination of barium is the reciprocal of the determination of sulphur or sulphuric acid. Precautions and directions given for the sulphur precipitation apply here also, with the exception that dilute sulphuric acid is used as the precipitating reagent in place of barium chloride.

The author found that precipitation of barium sulphate in a large volume of cold solution containing 10 cc. of concentrated hydrochloric acid per 1600 cc. of solution, by adding a slight excess of cold dilute sulphuric acid in a fine stream, exactly in the manner that barium chloride solution is added in the precipitation of sulphur, and allowing the precipitate to settle, at room temperature, for several hours (preferably over night), gives a precipitate that is pure and does not pass through the Gooch asbestos mat. We refer to the chapter on Sulphur for directions for filtering, washing, and ignition of the residue.

The addition of hydrochloric acid causes rapid settling of the barium sulphate. F. A. Gooch has shown that the precipitation should be conducted at temperatures over 75° C., preferably at 90° C. " Methods in Chemical Analysis," 1912, page 168.

VOLUMETRIC METHODS FOR THE DETERMINATION OF BARIUM

Titration of the Barium Salt with Dichromate

This method is of value for an approximation of the amount of barium present in a solution that may also contain calcium, strontium, and magnesium or the alkalies. It depends upon the reaction

$$2BaCl_2 + K_2Cr_2O_7 + H_2O = 2BaCrO_4 + 2KCl + 2HCl.$$

N/10 $K_2Cr_2O_7$ (precipitation purposes) contains 7.355 grams pure salt per liter.

Procedure. The solution containing the barium is treated with ammonia until it just smells of it. (If an excess of ammonia is present the solution is made faintly acid with acetic acid.) It is then heated to about 70° C. and the standard dichromate added, with stirring until all the barium is precipitated and the clear supernatant solution is a faint yellow color from the slight excess of the reagent. For accurate work it is advisable to titrate the precipitate formed by one of the methods given below. One cc. $K_2Cr_2O_7 = 0.00687$ gram Ba. (Note reaction given above.)

NOTE. An excess of potassium dichromate may be added, the precipitate filtered off, washed and the excess of dichromate determined as stated below.

Reduction of the Chromate with Ferrous Salt and Titration with Permanganate

Ferrous sulphate reacts with barium chromate as follows:

$$2BaCrO_4 + 6FeSO_4 + 8H_2SO_4 = 3Fe_2(SO_4)_3 + Cr_2(SO_4)_3 + 2BaSO_4 + 8H_2O.$$

An excess of ferrous salt solution is added and the excess determined by titration with N/10 $KMnO_4$ solution. Fe = $\frac{1}{3}$Ba.

Reagents. N/10 solution of $KMnO_4$. N/10 $FeSO_4$ (27.81 grams per liter) or $FeSO_4 \cdot (NH_4)_2SO_4$ (39.226 grams per liter). One cc. = 0.004579 Ba.

Procedure. The well-washed precipitate of barium chromate is dissolved in an excess of standard N/10 ferrous ammonium sulphate solution containing free sulphuric acid. The excess ferrous salt is titrated with standard N/10 potassium permanganate solution.

(Cc. N/10 ferrous solution minus cc. permanganate titration) multiplied by 0.004579 gives grams barium in the solution. Iron factor to barium is 0.8187.

Potassium Iodide Method

The procedure depends upon the reactions:

1. $2BaCrO_4 + 6KI + 16HCl = 2BaCl_2 + 2CrCl_3 + 6KCl + 8H_2O + 6I.$
2. $3I_2 + 6Na_2S_2O_3 = 6NaI + 3Na_2S_4O_6.$

Procedure. The precipitate, $BaCrO_4$, is dissolved in 50 to 100 cc. of dilute hydrochloric acid and about 2 grams of solid potassium iodide salt added and allowed to react about ten minutes. The liberated iodine is now titrated with N/10 thiosulphate. Near the end of the titration starch solution is added and followed by N/10 thiosulphate until the color disappears.

$$\text{One cc. N/10 } Na_2S_2O_3 = 0.004579 \text{ gram Ba.}$$

Titration of Barium Carbonate with Standard Acid

To the well-washed barium carbonate, $BaCO_3$, an excess N/10 H_2SO_4 is added and the excess acid determined.

$$\text{One cc. N/10 acid} = 0.00687 \text{ gram Ba.}$$

ANALYSIS OF BARYTES AND WITHERITE

Barytes or heavy spar is a variety of native barium sulphate, and witherite a native barium carbonate. These minerals are typical examples of barium-bearing ores. The analysis may involve the determination of barium and calcium sulphates or carbonates, magnesia, iron and aluminum oxides and moisture. Traces of lead, copper, and zinc may be present, as well as sulphide, sulphur and fluorine in fluorspar. The following is an approximate composition of a high-grade sample:

$$BaSO_4 = 96\%, \; CaCO_3 = 1.5\%, \; MgCO_3 = 0.3\%, \; SiO_2 = 0.5\%, \; Al_2O_3 = 0.5\%,$$
$$Fe_2O_3 = 0.2\%, \; H_2O = 0.5\%.$$

For complete analysis treat as directed under preparation of the sample.

Procedure for Commercial Valuation of the Ore

Total BaO as Barium Sulphate

One gram of the finely pulverized sample is weighed into a platinum crucible (35 cc. capacity), mixed with 8 grams of sodium carbonate, fused for twenty minutes over a Bunsen burner and twenty minutes over a Méker burner or blast lamp. The fusion is leached out with 200 cc. hot water in a 250 cc. beaker, the barium carbonate filtered off, washed thoroughly with hot dilute sodium carbonate solution (2 grams per liter) until no test for sulphate is obtained. The barium carbonate is dissolved from the paper with hot dilute hydrochloric acid, catching solution in a 600 cc. beaker and the paper thoroughly washed. The solution is neutralized with ammonium hydroxide, made slightly acid with hydrochloric acid (1–1.5 cc.), heated to boiling and the barium precipitated with hot ammonium sulphate solution (30 grams per liter). After standing on a steam plate for four hours, the barium sulphate is filtered off on a weighed platinum gooch crucible, ignited for 35 minutes and weighed.

The filtrate containing the SO_3 and washings from the barium carbonate is acidified with hydrochloric acid, 1–1.5 cc. excess heated to boiling, the sulphates precipitated with hot 10% barium chloride solution and proceeding as above.

[1] Standard Method of New Jersey Zinc Company.

Silica

One gram of the finely pulverized sample is weighed into a platinum crucible and fused with sodium carbonate as under Total BaO as Barium Sulphate. The melt is detached from the crucible, placed together with the crucible and cover in a 150 cc. platinum dish and covered with hot water. Leaching is carried on until disintegration of the fusion is complete. The residue, composed of mixed carbonates, silicates, etc., is filtered off, catching filtrate in a 400 cc. beaker, and washed thoroughly with hot water containing sodium carbonate. The residue on the paper is dissolved with hot dilute hydrochloric acid, catching in a separate 400 cc. beaker, and the paper thoroughly washed. This paper containing a portion of the silica is placed in a platinum crucible and retained.

The first filtrate, from the barium carbonate, is acidified and evaporated to dryness to dehydrate the silica. The second solution, containing the barium in solution as chloride is also evaporated to dryness. After dehydrating, the residues in both beakers are taken up with hydrochloric acid and filtered. They may be filtered through the same paper, washing the paper thoroughly before passing the second solution. Following the passage of the second solution, the paper is thoroughly washed, added to the original paper in the crucible, ignited and weighed. The contents of the crucible are treated with hydrofluoric and sulphuric acids, evaporated to dryness, ignited and weighed. The difference in weight is amount of silica present.

Iron and Alumina

The two filtrates from the silica determination are nearly neutralized with ammonium hydroxide (acidity 1 cc. hydrochloric acid), and combined. Ammonium sulphate solution is added to assure complete precipitation of all the barium, the beaker placed on a steam plate for two hours. The barium sulphate is then filtered off. The iron and alumina in the filtrate are precipitated with ammonium hydroxide, the precipitate filtered off, washed, ignited in a platinum crucible and weighed.

Lime

The filtrate from the precipitated iron and alumina is acidified slightly with hydrochloric acid, boiled down to a volume of less than 100 cc. and filtered if necessary. The solution is now made ammoniacal, heated to boiling, and 10 cc. of ammonium oxalate solution added. After standing for two hours in a warm place the precipitate of calcium oxalate is filtered off, washed with hot water, ignited and weighed.

Magnesia

The filtrate from the lime determination is boiled down to about 200 cc., slightly acidified, and 15 cc. of microcosmic salts (Saturated Solution) added. Ammonium hydroxide is added with 40 cc. in excess and the precipitate allowed to settle over night. The precipitate is then filtered off, dissolved with hydrochloric acid (1 part concentrated acid to four parts of water), diluted to 100 cc., 10 cc. of microcosmic salt solution added, followed by ammonium hydroxide with 40 cc. excess and allowed to stand over night. The precipitate of magnesium ammonium phosphate is filtered off, washed, carefully ignited and weighed.

Carbon Dioxide

The carbon dioxide is determined according to method given for carbonates under chapter on Carbon, page 111.   It is necessary to use a large sample, i.e., 5–10 grams, and for samples containing a small amount of carbonates a Geissler absorption bulb is preferable to the heavy Fleming bulb.

Fluorine

One gram of sample is placed in a lead bomb with 12 cc. of sulphuric acid, the bomb closed with glass plate in place and heated in an oil bath for 45 minutes at 165° C.   The etching on the glass plate is compared with etching using known amounts of fluorine as CaF_2 and the same kind of glass.

The glass plate is kept cool by circulating cold water.   The type of bomb and its connections are shown in figure 6a.

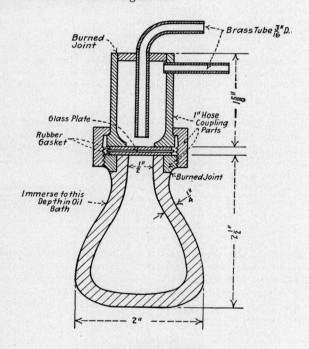

Loss on Ignition

One gram of the sample is weighed out into a platinum crucible dried at 110° C. for two hours and weighed.   This moisture is to be used for calculating all results to a dry basis.   The crucible is heated gently at first and then placed in a muffle furnace or over a blast lamp and ignited until it ceases to lose weight on reheating.   This loss is calculated to a basis of one gram of dry material.

Blanc Fixe

This sulphate of barium is generally marketed as paste and less commonly in the dry form. Both pulp and dry forms should contain not less than 97.5% $BaSO_4$ on the dry basis. The pulp (paste) should not contain over 30% H_2O. Blanc Fixe is used in photography, in coating paper and in paint.

Qualitative Tests. Suitability for photographic purposes. Spread a sample on a glass plate and apply a drop of 10% $AgNO_3$ solution. Set aside in a dark closet. No dark brown or black stain should be evident in five minutes.

Alumina and Iron. Heat a small portion with HNO_3, dilute and filter. Test the filtrate with NH_4OH. A gelatinous colorless precipitate = $Al(OH)_3$, a red precipitate = $Fe(OH)_3$ and possibly $Al(OH)_3$ with the iron.

Lead. Extract a small portion with ammonium acetate and test the extract with $K_2Cr_2O_7$. A yellow precipitate indicates $PbCrO_4$, showing the presence of $PbSO_4$.

Silica. Test about 0.5 g. with 10 cc. conc. H_2SO_4, complete solution shows the absence of SiO_2.

Organic Matter. Coloration of the acid in the silica test indicates the presence of organic matter.

Carbonates. Addition of acid will cause effervescence in presence of carbonate.

Phosphates. Extract a small portion with HNO_3 and test the extract with ammonium molybdate for phosphate.

Quantitative Analysis. The qualitative tests will indicate the presence of impurities. These can now be determined by the standard procedures. The following brief outline may be found useful as a guide for the method of examination.

Moisture. Determine loss on 2-gram sample dried at 105° C. for two hours.

Loss on Ignition. Ignite residue from above. Loss is due to escape of CO_2 and to organic matter and combined water.

Iron and Alumina. Digest a 5-gram sample with 150 cc. HCl (1 : 3). Evaporate extract to dryness, take up with 100 cc. dilute HCl and filter to remove SiO_2. Precipitate iron and alumina in filtrate in usual manner and determine. If P_2O_5 is present it will be present with the precipitate. If present it will be necessary to add a known quantity of iron as $FeCl_3$ to carry down all the P_2O_5. This iron and the P_2O_5, determined on a separate sample, must be deducted, to obtain the iron and alumina in the sample.

Phosphate. Extract a 2-gram sample with water by decantation. Digest the residue with a 10% solution of HNO_3 and filter. Precipitate P_2O_5 in the filtrate with ammonium molybdate in the usual manner. The yellow precipitate is dissolved in NH_4OH and P_2O_5 precipitated with magnesia mixture and determined by the standard procedure.

$$Mg_2P_2O_7 \times 0.6379 = P_2O_5 \qquad Mg_2P_2O_7 \times 2.7038 = Ba_3(PO_4)_2$$

Lead Sulphate. This may be determined in the residue from the iron and alumina determination by extraction with ammonium acetate and precipitation with dichromate reagent by usual procedure. $PbCrO_4 \times 0.9383 = PbSO_4$.

For Other Ingredients consult chapter on Paint Analysis, page 1188.

BERYLLIUM (GLUCINUM)

Gl, *at.wt.* 9.1; *sp.gr.* $1.85^{20°}$; *m.p.* $> 960°$ C.; *oxide,* GlO.

DETECTION

In the usual course of analysis glucinum will be precipitated by ammonia along with iron and aluminum hydroxides. Silica having been removed by evaporation to dryness of the acid solution of the substance, extraction of the residue with dilute hydrochloric acid and subsequent filtration; the members of the hydrogen sulphide group are precipitated from slightly acid solution by hydrogen sulphide. The filtrate is concentrated to about 30 cc., and about 2 grams of sodium peroxide are added to the cooled liquid, which is now heated to boiling and filtered. $Fe(OH)_3$ remains insoluble, if iron is present, while aluminum and glucinum dissolve. The filtrate is acidified with nitric acid, and ammonia then added in excess. If a precipitate forms, alumina or glucinum or both are indicated. Glucinum hydroxide and aluminum hydroxide cannot be distinguished by appearance; the solubility of the former in sodium bicarbonate solution makes it possible to separate the two. The precipitate is dissolved in acid and the solution made almost neutral with ammonia. Solid sodium bicarbonate is added in sufficient amount to make the solution contain 10% of the reagent and the mixture heated to boiling, then filtered. Alumina hydroxide remains on the filter paper and glucinum passes into the filtrate, in which it may be detected by diluting to ten volumes with water and boiling, whereupon glucinum hydroxide precipitates.

Glucinum hydroxide, $Gl(OH)_2$, is precipitated from neutral or acid solution by ammonia, insoluble in excess (distinction from $Al(OH)_3$). It is precipitated by sodium and potassium hydroxides, soluble in excess (separation from iron); if this solution is boiled $Gl(OH)_2$ is reprecipitated, $Al(OH)_3$ remains in solution. $Gl(OH)_2$ is soluble in an excess of ammonium carbonate, Al $(OH)_3$ is insoluble.

ESTIMATION

Glucinum occurs in the minerals beryl, euclase, davalite, chrysoberyl, helvite, leucophane, phencaite.

The oxide, GlO, is soluble in strong sulphuric acid. It is decomposed by fusion with potassium fluoride. The freshly precipitated hydroxide, $Gl(OH)_2$, is easily soluble in dilute acids, in alkalies and alkali carbonates and bicarbonates.

The methods of preparation and solution of the sample are the same as those described for the estimation of aluminum. For details of these procedures the analyst is referred to the chapter on this element.

Chapter contributed by W. W. Scott.

SEPARATIONS

Removal of Silica and Members of the Hydrogen Sulphide Group. See procedure given under "Detection."

Separation of Glucinum from Iron and Manganese. The acid solution is nearly neutralized with ammonia and then poured with constant stirring into an excess of a cold mixture of ammonium sulphide and carbonate. Iron and manganese are precipitated, whereas glucinum passes into the filtrate. (Zirconium and yttrium will be found with glucinum, if they are present in the material examined.)

Separation from Zirconium and Yttrium. The filtrate obtained from the separation of iron and manganese is boiled for an hour, the precipitate is filtered and washed, then dissolved in dilute hydrochloric acid. To this solution is added an excess of sodium hydroxide, zirconium and yttrium are precipitated, whereas glucinum remains in solution. After filtering, glucinum may be precipitated by boiling the diluted filtrate.

Separation from Aluminum, Chromium and Iron. The elements precipitated as hydroxides are ignited to oxides and fused with sodium carbonate for an hour or more. Upon leaching with water, aluminum and chromium dissolve, while iron and glucinum remain insoluble. The oxides of glucinum and iron may be separated by fusion with sodium acid sulphate, extracting with water and precipitating the iron with an excess of sodium hydroxide, glucinum remaining in solution.

Separation of Glucinum from Aluminum. The hydroxides of aluminum and glucinum are precipitated with ammonia and the precipitate treated with an excess of ammonium carbonate. $Gl(OH)_2$ dissolves, whereas $Al(OH)_3$ remains insoluble. See Detection, also Gravimetric Method for Determination of Glucinum.

NOTES. The mineral beryl contains approximately 14 per cent of beryllium oxide, BeO, and is the principal source of beryllium. Beryllium is used in alloys. The aluminum alloy has considerable tensile strength, is hard and light (sp.gr. 2.5 with 90% Al and 10% Be). Beryllium added to alloys of calcium and aluminum makes a tougher and more malleable product. (85% Al, 10% Be, 5% Cu, sp.gr. 2.8 with tensile strength equal to that of bronze.) Alloys of beryllium and copper are valuable for making scientific instruments on account of their electrical properties. Beryllium oxide is used in incandescent mantles. It acts as an accelerator for catalyzers. N.B. Ind. Eng. Chem., 16, 74 (Jan., 1924). Article on catalyzers by W. W. Scott.

GRAVIMETRIC DETERMINATION OF BERYLLIUM

The procedure recommended by Parsons and Barnes [1] depends upon the solubility of glucinum hydroxide in a 10% sodium bicarbonate solution, in the separation of this element from iron and aluminum hydroxide precipitate, with which it is commonly thrown out from solution. (Uranium, if present, also dissolves.)

Procedure. Silica and the members of the hydrogen sulphide group having been removed by the usual methods (See Detection), hydrogen sulphide is expelled by boiling, nitric acid is added in sufficient amount to oxidize iron (the hydrochloric acid solution turns yellow) and ammonium hydroxide added in slight excess. The precipitated hydroxides are allowed to coagulate by heating to boiling and, after settling a few minutes, filtered and washed with a 2% solution of ammonium acetate containing free ammonia.

Separation from Iron and Aluminum Hydroxide. The precipitate is dissolved in hydrochloric acid, the solution oxidized with nitric acid or hydrogen peroxide (C.P.), if necessary, and the free acid then neutralized with ammonia. To the cold solution are added 10 grams of sodium bicarbonate for each 100 cc. of liquid. The mixture is heated to boiling and boiled for one minute,[2] then cooled and filtered. The residue is washed with hot 10% solution of sodium bicarbonate. Iron and aluminum hydroxides remain on the filter and glucinum passes into the filtrate.

To recover occluded glucinum from the hydroxides of iron and alumina, the precipitate is dissolved in a few drops of hydrochloric acid, and the precipitation repeated. It is advisable to repeat this treatment a third time, adding the filtrates to the first portion containing the glucinum.

Precipitation of Glucinum. The combined filtrates from the alumina and iron hydroxides are acidified with strong hydrochloric acid, the beakers covered to prevent loss by spurting and the carbon dioxide completely removed by boiling. (CO_2 remaining in solution would form ammonium carbonate, on subsequent treatment with ammonia, which would dissolve glucinum.) A slight excess of ammonia is now added, the mixture again boiled and the precipitated glucinum hydroxide allowed to settle, then filtered and washed with a 2% solution of ammonium acetate containing free ammonia, until the chlorides are removed. After ignition the residue is weighed as glucinum oxide, GlO.

$$GlO \times 0.3626 = Gl.$$

[1] C. L. Parsons and S. K. Barnes, Jour. Am. Chem. Soc., **28**, 1589, 1906.
[2] Prolonged boiling would cause the loss of too much CO_2, so that $Al(OH)_3$ would be apt to pass into solution. The evolution of CO_2 may be mistaken for boiling.

Other Methods for Determining Beryllium (or Glucinum)

Gibson's Method.[1] The pulverized material is heated with ammonium fluoride. Much of the silica is volatilized as ammonium fluosilicate and the aluminum and beryllium are converted to fluorides, then to sulphates. Beryllium is now separated from aluminum by extraction with ammonium carbonate and precipitated from the extract by boiling.

Haven's Method.[2] The beryl is fused with caustic soda, and the fusion is dissolved in HCl. The hydroxides are now precipitated with NH_4OH and filtered and washed. The precipitate is dissolved in HCl and aluminum precipitated by saturation with HCl gas. (See notes below.) ($AlCl_3.6H_2O$) precipitated. Ammonium carbonate is now used to separate the beryllium by methods described.

NOTES.[3] $Be(OH)_2$ resembles $Al(OH)_3$ in being soluble in an excess of fixed alkali, but differs from it by the fact that $Be(OH)_2$ reprecipitates on boiling the solution, while aluminum remains in solution.

A saturated solution of sodium carbonate dissolves $Be(OH)_2$ but not $Al(OH)_3$ or $Fe(OH)_3$.

If ether is added to a solution or $BeCl_2$ and $AlCl_3$ and then HCl gas passed in, aluminum chloride is precipitated, while beryllium remains in solution.

Fusion of the oxides of aluminum and beryllium with Na_2CO_3 produces a water soluble aluminate of sodium, while beryllium remains insoluble.

Fusion with Na_2SIF_6 produces soluble Na_2BeF_4 and insoluble Na_3AlF_6.

[1] Jour. Chem. Soc., 63, 909 (1893).
[2] Amer. Journ. Sci. (4) 4, 111 (1897).
[3] Chemistry of the Rarer Elements by B. Smith Hopkins. The Analysis of Minerals and Ores of the Rarer Elements by W. R. Schoeller and A. R. Powell.

BISMUTH

Bi, *at.wt.* 209.0; *sp.gr.* 9.7474; *m.p.* 271°;[1] *b.p.* 1420° C.; *oxides,* Bi_2O_3, Bi_2O_5.

DETECTION

Bismuth is precipitated from its solution, containing free acid, by H_2S gas, as a brown sulphide, Bi_2S_3. The compound is insoluble in ammonium sulphide (separation from arsenic, antimony, and tin), but dissolves in hot dilute nitric acid (separation from mercury). The nitrate, treated with sulphuric acid and taken to SO_3 fumes, is converted to the sulphate and dissolves upon dilution with water (lead remains insoluble as $PbSO_4$). Bismuth is precipitated from this solution by addition of ammonium hydroxide, white $Bi(OH)_3$ being formed (copper and cadmium dissolve). If this hydroxide is dissolved with hydrochloric acid and then diluted with a large volume of water, the white, basic salt of bismuth oxychloride, BiOCl, is precipitated. The compound dissolves if sufficient hydrochloric acid is present. It is insoluble in tartaric acid (distinction from antimony).

Reducing Agents. Formaldehyde in alkaline solution, hypophosphorous acid, potassium or sodium stannite, reduce bismuth compounds to the metallic state. For example, a hot solution of sodium stannite poured onto the white precipitate of $Bi(OH)_3$ on the filter will give a black stain. The test is very delicate and enables the detection of small amounts of the compound.

$$3K_2SnO_2 + 2BiCl_3 + 6KOH = 2Bi + 3K_2SnO_3 + 6KCl + 3H_2O.$$

Blowpipe Test. A compound of bismuth heated on charcoal with a powdered mixture of carbon, potassium iodide and sulphur, will give a scarlet incrustation on the charcoal.

ESTIMATION

The determination of bismuth is required in complete analysis of ores of cobalt, nickel, copper, silver, lead, and tin, in which it is generally found in small quantities. In evaluation of bismuthite, bismuth ochre, etc. In the analysis of the minerals wolfram, molybdenite. It is determined in the residues from the refining of lead (the principal source of bismuth in the United States). In the analysis of alloys—antifriction metals, electric fuses, solders, stereotype metals, certain amalgams used for silvering mirrors (with or without lead or tin), and in bismuth compounds.

Preparation and Solution of the Sample

In dissolving the substance, the following facts must be kept in mind: nitric acid is the best solvent of the metal. Although it is soluble in hot sulphuric acid, it is only very slightly soluble in the cold acid. The metal is practically insoluble in hydrochloric acid, but readily dissolves in nitrohydrochloric acid.

Chapter contributed by W. W. Scott and J. J. Mulligan.

The hydroxides, oxides, and most of the bismuth salts are readily soluble in hydrochloric, nitric, and sulphuric acids.

Ores or Cinders. One gram of the finely pulverized ore or cinder (or larger amounts where the bismuth content is very low) is treated in a 400-cc. beaker with 5 cc. of bromine solution $(Br+KBr+H_2O)$,[1] followed by the cautious addition of about 15 cc. of HNO_3 (sp.gr. 1.42). When the violent action has ceased, which is apt to occur in sulphide ores, the mixture is taken to dryness on the steam bath, 10 cc. of strong HCl and 20 cc. of concentrated H_2SO_4 and the covered sample heated until SO_3 fumes are freely evolved. The cooled solution is diluted with 50 cc. of water and gently heated until only a white or light gray residue remains. The solution is filtered and the residue washed with dilute H_2SO_4 (1 : 10), to remove any adhering bismuth. Silica, the greater part of the lead (also $BaSO_4$) remain in the residue, whereas the bismuth, together with iron, alumina, copper, antimony, etc., are in the solution. Details of further treatment of the solution to effect a separation of bismuth are given under "Separations" and the procedures for determination of bismuth.

Alloys, Bearing Metal, etc. One gram of the borings, placed in a small beaker, is dissolved by adding 20 cc. of concentrated HCl and 5 cc. of strong HNO_3. The alloy will usually dissolve in the cold, unless considerable lead is present, in which case prolonged heating on the steam bath may be necessary. (A yellow or greenish-yellow color at this stage indicates the presence of copper.) Lead may now be removed either as a sulphate by taking to SO_3 fumes with H_2SO_4 or by precipitating as a chloride, in the presence of alcohol, according to directions given under Separations. The bismuth is determined in the filtrate from lead according to one of the procedures given under the quantitative methods.

Lead Bullion, Refined Lead. Ten to twenty-five grams of the lead, hammered or rolled out into thin sheets and cut into small pieces, are taken for analysis. The sample is dissolved by a mixture of 250 cc. of water and 40 cc. of strong nitric acid, in a large covered beaker, by warming gently, preferably on the steam bath. When the lead has dissolved, the beaker is removed from the heat and dilute ammonia (1 : 2) added to the warm solution, very cautiously and finally drop by drop until the free acid is neutralized and the liquid remains faintly opalescent, but with no visible precipitate. Now 1 cc. of dilute HCl (1 : 3) is added. The solution will clear for an instant and then a crystalline precipitate of bismuth oxychloride will form, if any considerable amount of bismuth is present. The beaker is now placed on the steam bath for an hour, during which time the bismuth oxychloride will separate out, together with a small amount of lead and with antimony if present in appreciable amounts. The further isolation and purification of bismuth is given under "Separations." In brief—antimony is removed by dissolving the precipitate in a small amount of hot dilute HCl (1 : 3), precipitating bismuth, traces of lead, and the antimony by H_2S, dissolving out the antimony sulphide with warm ammonium sulphide, dissolving the Bi_2S_3 and PbS in HNO_3 and reprecipitation of the bismuth according to the procedure given above. Bismuth is now determined as the oxychloride. Further details of this method are given under the gravimetric procedures for bismuth.

---

[1] Bromine solution is made by dissolving in water 75 grams of KBr, to which are added 50 grams of liquid bromine and the mixture diluted to 500 cc. with water.

SEPARATIONS

The following procedures are given in the order that would be followed in the complete analysis of an ore, in which all the constituents are sought. This general scheme, however, is not required for the majority of bismuth-bearing samples commonly met with in the commercial laboratory, direct precipitations of bismuth frequently being possible.

Separation of Bismuth from Members of Subsequent Groups, Fe, Cr, Al, Mn, Co, Ni, Zn, Mg, the Alkaline Earths and Alkalies, together with Rare Elements of these Groups. The solution should contain 5 to 7 cc. of concentrated hydrochloric acid (sp.gr. 1.19) for every 100 cc. of the sample. The elements of the hydrogen sulphide group are precipitated by saturating the solution with H_2S (Hg, Pb, Bi, Cu, Cd, As, Sb, Sn, Mo, Se, Te, Au, Pt). The members of subsequent groups remain in solution and pass into the filtrate.

Separation of Bismuth from Arsenic, Antimony, Tin, Molybdenum, Tellurium, Selenium. In presence of mercury, the soluble members of the hydrogen sulphide group are separated from the insoluble sulphides by digesting the precipitate above obtained with ammonium sulphide; in absence of mercury, however, which is generally the case, digestion of the sulphides with sodium hydroxide and sodium sulphide solution is preferred, the general procedure being followed. Mercury, lead, bismuth, copper, and cadmium remain in the residue, whereas the other members of the group dissolve.

Separation of Bismuth from Mercury. The insoluble sulphides, remaining from the above treatment with ammonium sulphide after being washed free of the soluble members of this group, are placed in a porcelain dish and boiled with dilute nitric acid (sp.gr. 1.2 to 1.3). The solution thus obtained is filtered, upon dilution, from the insoluble sulphide of mercury. A little of the lead may remain as $PbSO_4$, the solution may contain lead, bismuth, copper, and cadmium.

Separation of Bismuth from Lead. This is the most important procedure in the determination of bismuth as the separation is almost invariably necessary, as these elements commonly occur together.

Precipitating Lead as $PbSO_4$. This procedure is generally used in the process of a complete analysis of an ore containing lead and bismuth. The nitric acid solution of the sulphides, obtained upon removal of the soluble group and mercury by boiling the insoluble sulphides with dilute nitric acid, is treated with about 10 cc. of strong sulphuric acid, and taken to SO_3 fumes by heating. The cooled sulphate solution is diluted with water and the insoluble lead sulphate filtered off and washed with dilute sulphuric acid solution (1 : 20). Bismuth passes into solution, together with copper and cadmium, if also present in the original sample.

Precipitation of Lead as $PbCl_2$. This separation is used in the complete analysis of pig lead.

Separation of Bismuth from Copper and Cadmium. This separation is accomplished by precipitating bismuth as the oxychloride with hydrochloric acid, or as the carbonate by adding an excess of ammonium carbonate to the solution nearly neutralized by ammonia, or as the hydroxide by adding an excess of ammonia. Details of these procedures are given under the gravimetric methods for determining bismuth.

GRAVIMETRIC METHODS FOR THE DETERMINATION OF BISMUTH

Determining Bismuth by Precipitation and Weighing as the Basic Chloride, BiOCl

The determination depends upon the formation of the insoluble oxychloride, BiOCl, when a hydrochloric acid solution of bismuth is sufficiently diluted with water, the following reaction taking place, $BiCl_3 + H_2O = BiOCl + 2HCl$.

The procedure is recommended for the determination of bismuth in refined lead, bearing metal, and bismuth alloys. Copper, cadmium, and lead do not interfere; appreciable amounts of antimony and tin, however, should be removed by H_2S precipitation and subsequent treatment with Na_2S, and the residual sulphides dissolved in hot dilute nitric acid, according to directions given under "Separations."

Properties of BiOCl. *Mol.wt.*, 259.46; *sp.gr.*, $7.717^{15°}$; *m.p.*, red heat; *insol. in H_2O and in $H_2C_4H_4O_6$, soluble in acids. Appearance is white, quadratic crystalline form.*

Procedure. The solution of bismuth, freed from appreciable amounts of tin and antimony, is warmed gently and treated with sufficient ammonia to neutralize the greater part of the free acid. At this stage a precipitate is formed by the addition, which dissolves with difficulty; the last portion of the dilute ammonia (1 : 2) is added drop by drop, the solution is diluted to about 300 cc., and the remainder of the free acid neutralized with dilute ammonia added cautiously until a faint opalescence appears, but not enough to form an appreciable precipitate. One to 3 cc. of dilute hydrochloric acid (1 part HCl sp.gr. 1.19 to 3 parts H_2O) are now added, the mixture stirred and the bismuth oxychloride allowed to settle for an hour or so on the steam bath, then filtered hot by decanting off the clear solution through a weighed Gooch crucible. The precipitate is washed by decantation twice with hot water and finally washed into the Gooch, then dried at 100° C. and weighed as BiOCl.

$$BiOCl \times 0.8017 = Bi.$$

NOTE. Three cc. of dilute hydrochloric acid (or 1 cc. conc. HCl, sp.gr. 1.19) are sufficient to completely precipitate 1 gram of bismuth from solution.

9

Determination of Bismuth as the Oxide, Bi_2O_3

Preliminary Considerations. The determination of bismuth as the oxide requires the absence of hydrochloric acid or sulphuric acid from the solution of the element, since either of these acids invariably contaminates the final product. In presence of these acids, which is frequently the case, determination of bismuth by precipitation as Bi_2S_3 or by reduction to the metal and so weighing is generally recommended; a brief outline of the methods is given later; a solution of bismuth free from hydrochloric acid and practically free of sulphuric acid may be obtained by precipitating Bi_2S_3, together with CuS, CdS, and PbS, the amount of sulphuric acid formed by the reaction being negligible. Bismuth should be in a nitric acid solution, free from antimony and tin.

Two general conditions will be considered: 1. Solutions containing lead. Copper and cadmium may also be present. 2. Solutions free from lead. Copper and cadmium may be present.

1. Separation from Lead, Copper, and Cadmium, by Precipitation as Basic Nitrate.[1] Either the sulphuric or hydrochloric acid methods may be employed for effecting the separation of lead by precipitation. Furthermore advantage may be taken of the fact that bismuth nitrate is changed by the action of water into an insoluble basic salt, while lead, copper and cadmium do not undergo such a transformation.

Procedure. The bismuth nitrate solution is evaporated to syrupy consistency and hot water added with constant stirring with a glass rod. The solution is again evaporated to dryness, and the hot-water treatment repeated. Four such evaporations are generally sufficient to convert the bismuth nitrate completely into the basic salt; when this stage is reached the addition of water will fail to produce a turbidity. The solution is finally evaporated to dryness and, when free from nitric acid, is extracted with cold ammonium nitrate solution $(1.NH_4NO_3 : 500\ H_2O)$ to dissolve out the lead and other impurities. After allowing to stand some time with frequent stirring, the solution is filtered and the residue washed with ammonium nitrate solution, then dried.

Ignition to Bismuth Oxide. As much of the precipitate as possible is transferred to a weighed porcelain crucible, the filter is burned and the ash added to the main precipitate. This is now gently ignited over a Bunsen burner. Too high heating will cause the oxide to fuse and attack the glaze of the crucible.

Properties. $Bi(OH)_2NO_3$ *mol.wt.*, 304.03; *sp.gr.*, 4.928[15°]; *decomp.*, 260°; *insol. in* H_2O; *sol. in acids; hexagonal plates.*

$Bi_2O_3 = mol.\ wt.$, 464.0; *sp.gr.*, 8.8 to 9.0; *m.p.*, 820 to 860°; *insoluble in cold water and in alkalies, but soluble in acids; yellow tetragonal crystals.*

$$Bi_2O_3 \times 0.8965 = Bi.$$

2. Precipitation of Bismuth as the Subcarbonate or Hydroxide, Lead being Absent. Either of these procedures effects a separation of bismuth from copper and cadmium.

A. Procedure. Precipitation of the Subcarbonate. The solution is diluted to about 300 cc. and dilute ammonia added cautiously until a faint turbidity is obtained and then an excess of ammonium carbonate. The solution

[1] J. Löwe, Jour. prak. Chem., (1), **74**, 344, 1858.

is heated to boiling, the precipitate filtered off, washed with hot water, dried and ignited according to directions given in the bismuth subnitrate method. The residue is weighed as Bi_2O_3.

B. Procedure. Isolation of Bismuth by Precipitation as the Hydroxide.[1] The solution is taken to dryness and the residue treated with 5 cc. of nitric acid (1 : 4) and 25 cc. of water added. The resulting solution is poured, with constant stirring, into 25 cc. of concentrated ammonia and 50 cc. of 4% hydrogen peroxide. Upon settling of the bismuth hydroxide, the clear solution is filtered off and the residue is treated with more ammonia and peroxide. It is then filtered onto a filter paper, washed with hot, dilute ammonium hydroxide, (1 : 8), followed by hot water and washed free of any adhering copper or cadmium (no residue when a drop is evaporated on platinum foil). Re-solution in hot dilute nitric acid and reprecipitation may sometimes be necessary to obtain the pure product. The hydroxide may be dried, ignited and weighed as Bi_2O_3 according to directions already given on page 74.

Properties. $Bi_2O_3 \cdot CO_2 \cdot H_2O$, *mol.wt.*, 526.02; *sp.gr.*, 6.86; *decomp. by heat; insoluble in water, soluble in acids, insoluble in Na_2CO_3; white precipitate.*

$Bi(OH)_3$, *mol.wt.*, 259.02; *loses* $1\frac{1}{2}$ H_2O *at* 150°; *insol. in cold water and in alkalies; soluble in acids; white precipitate.*

Determination of Bismuth as the Sulphide, Bi_2S_3

The procedure is applicable to the determination of bismuth in a hydrochloric or sulphuric acid solution, freed from other members of this group.

Procedure. Bismuth sulphide is precipitated by passing H_2S into the slightly acid solution, preferably under pressure. When the precipitation is complete, the bismuth sulphide, Bi_2S_3, is filtered off into a weighed Gooch crucible, the precipitate washed with H_2S water, then with alcohol to remove the water, followed by carbon disulphide to dissolve out the precipitated sulphur, then alcohol to remove the disulphide, and finally with ether. After drying for fifteen to twenty minutes, the residue is weighed as Bi_2S_3. This weight multiplied by 0.8122 = Bi.

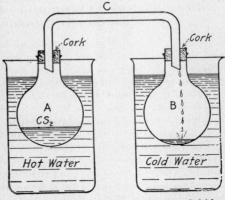

Fig. 7.—Purification of Carbon Disulphide.

NOTE. The carbon disulphide used should be freshly distilled. This may be accomplished by placing the carbon disulphide in a small flask (*A*, Fig. 7) connected by means of a glass tube (*C*) to a second flask (*B*), cork stoppers being used. The vessels are immersed in beakers of water, the container with the reagent being placed in hot water (60–80° C.) and the empty flask in cold water. The reagent quickly distills into the empty flask in pure form.

Properties of Bismuth Sulphide. Bi_2S_3, *mol.wt.*, 512.21; *sp.gr.*, 7–7.81; *decomposed by heat, solubility* = 0.000018g. *per* 100 cc. *cold* H_2O; *soluble in nitric acid; brown rhombic crystals.*

[1] P. Jannasch, Zeit. anorg. Chem., 8, 302, 1895.

Determination of Bismuth as the Metal

Reduction with Potassium Cyanide.[1] Bismuth precipitated as the carbonate and ignited to the oxide according to the procedure given, is fused in a weighed porcelain crucible with 5 times its weight of potassium cyanide over a low flame. The cooled melt is extracted with water, pouring the extracts through a filter that has been dried and weighed with the crucible. Bismuth is left undissolved as metallic bismuth. After washing with water, alcohol, and ether, the filter, with the metal and loosened pieces of porcelain glaze, is dried at 100° C. together with the crucible. These are then weighed and the increased weight taken as the amount of bismuth present in the sample.

Electrolytic Deposition of Bismuth

With samples containing less than 0.03 gram bismuth, the metal may be satisfactorily deposited by electrolysis of its sulphuric acid solution, lead having been removed previously by sulphuric acid by the standard procedure. The solution contains about 5 cc. of strong sulphuric acid per 100 cc. This is electrolyzed with a current of 0.6 to 0.7 ampere and about 2.7 to 3 volts.

VOLUMETRIC DETERMINATION OF BISMUTH

Determination of Bismuth by Precipitation as Oxalate and Titration by Potassium Permanganate [2]

Normal bismuth oxalate, produced by addition of oxalic acid to a nitric acid solution of the element, boiled with successive portions of water, is transformed to the basic oxalate. This may be titrated with potassium permanganate in presence of sulphuric acid.

Procedure. *Preparation of the Sample.* One gram of the finely ground sample is treated with 5 to 10 cc. of concentrated nitric acid and digested on the steam bath and finally evaporated to dryness, the residue is taken up with 5 cc. of nitric acid (sp.gr. 1.42)+25 cc. of water, and diluted to 100 cc.

Precipitation of the Oxalate. About 5 grams of ammonium oxalate or oxalic acid are added and the liquid boiled for about five minutes, the precipitate allowed to settle and the supernatant solution filtered off. The precipitate is boiled twice with 50-cc. portions of water and the washings poured through the same filter. If the filtrate still passes through acid, the washing is continued until the acid is removed and the washing passing through the filter is neutral. The bulk of the basic oxalate precipitate is placed in a beaker and that remaining on the filter paper is dissolved by adding 2 to 5 cc. of hydrochloric acid, 1 : 1, the solution being added to the bulk of the precipitate.

---

[1] Method by H. Rose, Pogg. Ann., **110**, p. 425.
Vanino and Treubert (Ber., **31** (1898), 1303), reduce bismuth by adding formaldehyde to its slightly acid solution and then making strongly alkaline with 10% NaOH solution and warming.
[2] The method is rapid and is sufficiently accurate for commercial work. Warwick and Kyle (C. N., **75**, 3).
Muir and Robbs, J. C. S., **41**, 1.

This is now warmed until it goes into solution and the liquid is diluted to 250 cc. with hot water. Dilute ammonia is now added until the free acid is neutralized; the resulting precipitate is taken up with dilute sulphuric acid, 1 : 4, added in slight excess. The resulting solution, warmed to 70°, is titrated with standard potassium permanganate.

One cc. KMnO₄ N/10 = 0.0104 gram Bi.

NOTE. Lead, copper, arsenic, iron, zinc, and tellurium do not interfere. Hydrochloric acid should not be used to dissolve the sample, as it interferes with the oxalate precipitation.

Cinchonine Potassium Iodide, Colorimetric Method [1]

This method is applicable for the determination of small amounts of bismuth, 0.00003 to 0.00015 gram, in ores and alloys. The procedure depends upon the fact that bismuth nitrate produces a crimson or orange color when its solution is added to a solution of cinchonine potassium iodide, the intensity of the color depending upon the amount of bismuth in the resulting product.

Special Reagents. *Cinchonine Potassium Iodide Solution.* Ten grams of cinchonine are dissolved by treating with the least amount of nitric acid that is necessary to form a viscous mass and taking up with about 100 cc. of water. The acid is added a drop at a time, as an excess must be avoided. Twenty grams of potassium iodide are dissolved separately and cinchonine solution added. The resulting mixture is diluted with water to 1000 cc. After allowing the reagent to stand forty-eight hours, any precipitate formed is filtered off and the clear product is ready for use. The reagent preserved in a glass-stoppered bottle keeps indefinitely. It should be filtered free of suspended matter before use.

Standard Bismuth Solution. One gram of metallic bismuth is dissolved in the least amount of dilute nitric acid (1 : 1) that is necessary to keep it in solution and diluted to 1000 cc., in a graduated flask. One hundred cc. of this solution is diluted to 1000 cc. One cc. of this diluted solution contains 0.0001 gram bismuth.

Procedure. *Isolation of Bismuth.* The solution is freed from lead by H₂SO₄, and from arsenic, antimony, and tin by precipitation of the sulphides and extraction with Na₂S solution. The residual sulphides are dissolved in hot dilute nitric acid, according to the standard methods of procedure. The free nitric acid is nearly neutralized by the cautious addition of dilute ammonia, the last portion being added drop by drop, until a faint cloudiness is evident, and then 10 to 15 cc. of 10% ammonium carbonate are added with constant stirring. The mixture is digested for about three hours on the steam bath, the clear solution decanted through a small filter, the residue washed by decantation once or twice with hot water containing ammonium carbonate and then on the filter twice with pure hot water.

Colorimetric Comparison

The residue of bismuth basic carbonate is dissolved in the least amount of dilute nitric acid necessary to effect solution and the filter washed free of bismuth with a little water containing a few drops of nitric acid. The solution is made up to a definite volume, 50 cc. or 100 cc. according to the bulk of

[1] Method of W. C. Ferguson.

precipitate dissolved. Two small beakers placed side by side may be used for the color comparison, a sheet of white paper or tile being placed under the beakers. Two 50-cc. Nessler tubes, however, are preferred. Three cc. of cinchonine solution are added to each container. From a burette the bismuth nitrate sample is run into one of these containers in just sufficient quantity to color the reagent a crimson or orange tint. The exact volume required to do this is noted and the equivalent amount of sample used calculated. (If no color is produced bismuth is absent.) The reagent in the adjacent beaker or Nessler tube is diluted to 5 to 7 cc., and into this is run, from a burette, the standard bismuth nitrate solution until the color exactly matches the sample. From the cc. of the standard required the amount of bismuth in the sample can readily be calculated.

Reaction. $3KI + C_{19}H_{22}N_2OKI + Bi(NO_3)_3 = C_{19}H_{22}N_2OKIBiI_3 + 3KNO_3.$

Precautions. The sensitiveness of the method is lost if the depth of color is too great. It is necessary, then, to add the sample to the cinchonine reagent in such quantity only as will produce a light crimson or orange color.

Solutions in the comparison tubes or beakers must not be overdiluted, since the bismuth salt formed by the reaction of the cinchonine reagent is soluble in water with the disappearance of color in too dilute solutions.

Comparison must be expeditiously made, as a precipitate is apt to form upon standing, and iodine will sometimes separate.

The order of addition must be observed; e.g., the bismuth solution is added to the cinchonine reagent, never the reverse.

Colorimetric Determination of Bismuth.   Bismuth Iodide Method [1]

Bismuth iodide gives an intense yellow, orange, or red color to its solution. The color is not destroyed by SO_2, as is that of free iodine. The intensity of the color varies as follows:

1 part of bismuth in   10,000 parts of water produces an orange-colored solution.
1 part of bismuth in   40,000 parts of water produces a light orange color.
1 part of bismuth in 100,000 parts of water produces a faint yellow color.

Reagents.   *Standard Bismuth Solution.* One gram of bismuth is dissolved in 3 cc. of strong nitric acid and with 2.8 cc. of water and made up to 100 cc. with glycerine. Glycerine is added to keep the BiI_3 in solution. Glycerine is not necessary for amounts of bismuth below 0.0075 gram per cc.

Potassium Iodide Solution. Five grams of potassium iodide dissolved in 5 cc. of water is diluted to 100 cc. with glycerine.

Procedure. The sample is dissolved with just sufficient nitric acid and water necessary to cause solution, 10 cc. of glycerine and 10 cc. of potassium iodide solution added and the sample diluted to 50 cc. Comparison is now made with 10 cc. of the standard bismuth solution to which has been added 10 cc. of potassium iodide and 30 cc. of water. It is advisable to have the standard stronger in bismuth than the sample and to draw out the standard from the comparison cylinder until the two colors match.

[1] T. C. Thresh, Pharm. Jour., 641, 1880.

Bismuth Determination in Lead Bullion[1]

Ten to twenty-five grams of the lead, hammered or rolled out and cut into small pieces, are taken for analysis. The sample is dissolved in a mixture of 200 cc. water and 50 cc. strong nitric acid, in a large covered beaker, and warmed gently on water or steam bath. When lead has dissolved, the beaker is removed and placed on cool surface and enough sulphuric acid (1–1) added to precipitate lead.

The lead sulphate is allowed to settle and the clear supernatant liquid is decanted into another beaker and held. To the residue of lead sulphate 10–20 cc. concentrated sulphuric acid is added and brought down to strong fumes on hot plate. After strong fuming, the portion containing lead sulphate is diluted with water. To the first clear decanted portion, 10 cc. sulphuric acid is added and this also evaporated down to fumes of sulphuric acid. Both portions are removed from hot plate and when cool add 50 cc. water and 3 to 5 grams of tartaric acid to each. Heat to dissolve tartaric acid and filter over asbestos pad, the clear portion first, and then follow with the one containing bulk of lead sulphate. The bulk of lead sulphate is washed by decantation three or four times with warm water before transferring to asbestos pad. When bismuth is higher than .30% in the bullion, the sulphate residue may be retreated with sulphuric acid, fumed and washed. The clear solution is allowed to stand for one hour and refiltered to ensure removal of all lead sulphate. The filtrate is then warmed and hydrogen sulphide gas passed filtered on a paper and washed with cold H_2S water. The sulphides are washed from filter back to precipitating beaker. The sulphides of Sb, Sn, Te, etc., are leached out with a 10% K_2S solution, which has been saturated with hydrogen sulphide, and allowed to stand in warm place and filtered over the original sulphide paper. After washing with warm water containing a few drops of K_2S solution, the precipitate is dissolved in nitric acid and a few drops of bromine to ensure solution of all sulphur. It is all important to remove Sb, Sn and Te from sulphide precipitate before going any further by repeating hydrogen sulphide precipitation.

The nitric acid solution of Bi, Cu, etc. is made faintly alkaline with ammonia and 1 gram ammonium carbonate added and the solution boiled for five to ten minutes when the bismuth is precipitated as basic salt. To ensure solution of the copper, a few drops of free ammonia are added with stirring before filtering. The bismuth precipitate is filtered on a tared gooch, washed with water, dried and ignited to Bi_2O_3 over Bunsen flame.

$$Bi_2O_3 \times .8966 = Bi.$$

If the bismuth precipitate is dark after precipitation with ammonia and ammonium carbonate, it may be due to tellurium. If so, the filtered precipitate is dried, ignited and fused with caustic potash and sulphur to put the tellurium in a soluble form and thus remove tellurium from insoluble bismuth sulphide; or redissolving in acid and reprecipitating as sulphides and washing the sulphides with K_2S solution as before mentioned.

This method is applicable to refined lead when larger portions are taken.

[1] J. J. Mulligan.

Bismuth in Alloys[1]

The alloy is dissolved in nitric acid, as little hydrochloric as possible, and 10 cc. sulphuric, and run down to strong SO_3 fumes and proceed as for ores.

Where bismuth is present with considerable tin, the cementation of the bismuth with pure iron wire to free it from tin and then treat residue same as for ores and mattes, seems to be the best means for obtaining a bismuth precipitation free of tin for control analyses.

In the presence of considerable copper, the bismuth can be precipitated as basic carbonate, filtered and the impure basic precipitate treated with nitric acid and evaporated to fumes of sulphuric, the procedure is then the same as for ores and matte.

Determination of Bismuth in Ores, Mattes[1]

A qualitative test for lead and insoluble bismuth compounds on all products to be analyzed for bismuth is an important step with the removal of lead as sulphate as soon as perfect solution of the sample is insured.

For products containing from $\frac{1}{2}\%$ to 25% bismuth, one gram portions are taken; for 50% or higher $\frac{1}{2}$ gram portions are weighed up.

The weighed portion is transferred to a 250 cc. beaker and 25 cc. H_2O and 10 cc. concentrated nitric acid added, and where sulphur is present a few drops of bromine are added cautiously after acid has been allowed to act for a sufficient time. After all sulphur has gone in solution, boil down to dryness on water bath or hot plate, 2 cc. of nitric acid and 1 cc. hydrochloric acid are added, then 10 cc. of sulphuric acid cautiously added, and the covered sample taken down on hot plate to strong fumes of SO_3. The cooled solution is diluted with 50 cc. water and 2 to 3 grams tartaric acid are added and the mixture brought to a boil. The residue containing insoluble lead sulphate, silica and other insolubles are filtered and washed with 2% solution of sulphuric acid to free residue of soluble bismuth.

The filtrate contains Bi, Cu, Sb, As, Te and a small amount of Pb may be in the solution. After standing for one-half hour or more, the solution is refiltered to remove any lead that may have gone through on first filtration.

The clear solution containing bismuth is saturated with hydrogen sulphide gas for one-half hour.

Filter sulphides on filter paper and wash with H_2S water. If much iron is present the sulphides are redissolved in nitric and sulphuric, taken to fumes, and the sulphide precipitation repeated.

The sulphides are leached with a 10% solution of K_2S saturated with H_2S gas to remove As, Sb, Sn, Te and Se. The remaining sulphides containing Cu, Cd, Bi are dissolved in nitric acid and a little bromine water, bromine boiled off and solution made slightly alkaline with ammonia water and 1 gram of ammonium carbonate added. The solution is boiled for five to ten minutes and 1 cc. of ammonia is added and the basic carbonate filtered on a tared Gooch, ignited and weighed as Bi_2O_3.

$$Bi_2O_3 \times .8966 = Bi.$$

[1] J. J. Mulligan.

BORON

B, *at.wt.* 10.82; $\begin{cases} \textit{amorp. sp.gr.} \; 2.45; \; \textit{m.p.} \; 2200°; \; \textit{b.p. sub}\text{'}\textit{imes.} \\ \textit{cryst.} \quad \textit{sp.gr.} \; 2.55; \; \textit{m.p.} \; 2500°; \; \textit{b.p.} \; 3500° \; \text{C}; \; \textit{oxide}, \text{B}_2\text{O}_3 \end{cases}$

DETECTION

Flame Test. Boric acid is displaced from its salts by nearly all acids, including even carbonic acid. Upon ignition, however, it in turn drives out other acids which are volatile at lower temperatures. A powdered borate, previously calcined, is moistened with sulphuric acid and a portion placed on the loop of a platinum wire is heated to expel the sulphuric acid,[1] then moistened with glycerine and placed in the colorless flame; a green color will be imparted to the flame. Copper salts should be removed with H_2S and barium as $BaSO_4$ if present, as these also color the flame green.

The flame test may be conveniently made by treating the powdered sample in a test-tube with sulphuric acid and alcohol (preferably methyl alcohol). A cork carrying a glass tube is inserted and the test-tube gently warmed. The escaping gas will burn with a green flame.

The test may be made by igniting the mixture of powder, alcohol, and sulphuric acid in an open porcelain dish. The green color will be seen in presence of a borate. The test is not as delicate as the one with the test-tube.

Borax Bead. $Na_2B_4O_7 \cdot 10H_2O$ fused in a platinum loop, swells to several times its original volume as the water of crystallization is being driven out, then contracts to a clear molten bead. If the bead is dipped into a weak solution of cobalt and plunged into the flame, until it again becomes molten, the bead upon cooling will be colored blue.

Turmeric Test. A few drops of acetic acid are added together with 2 or 3 drops of an alcoholic turmeric solution to an alcoholic extract of the sample, placed in a porcelain dish. The solution is diluted with water and then evaporated to dryness on the water bath. 1/1000 milligram of boric acid will produce a distinct color, 2/100 milligram will give a strong reddish-brown colored residue, which becomes bluish-black when treated with a drop of sodium hydroxide solution.

ESTIMATION

The determination of boron is required in the valuation of borax, $Na_2B_4O_7 \cdot 10H_2O$; boracite, $4MgB_4O_7 2MgOMgCl_2$; borocalcite, $CaB_4O_7 \cdot 6H_2O$; hydroboracite; boronatrocalcite, etc., the element being reported generally as the oxide, B_2O_3. The determination is required for obtaining the true value of commercial boric acid, in the analysis of fluxes and certain pigments. It is determined as a food-preservative in milk, meat, canned goods, etc. The element is determined in certain alloys of nickel, cobalt, zinc, chromium, tungsten, molybdenum and in the analysis of steel.

[1] Silicates should be mixed with potassium fluoride and potassium acid sulphate, $KHSO_4$, then held in the flame.

Chapter contributed by Wilfred W. Scott.

81

Preparation and Solution of the Sample

It will be recalled that crystalline boron is scarcely attacked by acids or alkaline solutions; the amorphous form, however, is soluble in concentrated nitric and sulphuric acids. Both forms fused with potassium hydroxide are converted to potassium metaborate. Boric acid is more readily soluble in pure water than in hydrochloric, nitric, sulphuric, or acetic acids, but still more soluble in tartaric acid (Herz, Chem. Zentr., 1903, **1**, 312). It is soluble in alcohol and volatile oils. Borax is insoluble in alcohol. With acids it becomes transposed to boric acid and the sodium salt of the acid.

Boric Oxide in Silicates, Enamel, etc. About 0.5 gram of the finely ground material is fused with five times its weight of sodium carbonate, the melt extracted with water and the extracts, containing the sodium salt of boric acid, evaporated to small volume. The greater part of the excess sodium carbonate is neutralized with hydrochloric acid and finally made acid with acetic acid (litmus paper test = red). Boric oxide is now determined by the distillation process according to the procedure given later in the chapter.

Boronatrocalcite, Borocalcite, Boracite, Calcium Borate. Ten grams of the powdered material is placed in a flask with a reflux condenser and about 50 cc. of normal hydrochloric acid added and the mixture boiled for half an hour. The contents of the flask, together with the washings, including those of the reflux condenser (CO_2-free water being used), are filtered into a 500-cc. flask and made to volume with CO_2-free water. Fifty cc. of this solution is titrated with half-normal sodium hydroxide, using paranitrophenol indicator. When a yellow color appears the hydrochloric acid has been neutralized. A second 50-cc. portion is now taken for analysis and the free hydrochloric acid neutralized with sodium hydroxide, using the amount of caustic required in the trial analysis (this time without an indicator). Boric acid is now determined by titration according to the procedure on p. 86.

Borax, Boric Acid. Ten grams of the material are dissolved in about 300 cc. of water (free from CO_2) and made to 500 cc. in a graduated flask, with pure water. One hundred-cc. portions are taken for analysis and the solution titrated, in presence of mannitol or glycerol, according to directions given under the volumetric procedures.

Boric Acid in Mineral Water. Water containing more than 0.1 gram boric acid per liter, about 200 cc. are evaporated to small volume, the precipitated salts are filtered off and washed. Boric acid passes into the filtrate and may be determined by the distillation method of Gooch given on p. 84.

With water containing traces of boric acid, 5 liters or more are evaporated to about one-tenth the original volume the precipitate filtered off and washed with hot water. The filtrate is evaporated down to a moist residue. If the residue is small, it is acidified with acetic acid and the boric acid determined by distillation, as stated on p. 84. If considerable residue is present, hydrochloric acid is added to acid reaction, and then the mixture digested with absolute alcohol in a corked flask for ten to fifteen hours, with occasional shaking. The solution is filtered, the residue washed with 95% alcohol, the filtrate diluted with water, 10 cc. of 10% sodium hydroxide solution added and the alcohol distilled off. A second alcoholic extraction is generally recommended. The final alkaline solution is taken to dryness and gently ignited. The residue is extracted with water, made acid with acetic acid and B_2O_3 determined by distillation.

Carbonates. The material is treated with sufficient acid (M. O. indicator) to liberate all the CO_2 and react with the combined alkali of boric and carbonic acid; it is boiled in a flask with reflux condenser to expel CO_2, ten to fifteen minutes, the solution exactly neutralized with sodium hydroxide, (M. O.), and the liberated boric acid titrated in presence of glycerol and phenolphthalein as usual.

Boric Acid in Milk, Butter, Meat and Other Foods

Milk.[1] One hundred cc. of milk is treated with 1 to 2 grams of sodium hydroxide, and evaporated to dryness in a platinum dish. The residue is thoroughly charred[2] by gently heating; at this stage care must be exercised or loss of boric acid will result; 20 cc. of water are added, the sample heated and hydrochloric acid added drop by drop until all but the carbon has dissolved. The mixture is washed into a 100-cc. flask with as little water as possible, 0.5 gram calcium chloride added, then a few drops of phenolphthalein indicator, then a 10% sodium hydroxide solution until a slight permanent pink color is obtained and finally 25 cc. of lime water. (All P_2O_5 is precipitated as calcium phosphate.) The liquid is made to 100 cc., mixed thoroughly, and then filtered through a dry filter. To 50 cc. of the filtrate, equivalent to 50 cc. of the milk taken, normal sulphuric acid is added until the pink color disappears, then methyl orange indicator is added, followed by more of the standard acid until the yellow color changes to a faint pink. Carbon dioxide is expelled and the liberated boric acid titrated in presence of glycerine, according to the procedure given for evaluation of borax and boric acid, under "Volumetric Determination of Boron."

Butter.[3] Twenty-five grams of butter are weighed out in a beaker and 25 cc. of a sugar sulphuric acid mixture added. (Mix = 6 grams sugar of milk, 4 cc. normal sulphuric acid per 100 cc. of solution.) The beaker is placed in the oven (100° C.) until the fat is melted and the mixture is thoroughly stirred. When the aqueous solution has settled, 20 cc. are pipetted out, phenolphthalein added, the solution brought to boiling and half-normal sodium hydroxide added until a faint pink color is obtained. Ten cc. of neutral glycerine are added and the titration carried on until a permanent pink color appears. The difference between the two titrations multiplied by the factor for equivalent boric acid gives the weight of boric acid in the portion taken.

The determination is not affected by the phosphoric or butyric acid or by the sugar of milk in the butter.

Meat.[4] Ten grams of the chopped meat are mixed in a mortar with 40 to 80 grams of anhydrous sodium sulphate, and dried in the water oven. The mass is powdered, then placed in a flask and 100 cc. of methyl alcohol added and allowed to stand for about twelve hours. The alcohol is distilled into a flask and saved. Fifty cc. more of alcohol are added to the residue and this again distilled into the first distillate. The distillates are made up to 150 cc., a

[1] R. T. Thomson, Glasgow City Anal. Soc. Repts., 1895, p. 3.
[2] The milk residue thoroughly charred will give a colorless solution upon extraction.
[3] H. Droop Richmond and J. B. P. Harrison, Analyst, **27**, 197.
[4] C. Fresenius and G. Popp, Chem. Centr., 1897, **2**, 69.

50-cc. portion diluted with 50 cc. of water and 50 cc. of neutral glycerine added with phenolphthalein indicator, and the boric acid titrated with twentieth-normal sodium hydroxide.

One cc. N/20 NaOH =0.0031 gram boric acid, H_3BO_3.

Boric acid in canned goods, sauces, cereals, etc., may be determined by evaporation of the substance with sodium hydroxide and incineration as in case of milk. The sodium hydroxide is neutralized and boric acid titrated as usual.

GRAVIMETRIC DETERMINATION OF BORON

The solubility of boron compounds prevents complete precipitation by any of the known reagents, hence most of the gravimetric methods are indirect.

Distillation as Methyl Borate and Fixation by Lime [1]

This excellent method, originally worked out by F. A. Gooch,[1] and later modified by Gooch and Jones,[2] depends upon the fact that the borates of alkaline earths and alkalies give up their boron in the form of the volatile methyl borate (b.p., 65° C.), when they are distilled with absolute methyl alcohol (acetone-free). The methyl borate passed over lime in presence of water is completely saponified, the liberated boric acid combining with the lime to form calcium borate, which may be dried, ignited, and weighed. The increase of the weight of the lime represents the B_2O_3 in the sample.

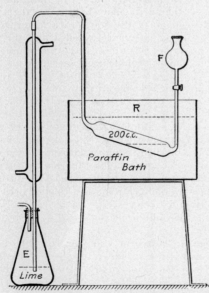

$$2B(OCH_3)_3 + CaO + 6H_2O$$
$$= 6CH_3OH + Ca(BO_2)_2 + 3H_2O.$$

Procedure. About 1 gram of pure calcium oxide is ignited to constant weight over a blast lamp and then transferred to the dry, Erlenmeyer receiving flask (Fig. 8). The crucible in which the lime was heated and weighed is set aside in a desiccator for later use.

0.2 gram or less of the alkali borates, obtained in solution by a procedure given under "Preparation of the Sample," is treated with a few drops of litmus (or lacmoid), solution and the free alkali neutralized with dilute HCl solution added drop by drop. A drop of dilute sodium hydroxide solution is added and

Fig. 8.—Distillation of Methyl Borate.

then a few drops of acetic acid. The slightly acid solution is transferred to the

[1] Proc. Am. Acad. of Arts and Sciences, **22**, 167–176 (1886). Anal. Chem., Treadwell-Hall, Vol. **2**.

[2] See note on p. 75.

pipette-shaped retort R, Fig. 8, by means of the attached funnel F, washing out the beaker and funnel with three 2- to 3-cc. portions of water. The stopcock of the funnel is closed, the apparatus is connected up as shown in the illustration, the paraffine bath, heated to not over 140° C., placed in position and the liquid in the retort distilled into the receiver containing the known amount of lime. When all the liquid has distilled over, the paraffine bath is lowered, the retort allowed to cool for a few minutes, 10 cc. of methyl alcohol (acetone-free) added to the residue in R and the contents again distilled by replacing the paraffine bath. The process is repeated three times with methyl alcohol. The contents of the retort (which are now alkaline), are made distinctly acid by addition of acetic acid, and three more distillations made with 100-cc. portions of methyl alcohol, as before. The paraffine bath is now removed, the receiving flask is stoppered, the contents thoroughly mixed by shaking, and set aside for an hour or more for complete saponification of the methyl borate. The contents are now poured into a large platinum dish and evaporated on the water bath at a temperature below the boiling-point of the alcohol. (Loss of boric acid will occur if the alcohol boils.) The adhering lime in the receiving flask is dissolved by wetting its entire surface with a few drops of dilute nitric acid (the flask being inclined and revolved to flow the acid over its sides). The contents are transferred to the dish with a little water and the evaporation repeated. No loss of boric acid will take place at this stage, the alcohol having been removed during the first evaporation. The residue is gently heated to destroy any calcium acetate that may have formed, the cooled borate and lime are taken up with a little water and transferred to the crucible in which the lime was heated and weighed. The material clinging to the dish is dissolved with a little nitric acid (or acetic acid), and washed into the crucible. The contents of the crucible are evaporated to dryness on the water bath, then heated very gently over a flame (the crucible being covered) and finally more strongly. The heating is continued until a constant weight is obtained. The increase of weight of the lime represents the amount of B_2O_3 in the sample.

NOTES. Gooch and Jones worked out a procedure which utilizes sodium tungstate as a retainer of the methyl borate, in place of the lime. This substance is definite in weight, not hydroscopic, soluble in water, and recoverable in its original weight after evaporation and ignition. "Methods in Chem. Anal.," p. 204, 1st Ed. By F. A. Gooch, John Wiley & Sons, Publishers.

The receiving flask has a cork stopper with a hole to accommodate the tube of the condenser and a slit to permit the escape of air from the flask.

Gooch recommends cooling of the receiving flask.

VOLUMETRIC DETERMINATION OF BORON

Titration of Boric Acid in Presence of Mannitol or Glycerol Evaluation of Borax

The method takes advantage of the fact that boric acid reacts neutral to methyl orange (or paranitrophenol), but is acid to phenolphthalein, and may be quantitatively titrated in the presence of mannitol or of glycerol, which prevent the hydrolization of sodium borate. If insufficient mannitol or glycerol are present the color change takes place too soon, the color fading upon adding more of these substances. The end-point is reached when the further addition of these reagents produces no fading of the color. In the procedure, the alkali is neutralized in presence of methyl orange (or paranitrophenol), and the liberated boric acid is now titrated.

Reactions. $Na_2B_4O_7 + 2HCl + 5H_2O = 2NaCl + 4H_3BO_3$
$H_3BO_3 + NaOH = NaBO_2 + 2H_2O$.

Procedure. One hundred cc. of the solution containing the borax, prepared according to directions under "Preparation and Solution of the Sample," equivalent to 2 grams of the substance, is taken for analysis.

A. Titration of Combined or Free Alkali. Methyl orange indicator is added and the solution is titrated with normal or half-normal sulphuric acid until the yellow color is replaced by an orange red. (With paranitrophenol the solution becomes colorless.) From this titration the combined alkali, together with any free alkali, is calculated. If free alkali is known to be absent (see note), the amount of borax may be calculated.

One cc. N. $H_2SO_4 = 0.031$ gram Na_2O, or $= 0.1911$ gram $Na_2B_4O_7 \cdot 10H_2O$, or $= 0.101$ gram $Na_2B_4O_7$.

B. Titration of Boric Acid. The liberated boric acid may now be titrated with caustic. This may be accomplished either on the above portion or on a fresh 100-cc. portion (free from methyl orange indicator), to which the amount of acid, required to neutralize the alkali, has been added. Fifty cc. of neutral glycerol or 1 gram of mannitol are added, followed by phenolphthalein indicator. Normal or half-normal sodium hydroxide is added from a burette until a change of color takes place. If methyl orange is present, the color, first becoming yellow, changes to an orange red. In absence of methyl orange the characteristic lavender or purplish pink of alkali phenolphthalein is obtained. More glycerol or mannitol is now added and if the color fades the titration is continued until the addition of these reagents no longer produces this fading of the end-point, From this titration boric acid is calculated and the equivalent borax determined.

One cc. N. NaOH $= 0.062$ gram H_3BO_3, equivalent to 0.0505 $Na_2B_4O_7$, or 0.0955 $Na_2B_4O_7 \cdot 10H_2O$.

Factors. Na_2O to $Na_2B_4O_7 = 3.2581$, reciprocal $= 0.3069$.
Na_2O to $Na_2B_4O_7 \cdot 10H_2O = 6.1638$, recip. $= 0.1622$.
Na_2O to $Na_2CO_3 = 1.7097$, recip. $= 0.5849$.
Na_2O to $Na_2CO_3 \cdot 10H_2O = 4.6155$, recip. $= 0.2167$.

NOTES. In borax (free from excess B_2O_3 or Na_2O), the acid titration is half the subsequent alkali titration (factor, acid to borax $= 0.1911$, alkali to borax $= 0.0955$).

If the acid titration exceeds this proportion, alkali other than that combined with boric acid is indicated; if the alkali titration is greater than twice the acid titration, free boric acid is indicated.

The glycerol should be made neutral with N/10 NaOH before use, in case it contains free fatty acids.

Mannitol is a solid and has some advantages over glycerol; it gives a sharper end-point, is less apt to contain free acids and does not appreciably alter the bulk of the solution to which it is added. (L. C. Jones, C. N., **80**, 65, 1899).

N.B. Paper on use of mannitol and glycerol in determining boric acid, by R. T. Thomson, J. S. C. I., **12**, 432.

Example. By actual test 2 grams $Na_2B_4O_7 \cdot 10H_2O$ required 10.66 cc. N. $H_2SO_4 =$ 10.66×0.1911 = 2.037 gram $Na_2B_4O_7 \cdot 10H_2O$. The liberated H_3BO_3 required 21.39 cc. of N. NaOH = 21.39×0.0955 = 2.043 gram $Na_2B_4O_7 \cdot 10H_2O$. The borax had lost a small amount of water of crystallization, hence the high results when calculated to $Na_2B_4O_7 \cdot 10H_2O$.

EVALUATION OF BORIC ACID

One hundred cc. of the solution, prepared as directed under "Preparation of the Sample," equivalent to 2 grams of the original material, is treated with 50 cc. of glycerol or 1 gram of mannitol, and the acid titrated with standard caustic, in presence of phenolphthalein indicator according to the procedure given in *B*, under "Evaluation of Borax."

One cc. normal acid contains 0.062 gram H_3BO_3, hence the cc. of caustic required multiplied by 0.062 = grams boric acid.

Examples. Two grams H_3BO_3 by actual test required 32.1 cc. N. NaOH = 32.1×.062 = 1.99 grams H_3BO_3.

Detection of Minute Amounts of Boron.

Robin's Test for Boron. To a few drops of the aqueous solution under examination (slightly acidified with HCl) are added two drops of a tincture of mimosa flowers, and the mixture evaporated to dryness on the water bath. The residue is treated with dilute ammonia water, whereupon in presence of boric acid, a rose pink to blood red color develops, according to the amount present. L. Robin claims that as little as 0.0001 milligram may be detected in presence of nitrates, chlorides, iodides, or calcium sulphate. Organic acids and sodium phosphate interfere. The reagent is prepared by extracting the mimosa flowers with ethyl alcohol. The extract is protected from the light.

Distillation Method as Modified by Chapin

Reagents[1]

1. *Paranithrophenol.*—One gram dissolved in 75 cc. of neutral ethyl alcohol and made up to 100 cc. with water.

2. *Phenolphthalein.*—One gram dissolved in 100 cc. of ethyl alcohol and made up to 200 cc. with water.

3. *Hydrochloric Acid, 0.1 Normal.*—The water should be boiled to remove carbonic acid.

4. *Hydrochloric Acid, 1.1 Strength.*—A dropping bulb should be filled with this acid when it is needed in accurate small amounts.

5. *Sodium Hydroxide, 0.5 and 0.1 Normal.*—These should be standardized as follows: Fuse pure boric acid in a platinum dish. While still warm break the melt up and place the fragments quickly in a weighing tube. Dissolve 1.75 grams in 250 cc. of hot,

[1] Bulletin 700, "The Analysis of Silicate and Carbonate Rocks," by W. F. Hillebrand. Department of the Interior, U. S. Geological Surveys.

[3] Am. Jour. Sci., 4th Ser., Vol. 14, 1903, p. 195.

recently boiled water, cool, and dilute the solution to 500 cc. This solution is 01. normal—that is, in presence of mannite or glycerol 1 cm.[3] is neutralized to the phenolphthalein end point by 1 cc. of 0.1 normal sodium hydroxide.

In standardizing the sodium hydroxide against this solution both indicators should be used, so that the end may be the same as that seen in actual titration. Follow exactly the directions given under *b*, below, for the final titraction, only assuming that the boric acid solution is exactly neutral to paranitrophenol—that is, free from mineral acid. When sodium hydroxide is standardized in this way the small amount of carbonate present does no harm.

6. *Mannite.*—This is preferable to glycerol, for it requires no special preparation, does not materially increase the bulk of the solution to be titrated, and gives an equally sharp end point.

7. *Methyl Alcohol.*—This should be distilled over lime after it has been heated for some hours in contact with the lime under a reflux condenser. The more nearly anhydrous the alcohol is the better.

8. *Calcium Chloride.*—This should be granular, anhydrous, and free from boron.

Apparatus

The main set up of the apparatus required is shown in Fig. 8a.

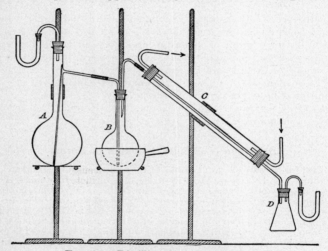

Fig. 8a.—Distilling Apparatus for Boron

A. Flask of 500 cc. capacity for methyl alcohol, having a U tube containing a little mercury as a safety trap, and a capillary "boiling tube" 3 mm. in bore, the bore closed at 1 cm. from the lower end.

B. Flask of 250 cc. capacity, supported in a casserole of water but not touching the bottom. The outlet tube of *B* should be not less than 12 cm. in vertical height and of rather wide bore. The end inside the flask should be jagged to allow condensed liquid to drop back instead of jumping up the tube.

C. Glass condenser, 40 cc. long, of tubing 3 cm. wide.

D. Receiving flask of 250 cc. capacity. Two or three of these flasks should be at hand, with necks of same size, so that the receiver can be changed without interrupting the distillation. The small U tube attached contains a little water to prevent the escape of any uncondensed methyl borate.

E. (Not shown.) A second glass condenser like *C*, connected on the one side to a flask exactly like *B*, which, however, has a stopper with only one hole and rests completely within a large bath. The flask carries also a "boiling tube." On the other side is a receiver like *D*, but without the U trap. This apparatus is used for distilling off the alcohol after the preliminary titration.

F. (Not shown.) Filter pump. At the end of the side tube is an elbow tube connected to a rubber stopper with two holes and of a size to fit the flask mentioned in connection with *E*. The second hole of the stopper is closed by a glass plug, which can be withdrawn to break the vacuum. Midway between pump and stopper is a small safety bottle, to prevent the tap water from sucking back the flask.

Procedure

Chapin used not more than half a gram of mineral powder for even very small amounts of boron, and not less unless the percentage exceeded 10. When the percentage is high it is best to so limit the weight of the sample that the of B_2O_3 shall not exceed 0.1 gram. If a flux is used it should be weighed to within a milligram of two; then the amount of acid required to take up the melt can be measured out at once and there is no danger of using too great an excess.

If the mineral is soluble in hydrochloric acid, transfer 1 gram of it to the flask B, without letting any adhere to the neck, and treat with not more than 5 cc. of 1 : 1 hydrochloric acid. Heat gently on a water bath until solution is complete.

If the mineral is not soluble, add to it exactly six times its weight of sodium carbonate or of an equimolecular mixture of sodium and potassium carbonates, mix, and fuse in the usual manner. Without removing from the crucible, decompose the melt with 1 : 1 hydrochloric acid in calculated amount added by degrees. While this is being done the crucible should rest in a casserole, and the lid should be kept in place as much as possible. Toward the end it may be necessary to heat a little, but care should be taken not to boil, for boric acid would be lost with the steam. Pour the solution into the flask B and rinse the crucible with a very little water.

Then add pure anhydrous calcium chloride, using about 1 gram for each cubic centimeter of solution and running it through a paper funnel to keep the neck of the flask clean. Twirl the flask a little to allow the chloride to take up the water, connect it with the rest of the apparatus, raise the casserole beneath it until the flask rests in the water but does not touch the bottom, and then begin the distillation of the alcohol from the flask A, taking care that the open end of the capillary "boiling tube" is free from alcohol and that the U tube attached to the receiver is trapped with water.

The decomposition flask, B, is not heated until about 25 cc. of alcohol has condensed in it. After that the water bath is heated by a small flame and the flask thus kept hot enough to prevent further condensation of alcohol. The distillation should not be so rapid as to permit escape of methyl borate from the system.

When a distillate of about 100 cc. has collected the receiver is exchanged for a fresh one and the collection of a second distillate is proceeded with. The contents of the trap tube are added to the first receiver, and a preliminary titration is made as follows: Add a few drops of paranitrophenol solution and run in the standard sodium hydroxide until the indicator shows that the free mineral acid is neutralized. Next, add 1 cc. of the phenolphthalein solution and continue the titration until the color of that indicator begins to appear. The end point will not be sharp, but the amount of alkali used between the two end points indicates approximately the amount of boric acid in the distillate. If the mineral is likely to contain more than 3 or 4% of B_2O_3 use the 0.5 normal solution, the object being to keep the distillate as free from the water as possible and thus facilitate the dehydration later on.

After completing the titration add to the distillate twice as much alkali as was used between the two end points, transfer the distillate to the second distilling apparatus, and boil off the alcohol. The addition of the indicated amount of alkali prevents loss of boron by converting unstable $NaBO_2$ to stable Na_3BO_3.

In the meantime the second distillate of 100 cc. will have collected. Unless the alcohol contained water or insufficient calcium chloride was used, this second distillate will contain the remainder of the boron. Therefore, before removing the receiver stop the distillation. Treat the second distillate exactly like the first and then add it to the first.

If the amount of alkali used for titrating the second distillate is less than 0.1 cc. of the 0.1 normal solution the distillation may be regarded as complete.

When the liquid from the alcohol is being distilled no longer boils in the steam bath, remove it to a casserole, rinsing the flask once or twice with water, and heat over a direct flame, rotating the casserole while doing so, until the little remaining alcohol has been boiled out. The residue should now be small —about 25 cc. if 0.1 normal alkali was used and much less if 0.5. If the volume is less than 25 cc. make up to that extent.

Return the solution to the flask and add drop by drop with constant twirling 1 : 1 hydrochloric acid from a dropping bulb, until the color of both indicators is discharged, taking care not to add a drop too much. Now put in the "boiling tube," heat the flask in the steam bath for a minute or two, then attach to the filter pump and boil under reduced pressure until the liquid is nearly cold and only an occasional bubble appears.

All carbon dioxide being thus removed, break the vacuum, cool if necessary under the tap, and proceed to the final titration.

First neutralize the excess of hydrochloric acid by running in carefully 0.5 normal sodium hydroxide until the yellow of paranitrophenol just appears. Bring back to acid reaction with 0.1 normal hydrochloric acid and then to the appearance of a faint yellow with 0.1 normal alkali. The solution is now exactly neutral as the indicators itself shows this color in a neutral solution. Make sure that the end point is reached exactly. One drop of 0.1 normal acid should discharge the color entirely.

Now add 1 gram of mannite, read the burette, and continue titrating to the end point of phenolphthalein. Add another gram of mannite, and if this causes a disappearance of the end color add more alkali until it reappears.

The number of cubic centimeters of alkali used multiplied by its equivalent in terms of B_2O_3 gives the amount of the latter contained in the solution. If the solution of alkali is exactly 0.1 normal this equivalent is 0.0035.

The changes of color that take place during the titration need mention. When the mannite (or glycerol) is added to the solution the yellow color due to the paranitrophenol disappears at once. This is due to the fact that the combination of these reagents with the boric acid forms an acid of sufficient strength to affect the indicators, although boric acid alone does not. As the titration proceeds the color reappears and grows stronger, becoming very intense just before the phenolphthalein end point appears. The latter will be noticed as a faint brownish tinge, and then another drop of the 0.1 normal alkali changes it to an intense brownish red, which is the real end point.

Notes

E. T. Allen and E. G. Ziez[1] tested the Chapin method very fully in its application to the determination of boron in glasses and regard it as far superior to other methods, even though it is subject to a slight but very uniform correction of 1 milligram or less, to be determined by a blank run. The correction seems to be due always in part to a boron content of the reagents used and in part to a titration error. The fact that such correction is unavoidable makes the method of uncertain value for determining the very small amounts of boron that rocks may be presumed to carry, but the constancy of results is so great that a consistent excess found over what the blank affords is strong evidence that boron is actually present.

Allen and Ziez found the method to be affected appreciably by relatively large amounts of arsenious acid but not by arsenic acid. The effect of the former can be eliminated by converting it to arsenic acid by oxidizing with hydrogen peroxide after making the solution distinctly alkaline with sodium hydroxide.

Allen and Ziez also found that relatively large amounts of fluorides effect the accuracy of the method but do not seriously impair its usefulness for ordinary work.

[1] "Analysis of Silicate and Carbonate Rocks," by W. T. Hillebrand, Bull. 700, U. S. Geol. Survey, 1919.

Glucose may be used in place of mannite in titrations of borax as shown by LeRoy S. Weatherby and H. H. Chesney (Ind. Eng. Chem., 18, 820, Aug., 1926). Though a larger quantity of glucose (about 10 times) is required than of mannitol, this is of no disadvantage, as a large background of white material is helpful in distinguishing the end-point. As commercial glucose (cerelose) may be obtained at a cost of a few cents a pound, while the price of mannitol is more than 200 times as great, the advantage of glucose is seen. (Glucose $3.24 per 100 lbs., mannitol $6.50 per lb., Aug., 1926, quotation.)

BROMINE

Br, *at.wt.* **79.92;** *sp.gr.* **3.1883°;** *m.p.* **−7.3°;** *b.p.* **58.7° C.;** *acids,* **HBr, HBrO, HBrO₃**

DETECTION

Silver Nitrate solution precipitates silver bromide, AgBr, light yellow, from solutions containing the bromine anion. The precipitate is insoluble in dilute nitric acid, but dissolves with difficulty in ammonium hydroxide and is practically insoluble in ammonium carbonate solution (distinction from AgCl).

Carbon Disulphide or Carbon Tetrachloride shaken with free bromine solution, or with a bromide to which a little chlorine water has been added, (a large excess of chlorine must be avoided, as this forms BrCl compound), will absorb the bromine and become a reddish-yellow color, or if much bromine is present, a brown to brownish-black. In the latter case a smaller sample should be taken to distinguish it from iodine.

Bromates are first reduced by a suitable reducing agent such as cold oxalic acid, sodium nitrite, hydrochloric acid, etc., and the liberated bromine tested as directed above. *Silver nitrate* added to bromates in solution precipitates AgBrO₃, which is decomposed by hydrochloric acid to bromine gas.

Barium Chloride precipitates Ba(BrO₃)₂, which is reduced readily to bromine as directed above.

Magenta Test for Bromine.[1] The test reagent is made by adding 10 cc. of 0.1% solution of magenta to 100 cc. of 5% solution of sulphurous acid and allowing to stand until colorless. This is the stock solution. Twenty-five cc. of this reagent is mixed with 25 cc. of glacial acetic acid and 1 cc. of sulphuric acid. Five cc. of this is used in the test.

Test. Five cc. of the magenta reagent is mixed with 1 cc. of the solution tested. Chlorine produces a yellow color. Bromine gives a reddish-violet coloration. The colored compound in each case may be taken up with chloroform or carbon tetrachloride and a colorimetric comparison made with a standard.

In halogen mixes, iodine is first eliminated by heating with an iron per salt. Bromine is now liberated by adding sulphuric acid and potassium chromate. A glass rod with a pendant drop of sodium hydroxide is held in the vapor to absorb bromine, and the drop then tested with the magenta reagent. After iodine and bromine are eliminated, chlorine may be tested by heating the substance with potassium permanganate, which liberates this halogen.

[1] G. Denigès and L. Chelle. Ann. Chim. anal., 1913, **18**, 11–15; The Analyst, 1913, 119.

Chapter contributed by Wilfred W. Scott.

ESTIMATION

Bromine never occurs free in nature. It is found chiefly combined with the alkalies and the alkaline earths, hence occurs in many saline springs and is a by-product of the salt industry. It is found in silician zinc ores, Chili saltpeter, in sea water (probably as $MgBr_2$), in marine plants. Traces occur in coal, hence in gas liquors.

The substance is used in metallurgy, the arts, and medicine. It is a valuable oxidizing agent for the laboratory.

Preparation and Solution of the Sample

The following facts regarding solubility should be remembered: The element bromine is very soluble in alcohol, ether, chloroform, carbon disulphide, carbon tetrachloride, concentrated hydrochloric acid and in potassium bromide solution. One hundred cc. of water at 0° C. is saturated with 4.17 grams of bromine, and at 50° C. with 3.49 grams. The presence of a number of salts increases its solubility in water, e.g., $BaCl_2$, $SrCl_2$, etc.

NOTE. The element is a dark, brownish-red, volatile liquid, giving off a dark reddish vapor with suffocating odor, irritating the mucous membrane (antidote dil. NH_4OH, ether), very corrosive. Acts violently on hydrogen, sulphur, phosphorus, arsenic, antimony, tin, the heavy metals, and on potassium, but has no action on sodium, even at 200° C. Bleaches indigo, litmus, and most organic coloring matter. It is a strong oxidizing agent. Bromine displaces iodine from its salts, but is displaced by chlorine from its combinations.

Bromides are soluble in water, with the exception of silver, mercury, lead, and cuprous bromides.

Bromates are soluble in water with the exception of barium and silver bromates and some basic bromates.

Decomposition of Organic Matter for Determination of Bromine. The substance is decomposed with nitric acid in presence of silver nitrate in a bomb combustion tube by the Carius method described in the chapter on Chlorine, under "Preparation and Solution of the Sample" The residue, containing the halides, is dissolved in warm ammonia water, and filtered, as stated. The filtrate and washings are acidified with nitric acid, heated to boiling and the silver bromide settled in the dark, then filtered through a weighed Gooch crucible, the washed precipitate dried at 130° C. and weighed as AgBr.

In presence of two or three halogens the lime method is recommended, as given in the chapter on chlorine, page 146.

Salts of Bromine. The ready solubility of bromides and bromates has been mentioned. A water extract is generally sufficient. Insoluble salts are decomposed by acidifying with dilute sulphuric acid and adding metallic zinc. The filtrate contains the halogens.

SEPARATIONS

Separation of Bromine from the Heavy Metals. Bromides of the heavy metals are transposed by boiling with sodium carbonate, the metals being precipitated as carbonates and sodium bromide remaining in solution.

Separation of Bromine from Silver (AgBr) and from Cyanides (AgCN). The silver salts are heated to fusion. The mass is now treated with an excess of zinc and sulphuric acid, the metallic silver and the paracyanogen filtered off and the bromine determined in the filtrate.

Separation of Bromine from Chlorine or from Iodine. Details of the procedure for determining the halogens in presence of one another is given in the chapter on Chlorine, page 154. Free bromine is liberated when the solution of its salt is treated with chlorine.

Separation of Bromine from Iodine.[1] The neutral solution containing the bromide and iodide is diluted to about 700 cc. and 2 to 3 cc. of dilute sulphuric acid, 1 : 1, added, together with about 10 cc. of 10% sodium nitrite, $NaNO_2$, solution. (Nitrous acid gas may be passed through the solution in place of adding sodium nitrite, if desired.)[2] The solution containing the halides is boiled until colorless and about twenty minutes longer, keeping the volume of solution above 600 cc. 0.5 gram KI may be decomposed and the iodine expelled from the bromide in half an hour. The bromine is precipitated from the residue remaining in the flask by addition of an excess of silver nitrate and determined as silver bromide.

The procedure for determining iodine is given in the chapter on this subject.

GRAVIMETRIC METHODS

Precipitation as Silver Bromide

The general directions for determination of hydrochloric acid and chlorides apply for determining hydrobromic acid and bromides.

I. Hydrobromic Acid and Bromides of the Alkalies and Alkaline Earths. Procedure. The bromide in cold solution is made slightly acid with nitric acid and then silver nitrate added slowly with constant stirring until a slight excess is present. The mixture is now heated to boiling and the precipitate settled in the dark, then filtered through a weighed Gooch crucible, and washed with water containing a little nitric acid and finally with pure water to remove the nitric acid. After ignition the silver bromide is cooled and weighed as AgBr.

$$AgBr \times 0.4256 = Br, \text{ or } \times 0.6337 = KBr.$$

II. Heavy Metals Present.

If heavy metals are present it is not always possible to precipitate silver bromide directly. The heavy metals may be removed by precipitation with ammonia, sodium hydroxide or carbonate and the bromide then determined in the filtrate as usual.

---

[1] F. A. Gooch and J. R. Ensign, Am. Jour. Sci., (3), xl, 145.
[2] Nitrous acid gas is generated by dropping dilute H_2SO_4, by means of a separatory funnel onto sodium nitrite in a flask.

VOLUMETRIC METHODS

Free hydrobromic acid may be titrated with standard alkali exactly as is described for the determination of hydrochloric acid in the chapter on Acids. One cc. normal caustic solution is equivalent to 0.08093 gram HBr.

Determination of Free Bromine. Potassium Iodide Method

The method depends upon the reaction $KI + Br = KBr + I$.

Procedure. A measured amount of the sample is added to an excess of potassium iodide, in a glass-stoppered bottle, holding the point of the delivering burette just above the potassium iodide solution. The stoppered bottle is then well shaken, and the liberated iodine titrated with standard thiosulphate solution.

One cc. of N/10 thiosulphate, $Na_2S_2O_3 = 0.007992$ gram Br.

Determination of Bromine in Soluble Bromides. Liberation of Bromine by Addition of Free Chlorine

When chlorine is added to a colorless solution of a soluble bromide, bromine is liberated, coloring the solution yellow. At boiling temperature the bromine is volatilized, the liquid becoming again colorless. When the bromide is completely decomposed and bromine expelled, further addition of chlorine produces no color reaction. $KBr + Cl = KCl + Br$.

Procedure. The solution containing the bromide is heated to boiling and standard chlorine water added from a burette (protected from the light by being covered with black paper), the tip of the burette being held just above the surface of the hot bromide solution to prevent loss of chlorine. The reagent is added in small portions until finally no yellow coloration is produced. From the value per cc. of the chlorine reagent the bromine content is readily calculated.

Standard Chlorine Water. The reagent is made by diluting 100 cc. of water saturated with chlorine to 500 cc. This solution is standardized against a known amount of pure potassium bromide (dried at 170° C.), the same amount of bromide being taken as is supposed to be present in the solution examined. The value per cc. of the reagent is thus established.

Silver=Thiocyanate=Ferric Alum Method. (Volhard)

The procedure is the same as that used for the determination of chlorine. The bromide solution is treated with an excess of tenth-normal silver nitrate solution, and the excess of this reagent determined by titration with ammonium thiocyanate, using ferric alum indicator. One cc. of the thiocyanate should be equivalent to 1. cc. of silver nitrate solution. The formation of the red ferric thiocyanate indicates the completed reaction. (Consult the procedure in the chapter on Chlorine, page 149.)

One cc. of N/10 $AgNO_3 = 0.007992$ gram Br.

Determination of Traces of Bromine

By means of the magenta reagent, described under "Detection," small amounts of bromine may be determined colorimetrically.

To 5 cc. of the solution is added 0.2 cc. of strong hydrochloric acid, 1 cc. of concentrated sulphuric acid, 1 cc. of the stock magenta reagent and 0.2 cc. of a 10% solution of potassium chromate, shaking the mixture with addition of each reagent, and without cooling, 1 cc. of chloroform is added. Comparison is made with a standard sample containing a known amount of bromide.[1]

Note. A solution containing 0.001 gram bromine per liter has a violet to reddish-violet color.

Determination of Bromates by Reduction with Arsenous Acid and Titration of the Excess [2]

Bromic acid may be reduced by arsenous acid in accordance with the reaction $3H_3AsO_3 + HBrO_3 = 3H_3AsO_4 + HBr$. In the process a considerable excess of arsenous acid is added, the excess titrated with iodine and the bromate calculated.

Procedure. The sample of bromate, dissolved in water, is treated with a considerable excess of N/10 arsenous oxide (dissolved in alkali hydrogen carbonate) reagent, the solution then acidified with 3 cc. to 7 cc. of dilute sulphuric acid (1 : 1) and diluted to a volume not exceeding 200 cc. After boiling for ten minutes, the free acid is neutralized with alkali hydrogen carbonate ($NaHCO_3$ or $KHCO_3$) and the excess of arsenite titrated with N/10 iodine.

Let x cc. equal the difference between the two titrations with N/10 iodine (i.e. of total arsenite minus excess arsenite) and w equal the weight of bromate desired, then

$$w = \left(\frac{x \text{ cc.} \times \text{mol. wt. } RBrO_3}{6 \times 10 \times 1000}\right) \text{milligrams.}$$

ANALYSIS OF CRUDE POTASSIUM BROMIDE AND COMMERCIAL BROMINE

Determination of Chlorine, Combined or Free

This is the principal impurity present and its estimation is concerned here. Andrews' modification of Bugarszk's method [3] is as follows:

[1] G. Denigés and L. Chelle, Ann. Chem. anal., 1913, 18–15; Analyst, 1913, p. 119.
By means of the magenta reagent it is possible to detect bromine in the ash of plants, beet root, spinach, etc. The organic substance may be decomposed by heating in a combustion tube. Filter paper moistened with the reagent and held in the fumes of the organic substances gives the characteristic test if bromine is present.
[2] Method of F. A. Gooch and J. C. Blake, Am. Jour. Sci., 14, Oct., 1902. Procedure communicated to the Editor by Prof. Gooch.
[3] Jour. Am. Chem. Soc., 1907, 29, 275–283; Zeits. anorg. Chem., 1895, 10, 387.

Procedure. The following amount of sample and reagents should be taken.

Approx. per cent Impurity if KCl Present is	Amount Substance to be Taken, Gram.	Iodate Solution 1/5 N. Required: cc.	2N. HNO₃ Required, cc.
Over 5	0.6	36	20
1.5 to 5	1.8	96	26
0.2 to 1.5	3.6	186	35

The mixture is gently heated to boiling in a long-necked Kjeldahl flask, inclined at an angle of 30°, potassium iodate solution added, then nitric acid and sufficient water to make the volume about 250 cc. The boiling is continued until bromine is expelled (test steam with 2% KI solution rendered faintly acid with hydrochloric acid). The mixture is boiled down to not below 90 cc. Now 1 to 1.5 cc. of 25% phosphorus acid are added and the mixture boiled for five minutes after all the iodine has been expelled. The colorless liquid is cooled, mixed with a slight excess of 1/20 or 1/50 normal silver nitrate solution (according to the proportion of chloride), the excess of silver nitrate then determined by titration with standard thiocyanate with ferric nitrate as indicator. (See procedure for silver-thiocyanate-ferric alum method of Volhard for determination of chlorine, page 149.)

Determination of Chlorine in Crude Bromine

Three grams of bromine (or more if less than 0.5% chlorine is present) in 50 cc. of 4% potassium iodide solution in a glass-stoppered flask (cooled in ice during hot weather) are shaken and then transferred to a Kjeldahl flask. Sixty cc. of 1/5 N. KIO₃ solution and 24 cc. 2N. HNO₃ introduced, the solution diluted to 250 cc. and chlorine determined as directed above.

All commercial bromine contains chlorine. The U. S. Bureau of Standards determines the specific gravity of the two.

Determination of Bromine in Mineral Waters in Presence of Iodine. Separation from Iodine

The method of Baughman and Skinner (U. S. Bureau of Chemistry) is as follows:

The neutral or slightly acid sample, which should contain not more than 0.1 g. bromine, or 10 g. total salts, is introduced into the distillation flask and adjusted to a volume of approximately 75 cc. One and a half to 2 grams of ferric sulphate are added, the liberated iodine distilled with steam into 100 cc. of potassium iodide solution (10 g. KI per 100 cc.). The potassium iodide solution may be titrated with sodium thiosulphate solution, to determine the iodine. The bromine is determined in the liquid remaining in the distillation flask.

Determination of Bromide in Mineral Waters and Brines[1]

Bromine occurs combined as bromide in natural and artificial brines, associated frequently with small amounts of iodide. Bromine may be obtained in the mother liquor or "bittern," a by-product in the manufacture of common salt. The following procedure for evaluation of these brines for their bromine content was developed by W. F. Baughman and W. W. Skinner, U. S. Bureau of Chemistry.

[1] J. Ind. Eng. Chem., 11, 954–959 (Oct., 1919).

Apparatus. Three tall form, 250 cc. glass stoppered Dreschsel gas washing bottles or cylinders are joined together in series, the first two are joined by welding together the outlet tube of the first and the intake tube of the second, and the second and third by rubber tubing, bringing the ends of the glass tubing in contact with each other. The drawing Fig. 8b shows the details of the apparatus.

Procedure. "Evaporate the sample of water or brine, which should not be acid (if necessary add a small amount of sodium carbonate), to dryness or nearly so. Charge the reaction cylinder by introducing first glass beads to a depth of about 1 in., then as much of the sample as can be scraped in, and finally enough glass beads to fill the cylinder half full. Make a solution of sodium sulphite and sodium carbonate of such a concentration that 25 cc. will contain 1 g. of sulphite and 0.2 g. of carbonate. Add 20 cc. of this solution to the first absorption cylinder, 5 cc. to the second, and dilute each to approximately 200 cc. Connect the three cylinders and draw through a slow current of air. Add 15 g. of chromic anhydride dissolved in 10 to 12 cc. of water to the reaction cylinder, followed by washings from the evaporat-

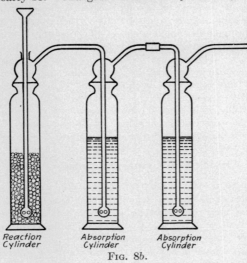

Reaction Cylinder Absorption Cylinder Absorption Cylinder

Fig. 8b.

ing dish which contained the sample, sufficient to bring the total volume added to about 25 cc. Aspirate until the contents of the reaction cylinder are in solution and thoroughly mixed, then discontinue, close the inlet tube with a small piece of rubber tubing and a clamp, and reduce the pressure in the apparatus slightly by sucking out some air in order to guard against any possible escape of bromine at the ground glass stopper. Allow to stand overnight, then aspirate with a rather strong current of air (about $\frac{1}{2}$ to $\frac{3}{4}$ l. per min.) for three hours, adding four 2 cc. portions of 3 per cent hydrogen peroxide at thirty minute intervals. Stop the aspiration and evaporate the contents of the two absorption cylinders nearly to dryness. Clean out the reaction cylinder and freshly charge with glass beads and 15 g. chromic anhydride. Into the first absorption cylinder, put 10 g. potassium iodide, dissolved in 200 cc. of water, and into the second 3 or 4 g. in a like amount of water. Connect the apparatus, draw through a slow current of air, and transfer the contents of the evaporating dish to the reaction cylinder by means of a small funnel, using 25 cc. of water. Aspirate with a rather strong current of air until all the bromine is evolved (about 1 hour) and titrate the potassium iodide solution with thiosulphate."

Chromic acid in concentrated solution liberates bromine from bromides quantitatively at room temperature.

Iodides should be removed if present in appreciable amount.

CADMIUM

Cd, *at.wt.* 112.41 *sp.gr.* 8.642; *m.p.* 320.9°; *b.p.* 778° C.; *oxide* CdO.

DETECTION

Cadmium is detected in the *wet way* by precipitation as the yellow sulphide by hydrogen sulphide from an acid solution. It is distinguished from arsenic, antimony and tin (stannic) by the insolubility of its sulphide in ammonium hydroxide or colorless ammonium sulphide; from tin (stannous) by its insolubility in yellow ammonium sulphide; and from mercury by its solubility in hot nitric acid. The separation of cadmium may be made from lead since cadmium sulphate is soluble in dilute sulphuric acid while lead sulphate is not; from bismuth since ammonium hydroxide precipitates bismuth hydroxide but holds cadmium in solution, and from copper by passing hydrogen sulphide into the solution containing potassium cyanide which prevents the precipitation of copper sulphide but not cadmium sulphide.

The detection of cadmium may be made in the *dry way* through the tube test. This test is carried out in the following manner. A piece of hard glass tubing of about 5 mm. bore is sealed at one end. From 200–400 milligrams of the fine dried ore is mixed with a reducing agent as dry powdered charcoal and introduced into the tube.[1] The tube is heated just above the mixture of ore and reducing agent and drawn out to a capillary of about 1 mm. diameter. The end of the tube containing the mixture is now heated in the blast lamp and the cadmium together with zinc, arsenic, antimony, etc., is volatilized and condensed in the capillary in the form of separated rings. The cadmium ring is detected from the others by introducing a little powdered sulphur into the tube and heating so that the sulphur vapor passes over the rings. The cadmium is converted to sulphide and appears red while hot and yellow while cold. Very small amounts of cadmium may be detected in this way and it is possible with experience to estimate, from the appearance of the ring, either metallic or sulphide the amount of cadmium present. See Fig. 8c page 102.

Cadmium. Brilliant Spectrum of Green and Blue Lines.

Blowpipe Tests. Heated on charcoal in the reducing flame, cadmium gives a brown incrustation which is volatile.

ESTIMATION

Cadmium occurs in small quantities in practically all zinc ores and is found in most spelters and commercial zinc materials as sheet zinc, zinc oxide, etc. In ores it occurs usually as the sulphide; the rare mineral greenockite being CdS. The metal cadmium is largely obtained as a by-product from zinc smelting. It is used in certain alloys, as trial plates for silver coinage and more recently in substitutes for tin base-bearing metals. Precipitated yellow cadmium sulphide is used as a paint pigment.

[1] All metals present in the ore must be in the oxidized state. Sulphide ores must be carefully roasted before using in this test. In the case of metallic substances, however, no reducing agent is necessary.

Chapter contributed by L. S. Holstein and L. A. Wilson.

Cadmium is determined, after separation from other elements, by weighing as the sulphide or as the metal following electrolysis. As it usually occurs in small quantities in the presence of large quantities of zinc, several precipitations with hydrogen sulphide are necessary and the methods must be followed in close detail to obtain accurate results. The determination of cadmium is necessary in case of spelters sold under specified rejection limits, in ores purchased and to be smelted for such spelter, and in other zinciferous material where cadmium is deleterious to the finished product. The methods to be used for the determination of cadmium in spelter is given under the chapter on zinc.

Preparation of Sample

Samples of metals, as spelter, cadmium metal, brass, etc., should be in the form of drillings, sawings or pourings, taken in a proper manner to be representative of the lot and of sufficient fineness to preclude against a non-representative sample being weighed for analysis. The samples of ore or fine material should be ground to pass a 100-mesh screen. Metallics, if also present, are kept separate from the fine material which passes through the screen, and in weighing out the sample, proportional amounts of each are taken.

SEPARATIONS

Removal of Silica. Evaporate with hydrochloric acid or sulphuric acid, and filter off the dehydrated silica, using suction.

Removal of Lead. Evaporate to fumes with sulphuric acid, cool, take up with water, warm until all soluble salts are dissolved and allow to stand until all lead sulphate settles. By using sulphuric acid to dehydrate the silica, lead and silica may be separated together.

Separation from Ammonium Sulphide Group, Except Zn, Alkaline Earths, and Alkalies. The solution from lead and silica separation containing 12 cc. sulphuric acid (1 : 1) per 100 cc. of solution is saturated cold with hydrogen sulphide, passing a steady stream for 20–30 minutes, and after the first 5 minutes adding a drop of ammonium hydroxide until zinc sulphide precipitates in quantity. It is necessary to add ammonium hydroxide to bring down zinc in order to assure the complete precipitation of cadmium. The precipitate of sulphides is filtered off and washed with cold water.

Removal of Arsenic, Antimony and Tin. The precipitate on the filter is washed with ammonium hydroxide or colorless ammonium sulphide, dissolving out the arsenic, antimony and tin (stannic). If tin in the stannous condition is found to be present, yellow ammonium sulphide must be used. Arsenious sulphide is practically insoluble in the hydrochloric acid used in dissolving the cadmium sulphide. Antimonious sulphide is also only slightly soluble, so that these sulphides remain behind in carrying out the analysis.

The sulphides of tin, however, are soluble in the hydrochloric acid and must be removed before the final precipitation of cadmium sulphide.

Removal of Bismuth, Copper and Mercury. Bismuth is not removed in the course of analysis as its sulphide is soluble in hydrochloric acid and hence it must be removed by precipitating with ammonium hydroxide before the final precipitation of cadmium is made. Copper sulphide is, however, practically insoluble in the hydrochloric acid used and remains behind when dissolving the first precipitations of cadmium sulphide, so that the use of potassium cyanide need seldom be resorted to mercuric sulphide is practically insoluble in cold hydrochloric acid (1 : 2) and is left behind in carrying out the analysis.

Separation from Zinc. Cadmium is separated from the accompanying zinc by successive precipitations with hydrogen sulphide, each time bringing down less zinc, until finally only cadmium is precipitated. In the presence of a large quantity of zinc it is not possible to precipitate all cadmium with the acidity required to prevent the precipitation of any zinc sulphide.

GRAVIMETRIC METHODS

Determination as Cadmium Sulphide

Procedure for Ores

A 10 gram sample of the finely pulverized ore is weighed out into a 400 cc. beaker, moistened with water and 50 cc. of aqua regia carefully added.[1] When violent action has stopped, the beaker is placed on a warm plate to complete decomposition. The cover glass and sides of beaker are washed down with water, 25 cc. of sulphuric acid (1 : 1) added and evaporation carried to fumes. Water (100 cc.) is added, boiled until all soluble salts are dissolved, and the residue filtered off, and washed, using suction.

A steady stream of hydrogen sulphide is passed through the filtrate, which should have a volume of approximately 300 cc. for 30–40 minutes. After all iron in solution has been reduced, ammonium hydroxide is added 1 cc. at a time until a heavy precipitation of zinc sulphide has taken place.[2] The precipitate is allowed to settle, the clear solution decanted and finally the bulk of the precipitate transferred to a 15 cc. paper, and washed with cold water. The sulphides on the paper are dissolved with hydrochloric acid (1 : 2) catching the solution in a clean beaker. Any precipitate adhering to the sides of the original beaker is also dissolved off and poured into the filter. After all zinc sulphide has been dissolved, the paper is washed three more times with the hydrochloric acid.[3] To the solution 15 cc. of sulphuric acid is added and evaporation carried just to fumes. Water is added (200 cc.) and hydrogen sulphide passed through as before. Ammonium hydroxide should be added one drop at a time, only to start the precipitation of cadmium sulphide. This precipitate is filtered off, redissolved as previously and a third precipitation made.[4] Before making the final precipitation, the solution should be allowed

---

[1] In the case of oxidized ores, hydrochloric acid alone may be used, although it is well to use a little nitric acid also.

[2] The solution should always be sufficiently acid so that no iron etc. precipitates.

[3] This strength of acid will leave on the paper as an insoluble residue all the As, Cu, and Hg, most of the Sb, and some of the Bi and Sn. The second precipitation of CdS should free the Cd of the rest of the Sb, but not the Bi or Sn.

[4] It is necessary to make three or even four precipitations of cadmium sulphide in order to free it completely of zinc.

to stand, and any lead sulphate filtered off.   The final precipitate of cadmium sulphide is filtered on a weighed Gooch crucible.

The cadmium sulphide on the Gooch crucible need only be washed once or twice, if at all, as it usually receives sufficient washing in the transfer to the crucible, and in the scrubbing and washing out of the beaker.   After drying at 110° C. for 1 hour, the crucible is cooled, weighed and the cadmium calculated from the difference in weights.

$$CdS \times .778 = Cd.$$

It is seldom that bismuth and tin will be encountered in making a determination for cadmium so that the procedure for removing these elements need rarely be used.   Cadmium sulphide precipitated from sulphuric acid is bright yellow to orange.   If the precipitate is brown in color, bismuth and tin should be looked for and removed.

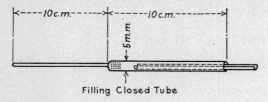

Filling Closed Tube

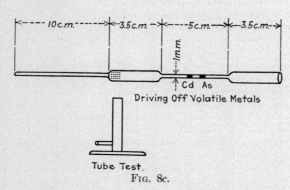

Driving Off Volatile Metals

Tube Test

FIG. 8c.

Determination as Sulphate

The final precipitate of cadmium sulphide is dissolved with hydrochloric acid, and evaporated to dryness in a weighed platinum crucible or dish.   A slight excess of dilute sulphuric acid is poured over the residue and evaporated over a steam bath or warm plate.   The excess of sulphuric acid is driven off by heating in an air bath, or in a muffle heated below a dull red heat.   For an air bath, the crucible or dish may be placed in a larger vessel, and this outer vessel heated to redness. $CdSO_4 \times 0.5392 = Cd.$

Electrolytic Determination of Cadmium

The final precipitate of CdS is dissolved and the cadmium determined electrolytically as given under cadmium in spelter, Chapter on Zinc, or if the amount of cadmium is large the electrolytic determination is best carried out with a potassium cyanide electrolyte.   The hydrochloric acid solution of cadmium, after separation of interfering elements is taken to fumes with sulphuric acid, a drop of phenolphthalein added for indicator, and a pure solution of sodium hydroxide added until a permanent red color is obtained. A strong solution of potassium cyanide is added drop by drop until the cadmium hydroxide just dissolves, avoiding any excess.   The solution is diluted to 100 cc. with water, electrolyzed cold using a gauze electrode with a current of 0.5–0.7 ampere and voltage of 4.8–5.0. At the end of 5–6 hours the current is increased to 1–1.2 amperes and electrolysis continued for an hour more. The electrode is removed from the solution the instant the current is broken and immediately washed with water, followed by alcohol and ether. After drying at 100° C., the electrode is cooled and weighed. Prolonged heating of the deposit should be avoided.

Rapid deposition can be effected by means of the rotating anode (600 revolutions per minute). The solution of cadmium sulphate containing 3 cc. of H_2SO_4 (1 : 10) per 150 cc. The solution, heated to boiling, is electrolyzed with a current of N.D.$_{100}$ =5 amperes, E.M.F. =8–9 volts. Fifteen minutes is sufficient for the deposition of .5 gram of cadmium.

NOTES. Before washing and discontinuing the current, it is advisable to add a little water to raise the level of the liquid and continue the electrolysis to ascertain whether the deposition is complete.

Traces of cadmium may be estimated in the above solution by saturating this with H_2S and comparing the yellow-colored colloidal cadmium sulphide solution with a known quantity of cadmium and the same amounts of potassium hydroxide and cyanide as in the solution tested.

VOLUMETRIC DETERMINATION OF CADMIUM

Titration of Cadmium Sulphide with Iodine.[2]

The titration of cadmium sulphide with standard iodine in a hydrochloric acid solution is the same as the procedure given for determination of sulphur by the evolution method, the following reaction taking place:

$$CdS+2HCl+I_2 =CdCl_2+2HI+S.$$

Procedure. Cadmium having been isolated as the sulphide according to the standard procedures given, the precipitate is washed and allowed to drain on the filter. The filter, together with the sulphide, is placed in a beaker or an Erlenmeyer flask, water added, and the whole shaken to break up the precipitate. A moderate quantity of hydrochloric acid is added and the solution titrated with standard N/5 or N/10 iodine solution. Towards the end a little starch solution is added and the titration continued until the excess of iodine colors the solution blue. If preferred, an excess of iodine solution may be added and the excess determined by a back-titration with standard thiosulphate solution.

One cc. N/10 iodine =0.00562 gram cadmium.

[1] Electro-Analysis, E. F. Smith. P. Blakiston's Son & Co. Pub.

[2] P. von Berg (Z. a. C., **26**, 23) transfers the precipitate and filter to a stoppered flask, expels the air with CO_2 and by boiling and then titrates in an hydrochloric acid solution. Experiments by the author have shown this caution to be unnecessary.

CALCIUM

Ca, *at.wt.* 40.07; *sp.gr.* 1.5446$^{29°}$; *m.p.* 810$°$[1] C.; *oxide*, CaO

DETECTION

In the usual course of qualitative and quantitative analysis calcium passes into the filtrates from the elements precipitated by hydrogen sulphide in acid and alkaline solutions (Ag, Hg', Hg", Pb, Cu, Cd, As, Sb, Sn, Fe, Cr, Al, Mn, Ni, Co, Zn, etc.), and is precipitated from an ammoniacal solution by ammonium carbonate as calcium carbonate, along with the carbonates of barium and strontium. The separation of calcium from barium and strontium is considered under Separations. The oxalate of calcium is the least soluble of the alkaline-earth group.[2] All, however, are soluble in mineral acids. Calcium oxalate may be precipitated from weak acetic acid solution by ammonium oxalate.

Flame Test. The flame of a Bunsen burner is colored yellowish red when a platinum wire containing calcium salt moistened with c ncentrated hyd ochloric acid is held in the flame.

Spectrum. An intense orange and green line with a less distinct violet line. Note chart of the spectra of the alkaline earths. Plate II.

See chapter on Barium under Separations—Preliminary Tests, page 58.

ESTIMATION

The determination of calcium is required in complete analyses of ores. It is of special importance in the analysis of mortar, cement, bleaching powder, plaster of Paris, certain paint pigments such as phosphorescent paint, CaS. The determination is required in the analysis of water.

Calcium occurs in the following substances: as carbonate in limestone, marble, chalk, Iceland spar, shells, coral, pearl. Together with magnesium it is found in dolomite. It occurs as sulphate in anhydrite, gypsum, alabaster, selenite; as silicate in the mineral wollastonite, $CaSiO_3$; as phosphate in phosphorite, $Ca_3(PO_4)_2$, also in bones and in apatite, $3Ca_3(PO_4)_2 \cdot CaF_2$; as fluoride in fluorspar, CaF_2. As oxalate it occurs in plant cells. It is found in nearly all mineral springs, artesian wells, and river waters, principally as bicarbonate of calcium, $CaHCO_3$.

Preparation and Solution of the Sample

The oxide, hydroxide, and salts of calcium are soluble in acids with the exception of gypsum and certain silicates which require fusion with sodium carbonate or bicarbonate followed by an hydrochloric acid extraction.

[1] Cir. 35 (2d Ed.) U. S. Bureau of Standards.
[2] Solubility: $CaC_2O_4 \cdot H_2O = 0.000554$ gram per 100 cc. H_2O. $BaC_2O_4 \cdot H_2O = 0.0093$ gram. $SrC_2O_4 \cdot H_2O = 0.0051$ gram. $MgC_2O_4 2H_2O = 0.07$ gram.
Van Nostrand's Chem. Annual—Olsen.

Chapter contributed by Wilfred W. Scott.

Solution of Limestones, Dolomites, Magnesites, Cements, Lime, etc.
One gram of the powdered material is digested in a 250-cc. beaker with
20 cc. of water, 5 cc. of concentrated hydrochloric acid, and 2 or 3 drops of
nitric acid (sp.gr. 1.42). The beaker is covered to prevent loss by effervescence.
When the violent action has subsided, the sample is placed on a hot plate and
boiled for a few minutes. The watch-glass is rinsed into the beaker and the
solution filtered. The residue is washed, dried and ignited in a platinum crucible, and then fused with a little sodium carbonate or bicarbonate. The cooled
fusion is dissolved in hot dilute hydrochloric acid, the liquid added to the main
solution and calcium determined by precipitation as calcium oxalate, after removal
of silica, iron, alumina, etc.

Solution of Gypsum, Plaster of Paris, and Sulphates of Lime, etc. The
treatment of the sample is similar to the one given above with the exception
that it is advisable to add a larger amount of strong hydrochloric acid, e.g.,
about 20 to 25 cc. If barium sulphate is present it is indicated by the clouding
of the solution, upon acidifying the water extract of the carbonate fusion.

Silicates. Solution of silicates is best obtained by direct fusion of 1 gram
of the powdered material with 4 to 5 grams of sodium carbonate, in a platinum crucible. The cooled melt is now covered with water and dissolved with
hydrochloric acid according to the standard procedure for carbonate fusions.
The hydrochloric acid solutions are taken to dryness and the silica dehydrated
in an oven at 110° C. for an hour and then the residue is extracted with dilute
hydrochloric acid and filtered. The filtrate contains iron, alumina, magnesium,
lime, etc.

Chlorides, Nitrates, and Other Water-soluble Salts. These are dissolved
in water slightly acidified with hydrochloric acid.

Sulphides, Pyrites Ore, etc. The ore should be oxidized with bromine or
by roasting, previous to the acid treatment.

SEPARATIONS

Removal of Silica. The solution obtained by one of the above procedures
is evaporated to dryness and the silica dehydrated at 110° C. for an hour. The
residue is now extracted with dilute hydrochloric acid. Silica remains insoluble
and may be filtered off. The solution contain. lime, together with iron, alumina,
magnesia, etc., as chlorides.

Removal of Iron and Alumina. The filtrate from the silica residue is
treated with a few drops of nitric acid and boiled to oxidize the iron. Ammonia
is now added cautiously until the solution just smells of it (a large excess over
that required to neutralize the acid and combine with iron and alumina,
will tend to dissolve $Al(OH)_3$). The precipitated hydroxides are allowed to
settle and then filtered hot through a rapid filter and washed with hot water.
Calcium, together with magnesium, is in solution and passes into the filtrate.

**Removal of Copper, Nickel, Cobalt, Manganese, Zinc, and Elements
Precipitated as Sulphides in Acid and Alkaline Solutions.** This separation
is required seldom in lime-bearing ores. In analysis of pyrites and certain
other ores, containing members of the hydrogen sulphide and ammonium sulphide groups, the removal of these impurities is necessary.

The solution from the residue of silica is made slightly ammoniacal and
H_2S passed into the solution to saturation (or ammonium sulphide may be

11

added). The precipitated sulphides are filtered off from the solution heated to boiling. The filtrate containing the calcium is boiled down to 50 to 75 cc. and the precipitated sulphur removed by filtration. Calcium is determined in the filtrate by precipitation with ammonium oxalate or oxalic acid according to directions given later.

Separation of Calcium from Magnesium and the Alkalies. In the presence of considerable amounts of calcium and comparatively small quantities of magnesium the oxalate method of precipitating calcium, in presence of ammonium chloride, is generally sufficient for precipitating calcium free from magnesium and the alkalies. In analysis of dolomite, $MgCO_3 \cdot CaCO_3$, and of samples containing comparatively large amounts of magnesium, a double precipitation of calcium is generally necessary for removal of occluded magnesium.

Separation of Calcium from Barium and from Strontium. The alkaline earths are converted to nitrates, all moisture expelled by heat, and calcium nitrate extracted from the insoluble nitrates of barium and strontium by a mixture of anhydrous ether and absolute alcohol, in equal parts, or by boiling the dry nitrates in amyl alcohol (b.p., 137.8° C.). Details of the procedure are given under Separations of the Alkaline Earths in the chapter on Barium, page 59.

Determination of Lime in Presence of Phosphates, Iron, and Alumina. Should phosphoric acid be present in the solution, calcium will be precipitated as a phosphate upon making the solution neutral or slightly alkaline with ammonia, and will remain with iron and alumina precipitates.

Precipitation of Calcium Oxalate in Presence of Iron and Alumina. The solution containing the phosphates freed from silica is oxidized by boiling with nitric acid as usual. Ammonia water is added to the cooled solution until a slight precipitate forms, and then citric acid is added in sufficient quantity to just dissolve the precipitate. If this does not readily occur, additional ammonia is added, followed by citric acid until the solution clears, then about 15 cc. of citric acid in excess. The solution is diluted to 200 cc. and heated to boiling. Calcium oxalate is now precipitated by addition of ammonium oxalate. Iron and alumina remain in solution.

Citric acid is made by dissolving 70 grams of the acid, $H_3C_6H_5O_7 \cdot H_2O$, in a liter of water.

Wagner's Solution. In place of citric acid, the following solution may be used. Twenty-five grams of citric acid and 1 gram of salicylic acid are dissolved in water and made to 1000 cc. Twenty-five to 50 cc. of this reagent is effective in preventing precipitation of iron and alumina.

Decomposition of Material. Though carbonates are easily dissolved in hydrochloric acid, sulphates and fluorides of calcium require fusion with sodium and potassium carbonate to effect decomposition. In case calcium fluoride is being decomposed, the addition of an equal weight or more of silica is necessary, and sodium or potassium hydroxide may be substituted for potassium carbonate. The fusion is leached with water to remove the mineral acids, and the residue, containing all of the calcium, is dissolved in hydrochloric acid. Calcium is now determined in the hydrochloric acid solution. If phosphate is present in the sample, it is not completely removed by the water leaching as sodium phosphate, as this reacts in the solution with calcium carbonate causing a partial conversion to calcium phosphate, which remains in the residue. In this case it is advisable to precipitate calcium oxalate from a slightly acetic acid solution, in which calcium phosphate will not precipitate, as it would if the solution was made ammoniacal.

GRAVIMETRIC DETERMINATION OF CALCIUM

Precipitation of Calcium Oxalate and Ignition to Calcium Oxide

Calcium oxalate is precipitated from feebly ammoniacal solutions or from solutions acidified with acetic, oxalic, citric, or salicylic acids, by means of ammonium oxalate. The presence of ammonium chloride hinders precipitation of magnesium and does not interfere with that of calcium. If, however, much magnesium (or sodium) is present it will contaminate the calcium precipitate so that a second precipitation is necessary to obtain a pure product. The compound formed from hot solutions is crystalline or granular and filters readily, whereas the flocculent precipitate formed in cold solutions does not. Calcium oxalate, $CaC_2O_4 \cdot H_2O$,[1] decomposes at red heat to CaO, in which form it is weighed.

Procedure. If the calcium determined is in the filtrate from previous groups, hydrogen sulphide is expelled by boiling and the precipitated sulphur filtered off, the solution having been concentrated to about 200 cc. The filtrate should contain sufficient ammonium chloride to hold magnesium in solution in presence of ammonium oxalate (i.e, about 10 grams NH_4Cl per 0.0015 gram MgO per 100 cc. of solution.) If not already present, the chloride is added in sufficient amount, and the solution diluted to about 400 cc.

Precipitation. The solution is heated to boiling and 10 cc. of acetic acid added to the neutral mixture. Fifteen cc. or more of a saturated solution of oxalic acid is added and after five minutes a slight excess of ammonia. The solution is allowed to cool an hour or so, the clear solution decanted through a 10-cm. filter and the precipitate washed three times by decantation and finally on the filter with dilute ammonia (1 : 10), or 1% ammonium oxalate.

To remove clinging impurities (Na or Mg) the precipitate is dissolved in dilute nitric acid (1 : 4) and the filtrate collected in the beaker in which the first precipitation was made. The solution is heated to boiling after addition of a few drops of oxalic acid and sufficient ammonium hydroxide to make the solution slightly alkaline. The precipitated oxalate is allowed to settle, filtered and washed as in the first precipitation, the oxalate adhering to the sides of the beaker being carefully "copped" out. The oxalate is ignited wet in a weighed crucible, the heat being low at first, until the filter has charred and then to the full heat of the Méker blast lamp. Fifteen minutes of blasting should be sufficient to obtain constant weight. If the precipitate is large a second ignition is advisable to insure the complete decomposition of the oxalate and carbonate to oxide.

The crucible is cooled in a desiccator and weighed as soon as possible.

Factors. $CaO \times 0.7146 = Ca$, or $\times 1.7847 = CaCO_3$, or $\times 2.8908 = Ca(HCO_3)_2$, or $\times 2.428 = CaSO_4$.

[1] Calcium oxalate dried at $100 = CaC_2O_4 \cdot H_2O$. Heated to $200°$ C. $= = CaC_2O_4$. At $500°$ C. the oxalate begins to decompose, free carbon is liberated, and calcium carbonate begins to form. At bright red heat carbon burns off and the carbonate is completely decomposed to the oxide and CO_2.

Calcium Phosphate. Fusion of the phosphate with sodium carbonate and silica does not effectively accomplish removal of P_2O_5 by leaching with water if calcium is present (also Ba and Sr) since the phosphate of calcium forms an insoluble phosphate, the reaction of calcium carbonate to phosphate and phosphate to carbonate being reversible. Under these conditions it is preferable to precipitate aluminum in a faintly acid solution by addition of sodium thiosulphate.

VOLUMETRIC DETERMINATION OF CALCIUM
Titration of the Oxalate with Permanganate

This procedure may be applied successfully in a great variety of instances on account of the readiness with which calcium oxalate may be separated. In the presence of iron, alumina, manganese, magnesia, etc., it is advisable to make a reprecipitation of calcium oxalate to free it from adhering contaminations.

The following reaction takes place when potassium permanganate is added to calcium oxalate in acid solution:

$$5CaC_2O_4 + 2KMnO_4 + 8H_2SO_4 = 5CaSO_4 + K_2SO_4 + 2MnSO_4 + 10CO_2 + 8H_2O.$$

Procedure. Calcium oxalate, obtained pure, by precipitation and washing according to directions given under the gravimetric determination of calcium, is washed into a flask through a perforation made in the filter paper, the filter is treated with a little warm, dilute sulphuric acid and the adhering oxalate dissolved and washed into the flask. About 25 cc. of dilute sulphuric acid, 1 : 1, is added and the solution diluted to 250 to 300 cc.

When the precipitate has dissolved, the solution warmed to 60 or 70° C. is titrated with standard potassium permanganate, added cautiously from a burette with constant agitation, until a faint permanent pink color is obtained.

One cc. N/10 $KMnO_4$ = 0.0020 gram Ca, or ×0.0028 = CaO.

Factors. Ca×1.3993 = CaO or ×2.4974 = $CaCO_3$ or ×3.3975 = $CaSO_4$ or ×2.581 = $Ca_3(PO_4)_2$.

NOTES. **Precipitation from Acetic Acid Solution.** This is recommended if the material contained phosphates. The solution should contain about 1 cc. free glacial acetic acid per two hundred cc. of solution. In presence of phosphate, iron and alumium cannot be removed by addition of ammonia as calcium would also precipitate as phosphate. Citric acid may also be used to prevent precipitation of iron.

Procedure. To a volume of about 75 cc. of solution containing 0.1 to 0.15 g. Ca add 10 cc. acetic acid (glacial), heat to boiling and add slowly 10 cc. of a saturated solution of ammonium oxalate. Now add a slight excess of ammonia and make slightly acid with acetic acid (about 0.5 cc. per 100 cc. solution). If phosphate is present this last acidification is necessary, likewise in presence of aluminum.

Available Lime

This is the usual method followed in cyanide mills for determining the per cent which will dissolve of the lime added to the mill solution, and therefore the amount to add to maintain the desired protective alkalinity.

Procedure. One gram of the finely ground sample is placed in a glass stoppered bottle which has been previously marked to hold 500 cc. Thirty grams of sugar and about 300 cc. of water are added and the bottle shaken vigorously. The solution is diluted to 500 cc. and shaken at ten minute intervals for 1½ to 2 hours. Then the insoluble material is allowed to settle, part of the solution filtered through a coarse filter paper, 50 cc. of the filtrate drawn out with a pipette into an Erlenmeyer flask, two drops of phenolphthalein indicator added, and the solution titrated with N/10 oxalic acid solution till the pink color disappears. The lime is reported as per cent available CaO.

1 cc. N/10 Oxalic Acid = .0028 grams CaO.

Standard N/10 Oxalic Acid Solution. Dissolve 6.303 grams $H_2C_2O_4.2H_2O$ in distilled water and dilute to 1000 cc. The solution ordinarily need not be standardized as the weight of the oxalic acid is constant.

To standardize take 30 cc. of the solution, add 100 cc. of water and 5 cc. of conc. H_2SO_4, heat to 70° and titrate to a permanent pink with N/10 $KMnO_4$.

Rapid Iodine Method for Calcium Oxide in Presence of Calcium Carbonate

The method worked out by John C. Bailar, Great Western Sugar Company, is based on the fact that a solution of iodine reacts with calcium hydroxide, but does not react with calcium carbonate. The method is used in the evaluation of lime.

Reagents. *Iodine Solution.* A standard solution is made by dissolving 90 grams of potassium iodide and 45.27 grams of iodine in the least quantity of water necessary to effect solution, and diluting with water to one liter. 1 cc. is equivalent to 0.01 g. CaO.

Thiosulphate Solution. The reagent is made by dissolving 44.27 grams of thiosulphate of sodium in water and diluting to one liter. Two cc. of this solution is equivalent to one cc. of the iodine solution.

Standardization. A definite weight of 0.5 to 1.0 gram of pure arsenous oxide (As_2O_3) is dissolved in 10% sodium hydroxide solution and the resulting product acidified with hydrochloric or sulphuric acid. This solution is now neutralized with sodium bicarbonate and 4 to 5 grams added in excess. Starch indicator is now added and the arsenite titrated with the standard iodine solution. Since one gram of As_2O_3 is equivalent to 0.5656 gram of CaO, the weight of the arsenic taken multiplied by 0.5656, divided by the cc. of iodine required, gives the equivalent lime per cc. of the standard iodine reagent. Use this factor in the iodine titrations of lime.

The thiosulphate may be standardized against a definite volume of the iodine reagent and its equivalent value established in terms of the standard iodine solution. See Notes below.

Procedure. The sample of lime (one gram is recommended) is slacked by adding boiling water (5 to 10 minutes is ample to accomplish this). An excess of iodine is added (see Notes) and the mixture stirred occasionally until the lime is all in solution. Insoluble silica is generally present but can easily be distinguished from the milky appearing lime. When the solution of the lime is complete (1 to 10 minutes will accomplish this), the excess of iodine is titrated with the standard thiosulphate. The excess thus determined is subtracted from the total iodine added and the equivalent CaO, to the combined iodine, calculated from the CaO factor of iodine.

NOTES. Any substance which liberates iodine quantitatively from a solution of potassium iodide can be used for the standardization of sodium thiosulphate. Among such substances are potassium permanganate, potassium dichromate, potassium iodate, potassium bi-iodate, metallic copper oxidized to a cupric salt, etc. In using any of the above reagents first acidify the solution of an excess of potassium iodide with hydrochloric acid (strongly acid if dichromate is used, end point in this case is green in place of colorless) and then add the permanganate or dichromate or other reagent desired and titrate the liberated iodine in presence of starch indicator.

Lime Equivalents.   1 g. $KMnO_4$ 0.8870 g. CaO
                        1 g. $K_2Cr_2O_7$ 0.5718 g. CaO
                        1 g. KIO_3 0.7860 g. CaO
                        1 g. KIO_3HIO_3 0.8602 g. CaO
                        1 g. Cu   0.4410 g. CaO

An excess of 5 cc. of iodine is recommended in the lime determination. To the same amount of water used in the analysis add 5 cc. of the iodine and use this as a standard for color comparison in running in the necessary excess of iodine in the sample.

STANDARD METHODS OF TESTING GYPSUM AND GYPSUM PRODUCTS

ADOPTED BY A. S. T. M. IN AMENDED FORM, 1923

Free Water. Not less than 1 lb. of the entire sample as received shall be weighed, spread out in a thin layer in suitable vessel, placed in a drying oven, and dried at 45° C. for 2 hours. It shall then be cooled in an atmosphere free from moisture, and weighed again. The loss of weight corresponds to the free water, and shall be calculated to percentage of sample as received.

Preparation of Sample. Dry sample as in Section 1, and reduce about 10 g. until it all passes a 100-mesh sieve, using extreme care not to unduly expose the material to the action of moisture or to overheating. The sample shall be kept in an air-tight container until ready for use.

Combined Water. Place 1 g. of the sample in a covered crucible and dry at 215 to 230° C. to constant weight. Calculate the loss of weight to percentage of sample as received and report as combined water.

Carbon Dioxide. Place the residue obtained after drying, as described above, in a suitable flask and dissolve it in dilute HCl (not stronger than 1 : 4) in such a way that the gas evolved, after being freed from H_2O vapor by calcium chloride or sulfuric acid, can be collected in either soda lime or caustic potash and weighed. The solution should be boiled for one minute, and a current of CO_2-free air kept passing through the apparatus for 30 minutes. The increase of weight of the soda lime or caustic potash corresponds to the weight of carbon dioxide, which is to be calculated to percentage of sample as received.

Silica and Insoluble. Place 0.5 g. of the sample prepared as described above in a porcelain casserole. Add about 25 cc. of 1 : 5 HCl, and evaporate to apparent dryness on a hot plate. Cool and add enough concentrated HCl to wet thoroughly. Add about 10 cc. of water, boil, filter, and wash. Put the filtrate back in the same casserole. Evaporate it to dryness and heat to about 120° C. for one hour. Cool. Add enough concentrated HCl to wet thoroughly. Add about 25 cc. of water, boil, filter, and wash. Transfer the two papers containing the two precipitates to the same crucible, ignite, and weigh. Calculate this weight to percentage of sample as received.

Iron and Alumina. To the filtrate obtained above. add a few drops of HNO_3 and boil to insure oxidation of the iron. Add 2 g. of NH_4Cl previously dissolved in water. Make alkaline with NH_4OH. Digest hot for a few minutes until the precipitate coagulates. Filter, wash, ignite the precipitate and weigh as $Fe_2O_3 + Al_2O_3$. Calculate this weight to percentage of sample as received. This precipitate may be further treated to separate the two oxides, but this is generally unnecessary.

Lime. (*a*) To the filtrate obtained above, add 5 g. of $(NH_4)_2C_2O_4$ dissolved in water. Digest hot for $\frac{1}{2}$ hour, making sure that the solution is always alkaline with NH_4OH. Filter, wash and ignite in a platinum crucible over a strong blast to constant weight. Calculate this weight to percentage of sample as received.

(b) Alternative Method. To the filtrate obtained from the iron and alumina precipitates, add 5 g. of $(NH_4)_2C_2O_4$ dissolved in water. Digest hot for $\frac{1}{2}$ hour, making sure that the solution is always alkaline with NH_4OH. Filter and wash. Transfer the precipitate to a beaker, and wash the filter paper with hot dilute H_2SO_4, catching the washings in the same beaker. Heat gently to complete solution, adding more H_2SO_4 if necessary. While still warm titrate with a solution of $KMnO_4$ containing 5.6339 g. per liter, until the pink color is permanent. The number of cubic centimeters of $KMnO_4$ used gives directly the percentage of lime in the dried sample. Recalculate to percentage of sample as received.

Magnesia. To the filtrate obtained from calcium precipitate, add enough water to give a total volume of about 600 cc. Cool. Add 10 cc. of NH_4OH and 5 g. $NaNH_4HPO_4$ dissolved in water. Stir until precipitate begins to form. Let stand over night. Filter, wash with a 2.5 per cent by weight solution of NH_4NO_3. Ignite and weigh. Multiply this weight by 40/111 to find the weight of MgO. Then calculate to percentage of sample as received.

Sulfur Trioxide. Dissolve 0.5 g. of the sample prepared as described in Section 3 in 50 cc. of 1 : 5 HCl. Boil. Add 100 cc. of boiling water, and continue boiling for 5 minutes. Filter immediately and wash thoroughly with hot water. Boil, and while boiling, add slowly 20 cc. of a boiling 10 per cent solution of $BaCl_2$. Digest hot for one hour, or until precipitate settles. Filter and wash. Dry carefully. Ignite over Bunsen burner at lowest heat possible until filter paper is burned off. Ignite at bright red heat for 15 minutes, and weigh. Multiply this weight by 80/233 to determine the weight of SO_3. Then calculate to percentage of sample as received.

Sodium Chloride. Dissolve in boiling water a 1-g. sample (prepared as described in Section 3), put on the filter and wash with 250 cc. of boiling water, then titrate the filtrate. Add two or three drops of potassium chromate solution and titrate with an N/20 solution of silver nitrate. Each cubic centimeter of silver nitrate solution = 0.002923 g. of sodium chloride. Calculate to percentage of sample tested.

Calculation of Results. By the methods given above, the results are obtained and reported in the following form:

PER CENT.

Free water...	
Combined water....................................	
Silica and insoluble, SiO_2..............................	
Iron and alumina, R_2O_3.................................	
Lime, CaO	
Magnesia, MgO	
Sulfur Trioxide, SO_3.................................	
Carbon Dioxide, CO_2...............................	
Sodium Chloride, NaCl...............................	

$$100.00 \pm$$

NOTE

Since it is frequently advisable to recalculate these results, that they may be more enlightening, the following method is submitted for consideration:

(a) Multiply percentage of MgO by 84/40 to find percentage of $MgCO_3$.

(b) Multiply the percentage of MgO by 44/40 to find the percentage of CO_2 as $MgCO_3$.

(c) Deduct CO_2 as $MgCO_3$ from the CO_2 determined.

(d) Multiply the CO_2 remaining by 100/44 to find percentage of $CaCO_3$.

(e) Add together the percentage of SiO_2, R_2O_3, $MgCO_3$, and $CaCO_3$, and report in the aggregate.

(f) Multiply the percentage of $CaCO_3$ by 56/100 to find the percentage of CaO as $CaCO_3$.

(g) From the total percentage of CaO deduct the percentage of CaO as $CaCO_3$. The remainder may be called " available CaO."

(h) The " available CaO " should bear to the SO_3 a ratio of 7 to 10. Determine which (if either) is in excess.

(i) If the CaO is in excess, multiply the SO_3 by 7/10, and subtract the result from the " available CaO." The remainder is reported as " excess CaO."

(j) If the SO_3 is in excess, multiply the " available CaO " by 10/7 and subtract the result from the SO_3. The remainder is reported as " excess SO_3."

(k) Add together the " available CaO" and the SO_3, and subtract the " excess CaO " or " excess SO_3." The remainder is $CaSO_4$.

(l) If the $CaSO_4$ is present as $CaSO_4 \cdot \frac{1}{2}H_2O$, the percentage of $CaSO_4$ should bear to the percentage of combined water a ratio of 136 to 9. Determine which (if either) is in excess.

(m) If the $CaSO_4$ is in excess, some of it is present in the anhydrous form. Multiply the percentage of combined water by 136/9 to find the percentage of $CaSO_4$ as $CaSO_4 \cdot \frac{1}{2}H_2O$. The difference between the total $CaSO_4$ and the percentage of $CaSO_4$ as $CaSO_4 \cdot \frac{1}{2}H_2O$ is the $CaSO_4$ in the anhydrous form.

(n) If the water is in excess, some of the $CaSO_4$ is present as gypsum. Let x = percentage of $CaSO_4 \cdot \frac{1}{2}H_2O$, and y = percentage of $CaSO_4 \cdot 2H_2O$. Then $x + y$ = percentage of $CaSO_4$ (as found in k) + percentage of water.

$$\frac{9x}{145} + \frac{36y}{172} = \text{percentage of combined water.}$$

Solve these equations for x and y. Report x as percentage of " calcined gypsum," $CaSO_4 \cdot \frac{1}{2}H_2O$. Report y as percentage of gypsum, $CaSO_4 \cdot 2H_2O$.

Having made these calculations, the result may be reported as follows:

Per Cent.

Gypsum, $CaSO_4 + 2H_2O$ ....
Calcined gypsum, $CaSO_4 + \frac{1}{2}H_2O$ ....
Anhydrite, $CaSO_4$ ....
Excess CaO ⎱
    or   ⎰ ....
Excess SO_3 ⎰
Sodium chloride, NaCl ....
Other ingredients ....

100.00 ±

The presence of the different forms of $CaSO_4$ may be corroborated by a microscopic examination.

Other Gravimetric Methods

Calcium may be converted to carbonate, sulphate, fluoride, tungstate and so weighed. The oxide, obtained by ignition of the oxalate may be converted to sulphate by moistening with a few drops of H_2SO_4, then adding an excess of NH_4OH and igniting to expel excess of sulphate and NH_3.

Calcium sulphate may be precipitated by adding an excess of H_2SO_4 and then 95% alcohol (two to four times the total volume of the solution). The precipitate is washed with alcohol and then ignited to constant weight. (Fresenius.)

Calcium tungstate by Saint Sernin's method is precipitated by adding ammonia until the solution is alkaline and then an excess of 20% solution of sodium tungstate. The precipitate is best filtered into a weighed Gooch crucible and washed with ammonia (1 : 10 sol.) then dried at 100° C. and weighed as $CaWO_4$.

CARBON

C,[1] *at.wt. 12.0; sp.gr. amorp. 1.75–2.10; cryst.: graphite, 2.25; diamond, 3.47–3.5585; m.p. sublimes at 3500° C.; oxides, CO and CO_2*

DETECTION

Element. Carbon is recognized by its appearance and by its inertness towards general reagents. It is seen in the charring of organic matter when heated or when acted upon by hot concentrated sulphuric acid.

Upon combustion with oxygen or by oxidation with chromic and sulphuric acids, carbon dioxide is formed. The gas passed into lime water forms a white precipitate, $CaCO_3$. Penfield fuses the substance with precipitated and washed $PbCrO_4$ in a hard glass tube closed at one end. The CO_2 is tested at the mouth of the tube.[2]

Carbon Dioxide. Carbonates. CO_2 in Gas. A white precipitate with lime water, baryta water, ammoniacal solutions of calcium, or barium chlorides, or lead acetate (basic), carbonates of the metals are formed.

Carbonates. Action of mineral acids cause effervescence, CO_2 being evolved. The gas is odorless (distinction from SO_2, H_2S, and N_2O_3) and is colorless (distinction from N_2O_3). The gas absorbed in the reagents above mentioned produces a white precipitate. The test is best made by placing the powdered material in a large test-tube with a stopper carrying a funnel and delivery tube as shown in the illustration, Fig. 9. For small amounts of combined CO_2, warming of the test-tube may be necessary. Sulphuric or phosphoric acid should be used to liberate the gas, which is conducted into the reagent used for the test.

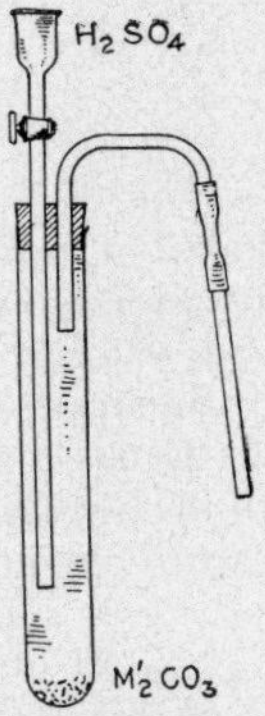

Distinction between Soluble Carbonates and Bicarbonates. The solution of the former is alkaline to phenolphthalein indicator (pink). Bicarbonate solutions remain colorless with this indicator. Normal carbonates precipitate magnesium carbonate when added to magnesium sulphate solution; bicarbonates cause no precipitation.

FIG. 9.—Test for Carbonate.

Free Carbonic Acid in Water in Presence of Bicarbonates. 0.5 cc. of rosolic acid (1 part acid in 500 parts of 80% alcohol), produces a red color with bicarbonates in absence of free CO_2, and a colorless or faintly yellow solution when free CO_2 is present.

Carbon Monoxide. The gas burns with a pale blue flame and is not absorbed by potassium hydroxide or lime water (distinction from CO_2). It is oxidized to CO_2 and so detected. With hot, concentrated potassium hydroxide, potassium formate is produced.

The gas is detected in the blood by means of the absorption spectrum.

---

[1] Van Nostrand's Chemical Annual, Olsen.

Chapter contributed by Wilfred W. Scott.

109

ESTIMATION

The element occurs free in nature in the crystalline forms, diamond and graphite, and in the amorphous form, charcoal, coke, etc. It occurs in iron, steel, and in certain alloys. Its estimation in these metals is generally required. Carbon is determined in the analysis of organic compounds in which it is invariably combined and may also be present as free carbon (asphaltum).

Combined as a carbonate it occurs in a large number of substances, among which are found calcite, marble, limestone, dolomite, magnesite, strontianite, witherite, spatic iron ore. It occurs as the dioxide in the air, in water ($H_2O \cdot CO_2$) and in flue gas. Carbon dioxide is the active constituent of baking powders ($NaHCO_3$).

Preparation of the Sample

Iron, Steel, and Alloys. The subject is discussed in the chapter on Iron and Steel, Volume II.

Organic Matter. It is advisable to fuse this in a nickel or iron crucible with sodium peroxide. The carbonate thus formed may be determined as usual. The organic substance may be oxidized directly in the combustion furnace.

Carbonates. Limestone, Dolomite, Cement, Alkali Carbonates and Bicarbonates. The powdered material is decomposed by addition of an acid as directed in the methods given later.

Separation of Carbon from Other Substances

The element is generally determined as carbon dioxide, in which form it is liberated from most of the combinations in which it occurs, free from other substances by ignition in a current of oxygen, or by oxidation with chromic acid as directed later.

Separation of Carbon in Iron and Steel. Cupric Potassium Chloride Method. 0.5 to 2 grams of the drillings are treated with 100 to 200 cc. of cupric potassium chloride solution and 10 cc. of hydrochloric acid (1.19). This mixture dissolves the iron according to the reaction

$$Fe + CuCl_2 = FeCl_2 + Cu \quad \text{and} \quad Cu + CuCl_2 = Cu_2Cl_2 + \text{carbon as a residue.}$$

The solution should be stirred frequently to hasten the solution of the iron. It is advisable to keep the temperature of the solution at about 50° C. When the iron and copper have dissolved the carbon is filtered off into a perforated platinum boat or crucible, as directed under the methods. It is now oxidized to CO_2 and so determined.

NOTE. The cupric potassium chloride solution is prepared by dissolving 150 parts of potassium chloride and 170 parts of crystallized cupric chloride in water and crystallizing out the double salt. Three hundred grams of this salt are dissolved in 1000 cc. The solution may be used several times by chlorinating the dirty brown filtrate from the carbonaceous residue. The cuprous chloride formed during the solution of the steel is converted again to cupric chloride, and the chlorinated double salt is even more energetic in its solvent action than the freshly made reagent. (Blair.)

GRAVIMETRIC METHODS FOR DETERMINATION OF CARBON [1]

The determination of carbon by combustion with oxygen is made in two general classes of substances: *A.* Steel, iron and in certain alloys. *B.* Organic compounds. Carbon in steel and alloys is considered in two forms: carbide or combined carbon, and graphitic carbon. In organic substances carbon occurs principally combined with hydrogen, oxygen, and nitrogen. For the present we will consider procedures for the determination of carbon in steel and alloys.

The most accurate procedure for determination of carbon in steel, alloys, and in other materials containing the substance combined or free is by combustion with oxygen in a furnace heated by gas or electricity; the carbon dioxide formed being absorbed in caustic, and weighed.

Apparatus. Combustion Furnace. Although the gas furnace has been used more commonly on account of gas being more available than electricity, the extension of generating electric plants makes it possible to use electric furnaces, and these are gradually displacing those heated by gas, as they are more compact, easily manipulated and comparatively simple in structure.

A simple electric furnace may be made by wrapping a silica tube with a thin covering of asbestos paper, which has been moistened with water. On drying the paper will cling to the tube. A spiral coil of nichrome wire (Driver and Harris) is wound around this core. On a 2-foot length of tube two 45-foot lengths of No. 18 wire, connected in parallel, will heat the tube to bright redness, attaching the terminals to an ordinary light socket. The coils should be covered with $\frac{1}{4}$-in. coating of alundum cement. For appearance' sake as well as for protection, the tube is placed in a large cylinder of sheet iron, packed around with asbestos, and is held in position by circular asbestos boards placed at the ends of the large cylinder. The cylinder is mounted on a stand.

Absorption Apparatus. A large number of forms are for sale. The Geissler and Liebig bulbs have been popular (Figs. 11 and 12), but are now being displaced by forms that have less surface exposed, that are more easily cleaned and less fragile, such as Gerhardt's, Vanier's and Fleming's apparatus (Figs. 13, 14 and 16). The Vanier and the Fleming absorption apparatus are especially to be recommended, on account of their capacity, compactness, efficiency, in handling gases passing at a rapid rate, and their simplicity of form.

[1] **The Wet Method.** Oxidation of carbon in iron and steel is accomplished by means of boiling chromic-sulphuric acids, containing copper sulphate (35 cc. chromic acid made by dissolving 720 grams of chromic acid with 700 cc. of water, 150 cc. of 20% copper sulphate and 200 cc. of strong sulphuric acid); the evolved carbon dioxide is absorbed in soda lime. The method is of value in cases where a combustion furnace is not available. The procedure, however, requires considerably more time and manipulation and can not compare with the standard combustion methods described.

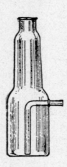

FIG. 10. — Hutchinson - Weirich Combustion Bulb for the Rapid Absorption of CO_2 in Carbon Determinations in Iron and Steel.

Procedure for Determining Carbon by Combustion. [1]

The greatest value of this rapid method is realized when it is used to follow a bath of steel in the open-hearth furnace preliminary to tapping. It abolishe

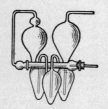   

FIG. 11.—Geissler Bulb.   FIG. 12.—Liebig Bulb.   FIG. 13.—Gerhardt Bulb.

completely the unreliable and dangerous color carbon. By this method abso lutely accurate results can be reported to the open hearth ten minutes afte the drillings are received.

In principle this method is not new; in manipulation it is new. Hereto fore chemists have been laboring under the impression that the flow of gas during a combustion must not exceed a certain snail-like pace. This false impression has beer injected into the minds of chemists by a few who were supposed to have investigated the matter. The truth i that the faster oxygen is fed to burning steel the mor complete the combustion will be. The rate of current i limited by the efficiency of the apparatus used to absorl the evolved carbon dioxide.

FIG. 14.—Vanier Bottle.

The Apparatus Described. The combustion train i shown complete in Fig. 15. The oxygen is delivered t the train through a regulating and reducing valve such a is used for welding. The regulating valve is not essen tial, yet any chemist who uses one will appreciate its con venience, especially in this method. Its convenience will b explained later. K is a mercury pressure gauge. It serve as a guide during the combustion and is an essential piec of apparatus. The graduated column is 6 ins. high and i divided into eighths. P is a washing bottle containing caustic potash solution Filled to the mark indicated with 50% solution it will serve for at least 1000 com bustions. It is used solely to indicate the flow of gas, not to purify it. If th chemist desires he may omit this from the train. T is a calcium chloride jar It is filled to the mark indicated with finely divided calcium chloride, about pea size retaining all the dust. A layer of asbestos is formed over the chloride and th remaining space filled with soda lime. The glass tubing leading from the jar i loosely packed for a distance of several inches with asbestos. This prevents any soda lime dust being carried into the combustion tube. G is a mercury valve lik that used in Johnson's train. It is used solely to maintain an atmosphere o

[1] By Wm. R. Fleming, Metallurgist, Andrews Steel Company, Newport, Ky. Iro Age, Jan. 1, 1924.

pure oxygen in the purifying train, a condition essential to accurate results. It is not used to prevent carbon dioxide backing into the purifying train, of which there is not the remotest possibility.

The combustion tube is the ordinary fused silica tube glazed on the inside only. The tube is 30 ins. long with inside diameter of from $\frac{7}{8}$ to 1 in. One tube of 30 ins. will serve twice as long as one of 24 ins. It is loosely packed with asbestos for a distance of 6 ins. at the exit end, and 3 ins. is allowed to project from the furnace. For about the first 100 combustions, the combustion boat is pushed close against the asbestos. The portion of the tube immediately above this will become coated with iron oxide. The asbestos is then moved up so that it covers this portion of the tube and a fresh area exposed to the spraying oxide. In this manner one tube can be made to serve 600 combustions or even more. Both platinum and nickel cylinders have been used inside the tube to protect it from the spraying oxide, but it is doubtful whether this practice pays. These cylinders are not used in this laboratory because it is believed that they delay incipient combustion for at least thirty seconds.

The Furnace and Combustion Tube. The furnace used is one of the ordinary resistance type. It is constantly maintained at a temperature of 1000° C. This temperature is verified daily by the use of a pyrometer. Many claim to be expert at judging temperatures, but none are expert enough to be without a pyrometer. The two-way stop following the combustion tube will be found very convenient when it is not desirable to pass the current through the jars Z and O.

Z is filled with 20-mesh zinc. Once filled it will serve for several thousand combustions. As a matter of fact it is included in this train as a filter. If nickel boats and aluminum are used the chemist may omit this zinc jar from the train, for with all ordinary grades of steel it serves no purpose.

O is the phosphoric anhydride jar. A little asbestos is placed in the lower part just above the lower stopper. The remaining space in the jar is completely filled with phosphoric anhydride. The upper stopper is packed tightly enough to prevent any powder being swept into the weighing apparatus. As the anhydride liquefies it passes down into the lower stopper, where it can be removed conveniently without disturbing the anhydride above it. Likewise

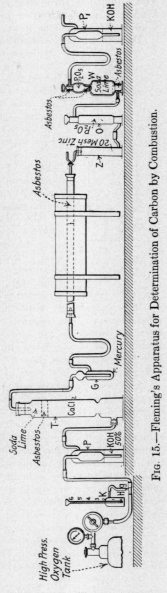

FIG. 15.—Fleming's Apparatus for Determination of Carbon by Combustion.

the anhydride can be replenished by removing only the upper stopper. The jar need not be washed oftener than once in 500 combustions. When filled with anhydride, fresh reagent need not be added for at least 150 combustions. After each combustion the jar should be given a few sharp taps with the hand to prevent canals being formed.

Details of the Absorption Apparatus. The absorption apparatus, shown in detail in Fig. 16, has been modified slightly at the suggestion of Henry G. Martin, of the Railway Steel Spring Company, Chicago Heights, Ill. This apparatus is no more efficient than the old style, but it is much more convenient and less troublesome. In the old-style tube the anhydride would liquefy after several days and require replenishing. To overcome this, Mr. Martin suggested using separate chambers for the anhydride and soda lime, so that communication could be broken when the tube was standing idle. The tube shown, Fig. 16, is Fleming's modification of Mr. Martin's suggestion. When properly filled this tube will serve for at least 70 combustions when operating on 1.5 grams of sample containing 1.03% carbon.

FIG. 16.—Fleming Absorption Apparatus.

The anhydride in the upper chamber serves for at least 300 combustions. Soda lime, placed in the lower tube in alternate layers ($\frac{1}{8}$ in.) of the different meshes, has proven a very convenient and desirable reagent. The 12-mesh soda lime for nitrogen can also be used with excellent results. If this is employed, part of it should be ground to about 60-mesh and alternate layers of fine and coarse used.

It is exceedingly important that the tube be loaded with alternate layers of coarse and fine reagent, for, if the 12-mesh reagent is transferred directly from the bottle to the absorption tube, the latter will fail to be effective for more than 30 combustions and in some cases less. The reason for this is evident. The lower stopper is packed loosely with asbestos, also the lower portion of the soda lime chamber just above the stopper. Beginning with a layer of 12-mesh soda lime, the entire chamber is filled with alternate layers of fine and coarse reagent. The small diameter of the anhydride chamber is packed with asbestos and the remaining space filled with phosphoric anhydride. Finally, the upper stopper is packed with asbestos. The anhydride chamber, filled as indicated, will not require refilling for at least 300 combustions. It is not necessary to turn the chamber to break communication while the tubes are idle, for the packing of the small diameter with asbestos prevents the absorption of moisture from the soda lime. The tubes must be used in pairs, so that one serves as a tare in weighing the other. A pair of tubes assures the operator of at least 140 combustions. A glass or rubber tubing about 12 ins. long serves as a guard for the absorption tube. It connects the bottle, P_1, which is used to indicate flow of gas.

The use of clay boats has been abandoned in favor of nickel boats filled with "carbon free" alundum.[1] These are greatly superior to clay boats in every conceivable way. The boat with alundum should always be burned in oxygen at 1000° before using. The boats are formed out of 22-gauge pure sheet nickel. One boat will serve for about 100 to 150 combustions, some more, some less.

[1] Carbon content is less than 0.003%. Sold by Norton Company, Worcester, Mass.

Details of the Analysis. The furnace being at 1000°, the two freshly prepared absorption tubes are placed in the train and oxygen run through at the rate of 300 cc. per minute for fifteen minutes. This insures the displacement of all air from the purifying train as well as the absorption tubes. Remove one absorption tube from the train and turn on the oxygen until the mercury stands at about 2 ins. The rate of current is then measured by inverting a graduated cylinder filled with water. Several trials will establish a rate of about 325 cc. per minute. Note the reading of the column of mercury at this rate and subsequently, when using the same absorption tube, maintain this same pressure in the train and the rate of flow will be 325 cc., the rate during all combustions. Shut off the oxygen and, when it comes to a slow bubbling through P_1, close the upper stopper of the absorption tube. Disconnect it from the train, but do not close the lower stopper for about five seconds after disconnection. Weigh against its mate as a tare. It is now ready for the first combustion.

Weigh 1.5 grams of drillings, preferably thin, curly drillings from a twist drill, and spread out in the nickel boat which is half filled with alundum. Place the absorption tube in the train and place its mate beside it. With the oxygen flowing about 100 cc. per minute, the drillings are pushed into the combustion tube. The current is immediately run up to the desired pressure, which gives 325 cc. per minute. The regulator will do the rest. It will feed the oxygen automatically to the burning steel. As a rule the drillings are entirely burned one and one-half minutes after insertion. Continue the flow of oxygen for three and one-half minutes more (five minutes, total time) and disconnect as before the absorption tube. Weigh immediately. The result will be accurate and reliable. Whether determining carbon in a standard steel, where the greatest accuracy is required, or in a bath test, the time required is always five minutes.

The weight of the boat, plus refractory lining, should be kept as low as possible, so as not to introduce too much cold material into the combustion tube. The boats used in this laboratory are $\frac{1}{2}$ in. wide, $\frac{1}{4}$ in. deep and 3 ins. long. Sheet nickel varies in percentage of carbon. As a rule, a nickel boat must be ignited in oxygen at 1000° for one to two hours.

There seems to be a difference of opinion concerning the physical condition of the steel after burning, some chemists believing that inaccurate results are obtained if the drillings have fused during combustion. Others maintain that complete fusion of the drillings is essential to accurate result. If drillings which happen to be a little thick are used, low results are obtained unless these are *perfectly fused.*

Graphitic Carbon

In Iron and Steel. The sample of 1 gram of pig iron or 10 grams of steel is treated with 15 cc. of nitric acid (sp.gr. 1.2), per gram of sample taken. When all the iron has dissolved, the graphite is allowed to settle and the supernatant liquid decanted onto an ignited asbestos filter, using either a perforated boat, Fig. 17, or a filtering tube. The residue is transferred to the filter, and washed thoroughly with hot water. It is treated with hot caustic solution (sp.gr., 1.1), washed thoroughly again with hot water, then with a little dilute hydrochloric acid, and finally with hot water. The carbon is now burned by one of the procedures given—the oxidation in the combustion furnace being recommended. The CO_2 is absorbed in caustic and estimated according to the standard procedure given for carbon.

$$CO_2 \times 0.2727 = \text{graphitic carbon.}$$

The perforated boat, shown in the cut, fits snugly into the receptacle below. Sufficient asbestos is poured into the boat to form a film over the bottom. A seal is made around the boat with additional asbestos, the apparatus having been inserted in a rubber stopper in the neck of a suction flask and suction applied.

The apparatus is recommended by Blair for combustion of graphitic carbon or of total carbon liberated from iron or steel by the cupric potassium chloride method. The boat may be placed directly in the combustion tube and the carbon oxidized as usual.

FIG. 17.—Boat and Holder for Carbon Determination.

The Shimer Combustion Apparatus [1]

The apparatus, designed for the rapid determination of iron and steel, is in general applicable to the same class of chemical operations as is the combustion tube of platinum, silica, or porcelain. It offers the advantage of neatness, reduction in the number of parts to be handled, diminished consumption of gas, and increased ease of manipulation. The simplified form, shown in the cut, Fig. 18, enables the use of the standard form of platinum crucible, A, with its inner wall ground to fit a tapered nickel, water-jacketed stopper, B. The rubber jacket of the original type is eliminated and a detachable nickel reinforcing ring, C, at the top of the crucible serves the double purpose of completing the security of the seal and as a support for the apparatus.

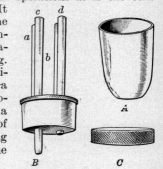

FIG. 18.—Shimer Combustion Apparatus, Simplified Form.

Water is circulated through the stopper through the tubes c and d. The current of oxygen passes through a into the crucible, oxidizing the material within the crucible, which is heated to red heat with a burner placed below it. The carbon dioxide formed passes through b to the absorption train. The remainder of the apparatus for the determination is the same as is used with the combustion tube. An asbestos shield protects the upper portion of the outfit, the crucible fitting snugly in a hole in the asbestos board.

Combined Carbon

Indirect Method. The excess of carbon remaining when the graphitic carbon is subtracted from total carbon (in iron and steel), is calculated as combined carbon. This difference method is generally accepted as being the most accurate for estimation of combined carbon.

[1] Courtesy of Baker Platinum Works.

Stetser and Norton Combustion Train for Carbon Determinations

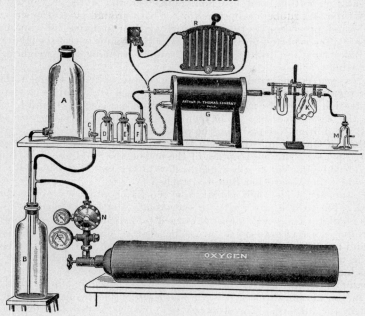

Fig. 18a.—Stetser and Norton Combustion Train for Carbon Determinations.[1]

A   Aspirator Bottle, 8 liters capacity, graduation interval 250 ml., with one hole rubber stopper.
B   Bottle, narrow mouth, of green glass, 2 gallons capacity, with two hole rubber stopper.
    Glass Tubing, 6 mm. outside diameter, for connections.
    Glass T-tube, 3 mm. bore.
C   Glass Stopcock, 2 mm. bore; used to prevent the gas from flowing through the train when turned
    on at the regulator.
D   Bottle, wide mouth, 8 oz. capacity, with two hole rubber stopper; to be used empty as a safety.
E   Ditto; to be filled one-third full with concentrated Sulphuric Acid.
F   Ditto; to be filled with Ascarite.
    Combustion Tube, of Silica, glazed, 30 inches long by ⅞ inch diameter.
G   Combustion Tube Furnace.
R   Rheostat, for controlling the temperature.
J   Vanier Zinc Drying Tube; to be filled with 30 mesh Zinc.
K   Vanier Sulphuric Acid Bulb; to be filled with Sulphuric Acid.
M   Stetser and Norton Modification of the Midvale Absorption Bulb.

Combustion Method

The train illustrated in Fig. 18a was described by J. B. Stetser and R. H. Norton in *"Combustion Train for Carbon Determinations," The Iron Age,* Vol. 102, No. 8. The special features of this outfit are rapidity with which combustions can be made, simplicity of the absorption bulb, and the use of Ascarite, a special sodium hydrate asbestos absorbent mixture according to the formula of Mr. J. B. Stetser. It is a dryer as well as an absorber.

The absorption bulb is a modification of the Midvale Absorption Bulb designed by Mr. H. L. Fevert, of the Midvale Steel Company, and with one

[1] By courtesy of Arthur H. Thomas Co.

filling can be used for at least 400 determinations using one-half factor weight with carbon 0.50%. The bulb measures 45 mm. in diameter at its widest portion at the bottom, and 120 mm. high over all. It is provided with a capillary tube and rubber stopper at top, or with ground glass stopper.

The usual procedure is to allow three or four minutes for combustion. Results are often reported in six minutes after the sample enters the laboratory: this includes time for drilling, weighing sample, running test, weighing bulb and reporting result.

The oxygen is delivered from a high pressure cylinder through the gas pressure regulator, and the authors also recommend that bottles as shown in the illustration be used to measure the amount of gas consumed and to supply rapidly the extra quantity of oxygen required during the burning of the drillings, in addition to equalizing the pressure.

Method of Operation.—The train is set up free from leaks, and the stopcocks are opened with the exception of the one on the Sulphuric Acid Bulb K. This is opened sufficient to allow gas to flow at the rate of 200 to 250 ml. per minute when the Absorption Bulb is attached.

The stopcock C is then closed, the remaining stopcocks being left as adjusted. The train is now ready for operation. Bottle B is filled with water. The gas regulator is opened, allowing oxygen to displace the water in bottle B, which water is driven into bottle A. When bottle A is filled, the regulator is closed and the train is ready for combustion. The exit end of the combustion tube, the usual glazed silica tube being recommended, is packed with some asbestos burned in a current of oxygen prior to using.

A sample of one-half factor weight properly prepared and weighed is placed in the alundum boat, RR Alundum protection being used, and then inserted in the furnace. The stopcock C is opened, and if the furnace is at 1800° C. or over, and the sample of drillings fine and uncovered, it should begin to burn in 20 seconds. The burning should take 40 to 60 seconds additional, consuming 500 ml. of oxygen. An additional 500 ml. of oxygen is turned on to wash out all CO_2, and the bulb is then ready to be weighed.

If the sample is covered by a lid or RR Alundum, or if the drillings are coarse, or the furnace is below 1800° C., the combustion may be delayed as much as two minutes. The actual burning will, as before, take from 40 to 60 seconds. In any event, 500 ml. of gas must be passed through the apparatus after the steel has ceased burning. The point at which the burning of the sample begins may be determined by the increased rate at which gas passes through the liquid in bottle E, due to the rapid absorption of oxygen by the burning steel. A similar decrease marks the end of the burning period.

Standardization of the Stetser and Norton Absorption Bulb.—A freshly filled bulb should be run on the train for an hour and then weighed. When the bulb has reached a constant weight, the train is checked by running a government standard

A bulb once filled and standardized will last for several weeks and is sufficient for from three to four hundred determinations. On account of the difference in color between the used and unused portions of the absorbent, it is possible to determine the moment a bulb can be discarded.

NOTES. In chromium, tungsten and titanium steels a temperature of 1500° C. is necessary to oxidize the carbon by direct combustion for thirty minutes. (J. R. Cain and H. E. Cleaves, J. Wash. Acad. Sci., **194**, 4, 393–397.)

DETERMINATION OF CARBON IN ORGANIC SUBSTANCES

Combustion of Organic Substances Free of Nitrogen, Halogens, Sulphur, and the Metals

The following modification of the procedure described for determination of carbon in iron and steel is applicable to the determination of carbon in organic substances free from the substances mentioned above.

Apparatus. This is practically the same as that shown in Fig. 19, with the exceptions that copper plugs may be used to advantage in place of the platinum plugs. In the absorption end of the train a calcium chloride tube is preferred. The calcium chloride should have been saturated with dry CO_2 gas, the excess of which has been removed by a current of pure air. This tube

Platinum   Boat.   Platinum   Copper Oxide   Platinum

FIG. 19.—Diagrammatic Sketch of Combustion Tube.

is weighed as well as the potash bulb, the calcium chloride retaining the water formed by the combustion of the hydrogen of the organic substance, which is thus determined.

The organic substance, if a solid, is introduced into the combustion boat directly; if it is a liquid, it is held in a bulb blown in a capillary tube. One end of the tube is sealed and a bulb blown. When cool, the tube is weighed, and the material then introduced by first warming the bulb and then inserting the open end of the tube into the liquid to be examined. By cooling the bulb, liquid is drawn into the tube. The end is wiped off, and the liquid expelled from the capillary by gently heating this portion. The end is now sealed if the liquid is volatile, otherwise it is left open, and the tube is weighed. The increased weight is due to the organic substance. The tip of the capillary is now broken, if sealing was necessary, by means of a file. The tube containing the sample is placed in the boat, the open end of the capillary pointing toward the open end of the combustion tube. After connecting up the apparatus, the copper oxide end of the tube is heated to redness and oxygen slowly passed through the tube at such a rate that the bubbles in the potash bulb can be readily counted. The entire tube is now heated and remaining operation is the same as has been described for iron and steel combustion.

The gain of weight of the calcium chloride tube is due to water formed by the combustion of hydrogen of the compound, that of the potash bulb to the carbon.

$$H_2O \times 0.1121 = H,$$
$$CO_2 \times 0.2727 = C.$$

Carbon in Soils. One to 3 grams of 60-mesh sample is treated with a solution of 3.3 grams $CrO_3 + 10$ cc. H_2O and 50 cc. conc. H_2SO_4 (1.84). The evolved CO_2 is absorbed in standard caustic and titrated with acid, phenolphthalein and methyl orange being used as indicators. (J. Ind. Eng. Chem., 1914, **6**, 843–846.)

NOTE. The oxygen gas should be free from hydrogen. A preheater, placed before the purifying tubes of the train, causes the combustion of the hydrogen and the absorption of the water formed before the gas enters the combustion tube.

Determination of Carbon and Hydrogen in Nitrogenous Substances

A modification of the first procedure described for determinations of carbon and hydrogen in organic substances must be made, since from substances containing nitrogen, nitroso and nitro compounds, oxides of nitrogen are formed which would be absorbed in the calcium chloride and potash bulbs, giving high results for hydrogen and carbon. To overcome this difficulty, a copper spiral, that has been reduced (See note below) is placed in the front end of the combustion tube (to the right in Fig. 19) to reduce the oxides of nitrogen to nitrogen.

NOTE. Reduction of copper spiral may be accomplished as follows: The copper spiral is prepared by rolling together a piece of copper gauze about 10 centimeters wide, making it as large as will conveniently pass into the combustion tube. The spiral is heated till it glows by holding it in a large gas flame, and while still hot it is dropped into a test-tube containing 1 or 2 cc. of methyl alcohol or ether. This quickly boils away, igniting at the end of the tube. The copper is completely reduced to bright metallic copper. The spiral is taken out with a pair of crucible tongs and dried by quickly passing it through a flame a few times, and while it is still warm it is introduced into the front of the combustion tube.

The substance is introduced into the tube and the connections made. The copper oxide spiral, that was pushed after the boat, is heated, and then the reduced spiral (right end of tube). The oxide near the boat, and finally the entire tube is heated to a red heat. When the bubbles cease to show in the potash bulb, the stopcock is opened to the oxygen-purifying train and a slow flow of oxygen turned on, the gas allowed to pass through the tube until it can be detected with a glowing splinter at the exit of the absorption end of the apparatus.

If the substance is difficult to burn, it is mixed with freshly ignited (cold) copper oxide, which assists combustion.

The remainder of the operation is the same as has been described.

Organic Substances Containing Halogens

The procedure is the same as that described for nitrogenous substances with the exception that a silver spiral is used in place of the reduced copper spiral. The heating of this spiral should be between 180 and 200° C. (not over 200°).

Organic Substances Containing Sulphur

These are best ignited with sodium peroxide and the carbonate formed is determined by the procedure given for carbon dioxide in carbonates.

The Wet Combustion Process for Determination of Carbon

The method depends upon the oxidation of carbon to carbon dioxide when the powdered material is digested with a mixture of concentrated sulphuric acid and chromic acid, or potassium dichromate, or permanganate. The procedure is applicable to oxidation of free carbon, carbon combined in organic substances and in certain instances to carbon combined with metals, where the substance may be decomposed by the action of the acids.[1] It is of value in determination of carbonates in presence of sulphides, sulphites, thiosulphates and nitrites, which would vitiate results were they not oxidized to more stable forms, from passing into the potash bulb with the carbon dioxide.

Apparatus. The apparatus is identical with that used for determining carbon dioxide in carbonates, Fig. 20, with the exception that in place of the acid bulb nearest the decomposition flask two bulbs are placed. The first of these contains a strong solution of chromic and sulphuric acids, the second is filled with glass beads moistened with chromic acid solution. Following this is the drying bulb containing concentrated sulphuric acid and finally the absorption apparatus, as shown in the illustration.

Procedure. 0.2 to 1 gram of the powdered material, fine drillings, free carbon, or organic substance is placed in the decomposition flask. If the material is apt to pack it is advisable to mix with it pure ignited sea-sand to prevent this. Five to 10 grams of granular potassium dichromate are added and the apparatus swept free of carbon dioxide by passing purified air through it before attaching the absorption apparatus. The potash bulb is now weighed, using a counterbalance bulb and following the precautions given in the dry-combustion method. The bulb is attached to the train.

Oxidation. Concentrated sulphuric acid placed in the acid funnel, attached to the decomposition flask, is allowed to flow down on the sample until the funnel is almost empty; the stop-cock is then closed. A flame is placed under the flask, when the vigorous action has ceased, and the material gently heated until the reaction is complete and the organic matter or carbon completely oxidized.

The apparatus is now swept free of residual CO_2 by applying suction, the gas being completely absorbed by the potash, or the soda lime reagent.

The increase of weight of the absorption bulb is due to carbon dioxide.

$$CO_2 \times 0.2727 = C.$$

NOTE. The following additional purifiers are frequently advisable: (a) an absorption bulb containing silver sulphate to absorb chlorine and vapors from sulphur compounds; (b) a capillary tube of silica or platinum heated to a dull redness to oxidize any hydrocarbons, carbon monoxide, etc., that may be evolved and imperfectly oxidized by the chromic acid.

DETERMINATION OF CARBON DIOXIDE IN CARBONATES

The method is applicable for determination of carbon dioxide in limestone, dolomite, magnesite, strontianite, witherite, spatic iron ore, carbonates of sodium, and potassium, bicarbonates in baking powder, carbon in materials readily oxidized to CO_2 chromic sulphuric acid mixture. The procedure depends upon the evolution of carbon dioxide by a less volatile acid, or the oxidation of carbon. The CO_2 is absorbed in caustic and weighed.

Apparatus. The illustration shows the apparatus found suitable for this determination. It is Knorr's apparatus slightly modified. The absorption bulb or bottle should be one that will effectively absorb carbon dioxide entering

[1] Not applicable for determining carbon in ferro-silicon, ferro-chrome or tungsten steel.

at a rapid rate. The Vanier or the Fleming forms is satisfactory for this purpose.

Procedure. A sample weighing 0.5 to 2 grams, according to the carbon dioxide content, is placed in the dry decomposition flask (*C*). The flask is closed by inserting the funnel tube (*B*) fitted with the soda lime tube (*A*), and connected by means of a condenser to the train for removing impurities from carbon dioxide, leading to the absorption bulb, as shown in Fig. 20.

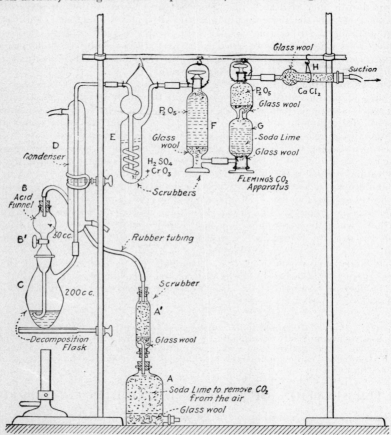

FIG. 20.—Apparatus for Determining Carbon Dioxide.

The apparatus is swept out with a current of **dry,** purified air before attaching the weighed absorption bottle. This is accomplished by applying gentle suction at the end of the purifying train. The absorption apparatus is now attached (Fleming absorption apparatus is shown in the illustration). The tube (*B*) is nearly filled with dilute sulphuric acid (1 : 3), the stop-cock (*B'*) being closed. The soda lime tube is now inserted into place as shown in the cut. The acid in (*B*) is now allowed to run slowly down on the sample at a rate that evolves gas not too rapidly to be absorbed; 1 to 2 cc. of acid being retained in (B) to act as a seal, the stop-cock (B') being then closed.

When the violent action has ceased, the solution in (C) is heated to boiling and boiled for about three minutes. If the sample is baking powder, or contains organic matter, the decomposition flask is protected from excessive heat by placing a casserole of hot water under it. This prevents charring of the starch or organic matter, which would be apt to occur if the direct flame was used. Gentle suction is now applied to the absorption end of the apparatus and the stop-cock (B') opened, allowing the remainder of the acid to flow into the flask (C) and admitting a current of air, purified by passing through the soda lime in (A). The suction should be gentle at first, and then the speed of the flow increased to the full capacity of the absorption bottle. A fairly rapid current is preferred to the old-time procedure of bubbling the gas through the apparatus at a snail-like pace, but discretion should be used in avoiding a too rapid flow.

In the analysis of baking powders, where foaming is apt to occur, the decomposition flask should be of sufficient capacity to prevent foaming over. A small flask is generally to be preferred for obvious reasons. By gently heating to boiling during the passage of the air, steam assists in expelling any residual CO_2 in the flask. When the passage of air is rapid, this boiling should be discontinued.

The increase of weight of the absorption bottle is due to the carbon dioxide of the sample. This procedure gives total CO_2.

Determination of Carbon Dioxide by Measuring the Gas

Fairly accurate results may be obtained by measuring the gas evolved. A large cylindrical tube having a capacity of about 1100 cc. is used. The tube is graduated from 1000 cc. to 0 at the upper portion of the cylinder; a space of about 100 cc. remains at the upper portion. A tube extending from a little above the 0 graduation to the bottom of the cylinder carries out the water as the gas is admitted.

To make the run, the cylinder is filled to the mark 0 with saturated salt solution.[1] It is now connected to a condenser.[2] Twenty-five cc. of saturated salt solution are admitted to the decomposition flask, and the generated gas measured by the water displacement in the tube described. Calculations are made after reduction to standard conditions. 5.1 cc. CO_2 at 0° C. and 760 mm. weigh 0.01 gram.

Residual Carbon Dioxide

This is the CO_2 remaining after baking powder has been treated with water and the evolved CO_2 expelled by warming.

The procedure recommended by the U. S. Department of Agriculture is as follows:[3]

[1] H. W. Brubacker, Jour. Ind. Eng. Chem., 1915, **7**, 432.
[2] The nitrometer may be used in place of the cylinder and atmospheric conditions obtained as usual. Formula for reduction to 760 millimeters and 0° C.:

$$V' = V \frac{(P-w)}{760(1+.00367t)}.$$

[3] Bureau of Chem. Bulletin No. 107.

Weigh 2 grams of baking powder into a flask suitable for the subsequent determination of carbonic acid, add 20 cc. of cold water, and allow to stand twenty minutes. Place the flask in a metal drying cell surrounded by boiling water and heat, with occasional shaking, for twenty minutes.

To complete the reaction and drive off the last traces of gas from the semi-solid mass, heat quickly to boiling and boil for one minute. Aspirate until the air in the flask is thoroughly changed, and determine the residual carbon dioxide by absorption, as described under total carbonic acid.

The process described, based on the methods of McGill and Catlin, imitate, as far as practicable, the conditions encountered in baking, but in such a manner that concordant results may be readily obtained on the same sample, and comparable results on different samples.

Available Carbon Dioxide

The residual is subtracted from the total, and the difference taken as available CO_2.

Determination of Carbon Dioxide by Loss of Weight

An approximate estimation of the carbon dioxide in carbonates—baking powders, bicarbonate of soda, limestone, etc., may be obtained by the loss of weight of the material when treated with a known weight of acid.

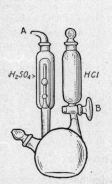
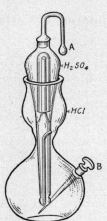

FIG. 21.—Schroetter's Alkalimeter.   FIG. 22.—Mohr's Alkalimeter.

Various forms of apparatus are used for this determination. The Schroetter and Mohr types are shown, Figs. 21 and 22.

About 0.5 to 1.0 gram of sample is taken and placed in the bottom of the flask, dilute hydrochloric and strong sulphuric acids then placed in the bulbs as indicated in the illustrations. The apparatus is weighed as it is thus charged. The hydrochloric acid is now allowed to flow down on the carbonate and the stopper closed. The evolved gas passes through the strong sulphuric acid, which absorbs the moisture. After the vigorous action has subsided the apparatus is placed over a low flame and the solution heated to boiling and boiled very gently for about three minutes. CO₂-free air is aspirated through the solution to expel the last traces of CO₂, by applying gentle suction at *a* and opening *b*, the air being purified by passing through soda lime. The apparatus is again weighed and the loss of weight taken as the CO_2 of the material.

Available CO_2 in baking powder may be determined by substituting water in place of hydrochloric acid.

Determination of CO₂ in Carbonates —

Hydrometer Method of Barker [1]

No mechanical balance or scale is required by the following procedure.

The method depends upon the principle of the hydrometer, following the law that when an object is immersed in a liquid it is buoyed up by a force equal to the weight of the liquid displaced by the object. The carbon dioxide set free from the sample decreases the weight; and the rise of the graduation scale tube above the water records the percentage of carbonates from which the gas was released. The procedure is suitable for determining the comparative strengths of baking powders, for rapid tests of the quality of limestone and for estimation of carbon dioxide of carbonates in general.

Procedure. To analyze a sample for carbonates measure out 40 cc. of hydrochloric acid (sp.gr. 1.15), using a small graduate; pour this into the acid reservoir through the opening *A*. With graduated stem disconnected hang a 10 gram weight at *B*. The hydrometer should then float in a cylinder of water and be immersed to some point at *C*. Remove the 10 gram weight and introduce pulverized limestone, or other substance that is being tested, until the instrument is immersed to exactly the same point that it occupied with the suspended weight. The reservoir will now contain 10 grams of sample. Connect up the graduated stem and add water, a drop at a time, through the funnel-shaped top, until immersed to the zero point. Raise the hydrometer out of the water and open the stopcock *D* until the acid drops slowly into the reaction chamber, decomposing the carbonate. As the reaction proceeds the instrument rises slowly and at the conclusion the point on stem at the surface of the water gives the per

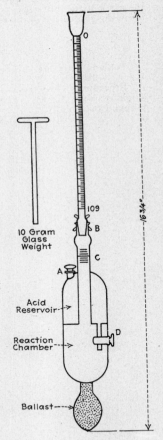

Fig. 22a.— Barker's Hydrometer.

[1] J. F. Barker, Jour. Ind. Eng. Chem., IX, 8, 786–787 (1917).

cent of calcium carbonate equivalent to the carbon dioxide in the sample. This figure is the calcium carbonate equivalent so often mentioned in connection with limestone analyses.

A Fahrenheit thermometer accompanies each instrument and is hung inside the floating cylinder. Its reading is taken before and after each determination to allow for any error due to change in temperature. To the figure for calcium carbonate equivalent add 0.5 for each degree rise, or subtract 0.5 for each degree fall in temperature between the two readings. This temperature change need seldom amount to more than a fraction of a degree.

NOTES. The limestone may be weighed to an accuracy of 0.02 gram. The weight of CO_2 remaining in the apparatus tends to offset the loss due to moisture escaping with the gas, but the difference, together with any other sources of error has been accounted for in the graduation of the reading stem.

Determination of Percarbonates

Percarbonates are decomposed by dilute sulphuric acid according to the reaction $K_2C_2O_6 + 2H_2SO_4 = 2KHSO_4 + 2CO_2 + H_2O_2$.

Procedure. Two tenths of a gram of the salt is added to about 300 cc. of cold dilute sulphuric acid (1 : 30). The liberated hydrogen peroxide is titrated with potassium permanganate.

$$1 \text{ cc. N/10 KMnO}_4 = 0.00991 \text{ g. K}_2C_2O_6.$$

VOLUMETRIC METHODS FOR THE DETERMINATION OF CARBON

Total Carbon.   Absorption of Carbon Dioxide in Barium Hydroxide

The carbon dioxide evolved by oxidation of the material by dry combustion with oxygen or by oxidation with chromic sulphuric acid mixture is absorbed in barium hydroxide, free from carbonate, and the precipitated barium carbonate titrated with standard hydrochloric acid.

Procedure. The essential difference in this method from those already described under the gravimetric methods is in the fact that a perfectly clear saturated solution of barium hydroxide is used for absorption of the carbon dioxide in place of caustic potash. Considerable care must be exercised to prevent contaminating the reagent with carbonate. The solution is drawn by suction through a siphon, dipping below the surface of the reagent, into the absorption tube, which should be of such construction that the material may readily be poured out. Details of the procedure are given in the chapter on Iron and Steel, Volume II, under carbon determination. The precipitate barium carbonate is filtered off and washed by a special procedure, and then titrated with standard hydrochloric acid.

$$1 \text{ cc. } 0.1 \text{ N HCl} = 0.0006 \text{ g.}$$

Determination of Carbon by Measurement of the Volume of Carbon Dioxide Evolved by Oxidation of Carbon, or by the Decomposition of Carbonates with Acid.

Description of the Scheibler and Dietrich Process and that of Lunge and Marchlewski are given in Mellor's work on "Inorganic Analysis," pp. 555–559, 1st Ed. A modification of Wiborg's method is described in Blair, "Chemical Analysis of Iron," pp. 146–149, 7th Ed.

Determination of Carbon Dioxide in a Gas Mixture.

See Gas Analysis.

Direct Colorimetric Method for Determination of Combined Carbon

The procedure is of value to the steel laboratory where a large number of daily determinations of combined carbon are required. By this method over a hundred determinations a day may be made by an experienced manipulator. The method depends upon the color produced by combined carbon dissolved in nitric acid, the depth of color increasing with the combined carbon content of the material. Comparison is made with a standard sample of iron or steel, which is of the same kind and in the same physical condition as the material tested.[1] That is to say, a Bessemer steel should be compared with a Bessemer standard, open hearth with open hearth, crucible steel with crucible steel, the standards containing approximately the same amounts of carbon, and as nearly as possible the same chemical composition. The samples should be taken from the original bar which has not been reheated, hammered, or rolled. Copper,

 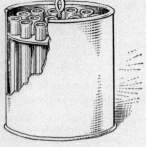

Fig. 23.   Fig. 24.

Hot Water Racks for Test Tubes.   Color Carbon Determination.

cobalt, and chromium will interfere with the test; the other elements have very little effect.

Procedure. One standard sample of 0.2 gram and the same amount of sample drillings are taken for analysis. The weighings are conveniently made in brass or aluminum pans, boat-shaped to enable the drillings to be dumped into test-tubes. A counterpoise, weighing the same as the boat, is placed on the opposite pan, together with the 0.2 gram weight. A magnetized knife will assist in removing the excess of material. The weighed sample is brushed into a test-tube 6 ins. long (150 mm.) $\frac{5}{8}$ in. (16 mm.) in diameter. (Each test-tube has a label near the open end to distinguish the sample.) A rack or a 600-cc. beaker may be employed for holding the test-tubes during the weighing. After the batch is ready the tubes are transferred to a perforated rack (Figs. 23 or 24) and this then stood in the water bath filled with cold water.

The proper amount of nitric acid (sp.gr. 1.2; e.g., 1 conc. HNO_3 : 1 H_2O), from a burette, is now added to each test-tube.

3 cc. HNO_3 for 0.3% C.	6 cc. HNO_3 for 0.8 to 1% C.
4 cc. HNO_3 for 0.3 to 0.5%.	7 cc. HNO_3 for over 1% C. steel[1]
5 cc. HNO_3 for 0.5 to 0.8% C.	

[1] Blair, "The Chemical Analysis of Iron."

The depth of color produced by the acid will give an idea of the amount required. One cc. of acid is added at a time until the depth of color is correct. This requires experience gained from observation of the color produced by standard samples. The acid is added slowly to the coarse drillings. Insufficient acid gives a darker tinted solution than it properly should be. The nitric acid should be free from chlorine and hydrochloric acid, since these produce a yellow color. (Cl and $FeCl_3$ are yellow.)

A glass bulb or a small funnel is placed in each test-tube and the water in the bath then heated to boiling and boiled until all the carbonaceous matter has dissolved, the tubes being shaken from time to time to prevent formation of a film of oxide. Low-carbon steels require about twenty minutes, whereas steels of over 1% carbon require about forty-five minutes. (Blair.) As soon as the bubbles cease and the brownish flocculent matter disappears, the rack is removed from the bath and placed in a casserole of cold water. (Prolonged heating and strong light each causes fading of the color due to combined carbon.)

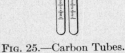

FIG. 25.—Carbon Tubes. FIG. 26.—Color Comparator or Camera.

Color Comparison. This is made in graduated, clear, colorless, glass cylinders called carbon tubes. The form shown in Fig. 25 was found by the writer [1] to be the most satisfactory type for a steel-works laboratory where rapidity of manipulation was essential. The bend at the upper portion of the tube facilitates mixing of the solution upon dilution with water, the tube being tilted back and forth until the solution is homogeneous, the bend preventing the liquid from spilling. The dilution should be at least twice that of the amount of nitric acid used, as this amount of water is necessary to destroy the color due to ferric nitrate.

The standard is poured into the carbon tube and the rinsings from the test-tube added. The solution is diluted to a convenient multiple in cc. of the carbon content. For example, 0.45% carbon sample may be diluted to 9 cc., then each cc. will represent 0.05% carbon. The sample is placed in a second tube of exactly the same diameter, wall thickness, and form. If the solution of the sample is darker than the standard, water is added little by little,

[1] W. W. Scott.

followed by mixing, until the shade matches the standard. If the standard, on the other hand, is darker than the sample, a greater dilution of the standard is necessary, the cc. again representing a multiple of the carbon content. For example dilution of the .45% carbon sample to 15 cc. makes each cc. to represent 0.03 carbon. (It is frequently advisable to take a standard of lower carbon content in place of greater dilution of the standard.)

Example. Suppose in the first case the dilution of the sample was 15 cc. in order to match the standard, then $15 \times 0.05 = 0.75\%$ carbon. Six cc. dilution case $2 = 6 \times 0.03 = 0.18\%$ carbon.

The color comparison can be best made in a "camera," a long box with one end closed by a ground-glass screen, Fig. 26. Parallel to the screen and near it, two holes through the top of the box admit the test-tubes. The inner walls of the camera are blackened to prevent reflection of light. If a camera is not available, the tubes may be held side by side and compared against a sheet of white paper held as a background.

CARBON DETERMINATION IN STEEL

Beneker's Modification of Eggertz's Method for Determining Carbon in Steel

By this procedure interference of the color due to iron is eliminated.

.2 g. of the sample and standard (which does not need to be very similar in carbon content to the sample) are weighed into two test tubes, treated with .7 cc. − 10 cc. of nitric acid of the usual dilution, say 3 parts water to 2 parts acid, and warmed until dissolved and perfectly clear. When cold, $\frac{1}{2}$ cc. of 85% phosphoric acid is added to each tube, transferred to the comparison tubes and read. When this method is used on low carbon steel the bleaching action is very pronounced, because in that case the iron color bleached out represents the major part of the original color. Dilution of standard or sample introduces no error, because of the absence of the interfering iron color.

ANALYSIS OF GRAPHITE

Determination of Carbon

The procedure for determining carbon in graphite is the same as that described for determination of carbon in difficultly combustible organic substances.

The material is broken down in a steel mortar and powdered in an agate mortar. About 0.2 gram is taken for the determination and mixed with copper oxide to assist the combustion, then placed in the boat and the combustion of the carbon carried on according to the standard method in the combustion tube.

$$CO_2 \times 0.2727 = C.$$

CYANOGEN

DETECTION

Traces of Hydrocyanic Acid. To the solution to be tested are added two drops of 10% NaOH and the mixture is evaporated to near dryness. After cooling, a drop of 2% ferrous sulphate is added and the sample allowed to stand in the cold for about fifteen minutes. Now 2–3 drops of strong HCl are added and the solution warmed gently, then cooled. The solution is a blue green if HCN was present in the original sample. 0.000002 g. HCN may be detected.

NOTE. Test for Cyanide. This depends upon the solvent action of HCN upon freshly precipitated HgO in presence of KOH. The filtrate is tested for mercury in an acid solution by addition of H_2S. (Hood.)

VOLUMETRIC DETERMINATION OF HYDROCYANIC ACID[1]

The method depends upon the decolorization of the blue ammoniacal solutions of cupric salts by a soluble cyanide, the reduction to cuprous condition being available for an accurate quantitative estimation of the cyanide.

Standard Copper Sulphate. Twenty-five grams of copper sulphate, $CuSO_4 \cdot 5H_2O$ are dissolved in a 1000-cc. flask with 500 cc. of distilled water and ammonium hydroxide added until the precipitate that first forms dissolves and a deep blue solution is obtained. Water is now added to make the volume exactly 1000 cc. The cupric solution is standardized by running a portion into a solution containing 0.5 gram pure potassium cyanide, KCN, per 100 cc. of water and 5 cc. of ammonium hydroxide until a faint blue color is evident. Chlorides do not interfere.

Procedure. 0.5 gram of the soluble cyanide is dissolved in 100 cc. of water and 5 cc. strong ammonium hydroxide added. The standard cupric sulphate solution is now added until the blue color is obtained. The cc. required multiplied by the factor of the copper salt in terms of the salt sought gives the weight of that salt in the sample.

Liebig's Method for Determination of Hydrocyanic Acid. Soluble Cyanides[2]

Silver nitrate reacts with an alkali cyanide in neutral or alkaline solution as follows: $AgNO_3 + 2KCN = Ag(CN)_2K + KNO_3$. The potassium silver cyanide is soluble, hence the precipitate that first forms immediately dissolves on stirring as long as the cyanide is present in excess or in sufficient quantity to react according to the equation. A drop of the silver salt in excess will produce a permanent turbidity, owing to the following reaction:

$$Ag(CN)_2K + AgNO_3 = 2AgCN + KNO_3, \text{ the insoluble AgCN being formed.}$$

[1] J. McDowell, C. N., 1904, p. 221.
[2] Ann. d. Chem. und Pharm., **77**, p. 102.

Procedure. The alkali cyanide contained in a beaker placed over a sheet of black glazed paper, is treated with 4 to 5 cc. of 10% KOH solution and diluted to 100 cc. The liquid is now titrated with standard silver nitrate, with constant stirring, until a faint permanent turbidity is obtained.

One cc. N/10 $AgNO_3 = 0.013022$ gram KCN.

Determination of Cyanide by Volhard's Method

The method involves Volhard's method for determining halogens, the procedure depending upon the fact that the silver salts of cyanides are insoluble in dilute cold nitric acid solutions.

The neutral cyanide solution is treated with an excess of silver nitrate reagent, and slightly acidified with nitric acid and diluted to a definite volume in a measuring flask. A portion of the solution is now filtered through a dry filter, and a convenient aliquot portion of this is titrated with standard thiocyanate solution, using ferric alum as indicator (see page 149) to determine the silver nitrate present. From this titrate the excess silver nitrate reagent added and that combined with the cyanide ascertained.

1 cc. N/10 $AgNO_3 = 0.006511$ g. KCN or 0.002601 g. CN.

Determination of Cyanide and Thiocyanate in Presence of One Another

The cyanide is determined preferably by Liebig's method, page 131 and the cc. of $AgNO_3$ required recorded. To the alkaline or neutral solution is added an excess of standard $AgNO_3$ and the solution acidified with HNO_3, then made to definite volume and a portion filtered. The silver nitrate in a convenient amount of the filtrate, an aliquot portion of the whole, is determined by titrating with standard thiocyanate solution, using ferric alum indicator (page 149). From this titration the excess of $AgNO_3$ added is determined, and the amount of reagent required for the thiocyanate is known.

1 cc. N/10 $AgNO_3 = 0.006511$ g. KCN

or 0.002601 g. CN or 0.005808 g. CNS.

In place of the above method the following may be used: One portion of the solution is treated with nitric acid and the thiocyanic acid oxidized to sulphate. By adding $Ba(NO_3)_2$ solution $BaSO_4$ is precipitated and the equivalent thiocyanic acid calculated, also the equivalent N/10 $AgNO_3$ that would be required to precipitate this. In another portion an excess of N/10 $AgNO_3$ is added and the thiocyanate and cyanide precipitated. The excess of the silver nitrate may now be determined by Volhard's method described above and the amount required by CN and CNS thus ascertained. The amount of reagent required for CNS is subtracted from this total and that required by CN thus obtained. The factors are given above.

For determination of cyanide in presence of the halogens see the chapter on chlorine, page 154.

COMPLEX COMPOUNDS—FERRO AND FERRI CYANIDES

Hydroferrocyanic Acid

On gram of the hydroferrocyanide in 100 cc. of water acidified with 10 cc·
of sulphuric acid is titrated in a casserole with standard potassium permanganate to a permanent pink color.   The end-point is poor, so that it is advisable
to standardize the permanganate against pure potassium ferrocyanide.

Reaction: $2H_4Fe(CN)_6 + O = H_2O + 2H_3Fe(CN)_6$

One cc. N $KMnO_4 = 0.3683$ gram $K_4Fe(CN)_6$

Hydroferricyanic Acid

Ten grams of hydroferricyanide are dissolved in water, the solution made
alkaline with KOH and heated to boiling and an excess of ferrous sulphate
solution added.   The yellowish brown ferrid hydroxide turns black with excess
of ferrous salt.   The solution is diluted to exactly 500 cc. and 50 cc. of a
filtered portion titrated with potassium permangate.

One cc. N $KMnO_4 = .3292$ gram $K_3Fe(CN)_6$.

13

The illustration below shows a convenient form of cabinet for electrolytic determinations. The cabinet can be obtained from the Denver Fire Clay Co., Denver, or from Braun Corporation, Los Angeles. The anode may be rotated or left stationary as desired.

FIG. 26a'. Electrolytic Analysis Cabinet.

In the determination of copper in alloys, 0.5 g. alloy is dissolved in 10 cc. HNO_3 in a 250 cc. beaker, the solution diluted to 100 cc. is treated with 5 cc. H_2SO_4 and electrolyzed at $1\frac{1}{2}$–2 amperes. Do not use greater current. 0.2 g. Cu deposits readily in 1 hour with the rotating anode. Turn switch, after washing the deposit with water and alcohol, to "Off" position, otherwise the current will be broken through remaining units. Each unit has its own switch.

In determining copper in ores, iron is removed by precipitation with NH_4OH.(30% Fe carries down 0.3% Cu on 0.5 g. sample), lead is removed as sulphate and the filtrate electrolyzed for copper.

Copper is removed from the electrode by means of HNO_3.

Fig. 26a' by courtesy of the Denver Fire Clay Company.

CERIUM AND THE OTHER RARE-EARTH METALS

Atomic number	Name	Symbol	Atomic weight	Colour of sesquioxide	Absorption spectrum	Solubility of double sodium sulphates
57	Lanthanum	La	139.0	White	None	Cerium group: Double sodium sulphates insoluble in saturated sodium sulphate solution.
58	Cerium	Ce	140.25	White	None	
59	Praseo-dymium	Pr	140.9	Greenish-yellow	Strong	
60	Neodymium	Nd	144.3	Lilac	Strong	
61	. . . .	. . .	. . . .	. . . .	. . . .	
62	Samarium	Sa	150.4	Yellowish-white	Strong	
63	Europium	Eu	152.0	Pale pink	Faint	Terbium group: Double sodium sulphates sparingly soluble in saturated sodium sulphate solution.
64	Gadolinium	Gd	157.3	White	In the Ultra-violet	
65	Terbium	Tb	159.2	White		
39	Yttrium	Y	88.7	White	None	Yttrium group: Double sodium sulphates readily soluble in saturated sodium sulphate solution.
66	Dysprosium	Dy	162.5	White	Strong	
67	Holmium	Ho	163.5	Pale yellow	Strong	
68	Erbium	Er	167.7	Rose	Strong	
69	Thulium	Tm	168.5	White	Moderate	
70	Ytterbium	Yb	173.5	White	None	
71	Lutecium	Lu	175.0	White	None	

All the metals in the above list have now been characterised with certainty as elements. The results of X-ray analyses of the elements indicate that another rare-earth metal, and only one, of atomic number 61, still remains to be discovered. Element 72 or celtium, discovered by Urbain in 1911, is not a rare-earth element but an analogue of zirconium; it appears that Coster and Hevesy's hafnium (1923) is identical with celtium. When a mixture of praseodymium and neodymium is referred to, the old term didymium (symbol Di) is often used for the sake of brevity.

Only a few of the rare-earth metals have so far been prepared in the pure state, but an alloy of the cerium metals, prepared from the oxides which form a by-product in the extraction of thoria from monazite, is used under the name of Mischmetall as a reducing agent for preparing a number of metals from their oxides. Cerium (and, to a very limited extent, didymium) nitrate is used in the gas-mantle industry. Cerium or Mischmetall forms the basis of the pyrophoric alloys for cigarette and gas lighters. Cerium glass (Crookes'

W. R. Schoeller, Ph.D., and A. R. Powell.
Metallurgical chemists, London, England.

glass) containing ceria and various other oxides, whilst transparent to light rays, intercepts the ultra-violet and a large proportion of the heat rays. A small quantity of yttria is consumed in the manufacture of filaments for Nernst lamps. To a limited extent, cerium compounds are employed in medicine; they have been proposed for use as catalysts, in dyeing, tanning and photography.

TABLE OF THE MORE IMPORTANT RARE-EARTHS MINERALS

Name	Composition	Solvent
Monazite	Phosphate of the ceria earths and thoria.	H_2SO_4
Xenotime	Phosphate of the yttria earths.	H_2SO_4
Gadolinite	$2BeO \cdot FeO \cdot Y_2O_3 \cdot 2SiO_2$.	HCl or H_2SO_4
Cerite	Hydrated silicate of ceria earths, lime and ferrous oxide.	HCl or H_2SO_4
Euxenite/Polycrase	Titanocolumbates of rare earths (chiefly yttria) containing uranium.	$NaHSO_4$ or HF
Samarskite	Columbate of the yttria earths, lime and ferrous oxide, containing uranium.	$NaHSO_4$ or HF
Yttrotantalite	Tantalate of yttria, lime and ferrous oxide.	$NaHSO_4$ or HF

The only mineral of considerable commercial importance is monazite; this is due, not to its content of rare earths, but of thoria, monazite forming the raw material for the gas-mantle industry. Ceria and the other earths of the cerium sub-group are obtained in large quantities as by-products in the extraction of thoria from monazite. Yttria is extracted from gadolinite.

QUALITATIVE ANALYSIS

Detection of the Rare Earths in the Systematic Procedure for Qualitative Analysis. The rare earths are precipitated as hydroxides by ammonia and are therefore found, together with alumina, ferric hydroxide, etc., in the precipitate obtained in the ordinary course of analysis. The precipitate is dissolved in hydrochloric acid and the solution evaporated to dryness or carefully neutralised with ammonia. In the former case the residue is dissolved in 150 cc. of dilute hydrochloric acid (30 cc. of strong acid per litre) and the solution, warmed to 60° C., treated with 30 cc. of saturated oxalic acid solution. A white crystalline precipitate, obtained either at once or, in the case of small quantities, after standing over night in a warm place, consists of the oxalates of some or all of the following oxides: the rare earths, thoria, and scandia. The washed oxalate precipitate is brought into solution by digestion with fuming nitric acid of sp. gr. 1.5 and subsequent evaporation to dryness; the residue of nitrates is converted into chlorides by the usual evaporation to dryness with hydrochloric acid. The chlorides are dissolved in water and the solution boiled with sodium thiosulphate: thoria and scandia are precipitated. The filtered solution is treated with excess of ammonia: a white flocculent precipitate, which discolors on exposure to air if cerium is present, indicates the presence of the rare earths.

For further characterisation the precipitate is collected, washed with hot water, dissolved in hydrochloric acid, and the solution saturated with solid sodium sulphate: if a white crystalline precipitate forms, the cerium group is present. The filtrate is again treated with ammonia when the earths of the yttrium group are precipitated as hydroxides. The metals of the terbium group are found in both fractions.

The double sulphate precipitate of the cerium metals may be treated as follows: it is converted into hydroxides by digestion with caustic soda and the washed precipitate is dissolved in dilute nitric acid. The solution is boiled with potassium bromate and lump marble by which procedure the cerous nitrate is oxidised to ceric salt which is then hydrolysed with precipitation of a yellow basic salt whilst the nitrates of the other rare earths remain unchanged. The cerium precipitate is filtered off and dissolved in strong nitric acid; the solution, on addition of citric acid, ammonia, and hydrogen peroxide, gives a yellow to brown coloration characteristic of cerium. The filtrate from the cerium precipitate is treated with ammonia, the precipitate dissolved in nitric acid and the liquid evaporated to dryness: the dry residue is heated to 450° C. for half a minute, the cooled mass extracted with water and the liquid filtered. The more basic lanthanum is chiefly in the filtrate, the less basic didymium in the residue. Lanthanum is characterised by addition of a few drops of acetic acid, iodine solution in potassium iodide, and ammonia, drop by drop until the brown color of the iodine nearly disappears; a blue color, appearing on gentle warming, indicates lanthanum. The residue of basic didymium nitrate is ignited: the brown color of the resulting oxides, due to a higher oxide of praseodymium, proves that metal to be present; it is invariably associated with neodymium.

The precipitate of hydroxides of the yttria group, obtained above, may be dissolved in nitric acid and the solution tested for its absorption spectrum (erbia, etc.). Yttria is by far the most common earth of this group.

Detection of Cerium. If it is desired to prove the presence of cerium only, the procedure is the same as that outlined above up to the point where the ammonia precipitate of rare-earth hydroxides has been obtained in the filtrate from the thoria-scandia precipitation by thiosulphate. The ammonia precipitate is collected, washed, and dissolved in nitric acid with a little hydrogen peroxide, if necessary. The solution is taken to dryness and the residue dissolved in water. Cerium is detected in this solution in the presence of the other rare earths by colour reactions based on oxidation to the ceric state:

1. Boiling with lead peroxide and nitric acid causes the appearance of a deep yellow color.

2. The solution is treated with tartaric acid, made ammoniacal, and boiled with a few cc. of hydrogen peroxide: a dark brown coloration develops.

3. Ammonium acetate and hydrogen peroxide produce a yellow precipitate or colour.

4. In the presence of large quantities of the other rare earths a precipitate rich in cerium may be obtained by addition of hydrogen peroxide to the solution, followed by ammonia added drop by drop until a slight permanent precipitate forms: cerium will be concentrated in the precipitate, the colour of which will be orange to reddish-brown.

QUANTITATIVE ANALYSIS

Decomposition of Minerals. Concentrated sulphuric acid is the most generally applicable solvent; it can be used for practically all rare-earth minerals except those containing tantalic, columbic and titanic acids, which are best decomposed by fusion with sodium bisulphate (pyrosulphate). When the mass resulting from the decomposition by sulphuric acid is dissolved in water, it should be borne in mind that the sulphates of the rare-earth metals, whilst soluble in cold water, yield sparingly soluble compounds at higher temperatures; care should therefore be taken in keeping the temperature of the solution below 20° C. The use of ice-cold water is recommended as this counteracts the rise in temperature resulting from the hydration of the strong acid; alternatively the cold pasty mass is simply stirred into a large excess of cold water. Again, the double sulphates of the cerium group and potassium are more insoluble than the corresponding sodium salts; sodium bisulphate is therefore to be preferred to the potassium salt as a flux. If the sodium bisulphate melt begins to solidify before decomposition is complete, it is allowed to cool somewhat and treated with 1 cc. of strong sulphuric acid; on renewed heating the bisulphate is regenerated. In the case of decomposition by hydrofluoric acid, the rare earths are obtained as a residue of insoluble fluorides which must be converted into sulphates by evaporation with strong sulphuric acid. Cerite and gadolinite are silicates decomposable by hydrochloric or sulphuric acid.

SEPARATIONS

1. **Separation of the Rare Earths (and Thoria) from Other Metals.** The oxalate precipitation outlined under " Qualitative Analysis " is also a quantitative method for separating the rare earths from base metals of the ammonia precipitate. The chloride solution of the ammonia precipitate, if acid, is evaporated to dryness and the residue dissolved in 0.3N hydrochoric acid so that the concentration of the oxides present is less than 1 gram per 60 cc. The solution is warmed and precipitated with a saturated solution of oxalic acid so that the final solution contains 3 grams of free acid per 100 cc. The oxalate precipitate, which is flocculent at first, soon becomes crystalline; it is allowed to stand over night in a warm place, filtered off, and washed with a 1% solution of oxalic acid. If the base metals are present in fairly large quantity, it is best to repeat the precipitation: the oxalates are digested with caustic soda, the hydroxides filtered off, washed and dissolved in nitric acid; the solution is evaporated to dryness and the residue treated as before. The solubility of the oxalates of the rare earths in water is of the order of 0.4 to 1 milligram of anhydrous oxalate per litre. They are more soluble in dilute mineral acids, but the solubility is reduced by the presence of free oxalic acid.

Another method of separating the rare earths and thoria from base metals of the ammonia group consists in precipitating them as fluorides from a weakly acid chloride solution. This method is less reliable and convenient than the oxalate precipitation.

2. **Separation of the Rare Earths from Thoria.** The oxalate precipitate obtained as just described is digested with fuming nitric acid and the solution evaporated to dryness. The mass is moistened with water and dried again; it is then dissolved in 100 cc. of water, 10 grams of ammonium nitrate are

added, the solution is heated to 60° C., and the thorium precipitated by addition of 20 cc. of hydrogen peroxide. The gelatinous precipitate is collected, washed with 2% ammonium nitrate solution, and dissolved in nitric acid. The solution is evaporated to dryness and the precipitation of the thorium repeated as before. The combined filtrates, which contain the rare earths, are treated with ammonia and the washed precipitate is redissolved in nitric acid; the solution is then ready for the separation of cerium from the other earths by one of the following methods.

3. **Separation of Cerium from the Rare Earth Metals.** The methods for the quantitative separation of ceria from the other earths are based on the fact that cerium alone can be oxidised to the quadrivalent ceric state, and ceric hydroxide being a weak base, its salts are easily hydrolysed.

(a) James and Pratt's method (Jour. Amer. Chem. Soc., 1911, 33, 1326) consists in boiling the neutral nitrate solution with potassium bromate. As soon as bromine vapours begin to be evolved two or three small lumps of marble are added to the solution which is boiled for some hours, water being added from time to time to replace that lost by evaporation. When a qualitative test by hydrogen peroxide in a small part of the filtered solution shows only a trace of cerium, the boiling is interrupted and the marble removed and washed. The precipitate is left to settle, filtered off, and washed with a 5% solution of ammonium nitrate. For very accurate work the precipitate is dissolved in nitric acid with the help of a little hydrogen peroxide, the solution evaporated to dryness and the whole process repeated. The combined filtrates containing the rare earths and a trace of cerium are boiled with more bromate and marble until the peroxide test for cerium is negative. The small precipitate so obtained requires purification by reprecipitation.

(b) In Brinton and James' method (Jour. Amer. Chem. Soc., 1919, 41, 1080) the cerium is oxidised by means of potassium bromate and the metal in its quadrivalent state precipitated as iodate. The cold solution of the rare-earth nitrates (50 cc.) containing not more than 0.15 gram of ceria is treated with 25 cc. of strong nitric acid and 0.5 gram of potassium bromate. This is followed by 10 to 15 times the quantity of potassium iodate (in the form of a solution containing 100 grams of iodate and 333 cc. of strong nitric acid per litre) required to precipitate ceric iodate $Ce(IO_4)_4$, added slowly with constant stirring. The bulky flocculent precipitate is left to settle in the cold and filtered off; the beaker and precipitate are washed once with a solution containing 8 grams of potassium iodate and 50 cc. of strong nitric acid per litre and the precipitate is rinsed back into the beaker, thoroughly stirred up with 50 cc. of the washing solution, and again filtered on the same paper. The precipitate is once more returned to the beaker with hot water and dissolved by boiling with not more than the necessary amount of strong nitric acid (20 to 25 cc.). The solution is again treated with potassium bromate (0.25-gram) and the same amount of potassium iodate as for the first precipitation. The precipitate is left to settle completely, filtered off on the paper previously used, returned to the beaker and stirred up with 50 cc. of washing solution, and finally washed several times on the filter with the same solution. Filter and precipitate are then boiled in the original beaker with 5 to 8 grams of oxalic acid and 50 cc. of water until iodine vapours cease to be evolved. After standing for several hours the cerous oxalate is collected, washed with cold water, ignited strongly in a platinum crucible and weighed as CeO_2.

ESTIMATION OF CERIUM

1. **Gravimetric.** In technical practice cerium is almost invariably estimated volumetrically as this procedure obviates its separation from the other rare earths. The only compound obtained in gravimetric work is the dioxide, CeO_2, which is formed when cerium salts of the volatile acids are strongly ignited. Thus in the preceding paragraph ceric iodate was converted into cerous oxalate which was ignited to dioxide. The routine process consists in obtaining the cerium in a nitric acid solution. For this purpose, ceric hydroxide as obtained in the bromate-marble separation from the other rare earths is dissolved in nitric acid and hydrogen peroxide and the solution is evaporated to dryness. The residue is dissolved in water and a few drops of nitric acid, and an excess of oxalic acid is added to the hot solution which is allowed to stand on the water bath until the precipitate becomes crystalline. The cerous oxalate is filtered off next day, washed with slightly acidulated water and ignited gradually at first, then more strongly over a blast burner. The residue is weighed as CeO_2. Factors: CeO_2 to $Ce = 0.8142$; CeO_2 to Ce_2O_3 $= 0.9536$. The oxalate for the purpose of gravimetric estimation should be precipitated in a solution free from hydrochloric acid and alkali metals or high results will be obtained.

2. **Volumetric.** The volumetric estimation of cerium is both quick and accurate and is therefore applied in commercial and industrial practice. The other rare earths need not be eliminated. Of the several processes published only Metzger's bismuthate method (Jour. Amer. Chem. Soc., 1909, 31,. 523) will be described here as it is easily carried out and reliable in the presence of thoria, zirconia and the rare earths.

Cerium and the rare earths are first separated from the base metals by precipitation as oxalates (see under " Separation from other Metals "); the precipitate is ignited to oxides and these are converted into sulphates by heating with 20 cc. of strong sulphuric acid. The cooled mass is poured into 100 cc. of ice-cold water and the solution treated with 2 grams of ammonium sulphate and 1 gram of sodium bismuthate, boiled for 5 minutes, cooled and filtered through a Gooch crucible on asbestos which has previously been washed with an acidified solution of potassium permanganate followed by water. The crucible is washed with about 100 cc. of 2% sulphuric acid, a small excess of ferrous sulphate solution is added to the filtrate and this excess is measured by N/20 permanganate. According to the equation

$$2Ce(SO_4)_2 + 2FeSO_4 = Ce_2(SO_4)_3 + Fe_2(SO_4)_3,$$

140.25 parts of cerium are equivalent to 55.84 parts of iron; hence the iron factor of the permanganate multiplied by 2.5116 gives the cerium factor.

ESTIMATION OF THE OTHER RARE EARTHS

1. Estimation of the Sum of the Rare Earths Other than Ceria. This is preceded by the separation of ceria by one of the methods described. If the bromate-marble method has been used, the solution is acidified with 5 cc. of nitric acid per 100 cc. In the case of the iodate-bromate method the combined filtrates are diluted somewhat and precipitated with caustic soda. The precipitate is filtered off, well washed, and dissolved in nitric acid. The solution is evaporated to dryness and the residue taken up in water containing 5 cc. of nitric acid per 100 cc.

Either solution is treated whilst hot with an excess of saturated oxalic acid solution. After standing over night the precipitate is collected and washed with acidulated water. For accurate work reprecipitation is recommended. The precipitate is rinsed off the paper and evaporated twice to dryness with fuming nitric acid. The residue is dissolved in 5% nitric acid and the oxalate precipitation repeated as before. The washed precipitate is dried and ignited strongly, first in the air then in hydrogen, cooled, and weighed as sesquioxides. If the reduction is omitted, a colored residue and high result will be obtained caused by higher oxides of praseodymium and terbium. The rare earths should be weighed without delay in covered crucibles as they absorb carbon dioxide and water like lime.

2. Estimation of the Cerium and Yttrium Groups. The weighed sesquioxides obtained in the preceding paragraph are converted into sulphates by heating with strong sulphuric acid and the mass is heated to 450° C. to expel the excess of acid. After fumes cease to be evolved, the mass is allowed to cool and dissolved in ice-cold water to give a nearly saturated solution. An equal bulk of water is added followed by powdered sodium sulphate until the solution is saturated with the salt. The precipitated double sulphates of sodium and the cerium metals are filtered off next day and washed with a saturated solution of sodium sulphate; they are digested with hot caustic soda, the hydroxides filtered off, dissolved in hydrochloric acid and the solution evaporated to dryness. The residue is dissolved in the minimum of water, an equal bulk of water added and the precipitation with sodium sulphate is repeated. The combined filtrates are treated with ammonia, the precipitate is dissolved in nitric acid and the yttria earths precipitated by addition of oxalic acid. The precipitate after ignition is weighed as " oxides of the yttrium group." The double sulphate precipitate is once more converted into hydroxides, which are dissolved in nitric acid and converted into oxalates as before. The precipitate is ignited in air, then in hydrogen and weighed as " oxides of the cerium group."

The above separation is more or less approximate. The mixture of cerium group oxides contains all the lanthana, praseodymia, neodymia and nearly all the samaria as well as the greater part of the terbia group and a minute quantity of the yttria earths; the remainder of the terbium group is not precipitated by sodium sulphate and remains with the yttria group. The separate estimation of the terbium group is neither feasible by ordinary analytical methods nor of any practical importance.

3. Estimation of the Individual Earths. The separation of a mixture of the rare earths into its individual constituents is not a practical proposition. It can be accomplished by fractional crystallisation but the process requires large quantities of material and the fractionations must be repeated many hundreds of times. As a general rule the solubility of the salts of the rare earths increases with their atomic weight, whereas the basicity of the oxides decreases .

TECHNICAL METHODS

1. Estimation of Cerium in Minerals. This is effected in the following manner by means of the methods described in this chapter: The ore is decomposed by heating with sulphuric acid and the sulphates are dissolved in ice-cold water. The solution is precipitated with oxalic acid, the oxalate precipitate strongly ignited and the oxides again converted into sulphates. These are brought into solution as before and the cerium is estimated volumetrically by the bismuthate method.

2. Colorimetric Estimation of Ceria in Thoria, Thorium Nitrate, etc. (Benz, Z. angew. Chem., 1902, 16, 300.) The oxide is fused with sodium bisulphate, the melt leached with cold water and the sulphates converted into nitrates by precipitation with ammonia, filtration and solution in nitric acid. The liquid is evaporated to dryness, the residue dissolved in water, the solution filtered to remove filter fibres and the filtrate heated with a known excess of citric acid and hydrogen peroxide. The solution is neutralised with ammonia and an aliquot part transferred to a colorimeter tube. The standard is made by mixing cerium nitrate solution (1 cc. = 0.0005 gram of ceria) with the same quantities of pure thorium nitrate, citric acid, hydrogen peroxide and ammonia as are contained in the test. Equality of tints is produced by withdrawal of a suitable amount of liquid from the solution showing the deeper colour.

Cerium in thorium nitrate is detected by heating a solution of 5 grams of the salt with ammonia and hydrogen peroxide. A yellow colour developing in the precipitate proves the presence of cerium. The quantitative estimation is carried out colorimetrically as described above.

3. Detection and Estimation of Didymium in Cerium Salts. (Dede, Chem. Zeit., 1923, 47, 82.) When the light from mercury vapour is made to pass through crystals or solutions of didymium salts, the absorption band covers the two yellow lines of mercury, and the solution or crystals appear blue. The reaction is sensitive: thus crystals of cerium nitrate, in which the didymium band can hardly be detected in a 20 mm. layer of 5 per cent. solution, appear distinctly blue in mercury vapour light. In dilute solutions the absorption of the mercury lines is proportional to the concentration of the didymium, which can therefore be estimated colorimetrically without a spectral photometer.

INDIUM, SCANDIUM, THALLIUM

INDIUM

At. wt.   *Symbol : In.*

Occurrence. Indium is a rare element found in many deposits of zinc blende, in some tungsten ores, in most tin ores and, sometimes, in pyrites, siderite and galena. It is often found concentrated as an indium-gallium alloy of low melting point in the residues from zinc retorts.

Detection. For the detection of indium in an ore, *e.g.*, in zinc blende, the mineral is dissolved in hydrochloric acid with the addition, if necessary, of a little nitric acid, the excess of which is expelled by boiling with hydrochloric acid. Digestion of the filtered solution with metallic zinc precipitates all the indium together with lead, copper, cadmium, etc. The precipitate is dissolved in nitric acid and the solution evaporated with sulphuric acid to fumes. The cold residue is extracted with water and the lead sulphate, etc., filtered off. The filtrate is treated with an excess of ammonia, boiled and filtered. The precipitate is dissolved in the minimum of hydrochloric acid, the solution neutralised with ammonia, an excess of sodium bisulphite added and boiling continued for some time. A white microcrystalline precipitate indicates indium. As a confirmatory test the precipitate is dissolved in a few drops of hydrochloric acid and a platinum wire is dipped into the solution and held in the Bunsen flame. A bright blue colour showing two characteristic bright blue lines (λ 4511.55 and λ 4101.95) when viewed through the spectroscope confirms the presence of indium.

Indium oxide is a yellow powder soluble in acids and readily reduced on charcoal to metallic globules with a yellow incrustation. Indium salts give a yellow precipitate with hydrogen sulphide in neutral, alkaline or acetic acid solutions and a white gelatinous precipitate with ammonia insoluble in excess but soluble in cold caustic alkali and re-precipitated on boiling.

Quantitative Analysis. For the quantitative determination of indium in zinc blende and retort residues as much as 100 grams may have to be taken. The procedure follows the same lines as that described above with the usual precautions to render the separations quantitative. As the bisulphite precipitate obtained contains a little iron, it is dissolved in hydrochloric acid and the solution treated with sodium acetate and a little α-nitroso-β-naphthol dissolved in 50% acetic acid. The precipitate is filtered off, washed with cold water and the indium recovered from the filtrate by boiling with a very slight excess of ammonia. The indium hydroxide is collected, washed well with 5% ammonium nitrate solution, ignited wet at a temperature below 800° and the residue weighed as In_2O_3. Factor to indium: 0.8271.

Chapter by W. R. Schoeller, Ph.D., and A. R. Powell, metallurgical chemists, London, England.

SCANDIUM

At. wt.   *Symbol : Sc.*

Occurrence. Scandium is very widely distributed in minute quantities in almost all rocks but it is found in appreciable quantity only in very few minerals. Mica, cassiterite and wolframite from some localities, euxenite and keilhauite contain a few tenths per cent. of scandia; wiikite contains a little more than 1% and thortveitite, the only mineral in which scandium is an essential constituent, over 30%.

Detection. For the detection of scandia in a mineral the solution obtained as described later is treated as usual for the separation of the ammonia group. The precipitate is dissolved in hydrochloric acid, the solution neutralised and boiled with sodium thiosulphate and the precipitate of basic scandium thiosulphate redissolved in hydrochloric acid. After filtering off the sulphur, scandium (and thorium) oxalates are precipitated by addition of oxalic acid. The precipitate is collected, washed and digested with fuming nitric acid. The solution is evaporated to dryness, the residue dissolved in a little water, the solution poured into a 20% solution of ammonium tartrate and the mixture boiled with ammonia; the gradual separation of a crystalline precipitate indicates the presence of scandium.

In the examination of mixtures of the rare earths, which often contain scandia, the mixed oxides obtained from the oxalates by ignition (see Rare Earth chapter) are dissolved in hydrochloric acid and the solution is boiled with sodium silicofluoride; a heavy gelatinous precipitate of scandium fluoride is obtained, free from all but traces of rare earths.

Minute quantities of scandia are readily detected in minerals by examination of the spark spectrum. The scandium must first be concentrated together with any rare earths by precipitation with oxalic acid. The precipitate is ignited to oxides and these are dissolved in hydrochloric acid; the filtrate from the oxalates should be examined also for scandia. The most intense lines of this element lie between 3500 and 3700 Å.; they are 3572.72, 3613.98, 3630.90, 3642.96.

Quantitative Analysis. Minerals containing rare earths are decomposed by treatment with hydrochloric or sulphuric acids or by fusion with sodium pyrosulphate (see Rare Earth chapter). Euxenite, and similar minerals containing titanium, tantalum and columbium, may be decomposed by moistening the finely ground sample with a little water, adding 40% hydrofluoric acid, and, after the violent action has subsided, evaporating to dryness on the water-bath. The residue is extracted with boiling water and the insoluble material, which contains all the scandium, is filtered off, using a rubber funnel, and dissolved by heating with strong sulphuric acid. Wolfram, which sometimes contains scandia, is decomposed by digesting the slimed mineral with aqua regia until the residue is a pale yellow colour.

The solution obtained by any of the above methods is saturated with hydrogen sulphide to remove heavy metals and the filtrate is boiled to expel excess of the gas and oxidised with nitric acid. The hot solution is treated with a slight excess of ammonia, the precipitate collected, well washed with 2% ammonium nitrate solution and dissolved in the minimum of 1 : 1 nitric acid, and the solution (50 to 60 cc.) treated hot with 20 cc. of a saturated solution of oxalic acid added slowly with vigorous stirring. After standing over night the precipitate is filtered off, washed with a dilute solution of oxalic acid, rinsed back into the original beaker and digested with fuming nitric acid until completely dissolved. The nitrates are converted into chlorides by two evaporations with hydrochloric acid to dryness. The residue is dissolved in 200 cc. of hot water and the solution boiled with 10 grams of sodium thiosulphate for one hour. The precipitate is filtered off, washed with hot water and digested with hydrochloric acid. The insoluble residue of sulphur is removed by filtration and well washed with hot water. The filtrate is evaporated to dryness, the residue dissolved in water and the thiosulphate precipitation repeated using 5 grams of this salt in a bulk of 150 cc. The second precipitate is treated as before and the filtered solution neutralised with ammonia. One drop of 1 : 1 hydrochloric acid is added and the solution (50 cc.) is allowed to drop slowly into 50 cc. of a boiling 20% solution of ammonium tartrate. Boiling is continued, with the addition of a few cc. of ammonia from time to time, for 40 minutes. After standing until cold, the solution is filtered; the precipitate is washed with cold 5% ammonium tartrate solution and ignited wet in a platinum crucible. The residue is weighed as scandia.

In the case of wolfram, the solution obtained by digesting the mineral with aqua regia is boiled with 5 grams of sodium silicofluoride for 1 hour. The precipitate of scandium fluoride is collected, washed with hot water and heated in a platinum dish with strong sulphuric acid until copious fumes are evolved. After cooling, the residue is dissolved in water and the scandium is precipitated as scandium ammonium tartrate as described above.

THALLIUM

At. wt.   *Symbol : Tl.*

Occurrence. Thallium occurs in quantity only in a few rare minerals such as crookesite, $(Cu, Tl, Ag)_2Se$, and lorandite, $TlAsS_2$. It occurs in small quantities however associated with the alkali metals, zinc, iron and lead, and is usually recovered from the flue dust from burning iron pyrites.

Detection. Thallium forms two series of salts. The reactions of thallous salts recall those of lead: thus thallous chloride is a white crystalline sparingly soluble compound and thallous iodide is a yellow crystalline compound almost insoluble in water. Thallous sulphide is produced as a brown precipitate by addition of sodium sulphide to an alkaline or acetic acid solution of a thallous salt. Thallous sulphate and hydroxide are both soluble like the corresponding potassium salts whilst the chloroplatinate and cobaltinitrite are insoluble yellow and red compounds respectively. Thallous chromate like lead chromate is insoluble in water and dilute acetic acid.

Thallic salts are obtained by oxidising thallous salts with chlorine, permanganate and similar oxidising agents; they are all readily hydrolysed yielding brown basic salts and give a brown gelatinous precipitate with ammonia, resembling ferric hydroxide.

In the usual course of qualitative analysis thallium passes into the zinc-nickel group, being precipitated with ammonium sulphide in ammoniacal solution. The most satisfactory procedure is to evaporate the solution to be tested with sulphuric acid to remove the lead and then saturate the filtrate with hydrogen sulphide. After filtration the excess of the precipitant is removed by boiling and the solution oxidised with nitric acid. Iron, aluminum, the zinc-nickel group and the alkaline earths are removed by boiling with sodium carbonate and the filtrate is treated with ammonium sulphide. If a brown precipitate is formed, it is dissolved in a little dilute sulphuric acid and the solution, after boiling, is treated with potassium iodide. A yellow crystalline precipitate proves the presence of thallium. A little of this substance held in the loop of a platinum wire in the Bunsen flame gives a characteristic green flame which, viewed through the spectroscope, shows a broad green line at λ 5350.7. For the isolation of minute quantities of thallium from a solution free from heavy metals a mixture of potassium iodide and antimony trichloride is added, when a characteristic orange-red micro-crystalline precipitate of $3TlI \cdot 2SbI_3$ is formed.

Separation from Other Metals and Gravimetric Determination. Practically all materials containing thallium yield the metal to treatment with hydrochloric or nitric acids or to *aqua regia*. The solution obtained in all cases is evaporated with strong sulphuric acid until copious fumes are evolved; after cooling, the residue is dissolved in water and the solution filtered for the removal of lead sulphate. If the amount of thallium present in the original material is very small, a large amount should be taken for the analysis; after removal of the nitric acid used in dissolving the substance the solution is boiled with several strips of zinc until effervescence ceases and basic salts begin to be precipitated. A few cc. of hydrochloric acid are added and the solution is filtered whilst a feeble evolution of hydrogen is maintained from the zinc. The precipitate is washed, digested with hot 20% sulphuric acid and the solution containing the thallium filtered.

The sulphate solution obtained by either of the above methods is neutralised with sodium carbonate, 5 grams of potassium cyanide and a further 2 grams of sodium carbonate are added and the mixture warmed on the water-bath. The precipitate is filtered off and washed with a 1% solution of sodium carbonate. To the filtrate, which contains all the thallium, are added a few cc. of colourless ammonium sulphide : the precipitated thallous sulphide is collected on a filter, washed with a very dilute solution of ammonium sulphide and dissolved in the minimum of hot 10% sulphuric acid. After boiling to expel hydrogen sulphide, the solution is neutralised with sodium carbonate, diluted to 100 cc. and treated at 80° C. with a 10% solution of potassium iodide drop by drop until no further precipitate forms. One gram of the solid salt is added in excess and the assay set aside over night. The precipitate is collected in a tared Gooch crucible, washed first with the minimum quantity of a cold 1% solution of potassium iodide, then with 82% alcohol until the washings cease to react for iodide, dried at 110° C. and weighed as TlI. Factor to thallium: 0.6165.

Volumetric Method. The sulphide obtained as described above is dissolved in 4 cc. of hydrochloric acid and a little water. The solution is boiled to expel hydrogen sulphide, cooled, diluted to 60 cc. and titrated with N/10 potassium permanganate standardised against pure thallous chloride. (Hawky, Jour. Amer. Chem. Soc., 1907, 29, 300.)

Electrolytic Method. Thallous sulphide is dissolved in nitric acid and the filtered solution is evaporated to dryness on the water-bath. The residue is dissolved in 100 cc. of water and transferred to a platinum dish with roughened surface to act as anode. Ten cc. of absolute alcohol or acetone are added and the solution is electrolysed for 6 to 8 hours at 60° C. with a disc of iridoplatinum, rotating at 300 revolutions per minute, as cathode. A current of 0.5 amp. at 2.5–3 volts is used. The thallic oxide deposited on the anode is washed with distilled water, dried at 160°, and weighed. Factor to thallium: 0.8947.

Uses. The chief use of thallium appears to be in the manufacture of artificial stones and optical glass of very high refracting power. In the analysis of these products the thallium is rendered soluble by fusion with sodium carbonate followed by solution of the melt in water and evaporation with sulphuric acid. The resulting mass is dissolved in water, the solution filtered from insoluble matter and the thallium determined in the filtrate as described above.

CHLORINE

Cl_2, *at.wt.* **35.46;** *D.* (*air*), **2.491;** *m.p.* **−101.5°;**[1] *b.p.* **−33.6° C.;** *oxides,* Cl_2O, ClO_2, Cl_2O_7.

DETECTION

Free Chlorine. The yellow gas is recognized by its characteristic odor. It liberates iodine from iodides; it bleaches litmus, indigo, and many organic coloring substances.

Chlorides. **Silver Nitrate Test.** In absence of bromides and iodides, which also form insoluble silver salts, silver nitrate precipitates from solutions containing chlorides white, curdy, silver chloride, AgCl (opalescent with traces), soluble in NH_4OH (AgBr slowly soluble, AgI difficultly soluble), also soluble in concentrated ammonium carbonate (AgBr is very slightly soluble; AgI is insoluble). Silver chloride turns dark upon exposure to light.

Free Hydrochloric Acid. **Manganese Dioxide, Potassium Permanganate,** and certain oxidizing agents liberate free chlorine gas when added to solutions containing free hydrochloric acid. The gas passed into potassium iodide liberates free iodine, which produces a blue solution with starch.

Concentrated Sulphuric Acid added to chlorides and heated liberates HCl gas, which produces a white fume in presence of ammonium hydroxide.

Detection in Presence of Cyanate, Cyanide, Thiocyanate. An excess of silver nitrate is added to the solution, the precipitate filtered off and boiled with concentrated nitric acid to oxidize the cyanogen compounds and the white precipitate, silver chloride, subjected to the tests under chlorides to confirm the compound.

Detection in Presence of Bromide and Iodide. About 10 cc. of the solution is neutralized in a casserole with acetic acid, adding about 1 to 2 cc. in excess, and then diluting to about 6 volumes with water. About half a gram of potassium persulphate, $K_2S_2O_8$, is added and the solution heated. Iodine is liberated and may be detected by shaking the solution with carbon disulphide, which is colored blue by this element. Iodine is expelled by boiling, the potassium persulphate being repeatedly added until the solution is colorless. Bromine is liberated by adding 2 or 3 cc. of dilute sulphuric acid and additional persulphate. A yellowish-red color is produced by this element. Carbon disulphide absorbs bromine, becoming colored yellowish red. Bromine is expelled with additional persulphate and by boiling. The volume of the solution should be kept to about 60 cc., distilled water being added to replace that which is expelled by boiling. When bromine is driven out of the solution, the silver nitrate test for chlorides is made. A white, curdy precipitate, soluble in ammonium hydroxide and reprecipitated upon acidifying with nitric acid, is produced, if chlorides are present.

Chapter contributed by Wilfred W. Scott.

If Chlorates are Present. The halogens are precipitated with silver nitrate, the precipitate dissolved with zinc and sulphuric acid and the solution treated as directed in the preceding paragraph.

Test for Hypochlorite. Potassium hypochlorite, KClO, shaken with mercury forms the yellowish-red compound Hg_2OCl_2,[1] which does not form with the other potassium salts of chlorine, i.e., KCl, $KClO_2$, $KClO_3$, $KClO_4$.

Hypochlorites decolorize indigo, but do not decolorize potassium permanganate solutions. If arsenious acid is present, indigo is not decolorized until all of the arsenious acid has been oxidized to the arsenic form.

Tests for Chlorite. Potassium permanganate solution is decolorized by chlorites. (The solution should be dilute.)

A solution of indigo is decolorized, even in presence of arsenious acid (distinction from hypochlorites).

Detection of Chlorate. The dry salt heated with concentrated sulphuric acid detonates and evolves yellow fumes.

Chlorates liberate chlorine from hydrochloric acid.

Perchlorate. The solution is boiled with hydrochloric acid to decompose hypochlorites, chlorites and chlorates. Chlorides are removed by precipitation with silver nitrate, the filtrate evaporated to dryness, the residue fused with sodium carbonate to decompose the perchlorate to form the chloride, which may now be tested as usual.

ESTIMATION

The determination of chlorine is required in a large number of substances. It occurs combined as a chloride mainly with sodium, potassium and magnesium. Rock salt, NaCl, sylvine, KCl, carnallite, $KCl \cdot MgCl_2 \cdot 6H_2O$, matlockite, $PbCl_2 \cdot PbO$; horn silver, AgCl, atacamite, $CuCl_2 \cdot 3Cu(OH)_2$, are forms in which it is found in nature. Chlorine is determined in the evaluation of bleaching powder. It is estimated in the analysis of water.

Preparation and Solution of the Sample

In dissolving the sample the following facts should be borne in mind: Although chlorides are nearly all soluble in water, silver chloride is practically insoluble (100 cc. dissolves 0.000152 gram at 20° C.); mercurous chloride is nearly as insoluble as silver chloride (0.00031 gram); lead chloride requires heat to bring it into solution (in cold water only 0.673 gram soluble per 100 cc. of water). Chlorides of antimony, tin, and bismuth require free acid to keep them in solution. Hydrochloric acid increases the solubility of silver, mercury, lead, antimony, bismuth, copper (Cu'), gold and platinum, but decreases the solubility of cadmium, copper (Cu''), nickel, cobalt, manganese, barium, calcium, strontium, magnesium, thorium, sodium, potassium and ammonium chlorides.

Chlorine gas is most readily dissolved in water at 10° C. (1 vol. H_2O dissolves 3.095 vols. Cl). Boiling completely removes chlorine from water.

Hypochlorites, chlorites, chlorates, and perchlorates are soluble in water.

The chlorine may be present either combined or free. In the combined state it may be present as free hydrochloric acid or as a water-soluble or insoluble salt.

14   [1] Prescott and Johnson, Qual. Chem. Anal.   D. Van Nostrand Co.

Water-soluble Chlorides. Chlorides of the alkali or alkaline earth groups may be treated directly with silver nitrate upon making slightly acid with nitric acid, the chlorine being determined either gravimetrically or volumetrically according to one of the procedures given later. It is convenient to work with samples containing 0.01 gram to 1 gram of Cl. The sample is dissolved in about 150 cc. of water, made acid with nitric acid with about 5 to 10 cc. in excess of the point of neutralization, should the sample be alkaline. Then the chlorine combined as chloride is determined as directed later.

If the water solution contains a chloride of a heavy metal which forms basic salts (e.g., stannic, ferric, etc., solutions), or which may tend to reduce the silver solution, it is necessary to remove these by precipitation with ammonium hydroxide, or by sodium hydroxide, or potassium carbonate solution. The salt is dissolved in water and acidified with HNO_3, adding about 10 cc. in excess, for about 150 cc. of solution. (This excess HNO_3 should be sufficient to oxidize substances which would tend to reduce the silver reagent; e.g., $FeSO_4$, etc.) Ammonia solution (free from chloride) is added in sufficient quantity to precipitate the heavy metals iron, manganese, aluminum, etc. The mixture is filtered and the residue washed several times with distilled water. Chlorine is determined in the filtrate by acidifying with HNO_3 as directed above.

Water-insoluble Chlorides. The chloride may frequently be decomposed by boiling with sodium carbonate solution. Many of the minerals, however, require fusion with sodium carbonate to prepare them for solution; e.g., apatite, sodalite, etc. Silver chloride may also be decomposed by fusion.

Silver Chloride. The sample is mixed with about three times its weight of Na_2CO_3 and fused in a porcelain crucible until the mass has sintered together. The soluble chloride, NaCl, is leached out with water, leaving the water-insoluble carbonate of silver, which may be filtered off. The filtrate is acidified with HNO_3 and chlorine determined as usual.

Chlorine in Rocks. The finely ground material is fused with about five times its weight of potassium carbonate. The melt is extracted with hot water, cooled and the solution acidified with nitric acid (methyl orange indicator), and the solution allowed to stand several hours (preferably over night). If silicic acid precipitates, the solution is treated with ammonia and boiled, filtered and the filter washed with hot water. The cooled filtrate is acidified with nitric acid and chlorine determined as usual. If silicic acid does not separate, the addition of ammonia may be omitted and chlorine determined in the solution.

Free Chlorine. Free chlorine may be determined volumetrically according to the procedure given under this section. If it is desired to determine this gravimetrically, a definite amount of the chlorine water is transferred by means of a pipette to a flask containing ammonia solution and the mixture heated to boiling. The cooled solution is acidified with nitric acid and the chloride precipitated with silver nitrate according to the standard procedure given on page 151.

NOTE. Free chlorine cannot be precipitated directly, as the following reaction takes place: $6Cl+6AgNO_3+3H_2O=5AgCl+AgClO_3+6HNO_3$.
Reaction of chlorine with ammonia: $2Cl+2NH_4OH=NH_4Cl+NH_4OCl+H_2O$. When the solution is boiled, NH_4OCl breaks down, e.g, $3NH_4OCl+2NH_3=3NH_4Cl+N_2+3H_2O$.

Chlorine in Ores and Cinders. One hundred grams of the finely ground ore or cinder are placed in a 500-cc. flask, containing 300 cc. of strong sulphuric acid (Cl-free). The flask is shaken to mix the sample with the acid and then connected with an absorption apparatus, containing distilled water or dilute caustic solution. The sample is gradually heated, the distillation flask resting upon a sand bath. After two hours, which is sufficient to expel all the chlorine as hydrochloric acid, the contents of the absorption tubes are filtered, if free sulphur is present (sulphide ores), nitric acid added and the filtrate brought to boiling to oxidize any SO_2 that may be present. Chlorine is precipitated according to the standard procedure on page 148.

During the run the distilling flask should be shaken occasionally to prevent caking. Suction applied at the absorption end of the apparatus and a current of air swept through the system aids in carrying over the HCl into the water or NaOH.

Determination of Halogens in Organic Compounds. Method of Carius [1]

Organic compounds may be decomposed by heating with strong nitric acid at high temperatures under pressure. If this heating is conducted in the presence of silver nitrate, the halogen hydride, formed by the action of nitric acid on the organic compound, is converted to the silver halide. This is weighed, or the excess $AgNO_3$ titrated (p. 149). Arsenic, phosphorus, and sulphur are oxidized to arsenic, phosphoric, and sulphuric acids, the metals present being converted to nitrates.

Procedure. About 0.5 to 1 gram of powdered silver nitrate is introduced, by means of a glazed paper funnel, into a heavy-walled, bomb-glass tube, which is sealed at one end and is 50 cm. long, 2 cm. in diameter and about 2 mm. thickness of wall. About 30 cc. of strong nitric acid (96%), free from chlorine, are introduced by means of a long-stemmed funnel, to avoid wetting the upper portion of the tubing. About 0.1 gram of the organic substance, contained in a small bore, thin wall, glass tube closed at one end (4–5 cm. long), is introduced into the bomb tube, inclined to one side. The small tube should float in the nitric acid, as it is important that the material should not come in contact with nitric acid until the bomb has been sealed, as loss of halogen is apt to occur with open tubes. The upper end of the bomb is softened in the blast-lamp flame, drawn out to a thick-walled capillary tube and fused.

When cold, the bomb is wrapped in asbestos paper, shoved into an iron tube of a bomb furnace and the heat turned on. The heating is so regulated that the temperature is raised to 200 ° C. in three hours. If a higher temperature is necessary, the heating should be such as to cause a rise of 50° C. in three hours. Substances of the aromatic series require eight to ten hours heating at 250 to 300° C., while aliphatic substances may be decomposed at 200° C. in about four hours.[2] Occasionally it is necessary to relieve the pressure in a tube after heating to 200° C., before taking to a higher temperature, by softening the tip of the cooled bomb in a flame, allowing the accumulated gas to blow out, resealing and again heating to the desired temperature. Evidence of crystals or drops of oil in the glass tube indicate incomplete decomposition. When the bomb is cooled, it is removed by

[1] Ann. d. Chem. u. Pharm. (1865), **136**, p. 129.
[2] Treadwell and Hall, Anal, Chem., J. Wiley & Son. P. C. R. Kingscott and R. S. G. Knight, Methods of Quant. Org. Anal. Longmans, Green & Co. (1914), Clowes and Coleman, Quant. Chem. Anal., P. Blakiston's Son & Co., 1900.

taking out the iron sheath from the furnace and inclining it so that the glass capillary tip slides partly out of the tube. (The eyes should be protected by goggles.) The point of the capillary is held in the flame until the tip softens and the gas pressure is released by blowing through a passage in the softened glass. When the gas has escaped, a scratch with a file is made below the capillary and the tip broken off by touching the scratch with a hot glass rod. The contents of the bomb are poured out into a beaker, the tube washed out with water and the combined solution made to about 300 cc. This is heated to boiling and then allowed to cool. The halide precipitate is filtered through a Gooch crucible, then dried and weighed. or by titrating the excess $AgNO_3$ by Volhard's method, the halide may be estimated.

If pieces of glass should be present, the precipitates, AgCl or AgBr, are dissolved, in ammonium hydroxide, filtered and reprecipitated by acidifying with nitric acid. AgI may be dissolved by means of dilute sulphuric acid and zinc. The excess zinc is removed, the glass washed free of iodine, dried and weighed and its weight subtracted from the original impure AgI, giving the weight of the pure silver iodide.

Lime Method for Determination of Halogens in Organic Matter

A layer of lime (free from chloride), about 6 cm. long, is introduced into a difficultly fusible glass tube, closed at one end (35 cm. long and with 1 cm. bore), followed by 0.5 gram of the substance, and 6 cm. more of the lime. The substance is thoroughly mixed by means of a copper wire with a spiral end. The tube is nearly filled with lime, and in a horizontal position, gently tapped to cause the lime to settle and form a channel above the layer. The tube is placed in a small carbon combustion furnace. The heat is turned on, so that the front end of the tube is heated to dull redness and then the end containing the substance. When the organic matter has been decomposed, the tube is cooled and the contents transferred to a beaker and the lime dissolved in dilute nitric acid (Cl-free). The carbon is filtered off and the halogen determined as usual in the filtrate.

Should a sulphate be present in the mixture, organic matter will reduce it to a sulphide, so that AgS will be precipitated along with the halides. To prevent this, hydrogen peroxide is added to the solution which should be slightly alkaline. The mixture is boiled to remove the excess of H_2O_2 and is then acidified with nitric acid, the solution filtered and the halide determined in the filtrate.

With substances rich in nitrogen, some soluble cyanide is apt to form. The silver precipitate containing the halides and the cyanide is heated to fusion. The residue is now treated with zinc and sulphuric acid, the metallic silver and the paracyanogen filtered off and the halides determined in the filtrate.

Sodium Peroxide Method

Organic compounds may be decomposed by sodium peroxide in an open crucible without recourse to a sealed tube, as is required by the Carius method. The following is the procedure outlined by Pringsheim.[1]

About 0.2 gram of substance in a small steel crucible is treated with a calculated quantity of sodium peroxide.[2] The crucible should be only two-thirds of its height full; this is put in a large porcelain crucible, in which a little cold water is carefully placed, so that the steel crucible stands out 1 to 2 cm. This latter crucible is covered with its own cover, in which is a hole through which an iron wire heated to redness can be introduced with the object of starting the combustion. As soon as the combustion is completed the whole is plunged into the water in the larger crucible. The porcelain crucible is covered with a watch-glass and heated gently until the whole mass is dissolved. This point is recognized when no more bubbles are given off and when there are no more particles of carbon which have escaped combustion. The steel crucible is then removed and washed carefully; the solution is filtered and treated with an excess of sulphurous acid (to neutralize the alkaline liquid, and to reduce the oxidized products: bromic, iodic acids, etc.). The solution is acidulated with nitric acid, then made to a volume of about 500 cc., and the halogens precipitated with silver nitrate and the precipitate washed, dried and weighed as usual.

Chlorine and Chlorides in Gas. The gas is bubbled through dilute sodium hydroxide contained in one or more cylinders, gas wash bottle type, measuring the gas by means of a dry meter, placed after the cylinders. The meters are protected from moist gas by passing this through sulphuric acid and an asbestos filter, loosely packed. Aliquot portions of the sodium hydroxide are now examined for chlorine by acidifying with nitric acid and adding silver nitrate. If only traces are present the turbidity of the solution is compared with standards made up with known amounts of sodium chloride dissolved in water. The comparisons may be made conveniently in Nessler tubes. To different quantities of the standard made up to a convenient volume, silver nitrate reagent is added and the solution diluted to 50 or 100 cc. The unknown, placed in a Nessler tube, is treated with nitric acid and silver nitrate and matched with the standards, after dilution to the same volume adopted for the standards.

[1] C. N., 1905, **91**, 2372, 215.
[2] Charge of sodium peroxide is judged as follows:

Per cent C and O in material.	Amount of sugar to add.	Amount of Na$_2$O$_2$ required.
Over 75	0	18 times wt. of sub.
30 to 75	0	16 times wt. of sub.
25 to 50	$\frac{1}{2}$ the wt. of sub.	16 times wt. of sub.
Below 25	An equal weight	16 times wt. of sub.

SEPARATIONS [1]

Separation of Chlorine and the Halides from the Heavy Metals. Halides of the heavy metals are transposed by boiling their solutions with sodium carbonate, the heavy metals being precipitated as carbonates and the halides going into solution as sodium salts.

Separation of Halides from Silver and from Silver Cyanide. The silver salt is treated with an excess of zinc and sulphuric acid, the metallic silver and the paracyanogen filtered off, and the halides determined in the filtrate.

Separation of the Halides from One Another. Separation of Chlorine from Iodine. The method depends upon the fact that nitrous acid sets iodine free from dilute solutions containing a mixture of halogen salts, bromides and chlorides being unaffected.

The solution of the chloride and the iodide in an Erlenmeyer flask is diluted to 400 cc. and 10 cc. of dilute sulphuric acid, 1 : 1, are added. The gas from 2 grams of sodium nitrite is passed into the solution at the rate of about five bubbles per second. [2] (Pure sodium or potassium nitrite may be added directly to the solution in the flask.) The liberated iodine is now completely expelled by boiling until the evolving steam no longer reacts upon litmus paper. Should a determination of iodine be desired the evolved gas is absorbed in a hydrogen peroxide sodium hydroxide solution according to the procedure described under iodine.

The contents of the flask are treated with silver nitrate and the precipitated silver chloride determined as usual.

Separation of Chlorine and Bromine from Iodine. The procedure is similar to the separation of chlorine from iodine with the exception that a more dilute solution is necessary to prevent the volatilization of bromine with the iodine.

The neutral solution containing the halogens is diluted to about 700 cc. and about 2 to 3 cc. of dilute sulphuric acid, 1 : 1, are added and a sufficient amount of pure sodium nitrite introduced or nitrous acid gas passed into the solution as directed above. The solution is boiled until colorless and until the evolved steam no longer acts upon litmus paper. About twenty minutes' boiling after the color of iodine has disappeared from the flask will completely eliminate iodine; in this case, however, water should be added to the flask to replace that evaporated before the solution has been reduced to a volume of less than 600 cc.

For determination of bromine in the residue remaining in the flask, see the chapter on this subject, page 93 and page 154.

[1] Attention is called to "Methods in Chemical Analysis," by F. A. Gooch for useful information on the separation of the halogens.

[2] Nitrous acid is generated by addition of dilute H_2SO_4 to $NaNO_2$, the acid being added drop by drop through a thistle tube with glass stop-cock.

GRAVIMETRIC
Silver Chloride Method

The procedure is based on the insolubility of silver chloride in dilute nitric acid solution, the following reaction taking place, M representing a monatomic element:

$$M.Cl + AgNO_3 = M.NO_3 + AgCl$$

From the equation it is evident that 35.45 grams of chlorine require 169.9 grams of silver nitrate. In practice it is best to add about 20% excess of the silver salt.

Equivalents: 1 gram Cl, 4.79 grams $AgNO_3$, 3.043 grams Ag.

Reagents. *Silver Nitrate Solution.* Make up a solution containing 4.8 grams $AgNO_3$ per 100 cc. of distilled water, or dissolve 3.05 grams of silver foil in 10 cc. of dilute nitric acid (1 : 1.6) and make up to 100 cc. 1 cc. of this reagent will precipitate 0.01 gram of chlorine, or 0.0404+ gram AgCl.

Dilute Nitric Acid. One vol. HNO_3 to 1.6 vols. H_2O (dist.).

Procedure. *Soluble Chlorides.* *Preparation of the Solution.*

1. Weigh 0.4 to 0.5 gram of the salt on a watch glass or in a weighing bottle and transfer to a beaker or an Erlenmeyer flask.

2. Dissolve in 100 cc. of water and add 2 cc. of dilute nitric acid.

3. **Precipitation.** Calculate roughly the cc. of the silver nitrate reagent that are required to precipitate the chlorine in the sample. If it is an unknown, consider the chlorine in the material to be about 50%. Run the determination in duplicate. The first will be a guide to the amount of silver nitrate solution required.

4. Add the silver nitrate from a burette, drop by drop, to approximately the quantity calculated to be necessary, stirring the solution during the addition. Allow the precipitate to settle and add a few more drops of the reagent, and continue the addition as long as a precipitate forms with the reagent. Now add about 20% in excess.

5. Heat to boiling, covering the beaker with a watch glass.

6. If the solution is still cloudy, stir vigorously. If the solution is in an Erlenmeyer flask, the mixture may be shaken. This will cause the finely divided silver salt to coagulate so that the solution will settle out clear. Avoid exposing the precipitate to strong light, as this will cause the exposed surface to decompose into the subchloride Ag_2Cl and liberate chlorine.

7. *Filtration.* Two processes are commonly practiced.

Filter Paper Method. Decant the clear solution into a filter. Test the filtrate to be sure all chlorine has been precipitated by adding a drop or so of silver nitrate solution. Wash the precipitate in the beaker twice by decantation and then transfer to the filter and wash until free of chlorides. The wash water should contain 1 cc. of nitric acid per 100 cc. of distilled water.

8. Dry the filter with its contents at 105° C. either in the funnel in which the operation was conducted or on a watch glass. It is advisable to protect the sample from dust by placing a large filter over the material.

9. Remove as much of the precipitate as possible from the filter, placing the silver chloride on a 4-inch square of glazed paper.

10. Ignite the filter allowing the ash to drop into a weighed porcelain crucible. Add a drop of nitric acid and a drop of hydrochloric acid to react

with any reduced silver.   Place the silver salt (on the glazed paper) in the crucible and gently ignite until the chloride begins to melt.   If the AgCl appears dark, moisten with HCl and again apply heat to expel the free acid.

11. Cool in a desiccator.

12. Weigh as silver chloride.   (wt. crucible + AgCl) − wt. crucible = wt. AgCl.   The compound contains 24.74% of chlorine.

$$AgCl \times 0.2474 = Cl.$$

13. Calculate the per cent chlorine from the weight of sample taken.

Gooch Crucible Method.   8[a].   Prepare a Gooch crucible filter with a moderately thick asbestos mat and wash thoroughly with distilled water containing 1 cc. nitric acid per 100 cc. of water.

9[a].   Dry the crucible with mat in an oven at 100° C., then heat gently over a free blue flame.   Cool in desiccator and weigh.

10[a].   Wash the silver chloride, first by decantation, then transfer to the Gooch crucible, which has been placed in position on a suction flask.   (Consult directions for the preparation of a Gooch and the method of filtration under " Laboratory Apparatus and Manipulation " in the introductory chapter.) Gentle suction is applied and the precipitate is washed free of chloride by repeated additions of the wash water containing the nitric acid.

11[a].   Place the crucible in an oven for 15 minutes or more and dry at 105° C. Now place over a free blue flame and heat until the silver begins to melt. Cool in a desiccator.

12[a].   Weigh.   The weight in excess of that due to the crucible is due to silver chloride.

13[a].   Calculate percent chlorine as stated under the " filter paper method," 12 and 13.

NOTE.   The silver chloride may be removed from the crucible by adding a piece of zinc and dilute sulphuric acid to the residue.   AgCl is soluble in ammonia.

VOLUMETRIC METHODS

Determination of Chlorine in Acid Solution, Silver Thiocyanate Ferric Alum Method

The method, devised by Volhard,[1] is applicable to titration of chlorine in acid solutions, a condition frequently occurring in analysis, where the Silver-Chromate Method of Mohr cannot be used. The method is based on the fact that when solutions of silver and an alkali thiocyanate are mixed in presence of a ferric salt, the thiocyanate has a selective action towards silver, combining with this to form thiocyanate of silver, any excess of that required by the silver reacting with the ferric salt to form the reddish-brown ferric thiocyanate, which color serves as an indication of the completion of the reaction. An excess of silver nitrate is added to the nitric acid solution containing the chloride, AgCl filtered off, and the excess of silver titrated with the thiocyanate in presence of the ferric salt.

Copper (up to 70%), arsenic, antimony, cadmium, bismuth, lead, iron, zinc, manganese, cobalt, and nickel, do not interfere, unless the proportion of the latter metals is such as to interfere by intensity of the color of their ions.

Preparation of Special Reagents. *N/10 Ammonium or Potassium Thiocyanate Solution.* About 8 grams of ammonium or 10 grams of potassium salt are dissolved in water and diluted to one liter. The solution is adjusted by titration against the N/10 silver nitrate solution. It is advisable to have 1 cc. of the thiocyanate equivalent to 1 cc. of the silver nitrate solution. Owing to the deliquescence of the thiocyanates the exact amount for an N/10 solution cannot be weighed.

N/10 Silver Nitrate. This solution contains 10.788 grams Ag or 16.989 grams $AgNO_3$ per liter. The silver nitrate salt, dried at 120° C., or pure metallic silver may be taken, the required weight of the latter being dissolved in nitric acid and made to volume, or 17.1 grams of the salt dissolved in distilled water and made to 1000 cc. The solution is adjusted to exact decinormal strength by standardizing against an N/10 sodium chloride solution, containing 5.846 grams of pure NaCl per liter.

Ferric Indicator. Saturated solution of ferric ammonium alum. Should this not be available, $FeSO_4$ may be oxidized with nitric acid, and the solution evaporated with an excess of H_2SO_4 to expel the nitrous fumes. A 10% solution is desired. Five cc. of either of these reagents are taken for each titration.

Pure Nitric Acid. This should be free from the lower oxides of nitrogen. Pure nitric acid is diluted to contain about 50% HNO_3, and boiled until perfectly colorless. The reagent should be kept in the dark. Dilute nitric acid does not interfere with the method.

Procedure. To the solution, containing 0.003 to 0.35 gram chlorine, in combination as a chloride, is added sufficient of the pure HNO_3 to make the solution acid and about 5 cc. in excess. To the solution, diluted to about 150 cc., is added an excess of standard silver nitrate reagent. The precipitated AgCl

[1] Liebig's Ann. d. Chem., **190**, 1; Sutton, " Volumetric Analysis," 10 Ed.   Z. Anorg. Chem., **63**, 330, 1909.

is filtered off and washed free of silver nitrate. The filtrate and washings are combined and titrated with standard thiocyanate.[1]

The filtrate from the precipitated chloride is treated with 5 cc. of the ferric solution,[2] and the excess silver determined by addition of the thiocyanate until a permanent reddish-brown color is produced. Each addition of the reagent will produce a temporary reddish-brown color, which immediately fades as long as silver uncombined as thiocyanate remains. The trace of excess produces ferric thiocyanate, the reddish-brown color of this compound being best seen against a white background. From this titration the amount of silver nitrate used by the chloride is ascertained.

One cc. $N/10$ $AgNO_3 = 0.00355$ gram Cl or 0.00585 gram NaCl.

Volumetric Determination of Chlorine in a Neutral Solution, Silver Chromate Method

The method, worked out by Fr. Mohr, is applicable for determination of chlorine in water or in neutral solutions containing small amounts of chlorine; the element should be present combined as a soluble chloride. Advantage is taken of the fact that silver combines with chlorine in presence of a chromate, Ag_2CrO_4 being decomposed as follows: $Ag_2CrO_4 + 2NaCl = 2AgCl + Na_2CrO_4$. When all the chlorine has gone into combination as AgCl, an excess of K_2CrO_4 immediately forms the red Ag_2CrO_4, which shows the reaction of $AgNO_3$ with the chloride to be complete.

Reagents. *Tenth Normal Silver Nitrate Solution.* Theoretically 16.989 grams $AgNO_3$ per liter are required. In practice 17.1 grams of the salt are dissolved per 1000 cc. and the solution adjusted against an $N/10$ NaCl solution containing 5.846 grams NaCl per liter.

Potassium Chromate. Saturated solution.

Procedure. To the neutral solution are added 2 or 3 drops of the potassium chromate solution. A glass cell [3] (or a 50-cc. beaker) is filled to about 1 cm. in depth with water tinted to the same color as the solution being titrated. The cell is placed on a clear glass plate half covering the casserole containing the sample. The standard silver solution is now added to the chloride solution from a burette until a faint blood-red tinge is produced, the red change being easily detected by looking through the blank, colored cell.

One cc. $N/10$ $K_2CrO_4 = 0.003546$ gram Cl.

NOTES. Chlorides having an acid reaction ($AlCl_3$) are treated with an excess of neutral solution of sodium acetate and then titrated with silver nitrate.

Elements whose ions form colored solution with chlorine are precipitated from the solution by sodium hydroxide or potassium carbonate, and the filtrate, faintly acidified with acetic acid, is titrated as usual.

---

[1] Time is saved by filtering, through a dry filter paper, only a portion of the mixture made to a definite volume, and titrating an aliquot portion. The first 10–15 cc. of the filtrate are rejected.

[2] Upon addition of the ferric solution no color should develop. If a reddish or yellowish color results, more nitric acid is required to destroy this. The amount of nitric acid does not affect results when within reasonable limits.

[3] Depré, Analyst, **5**, 123; also, Systematic Handbook of "Volumetric Analysis," F. A. Sutton.

Free hydrochloric acid is neutralized with ammonium hydroxide and titrated.

It is advisable to titrate the sample under the same conditions as those observed during standardization. The solution should be kept to small bulk and low temperature for accuracy on account of the solubility of the silver chromate.

Free chlorine should be converted to a chloride before titration. This may be accomplished, as stated under preparation of the sample, by boiling with ammonium hydroxide. Free chlorine may be determined by sweeping the gas, by means of a current of air, into a solution containing potassium iodide, the liberated iodine titrated by N/10 thiosulphate, $Na_2S_2O_3$, and the equivalent chlorine estimated.

Volumetric Determination of Free Chlorine

The determination depends upon the reaction $Cl+KI = KCl+I$. The iodine liberated by the chlorine is titrated with $Na_2S_2O_3$ and the equivalent Cl calculated.

Procedure. A measured amount of the chlorine water is added to a solution of potassium iodide in a glass-stoppered bottle by means of a pipette, the delivery tip of which is just above the surface of the iodide solution. The bottle is then closed and the contents vigorously shaken. The liberated iodine is titrated with tenth-normal sodium thiosulphate ($2Na_2S_2O_3+I_2 = 2NaI+Na_2S_4O_6$). When the yellow color of the iodine has become faint, a little starch solution is added and the titration completed to the fading out of the blue color.

One cc. N/10 $Na_2S_2O_3 = 0.003546$ gram Cl.

Determination of Hypochlorous Acid in the Presence of Chlorine

The determination depends upon the reactions:

$$2KI+HOCl = KCl+KOH+I_2 \text{ and } 2KI+Cl_2 = 2KCl+I_2.$$

The alkali liberated by hypochlorous acid and the total iodine are determined and the calculations made for each of the constituents.

Procedure. A measured volume of N/10 HCl is added to a potassium iodide solution. To this the sample containing the hypochlorous acid and chlorine are added. The liberated iodine is titrated with N/10 $Na_2S_2O_3$. (The addition of starch is omitted.) The colorless solution is treated with methyl orange indicator and the excess of hydrochloric acid is titrated with N/10 NaOH. The potassium hydroxide, produced by the action of the hypochlorous acid upon the iodide, requires half as much acid for neutralization as the volume of thiosulphate required by the iodine set free by the hypochlorous acid.

Calculation. The cc. back titration with NaOH are subtracted from the total cc. of HCl taken = cc. HCl required by NaOH liberated by HOCl = A. Then $2A$ cc. = cc. $Na_2S_2O_3$ required by the I liberated by HOCl. Cc. $A \times 0.005247$ = gram HOCl. The total $Na_2S_2O_3$ titration minus $2A$ cc. (due to the iodine liberated by HOCl) = cc. $Na_2S_2O_3$ that are required by the iodine liberated by chlorine. The cc. thus required multiplied by 0.003546 = grams chlorine in the sample taken.

[1] Six parts $AgCrO_4$, dissolve in 100,000 parts H_2O at 15.5°.—W. G. Young, Analyst, **18**, 125.

Gravimetric Determination of Chloric Acid, $HClO_3$, or Chlorates, by Reduction to Chloride and Precipitation as Silver Chloride

Reduction of the Chlorate. Among the methods of reduction of chlorates the following deserve special mention: 1. *Reduction with Sulphurous Acid.*[1] 2. Ferrous sulphate. 3. Zinc.

1. About 0.2 to 0.5 gram of the salt is dissolved in 100 cc. of distilled water and either SO_2 gas passed into the solution or sulphurous acid in solution added in excess. The solution is now boiled to expel SO_2 and the chloride precipitated as AgCl in presence of free nitric acid.

2. The sample in 100 cc. of distilled water is treated with 50 cc. of crystallized ferrous sulphate (10% solution), heated to boiling, with constant stirring, and then boiled for fifteen minutes. Nitric acid is added to the cooled solution, until the deposited basic ferric salt is dissolved. The chloride is now precipitated as AgCl, as usual.

3. The dilute chlorate solution is treated with acetic acid until it reacts distinctly acid. An excess of powdered zinc is now added and the solution boiled for an hour. Nitric acid is added to the cooled solution in sufficient quantity to dissolve the zinc remaining. The solution is filtered, if necessary, and the chloride precipitated as usual.

Factors. $AgCl \times 0.855 = KClO_3$, or $\times 0.2474 = Cl$.

NOTE. In absence of cyanides, carbonates and acids decomposed and volatilized by hydrochloric acid, or oxides, hydroxides and substances other than chlorates that may be decomposed or acted upon by this acid, evaporation of the salt with HCl and ignition of the residue, or addition of an excess of ammonium chloride,[2] and subsequent heating will give a residue of chloride, which may be determined as usual and the equivalent chlorate calculated. Method by L. Blangey.

The methods may be used in determining chlorates in presence of perchlorates, only the former being reduced to chlorides. Outline of the procedure is given later.

Gravimetric Determination of Perchloric Acid by Reduction to Chloride

A perchlorate ignited with about four times its weight of ammonium chloride in a platinum dish may be decomposed to chloride. A second treatment is usually necessary to change the salt completely. Platinum appears to act as a catalyser, so must be added in solution if a porcelain crucible is used.

Procedure. About 0.2 to 0.5 gram of potassium perchlorate is intimately mixed with about 2 grams of ammonium chloride in a platinum crucible, the latter then covered with a watch-glass and the charge ignited gently for one and a half to two hours, the temperature being below the fusing-point of the residual chloride (otherwise the platinum would be attacked). A second addition of ammonium chloride is made and the mix again heated as before. The resulting chloride may now be determined as usual.

Factors. $AgCl \times 0.9667 = KClO_4$, $\times 0.2474 = Cl$.

[1] Blattner and Brassuer, Chem. Zeit. Rep., 1900, **24**, 793.
[2] Perchlorates are decomposed by ignition with NH_4Cl in presence of platinum.

Determination of Chlorates and Perchlorates in Presence of One Another

(1) A portion of the sample is treated with about twelve times its weight of ammonium chloride in a platinum dish (or in a porcelain dish with the addition of 1 cc. of hydroplatinic acid), and the mixture heated according to the procedure given for perchloric acid (page 152). The resulting chloride is determined as usual. This is the total chlorine in the sample.

(2) In a second portion the chlorate is reduced by means of SO_2 or $FeSO_4$, according to directions given for determination of chloric acid, and chlorine determined. The chlorine of this portion is subtracted from the total chlorine, the difference multiplied by $3.9075 = KClO_4$. The chlorine of the second portion multiplied by $3.4563 = KClO_3$, or AgCl in (2) subtracted from AgCl of (1) and the difference multiplied by $0.9667 = KClO_4$. AgCl of (2) multiplied by $0.855 = KClO_3$.

Determination of Hydrochloric, Chloric, and Perchloric Acids in the Presence of One Another

(1) **Total Chlorine.** If the determination is made in the valuation of niter a 5-gram sample is fused with about three times its weight of alkali carbonate [1] or calcium hydroxide,[2] in a platinum dish, whereby all the chlorine compounds are converted to chlorides. If the compounds are present as alkali salts, fusion with ammonium chloride in a platinum dish may be made and the total chlorides determined after dissolving the residue in nitric acid.

(2) **Chloride and Chlorate.** If the estimation is being made in niter, 5 grams of the salt are treated with 10 grams of zinc dust (Cl-free) in presence of 150 cc. of 1% acetic acid. The solution is boiled for half an hour, filtered, and the chloride determined. In a mixture of alkali salts of hydrochloric, chloric, and perchloric acids, reduction may be accomplished by passing in SO_2 gas or by adding ferrous sulphate and boiling according to directions given for the determination of chlorate. The chloride now present in the residue is due to the reduced chlorate and to the original chloride of the sample.

(3) The chloride of the sample is determined by acidifying the salt with nitric acid (cold) and precipitating as AgCl.

Perchlorate. The chloride and chlorate in terms of chlorine are subtracted from total chlorine of (1) and multiplied by the factor for the salt desired.

Chlorate. The chlorine of (3) is subtracted from chlorine of (2) and multiplied by the factor for the compound desired.

Chloride. The AgCl of (3) is multiplied by the appropriate factor.

Factors. $AgCl \times 0.2474 = Cl$, or $\times 0.2544 = HCl$, or $\times 0.4078 = NaCl$, or $\times 0.5202 = KCl$.

$AgCl \times 0.855 = KClO_3$, or $\times 0.9667 = KClO_4$.

$Cl \times 3.4563 = KClO_3$, or $\times 3.9075 = KClO_4$, or $\times 2.1027 = KCl$, or $\times 3.0028 = NaClO_3$, or $\times 3.4535 = NaClO_4$, or $\times 1.6486 = NaCl$.

[1] Mennick, Chem. Zeit. Rep., 1898, **22**, 117.
[2] Blattner and Brasseur, Chem. Zeit. Rep., 1900, **24**, 793.

Determination of Chlorine, Bromine, and Iodine in the Presence of Each Other

The procedure is Bekk's modification of Baubigny's method.[1]

Procedure. The halogens are precipitated with an excess of silver nitrate, filtered onto asbestos or glass wool, washed, dried, and weighed as total halogens as silver salts. A second portion is precipitated and the moist, washed silver salts (0.3 to 0.4 gram) are treated with a solution of 2 grams of potassium dichromate in 30 cc. of concentrated sulphuric acid at 95° C., and digested for thirty minutes. By this procedure the iodine is oxidized to hydriodic acid (HIO_3) and chlorine together with bromine is liberated in form of the free halogen. Toward the end of the reaction a stream of air is led through the solution to remove any chlorine and bromine. This is now diluted to 300 to 400 cc., filtered, and the hydriodic acid reduced by adding, drop by drop, with constant stirring, a concentrated solution of sodium sulphite, Na_2SO_3, until a faint odor of SO_2 remains after standing ten minutes. (Under certain conditions an excess may result in a partial reduction of the silver iodide.) The precipitated silver salt is filtered, washed with hot, dilute nitric acid, dried and weighed as AgI. The filtrate containing the silver, formerly with the chlorine and bromine, is treated with potassium iodide in sufficient amount completely to precipitate the silver as AgI. This is filtered, washed and weighed. From the three weights the chlorine, bromine and iodine can be easily calculated.

NOTE. Bekk claims an accuracy within less than 0.15%.

Determination of Free Hydrochloric Acid

In absence of other free acids, hydrochloric acid may be accurately determined by titration with standard alkali. Details for the volumetric analysis of muriatic acid in presence of commonly occurring impurities are given in Volume II in the chapter on Acids.

1 cc. N/1 NaOH = 0.03647 g. HCl.

Determination of Chloride and Cyanide in Presence of One Another

The cyanide is determined by Liebig's method described on page 131. To the neutral solution is added sufficient N/10 silver nitrate to combine with all of the cyanide and chloride present and an excess. The solution is acidified with nitric acid and diluted to a definite volume and a portion filtered through a dry filter. A portion of the filtrate, an aliquot of the whole is titrated with standard thiocyanate solution (page 149) using ferric alum indicator and the excess of the $AgNO_3$ added thus ascertained. From this the amount combined with the CN and Cl is known. The equivalent required by the cyanide is deducted, the difference being due to the chloride present in the solution.

1 cc. N/10 $AgNO_3$ = 0.005203 g. CN, or 0.013022 g. KCN, or 0.003546 g. Cl or 0.005846 g. NaCl, or 0.007456 g. KCl.

Determination of Chloride, Cyanide and Thiocyanate in Presence of One Another

The cyanide is determined by the method of Liebig described on page 131, and the equivalent $AgNO_3$ required recorded = A.

[1] Julius Bekk, Chem. Ztg., **39**, 405–6 (1915). C. A., **9**, 2042, (1915).

An excess of N/10 $AgNO_3$ over that required by CN, CNS and Cl is added and the solution acidified with nitric acid. After making to a definite volume, the solution is filtered through a dry filter, the residue being saved. A portion of the filtrate, an aliquot of the whole solution, is titrated with standard thiocyanate solution using ferric alum indicator (see page 149), the excess of $AgNO_3$ is calculated. The amount combining with CN, CNS and Cl is now known = B.

The silver salts on the filter paper are washed with water and transferred by means of strong hydrochloric acid to a flask and boiled for an hour. The cyanide and thiocyanate are decomposed and dissolve, while the silver chloride remains unchanged. The sulphuric acid formed by oxidation of the thiocyanate is precipitated by barium nitrate (as $BaSO_4$). Without removing the precipitates AgCl and $BaSO_4$ the silver nitrate in this solution is determined by Volhard's method (page 149) and the $AgNO_3$ required by thiocyanic acid and cyanide thus ascertained = C.

By deducting the $AgNO_3$ of (A) from (C) the silver nitrate required by thiocyanic acid is determined.

Deducting the $AgNO_3$ required by CN and CNS (C) from the total $AgNO_3$ required by CN, CNS and Cl is (B) the amount required by chlorine is obtained.

We now have the silver nitrate equivalent of Cl, CN, CNS.

N/10 $AgNO_3$ = 0.005203 g. CN, or 0.005808 g. CNS or 0.003546 g. Cl.

NOTE. In the analysis of compounds containing hypochlorites and chlorides, the conversion of hypochlorites to chlorides by heating with hydrogen peroxide is a great convenience.

For instances in the analysis of bleach liquors, washes, etc., the (OCl) and Cl may be very easily and quickly determined by titrating an aliquot with As_2O_3 and then a similar aliquot with $AgNO_3$ after converting all the OCl to Cl by warming with H_2O_2.

EVALUATION OF BLEACHING POWDER, CHLORIDE OF LIME, FOR AVAILABLE CHLORINE

When chloride of lime is treated with water, it is resolved into calcium hypochlorite, $Ca(OCl)_2$, and calcium chloride, $CaCl_2$. The calcium hypochlorite constitutes the bleaching agent. The technical analysis is confined to the determination of available chlorine, which is expressed as percentage by weight of the bleaching powder.

Procedure. Ten grams of the sample are washed into a mortar and ground with water, the residue allowed to settle and the supernatant liquor poured into a liter flask. The residue is repeatedly ground and extracted with water until the whole of the chloride is transferred to the flask. The combined extracts are made up to 1000 cc.

To 50-cc. portions (0.5 gram) of the solution, 3 to 4 grams of solid potassium iodide and 100 cc. of water are added and the solution acidified with acetic acid. Iodine equivalent to the available chlorine is liberated. This is titrated with N/10 arsenious acid.

One cc. N/10 arsenious acid = 0.003546 gram Cl. This multiplied by 200 = %Cl.

In France the strength is given in Gay-Lussac degrees, e.g., liters of gas evolved by 2 kilograms of bleaching powder, 0° C. and 760 mm. 100° = 31.78% Cl.

CHROMIUM

Cr, *at.wt.* **52.0;** *sp.gr.* **6.92;** *m.p.* **1520°;** *b.p.* **2200° C;** *oxides,* **CrO$_2$;**
Cr$_2$O$_3$, CrO$_3$.

DETECTION

Chromium is precipitated by hydrogen sulphide and ammonium hydroxide
as bluish-green, Cr(OH)$_3$, along with the hydroxides of iron and aluminum
(members of previous groups having been removed). The chromic compound
is oxidized to chromate by action of chlorine, bromine, sodium peroxide, or
hydrogen peroxide added to the substance containing an excess of caustic alkali.
The chromate dissolves and is thus separated from iron, which remains insoluble as Fe(OH)$_3$. The alkali chromates color the solution yellow.

Barium acetate or chloride added to a neutral or slightly acetic acid
solution of a chromate precipitates yellow barium chromate, BaCrO$_4$. Addition
of ammonium acetate to neutralize any free inorganic acid aids the reaction.

Lead acetate produces a yellow precipitate with chromates, in neutral
or acetic acid solutions.

Mercurous nitrate or silver nitrate gives red precipitates with chromates.

Hydrogen peroxide added to a chromate and heated with an acid, such
as sulphuric, nitric, or hydrochloric, will form a greenish-blue colored solution.
Chromates are reduced by hydrogen peroxide in acid solution, the action being
reversed in alkaline solution.

**Reducing agents, hydrogen sulphide, sulphurous acid, ferrous salts,
alcohol** form green chromic salts when added to chromates in acid solution.

Ether shaken with a chromate to which nitric acid and hydrogen peroxide
are added, is colored a transient blue. Oxygen is given off as the color fades.

$$HCrO_4 + 3HNO_3 = Cr(NO_3)_3 + 2H_2O + O_2.$$

Diphenyl carbazide test. To 5 cc. of the solution containing chromium as
chromate, 2 drops of hydrochloric or acetic acid are added, and 1 drop of an acetic
acid solution of diphenyl carbazide (0.2 gram CO (NH·NH·C$_6$H$_5$)$_2$ is dissolved
in 5 cc. glacial acetic acid and diluted to 20 cc. with ethyl alcohol). A violet
pink color is produced in presence of a chromate. Less than 0.0000001 gram
chromium may be detected.

Chromic salts are bluish green; chromic acid is red; chromates, yellow;
bichromates, red; chrome alum, violet.

The powdered mineral, containing chromium, when fused with sodium
carbonate and nitrate, produces a yellow colored mass.

ESTIMATION

Among the substances in which chromium is determined are the following:
Chrome iron or chromite, Cr$_2$O$_3$·FeOMgO; crocoisite, PbCrO$_4$; slags; chromic
oxide, chrome green, in pigments; chromates and dichromates; chrome steel
and ferro-chrome.

Chapter contributed by Wilfred W. Scott.

Preparation and Solution of the Sample

Although powdered metallic chromium is soluble in dilute hydrochloric or sulphuric acid, it is only slightly soluble in dilute or concentrated nitric acid. It is practically insoluble in aqua regia and in concentrated sulphuric acid. Chrome iron ore is difficult to dissolve. It is important to have the material in finely powdered form to effect a rapid and complete solution of the sample. An agate mortar may be used to advantage in the final pulverizing of the substance.

General Procedures for Decomposition of Refractory Materials Containing Chromium. The following fluxes may be used:

A. Fusion with $KHSO_4$ and extraction with hot dilute HCl. The residue fused with Na_2CO_3 and $KClO_3$, 3 : 1, or fusion with soda lime and $KClO_3$, 3 : 1.

B. Fusion with $NaHSO_4$ and NaF, 2 : 1.

C. Fusion with magnesia or lime and sodium or potassium carbonates, 4 : 1.

E. Fusion with Na_2O_2, or NaOH and KNO_3, or NaOH and Na_2O_2. Nickel, iron, copper, or silver crucibles should be used for *E*. Platinum may be used for *A*, *B*, or *C*.

Special Procedures. Materials High in Silica. The finely ground sample, 1 to 5 grams, is placed in a platinum dish and mixed with 2 to 5 cc. concentrated sulphuric acid (1.84), and 10 to 50 cc. of strong hydrofluoric acid added. The solution is evaporated to small volume on the steam bath and to SO_3 fumes on the hot plate. Sodium carbonate is added in sufficient amount to react with the free acid, and then an excess of 5 to 10 grams added and the mixture heated to fusion and kept in molten condition for half an hour. From time to time a crystal of potassium nitrate is added to the center of the molten mass until 1 to 2 grams are added. (*Caution.* Platinum is attacked by KNO_3, hence avoid adding a large amount at any one time.) Chromium and aluminum go into solution in the flux, but iron is thrown out as $Fe(OH)_3$. The cooled fusion is extracted with hot water and filtered from the iron residue. Chromium is in solution together with aluminum. If much iron is present it should be dissolved in a little hydrochloric acid and the solution poured into boiling 10% solution of potassium hydroxide, the cooled solution+$Fe(OH)_3$ precipitate is treated with hydrogen peroxide or sodium peroxide to oxidize any chromium that may have been occluded by the iron in the first precipitate. The mixture is again filtered and the combined filtrates examined for chromium.

Sodium Peroxide Fusion. Chrome Iron Ores. One to two grams of finely pulverized ore are placed in a nickel or iron crucible of 50 to 75 cc. capacity and mixed with 5 to 10 grams of yellow sodium peroxide. (Fresh peroxide is best). The mass is gently heated over a Bunsen burner until it melts. The fusion is kept at a low red heat for about fifteen minutes. About 5 grams more of the Na_2O_2 are added and the fusion heated for about ten minutes more. The cooled fusion is dissolved in a casserole with 100 cc. to 150 cc. of water, more peroxide being added to this solution if it appears purple. The excess of peroxide is decomposed by boiling the solution, and to the caustic solution free from peroxide is added 10 to 15 grams of ammonium carbonate or a sufficient quantity of the salt to neutralize four-fifths of the sodium hydroxide present in the solution, as the strong caustic would otherwise dissolve the filter. The solution is now filtered. The insoluble matter is treated on the filter with dilute sulphuric

15

acid, 1 : 4. If a portion remains insoluble, it is an indication of incomplete decomposition of the ore, and this residue is again fused with peroxide and treated as above. The combined filtrates contain the chromium.

Since chromates are reduced in presence of free acid and peroxide, the latter should be expelled before making the solution acid.[1]

If the chromate is to be precipitated as $BaCrO_4$ or $PbCrO_4$, the solution should be acidified with hydrochloric acid. If the reduced solution is to be titrated with potassium permanganate, it is best to use sulphuric acid in neutralizing the caustic solution. Further directions will be given under the method chosen.

Method for Solution of Iron and Steel. See methods at close of chapter.

SEPARATIONS

Chromium, Iron, and Aluminum. If chromium has been fused with sodium peroxide or carbonate containing a little potassium nitrate, and the fusion extracted with boiling water, most of the chromium goes into solution as a chromate, together with alumina, but some of the chromium is occluded by $Fe(OH)_3$. If the amount of the iron precipitate is appreciable, and warrants the recovery of occluded chromium, it is dissolved in hydrochloric acid and the iron reprecipitated by pouring into a solution of strong sodium hydroxide. Before filtering off the iron hydroxide, a little H_2O_2 is added to oxidize the Cr_2O_3, if accidentally present, and the solution boiled and filtered. The combined filtrates will contain all of the chromium and aluminum.

If chromium is present as a chromic salt, instead of a chromate, it is oxidized to the higher form, by adding peroxide (H_2O_2 or Na_2O_2) to the alkaline solution. Bromine added to this solution or chlorine gas passed in will accomplish complete oxidation.[2] It must be remembered that in acid solutions hydrogen peroxide, sodium peroxide, or nitrites will cause reduction of chromates to chromic salts (exception, see method for solution of steel), so that these should be boiled out of the alkaline solution before making decidedly acid with hydrochloric or sulphuric acids. Since these are difficult, if not impossible, to completely expel from an alkaline solution, after boiling the strongly alkaline solution, dilute sulphuric acid is added until the solution acquires a permanent brown color (nearly acid), acid potassium sulphate, $KHSO_4$, is added, and the boiling continued.[3] This will decompose the bromates and expel bromine, etc., but will not cause the reduction of the chromate, as would a strong acid solution.

Separation of Chromium from Aluminum. This separation is necessary if chromium is to be precipitated as $Cr(OH)_3$. The sodium chromate and aluminate solutions are made slightly acid with nitric acid and then faintly alkaline with ammonium hydroxide, $Al(OH)_3$ is precipitated and chromium remains in solution as a chromate.

---

[1] See Separations.

[2] Br may be added and then NaOH to oxidize Cr and precipitate $Fe(OH)_3$.
Chromic oxide and most of its compounds, except chrome iron stone, may be decomposed by conc. $HNO_3 + KClO_3$ (added in small portions). M. Gröger, Zeitsch. anorg. Chem., **81**, 233–242, 1913.

[3] $KHSO_4$ will not cause reduction of chromates. A. Kurtenacker, Zeitsch. anal. Chem., **52**, 401–407, 1913. The Analyst, **38**, 449, page 387.

GRAVIMETRIC METHODS FOR THE DETERMINATION OF CHROMIUM

Precipitation of Chromic Hydroxide and Ignition to Cr_2O_3[1]

Chromium present as a chromic salt in solution, free from iron and aluminum or elements precipitated as hydroxides, is thrown out of solution by NH_4OH as $Cr(OH)_3$, the precipitate ignited to the oxide, Cr_2O_3,[2] and so weighed. The presence of hydrochloric acid or sulphuric acid does not interfere.

Reduction. If the chromium is already present as the chromic salt, free from iron and alumina, it may be precipitated directly as the hydroxide by addition of ammonia; otherwise, if present as the chromate, as is the case when a separation from iron and alumina has been necessary, and in cases where the chromium has been brought into solution by fusion with an oxidizing reagent, reduction is necessary. This is accomplished by passing SO_2 or H_2S into the slightly acid solution of the chromate, or by adding alcohol to the hydrochloric acid solution and boiling until the solution appears a deep grass green. Twenty cc. of alcohol for every 0.1 gram of Cr has been found to be ample for this reduction. The SO_2 or H_2S should be expelled from solution by boiling, in case either has been used for reduction of the chromate.

Precipitation. *Ammonium hydroxide or ammonium sulphide are added in slight excess and the solution boiled for about ten minutes.* The solution should be slightly alkaline (litmus), otherwise a few drops of ammonia should be added, but not a large excess; the solution will then settle out clear. A cloudy solution results from prolonged boiling when the solution has become acid; on the other hand, a large excess of ammonia will prevent complete precipitation of chromium and the filtrate will be colored pink or violet. *The chromic hydroxide is filtered off on S and S 589 filter paper.* Since the precipitate is apt to be gelatinous it is advisable to wash two or three times by decantation and several times on the paper. The well-drained precipitate and filter is ignited wet in a porcelain or platinum crucible, first over a low flame until the paper has been charred, then over a strong gas flame for about thirty minutes, and finally a blast heat for five minutes. The green residue is weighed as Cr_2O_3.[1]

$$Cr_2O_3 \times 0.6846 = Cr.$$

[1] It is advisable to take such a weight of sample that the ignited Cr_2O_3 does not exceed 0.5 gram in weight.
[2] Cr_2O_3, *mol.wt.*, 152; *sp.gr.*, 5.04; *m.p.*, 2059°; insol. in H_2O, slightly sol. in acids, dark green hexagonal.

Determination of Chromium as Barium Chromate

Chromium, present as a chromate, is precipitated from a neutral or faintly acetic acid solution of an alkali chromate by addition of barium acetate or chloride. The $BaCrO_4$ is gently ignited and weighed. The solution should be free from sulphuric acid or sulphates.

Procedure. The alkali chromate solution is neutralized with nitric acid or ammonia as the case may require, precautions for avoiding reduction having been observed as indicated under Preparation and Solution of the Sample. 10 cc. of $\frac{1}{2}$ N. $BaCl_2$ or $Ba(C_2H_3O_2)_2$ (approx. 10% sol.) are added to the boiling solution for each 0.1 gram of chromium present. The reagent should be added in a fine stream or drop by drop to prevent occlusion of the reagent by the precipitate. The precipitated chromate is allowed to settle on the steam bath for two or three hours and then filtered into a weighed Gooch crucible and washed with 10% alcohol solution. The precipitate is dried for an hour in the oven, then placed in an asbestos ring suspended in a large crucible with cover and thus heated over a low flame, gradually increasing the heat until the outer crucible becomes a dull red. The cover is removed and the heating continued for five minutes, or until the precipitate appears a uniform yellow throughout. High heating should be avoided. The cooled residue is weighed as $BaCrO_4$.

$$BaCrO_4 \times 0.2055 = Cr,$$

$$BaCrO_4 \times 0.3002 = Cr_2O_3.$$

$$BaCrO_4 \times 0.7666 = K_2CrO_4.$$

$$BaCrO_4 \times 0.5807 = K_2Cr_2O_7.$$

NOTES. If the precipitate on the sides of the crucible appears green, it is ignited until the green color disappears.

If sulphates are present, $BaSO_4$ will be precipitated, hence this method could not be used. In this case either reduction to the chromic salt and precipitation of chromium as $Cr(OH)_3$ or a volumetric procedure should be followed.

Oxidize chromium with an excess of hydrogen peroxide in alkaline solution, reduce in acid solution with ferrous sulphate and titrate with permanganate. Decomposition of hydrogen peroxide is accelerated by heat and by presence of sodium sulphate or ferric salts. Salts of nickel cobalt, or manganese, decompose H_2O_2 energetically and lower results are obtained. F. Bourin and A. Senechal. Compt. rend., **157**, 1528–31.

Mercurous Nitrate Method. To the chromate solution (containing 0.2–0.5 g. Ccr) heated to boiling is added 2 grams of Na_2CO_3 and then a saturated solution of pure mercurous nitrate in 1/10 sol. conc. HNO_3 (free from nitrous oxides), added in slight excess. The mixture is boiled until the brown precipitate changes to the orange crystalline form. The precipitate is filtered and washed with hot water, then ignited three hours over a Meeker burner until the weight is constant.

[1] If the filtrate appears yellow, chromate is indicated, the solution should be reduced and the chromium precipitated as $Cr(OH)_3$. If the filtrate is pink, it should be boiled until it appears green and $Cr(OH)_3$ precipitates. These precipitates should be included in the above calculation for chromium.

$BaCrO_4$, *mol.wt.*, 253.47; *sp.gr.*, 4.498; solubility per 100 cc. H_2O, 0.00038[18°] and 0.0043 hot. Soluble in HCl and in HNO_3; yellow rhombic plates.

VOLUMETRIC METHODS FOR THE DETERMINATION OF CHROMIUM

Potassium Iodide Method for Determination of Chromium

Chromium present as a chromate is reduced in acid solution by addition of potassium iodide and the liberated iodine titrated by standard sodium thiosulphate. The method depends upon the following reactions:

$$(a)\ 2CrO_3 + 6KI = Cr_2O_3 + 3K_2O + 6I.$$

$$(b)\ I_2 + 2Na_2S_2O_3 = 2NaI + Na_2S_4O_6.$$

The presence of large quantities of Ca, Ba, Sr, Mg, Zn, Cd, Al, Ni, Co, H_2SO_4, HCl, does not interfere.[1]

Procedure. The alkali chromate solution containing not over 0.17 gram Cr [2] and free from Fe_2O_3, is made nearly acid with H_2SO_4, boiled with 20 cc. of 30% potassium acid sulphate to decompose bromates or expel Br, Cl, or H_2O_2' as the case may require, more $KHSO_4$ being added if necessary. If the solution is not acid it is made so with sulphuric acid and 5 cc. of the acid per 100 cc. of solution is added in excess.[3] About 2 grams of solid potassium iodide are added and, after five minutes, the liberated iodine is titrated with N/10 $Na_2S_2O_3$ solution. When the green color of the reduced chromate begins to predominate over the free iodine color (brownish red) a little starch solution is added and the titration with the thiosulphate continued until the blue color of the starch compound is just destroyed, care being taken not to confuse the green color of the reduced chromium with the blue of the starch.

One cc. of N/10 $Na_2S_2O_3$[4] = 0.001733 gram Cr.

Determination of Chromium by Reduction of the Chromate with Ferrous Salts

The procedure may be used for the determination of chromium in presence of ferric iron and alumina. Hydrochloric or sulphuric acids do not interfere. If hydrochloric acid is present in solution, the $K_2Cr_2O_7$ back titration should be made. In presence of H_2SO_4 either $KMnO_4$ or $K_2Cr_2O_7$ titrations may be made. The method depends upon the reduction of soluble chromates by ferrous salts, the excess being determined by titration.

Reactions. $a.\ 2CrO_3 + 6FeO + xsFeO = Cr_2O_3 + 3Fe_2O_3 + xsFeO.$

$b.$ $xsFeO$ is oxidized by standard oxidizing reagent to Fe_2O_3.

---

[1] Sodium peroxide is generally used for oxidation of chromium. The solution is now neutralized with acid, the iodide added and 10 cc. of strong HCl. The liberated iodine is immediately titrated.

[2] If desired, stronger solution of titration reagents may be used, and consequently a larger sample taken. A normal sol. of $Na_2S_2O_3$ may be used to advantage with 1 gram samples of chromium salts or hydrates, where Cr exceeds 10%.

[3] Sutton recommends for every 0.5 gram $K_2Cr_2O_7$ present to add .5 gram KI and 1.8 gram H_2SO_4 per 100 cc. of solution. If more $K_2Cr_2O_7$ is present, increase the KI and H_2SO_4, but not the water.

[4] If desired, a normal solution of thiosulphate may be used with one gram sample of chromium salts or hydroxides, when the chromium present exceeds 10 per cent.

Procedure. *Reduction.* The sample, containing not over 0.17 gram chromium present as a chromate, is boiled to expel oxidizing reagents according to the method described under the potassium iodide procedure for chromium. The solution is made acid, if not already so, and about 5 cc. conc. H_2SO_4 per 100 cc. of solution, added in excess. Tenth normal ferrous ammonium sulphate solution containing free sulphuric acid is added until the solution changes from yellow through olive green to deep grass green. For every 0.1 gram of chromium about 65 to 70 cc. of N/10 ferrous salt solution should be added. After five minutes, the excess of this reducing reagent is titrated either with permanganate or with dichromate as directed below.

Potassium Permanganate Titration. To be used in presence of free sulphuric acid, free hydrochloric acid being absent.

Tenth-normal potassium permanganate solution is run into the reduced chromate until the green color gives place to a violet tinge. At the end-point the solution appears to darken slightly. A little practice enables one to get this with accuracy. A slight excess of permanganate gives the solution a pinkish color, readily distinguishable in the green. Addition of 3 to 4 cc. syrupy phosphoric acid gives a sharper end-point. The color should hold one minute.

Potassium Dichromate Titration. N/10 $K_2Cr_2O_7$[1] is run into the solution until a drop of the sample placed on a white glazed surface with a drop of potassium ferricyanide reagent no longer gives a blue color.

Calculation. From the total ferrous ammonium sulphate added, subtract the cc. of back titration (the reagents being exactly N/10), the difference gives the cc. of ferrous salt required for chromium reduction. If reagents are not N/10, multiply cc. titrations by factor converting to N/10.

Cc. ferrous ammonium sulphate $\times 0.001733 =$ Cr.

$$Cr_2O_3 + 3O = Cr_2O_6. \quad \therefore \ Cr = 1\tfrac{1}{2}O \ \text{or} \ = 3H; \ \text{hence} \ \tfrac{1}{3} \ \text{mol. wt. Cr per liter} = N \ \text{sol.}$$

Determination of Small Amounts of Chromium [2]

Advantage may be taken of the color produced by chromates in solution in determining small amounts, the depth of color depending upon the amount of chromate in solution. The method possesses the usual disadvantage of colorimetric procedures in that there is always room for doubt as to whether the element sought is entirely responsible for the color of the solution.

Procedure. The solution containing the sample is nearly neutralized with sodium carbonate, the reagent being added until a slight cloudiness results. The solution is now cleared with a few drops of sulphuric acid, and then sufficient excess of a strong solution of sodium thiosulphate added to precipitate aluminum, chromium, manganese, etc. The precipitate is filtered off, dissolved in the least amount of dilute nitric acid, then filtered from the precipitated sulphur and diluted to 300 to 400 cc. Chromium is now oxidized by adding 10 cc. of 0.2% silver nitrate solution, about 10 grams each of ammonium nitrate and persulphate. After boiling for about twenty minutes, sufficient hydrochloric acid is added to decompose any permanganate present and to precip-

---

[1] If desired, a larger sample may be taken and N/5 or N solutions used in titration. It is advisable to titrate chromium salts, e.g., over 1.0% Cr, with normal solutions, so that one gram sample may be taken for analysis.

[2] M. Dittrich, Zeitsch. anorg. Chem., **80**, 171–174, 1913.

itate the silver, and a few cc. added in excess. The solution is again boiled for about ten minutes and then filtered. The filtrate is treated with a little sodium phosphate to repress the color of traces of iron that may be present and made to a definite volume.

The solution may now be compared with a standard solution containing the same amounts of acids, manganese, alumina, etc., as are present in the sample, tenth normal potassium dichromate being run into this standard solution until its color matches that of the sample. The burette reading is taken and the chromium calculated.

One cc. of N/10 K₂Cr₂O₇ =0.00173 gram Cr.

NOTES. Prolonged boiling after addition of hydrochloric acid to the solution of the chromate will cause its reduction. A green tint usually indicates that the chromate has been reduced.

The test may be carried on in the presence of sulphuric, hydrochloric, phosphoric, hydrofluoric, and nitric acids. Alumina, manganese, and small amounts of iron do not interfere.

Organic matter should be destroyed by either calcining the sample or by oxidation by taking to fumes with sulphuric acid. The presence of this prevents precipitation of chromium.

Colorimetric Estimation of Small Amounts of Chromium with Diphenyl Carbazide

Diphenyl carbazide, $CO(NH \cdot NH \cdot C_6H_5)_2$, gives a violet color with chromic salts or chromates in acid solution, the intensity of the color being proportional to the concentration of the chromate. Less than 0.0001 milligram of chromium may be detected by this reagent. The following procedure may be used for determining traces of the element:

1 to 2 grams of the substance, which has been brought into solution with water or acid, is treated with an excess of sodium peroxide to oxidize chromium, and the solution filtered. The filtrate, concentrated to 75–80 cc. is acidified with hydrochloric, so that there is present about 5 cc. of free concentrated acid (sp. gr. 1.19) per 100 cc. of Solution. The Solution transferred to a Nessler tube, is treated with 1 cc. of the reagent, and the color compared with standards containing the same reagents as the sample examined. A colorimeter may be used and comparison made with a standard according to details given for the colorimetric comparison of traces of lead or titanium.

Preparation of Diphenyl Carbazide Reagent. One tenth of a gram of the compound is dissolved in 10 cc. of glacial acetic acid and diluted to 100 cc. with ethyl alcohol.

Diphenyl carbazide may be made by heating a mixture of 15 grams of urea with 50 grams of phenyl hydrazine four hours, finishing at 155° C. The solid product is crystallized three times with alcohol. A light straw-colored product is obtained. A white product is obtained if the urea is cut down to 5 grams, the yield, however, is only 25 per cent of that obtained by the first method and the compound possesses no advantages.

Determination of Chromium in a Soluble Chromate

To a concentrated solution of potassium iodide is added a known amount of the soluble chromate dissolved in a little water. The liberated iodine is now titrated with standard thiosulphate reagent.

$$N/10 \ Na_2S_2O_3 = 0.001733 \text{ g. Cr.}$$

Determination of Chromium in Chromite

About 0.2 g. of the powdered ore is fused with ten times its weight of Na_2O_2 in a porcelain crucible, placed inside a larger crucible. The melt is dissolved in water and the $Fe(OH)_3$ filtered off. The filtrate is evaporated to dryness, the residue taken up with as little water as possible and about two grams of KI added. The solution is now diluted to about 300 cc. and the liberated iodine titrated with standard thiosulphate.

$$N/10 \ Na_2S_2O_3 = 0.001733 \text{ g. Cr.}$$

Rapid Method for Determination of Chromium in Steel[1]

Three grams of steel are dissolved in 100 cc. sulphuric acid 1 to 4. If it does not dissolve rapidly a crystal of $(NH_4)_2(SO_4)_2$ will hasten it. After solution add conc. HNO_3 until iron is oxidized. Boil for one minute, add 10 cc. 5% $AgNO_3$, remove from flame and add crystals of Am. Persulphate until all the chromium is oxidized as shown by the manganese being converted to Permanganate. Boil until persulphate is broken up and then add 5% sod. chloride until permanganate is destroyed as shown by the disappearance of the red color. An excess of salt sol. should be avoided. Cool thoroughly and titrate with ferous sulphate, determining the excess with permanganate. These solutions are standardized on a known sample of chrome steel.

Two determinations run simultaneously will check within .02%.

[1] Method by A. I. Appelbaum, the chemist analyst, J. T. Baker Chem. Co.

DICHROMATE=DIPHENYLAMINE METHOD FOR DETERMIN= ING CHROMIUM IN IRON ORES AND ALLOYS

The method takes advantage of Knop's reaction with diphenylamine (Journal of the American Chemical Society, Feb., 1924) in titration of iron with potassium dichromate, here chromate is titrated with a solution of iron. The procedure is applicable to the determination of chromium in ores, ferrochrome, chrome steels and soluble chromates.

Reagents

Potassium Dichromate. 0.1 N solution.
Ferrous Ammonium Sulphate or Ferrous Sulphate. 0.1 N solution.
Phosphoric-Sulphuric Acid. 150 cc. sulphuric acid (d. 1.84) and 150 cc. phosphoric acid (d. 1.70) diluted to 1000 cc. with water.
Diphenylamine Indicator. 1 g. of the reagent dissolved in 100 cc. of sulphuric acid (d. 1.84). Use 4 drops (0.2 cc.). Deduct 0.1 cc. blank.
Sodium Peroxide. Fresh powder.

Procedure

The amount of the sample should be such as contains between 0.002 to 0.08 g. chromium. The finely powdered material is fused with ten times its weight of sodium peroxide in a nickel or iron crucible. (It appears unnecessary to heat to molten condition, provided the mass sinters.) After heating at dull red heat for ten minutes, the crucible is cooled and then upset in a 400 cc. beaker containing about 100 cc. of water. (The beaker should be immediately covered as the reaction is violent.) The crucible is washed out and removed, and the solution boiled to expel the peroxide. The solution is cooled and dilute sulphuric acid added until the alkali is neutralized and the solution is slightly acid. (Iron hydroxide dissolves, but manganese dioxide remains in suspension.[1]) If manganese is present, it is removed by filtration. To the filtrate 15 cc. of phosphoric-sulphuric acid mixture is now added and, from a burette, a measured excess of standard ferrous ammonium sulphate. (With an excess of ferrous salt the solution turns green.) Four drops (0.2 cc.) of diphenylamine indicator are now added and the excess of ferrous salt titrated with standard potassium dichromate. The green color changes to a blue green and then to an intense blue or violet color. (If the end-point is overrun, titrate back with ferrous sulphate to a green color and repeat the dichromate titration. Convert the reagents to exact equivalents, i.e., terms of 0.1 N solution.) The difference between the cc. of ferrous solution and the dichromate reagent multiplied by the chromium equivalent represents the chromium in the sample. 1 cc. 0.N solution is equivalent to 0.00173 g. chromium.

NOTES. 1. If it is desired to filter off the iron and manganese precipitates, it will be necessary to filter through asbestos or to neutralize the caustic with ammonium carbonate (1.5 times as much as peroxide used in fusion), boil and filter through paper. The ferric hydroxide occludes chromium, hence solution of the iron with acid and reprecipitation is necessary to recover chromium.
2. If the iron precipitate has been dissolved and much manganese is present, the precipitate may be filtered off. Manganese dioxide does not occlude an appreciable amount of chromium.
3. Chrome steels may be dissolved with acid followed by treatment with permanganate to oxidize the iron.

5. It is evident that neither ferric salt nor dichromate alone produce the blue color, but an excess of dichromate in presence of ferric iron. If much chromium is present, the end point may be overrun owing to the depth of color, the excess of dichromate causing a greenish blue color to reappear. Back titration with ferrous sulfate will restore the blue or violet color, and an excess will change the color to blue green.

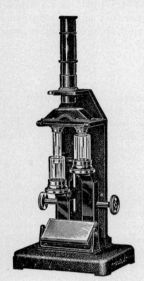 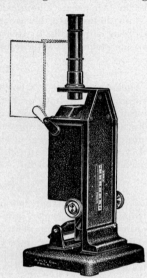

By courtesy of Arthur H. Thomas Company, Philadelphia, Pa.

FIG. 26a.—Colorimeter.

DIRECTIONS FOR THE USE OF A DUBOSCQ TYPE OF COLORIMETER

The mirror is turned so that the two halves of the field appear to be equally illuminated with the cups clean and empty. The solutions are then poured into the cups. The cup containing the standard solution is then lowered to a definite thickness of the standard solution between the bottom of the cup and the end of the plunger. With this movement the half of the field corresponding to the standard solution is seen to darken, while the other half remains luminous and colorless. If the cup containing the unknown solution is now moved in its turn, the two halves of the field are brought to the same intensity, after which the height at which the two liquid columns display this equal absorptive power is read by means of this scale. The proportion of coloring matter in two solutions is inversely proportional to the heights of the two columns necessary to obtain the same intensity of illumination, thus if the standard tube is set at 10 mm., and the solution under examination is the same intensity of color at 20 mm., the latter is just one-half the concentration of the standard. This is usually expressed by the formula:

$$\frac{Color \ of \ test \ solution}{Color \ of \ standard \ solution} = \frac{Height \ of \ standard \ solution}{Height \ of \ solution \ to \ be \ tested}$$

If, therefore, the scale reading is 20 mm. for the standard, and 15 mm. for the solution to be tested, the formula reads: $\frac{20}{15} = 1.33$.

If, for example, the standard solution contains 4 ml. of coloring matter in 100 ml., the solution under test will be found to contain 4 x 1.33 = 5.32 ml. in 100 ml.

DETERMINATION OF CHROMIUM IN STEEL—
CHLORATE METHOD

The following method is recommended for routine by the A.S.T.M.

Reagents

Nitric Acid. 1000 cc. HNO_3 (d. 1.42) and 1200 cc. water.

Potassium-Ferricyanide Indicator. 0.1 g. salt per 50 cc. water. Prepare fresh each day the indicator is used.

Potassium Dichromate. 5 grams of salt per liter. Adjust to 0.1 N.

Potassium Permanganate. 2 grams of salt per liter, standardized against sodium oxalate. The solution adjusted so that 1 cc. is equal to 0.001 g. chromium (1 cc.=.1% per 1 g. sample). 1 g. $Na_2C_2O_4$=0.2584 g. Cr; hence 0.001 g. Cr=0.00387 g. $Na_2C_2O_4$ and 0.1548 g. $Na_2C_2O_4$ should require 40 cc. titration.

Ferrous Sulphate. 25 grams of ferrous ammonium sulphate dissolved in 900 cc. water and 100 cc. (1 : 1) H_2SO_4 added.

Procedure

A sample of 1 gram of steel in a 300 cc. Erlenmeyer flask is dissolved in 30 cc. of the nitric acid and the solution evaporated rapidly to about 15 cc. 50 cc. of strong HNO_3 (d. 1.42) are added and 1 g. of sodium or potassium chlorate. The solution is boiled down to about half the volume. Either of the following procedures are recommended.

1. The solution is diluted with 100 cc. of water, the manganese filtered off and washed with hot water. The filtrate containing the chromium is cooled, diluted to 600 cc., an excess of the ferrous sulphate solution added and the excess titrated with potassium dichromate reagent, using the spot test with potassium-ferricyanide indicator, the spot no longer showing a blue color of ferrous iron.

NOTE. In place of the outside indicator, the editor, W. W. Scott, recommends the use of the internal indicator, diphenylamine (0.2 cc. of a solution l. g. salt per 100 cc. strong H_2SO_4). See method, page 164a.

2. 10 cc. of hydrochloric acid (1 : 1) are added and the solution boiled until the manganese dissolves. The cooled solution is diluted to 300 cc. An excess of ferrous sulphate is added and the excess titrated with standard potassium permanganate solution to a faint pink color.

NOTES. The ferrous ammonium sulphate should be compared with the dichromate or permanganate reagents the day it is used and the ratio of the reagents established.

Oxidation may be effected of the manganese with potassium permanganate added to a boiling solution of the steel dissolved in sulphuric acid (in volume of about 150 cc.), a large excess of the permanganate being avoided, as it is claimed that MnO_2 carries down Cr. The A.S.T.M. method now recommends either of two methods: (a) precipitation of iron as hydroxide with ammonia, filtering off a portion and determining Cr in the filtrate, or method 2 above. The editor finds that the $Fe(OH)_3$ is largely responsible for holding up chromium, and advises against this procedure (a).

COBALT

Co, *at.wt.* 58.97; *sp.gr.* 8.7918; *m.p.* 1478°; *b.p.* *unknown*; *Oxides*, Co_3O_4, Co_2O_3, CoO, CoO_2.

DETECTION

After the removal of the elements precipitated by hydrogen sulphide from acid solution, a little nitric acid is added to the solution to oxidize to the ferric state any ferrous salts which may be present, and ammonia is added until its odor is distinctly perceptible, to precipitate iron, aluminum and chromium.[1] This precipitate is removed by filtration and hydrogen sulphide passed through the ammoniacal solution to precipitate cobalt, nickel, manganese and zinc. After collecting this precipitate it is washed thoroughly with cold hydrochloric acid of approximately 1.035 specific gravity, to remove manganese and zinc. A small quantity of the residue is fused with borax in the loop of a platinum wire. A blue color in the cold bead indicates cobalt. This test is masked in the presence of large quantities of nickel. In this case the residue is dissolved in hydrochloric acid to which a few drops of nitric acid have been added and the solution evaporated to dryness. The residue is redissolved in water, acidified with hydrochloric acid and the cobalt precipitated with a hot solution of nitroso-beta-naphthol in 50% acetic acid. A brick red precipitate indicates cobalt.

Potassium sulphocyanate, KCNS, produces a red color with cobalt. If to this neutral or faintly acid solution are added twice its volume of alcohol and four times its volume of ether and the solution shaken the ether layer is colored blue by cobalt. If iron is present a solution of sodium thio-sulphate, $Na_2S_2O_3$, is added until the red color disappears, the solution filtered and then treated with the alcohol-ether mixture.

Potassium Nitrite, KNO_2, added to a neutral or slightly acid solution containing acetic acid, will precipitate cobalt as a yellow complex nitrite having the formula $K_3Co(NO_2)_6$.

A solution of *dicyandiamidine sulphate* and sodium hydroxide added to a cobalt solution to which ammonia has been added until the odor is distinctly discernible, and containing from 10 to 20 cc. of 10% sugar solution, will change the color of the solution to red or reddish violet. If large quantities of nickel are present the color will be yellow or reddish yellow, after which the nickel will separate out in brilliant crystals, leaving the cobalt in solution, coloring it as described above.

A concentrated solution of *ammonium sulphocyanate* added to a cobaltous solution colors it blue. On dilution this becomes pink. Amyl alcohol or a mixture of amyl alcohol and ether 1 : 1, added to this and shaken, extracts this blue compound. Iron sulphocyanate, $Fe(CNS)_3$, likewise colors the ether-

---

[1] If a relatively large amount of iron is present, the basic acetate method of separation is necessary, as iron occludes cobalt.

Chapter contributed by W. L. Savell.

alcohol extract red, which may mask the cobalt blue. By addition of sodium carbonate solution ferric hydroxide precipitates, while the cobalt color will remain after this treatment.

ESTIMATION

Cobalt is usually estimated as metal; either reduced by hydrogen from the ignited oxide or reduced by electrolysis from an ammoniacal solution of its salts. Sometimes, however, it is estimated as oxide; usually as Co_3O_4. The reduction of the oxide by hydrogen may be carried out in conjunction with any process giving an oxide, hydroxide, carbonate, nitrate, chloride or an organic compound, as a final product.

The reduction of the metal, in solution, by electrolysis, must be accomplished in a strongly ammoniacal solution free from copper and nickel, as these metals are deposited with the cobalt on the cathode. When desirable the copper and nickel may be estimated after the electrolysis by dissolving the deposit from the cathode and proceeding in the usual manner.

Preparation and Solution of the Sample

General Procedure for Ores. The ores containing cobalt vary so widely in their chemical nature that it is difficult to lay down a method for treating all ores. However, as the principal ores contain the cobalt as a sulphide or arsenide the same general methods may be used in the majority of cases. In all cases it is necessary to prepare the sample for treatment by grinding finely. Usually either of the above ores may be brought into solution by heating with strong nitric acid or a mixture of nitric and hydrochloric acids, except silver-bearing ores, which may usually be dissolved in a mixture of nitric and sulphuric acids.

While it is desirable to use no more acid than is necessary to bring the sample into solution, an excess will not interfere, as it may be driven off by evaporation and in the event of determining the cobalt electrolytically it is essential that the solution be free from nitric acid, so that this evaporation becomes part of the procedure.

In the case of especially refractory ores or oxides of cobalt or nickel, a fusion with potassium bisulphate will usually be found sufficient as a preliminary treatment to enable it to be brought into solution. Under certain conditions, however, it has been found necessary to fuse the ore with sodium peroxide in a silver crucible, dissolving the cobalt oxide formed in hydrochloric acid. In somewhat less refractory ores of a silicious nature a preliminary fusion with a mixture of sodium carbonate and potassium carbonate with subsequent solution in hydrochloric acid or sulphuric acid, if the ore is a silver-bearing one, will be found satisfactory.

Cobalt Oxides. Cobalt oxide, gray or black, may be fused with potassium bisulphate, and the melt leached with water; or they may be treated with sulphuric acid, in which they dissolve slowly; or with hydrochloric acid, in which they dissolve more rapidly.

Metallic Cobalt, Nickel and Cobalt Alloys. Metallic cobalt dissolves readily in nitric acid, as do nickel and the ordinary cobalt alloys. There are some alloys of cobalt, however, which require fusion with sodium peroxide before they become amenable to further treatment. Among these are certain cobalt-chromium alloys.

SEPARATIONS

Separation of the Ammonium Sulphide Group Containing Cobalt from the Hydrogen Sulphide Group—Mercury, Lead, Bismuth, Copper, Cadmium Arsenic, Antimony, Tin, Gold, Molybdenum, etc.

Hydrogen sulphide passed into a hydrochloric acid solution containing from 5 to 7 cc. of concentrated hydrochloric acid per 100 cc. of solution, precipitates only the members of that group and silver, whereas the members of the subsequent groups remain in solution. If the solution is too acid, lead and cadmium are not completely precipitated.

Separation of the Ammonium Sulphide Group from the Alkaline Earths and Alkalies. Ammonium sulphide, free from carbonate, added to a neutral solution containing the above elements in the presence of ammonium chloride, precipitates only the members of this group; the alkaline earths metals, magnesium and the alkalies remain in solution. A second precipitation should be made if large quantities of the alkaline earths or alkalies are present.

Separation of Cobalt and Nickel from Manganese. The solution of the chlorides or sulphates of cobalt or nickel is treated with an excess of sodium carbonate and then made strongly acid with acetic acid. About 5 grams of sodium acetate for each gram of cobalt or nickel present is now added, the solution diluted to 200 cc. and heated to about 80° C. and saturated with hydrogen sulphide. Cobalt and nickel are precipitated as sulphides and the manganese remains in solution. The filtrate is concentrated, and colorless ammonium sulphide added when the cobalt and nickel that may have passed in solution from the hydrogen sulphide treatment, will be precipitated. The treatment should be repeated with the second filtrate to ensure complete precipitation of the cobalt and nickel.

Separation of Cobalt from Nickel. Among a number of methods for effecting this separation the following give good results:

A. Nickel is removed from the solution by precipitation with dimethylglyoxime. The details of the procedure may be found in the gravimetric methods for the determination of nickel. Cobalt remains in solution.

B. Cobalt is precipitated by nitroso-beta-naphthol, leaving nickel in solution. Details of the procedure are given under gravimetric methods for determination of cobalt.

C. Cobalt is precipitated as potassium cobalti-nitrite, nickel remaining in solution. Details of the procedure are given under gravimetric methods for the determination of cobalt.

Separation of Cobalt from Zinc. Zinc is precipitated from weak acetic or formic acid solution by hydrogen sulphide as zinc sulphide. Cobalt, nickel and manganese remain in solution. The details of the procedure are given under the methods of determination of zinc.

GRAVIMETRIC METHODS FOR THE DETERMINATION OF COBALT

Precipitation of Cobalt by Potassium Nitrite

Cobalt may be precipitated from a solution made slightly acid with an excess of acetic acid by adding a hot solution of potassium nitrite. The cobalt is precipitated as potassium cobalti-nitrite, $K_3Co(NO_2)_6$, very completely, after standing for a period of six hours in a warm place. This method has the advantage of making possible the separation of cobalt from nickel and iron, although it has the one disadvantage, for commercial purposes, of requiring a long time to complete the determination.

Procedure. After bringing the material into solution and separating the silica and members of the first and second groups in the usual manner, the solution is boiled to eliminate hydrogen sulphide. Oxidize the iron present with a little hydrogen peroxide and evaporate the solution to a syrup. Take up in a little water and neutralize with a practically saturated solution of sodium carbonate. Render slightly acid with acetic acid and add an excess of 1 : 1 acetic acid. Heat to almost boiling and add solution of 50% potassium nitrite containing 100 cc. of glacial acetic acid per liter, also heated to nearly boiling. This solution should be added slowly to the solution of the sample which should be agitated, preferably by rotating gently while the addition is being made. The sides of the beaker should be washed down with a 1% solution of potassium nitrite containing 1 cc. of glacial acetic acid per liter. Allow to stand for at least six hours and if possible overnight. Filter through thick Swedish filter paper without previous wetting. As this precipitate shows a very decided tendency to creep, considerable care is required to keep it well down in the apex of the filter-paper cone. Wash about ten times with the warm nitrite solution mentioned above. Transfer to a beaker by removing the filter paper from the funnel and opening it into the beaker with the outside of the paper against the glass. This leaves it in a convenient position for washing. The bulk of the precipitate is washed off with 10 cc. of 1 : 1 sulphuric acid, heated to about 80° C. This should leave only a slight film of precipitate on the paper. Keep the solution in the beaker at about 80° C. to assist in dissolving the precipitate and wash the paper with the hot sulphuric acid solution five times, using about 10 cc. each time. Gradually withdraw the filter paper from the top of the beaker during the washing operation. Give the paper a final wash with hot water and squeeze the last drops from it into the beaker. Evaporate and allow to fume strongly for ten minutes. Set the beaker in a cooling trough and add water until the volume is about doubled. Neutralize and make slightly ammoniacal and then add an excess of 50 cc. of strong ammonia and electrolyze as described under Electrolysis in Reduction of Cobalt by Electrolysis, page 170.

Precipitation of Cobalt by Nitroso=beta=Naphthol [1]

Nitroso-beta-naphthol, $C_{10}H_6(NOH)$, added to a hydrochloric acid solution of cobalt, precipitates cobalti-nitroso-beta-naphthol, $Co(C_{10}H_6O(NO))_3$; nickel, if present, remains, in solution. The method is especially suitable for the determination of small amounts of cobalt in the presence of comparatively large

[1] Burgess, Z. Angew., 1896, 596.

amounts of nickel. The cobalt precipitate is voluminous, so that the sample taken for the determination should not contain over 0.1 gram of cobalt. The reagent will also precipitate copper and iron completely from solution, and silver, bismuth, chromium and tin partially; but mercury, lead, cadmium, arsenic, antimony, aluminum, manganese, nickel, glucinum, calcium and magnesium remain in solution.

Procedure. To the solution containing the cobalt is added a freshly prepared hot solution of nitroso-beta-naphthol, in 50% acetic acid, as long as a precipitate is produced. After allowing it to settle, more of the reagent is added to insure complete precipitation of the cobalt. The compound is allowed to settle for two of three hours, the clear solution decanted through a filter and the precipitate washed by decantation with cold water, then with warm 12% hydrochloric acid solution to remove the nickel, and finally with hot water until free of acid.

The brick-red precipitate is dried, then ignited in a weighed platinum crucible (Rose crucible), first over a low flame and finally at a white heat, the crucible being covered by a platinum cover (Rose crucible type) with a platinum tube, through which is passed a slow current of oxygen. The residue is weighed as Co_3O_4. The oxide may be reduced in a current of hydrogen and weighed as metallic cobalt. **Ignited in the presence of CO_2** the oxide CoO is formed.

Edler finds it unnecessary to ignite the cobalt in a current of oxygen. Ignition in air is best.

Precipitation of Cobalt by Electrolysis [1]

Metallic cobalt is readily deposited from an ammoniacal solution of the sulphate, but in the presence of copper and nickel these are also completely precipitated on the cathode; so, in case it is desired to determine the cobalt alone it is necessary to separate these metals from the solution before electrolysis or to determine them separately after electrolysis in a solution of the metallic deposit. In practice the copper is usually separated before electrolysis and the nickel, if determined separately, is estimated afterward by one of the methods given under Nickel, the cathode deposit being dissolved for this purpose.

Procedure. After preparation and solution of the sample the usual separations with hydrogen sulphide in acid solution are made if necessary. In most cases it is necessary to pass hydrogen sulphide through the warmed solution for at least one hour to insure the complete precipitation of arsenic. Filter and boil to expel hydrogen sulphide. Add 5 cc. hydrogen peroxide to insure oxidation of iron compounds to ferric state and add ammonium hydroxide until slightly alkaline to litmus. Filter off ferric hydroxide and wash with water containing a small quantity of ammonium hydroxide. Redissolve and reprecipitate this ferric hydroxide in the above manner, using a little hydrogen peroxide in each instance, until the last traces of cobalt have been removed from it, keeping the filtrates, which should be as small as possible, to add to the main filtrate. If much iron is present this is best removed as the basic acetate.

Electrolysis. If the treatment of the iron precipitate has made a large volume of solution this may be reduced by evaporation, after which 50 cc. of strong ammonia are added and the solution electrolyzed, using direct current of 2 volts and 0.5 ampere per square decimeter. The electrodes should be platinum, the anode a spiral wire and the cathode either a hollow cylinder or a cylindrical gauze. By agitating the solution, raising the voltage and the cur-

[1] Low, "Technical Methods of Analysis."

rent density, the rate of deposition may be increased. In a properly agitated solution the deposition may be completed in forty-five minutes.

The current should not be cut off until the solution is tested to determine if the electrolysis is complete. This is done by mixing a drop or two of the solution from the end of a stirring rod with a few drops of ammonium sulphide. If the electrolysis is complete the mixture will remain colorless, but if some cobalt still remains in the solution the mixture will be darkened. After the electrolysis is complete the cathode is carefully removed from the solution and dipped into a beaker of clean water, after which it is washed with alcohol, preferably ethyl alcohol.

If a large number of electrolytic determinations are to be made, it is convenient to have a wide-mouthed bottle with a well-ground-in glass stopper or a cork stopper for holding the alcohol for the preliminary washing. The mouth should be large enough to receive the cathode without pouring out the alcohol. The cathode may be lowered into the alcohol in this bottle, which should only be partly filled, and then rinsed again by pouring fresh alcohol over it and allowing it to drain into the wide-mouth bottle. This allows a great many cathodes to be washed with a comparatively small quantity of alcohol. Directly after the final washing with alcohol the cathode is passed through the flame of a Bunsen burner and the alcohol ignited. After this is entirely burned off the cathode is placed in a desiccator to cool and when cool is weighed. The increase in weight of the cathode is the weight of cobalt in the sample if the solution had been free from nickel before electrolysis. If the nickel remained in the solution the increase in weight of the cathode represents the cobalt and nickel in the sample. If it is desired to determine the cobalt and nickel together the increase in weight of the cathode is divided by the weight of the sample and multiplied by 100 to obtain the percentage. If it is desired to obtain the percentage of cobalt separately, the plate is dissolved from the cathode in a few cc. of nitric acid and the nickel determined in the resulting solution by precipitation with dimethyl-glyoxime as described in the chapter on Nickel, after which the cobalt is found by difference.

Cobalt in Cobalt Oxide [1]

One gram of finely ground cobalt oxide is either fused with 10 grams of potassium bisulphate or heated with 20% sulphuric acid until dissolved. If the fusion method is used the melt is extracted with water and acidified with sulphuric acid. Arsenic and copper are precipitated by passing hydrogen sulphide through the warmed solution, which should be diluted to about 200 cc. for about one hour. These are removed by filtration and the cobalt determinated by one of the above methods. The following procedure is one of the most satisfactory:

Procedure. If it is desired to determine the nickel separately, as is usually the case, this is first precipitated with dimethylglyoxime as described in the chapter on Nickel, after boiling the solution to expel hydrogen sulphide. It is then evaporated to fumes of sulphur trioxide and taken up with twice its volume of water. The free acid is neutralized with ammonium hydroxide and an excess of 50 cc. of strong ammonium hydroxide added. The solution is made up to 250 cc. and electrolyzed as under Precipitation of Cobalt by Electrolysis.

[1] R. W. Landrum, Proc. Am. Ceramic Soc., **12**, 1910.

Cobalt in Metallic Cobalt and Ferro=cobalt

Cobalt is usually determined in metallic cobalt and ferro-cobalt by electrolysis, after separation of the elements precipitated by hydrogen sulphide in acid solution and elimination of iron, if present in large quantities. In case it is desired to estimate nickel separately it is precipitated by dimethylglyoxime as described in the chapter on Nickel, before electrolysis, taking the solution down to sulphur trioxide fumes, diluting with water and adding ammonium hydroxide in excess and electrolyzing. In case the solution is electrolyzed before separating the nickel the determination of this element may be made in the solution of the electrolytic deposit dissolved in acid, the cobalt then found by difference.

Procedure. Dissolve 1 gram of well-mixed drillings in the least possible quantity of nitric acid and add 20 cc. of 1 : 1 sulphuric acid. Evaporate to fumes of sulphur trioxide and allow to fume strongly for ten minutes. This insures the complete elimination of nitrates, which would interfere subsequently with the electrolysis. Cool and dilute carefully with 20 cc. of water. Heat the solution to nearly boiling and pass in hydrogen sulphide for one hour to precipitate copper and arsenic. Filter and boil the solution to expel the last traces of hydrogen sulphide. Add 2 cc. of hydrogen peroxide to oxidize ferrous compounds to ferric state, and add ammonium hydroxide until slightly alkaline to litmus paper and heat to boiling. Filter off the ferric hydroxide and wash with water containing a small quantity of ammonium hydroxide. Redissolve the precipitate in a little 1 : 1 sulphuric acid, adding a little hydrogen peroxide to keep the iron in the ferric state, and reprecipitate in the same manner as that described above. In presence of comparatively large amounts of iron the basic acetate separation of iron is necessary, as $Fe(OH)_3$ occludes cobalt and nickel. The filtrates from these precipitations are added to the main one.

In determining the cobalt in metallic cobalt it is not necessary to filter off the iron precipitate, if this is small, as it has been found by W. L. Rigg, of Deloro, Ontario, that this precipitate does not interfere with the accuracy of the determination. The iron content may be up to 5% without interfering seriously with the electrolysis.

The solution is made ammoniacal with 50 cc. of strong ammonium hydroxide and electrolyzed as described above.

Cobalt in Metallic Nickel

The cobalt in metallic nickel may be determined by precipitation with potassium nitrite from a solution of the sample containing an excess of acetic acid. The precipitate is filtered off and dissolved in hot sulphuric acid solution, after which the solution is evaporated to fumes of sulphur trioxide and carefully diluted. The excess of acid is neutralized and made strongly ammoniacal with ammonium hydroxide. The solution is then electrolyzed as previously described.

Procedure. Dissolve 5 grams of thoroughly mixed drillings in a minimum quantity of nitric acid. Evaporate to a syrup. Care must be exercised at this point to prevent evaporating too far and decomposing the nitrates. Dissolve in 50 cc. of water. Neutralize with a practically saturated solution of sodium carbonate. For this purpose a dropping bottle is very convenient. Render slightly acid with acetic acid and add an excess of 10 cc. of 1 : 1 acetic acid. Heat to almost boiling and add 10 cc. of a 50% solution of potassium nitrite to which has been added 10 cc. of glacial acetic acid per 100 cc. of solution. This solution must also be nearly boiling and should be added while gently rotating the nickel solution. Wash down the sides of the beaker with a 1% solution of potassium nitrite containing 1 cc. glacial acetic acid per liter. Allow to stand for at least six hours and preferably overnight. Filter through a thick, 9-cm. filter paper without previous wetting. Considerable care is required to keep the precipitate well down in the apex of the filter paper cone, as it creeps very badly. Wash about ten times with the warm nitrite solution mentioned above. Lift gently from the funnel and open the filter paper into a beaker. Lay the paper against the side of the beaker with the outside against the glass. This leaves the paper adhering to the side of the beaker in a most convenient position for washing. Wash down as much of the precipitate as possible with about 10 cc. of 1 : 1 sulphuric acid solution, heated to about 80° C. This should leave only a slight film of precipitate on the paper. Keep the solution at about 80° C. and wash the paper five times with the warm sulphuric acid solution, using about 10 cc. each time, gradually withdrawing the paper from the top of the beaker. Give a final wash with hot water and squeeze the last drops from the filter paper into the beaker. Evaporate and allow to fume strongly for ten minutes. Add water in a cooling trough until the volume is about doubled. Neutralize with ammonium hydroxide and add an excess of 50 cc. of strong ammonium hydroxide and electrolyze as described in Precipitation of Cobalt by Electrolysis.

Cobalt in Ores and Enamels [1]

The determination of cobalt in ores and enamels is usually made by a slight variation of the above methods. The silica is separated in the usual manner by taking down to dryness with hydrochloric acid and the warmed solution is treated with hydrogen sulphide to precipitate sulphides insoluble in acid solution. Aluminum, chromium and iron are precipitated by adding ammonium hydroxide to the oxidized solution. In the enamel industry it has been the practice to follow R. W. Landrum's method, in which the cobalt, manganese and nickel are precipitated together as sulphides and filtered off. The manganese is dissolved from this precipitate with cold hydrogen sulphide water acidified with one-fifth its volume of hydrochloric acid (sp.gr. 1.11). The residue of cobalt sulphide is burned in a porcelain crucible, dissolved in aqua regia and evaporated with hydrochloric acid. The platinum and copper, if they are present, are thrown down by passing hydrogen sulphide through the solution. The filtrate is made ammoniacal and the cobalt is precipitated with hydrogen sulphide. This is filtered off and washed with water containing a small quantity of ammonium sulphide. The precipitate is either ignited and weighed as oxide or reduced in hydrogen to metallic cobalt, taking care to cool it thoroughly in an atmosphere of hydrogen before allowing it to come into contact with the atmosphere of the room, as finely divided cobalt is decidedly pyrophoric and oxidizes readily, particularly if reduced at a low temperature.

Instead of igniting the sulphide precipitate it may be dissolved in hot 1 : 1 sulphuric acid solution with the aid of a little nitric acid and treated as described under Precipitation of Cobalt by Electrolysis.

[1] R. W. Landrum, Trans. Am. Cer. Soc., **12**, 1910.

Cobalt in Steel

This determination is a modification of the nitroso-beta-naphthol method already described, as worked out in the laboratory of the Firth Stirling Steel Company, McKeesport, Pa. The procedure as described by Mr. Giles, Chief Chemist, is as follows.

Two grams of the sample are weighed into a 500-cc. Erlenmeyer flask and dissolved in 50 cc. of concentrated hydrochloric acid. When the sample is completely decomposed 10 cc. of concentrated nitric acid are added to oxidize the iron, tungsten, etc. The solution is evaporated to 10 cc.; 50 cc. of water are added; the contents of the flask are then transferred to a 500-cc. volumetric flask and cooled to room temperature. A fresh solution of zinc oxide is added in slight excess, the contents of the flask diluted to the mark, well mixed, transferred back to the original Erlenmeyer flask and allowed to settle. Filter 250 cc. (equivalent to 1 gram of the sample) through a dry filter paper, transfer it to a 500-cc. flask, then add 6 cc. of concentrated hydrochloric acid.

The solution, which should now be between 300 and 350 cc. in volume, is heated to boiling and 10 cc. of freshly prepared solution of nitroso-beta-naphthol (1 gram of salt to 10 cc. glacial acetic acid) are added for each 0.025 gram of cobalt present. Continue to heat for two minutes, remove from plate, shake well, and set aside until the bright red precipitate settles, which will only take a few minutes. Filter the hot solution and wash the flask out with hot 1 : 1 hydrochloric acid and then wash the flask out with hot acid of the same strength. Wash the paper alternately with hot (1 : 1) hydrochloric acid and hot water until it has been washed five times with the acid, then wash ten times with hot water. The precipitate is transferred to a quartz or porcelain crucible, heated gently to expel the carbonaceous matter, then at a high temperature until ignition is complete. After cooling the crucible is weighed and the weight of the residue (Co_3O_4) is multiplied by 0.734 to obtain the percentage of cobalt present. If desired the Co_2O_4 may be reduced in hydrogen and weighed as metal.

Sulphide Pyrophosphate Method (**Dufty**). Cobalt is precipitated as sulphide and this is converted to ammonium cobalt phosphate. The precipitate is filtered and washed, then ignited and the cobalt weighed as pyrophosphate. Traces of cobalt passing into the filtrate are recovered by precipitation as sulphide, this is filtered off, ignited to oxide and added to results obtained. (Oxide is Co_3O_4.)

Volumetric Methods for the Determination of Cobalt

Perborate Method. Nickel and cobalt are oxidized to trivalent oxide or hydroxides by various oxidizing agents but neither the cobalt and nickel salts of the trivalent metals are stable. Both the trivalent hydroxides liberate iodine when treated with acids and potassium iodide and form the nickelous or cobaltous salts, e.g.,

$$Co(OH)_2 + KI + 3HCl = CoCl_2 + I + KCl + 3H_2O.$$

Cobalt is more readily oxidized than nickel and when oxidized, forms the more stable compounds. Cobaltic hydroxide may be formed by treating the sulphate with potassium hydroxide and hydrogen peroxide while nickel is not affected even on boiling with these reagents. This procedure, however, has not been satisfactory as a method of analysis on account of the difficulty of removing the excess hydrogen peroxide which persists in the precipitate even after boiling and prolonged washing.

After some rather extended tests, W. D. Engle and R. G. Gustavson have found that the differential oxidation may be very satisfactorily accomplished by means of sodium perborate in the presence of potassium hydroxide and the excess of this oxidizing agent readily decomposes on boiling. Advantage is then taken of the reaction stated above and the iodine liberated is titrated with a standard sodium thiosulphate solution.

Procedure. The ore, metal or other material may be dissolved with acids or brought into solution by any of the usual methods and the metals of the Cu and Fe groups including manganese removed in the usual way. The solution so obtained may contain nickel, cobalt, zinc and the metals of the alkaline and alkaline earth groups, but must be free from any substances capable of liberating iodine from an acid solution of potassium iodide. In some cases, it is desirable to precipitate Fe by means of zinc oxide or zinc hydroxide emulsion, making up the mixture so obtained containing the emulsion to a certain volume and after allowing the precipitate to settle, filtering aliquot portions and treating as described. This solution with a volume of about 100 cc. is made acidified with dilute H_2SO_4 using about 5 cc.'s excess (1 : 5) and then adding one or two grams of dry sodium perborate. Sodium hydroxide is added to a strong alkaline reaction. The mixture is boiled for ten minutes to decompose the excess of perborate. This solution is then cooled to room temperature and after one gram of potassium iodide is added, acidified with dilute H_2SO_4 the potassium iodide slowly reduces the cobaltic hydroxide liberating iodine. Any attempt to hasten the determination at this point will cause low results as cobaltic hydroxide itself is not readily soluble and must be reduced to the cobaltus form before going into solution. This reduction consequently takes place very slowly and care must be exercised to make certain that all the precipitate has been entirely dissolved before the titration is attempted.

After the complete solution of the precipitate, the liberated iodine is titrated with standard solution of sodium thiosulphate, using starch paste as an indicator in the usual manner.

As it is somewhat difficult to obtain cobalt compounds of known cobalt content, the sodium thiosulphate solution may be standardized by means of potassium bichromate in the usual manner and the factor for cobalt calculated from this determination.

Since one $K_2Cr_2O_7 = 6I = 6Co$, then the potassium bichromate factor of the solution multiplied by 1.2027 gives its cobalt factor.

General Procedure for Determining Nickel and Cobalt in Ores—Method of Schoeller and Powell [1]

1. *The ore is free from copper, manganese, lime, and magnesia; presence of arsenic immaterial.* Dissolve in nitric or hydrochloric acid or both, but convert metals into nitrates; take almost to dryness, avoiding separation of basic salts. Add tartaric acid (10 times the weight of trivalent metals) dissolved in a minimum of water. Add 50 cc. of strong ammonia whilst cooling, and 3 to 5 gms. of solid potassium iodide according to the quantity of nickel or cobalt present. Allow to stand, stoppered, with occasional shaking, for fifteen minutes, or longer if cobalt predominates. Filter on loose paper, wash with ammoniacal iodide solution (4 per cent of potassium iodide dissolved in 4 vol. strong ammonia: 1 vol. water). Dissolve precipitate in dilute hydrochloric acid, filter off gangue, precipitate cobalt as phosphate, titrate nickel in filtrate.

2. *The ore contains copper, otherwise the same as 1.* Proceed as above to the point where the iodide precipitate is dissolved in dilute hydrochloric acid. Decolorize the liquid with sulphurous acid, heat just to boiling, allow to cool, filter off insoluble gangue and cuprous iodide, apply phosphate separation.

3. *General procedure for complex ores free from manganese.* Dissolve in a suitable acid, precipitate heavy metals with hydrogen sulphide. To the filtrate add ammonium chloride and a slight excess of ammonia; saturate with hydrogen sulphide, taking care to precipitate the whole of the nickel. Dissolve the precipitate in *aqua regia*, evaporate almost to dryness, etc., as under 1.

4. *The ore contains manganese.* (a) *Subordinate amounts.* Apply one of the preceding methods; titrate cobalt ammonium phosphate with N/5 acid (not hydrochloric), or dissolve the weighed cobalt pyrophosphate in sulphuric acid. Determine the manganese colorimetrically with persulphate and silver nitrate in an aliquot portion of the cobalt solution, obtaining cobalt by difference.

(b) *Moderate to large amounts.* Dissolve the mixed iodide precipitate in dilute acid, and precipitate nickel and cobalt sulphides from an acetic solution. Or neutralize the acid solution and precipitate cobalt and nickel as xanthates. Either precipitate is dissolved in *aqua regia* and the two metals separated as before.

(c) With nickel ores free from cobalt, manganese does not interfere in the cyanide titration in the presence of citrate.

[1] By W. R. Schoeller and A. R. Powell, The Analyst, Aug., 1919.

COPPER

Cu, *at.wt.* 63.57; *sp.gr.* 8.89$^{20°}$; *m.p.* 1083 (*in air* 1065); *b.p.* 2310; *oxides* Cu$_2$O *and* CuO.

DETECTION

Copper is precipitated in an acid solution by H$_2$S gas, along with the other members of the hydrogen sulphide group. The insolubility of its sulphide in sodium sulphide is a means of separating copper from arsenic, antimony, and tin. The sulphide dissolves in nitric acid (separation from mercury) along with lead, bismuth, and cadmium. Lead is precipitated as PbSO$_4$ by sulphuric acid and bismuth as the hydroxide, Bi(OH)$_3$, upon adding ammonium hydroxide. Copper passes into the filtrate, coloring this solution blue,

$$Cu(OH)_2 \cdot 2NH_4OH \cdot (NH_4)_2SO_4.$$

Flame Test. Substances containing copper (sulphides oxidized by roasting), when moistened with hydrochloric acid and heated on a platinum wire in the flame, give a blue color in the reducing flame and a green tinge to the oxidizing flame.

Wet Tests. Nitric acid dissolves the metal or the oxides (sulphides should be roasted), forming a green or bluish-green solution. Ammonium hydroxide added to this solution will precipitate a pale blue compound, which dissolves in excess with the formation of a blue solution. (Nickel also gives a blue color.)

Hydrogen sulphide passed into a copper solution which is free of SO$_2$ or an oxidizing agent, but somewhat acid with a mineral acid, precipitates at once brownish black CuS or Cu$_2$S (distinction from nickel, cobalt and zinc), which is difficulty soluble in strong, hot HCl (distinction from antimony), insoluble in fixed alkaline polysulphides (distinction from gold and platinum), soluble in alkaline cyanides (distinction from lead, bismuth, cadmium, mercury and silver), soluble in nitric acid (distinction from sulphide of mercury), with production of a bluish solution (distinction from all other metals except silver).

Potassium ferrocyanide precipitates from an acid or neutral solution of a cupric salt reddish-brown cupric ferrocyanide, which can be confused only with similarly colored precipitates from molybdenum or uranium solutions.

Cupric salts in halogen acid solution are reduced to colorless cuprous compounds by metallic copper, stannous chloride and sulphurous acid; and in alkaline solution are reduced by grape sugar, arsenious or sulphurous acids.

ESTIMATION

The estimation of copper is required in the following substances: *In ores*[1]

[1] Ores, copper pyrites, chalcopyrite, Cu$_2$S·Fe$_2$S$_3$; copper glance chalcocite, Cu$_2$S (gray to bluish-black); malachite, CuCO$_3$·CuOH·H$_2$O (green); azurite, 2CuCO$_3$·-CuOH$_2$O (blue); cuprite, red copper, Cu$_2$O; malaconite, CuO (black); dioptase, CuOSiO$_2$H$_2$O (green vitreous); atacamite, CuCl$_2$3Cu(OH)$_2$; bornite, Cu$_3$FeS$_3$; brochantite, 4CuO·SO$_3$·3H$_2$O; corellite, CuS; crysocolla, CuO·SiO$_2$·2H$_2$O; tetrahedrite, 4Cu$_2$S·Sb$_2$S$_3$; termantite, 4Cu$_2$S·As$_2$S$_3$; olirenite, 4CuO·As$_2$O$_5$·H$_2$O.

Chapter contributed by W. W. Scott and W. G. Derby.

177

of copper, in which it occurs as native copper or combined as sulphide, oxide, carbonate, chloride, silicate and basic sulphate. In furnace slags, mattes, concentrates, blister copper, bottoms. The determination of copper is required in the analysis of alloys containing copper,[1] brass, bronze, etc. It is occasionally looked for as an undesirable impurity in food products. It is determined in salts of copper, in insecticides, germicides, etc.

Preparation and Solution of the Sample

Hydrochloric and sulphuric acids are effective in dissolving metallic copper only in presence of an oxidizing agent; nitric acid is the most active solvent. The oxides of copper may be dissolved in hydrochloric or sulphuric acid, but nitric acid is commonly used. Some refractory furnace products are with most certainty decomposed by treatment with hydrofluoric and fuming sulphuric acid followed by a bisulphate fusion.

Ores. If the ore consists practically of a single mineral, the fineness of the sample need not exceed 80 mesh. If the ore is a mixture of minerals, lean and rich in copper, the laboratory sample should pass a 120-mesh sieve.

Metallic particles or masses are separated at some stage in the process of sampling and made into a separate sample. If the metallic portion is a small percentage of the total sample and consists of particles, the copper value of which is known to vary by a few percent, no attempt is made to refine the sample of such, but a large portion, 10–100 grams, is taken for analysis and the copper determined in an aliquot part of the solution. If the metallic masses are a large percentage of the sample, large of size, or consisting of particles differing widely in copper content, a weighed amount of 1 to 50 lbs. is melted in a graphite crucible, with addition of suitable fluxes, such as powdered silica or lime, if necessary. Separate samples are made of the weighed products of the fusion and the copper content of the material before melting calculated from their analyses. The amount of sample taken for assay depends upon the richness of the ore, homogeneity of the material and the commercial importance of the determination. As a rule in the assay for purchase and sale, a 1 gram test portion is taken of finely pulverized samples containing over 20% copper, 2 to 3 grams of 30% ores and 5 grams of ores containing less than 10% copper.

Sulphide Ores and Matte. One to five grams of the sample are dissolved by adding 10–20 cc. dilute nitric acid (1.2 sp.gr.), or 10 cc. strong nitric acid saturated with potassium chlorate, and allowing the mixture to stand in a warm place for about 15 minutes before applying heat. Decomposition is completed by evaporating to small volume in a casserole and continuing to dryness after addition of 10 cc. hydrochloric acid, or by adding 5–10 cc. 50% sulphuric acid, to the assay in a flask or tall beaker and heating to fumes. Continuing the first procedure, the residue is taken up by warming with 20 cc. 10% H_2SO_4, diluting, boiling and filtering the residue of silica, lead sulphate and silver chloride from the copper solution. Continuing the second procedure, the mass of anhydrous salts are dissolved with water, silver dissolved by just sufficient NaCl solution if the determination is to be by the electrolytic method, and the solution filtered; after it has reached room temperature when the per cent of silver present is high.

Oxidized Ores with the exclusion of details relating to the oxidation of sulphur are brought into solution by the same method of treatment as sulphide ores.

NOTES. The sulphur that appears upon adding acid to the ore, with proper precautions, should be yellow. If it is dark and opaque, the solution has been over-heated, and some of the ore has been occluded. It is advisable in this case to remove the globule of sulphur and oxidize it separately with bromine and nitric acid, then boil out the bromine and add the solution to the rest of the sample.

Sulphide ores may be treated according to the procedure recommended for iron pyrites in the chapter on Sulphur, the ore being decomposed with a mixture of bromine and carbon tetrachloride, 2 : 3, followed by nitric acid and then sulphuric acid.

Treatment of Matte Slag. Only by quick quenching of the molten slag is decomposition of the sample by acids made possible, without preliminary treatment with hydrofluoric acid. As a rule lime slags are readily decomposed by mixed acids. Extremely acid, and high iron slags are apt to be refractory and are decomposed with most certainty by treatment with hydrofluoric acid followed by fusion with potassium bisulphate.

The following scheme[1] of attack, which also can be applied to silicious ores, with skilful manipulation gives very satisfactory results:

One gram of the 100 mesh fine slag is placed in a 250 cc. beaker of Jena glass, moistened with water, mixed with 3 cc. of sulphuric acid (sp.gr. 1.54), and then, while the particles of the slag are in suspension through rotary movement of the beaker, 15 cc. hydrochloric acid are added. The silica is gelatinized in 2 or 3 minutes by heating the beaker over a free flame. One cc. nitric acid followed by a few drops of hydrofluoric acid are added, and the heating continued in a hood until the material is nearly dry, and then to strong sulphuric acid fumes on a hot plate. When cool, 4 cc. of sulphuric acid are added.

The remainder of the procedure depends upon the method that is to be followed in the determination of copper. If the electrolytic method is preferred, 3 cc. of nitric acid are added; the mass heated until solution is effected, the liquid diluted to 175 cc. with cold, distilled water, and copper plated out in 20–35 minutes, using a rotating anode and $2\frac{1}{2}$ amperes current.

If the iodide method is to be followed, without addition of other acid than sulphuric, the mass is again heated to fumes. When cooled, 25–30 cc. water and 5 cc. hydrochloric acid are added and the liquid boiled until clear. After addition of 40 cc. saturated solution of sodium acetate, $4\frac{1}{2}\%$ solution of sodium fluoride is added until the color of ferric acetate is discharged, and then an excess of 10 cc. When cold, titration is commenced, using a thiosulphate solution with a copper equivalent of 0.0005 g. per cc.

The following quick method has been systematically and satisfactorily checked for a long period by a hydrofluoric acid-bisulphate fusion method, by which copper, precipitated as a sulphide, is ignited, the oxide dissolved in nitric acid and copper determined by electrolysis.

Three grams of the 100 mesh fine sample are placed in an 800 cc. resistance beaker. The slag is spread over the bottom of the beaker, and while in motion 5 cc. of sulphuric acid are added rapidly to prevent the slag gathering into a mass. After addition of 40 cc. hydrochloric acid, the beaker is heated over a bare flame for about 3 minutes until the silica has gelatinized. To the hot

[1] White, Chemist Analyst, July, 1912.

solution nitric acid is added, drop by drop, until the liquid becomes dark brown. To the liquid, while in a state of agitation, 1–2 cc. hydrofluoric acid are added and the mixture boiled until the solution is complete. The liquid is diluted to 400 cc. and saturated with hydrogen sulphide and the precipitate filtered and washed as usual. The copper sulphide is ignited in a silica crucible; the residue, if washing of the precipitate has been thorough, can be brushed into a 250 cc. beaker and dissolved with a few cc. of nitric acid. After boiling gently to expel nitrogen gases, the free acid is neutralized with ammonia, and the solution then acidified with a slight excess of acetic acid. The cold solution is titrated by the iodide method, using a thiosulphate solution having a copper equivalent of about 0.0005 g. per 1 cc.

Metals. A casting of a copper alloy and even of refined copper is not homogeneous, and the zones of segregation of the constituents of the alloy (usually roughly parallel to the cooling surfaces) are the more sharply defined as the conditions which favor diffusion of the eutectic prevail, therefore, unless the casting be quite thin and quickly cooled, a satisfactorily representative sample of it cannot be obtained from a single drill hole. A single casting may be sampled by complete cross-sectional cuts by a suitable saw or by a series of drill holes located in such a manner as to amount substantially to one or more cross-sectional cuts. Steel is usually present as a contaminant of the drill or saw shavings from refined copper and the tougher alloys and should be removed by a magnet. Crude copper, such as blister or black copper, is sampled by drilling one hole in each piece of a definite fraction of the total pieces of the average lot. The position of the hole in successive pieces is changed to conform with a pattern or " templet " which will cover a quarter, or half, or the complete top surface of the average piece, the " templet " is divided into squares, preferably about 1 inch on a side, and in the centre of each square the $\frac{1}{2}$-inch hole is drilled. The drillings are ground to pass a 20-mesh screen and the sample then withdrawn by means of a riffle sampler.

Sampling by splashing a molten stream with a wooden paddle and by slowly pouring the metal into water are methods frequently practiced. The size of the particles, the degree of homogeneity and the limit of accuracy of result required are factors which determine whether one or more grams of the sample should be taken for analysis.

Iron Ores and Iron Ore Briquettes. A 5-gram sample of the finely divided material is fused in a large platinum dish with 40 grams of pure potassium bisulphate. If the ore is high in sulphur, it should be roasted by heating to redness in a silica or porcelain crucible before placing in the platinum dish and mixing with the bisulphate.

The cooled fusion is broken up into small pieces and placed in an 800-cc. beaker with clock-glass cover. Three hundred cc. of hot water and 25 cc. of strong hydrochloric acid are added and the fusion is boiled until it passes into solution. If an appreciable residue remains, the solution is filtered, the residue fused with additional bisulphate, then dissolved in hot dilute acid and the filtrate added to the first solution. Silica and barium sulphate remain in the residue.

The solution is now reduced and copper precipitated according to directions given under "Separation of Copper by Precipitation in Metallic Form by a more Positive Element," aluminum powder being preferably used.

The precipitated copper is filtered free from iron and other commonly occurring impurities, then dissolved by pouring on the precipitated metal 30 cc. of hot dilute nitric acid, 1 : 1, followed by 10 cc. of bromine water and then 10 cc. of hot water. The filter paper is removed, ignited and the ash added to the copper solution. The whole solution is now evaporated to small volume and determined, preferably, by the "Potassium Iodide" method as described under the volumetric procedures.

Steel, Cast Iron, and Alloy Steels.[1] From 3 to 5 grams of steel, depending upon the amount of copper present, are dissolved in a mixture of 60 cc. of water and 7 cc. of sulphuric acid (sp.gr. 1.84) in a 250-cc. beaker. After all action has ceased, a strip of sheet aluminum, $1\frac{1}{2}$ ins. square, bent so that it will stand upright in the beaker, is placed in the solution.

After boiling the solution for twenty to twenty-five minutes, which is sufficient to precipitate all of the copper in the sample, the beaker is removed from the heat and the cover and the sides washed down with cold water. The liquid is decanted through an 11-cm. filter, the precipitate washed three times with water, then placed with the filter in a 100-cc. beaker, and 8 cc. of concentrated nitric acid and 15 cc. of water are poured over the aluminum and the solution heated to boiling. This hot solution is poured over the precipitate and filter in the 100-cc. beaker, and boiled until the paper becomes a fine pulp, only a few minutes being required. The solution is filtered, the residue washed several times with hot water and copper determined in the filtrate by the electrolytic or iodide methods.

SEPARATIONS

Separation of Copper as Cuprous Thiocyanate. Isolation of copper from solutions containing iron, nickel, cobalt, zinc, cadmium, arsenic, antimony and tin may be accomplished by this method. When much arsenic is present, precipitation should be from a solution in which hydrochloric is the only free strong acid. Unless previously removed from the solution, lead, mercury, tellurium and the precious metals will contaminate the precipitate. Selenium may be a contaminant when present in considerable quantity, sometimes when the only free acid is sulphuric, always when hydrochloric acid is present.

Cuprous thiocyanate, besides being the medium of separation of copper from interfering elements preliminary to its determination by the standard electrolytic, iodide or cyanide methods, is the basis of a number of other more or less useful gravimetric and volumetric methods of determining copper. The details of the procedure of procuring the precipitate vary to some extent with its object.

When the intention is to secure in the precipitate substantially all the copper from the solution, the procedure in the main is as follows: To the cold, concentrated and very slightly acid copper solution which must be free of any oxidizing agent, sulphur dioxide, gaseous or in solution, or a solution of an alkaline bisulphite or metabisulphite is added somewhat in excess of the quantity theoretically required to reduce all the copper and ferric iron present, or until saturation of the liquid with SO_2.

[1] W. B. Price, Jour. Ind. Eng. Chem., Vol. **6**, No. 2, p. 170.

Cool the liquid if hot through the formation of H_2SO_3, and add then with constant stirring a solution of alkali thiocyanate of about normal strength until precipitation ceases. Reaction is $2CuSO_4 + 2KCNS + H_2SO_3 = 2Cu-CNS + 2H_2SO_4 + K_2SO_4$. It is common practice to continue introduction of SO_2 throughout precipitation. Let the precipitate stand until it is white and the liquid above it is clear. The presence of $FeSO_4$ accelerates conversion of cupric to cuprous thiocyanate when sulphuric is the only free acid. Filter with the aid of reduced pressure through doubled filter papers of tight texture. Wash with cold water until a washing is obtained which gives no or very slight indication of thiocyanate when tsted with a ferric salt.

The precipitate is inclined to float and creep with the capillary film of fluid, so precautions must be taken in its manipulation to avoid loss of copper on account of this characteristic.

The collected precipitate may now be treated in several ways which will produce a solution fit for the determination of copper by one of the standard methods. (a) Transfer the filter and precipitate to a porcelain or silica crucible which is very much larger than the volume of the wet precipitate, dry slowly in an oven or muffle and finally incinerate. Some operators made the final washing with 20% alcohol to facilitate drying or with a weak solution of an alkali nitrate to aid incineration. Dissolve the residue in the crucible with hot, strong nitric acid. (b) The point of the filter is punctured and the precipitate washed with as little water as possible into a flask or tall beaker. The filter is finally cleansed of adherent precipitate by washing with dilute nitric acid. The filter is dried, incinerated and dissolved separately or added to the main precipitate before its decomposition. Add 15 cc. strong nitric acid for each gram or fraction of a gram of copper present and let the covered beaker or funnel-closed flask stand in a warm place until the precipitate is dissolved, then boil until solution is free of the nitrogen gases. It is the practice of some to add now 10–15 cc. sulphuric acid and evaporate to fumes. Because evolution of gas during dissolution of the precipitate is profuse, care must be taken to expel the gas slowly to prevent boiling over.

The method as described is rather tedious in its operation and therefore is commonly carried out with modifications which may include: Employing a mixture of the precipitating and reducing agents (for instance, 25 grams alkali thiocyanate with an equal or larger amount of alkali bisulphite in a liter solution); quickening settlement of the precipitate by letting the assay stand in a warm place or even bringing the fluid to boiling; filtering through asbestos, alundum or porous stoneware; washing with hot water. Some of these modifications yield a certain amount of copper to the filtrate because of the appreciable solubility of cuprous thiocyanate in warm or hot water. Consistent with the value of the determination, the amount so dissolved is disregarded, accounted for by an empirical correction, or recovered by expelling SO_2 and concentrating filtrate and washings by boiling, precipitating with H_2S and determining copper, usually by a colorimetric method.

Separation of Copper by Precipitation in Metallic Form by a More Positive Element. Metallic aluminum or zinc is more commonly used in this procedure. A strip of pure aluminum or zinc, placed in the neutral or slightly acid solution, causes the complete deposition of copper. To obtain quick precipitation, a sheet of aluminum, 2.5 by 14 cm., is bent to form a triangle, placed in a covered 150 cc. beaker containing the copper solution, which should be not much over 75 cc. in volume and should hold about 10% free H_2SO_4. Boil 7-10 minutes. (Low, "Technical Methods of Analysis.") The copper is removed mechanically from the displacing metal and dissolved in nitric acid and then estimated, or the aluminum may be dissolved with the copper.

A method of precipitation by means of powdered aluminum is recommended especially for separation of copper from large amounts of iron, iron ores and iron ore briquettes. The solution of the bisulphate fusion of the iron ore is heated until bubbles appear over the bottom of the containing beaker. Aluminum powder is now added in small portions at a time, in sufficient quantity to reduce the iron, the solution becoming colorless. The solution is now heated until the aluminum completely dissolves. Metallic copper is precipitated. It is advisable to add 25 cc. of water saturated with H_2S gas to precipitate traces of copper in solution. The solution is filtered while hot through a close thick filter, and washed six times, keeping the residue covered with water to prevent oxidation by air. The copper is now dissolved in hot dilute nitric acid, evaporated to small volume and determined by the procedure preferred. The potassium iodide method gives excellent results.

Separation of Copper from Members of the Ammonium Sulphide and Subsequent Groups by Precipitation as Copper Sulphide in Acid Solution. The solution containing free hydrochloric or sulphuric acid is saturated with H_2S gas,[1] the precipitated copper sulphide (together with the members of the group), is filtered and washed, first with water containing H_2S and finally with a little pure water. The residue is dissolved in nitric acid and the resulting solution examined for copper.

Removal of Silver. This element is precipitated as the insoluble chloride, AgCl, by addition of hydrochloric acid, and may be removed by filtration, copper passing into the filtrate.

Removal of Bismuth. Upon adding ammonium hydroxide to a solution containing copper and bismuth the latter is precipitated as $Bi(OH)_3$ and may be removed by filtration. Copper passes into the filtrate as the double ammonium salt. Ammonium carbonate or potassium cyanide may be used instead of ammonium hydroxide.

Removal of Lead. Lead is precipitated by sulphuric acid as $PbSO_4$ and may be removed by filtration, copper passing into the filtrate.

Removal of Mercury. The sulphide of mercury remains undissolved when the precipitated sulphides are treated with dilute nitric acid, copper sulphide dissolving readily.

Removal of Selenium and Tellurium. Selenium can be eliminated from a copper solution by evaporating several times to dryness with hydrochloric acid. Saturation with SO_2 of a slightly acid sulphate of copper solution which contains about twice as much silver as selenium and boiling then to expel most of the gas, will precipitate selenium and tellurium free of copper, also nearly all the silver. Precipitation of the remainder of the silver as AgCl helps to retain the fine precipitate on the filter.

Removal of Arsenic, Antimony, Tin, Selenium and Tellurium. Copper in minor quantity may be separated from these elements by passage of H_2S into the solution made slightly alkaline with sodium hydrate; after addition of about a gram of tartaric acid if iron is present. Copper sulphide remains insoluble also when the mixed sulphides precipitated from an acid solution are treated with a hot mixture of sodium sulphide and hydrate. These elements in minor quantity, whose presence in the electrolyte for copper determination is an objectionable impurity, are removed very satisfactorily by adding enough ferric iron to the solution to make the total iron present about twenty times that of the combined impurity, making ammoniacal and filtering after settling. Some copper is retained in the precipitate. Moderate amounts of arsenic may be eliminated as arsenious fluoride by the method employed to expel silica from an assay. Oxidizing agents must not be present.

In an alloy tin and antimony may be precipitated as oxides by evaporation of the solution of the alloy with strong nitric acid. A slight amount of copper may remain insoluble.

Separation from Cadmium. The sulphides in a solution of dilute sulphuric acid, 1 : 4, are boiled and H_2S gas passed in for twenty minutes, the solution being kept at boiling temperature. Cadmium sulphide dissolves while copper sulphide remains unaffected. The solution is filtered hot, the air above the filter being displaced by CO_2 to prevent oxidation. Traces of cadmium are removed by repeating the operation. (Method by A. W. Hofmann.)

GRAVIMETRIC METHODS
Methods of Isolating Copper

Separation of Copper as Cuprous Thiocyanate. Isolation of copper from solutions containing iron, nickel, cobalt, zinc, cadmium, arsenic, antimony and tin may be accomplished by this method. When much arsenic is present, precipitation should be from a solution in which hydrochloric is the only free strong acid. Unless previously removed from the solution, lead, mercury, tellurium and the precious metals will contaminate the precipitate. Selenium may be a contaminant when present in considerable quantity, sometimes when the only free acid is sulphuric, always when hydrochloric acid is present.

Cuprous thiocyanate, besides being the medium of separation of copper from interfering elements preliminary to its determination by the standard electrolytic, iodide or cyanide methods, is the basis of a number of other more or less useful gravimetric and volumetric methods of determining copper. The details of the procedure of procuring the precipitate vary to some extent with its object. Low grade copper ores may be conveniently determined by this method.[1]

Procedure. To the cold, concentrated and very slightly acid copper solution (which must be free of any oxidizing agent) sulphur dioxide, gaseous or in solution, or a solution of an alkaline bisulphite or metabisulphite is added somewhat in excess of the quantity theoretically required to reduce all the copper and ferric iron present.

The liquid is cooled if hot, and then, with constant stirring, a solution of alkali thiocyanate of about normal strength is added until precipitation ceases. Reaction: $2CuSO_4 + 2KCNS + H_2SO_3 + H_2O = 2CuCNS + 2H_2SO_4 + K_2SO_4$. It is common practice to continue introduction of SO_2 throughout precipitation. The precipitate is allowed to stand until it is white and the liquid above it is clear. The presence of $FeSO_4$ accelerates conversion of cupric to cuprous thiocyanate when sulphuric is the only free acid. The precipitate is filtered off with the aid of reduced pressure through doubled filter papers of tight texture and washed with cold water until a washing is obtained which gives but very slight indication of thiocyanate when tested with a ferric salt.

The precipitate is inclined to float and creep with the capillary film of fluid, so precautions must be taken in its manipulation to avoid loss of copper on account of this characteristic.

The collected precipitate may now be treated in several ways which will produce a solution fit for the determination of copper by one of the standard methods. (a) The filter and precipitate are transferred to a porcelain or silica crucible which is very much larger than the volume of the wet precipitate, dried slowly in an oven or muffle and finally incinerated. Some operators made the final washing with 20% alcohol to facilitate drying or with a weak solution of an alkali nitrate to aid incineration. The residue is dissolved in the crucible with hot, strong nitric acid. (b) The point of the filter is punctured and the precipitate washed with as little water as possible into a flask or tall beaker. The filter is finally cleansed of adherent precipitate by washing with dilute nitric acid. The filter is dried, incinerated and dissolved separately

---

[1] With high grade ores or copper bullion a trace of copper (usually less than 0.0005 g. Cu) will pass into the filtrate when the copper precipitated amounts to 0.5 gram Cu. The loss with low grade copper ores is negligible.

or added to the main precipitate before its decomposition. 15 cc. strong nitric acid is added for each gram or fraction of a gram of copper present and the covered beaker or funnel-closed flask allowed to stand in a warm place until the precip.tate is dissolved, then boil until solution is free of the nitrogen gases. It is the practice of some to add now 10–15 cc. of sulphuric acid and evaporate to fumes. Because evolution of gas during dissolution of the precipitate is profuse, care must be taken to expel the gas slowly to prevent boiling over.

Separation of Copper by Precipitation in Metallic Form by a More Positive Element. Metallic aluminum or zinc is more commonly used in this procedure. A strip of pure aluminum or zinc, placed in the neutral or slightly acid solution, causes the complete deposition of copper. To obtain quick precipitation, a sheet of aluminum, 2.5 by 14 cm., is bent to raise the metal from the bottom of the beaker, placed in a covered 150 cc. beaker containing the copper solution, which should be not much over 75 cc. in volume and should hold about 10% of free H_2SO_4. Boil 7–10 minutes. The copper is removed mechanically from the displacing metal and dissolved in nitric acid and then estimated.

A method of precipitation by means of powdered aluminum is recommended especially for separation of copper from large amounts of iron, iron ores and iron ore briquettes. The solution of the bisulphate fusion of the iron ore is heated until bubbles appear over the bottom of the containing beaker. Aluminum powder is now added in small portions at a time, in sufficient quantity to reduce the iron, the solution becoming colorless. The solution is now heated until the aluminum completely dissolves. Metallic copper is precipitated. It is advisable to add 25 cc. of water saturated with H_2S gas to precipitate traces of copper in solution. The solution is filtered while hot through a close thick filter, and washed six times, keeping the residue covered with water to prevent oxidation by air. The copper is now dissolved in hot dilute nitric acid, evaporated to small volume and determined by the procedure preferred. The potassium iodide method gives excellent results.

Occasionally the aluminum lies inert in the solution. If this occurs, two or three drops of hydrochloric acid (do not use much) will start a vigorous action and cause a rapid precipitation of metallic copper.

Deposition of Metallic Copper by Electrolysis

The electrolytic method of determining copper is the most accurate of the gravimetric methods. This deposition may conveniently be made from acid solutions containing free nitric or sulphuric acid) or from an ammoniacal solution.

FIG. 27.—Terminal Case Showing Battery of Electrodes for Electrolytic Deposition of Copper.

The end sought by this method is to plate out all, except a trace, of the copper in the form of an evenly distributed, firmly adherent, very finely crystalline deposit, which is free from a weighable amount of impurity.

17

In ores, mattes, alloys (from which lead has been removed as the sulphate by taking the solution to fumes with sulphuric acid) deposition by electrolysis, from a solution containing free sulphuric acid, is convenient. On the other hand, deposition from a nitric acid solution is advantageous under conditions where this reagent has been used as a solvent and evaporation with sulphuric acid is unnecessary. This is the case in the analysis of certain alloys and the determination of copper from which impurities have been largely removed. A chloride in an acid solution gives rise to a spongy deposit of copper, and endangers a solvent action on the anode and deposition of platinum on the cathode.

Conditions other than the presence of precipitable impurities, which affect the character of the deposit are—quantity and concentration of copper, size and shape of electrodes, current density, uniformity of distribution of current to the cathode, volume, temperature and rate of circulation of the electrolyte, and concentration of oxidizing agents such as nitric acid and ferric salts. Inasmuch as the change of one condition limits or makes possible or necessary a modification of others, a large number of practicable combinations of conditions are possible. For discussion of these conditions reference is made to articles by Blasdale and Cruess, Jour. Am. Chem. Soc. Oct. 1910, 1264; and by Richards and Bisbee, Jour. Am. Chem. Soc., May, 1904, 530.

By the feature of rate of deposition, electrolytic methods may be classified as "slow" or "rapid." The slow methods, with 12 to 24 hour periods of electrolysis, are practiced when extreme accuracy is required, or when the distribution of laboratory labor and time allowed for completion of the assays permit their economical employment. The electrolyte is a solution of sulphate salts of the metals present, ammonium sulphate or nitrate, and a quantity of free nitric acid, which varies with the amount of copper and ferric salts present, and the current density employed. The oxidizing effect of nitric acid is intensified by the presence of ferric ions.[1] Electrolysis is carried out at room temperature, at current densities varying from $ND/100$, 0.15 to 0.5 amperes; and deposition on plain, corrugated, slit or perforated platinum cylinders from 0.75 to 2 in. diameter having 50 to 200 cm. depositing surface. A perforated cylinder permits freedom of circulation between the two surfaces of the electrode, the most even distribution of current density, and produces the most uniform coating of the foil. On account of the effect on the character of the deposit by oxygen lodging in regions of the cathode where the current density and circulation is least, the anode should be of such a form that all the gas liberated will be in the zone of maximum circulation. To procure uniform behavior under given conditions the size and shape of the electrolytic beaker should be such as to present the smallest practicable volume of electrolyte between the outer surface of the cylinder and the inside of the beaker. An unclosed seam or rivetted joint in a negative electrode will hold tenaciously salts which require extreme care to remove. It is probable that such recesses retain traces of the electrolyte underneath the coating of copper.

Rapid methods have a tendency to procure high results, resolution and mechanical loss through misting having been prevented. Deposition is hastened by increasing the rate of circulation and the current density. Circulation is

[1] Larison, Eng. and Min. Jour. 84, 442. Fairlie and Boone, Elect. and Met. Ind. 6, 58.)

promoted by the use of the gauze cathode,[1] by rotating either cathode,[2] or by placing the vessel, containing the solution and electrodes, in a field of electromagnetic force.[3] Quick deposition of a quality satisfactory for some classes of work is brought about by increase of current density upon an electrolyte heated to 50° to 80° C. In all the quick methods, the progress of electrolysis should be watched, and the cathode removed as soon as completion of deposition is detected by the evolution of gas about its surface. The completion of action is ascertained with greater certainty by addition of water to the electrolyte and observing whether the newly exposed surface of the cathode remains bright. When the electrolyte is hot or has a high acid content, detachment of the cathode should be preceded by removal of the electrolyte and simultaneously washing the cathode without interruption of the current. A syphon may be employed, water being added as the liquid drains from the beaker until the acid is removed.

RAPID METHODS

Rapid Deposition of Copper—Solenoid Method of Heath [4]

The solenoid is made by winding 500 turns of No. 13 B and S gauge magnet wire upon a copper cylinder $2\frac{3}{4}$ in. in diameter, $3\frac{1}{4}$ in. high, $\frac{3}{32}$ in. thickness of metal. The cylinder is brazed water tight at the bottom to a $5\frac{1}{2}$ in. disc of $\frac{3}{32}$ in. soft steel. In this disc is a 1-in. hole for the insertion of a rubber plug, through which glass tubes may be inserted for inlet and outlet of air or water to cool the electrolytic beaker. A steel disc of the same size as the bottom and with an opening to fit is brazed to the top of the cylinder. The solenoid thus made is suitable for a 300 cc. lipless beaker $4\frac{7}{8}$ in. high and $2\frac{1}{4}$ in. diameter. The solenoid coil may be in series in the electrolytic line or excited separately.

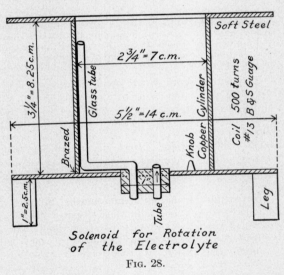

Solenoid for Rotation of the Electrolyte

FIG. 28.

The negative electrode is of gauze 40 meshes per linear inch, with a depositing surface of 100 cm. and is slit to permit quick removal from the electrolyte.

[1] Stoddard, Jour. Am. Chem. Soc., 1909, 385. Price and Humphreys, Jour. Soc. Chem. Ind., 1910, 307.
[2] Eng. and Min. Jour., 89, 89, 1910.
[3] Frary, Jour. Am. Chem. Soc., Nov., 1907, 1592. Heath, Jour. Ind. Eng. Chem., Feb., 1911, 74.
[4] Heath, Jour. Ind. Eng. Chem., Feb., 1911, 76.

Procedure. Five grams of the thoroughly cleaned copper sample are dissolved in the covered electrolytic beaker on a steam plate with 40 cc. of stock acid solution composed of 7 parts (1.42 sp.gr.) nitric acid, 10 parts sulphuric acid (1.84 sp.gr.) and 25 parts by volume of water. The temperature during the solution is kept just below the boiling point, 50 cc. of the stock solution is used for copper containing 0.03 to 0.1 per cent of arsenic, 60 cc. for material containing 0.11 to 0.5 per cent arsenic. The electrolyte is diluted to 120 cc. A current of 4.5 amperes is used for the electrolysis and the same amount employed to excite the solenoid. During the deposition a double pair of watch glasses cover tightly the beaker until the color of the electrolyte fades out, when they are rinsed and removed. Twenty minutes later and thereafter at 5 minute intervals, a test for completion of deposition is made by withdrawal of 1 cc. onto a porcelain tile and treating with a few drops of freshly prepared hydrogen sulphide water. This test will detect the presence of 0.000005 g. copper or more remaining in the solution. The determination is complete in two and a half hours. Extremely accurate results are obtained when the electrolyte is kept very cold by circulation of water about it and when the cathode is withdrawn within 5 minutes after completion of deposition.

In the assay of casting copper, in case the deposit is evidently impure, the cathode may be stripped by treatment with 50 cc. of the stock solvent and then replated under the conditions described.

NOTES. The advantage of the solenoid over any mechanical device for the rotation of electrodes is due to the prevention of loss by spraying from the anode, as the beaker can be covered with a double pair of watch glasses.

Results range from 0.003 to 0.01 per cent higher than the author's slow method of assay of refined copper, and is due to platinum from the anode, which is corroded by the influence of heat, nascent nitrous acid and high current.

Deposition from Nitric Acid Solution. The solution should not contain over 2–3 cc. of free concentrated nitric acid. If more than this is present, the solution is evaporated to expel most of the acid, the remainder neutralized with ammonia and the requisite amount of nitric acid added. The solution is diluted to 100 cc., warmed to 50° or 60° C. and electrolyzed with a current of 1 ampere and 2–2.5 volts. Two hours are sufficient to deposit 0.3 gram copper. Since the hot acid acts vigorously on copper, it is necessary to wash out the acid from the beaker before breaking the current. (See method for copper in alloys, page 207.)

Deposition from an Ammoniacal Solution. Ammonium hydroxide is added to the solution containing copper until the precipitate, first formed, dissolves. Twenty to twenty-five cc. of ammonium hydroxide (sp.gr. 0.96) are required for 0.5 gram copper or 30-35 cc. for 1 gram. Three to four grams of ammonium nitrate are added and the solution electrolyzed with a current of $ND/100 = 2$ amperes. The electrodes are washed, without breaking the current, until the ammonia and nitrate are removed.

Lead, bismuth, mercury, cadmium, zinc and nickel should be absent from the ammoniacal solution. Arsenic is not deposited. Unless a very pure platinum anode is used, platinum may contaminate the deposit appreciably. Jena or other brand of zinc borate resistance glass should not be used for the electrolytic beaker.

SLOW METHODS

Electrolytic Determination of Copper in Blister Copper

The sample should be no coarser than 20 mesh. Because fine particles are comparatively poor in copper, extreme care must be taken in drawing the portion for analysis to preserve the ratio of the coarse to fine. Some analysts, to avoid sampling error, sieve the coarse from the 40 or 60 mesh fine and either make a separate analysis of each weighed product, or weigh into a single test the due proportion of each. Others draw a large portion, by means of a riffle (Fig. 29) or similar sampling device and from its solution in a volumetric flask, pipette an aliquot part equivalent to one or more grams.

By the small portion method insoluble matter must be removed by filtration. When the sample contains an insignificant quantity of insoluble matter, the practice is to deposit the silver with the copper and make a correction for its presence in accordance with the result of the silver assay of the sample.

By the large portion method, insoluble matter and silver, as silver chloride, is removed from the electrolyte by sedimentation in the volumetric flask.

Procedure. Small Portion Method. The coarse and fine portions are quartered down to convenient amounts and from these a 5-gram composite weighed, which contains the coarse and fine portions in ratio of their percentage weights. The sample is placed in a 350-cc. tall-form beaker, without lip and with flaring rim. Fifty cc. of chlorine-free, stock acid solution (15 parts nitric and 5 parts sulphuric acids) are added, the beaker covered with a funnel (stem up), which just fits in the rim, and the mixture heated gently at first and finally to boiling. When the sample has dissolved, 5 cc. saturated solution of ammonium nitrate are added and the sample

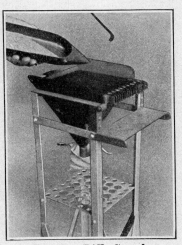

Fig. 29.—Riffle Sampler.

diluted to 200 cc. with water.

When the electrolyte has cooled to room temperature the electrodes are introduced, the beaker covered with split watch glasses and electrolysis started with a current of .05 ampere and continued until the appearance of the foil indicates that the silver has deposited. The current is then raised to ND/100 = .75 ampere and this continued for twenty to twenty-two hours, or until the appearance of gas about the negative electrode indicates that deposition of the copper is practically complete. For the unexperienced a simple method is to add a little water to the electrolyte without breaking the current and after 15 minutes to observe whether any deposition or copper takes place on the freshly exposed surface. The watch glasses and electrode stems should be rinsed when the electrolysis has continued 15–16 hours.

Procedure. Large Portion Method.[1] The sample is quartered by a riffle sampler (see Fig. 29) to an amount very close to 80 grams. This quantity is weighed and transferred by a paper chute into a 2000 cc. flask, which has been calibrated by the method of repeated delivery at constant temperature, of a 50 cc. overflow, dividing pipette (see Fig. 30). The liquid employed in calibrating is a copper solution of the same composition as that for which the flask is to be used. A cold mixture of 80 cc. sulphuric acid (sp.gr. 1.82) and 200 cc. nitric acid (1.42) with 500 cc. of water is added. A standard solution of sodium chloride is added in sufficient quantity to precipitate the silver, care being taken to add less than 20% excess. A bulbed condenser tube is placed in the neck before putting the flask on a hot plate.

The solution is gradually heated to boiling and when the solution is nearly complete, boiled gently for one hour. This generally completely dissolves the copper present. Residues of lead, tin, silver, or silica if present in appreciable amounts are separated at this point by filtration.

When the solution in the flask has cooled for half an hour, water is added to a little above the 2000-cc. mark, giving the flask a rotary motion, during the addition, to mix the solution. The flask is placed in a large tank, Fig. 30,

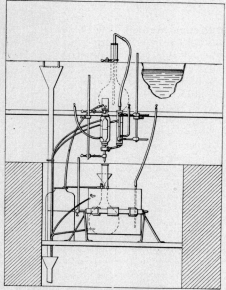

FIG. 30.—Constant Temperature Bath and Dividing Pipette.

containing water and allowed to remain until it becomes of the same temperature as the water and very close to that of the room. The solution is then made exactly to the mark and allowed to settle, after thorough mixing, by placing the flask again in the water tank.

[1] W. C. Ferguson, Jour. Ind. and Eng. Chem., May, 1910.

Electrolysis. Portions equivalent to 2 grams of sample are measured out by means of a dividing pipette, with water-jacket through which the tank-water flows. The solution is run into glasses, hydrometer-jar in shape, with concave bottoms, height of glass, $6\frac{1}{2}$ ins., diameter $2\frac{1}{6}$ ins., Fig. 31. Each portion is treated with 5 cc. of a saturated solution of ammonium nitrate and diluted to 125 cc. with water. (NH_4NO_3 or $(NH_4)_2SO_4$ delays deposition of As and Sb until electrolyte is freed from Cu.) The electrolyte, at this stage, contains about 3.7 cc. of nitric acid.

Fig. 31.

The copper is deposited by electrolysis, using a current of .33 ampere per 100 sq.cm., which is kept constant until deposition is complete, about twenty hours. It is advisable to begin the electrolysis in the evening, 5 P.M. The following morning, the inside of the jar, the rods of the electrodes, and the split watch-glasses which cover the jar are rinsed with a spray of water into the glass and the run continued for two or three hours. Each electrode is quickly detached from the binding posts, the cathode plunged into cold water, then successively into three jars of 95% alcohol, shaken free of adherent drops and dried by revolving rapidly over a Bunsen flame for a few seconds after ignition of the film of alcohol.

The weighing of foil plus the deposit is made with as little delay as possible.

Determination of the Copper Remaining in the Electrolytes. Since the exhausted electrolyte seldom contains over 0.01% copper, this residual copper can be closely estimated by observation of the depth of the sulphide precipitate. Should circumstances be such that the quantity be more than can be estimated by the appearance of the sulphide precipitate and a determination without rerun is necessary, the precipitate is filtered, incinerated, dissolved with a little hot HNO_3, made ammoniacal and after settling, filtered through asbestos. The color of the solution is compared with a standard solution treated with the same amount of reagents as the sample, care being taken that similar conditions prevail when making comparison.

Notes and Precautions

Character of the Deposits. The ideal deposit is of a salmon-pink color, silky in texture and luster, smooth and tightly adherent. A slightly spongy and coarsely crystalline deposit, although good in color and perfectly adherent, will invariably give high results. A loosely adherent deposit caused by either too rapid a deposition at the commencement or too low a current density at some period of the electrolysis, usually shows a red tint and may give a high result on account of oxidation or a low result because of detachment of particles. A darkly shaded deposit indicates the presence of impurity in greater or less extent. If it is impossible to complete the electrolysis without this appearance the electrolyte should be purified. Impurities such as arsenic, antimony, bismuth, selenium and tellurium may occur in the blister copper.

A dark colored, but perfectly adherent deposit is dissolved very slowly from the foil, in a covered electrolytic jar, by gently heating for several hours with about 60–70 cc. of a solution containing 2 cc. sulphuric and 5 cc. nitric acids. When the solution is complete the temperature is raised to expel dissolved gases. Five cc. saturated ammonium nitrate solution is added and the electrolyte diluted to 125 cc. When cooled to room temperature, electrolysis is carried out under the same conditions as that of the first deposit and on the same foil, if arsenic or antimony is the interfering impurity; on a fresh foil if selenium or tellurium has been the contaminating element. The undeposited copper is determined colorimetrically in the mixture of the first and final electrolytes and added to the weight of the copper deposited.

If the sample contains a large percentage of arsenic or antimony, a portion representing 2 grams is drawn from a pipette into a Kjeldahl flask, 10 cc. of sulphuric acid added, and the liquid evaporated to fumes to expel nitric acid. From this solution cuprous thiocyanate is precipitated according to the method described on page 153. The funnel containing the filter is placed in a 500 cc. flask with long neck, the filter is punctured and the precipitate washed into the flask with the least quantity of water possible, the adherent precipitate is dissolved from the filter with warm dilute nitric acid, added cautiously to avoid violent evolution of gases from the dissolving precipitate in the flask. The washed filter is incinerated and the solution of its ash by nitric acid added to the electrolyte after completion of electrolysis. When solution of the precipitate is complete, the liquid is boiled to small volume, neutralized, and 5 cc. ammonium nitrate solution and 3 cc. excess free nitric acid added. The liquid is transfered to an electrolytic jar and electrolysis carried out in the manner already described.

The amounts of bismuth, arsenic, antimony, selenium or tellurium usually found in blister copper may be separated together with iron present by addition of ammonia to a pipetted portion. The filtered precipitate is purified of copper by solution with nitric acid and reprecipitation. The combined filtrates are neutralized, $3\frac{1}{2}$ cc. of free nitric acid added and the solution electrolyzed under the conditions already described. The nitric acid solution of the incinerated filter, carrying the iron, etc., is added to the electrolyte, after electrolysis is complete, for determination as undeposited copper. The undeposited copper is determined colorimetrically by one of the procedures outlined on pages 165, 166 or 167.

The deposited copper is never absolutely pure. The total impurities seldom exceed 0.03%. Ag from 0.000 to 0.18%; As from 0.000 to 0.003%; Sb from 0.000 to 0.004%; Se and Te from 0.001 to 0.027%; Bi from 0.000 to 0.0003%. Periodical complete analyses may be made and corrections applied to the analysis when exceedingly accurate percentages are required.

Too low a current density or excessive oxidizing power of the electrolyte may produce high results, due to the oxidation of the deposited copper. Too high a current density or a deficiency of oxidizing power in the electrolyte, by causing a deposition of impurities, will give high results.

The electrodes used by the Nichols Copper Co. are straight platinum wires for the positive ends and cylinders $1\frac{3}{4}$ in. long, 1 in. in diameter of 0.004 in. iridoplatinum foil, $11\frac{1}{2}$ sq. in. depositing surface, for the cathodes.

A uniform current is essential.

The nitric acid used should be free of iodic acid.

The presence of oxide of nitrogen gases, or a chloride in an acid solution, will cause a coarsely crystalline or brittle deposit, under conditions which in their absence would produce a good plating. The deposit moreover may contain platinum from the anode if the electrolyte contains a chloride salt.

Silver may be deposited with the copper and correction made for its presence from the result of a separate assay. Copper deposits in poor form, unless the silver be first plated out at a very low current density, about 0.03 Amp. ND_{100}.

Solid matter, unless removed, will contaminate the deposit mechanically.

Arsenic, antimony, selenium or tellurium have an influence on the physical character of the deposit which may affect the copper result beyond the sum of such impurities deposited.

Whether impurities are deposited or not, appreciably high results are obtained by continuing electrolysis for some time after the electrolyte has become impoverished of copper.

Overheating of the copper deposit, in the process of ignition of the alcohol clinging to the cathode, will cause oxidation of the copper. As much as possible of the alcohol must be shaken off before passing the electrode rapidly through the flame. It is advisable to weigh the copper shortly after deposition, as prolonged contact with air is undesirable, if extreme accuracy is desired.

The copper deposits may be removed by plunging the electrode, for a few moments, in hot nitric acid. After washing with water, the foil is ignited to a cherry red in a direct colorless flame. The ignition removes any grease which would be objectionable, that may contaminate the platinum. Alcohol frequently contains oily matter which will cling to the electrode in spite of the rapid ignition for drying the deposit.

VOLUMETRIC METHODS FOR THE DETERMINATION OF COPPER

Potassium Iodide Method

The procedure depends upon the fact that cupric salts when treated with potassium iodide liberate iodine, the cuprous iodide formed being insoluble in dilute acetic acid.

Reactions. $2CuSO_4 + 4KI = Cu_2I_2 + 2K_2SO_4 + I_2,$
or $2Cu(C_2H_3O_2)_2 + 4KI = Cu_2I_2 \downarrow + 4KC_2H_3O_2 + I_2.$

The liberated iodine is titrated with standard thiosulphate.

$$2Na_2S_2O_3 + I_2 = Na_2S_4O_6 + 2NaI.$$

The method is extremely accurate.   Only a few elements interfere, such as selenium, trivalent arsenic, antimony and iron.   Lead, mercury and silver increase the consumption of potassium iodide, but otherwise interfere only because of the color of their iodides.   No free iodine is consumed by these.

N/10 solution contains 24.82 grams $Na_2S_2O_3.5H_2O$ per liter.   According to the reaction above, 1 cc. of this will be equivalent to 0.006357 g. Cu.

If a solution is desired so that 1 cc. = 0.001 g. Cu, dissolve 3.92 grams of the $Na_2S_2O_3.5H_2O$ crystals per liter (248.2 ÷ 63.57).   If a 0.5 gram sample is to be taken, it is convenient to have the reagent of such strength that 1 cc. is equivalent to 0.005 g. Cu so that 1 cc. is equivalent to 1% Cu.   3.92 × 5 = 19.6 *grams* of thiosulphate crystals are required.   The addition of 2–4 grams NaOH per liter is claimed to preserve the solutions.

Standardization.   Copper Method.   If either a N/10 reagent or the one containing 19.6 grams of the salt are being standardized, weigh 0.2–0.25 gram of pure copper (electrolytic) (not over 0.1 gram should be taken for the weakest solution)   Dissolve the copper in about 5 cc. of nitric acid (in a flask or a covered beaker), boil to expel brown fumes, dilute to about 50 cc., add ammonia in slight excess (10–12 cc. is generally sufficient), boil until the odor of ammonia is faint and add 5 cc. glacial acetic acid, or sufficient to make the solution acid (test with litmus) and again boil for about one minute.   Cool and add to the cold solution 6 cc. of a 50% potassium iodide solution or about 3–4 grams of the solid, and titrate with the standard thiosulphate reagent.   The iodine that is liberated (see reaction above) colors the solution brownish; on addition of the thiosulphate the color fades; when the brown color has become faint, add starch solution and complete the titration adding the thiosulphate until the blue color (lilac at end) fades out with an added drop of the reagent.

Divide the weight of copper taken by the cc. of reagent required and the result is the weight of copper represented by one cc. of the reagent.   An exactly N/10 solution is equivalent to 0.006357 g. Cu.

Standardization with Permanganate.   The reagent may be standardized against N/10 potassium permanganate or a permanganate solution whose iron equivalent is known.   To about 40 cc. of the N/10 permanganate solution add 6 cc. of the potassium iodide (50%) reagent and titrate the liberated iodine in presence of acetic acid exactly as is described above.   1 cc. of N/10 permanganate should equal 1 cc. of the N/10 thiosulphate.   Establish its normality by converting the permanganate to exact cc. in normality equivalent and dividing the cc. by the cc. of thiosulphate required. If preferred, multiply the iron equivalent by $\dfrac{63.57}{55.84} = 1.139$ to get the copper equivalent.

Procedure for Copper in Ores

Solution of the Sample. Weigh 0.5 gram of the powdered ore (if a N/10 thiosulphate reagent is to be used, the factor weight 0.636 gram of ore is convenient so that 1 cc. of the reagent is equivalent to 1% Cu). Dissolve in 10 cc. hydrochloric acid and 5 cc. nitric, heating gently to effect solution, adding more of the acids if necessary. Add 10 cc. of sulphuric acid and evaporate to fumes. Heat until any free sulphur that precipitates disappears. Allow to cool. Dilute with 30–40 cc. of water, heat to boiling and keep hot until ferric sulphate has dissolved (copper will all be in solution). Filter into a small beaker and wash the residue at least six times with hot water, using small portions at a time. The volume of the filtrate need not exceed 75–100 cc. This contains all of the copper.

Precipitation of Metallic Copper. Place in the beaker containing the copper solution a piece of heavy sheet aluminum (1.5 inches square) or a bent strip of aluminum bent in form of a triangle, or add 1–2 grams of pure granulated aluminum. If a sheet is used, bend the corners at right angles so it will not lie flat in the beaker. (The sheet may be used repeatedly as it is not attacked to any great extent by the nitric acid subsequently used). Heat the solution to boiling (beaker covered) and keep at this temperature for about 10 minutes. All of the copper will precipitate an equivalent amount of aluminum dissolving. If the action is sluggish, add 2 or 3 drops of hydrochloric acid. Wash down the cover and sides of the beaker with hydrogen sulphide water. (This precipitates any trace of copper in the solution and prevents oxidation of the copper.)

Isolation of Copper. Decant the solution through a filter and rinse the metallic copper into the filter with a jet of the hydrogen sulphide water, leaving the aluminum as clean as possible in the beaker. Save this. Wash the precipitate 6 times with hydrogen sulphide wash water, allowing to drain, but following up immediately with more solution, until the washing is complete. (Copper will oxidize if allowed to stand exposed to the air. Hence the washing should be completed as soon as possible.)

Solution of the Copper. Punch a hole in the filter and wash the copper into a beaker with a jet of water, using as little as possible. (If much copper is present, open the filter on a watch glass and wash the precipitate into a beaker. Again fold the filter and place in the funnel over the beaker.) Pour 5 cc. of strong nitric acid over the aluminum which still contains a little of the copper. When all of this has dissolved, pour the nitric acid solution over the filter, catching the acid in the beaker containing the bulk of the copper. (Removing the beaker, place the one containing the foil under the filter.) Cover the beaker containing the copper and boil until the metal is in solution and again place under the filter funnel. Wad the filter paper loosely in the funnel. Pour over this filter 5–10 cc. of bromine water, catching the solution in the beaker containing the copper; this should impart a yellow color to the copper solution. Wash the filter 6 times with water, pouring the water first into the beaker with the aluminum and from this over the wadded filter.

Concentration. Boil the combined solution down to about 25 cc. The bromine will be expelled. Add a small excess of ammonia; the free acid is neutralized and the solution smells of ammonia. Again boil to expel the excess of ammonia. The solution should still smell faintly of NH_3. Add glacial acetic acid until slightly acid (litmus test); 5 cc. should be sufficient. Cool thoroughly.

Potassium Iodide Treatment and Titration with Thiosulphate. To the cold solution add 5–6 cc. of 50% solution of potassium iodide and titrate the liberated iodine with standard thiosulphate until the brown color has become faint, add starch and complete the titration.

$$1 \text{ cc. } N/10 \text{ } Na_2S_2O_3 = 0.00636 \text{ g. Cu.}$$

NOTE. Since a yellow color may be due to other causes than to free iodine, it is advisable to make a preliminary titration by adding the starch indicator before starting titration with thiosulphate. In a check run the starch is added upon neutralizing the greater part of the free iodine. This precaution prevents overrunning the endpoint, which one is apt to do when depending upon a color change of iodine brown to yellowish brown due to a trace of free iodine.

Short Iodide Method for Copper

The method takes advantage of the repression of ionization of iron by addition of potassium fluoride to a neutral or acetic acid solution containing the ferric salt. Ag, As, Bi, Cd, Co, Fe, Hg, Mn, Mo, Ni, Pb, Sn, U, Zn do not interfere. Cr forms an insoluble sulphate, which holds Cu, hence avoid using H_2SO_4. V interferes.

Procedure

Decomposition. 0.5 to 1.0 gram of the material is decomposed by 10–15 cc. hydrochloric acid and 5 cc. nitric acid and warming. When the action has ceased, 5 cc. of sulphuric acid are added and the solution taken to strong fumes. (In presence of chromium the sulphuric acid is omitted, the solution being taken to near dryness, 10 cc. more HCl is added and the evaporation repeated to expel HNO_3.) 30 cc. water are now added to the cooled solution and the solution boiled to dissolve any salt that has crystallized out. 5 cc. bromine water are added to oxidize any arsenic or antimony that may be present and the excess expelled by boiling.

Ammonium hydroxide is now added until the solution shows an alkaline reaction, the odor of ammonia being evident in the solution. The mixture is now acidified with glacial acetic acid (5–10 cc. should make the solution decidedly acid). Two grams (approximately 2 cc.) of solid potassium fluoride are now added to precipitate iron, additional fluoride may be necessary if the supernatant solution is colored by iron. The solution is boiled for a minute or so and then cooled under the tap.

To the cold solution 5 cc. of a 50% solution (or its equivalent of 2 g. of KI) of potassium iodide are added, together with 5 cc. of starch solution.

The liberated iodine is now titrated with 0.1 N thiosulphate, to the point where one drop destroys the blue color. It is advisable to add 1–2 cc. more of potassium iodide, and if a blue color results, to add more of the thiosulphate until the color is destroyed.

$$1 \text{ cc. } 0.1 \text{ N } Na_2S_2O_3 = 0.00636 \text{ g. Cu.}$$

NOTE. The additional iodide is occasionally necessary owing to its consumption by Co, Ni, U, Mo, Pb, etc.

Potassium Cyanide Method

This procedure is largely employed on account of its simplicity, although it does not possess the degree of accuracy of the Iodide Method. The procedure depends upon the decoloration of an ammoniacal copper solution by potassium cyanide.

The operations of the standardization of potassium cyanide and of making the assay should be as near alike as possible. If iron is present in the assay it should be added to the standard copper solution titrated, in order to become accustomed to the end-point in its presence.

Silver, nickel, cobalt, cadmium, and zinc interfere and should be removed if present in appreciable quantities. Precipitation of metallic copper by aluminum powder, as directed under Separations, is recommended as a procedure for iron ores and briquettes. In presence of smaller amounts of iron, the titration may be made in presence of iron suspended in the solution. It is not advisable to filter off this precipitate, as it invariably occludes copper. With practice, the shade of color the iron precipitate assumes at the end of the reaction serves as an indicator, so that the operator is assisted rather than retarded by its presence.[1]

$$2Cu(NH_3)_4SO_4 \cdot H_2O + 7KCN =$$
$$K_3NH_4Cu_2(CN)_6 + NH_4CNO + 2K_2SO_4 + 6NH_3 + H_2O.$$

Standard Potassium Cyanide Solution. Thirty-five grams of the salt are dissolved in water, then diluted to 1000 cc.

Standardization. 0.5 gram of pure copper is dissolved in a flask by warming with 10 cc. of dilute nitric acid (sp.gr. 1.2), the nitrous fumes expelled by boiling, the solution neutralized, diluted and titrated as directed under Procedure. If iron is present in the samples titrated, it is advisable to add iron to the standard copper solution as directed above.

$$\frac{0.5}{cc.\ KCN\ solution} = wt.\ Cu\ per\ cc.\ of\ standard\ KCN.$$

Procedure. The solution containing the copper is neutralized with sodium carbonate or hydroxide, the reagent being added until a slight precipitate forms. One cc. of ammonium hydroxide is now added and the solution titrated with standard potassium cyanide solution. The blue color changes to a pale pink; finally a colorless solution is obtained. In presence of iron, when the copper is in excess of the cyanide, the iron precipitate possesses a purplish-brown color, but, as this excess lessens, the color becomes lighter until it is finally an orange brown, the solution appearing nearly colorless. The reagent should be added from a burette drop by drop as the end-point is approached.

Cc. KCN × factor per cc. = weight Cu in assay.

[1] Sutton, "Volumetric Analysis." Davies, C. N., **58**, 131. J. J. and C. Beringer, C. N., **49**, 3. Dr. Steinbeck, Z. a. C., 8, 1; C. N., **19**, 181.

VOLUMETRIC METHODS BASED UPON THE PRECIPITATION COPPER AS CUPROUS THIOCYANATE

Iodide Method[1]

A thiocyanate salt is oxidized by a series of reactions which when completed has this form: $4CuCNS+7KIO_3+14HCl=4CuSO_4+7KCl+7ICl+4HCN +5H_2O$. Other cyanate than copper, lead and antimony must be absent from the precipitate of cuprous thiocyanate. This precipitate together with the filter paper is placed in a 250 cc. glass stoppered bottle and 5 cc. chloroform, 20 cc. water and 30 cc. concentrated hydrochloric acid added.

A certain amount of standard potassium iodate solution (11.784 grams KIO_3 per liter, 1 cc. = 2 milligrams copper) is added from a burette. When the bottle is thoroughly shaken, a violet color appears in the stratum of chloroform which may increase in depth of tint and then fade with each addition of iodate solution. Disappearance of color determines the end point. The KIO_3 solution is very stable. The same layer of chloroform can be used for successive determinations by decanting only the liquor carrying the paper pulp. The immediate coloring of the chloroform on addition of a precipitate to the bottle is due to the reaction with residual ICl and has no harmful effect on the determination.

Permanganate Method[2]

Although this method is based upon these reactions: $CuCNS+NaOH =CuOH+NaCNS$ and $5HCNS+6KMnO_4+4H_2SO_4=3K_2SO_4+6MnSO_4 +5HCN+4H_2O$, solution of the copper salt by the method of manipulation and possibly incompletion of oxidation makes necessary the use of an empirical factor. Theoretically the iron value multiplied by 0.1897 gives the copper value of permanganate. By this method the Fe to Cu factor may be from 0.192 to 0.2. The thiocyanate precipitate is decomposed on the filter with boiling 8% NaOH. The filter is washed with hot water. After making the filtrate acid with dilute H_2SO_4, titration of the solution to permanent tint is made with standard $KMnO_4$ solution having a strength of 1 cc. equivalent to about 1 mgm. copper.

[1] Jamieson, Levy and Wells, Jo. Am. Chem. Soc., May, 1908, p. 760. Price and Meade, "Technical Analysis of Brass," Sec. Ed., p. 84.

[2] Low, "Technical Methods of Analysis." Chapter on Copper has table of empirical factors, compiled by G. A. Guess.

Demorest's Method [1]

Precipitation of cuprous thiocyanate is from a nearly boiling 75–100 cc. solution containing ammonium sulphate about 3 grams ammonium tartrate and 1 cc. free sulphuric acid. 5% solution of sodium sulphite is first added then slowly, with stirring 5% solution of potassium thiocyanate. The coagulated precipitate is collected from the hot liquid in an asbestos lined Gooch crucible and washed thoroughly. The precipitate is decomposed in the crucible by pouring on it hot, 10% NaOH. The residue of CuOH in the crucible is washed well with hot water.

The alkaline filtrate is warmed to 50° C. and a few cc. of standard permanganate solution introduced. At 10 or 5 cc. intervals during the titration, a drop of the green solution is placed on a parafined white plate together with a drop of strong hydrochloric acid. When the red color of this test drop becomes faint on addition of a drop of a 10% solution of $FeCl_3$, 30 cc. of 50% H_2SO_4 are added and the flask shaken until all the MnO_2 is dissolved. The characteristic end point will appear on addition of very little more permanganate. This method has the merit of not requiring an empirical factor, the copper value of the permanganate being 0.1897 times that of the iron.

Volhard's Method

The copper is precipitated with a standard solution of alkali thiocyanate. After filtering and washing the precipitate free of reagent, the excess is titrated with standard silver nitrate solution.

Garrigue's Method

The precipitate of cuprous thiocyanate is collected in a Gooch crucible, decomposed in a casserole with an excess of hot, standard NaOH. The cuprous hydrate formed is filtered, washed and the excess alkali titrated with standard HCl. The reactions are: $CuCNS + NaOH = CuOH + NaCNS$ and $NaOH + HCl = NaCl + H_2O$.

[1] Jo. Ind. and Eng. Chem., Mar., 1913, p. 215.

COLORIMETRIC DETERMINATION OF SMALL AMOUNTS

OF COPPER
Potassium Ethyl Xanthate Method

The method is based upon the fact that potassium ethyl xanthate produces a yellow-colored compound with copper. The reagent added to a solution containing traces of copper will produce a yellow color varying in intensity in direct proportion to the amount of copper present. Larger amounts of copper with the reagent produce a bright yellow precipitate of copper xanthate. Small quantities of iron, lead, nickel, cobalt, zinc, or manganese do not interfere. The procedure is especially valuable for determination of the purity of salts crystallized in copper pans.

Special Solutions. *Stock Solution of Copper Sulphate.* 3.9283 grams $CuSO_4 \cdot 5H_2O$ are dissolved in water and made up to a volume of 1000 cc. One cc. is equivalent to 0.001 gram Cu.

Standard Copper Sulphate. Ten cc. of the stock solution are diluted to 1000 cc. with distilled water. One cc. =0.00001 gram Cu.

Potassium Ethyl Xanthate Solution. One gram of the salt is dissolved in 1000 cc. of water. The solution is kept in an amber-colored glass-stoppered bottle.

Procedure. Five grams of the substance are dissolved in 90 cc. of water (see note) and the solution poured into 100-cc. Nessler tube; 10 cc. of the potassium xanthate reagent are added and the solution mixed by means of a glass plunger. To a similar tube containing 50 or 60 cc. of water are added 10 cc. of the xanthate reagent and then gradually drop by drop the standard copper solution from a 10-cc. burette (graduated in $\frac{1}{10}$ cc.) until the colors in both tubes match.

If a =grams of the substance taken for analysis, b =number of cc. standard copper solution required to match the sample; then $b \times 0.00001 \times 100 \div a = \%$ Cu.

NOTES. The amount of the substance to be taken varies according to its copper content. The greater the copper contamination of the salt, the less sample required. The solution should be neutral or only very slightly acid.

In place of the Nessler tubes the special colorimetric apparatus described under Titanium and under Lead may be used. A very weak copper standard will be required for the comparison tube.

If the substance is insoluble in water the copper is rendered soluble by treatment with nitric acid. Hydrochloric acid is added and the nitric expelled by evaporation. The substance is taken up with water and the insoluble residue filtered off.

Starch and organic matter are destroyed by addition of 10 cc. 10% sodium hydroxide +10 cc. of saturated sodium nitrate solution, then evaporating to dryness and igniting. Hydrochloric acid is now added to expel the nitric acid as directed above.

In dealing with a flotation concentrate containing oil, the sample should be taken to fumes with nitric acid, otherwise the color is apt to be green instead of blue.—C. Y. Pfoutz.

Ferrocyanide Method for Determination of Small Amounts of Copper

By this colorimetric method it is possible to detect one part of copper in 2,500,000 parts of water. The procedure depends upon the purplish to chocolate-brown color produced by potassium ferrocyanide and copper in dilute solutions. The procedure is applicable to the determination of copper in water and may be used in presence of a number of elements that occur in slags. Iron also produces a colored compound with ferrocyanide (1 part Fe detected in 13 million parts H_2O), so this element must be removed from the solution before testing for copper.

Solutions. *Standard Copper Solution.* 0.393 gram $CuSO_4 \cdot 5H_2O$ per liter. 1 cc. = 0.0001 gram Cu.

Ammonium Nitrate. 100 grams of the salt per liter.

Potassium Ferrocyanide. Four grams of the salt per 100 cc. of solution.

Procedure. A volume of 5 to 20 drops of potassium ferrocyanide, according to the amount of copper present in the solution, is placed in a tall, clear, glass cylinder or Nessler tube of 150 cc. capacity, 5 cc. of ammonium nitrate solution added and then the whole or an aliquot portion of the neutral [1] solution of the assay. The mixture is diluted to 150 cc. The same amount of ferrocyanide and ammonium nitrate solutions are poured into the comparison cylinder, placed side by side with the one containing the sample, on a white tile or sheet of white paper. The standard copper solution is now run from a burette into the comparison cylinder, stirring during the addition, until the color matches that of the assay. The number of cc. required multiplied by 0.0001 gives the weight of copper in the sample contained in the adjacent cylinder.

$$\frac{\text{Amount of Cu} \times 100}{\text{Wt. of sample compared}} = \% \text{ Cu in the sample.}$$

NOTES. The solution must be neutral, as the copper compound is soluble in ammonium hydroxide and is decomposed by the fixed alkalies. If the solution contains free alkalies, it is made slightly acid and then the acid neutralized with ammonia, added in slight excess. This is boiled to expel the excess of ammonia, and then tested according to the directions under "Procedure." Solutions containing free acids are neutralized with ammonia.

Iron may be removed by precipitation with ammonia. As this hydroxide occludes copper, the precipitate should be dissolved and reprecipitated to recover the occluded copper.

Determination of copper in water is accomplished by evaporating a quantity of water to dryness, taking up the residue with a little water containing 1 cc. nitric acid, the residue having been ignited to destroy organic matter, precipitating iron with ammonia, as directed above, and determining copper in the filtrate.

The colorimeter used in determination of traces of lead and for the colorimetric determination of titanium may be employed in place of the Nessler tubes.

Ammonia Method for Determining Small Amounts of Copper

In the absence of organic matter, nickel and elements giving a precipitate with ammonia, copper to an upper limit of 10 milligrams can be determined by comparison of the depth of the blue tint of its ammonium solution with a temporary or permanent standard copper solution of equal volume. Because copper in ammoniacal solution combines with the cellulose of filter paper, clarification of such a solution should be through asbestos. Permanent standard solution of copper sulphate, free of nitrate, if kept cool and away from the direct sunlight, lasts for a long time.[1]

Hydrogen Sulphide Method

In the absence of elements precipitated by hydrogen sulphide, copper to the limit of about 1 milligram, in a solution not too strongly acid with sulphuric or hydrochloric acid, may be determined by comparison of its sulphide with that of a known quantity of copper in equal volume and similarly treated. The liquid should be cold and the passage of the hydrogen sulphide stopped before the compound coagulates.

NOTE. Either the ammonia or the hydrogen sulphide method is applicable to the determination of the copper not deposited in the operation of the electrolytic method.

DETERMINATION OF IMPURITIES IN BLISTER AND REFINED COPPER

Introduction. In the complete analysis of copper the following impurities are generally estimated: silver, gold, lead, bismuth, arsenic, antimony, selenium, tellurium, iron, zinc, cobalt, nickel, oxygen, sulphur, and less commonly, tin and phosphorus. In high grades of blister and in refined copper the percentage of these impurities is very low, the blister copper usually averaging over 99.0% copper with silver and the refined copper over 99.93% of the metal. The principal impurity in the refined element is oxygen, which may be present to the extent of .02 to .15%, the remaining impurities being in the third decimal place. From this it is readily seen that large samples are required for the accurate determination of these constituents. The amount of sample taken in blister copper depends upon the grade of copper analyzed. The impurities in this vary from tenths of a per cent to thousandths, as the metal from one locality may contain quite appreciable amounts of a constituent, which may be present only in extremely small quantities or not at all in copper from a different section. In usual practice it is customary to take from 10 to 50 grams of blister and 50 to 500 grams of refined copper for analysis, depending upon the purity of the material. If a larger sample than 50 grams is taken, it is necessary to divide the material into several lots, and, after removal of the bulk of copper and isolation of the impurities, to combine the filtrates or residues containing the constituents sought.

In the procedures the smallest amount of sample, 10 grams, is taken as the basis of calculation for amounts of reagents used. For larger samples, in the initial treatment for removal of copper, proportionately larger amounts of the reagents are required, i.e., multiples of from 2 to 5 times the amount stated. A 50-gram sample is the largest amount of material handled in one lot.

Scrupulous care must be exercised throughout the analysis to prevent contamination of the sample or reagents, and to avoid loss of constituents. The reagents used should be free from the substance sought or from interfering substances. It is the practice to carry blank tests of the reagents through under conditions similar to a regular analysis for iron, lead, zinc, arsenic antimony and sulphur.

It is found best to determine the impurities in several portions, i.e., gold and silver by assay; bismuth and iron in one portion; lead, zinc, cobalt, and nickel in a second; arsenic, antimony, selenium, and tellurium in a third; and separate portions for sulphur, oxygen, phosphorus and tin, when these are occasionally required.

Determination of Bismuth and Iron

Separation of Copper. *Amount of Sample.* Blister copper 10 to 25 grams refined copper 100 to 500 grams. The drillings are dissolved in a large beaker in 40 cc. of nitric acid per 10-gram sample and the free acid expelled by boiling The solution should not become basic during the evaporation. Water is added to make the volume 130 cc. per 10 grams or proportionately more for larger samples. Ammonia is now added in sufficient excess to hold the copper in solution and 5 cc. of saturated ammonium carbonate solution and the sample diluted to 200 cc. (25 cc. $(NH_4)_2CO_3$ per 50 grams, and dilution to 1000 cc.) The beaker is placed on the steam bath for several hours, preferably over night. The solution is filtered hot (to avoid crystallization of the copper salt), the first 100 cc. being refiltered, and the residue washed with hot water containing a little ammonia. By this procedure the copper passes into the filtrate and bismuth and iron remain in the residue on the filter.

Separation of Iron and Bismuth. The precipitate is dissolved in warm, dilute hydrochloric acid (1 : 3), ammonia added to the solution in sufficient amount to almost neutralize the acid and the solution then saturated with hydrogen sulphide. After settling some time, the precipitate containing bismuth sulphide is filtered off, iron passing into the solution.

Determination of Iron. Hydrogen sulphide is expelled by boiling the filtrate, and iron oxidized by addition of hydrogen peroxide, or potassium chlorate (nitric acid should not be used). The solution is evaporated to dryness and iron then determined in the residue by the stannous chloride method, details of which may be found in the chapter on Iron, page 257.

Determination of Bismuth. The sulphides remaining on the filter are dissolved in nitric acid, the solution evaporated with sulphuric acid to SO_3 fumes to expel nitric acid, the concentrate diluted with water, and lead filtered off. Bismuth is precipitated in the filtrate by addition of ammonia in slight excess, followed by 10 cc. of a saturated solution of ammonium carbonate, and boiling. The precipitate is settled for several hours or over night if preferred, and then separated by filtration. This is now dissolved in the least amount of nitric acid, added to the filter drop by drop from a burette and bismuth determined in the solution by the cinchonine iodide method, given in detail in the chapter on Bismuth, page 77.

NOTES. An excess of nitric acid, or the presence of cadmium, lead, silver, or hydrochloric acid interferes with the colorimetric procedure.

In analysis of refined copper several 50-gram portions are taken for analysis, ten such portions on a 500-gram sample; the filtrates, obtained upon dissolving the residue freed from copper, are combined and bismuth and iron determined on this combined solution.

Determination of Lead, Zinc, Nickel, and Cobalt

Removal of Copper. Ten to 25 grams of blister copper, and 100 to 250 grams of refined copper in 25-gram portions are taken for analysis. The metal is dissolved in nitric acid (40 cc. per 10 grams) and the solution boiled until a faint green precipitate begins to appear on the surface of the solution. The free acid being expelled, the solution is made faintly acid by adding 1 to 2 cc. of nitric acid, the solution diluted 300 to 700 cc., according to the amount of copper taken, and then electrolyzed with a current of 1.5 to 2 amperes for thirty-six hours, with a spiral anode and a cathode with about 160 cm. depositing surface. The solution should remain slightly acid throughout the electrolysis, otherwise cobalt, nickel, and zinc may be precipitated as hydroxides from a neutral solution. When the copper is nearly removed, the electrodes are disconnected, and removed.

The solution is concentrated by boiling, a few crystals of oxalic acid added, and the anode (which may be coated with PbO_2) immersed in the hot solution for a few minutes, then rinsed off into the solution.

Separation of Lead. The solution is evaporated to small volume, about 40 cc. of dilute sulphuric acid (1 : 1) are added and the mixture evaporated to SO_3 fumes. The cooled concentrate is diluted with 100 cc. of water and again evaporated to fumes. About 300 cc. of water added and when the soluble salts have dissolved, the solution is filtered and the residue, $PbSO_4$, washed. The filtrate contains Zn, Ni, Co, etc.

Determination of Lead. The residue, $PbSO_4$, is dissolved by successive treatments with ammonium acetate and hot water, the lead precipitated from the solution, made slightly acid with acetic acid, by adding a slight excess of potassium chromate and the element determined as lead chromate according to the standard procedure for lead. See page 274 in the chapter on Lead.

Removal of the Hydrogen Sulphide Group. The filtrate from the lead sulphate is saturated with H_2S and filtered. The filtrate contains zinc, cobalt, and nickel. To recover any occluded zinc, the precipitate is dissolved in nitric acid, taken to fumes with sulphuric acid, diluted to about 200 cc., and again treated with H_2S. The filtrate from this precipitate is combined with the first portion. The precipitate is rejected.

Removal of Iron. This, if present, will be found in the filtrate. The H_2S is expelled by boiling and the solution concentrated to 400 cc. after adding 5 cc. of H_2O_2 to oxidize the iron. Five grams of ammonium sulphate are added, the solution made strongly ammoniacal, and filtered. Iron is precipitated as $Fe(OH)_3$ and is thus removed. If much iron is present, a double precipitation is advisable to recover any occluded zinc, nickel, or cobalt, and the filtrates combined.

Determination of Zinc. The filtrate from iron is concentrated to 400 cc., then made neutral to litmus by cautious addition of dilute sulphuric acid, drop by drop, and then faintly acid with 3 drops in excess. Zinc is now precipitated as the sulphide by saturating the solution with H_2S and allowing to stand over night. The sulphide is filtered off. The filtrate contains cobalt and nickel.

Zinc sulphide is dissolved in hot dilute HCl (1 : 2) and a few crystals of $KClO_3$. The solution is evaporated to dryness, the residue taken up water containing a few drops of HCl and the extract filtered. (To remove any SiO_2 dissolved from the beakers.) Zinc carbonate is now precipitated (in a beaker of glass, which does not contain zinc) from the filtrate by addition of sodium carbonate, and ignited to the oxide ZnO.

$$ZnO \times 0.8034 = Zn.$$

Determination of Nickel and Cobalt. The filtrate from the zinc sulphide is examined for nickel and cobalt. About 0.5 cc. of sulphuric acid is added. H_2S is expelled by boiling, and 2 cc. of H_2O_2 added. The solution is concentrated to about 400 cc. (this should be free from nitric acid), treated with about 25 cc. of ammonium hydroxide, and electrolyzed over night with a current of 0.5 amperes. Nickel and cobalt, if present, are deposited on the cathode as metals and so determined. For greater details, consult the chapter on Nickel under the method by electrolysis.

Determination of Arsenic, Antimony, Selenium, and Tellurium

Separation of Copper. Ten to 50 grams of blister copper and 100 to 500 grams of refined copper are required for the determination. (For 500-grams sample, 5 lots of 100 grams are taken.) The drillings are dissolved in nitric acid (40 cc. per 10 grams) and the solution boiled until a light-green precipitate appears on the surface. The liquor is diluted to 500 cc., and 5 cc. of ferric nitrate containing 3% of iron are added. A basic acetate precipitate is now made, weak sodium hydroxide being added to neutralize the free acid, but not in sufficient amount to produce a permanent precipitate. If the end-point is overrun, nitric acid is added drop by drop until the solution clears. The solution is diluted to about 800 cc., 20 cc. of a saturated solution of sodium acetate added, the liquor brought to boiling and filtered hot through a large creased filter paper, the first portion of the filtrate being poured back on the filter. The residue is washed twice with hot water to remove the copper. Five cc. additional iron are added to the filtrate and a second basic acetate precipitation made, a separate filter being used. The precipitates are dissolved in the least amount of nitric acid necessary and the solutions combined. The liquor is concentrated to 150 cc., a pinch of potassium chlorate added, and the concentration continued until the volume has been reduced to about 30 cc. An equal volume of strong hydrochloric acid is added and a second pinch of chlorate and the evaporation repeated to eliminate all traces of nitric acid.

The evaporation is best conducted in a casserole, resting in the circular opening of an asbestos board, in order that the sides of the vessel may be protected from the flame.

Separation and Determination of Arsenic. The solution is transferred to a distillation flask, arsenic reduced with ferrous chloride, and distilled according to the standard procedure for this element, p. 37.[1] In this distillate arsenic is determined volumetrically.[2] (See chapter on subject.) Antimony, selenium and tellurium remain in the flask.

Separation and Determination of Antimony. Twenty-five cc. of a saturated solution of zinc chloride are added to the liquor remaining in the distilling

---

[1] The concentration should not be carried below 30 cc.

[2] Arsenic may be precipitated by H_2S, the sulphide dissolved in NH_4OH, the filtrate taken to dryness, HNO_3 added and the evaporation repeated. Arsenic now is determined by precipitation with $AgNO_3$ and titration of the silver with KCNS in presence of a ferric salt. $Ag \times 0.2316 = As$.

flask after the elimination of arsenic. The antimony is now distilled, strong hydrochloric acid being introduced in the distilling flask drop by drop by means of a separatory funnel, to replace the solution distilled, the volume in the flask being kept as low as possible, avoiding crystallization. When the antimony has been completely eliminated, the contents of the distilling flask is poured out while still hot, and, together with the rinsings of the flask, placed aside for the subsequent determination of selenium and tellurium.

The distillate is neutralized with ammonia, then made slightly acid with HCl and antimony precipitated with H_2S. Most of the selenium and tellurium remain in the flask. Some of the selenium, however, distills with the antimony, hence this must be recovered from the antimony sulphide precipitate and at the same time this must be purified.

The precipitate is dissolved in dilute HCl (1 : 2), containing a little bromine to oxidize the sulphur. The solution is filtered free from sulphur and the filter washed with a little dilute HCl. The filtrate should contain one-third its volume of strong HCl. Selenium is now precipitated by passing in SO_2 gas to saturation and bringing the solution to boiling. The precipitate is allowed to settle several hours and then filtered through a tared Gooch crucible. (To this is added the selenium and tellurium later obtained from the residue of the flask.) The filtrate contains antimony.

After boiling out the SO_2, the filtrate is first neutralized with ammonia, then made slightly acid with hydrochloric acid and antimony precipitated as the sulphide by saturating the solution with H_2S, allowing the precipitate to settle, resaturating with H_2S and again allowing to settle. The filtered, washed precipitate is dissolved with sodium sulphide, and 10 cc. of 25% potassium cyanide (poison) added to the filtrate, together with 2 cc. of 25% sodium hydroxide.

The solution is now electrolyzed hot (90° C.) for an hour with a current of 0.5 ampere and antimony deposited as the metal on the cathode. This is quickly removed and washed by dipping it successively into a beaker of cold water, three of hot water and one of 95% alcohol. The foil is dried at 100° C., and then weighed, on cooling, as usual. Antimony is now removed by immersing the cathode in boiling nitric acid containing tartaric acid, and washing as before. The loss of weight of the foil is taken as antimony.

NOTE. It is advisable to test the electrolyte for antimony by acidifying the solution with oxalic acid (Hood). A reddish coloration indicates the incomplete removal of the element.

Determination of Selenium and Tellurium. The solution from the distillation flask is nearly neutralized with ammonia and saturated with H_2S. The precipitate is filtered off and dissolved in equal parts of nitric acid (sp.gr. 1.42) potassium bromide bromine solution (20 cc. Br added to a saturated solution of KBr and diluted to 200 cc.). The liquor is diluted to 400 cc., 5 cc. of ferric nitrate (3% Fe''') solution added, and sufficient ammonia to make the solution decidedly alkaline. The precipitate contains, besides the iron, all of the selenium and tellurium, whereas any copper that may have been present is removed. The precipitate, washed, is dissolved in hydrochloric acid, the free acid nearly neutralized and H_2S passed in to saturation. The precipitate is filtered off, washed, and dissolved in the nitric acid potassium bromide and bromine mixture stated above. The solution is filtered and then sufficient hydrochloric acid added to make the solution contain about one-third its volume of strong HCl. Selenium and tellurium are precipitated from this solution by passing in SO_2 to saturation, and boiling for a minute or so. The precipitate is now filtered into the crucible containing the selenium obtained in the purification of the antimony precipitate. After washing with hot water and once with 95% alcohol, the residue is dried at 100° C, for an hour and weighed as selenium and tellurium. Solution should stand three hours at least, or overnight, before filtering.

NOTE. The precipitate of selenium and tellurium may contain gold, which should be determined by assay.

Determination of Oxygen

This determination is required only in refined copper. The method depends upon the reduction with hydrogen of cuprous oxide heated to redness; the water formed by the reaction being the measure of the oxygen.

Apparatus. The combustion-furnace is the same as that used for the determination of carbon. As it is necessary that the hydrogen be absolutely free from oxygen and moisture, the gas is passed through a preheater consisting of a platinum or silica tube of small bore heated to redness by a flame or an electrical device. The gas is then passed through a tube containing calcium chloride and finally through a P_2O_5 bulb containing the anhydride. In this purified form it enters the combustion-tube. The product of combustion, water, is absorbed in a tared bulb by P_2O_5, to which is attached a tube of calcium chloride.

Procedure. The sample, which has been drilled with considerable care to avoid overheating, is dried under partial vacuum in a desiccator after warming to below 70° C. for a few minutes.

One hundred grams are taken for analysis and placed in the combustion tube, the drillings being held in a large boat. Purified hydrogen is rapidly passed through the tube for half an hour to sweep out the air, the tube being cold. The tared P_2O_5 bulb and the calcium chloride tube are now attached. The heat is turned on to bring the sample to cherry red heat, 900° C., and the current of hydrogen passed slowly over the sample for several hours.

The increase of weight of the P_2O_5 bulb $=H_2O$.

$$H_2O \times 0.8881 = O. \quad O \times 4.9687 = CuO.$$

Determination of Sulphur

This determination is rarely required in refined copper.

Twenty grams of blister, unrefined or cement copper, placed in a casserole, are treated cold with 50 cc. bromine-potassium bromide mixture (see under Determination of Selenium and Tellurium). After standing at least ten minutes, 100 cc. of strong nitric acid are added. After another ten minutes the casserole is placed on the steam bath and the solution evaporated to small volume. This is taken up with 25 cc. of strong hydrochloric acid and evaporated to a pasty mass. The treatment is repeated to ensure the decomposition of nitrates and to expel nitric acid. It is now taken up with 5 cc. of hydrochloric acid, diluted with water and sulphuric acid precipitated as $BaSO_4$, according to the standard procedure for sulphur.

$$BaSO_4 \times 0.1374 = S.$$

Determination of Phosphorus

This determination is seldom required, and then only in low-grade copper and copper scrap containing phosphor bronze. The sample, dissolved in nitric acid, is treated with ferric nitrate and the basic acetate precipitation made as has been described for the determination of arsenic, etc. The precipitate is dissolved in HCl, this solution then made strongly ammoniacal, and saturated with H_2S, and filtered. The filtrate containing the arsenic and phosphoric acid is acidified, arsenic sulphide and sulphur filtered off, and phosphoric acid determined in the filtrate by precipitation with magnesia mixture as usual. See chapter on Phosphorus.

$$Mg_2P_2O_7 \times .2787 = P.$$

DETERMINATION OF COPPER IN REFINED COPPER

In determining the quality of copper for electrical purposes each hundredth of a percent above 99.90 has its significance. The methods employed are the electrolytic and the hydrogen reduction methods. Silver present is rated as copper.

Electrolytic Method.[1] The sample, consisting of unground drillings, should be untarnished, free of grease or oil, and cleaned of particles of iron by use of a good magnet.

Procedure. A catch weight of about 5 grams is taken, each piece being examined for dust, particles from the drill and surface oxidation before it is placed on the balance pan. Solution is effected in a special 400 cc. beaker which has hipped sides to support a series of watch glasses, the lower hip at the 125 cc. mark, the upper at 350 cc. (Fig. 32.)

Fig. 32.

The drillings are treated with 50 cc. of a stock solution (10.5 p rts nitric acid and 4.5 parts of sulphuric). The watch-glass traps are put in place to retain the copper which is always entrained in the nitrogen peroxide fumes. Except that the current is maintained at .75 ampere throughout the period of electrolysis, the conditions are the same as have been described for the determination of copper by the " Small Portion Method."

Hydrogen Reduction Method. This method is applicable to the determination of copper in grades of refined copper which are characterized by a metallic impurity content which is constant and less than 0.01 per cent. The apparatus consists of a combustion furnace, preferably electrolytically heated, the temperature of which can be kept constant at about 950° C.; a silica tube of $\frac{3}{4}$-in. bore, one end of which is connected with a large Peligot tube containing concentrated sulphuric acid, the other end is connected by a rubber plug and flexible tube with a source of purified hydrogen; porcelain combustion boats 95 mm. long, 18 mm. wide and 10 mm. deep.

Proce ure. A catch weight of about 25.1 grams of drillings is placed in the combustion boat, and the boat inserted in the silica tube. After passing hydrogen for half an hour through the cold tube, the temperature is raised to 950° C. and so maintained for two hours. If the furnace is of a type, which will permit the removal of the tube without disconnecting the train,[2] the tube is taken from the

[1] Ferguson, Jour. Ind. and Eng. Chem., May, 1910.
[2] Electric Heating Apparatus Co., New York.

furnace without interruption of the stream of hydrogen and cooled by a jet of cold air. When cold, the mass of copper, the particles of which are cemented, is taken from the boat and weighed.

NOTE. If the sample is allowed to become molten, the boat and tube will be coated with a film of copper.

A convenient and efficient type of combustion furnace, hinged design, is shown in Fig. 32b. This furnace may be purchased from the Electric Heating Apparatus Company, New York City.

FIG. 32a.—Combustion Furnace, Hinged Design, Type 70—Shown with one "Spare" Unit. Height to center, $9\frac{1}{2}''$.

By courtesy of the Electric Heating Apparatus Company, New York City.

CHLORINE IN CEMENT COPPER AND COPPER ORES

If the material contains very little silver the following method is applicable in laboratories equipped with apparatus for furnace assaying.

Ten grams of the finely ground sample placed in an 800 cc. beaker are treated with 600 cc. water, 100 cc. nitric acid (free from iodic acid) and the mixture brought to boiling by gentle heating. After filtration and thorough washing, the insoluble residue is treated repeatedly with additional water and acid, of the above proportion, until a test of the filtrate with silver nitrate indicates complete extraction of the soluble chloride. The combined filtrates are treated with a slight excess of silver nitrate and chloride of silver precipitated and determined in the usual way.

On a separate 10 gram sample an assay of silver is made and the equivalent weight of chloride calculated. This equivalent is added to the weight of silver chloride obtained in the extract. The percent of chlorine is calculated from this result by the formula.

$$\frac{\text{Weight of AgCl}\times.2474\times100}{10}=\text{gram chlorine.}$$

DETERMINATION OF COPPER IN BLUE VITROL

This is best determined on a 2 gram sample of the finely powdered dry salt or a catch weight of approximately 2 grams if the salt is moist. Copper is deposited electrolytically, the electrolyte being diluted to 130 cc. and containing 4 cc. of nitric acid and 5 cc. saturated solution of ammonium nitrate. A current of .18 amperes and an electrode of $11\frac{1}{2}$ sq. in. depositing surface are used. If the salt contains insoluble matter consisting wholly of basic salts, complete solution is brought about by gently boiling after adding 4 cc. nitric acid and 25 cc. of hot water to the salt. If the insoluble matter shows a tendency to remain in suspension, the presence of arsenic or antimony is indicated. In this case the impurities are precipitated along with ferric hydroxide as has been previously described under the notes on the electrolytic determination of copper in blister copper.

DETERMINATION OF COPPER AND LEAD IN BRASS [1]

One gram of the alloy is dissolved in 8 cc. nitric acid and the nitrous fumes are boiled off; if tin is present, 40 cc. of boiling water are added, the metastannic acid allowed to settle on the hot plate for fifteen minutes and filtered off. (Method for tin is accurate only for wrought brass; high iron or antimony interfere).

The filtrate from the tin is electrolyzed for copper and lead. If the lead is less than 0.75 per cent, an ordinary sandblasted, spiral anode is used; if the amount of lead is 0.75 to 5 per cent a sandblasted gauze cylinder is necessary. For amounts of lead over 5 per cent either a smaller sample is taken or the greater part of the lead is precipitated as lead sulphate and the small amount of lead passing into the filtrate is recovered by electrolysis, using $\frac{1}{4}$ ampere current per solution, after adding 3 cc. of nitric acid. For lead under 0.5 per cent; 5 cc. of 1 : 1 sulphuric acid are stirred in, after the current has been passing for at least ten minutes. If the lead is high the sulphuric acid is added after the electrolysis has continued for at least an hour. Under these conditions no lead sulphate deposits from the solution and as long as the current passes, the sulphuric acid present does not attack the PbO_2 deposited. After the sulphuric acid is added the current is raised to $\frac{1}{2}$ ampere per solution and the electrolysis continued overnight.

The lead peroxide is dried at 250°C. for half an hour. The factor 86.43 gives the equivalent per cent lead. (Factor determined from the average of a large number of tests made on pure lead. The factor is best obtained under the conditions of the laboratory where the determinations are made, as it varies slightly with change of conditions.)

The copper on the cathode is washed, dried and weighed according to the usual standard procedure.

[1] Method of The National Brass and Copper Tube Company, communication by R. T. Roberts.

METHODS OF DETERMINING THE COMBINATIONS OF COPPER IN ORES AND FURNACE PRODUCTS

Sulphurous Acid Method[1]

The method is based on results which show that cuprite, melaconite, malaconite, chrysocolla, and metallic copper, when finely pulverized, are readily and completely soluble in sulphurous acid. Copper sulphides, chalcocite and chalcopyrite, are not attacked, no matter how finely pulverized or how long the period of contact. Metallic iron in quantities ordinarily found in pulverized mineral, even up to 3%, dissolves and affects the determination not at all, provided there is a strong excess of H_2SO_3. The essential conditions of the method are: (a) Fine pulverization of the sample in order completely to liberate the particles of copper minerals from the gangue; (b) The powdered sample must be kept in suspension during the period of lixiviation. Most ores give recovery in a half hour's contact and refractory ores yield in less than two hours.

In general, a solution containing 3% SO_2 should be used, but a much weaker one, as low as 0.75%, will suffice in the case of some ores.

NOTE. For the manufacture of sulphurous acid, an absorption tower of 1 in. dia. and 4 ft. long glass tubing filled with broken hard-burned fire clay is set at an angle of 75° between two 3 to 5 gal. bottles, one about 5 ft. above the other. The tube is open at the top and sealed at the bottom with a plug of sealing wax, through which two small glass tubes extend. The upper bottle contains cold distilled water which is siphoned into the upper end of the absorption tower, the flow being regulated by a stopcock. A 6 to 50 lb. cylinder of sulphur dioxide is connected to one of the glass tubes extending into the absorption tower. On opening the valve of the cylinder the issuing SO_2 is gasified and passes to the tower where it is absorbed by water from the upper bottle, converted into SO_2 solution of the desired strength and caught in the lower bottle. A gas generator may be employed instead of cylinder of SO_2. A 3% solution of SO_2 may be produced by this apparatus at the rate of 3 liters per hour.

The procedure is as follows: Place 2 grams of pulp of 100 to 150 mesh fineness in a bottle, add 100 cc. 3% solution of SO_2, seal the bottle and agitate by rolling $\frac{1}{2}$ to 3 hours. Filter, wash the residue with SO_2 solution and add the washings to the filtrate which contain in solution all oxides, carbonates and silicates of copper and all the metallic copper. To this solution and 5 to 10 cc. nitric acid, evaporate to 20 cc., dilute with distilled water and determine copper by the electrolytic or other method suitable to the quantity.

The residue from the filtration contains the unaltered and undissolved sulphides, the copper in which is determined by the method suitable to the grade of ore.

The Silver Sulphate=Sulphuric Method[2]

The method as described is especially adapted to the determination of metallic copper, cuprous oxide and cupric oxide in the raw material used for the manufacture of marine paints, but can with obvious modifications be applied

[1] Van Barneveld and Leaver, Met. and Chem. Eng., Feb. 15, 1918, pp. 204–206. Eng. and Min. Jour., March 23, 1918, pp. 553–4.
[2] Communicated by E. F. Fitzpatrick, chemist, Nichols Copper Co.

to the differentiation of sulphide and metallic from oxidized copper in ores. Particles of iron from the grinding plates do not interfere. The sample should be no coarser than 150 mesh.

According to the importance of close valuation of any constituent, 2, 5, or 10 grams of the sample are placed in an 800 cc. beaker together with 300 cc. of a neutral, saturated solution of silver sulphate. Boil gently for 10 minutes, decant the solution on to a 15 cm. thick filter paper of close texture and wash the residue in the beaker by decantation. These operations are repeated with addition each time of more silver sulphate, until a NH_4OH test of a filtrate shows no copper. When extraction of all metallic or sulphide copper has been accomplished, the entire residue is transferred to the filter and washed with hot water until an HCl test of the washing shows the presence of no silver. After evaporation to convenient volume and precipitation of silver from the hot liquid by dilute HCl (cautiously added to ensure no excess in case determination is to be by the electrolytic method), removal of the precipitate by filtration only after the liquid has become cold, copper is determined in the whole or an aliquot part by the electrolytic or iodide method.

The residue is washed (with care not to break the paper) into a beaker or flask and boiled for 5 minutes with 200 cc. of 10% H_2SO_4 to bring into solution all the copper combined as CuO and half that combined as Cu_2O. Decant the solution upon the filter used for the silver sulphate leach to decompose particles of oxides which may be retained upon the paper. Wash thoroughly with cold water.

The copper on the filter, precipitated by the reaction $Cu_2O + H_2SO_4 = Cu + CuSO_4$, is dissolved with dilute nitric acid (about 12% HNO_3), solution filtered and freed from silver by the same method and with the same precautions as were employed on the filtrate from the treatment with 10% H_2SO_4. According to the amount of sample taken or the evident quantity of copper in solution, copper is determined, by any method suitable to the quantity, in the whole or an aliquot part. This copper in terms of per cent content in the sample multiplied by two gives the copper combined as cuprous oxide, and multiplied by 1.1258 gives per cent Cu_2O.

From the per cent of copper in the sample dissolved by 10% H_2SO_4 and determined by any method suitable to the quantity, is subtracted that precipitated by the same operation to obtain by difference copper combined as cupric oxide. This multiplied by 1.2518 gives the per cent CuO.

Phosphoric Acid=Ammonium Chloride Method[1]

This method depends upon the solubility of carbonates and silicates and insolubility of sulphides of copper in 15% phosphoric acid; also upon the solubility of metallic copper in a solution of ammonium chloride. Metallic iron does not interfere.

[1] Cremer, Met. and Chem. Eng., June 15, 1918, p. 645.

One gram of the fine sample is placed in a 500 cc. flat bottom flask, covered with 20 cc. of 15% H_3PO_4 and an equal amount of a 20% NH_4Cl solution and boiled gently for 10 minutes. Because H_3PO_4 gives to the solution a yellow tint which interferes with a sharp end point in the determination of copper by the cyanide method, a pinch of burned lime is added to the solution after it has cooled somewhat. The flask is thoroughly shaken, 25 cc. strong ammonia added and the solution boiled again for some time. The ammoniacal copper solution is filtered from the residue, allowed to cool and titrated by the cyanide method. The method as described is adapted to ores containing up to 3% of oxidized copper.

Caustic Soda=Sodium Tartrate Method[1]

This method is based on the permanence of sulphides, the rapid yielding to solution of cupric oxide and the decomposition of cuprous oxide ores on treatment with a mixture of caustic soda and sodium tartrate; also upon the solubility of the more refractory form of oxide of copper by a mixture of ammonium hydrate and sulphate. The method is subject to some error because washing with the ammonia solution dissolves some copper from the chalcocite, and because of the difficulty of washing out all of the dissolved copper from the gelatinized spongy mass on the filter when chrysocolla is a constituent of the sample. These errors tend to balance each other when chalcocite and chrysocolla are both present and the resultant error in assays of ores containing less than 5% copper is not serious.

The procedure is to add 20 cc. of a solution of sodium hydrate and tartrate to 2 grams of pulp and boil gently for 5 to 10 minutes with occasional shaking of the beaker. While hot add 25 cc. 20% solution of ammonium sulphate, heat for 10 minutes, filter, wash several times with a hot mixture of ammonium hydrate and sulphate and finish the washing with hot water. Neutralize the filtrate with dilute H_2SO_4, add $2\frac{1}{2}$ cc. concentrated HNO_3 and determine copper by the electrolytic method.

NOTE. The caustic soda-sodium hydrate stock solution consists of 100 grams sodium hydrate and 50 of sodium tartrate dissolved in 1000 cc. distilled water. The ammonium hydroxide and sulphate washing solution is a mixture of 100 cc. ammonium hydrate, 100 grams ammonium sulphate and 1000 cc. distilled water.

[1] Hunt and Thurston, Col. School of Mines Meg., Sept., 1917, pp. 157-8.

Sulphuric Acid=Mercury Method[1]

This method depends upon the dissolution of oxide copper by dilute sulphuric acid and amalgamation of metallic copper, native and that resulting from the decomposition of cuprous oxide, thus separating intact sulphide copper.

According to the copper content, 1 to 3 grams of the sample are heated at 80° to 90° C. for 30 to 45 minutes in a 150 cc. casserole with 50 cc. of 4% H_2SO_4, care being taken to avoid boiling. Cool then to room temperature and add 4 to 5 cc. clean mercury. Rub the mercury about the casserole with the finger or other object for 3 or 5 minutes until it is certain that all the particles of ore have come in contact with the mercury. Pour the supernatant solution, containing the sulphide minerals suspended in it into a beaker, taking care to retain the mercury in a single globule in the casserole. Wash the last traces of the ore into the beaker by means of a fine stream of water. The solution is now filtered and the copper determined by any suitable method.

The copper in the residue represents the sulphide, while that in the filtrate represents the so-called "oxide" or sulphuric acid soluble copper.

Copper in Brass—Short Iodide Method

Determination of copper in brass may be accomplished by the short iodide method given on page 193b. The tin should remain in the solution since the tin oxide carries down a small amount of copper, which would be lost if the tin were removed by filtration as is prescribed for wrought brass using the electrolytic method. The sample brought into solution by dilute nitric acid (1 : 1), the factor weight 0.6357 requires about 10 cc. HNO_3, and 5 cc. strong sulphuric acid are added and the sample evaporated to strong fumes. Any copper that may have been occluded by tin passes into solution upon dilution with water, lead will remain as a white precipitate, but does not interfere and should be left in the solution. Addition of bromine water is generally necessary, and in any case will do no harm, the bromine being subsequently removed by boiling the solution.

Copper in Babbitt Metal and Alloys High in Tin— Short Iodide Method

High tin alloys may be decomposed, generally, by strong sulphuric acid. This prevents formation of tin oxide obtained by the nitric acid method. If much copper is present, dilute nitric acid decomposition must be used, since copper is more readily attacked by nitric acid. (Copper is attacked by strong sulphuric acid, but is insoluble in dilute sulphuric acid or hydrochloric acid.) Tin oxide occludes copper, as has been stated under the iodide method for brass, treatment with sulphuric acid, as prescribed, frees any occluded copper. Since tin and lead do not interfere in the method, their removal is not necessary nor advisable.

[1] Maier, Eng. and Min. Jour., Feb. 23, 1819, pp. 372–3.

The editor (W. W. Scott) has taken the liberty to revise portions of this chapter.

FLUORINE

F', at.wt. 19; D (air) 1.31$^{15°}$, sp.gr. (−187°) 1.14; m.p. −223; b.p. −187° C; acids, HF, H$_2$SiF$_6$.

DETECTION

Fluorine is the most active element known, and is by far the most active of the halogens, displacing chlorine, bromine, and iodine from their combinations.

Etching Test. The procedure depends upon the corrosive action of hydrofluoric acid on glass, the acid being liberated from fluorides by means of hot concentrated sulphuric acid. This test is applicable to fluorides that are decomposed by sulphuric acid. The reactions taking place may be represented as follows:

I. $CaF_2 + H_2SO_4 = CaSO_4 + 2HF.$

II. $SiO_2 + 4HF = 2H_2O + SiF_4.$

The test may be carried out in the apparatus shown in the illustration, Fig. 33. A clear, polished glass plate 2 ins. square, free from scratches, is warmed and molten wax allowed to flow over one side of the plate, the excess of wax being drained off. A small mark is made through the wax, exposing the surface of the plate, care being exercised not to scratch the glass. If the test is to be quantitative, the marks should be of uniform length and width. The powdered material is placed in a large platinum crucible (*B*) (a lead crucible will do); sufficient concentrated sulphuric acid is added to cover the sample. The plate (*D*) with the wax side down is placed over the crucible and pressed firmly down. To prevent the wax from melting, a condenser (*C*), with flowing water, cools the plate. An Erlenmeyer flask (*C*) is an effective and simple form of condenser, though a metallic cylinder is a better conductor of heat. A little water placed on the plate makes better contact with the condenser. As a further protection a wide collar of asbestos board (*E*) may be placed as shown in the figure. In quantitative work, where a careful regulation of heat is necessary, the crucible is placed in a casserole with concentrated sulphuric acid or in a sand bath, containing a thermometer to register the temperature. The run is best conducted at a temperature of 200° C. (not over 210°—H$_2$SO$_4$ fumes).

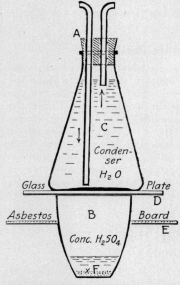

FIG. 33.—Etching Test for Fluorine.

Chapter contributed by Wilfred W. Scott.

After an hour the wax is removed with hot water and the plate wiped clean, and examined by reflected light for etching. A test is positive when the mark can be seen from both sides of the glass. Breathing over the etched surface intensifies the mark.

Treatment of Fluo-Silicates not Attacked by Sulphuric Acid. The powdered material is mixed with about eight times its weight of sodium carbonate and fused in a platinum crucible. The cooled melt is extracted with water. Calcium fluoride is thrown out from the filtrate, according to directions under Preparation and Solution of the Sample. The fluoride may now be tested as directed in the etching test or as follows by the hanging drop test.

The Hanging Drop Test. The test depends upon the reaction $3SiF_4+3H_2O = 2H_2SiF_6+H_2SiO_3$.

If the material contains carbonates, it is calcined to expel carbon dioxide. Half a gram of the powdered dry material is mixed with 0.1 gram dried precipitated silica and placed in a test-tube, Fig. 34, about 5 cm. long by 1 cm. in diameter. A one-hole rubber stopper fits in the tube. A short glass tube, closed at the upper end, passes through the stopper extending about 3 mm. below. Two or three drops of water are placed in this small tube by means of a pipette, nearly filling it. Two cc. of concentrated sulphuric acid are added to the sample in the test-tube and this immediately closed by inserting the stopper carrying the hanging drop tube, exercising care to avoid dislodging the drop of water. The test-tube is placed in a beaker of boiling water and kept there for thirty minutes. If an appreciable quantity of fluorine is present a heavy gelatinous ring of silicic acid will be found at the end of the hanging drop tube in the stopper.

It is important to have material, test-tube, and rubber stopper dry, so that the deposition may occur as stated.[1]

FIG. 34.
Hanging Drop Test for Fluorine.

NOTE. Dr. Olsen[2] makes the test by heating the sample in a small Erlenmeyer flask, with concentrated sulphuric acid. A watch-crystal with a drop of water suspended on its curved surface is placed over the mouth of the flask. A spot etch is obtained in presence of fluorine.

Black Filter Paper Test. According to Browning,[3] small amounts of fluorine may be detected by the converse method for detection of silicates and fluosilicates (See silicon). The fluoride is placed with a suitable amount of silica, in a small lead cup, 1 cm. in diameter and depth (Fig. 35); a few drops of concentrated sulphuric are added; the cup is covered by a flat piece of lead with a small hole in the center; upon the cover is placed a piece of moistened black filter paper and upon this a small pad of moistened filter paper. The cup is heated on the steam bath for ten or fifteen minutes. A white deposit will

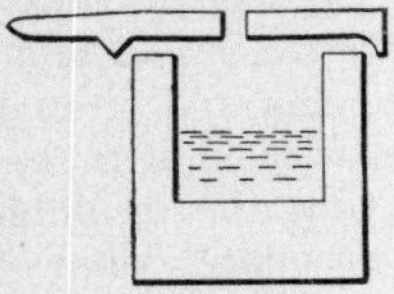

FIG. 35.

[1] C. D. Howard; Jour. Am. Chem. Soc., 1906, **28**, 1238-1239. C. N., 1906, **30**. 420.

[2] Communicated to the author by J. C. Olsen.

[3] P. E. Browning, Am. Jour. Sci. (4), **32**, 249. "Methods in Chemical Analysis," by F. A. Gooch.

be found on the under side of the black filter paper, over the opening in the cover, if fluorine is present in an appreciable amount. (0.001 gram CaF_2 or above, and 0.005 gram Na_3AlF_6 will give the test.)

ESTIMATION

The determination of fluorine in the evaluation of minerals used for the production of hydrofluoric acid is of technical importance. The demand for elimination of the use of fluorides for preservatives of food makes its estimation in small amounts of importance.

Fluorine occurs only combined. It is found abundantly combined with lime in the mineral fluorspar, CaF_2. It occurs as cryolite, Na_3AlF_6; apatite, $3Ca_3(PO_4)_2CaF_2$. It is found in mineral springs, ashes of plants, in bones, and in the teeth (CaF_2). It occurs sparingly, with aluminum and silicon, in topaz, and with cerium and yttrium in fluocerite, yttrocerite, also in wavellite, wagnerite, etc.

Preparation and Solution of the Sample

Fluorides of the alkalies, and of silver and mercury, are readily soluble; copper, lead, zinc, and iron fluorides are sparingly soluble; the alkaline earth fluorides dissolve in 100 cc. H_2O as follows: $BaF_2 = 0.163$ gram, $SrF_2 = 0.012$ gram, $CaF_2 = 0.0016$ gram.

Fluosilicates of potassium, sodium, and barium are slightly soluble in water and practically insoluble if sufficient alcohol is added.

Organic Substances.[1] These are best decomposed by the lime method, the details of which are given in the chapter on chlorine under the section for the preparation and solution of the sample, p. 146. For fluorides in organic matter it is advisable to decompose the substance in a seamless nickel tube, 40 mm. long by 4–5 mm. bore. The end of the tube is sealed with silver solder. The lime used should be soluble in acetic acid. The tube is heated to yellow heat for two hours. The lime is then extracted with acetic acid and fluorine determined as calcium fluoride.

Silicious Ores and Slags. 0.5 to 1.0 gram of material is fused in a crucible with ten times its weight of sodium and potassium carbonates (1 : 1) and poured into an iron mould. If a porcelain crucible has been used, this is broken up and added to the cooled fusion. The mass is digested with about 200 cc. of hot water for an hour, the mass having been broken up into small lumps, (Kneeland recommends using an agate-ware casserole as diminishing the liability of subsequent bumping)[2] then boiled briskly for ten minutes longer and filtered, the solution being caught in a large beaker. The residue is washed with hot water, followed by a hot solution of ammonium carbonate and the insoluble material rejected. The silica is removed with ammonium carbonate, followed by the zinc oxide treatment of the second filtrate, as described under the section of Separations. In presence of appreciable amounts of fluorides, the gravimetric precipitation of fluorine as calcium fluoride is recommended.

[1] H. Meyer and A. Hub, Monatsch. für Chem., 1910, **31**, 933–938. C. N., 1910, **35**, 489.
[2] E. Kneeland, Eng. and Min. Jour., **80**, 1212. A. H. Low, "Technical Methods of Ore Analysis."

Calcium Fluoride. The product is best decomposed by fusion with sodium and potassium carbonates, after mixing the fluoride with 2.5 times as much silicic acid, followed by ten times its weight of carbonates. Most of the silicic acid and all the fluorine will be changed to soluble alkali salts, while the calcium will be left as insoluble calcium carbonate. The mixture must be heated gradually to prevent the contents of the crucible from running over by the rapid evolution of carbon dioxide. The thin liquid fusion soon thickens to a pasty mass. The reaction is complete when there is no further evolution of carbon dioxide. The fused mass is now extracted with hot water as indicated above, and the soluble fluoride filtered from the calcium carbonate residue. Silicic acid is removed from the filtrate by addition of ammonium carbonate. Traces of silicic acid are removed from the filtrate taken to near dryness, after neutralizing the alkali with dilute hydrochloric acid (phenolphthalein indicator), by the zinc oxide emulsion method given under Separations. Fluorine is precipitated as calcium fluoride, according to the procedure given later on page 216.

Soluble Fluorides. See page 223b, 226.

Hydrofluoric Acid. See page 216. Also Vol. II Acids.

Valuation of Fluorspar. Perchloric Acid Method. One gram of fluorspar is treated with 15 cc. perchloric acid and 15 cc. of water in a suitable distillation flask and heated in an oil bath until the residue is almost dry. The distillation is continued with 10 cc. and finally 5 cc. portions of perchloric acid and equal amounts of water. Hydrofluoric acid may be determined as lead chlorofluoride in the distillate and water soluble residue analyzed for metals. If a residue analysis is desired treat first with H F evaporate and follow with perchloric acid. The residue is soluble in water or dilute HCl.

SEPARATIONS

Removal of Silicic Acid from Fluorides. This separation is frequently required, especially in samples where the sodium and potassium carbonate fusion has been required for decomposition of fluosilicates, or calcium fluoride mixed with silicic acid. (See Preparation and Solution of the Sample.)

To the alkaline solution about 5 to 10 grams of ammonium carbonate are added, the solution boiled for five minutes and allowed to stand in the cold for two or three hours. (Treadwell and Hall recommend heating to 40° C., and allowing to stand over night.) The precipitate is filtered off and washed with ammonium carbonate solution. The fluoride passes into the filtrate, while practically all of the silicic acid remains on the filter.

Small amounts of silica in the filtrate are removed by evaporating the solution to near dryness on the water bath, then neutralizing the carbonate with dilute hydrochloric acid (phenolphthalein indicator) added to the residue taken up with a little water. Upon boiling the pink color is restored, the solution then cooled and acid again added to discharge the color; this is repeated until finally the addition of 1–2 cc. of 2 N. HCl is sufficient to discharge the color. Four to 5 cc. of ammoniacal zinc oxide solution (moist ZnO dissolved in NH₄OH—Low recommends 20 cc. of an emulsion of ZnO in NH₄OH) is added and the mixture boiled until ammonia has been completely expelled. The precipitate of zinc silicate and oxide is filtered and washed with water. The fluoride is determined in the filtrate by precipitation with calcium chloride as directed later.

19

Separation of Hydrofluoric and Phosphoric Acids. The method of Rose modified by Treadwell and Koch,[1] takes advantage of the fact that silver phosphate is insoluble in water, whereas silver fluoride is soluble. The alkaline solution of the salts of the acids (solution of the sodium carbonate fusions) is carefully neutralized with nitric acid and transferred to a 300-cc. calibrated flask. A slight excess of silver nitrate solution is added, and the mixture made to volume and thoroughly shaken. After settling, the solution is filtered through a dry filter, the first 10 to 15 cc. being rejected; 225 cc. of this filtrate is again transferred to a 300-cc. calibrated flask, the excess of silver precipitated by adding sodium chloride solution, and after diluting to the mark and shaking, the precipitate is again allowed to settle; 200 cc. of this solution is taken for analysis, after filtering as previously directed. This sample represents 50% of the original sample taken. Fluorine is now determined by one of the procedures outlined.

Separation of Hydrofluoric and Hydrochloric Acids. The solution containing hydrofluoric and hydrochloric acids, in a platinum dish, is treated with nitric acid and silver nitrate. The chloride is precipitated as the silver salt, whereas the fluorine remains in solution and may be filtered off through a glass funnel coated with paraffine or wax, or a hard rubber funnel. In presence of phosphoric acid, silver nitrate added to the solution will precipitate the phosphate as well as the chloride, whereas the fluoride remains in solution. The phosphate may be dissolved out from the chloride by means of dilute nitric acid.

Separation of Hydrofluoric and Boric Acids. An excess of calcium chloride is added to the boiling alkali salt solutions of the two acids. The precipitate is filtered off and washed with hot water. The residue, consisting of calcium fluoride, borate and carbonate, is gently ignited and then treated with dilute acetic acid, taken to dryness, and the residue taken up with acetic acid and water. Calcium acetate and borate are dissolved, whereas the fluoride remains insoluble and may be filtered off and determined.

GRAVIMETRIC METHODS FOR THE DETERMINATION OF FLUORINE

Precipitation as Calcium Fluoride

The method utilizes the insolubility of calcium fluoride in dilute acetic acid in its separation from calcium carbonate, the presence of which facilitates filtration of the slimy fluoride. The reaction for precipitation is as follows:

$$2NaF + CaCl_2 = CaF_2 + 2NaCl.$$

Procedure. Solution of the sample and the removal of silica having been accomplished according to procedures given under Preparation and Solution of the Sample, and Separations, the solution is neutralized, if acid, by the addition of sodium carbonate in slight excess; if basic, by addition of hydrochloric acid in excess, followed by sodium carbonate. To this solution, faintly basic, 1 cc. of twice normal sodium carbonate reagent is added, followed by sufficient

[1] Z. anal. Chem., **43**, 469, 1904. "Analytical Chemistry," Vol. 2, by Treadwell and Hall. John Wiley and Sons.

calcium chloride solution to precipitate completely the fluoride and the excess of carbonate, i.e., until no more precipitate forms, and then 2–3 cc. in excess. After the precipitate has settled, it is filtered and washed with hot water. (The filtrate should be tested for fluoride and carbonate with additional calcium chloride.) The precipitate of calcium fluoride and carbonate is dried and transferred to a platinum dish, the ash of the filter, burned separately, is added and the material ignited. After cooling, an excess of dilute acetic acid is added, and the mixture evaporated to dryness on the water bath. The lime is converted to calcium acetate, while the fluoride remains unaffected. The residue is taken up with a little water, filtered and washed with small portions of hot water, by which procedure calcium acetate is removed, while calcium fluoride remains on the filter.[1] The residue is dried, separated from the filter and ignited. This, together with the ash of the filter, is weighed as calcium fluoride, CaF_2.

To confirm the result, the residue is treated with a slight excess of sulphuric acid and taken to fumes in a platinum dish. The adhering acid is removed as usual by heating with ammonium carbonate, and the ignited residue weighed as calcium sulphate. One gram of calcium fluoride should yield 1.7436 grams of calcium sulphate.[2]

$$CaO \times 1.3924 = CaF_2, \text{ or } \times 0.677 = F.$$

Factors. $CaF_2 \times 0.4867 = F$, or $\times 0.5126 = HF$, or $\times 1.0757 = NaF$. $CaSO_4 \times 0.5735 = CaF_2$.

Precipitation of Fluorine as Lead Chlorofluoride

The method, worked out by Starck,[3] takes advantage of the double halide formed by action of lead chloride upon a soluble fluoride. The compound, $PbFCl$, is about fourteen times the weight of the fluorine it contains. Unfortunately, the compound is quite appreciably soluble in water,[4] so that a loss occurs if pure water is used for washing the precipitate. The method is limited to the determination of soluble fluorides.

The sample, made neutral, is treated with a large excess of a cold saturated solution (200 cc. $PbCl_2$ per 0.1 gram NaF in 50 cc. solution) of lead chloride, the precipitate, settled over night, is filtered off in a weighed Gooch crucible, washed several times with a saturated solution of lead chlorofluoride, and finally two or three times with ice-cold water. The compound is dried two hours at 140–150° C., and weighed as $PbFCl$.

[1] The results are slightly low, owing to the solubility of calcium fluoride: 100 cc. H_2O dissolves 0.0016 gram CaF_2; 100 cc. 1.5 N. $HC_2H_3O_2$ dissolves 0.011 gram.

[2] Low recommends disintegration of the fluoride with sulphuric acid, diluting the mixture with water, boiling with ammonium chloride, and then with ammonium hydroxide and hydrogen peroxide. Calcium oxalate is now precipitated from the filtrate and CaO determined by titration with standard permanganate according to the usual procedure for determination of lime.

[3] Z. anorg. Chem., **70**, 173 (1911); Chem. Abs., **5**, 2049 (1911).

[4] One hundred grams H_2O at 18° C. dissolves 0.0325 gram $PbFCl$ and 0.1081 gram at 100° C.

VOLUMETRIC METHODS FOR THE DETERMINATION OF FLUORINE

Volumetric Determination of Fluorine—Formation of Silicon Tetrafluoride and Absorption of the Evolved Gas in Water. Offerman's Method [1]

Silicon tetrafluoride is formed by the action of sulphuric acid upon a fluoride in presence of silica, the evolved gas is received in water and the resulting compound titrated with standard potassium hydroxide. The following reactions take place:

A. $3SiF_4 + 2H_2O = 2H_2SiF_6 + SiO_2$.

B. $H_2SiF_6 + 6KOH = 6KF + SiO_2 + 4H_2O$.

The method is suitable for determining fluorine in fluorspar in evaluation of this mineral.

Procedure. The powdered sample, containing the equivalent of 0.1–0.2 gram calcium fluoride, is mixed with about ten times its weight of pulverized quartz (previously ignited and kept in a desiccator), placed in the decomposition flask F, shown in Fig. 36, and about 1 gram of anhydrous copper sulphate

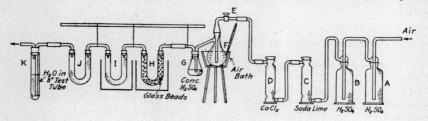

FIG. 36.

added, followed by 25 cc. of concentrated sulphuric acid. The stopcock E is closed and the air bath heated gradually till in one-half hour the temperature has risen to 220°. The cock E is now opened and air slowly forced through the apparatus (by means of water pump) at the rate of about three bubbles per second, the temperature being kept at 220°, and the flask containing the sample occasionally shaken. When the bubbles of silicon tetrafluoride have disappeared from F, the flame is removed, but the air current continued for half an hour longer. The solution in the receiving flask is now titrated with 0.1 N. KOH.

NOTES. The apparatus shown in the cut is the form recommended by Adolph, and the details of procedure are essentially his. This method is preferred to that of Penfield,[2] in which an alcoholic solution of potassium chloride is used to absorb the tetrafluoride, and the liberated hydrochloric acid titrated with the standard alkali in presence of cochineal indicator.

The results obtained by this method are generally low, but the procedure is useful for rapid valuation of fluorspar.

[1] Z. angew. Chem., **3**, 615 (1890). Wm. H. Adolph, Jour. Am. Chem. Soc., **37**, 11, 2500 (1915).

[2] Am. Chem. Jour., **1**, 27 (1879).

The run having been made as directed, the solution in tube " K " is poured into a beaker, an excess of standard potassium hydroxide added and the excess alkali titrated with standard sulphuric acid, to the boiling solution. Norris prefers the use of litmus indicator to phenolphthalein, claiming that the end point is sharper.

It is found advisable to use mercury in the tube "K" as a trap, thus preventing the stoppage of the delivery tube by crystallization. The gas readily passes up through the mercury and is absorbed in the supernatant solution.

In place of using N/10 solutions the potassium hydroxide may be made of such strength that 1 cc. will equal 1% fluorine with 0.5 g. sample taken and the acid made to a corresponding strength.

NOTES. The following suggestions for the method are made by W. V. Norris, Colorado School of Mines.

It is especially necessary that all apparatus be dry, as the least amount of moisture will make the results run low. For this reason it is better to use phosphoric anhydride in washing bottle "B" instead of sulphuric acid.

The sulphuric acid used should be previously treated as follows:—Heat 500 cc. of acid to white fumes, cool, warm gently but not to fumes, then again cool in a desiccator until ready for use. This will produce an acid that will be an efficient dehydrator and will give off no free sulphur trioxide.

It is advisable to use a large excess of silica in the generator apparatus, preferably ten times the weight of the sample taken.

The copper sulphate must be anhydrous, and can best be obtained by heating very thin layers of the pure blue crystals in an oven for about five hours at 215° C.

Adhere strictly to the directions of keeping the temperature in the flask at 220° C., as that temperature will give the maximum recovery.

The bottles "A," "B," "C" and "D" are for the purpose of thoroughly drying the air. "G" contains strong sulphuric acid, prepared as suggested above. "H" is filled with glass beads, to remove sulphuric acid spray; "I" and "J" are empty tubes which should be thoroughly dry. The gas is completely absorbed in tube "K."

Colorimetric Determination of Fluorine—Method of Steiger [1] and Merwin [2]

The method is based on the bleaching action of fluorine upon the yellow color produced by oxidizing a solution of titanium with hydrogen peroxide. A known amount of titanium in solution is mixed with definite volume of the solution containing the fluorine and the tint compared with a standard solution containing an equivalent amount of titanium. The extent of bleaching enables the computation of the fluorine present. The method is applicable to determination of fluorine in amounts ranging from 0.00005 to 0.01 gram. Merwin has shown that large amounts of alkali sulphates have a bleaching action similar to fluorine. Addition of free acid, or rise of temperature, intensifies the color lost by bleaching. Aluminum sulphate has no marked effect on standard solutions, or on solutions bleached by alkali sulphates, but it restores the color to a considerable degree to solutions bleached by fluorine. Ferric sulphate has a similar effect. Phosphoric acid bleaches a standard solution. Silica has little

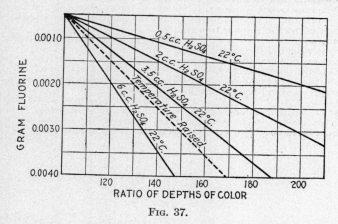

Fig. 37.

effect. According to Merwin an accuracy of 0.002 gram may be expected, an error which is half that of the most reliable gravimetric method.

Reagents. *Standard Titanium Solution.* An intimate mixture of 1 gram of TiO_2 and 3 grams of ammonium persulphate is heated until the vigorous action has ceased, and the ammonium sulphate is expelled. The residue is treated with 20 cc. of strong sulphuric acid, heated to fuming and, when cold, poured into about 800 cc. of cold water. When the suspended salt has dissolved, 57.5 cc. of strong sulphuric acid are added, and the solution made up to 1000 cc. (50 cc. or more of the solution should be analyzed for TiO_2). *One cc. will contain* 0.001 *gram* TiO_2.

Standard Fluorine Solution. 2.21 grams of sodium fluoride, which has been purified by recrystallizing, washing, and igniting strongly, is dissolved in 1000 cc. of water. One cc. will contain 0.001 gram fluorine.

[1] G. Steiger, Jour. Am. Chem. Soc., **30**, 219, 1908.
[2] H. E. Merwin, Am. Jour. Sci. (4), **28**, 119, 1909. Chem. Abs., 3, 2919 (1909). J. W. Millor, "A Treatise on Quantitative Inorganic Analysis." Chas. Griffin & Co.

Sulphuric Acid.   95.5% solution, sp.gr., 1.84.

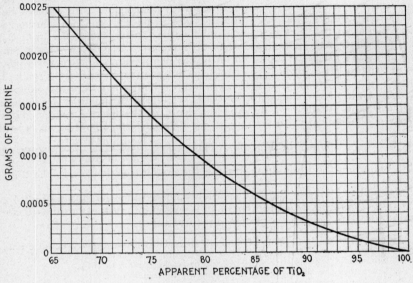

FIG. 38.

Hydrogen Peroxide.   Ordinary strength.

Standard Colored Solution.   The solution used in determining fluorine in materials fused with alkali carbonates contains 10 cc. of the titanium solution, 4 cc. of hydrogen peroxide, and 4 cc. of concentrated sulphuric acid.

Apparatus.   *Nessler Tubes* 6 cm. long, 2.7 cc. in diameter are recommended by the authors. Colorimeters may be used in place of Nessler tubes. A very suitable type for this purpose is shown on page 283, Fig. 43.

Procedure.   Two grams of the powdered sample are fused with 8 grams of mixed sodium and potassium carbonates, the fusion taken up with hot water, and when leached, 3 to 4 grams of ammonium carbonate added. The mix is warmed for a few minutes and then heated on the water bath till the ammonium carbonate is decomposed and the bulk of liquid is small. Silica, ferric oxide, and alumina oxide are thrown down and are removed by filtration. The filtrate, which should not exceed 75 cc., is treated with 4 cc. of hydrogen peroxide, and then 10 cc. of standard titanium solution cautiously added (H_2O_2 prevents precipitation of TiO_2 by the alkali carbonates), followed by 4 cc. of strong sulphuric acid to neutralize the alkali carbonates. The solution, neutral or slightly acid, acquires a light orange tint. A little sodium carbonate is added in just sufficient amount to discharge the color, and then a drop or so of acid to again restore it. The amount of excess acid now required depends upon the amount of fluorine present in the solution. For amounts of fluorine less than 0.0025 gram (0.125% of sample), 3 cc. of acid are added. For amounts of 0.0025 to 0.012 gram fluorine, 12 cc. of acid are added. The solution is diluted to 100 cc.

Comparison.   The test solution is now compared with the standard solution containing 10 cc. titanium reagent, and the same amount of acid and hydrogen peroxide as in the test sample, in a volume of 100 cc. If Nessler tubes are used, these are held over a white surface illuminated with diffused light. In the absence of a bleaching substance, such as fluorine, the two solutions will have the same tint, but in presence of fluorine the bleaching effect will cause the test solution to appear paler than the standard. The depths of the liquids are adjusted so that the tubes will have the same intensity of color when moved from right to left or reversed. Should the left eye perceive a darker shade, the tube on the left will appear uniformly darker whether it be the test sample or the standard. The comparative depths of the liquids in the tubes are measured and the ratio obtained by dividing the depth of the fluorine solution by the depth of the standard and multiplying by 100. Reference may be made to the plotted curve shown in Fig. 38. The ratio $\dfrac{\text{Depth of F Sol.}}{\text{Depth of Standard}} \times 100 =$ the abscissa, while the ordinate represents the amount of fluorine in the 2-gram sample.

Example. Suppose the test solution = 3.6 cc. and the standard = 4.5 cc., the ratio then = 80, from the curve it is evident that the fluorine = 0.00095 gram or 0.0475%, since a 2-gram sample was taken.

NOTES. 1. The destruction of ammonium carbonate is necessary because ammonium sulphate bleaches the final solution and should be absent.

2. Changes of temperature of 50° C. intensify the color 5 to 15%.

3. Increasing the acidity tends to restore the bleached color.

4. The same ratios are obtained by dividing the final volume of the standard by the volume of the test in cases in which a colorimeter is used which requires the standard to be diluted.

According to Merwin, however, the bleaching effect of alkali sulphates, which are present, will make the ratio much higher than it would be if they were absent. (The sulphates alone give a ratio of 125.) This ratio should be determined on two 8-gram portions of the alkali carbonate mixture used in the fusion and the correction made accordingly. If this ratio is represented by m and r is the ratio of the two solutions, then $(r-m) \div 23{,}000 =$ g. F for amounts of fluorine not exceeding .0025 g. (3 cc. H_2SO_4, 22° C., 4 cc. H_2O_2, .01 g. TiO_2). If the fluorine amounts to .0025 to .012 g., then 12 cc. H_2SO_4 is added and the formula required is $(r-m-3) \div 6300 =$ g. F (m blank should be determined and should not much exceed 108). In absence of sulphates the following formulas are given—

(a) $(r-100) \div 70{,}000 =$ g. F, with .5 cc. H_2SO_4, limits of F = .00005 to .001 g.

(b) $(r-100) \div 22{,}000 =$ g. F, with 3.5 cc. H_2SO_4, limits of F = .001 to .004 g.

Example (a) if $r = 142$, then $(142-100) \div 70{,}000 = .0006$ g. F.

THE VOLUMETRIC DETERMINATION OF FLUORINE [1]

Two procedures are suggested—a rapid method depending upon the estimation of fluorine from the percentage of calcium present with fluorine, the calcium combined with commonly occurring substances being extracted by glacial acetic acid; and a procedure that depends upon separation of fluorine from its combination by converting it to soluble alkali salt and reprecipitating it from solution by addition of a known amount of calcium salt, the excess of the calcium being converted to oxalate and so determined, the amount combined with fluorine being thus estimated and the equivalent fluorine calculated.

Reagents

Calcium Acetate (0.25 N solution). 12.51 grams of pure calcium carbonate are dissolved in 500 cc. of water and 75 cc. glacial acetic acid (large beaker necessary) and the acetate formed is placed in a graduated flask and diluted to 1000 cc. The solution is standardized by precipitation of the calcium oxalate in an aliquot portion (40 cc.) and titration with standard potassium permanganate. Exact normality is recorded.

1 cc. 0.25 N solution = 0.005 gram calcium.

Potassium Permanganate (0.25 N solution). 7.91 grams of pure crystals ($KMnO_4$) per liter are standardized against 0.67 gram of pure sodium oxalate equivalent to 40 cc. 0.25 N solution. Exact normality is recorded.

Sodium Oxalate (0.25 N solution containing 16.75 grams of the salt per liter). Solution is best effected in hot water. (Solubility, 3.22 grams per 100 cc. at 15° C.)

Preliminary Procedure

(a) Minerals Containing Phosphates or Sulfates. One gram of the finely powdered mineral, ore, or calcium fluoride salt is extracted with 50 cc. of dilute acetic acid (1 part glacial, 10 parts water) by gently warming for 15 to 20 minutes with stirring. The residue, transferred to a small, ashless filter, is washed with about 50 cc. of water making the total extract 100 cc. (Save for calcium determination, if desired.) The filter and residue are dried rapidly by spreading out on a watch glass. The fluoride is carefully transferred onto a sheet of glazed paper, the filter ignited, and the ash added to the fluoride. The residue is fused as directed under Fusion.

NOTE. For exact work an allowance has to be made for the solubility of the calcium fluoride. The following solubilities were found, 0.5-gram samples of material being taken and treated with 100 cc. of acetic acid of the strength stated:

Acid	H_2O	CaF_2	$CaCO_3$	Ca_3PO_4	$CaSO_4$
1 part	2 parts	0.0103	Very soluble	0.240	0.084
1 part	10 parts	0.0144	Very soluble	0.276	0.170

(b) Phosphates and Sulfates Absent. No acetic acid extraction is necessary.

(c) Sulfides Present. Sulfur as sulfide occurs generally combined with iron, copper, cobalt, etc. No special procedure is necessary here, as the sulfide is oxidized later.

[1] By Wilfred W. Scott. Reprinted from Industrial and Engineering Chemistry, Vol. 16, No. 7, page 703. July, 1924.

Fusion. Five grams of sodium carbonate and 10 grams of potassium hydroxide, placed in a 50 to 60-cc. silver or iron crucible, are brought to quiet fusion, and allowed to cool until a crust forms over the melt. Half a gram of the fluoride sample is intimately mixed with 0.5 to 1 gram of powdered silica prepared as outlined above (powdered sand free of fluorine will do), and placed in the crucible over the fusion. The crucible is covered and heat applied to bring the contents of the crucible to molten temperature. (High heat is not necessary.) Complete decomposition is effected in half to three-quarters of an hour. The crucible should be agitated frequently during fusion to mix the contents.

NOTE. Calcium fluoride is not so easily decomposed as many existing methods indicate. Hydrochloric acid apparently dissolves the mineral, but on dilution calcium fluoride precipitates. Sulfuric acid and potassium acid sulfate fusion is far from satisfactory; platinum is required and a loss due to bumping is liable to occur. Complete decomposition by acid treatment is frequently doubtful. The alkali fusion appears to be the best method for decomposing the fluoride.

If the mass is in molten condition, it may be poured in the lid of the crucible; if too viscous to pour, it is spread over the inner surface of the crucible by rotating the crucible over the flame. The material is now disintegrated and removed from the crucible and lid by action of about 200 cc. of hot water in a 500-cc. beaker. Ten cubic centimeters of hydrogen peroxide are added and the solution is boiled for about 5 minutes.

NOTE. Fusions made in silver disintegrate more readily than those made in iron. Calcium carbonate tends to adhere to the walls of the crucible.
Boiling the solution expels the excess of peroxide, which interferes in the oxalate precipitation of calcium, if left in solution. Sulfides, iron, and other oxidizable materials are oxidized by the peroxide.

The solution containing the excess of sodium carbonate, potassium hydroxide, alkali fluoride, and the greater portion of the silica is filtered off; the residue, containing calcium carbonate (or phosphate if present in the fused material), silver, iron, some silica (10 to 15% of total), etc., is washed with hot water (10 times) and the washings are combined with the first filtrate. The residue is used for Procedure A, the filtrate for Procedure B.

NOTE. Should phosphates be present in the material, the greater portion will remain in the residue, and a small amount will pass into the filtrate as sodium salt.

$$Ca_3(PO_4)_2 + 3Na_2CO_3 \rightleftharpoons 3CaCO_3 + 2Na_3PO_4.$$

Procedure A—Determination of Calcium and Equivalent Fluorine

The residue washed into a beaker is dissolved in hydrochloric acid (200 cc. of water, 20 cc. HCl). If any gritty material remains, it is advisable to fuse this with about 2 grams of sodium carbonate and 3 to 5 grams of potassium hydroxide, repeating the extraction with water; the residue is dissolved in hydrochloric acid and added to A and the water extract to B. The free acid is neutralized with ammonia, the solution heated and filtered, and the residue washed. Calcium passes into solution, iron (and silver) remains on the filter. (The crucible should be rinsed out with dilute hydrochloric acid, as calcium carbonate may adhere to the walls of the vessel.)

NOTE. A small amount of calcium is liable to be occluded by the hydroxide of iron. If this is present in appreciable amount, it is necessary to dissolve this in hydrochloric acid, reprecipitate with ammonia, and filter, adding the filtrate to the main portion containing calcium.

Calcium is now precipitated from the filtrate by adding 0.25 N sodium oxalate. About 60 cc. are necessary for 0.5 gram of calcium fluoride (fluorspar). After heating to crystallize the oxalate, the calcium is filtered off, washed with water (6 times), dissolved in water containing sulfuric acid (200 cc. $H_2O + 10$ cc. H_2SO_4), and titrated hot with 0.25 N potassium permanganate.

1 cc. 0.25 N $KMnO_4 = 0.005$ gram Ca and 0.00474 gram F.

NOTE. If the mineral was extracted with dilute acetic acid (1 : 10) to remove calcium phosphate, carbonate, or sulfate, an allowance should be made for the solubility of calcium fluoride of approximately 0.014 gram CaF_2 per 100 cc. extract at 18° C.

If total calcium is desired, the calcium in the extract should be determined and added to the calcium of the fluoride.

Procedure B—Determination of Fluorine. Calcium Acetate Method

The alkaline filtrate (water extract of the fusion) contains the fluorine, sodium and potassium salts, and silicic acid.

The filtrate is heated to near boiling and sufficient 0.25 N calcium acetate reagent is added to precipitate all the fluorine and about 5 to 10 cc. excess (60 cc. per 0.5 gram CaF_2). Glacial acetic acid is now added until faintly acid (if the solution is alkaline, litmus paper test) and then an excess of 1 cc. per 100 cc. of solution. The heating is continued for about 5 minutes.

NOTE. Upon addition of the calcium acetate, calcium carbonate also precipitates with calcium fluoride. When the solution becomes acid, the carbonate dissolves. If the acidity is correct, the precipitate settles readily and is easily filtered. Should it be finely divided and remain in suspension, the addition of sufficient potassium or sodium hydroxide to give an alkaline reaction will coagulate and settle the precipitate.

The solution and precipitate are transferred to a 500-cc. (or larger) graduated volumetric flask and, after cooling (18° C.), made to volume, then transferred to a large beaker and the precipitate allowed to settle for a few minutes. An aliquot portion of the clear solution is decanted through a filter, the first 5 to 10 cc. being rejected (several filters may be used to hasten filtration, if slow). A measured volume of the filtrate is now taken for the determination of excess calcium.

Precipitation and Titration of Calcium. Sufficient 0.25 N sodium oxalate solution is added to precipitate the calcium. It is safe to use as much oxalate as the aliquot requires in case no calcium was removed by fluorine—i.e., if one-half the total solution represents the aliquot, then 30 cc. of oxalate are added. The author prefers to precipitate the calcium from a weak acetic acid solution (about 0.5 cc. free glacial acetic acid per 100 cc.). This is the acidity of the solution obtained on adding calcium acetate and acetic acid, as directed, no alkali being added, as suggested, for settling stubborn calcium fluoride precipitates.

The calcium oxalate is coagulated by heating, then filtered, washed, and titrated with 0.25 N potassium permanganate in a hot solution containing sulfuric acid. The oxalate is best dissolved from the filter by hot water containing sulfuric acid.

1 cc. 0.25 N potassium permanganate = 0.005 gram calcium.

Calculation. If $A = $ cc. 0.25 N calcium acetate,

$\quad\quad\quad\quad B = $ cc. 0.25 N potassium permanganate,

$\quad\quad\quad\quad X = $ factor for converting the aliquot portion of solutio

$\quad\quad\quad\quad\quad\quad$ taken in the calcium determination to total solutior

Then A cc. $- XB$ cc. $= $ cc. 0.25 N calcium acetate required by fluorine.

The difference multiplied by $0.005 = $ calcium combined with fluorine, o
multiplied by $0.006 = $ equivalent fluorine (Ca $\times 1.2 = $ F). (See Discussion.)

CORRECTION. Owing to the slight solubility of calcium fluoride and possibly t
the formation of a complex compound, calcium fluoride with a fluosilicate, a correctiv
factor seems to be necessary, the ratio of calcium to fluorine being 40 : 48, rather tha
the ratio represented in the formula CaF_2.

Procedure C—Determination in Alkali Fluorides

Decomposition. 0.5 to 1 gram of the alkali fluoride is dissolved in abou
100 cc. of hot water.

Precipitation. The fluorine is precipitated by adding, from a buret, ,
known amount of 0.25 N calcium acetate in sufficient amount to precipitat
all the fluorine, and then 5 to 10 cc. in excess. If the solution has not becom
acid by addition of the reagent, make it so by adding acetic acid. The solutior
and precipitate are transferred to a 250-cc. graduated flask, and after coolin;
are made to volume and well mixed. An aliquot portion is now filtered througl
a fine-mesh filter (rejecting the first 5 to 6 cc.). A measured portion (ha\
the original total is recommended) is heated to boiling, and calcium is precipi
tated by adding an excess of sodium oxalate. The solution is neutralizec
with ammonia and the calcium oxalate filtered off, washed, and titrated witl
0.25 N potassium permanganate, according to the standard procedure. (Se
Precipitation and Titration of Calcium.)

Calculations of Fluorine.

If $A = $ total cc. of 0.25 N calcium acetate,

$\quad\quad B = $ cc. of 0.25 N potassium permanganate required by the calcium i
$\quad\quad\quad$ half the total volume.

Then $A - 2B = $ per cent fluorine per half gram sample or $A - 2B \times 0.00$
$\times 0.948 = $ gram fluorine.

NOTES. It appears that the compound formed by addition of calcium acetate t
the soluble fluoride is CaF_2; it is thus possible to use the conversion factor 0.948 fo
converting the calcium, combined with fluoride, to its equivalent fluoride.

The method does not distinguish fluorine combined as a fluosilicate from fluorin
combined as a fluoride.

The best method for decomposing fluorspar or calcium fluoride was found to be b;
fusion with sodium and potassium carbonates, sodium carbonate and potassium hy
droxide, or sodium or potassium carbonate and sodium hydroxide. The presence o
silica is necessary.

CaF_2 is precipitated by adding the calcium acetate reagent to the alkaline solutior
of the fluoride. The calcium carbonate, which also forms, redissolves as soon as the
solution becomes acid. A large amount of acid is to be avoided, as this liberates silici
acid, which prevents the settling of the fluoride. When the solution is first acidifiec
and the calcium fluoride is then precipitated, the compound settles badly and is difficul
to filter. Should the fluoride be difficult to settle, it is preferable to make the solutior
alkaline by addition of sodium or potassium hydroxide, rather than to add an insolubl
substance to carry down the flocculent material. The alkali treatment coagulates the
fluoride (probably dissolving silicic acid) and causes rapid settling.

VALUATION OF FLUORSPAR

The following procedure, worked out by Dr. Bidtel,[1] meets the commercial requirements for the valuation of fluorspar. The determinations usually required are calcium fluoride, silica, and calcium carbonate; in some particular cases lead, iron, zinc, and sulphur.

Procedure. Calcium Carbonate. One gram of the finely powdered sample is placed in a small Erlenmeyer flask, 10 cc. of 10% acetic acid are added, a short-stemmed funnel inserted in the neck of the flask as a splash trap, and the mixture heated for an hour on a water bath, agitating from time to time. The calcium carbonate is decomposed and may be dissolved out as the soluble acetate, whereas the fluoride and silica are practically unaffected. The solution is filtered through a 7-cm. ashless filter, the residue washed with warm water four times, and the filter burned off in a weighed platinum crucible at as low a temperature as possible. The loss of weight minus 0.0015 gram (the amount of calcium fluoride soluble in acetic acid under the conditions named) is reported as *calcium carbonate*.

Silica. The residue in the platinum crucible is mixed with about 1 gram of yellow mercuric oxide, in form of emulsion in water (to oxidize any sulphide that may be present); any hard lumps that may have formed are broken up, the mixture evaporated to dryness and heated to dull redness, then cooled and weighed. About 2 cc. of hydrofluoric acid are added and the mixture evaporated to dryness. This is repeated twice to ensure complete expulsion of silica (as SiF_4). A few drops of hydrofluoric acid are then added, together with some macerated filter paper, and a few drops of ammonium hydroxide to precipitate the iron. The solution is evaporated to dryness, heated to dull redness, cooled and weighed. The loss of weight is reported as *silica*.

Calcium Fluoride. The residue is treated with 2 cc. of hydrofluoric acid and 10 drops of nitric acid (to decompose the oxides), the crucible covered and placed on a moderately warm water bath for thirty minutes, the lid then removed and the sample taken to dryness. The evaporation with hydrofluoric acid is repeated to ensure the transposition of the nitrates to fluorides, and if the residue is still colored, hydrofluoric acid again added and the mixture taken to dryness a third time; then a few drops of hydrofluoric acid are added and 10 cc. of ammonium acetate solution (the acetate solution is made by neutralizing 400 cc. of 80% acetic acid with strong ammonia, adding 20 grams of citric acid and making the mixture up to 1000 cc. with strong ammonium hydroxide). The mixture is digested for thirty minutes on a boiling water bath, then filtered and washed with hot water containing a small amount of ammonium acetate, and finally with pure hot water. (Several washings by decantation are advisable.) The residue is ignited in the same crucible and weighed as *calcium fluoride*. An addition of 0.0022 gram should be made to compensate for loss of CaF_2.

Pure calcium fluoride is white. To test the purity of the residue, 2 cc. of sulphuric acid are added and the material taken to fumes to decompose the fluoride; 1 cc. of additional sulphuric acid is added and the excess of acid expelled by heating. The residue is weighed as calcium sulphate. This is now fused with sodium carbonate, and the fusion treated with hydrochloric acid in excess. If barium is present the solution will be cloudy ($=BaSO_4$).

[1] Dr. E. Bidtel, Jour. Ind. Eng. Chem., Vol. 4, No. 3, March, 1912.

ANALYSIS OF SODIUM FLUORIDE

Preparation of the Sample and Insoluble Residue. Ten grams of the sample are dissolved in 250 cc. of water in a beaker, and boiled for five minutes, then filtered into a liter flask through an ashless filter; the residue is washed with several portions of water and ignited. This is weighed as insoluble residue. The filtrate and washings are made to 1000 cc. with distilled water.

Sodium Fluoride. Fifty cc. of the solution equivalent to 0.5 gram of sample are diluted to 200 cc. in a beaker, 0.5 gram sodium carbonate is added and the mixture boiled. An excess of calcium chloride solution is now added slowly and boiled for about five minutes. A small amount of paper pulp is added to prevent the precipitate from running through the filter, the precipitate allowed to settle and then filtered, using a 9-cm. S. & S. 590, or B. & A. grade A, filter paper. The fluoride is washed twice by decantation, and four or five times on the filter with small portions of hot water. The final washings should be practically free of chlorine.

The residue is ignited in a platinum dish, then treated with 25 cc. of acetic acid, and taken to dryness. This treatment is repeated and the residue taken up with a little hot water and filtered. The calcium fluoride is washed free of calcium acetate with small portions of water, remembering that CaF_2 is slightly soluble in water. The ignited residue is weighed as CaF_2.

$$CaF_2 \times 1.0757 = NaF.$$

Sodium Sulphate. To the filtrate from calcium fluoride is added 10 cc. hydrochloric acid and then a hot solution of barium chloride. The $BaSO_4$ is allowed to settle, filtered, washed, dried, ignited, and weighed as usual.

$$BaSO_4 \times 0.6086 = Na_2SO_4.$$

Sodium Carbonate. Sodium carbonate is determined on a 5-gram sample by the usual method for carbon dioxide as described in the chapter on Carbon.

Approximate results may be obtained by adding a small excess of normal sulphuric acid to 5 grams of the fluoride in a platinum dish, boiling off the carbon dioxide, and titrating the excess of acid with normal caustic, using phenolphthalein indicator.

$$\text{One cc. N. } H_2SO_4 = 0.053 \text{ gram } Na_2CO_3.$$

$$H_2SO_4 \times 1.0816 = Na_2CO_3.$$

Sodium Chloride. Fifty cc. of the sample is titrated with N/10 $AgNO_3$ solution.

Silica. This is probably present as sodium fluoride and silicate. One gram of the sample is dissolved in the least amount of water and a small excess of hydrofluoric acid added to convert the silicate to silico-fluoride, then an equal volume of alcohol. After allowing to stand for an hour, the precipitate is filtered, washed with 50% alcohol until free of acid and the filter and fluoride are placed in a beaker with 100 cc. of water, boiled and titrated with N/10 NaOH.

$$\text{One cc. } N/10 \text{ NaOH} = 0.0015 \text{ gram } SiO_2 \text{ or } 0.0047 \text{ gram } Na_2SiF_6.$$

Volatile Matter and Moisture. One-gram sample is heated to dull redness to constant weight. Loss of weight is due to moisture and volatile products.

Determination of Traces of Fluorine

An approximate estimation of traces of fluorine may be made by utilizing the method outlined for detection of this element. By varying the amounts of substance tested, an etch is obtained that is comparable with one of a set of standard etches, obtained with known amounts of fluorine in form of calcium fluoride, added to the same class of material examined.

The conditions in obtaining the standard etches and those of the tests should be the same. This applies to the temperature of the paraffine bath, duration of the run, size of mark exposing the surface of the test-plate, and the general mode of procedure.

One gram of sample is placed in a lead bomb with 12 cc. of sulphuric acid, the bomb closed with glass plate in place and heated in an oil bath for 45 minutes at 165° C. The etching on the glass plate is compared with etching using known amounts of fluorine as CaF_2 and the same kind of glass.

The glass plate is kept cool by circulating cold water. The type of bomb and its connections are shown in Figure 38a.

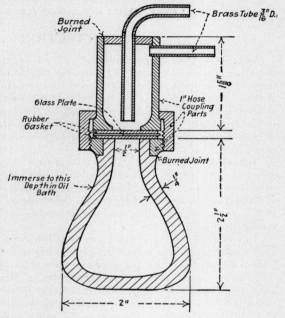

Fig. 38a.

NOTE. The importance of regulating the temperature may be seen by the results obtained by Woodman and Talbot. With a temperature of 79–80° C., one part of fluorine may be detected in 25 to 100 thousand parts of material; by raising the temperature to 136° C., the delicacy of the procedure is increased to one part of fluorine in 1 to 5 million parts. The limit of delicacy is apparently reached at 213–218° C. (i.e., 1 part F per 25 million).

A metal condenser, such as is recommended for mercury determinations, may be used and the oil or paraffine bath substituted by an electric heater automatically controlled.

Crisco is claimed to be better than paraffine, as this does not give off any unpleasant fumes when heated.

GOLD

Au, *at.wt.* **197.2**; *sp.gr.* **19.33**; *m.p.* **1063**; *b.p.* **2530° C**; *oxides,* **Au₂O, Au₂O₃**

DETECTION

Because of the limited application and tediousness of wet methods, the detection of a small quantity (2 parts per million or less) of gold in a mineral or base met l is most positively carried out by furnace methods of assaying. Wet methods of detection of traces of gold can be applied only to solutions free of colored salts and elements precipitated by the reagents employed. As a rule, in the treatment of an unknown substance, advantage is taken of the solubility of most metals and their compounds, and insolubility of gold by one of the mineral acids.

Detection of Gold in Alloys. In metals or alloys which produce colorless solutions with dilute nitric acid, gold, in the absence of other insoluble matter, exhibits itself as a black or brownish residue which settles readily, and from which the liquid can be separated by careful decantation. If unassociated with metals of the platinum group, this residue will become yellowish brown on heating with strong nitric acid.

In copper, nickel and such alloys, which leave a residue of sulphur, carbon or silicious matter on treatment with dilute nitric acid, the solution is filtered through double ashless filters and the filter and residue incinerated in a porcelain crucible. The residue, which may require pulverizing, is digested for a few minutes with aqua regia, and the dilute, filtered solution evaporated to dryness by heating below 200° F. Just as soon as dry, the mass is moistened with the least quantity of hydrochloric acid and the purple of Cassius test applied to its water solution in a small volume. This test is made by adding a solution of stannous chloride, containing stannic chloride. In strongly acid and concentrated gold solutions a precipitate of brown metallic gold is obtained. If the solution is but slightly acid and dilute, a reddish purple color is produced by colloidal gold and the stannic acid. The tint fades on standing. Addition of ammonia produces a red coloration.

This test applied to 1 part of gold in 600,000 of solution will impart a perceptible shade; to double this quantity, a mauve color. When gold is present in somewhat greater proportion a flocculent precipitate will form.

Test for Gold in Minerals. From minerals, in which the metal exists in unalloyed, or uncombined state, gold may be extracted by iodine in potassium iodide solution, or by chlorine or bromine water. All minerals containing sulphides should be roasted. In natural or roasted state the sample should be ve y finely pulverized, and usually yields the gold best if first digested with nitric acid and washed free of soluble salts. The sample in a flask is covered with bromine water, the flask closed with a plug and shaken frequently during a period of three

Chapter contributed by W. G. Derby.

228

or four hours. The purple of Cassius test is applied to the extract, removed by decantation after concentration.

If it is evident that base metals are present in the bromine water extract in quantity sufficient to mask the purple of Cassius test, hydrogen peroxide is added to the concentrated liquid, slightly alkaline with sodium or potassium hydroxide or carbonate.[1] After boiling the solution until hydrogen peroxide is removed, precipitated hydroxides or carbonates are dissolved by hydrochloric acid. Gold in exceedingly small quantity exhibits itself as a light-brown residue on a fine filter. This indication should be confirmed by a purple of Cassius test on the aqua regia solution of the residue; the test carried out in the same manner as on the residue from a solution of a metal.

Benzidine Acetate Tests. Maletesta and Nola[2] make use of benzidine acetate (1 gram benzidine dissolved in 10 cc. acetic acid and 50 cc. water) as a reagent in the detection of gold and platinum in quite dilute solutions. Gold gives a blue coloration which gradually changes to violet. The coloration is green in the presence of free acetic acid, changing to blue with addition of benzidine in excess. Platinum gives a blue flocculent precipitate, the formation of which is promoted by heating. Free mineral acids have no influence on the gold and retard the platinum reaction only in the cold. Since ferric salts give a blue coloration, stable only in excess of benzidine, their absence must be assured before application of the test for the precious metals. The limit of sensitiveness of the test is 35 parts for gold and 125 parts for platinum per 10,000,000.

Phenylhydrazine Acetate Test. E. Pozzi Escot[3] adds phenylhydrazine acetate to a very dilute gold solution which contains an excess of an organic acid (formic or citric). A violet coloration, permanent for several hours, is imparted. The depth of color is proportional to the quantity when the gold is present in less amount than one part in 500,000.

ESTIMATION

Solubility

Gold in massive form is practically insoluble in pure nitric, sulphuric or hydrochloric acids, but in the presence of oxidizing agents, is attacked appreciably by sulphuric, and actively by hydrochloric acid. Gold is found in minute quantity in the nitric acid[4] solution of its alloys and in such as contain selenium, the amount may be a large part of the total present.

Gold is attacked energetically by aqua regia. Large amounts of gold are dissolved with requirement of least attention when the proportion of hydrochloric acid is several times that of the aqua regia formula, ($3HCl : 1HNO_3$).

Gold is dissolved by solutions of chlorine or bromine, by alkaline thiosulphates; in the presence of free oxygen by iodine in potassium iodide solution, by soluble cyanides, by fused potassium or sodium hydroxide; by fused potassium or sodium nitrate or sulphide. In a finely divided state, it is dissolved by a solution of potassium or sodium hydroxide.

Gold alloys quickly with molten lead. When in the form of bright, untarnished particles it alloys readily with mercury.

[1] Vanino and Seeman, Berichte, **32**, 1968; Rossler, Zeit. Anal. Chem., **49**, 733.
[2] Bull. Chim. Farm, **52**, 461; Chem. Abs., April 20, 1397, 1914.
[3] Am. Chim. Anal. Appl., 1907, **12**, 90; J.S.C.I., June 15, 1907, 645.
[4] Dewey, J.A.C.S., March, 1910, 318; E. Keller, Bull. Am. Inst. Min. Eng., **67**, 681.

GRAVIMETRIC METHODS

Gold is always weighed in metallic state, and is determined most accurately in the form of the mass obtained by dilute nitric acid treatment of the silver alloy resulting from the operation of cupellation in the method of assaying by furnace processes. On account of tediousness in making complete separation from associated metals, and of uncertainty in collection of the product in a form suitable for accurate weighing, direct precipitation methods are never used for the valuation of gold-bearing material, but may be applied to the estimation of gold in plating baths, the Wohlwill parting electrolyte and solutions of similar type.

Precipitation of Gold. From such solutions of auric chloride, slightly acid with hydrochloric, freed of oxidizing agents by evaporation and displacement with hydrochloric acid, and containing but little of the salts of the alkalis or alkali earths, gold is separated from other than occluded platinum and palladium by precipitation with oxalic acid, ferrous sulphate, or hydrazine hydrochloride. The reactions are hastened by heat. When salts of the alkalis or earths are present, equally good separation and more complete precipitation can be obtained by addition of excess of sodium peroxide, boiling vigorously for a few minutes and then acidifying with hydrochloric acid. The precipitated metal is collected on an ashless filter paper, and after drying, weighed.

Gold precipitated from a very weak solution is in such fine form that it is not wholly retained by the finest paper.

Wet Gold Assay of Minerals

A wet gold assay, suitable for prospector's use,[1] is carried out by covering one assay ton (29.17 grams), of the finely pulverized natural or roasted ore, in a porcelain mortar, with 50 cc. of a solution of 2 parts of iodine and 4 parts potassium iodide in 100 cc. of water. Sulphide ores should be roasted and digested with nitric acid before treatment with the iodine solution. Similar treatment is advantageously applied to all ores. The ore is ground in contact with the iodine solution and additions of the halogen are made whenever the liquid becomes colorless. The solution is then allowed to stand at least an hour. To the filtrate and washings from the pulp, in a glass-stoppered bottle or flask, are added 5 grams of gold free mercury. The liquid is shaken vigorously with the mercury until clear. The mercury is then transferred to a small porcelain casserole, washed with clean water and dissolved by warming carefully with 10 cc. nitric acid. The gold mass is washed free of nitrate of mercury by decantation, dried and annealed by heating in a casserole over a Bunsen flame, and the metal weighed. Each milligram represents an ounce per ton. Results obtained by this method of assaying are usually more than 50 per cent of the actual gold content.

Electrolytic Method. The gold content of a cyanide plating bath containing no potassium ferrocyanide may be estimated by electrolysis.[2]

Procedure. A measured quantity, 25 to 50 cc. in a tared platinum dish, is diluted to 1 cm. of the rim of the dish and using a carbon or platinum anode, elec-

[1] De Luce, Min. Sci. Press, **100,** 895; Hawson, Min. Sci. Press, **100,** 936; Davis, Mines and Minerals, Oct., 1910, Feb., 1911; Austen, Inst. of Min. and Met., May 31, 1911.

[2] Electro Deposition of Metals, Langbein.

trolyzed for about three hours at a current density $ND_{100} = 0.067$ amp. (.0.0043 per square inch). Completion of deposition is recognized by the lack of any deposit within fifteen minutes, on a platinum strip suspended on the rim of the dish. The dish plus gold deposit is washed, rinsed with alcohol, dried at 212° and when cold weighed.

The following is a summary of the conditions of deposition of gold in compact form as described by Classen[1] 3 grams potassium cyanide were added to a gold chloride solution containing 0.0545 grams of gold in 120 cc. This solution heated to about 55° C. when electrolyzed at a current density of $ND_{100} = 0.38$ amp. (0.024 amp. per square inch) with a potential difference of 2.7–4.0 volts, deposited its gold content in one and a half hours. Time required for deposition is tripled if the electrolyte is at room temperature.

Miller[2] deposited 0.1236 gram of gold in two and a quarter hours from 125 cc. of electrolyte at 50° C. containing 1 gram potassium cyanide by a current of $ND_{100} = 0.03$ amp. (0.002 amp. per square inch) and 2.5 volts.

Perkin and Preble[3] use an electrolyte containing ammonium thiocyanate in place of potassium or sodium cyanide.

Gold is removed from the platinum electrode by warming with a solution of chromic anhydride in a saturated salt solution,[4] or with a solution of potassium cyanide containing some oxidizing agent as hydrogen peroxide, sodium peroxide or alkali persulphate.[5]

VOLUMETRIC METHODS

These methods are applicable to the determination of the strength of chloride of gold solutions used in photography, electro gilding, and as electrolyte in the Wohlwill parting process.

Preparation of the Sample. Nitric acid or nitrates in the solutions should be removed by repeated evaporations to syrup with addition of hydrochloric acid saturated with chlorine. Free chlorine or bromine should be removed by addition of ammonia to formation of permanent precipitate, then making the solution very slightly acid with hydrochloric acid and heating until the precipitate of fulminating gold dissolves. The gold solution should contain but little free hydrochloric acid, an excessive amount of which may be removed by ammonia.

Permanganate Method

Weak gold solutions should be concentrated whenever possible. The permanganate method,[6] which is not applicable when the sample contains organic matter, depends upon the titration, after complete precipitation of gold, of the unoxidized portion of a measured quantity of an added reagent of a known gold precipitating value. The reagent may be ammonium or potassium oxalate, ferrous sulphate or ferrous ammonium sulphate in solutions varying from 5 to 25 milligrams gold precipitating value and is titrated with a permanganate solution of approximately equal oxidizing strength. One part of gold requires for precipitation 1.08 of ammonium oxalate, 1.40 of potassium oxalate, 4.22 of ferrous sulphate, 5.96

[1] Classen, "Quantitative Chemical Analysis by Electricity," Classen-Boltwood.
[2] J.A.C.S., Oct., 1904, 1255.
[3] Elec. Chem. and Met. Ind., **3**, 490.
[4] Classen-Boltwood, "Quantitative Chemical Analysis by Electricity."
[5] Rose, "Met. of Gold," 5th Ed., 469.
[6] Bull. Chim. Farmac., 1894, XXX, III, 35; Oestr. Zeit. f. Berg. und Hut., 182, 1880; Sutton, "Volumetric Analysis," 10th Ed.; E. A. Smith, "Sampling and Assaying of Precious Metals"; Min. Eng. World, **37**, 853.

parts ferrous ammonium sulphate, each in crystalline form. The most satisfactory precipitations are made with the iron salts. The standard solution of either should contain about 0.1 per cent of sulphuric acid. One part of gold, in solution as auric chloride, has an oxidizing value equivalent to 0.4808 part of potassium permanganate.

The precipitating value of 0.2548 gram of dry Sorenson's sodium oxalate is 250 milligrams of gold, and by titrating a solution of this amount of oxalate in 250 cc. of water, aciduated with a few drops of sulphuric acid, the oxidizing value of the permanganate solution is obtained in terms of gold.

The value of the precipitating reagent and relative oxidizing value of the permanganate solution can be checked very accurately by adding a measured quantity of the reagent to an excess of gold chloride, filtering, washing thoroughly, incinerating and weighing the precipitate obtained in a tared porcelain crucible.

Procedure. In carrying out the determination of a gold solution, a measured or weighed portion is freed of oxidizing agents, a measured amount of the standard precipitating reagent added in slight excess of the amount required to decolorize the solution, and digestion on a steam bath or hot plate continued until the gold settles out, leaving a clear liquid. A few drops of sulphuric acid may be then added and, without filtering, titration performed. The gold value of the quantity of reagent added, minus that found of the excess of reagent, is the gold content of the amount of the sample taken.

Iodide Method

Small quantities of gold are determined by Gooch and Morley's iodide method.[1] A measured or weighed portion of the gold solution is treated, as has been described for removal of oxidizing agents, with an excess of free hydrochloric acid. Potassium iodide solution is run into the cold liquid until the gold precipitated as aurous iodide is completely dissolved. Starch solution is then added, and the amount of N/1000 thiosulphate required to decolorize the liquid noted. From this amount is deducted the amount of N/1000 iodide required to just produce a perceptible rose tint in the liquid.

The reactions involved are $AuCl_3 + 3KI = AuI + I_2 + 3KCl$ and $I_2 + 2Na_2S_2O_3 = 2NaI + Na_2S_4O_3$.

The gold value of the N/1000 solution of sodium thiosulphate should be determined by performance of the operations of the method on a known quantity of gold, similar in amount and contained in a volume of solution approximately equal to that of the analysis.

Lenher's Method. By Lenher's method[2] of determining gold in solutions free of oxidizing agents, sulphurous acid of a reducing strength of 2–5 milligrams gold per cc. is used as the reagent. The sulphurous acid requires frequent standardizing by means of standard iodine or potassium iodide to which a definite amount of standard permanganate has been added or by a gold solution of known strength. Using starch as indicator, the iodine liberated by addition of potassium iodide can be titrated by sulphurous acid. Bromine liberated by potassium bromide according to the equation, $AuCl_3 + 2KBr = AuCl + 2KCl + Br_2$, can be titrated by sulphurous acid. Excess of magnesium or sodium chloride gives to auric chloride a yellow color which by sulphurous acid can be titrated to the colorless or aurous state. These alkaline salts do not interfere in the potassium bromide or iodide reactions.

[1] Amer. Jour. Sci., Oct., 1899, 261; Min. and Eng. World, **37**, 853; Vol. Am., Sutton, 10th Ed.; "Assaying of Precious Metal," E. A. Smith.
[2] Jour. Am. Chem. Soc., June, 1913, **735**.

COLORIMETRIC METHODS

Practical application of these methods is made in the estimation of gold in the liquors produced in the treatment of ores by the cyanide process.

Prister's Method

By Prister's method [1] a slight excess of copper solution is added to a 100 to 200-cc. portion of a cyanide solution in which the cyanide has been decomposed by boiling several minutes after acidifying with hydrochloric acid. Assurance of the presence of an excess of copper is made by spot test with a solution of potassium ferrocyanide.

The copper solution is made by boiling for ten minutes in contact with copper shavings, a solution of 1 part blue vitriol and 2 parts salt in 10 parts of water, and adding a little acetic acid on cooling. A few drops of a 1 to 2 % sodium sulphide solution are added, the liquid boiled for five minutes, the precipitate allowed to settle, and liquid separated by decantation on to a filter. The precipitate in the beaker and on the filter is dissolved with $2\frac{1}{2}$ to 3 cc. of a 3 to 5% solution of potassium cyanide to which a few drops of potassium hydrate solution has been added.

Gold is precipitated from this cyanide solution (which may be turbid), by addition of 1 to 2 grams of zinc dust and warming to 100° F. for half an hour. Liquid is separated by decantation through a filter. The residue on the filter and in the beaker is first treated with hydrochloric acid to dissolve zinc, then with 10 cc. aqua regia, the reagent being passed several times through the filter. Stannous chloride solution is then added to the liquid diluted to 20 cc. Comparison of the coloration produced is made with that from a standard solution of gold treated in the same manner.

Cassel's Method. By Cassel's method [2] 0.5 gram potassium bromate is mixed with 10 to 50 cc. of the cyanide solution and concentrated sulphuric acid added gradually with constant agitation until reaction commences. When the reaction stops, saturated solution of stannous chloride is added dropwise until the liquid is just colorless. The tint produced is compared with that from a standard gold solution treated in the same manner.

Moir's Method. By Moir's method [3] a measured quantity of the cyanide solution is oxidized by addition of 1 to 2 grams of sodium peroxide and boiling. If sufficient sodium peroxide is present, the brown spot produced by addition of a few drops of lead acetate will immediately dissolve. The lead-aluminum couple formed by addition of aluminum powder precipitates gold which is filtered off. To the aqua regia solution of the precipitate a solution of stannous chloride is added drop by drop until the liquid is dissolved. The purple of Cassius tint developed is compared with permanent standards composed of mixtures of solutions of copper sulphate and cobalt nitrate which have been adjusted to shades corresponding to those produced by known amounts of gold treated according to the method described.

Bettel [4] filters suspended matter from the cyanide solution, adds a measured quantity of a strong solution of potassium cyanide which contains some cuprous cyanide and precipitates gold by the copper zinc couple produced by addition of a measured quantity of zinc fume. The remainder of the method is the same as Prister's.

[1] Proc. Chem. Met. and Min. Soc. of So. Af., IV, 235, 1904.
[2] Eng. and Min. Journal, Oct. 31, 1903.
[3] Proc. Chem. Met. and Min. Soc. of So. Af., Sept., 1913.
[4] Min. World, **33**, 102 and **35**, 987.

Dowsett's [1] factory test of barren cyanide solutions is capable of detecting variation in gold value of 1 cent per ton in solutions varying from one cent to about 15 cents per ton. To 500 cc. of the sample in a bottle with slight shoulder are added 10-15 cc. saturated sodium cyanide solution, 2 or 3 drops saturated lead nitrate solution and 1-2 grams 200-mesh fine zinc dust. The stoppered bottle is shaken violently until the precipitate settles rapidly. Inverting the bottle allows the precipitate to settle into a casserole. Clear liquid is removed by decantation. Zinc is dissolved by hydrochloric acid added drop by drop until reaction ceases. A few drops excess hydrochloric acid and 3-5 drops dilute nitric acid (sp.gr. 1.18) are added and the liquid concentrated to 1-2 cc. The solution is transferred to a $\frac{1}{2}$-in. diameter test-tube, about 1 cc. of stannous chloride reagent added and grade of cyanide solution estimated by the tint obtained after one or two minutes standing. 1/1000 oz. gold per ton of original cyanide solution gives a very slight coloration; 15/10000 a slight yellow; 1/500 a slight pinkish yellow; 3/1000 a strong pink; 1/250 the purple of Cassius. Too much nitric acid hinders the production and the presence of mercury causes modification of the color. No more lead nitrate should be used than is sufficient to produce a rapidly settling precipitate. The stannous chloride reagent is a water solution containing about $12\frac{1}{2}\%$ crystals and 10% concentrated hydrochloric acid.

PREPARATION OF PROOF GOLD

Commercial gold may contain arsenic, antimony, selenium, tellurium, copper, lead, mercury, silver, zinc, palladium, platinum and other metals of the platinum group. The method of making pure gold depends to a certain extent upon the character and quantity of impurities.[2] The method described assumes the raw material to be of extreme impurity. The metal is treated in 10-g am portions.

When the metal contains silver its solution is effected most quickly by rolling extremely thin and annealing before treatment with acids.

The strips, in a covered No. 6 casserole on a steam bath, are dissolved with a mixture of 5 cc. nitric and 50 cc. hydrochloric acid. If but little silver is present the quantity of hydrochloric acid may be decreased to 25 cc. The solution is evaporated to dryness and the casserole gently heated over a Bunsen flame until all the gold is reduced to metal.

Digestion with ammonia will dissolve most of the silver and copper. After decanting the ammoniacal solution and washing with water, the gold is digested with hot nitric acid. If the solution is wine colored the digestion is continued for several hours, and reheated with fresh portions of acid until the absence of color indicates removal of palladium. The gold is now dissolved with 5 cc. of nitric and 15 to 20 cc. hydrochloric acids, evaporated to dryness, residue moistened with the least quantity of hydrochloric acid, dissolved with about 800 cc. water and liquid transferred to a 1000 cc. beaker. After the faint cloud of silver chloride settles to the bottom of the beaker, the clear liquid only is siphoned to another beaker, and allowed to stand another period of several days if it appears at all cloudy. The clear liquid is now siphoned into a 1000-cc. flask and sulphur dioxide gas passed until the gold is practically all precipitated. The gold is allowed to

---

[1] Trans. I.M.M., 1912-13, 190; Met. and Chem. Eng., July, 1914.
[2] Eng. and Min. Jour., **68,** 785, 1899; "Metallurgy of Gold," Rose, 5th Ed.; Min. and Sci. Press, Nov. 14, 1903; "Manual of Fire Assaying," Fulton; "Assaying of Precious Metals," Smith.

settle, digested with hot nitric acid for a few minutes, washed by decantation several times, redissolved with aqua regia, solution transferred to a casserole, and nitric acid expelled by repeated evaporation to syrup with addition of hydrochloric acid. The product of the second evaporation is moistened with the least quantity of hydrochloric acid, dissolved with water and solution transferred to a 1000-cc. beaker or Erlenmeyer flask. To the liquid of about 500-cc. volume is added 11 grams of ammonium oxalate crystals. The beaker is permitted to remain on a steam bath until reaction is complete. The spongy mass of gold is now washed with hot water by decantation until free of salts.

The gold is dried, melted in a clay crucible which has previously been thinly glazed with borax glass and poured out into a mold of charcoal, graphite and clay or iron polished with graphite.

The ingot, which will have a volume of half a cubic centimeter, is cleaned by paring with a knife and rolled or hammered into a thin sheet. The rolls or hammer should be clean, bright and free of grease.

The gold, cut into convenient strips, is digested for several hours with hydrochloric acid and finally washed thoroughly with distilled water.

The dried gold thus prepared may be considered 1000 fine.

FURNACE METHODS OF ASSAY FOR GOLD

Details of assay for gold and silver by furnace methods will be found following the chapter on silver.

IODINE

I, *at.wt.* 126.93; *sp.gr.* 4.948$^{17°}$; *m.p.* 113.5°;[1] *b.p.* 184.4° C; *acids,* **HI, HIO, HIO$_3$, HIO$_4$.**

DETECTION

The element may be recognized by its physical properties. It is a grayish black, crystalline solid, with metallic luster, brownish-red in thin layers. It vaporizes at ordinary temperatures with characteristic odor. Upon gently heating the element the vapor is evident, appearing a deep blue when unmixed with other gases, and violet when mixed with air. It colors the skin brown. Chemically it behaves very similarly to chlorine and bromine.

Free iodine colors water yellow to black, carbon disulphide violet, ether or chloroform a reddish color, cold starch solution blue.

Tannin interferes with the usual tests for iodine, unless ferric chloride is present.

Iodide. The dry powder, heated with concentrated sulphuric acid, evolves violet fumes of iodine. Iodine is liberated from iodides by solutions of As^5, Sb^5, Bi^5, Cu'', Fe''', Cr^6, H$_3$Fe(CN)$_6$, HNO$_2$, Cl, Br, H$_2$O$_2$, ozone.

Insoluble iodides may be transposed by treatment with H$_2$S, the filtered solution being tested for the halogen.

Iodate. The acidulated solution is reduced by cold solution of SO$_2$, or K$_4$Fe(CN)$_6$, (acidulated with dilute H$_2$SO$_4$), or by Cu$_2$Cl$_2$, H$_3$AsO$_3$, FeSO$_4$, etc. An iodate in nitric acid may be detected by diluting the acid with water, adding starch solution, then hydrogen sulphide water, drop by drop, a blue zone forming in presence of the substance.

ESTIMATION

The element is found free in some mineral waters; combined as iodides and iodates in sea water; in ashes of sea plants; small quantities in a number of minerals, especially in Chili saltpeter as sodium iodate, hence in the mother liquor from the Chilian niter works from which iodine is principally produced. Sea-weed ash (drift kelp, Laminaria digitata and L. stenophylla) is an important source of iodine.

Free iodine, potassium iodide, iodoform, are the principal commercial products.

Preparation and Solution of the Sample

In dissolving the substance it will be recalled that free iodine is soluble in alcohol, ether, chloroform, glycerole, benzole, carbon disulphide, solutions of soluble iodides. One hundred cc. of water at 11° C. is saturated with 0.0182 gram iodine, at 55° with 0.092 gram.

Chapter contributed by Wilfred W. Scott.

IODINE

Iodides of silver, copper (cuprous), mercury (mercurous), and lead are insoluble, also TlI, PdI$_2$. Iodides of other metals are soluble; those of bismuth, tin, and antimony, require a little acid to hold them in solution.

Iodates of silver, barium, lead, mercury, bismuth, tin, iron, chromium require more than 500 parts of water at 15° C. to hold them in solution. Iodates of copper, aluminum, cobalt, nickel, manganese, zinc, calcium, strontium, magnesium, sodium, and potassium are more soluble. One hundred cc. of cold water dissolves 0.00385 gram AgIO$_3$ and 0.000035 gram AgI at ordinary temperatures.

Free Iodine (Commercial Crystals). Iodine is best brought into solution in a strong solution of potassium iodide according to the procedure described for standardization of sodium thiosulphate under Volumetric Methods. The iodine is now best determined volumetrically by titration with standard thiosulphate or arsenic.

Iodine or Iodides in Water. The sample of water is evaporated to about one-fourth its volume and then made strongly alkaline with sodium carbonate. The precipitated calcium and magnesium carbonates are filtered off and washed. The filtrate containing the halogens is evaporated until the salts begin to crystallize out. The hot concentrated solution is poured into three volumes of absolute alcohol and the resulting solution again filtered. The residue is washed four or five times with 95% alcohol. All of the bromine and iodine pass into the solution, whereas a large part of chlorine as sodium chloride remains insoluble and is filtered off. About half a cc. of 50% potassium hydroxide is added and a greater part of the alcohol distilled off with a current of air. The residue is concentrated to crystallization and again poured into three times its volume of absolute alcohol and filtered as above directed. This time only one or two drops of potassium solution is added and the procedure repeated several times. The final filtrate is freed from alcohol by evaporation, the solution taken to dryness and gently ignited, then taken up with a little water and filtered. Iodine is determined in the filtrate, preferably by the volumetric procedure III, decomposition with nitrous acid, described under Volumetric Methods, p. 242.

Organic Substances. If only an iodide is present, the Carius method is followed; in presence of other halogens, the "lime method" is preferred. Details of these methods are given in the chapter on Chlorine under Preparation and Solution of the Sample, p. 145.

Silver iodide cannot be separated from the glass of the combustion-tube by solution with ammonium hydroxide as is the chloride or bromide of silver. The compound, together with the glass, is collected upon a filter paper, and washed with dilute nitric acid, followed by alcohol; then dried at 100° C. After removing most of the iodide and the glass, the filter is ignited in a weighed porcelain crucible, the main bulk of the material then added, the substance fused and weighed as AgI+glass. The mass is then covered with dilute sulphuric acid and a piece of pure zinc added. After several hours (preferably over night) the excess zinc is carefully removed and the iodine solution decanted from the glass and metallic silver, and the residue washed by decantation. The silver is now dissolved in hot dilute nitric acid, then filtered from the residue of glass through a small filter. The glass and filter are ignited and weighed. The difference between the two weighings is due to silver iodide.

Minerals. Phosphates. The substance is decomposed by digestion with 1 : 1 sulphuric acid in a flask through which a current of air passes to sweep out the iodine vapor into a solution of potassium hydroxide, the sample being boiled until all the iodine vapors have been driven into the caustic. Iodates are converted to iodides by reduction with sulphurous acid.

With the iodine content below 0.02%, a 50 to 100-gram sample should be taken.

SEPARATIONS

Separation of Iodine from the Heavy Metals. The heavy metals are precipitated as carbonates by boiling with solutions of alkali carbonates, the soluble alkali iodide being formed.

Iodine is liberated from combination by nitrous acid.

Silver iodide may be decomposed by warming with metallic zinc and sulphuric acid.

Separation of Iodine from Bromine or from Chlorine. Details of separation and estimation of the halides in presence of one another are given in the chapter on Chlorine. Advantage is taken of the action of nitrous acid on dilute solutions, free iodine being liberated, while bromides and chlorides are not acted upon.

The solution containing the halogens is place in a large, round-bottom flask and diluted to about 700 cc. Through a two-holed stopper a glass tube passes to the bottom of the flask; through this tube steam is conducted to assist the volatilization of iodine. A second short tube connected to the absorption apparatus conducts the evolved vapor from the flask into a 5% caustic soda solution containing an equal volume of hydrogen peroxide (about 50 cc. of each). The absorption system may be made by connecting two Erlenmeyer flasks in series, the inlet tubes dipping below the solutions in the flasks. It is advisable to cool the receivers with ice.

Two to 3 cc. of dilute sulphuric acid (1 : 1) and 25 cc. of 10% sodium nitrite solution are added to the liquid containing the halogens, the apparatus is immediately connected, and the contents of the large flask heated to boiling, conducting steam into it at the same time. The iodine vapor is gradually driven over into the cooled receiving flasks.

When the solution in the large flask has become colorless it is boiled for half an hour longer. The steam is now shut off, the flask disconnected from the receiving flasks and the heat turned off. The contents of the receiving flasks are combined with the washing from the connecting tubes and the solution heated to boiling to expel, completely, hydrogen peroxide. The cooled liquid is acidified with a little sulphuric acid and the solution decolorized with a few drops of sulphurous acid. Iodine is now precepitated as silver iodide by adding an excess of silver nitrate and a little nitric acid and boiling the mixture to coagulate the precipitate. AgI is determined as directed on page 239.

Chlorine and bromine remain in the large flask in combined form and may be determined in this solution if desired.

Separation of Iodine from Chlorine and Bromine by Precipitation as Palladous Iodide. The solution containing the halogens is acidified with hydrochloric acid, and palladous chloride solution added to the complete precipitation of the iodide. The compound is allowed to settle in a warm place for twenty-four hours or more and then filtered and washed free of the other halogens. It may now be dried and weighed as palladous iodide, PdI_2, or ignited in a current of hydrogen, then weighed as metallic palladium and the equivalent iodine calculated. See Gravimetric methods.

GRAVIMETRIC METHODS
Precipitation as Silver Iodide

The procedure is practically the same as that described for determining chlorine.

Silver nitrate solution is added to the iodide solution, slightly acidified with nitric acid. The precipitate is filtered into a weighed Gooch crucible, then washed, dried, gently ignited, and weighed as silver iodide.

$$AgI \times 0.5406 = I \text{ or } \times 0.7071 = KI.$$

NOTE. If filter paper is used in place of a Gooch crucible, the precipitate is removed and the filter ignited separately. A few drops of nitric and hydrochloric acid are added, the acids expelled by heat and the residue weighed as AgCl. This, multiplied by 1.638 = AgI. The result is added to the weight of the silver iodide, which is ignited and weighed separately.

Determination of Iodine as Palladous Iodide

This method is applicable for the direct determination of iodine in iodides in presence of other halogens.

The method of isolation of iodine as the palladous salt has been given under Separations. The salt dried at 100° C. is weighed as PdI_2.

$$PdI_2 \times 0.704 = I.$$

PdI_2 ignited in a current of hydrogen is changed to metallic palladium.

$$Pd \times 2.379 = I.$$

VOLUMETRIC METHODS
Determination of Hydriodic Acid—Soluble Iodides

Free hydriodic acid cannot be determined by the usual alkalimetric methods for acids. The procedures for its estimation, free or combined as a soluble salt, depends upon the liberation of iodine and its titration with standard sodium thiosulphate, in neutral or slightly acid solution; or by means of standard arsenious acid, in presence of an excess of sodium bicarbonate in a neutral solution. The following equations represent the reactions that take place:

I. Thiosulphate. $2Na_2S_2O_3 + I_2 = 2NaI + Na_2S_4O_6.$

II. Arsenite. $Na_3AsO_3 + I_2 + H_2O = Na_3AsO_4 + 2HI.$

The free acid formed in the second reaction is neutralized and the reversible reaction thus prevented:

$$HI + NaHCO_3 = NaI + H_2O + CO_2.$$

The presence of a free alkali is not permissible, as the hydroxyl ion would react with iodine to form iodide, hypoiodite and finally iodate, hence sodium or potassium carbonates cannot be used. Alkali bicarbonates, however, do not react with iodine.

Standard Solutions. *Tenth Normal Sodium Thiosulphate.* From the reaction above it is evident that 1 gram molecule of thiosulphate is equivalent to 1 atom iodine = 1 atom hydrogen, hence a tenth normal solution is equal to one-tenth the molecular weight of the salt per liter, e.g., 24.82 grams $Na_2S_2O_3 \cdot 5H_2O$; generally a slight excess is taken—25 grams of the crystallized salt. It is advisable to make up 5 to 10 liters of the solution, taking 125 to 250 grams sodium thiosulphate crystals and making up to volume with distilled water, boiled free of carbon dioxide. The solution is allowed to stand a week to ten days, and then standardized against pure, resublimed iodine.

About 0.5 gram of the purified iodine is placed in a weighing bottle containing a known amount of saturated potassium iodide solution (2 to 3 grams of KI free from KIO_3 dissolved in about $\frac{1}{2}$ cc. of H_2O), the increased weight of the bottle, due to the iodine, being noted. The bottle and iodine are placed in a beaker containing about 200 cc. of 1% potassium iodide solution (1 gram KI per 200 cc.), the stopper removed with a glass fork and the iodine titrated with the thiosulphate to be standardized.

Calculation. The weight of the iodine taken, divided by the cc. thiosulphate required, gives the value of 1 cc. of the reagent; this result divided by 0.012692 gives the normality factor.

Note. The thiosulphate solution may be standardized against iodine, which has been liberated from potassium iodide in presence of hydrochloric acid by a known amount of standard potassium bi-iodate, a salt which may be obtained exceedingly pure.

$$KIO_3 \cdot HIO_3 + 10KI + 11HCl = 11KCl + 6H_2O + 6I_2.$$

A tenth normal solution contains 3.2496 grams of the pure salt per liter. (One cc. of this will liberate 0.012692 gram of iodine from potassium iodide.) The purity of the salt should be established by standardizing against thiosulphate, which has been freshly tested against pure resublimed iodine.

About 5 grams of potassium iodide (free from iodate) are dissolved in the least amount of water that is necessary to effect solution, and 10 cc. of dilute hydrochloric acid (1 : 2) are added, and then 50 cc. of the standard bi-iodate solution. The solution is diluted to about 250 cc. and the liberated iodine titrated with the thiosulphate reagent; 50 cc. will be required if the reagents are exactly tenth normal.

Tenth Normal Arsenite. From the second reaction above it is evident that As_2O_3 is equivalent to $2I_2$, e.g., to 4H, hence $\frac{1}{4}$ the gram molecular weight of arsenious oxide per liter will give a normal solution: $198 \div 4 = 49.5$.

4.95 grams of pure arsenious oxide is dissolved in a little 20% sodium hydroxide solution, the excess of the alkali is neutralized with dilute sulphuric acid, using phenolphthalein indicator, the solution being just decolorized. Five hundred cc. of distilled water containing about 25 grams of sodium bicarbonate are added. If a pink color develops, this is destroyed with a few drops of weak sulphuric acid. The solution is now made to volume, 1000 cc. The reagent is standardized against a measured amount of pure iodine. The oxide may be dissolved directly in sodium bicarbonate solution.

NOTE. Commercial arsenious oxide is purified by dissolving in hot hydrochloric acid, filtering the hot saturated solution, cooling, decanting off the mother liquor, washing the deposited oxide with water, drying and finally subliming.

Starch Solution. Five grams of soluble starch are dissolved in cold water, the solution poured into 2 liters of hot water and boiled for a few minutes. The reagent is kept in a glass-stoppered bottle.

The addition of a few cc. of 5% NaOH, then heating to boiling and filtering will preserve the starch.

Decomposition of the Iodide by Ferric Salts

The method takes advantage of the following reaction:

$$Fe_2(SO_4)_3 + 2KI = K_2SO_4 + I_2 + 2FeSO_4.$$

The procedure enables a separation from bromides, as these are not acted upon by ferric salts.

Procedure. To the iodide in a distillation flask is added an excess of ferric ammonium alum, the solution acidified with sulphuric acid, then heated to boiling, and the iodine distilled into a solution of potassium iodide. The free iodine in the distillate is titrated with standard thiosulphate, or by arsenious acid in presence of an excess of sodium bicarbonate.

The reagent is added from a burette until the titrated solution becomes a pale yellow color. About 5 cc. of starch solution are now added and the titration continued until the blue color of the starch fades and the solution becomes colorless.

One cc. of tenth normal reagent = 0.012692 gram iodine, equivalent to 0.012793 gram HI, or 0.016602 gram KI.

Decomposition with Potassium Iodate [1]

The reaction with potassium iodate is as follows:

$$5KI + KIO_3 + 6HCl = 6KCl + 3H_2O + 3I_2.$$

It is evident that $\frac{5}{6}$ of the titration for iodine would be equal to the iodine of the iodide, hence 1 cc. of tenth normal thiosulphate is equivalent to 0.012692 $\times \frac{5}{6} = 0.01058$ gram iodine due to the iodide. The procedure is as follows:

Procedure. A known amount of tenth normal potassium iodate is added to the iodide solution, in sufficient amount to liberate all of the iodine, combined as iodide, and several cc. in excess. Hydrochloric acid and a piece of calcite are added. The mixture is boiled until all of the liberated iodine has been expelled. To the cooled solution 2 or 3 grams of potassium iodide are added and the liberated iodine, corresponding to the excess of iodate in the solution, is titrated with standard thiosulphate. If 1 cc. of thiosulphate is equal to 1 cc. of the iodate, then the total cc. of the iodate used, minus the cc. thio-

[1] H. Dietz and B. M. Margosches, Chem. Ztg., **2**, 1191, 1904. Treadwell and Hall, "Analytical Chemistry," Vol. **2**.

sulphate required in the titration gives a difference due to the volume of iodate required to react with the iodide of the sample.

One cc. of N/10 KIO$_3$ = 0.01058 gram I in KI.

NOTE.   Tenth normal potassium iodate contains 3.5675 grams KIO$_3$ per 1000 cc.

Decomposition of the Iodide with Nitrous Acid (Fresenius)

Nitrous acid reacts with an iodide as follows:

$$2HNO_2 + 2HI = 2NO + 2H_2O + I_2.$$

Since neither hydrochloric nor hydrobromic acids are attacked by nitrous acid, the method is applicable to determining iodine in presence of chlorine and bromine; hence is useful for determining small amounts of iodine in mineral waters containing comparatively large amounts of the other halogens.

Nitrous Acid. The reagent is prepared by passing the gas into strong sulphuric acid until saturated.

Procedure. The neutral or slightly alkaline solution of the iodide is placed in a glass-stoppered separatory funnel, Fig. 39, and slightly acidified with dilute sulphuric acid. A little freshly distilled colorless carbon disulphide (or chloroform) is added, then 10 drops of nitrous acid reagent. The mixture is well shaken, the disulphide allowed to settle, drawn off from the supernatant solution and saved for analysis. The liquor in the funnel is again extracted with a fresh portion of disulphide and if it becomes discolored it is drawn off and added to the first extract. If the extracted aqueous solution appears yellow, it must be again treated with additional carbon disulphide until all the iodine has been removed (e.g., until additional CS$_2$ is no longer colored when shaken with the solution). The combined extracts are washed with three or four portions of water, then transferred to the filter and again washed until free from acid. A hole is made in the filter and the disulphide allowed to run into a small beaker and the filter washed down with about 5 cc. of water. Three cc. of 5% sodium bicarbonate are added and the iodine titrated with N/20 or N/50 standard thiosulphate, the reagent being added until the reddish-violet carbon disulphide becomes colorless.

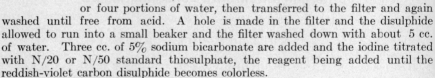

FIG. 39.

The sodium thiosulphate used is standardized against a known amount of pure potassium iodide treated in the manner described above.

One cc. N/20 Na$_2$S$_2$O$_3$ = .00635 gram I, 1 cc. N/50 Na$_2$S$_2$O$_3$ = .002538 gram I.

Liberation of Iodine by Means of Hydrogen Peroxide and Phosphoric Acid [1]

Principle. Iodine is liberated from an iodide by addition of hydrogen peroxide to the solution acidified with phosphoric acid, the iodine distilled into potassium iodide and titrated with thiosulphate.

Procedure. Fifty cc. of the iodide solution are mixed with 5 cc. of pure phosphoric acid and 10 to 20 cc. hydrogen peroxide added, the mixture being placed in a round-bottomed flask, connected with a short condenser, delivering into two absorption vessels containing a 10% solution of potassium iodide. A current of air is drawn through the apparatus, and the contents of the flask gradually heated to boiling. The iodine is absorbed in the potassium iodide solution and titrated as usual with standard sodium thiosulphate. Twenty minutes' heating is generally sufficient.

One cc. $Na_2S_2O_3 = 0.012692$ gram I, or 0.016602 gram KI.

NOTE. Iodine in urine may be determined by evaporating to 1/10 its volume. After adding an excess of sodium hydroxide, the mixture is taken to dryness and gently ignited. The ash may be used for the iodine determination.

Oxidation of Combined Iodine with Chlorine. (Mohr's Modification of Dupré's Method)[2]

When a solution of potassium iodide is treated with successive amounts of chlorine water, iodine is liberated, which reacts with an excess of chlorine with formation of chloride of iodine (ICl) and with greater excess the pentachloride (ICl_5) which is changed in presence of water to iodic acid (HIO_3).

Procedure. The weighed iodide compound is brought into a stoppered flask, and chlorine water delivered from a large burette until all yellow color has disappeared. A drop of the mixture brought in contact with a drop of starch solution should produce no blue color. Sodium bicarbonate is now added until the mixture is slightly alkaline, followed by an excess of potassium iodide and 4 to 5 cc. of starch reagent. Standard thiosulphate is now added until the blue color is removed. The excess of chlorine water is thus ascertained. From the value of the chlorine reagent the iodine of the sample may readily be calculated.

The chlorine water is standardized by running 25 to 50 cc. of the reagent into potassium iodide solution (see procedure for bromides, p. 95), and titrating the liberated iodine with standard sodium thiosulphate. The value of the reagent in terms of thiosulphate are thus ascertained and from this the value per cc. in terms of iodine.

OTHER METHODS

Volhard's Method for Determining Iodides

This procedure is very similar to those for determining chlorine or bromine, with the exception that silver iodide formed will occlude both the iodide solu-

---

[1] E. Winterstein and E. Herzfeld, Zeit. Physiol. Chem., **63**, 49–51, 1909. Chem. Zentralbl., (1), 473–474, 1910.
[2] Sutton, "Volumetric Analysis," 10th Ed.

tion and silver nitrate unless the additions of the silver salt are made in small portions with vigorous shaking.

Standard silver nitrate is added to the solution in a glass-stoppered flask, shaking vigorously with each addition. As long as the solution appears milky the precipitation is incomplete. When the silver iodide is coagulated and the supernatant liquid appears colorless, ferric alum solution is added, and the excess of silver nitrate titrated with potassium sulphocyanate until the characteristic reddish end-point is obtained.

The iodine is calculated from the amount of silver nitrate required. E.g., total $AgNO_3$ added, minus excess determined by $KCNS = cc$. $AgNO_3$ required by the iodine.

NOTE. The ferric salt oxidizes hydriodic acid with separation of iodine, whereas the silver iodide is not acted upon, hence the indicator is added after all the iodine has combined with silver.

Determination of Iodates

The procedure is the reciprocal of the one for determination of iodide by means of an iodate:

Reaction. $KIO_3 + 5KI + 6HCl = 6KCl + 3H_2O + 3I_2$.

Procedure. The solution containing the iodate is allowed to run into an excess of potassium iodide solution containing hydrochloric acid. The liberated iodine is titrated with sodium thiosulphate as usual.

One cc. $N/10$ $Na_2S_2O_3 = 0.002932$ gram HIO_3, or 0.003567 gram KIO_3.

Determination of Periodates

The procedure is the same as that described for iodates, the reaction in this case, however, being as follows:

$$KIO_4 + 7KI + 8HCl = 8KCl + 4H_2O + 4I_2.$$

From the equation it is evident that 1 gram molecule of the iodate is equivalent to 8 atoms of iodine $= 8$ atoms of hydrogen, hence $\frac{1}{8}$ the molecular weight per liter of solution would equal a normal solution. Therefore, 1 cc. of a tenth normal solution would contain $0.019193 \div 8 = 0.002399$ gram HIO_4.

One cc. $N/10$ $Na_2S_2O_3 = 0.002399$ gram HIO_4, or $= 0.002849$ gram $HIO_4 \cdot 2H_2O$, or $= 0.002875$ gram KIO_4.

Determination of Iodates and Periodates in a Mixture of the Two

The procedure depends upon the fact that an iodate does not react with potassium iodide in neutral or slightly alkaline solutions, whereas a periodate undergoes the following reactions:

$$KIO_4 + 2KI + H_2O = 2KOH + KIO_3 + I_2.$$

Procedure. The sample, dissolved in water, is divided into two equal portions.

A. To one portion a drop of phenolphthalein indicator is added and the solution made just faintly alkaline by addition of alkali to acid solutions or hydrochloric acid to alkaline solution, as the case may require. Ten cc. of cold saturated solution of sodium bicarbonate are added and an excess of potassium iodide. The liberated iodine is titrated with tenth normal arsenious acid.[1] ($Na_2S_2O_3$ will not do in this case, as the solution is alkaline.)

One cc. N/10 As_2O_3 = 0.0115 gram KIO_4.

B. To the other portion potassium iodide is added in excess and the solution made distinctly acid. The liberated iodine is titrated with standard sodium thiosulphate. (As_2O_3 will not do.)

Calculation. In the acid solution, B, both iodates and periodates are titrated, whereas in the alkaline solution, A, only the periodates are affected. From the reactions for periodates it is evident that 1 cc. $Na_2S_2O_3$ = 4 cc. As_2O_3 for the periodate titration, hence

Cc. $Na_2S_2O_3$ − cc. $As_2O_3 \times 4$ = cc. thiosulphate due to KIO_3.

The difference, multiplied by 0.003567 = grams KIO_3 in the sample.

Determination of Iodine in Mineral Waters and Brines

The following procedure is given by W. F. Baughman and W. W. Skinner.[2] Take such a quantity of the brine or water as will contain not more than 0.1 g. iodine as iodide or more than 10 g. total salts. Adjust the volume to 100 cc. to 150 cc. and boil it with a sufficient amount of sodium hydroxide and sodium carbonate to precipitate the calcium and magnesium. Filter off the precipitate and wash with hot water. Introduce the filtrate into an Erlenmeyer flask, adjust the volume to about 100 cc., neutralize with dilute sulphuric acid, and add 1 cc. of a solution of sodium hydroxide (4 g. per 100 cc.). Heat to boiling, add an excess of potassium permanganate, continue the heating until the precipitate begins to coagulate, and then allow to cool. Add sufficient alcohol to cause the permanganate color to disappear and allow the precipitate to settle on the steam bath. Filter and wash with hot water. After cooling, add one or two grams of potassium iodide, acidify with hydrochloric acid, and titrate with standard thiosulphate. The number of cc. required, divided by 6, represents the number of cc. required by the iodine in the sample.

[1] In alkaline solutions the arsenious acid titration must be made, whereas in acid solutions potassium thiosulphate is used.

[2] Bureau of Chemistry, Dept. Agriculture, Washington, D. C. Ref. Jour. Ind. Eng. Chem., Vol. 11, No. 6, page 563 (June, 1919).

IRON

Fe, at.wt. 55.84; sp.gr. 7.85–7.88; m.p. pure 1530°,[1] wrought 1600°,[2] white pig 1075°,[1] gray pig 1275°,[1] steel 1375°; [1] b.p. 2450° [1] oxides FeO, Fe_2O_3, Fe_3O_4.

DETECTION

Ferric Iron. The yellow to red color in rocks, minerals, and soils is generally due to the presence of iron.

Hydrochloric acid solutions of iron as ferric chloride are colored yellow.

Potassium or ammonium sulphocyanate produces a red color with solutions containing ferric iron. Nitric acid and chloric acid also produce a red color with potassium or ammonium sulphocyanate. This color, however, is destroyed by heat, which is not the case with the iron compound. The red color of ferric iron with the cyanate is destroyed by mercuric chloride and by phosphates, borates, certain organic acids, and their salts, e.g., acetic, oxalic, tartaric, citric, racemic, malic, succinic, etc.

Potassium ferrocyanide, $K_4Fe(CN)_6$, produces a deep blue color with ferric salts.

Salicylic acid added to the solution of a ferric salt containing no free mineral acid gives a violet color. Useful for detecting iron in alum and similar products.

Ferrous Iron. Potassium Ferricyanide, $K_3Fe(CN)_6$, gives a blue color with solutions of ferrous salts.

Distinction between Ferrous and Ferric Salts.

KCNS gives red color with Fe''' and no color with Fe''.

$K_3Fe(CN)_6$ gives a blue color with Fe'' and a brown or green with Fe'''.

NH_4OH, NaOH or KOH precipitates red, $Fe(OH)_3$ with Fe''' and white, $Fe(OH)_2$ with Fe'' turning green in presence of air due to oxidation.[3]

Sodium peroxide produces a reddish-brown precipitate of $Fe(OH)_3$ with either ferrous or ferric salt solutions, the former being oxidized to the higher valence by the peroxide. Chromium and aluminum remain in solution, if present in the sample.

ESTIMATION

Iron is so widely diffused in nature that its determination is necessary in practically all complete analyses of ores, rocks, minerals, etc. It is especially important in the evaluation of iron ores for the manufacture of iron and steel. Among the ores of iron the following are more common:

Oxides. Red hematite, Fe_2O_3; brown hematite, $2Fe_2O_3\cdot3H_2O$; black magnetite or magnetic iron ore, Fe_3O_4. Ferric oxide with varying amounts of water

---

[1] Circular 35 (2d Ed.) U. S. Bureau of Standards.
[2] D. Van Nostrand's Chemical Annual.—Olsen.
[3] The green salt is a hydrate of Fe_3O_4. The white precipitate can be obtained in absence of air or by using H_2SO_3 to take up oxygen in solution.

Chapter contributed by Wilfred W. Scott.

forms the substances known as hematite, göthite, limonite, yellow ochre, bog iron ore.

Sulphide. Iron pyrites or "fool's gold," FeS_2; pyrrhotite, FeS.

Carbonates. Spatic iron ore, $FeCO_3$; combined with clay in clay ironstone with bituminous material as "black band."

Iron is determined in the cinders and in iron ore briquettes from burned iron pyrites, by-products of sulphuric acid.

It is found as an impurity in a large number of commercial salts and in the mineral acids.

Preparation and Solution of the Sample

The material should be carefully sampled and quartered down according to the general procedure for sampling. Ores should be ground to pass an 80-mesh sieve. In analysis of metals, both the coarse and fine drillings are taken.

The following facts regarding solubility should be remembered: The *element* is soluble in hydrochloric acid and in dilute sulphuric acid, forming ferrous salts with liberation of hydrogen. It is insoluble in concentrated, cold sulphuric acid, but is attacked by the hot acid, forming ferric sulphate with liberation of SO_2. Moderately dilute, hot nitric acid forms ferric nitrate and nitrous oxide; the cold acid gives ferrous nitrate and ammonium nitrate or nitrous oxide or hydrogen. Cold, concentrated nitric acid forms "passive iron," which remains insoluble in the acid. The *oxides of iron* are readily soluble in hydrochloric acid, if not too strongly ignited, but upon strong ignition the higher oxides dissolve with extreme difficulty. They are readily soluble, however, by fusion with acid potassium sulphate followed by an acid extraction. *Silicates* are best dissolved by hot hydrochloric acid containing a few drops of hydrofluoric acid or by fusion with sodium and potassium carbonates, followed by hot hydrochloric acid.

Soluble Iron Salts. Water solutions are acidified with HCl or H_2SO_4, so as to contain about 3% of free acid.

Ores. The samples should be pulverized to pass an 80- to 100-mesh sieve.

Sulphides, Ores Containing Organic Matter. One- to 5-gram samples should be roasted in a porcelain crucible over a Bunsen flame for about half an hour, until oxidized. The oxide is now dissolved as directed in the following procedure.

Oxides, Including Red and Brown Hematites, Magnetic Iron Ore, Spatose Iron Ore, Roasted Pyrites, and Iron Ore Briquettes. One to 5 grams of the ore, placed in a 400-cc. beaker, is dissolved by adding twenty times its weight of strong hydrochloric acid with a few drops of 5% stannous chloride solution. Addition of 4 or 5 drops of HF is advantageous if small amounts of silica are present. The solution is covered with a watch-glass and heated to 80 or 90° C. until solution is complete. Addition of more stannous chloride may be necessary, as this greatly assists solution. An excess sufficient to completely decolorize the solution necessitates reoxidation with hydrogen peroxide, hence should be avoided. If a colored residue remains, it should be filtered off, ignited and fused with a mixture of Na_2CO_3 and K_2CO_3 in a platinum crucible. The fusion dissolved in dilute HCl is added to the main filtrate.

NOTE. The ore placed in a porcelain boat in a red-hot combustion tube may be reduced with hydrogen (taking precaution first to sweep out oxygen with CO_2) and after cooling in an atmosphere of hydrogen the reduced iron may be dissolved in acid and titrated.

Iron Silicates. One to 5 grams of the material, placed in a deep platinum crucible, is treated with ten times its weight of 60% HF and 3 to 4 drops of conc. H_2SO_4. The mixture is evaporated to near dryness on the steam bath and taken up with dilute sulphuric acid or hydrochloric acid. The latter acid is the best solvent for iron.

Fusion with Potassium Bisulphate. The sample is mixed with ten times its weight of the powdered bisulphate and 2–3 cc. of concentrated sulphuric acid added. A porcelain or silica dish will do for this fusion. The fusion should be made over a moderate flame and cooled as soon as the molten liquid becomes clear. Complete expulsion of SO_3 should be avoided. It may be necessary to cool and add more conc. sulphuric acid to effect solution. Iron and alumina completely dissolve, but silica remains undissolved. The melt is best cooled by pouring it on a large platinum lid.

Fusion with Carbonates of Sodium and Potassium. The residues insoluble in hydrochloric acid are fused with 5 parts by weight of the fusion mixture $(Na_2CO_3 + K_2CO_3)$ in a platinum crucible. The Méker blast will be necessary. When the effervescence has ceased and the melt has become clear, the crucible is removed from the flame, a platinum wire inserted and the melt cooled. Upon gently reheating, the fuse may be readily removed by the wire in a convenient form for solution in dilute hydrochloric acid.

The bisulphate fusion is recommended for fusion of residues high in iron and alumina. It is an excellent solvent for ignited oxides of these elements. The carbonate fusions are adapted to residues containing silica.

SEPARATIONS

General Procedure. In the usual course of analysis silica is removed by evaporating the acid solution to dryness, taking up with water and filtering. Mercury, lead, bismuth, copper, cadmium, arsenic, antimony, tin, molybdenum and other elements precipitated from an acid solution as sulphides are removed as such by filtration and iron, after oxidation to the ferric state, is precipitated as $Fe(OH)_3$. In the majority of cases it may now be determined by titration.

Ether Method for Removing Iron from a Solution. Ferric chloride dissolved in HCl (sp.gr. 1.1) is more soluble in ether than in this acid. Advantage is taken of this fact when it is desired to remove a greater portion of the iron in determining copper, nickel, cobalt, chromium, vanadium and sulphur (as H_2SO_4) in steel. The hydrochloric acid solution of iron, etc., is evaporated to a syrupy consistency and then taken up with HCl (sp.gr. 1.1) and transferred by means of more of the acid to a separatory funnel. The cold acid solution is now extracted several times by shaking with ether, each time allowing the ether carrying the iron to separate before drawing off the lower layer for re-extraction. Three extractions are generally sufficient for removing the iron.

Since alkali salts cause trouble by crystallizing and clogging the borings of the stopcock, the use of alkalies should be avoided when this method of separation is used. The iron may be extracted from the ether by shaking this with water and drawing off the lower water layer. Since heat is generated by the mixing of ether and the ferric chloride-hydrochloric acid solution, cooling the mixture under the tap during mixing may be necessary. This heating is reduced by using a mixture of ether and hydrochloric acid. Strong hydrochloric acid (sp. gr. 1.19) is saturated with ether, an excess of ether separating out as an upper layer. 100 cc. of the acid will absorb 150 cc. ether. (Dilute hydrochloric acid (sp. gr. 1.1) absorbs only 30 cc. ether.)

GRAVIMETRIC METHODS FOR THE DETERMINATION OF IRON

The gravimetric determination of iron may be made from solutions practically free from other metals. A number of elements such as phosphorus, arsenic, molybdenum, tungsten, and vanadium, in neutral or slightly alkaline solutions, form fairly stable compounds with iron, whereas others, such as lead, copper, nickel, cobalt, sodium, and potassium may be occluded in the ferric hydrate precipitate and are removed only with considerable difficulty. Aluminum, chromium, and several of the rare earths are precipitated with iron, if present. These facts taken into consideration, the volumetric methods are generally preferred as being more rapid and trustworthy.

Determination of Iron as Fe_2O_3

Iron is precipitated as the hydroxide and ignited to the oxide, Fe_2O_3, in which form it is weighed.

Reactions. $FeCl_3 + 3NH_4OH = Fe(OH)_3 + 3NH_4Cl.$
$2Fe(OH)_3 + heat = Fe_2O_3 + 3H_2O.$

Procedure. One-gram sample or a larger amount of material if the iron content is low, is brought into solution with hydrochloric acid, aqua regia, or by fusion with potassium carbonate or potassium acid sulphate, as the case may require. Silica is filtered off and the acid solution treated with H_2S if members of that group are present. The filtrate is boiled to expel H_2S and the iron oxidized to ferric condition by boiling with 5 cc. concentrated nitric acid.

Absence of Aluminum and Chromium. About 1 gram of ammonium chloride salt or its equivalent in solution is added, the volume made to about 200 cc. and ammonium hydroxide added in slight excess to precipitate $Fe(OH)_3$. The solution is boiled for about five minutes, then filtered through an ashless filter.

If Aluminum and Chromium are Present. In place of ammonium hydroxide powdered sodium peroxide is added in small portions until the precipitate first formed clears, the solution being cold and nearly neutral. It is diluted to about 300 cc. and boiled ten to fifteen minutes to precipitate the iron. Aluminum and chromium are in solution. (Mn will precipitate with Fe, if present.) The precipitate is filtered onto a rapid filter and washed with hot water.

Second Precipitation. In either case dissolve the precipitate with the least amount of hot dilute hydrochloric acid and wash the paper free of iron. Add a few cc. of 10% ammonium chloride solution and reprecipitate the hydroxide of iron by adding an excess of ammonium hydroxide, the volume of the solution being about 200 cc. Washing the precipitate by decantation is advisable. Three such washings, 100-cc. portions, followed by two or three on the filter paper, will remove all impurities.

Ignition. The precipitate is ignited wet over a low flame, gradually increasing the heat. Blasting is not recommended, as the magnetic oxide of iron, Fe_3O_4, will form with high heating. The oxide heated gently appears a reddish-brown. Higher heat gives the black oxide, Fe_3O_4. Twenty minutes' ignition, at red heat, is sufficient.

The crucible, cooled in a desiccator, is weighed and Fe_2O_3 obtained.

Factors. $Fe_2O_3 \times 0.6994 = Fe.$
$Fe_2O_3 \times 0.8998 = FeO.$

Precipitation of Iron with "Cupferron," Amino nitrosophenyl=hydroxylamine [1]

By this procedure iron may be precipitated directly in acid solution in presence of a number of elements. Mercury, lead, bismuth, tin, and silver may be partially precipitated. Copper precipitates with iron, but may be easily removed by dissolving it out with ammonia. The method is especially adapted for separation of iron from aluminum, nickel, cobalt, chromium, cadmium, and zinc.

Procedure. The solution containing the iron is made up to 100 cc. and 20 cc. of concentrated hydrochloric acid added. To this cool solution (room temperature) Baudisch's reagent, cupferron, is slowly added with constant stirring, until no further precipitation of iron takes place, and crystals of the reagent appear. The iron precipitate is a reddish-brown. Copper gives a grayish-white flocculent compound. An excess of the reagent equal to one-fifth of the volume of the solution is now added, the precipitate allowed to settle for about fifteen minutes, then poured into a filter paper and washed, first with 2N. HCl, followed by water, then with ammonia and finally with water. The drained precipitate is slowly ignited in a porcelain or platinum crucible and the residue weighed as Fe_2O_3

$$Fe_2O_3 \times 0.6994 = Fe.$$

NOTES. Baudisch's reagent, amino nitrosophenyl-hydroxylamine (cupferron), is made by dissolving 6 grams of the salt in water and diluting to 100 cc. The reagent keeps for a week if protected from the light. It decomposes in the light, forming nitrobenzine. Turbid solutions should be filtered.

The precipitates of copper or iron are but slowly attacked by twice normal hydrochloric acid in the cold, but decomposed by hot acid, hence the solution and reagent should be cold.

Cold, dilute potassium carbonate solution, or ammonium hydroxide, have no action on the iron precipitate; the copper compound dissolves readily in ammonia. Alkaline hydroxide causes rapid decomposition.

The precipitation is best made in comparatively strong acid solutions (HCl, H_2SO_4, or acetic acid).

[1] O. Baudisch, Chem. Ztg., **33**, 1298, 1905. Ibid., **35**, 913, 1911. O. Baudisch and V. L. King, Jour. Ind. Eng. Chem., **3**, 627, 1911.

VOLUMETRIC DETERMINATION OF IRON IN ORES AND METALLURGICAL PRODUCTS

General Considerations. Two general procedures are commonly employed in the determination of iron.

A. Oxidation of ferrous to ferric condition by standard oxidizing agents.

B. Reduction of ferric iron to ferrous condition.

The sample is dissolved as directed under Preparation and Solution of the Sample.

Determination of Iron by Oxidation Methods

Some modification of either the dichromate or permanganate methods is commonly employed in the determination of iron by oxidation. To accomplish this quantitatively, the iron must be reduced to its ferrous condition. This may be accomplished in the following ways:

1. Reduction by Hydrogen Sulphide. During the course of a complete analysis of an ore, H_2S is passed into the acid solution to precipitate the members of that group (Hg, Pb, Bi, Cu, Cd, As, Sb, Sn, Pt, Au, Se, etc.). The filtrate contains iron in the reduced condition suitable for titration with either dichromate or permanganate, the excess of H_2S having been boiled off. If the expulsion of H_2S is conducted in an Erlenmeyer flask there is little chance for reoxidation of the iron during the boiling. Reduction by H_2S is very effective and is frequently advisable. This is the case when titanium is present, since this is not reduced by H_2S, but by methods given below. Arsenic, antimony, copper, and platinum, which, if present would interfere, are removed by this treatment.

Reaction. $2FeCl_3 + H_2S = 2FeCl_2 + 2HCl + S.$

2. Reduction with Stannous Chloride. $SnCl_2$ solution acts readily in a hydrochloric acid solution of the ore; the reduction of the iron is easily noted by the disappearance of the yellow color. The excess of the reagent is oxidized to $SnCl_4$ by addition of $HgCl_2$.

Reactions.   1. $2FeCl_3 + SnCl_2 = 2FeCl_2 + SnCl_4.$
2. Excess $SnCl_2 + 2HgCl_2 = SnCl_4 + 2HgCl$ precipitated.

An excess of $SnCl_2$ is advisable, but a large excess is to be avoided, as a secondary reaction would take place, as follows: $2SnCl_2 + 2HgCl_2 = 2SnCl_4 + 2Hg.$ This reaction is indicated by the darkening of the solution upon the addition of $HgCl_2$. Precipitation of metallic mercury would vitiate results. The solution should be cooled before addition of mercuric chloride. About 15–20 cc. of saturated mercuric chloride, $HgCl_2$, solution should be sufficient.

3. Reduction by a Metal such as Test Lead, Zinc, Magnesium, Cadmium, or Aluminum, in Presence of Either Hydrochloric Acid or Sulphuric Acid. The former acid is preferred with the dichromate titration, and the latter with the permanganate. Two methods of metallic reduction are in common use— reduction by means of test lead, and reduction with amalgamated zinc by means of the Jones reductor.

(*a*) **Reduction with Test Lead.** By this method copper is precipitated from solution and small amounts of arsenic and antimony expelled. Sufficient test lead is added to the acid ferric solution to completely cover the bottom of the beaker. The solution is covered and boiled vigorously until the yellow color has completely disappeared, and the solution is colorless. The reduced iron solution, cooled, is decanted into a 600-cc. beaker, the remaining iron washed out from the lead mat by several decantations with water; two or three 50-cc. portions of water should be sufficient; the washings are added to the first portion. If the solution becomes slightly colored, a few drops of stannous chloride, $SnCl_2$, solution are added, followed by 10 cc. mercuric chloride, $HgCl_2$, solution. The sample is now ready for titration.

(*b*) **Reduction with Zinc, Using the Jones Reductor.** The acid solution of iron, preferably sulphuric acid, is passed through a column of amalgamated zinc.[1] The hydrogen evolved in presence of the zinc reduces the ferric iron to ferrous condition. The procedure is described in detail under the Permanganate Method for Determination of Iron, page 254. Titanium if present will also be reduced.

4. Reduction with Sulphurous Acid, Sodium Sulphite or Metabisulphite. SO_2 gas is passed into a neutral solution of iron, since iron is not reduced readily in an acid solution by this method. The excess SO_2 is expelled by acidifying the solution and boiling.

5. Reduction with potassium iodide, the liberated iodine being expelled by heat.

In the solution of the ore with stannous chloride and hydrochloric acid, if an excess of the former has been accidentally added, it will be necessary to oxidize the iron before reduction. This may be accomplished by addition of hydrogen peroxide until the yellow color of ferric chloride appears (or by addition of $KMnO_4$ solution), the excess H_2O_2 may be removed by boiling. The iron may now be reduced by one of the above methods.

---

[1] Amalgamated zinc is best prepared by dissolving 5 grams of mercury in 25 cc. of concentrated nitric acid with an equal volume of water, 250 cc. of water are added and the solution poured into 500 grams of shot zinc, 20-mesh. When thoroughly amalgamated the solution is poured off, and the zinc dried.

Volumetric Determination of Iron by Oxidation with Potassium Dichromate

Principle. This method depends upon the quantitative oxidation of ferrous salts in cold acid solution (HCl or H_2SO_4) to ferric condition by potassium dichromate, the following reaction taking place:

$$6FeCl_2 + K_2Cr_2O_7 + 14HCl = 6FeCl_3 + 2CrCl_3 + 2KCl + 7H_2O.$$

Potassium ferricyanide is used as an outside indicator. This reagent produces a blue compound with ferrous salts and a yellowish-brown with ferric. The chromic salt formed by the reaction with iron colors the solution green.

Reagents Required. Standard Potassium Dichromate. When oxygen reacts with ferrous salts, the following reaction takes place:

$$6FeCl_2 + 6HCl + 3O = 6FeCl_3 + 3H_2O.$$

Comparing this reaction with that of dichromate, it is evident that a normal solution of dichromate contains one-sixth of the molecular weight of $K_2Cr_2O_7$ per liter, namely, 49.033 grams. For general use it is convenient to have two strengths of this solution, N/5 for ores high in iron and N/10 for products containing smaller amounts.

Standardization. For N/5 solution 9.807 grams of the recrystallized dehydrated salt are dissolved and made up to one liter; N/10 potassium dichromate contains 4.903 grams of the pure salt per liter. It is advisable to allow the solution to stand a few hours before standardization. The Sibley iron ore furnished by the U. S. Bureau of Standards, Washington, D. C., is recommended as the ultimate standard. Other ores uniform in iron may be standardized against the Sibley ore and used as standards. For accurate work it is desirable to use a chamber burette with graduations from 75 to 90 cc. in tenths and from 90 to 100 in twentieths of a cc. A titration of 90 to 100 cc. of the dichromate would require 0.9 to 1.1 gram of iron for a fifth normal solution and half this amount for a tenth normal solution of dichromate. The ore is best dissolved in strong HCl, adding a few drops of stannous chloride solution and heating just below boiling. In case of an ore or iron ore briquette, containing silica in an appreciable amount, a carbonate fusion of the residue may be necessary. Reduction and titration of the ore is done exactly as prescribed later.

The equivalent iron in the ore divided by the cc. titration required for complete oxidation gives the value in terms of grams per cc., e.g., 1.4 gram of ore containing 69.2% Fe required a titration of 95 cc. $K_2Cr_2O_7$ solution, then,

$$1 \text{ cc.} = \frac{(69.2 \times 1.4)}{100} \div 95 = 0.0102 \text{ gram Fe.}$$

Stannous Chloride. Sixty grams of the crystallized salt dissolved in 600 cc. of strong HCl and made up to one liter. The solution should be kept well stoppered.

Mercuric Chloride. Saturated solution of $HgCl_2$ (60 to 100 grams per liter).

Potassium Ferricyanide, $K_3Fe(CN)_6$. The salt should be free of ferrocyanide, as this produces a blue color with ferric salts, which would destroy the end-point. It is advisable to wash off the salt before using. A crystal the size of a pinhead dissolved in 50 cc. of water is sufficient for a series of determinations. The solution should be made up fresh for each set of determinations.

Apparatus. *Chamber burette.* This should read from 75 to 90 cc. in tenths and from 90 to 100 cc. in twentieths of a cc.

Test Plate. The usual porcelain test-plate with depressions may be replaced by a very simple and efficient test-sheet made by dipping a white sheet of paper in paraffin. The indicator does not cling to this surface, the drops assuming a spherical form, which renders the detection of the end-point more delicate.

Procedure. **Iron Ores.** The amount of sample taken should be such that the actual iron present would weigh between 0.9 to 1.1 gram. This weight can be estimated by dividing 95 by the approximate percentage of iron present, e.g., for 50% Fe ore take $\frac{95}{50} = 1.9$ gram; 95% iron material would require 1 gram, whereas 20% Fe ore would require 4.75 grams.

For samples containing less than 20% Fe it is advisable to use N/10 $K_2Cr_2O_7$ solution.

The sample should be finely ground (80-mesh).

Solution. The hydrochloric acid method for solution of the oxidized ore with subsequent carbonate fusion of the residue is recommended as being suitable for iron ores, briquettes, and materials high in iron.

Reduction. H_2S reduction is recommended in ores containing arsenic or titanium. $SnCl_2$ in very slight excess, followed by mercuric chloride, $HgCl_2$, gives excellent results in absence of other reducible salts of elements, Cu, As, etc.

Test Lead. The easy manipulation and efficiency of this method of reduction makes it applicable for a large variety of conditions. The acid solution preferably, HCl, is diluted to about 150 to 200 cc., containing 15 to 20 cc. concentrated hydrochloric acid (sp.gr. 1.19). Sufficient test lead is added to cover the bottom of a No. 4 beaker. The solution covered is boiled vigorously until it becomes colorless. Copper, if present, is precipitated as metallic copper, and small amounts of arsenic and antimony eliminated from the solution during the reduction of the iron. The cooled solution is poured into a 600-cc. beaker and the mat of lead remaining in the No. 4 beaker washed free of iron, two or three 50-cc. washings being sufficient. The main solution and washings are combined for titration. If the solution is slightly colored, due to reoxidation of iron, a few drops of stannous chloride solution are added to reduce it, followed by an excess of $HgCl_2$ solution, 20 to 25 cc., and allowed to stand five minutes.

Titration. The standard potassium dichromate is run into the solution to within 5 to 10 cc. of the end-point, this having been ascertained on a portion of the sample. The dichromate is run in slowly near the end-reaction, and finally drop by drop until a drop of the solution mixed with a drop of potassium ferricyanide solution produces no blue color during thirty seconds. A paraffined surface is excellent for this test.

Cc. $K_2Cr_2O_7$ multiplied by value per cc. $=$ Fe present in sample. $\% = \dfrac{Fe \times 100}{\text{wt. taken}}$.

NOTES. If $SnCl_2$ solution has been used for reduction of the iron, it is necessary to add the $HgCl_2$ rapidly to a cold solution, as slow addition to a warm solution is apt to precipitate metallic mercury.

In case an excess of dichromate has been added in the titration, as often occurs, back titration may be made with ferrous ammonium sulphate $(NH_4)_2SO_4 \cdot FeSO_4 \cdot 6H_2O$.

N/10 solution of this reagent may be prepared by dissolving 9.81 grams of the clear crystals in about 100 cc. of water, adding 5 cc. of concentrated H_2SO_4 and making to 250 cc. The solution should be standardized against the dichromate solution to get the equivalent values, by running the dichromate directly into the ferrous solution.

The ferricyanide indicator should be made up fresh each time it is required.

Large amounts of manganese in the iron solution titrated cause a brown coloration, which masks the end-point. Nickel and cobalt, present in large amounts, are objectionab'e for the same reason. This interference may be overcome by using very dilute acid solutions of ferricyanide indicator, so that the insoluble ferricyanide of these metals will not form.

Dichromate Method for Iron with Diphenylamine Indicator

The disadvantage of the dichromate method in requiring an outside indicator (potassium ferricyanide on a spot plate) is overcome by the procedure outlined by J. Knop (The Journal of the American Chemical Society, 46, 263 (Feb., 1924)) in which diphenylamine is used as an internal indicator, the end point being a violet blue or blue-black.

Reagents

0.1 N Potassium Dichromate solution.

Sulphuric-Phosphoric Acid Mixture. 150 cc. sulphuric acid (d. 1.84), 150 cc. phosphoric acid (d. 1.7), diluted to 1000 cc., 15 cc. used.

Diphenylamine solution. 1 g. diphenylamine dissolved in 100 cc. of strong sulphuric acid. 3 drops used as indicator. (A color change to brown does not impair the efficiency of the indicator.)

Procedure for Ores

Half to 1 gram of the ore is digested with 20 cc. of strong hydrochloric acid at 70–80° C., the solution diluted with an equal volume of water and filtered. The insoluble residue is fused with sodium carbonate in a platinum crucible,[1] the melt dissolved in dilute hydrochloric acid, the iron precipitated with ammonium hydroxide and filtered off. The ferric hydroxide on the filter is dissolved with a few cc. of dilute HCl and the solution added to the main solution containing the iron.

The iron is now reduced by addition of stannous chloride solution (60 g. $SnCl_2$ in 600 cc. HCl and 400 cc. H_2O) added cautiously until the color of iron is no longer evident (other reducible elements must be absent, i.e., Cu, As, etc.), the excess of stannous chloride is overcome with mercuric chloride solution (saturated solution) as usual.

15 cc. of sulphuric-phosphoric acid mixture are added and 3 drops of diphenylamine indicator, the solution is diluted to 150–200 cc. and titrated with 0.1 N potassium dichromate solution.

Near the end the green color of the solution deepens to a blue-green or in presence of a large amount of iron to a grayish blue. The dichromate is now added dropwise until the color changes to an intense violet blue.

$$1 \text{ cc. } 0.1 \text{ N } K_2Cr_2O_7 = 0.0056 \text{ g. Fe.}$$

NOTES. The method has the advantage of having no fading end point as is obtained in the permanganate titrations in presence of mercurous chloride. The method permits back titration with standard ferrous solution. Organic substances do not interfere as they do in permanganate titrations. Zinc, aluminum, manganese, nickel, cobalt and chromium do not interfere. Copper present in quantities less than 1 mg. does not interfere, in larger quantities it lowers results as it assists oxidation of iron by air. Trivalent arsenic raises results as it is oxidized by dichromate to pentavalent form.

[1] The residue may be fused with $KHSO_4$ in glass in place of Pt.

Potassium Permanganate Method for Determination of Iron

Introduction. The method depends upon the quantitative oxidation of ferrous salts to the ferric condition when potassium permanganate is added to their cold solution, the following reaction taking place:

$$10FeSO_4 + 2KMnO_4 + 8H_2SO_4 = 5Fe_2(SO_4)_3 + K_2SO_4 + 2MnSO_4 + 8H_2O.$$

Hydrochloric acid in presence of iron salts has a secondary reaction upon the permanganate, e.g.,

$$2KMnO_4 + 16HCl = 2KCl + 2MnCl_2 + 8H_2O + 10Cl.$$

This reaction may be prevented by addition of large amounts of zinc or manganous sulphates together with an excess of phospnoric acid or by large dilution. See note on page 257. The solution is diluted and reduced with zinc and titrated as directed.

The reduction of ferric sulphate is best accomplished by passing the solution through a column of amalgamated zinc in the Jones reductor. In presence of titanium, reduction is accomplished by H_2S in a hydrochloric acid solution of the iron.

Since potassium permanganate enters into reaction with acid solutions of antimony, tin, platinum, copper and mercury, when present in their lower state of oxidation, (also with manganese in neutral solutions) and with SO_2, H_2S, N_2O, ferrocyanides and with most soluble organic bodies, these must be absent from the iron solution titrated.

Potassium permanganate produces an intense pink color in solution, so that it acts as its own indicator.

Solutions Required. *Standard Permanganate Solutions.* As in case of potassium dichromate, it is convenient to have two standard solutions, N/5 and N/10.

From the reaction given above it is evident that 2 $KMnO_4$ are equivalent to 5 oxygens, e.g., $2KMnO_4 = K_2O + 2MnO + 5O$, hence a normal solution would contain one-fifth of the molecular weight of $KMnO_4 = 31.6$ grams of the pure salt.

Since commercial potassium permanganate is seldom pure, it is necessary to determine its exact value by standardization. This is commonly accomplished by any of the following methods:

(a) By a standard electrolytic iron solution.

(b) By ferrous salt solution, e.g., $(NH_4)_2SO_4 \cdot FeSO_4 \cdot 6H_2O$.

(c) By oxalic acid or an oxalate.

Reaction. $2KMnO_4 + 5Na_2C_2O_4 + 8H_2SO_4$
$$= K_2SO_4 + 2MnSO_4 + 5Na_2SO_4 + 10CO_2 + 8H_2O.$$

Standardization of $KMnO_4$ against *sodium oxalate* is recommended as the most accurate procedure.

To standardize 0.2 g. of pure sodium oxalate, equivalent to $0.2 \div .0067 = 29.9$ cc. 0.1 N solution (or if preferred 0.67 g. ≈ 100 cc. 0.1 N solution), is dissolved in 100 cc. hot water and 5–10 cc. H_2SO_4 (1.84) and titrated with the permanganate solution until a faint pink color, that persists, is obtained. The oxalate equivalent in cc. divided by cc. $KMnO_4$ required = normality of $KMnO_4$, in terms of N/10 solution.

Procedure for the Determination of Iron by the Jones Reductor

Preparation of Sample. Such an *amount* of the sample is taken that the iron content is between two- and three-tenths of a gram (0.2 to 0.3 gram). If hydrochloric acid has been required to effect solution, or hydrochloric acid and nitric acid (25 cc. : 1 cc.), as in case of iron and steel, 4 to 5 cc. conc. sulphuric are added, and the solution evaporated to small bulk on the steam bath and to SO_3 fumes to remove hydrochloric acid. The iron is taken up with about 50 cc. dilute sulphuric acid, 1 : 4, heating if necessary, and filtering if an insoluble residue remains.

Preparation of the Reductor. Cleaning out the apparatus. See Fig. 40. The stop-cock of the reductor is closed, a heavy-walled flask or bottle is put into position at the bottom, and 50 cc. of dilute sulphuric acid poured into the funnel. The cock is opened and the acid allowed to flow slowly through the zinc in the tube, applying a gentle suction. Before the acid has drained out of the funnel, 50 cc. of water are added, followed by 50 cc. more of dilute sulphuric acid and 50 cc. of water in turn. The stop-cock is turned off before the water has drained completely from the funnel so that the zinc is always covered by a solution of acid or water.

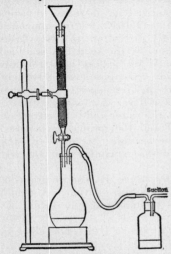

FIG. 40.—Jones Reductor.

Determination of the Blank. Fifty cc. of dilute sulphuric acid, 1 : 4, are passed through the reductor, followed by 250 cc. of distilled water, according to the directions given above. The acid solution in the flask is then titrated with N/10 $KMnO_4$ solution. If more than 3 or 4 drops of the permanganate are required, the operation must be repeated until the blank titration does not exceed this amount. The final blank obtained should be deducted from the regular determinations for iron. The end-point of the titration is a faint pink, persisting for one minute.

Reduction and Titration of the Iron Solution. The sample is diluted to 200 cc., and, when cold, is run into the funnel, the stop-cock opened and the solution drawn slowly through the column of zinc into the flask, about four minutes being required for 200 cc. of solution. Before the funnel has completely drained, rinsings of the vessel which contained the sample are added; two 50-cc. portions are sufficient, followed by about 50 cc. of water. The stop-cock is closed before the solutions have completely drained from the funnel.

Titration. The flask is removed and tenth normal solution of permanganate added until a faint pink color, persisting one minute, is obtained. The blank is deducted from the cc. reading of the burette.

One cc. N/10 $KMnO_4$ = .005584 gram Fe; or .007984 gram Fe_2O_3.

Zimmermann-Reinhard reagent to overcome action of HCl on $KMnO_4$—10 cc. of a solution, 1 liter of which contains 2 mols. of H_2SO_4, 2 mols. of H_3PO_4 and 0.3 mols. of $MnSO_4$ will take care of 10 cc. of 12 N. HCl for 600 cc. of solution.

OPTIONAL PERMANGANATE METHOD FOR IRON IN ORES

1. **Solution of the Sample.** 0.5 gram of ore. 8-oz. flask. With oxidized ores add 10–15 cc. of HCl and warm gently until the iron oxide is dissolved; then if sulphides are also present add 5 cc. of HNO_3 to decompose them also. With straight sulphides use 10 cc. of HCl and 5 cc. of HNO_3. When decomposition is complete add 5 cc. of H_2SO_4 and boil over a free flame nearly to dryness.

Refractory Ores. Certain silicates, oxides of iron, furnace slags, etc., do not decompose with acid treatment. Decomposition may be accomplished by fusion methods followed by solution of the fused mass with water and hydrochloric acid. Fusions with Na_2CO_3 and K_2CO_3 are made in a nickel or platinum crucible and are recommended for materials high in silica. Fusions with an acid flux, K_2SO_4 (+5 cc. H_2SO_4 d. 1.84) or $KHSO_4$ are recommended for oxides. These latter fusions are conveniently carried out in pyrex flasks held by a heavy wire clamp. About 10–15 grams of the solid flux is added to the ore in the flask and the fusion completed by heating until the mix becomes transparent. Oxides which do not readily decompose may be brought into solution, frequently, by adding a little piece of filter paper to the molten mass in the flask. The carbon thus furnished reduces the oxides, effecting decomposition. If decomposition is incomplete, the water and acid extraction is made, the soluble constituents decanted off and the residue again fused with more $KHSO_4$ and filter paper. It may be necessary to follow $KHSO_4$ fusion by the Na_2CO_3 fusion, on the insoluble residue remaining from the acid extraction of the $KHSO_4$ mass.

2. **Reduction.** After cooling, add 30 cc. water, 10 cc. HCl and 6 grams of 20 mesh granulated zinc. Not necessary to get salts into solution. Now add 3 cc. of a 4% copper sulphate solution. Allow to stand until the action has become feeble.

3. Add 50 cc. of cold water and then 10 cc. of strong H_2SO_4 and allow to stand until the zinc is nearly all dissolved.

4. **Filtration from Insoluble Gangue and Excess Zinc.** Prepare a filter by placing a rather thick wad of absorbent cotton in a funnel and wetting it into place. Place a battery jar, or a liter beaker containing about an inch of cold water, under the funnel. Have the beaker marked at the 700 cc. point.

5. When the zinc in the flask has nearly all dissolved, filter the liquid through the absorbent cotton and wash out the flask at least 10 times with cold water, pouring through the filter. Use the wash bottle reversed to save time, and use enough water for each wash to completely cover the absorbent cotton. Allow to drain between washes. Continue the washing until the filtrate reaches the 700 cc. mark on the beaker.

6. **Titration of the Sample.** Titrate at once with standard permanganate to a very faint pink tinge and take reading.

7. A blank should previously be run on the zinc to determine any correction (usually due to a little iron) necessary. Deduct this correction from the above reading.

8. Multiply the cc. of permanganate used by the factor for iron.

9. **Standardization of Permanganate.** See procedure on the following page.

Diphenylamine Indicator in Permanganate Titrations of Iron.[1] As in case of titrations of iron with potassium dichromate, diphenylamine may be used in potassium permanganate titrations. The advantage of the indicator is in the fact that titrations may be made in presence of considerable hydrochloric acid and in the presence of tin and mercury salts without a fading endpoint, obtained, when permanganate is used alone. The blue color is more intense than the pink of potassium permanganate.

Procedure

Decomposition and Reduction of Sample. The procedure given on page 257a applies also to this method.

Titration. About 15 cc. of phosphoric-sulphuric acid mixture are added and the solution is diluted to about 100 cc. 0.2 cc. (4 drops) of diphenylamine indicator are added and the titration now made with the standard potassium permanganate solution. The color becomes green, deepening to a blue-green or grayish blue. The reagent is now added " dropwise " until an intense violet blue or dark blue color is obtained.

$$\text{1 cc. 0.1 N KMnO}_4 \text{ 0.005584 g. Fe.}$$

NOTES. If 0.5584 gram of sample is taken then 1 cc. of 0.1 N potassium permanganate is equivalent to one per cent of iron.

The phosphoric-sulphuric acid mixture prevents the yellow color of iron from producing a green with the blue end-point. A larger amount of indicator than is recommended should not be used, as this necessitates a blank being deducted for the action of the oxidizing agent on the indicator.

Standardization of Potassium Permanganate-Oxalate Method. Weigh 0.2 to 0.3 gram of pure sodium oxalate, place in a beaker and add 200 cc. of water and about 5 cc. strong sulphuric acid. Heat to boiling and titrate with the potassium permanganate solution. The reaction starts slowly, but with the progress of the titration the action becomes vigorous. Towards the end of the reaction the pink color fades less rapidly and finally a permanent pink color s obtained with one drop of the reagent.

Calculation of Normality. Since 67 grams of sodium oxalate per 1000 cc. f solution is a normal solution the cc. equivalent of the amount taken is btained by dividing by 0.067. This value divided by the titration of the oxalate with the reagent being standardized will give the normality of the reagent.

Example. Suppose 0.268 gram of sodium oxalate required a titration of 50 cc. of the permanganate solution. Since 67 g. is equivalent to 1000 cc. N solution of sodium oxalate, then 0.268 is equivalent to

$$0.268 \times 1000 \text{ divided by } 67 = 4 \text{ cc. N sol.}$$
$$4 \text{ divided by } 50 = 0.08 \text{ N.}$$

[1] Wilfred W. Scott, Jour. Am. Chem. Soc., **46**, 1396, (June, 1924).

Determination of Ferric Iron by Titration with Titanous Salt Solution

Method of Knecht and Hibbert[1] (modified by Thornton and Chapman[2])

Introduction. The iron solution, which may contain hydrochloric acid, sulphuric acid or hydrofluoric acid (provided boric acid be added in considerable excess) — but not nitric acid, is titrated while cold (15–20° C.) with a standard solution of titanous sulphate,[3] using ammonium thiocyanate as indicator, until the red color of the ferric thiocyanate just disappears.

Any substance that might reduce the ferric iron or oxidize the titanous salt should, of course, be absent. Many forms of organic matter, however, do not interfere. Copper is quantitatively precipitated as cuprous thiocyanate, and certain other metals of the hydrogen sulphide group are not without effect. Platinum salts, chromic acid (not chromic salts) and vanadic acid exert an oxidizing action on the titanium compound.

Reagents. Titanous Sulphate Solution (1/10 Normal). The titanium solution is prepared by mixing 100 cc. of a 20% solution of titanous sulphate with 80 cc. of sulphuric acid (1 : 1) and making up the volume to 1 liter with cold water.

Ammonium Thiyanate Solution. One hundred grams of the crystals are dissolved in water, filtered and the clear solution diluted to 1 liter.

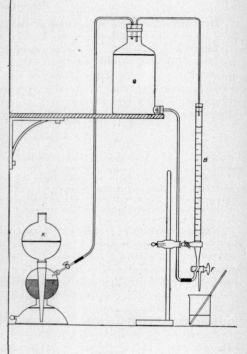

Fig. 40a.—Apparatus for Storing and Using Titanous Salt.

Ferric Ammonium Sulphate Solution.[4] Forty eight grams of ferric alum crystals are dissolved in water, the solution acidified with 50 cc. of concentrated sulphuric acid and after filtration the volume made up to 1 liter.

[1] E. Knecht and E. Hibbert, "New Reduction Methods in Volumetric Analysis," 1918, pp. 12, 49, 51 and 53.

[2] Wm. M. Thornton, Jr. and J. E. Chapman, J. Am. Chem. Soc., **43**, 91 (1921).

[3] Knecht and Hibbert employ titanous chloride, which for most purposes is equally efficient. Either salt, in the form of a 20% solution, may be obtained from The La Motte Chemical Products Co., 13 W. Saratoga St., Baltimore, Md.

[4] The ferric alum solution will serve for a "back titer" in any analysis, and it will enable the operator to quickly detect any change in the oxygen consuming capacity of the titanous salt solution.

Apparatus. The titanous sulphate solution, prepared as described above, is charged into the storage bottle S, Fig. 40a, the volume being so regulated as to fill the container to its neck. The stopcock F is then turned so that the liquid rises in the burette B and on up until hydrostatic equilibrium has been reached. With the hydrogen supply turned on from the Kipp generator K, the burette is allowed to empty itself by properly manipulating the cock F, and a current of pure hydrogen is maintained through the apparatus till it is reasonably certain that all air from within the system has been displaced. On being refilled the burette will be ready for service.

Standardization. One and five tenths grams of ferrous ammonium sulphate (C. P. crystals) are weighed with exactness into a 500 cc. Erlenmeyer flask and dissolved in about 150 cc. of water. Ten cc. of sulphuric acid (1 : 1) are added and the solution titrated at once with potassium permanganate, which has previously been standardized against Bureau of Standards sodium oxalate, until the end-point is just reached. After boiling for 10 minutes, the liquid is cooled at the tap to about 15° C. and titrated carefully with titanous sulphate in the presence of 10 cc. of the ammonium thiocyanate solution to complete loss of color. Another 1.5 gram portion of the same ferrous ammonium sulphate is titrated with titanous sulphate—omitting, however, the preliminary oxidation with permanganate. If the second small titer be deducted from the first titer, the difference will give the iron value of the titanium solution against sodium oxalate; while the main titer (uncorrected) will serve as a basis on which to calculate the standard against ferrous ammonium sulphate. With specimens of ferrous ammonium sulphate of good quality, the results will be found to be in close agreement; but in case there is any considerable discrepancy, preference should be given to the sodium oxalate value.

Procedure. Ferric Iron. The solution, which may conveniently contain any amount of iron up to 0.28 gram and whose initial volume should not at the most exceed 500 cc., is titrated with titanous sulphate—using the same quantities of reagents and otherwise as prescribed above.

Total Iron. To determine the entire amount of iron present, notwithstanding its state of oxidation, the whole must be gotten into the ferric condition. This may be accomplished in one of three ways: (1) the solution is treated with hydrogen peroxide in ammoniacal solution and boiled to decompose the excess of the peroxide and finally acidified with hydrochloric acid; (2) the sulphuric acid solution is titrated with potassium permanganate (the exact strength of which need not be known) as detailed under standardization; (3) the solution is oxidized with potassium chlorate and hydrochloric acid and the excess chlorate removed by evaporation. The second method is generally to be preferred. Large amounts of hydrochloric acid may be removed by evaporation on the water bath with sulphuric acid in excess; but with small concentrations of said acid the evil effect upon permanganate may be offset by adding a sufficient quantity of the Zimmermann-Reinhardt preventive reagent. 67 g. $MnSO_4.4H_2O$ + 500 cc. H_2O + 138 cc. H_3PO_4 (sp. gr. 1.7) + 130 cc. H_2SO_4 (sp. gr. 1.82). Mixture diluted to 1000 cc.

Stannous Chloride Method for Determination of Ferric Iron

The procedure is based upon the reduction of the yellow ferric chloride to the colorless ferrous salt by stannous chloride, the following reaction taking place:

$$2FeCl_3 + SnCl_2 = 2FeCl_2 + SnCl_4.$$

The method is of value in estimating the quantity of ferric iron in presence of ferrous, where the two forms are to be determined. In order to obtain the total iron the ferrous is oxidized by adding a few crystals of potassium chlorate and taking to dryness to expel chlorine, and then titrated with stannous chloride.

The accuracy of the method depends upon the uniformity of conditions of temperature, concentration, etc., of making the run with the sample and of standardizing the stannous chloride. The solution should be free from other oxidizing agents, or from salts that give colored solutions.

The amount of iron in terms of ferric oxide that can be estimated by this procedure ranges from 0.002 gram to 0.05 gram.

Reagents. *Stannous Chloride Solution.* The reagent is prepared by dissolving 2 grams of stannous chloride crystals in hot concentrated hydrochloric acid and making up to 1 liter. The solution should be kept in a dark bottle to which the titrating burette is attached in such a way that the liquid may be siphoned out into this, as shown in the illustration, Fig. 41. The air entering the bottle passes through phosphorous or pyrogallic acid to remove the oxygen. In this way, protected from the air, the reagent will keep nearly constant for several weeks. It is advisable, however, to restandardize the solution about every ten to fifteen days. One cc. will be equivalent to about 0.001 gram Fe.

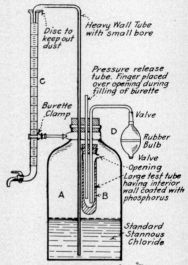

FIG. 41.—Apparatus for Stannous Chloride Titration of Iron.

Standard Iron Solution. 8.6322 grams of ferric ammonia alum is dissolved in dilute hydrochloric acid and made up to one liter. The iron is determined in 100-cc. portions by the dichromate method. One cc. will contain about 0.001 gram Fe.

Procedure. To the sample in a casserole is added 25 cc. of concentrated hydrochloric acid and an equal volume of water. The resulting solution is heated to boiling and quickly titrated with the stannous chloride reagent, until the yellow color fades out and the solution becomes colorless.

NOTE. The titration should be done quickly, as the iron will reoxidize on standing and the solution again become yellow. The true end-point is the first change to a colorless solution.

COLORIMETRIC METHODS FOR THE DETERMINATION OF SMALL AMOUNTS OF IRON

Iron Traces. Sulphocyanate (Thiocyanate) Method [1]

Introduction. By this method 1 part of iron may be detected in 50 million parts of water. The presence of free mineral acid increases the sensitiveness of the method, so that it is especially applicable to the determination of small amounts of iron in mineral acids. It is available in presence of many of the ordinary metals and in presence of organic matter. Silver, copper, cobalt, mercuric chloride, however, interfere.

Nitric acid gives a color with sulphocyanates that may be mistaken for iron.

This method, like the stannous chloride method, determines only the ferric iron. It is based on the fact that ferric iron and an alkali sulphocyanate, ammonium or potassium sulphocyanates, in an acid solution gives a red color, the intensity of which is proportional to the quantity of iron present. The color is due to the formation of the compound, $Fe(CNS)_3 \cdot 9KCNS \cdot 4H_2O$.

Reagents Required. *Standard Iron Solution.* A ferric solution, the iron content of which has been determined, is diluted and divided so as to obtain 0.0004 gram Fe. This is made up to 2 liters with water containing 200 cc. of iron-free, C.P. H_2SO_4. One hundred cc. of this solution, together with 10 cc. of normal ammonium sulphocyanate solution, is used as a standard. One hundred cc. contains 0.00002 gram Fe.

Normal sulphocyanate contains 76.1 grams of NH_4CNS per liter.

Procedure. The weighed sample, 1 to 10 grams, or more if necessary, is dissolved in dilute H_2SO_4 and oxidized by adding dilute permanganate, $KMnO_4$, solution drop by drop until a faint pink color is obtained. The sample is diluted to exactly 100 cc. and is poured into a burette graduated to $\frac{1}{10}$ cc. Two colorless glass cylinders of the 100-cc., Nessler type are used for comparison of standard and sample. Into one cylinder is poured 100 cc. of the standard solution, made as directed above. Into the second cylinder containing 10 cc. of sulphuric acid with 10 cc. ammonium sulphocyanate, NH_4CNS, diluted to 60 or 70 cc., the sample is run from the burette until the depth of the color thus produced on dilution to 100 cc. exactly matches the standard. From the number of cc. used the weight of the sample is calculated. One hundred cc. of the standard contains 0.00002 gram Fe.

Dividing the weight of iron in the standard by the weight of sample used and multiplying by 100 gives the per cent of iron in the sample.

NOTES. If other metals are present, that form two series of salts, they must be in the higher state of oxidation, or the color is destroyed. (Sutton.) Oxalic acid, if present, destroys the color. Oxidation with $KMnO_4$ or $KClO_3$ with subsequent removal of Cl_2 prevents this interference. (Lunge, C. N., **73**, 250.)

Chlorides of the alkaline earths retard or prevent the sulphocyanate reaction. (Weber, C. N., **47**, 165.)

The colorimeter used for the determination of minute quantities of lead would serve admirably for the determination of traces of iron by the sulphocyanate method.

Acids, hydrochloric or sulphuric (diluted), may be added directly to the ammonium sulphocyanate solution.

[1] Thomson, J. C. S., 493, 1885, and C. N., **51**, 259. Kruss and Moraht, C. N., **64**, 255. Davies, C. N., **8**, 163.

Salicylic Acid Method for Determining Small Amounts of Iron

Salicylic acid produces an amethyst color with neutral solutions of ferric salts, the depth of the color being proportional to the concentration of the ferric iron in the solution. The reaction is useful in determining small amounts of iron in neutral salts, such as sodium, ammonium, or potassium alums, sulphates, or chlorides, zinc chloride, etc. Phosphates, fluorides, thiosulphates, sulphites, bisulphites and free mineral acids should be absent. The sample should not contain over 0.0002 gram iron, as the depth of color will then be too deep for colorimetric comparisons. As low as 0.00001 gram ferric iron may be detected. Ferrous iron produces no color with the reagents, hence the procedure serves for determining ferric iron in presence of ferrous.

The material is dissolved in 20 cc. of pure water, the sample filtered if cloudy, and transferred to a Nessler tube. Dilute potassium permanganate solution is added until a faint pink color is produced and then 5 cc. of a saturated solution of salicylic acid. (The reagent is filtered and the clear solution used.) Comparison is made with standard solutions containing known amounts of ferric iron, the standards containing the same reagents as the sample. If desired the standard iron solution (0.086 gram ferric ammonium alum, clear crystals, dissolved in water containing 2 cc. of dilute sulphuric acid and made to 1000 cc., each cc. contains approximately 0.00001 gram Fe''') is added from a burette to 5 cc. of salicylic acid diluted to 25 cc. in a Nessler tube, until the color of the standard matches the sample. A plunger is used to stir the liquids.

TECHNICAL ANALYSIS OF IRON AND STEEL

The elements carbon, manganese, phosphorus, sulphur, and silicon are invariable constituents of iron and steel, and are always included in an analysis. Copper and arsenic are sometimes found; aluminum, chromium, nickel, molybdenum, tin, titanium, tungsten, vanadium, and zinc occur in special alloy steels. Minute traces of oxygen, hydrogen, and of many other elementary constituents frequently are present, but are of so little importance that they are seldom considered in an analysis.

The importance of the subject has called for a special chapter on the subject of iron and steel. This may be found in volume II of this work. The individual determinations are given in chapters throughout volume I. We will deal with a few determinations here.

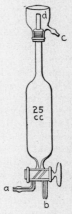

Fig. 42.

Preparation of the Sample

The sample of borings is taken from several portions of the piece by a drill, free from oil or grease, and stored in a heavy manilla envelope. For carbon determinations samples contaminated by oil or grease should be thoroughly washed with ether before making the determination.

Total Carbon

The determination is required for an accurate estimation of carbon where the color test indicates the carbon content outside the limits of requirement, or in cases where interfering substances are present. In material where the carbon content is of extreme importance, the color method is not used. Details of the procedure for determining carbon by direct combustion are given in the chapter on Carbon. The following procedure is recommended by the Bureau of Standards:

(a) **In Irons.** Two grams of iron are mixed with about twice the weight of purified ferric oxide. The mixture is placed in a platinum boat, which is lined with a suitable bed material, and is burned in a current of oxygen, as described below.

(b) **In Steels.** The method is the same as for irons with omission of the ferric oxide mixture.

Details of Direct Combustion Method. *Furnaces and Temperature of Burning.* Porcelain tubes wound with "nichrome" wire, provided with suitable heat insulation and electrically heated, are used, and readily give temperatures to 1100° C. Type FB 301 Hoskins tube furnace and the hinged type, Fig. 32b, are satisfactory. The temperature control is by means of an ammeter and rheostat in series with the furnace, with occasional check by a thermocouple.

Boats and Lining. Platinum boats provided with a long platinum wire for manipulation in the tube are mostly used; alundum ones occasionally. The bed or lining on which the steel rests is 90-mesh "RR alundum, alkali-free, specially prepared for carbon determination." A layer of this alundum is also placed in the bottom of the combustion tube to prevent the boat sticking to the glaze. A platinum cover for the boat is sometimes used, and is essential when the combustion is forced.

The nature and quality of the bed material are matters of great importance. Alumina as prepared from the sulphate or from alum may not be free from sulphate or alkali, both of which have given serious trouble at the Bureau. The alkali, if present, may not manifest itself by an alkaline reaction until after one or two combustions have been made, using the same bed material. Even the ordinary white "alundum" on the market carries a few hundredths of 1% of alkali. Iron oxide has been tried, and when pure should, apparently, give good service. As yet, however, it has been difficult to obtain or prepare acceptable material for use with steels. Quartz sand gives rise to a fusible slag, which melting before combustion is complete, incloses bubbles of carbon dioxide gas. This defect would probably inhere in any other material of an acid character. The presence in the silica bed after combustion of crystals which appear to be carborundum, have occasionally been noted.[1]

Purity of Oxygen. Blanks. Oxygen may be made electrolytically, whereby the content of this element is usually 99 to 99.5%, and sometimes higher. Even with this gas a slight blank is usually obtained. When running a blank, in addition to the usual precautions, the rate at which the oxygen is introduced should be the same as when burning a sample, and the time should be three to five times as long.

Method of Admitting Oxygen and Rate of Combustion. The furnace being at the proper temperature, the boat containing the sample is introduced. Oxygen is admitted either at once or after the boat has reached the temperature of the furnace, as the operator prefers, or as the nature of the steel may demand. The rate of flow of the oxygen varies with the absorption apparatus used and with the preference of the operator, and may be considerably more rapid when absorbing carbon dioxide in soda lime than in an alkaline solution. A rapid flow of oxygen also facilitates the burning of resistant samples. A continuous forward movement of the gas current is maintained at all times. The time for a determination varies, of necessity, with the nature of the sample and the rate of flow of the oxygen, ranging from ten to thirty minutes. The endeavor is to obtain a well-fused oxide. With all samples close packing in a small space is conducive to rapid combustion and to fusion of the resulting oxide.

Authorities differ as to the advisability of allowing the oxide of iron to fuse thoroughly. Even when fusion does take place additional carbon dioxide is obtained very frequently by grinding the oxide and reburning. Often more than one regrinding and reburning is necessary in order to reduce the amount of carbon dioxide obtained to that of the constant blank.

Oxides of sulphur have been found very difficult to eliminate from the gases leaving the tube. Lead peroxide ("nach Dennstedt") heated to 300° C. and zinc at room temperature appear to retain them best.

Attention is called to the inadmissibility of using dry agents of different

[1] Statement of Mr. George M. Berry, of the Halcomb Steel Co.

absorptive power in the same train, in positions where a difference could possibly affect results.

Weighing of Tubes. There is much greater difficulty in securing constant conditions when weighing absorption tubes than is usually considered to be the case. Electrical effects, caused by wiping as a preliminary to weighing, may occasionally cause errors in weight running into the milligrams. The use of counterpoises of equal volume and similar material and shape is recommended.

If tubes are weighed full of oxygen, care is necessary to secure a uniform atmosphere in them. Even though the attempt is made to keep the apparatus always full of oxygen, some air is admitted when the boat is pushed into the combustion tube, and a much longer time is required to displace this than is usually allowed, unless the flow of oxygen during aspiration is rapid. The same is true if the tubes are weighed full of air by displacing the oxygen left in them after the steel is burned. Another source of error may arise from the air admitted when putting the boat into the tube, if this air contains much carbon dioxide, as is the case when a gas furnace is used. The boat is usually pushed at once into the hot furnace, and as combustion begins almost immediately, there is no opportunity for displacing this air before the steel begins to burn.

Graphite in Iron

Two grams of iron are dissolved in nitric acid (sp.gr. 1.20), using 35 cc. and heating very gently. The residue is collected on an asbestos felt, washed with hot water, then with a hot solution of potassium hydroxide (sp.gr. 1.10), followed by dilute hydrochloric acid and finally by hot water. After drying at 100° C., the graphite is burned in the same manner as the total carbon, but without admixture of ferric oxide.

Standard Specifications for Carbon

Material	Per Cent Carbon
Automobile Carbon Steel	0.05–0.15 up to 0.95–1.15
Boiler and Fire Box Steel Stationary Service	not over 0.25
Carbon Steel Bars, Auto Springs	0.45–0.65
Carbon Steel Bars, Railway Springs	0.90–1.15
Carbon Steel Rails	{ Bessemer 0.37–0.55 { Open Hearth 0.50–0.75
Carbon Tool Steel	0.05±
Carbon Tool Steel High Speed	0.05±
Cold Drawn Bessemer Steel Automatic Screw Stock	0.08–0.16
Fire Box Steel	0.12–0.30
Forge Welding Steel Plates Flange	0.2
Forge Welding Steel Plates Structural	0.2
Lap Welded Seamless Boiler Tubes	0.08–0.18
Low Carbon Steel Splice Bars	not over 0.3
Medium Carbon Steel Splice Bars and Track Bolts	not under 0.45
High Carbon Steel Splice Bars	not over 0.60
Rivet Steel for Buildings, Cars and Ships	no specification stated
Steel Castings	not over 0.45
Quenched and Tempered Steel, Axles, Shafts and Forgings	0.25–0.70
Wrought Solid Carbon Steel Wheels and Tires	{ Acid 0.60–0.80 { Basic 0.65–0.85

Manganese in Iron and Steel

Ammonium Persulphate Method

Small amounts of manganese may be determined colorimetrically by the persulphate method, provided the sample does not contain over 1.5% of manganese.

Reaction. $2Mn(NO_3)_2+5(NH_4)_2S_2O_8+8H_2O$
$$=5(NH_4)_2SO_4+5H_2SO_4+4HNO_3+2HMnO_4.$$

0.1 to 0.2 gram of steel, according to the amount of manganese in the sample, is placed in a 10 in. test-tube and 10 cc. of nitric acid (sp.gr. 1.2) are added. The sample is heated in a water bath until the nitrous fumes are driven off and the steel is completely in solution. 15 cc. of $AgNO_3$ sol. (1.5 g. per l.) are added to the cooled sample, followed immediately with about 1 gram of ammonium persulphate crystals. The solution is warmed (80 to 90° C.) until the color commences to develop, and then for half a minute longer, and then placed in a beaker of cold water until the solution is cold. Comparison is now made with a standard steel treated in the same way, the comparison being made exactly as indicated for determining carbon by the color method. See chapter on Carbon.

Example. If the standard, containing 0.6% Mn, is diluted to 15 cc., each cc. = 0.04% Mn. If the sample required a dilution of 20 cc. to match the standard, then 0.04 × 20 = 0.8% Mn.

NOTE. If preferred, the sample may be titrated with standard sodium arsenate, one cc. of which is equivalent to 0.1% on basis of 0.1 gram sample.

Lead Oxide Method (Deshey)

Oxidation of the manganese in the steel is effected in a nitric acid solution by addition of red lead (or by lead peroxide); the lead peroxide, formed, oxidizes the manganese nitrate to permanganic acid. The solution is now titrated with standard sodium arsenite, the following reaction taking place:

$$2HMnO_4+5Na_3AsO_3+4HNO_3=5Na_3AsO_4+3H_2O+2Mn(NO_3)_2.$$

0.5 gram of steel is placed in a 150 cc. beaker and dissolved with about 30 cc. of nitric acid (sp.gr. 1.12). After violent action has subsided, the beaker is placed on a hot plate and when the iron has dissolved, 20 cc. of water added. The manganese is now oxidized by adding red lead in small portions at a time, until the solution appears brown with a pinkish purple foam on the surface. The solution is diluted with hot water until the volume is about 100 cc. and then boiled for a few minutes. It is now placed in a dark closet to cool. (A fresh batch of samples may be started in the meantime.) The solution is carefully decanted off from the peroxide, and with the washings of the peroxide residue, titrated with standard sodium arsenite to the yellowish green end-point. The sodium arsenite is made by dissolving 4.96 grams of pure arsenous acid together with 25 grams of sodium carbonate in 200 cc. of hot water and the solution diluted to 2500 cc. The arsenite is standardized against a steel sample of known manganese content, or against standard permanganate solution.

Bismuthate Method for Determining Manganese

(See also A. S. T. M. methods in Vol. 2)

This is the most accurate method for determining manganese in iron and steel. The procedure is as follows:

Procedure. One gram of drillings is dissolved in 50 cc. of nitric acid (sp.gr. 1.135) in a 200 cc. Erlenmeyer flask. Irons should be filtered. The solution is cooled, about 0.5 gram of sodium bismuthate is added, and it is then heated until the pink color has disappeared. Any manganese dioxide separating is dissolved in a slight excess of a solution of ferrous sulphate or sodium sulphite. The solution is boiled till free from nitrous fumes. After cooling to 15° C., a slight excess of bismuthate is added and the flask is shaken vigorously for a few minutes. Then 50 cc. of 3% nitric acid is added and the solution is filtered through asbestos. A measured excess of ferrous sulphate is run in and the excess titrated against permanganate solution which has been compared with the iron solution on the same day. A great many steels now carry small amounts of chromium as impurity. In such cases titration against arsenite solution is recommended, or removal of the chromium by zinc oxide and subsequent determination of the manganese by the bismuthate method.

Permanganate solutions are standardized against sodium oxalate.

Standard Specifications for Manganese

Material	Per Cent Manganese
Automobile Carbon Steels	0.25–0.60
Boiler and Fire Box Steel Stationary Service	0.30–0.60
Carbon Steel Bars Auto Springs	0.60–0.80
Carbon Steel Bars R. R. Springs	max. 0.5
Carbon Steel Rails	{ Bessemer 0.8–1.10 { Open Hearth 0.6–0.9
Carbon Tool Steel	0.10±
Carbon Tool Steel High Speed	0.15±
Cold Drawn Bessemer Steel Automatic Screw Stock	0.60–0.80
Fire Box Steel	0.30–0.60
Forge Welding Flange and Structural Steel Plates	0.35–0.60
Lap Welded Seamless Boiler Tubes	0.30–0.60
Open Hearth Steel Girders	0.60–0.90
Structural Nickel Steel	0.70. Ni not under 3.25%
Quenched High Carbon Steel	not over 0.80
Wrought Solid Carbon Steel Wheels and Tires { Acid.... { Basic....	0.55–0.80

Determination of Phosphorus

See A. S. T. M. methods in Vol. 2

(a) **Preparation of Solution and Precipitation of Phosphorus.** Two grams of sample are dissolved in nitric acid (sp.gr. 1.135) and the solution is boiled until brown fumes no longer come off. Ten cc. of permanganate solution (15 grams to 1 liter) are added, and the boiling is continued. Sodium sulphite solution is added to dissolve the oxide of manganese, and the solution is again boiled and then filtered. With irons the insoluble residue should be tested for phosphorus. After cooling the filtrate, 40 cc. of ammonia (sp.gr. 0.96) are added, the solution is agitated, and when the temperature is at 40° C., 40 cc. of molybdate solution [1] are added and the solution is shaken vigorously for five minutes. After settling out, the yellow precipitate is treated according to one of the following methods, b or c:

(b) **Alkalimetric Method.** The precipitate is washed with 1% nitric acid solution followed by 1% potassium nitrate solution until the washings are no longer acid. The precipitate is dissolved in a measured excess of standardized sodium hydroxide solution and titrated back with standardized nitric acid using phenolphthalein. The solutions are standardized against a steel with a known amount of phosphorus.

(c) **Molybdate Reduction Method.** The precipitate is washed ten to fifteen times with acid ammonium sulphate (prepared according to Blair) or until the washings no longer react for iron or molybdenum. It is dissolved in 25 cc. of ammonia (5 cc. ammonia of 0.90 sp.gr. to 20 cc. water). The filter is washed well with water and 10 cc. of strong sulphuric acid added to the filtrate, which is run through the reductor at once and titrated against a N/30 permanganate solution which has been standardized against sodium oxalate.

Standard Specifications for Upper Limits of Phosphorus

Material	Per Cent Limit of Phosphorus
Automobile Carbon Steels	0.045
Boiler and Fire Box Steel, Stationary Service { Acid { Basic	} not over 0.04
Carbon Steel Bars Auto Springs	0.045
Carbon Steel Bars R. R. Springs	0.05
Carbon Steel Rails	max. 0.10
Carbon Tool Steel	0.005
Carbon Tool Steel High Speed	0.01 or more
Cold Drawn Automatic Screw Stock	0.09–0.13
Fire Box Steel	0.05
Forge Welding Flange and Structural Steel Plates	0.04–0.06
Lap Welded Seamless Boiler Tubes	0.04
Low, Medium and High Carbon Splice Bars	0.04
Rivet Steel for Ships	{ Acid 0.06 { Basic 0.04
Open Hearth Steel Girders	0.04
Steel Castings	{ Acid 0.07 { Basic 0.06
Structural Steel for Cars and Ships	{ Acid 0.06 { Basic 0.04
Structural Steel for Buildings	Bessemer 0.10
Quenched and Tempered Steel Axles, Shafts and Forgings	0.05
Wrought Solid Carbon Steel Wheels and Tires	0.05

Gravimetric Method. Sulphur by Oxidation

Five grams of iron or steel are dissolved in a 400-cc. Erlenmeyer flask, using 50 cc. of strong nitric acid. A little sodium carbonate is added, the solution is evaporated to dryness, and the residue baked for an hour on the hot plate. To the flask 30 cc. of strong hydrochloric acid are added, and the evaporation and baking are repeated. After solution of the iron in another 30 cc. of strong hydrochloric acid and evaporation to a sirupy consistency, 2 to 4 cc. of the same acid are added, followed by 30 to 40 cc. of hot water. The solution is then filtered and the residue washed with hot water. The sulphur is precipitated in the cold filtrate (about 100 cc.) with 10 cc. of a 10% solution of barium chloride. After forty-eight hours the precipitate is collected on a paper filter, washed first with hot acid (containing 10 cc. of concentrated hydrochloric acid and 1 gram of barium chloride to the liter) until free from iron and then with hot water till free from chloride; or, first with cold water, then with 25 cc. of water containing 2 cc. of concentrated hydrochloric acid to the liter. The washings are kept separate from the main filtrate and are evaporated to recover dissolved barium sulphate.

With iron the paper containing the insoluble residue above mentioned is put into a platinum crucible, covered with sodium carbonate free from sulphur, and charred without allowing the carbonate to melt. The crucible should be covered during this operation. Sodium nitrate is then mixed in and the mass fused with the cover off. An alcohol flame is used throughout. The melt is dissolved in water and evaporated with hydrochloric acid in excess to dryness in porcelain. The evaporation with water and hydrochloric acid is repeated to insure removal of nitrates. The residue is extracted with a few drops of hydrochloric acid and water, the insoluble matter is filtered off, and barium chloride is added to the filtrate. The barium sulphate obtained is added to the main portion.

Careful blanks are run with all reagents.

NOTE. For volumetric method see chapter on sulphur.

Standard Specifications for Limits of Sulphur

Material	Percentage Limit of Sulphur
Automobile Carbon Steels	0.050
Boiler and Fire Box Steel	0.040
Carbon Steel Bars for Auto Springs	0.045
Carbon Steel Bars for R. R. Springs	0.050
Cold Drawn Bessemer Steel Automatic Screw Stock	0.075–0.150
Carbon Tool Steel	0.005±
Carbon Tool Steel High Speed	0.01 or more as specified
Fire Box Steel	0.050
Forge Welding Flange and Structural Steel Plates	0.050
Lap Welded Seamless Boiler Tubes	0.045
Rivet Steel for Ships	0.045
Steel Castings	0.060
Structural Steel for Buildings, Bridges, Cars, Ships	0.045–0.060
Wrought Solid Carbon Steel Wheels and Tires	0.050

Determination of Silicon

One gram of pig iron, cast iron, and high silicon iron, or 5 grams of steel, wrought iron, and low silicon iron are taken for analysis. (By taking multiples of the factor weight 0.4693, SiO_2 to Si, the final calculation is simplified.) The sample is placed in a 250-cc. beaker and 20 to 50 cc. of dilute nitric acid added. If the action is violent, cooling the beaker in water is advisable. When the reaction subsides, 20 cc. of dilute sulphuric acid, 1 : 1, are added, the mixture placed on the hot plate and evaporated to dense white fumes The residue is taken up with 150 cc. of water containing 2 to 5 cc. of sulphuric acid and heated until the iron completely dissolves.

The solution is filtered and the silica residue washed first with hot dilute hydrochloric acid, sp.gr. 1.1, and then with hot water added in small portions to remove the iron sulphate. The residue is now ignited and weighed as silica.

If there is any doubt as to the purity of the silica, moisten the residue (in a platinum crucible) with strong sulphuric acid and add a few cc. of hydrofluoric acid (crucible cover full), evaporate to dryness, ignite and weigh. The loss of weight is due to silica.

NOTE. If the ash is colored by iron oxide, silica is determined by difference after expelling the silica by adding 4 to 5 cc. of hydrofluoric acid and a few drops of sulphuric, taking to dryness and igniting the residue.

The following acid mixtures are recommended by the U. P. Ry. For steel, wrought iron and low silicon iron, 8 parts by volume of HNO_3, sp.gr. 1.42; 4 parts of conc. H_2SO_4, sp.gr. 1.84; 6 parts HCl, sp.gr. 1.2 and 15 parts by volume of water. For dissolving pig iron, cast iron and high silicon iron, a mixture of 8 parts by volume of strong nitric acid and 5 parts of strong sulphuric acid, diluted with 17 volumes of water, is used.

Rapid Method for Determining Silicon in Foundry Work. Liquid iron, dropped into cold water from a ladle 3 ft. above the water, will form shot shaped according to forms resulting from its chemical constitution, silicon being an important factor. Round shot, concave upper surface, 1 to 3 in. in diameter, indicate over 2% silicon. Flat or irregular shot indicate low silicon. Shot with elongated tails indicate very low silicon.

Other Methods for Determination of the Less Common Elements in Steel. Other elements more commonly sought in alloy steel are copper, nickel, chromium, vanadium, tungsten, titanium and molybdenum. Methods for estimation of these elements are given in the chapters dealing with these substances. A.S.T.M. Methods for Steel may be found in Vol. 2.

Standard Specifications for Silicon

Material	Per Cent Silicon
Boiler and Fire Box Plates	not over 0.03
Carbon Steel Bars Auto Springs	1.80–2.10
Carbon Steel Bars R. R. Springs	0.25–0.50
Carbon Steel Rails Bessemer and Open Hearth	0.20
Carbon Tool Steel	0.15±
Carbon Tool Steel High Speed	0.15±
Cylinder Grade Pig Iron	1.25–1.60
Floor Grade Pig Iron	2.25–2.75
Open Hearth Steel Girders	0.20
Wrought Solid Carbon Steel Wheels and Tires	{ Acid 0.15–0.35 { Basic 0.10–0.30

NOTE. Tool Steels may contain 0.5±Cr, 0.75±W, 0.25±V, 0.5±Co and other elements.

LEAD

Pb, *at.wt.* 207.2; *sp.gr.* 11.34; *m.p.* 327°; *b.p.* 1525° C; *oxides,* PbO, PbO$_2$, Pb$_3$O$_4$.

DETECTION

Hydrochloric acid precipitates lead incompletely from its cold solution as white PbCl$_2$, soluble in hot water by which means it is separated from mercurous chloride and silver chloride. PbCl$_2$ forms needle-like crystals upon cooling the extract.

Hydrogen sulphide precipitates black PbS from slightly acid solutions along with the other elements of the group.[1] Yellow ammonium sulphide, sodium sulphide and the fixed alkalies dissolve out arsenic, antimony and tin. The sulphide of lead, together with bismuth, copper and cadmium, dissolve in hot dilute nitric acid, leaving mercuric sulphide insoluble. The extract evaporated to dryness and then to SO$_3$ fumes, after addition of sulphuric acid, expels nitric acid. Upon adding water to the residue and boiling with a little additional sulphuric acid the sulphates of bismuth, copper and cadmium are dissolved out, lead sulphate remaining as a white residue.

Lead may be further confirmed by dissolving the sulphate in ammonium acetate (barium sulphate is very slightly soluble,) and precipitating the yellow chromate, PbCrO$_4$, by addition of potassium dichromate solution.

ESTIMATION

The determination of lead is required in valuation of its ores—galena, PbS; anglesite PbSO$_4$; cerussite, PbCO$_3$; krokoite, PbCrO$_4$; pyromorphite, 3Pb$_3$P$_2$O$_8$·PbCl$_2$. It is determined in lead mattes; certain slags; drosses from hard lead; cupel bottoms; skimmings; lead insecticides (arsenate of lead); paint pigments such as white lead, red lead, yellow and red chromates, etc. It is determined in alloys such as solder, type metal, bell metal, etc. The estimation is necessary in the complete analysis of a large number of ores, especially in minerals of antimony and arsenic. Traces of lead are determined in certain food products where its presence is undesirable.

Preparation and Solution of the Sample

In dissolving lead, its alloys, or ores the following facts will be recalled. Hot, dilute nitric acid is the best solvent of the metal. Lead nitrate is insol-

---

[1] Lead precipitates best from solutions containing 1 cc. of concentrated free hydrochloric acid (sp.gr. 1.19) for each 100 cc. of solution. The sulphide is appreciably soluble if the acidity is increased to 3 cc. HCl per 100 of solution.

Chapter contributed by Wilfred W. Scott.

uble in concentrated nitric acid, but dissolves readily upon dilution with water. The metal is insoluble in dilute sulphuric acid, but dissolves in the hot, concentrated acid. Although not soluble in dilute hydrochloric, it dissolves in the hot, concentrated acid, especially in presence of the halogens chlorine, bromine and iodine. The metal is soluble in glacial acetic acid. The salts are soluble in hot, dilute nitric acid. In dissolving sulphide ores it should be kept in mind that strong nitric acid will form some lead sulphate which will be precipitated upon dilution of the solution. Oxidation is less apt to occur with the dilute acid. Silicates and slags require fusion with sodium carbonate and potassium carbonate. The cooled mass may then be extracted with hot water to remove silica and the residue containing the carbonates of the heavy metals dissolved in dilute nitric acid. Lead salts are soluble in ammonium acetate.

Ores, Minerals of Lead, etc. In decomposing lead ores it is well to start with hydrochloric acid, about 20 cc. per gram sample. The pear-shaped flask is convenient for this work. Warm gently but do not boil, add more acid if necessary. If the ore contains galena, PbS, some H_2S is expelled. There is generally a liberation of free sulphur, which floats in globules on the surface of the solution. After the reaction with hydrochloric acid subsides, nitric acid is added and the solution boiled, then follows the addition of sulphuric acid and the solution is evaporated to strong fumes. This expels nitric and hydrochloric acids. On dilution of the solution lead sulphate remains insoluble, with silica and certain insoluble substances, such as barium sulphate, etc. The residue should not contain black particles of undecomposed ore as these are very likely to contain lead. Addition of HF assists decomposition of silicates.

It may be advisable, in certain cases, to open up the ore with nitric acid or aqua regia, followed by sulphuric acid and hydrofluoric acid.

Iron Pyrites and Ores with Large Amounts of Impurities with Small Amounts of Lead. Ten grams of the sample or more, if lead is present in very small amounts (less than 0.1%), are taken for analysis, and 50 cc. of a mixture of potassium bromide and bromine solution added (75 grams of KBr dissolved in 400 cc. of water and 50 cc. of bromine added). After ten to fifteen minutes about 50 cc. of concentrated nitric acid are added and after the violent reaction has ceased 25 to 30 cc. of concentrated hydrochloric acid and the solution is evaporated on the hot plate to near dryness. Fifty cc. of C.P. (lead free) concentrated sulphuric acid is now added and the sample taken to fumes of SO_3 on a sand bath. After cooling, the concentrate is diluted to 500 cc. with water, about 5 cc. of strong sulphuric acid added, the solution heated to boiling and cooled. The precipitate is filtered by decantation onto a fine-grained filter (quality of an S. & S. 590 or B. & A. grade A), the residue boiled with more water containing H_2SO_4 and again decanted. This is repeated until all the iron sulphate is removed. (The filtrates should be kept several hours to see whether any of the lead has passed through the paper in a colloidal condition.) The precipitate is finally poured on the filter and washed with 2% H_2SO_4. Impure residues are extracted for lead with ammonium acetate.

Solution of Lead Alloys. As a rule these are best decomposed by treating 0.5 to 1.0 gram of the material, or more as the case may require, with a hot solution of nitric acid, 1 : 1, and evaporating the solution to low bulk, but not to dryness. Hot water is now added and the material boiled and the soluble portion filtered off. The insoluble material is digested with concentrated hydrochloric acid to which a little bromine has been added.

SEPARATIONS

Separation of Lead as Sulphate. Lead is most frequently separated from other metals by precipi ation as sulphate, $PbSO_4$, according to the details given under " Preparation and Solution of the Sample." In the presence of much bismuth or iron it is necessary to wash the precipitate with a 10% sulphuric acid solution to ke p the bismuth in solution and to prevent the formation of the difficultly soluble basic ferric sulphate. In absence of appreciable amounts of these elements the lead sulphate is more completely separated by adding to the dilute sulphuric acid solution an equal volume of alcohol, filtering and washing the residue with 50% alcohol.

Separation of Lead from Barium. In the analysis of minerals containing barium, the insoluble sulphate, $BaSO_4$, will be precipitated with lead. Since barium sulphate is slightly soluble in ammonium acetate it will contaminate the lead in the subsequent extraction by this reagent. The presence, however, of a little sulphuric acid, renders this solubility practically neglig ble. The sulphuric acid should not exceed 1–2% in the ammonium acetate reagent, as lead sulphate will precipitate if sufficient sulphuric acid is added to the acetate extract. (Lead sulphate is precipitated almost completely if the acetate solution contains 10% sulphuric acid.)

Lead may be separated from barium sulphate by digesting the mixed sulphates with ammonium carbonate solution, whereby the lead sulphate is transposed to lead carbonate and ammonium sulphate, while barium sulphate is not changed. The soluble ammonium sulphate may be washed out with ammonium solution followed by water. Since lead carbonate is slightly soluble in the ammonium salt, the filtrate is treated with hydrogen sulphide and the dissolved lead recovered as PbS. The residue containing lead carbonate and barium sulphate is treated with dilute nitric or acetic acid. Lead passes into solution, while barium sulphate remains insoluble.

Extraction of Lead from the Impure Sulphate by Ammonium Acetate. The filter containing the impure sulphate, obtained by one of the procedures for solution of the sample, is placed in a casserole and extracted with about 50 cc. of hot, slightly ammoniacal ammonium acetate, the stronger the acetate the better. The clear liquid is decanted through a filter and the extraction repeated until the residue is free from lead (i.e., no test is obtained for lead with $K_2Cr_2O_7$). A very effective method of extraction is by adding solid ammonium acetate directly to the sample on a filter and pouring over it a hot solution of ammonium acetate. The filtrate containing the pure lead acetate solution may now be examined by one of the following procedures.

Lead sulphate containing arsenic should be dissolved in ammonium acetate, the extract made alkaline and lead precipitated as PbS. Arsenic remains in solution.

GRAVIMETRIC METHODS

Determination of Lead as the Sulphate, PbSO$_4$

Procedure. The sample having been dissolved according to a method outlined, the lead precipitated as PbSO$_4$ by addition of an excess of sulphuric acid, and taking to SO$_3$ fumes, the lead sulphate is filtered off, upon cooling and diluting the sample. The PbSO$_4$ is washed with water containing 10% H$_2$SO$_4$ until free from soluble impurities. If insoluble sulphates or silica are present the lead must be purified. If such impurities are known to be absent (alloys), the sulphate may be filtered directly onto an asbestos mat in a tared Gooch crucible, dried, then ignited to dull red heat, cooled and finally weighed as PbSO$_4$. In the analysis of ores, however, it is generally advisable to purify the sulphate.

Purification of Lead Sulphate. Details of the procedure have been given under Separations—Extraction of Lead from the Impure Sulphate. The lead sulphate having been brought into solution by extraction with strong ammonium acetate solution, the excess acetic acid is volatilized by evaporation, the residue cooled and diluted with water. An excess of sulphuric acid is added and the precipitated sulphate is filtered off, washed with dilute sulphuric acid and alcohol, dried at about 110° C., or if preferred by ignition at dull red heat, and weighed.

PbSO$_4 \times 0.6831 =$ Pb. Pb multiplied by 100 and divided by weight of sample taken equals per cent.

NOTES. Lead sulphate may be precipitated from ammonium acetate solution by adding sulphuric acid until the solution contains approximately 10% H$_2$SO$_4$.

An acetate extraction may not be necessary, as is generally the case in the analysis of alloys. In analysis of ores, however, PbSO$_4$ may be contaminated by sulphates of the alkaline earths and by silica. The difficultly soluble oxides of iron and alumina may also be present.

If arsenic is in the sulphate it will pass into the filtrate with the lead.

Determination of Lead as the Chromate, $PbCrO_4$

This excellent method is applicable to a large class of materials and is of special value in precipitation of lead from an acetic acid solution, the method depending upon the insolubility of lead chromate in weak acetic acid.

Procedure. The solution of the sample, precipitation of the lead as the sulphate and extraction of lead with ammonium acetate have been given in detail.

The filtrate, containing all the lead in solution as the acetate, is acidified slightly with acetic acid and heated to boiling. Lead is precipitated by addition of potassium dichromate solution in excess (10 cc. of 5% $K_2Cr_2O_7$ solution are generally sufficient). The solution is boiled until the yellow precipitate turns to a shade of orange or red.[1] The precipitate is allowed to settle until the supernatant solution is clear. (This should appear yellow with the excess of dichromate reagent.) The $PbCrO_4$ is filtered onto an asbestos mat in a tared Gooch crucible, washed with water, dried in an oven at about 110° C. and the cooled compound weighed as $PbCrO_4$.

$$PbCrO_4 \times 0.641 = Pb. \qquad \frac{Pb. \times 100}{Wt. \text{ of sample}} = \text{per cent Pb.}$$

NOTES. Impurities, such as iron, copper, cadmium, etc., in the acetate solution of lead seriously interfere in the chromate precipitation. These should be leached out with water containing a little sulphuric acid before extracting the lead sulphate with ammonium acetate. See remarks under section on Traces of Lead.

If a standard solution of potassium dichromate is used in the precipitation of lead the excess of the reagent, upon filtering of the precipitate, may be titrated and the lead determined volumetrically. A known amount of dichromate solution (added from a burette) sufficient to precipitate all the lead and about one-third of the volume in excess is added to the hot solution. After boiling about two minutes the precipitate is filtered off quickly and washed several times with hot water. The filtrate, or an aliquot part of it, is made acid with 5 cc. concentrated sulphuric acid and titrated with standard ferrous sulphate at about 60° C., using potassium ferricyanide as an outside indicator; the end-point is a blue color produced by the slight excess of the ferrous salt reacting with the indicator. The excess of dichromate may be determined by adding 3 to 4 grams of solid potassium iodide, KI, to the solution diluted to about 500 cc. with water to which 15 cc. of concentrated sulphuric acid has been added. The liberated iodine is titrated with standard thiosulphate, the end-point being colorless, with starch solution internal indicator, changing from blue. Bi, Sb, Ba, Sr and Ca interfere slightly.

One cc. N/10 $K_2Cr_2O_7$ = 0.010355 gram Pb. One cc. N/5 $K_2Cr_2O_7$ = 0.02071 gram Pb.

---

[1] The yellow precipitate gives high results, since it is difficult to wash. The crystalline orange or red compound may be quickly filtered and washed.

Determination of Lead as the Molybdate, $PbMoO_4$

This method is rapid and has the following advantages:
   a. The sulphation of lead is avoided.   *b.* The acetate extraction is eliminated.   *c.* The precipitate may be ignited.   *d.* The ratio of lead to its molybdate compound is greater than either lead to $PbSO_4$ or to $PbCrO_4$, lessening the chance of error through weighing.

Cobalt, calcium, strontium and barium have little effect in presence of ammonium acetate.   In absence of this salt they interfere slightly.

Procedure.   The ore or alloy is decomposed with nitric acid or aqua regia as the case may require.   (Silica if present is eliminated by taking to dryness, dehydrating, taking up with dilute nitric acid and filtering.)   To the clear liquid 2 g. ammonium chloride is added and then sufficient ammonium acetate to destroy the excess of free nitric acid, i.e., 2 g. per each cc. of free HNO_3 present.

Lead is now precipitated by adding 40 cc. of ammonium molybdate per each 0.1 g. of lead present (4 g. per liter + 10 cc. acetic acid), stirring the mixture during the addition.   After boiling for two or three minutes the precipitated lead molybdate is allowed to settle, then filtered and washed with small portions of hot water and ignited over a Bunsen burner.

The cooled residue is weighed as $PbMoO_4$. $PbMoO_4 \times 0.5642 = Pb$.

NOTES.   If antimony or other members of the group are present in the original sample, it is advisable to dissolve the residue in HCl and reprecipitate the lead with molybdate reagent.

If lead is in the form of the sulphide, as may be the case in a complete analysis of a substance, it is decomposed with hot dilute HNO_3 and precipitated as $PbMoO_4$.

Galena is best decomposed by treating with hydrochloric acid to expel sulphur as oxidation of sulphur to sulphate is not desirable in this method. If lead sulphate has formed due to oxidation of sulphur, it is advisable to treat any residue, remaining from the acid extraction, with ammonium acetate, adding the acetate extract to the solution containing the lead.

The sample evaporated to dryness is treated with about 10 cc. of dilute nitric acid (1 : 1) and a little water then heated and filtered.   About 2 grams of ammonium chloride are added and for each cc. of free nitric acid (d. 1.42) two grams of ammonium acetate are necessary (total 10–15 g.).   The precipitate generally appears a light canary yellow or yellowish white.

Washing the precipitate with water containing a little ammonium nitrate prevents the formation of colloidal lead molybdate, which would pass through the filter paper.   (Use about 2 g. nitrate per 100 cc.)

The pulp used is paper pulp made by breaking up ashless filter paper and agitating it thoroughly with hot water in a flask.   The Editor prefers omitting the pulp and using a fine-grained ashless filter paper, the washing of the precipitate being conducted with wash water containing ammonium nitrate.

Electrolytic Determination of Lead as the Peroxide, PbO_2

An electric current passed through a solution of lead containing sufficient free nitric acid will deposit all the lead on the anode as lead peroxide. The method is excellent for analysis of lead alloys.

Procedure. The sample containing not over 0.5 gram lead is brought into solution by heating with dilute nitric acid, 1 : 1. The solution is washed into a large platinum dish with unpolished inner surface. Twenty to 25 cc. concentrated nitric acid (sp.gr. 1.4) are added and the solution diluted to about 150 cc.

The sample is electrolyzed in the cold with 0.5 to 1 ampere current and 2 to 2.5 volts, the platinum dish forming the anode of the circuit, a spiral platinum wire or a platinum crucible dipped into the solution being the cathode. Three hours are generally sufficient for the deposition of 0.5 gram Pb. Overnight is advisable, a current of 0.05 ampere being used.

A rapid deposition of the lead may be obtained by heating the solution to 60 to 65° C. and electrolyzing with a current $ND_{100} = 1.5$ to 1.7 amperes, the E.M.F. varying within wide limits. Stirring the solution with a rotating cathode aids in the rapid deposition of the PbO_2.

To ascertain whether all the lead has been removed from the solution, more water is added so as to cover a fresh portion of the dish with water. The electrolysis is complete if no fresh deposition of the peroxide takes place after half an hour.

The water is siphoned off while more water is being added until the acid is removed, the current is then broken, the dish emptied of water and the deposits dried at 180° C. and weighed as PbO_2.

The deposit of lead peroxide gently ignited forms lead oxide, PbO, a procedure recommended by W. C. May,[1] confirmed by Treadwell and Hall as giving more accurate results than the peroxide, PbO_2.

$$PbO_2 \times 0.8662 = Pb.$$

$$PbO \times 0.9283 = Pb.$$

NOTE. The deposits of lead oxide or peroxide may be removed by dissolving off with warm dilute nitric acid.

For volumetric procedure-titration of the peroxide PbO_2 see page 278.

[1] Am. Jour. Sci. and Arts (3), **6**, 255.

VOLUMETRIC METHODS

Volumetric Ferrocyanide Method for the Determination of Lead

Although the gravimetric methods for the determination of lead are considered the more accurate, yet the volumetric procedures may be frequently used with advantage. The ferrocyanide method has been pronounced by Irving C. Bull[2] to be the best of the procedures in common use, the results being accurate.

Procedure. Lead sulphate is obtained according to the method outlined under Preparation and Solution of the Sample. The lead sulphate is transferred to a small beaker and gently boiled with 10 to 15 cc. of a saturated solution of ammonium carbonate, the liquid having been added cold and brought up to boiling. After cooling, the precipitate is filtered onto the original filter paper from which the lead sulphate was removed. The lead carbonate is washed free of alkali with cold water. The filter with the precipitate is dropped into a flask containing a hot mixture of 5 cc. of glacial acetic acid with 25 cc. of water. The lead carbonate is decomposed by boiling and the solution diluted to 150 cc.

Titration. The sample warmed to 60° C. is titrated with a standard solution of potassium ferrocyanide, using a saturated solution of uranium acetate. as an outside indicator. The excess of ferrocyanide produces a brown color with the uranium acetate drop on the tile.

Free ammonia must be absent, as it reacts with uranium acetate and gives low results. NH_4OH precipitates reddish brown, gelatinous uranous hydroxide, $U(OH)_4$.

The bulk of solution to be titrated should be as near as possible to 100 cc., including 10 cc. of 50% acetic acid.

One per cent potassium ferrocyanide reagent is used in the titration. This reagent is standardized against a known amount of lead in solution as an acetate.

A correction of 0.8 cc. is generally necessary on account of the indicator. This is determined by a blank titration.

Antimony, bismuth, barium, strontium and calcium interfere only to a very slight extent, the error being negligible.

[1] C. N., 2253, **87**, 1903.

The Permanganate Method for Lead

By Albert H. Low

The following method for the determination of lead in ores has proved very satisfactory in the great majority of cases. It depends upon the separation of the lead as sulphate, the conversion of the sulphate to carbonate, the solution of the carbonate in acetic acid, followed by the precipitation of the lead as oxalate. The lead oxalate is then decomposed in dilute sulphuric acid and the separated oxalic acid titrated with standard permanganate.

Ordinary constituents of lead ores do not interfere, with the exception of lime. As high as 10% of CaO in an ore, however, is without effect. Barium interferes only by forming a combination with lead that resists the reactions, with consequent low results. The remedy is easy and is described below.

Procedure. Decompose 0.5 gram of the ore in an 8-oz., pear-shaped flask, such as is commonly called a "copper-flask." The treatment may usually be a very gentle boiling with 10 cc. of hydrochloric acid for a short time, then adding 5 cc. of nitric acid and continuing the gentle boiling until decomposition is complete. Now add 6 cc. of sulphuric acid and boil over a free flame to strong fumes. Allow to cool.

Add 100 c.c of cold water and 5 cc. of sulphuric acid and heat to boiling. Remove from the heat, add 10 cc. of alcohol (cautiously) and cool under the tap.

Fold a 9-cm. filter with particular care to creasing the fold that will come next to the precipitate as thin as possible, so it will lie flat and not easily allow material to get under the edge. Filter the mixture through this. Return the first portions of the filtrate if not clear. Wash 6 times with cold water containing 10% of alcohol. Any trace of lead sulphate remaining in the flask will be recovered subsequently.

With a jet of hot water, using as little as possible, rinse the precipitate from the filter, through a short funnel, back into the flask. (In the known or assumed presence of barium, interpolate the following short procedure: Add 10 cc. of hydrochloric acid and boil over a free flame almost to dryness. Allow to cool, add 20 cc. of water and a few drops of ammonia, sufficient to neutralize the acid.) Place the flask again under the original funnel and pour through the filter 10 cc. of a cold saturated solution of ammonium carbonate. Remove the flask and heat the contents just to boiling, then cool completely under the tap. Pour the cold mixture through the original filter. Wash out the flask well with cold water, pouring through the filter, and then wash filter and precipitate 10 times with cold water containing about 5% of the ammonium carbonate solution. Reject the filtrate.

Again using a jet of hot water, wash the precipitate from the filter into a small beaker. Add 5–6 cc. of glacial acetic acid and heat to boiling. Replace the flask under the funnel and pour the hot acid mixture through the filter. Wash out the beaker with hot water and then wash the filter 10 times with hot water slightly acidulated with acetic acid. (Small amounts of lead carbonate may be dissolved directly upon the filter, without previous transference to a beaker.)

Add to the filtrate 10 cc. of a cold saturated solution of oxalic acid, heat to boiling and then cool completely under the tap. Be particular to get as cold as possible. Now filter the lead oxalate through a 9-cm. filter. Using cold water, wash out the flask thoroughly and then wash filter and precipitate 10 times.

Place about 25 cc. of cold water in the flask, add 5–6 cc. of sulphuric acid and then about 100 cc. of hot water. Drop the filter and precipitate into this. Wipe out any lead oxalate adhering in the funnel with a small piece of dry filter paper and drop into the flask. Heat the acid mixture nearly to boiling and then titrate it with standard potassium permanganate solution to a faint pink tinge. Calculate the result from the known lead value of the permanganate.

The permanganate commonly used for iron titrations will serve, although rather strong for lead. Theoretically, 1.857 times the iron factor will give the lead factor. Owing to slight losses of lead an empirical factor must be used. This is 1.879 times the iron factor. Based on this factor and on 0.5 gram of ore taken for assay, 1 cc. of a permanganate solution containing 1.544 grams per liter will equal 1% lead. It may be standardized directly on lead as follows: Convert about 0.250 gram of pure lead foil to sulphate by boiling with 6 cc. of sulphuric acid. Continue according to the above entire process. Finally, divide the percentage value of the lead taken by the cc. of permanganate required, to obtain the percentage value of 1 cc. in lead. A comparison of this figure with the iron value of the permanganate may be made, to check the conversion factor given above. The personal equation may cause a slight difference.

NOTES. Metallic lead is converted to lead sulphate by boiling with strong sulphuric acid. The reaction takes place with the hot concentrated acid, the metal changing to the white lead sulphate solid, soluble in large excess of sulphuric acid; this is unnecessary, as decomposition is complete with the amount stated.

Conversion of lead sulphate to carbonate before changing to acetate appears at first thought to be an unnecessary step, but experience has shown that a direct conversion of sulphate to acetate by dissolving in ammonium acetate leaves sufficient sulphate in the solution to cause low results, as much as 10 per cent of lead apparently escaping subsequent conversion to oxalate. In the procedure for converting the lead to carbonate any small amount of lead sulphate remaining does no harm. It frequently occurs that the carbonate formed does not completely dissolve in acetic acid. If a cloudy solution is obtained, a few drops of ammonia will furnish enough ammonium acetate to dissolve the small amount of sulphate remaining. The precipitation of lead oxalate is not interfered with by the ammonium salt present, a large amount of which, however, should be avoided. Ammonium oxalate may be added in place of oxalic acid.

In the molybdate method, the procedure given for the permanganate method may be followed to the point where the oxalate or oxalic acid is added. This step is omitted and the lead acetate solution is titrated with ammonium molybdate as directed under this procedure.

Volumetric Determination of Lead by the Molybdate Method

Lead is precipitated from an acetic acid solution by a standard solution of ammonium molybdate, the termination of the reaction being recognized by the yellow color produced by the excess of reagent when a drop of the mixture comes in contact with a drop of tannin solution, used as an outside indicator. The method is rapid, but is not as accurate as the chromate-iodide method following.

Reagents

Ammonium Molybdate. 4.26 grams of the salt are dissolved in water and diluted to 1000 cc. On a half gram sample basis 1 cc. of the reagent is equivalent to about 1 per cent lead.

Standardization. Dissolve 0.2 g. of pure lead foil in 5 cc. of strong sulphuric acid by boiling gently in a 250-cc. pear-shaped flask. When the lead has decomposed to sulphate, dilute (on cooling) with water and filter off the PbSO$_4$. Now follow the details of the procedure used in the method given below, after isolating the lead as sulphate. Note the cc. of molybdate reagent required and divided this into 0.2 to get the equivalent value in terms of lead per cc. of reagent. 1 cc. should be equivalent, approximately, to 0.005 g. Pb.

Tannin Indicator. 0.1 g. tannic acid per 20 cc. water. The reagent should be prepared fresh for each day's analysis.

Procedure

Decomposition. Follow the usual procedure recommended for decomposing lead ores, using HCl, HNO$_3$ and finally H$_2$SO$_4$. Evaporate to strong sulphuric fumes, and take up with water. Filter off the lead sulphate and wash with a 10 per cent sulphuric acid solution, to remove sulphates of the metals, and finally with water containing a little alcohol, remembering the PbSO$_4$ is slightly soluble in water.

The lead sulphate is now brought into solution as lead acetate by extraction with ammonium acetate slightly acidified with acetic acid. (Use a strong solution of the reagent.) In absence of calcium and barium, the writer prefers to convert the lead sulphate to carbonate by boiling with ammonium carbonate solution, according to the permanganate method for lead (p. 277), and then to the acetate by dissolving the lead carbonate in dilute acetic acid. Thus sulphates are eliminated. The addition of a few drops of ammonia to the acetic acid solution insures the solubility of the lead. (PbSO$_4$ may be present in small amount and does not readily dissolve in acetic acid.) The results are more concordant in absence of sulphate. The acetate solution of lead is now titrated with the standard molybdic solution.

Titration. The solution is divided into two portions, one being kept in reserve. To one is added the standard solution of ammonium molybdate, from a burette, until a drop of the titrated solution, brought in contact with a drop of tannin indicator, on a white tile, or paraffined surface, gives a brown or yellow color. The reserve solution is now added in portions, the titration being continued, until the last portion has been used and the brown color obtained. This precaution avoids over running the end point.

The Chromate=Iodide Method for the Volumetric Determination of Lead

The method depends upon the action of chromates on potassium iodide with a resulting liberation of an amount of free iodine in direct ratio to the chromate present, which in turn is a measure of the amount of lead isolated as lead chromate. The liberated iodine is determined by titration with a standard solution of thiosulphate. The discussion of the reactions that take place, with equations, are given in the notes that follow the directions of the method.

Solutions Required

Ammonium Acetate Extraction Solution. A saturated solution of ammonium acetate, filtered to remove foreign matter if present, is diluted with twice its volume of distilled water and 30 cc. of 80 per cent acetic acid is added per liter of solution.

Hydrochloric Acid Mixture. To a liter of saturated salt solution, filtered if necessary, is added 150 cc. of distilled water and 100 cc. of concentrated hydrochloric acid.

Potassium Dichromate. Saturated solution, filtered if not clear.

Starch Solution.

Procedure. 1. *Solution of the Sample.* Half a gram of the finely divided material (if the factor weight 0.6907 g. is taken, 1 cc. N/10 reagent in final titration is equivalent to about 1% Pb) is dissolved in a beaker or a flask (Low's type) by adding 20 cc. of strong hydrochloric acid and heating gently until the action subsides. If the decomposition is incomplete, about 5 cc. of nitric acid are added and the heating continued.

2. About 5 cc. of sulphuric acid are added and the solution evaporated to strong fumes. After cooling, about 50 cc. of water are added and the solution is boiled to dissolve the soluble salts. If the ore is low grade, 5–10 cc. of ethyl alcohol are now added, the precipitate is allowed to settle and then washed by decantation three or four times with 1 : 15 sulphuric acid (i.e., about 10% solution), and finally transferred to the filter with the dilute acid and washed once with pure cold water.

3. By means of a fine jet from a wash bottle (500 cc.) filled with ammonium acetate extraction reagent, heated to near boiling, the precipitate is transferred to the beaker or flask in which the precipitation was made. This may be done by carefully spreading out the filter in the funnel or by breaking the filter and washing the paper free of the lead sulphate with a fine stream of the reagent. If the precipitate does not go into solution, more of the acetate is added and heat gently applied until it dissolves. The solution is now diluted to 150 cc., heated to boiling and 10 cc. of the saturated dichromate solution added and the boiling continued ten minutes. The yellow color of the lead chromate precipitate changes to red. This is important to obtain a precipitate of definite composition.

4. The precipitate is filtered, the containing vessel washed out with hot dilute sodium acetate wash solution (50 cc. of the extraction solution diluted to 1000 cc.) and the precipitate washed ten times with the reagent.

5. The original beaker or flask is now placed under the funnel and the lead chromate is dissolved on the filter by adding cold dilute hydrochloric acid mixture, stirring up the precipitate with a jet of the reagent, adding the

acid until all of the chromate has dissolved and the color has been completely removed from the filter. At least 50 cc. of the reagent should be used.

6. For low grade ores the entire solution is taken and treated with potassium iodide solution; in case of high grade ores, about half the solution is set aside in reserve, and upon completing the titration of the first portion, the reserve is added and the titration completed. This precaution is taken because a loss of iodine is apt to occur if much iodine is liberated at one time, free iodine being apt to escape as vapor from the easily saturated solution. (The solution is a poor solvent of iodine.) To the solution are added 5 cc. of 25% potassium iodide and the liberated iodine is titrated with N/10 sodium thiosulphate until the iodine color begins to fade; starch solution is now added in sufficient quantity to produce a distinct blue color and the titration continued until the blue color changes to pale green.

A background of white assists in recognition of the end-point. A sheet of white paper placed under the beaker will do, if the base of the stand is not already white.

Standardization of the Thiosulphate. This is best standardized against metallic lead. 0.6907 gram of pure lead should require 100 cc. of N/10 thiosulphate. 0.2 gram of lead is taken or a fraction of the factor weight (0.6907). The lead foil is dissolved in 5 cc. of sulphuric acid by bringing to vigorous boiling, upon cooling the residue is taken up with water and treated exactly according to the method outlined in steps 2–6 of the regular procedure. 1 cc. of N/10 thiosulphate is equivalent approximately to 0.0069 gram of lead.

NOTES. If barium is present in the sample, the residue left from the acetate extraction may contain lead. This is treated with about 10 cc. of strong hydrochloric acid, evaporated to dryness, 25 cc. of the acetate reagent added, the mixture boiled, filtered and the residue washed. The filtrate contains the lead that remained with the residue.

In considering the reactions that take place it must be remembered that it is the combined chromate radical that is responsible for the liberated iodine. The graphical formula represents what takes place:

$$2PbCrO_4 \text{ (or } (PbO)_2.Cr_2O_3.3O)+6KI=3K_2O+6I, \text{ etc.}$$

The chromate first formed is probably the normal salt $PbCrO_4$; this however is unstable and changes with boiling to a stable acid chromate of lead under the conditions of the procedure, hence $\frac{1}{2}$ the atomic weight of lead does not give the exact normal equivalent but the true value is an arbitrary factor, obtained by standardizing the thiosulphate against metallic lead as recommended.

It is evident that Pb is equivalent to 3I. Therefore a normal equivalent of Pb is $\frac{1}{3}$ of its atomic weight, 207.2 divided by 3=69.07, hence 1 cc. of a N/10 solution will titrate iodine equivalent to .006907 g. Pb.

Since Fe equivalent is 55.84, Fe to Pb = 69.07 divided by 55.84 = 1.237, factor of iron to lead.

Example in Standardization of Sodium Thiosulphate

If 0.2035 g. of lead required 30.05 cc. of thiosulphate solution, then 1 cc. would be equivalent to 0.00677 g. lead.

Volumetric Chromate=Ferrous=Diphenylamine Method for Determining Lead

The procedure depends upon isolating lead as chromate and titrating hexavalent chromium with standard ferrous salt after dissolving lead chromate in hydrochloric acid. As in the chromate-iodide method reduction of chromium from hexavalent to trivalent form makes the lead equivalent $\frac{1}{3}$ its atomic weight.

Reactions $PbCrO_4 + 2HCl \rightarrow PbCl_2 + H_2CrO_4$
$$H_2CrO_4 + 3FeCl_2 + 6HCl \rightarrow CrCl_3 + 3FeCl_3 + 4H_2O.$$

Reagents—*Ferrous Salt Solution.* 0.1 N solution of ferrous ammonium sulphate or ferrous sulphate.

Potassium Dichromate or Permanganate—0.1 N solution.

Phosphoric-Sulphuric Acid. 1 volume 85% H_3PO_4 mixed with 1 volume H_2SO_4 (sp.gr. 1.84).

Diphenylamine Indicator. 1 gram of salt dissolved in 100 cc. strong sulphuric acid.

Hydrochloric-Salt Solution. See under Procedure.

Procedure

Decomposition of Sample. The ore is dissolved in HCl and HNO_3 according to standard procedure for lead ores. Alloys are treated with HNO_3. H_2SO_4 is now added and the HCl and HNO_3 expelled by taking to sulphuric acid fumes. Pure lead is decomposed by direct treatment with strong H_2SO_4 and heating. The residue is treated with water and a little alcohol and the lead sulphate filtered off and washed with 10% alcohol solution. The lead sulphate is now dissolved in a saturated solution of ammonium or sodium acetate slightly acidified with acetic acid. The extract is diluted to about 150–200 cc. with water and lead precipitated as chromate ($PbCrO_4$) by addition of a saturated solution of potassium dichromate (10–15 cc. is ample).

Lead Chromate. The crystalline product is obtained by boiling the precipitate until the yellow color changes to orange red. After cooling under tap water, the precipitate is filtered and washed with a 5% solution of ammonium acetate, until free from the excess of chromate reagent (the wash water passing through the precipitate colorless).

Solution and Titration. The lead chromate is now dissolved by adding hydrochloric acid-salt solution (a mixture of 1000 cc. saturated solution of NaCl, 120 cc. H_2O and 100 cc. HCl, d. 1.19), 50 to 100 cc. being used. All yellow color being washed from the filter. The chromate solution is diluted to 150–200 cc., 10 cc. phosphoric-sulphuric acid mixture added and 4–6 drops of diphenylamine indicator. The 0.1 N ferrous solution is now added until the color of the chromate solution changes to green. The excess of ferrous solution is determined by titration with standard dichromate or permanganate, the green color changing to a deep blue.

1 cc. of 0.1 N ferrous solution is equivalent to 0.0069 g. Pb.

NOTES. In presence of a large amount of HCl the end point is a change of green to a dark green. The action of HCl may be overcome by adding 1 g. ammonium acetate for each cc. HCl present, the end point is now the characteristic blue mentioned under Procedure.

DETERMINATION OF SMALL AMOUNTS OF LEAD

The determination of minute quantities of lead is required in baking powders canned goods and like products in which small amounts of lead are objectionable. Traces of lead ranging from 5 to 100 parts per million (0.0005 to 0.01% Pb) are best determined colorimetrically on 0.5 to 1 gram samples; larger amounts of lead should be determined gravimetrically.

Gravimetric Methods for Determining Traces of Lead

The determination of extremely small amounts of lead cannot be accomplished by the usual methods of precipitation, as the lead compounds remain in solution in a colloidal state. The addition, however, of certain substances, which form amorphous precipitates with the reagents used for throwing out lead causes the removal of lead from the solution by occlusion. For example, the addition of a sufficient quantity of a soluble salt of mercury, copper, or arsenic to a solution containing a trace of lead, and then saturating the solution with H_2S, will cause the complete removal of lead from the solution. Iron and alumina thrown out of the solution as hydroxides will carry down small amounts of lead, and completely remove it from the solution, if they are present in sufficient quantity. Lead may be extracted from finely pulverized substances by means of hot ammonium acetate and precipitated from the extract as lead sulphide. Advantage may be taken of these facts in determining traces of lead in presence of large amounts of other substances.

Amount of the Sample. It is advisable to have the final isolated lead compound over 0.01 gram in weight, hence, in a sample containing 10 parts of lead per million, 800 to 1000 grams of the material should be taken, since a kilogram of the material would contain 0.01 gram, Pb or 0.0156 gram $PbCrO_4$, or 0.01464 gram $PbSO_4$, or 0.0177+ gram $PbMoO_4$. Large samples should be divided into several portions of 100 to 250 grams each, the lead isolated in each, and the final extracts, containing the lead, combined. For the given amount of occluding agent, stated in the procedure, the treated portion should contain not over 0.01 gram lead.

I.—Extraction of Lead with Ammonium Acetate and Subsequent Precipitation

It is frequently desirable to extract the lead from the mass of material and precipitate it from the liquor thus obtained. The procedure worked out by the writer is applicable to determining traces of lead in aluminum salts, but with modifications may be applied to a wide range of substances.

Extraction of Lead. The desired weight of finely powdered substance, in 100-gram portions, is placed in 6-inch porcelain casseroles (1000 cc. capacity). To each portion are added, with vigorous stirring, 500 cc. of *lead-free*, boiling hot ammonium acetate solution (33%).[1] The reaction is apt to be energetic, so that

---

[1] The reagent must be boiling, when added, to obtain best results. Experiments have shown that considerable alumina and iron dissolve if the proportion of the reagent falls much below 5 cc. of 33% acetate per gram of sample. With twice this amount of reagent the extract is free from iron and alumina. Small amounts of alumina and iron, however, do not interfere in the lead determination.

care must be exercised to avoid boiling over. The residue from aluminum salts is crystalline and may be separated from the extract very readily by filtering through two filter papers in a large Büchner funnel and applying suction.[1] The residue is tamped down to squeeze out the adhering extract and washed with 100 cc. more of hot ammonium acetate followed by 100 to 200 cc. of hot water, again tamped down and sucked as dry as possible. The lead extracts are now combined and lead precipitated as sulphide.

The reagent is made by dissolving one part of lead-free ammonium acetate in two parts of distilled water. The purity of the reagent should be tested.

Precipitation of Small Amounts of Lead. To the solution containing lead is added 2–3 cc. of a 10% copper sulphate or cadmium sulphate reagent. Hydrogen sulphide is passed into the liquor until it is saturated. The copper or cadmium sulphide assists the settling of lead sulphide. Gently warming on the steam bath for half an hour coagulates the precipitate and facilitates settling. The liquor is decanted through a double filter in a small Büchner funnel and the residue washed onto the filter with water saturated with H_2S gas.

The precipitate is washed several times with ammonium sulphide to remove sulphides of the arsenic group and the residue then dissolved in a hot mixture of hydrochloric and nitric acids (1 part HCl. 5 parts HNO_3 and 15 parts H_2O). Ten cc. of strong sulphuric acid are added to the solution, and the mixture is evaporated to SO_3 fumes *but not to dryness*. The residue is taken up with 100–125 cc. of water containing 2 cc. of sulphuric acid and boiled to dissolve the soluble salts of iron, alumina, copper, etc. After cooling, one-third the volume of 95% alcohol is added (30–40 cc.), the lead sulphate allowed to settle for an hour or more, then filtered and washed several times with 30% alcohol. The residue is extracted with hot ammonium acetate and lead chromate precipitated from the filtrate,[2] made slightly acid with acetic acid, by adding 10 cc. of potassium dichromate reagent and boiling, according to the standard procedure. (Page 274.)

$$PbCrO_4 \times 0.641 = Pb.$$

II.—Precipitation of Lead by Occlusion with Iron Hydroxide

Wilkie found[3] that ferric hydroxide has the property of occluding lead, five parts of $Fe(OH)_3$ removing one part of lead from solution. Advantage is taken of this property of iron hydroxide in precipitating small amounts of lead.[4]

Procedure. The required amount of material is weighed out in 50-gram lots and brushed into No. 8 beakers. If the material contains organic matter, it is treated with 200-cc. portions of concentrated hydrochloric acid, the mixture heated just below the boiling-point of HCl solution, and potassium chlorate added, a few crystals at a time, until the organic matter is decomposed (hood). If the material dissolves in water, the water solution is treated with 5 cc. of concentrated hydrochloric acid and a few crystals of potassium chlorate and the liquor boiled.

[1] 200 to 300 grams of material may be handled in a 6-inch Büchner funnel.
[2] Should lead chromate fail to precipitate, the solution should be treated with H_2S to complete saturation, the sulphide collected on a filter, then dissolved in acid and the procedure described above repeated. If the solution still remains clear, the absence of lead is confirmed.
[3] J. M. Wilkie C. N., 2597, 117, 1909.
[4] Occlusion of lead by zinc sulphide, precipitated by H_2S from a formic acid solution, is suggested; iron and alumina would not interfere.

Addition of Ferric Iron. If sufficient iron is not already present, ferric chloride is added in such quantity that the iron content of the sample will be from twenty to fifty times that of the lead (larger amounts of iron will do no harm) present in the solution. Five to 10 cc. of concentrated nitric acid are added and the sample boiled for ten to fifteen minutes.

Precipitation of Iron and Lead. If alumina is present, iron is precipitated by addition of a large excess of potassium hydroxide, the alumina going into solution as potassium aluminate. In absence of alumina, ammonia may be used to precipitate the ferric hydroxide. Lead is completely occluded by the precipitate and carried down. The solution is filtered hot through Baker and Adamson's fast filters, threefold. The filtering must be rapid and the liquid kept hot to prevent clogging of the filters.

Separation of Lead from Iron. The precipitate is dissolved in hot hydrochloric acid (free from lead). The solutions are combined, if several portions of the sample are taken. Concentrated sulphuric acid is added and the sample evaporated to small volume and heated until the white sulphuric acid fumes appear. The usual procedure is now followed for separation of the lead sulphate, acetate extraction of lead and final precipitation of lead chromate.

$$PbCrO_4 \times 0.641 = Pb.$$

NOTE. In place of using alcohol to decrease the solubility of lead sulphate, many prefer to add sulphuric acid so that the acidity of the solution will be 2–10% free H_2SO_4.

III. Modification of Seeker=Clayton Method for Traces of Lead in Baking Powder

One hundred grams of baking powder are treated with 25 cc. of water followed by 75 cc. of strong hydrochloric acid added in small portions to avoid excess frothing. The mixture is heated until the starch has decomposed (iodine test gives blue color with starch), the solution becoming clear and turning yellow. The free acid is neutralized and when the solution is cold, 400 cc. of lead-free ammonium citrate, saturated with H_2S, are added. Additional H_2S is passed into the slightly alkaline solution, the sulphides of iron and lead allowed to settle, the clear supernatant liquor decanted off, the sulphides collected on a filter and washed. The precipitate is dissolved in nitric acid, lead separated as a sulphate, extracted with acetate and precipitated as dichromate according to the procedure recommended under the acetate extraction method I.

COLORIMETRIC ESTIMATION OF SMALL AMOUNTS OF LEAD

Introduction. Estimation of small amounts of lead by the intensity of the brown coloration produced by the sulphide in colloidal solution was first proposed by Pelouze. The procedure was modified by Warington and by Wilkie to overcome the color produced by accompanying impurities, among these, of iron, which is almost invariably associated with lead. The method is useful in determining traces of lead in drinking water, in food products, baking powders, canned goods, phosphates, alums, acids such as sulphuric, hydrochloric, citric, tartaric and the like. By this procedure on a gram sample one part of lead per million may be detected and as high as 50 parts may be estimated. For larger amounts of lead, a smaller sample must be taken. Nickel, arsenic, antimony, silver, zinc, tin, iron, and alumina, present in amounts such as commonly occur in these materials, do not interfere.

In order to obtain accurate results it is necessary to have the solutions under comparison possess the same general character. " It must be remembered that the tint depends to a large extent on the size of the colloidal particles of lead, which in turn depend upon the nature of the salts in the solution and upon the way that the solution has been prepared." Vigorous agitation, salts of the alkalies and alkaline earths tend to coagulate the colloidal sulphide.

Reagents Required. *Standard Lead Solution.* A convenient solution may be made by dissolving 0.1831 gram of lead acetate, $Pb(C_2H_3O_2)_2 \cdot 3H_2O$ in 100 cc. of water, clearing any cloudiness with a few drops of acetic acid and diluting to 1000 cc. If 10 cc. of this solution is diluted to 1000 cc. each cc. will contain an equivalent of 0.000001 gram Pb.

Harcourt suggests a permanent standard made by mixing ferric, copper and cobalt salts. For example 12 grams of $FeCl_3$ together with 8 grams of $CuCl_2$ and 4 grams of $Co(NO_3)_2$ are dissolved in water, 400 cc. of hydrochloric acid added and the solution diluted to 4000 cc. 150 cc. of this solution together with 115 cc. of hydrochloric acid (1 : 2) diluted to 2000 cc. will give a shade comparable to that produced by the standard lead solution above, when treated with the sulphide reagent. The exact value per cc. may be obtained by comparison with the lead standard.

Alkaline Tartrate Solution. Twenty-five grams of C.P. sodium potassium tartrate, $NaKC_4H_4O_6 \cdot 4H_2O$, is dissolved in 50 cc. of water. A little ammonia is added and then sodium sulphide solution. After settling some time the reagent is filtered. The filtrate is acidified with hydrochloric acid, boiled free of H_2S and again made ammoniacal and diluted to 100 cc.

Ammonium Citrate Solution. Ammonium citrate solution is prepared in the same way as the tartrate solution above, 25 grams of the salt being dissolved in 50 cc. of water.

Potassium Cyanide. Ten per cent solution. The salt should be lead-free.

Sodium Sulphide. Ten per cent solution, made from colorless crystals. Sodium sulphide may be made by saturating a strong solution of sodium hydroxide with hydrogen sulphide gas, and then adding an equal volume of the sodium hydroxide. The solution is diluted to required volume, allowed to stand several days, and filtered.

Sodium metabisulphite. Solid salt of $Na_2S_2O_5$.

Apparatus. The color comparison may be made in Nessler tubes, or in a colorimeter. The Campbell and Hurley modification of the Kennicott-Sargent colorimeter is excellent for this purpose,[1] Fig. 43. The colorimeter is simple in construction and operation.

The tubes for holding the solutions to be compared are those of one of the well-known colorimeters, in which the unknown solution is placed in the left-hand tube while the color is matched by raising or lowering the level of a standard solution in the right-hand tube by means of a glass plunger working in an attached reservoir.

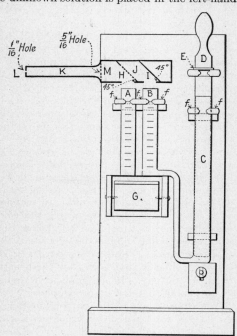

The accompanying d i a g r a m shows the essential features of construction of the colorimeter employed in the tests described below. The unknown solution is placed in the left-hand tube A, which is 19 cm. long, 3 cm. in diameter, and graduated for 15 cm. The standard solution is placed in the right-hand tube B, which is the same size as A, the graduated portion being divided into 100 divisions of 1.5 mm. each. The tube B is permanently connected by a glass tube with the reservoir C in which the glass plunger D works, so that the level of the liquid in B can be readily controlled by raising or lowering the plunger. As the tube B and reservoir C are made in one piece, the liquid used for the standard solution comes in contact with glass only, thus preventing any possibility of

FIG. 43.—Hurley's Colorimeter.

chemical change due to contact with the container. The plunger is provided with a rubber collar E, so placed as to prevent the plunger from accidentally striking and breaking the bottom of the reservoir. The tubes A and B, with the connecting reservoir, rest on wooden supports, the one under A and B being provided with holes for the passage of the light, and are held in position by spring clips F F. This arrangement allows the glass parts to be readily removed for cleaning and filling. The light for illuminating the solution is reflected upward through the tubes A and B by means of the adjustable mirror G. The best results are obtained by facing the colorimeter toward a north window in order to get reflected skylight through the tubes, care being taken to avoid light reflected from adjacent objects. The black wooden back of the colorimeter serves the double purpose of a support for the parts of the instrument and of a screen, as it is interposed between the color tubes and the source of light.

The light, passing upward through the tubes A and B, impinges on the two mirrors H and I cemented to brass plates sliding in grooves cut at an angle of 45° in the sides of the wooden box J. This box is supplied with a loosely-fitting cover, thus allowing easy access for the purpose of removing and cleaning the

[1] Jour. Am. Chem. Soc., **33**, 1112, July, 1911.

mirrors. The mirror H is cut vertically and cemented in such a position as to reflect one-half of the circular field of light coming through the tube A. The light passing upward through B is reflected horizontally by the mirror I, through a hole in the brass plate supporting the mirror H. One-half of the circular field of light from the tube B is cut off by the mirror H, the vertical edge of which acts as a dividing line between the two halves of the circular field. The image of one-half of the tube B is then observed in juxtaposition to the opposite half of the image of the tube A.

The juxtaposed images are observed through a tube K, 2.5 cm. in diameter and 16 cm. long, lined with black felt and provided with an eye-piece having a hole 1.5 mm. in diameter. At the point M in the tube K is placed a diaphragm having an aperture 8 mm. in diameter. All parts inside the box J except the mirrors are painted black so that no light except that coming through the tubes A and B passes through the tube K. By having the apertures in the eye-piece and diaphragm properly proportioned only the image of the bottoms of the tubes A and B can be seen, thus preventing interference of light reflected from the vertical sides of the tubes A and B.

A person looking through the eye-piece observes a single circular field divided vertically by an almost imperceptible line when the two solutions are of the same intensity. By manipulating the plunger D, the level of the liquid in B can be easily raised or lowered, thus causing the right half of the image to assume a darker or lighter shade at will. In matching colors with an ascending column in B, that is, gradually deepening the color of the right half of the field, the usual tendency is to stop a little below the true reading while in a comparison with a descending column the opposite is the case.

Procedure. If lead is between 10 to 50 parts per million a 1-gram sample is taken. If it is above or below these extremes the amount of sample is regulated accordingly. In materials containing organic matter it is not advisable to take more than a 1-gram sample.

Substances containing organic matter, such as starch in baking powder, should be decomposed by fusion with sodium peroxide, sodium or potassium sulphate containing a few drops of sulphuric acid. A Kjeldahl digestion with concentrated sulphuric acid and potassium bisulphate may occasionally be advisable. Sulphuric acid discolored by organic matter should be mixed with 4 to 5 grams of potassium bisulphate, taken to fumes and then diluted with water. The material may be extracted with ammonium acetate and lead determined in the extract. See notes.

To the solution containing the sample are added 10 cc. of tartrate solution (or 20 cc. of citrate solution with phosphates of lime, etc.), 10 cc. of hydrochloric acid and the mixture brought to boiling. Small amounts of ferric iron are now reduced by adding 0.5 gram sodium metabisulphite. Sufficient ammonium hydroxide is added to neutralize the free acid and 5 cc. in excess; then 3 cc. potassium cyanide (to repress any copper color that may be present to reduce higher oxides), and the mixture heated until the solution becomes colorless. The entire solution or an aliquot portion is placed in the comparison cylinder, and diluted to nearly 100 cc. If the Kennicott-Sargent apparatus is used the standard color solution is forced into the adjacent cylinder, until the color in this cylinder matches the one containing the sample. The number of cc. of the standard is noted. This blank is due to the slight color that the solutions of the samples invariably have. Four drops of the sulphide reagent are added to the sample and this is mixed by means of a plunger, avoiding any more agitation than is absolutely necessary to make the solution homogeneous. After one minute the comparison is again made, the colored standard being forced into the cylinder until its color matches the sample. It is advisable to take several readings with ascending and descending column of standard reagent, taking the average as the true reading.

Calculation. Suppose the standard $= 0.000001$ gram Pb per cc., blank $= 5$ cc., total reading $= 22$ cc., one gram of sample being taken for analysis. Then $22 - 5 = 17$ cc. $= 0.0017\%$ Pb or 17 parts per million.

NOTES. Iron must be completely reduced before adding ammonium hydroxide and potassium cyanide.

Allen's method of reducing iron with sodium metabisulphite is excellent. The salt may be made by passing SO_2 into a saturated solution of sodium carbonate at boiling temperature, until the liquor is just acid to methyl orange. The water evaporated during the treatment is replaced during the action. $Na_2S_2O_5$ separates and may be filtered off and the water removed by centrifuging.

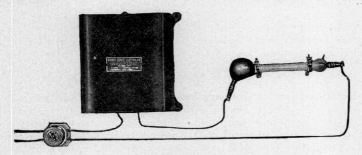

FIG. 44.—Cooper Hewitt Mercury Light.

The Cooper Hewitt Mercury light is excellent for colorimetric lead determinations, where an artificial light is desired. The yellow shades appear yellowish-green and may be matched more readily than the yellows obtained by daylight.

The illustration, Fig. 44, shows the type of light recommended for this work.

If a separation from iron is desired, the lead may be extracted with ammonium acetate solution. Ten grams of the powdered material are mixed with 75 cc. of a 33% ammonium acetate solution [1] (25 grams of the salt dissolved in 50 cc. H_2O), the reagent being added boiling hot. The mixture is diluted to 500 cc., a portion filtered, and the determination made on an aliquot part of the total, following the directions above.

NOTES. Nickel up to 0.1%, arsenic up to 0.2%, zinc 0.2%, antimony 0.05%, copper 0.25%, iron up to 1.0%, aluminum 10%, tin up to 1.4% do not interfere with the colorimetric determination of lead.

[1] The ammonium acetate should be free from lead.

STANDARD METHODS OF CHEMICAL ANALYSIS OF PIG LEAD [1]

A. Determination of Silver

Weigh 100 grams of the sample into a 3-inch scorifier and heat in a muffle furnace until the assay " covers." Pour into an iron mold and allow to cool. Free the resulting lead button from PbO, scorify again and pour as before. The button last obtained should not weigh over 20 grams and can be cupelled directly. Weigh the silver button obtained upon an assay balance.

NOTE. If the amount of silver is large, the button should be examined for gold, in the usual manner.

B. Determination of Bismuth

Solutions Required

Sodium Carbonate. Dissolve 100 grams Na_2CO_3 in a liter of distilled water.

Ammonium Carbonate. Make a half saturated solution.

Alkaline Sulphide Wash Solution. Dissolve 200 grams of KOH in a liter of distilled water and mix one part of this solution with 4 parts of H_2S water.

Method

(a) **Determination for Ordinary Amounts of Bismuth.** Dissolve 20 grams of the sample in a 400-cc. beaker with 100 cc. of HNO_3 (1 : 4), with the aid of heat. When solution is complete, add dilute ammonia (1 : 2) with constant stirring, drop by drop from a burette, until a faint opalescence appears. If an actual precipitate is formed, redissolve by the addition of a small amount of HNO_3 (1 : 4) and repeat the addition of ammonia. Now add 5 cc. of HCl (1 : 9), fill the beaker with hot water, bring to boiling, and allow to stand on a steam bath for two hours. The assay, while standing, must not reach the boiling temperature. Filter through a 7-cm. paper. Transfer the precipitate completely to the paper by means of a " policeman " and wash twice with hot water. Carefully examine the filtrate, washings, and any decanted liquid and reject if clear. Dissolve the precipitate by dropping around the edges, from a 5-cc. pipette, 5 cc. of boiling HCl (1 : 9), receiving the solution in the original beaker. Wash the paper thoroughly with hot water, fill the beaker with water, bring to boiling, and allow to stand as before. Filter the bismuth oxychloride upon a weighed Gooch crucible, wash thoroughly with water, once with alcohol, once with ether, and dry upon the hot plate. Cool and weigh.

NOTE. If time permits, it is convenient to allow the assay to stand over night. In that case, the precipitate of bismuth oxychloride generally settles so completely that the clear supernatant solution can be decanted.

(b) **Determination for Amounts of Bismuth Smaller Than Can Be Determined by (a).** Dissolve 100 grams of the sample in 500 cc. of dilute HNO_3 (1 : 4). When solution is complete, allow to cool and add Na_2CO_3 solution little by little until a slight permanent precipitate has formed. Then add 50 cc. of the Na_2CO_3 solution, bring to boiling, allow to stand warm until the

---

[1] Reprinted by permission from the 1921 Book of the A. S. T. M. Standards of the American Society for Testing Materials.

supernatant liquor is clear again, filter and reject the filtrate. Dissolve the precipitate without washing by slowly pouring hot HNO_3 (1 : 4) around the edges of the filter paper, using no more acid than necessary. Wash the paper once with hot water and determine bismuth in the filtrate as described in (a).

NOTE. When the sample contains a small amount of bismuth, it is often difficult to ascertain when the correct amount of ammonia has been added to the nitrate solution. In this case, place a small piece of litmus paper in the solution and add the ammonia very gradually until the litmus paper indicates a neutral reaction.

(c) **Determination of Bismuth in Samples Containing Appreciable Amounts of Tin and Antimony.** Use a piece of litmus paper as in Note under (b), and after the first precipitation of bismuth oxychloride has been filtered off and washed (see (a)), proceed as follows: Dissolve the bismuth by dropping around the edges of the paper 10 cc. of boiling HCl (1 : 2), receiving the solution in the original beaker. Discard the paper after washing. Dilute the solution to about 200 cc. with fresh H_2S water and then pass H_2S gas through the hot solution for fifteen minutes. Filter and wash with hot water. Remove any tin or antimony present by washing three times with alkaline sulphide wash solution. Wash the precipitate again with hot water, place it, together with the filter paper, in a 100-cc. beaker, add 20 cc. of HNO_3 (1 : 4), boil until sulphides are completely dissolved and the paper well pulped. Filter the solution, receiving the filtrate in the original beaker, and wash well. Determine bismuth in the filtrate as in (a).

NOTES. When the sample contains comparatively large amounts of tin or antimony, the residue left after the solution of the lead in the nitric acid obscures the opalescence found upon the addition of ammonia.

If the original sample contains more than 0.25 per cent of bismuth, it is preferable to use only a 10-gram charge.

C. Determination of Arsenic

Method

Dissolve 111.11 grams of the sample in 550 cc. of HNO_3 (1 : 4). When solution is complete, wash into a graduated liter flask. Add 75 cc. of H_2SO_4 (1 : 1), cool, and make up to the mark with water. Transfer to a large beaker, rinsing out the flask with 25 cc. of water. Mix thoroughly, allow to settle and filter off 900 cc., equivalent to a 100-gram charge. Evaporate in a large porcelain dish until only enough H_2SO_4 is left to moisten the residue. When cool, wash into a small distilling flask with 60 cc. of HCl (specific gravity 1.20) and 20 cc. of water, cleaning the dish carefully. Add 10 grams of ferrous sulphate and distill, boiling to as small volume as possible. When cool, add 50 cc. more HCl (specific gravity 1.20) and redistill. Pass H_2S gas through the cold distillate for 45 minutes. Filter, and weigh the As_2S_3 on a Gooch crucible, washing in cold water, alcohol and CS_2. After drying and weighing, redissolve with $(NH_4)_2CO_3$ solution and reweigh the Gooch crucible, calculating the loss in weight to arsenic.

NOTE. The 25 cc. of water added is equivalent to the volume of precipitated lead sulphate.

D. Determination of Remaining Metals

Solutions Required

Tartaric Acid. Dissolve 50 grams of tartaric acid in 250 cc. of distilled water to which has been added 250 cc. HCl (specific gravity 1.20).

Method

Dissolve 222.23 grams of the sample in 1100 cc. of HNO_3 (1 : 4), using a 1300-cc. beaker. When solution is complete, examine for color and turbidity. If clear, wash the solution at once into a 2000-cc. graduated flask. In case of a residue (Note 1), however, dilute to about 1100 cc. and allow to stand until the supernatant liquor is clear. Decant as much as possible into a 2000-cc. flask, filter the remainder and receive the filtrate in the same flask. Wash the precipitate well and then place it, together with the filter paper, in a 100-cc. beaker and add 20 cc. of the tartaric acid mixture. Heat to boiling and when the paper is well pulped allow to digest warm for 30 minutes. Now add 50 cc. of hot water, filter and wash. (Note 2.) Carefully dry the residue and ignite. If any appreciable residue remains, brush it into a small silver dish containing 1 gram of molten KOH. Fuse for 5 minutes, and after cooling dissolve in as little hot water as possible and add to the above-mentioned tartaric acid filtrate. Render this solution just alkaline with ammonia and then just acid with HCl, and saturate hot with H_2S gas. After digesting for 30 minutes on the steam bath, pass H_2S through the solution again for 15 minutes. Filter and wash with slightly acidified H_2S water. Reject the filtrate. Wash the sulphides from the paper into the original beaker and add 5 cc. of KOH solution (1 : 5) for every 25 cc. volume present. Digest hot for 5 minutes and filter through the original paper into a small flask graduated to 110 cc. After washing with H_2S water containing a little of the KOH solution, cool the filtrate and make up to the mark. Mix and reserve 100 cc. as alkaline sulphide solution No. 1. The precipitate may be discarded.

Add slowly to the main solution in the 2000-cc. flask 150 cc. of H_2SO_4 (1 : 1). After cooling and filling up to the mark, pour into a clean 3-liter flask provided with a rubber stopper. Rinse the flask out with 50 cc. of water, which is equivalent to the volume of lead sulphate present and is added to the portion. After mixing thoroughly by shaking, allow the precipitate to settle and filter off 1800 cc. of the liquid. This is equivalent to a 200-gram charge. Place this in a No. 9 porcelain evaporating dish and evaporate, first over a free flame and later on the hot plate until only enough H_2SO_4 is left to moisten the residue remaining. Add 50 cc. of water and, after digesting warm for a short time, wash the solution into a 250-cc. beaker, cleaning the dish carefully. Allow the solution to digest on the steam bath for 4 or 5 hours. (Note 3.) Then filter, wash and evaporate the filtrate to 200–250 cc. Place any residue (Note 4), together with the filter paper, in a 100-cc. beaker and treat with 20 cc. of the tartaric acid mixture. Boil for 5 minutes, dilute with 50 cc. of hot water and filter. Make the filtrate alkaline with ammonia and just acid with HCl, and obtain the tin, antimony and arsenic as previously described, reserving the whole of the alkaline sulphide solution as solution No. 2. Reject the sulphide residue.

To filtrate from the lead sulphate, add ammonia until the neutral point is reached, and then for every 50 cc. of the solution present add 2 cc. of HCl (specific gravity 1.20). Pass H_2S gas into the hot solution until saturated, digest for 30 minutes on the steam bath and again pass H_2S gas into the solution. Filter and wash with H_2S water slightly acidified. (Note 5.) Separate tin, antimony and arsenic in the precipitate with KOH solution as usual, obtaining an alkaline sulphide solution No. 3. (Note 6.)

To sum up, three alkaline sulphide solutions have been obtained, containing tin, antimony and arsenic, a precipitate of metallic sulphides containing copper, lead, etc., and a solution containing iron, zinc, nickel, etc.

NOTES. 1. A residue indicates the presence of antimony, tin possibly arsenic, or sulphur as lead sulphate.

2. It has been found that even this treatment occasionally fails to dissolve stannic acid completely.

3. If it is preferred, allow solution to stand over night to insure the complete solution of all soluble salts.

4. Any residue of lead sulphate may contain some tin, antimony, or possibly arsenic.

5. The filtrate will contain any iron, zinc, nickel, cobalt and manganese; while in the precipitate will be found any copper, cadmium, lead, silver, bismuth, tin, antimony and arsenic.

6. When separating the sulphides of arsenic, antimony and tin from sulphides of copper, lead, etc., it is necessary to wash all the sulphides back into the beaker in which they were precipitated. These sulphides sometimes cling so tenaciously to the paper that in dislodging them more water than the 25 cc. specified is required. In this case allow the sulphides to settle and then decant the clear supernatant liquor through the filter until the volume is reduced to 25 cc. Before rejecting decanted fluid always test with H_2S water.

In washing sulphide precipitates with water, much trouble is experienced from the tendency of the precipitate to pass through the filter in the colloidal form. This is particularly true in washing sulphides that have been digested with KOH. Time and trouble will be saved by washing all the sulphides precipitated from mineral acid solutions with H_2S water containing a little of the acid in which they were precipitated. The same is true of sulphides precipitated in or filtered from alkaline solution.

When working with alkaline solutions in which tin is to be determined, avoid the use of Jena or other glass that contains zinc. The zinc content of the glass may influence the result.

Determination of Antimony and Tin

Solutions Required

Potassium Iodide. Dissolve 100 grams in a liter of distilled water.

Standard Sodium Thiosulphate. Dissolve 24.8 grams of $Na_2S_2O_3.5H_2O$ in 1000 cc. of distilled water, and allow to stand for 24 hours. Standardize against Antimony Metal, c. p., using same quantity of reagents and same procedure as under method. Each cubic centimeter is equivalent to approximately 0.006 gram of antimony.

Method

Wash the alkaline sulphide solutions Nos. 1, 2, and 3 into a 600-cc. beaker, and acidify with 5 cc. of HNO_3 (specific gravity 1.42) and 20 cc. of HCl (specific gravity 1.20). Evaporate the solution to dryness on the steam bath. Dissolve the residue in 200 cc. of water, add 10 grams of oxalic acid and 10 grams ammonium oxalate and heat the solution until it is clear. Then pass H_2S gas through the boiling solution for 45 minutes. Filter off the precipitate, consisting of arsenic and antimony sulphides, and wash with hot water. Determine tin electrolytically in the filtrate, continuing the electrolysis until all the oxalic acid is decomposed and the solution becomes alkaline. Dissolve the deposit on the cathode with a small amount of HCl and examine qualitatively for tin.

Dissolve the sulphides of arsenic and antimony in KOH as usual, collecting the filtrate in a 500-cc. Erlenmeyer flask. Add 50 cc. of HCl (specific gravity 1.20) and boil the solution until about 30 cc. are left. Expel the arsenic as chloride. Now oxidize the solution with a pinch of $KClO_3$ and boil until no more chlorine remains. Cool and add 5 cc. potassium iodide solution. Titrate the liberated iodine with N/10 sodium thiosulphate solution, using carbon disulphide as an indicator.

Determination of Copper and Cadmium

Solutions Required

Sodium Chloride (Solution No. 1). Dissolve 1 gram NaCl in 100 grams distilled water.

Sodium Chloride (Solution No. 2). Dissolve 10 grams NaCl in 100 grams distilled water.

Standard Potassium Cyanide. Dissolve 2 grams of KCN in a liter of distilled water and standardize against a known amount of copper as treated in the analysis.

Sodium Carbonate (Solution No. 1). Dissolve 50 grams Na_2CO_3 in 1 liter of distilled water.

Sodium Carbonate (Solution No. 2). Dissolve Na_2CO_3 in distilled water to saturation.

Method

(a) **Determination Where Copper Exceeds 0.0025 Per Cent.** Place the filter containing the sulphides in a 100-cc. beaker and add 20 cc. of HNO_3 (1 : 4). Heat with occasional stirring until the paper is thoroughly pulped and the sulphides are completely dissolved. Filter into a 250-cc. beaker. Dry the residue, which generally contains a small amount of copper, ignite in a porcelain crucible, boil with 5 cc. of HNO_3 (1 : 1), and wash into the main portion, keeping the volume below 100 cc. Render it strongly alkaline with ammonia, and add 5 grams of potassium cyanide, then saturate it in the cold with H_2S gas. (Note 1.) Filter the solution and evaporate the filtrate to a volume of 20 to 30 cc. in a 4-inch porcelain casserole. Boil until solution is complete. Add 20 cc. of H_2SO_4 (1 : 1) and evaporate the solution under a hood until dense fumes of H_2SO_4 escape. Cool, dilute, and warm until copper sulphate is all dissolved. Now filter, if necessary, into a 200-cc. beaker, render just alkaline with ammonia, make acid by the addition of 3 cc. of HNO_3 per 100 cc. solution, and electrolyze for copper.

Dissolve the precipitate of sulphides in the usual manner with 20 cc. of HNO_3 (1 : 4). Add 1 cc. NaCl solution No. 1 to the solution, still containing the pulped filter, and digest for one-half hour. Filter off the AgCl, wash and reject. Make the filtrate, not exceeding 100 cc., alkaline with a slight excess of Na_2CO_3 and add 5 grams of KCN. Digest for 1 hour. Filter and wash with Na_2CO_3 solution No. 1. Reject the precipitate of bismuth. Now add a few cubic centimeters of ammonium sulphide solution to the filtrate to precipitate any cadmium as yellow cadmium sulphide. Filter upon a weighed Gooch crucible and weigh as cadmium sulphide. (Note 2.)

NOTES. 1. Copper remains in the solution while lead, silver and bismuth and cadmium are precipitated.

2. If an appreciable amount of cadmium sulphide is found, it should be converted

to and weighed as cadmium sulphate, according to method for cadmium under Standard Methods of Chemical Analysis of Spelter (see page 130) of the American Society for Testing Materials.

(b) Determination Where Copper Is Less Than 0.0025 Per Cent. Place the paper containing the sulphides in a porcelain crucible, dry carefully and ignite. When the carbon has been all burned off, cool and dissolve the residue in 5 to 10 cc. of HNO_3 (1 : 1). After evaporating to a volume of 1 to 2 cc., add 1 cc. of H_2SO_4 (1 : 1). Then evaporate the solution until fumes appear, cool, dilute, add a few drops of NaCl solution No. 2, and filter off the lead sulphate and silver chloride. Again evaporate the filtrate until fumes of H_2SO_4 appear, and when cold, dilute and neutralize with Na_2CO_3 solution No. 2. Then add about six drops of concentrated ammonia and titrate the solution with standard KCN solution until the blue color is discharged.

The cadmium can be obtained by making this solution, titrated for copper, strongly alkaline, diluting a little, and adding 5 grams of KCN. Saturate the solution cold with H_2S gas, filter, discard the filtrate and treat the precipitate for cadmium as described in Method (a).

Determination of Iron

Method

Evaporate the filtrate containing iron, zinc, etc., to 100 cc. and oxidize with a few drops of HNO_3. Separate the iron with ammonia as usual, making two separations, and receive the filtrate in a 500-cc. Erlenmeyer flask. Redissolve the iron hydroxide with hot HCl (1 : 1) or dilute H_2SO_4 and determine the iron volumetrically by any of the standard methods.

Determination of Zinc in Pig Lead

Solutions Required

Nitric Acid (1 : 4). Mix 200 cc. of HNO_3 (sp. gr. 1.42) with 800 cc. of distilled water.

Sulphuric Acid (1 : 1). Carefully pour, with stirring, 500 cc. of H_2SO_4 (sp. gr. 1.84) into 500 cc. of distilled water.

Acidulated Hydrogen Sulphide Water. Add 20 cc. of HCl (sp. gr. 1.19) to 1000 cc. of distilled water and saturate with hydrogen sulfide.

Ammonium Thiocyanate Solution (2%). Dissolve 20 g. of NH_4CNS in 1000 cc. of distilled water.

Hydrochloric Acid (1 : 3). Mix 100 cc. of HCl (sp. gr. 1.19) and 300 cc. of distilled water.

Standard Zinc Solution (0.1 mg. of zinc per cc.). Dissolve exactly 0.1 g. of U. S. Bureau of Standards pure zinc in 5 cc. of HCl (sp. gr. 1.19) and dilute to exactly 1000 cc. with distilled water.

Potassium Ferrocyanide Solution. Dissolve 34.8 gr. of $K_4Fe(Cn)_6.3H_2O$ in 1000 cc. of distilled water.

Method

Dissolve 222.23 g. of the sample in 1100 cc. of HNO_3 (1 : 4), using a 1300-cc. beaker. When the lead is dissolved, transfer the solution to a 2000-cc. graduated flask and add slowly 150 cc. of H_2SO_4 (1 : 1). Cool, fill the flask to the mark and then pour the solution into a clean 3000-cc. flask provided with a rubber stopper. Rinse the measuring flask with exactly 50 cc. of water, which is equivalent to the volume of lead sulfate which is present. Mix the solution thoroughly by shaking, allow the precipitate to settle and filter through a dry filter until 1800 cc. of filtrate has been obtained.

Place exactly 1800 cc. of filtrate (equivalent to a 200-g. charge) in a No. 9 porcelain evaporating dish and evaporate the solution to approximately 100 cc. Transfer the solution to a 600-cc. beaker, neutralize with ammonia, and then add 5 cc. of HCl (sp. gr. 1.19) for every 100 cc. of solution. Warm the solution and pass in a rapid current of hydrogen sulphide until it is saturated. Digest for 30 minutes on the steam bath, add an equal volume of water and again saturate with hydrogen sulfide. Filter and wash with acidulated H_2S water.

Discard the precipitate and evaporate the filtrate in glassware containing no zinc (such as Pyrex) until the volume of the solution is approximately 100 cc.

Neutralize the solution with ammonium hydroxide, add 5 g. of citric acid, and warm until the acid is dissolved. Add small portions of calcium carbonate to the hot citric acid solution until about 1 g. of calcium citrate has separated and then pass in a rapid current of H_2S as the solution is allowed to cool. Allow the solution to stand for from 2 to 4 hours, part of the time on a water bath, until the supernatant liquid is clear.

Collect the precipitate on a filter, wash with a 2% solution of ammonium thiocyanate and then dissolve the precipitate in hot dilute hydrochloric acid (1 : 3). If the solution has a reddish color (due to iron), the zinc must be reprecipitated as above. If the solution is clear, evaporate it to dryness on the steam bath, take up the residue in 3 cc. of HCl (sp. gr. 1.19), add 20 cc. of water and filter if not perfectly clear.

Transfer the solution (Note 2) to a 50-cc. Nessler jar and dilute to 45 cc. Prepare other Nessler jars containing 3 cc. of HCl (sp. gr. 1.19), definite volumes of standard zinc solution, and diluted to 45 cc. Add 5 cc. of potassium ferrocyanide solution to each jar, mix quickly, and compare the turbidities by viewing longitudinally as the jars are held over a sheet of fine print. Add more of the standard zinc solution from a burette to the jar which approximates the turbidity of the unknown most closely until the turbidities match each other and calculate the percentage of zinc on the basis of a 200-g. sample or the aliquot portion taken.

NOTES. 1. All glassware that contains zinc must be avoided and in umpire work a blank test should be carried along with the test.

2. The whole solution can be used if the lead contains no more than 0.002 per cent of zinc. If more zinc is present, it is best to take such an aliquot portion of the solution as will give approximately 4 mg. of zinc and then to add enough HCl to provide 3 cc.

3. For further details concerning the turbidmetric test, consult the "Determination of Small Quantities of Zinc" by Mr. Bodansky, in the Journal of Industrial and Engineering Chemistry, Vol. 13, pp. 696–697 (1921).

4. The addition of calcium carbonate with the formation of a precipitate of calcium citrate serves the purpose of giving a clear filtrate, and prevents the loss of colloidal sulphide.

Determination of Nickel and Cobalt

Solutions Required

Hydrogen Sulphide Wash Water. To each 100 cc. of hydrogen sulphide water add 20 cc. neutral ammonium acetate.

Method

Render the filtrate just alkaline with ammonia and saturate with H_2S. Heat to boiling and then make just acid with acetic acid, add 20 cc. of neutral ammonium acetate solution, and boil until the sulphides of nickel and cobalt separate out. Filter and wash with warm H_2S wash water. Dry the precipitate and paper in a porcelain crucible and carefully ignite. If there is an appreciable amount of residue after ignition, dissolve by boiling with 10 cc. of aqua regia, wash into a 250-cc. beaker, add 10 cc. of H_2SO_4 (1 : 1), evaporate until fumes appear, cool, dilute to 200 cc., make alkaline with ammonia, and add 15 cc. of concentrated ammonia. Then electrolyze the solution and weigh the nickel and cobalt as such.

NOTES. If the amount of nickel and cobalt is small, it can be weighed as oxide.

If the filtrate from the nickel and cobalt sulphides shows a brown color, it indicates that the precipitation has not been complete. In this case render the solution ammoniacal and repeat the above process.

Determination of Lead in Basic Carbonate of Lead

(Corroded White Lead)

Basic carbonate white lead ($2PbCO_3.Pb(OH)_2$) contains approximately 80 per cent metallic lead and 20 per cent carbonic acid and combined water, with traces of silver, antimony, lead, and other metals. The analysis of basic carbonate white lead can best be carried out by Walker's method.[1]

Analysis

Total Lead. " Weigh 1 gram of the sample, moisten with water, dissolve in acetic acid, filter, wash, ignite, and weigh the insoluble impurities. To the filtrate from the insoluble matter add 25 cc. of sulphuric acid (1 : 1), evaporate and heat until the acetic acid is driven off; cool, dilute to 200 cc. with water, add 20 cc. of ethyl alcohol, allow to stand for two hours, filter on a Gooch crucible, wash with 1 per cent sulphuric acid, ignite, and weigh as lead sulphate. Calculate to total lead ($PbSO_4 \times 0.68292 = Pb$) or calculate to basic carbonate of lead (white lead) by multiplying the weight of lead sulphate by 0.85258.

" The filtrate from the lead sulphate may be used to test for other metals, though white lead is only rarely adulterated with soluble substances; test, however, for zinc, which may be present as zinc oxide.

" Instead of determining the total lead as sulphate it may be determined as lead chromate by precipitating the hot acetic acid solution with potassium bichromate, filtering on a Gooch crucible, igniting at a low temperature, and weighing as lead chromate."

[1] P. H. Walker, Bureau of Chemistry Bulletin No. 109, revised, U. S. Dept. of Agriculture, pp. 21 and 22.

The Determination of True Red Lead [1]

Solutions Required

1. "*Red Lead Solution.*"—Weigh out into a 1300-cc. beaker, preferably Pyrex glass, 600 g. C.P. "Tested Purity" crystals sodium acetate (360 g. C.P. dried) and 48 g. C.P. potassium iodide. Make up a solution of 25% acetic acid by mixing 150 cc. C.P. glacial acetic acid with 450 cc. distilled water; 184 cc. 80% acetic acid with 416 cc. distilled water. Now pour about 500 cc. of this 25% acid into the 1300-cc. beaker above mentioned, reserving the remainder. Warm the beaker on the steam bath, stirring occasionally, until a clear solution is obtained. Cool this solution to room temperature and pour it into a 1-liter graduated flask. Then add enough of the reserved 25% acetic acid to make exactly 1000 cc. and mix thoroughly.

2. *1/20 N Sodium Thiosulphate Solution.*—Weigh out into a large beaker 25 g. C.P. crystals sodium thiosulphate, add 800–1000 cc. distilled water and stir until dissolved. Wash this solution into a 2-liter graduated flask, make up to the mark with distilled water and mix thoroughly.

Each 1 cc. of this solution is equivalent to about 1.74% true red lead (using 1 g. charge) but its exact value should be determined by standardizing against pure iodine, or better, against a standard sample of red lead, the "true red lead" content of which is accurately known. The strength of this solution gradually decreases upon standing. It is a good plan, therefore, when analyzing a sample of red lead, to run a parallel determination with the standard red lead, thus ascertaining the exact strength of the thiosulphate solution. This procedure consumes but little extra time and will often prevent error.

3. *Starch Indicator Solution.*—Weigh out 3 g. of ordinary starch (better soluble starch) into a small beaker, add about 50 cc. cold distilled water and mix to a thin paste. Measure into an 800-cc. beaker 500 cc. cold distilled water, heat to boiling, and pour into it slowly, with constant stirring, the previously prepared starch paste. Then boil for 2 minutes, stirring, and add 5–6 drops of oil of cassia. Allow to stand over night and decant the clear supernatant liquid into a stock-bottle. When prepared in this way the solution will keep indefinitely.

Method. Weigh 1 g. of sample into a 300-cc. Erlenmeyer flask. Add 25 cc. of distilled water and whirl the mixture into a smooth paste. Then add 30 cc. of the red lead solution and continue whirling until nearly all the red lead has been dissolved. Titrate off once with the 1/20 N thiosulphate, adding the latter rather slowly and keeping the liquid constantly in motion by whirling the flask. When the color of the assay has been reduced to a light yellow, examine it carefully for undissolved particles of red lead. If present, they can often be dissolved by shaking the flask for a short time, but if they dissolve too slowly, they should be crushed by rubbing with a glass rod flattened on end until completely dissolved. The rod should then be removed and washed with a few cc. of water. After the addition of thiosulphate has reduced the color of the assay to a very pale lemon tint and care has been taken to see that solution of the red lead is complete, add 2 cc. of the starch indicator solution. The assay should then turn blue. Now

---

[1] Contributed by F. E. Dodge, National Lead Co.

finish the titrating by adding the thiosulphate solution, drop by drop, shaking the flask very thoroughly after each addition, until the blue color disappears.

Calculation. Suppose, for instance, that a red lead sample, by this method, requires 24.8 cc. thiosulphate, standard red lead requires 26.5 cc. thiosulphate.

Then the per cent " true red lead "

$$\frac{92.00}{26.5} \times 24.8 = 86.1\% \text{ "true red lead,"}$$

92.00% Pb 304 being the true red lead contents of our standard.

NOTES. Before analyzing a course sample, "glassmakers'" red lead, for instance, it is necessary to grind the sample, thus rendering it more readily soluble. As a rule, it is not necessary to secure complete solution of the sample before beginning to titrate, for it will generally be found, by the time the titration is nearly finished, that the sample has entirely dissolved. Occasionally, however, a sample dissolves very slowly. In this case titrate until nearly all the free iodine has been used up. Then warm gently, not over 40–50° C., agitate until the solution is complete and finish the titration as usual. The loss of iodine, owing to its small concentration, will be inappreciable, provided that the heating is not too prolonged.

Never attempt to hasten solution of the sample by warming if much free iodine is present, as this will cause loss of iodine and, consequently, too low a result.

The temperature of the "red lead solution," when used, should be the same as normal room temperature, say 20°–25° C. (68–77° F.). If colder than this, lead iodide is apt to separate out. In this case, however, it may be redissolved by warming gently or adding a small amount of warm water.

The red lead solution should be kept in a moderately cool, dark place, but even then, however, it may gradually decompose, with liberation of iodine. The error thereby introduced in a determination is inappreciable, if the thiosulphate is freshly standardized, before use, against a standard sample of red lead, as recommended.

The main consideration, in this method, is to see that the determination is run in exactly the same way and under exactly the same conditions as the standardization of the thiosulphate, by means of the standard sample of red lead, then any small errors in the determination will be offset by similar errors in the standardization.

Theoretical factor for $N/10$ $Na_2S_2O_3$ to Pb_3O_4—.03425 per cc.
$N/20$ $Na_2S_2O_3$ to Pb_3O_4—.017125 per cc.

Modifications for Paste Red Lead or Red Lead from a Mixture with Linseed Oil

A modification of the method of solution in carrying out the determination for true red lead consists in the dispersion of the sample of red lead in a mixture of 7 parts chloroform and 3 parts glacial acetic acid before the addition of the regular "red lead solution." This is an excellent procedure to adopt, since it shortens the time of solution and titration of the sample remarkably. In following the regular procedure in the standard A. S. T. M. method, the red lead solution acts on the sample of red lead very slowly, and in fact the entire sample is not decomposed with resulting liberation of iodine until the titration is nearly completed, the entire operation taking from 10 to 15 minutes. With the modification proposed, the sample of red lead after dispersion in the chloroform and glacial acetic acid mixture reacts readily and completely almost at once with the red lead solution, and the entire operation of solution and titration takes only from 3 to 4 minutes. The use of this modification is particularly desirable with red lead after extraction from paste or paint.

NOTE. It would appear from this that the chloroform and acetic acid mixture is added before the 25 cc. of distilled water.

Determination of Lead Peroxide in Minium (Red Lead)

Lead per oxide is reduced by oxalic acid according to the reaction:

$$PbO_2 + H_2C_2O_4 = PbO + 2CO_2 + H_2O.$$

The peroxide reduced by an excess of oxalic acid is determined by titrating this excess with standard potassium permanganate.

Procedure.   About 0.2 g. minium is heated with 25 cc. of 2N HNO_3 (one volume of HNO_3, sp.gr. 1.2, to two volumes H_2O) and then 50 cc. N/5 oxalic acid added.   The solution is heated to boiling and the excess oxalic acid titrated with N/5 potassium permanganate.   1 cc. N/5 $KMnO_4$ = .0239 g. PbO_2.

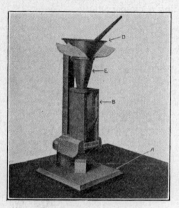

Fig. 44a.   Modified form of Scott Volumeter for Testing Lead Pigments.

Optional Method of Diehl Modified by Topf [1]

The method depends of the liberation of iodine from potassium iodide by lead peroxide, which in turn is reduced as shown in the reaction below.   The action is carried on in an acetic acid solution in presence of considerable alkali acetate.

$$PbO_2 + 4HI = I_2 + PbI_2 + 2H_2O.$$

Procedure.   About 0.5 gram of the material is dissolved by adding 1.2 g. of potassium iodide and 10 g. of sodium acetate in 5 cc. of 5 per cent acetic acid.   The solution is now diluted to about 25 cc. and the liberated iodine titrated with 0.1 N sodium thiosulphate.

1 cc. 0.1 N $Na_2S_2O_3$ titrates iodine equivalent to 0.01036 g. Pb.

NOTES.   The solution should be perfectly clear and no PbI_2 precipitate present. If present 3–4 g. sodium acetate are added and a few cc. of water.   The mixture is shaken until clear and then diluted to 25 cc.

[1] Diehl, Dingl. polyt. J., **246**, 196.   Topf, Z. anal. Chem., **26**, 296 (1887).

MAGNESIUM

Mg, *at.wt.* **24.32;** *sp.gr.* **1.69–1.75; m.p. 651°** [1]**;** *b.p.* **1120° C.** [2]**; oxide MgO.**

DETECTION

In the usual course of analysis magnesium is found in the filtrate from the precipitated carbonates of barium, calcium, and strontium. The general procedure for removal of the preceding groups may be found in the section on Separations given on the following page, 292. Magnesium is precipitated as white magnesium ammonium phosphate, $MgNH_4PO_4$, by an alkali phosphate, Na_2HPO_4, $NaNH_4HPO_4$, etc., in presence of ammonium chloride and free ammonia. The precipitate forms slowly in dilute solution. This is hastened by agitation and by rubbing the sides of the beaker during the stirring with a glass rod. Crystals soon appear on the sides of the beaker in the path of contact, and finally in the solution.

Baryta or lime water added to a solution containing magnesium produces a white precipitate of magnesium hydroxide.

Both the phosphate and the hydroxide of magnesium are soluble in acids.

ESTIMATION

The element is determined in the complete analysis of a large number of substances; in the analysis of ores, minerals, rocks, soils, cements, water, etc. The following are the more important ores in which the element occurs: Magnesite, $MgCO_3$; dolomite, $CaCO_3 \cdot MgCO_3$; kieserite, $MgSO_4 \cdot H_2O$; kainite, $MgSO_4 \cdot KCl \cdot 6H_2O$; carnallite, $MgCl_2 \cdot KCl \cdot 6H_2O$; in the silicates, enstatite, $MgSiO_3$; talc, $H_2Mg_3(SiO_2)_4$; meerschaum, forsterite, Mg_2SiO_4; titanate, $MgTiO_3$; olivine, $Mg_2SiO_4 \cdot Fe_2SiO_4$; serpentine, $H_4Mg_3Si_2O_4$. It occurs as boracite, $4MgB_4O_7 \cdot 2MgO \cdot MgCl_2$. It is found in sea-water, and in certain mineral waters. It occurs as a phosphate and carbonate in the vegetable and animal kingdoms, especially in seeds and bones.

Preparation and Solution of the Sample

In solution of the material it will be recalled that the metal is soluble in acids and is also attacked by the acid alkali carbonates. It is soluble in ammonium salts. The oxide, hydroxide, and the salts of magnesium are soluble in acids. Combined in silicates, however, the substance requires fusion with alkali carbonates to bring it into solution.

General Procedure for Ores. One gram of the ore is treated with 20 cc. of strong hydrochloric acid and heated gently until the material is decomposed. If sulphides are present, 5 to 10 cc. of strong nitric acid are added and the material decomposed by the mixed acids. If silicates are present and the decompo-

[1] Circular 35 (2d Ed.), U. S. Bureau of Standards.

Chapter contributed by Wilfred W. Scott.

sition is not complete by the acid treatment, the insoluble material is decomposed by fusion with sodium carbonate, or the entire sample may be fused with the alkali carbonate, the fusion is dissolved in hydrochloric acid and taken to dryness. Silica is dehydrated as usual by heating the residue from the evaporated solution. This is taken up with 50 cc. of water containing about 5 cc. strong hydrochloric acid, the silica filtered off and, after removal of the interfering substances according to procedures given under the next section on Separations, magnesium is determined as directed in the sections on Methods.

SEPARATIONS

Removal of Members of the Hydrogen Sulphide Group. Copper, Lead, Bismuth, Cadmium, Arsenic, etc. The filtrate from silica [1] is diluted to about 200 cc. and hydrogen sulphide gas passed in until the members of this group are completely precipitated. The sulphides are filtered off and washed with H_2S water and the filtrate and washings concentrated by boiling. This treatment is seldom necessary in analysis of many silicates and carbonates in which these elements are absent.

Removal of Iron, Aluminum, Manganese, Zinc, etc. The concentrated filtrate from the hydrogen sulphide group, or in case the treatment with hydrogen sulphide was not required, the filtrate from silica, is boiled with a few cc. of nitric acid to oxidize the iron (solution turns yellow), about 5 cc. of concentrated hydrochloric acid added, and if manganese is present, 15 to 20 cc. of a saturated solution of bromine water, and the solution made alkaline to precipitate iron, aluminum, manganese. If zinc, cobalt, and nickel are present, these are best removed as sulphides by passing hydrogen sulphide into the ammoniacal solution under pressure. (See Fig. 3 and Fig. 4, pages 42 and 43.)

Separation of Magnesium from the Alkaline Earths. The alkaline earths are precipitated either as oxalates, recommended when considerable calcium is present, or as sulphates, recommended in presence of a large proportion of barium, the magnesium salts being soluble. Magnesium is precipitated from the filtrates as a phosphate, according to directions given later. Details of the separation of magnesium from the alkaline earths may be found in the chapter on Barium, page 59.

An excellent procedure for the separation by means of sulphuric acid is to evaporate the solution to dryness, concentrating first in a porcelain dish and finally to dryness in a platinum dish, and then adding about 50 cc. of 80% alcohol and sufficient sulphuric acid to combine with the alkaline earths and magnesium, with slight excess. This precipitates barium, strontium, and calcium as sulphates, while the greater part of the magnesium is in solution. After settling, the precipitate is filtered and washed free of sulphuric acid by means of absolute alcohol, then with 40% alcohol to remove any magnesium sulphate remaining with the precipitate. Magnesium is determined in the filtrate by expelling the alcohol by evaporation, and then precipitating as magnesium ammonium phosphate according to directions given for the determination of this element.

NOTE. Magnesium is prevented from precipitation as a hydroxide by the presence of ammonium salts. See note, bottom of page 8.

[1] See previous paragraph.

GRAVIMETRIC DETERMINATION OF MAGNESIUM

Precipitation of Magnesium by a Soluble Phosphate as Ammonium Magnesium Phosphate

Magnesium is determined in the filtrate from calcium oxalate by the addition of sodium ammonium phosphate to a hot slightly acid or neutral solution followed by a definite amount of ammonia. The practice of precipitating magnesium from a cold solution necessitates a double precipitation as the composition of the phosphate is considerably modified by that of the solution in which the precipitation takes place, so that it is necessary to adjust conditions by having a definite amount of ammonia, ammonium salts and phosphate for the approximate amount of magnesium present.[1] Accurate results are obtained by precipitation of the compound from a hot solution by the method of B. Schmitz,[2] by addition of the soluble phosphate to a slightly acid solution and then making ammoniacal, or that of W. Gibbs,[3] by precipitation of the amorphous magnesium hydrogen phosphate in a neutral solution and transforming the precipitate to magnesium ammonium phosphate by addition of ammonia to the hot solution. Upon ignition of the precipitate, magnesium pyrophosphate ($Mg_2P_2O_7$) is formed.

Reactions.

A. $Na_2NH_4PO_4 + MgCl_2 = 2NaCl + MgNH_4PO_4$ (B. Schmitz).

B. $NaHNH_4PO_4 + MgCl_2 = NaCl + NH_4Cl + MgHPO_4$ and

$MgHPO_4 + NH_3 = MgNH_4PO_4$ (W. Gibbs).

Decomposition with Heat.

$2MgNH_4PO_4 = 2NH_3 + H_2O + Mg_2P_2O_7$.

The following procedure gives accurate results.

Procedure. The neutral or slightly acid solution, containing magnesium in presence of ammonium salts, is heated to boiling and treated, drop by drop, with an excess of sodium or ammonium phosphate, or microcosmic salt (10% solutions), stirring constantly during the addition. Then ammonium hydroxide (sp.gr. 0.96) is added, its volume measuring one-third that of the magnesium solution. The crystalline precipitate is allowed to cool and settle for two hours or more. The supernatant liquid is filtered off, the precipitate washed by decantation two or three times, then transferred to the filter, using dilute ammonia water (2%). The precipitate is dried and then transferred as completely as possible to a weighed platinum crucible, the ash of the filter paper, burned separately, is added and the compound heated gently at first, the crucible being covered until the ammonia is driven off, and then more strongly until the mass is snow white. The residue is cooled in a desiccator and weighed as $Mg_2P_2O_7$. The ammonium magnesium phosphate may be filtered directly into a weighed Gooch crucible and ignited, thus avoiding the carbon of the filter paper, and shortening the period of ignition, less heat being required to obtain the white magnesium pyrophosphate.

[1] F. A. Gooch and M. Austin, Am. Jour. Sci. (4), **7**, 187, 1899. W. Gibbs, C. N. **28**, 51, 1873. H. Struve, Zeit, anal. Chem., **36**, 289, 1897.
[2] Z. anal. Chem., **512**, 1906.   [3] Am. Jour. Sci. (3), **5**, 114, 1873.

Factors.[1] $Mg_2P_2O_7 \times 0.3621 = MgO$ or $0.2184 = Mg$ or $\times 0.7572$
$\qquad = MgCO_3$ or $\times 1.0811 = MgSO_4$ or $\times 2.2143 = MgSO_4 \cdot 7H_2O$.

Notes on Magnesium

The ignition is conducted gently at first to gradually oxidize the carbon that the precipitate contains. With rapid ignition the particles are inclosed in the mass in a form that it is almost impossible to completely oxidize, so that the final residue is gray instead of white. L. L. de Koninck [2] considers that the blackening of the precipitate is frequently due to the presence of organic bases in commercial ammonia and its salts, rather than to the fibers of filter paper occluded in the mass. With caution, the filter and residue may be ignited wet, the heat being low until the filter completely chars and then being increased, with the cover removed, until the residue is white.

Impurities. The precipitate may contain traces of lime that remained soluble in ammonium oxalate. This may be determined by dissolving the pyrophosphate in dilute sulphuric acid followed by addition of 9 to 10 volumes of absolute alcohol. Calcium sulphate, $CaSO_4$, precipitates and settles out on standing several hours. It may be filtered off, dissolved in hydrochloric acid and precipitated as oxalate in the usual way and so determined.

A residue remaining after treating the pyrophosphate with acid is generally SiO_2.

The presence of manganese may be detected by dissolving the magnesium pyrophosphate, $Mg_2P_2O_7$, in nitric acid and oxidizing with sodium bismuthate. (See method under Manganese.)

Properties of Ammonium Magnesium Phosphate. Readily soluble in dilute acids. One hundred cc. of pure water at $10°$ C. will dissolve 0.0065 gram. The presence of ammonia greatly decreases the solubility of the salt, e.g., 2.5% ammonia decreases the solubility to 0.00006 gram MgO per 100 cc. The presence of ammonium salts increase the solubility of the precipitate, e.g., 1 gram of ammonium chloride will increase the solubility to 0.0013 gram MgO.[3]

VOLUMETRIC DETERMINATION OF MAGNESIUM

Titration of the Ammonium Magnesium Phosphate with Standard Acid

The procedure known as Handy's volumetric method for magnesium,[4] depends upon the reaction $MgNH_4PO_4 + H_2SO_4 = MgSO_4 + NH_4H_2PO_4$. An excess of standard sulphuric acid is added to the precipitate and the excess of acid titrated back with standard ammonium hydroxide.

Procedure. The method of precipitation of the magnesium ammonium phosphate is the same as has been described under the gravimetric method. The precipitate is washed several times by decantation with 10% ammonium hydroxide solution (1 part NH_4OH, sp.gr. 0.90 to 9 parts water), and finally on the filter. After draining, the filter is opened out, the moisture removed as much as possible by means of dry filter papers. The residue may be dried in the room for about forty-five minutes or in the air oven at 50 to 60° C. for fifteen to twenty minutes.[5] When the filter has dried, ammonia will have been expelled. The substance is placed in a dry beaker, N/10 sulphuric acid added in excess (methyl orange indicator), the solution diluted to 100 cc. and the excess of acid titrated with N/10 sodium hydroxide.

One cc. N/10 $H_2SO_4 = 0.002$ gram MgO.

[1] Based on atomic weights of 1916.
[2] Zeit. analy. Chem., **29**, 165, 1890.
[3] Mellor, "Quantitative Inorganic Analysis," J. B. Lippincott Co., Pub.
[4] James Otis Handy, Jour. Am. Chem. Soc., **22**, 31.
[5] Low, "Technical Methods of Ore Analysis," Wiley & Sons, Pub.

MANGANESE

Mn, *at.wt.* 54.93; *sp.gr.* 7.42 [1]; *m.p.* 1260° [2]; *b.p.* 1900° C [1]; *oxides,* MnO, Mn_2O_3, (Mn_3O_4 *ignition in air*), MnO_2, MnO_3, Mn_2O_7.

DETECTION

In the usual course of analysis manganese is found in the filtrate from the hydroxides of iron, aluminum and chromium, the previous groups having been removed with hydrochloric acid, hydrogen sulphide and ammonium hydroxide in presence of ammonium chloride. Manganese, cobalt, nickel and zinc are precipitated as sulphides in an ammoniacal solution. The sulphides of manganese and zinc are dissolved by cold dilute hydrochloric acid, H_2S expelled by boiling and manganese precipitated as the hydroxide by addition of potassium hydroxide in sufficient amount to dissolve the zinc (sodium zincate). Manganese is now confirmed by dissolving this precipitate in nitric acid and adding red lead or lead peroxide to the strong nitric acid solution. A violet-colored solution is produced in presence of manganese. Chlorides should be absent.

Manganese in soils, minerals, vegetables, etc., is detected by incinerating the substance, treating the ash with nitric acid and taking to dryness, the residue is taken up with water and the mixture filtered. To the filtrate is added a few cc. of 40% ammonium persulphate and a little 2% silver nitrate solution. A pink color is produced in presence of manganese.

Manganese compounds heated with borax in the oxidizing flame produce an amethyst red color. The color is destroyed in the reducing flame.

Fused with sodium carbonate and nitrate on a platinum foil manganese compounds produce a green-colored fusion ("robin egg blue").

ESTIMATION

Manganese may be determined accurately gravimetrically or volumetrically. The former methods may be used for high-grade manganese ores, the latter are generally preferred for determining manganese in steel and in alloys and are applicable to a wide range of substances.

The most important ore of manganese is pyrolusite, MnO_2. Other ores are braunite, Mn_2O_3; hausmannite, Mn_3O_4; manganite, $Mn_2O_3 \cdot H_2O$; albanite, MnS; haurite, MnS_2; dialogite, $MnCO_3$; rhodonite, $MnSiO_3$.

Speigeleisen or ferromanganese is an important alloy for the steel industry. In addition to the requirement of the element in the analysis of the above substances it is determined in certain paint pigments—green and violet manganous oxides, in dryers of oils, etc. It occurs in a number of alloys.

[1] Van Nostrand's Chem. Annual—Olsen.
Chapter contributed by Wilfred W. Scott.

Preparation and Solution of the Sample

In dissolving the sample the following facts will be recalled: The metal dissolves in dilute acids, forming manganese salts. The oxides and hydroxides of manganese are soluble in hot hydrochloric acid. Manganous oxide is soluble in nitric or in sulphuric acid; the dioxide is insoluble in dilute or concentrated nitric acid, but is soluble in hot concentrated sulphuric acid.

Ores of Manganese. A sample of powdered ore weighing 1 gram is brought into solution by digesting with 25 to 50 cc. of strong hydrochloric acid for fifteen to thirty minutes on the steam bath. If much silica is present 5 to 10 cc. hydrofluoric acid will assist solution. Five cc. of sulphuric acid are added and the mixture evaporated and heated until fumes of sulphur trioxide are evolved. The residue is taken up with a little water and warmed until the sulphates have dissolved. If decomposition is incomplete and a colored residue remains, this is filtered off, ignited in a platinum dish and fused with a little potassium bisulphate. The fusion is dissolved in water containing a little nitric acid and the solution added to the bulk of the sample.

If manganese is to be determined volumetrically the removal of iron is not necessary. If, however, a gravimetric procedure is to be followed, iron and alumina are removed by the basic acetate method given under separations and manganese precipitated in the filtrate. In presence of small amounts of iron and alumina, precipitation with ammonia in presence of ammonium chloride will remove these elements without appreciable loss of manganese, a double precipitation being usually advisable. For volumetric procedures in ores containing over 2% manganese an aliquot portion of the sample is taken for the determination. The portion should not contain over 0.01 gram of manganese.

Sulphide Ores—Pyrites, etc. The sample is either roasted to oxidize the sulphide and then dissolved in hydrochloric acid as above stated or it is treated according to the procedure given for iron pyrites under sulphur.

Slags. These may be decomposed with hydrofluoric and hydrochloric acid with final expulsion of these acids with sulphuric acid. Manganese is best determined in the extract by a volumetric method.

Iron Ores. The treatment is the same as that recommended for ores of manganese. The residue remaining upon evaporation with sulphuric acid is dissolved in a little water and about 30 cc. of nitric acid (sp.gr. 1.135) added. Manganese is now determined by the bismuthate method.

Alloys. Manganese Alloys. One gram of ferromanganese is dissolved in 50 cc. of dilute nitric acid (sp.gr. 1.135) and oxidized with sodium bismuthate with boiling. The cooled solution is diluted to 500 cc. and 10 to 25 cc. is treated with about 30 cc. of dilute nitric acid and manganese determined by the bismuthate method. The amount of sample taken is governed by the manganese content. This should not exceed 0.01 gram of the element if the volumetric procedure is to be followed.

Manganese Bronze. Five grams of drillings are dissolved in dilute nitric acid (1.2), in a large beaker, using only sufficient acid to cause solution. If much free acid is present evaporation to small volume is necessary to expel the nitric acid. The concentrate is diluted to 200 cc. and hydrogen sulphide passed in to precipitate copper. The solution is diluted to 250 cc. and 50 cc. filtered off (=1 gram). The H_2S gas is expelled by boiling, the solution being concentrated to about 15 cc. Twenty-five cc. of nitric acid are added and manganese precipitated by adding potassium chlorate in small portions. The chlorine is boiled off and the precipitate filtered onto asbestos and washed with concentrated nitric acid. This is now determined volumetrically by treating with an excess of ferrous sulphate of known strength and titrating the excess with standard permanganate.

Ferro-titanium Alloy. This is best decomposed by fusion with sodium carbonate, to which a pinch of sodium peroxide has been added. The fusion is extracted with water and the residue containing iron, manganese and nickel filtered onto asbestos. Manganese is dissolved in 25 to 30 cc. of nitric acid by treating with SO_2 gas or hydrogen peroxide and manganese determined by the bismuthate method.

Ferro-chromium, Metallic Chromium. These are best decomposed by fusion with sodium peroxide (five times the weight of sample taken), the fusion being made in a nickel crucible. The treatment is now the same as that recommended for ferro-titanium.

Ferro-aluminum, Vanadium Alloys. The method used for steel is suitable to either of these substances.

Molybdenum Alloys. The alloy is decomposed with hydrochloric acid, and iron separated by the basic acetate method, a large excess of acetate being used. Manganese is precipitated as the dioxide by means of bromine and ammonia by the detailed procedure given later. Manganese is dissolved in nitric acid after reduction in the acid solution by addition of a little sodium thiosulphate or SO_2 gas. It is now oxidized to permanganate by means of red lead and determined either colorimetrically or by titration with a standard solution of sodium arsenite.

Tungsten Alloys. These are best decomposed by treating 1 gram of the substance with 5 to 10 cc. of hydrofluoric acid and a few cc. of strong nitric acid and digesting until the solution is complete. The hydrofluoric acid is expelled [1] by taking to dryness, a few drops of sulphuric acid having been added. The residue is taken up with water and boiled with SO_2 water. The solution is made to definite volume and manganese determined volumetrically on an aliquot portion.

Silicon Alloys. One gram of the alloy is treated with 50 cc. of dilute nitric acid (sp.gr. 1.2) and 5 cc. of hydrofluoric acid. The graphite is filtered off and the hot solution treated with sodium bismuthate and kept boiling for about fifteen minutes after the manganese dioxide has been precipitated. The bismuthate method for estimating manganese is recommended.

Iron and Steel. 0.5 to 1 gram of steel is dissolved by heating with 30 to 50 cc. of dilute nitric acid (1.135). The volumetric method by oxidation with sodium bismuthate is generally recommended, no separations of other substances being required, as manganese may be determined directly in the sample.

Pig Iron. One gram of the drillings is dissolved in 30 cc. of dilute nitric acid (1.135 sp.gr.), and as soon as the action has ceased the sample is filtered through a 7-cm. filter and the residue washed with 30 cc. more of the acid. The filtrate containing the manganese is now treated according to the procedure for steel.

---

[1] Brearly and Ibbotson state that although neither tungsten nor hydrofluoric acid interfere with the bismuthate method of determining manganese, the two combined lead to erratic results, hence the removal of hydrofluoric acid is necessary.

SEPARATIONS

This section includes methods of special separations of manganese from elements that may interfere in its determination. As is frequently the case, isolation of manganese is not necessary, since it may be determined volumetrically in presence of a number of elements, which would interfere in its gravimetric determination. The analyst should be sufficiently familiar with the material to avoid needless manipulations, which not only waste time, but frequently lead to inaccurate results.

Removal of Elements of the Hydrogen Sulphide Group. This separation may be required in the analysis of certain alloys where a separation of manganese from copper is required.

The acid solution containing about 4% of free hydrochloric acid (sp.gr. 1.2), is saturated with hydrogen sulphide and the sulphides filtered off. Manganese passes into the filtrate. This treatment will effect a separation of manganese from mercury, lead, bismuth, cadmium, copper, arsenic, antimony, tin and the less common elements of the group.

Separation of Manganese from the Alkaline Earths and the Alkalies. The separation is occasionally required in the analysis of clays, limestone, dolomite, etc. It is required in the complete analysis of ores. In the usual course of a complete analysis of a substance, the filtrate from the hydrogen sulphide group is boiled free of H$_2$S and is treated with a few cc. of nitric acid to oxidize the iron. The solution is made slightly ammoniacal with ammonia, in presence of ammonium chloride, whereby iron, aluminum and chromium are precipitated as hydroxides. The filtrate is treated with hydrogen sulphide or colorless ammonium sulphide, whereby manganese, nickel, cobalt and zinc are thrown out as sulphides and the alkaline earths and alkalies remain in solution.

Separation of Manganese from Nickel and Cobalt

The free acid of the sulphate or chloride solution of the elements is neutralized with sodium carbonate and a slight excess added. It is now made strongly acid with acetic acid and 5 grams of ammonium acetate added for every gram of nickel and cobalt present. The solution is now diluted to about 200 cc. and saturated with hydrogen sulphide, whereby nickel and cobalt are precipitated as sulphides and manganese remains in solution.

Separation of Manganese from Iron and Aluminum, Basic Acetate Method

The procedure effects a separation of iron, aluminum, titanium, zirconium and vanadium from manganese, zinc, cobalt and nickel.

The separation depends upon the fact that solutions of acetates of iron, aluminum, titanium, zirconium and vanadium are decomposed when heated and the insoluble basic acetates precipitated, whereas the acetates of manganese, zinc, cobalt and nickel remain undecomposed when boiled for a short time.

$$Fe(C_2H_3O_2)_3 + 2HOH = 2HC_2H_3O_2 + Fe(OH)_2 \cdot C_2H_3O_2.$$

The solvent action of the liberated acetic acid is prevented by the addition of sodium acetate which checks ionization of the acid. The method requires care and is somewhat tedious, but the results attained are excellent.

Procedure. To the cooled acid solution of the chlorides is added a concentrated aqueous solution of sodium carbonate from a burette with constant stirring until the precipitate that forms dissolves slowly. A dilute solution of the carbonate is now added until a slight permanent opalescence is obtained. With the weak reagent and careful addition of the carbonate drop by drop the proper neutralization of the free acid is obtained. With considerable iron present the solution appears a dark red color, fading to colorless as the quantity of iron decreases to a mere trace in the solution. Three cc. of acetic acid (sp.gr. 1.044) are added to dissolve the slight precipitate. The more perfect the neutralization before heating the less amount of reagent required for precipitating iron—an excess of reagent does no harm. If this does not clear the solution in two minutes, more acetic acid is added a drop at a time until the solution clears, allowing a minute or so for the reaction to take place with each addition. The solution is diluted to about 500 cc. and heated to boiling and 6 cc. of a 30% sodium acetate solution added. The solution is boiled for one minute and removed from the flame. (Longer boiling will form a gelatinous precipitate, difficult to wash and filter.) The precipitate is allowed to settle for a minute or so, then filtered, while the liquid is hot, through a rapid filter and washed with hot, 5% sodium acetate solution three times. The apex of the filter is punctured with a glass stirring rod and the precipitate washed into the original beaker in which the precipitation was made with a fine stream of hot, 1 : 1 hydrochloric acid solution from a wash bottle. (Dilute HNO_3 may be used in place of HCl.)

A second precipitation with neutralization of the acid and addition of sodium acetate is made exactly as directed above. It is advisable to evaporate the solution to small volume to expel most of the free mineral acid before addition of Na_2CO_3 to avoid large quantities of this reagent. The filtrates contain manganese, zinc, cobalt and nickel; the precipitate iron. aluminum, titanium, zirconium, vanadium.

Materials Insoluble in Acid

Silico-Manganese, Ferro-Chrome, Ferro-Tungsten. These alloys are best decomposed by fusion with sodium carbonate and magnesium oxide, in proportion of 2 parts Na_2CO_3 and 1 part MgO. The fused mass is broken down in a mortar with water, the paste transferred to a beaker and then water added to make the solution to about 150 cc. Sodium peroxide is added and the solution boiled. The precipitate MnO_2 is allowed to settle and is then washed by decantation. The residue may contain impurities such as SiO_2, etc., from which separations may be necessary.

Manganese in Cast Iron. 1 gram of sample is treated with 25 cc. of dilute nitric acid (1 : 3) and when all action has ceased the solution is filtered and the residue washed with 30 cc. more of the acid. The filtrate is now examined for manganese according to the method for steel.

The carbide in white irons will prevent the quantitative conversion of manganese to permanganate. Several treatments with sodium bismuthate may be necessary.

Ferro-Silicon. Decompose the sample with hydrofluoric and sulphuric acids and evaporate to fumes. (Use platinum dish.) Take up with dilute nitric acid and proceed as in steels.

Separation of Manganese as Manganese Dioxide. See page 300a.

GRAVIMETRIC METHODS

Determination of Manganese as Pyrophosphate

Manganese is precipitated as ammonium manganese phosphate, NH_4MnPO_4, and then ignited to pyrophosphate, $Mn_2P_2O_7$. The method is known as Gibbs' Phosphate Process.[1]

Procedure. The cold solution of manganese chloride[2] obtained as directed in previous sections, should be diluted so as to contain not over 0.1 gram of manganese oxide equivalent per 100 cc. of solution. A cold saturated solution of ammonium sodium phosphate (microcosmic salt, 170 grams per liter; 9 cc. precipitates an equivalent of 0.1 gram of the oxide) is now added in slight excess. The solution is made strongly ammoniacal and heated to boiling, the boiling being continued until the precipitate becomes crystalline. After allowing to settle until cold, the precipitate is filtered off (the filtrate being tested with more of the precipitating reagent to assure that an excess had been added), and dissolved in a little dilute hydrochloric or sulphuric acid.

Reprecipitation of the phosphate. The free acid is neutralized with ammonia added in slight excess until the odor is quite distinct, the solution heated to boiling, and a few cc. of additional phosphate reagent added. The crystalline precipitate is filtered into a weighed Gooch crucible, washed free of chlorides with very dilute ammonia ($AgNO_3+HNO_3$ test), dried and ignited to the pyrophosphate. The ignition is conducted, as in case of magnesium, by heating first over a low flame and gradually increasing the heat to the full power of the burner. The final residue will appear white or a pale pink.

$$Mn_2P_2O_7 \times 0.4996 = MnO,$$

$$Mn_2P_2O_7 \times 0.3869 = Mn.$$

NOTES. Zinc, nickel, copper and other elements precipitated as phosphates should be absent from the solution. The separation from iron is generally made by the basic acetate method and manganese precipitated from the filtrate, free of other elements, as the peroxide MnO_2, by means of bromine added to the ammoniacal solution. Other oxidizing reagents may be used, as has been stated. The dioxide is dissolved in strong hydrochloric acid and the above precipitation effected.

[1] Gibbs, C. N., **17**, 195, 1868.
[2] Some analysts prefer to add the phosphate reagent to the strongly ammoniacal solution, boiling hot.

Determination of Manganese as the Dioxide, MnO_2

The procedure is of special value in the complete analysis of ores where a basic acetate separation of iron and aluminum has been made, and a gravimetric estimation of other constituents in the solution are desired.

The procedure depends upon the principle that manganese in a dilute solution of manganous salt is oxidized to manganese dioxide and so precipitated, when boiled with bromine or certain other oxidizing agents:

$$MnCl_2 + Br_2 + 2H_2O = MnO_2 + 2HCl + 2HBr.$$

The free acid formed by the reaction must be neutralized either by ammonia or by the presence of a salt of a weak acid such as sodium acetate, otherwise the precipitation of manganese will be incomplete. In presence of ammonium salts much of the bromine is used up reacting with ammonia,

$$MnCl_2 + Br_2 + 3NH_3 + 2H_2O = MnO_2 + 2NH_4Cl + NH_4Br + HBr.$$

At the same time an acid is formed, which reacts with the free ammonia. It is necessary to have the solution ammoniacal throughout the reaction to prevent resolution of the manganese.

Procedure. To the solution containing manganese is added 4 to 5 grams of sodium acetate (unless already present in excess), the solution being diluted to about 200 cc. Bromine water is added until a distinct color of bromine is evident. The mixture is boiled and kept boiling for ten to fifteen minutes, additional bromine being added in small portions. The precipitate is allowed to settle and filtered off. The filtrate is boiled with additional bromine to ascertain whether the manganese has been completely removed from the solution.

If ammonia is present, as is frequently the case, it is advisable to add more of the reagent from time to time, the solution having a distinct odor of ammonia after the last portion of bromine has been added. When large amounts of manganese are present, several separations may be required to remove the element from the subsequent filtrates.

The precipitated dioxide may be dissolved in sulphuric acid and manganese determined volumetrically or gravimetrically.

It may be ignited directly and weighed as Mn_3O_4.

It may be evaporated with sulphuric acid and manganese determined as $MnSO_4$.

Manganates of zinc or calcium will be precipitated if present in large amounts.

Manganese may also be precipitated by ammonium persulphate in an ammoniacal solution, potassium chlorate and chloride of lime in presence of zinc chloride in a neutral solution.[1]

[1] J. Pattinson's Method, Jour. Chem. Soc., **35**, 365, 1899.

VOLUMETRIC METHODS
Volhard's Method for Manganese [1]

The method is based on the principle that when potassium permanganate is added to a neutral manganese salt all of the manganese is oxidized and precipitated. When this stage is reached any excess of permanganate is immediately evident by the color produced. The calculation of results may be based on the reaction,

$$3MnSO_4 + 2KMnO_4 + 2H_2O = 5MnO_2 + K_2SO_4 + 2H_2SO_4,$$

or

$$5ZnSO_4 + 6MnSO_4 + 4KMnO_4 + 14H_2O = 4KHSO_4 + 7H_2SO_4 + 5ZnH_2 \cdot 2MnO_3,$$

the ratio in either case being $2KMnO_4 = 3Mn.$

Procedure. The material decomposed with hydrochloric and nitric acid and taken to fumes with sulphuric acid, as stated for the preparation of the sample, is cooled and boiled with 25 cc. of water until the anhydrous ferric sulphate has dissolved and continue as follows: Transfer the mixture to a 500-cc. graduated flask and add an emulsion of zinc oxide in slight excess to precipitate the iron (C.P. $ZnSO_4$ precipitated by KOH added to slight alkalinity. The washed precipitate is kept in a stoppered bottle with sufficient water to form an emulsion).

Agitate the flask to facilitate the precipitation and see that a slight excess of zinc oxide remains when the reaction is complete. Now dilute the contents of the flask up to the mark with cold water, mix thoroughly and allow to stand a short time and partially settle. By means of a graduated pipette draw off 100 cc. of the clear supernatant liquid and transfer it to an 8-oz. flask. While the precipitate in the 500-cc. flask may appear large, it actually occupies but a very small space, and any error caused by it may consequently be neglected. Likewise the error in measurement due to change of temperature during the manipulation is insignificant. Heat the solution in the small flask to boiling, add two or three drops of nitric acid (which causes the subsequent precipitate to settle more quickly) and titrate with a standard solution of potassium permanganate. The permanganate causes a precipitate which clouds the liquid and it is therefore necessary to titrate cautiously and agitate the flask after each addition, and then allow the precipitate to settle sufficiently to observe whether or not the solution is colored pink. A little experience will enable one to judge by the volume of the precipitate formed, about how rapidly to run in the permanganate. The final pink tinge, indicating the end of the reaction, is best observed by holding the flask against a white background and observing the upper edges of the liquid. When this point is attained, bring the contents of the flask nearly to a boil once more and again observe if the pink tint still persists, adding more permanganate if necessary. In making this end-test avoid actually boiling the liquid, as a continual destruction of the color may sometimes thus be effected and the true end-point considerably passed. When the color thus remains permanent the operation is ended. Observe the number of cc. of permanganate solution used and calculate the result.

It is customary to use the same permanganate solution for both iron and manganese. Having determined the factor for iron, this may be multiplied by 0.2952 to obtain the factor for manganese. It will be observed that $2KMnO_4$ are required for 3Mn, and in the reaction for iron that $2KMnO_4$ are required for 10Fe.

[1] Applicable for high grade ores.

VOLUMETRIC DETERMINATION OF MANGANESE
Bismuthate Method

The determination of manganese by the bismuthate method is generally conceded to be the most accurate analytical procedure for determination of this element in iron and steel. It is simple and rapid, and generally can be accomplished without a previous separation being necessary. The principle of the process depends upon the fact that under certain conditions bivalent manganese can be quantitatively oxidized to permanganic acid by sodium bismuthate. This permanganic acid can now be titrated by a standard reducing agent such as sodium thiosulphate or arsenous acid, or ferous sulphate.

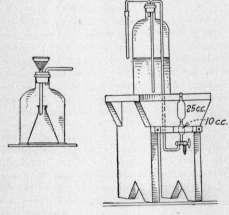

The application of the bismuthate method to the determination of large amounts of manganese has been made by T. R. Cunningham and R. W. Coltman (Ind. and Eng. Chem., **16**, 58, Jan., 1924). They recommend the following procedures:

Fig. 45.   Fig. 46.

Determination of Manganese in Manganese Ores

Descriptions of the application of the bismuthate method to the determination of manganese in manganese ores, ferro-manganese, and manganese metal are given in the following paragraphs. The adaptation of the method to other products high in manganese is easily made after the operator becomes familiar with the conditions necessary.

Reagents Required. (1) Concentrated nitric acid (specific gravity 1.42), freed from nitrous oxide by passing air through the solution for half an hour. If air from a compressor is used, it must be freed from oil and dust.

(2) Nitric acid of specific gravity 1.135, made by adding 500 cc. of the above acid to 1300 cc. of water.

(3) Dilute nitric for washing, prepared by adding 30 cc. of the concentrated acid described above to 1 liter of water.

(4) Sodium bismuthate. This reagent generally contains approximately 80% of $NaBiO_3$. If there is any doubt as to its oxidizing power, it may be tested as follows:

One-half gram is shaken up with 4 grams of potassium iodide and a little water in a stoppered flask. Fifteen cubic centimeters of hydrochloric acid (specific gravity 1.20) are added and the solution is allowed to stand in the dark, with occasional shaking, until the bismuthate has entirely decomposed. The solution is diluted to 300 cc. and titrated with 0.1 N sodium thiosulphate, starch being used as indicator.

$$1 \text{ cc. of } 0.1 \text{ N } Na_2S_2O_3 = 0.0140 \text{ gram } NaBiO_3.$$

(5) Ammonium persulphate (C. P.).

(6) Ferrous ammonium sulphate.

(7) Hydrogen peroxide (3%).

(8) Sulphurous acid. A freshly prepared solution of sulphur dioxide in water.

(9) Standardized 0.1 N potassium permanganate.

Solution of the Ore. Grind the sample to pass 100 mesh and dry at 120° C. for 1 hour. Dissolve 2.0000 grams (±0.0002 gram) by boiling with about 40 cc. of nitric acid (1 : 1) in a 400-cc. beaker covered with a clock glass, adding hydrogen peroxide in small portions until the violent evolution of oxygen is over and no black particles of ore remain. Rinse the clock glass and sides of the beaker with hot water, and add bismuthate in small (about 0.05 gram) portions, until a permanent precipitate of manganese dioxide forms. Any organic matter is thus destroyed and the excess of hydrogen peroxide removed. Boil for 2 or 3 minutes, and then add sulphurous acid drop by drop until the solution clears. Boil the solution for 5 minutes, and filter it into a 500-cc. volumetric flask. Wash the siliceous residue well with water and ignite it in platinum. Treat the residue with several drops of sulphuric acid (specific gravity 1.84) and sufficient hydrofluoric acid to dissolve the silica, and evaporate the solution until it fumes. Dissolve the residue in water and add to the solution in the flask. Bring the solution to a temperature of about 20° C. and make it up to the mark with water at this temperature.

With artificial manganese dioxide, where the weight of insoluble residue is insignificant, or in case the residue from an ore is flocculent and not sandy, the treatment with hydrofluoric and sulphuric acids may be omitted and the ore solution rinsed directly into the flask. Any barium in the ore will precipitate when the sulphuric acid solution of the residue is added to the main solution, but the barium sulphate does not interfere with the subsequent operations.

Trial Determination. As it is desirable that the back titration with permanganate be not too great (say, from 10 to 20 cc.), it is best, when ores of unknown manganese content are being analyzed, to make a rough preliminary determination of the amount of manganese present, so that the amount of ferrous salt added may not be excessive. For this purpose transfer 25.0 cc. of the well-mixed solution with a pipet to a 300-cc. Erlenmeyer flask and add 12 cc. of nitric acid (specific gravity 1.42) and about 13 cc. of water. Cool this solution, add 1.7 grams of bismuthate, agitate the mixture for 1 minute, dilute with 50 cc. of water, and filter through asbestos, washing the residue with dilute (3%) nitric acid. Add 2.5 grams (weighed) of the solid ferrous ammonium sulphate to the filtrate and titrate back with the 0.1 N permanganate. Calculate in the usual manner the weight of manganese present in the 25-cc. portion. As 100 cc. will be used in the actual determination, multiply the result by 4, and calculate approximately the amount of bismuthate necessary (26 grams for 1 gram for manganese) and also a weight of ferrous sulphate that will give from 10 to 20 cc. in the back titration. The bismuthate is weighed out roughly, and the necessary ferrous sulphate weighed out to within half a milligram just preceding the titration.

Final Determination. Transfer two 100-cc. aliquot portions with a pipet to 750-cc. Erlenmeyer flasks and add 50 cc. concentrated nitric acid and 50 cc. of water to each. (Total volume, 200 cc.) Cool the solutions (with ice, if possible) and add the calculated amount of bismuthate all at once. *Agitate the contents of the flask briskly for 1 minute*, add 200 cc. of cold water, and filter through asbestos, washing the residue with dilute nitric acid (solution No. 3) until the washings do not show the slightest trace of pink.

Place the weighed ferrous salt in a liter beaker and add the contents of the suction flask to it. Stir thoroughly until the permanganic acid has been decolorized and all the salt dissolved. Titrate at once with the standardized 0.1 N potassium permanganate to faint pink color.

Typical Analysis. Two grams of ore were dissolved and made up to 500 cc.; two 100-cc. aliquot portions were oxidized and 9 grams of ferrous salt added to reduce the permanganic acid formed.

Manganese in Manganese Ore

	Aliquot No. 1	Aliquot No. 2
Back titration, cc.	20.1	20.2
Equivalent of ferrous salt, cc.	228.6	228.6
Difference, cc.	208.5	208.4
Weight of manganese, gram	0.2291	0.2290
Percentage of manganese	57.26	57.24

Analysis of Ferromanganese

The weight of sample used is governed by the manganese content; for 80% ferromanganese from 0.25 to 0.30 gram may be used. As there is no difficulty in obtaining a uniform sample, it is preferable to weigh out individual portions of the 100-mesh sample for the determination, instead of using an aliquot part of a large sample. The following procedure as regards the quantity of acid and final volume presupposes the presence of about 0.2 gram of manganese.

When working with high-carbon ferromanganese, dissolve the sample in 60 cc. of nitric acid (specific gravity 1.42) and boil in a flask or covered beaker until nitrous fumes cease to be evolved. If the sample contains over 1% chromium, it is necessary to dilute the solution with an equal volume of water, filter, ignite the residue, and fuse it with a little sodium carbonate, dissolving the melt in the least possible quantity of nitric acid and adding it to the main solution.

Oxidize the carbon present by adding ammonium persulfate, a little at a time, until a total of from 2 to 2.5 grams has been introduced. Boil the solution about 10 minutes, and then add small amounts of bismuthate to the boiling liquid, until a precipitate of manganese dioxide has formed. Dissolve the precipitate by the addition of sulphurous acid, drop by drop, adding an excess of 1 cc. after the precipitate has dissolved to reduce any chromium to the trivalent form. After having boiled the solution for 5 minutes, bring it to a volume of 200 cc., and cool with ice. This cooling is essential in order that any chromium may not interfere, as it would be oxidized by the bismuthate if the solution were at elevated temperatures. The manganese is then oxidized with bismuthate and determined as described under " Analysis of Manganese Ore."

The following data are from typical determinations on a sample of standard high-carbon ferromanganese:

MANGANESE IN FERROMANGANESE

Weight of Sample G.	Ferrous Salt G.	KMnO$_4$ Equivalent of Ferrous Salt Cc.	Back Titration Cc.	0.1 N KMnO$_4$ Consumed Cc.	Per Cent Manganese
0.2500	7.5	191.3	13.4	177.9	78.15
0.2500	7.5	191.3	13.4	177.9	78.15
0.3000	9.0	229.2	15.7	213.5	78.19

Low-carbon ferromanganese is dissolved directly in nitric acid of 1.135 specific gravity (about 1 cc. for every milligram of manganese present). After preliminary oxidation and reduction of the solution, it is made up to the original volume with nitric acid (specific gravity 1.135) and cooled as a preliminary to the oxidation and determination of the manganese.

Analysis of Manganese Metal

Dissolve 0.2500 gram in 250 cc. of nitric acid (specific gravity 1.135) in a 750-cc. Erlenmeyer flask provided with a cut-off funnel or some similar device to prevent loss by spraying, and make a preliminary oxidation with bismuthate, subsequently reducing with sulphurous acid. If the carbon is high (4%), oxidation with ammonium persulphate should precede the bismuthate oxidation. Dilute the solution to a volume of 250 cc. by adding nitric acid (specific gravity 1.135), cool with ice, and finish the determination as previously outlined. About 6.5 grams of sodium bismuthate are necessary for the oxidation and 9 grams of ferrous salt will reduce the permanganic acid resulting from a 0.2500-gram sample of 95% manganese metal.

TABLE X—MANGANESE IN MANGANESE METAL

No.	Weight of Sample G.	Ferrous Salt G.	KMnO$_4$ Equivalent of Ferrous Salt Cc.	Back Titration Cc.	0.1 N KMnO$_4$ Consumed Cc.	Per Cent Manganese
1	0.2000	8.000	203.8	28.5	175.3	96.29
2	0.2500	9.000	229.2	10.0	219.2	96.32
3	0.2500	9.000	229.2	10.2	219.0	96.24

NOTES. The conditions necessary for securing complete oxidation of large quantities of manganese and for preventing the permanganic acid from undergoing any appreciable decomposition during the subsequent filtration are summarized below.

Concentration of Nitric Acid. The manganese should be present in a solution containing from 11% (specific gravity 1.062) to 22% (specific gravity 1.135) by weight of nitric acid. If the concentration of nitric acid falls much below 11%, the oxidation of the manganese will not be complete unless the time of shaking be increased to more than 1 minute.

Concentration of Manganese. A solution of permanganic acid containing about 0.05 gram of manganese per 100 cc. has the maximum stability, but the weight of manganese can be increased to 0.1 gram in 100 cc. without danger of any material decomposition occurring during the time required for filtering off the excess of bismuthate. When the concentration of manganese rises much above 0.10 gram per 100 cc., the rate of decomposition of the permanganic acid is unduly rapid.

Amount of Sodium Bismuthate Necessary. Approximately 26 grams of sodium bismuthate (79 per cent NaBiO$_3$) must be used for 1 gram of manganese.

Time of Oxidation. Shaking for 1 minute is sufficient to insure complete oxidation of the manganese to permanganic acid provided the foregoing conditions are adhered to.

Chlorides should be removed by taking the solution to fumes with H$_2$SO$_4$. The residue is dissolved in a small amount of water and the solution is evaporated to fumes a second time to insure the removal of every trace of chloride.

The only common metals that seriously interfere with the determination are cerium, cobalt, and sexivalent chromium. A method for the determination of cerium outlined by Metzger is exactly the same in principle as the bismuthate method for manganese. Any cerium present must therefore be separated as oxalate in acid solution, and the oxalic acid in the filtrate destroyed by evaporation with sulphuric and nitric acids as a preliminary to the determination of manganese.

The pink color produced by large amounts of cobalt interferes with the titration of permanganic acid. This can be overcome by separating the manganese from the bulk of the cobalt by precipitating it with sodium or potassium chlorate.

While trivalent chromium is in hot solution oxidized to the sexivalent state by bismuthate and by permanganic acid, the error caused by small amounts of trivalent chromium is not appreciable provided the solution is kept cold (10° C.), and is oxidized, filtered, and titrated as rapidly as possible. When more than a small percentage of chromium is present, it should be separated from the manganese by one of the several methods that have been proposed. Precipitation of the manganese from a nitric acid solution with sodium or potassium chlorate with subsequent filtration does not effect complete removal of chromium, but is useful in some cases. Fusion with sodium peroxide followed by filtration will give a complete separation, manganese remaining in the residue as oxide and chromium passing into the filtrate as sodium chromate. Watters precipitates chromium and ferric iron with zinc oxide and determines manganese in the filtrate, while Cain precipitates chromium and vanadium from a ferrous solution with cadmium carbonate and analyzes the filtrate for manganese. Sexivalent chromium interferes with the determination of manganese by the bismuthate method and must be reduced to the trivalent condition prior to the final oxidation with bismuthate.

Although any vanadium present is reduced by the ferrous sulphate added during the determination, it is re-oxidized by an equivalent amount of permanganic acid during the back titration, the manganese titration as a consequence being unaffected.

Provided accurate volumetric apparatus is available, it is usually preferable to make the manganese solution up to a definite volume and to work on an aliquot portion. The solution should always be given a preliminary oxidation with bismuthate before being made up to volume. Bismuthate is added to the hot solution in small portions until manganese dioxide precipitates, the liquid is boiled for 5 minutes, and then sulphurous acid is added drop by drop until the precipitate has dissolved. Sexivalent chromium should be reduced by SO_2 to trivalent form. The concentration of nitric acid to 1.135 sp.gr. and manganese to 0.0001 gram per cc. are required for best results.

Determination of Permanganic Acid

Ferrous ammonium sulphate is used to reduce the permanganic acid formed. It is added in solid form, weighed to ±0.5 mg., just before the determination is started. This salt is preferable to ferrous sulphate, as it goes into solution more readily. The manganese value of the ferrous salt must be known, and for this purpose 5.0000 grams (±0.5 mg.) are titrated with the standard permanganate used. This titration should be made in sulphuric acid solution. If the salt is kept in a well-corked bottle, the standard will suffer practically no change, and having once been well mixed and standardized may be used indefinitely. If preferred, a 0.1 N solution of ferrous ammonium sulphate may be used instead of the solid salt.

The excess of ferrous salt is now titrated with a standardized solution of potassium permanganate, preferably of about 0.1 N strength. The permanganate may be standardized against Bureau of Standards sodium oxalate, according to McBride's recommendations, by means of pure manganous oxalate as hereinafter described, or against pure anhydrous manganous sulphate. Owing to the difficulty of preparing the latter salt so that it is of theoretical composition, the first two methods of standardization are preferable. After the normality of the solution against sodium oxalate has been determined, the theoretical factor—viz., 1 cc. 0.1 N $KMnO_4$=0.001099 gram of manganese—may be used.

Correlation of Value of Permanganate Used for Titration with an Absolute Standard

In order to determine the value of the permanganate in terms of manganese, manganous sulphate prepared in the following manner may be used:

Potassium permanganate (the ordinary C. P. product) was reduced in the presence of sulphuric acid with a little less than the calculated amount of oxalic acid, and the manganese dioxide formed is filtered off and discarded. The solution of manganese sulphate is treated with ammonium carbonate and the resulting manganese carbonate filtered and washed free from sulphates by decantation. This precipitate is added to a boiling solution of oxalic acid, and the manganous oxalate formed filtered and washed free from acid with distilled water. This product is converted to the sulphate by heating to constant weight at 480° to 520° C., with an excess of sulphuric acid. An accurately weighed amount of the pure manganous sulphate (2.7536 grams of $MnSO_4$, equivalent to 1.0018 grams of manganese) is dissolved and made up to 1 liter with nitric acid (specific gravity 1.135). One hundred and 200-cc. portions of this solution are oxidized with sodium bismuthate, filtered, and reduced by addition of accurately weighed amounts of ferrous ammonium sulphate, the excess of which is determined by means of 0.1 N permanganate that had been standardized against sodium oxalate. The relation between the ferrous ammonium sulphate and the permanganate is carefully determined. The results that are obtained, confirm Blum's statement that either manganous sulphate or sodium oxalate may be used as the primary standard for the bismuthate method, or expressed in another form, that the manganese is quantitatively oxidized to the heptavalent state.

Ores. If mercury is to be determined by the dry procedure, the finely ground sample may be mixed directly with the flux and determined as directed later.

For opening up the ore for the volumetric method by Seamon see method at close of the chapter, page 312.

For decomposition of ores see also thiocyanate method, page 312a.

SEPARATIONS

Separation of Mercury from the Iron and Zinc Groups, or from the Alkaline Earths and the Alkalies. Mercury is precipitated as a sulphide from an acid solution of the mercuric salt by hydrogen sulphide, together with the members of the hydrogen sulphide group. Sufficient acid should be present to prevent the precipitation of zinc sulphide. Iron, aluminum, chromium, manganese, cobalt, nickel, zinc, the alkaline earths and the alkalies remain in solution.

Separation of Mercury from Arsenic, Antimony, and Tin. The sulphides obtained by passing hydrogen sulphide into an acid solution, preferably of the chlorides, are digested with yellow ammonium sulphide solution. Arsenic, antimony and tin dissolve, whereas mercury sulphide remains insoluble. Sulphides of the fixed alkalies dissolve mercury as well as arsenic, antimony and tin, so cannot be used in effecting a separation.

Separation from Lead, Bismuth, Copper and Cadmium. These elements remain with mercury upon removal of arsenic, antimony and tin as their sulphides are insoluble in ammonium sulphide. (CuS slightly soluble.) The precipitated sulphides are transferred to a porcelain dish and boiled with dilute nitric acid, sp.gr. 1.2 to 1.3. After diluting slightly with water the solution is filtered and the residue of mercuric sulphide washed with dilute nitric acid and finally with water. If much lead is present in the solution it is apt to contaminate the residue by a portion being oxidized to lead sulphate and remaining insoluble. In this case the residue is treated with aqua regia, the solution diluted and mercury chloride filtered from $PbSO_4$ and free sulphur. Mercury is best determined as HgS by the ammonium sulphide method described later. Traces of lead do not interfere, as lead is completely removed by remaining insoluble in potassium hydroxide, whereas mercury sulphide dissolves. See method.

Separation from Selenium and Tellurium. The mercury selenide or telluride is dissolved in aqua regia, chlorine water added and the solution diluted to 600 to 800 cc., phosphorous acid is added and the solution allowed to stand for some time; mercurous chloride is precipitated, selenium and tellurium remaining in solution. Selenium and tellurium will precipitate in hot concentrated solutions when treated with phosphorous acid, but not in dilute hydrochloric acid solutions.

Mercury in Organic Substances. The material is decomposed by heating in a closed tube with concentrated nitric acid, or by heating with 10% H_2SO_4 and sufficient $(NH_4)_2S_2O_8$, added in small portions until the organic matter is decomposed.

GRAVIMETRIC METHODS

Determination of Mercury by Precipitation with Ammonium Sulphide

The following method, suggested by Volhard, is generally applicable for determination of mercury. The element is precipitated by ammonium sulphide as HgS. The precipitate dissolved in caustic is again thrown out by addition of ammonium nitrate to the sulpho salt solution of mercury.

$$Hg(SNa)_2 + 2NH_4NO_3 = 2NaNO_3 + (NH_4)_2S + HgS.$$

Procedure. The acid solution of the mercuric salt is nearly neutralized by sodium carbonate, and is then heated with a slight excess of ammonium sulphide reagent, freshly prepared. Sodium hydroxide solution is added until the dark-colored liquid begins to lighten. The solution is now heated to boiling and more sodium hydroxide added until the liquid is clear. If lead is present it will remain undissolved and should be filtered off. Ammonium nitrate is now added to the solution in excess and the mixture boiled until the greater part of the ammonia has been expelled. The clear liquid is decanted from the precipitate through a weighed Gooch crucible and the precipitate washed by decantation with hot water and finally transferred to the crucible and washed two or three times more. The mercuric sulphide is dried at 110° C. and weighed as HgS.

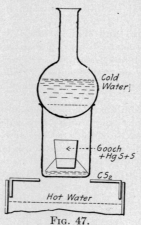

Cold Water

Gooch + Hg S + S

CS₂

Hot Water

Fig. 47.

$$HgS \times 0.8622 = Hg \quad \text{or} \quad \times 0.9307 = HgO.$$

Notes. Alumina and silica are apt to be present in caustic.

Free sulphur may be removed, if present, by boiling with sodium sulphite, $Na_2SO_3 + S = Na_2S_2O_3$. The sulphur may be extracted with carbon disulphide. The Gooch crucible is placed upon a glass tripod in a beaker, containing carbon disulphide, and a round-bottomed flask filled with cold water is placed over the mouth of the beaker to serve as a condenser, Fig. 47. By gently heating over a water bath for an hour the sulphur is completely extracted from the sulphide. Carbon disulphide is removed from the precipitate by washing once with alcohol followed by ether. The residue is now dried and weighed.

Determination of Mercury by Electrolysis

Mercury is readily deposited as a metal from slightly acid solutions of its salts.

Procedure. The neutral or slightly acid solution of mercuric or mercurous salt is diluted in a beaker to 150 cc. with water and 2 to 3 cc. of nitric acid added. The solution is electrolyzed with a current of 0.5 to 0.1 ampere, and an E.M.F. of 3.5 to 5 volts. A gauze cathode is recommended, or a platinum dish with dulled inner surface may be used. One gram of mercury may be deposited in about fifteen hours (or overnight). The time may be shortened to about three hours by increasing the current to 0.6 to 1 ampere.

The metal is washed with water without interrupting the current and then with alcohol. After removing the adhering alcohol with a filter paper, the cathode is placed in a desiccator containing fused potash and a small dish of mercury. The object of this mercury is to prevent loss of the deposit by vaporization.

The increased weight of the cathode is due to metallic mercury.

NOTES. In the electrolysis of mercuric chloride turbidity may be caused by formation of mercurous chloride by reduction, but this does no harm, as the reduction to metallic mercury follows.

Mercury may be electrolyzed from its sulpho solutions, obtained by dissolving its sulphide in concentrated sodium sulphide.

Determination of Mercury by the Holloway=Eschka Process Modified

When mercury sulphide is heated with iron filings metallic mercury is volatilized, iron sulphide being formed. The mercury vapor is condensed on a silver or gold plate. The use of iron for this reduction was suggested by Eschka and his method modified by Holloway. In ores containing arsenic the addition of zinc oxide is recommended. Erdmann and Marchand use lime for decomposing the mercury compound. The reactions may be represented as follows:

$$HgS + Fe = FeS + Hg \quad \text{or} \quad HgX + CaO = CaX + Hg + O.$$

Apparatus. This consists of a deep glazed porcelain crucible, the size depending upon the charge of the sample to be taken. Generally a 30-cc. crucible is used for a 2-gram sample with 4 grams of flux. The crucible is covered by a silver or gold plate that lies perfectly flat and fits snugly around the edges of the crucible. It may be necessary to grind the top of the receptacle on emery paper to obtain a perfectly level edge.

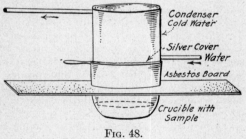

Condenser
Cold Water

Silver Cover
Water
Asbestos Board

Crucible with
Sample

FIG. 48.

The crucible is suspended in a hole through an asbestos board or quartz plate, to prevent the flame heating the upper portion of the vessel.

The lid of the crucible is kept cool by a cylindrical condenser of metal through which a stream of water passes. A small Erlenmeyer flask may be used, with a tube passing to the bottom of the flask through a rubber stopper, and a second tube just passing through the stopper.

Holloway has a weight placed on the metal condenser to hold the lid firmly against the crucible. The illustration (Fig. 48) shows the form of the apparatus set up for the run.

Procedure. The sample containing not over 0.1 gram of mercury is placed in the crucible with 5 to 10 grams of fine iron filings and intimately mixed. Additional filings are put over the charge. Sulphide ores containing arsenic are best mixed with about twice the weight of a flux of zinc oxide and sodium carbonate in the proportion 4 to 1, and about five times the weight of iron filings added.

The weighed silver cover is placed on the crucible and the apparatus set up as shown in the illustration, Fig. 48.

The bottom of the crucible is gradually heated with a small Méker flame until it glows slightly. Overheating should be avoided. The upper portion of the crucible should never become hot and the lid should remain cold. After heating for about thirty minutes the system is allowed to cool without disconnecting the condenser. The disk is now removed, dipped in alcohol and dried in a desiccator over fused potash or soda. The increase of weight of the dried disk is due to metallic mercury.

NOTES. If the sample contains less than 1% mercury, take 2 grams; if 1 to 2% mercury, take 1 gram; if the sample contains 2 to 5%, take 0.5-gram sample. If high in mercury, grind sample with sand and take an aliquot portion.

It is advisable to repeat the test with a clean foil to be sure that all the mercury has been driven out of the sample. The foil may be freed from mercury by heating.

VOLUMETRIC DETERMINATION OF MERCURY
Seamon's Volumetric Method [1]

Seamon's Volumetric Method. Weigh 0.5 gram of the finely ground ore into an Erlenmeyer flask of 125 cc. capacity. Add 5 cc. of strong hydrochloric acid and allow it to act for about ten minutes at a temperature of about 40° C., then add 3 cc. of strong nitric acid and allow the action to continue for about ten minutes longer. The mercury should now all be in solution. Now if lead be present, add 5 cc. of strong sulphuric acid; it may be omitted otherwise. Dilute with 15 cc. of water and then add ammonia cautiously until the liquid is slightly alkaline. Bismuth, if present, will be precipitated. Acidify faintly with nitric acid, filter, receiving the filtrate in a beaker, and wash thoroughly.

Add to the filtrate 1 cc. of strong nitric acid that has been made brownish in color by exposure to the light, and titrate with a standard solution of potassium iodide until a drop of the liquid brought into contact with a drop of starch liquor, on a spot-plate, shows a faint bluish tinge. It is a good plan to set aside about one-third of the mercury solution and add it in portions until the end-point is successively passed, finally rinsing in the last portion and titrating to the end-point very carefully.

Deduct 0.5 cc. from the burette reading and multiply the remaining cc. used by the percentage value of 1 cc. in mercury to obtain the percentage in the ore.

The standard potassium iodide solution should contain 8.3 grams of the salt per liter. Standardize against pure mercuric chloride. Dissolve a weighed amount of the salt in water, add 2 cc. of the discolored nitric acid and titrate as above. One cc. of standard solution will be found equivalent to about 0.005 gram of mercury, or about 1% on the basis of 0.5 gram of ore taken for assay.

The precipitate of red mercuric iodide which forms during the titration may not appear if the amount of mercury present is very small, but this failure to precipitate does not appear to affect the result.

Iron, copper, bismuth, antimony, and arsenic, when added separately to the ore, did not influence the results in Seamon's tests. Silver interferes. Duplicate results should check within 0.1 to 0.2 of 1%.

[1] "Manual for Assayers and Chemists," p. 112.

Volumetric Thiocyanate Method for Mercury [1]

A sample containing 0.1 to 0.5 g. Hg placed in a flask is decomposed by adding 10 cc. dilute H_2SO_4 (1 : 1) and about 0.5 g. $KMnO_4$ crystals. The mixture is agitated and heated to fumes. The solution cooled is diluted to 50 cc. with cold water, then boiled and the MnO_2 dissolved by adding a few crystals of oxalic acid (small portions at a time).

The solution is filtered, and any residue washed with dilute (1 : 10) H_2SO_4. The sulphide group are now precipitated with H_2S and filtered off. The precipitate, transferred to a casserole, is digested for some time with dilute HNO_3 (2 : 1), the solution then diluted with hot water and filtered and the HgS washed with dilute HNO_3 (1 : 1).

The HgS is transferred to a flask with a few cc. of hot water and then 5 cc. of strong H_2SO_4 and 0.5 g. $KMnO_4$ are added and the mixture heated to fumes. Oxalic acid crystals are added until the MnO_2 dissolves and the mixture again heated to fumes to destroy the excess of oxalic acid. The solution, cooled, is diluted to 100 cc. (It should now be clear.)

About 5 cc. of a saturated solution of ferric ammonium sulphate solution (acidified with HNO_3) are added and the solution titrated with 0.1 N thiocyanate solution.

1 cc. 0.1 N thiocyanate = 0.01003 g. Hg.

Solutions. Ferric Indicator. Make a saturated solution of ferric ammonium sulphate or ferric sulphate. Add sufficient nitric acid (freed from nitrous acid by heating) to clear the solution and produce a pale yellow color. 5 cc. of this solution (the Editor prefers less) is used in the test. Ferric nitrate may be used if the sulphate is not available.

Thiocyanate Reagent. A tenth normal solution may be made by dissolving 7.4 grams of NH_4CNS or 9.2 grams of $KCNS$ in water and diluting to a liter. The solution may be standardized against a standard silver solution, containing 0.01079 grams silver per cc.

40 cc. of the silver solution is measured into a beaker or Erlenmeyer flask and diluted to about 100 cc. The ferric indicator is added and the solution is titrated with the thiocyanate solution. Each addition of the thiocyanate will produce a temporary red color, which fades out as long as there is silver uncombined with thiocyanate. A drop in excess of the thiocyanate produces a permanent faint red color.

[1] By A. H. Low (Chemist-Analyst, 1919, 29, 13).

Mercury in Organic Matter. The compound is decomposed by the method of Carius by heating in a closed tube (see page 145) with strong nitric acid (d. 1.42). The acid solution is neutralized by addition of sodium hydroxide and sufficient excess of the alkali added to insure a slight excess. Pure potassium cyanide is now added in quantity sufficient to dissolve the mercuric oxide precipitate, and the solution saturated with H_2S gas. Ammonium acetate is added and the solution boiled until nearly all the NH_3 has been expelled. The precipitate is allowed to settle and then filtered off and washed with hot water, and then with hot dilute HCl and again with water. The precipitate is dried at 110° C. and weighed as mercuric sulphide HgS.

NOTE. Should free sulphur be present its removal is accomplished by extraction with pure CS_2, see page 310.

MOLYBDENUM

Mo, *at.wt.* **96.0;** *sp.gr.* **8.6 — 9.01;** *m.p.* **2500° C;** *oxides,* **Mo₂O₃, MoO₂, MoO₃**

DETECTION

Molybdenum appears in the hydrogen sulphide group, being precipitated by H₂S in acid solution as the sulphide. It passes into solution by digestion with ammonium sulphide or sodium sulphide along with arsenic, antimony, tin, gold and platinum. By addition of metallic zinc, antimony, together with tin, gold and platinum are precipitated as metals while molybdenum remains in solution. Arsenic, that has not volatilized as arsine, is expelled by evaporation. Nitric acid is now added and the solution taken to dryness. Molybdenum is extracted from the residue with ammonium hydroxide.

A dilute solution of ammonium molybdate treated with a soluble sulphide gives a blue solution.

Sodium thiosulphate added to a slightly acid solution of ammonium molybdate produces a blue precipitate with a supernatant blue solution. With more acid a brown precipitate is formed.

Sulphur dioxide produces a bluish-green precipitate if sufficient molybdenum is present, or a colored solution with small amounts. The reducing agents, stannous chloride, or zinc in acid solution, produce a play of colors when they react with molybdenum solutions, due to the formation of the lower oxides. The solution becomes blue, changing to green, brown and yellow.

Molybdenum present as molybdate is precipitated by *disodium phosphate* as yellow ammonium phosphomolybdate from a nitric acid solution. The precipitate is soluble in ammonium hydroxide.

A pinch of powdered mineral on a porcelain lid, moistened with a few drops of *strong sulphuric acid*, stirred and heated to fumes, then cooled, will produce a blue color when breathed upon. The color disappears on heating, but reappears on cooling. Water destroys the color.

Molybdenite is very similar to graphite in appearance. It is distinguished from it by the fact that nitric acid reacts with molybdenite, MoS₂, leaving a white residue, but has no action upon graphite. The blowpipe gives SO₂ with molybdenite and CO₂ with graphite.

ESTIMATION

The determination is required in the ores—molybdenite, MoS₂, (60% Mo); molybdite, MoO₃ (straw yellow); wulfenite, PbMoO₄ (yellow, bright red, olive green or colorless); Ilsemannite, MoO₃+MoO₂; powellite, CaMoO₄; pateraite, CoMoO₄; belonesite, MgMoO₄; eosite, lead-vanado-molybdate; achromatite,

---

Chapter contributed by Wilfred W. Scott, A. M. Smoot and J. A. Holladay.

lead molybdate and arsenate with tin oxide and lead chloride. Some iron and copper ores also contain molybdenum.

The metal is determined in certain self-hardening steels and alloys.

The reagents ammonium molybdate and the oxide-molybdic acid, MoO_3, are valuable for analytical purposes. Tests of their purity may be required.

Preparation and Solution of the Sample

In dissolving the substance the following facts should be kept in mind: The metal is easily soluble in aqua regia; soluble in hot concentrated sulphuric acid, soluble in dilute nitric acid, oxidized by excess to MoO_3. It is dissolved by fusion with sodium carbonate and potassium nitrate mixture. It is insoluble in hydrochloric, hydrofluoric and dilute sulphuric acids.

The oxide MoO_3 is but slightly soluble in acids and alkalies; MoO_2 is insoluble in hydrochloric and hydrofluoric acids. MoO_3, as ordinarily precipitated, is soluble in inorganic acids and in alkalies. The oxide sublimed is difficultly soluble.

Molybdates of the heavy metals are insoluble in water, the alkali molybdates are soluble.

Ores. Molybdenum ores are decomposed by fusion with a mixture of sodium carbonate and potassium nitrate, or with sodium peroxide, in an iron crucible, 0.5 gram of the sample being taken and 10 times its weight of fusion mixture. The melt is disintegrated with about 150 cc. of water, the alkali partly neutralized with $(NH_4)_2CO_3$ and filtered. The molybdenum is in the filtrate, the iron remains in the residue.

It is advisable to dissolve the residue in a little dilute HCl, pour this solution into a hot solution of an excess of NaOH and again filter off the iron hydroxide, adding the filtrate to the first lot.

The combined filtrates and washings are treated with about 5 cc. of a 50% tartaric acid solution or its equivalent in crystals. (This prevents W and V from separating out) and the solution saturated with H_2S. The thio-molybdate solution is made slightly acid with H_2SO_4 (1 : 2) MoS_3 precipitates. Further details are the same as those given on page 323.

For Acid Decomposition see procedure on page 322.

Steel and Iron. One to 2 grams of the drillings are dissolved, a mixture of 25 cc. HCl and 2 cc. HNO_3, additional HNO_3 (1 : 1) being added to oxidize the iron, if necessary. ($KClO_3$ crystals may be used.) A large excess of the oxidizing agent is to be avoided. The solution is evaporated to near dryness and the pasty residue taken up with about 25 cc. water and 10 cc. HCl, and gently heated. A yellow residue is due to WO_3. This is removed by filtration and washed. The filtrate contains the molybdenum. This is now treated according to the procedure given under "Separations" for removal of iron.

Separations

Separation of Molybdenum from Iron. The solution containing the molybdenum is treated very cautiously with 2 N.NaOH solution to neutralize the greater part of the free acid, but not with such an amount that would color the solution red. The yellow solution is heated to boiling.

In a separate vessel is placed 2 N.NaOH in sufficient quantity to combine with all the iron of the sample and about 50% excess (1 cc. of 2 N.NaOH = 0.1 g. Fe) 30 cc. should be sufficient. This solution is heated to boiling and to it is added the hot solution containing the molybdenum. The sample should be added very slowly, preferably through a special funnel with capillary tube, stirring the solution vigorously during the addition. With care a complete separation of iron, free from molybdenum, may be effected, the molybdenum remaining in solution. The mixture is transferred to a 500 cc. volumetric flask.

The volume is made up to exactly 500 cc. and the precipitate allowed to settle. A portion is now filtered off, the first 5–10 cc. being rejected and the following 250 cc. of filtrate is retained for analysis of molybdenum.

Separation from the Alkaline Earths. Fusion of the substance with sodium carbonate and extraction of the melt with water gives a solution of molybdenum, whereas the carbonates of barium, calcium and strontium remain undissolved.

Separation from Lead, Copper, Cadmium and Bismuth. The sulphides of the elements are treated with sodium hydroxide and sodium sulphide solution and are digested by gently heating in a pressure flask. Molybdenum dissolves, whereas lead, copper, cadmium and bismuth remain insoluble. If the solution of the above elements is taken, made strongly alkaline, and treated with H_2S, the sulphides of the latter elements are precipitated and molybdenum remains in solution. The precipitates are filtered off and the filtrate containing molybdenum is placed in the pressure flask, the solution made slightly acid with sulphuric acid and the mixture heated under pressure, until the liquid appears colorless, MoS_2 is precipitated and may be converted into the oxide as described later.

Separation from Vanadium is effected by a molybdenum sulphide precipitation in acid solution.

Separation from Arsenic. Arsenic, present in the higher state of oxidation, is precipitated by magnesia mixture, added to a slightly acid solution (5 cc. of concentrated hydrochloric acid per 100 cc. of solution for each 0.1 gram arsenic). The solution is neutralized with ammonia (methyl orange), and the arsenic salt filtered off. MoS_2 is now precipitated with H_2S in presence of free sulphuric acid in the pressure flask.

Separation from Phosphoric Acid. Phosphoric acid is precipitated from an ammoniacal solution as magnesium ammonium phosphate. Molybdenum may then be precipitated as the sulphide from the filtrate.

Separation from Titanium. The metals of the ammonium sulphide group are precipitated by adding ammonium hydroxide and ammonium sulphide. Molybdenum remains in solution and passes into the filtrate. H_2S is passed into the solution until it appears red; sulphuric acid is then added until the solution is acid, when molybdenum sulphide precipitates.

Separation from Tungsten. Molybdenum may be precipitated by H_2S as MoS_2 in presence of tartaric acid. Tungsten does not precipitate.

Ether Extraction Method. Ether extracts not only iron but also molybdenum (see p. 248). The ether is evaporated off on a steam bath (avoid a free flame, as ether is inflammable) and the solution taken to near dryness. 10 cc. of sulphuric acid are added and hydrochloric acid expelled by concentration to fumes. After cooling, 100 cc. of water are added and 2–3 grams of ammonium bisulphite, to reduce the iron. The solution is boiled to expel the excess of SO_3 and molybdenum is precipitated by H_3S in a pressure flask. After cooling slowly, the sulphide, MoS, is filtered off, washed and ignited and weighed as MoO_2.

GRAVIMETRIC METHODS FOR THE DETERMINATION OF MOLYBDENUM

Precipitation as Lead Molybdate

Preliminary Remarks. This method, suggested by Chatard, has been pronounced by Brearly and Ibbotson to be " one of the most stable processes found in analytical chemistry." " It is not interfered with by the presence of large amounts of acetic acid, lead acetate, or alkali salts (except sulphates). The paper need not be ignited separately and prolonged ignition at a much higher temperature than is necessary to destroy the paper does no harm. From faintly acid solution lead molybdate may be precipitated free from impurities in the presence of copper, cobalt, nickel, manganese, zinc, magnesium and mercury salts." It may be readily separated from iron and chromium. Barium, strontium, uranium, arsenic, cadmium and aluminum do not interfere if an excess of hydrochloric acid has been added to the solution followed by lead acetate and sufficient ammonium acetate to destroy the free mineral acid.

The method is not adapted to use with molybdenite, MoS_2, because of the sulphate that forms on oxidation.

Vanadium and tungsten, if present, must be removed.

Special Reagents. *Lead Acetate.* A 4% solution is made by dissolving 20 grams of the salt in 500 cc. of warm water. A few cc. of acetic acid are added to clear the solution.

Precipitation of Lead Molybdate. The solution acidified with acetic acid (5 cc. per 200 cc.) and free from iron, is heated to near boiling and the lead acetate reagent added slowly until no further precipitation occurs and then about 5% excess. (1 cc. of the 4% lead acetate reagent will precipitate about 0.01 gram of molybdenum.) The precipitate is allowed to settle a few minutes and filtered hot into a weighed Gooch crucible or into a filter paper. (Refiltering first portion if cloudy.)[1] The precipitate is washed with hot water until free of chlorides and the excess of the lead acetate.

The precipitate dried and ignited in a porcelain crucible at red heat for about twenty minutes is weighed as $PbMoO_4$.

$$PbMoO_4 \times 0.2615 = Mo. \quad PbMoO_4 \times 0.3923 = MoO_3.$$

$$Mo \times 3.8241 = PbMoO_4. \quad MoO_3 \times 2.5491 = PbMoO_4.$$

Determination of Molybdenum as the Oxide, MoO_3

Especially applicable where fusion with an alkali carbonate has been required.

Decomposition of Ore. One gram of the ore is fused with 4 grams of fusion mixture, $(Na_2CO_3 + K_2CO_3 + KNO_3)$, and the cooled melt extracted with hot water.

If *manganese is present*, indicated by a colored solution, it may be removed by reduction with alcohol, the manganese precipitate filtered off and washed with hot water, the solution evaporated to near dryness and taken up with water, upon addition of nitric acid as stated below.

The solution containing the alkaline molybdate is nearly neutralized by adding HNO_3, the amount necessary being determined by a blank, and to the cold, slightly alkaline solution, a faintly acid solution of mercurous nitrate is

[1] NOTE. Addition of ammonium nitrate to the solution tends to prevent formation of colloidal $PbMoO_4$. Paper pulp (ashless) may be added to assist rapid filtration.

added until no further precipitation occurs. The precipitate consists of mercurous molybdate and carbonate (chromium, vanadium, tungsten, arsenic and
phosphorus will also be precipitated if present). The solution containing the
precipitate is boiled and allowed to stand ten to fifteen minutes to settle, the black
precipitate is filtered off and washed with a dilute solution of mercurous nitrate.
The precipitate is dried, and as much as possible transferred to a watch-glass.
The residue on the filter is dissolved with hot dilute nitric acid, and the solution
received in a large weighed porcelain crucible. The solution is evaporated to
dryness on the water bath and the main portion of the precipitate added to this
residue, and the product heated cautiously over a low flame [1] until the mercury
has completely volatilized. The cooled residue is weighed as MoO_3.

$$MoO_3 \times 0.6667 = Mo.$$

Note. If Cr, V, W, As or P are present a separation must be effected. Molybdenum should be precipitated in an H_2SO_4 solution in a pressure flask as the sulphide
by H_2S as given in the following method, and arsenic if present removed by magnesia
mixture as indicated in the procedure for separation of arsenic from molybdenum.
If these impurities are present the molybdenum oxide may be fused with a very little
Na_2CO_3, and leached with hot water and the filtrate treated with H_2S as directed.

Precipitation of Molybdenum as the Sulphide by H_2S

A. **Precipitation from Acid Solution.** By this procedure molybdenum
is precipitated along with members of the hydrogen sulphide group, if present,
but free from elements of the following groups.

The cold molybdenum solution slightly acid with sulphuric acid (in presence
of Ba, Sr or Ca an HCl solution is necessary) is placed in a small pressure
flask and saturated with H_2S, the flask closed and heated on the water bath until
the precipitate has settled. The solution is cooled and filtered through a weighed
Gooch crucible.

B. **Precipitation from an Ammoniacal Solution.** By this procedure molybdenum is precipitated with antimony, arsenic, tin if present, but is free from
mercury, lead, bismuth, copper and cadmium.

Hydrogen sulphide is passed into the cold ammoniacal solution of molybdenum (in presence of tungsten or vanadium add tartaric acid) until it assumes
a bright red color, it is now acidified with dilute sulphuric acid, the precipitate
allowed to settle and the solution filtered through a weighed Gooch crucible.

In either case A or B the precipitate is washed into the Gooch crucible
with very dilute sulphuric acid followed by several washings with the acid and
then with alcohol until free from acid. The Gooch is placed within a larger
nickel crucible and covered with a porcelain lid. After drying at 100° C. it is
placed over a small flame and carefully heated until the odor of SO_2 can no longer
be detected. The cover is now removed and the open crucible heated to constant
weight. The residue consists of MoO_3.

$$MoO_3 \times 0.6667 = Mo.[1]$$

Note. Arsenic will contaminate the residue if present. The method for its
removal has been given.

[1] The oxide, MoO_3, sublimes at bright red heat. The volumetric method is more
reliable. See page 319.

VOLUMETRIC METHODS FOR THE DETERMINATION OF MOLYBDENUM OR MOLYBDIC ACID

The Iodometric Reduction Method [1]

Principle. When a mixture of molybdic acid and potassium iodide in presence of hydrochloric acid is boiled, the volume having defined limits, free iodine is liberated and expelled and the molybdic acid reduced to a definite lower oxide; by titrating with a standard oxidizing agent the molybdic acid is determined.

Reaction. $2MoO_3 + 4KI + 4HCl = 2MoO_2I + I_2 + 4KCl + 2H_2O$.

Reagents. N/10 solutions of iodine, sodium arsenite, potassium permanganate, sodium thiosulphate.

Analytical Procedure.[2] **Reduction.** The soluble molybdate in amount not exceeding an equivalent of 0.5 gram MoO_3 is placed in a 150-cc. Erlenmeyer flask, 20 to 25 cc. of hydrochloric acid (sp.gr. 1.2) added together with 0.2 to 0.6 gram potassium iodide. A short stemmed-funnel is placed in the neck of the flask to prevent mechanical loss during the boiling. The volume of the solution should be about 60 cc. The solution is boiled until the volume is reduced to exactly 25 cc. as determined by a mark on the flask. The residue is diluted immediately to a volume of 125 cc. and cooled. Either process A or B may now be followed.

A. **Reoxidation by Standard Iodine.** A solution of tartaric acid, equivalent to 1 gram of the solid, is now added, and the free acid nearly neutralized with sodium hydroxide solution (litmus or methyl orange indicator) and finally neutralized with sodium acid carbonate, $NaHCO_3$, added in excess. A measured amount of N/10 iodine is now run in. The solution is set aside in a dark closet for two hours, in order to cause complete oxidation, as the reaction is slow. The excess iodine is now titrated with N/10 sodium arsenite.

One cc. N/10 iodine = .0144 gram MoO_3 = .0096 gram Mo.

On long standing a small amount of iodate is apt to form. This is determined by making acid with dilute HCl and titrating with N/10 sodium thiosulphate.

B. **Reoxidation of the Residue by Standard Permanganate.** To the reduced solution about 0.5 gram of manganese sulphate in solution is added, followed by a measured amount of N/10 permanganate solution, added from a burette until the characteristic pink color appears. A measured amount of standard N/10 sodium arsenite, equivalent to the permanganate is then run in and about 3 grams of tartaric acid added. The acid is neutralized by acid sodium or potassium carbonate, the stopper and the sides of the flask rinsed into the main solution. The residual arsenite is now titrated by N/10 iodine, using starch indicator.

NOTES. Tartaric acid prevents precipitation during the subsequent neutralization with $NaHCO_3$. A and B.

The addition of manganese salt in B is to prevent the liberation of free chlorine by the action of $KMnO_4$ on HCl.

In addition to the oxidation of the lower oxides to molybdic acid, potassium permanganate added in B liberates free iodine from HI, it produces iodic acid, and forms the higher oxides of manganese. The standard arsenite, on the other hand, converts free iodine and the iodate to HI and reduces the higher oxides of manganese.

[1] F. A. Gooch and Charlotte Fairbanks, Am. Jour. Sc. (4), **2**, 160.

[2] F. A. Gooch and O. S. Pulman, Jr. Am. Jour. Sc. (4), **12**, 449.

Estimation by Reduction with Jones Reductor and Oxidation by Standard Permanganate Solution

Principle. The procedure depends upon the reduction of molybdic acid to Mo_2O_3 by passing its solution through a column of amalgamated zinc into a solution of ferric alum, and subsequent oxidation to MoO_3 by standard potassium permanganate solution.

Reactions. $2MoO_3 + 3Zn = Mo_2O_3 + 3ZnO.$

$$5Mo_2O_3 + 6KMnO_4 + 9H_2SO_4 = 10MoO_3 + 3K_2SO_4 + 6MnSO_4 + 9H_2O.$$

Reagents. *Potassium permanganate* approximately N/10 standardized against a standard molybdic acid solution.

10% solution of *ferric alum*.

2.5% solution of *sulphuric acid*.

Apparatus. *Jones Reductor*.

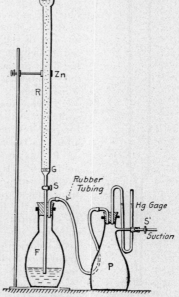

$R =$ reductor tube 50 cm. long, 2 cm. inside diameter. Smaller tube prolongation length 20 cm. inside diameter 0.5 cm.

$Zn =$ column of zinc 40 cm. long. Zn shot 8 mesh to sq.cm.;

$F =$ receiving flask;

$P =$ pressure regulator with gauge, set to give pressure in receiving flask of less than 20 cm. water;

$G =$ platinum cone or gauze with mat of fine glass wool 2 cm. thick;

The zinc in reductor should be protected from the air by covering with water, stop-cock S being closed when not in use.

Procedure. The receiving flask of the Jones reductor, Fig. 49, is charged with about 30 cc. of 10% ferric alum and 4 cc. of phosphoric acid. Through the 40-cm. column of amalgamated zinc in the reductor are passed in succession 100 cc. of dilute sulphuric acid (2.5% sol.), the molybdic acid in the form of ammonium molybdate dissolved in 10 cc. of water and acidified with 100 cc. of dilute sulphuric acid followed by 200 cc. more of the

FIG. 49.—Jones Reductor.

dilute sulphuric acid and 100 cc. of water. The reduced green molybdic acid upon coming in contact with the ferric alum solution produces a bright red color.

The solution is titrated with N/10 KMnO$_4$ solution

$$\text{One cc. of N/10 KMnO}_4 = \frac{.0144}{3} \text{ gram MoO}_3 = \frac{.0096}{3} \text{ gram molybdenum.}$$

[1] W. A. Noyes and Frohman, Jr. Am. Chem. Soc., **35**, 919. After a thorough examination of the various methods for determining molybdenum, Messrs. Smoot, Lundell and Holladay pronounce this method to be superior to any other.

Notes on Reductor Technique

Use of the following greatly simplified and correspondingly more rapid reductor technique is recommended by J. A. Holladay.

1. The reductor should contain a column of 20 x 30 mesh amalgamated zinc $\frac{3}{4}''$ in diameter and 10 inches long. If the molybdenum solution is given a preliminary reduction by heating with 2 grams of 20 mesh zinc for the purpose of precipitating copper (which is subsequently removed by filtration on asbestos), the length of the zinc column may safely and advantageously be reduced to 8''. The lengths specified apply only to a column $\frac{3}{4}''$ in diameter; the lengths for tubes of different diameters would have to be determined experimentally. The zinc should be as free from iron and as low in other impurities as possible.

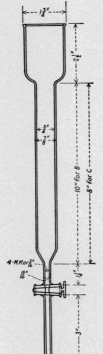

All Dimensions Approx.

FIG. 49a.—Reductor Tube.

2. If the reductor has been standing idle for longer than one day it is cleaned by passage of 50 cc. of 6% sulphuric acid, and 50 cc. of water, both at room temperature, a small amount of the water being left in the funnel which forms the reductor inlet. The acid and water are discarded after having been passed through the reductor. These operations are only necessary when the reductor has been standing idle for some time.

3. The required amount of ferric phosphate solution and sufficient water (about 50 cc.) to cause the tip of the reductor outlet tube to dip beneath the surface of the liquid are added to the flask.

4. The molybdenum solution (volume 50 to 100 cc., acidity 6% H_2SO_4) which may be at room temperature or slightly warm, is passed through the reductor rapidly. The total time required for the passage of the molybdenum solution and the liquids which follow it need not exceed from 1 to 3 minutes. No advantage results from having the solution hot—on the contrary, the greater action of the hot acid solution on the zinc is a disadvantage. No increase in accuracy is obtained by passing the solution through the reductor at a slower rate than that specified.

5. When the funnel which forms the inlet of the reductor is nearly but not entirely empty, 150 cc. of water at room temperature are passed through the reductor. In the case of samples containing relatively small amounts of molybdenum, 100 cc. of water is sufficient. In no case is the funnel permitted to become entirely empty and the stopcock is closed while some of the wash water remains above the surface of the zinc.

NOTE. It is unnecessary to have this solution hot as is sometimes recommended. This statement is confirmed by recent work of G. E. F. Lundell and H. B. Knowles, U. S. Bureau of Standards, Ind. Eng. Chem.

Method for Determining Molybdenum and Vanadium in a Mixture of their Acids

Principle of the Method. The procedure depends upon the fact that vanadic acid alone is reduced by SO_2[1] in a sulphuric acid solution, whereas both vanadic and molybdic acids are reduced by amalgamated zinc, in each case the reducing agents forming definite lower oxides which are readily oxidized to definite higher oxides by $KMnO_4$.

Reactions.

SO_2 Reduction:

    1. $V_2O_5 + SO_2 = V_2O_4 + SO_3$. (No action on MoO_3.)

Zn Reduction:

    2. $V_2O_5 + 3Zn = V_2O_2 + 3ZnO$.
    3. $2MoO_3 + 3Zn = Mo_2O_3 + 3ZnO$.

$KMnO_4$ Oxidation:

    4. $5V_2O_4 + 2KMnO_4 + 3H_2SO_4 = 5V_2O_5 + K_2SO_4 + MnSO_4 + 3H_2O$.
    5. $5V_2O_2 + 6KMnO_4 + 9H_2SO_4 = 5V_2O_5 + 3K_2SO_4 + 6MnSO_4 + 9H_2O$.
    6. $5Mo_2O_3 + 6KMnO_4 + 9H_2SO_4 = 10MoO_3 + 3K_2SO_4 + 6MnSO_4 + 9H_2O$.

From the reactions "4" and "5" it is seen that three times the amount of $KMnO_4$ is required to oxidize V_2O_2 to V_2O_5 as is required in the case of V_2O_4, hence—total cc. $KMnO_4$ required in oxidation of the zinc-reduced oxides minus three times the cc. $KMnO_4$ required in oxidizing the tetroxide of vanadium formed by the sulphur dioxide reduction = cc. $KMnO_4$ required to oxidize Mo_2O_3 to MoO_3. From these data molybdenum and vanadium may readily be calculated.

Method of Procedure. *A.* **Vanadic Acid.** The solution containing the vanadic and molybdic acids in a 250- to 300-cc. Erlenmeyer flask, is diluted to 75 cc. acidified with 2 to 3 cc. of strong sulphuric acid, heated to boiling and the vanadic acid reduced by a current of SO_2 passed into the solution until the clear blue color indicates the complete reduction of the vanadic acid to V_2O_4. The boiling is now continued and CO_2 passed into the flask to expel the last trace of SO_2.

Standard N/10 $KMnO_4$ is now run into the reduced solution to the characteristic faint pink. From reaction "4," vanadic acid may be calculated.

    One cc. N/10 $KMnO_4 = .0182$ gram $V_2O_5 = .0051$ gram vanadium.

B. **Molybdic Acid.** The reduction by Jones' reductor, and titration of the combined acids reduced by amalgamated zinc with N/10 potassium permanganate solution, is carried out exactly as described in the determination of molybdic acid alone. In this case 50 cc. of 10% ferric alum and 8 cc. of the phosphoric acid is placed in the receiving flask.

Calculation. Total permanganate titration in B minus three times the titration in A gives the permanganate required to oxidize Mo_2O_3 to MoO_3. From equation 6 the molybdic acid may now be calculated.

$$\text{One cc. N/10 } KMnO_4 = \frac{.0144}{3} \text{ gram } MoO_3 = \frac{.0096}{3} \text{ gram molybdenum.}$$

[1] Reduction of vanadium by SO_2 in presence of molybdenum, Graham Edgar, Am. Jour. Sc., (4) **25**, 332. No reduction of MoO_3 when 0.4 gram is present with 5 cc. H_2SO_4 in 25 cc. volume.

For theoretical considerations and data on accuracy of method see "Methods in Chemical Analysis," F. A. Gooch.

27

DETERMINATION OF MOLYBDENUM IN ORES[1]

The commercial ores of molybdenum are molybdenite, the native sulphide (MoS_2), and wulfenite, lead molybdate. Occasionally molybdenum ochre (the trioxide) may be met with, but this mainly occurs in very small proportion associated with molybdenite as an oxidation product of the sulphide.

Molybdenite is sold either as high grade selected mineral or as concentrates which are usually the product of flotation processes. The important determinations in molybdenite are molybdenum, arsenic, copper and phosphorus; sometimes bismuth is present and occasionally lead.

Owing to the complex nature of wulfenite the determination of molybdenum is more difficult and the number of impurities which may be sought for is much greater.

The following methods for the determination of molybdenum in commercial ores and concentrates are the result of co-operative work between the laboratories of the Electro Metallurgical Company and Ledoux & Company. It is believed that all important sources of error have been considered and eliminated so far as the usual commercial ores are concerned.

Determination of Molybdenum in Wulfenite or Molybdenite

One gram of the finely ground ore is dissolved by treatment with 15 cc. of nitric acid (sp.gr. 1.42) and 7 cc. of sulphuric acid (sp.gr. 1.84) at a temperature just short of boiling in a 150 cc. covered beaker. When practically complete decomposition has been effected, the liquid is evaporated until fumes of sulphur trioxide are freely expelled. After cooling addition is made of about 40 cc. of water, the solution is boiled to dissolve the molybdenum, cooled to room temperature and filtered into a 250 cc. beaker. The residue, consisting of lead sulphate, silica, etc., is washed with very dilute H_2SO_4. The residue rarely, if ever, contains Mo, nevertheless it should be examined to make sure that decomposition is complete. To this end it is digested with 15 cc. to 25 cc. of ammonium acetate solution (prepared by mixing 8 cc. of $(NH_4)OH$ (0.90) with 10 cc. H_2O and adding 7 cc. of 99% $C_2H_4O_2$) to remove all lead sulphate. The filter paper is washed with hot dilute acetic acid and with water. It is then ignited in a platinum crucible and the silica is removed by treatment with a drop of sulphuric acid and several cc. of hydrofluoric acid and evaporating to the expulsion of sulphuric acid. If an appreciable residue remains after this treatment, it is fused with potassium pyrosulphate and tested for molybdenum by means of tartaric acid and ammonium sulphide as described later.

To the solution containing all of the molybdenum there is added sufficient ferric sulphate to provide ten times as much iron as there is arsenic present: from 0.2 to 0.3 gram of iron is usually ample. The acid solution is then nearly neutralized with ammonia (addition of a sufficient amount to impart a red tint to the clear yellow solution is to be avoided), heated nearly to boiling and poured very slowly and with vigorous stirring into 75 cc. of warm ammonia solution (15 cc. $(NH_4)OH$ 1.90 sp.gr. 60 cc. H_2O) contained in a 250 cc. beaker. When it has settled the ferric hydroxide (which will carry down quantitatively all arsenic in the ore) is filtered and washed thoroughly with hot water; it is then dissolved in a slight excess of hot dilute (1 to 4) sulphuric acid and the resulting solution is again heated to boiling and poured into 75 cc. of warm

[1] By James A. Holladay and A. M. Smoot.

ammonia solution as before; the iron hydrate is again washed. The two filtrates, which will contain all the molybdenum, are collected in a 600 cc. beaker. It is essential that arsenic, which is usually present in these ores, be eliminated, and the method described furnishes a simple and effective way for accomplishing this. The addition of ferric sulphate would be omitted only in case arsenic is known to be absent or when the ore itself carries sufficient iron to take care of the arsenic.

To the combined ammoniacal filtrates there are added 2 grams of tartaric acid and when the acid has dissolved the liquid is saturated with hydrogen sulphide. The presence of tartaric acid is necessary to prevent precipitation of vanadium and tungsten along with the molybdenum. Both of these elements are ordinary constituents of wulfenite concentrates. Under these conditions the molybdenum remains in the solution as ammonium thiomolybdate, $(NH_4)_2MoS_4$, which imparts a deep red color to the solution. If a small precipitate of insoluble sulphides separates out, it is filtered off and washed with dilute ammonium sulphide solution; if the solution remains clear this step is omitted. Copper in the amounts usually present remains entirely in solution at this point and is reprecipitated with molybdenum when the solution is subsequently acidified. The thiomolybdate solution is then made slightly acid with sulphuric acid (1 to 2). Molybdenum is thus precipitated as trisulphide.

The cessation of effervescense on addition of more acid and the disappearance of the red color of the ammonium thiomolybdate mark the point where sufficient acid has been added to complete the reactions.

After heating for a short time, the precipitate is allowed to settle and filtered on an 11 cm. paper and washed thoroughly with hydrogen sulphide water containing a small amount of sulphuric acid.

The filtrate from the molybdenum sulphide sometimes contains appreciable amounts of molybdenum; addition is made to it of 15 cc. of nitric acid (sp.gr. 1.42) and the solution is evaporated to fumes of sulphur trioxide, more sulphuric acid being added if sufficient is not already present. After allowing it to cool, 5 cc. of concentrated nitric acid are added and the solution is again evaporated to fumes. The addition of nitric acid and evaporation to strong fumes is repeated once more to insure the destruction of all organic compounds. After allowing to cool, sufficient water is added to dissolve all salts, two grams of tartaric acid are added, and after addition of an excess of ammonia (sp.gr. 0.90) the warm liquid is thoroughly saturated with washed hydrogen sulphide and filtered. The filtrate is acidified with sulphuric acid (1 to 2) and if after standing for fifteen to thirty minutes in a warm place any molybdenum trisulphide has separated out it is filtered, washed well with hydrogen sulphide water containing a small amount of sulphuric acid, and combined with the main precipitate. The precipitation of the molybdenum as sulphide when carried out as described effects the separation and removal of tungsten, vanadium and chromium.

The molybdenum sulphide precipitate and paper, or precipitates and papers, are put into a 250 cc. beaker and treated with 6 cc. of sulphuric acid (sp.gr. 1.84) and 10 cc. of nitric acid (sp.gr. 1.42) and the liquid is cautiously boiled until dense fumes of sulphur trioxide are freely evolved. After allowing it to cool somewhat, 5 cc. of nitric acid (sp.gr. 1.42) are added and the evaporation is repeated. The evaporation with 5 cc. portions of strong nitric acid is repeated several times until the filter paper has been completely destroyed and every trace of yellow color, due to carbonaceous matter, has disappeared. When this has been accomplished the solution is fumed strongly for a short while, cooled, 5 cc. of water are added and the liquid is again taken to fumes of sulphur trioxide in order to insure the expulsion of every trace of nitric acid. When cool, addition is made of approximately 75 cc. of water and the contents of the beaker are boiled for a few minutes, which should give a perfectly clear solution. Five grams of granulated zinc (0.002% iron or under) are then added and the solution is warmed until most of it has dissolved; this results in partial reduction of molybdenum and complete precipitation of copper, which is usually present. The liquid is then filtered on an asbestos or "alundum" filter to remove the undissolved zinc and the copper.

All the molybdenum in the solution is then reduced from the sexivalent to the trivalent condition by passage through the zinc reductor into a solution of ferric sulphate containing phosphoric acid—the solution is then titrated with N/10 permanganate. The details of the reductor and the method of preparing it are given on page 319 and 320.

It is essential to run a blank determination, using the same quantity of zinc that was used to separate copper and the same volumes of water and acid—passing the solution through the reductor under the same conditions. With good zinc the correction may be about 0.2 cc.

Method for Molybdenum in Pure Molybdenite Concentrates

Molybdenum may be determined in the same way as in wulfenite, but owing to the higher Mo content of the material only 0.5 gram may be taken for the determination.

In case of very pure and high-grade molybdenite, which rarely contains any interfering element except traces of arsenic, the method may be simplified.

It suffices to treat 0.5 gram of the pulp with 10 cc. of strong HNO_3 and 7 cc. of H_2SO_4 and evapoate to fumes. A single treatment usually results in complete decomposition, but if the solution shows any particle of undecomposed material a further addition of 5 cc. of HNO_3 and a second evaporation to fumes may be necessary. In order to remove all HNO_3, the H_2SO_4 solution is diluted by washing down the sides of the beaker with 5 cc. of water and the evaporation to fumes is repeated.

After cooling, 35 cc. of water and ferric sulphate equivalent to about 0.1 gram of metallic iron are added. The solution is poured into an ammonia solution, the same as in the longer method, the iron hydrate carrying any arsenic is filtered off, dissolved in H_2SO_4 and reprecipitated. The united filtrates contain all the molybdenum; they are concentrated by evaporation to a volume of 150 cc. which also serves to expel the excess of ammonia. The solution is acidulated with H_2SO_4, reduced and titrated as in the longer method.

This method, of course, is not applicable in the presence of vanadium, tungsten or other impurities which might be reduced and reoxidized in the final operations, but it serves very well for the analysis of high grade molybdenite.

Determination of Commonly Occurring Substances in Molybdenum Ores

Determination of Arsenic

Two grams of the 100-mesh sample are treated with 20 cc. of nitric acid (sp.gr. 1.42) and 7 cc. of sulphuric acid (sp.gr. 1.84) in a 150 cc. beaker. The liquid is heated for a short time at a temperature just below the boiling point and is then evaporated until fumes of sulphuric anhydride are freely evolved. The residue is taken up with 40 cc. of water and boiled for a few minutes. Approximately one gram of cuprous chloride and five grams of ferrous ammonium sulphate are added, and transfer is made of the solution to the distillation flask of the apparatus shown on page 39, arsenic chapter, the beaker being rinsed with 40 cc. of hydrochloric acid (sp.gr. 1.20). It is important that the amount of water used be kept to a minimum so as to insure presence of sufficiently concentrated hydrochloric acid in the distillation flask to rapidly and completely volatilize the arsenic.

For distillation of the arsenic, the Knorr Arsenic and Antimony Distillation Apparatus (mentioned above see page 39) is very satisfactory. The rate at which arsenic distills over is greatly influenced by the size and form of apparatus employed; by use of the Knorr apparatus it is rapidly and completely volatilized. After having attached the flask, the inlet funnel is filled about half full of hydrochloric acid (sp.gr. 1.20). The outlet of the condenser tube is caused to dip just beneath the surface of 100 cc. of distilled water in a 350 cc. beaker containing a lump of ice. The solution in the distillation flask is heated to boiling; concentrated hydrochloric acid is introduced through the funnel tube drop by drop at a rate sufficient to replace the evaporation, or in other words, in amount sufficient to keep the level of liquid in the flask constant. It is necessary to keep a steady flame under the flask; if the solution in the beaker starts to be sucked back into the condenser tube, the apparatus is quickly raised so as to lift the lower end of the condenser out of the distillate. All of the arsenic will usually distill over in 30 or 35 minutes; at the end of this time the beaker holding the condensate is replaced by another one containing about 100 cc. of water and a lump of ice, and the distillation is continued for another 15 minutes. The second distillate is tested separately to determine whether it contains any arsenic.

By the above described procedure the arsenic is separated from any other elements likely to be present by volatilization as arsenious chloride, $AsCl_3$; it may be determined in the resulting solution by either gravimetric or volumetric methods. Where only occasional determinations have to be made, the gravimetric method is perhaps the more convenient. The hydrochloric acid distillate is warmed and saturated with hydrogen sulphide by passage of a rapid stream of gas. Filtration is made of the arsenious sulphide, As_2S_3, and it is washed with hydrogen sulphide water. The yellow precipitate is dissolved in the least necessary amount of ammonia (1 to 1) and the paper washed with hot water; the volume is kept as low as possible, the filtrate being received in a 150 cc. beaker. To the filtrate there are added 6–10 cc. of "perhydrol" (30% H_2O_2) for every 0.1 gram of arsenic supposed to be present; if "perhydrol" is not available, there are used instead 60–100 cc. of 3% hydrogen peroxide (free from phosphorus), but use of the former is preferable since it permits keeping the volume small. It is essential that enough hydrogen peroxide be used to oxidize the arsenic completely to the arsenic condition. Some commercial brands of hydrogen peroxide contain considerable amounts of phosphorus, making it always necessary to determine it and make a correction for the weight of magnesium pyrophosphate formed from it which would otherwise be counted as arsenic. The resulting solution is evaporated on a hot plate for some time, boiled for ten minutes with the cover glass on, and cooled in ice water. Ten cc. of "magnesia mixture" are added, followed by ammonia (sp.gr. 0.90) to the extent of one-third the volume of the solution, and the liquid is stirred vigorously for some time and then allowed to stand in a cool place for at least 15 hours. The magnesium ammonium arsenate is filtered on a weighed Gooch crucible and washed 10–15 times with small amounts of cold 2.5% ammonia water. It is then drained as completely as possible by suction, dried at 105° C., and heated gradually in an electric muffle furnace to a temperature of 400°–500° C. until no more ammonia is evolved. The temperature is then increased to 800°–900° C. and kept there for about 10 minutes. After having cooled the crucible in a desiccator and weighed it, the amount of arsenic is calculated from the weight of magnesium pyroarsenate, $Mg_2As_2O_7$, which contains 48.27% arsenic. A "blank" must be run on the reagents used and any arsenic found be deducted.

When more than occasional analyses have to be made, the following volumetric method can be used to advantage; the only superiority of the gravimetric process is that visual evidence of the presence of arsenic is furnished by the yellow arsenious sulphide. The distillate containing the arsenic in the form of arsenious chloride, As_2Cl_3, is made slightly ammonical, litmus paper being used as indicator, and the solution being kept cool. Sufficient sulphuric acid (1 to 1) to render the liquid slightly acid is next added. To the cold, slightly acid solution there is added 8–10 grams of sodium bicarbonate, 0.3 gram of potassium iodide and several cc. of starch solution (a water solution of Kahlbaum's "Lösliche Starke" or soluble starch is used) and the solution is titrated with a standard iodine solution each cc. of which is equal to 0.001 gram of arsenic. The titration is based on the reaction:

$$KH_2AsO_3 + I_2 + 2NaHCO_3 = KHAsO_4 + 2NaI + 2CO_2 + H_2O.$$

The reaction between arsenic and iodine is shown by the equation:

$$As_2O_3 + 2I_2 + 2H_2O = As_2O_5 + 4HI.$$

A "blank" must be run and deducted.

The iodine solution is prepared by dissolving 3.5 grams of pure resublimed iodine with 7 grams of potassium iodide and a very small amount of water. When the iodine has completely dissolved, the solution is diluted to 1 liter and mixed thoroughly; it is standardized against pure As_2O_3. This is accomplished by carefully weighing 0.0600 gram of pure resublimed As_2O_3 and transferring it to a 350 cc. beaker. Eight to ten grams of sodium bicarbonate and sufficient hot water to dissolve it are added and the solution is heated until the As_2O_3 has completely dissolved. The solution is then diluted to 250 cc., cooled, 0.3 gram of potassium iodide and 2–3 cc. of starch solution are added and the liquid is titrated with the iodine solution to the usual blue end point. The weight of As_2O_3 used (0.0600 gram) contains exactly 0.04545 gram of arsenic so the number of cc. used in the titration, divided into 0.04545, gives the arsenic value of one cc. of the iodine.

Determination of Copper

(a) **In Molybdenite.** Four grams of finely ground sample are treated with 35 cc. of nitric acid (sp.gr. 1.42) and 10 cc. of sulphuric acid (sp.gr. 1.84) in a 250 cc. beaker provided with a clock-glass cover. The solution is digested at a temperature somewhat below the boiling point until most of the molybdenite has dissolved. Subsequently the liquid is boiled until strong fumes of sulphur trioxide are expelled. After having allowed the beaker and its contents to cool, 50 cc. of water are added and the solution is boiled briskly for a few minutes and filtered on a 9 cm. paper into a 250 cc. beaker. The residue is washed thoroughly with hot water and discarded.

An excess of sodium hydroxide solution is added to the filtrate and the solution is boiled for several minutes and filtered on a 9 cm. paper. The precipitate of ferric hydroxide, etc., is washed with hot water to remove the molybdenum. Solution is then made of this precipitate, which will contain all of the copper, in 30 cc. of hot dilute (1 to 3) sulphuric acid, the paper being washed with about 50 cc. of hot water and the filtrate and washings being collected in a 150 cc. beaker.

The copper in the solution obtained as above described is completely precipitated in the metallic condition by placing a sheet of pure aluminum (conveniently bent into the form of a triangle 1 inch in height which stands on its edge in the beaker) in the beaker and boiling for about ten minutes. Complete precipitation is not obtained until the iron has been reduced, when the aluminum should appear clean and the precipitated copper be detached or only loosely adherent. After removal from the source of heat, the clock glass and sides of the beaker are rinsed with a jet of hydrogen sulphide water. The aluminum, the copper content of which should be accurately determined, is weighed before and after use and a correction is applied for the copper introduced from it.

Filtration is made of the solution on a very small filter paper, the copper being transferred to the paper and the aluminum being left as clean as possible in the beaker. The precipitate is washed with hot water and the filtrate and washings are discarded. Solution is then made of the copper in 5 cc. of hot nitric acid (1 to 1), this being accomplished by first dropping the acid over the aluminum and then pouring it on the copper. The aluminum and filter paper are both washed with hot water, the filtrate and washings being collected in a 100 cc. lipless beaker. After boiling the solution to expel oxides of nitrogen, 2 cc. of sulphuric acid (sp.gr. 1.84) are added and determination is made of the copper by electrolysis, a platinum gauze cathode and spiral anode being employed.

Determination of Copper

(b) **In Wulfenite.** Four (4.0000) grams of the 100-mesh sample are treated with approximately 30 cc. of 10% sodium hydroxide solution in a 250 cc. beaker provided with a clock-glass cover. The solution is boiled briskly for 10 minutes, which is usually sufficient to insure practically complete decomposition of the wulfenite. Addition is then made of about 70 cc. of hot water and the precipitate of iron oxide, etc., which will contain all of the copper, is allowed to settle. The supernatant liquid is decanted through a 9 cm. filter paper and the precipitate is then transferred to the paper and washed well with hot water. The filtrate and washings are discarded. As a result of these operations separation of the copper from practically all of the molybdenum and from the greater part of the lead is accomplished.

The paper containing the residue is returned to the original 250 cc. beaker and treated with 6 cc. of sulphuric acid (sp.gr. 1.84) and 10 cc. of nitric acid (sp.gr. 1.42). The contents of the beaker are cautiously heated until frothing is nearly over, and then boiled briskly until fumes of sulphuric anhydride are evolved. After having permitted the liquid to partly cool, 10 cc. of nitric acid (sp.gr. 1.42) are added and the evaporation to fumes is repeated. The cover glass and sides of the beaker are cleaned with a few cc. of water and the solution is once more evaporated to fumes in order to insure destruction of all carbonaceous matter from the filter paper and complete expulsion of nitric acid. In practice the total time consumed in making these evaporations is small.

Having allowed the beaker and its contents to cool somewhat, addition is made of 50 cc. of water and the solution is boiled for a short while, cooled, and filtered on a 9 cm. paper into a 150 cc. beaker. The precipitate of lead sulphate, etc., is washed with cold 10% sulphuric acid and discarded, the copper passing completely into the filtrate and washings, which should have a total volume of not more than 100 cc.

The copper in the solution obtained as above described is completely precipitated in the metallic condition by placing a sheet of pure aluminum (conveniently bent into the form of a triangle 1 inch in height which stands on its edge in the beaker) in the beaker and boiling for about ten minutes. Complete precipitation is not obtained until the iron has been reduced, when the aluminum should appear clean and the precipitated copper be detached or only loosely adherent. After removal from the source of heat, the clock-glass and sides of the beaker are rinsed with a jet of hydrogen sulphide water. The aluminum, the copper content of which must be accurately determined, is weighed before and after use and a correction is applied for the copper introduced into the solution from it.

Filtration is made of the solution on a very small filter paper, the copper being transferred to the paper and the aluminum being left as clean as possible in the beaker. The precipitate is washed with hot water and the filtrate and washings are discarded. Solution is then made of the copper in 5 cc. of hot nitric acid (1 to 1), this being accomplished by first dropping the acid over the aluminum and then pouring it on the copper. The aluminum and filter paper are both washed with hot water, the filtrate and washings being collected in a 100 cc. lipless beaker. After boiling the solution to expel oxides of nitrogen, 2 cc. of sulphuric acid (sp.gr. 1.84) are added and determination is made of the copper by electrolysis, a platinum gauze cathode and spiral anode being employed.

Determination of Phosphorus

(a) **Molybdenite.** Four grams of the finely ground sample are treated with 35 cc. of nitric acid (sp.gr. 1.42) and 10 cc. of sulphuric acid (sp.gr. 1.84) in a 250 cc. beaker provided with a cover glass. The liquid is heated at a temperature somewhat below the boiling point until decomposition of the mineral appears to have been secured, when it is boiled until strong fumes of sulphuric anhydride are given off. When the residue has cooled, 40 cc. of water are added and the solution is boiled for several minutes and filtered on a 9 cm. paper into a 300 cc. Erlenmeyer flask. The residue is washed well with hot water, ignited in a porcelain crucible and transferred to a platinum crucible and evaporated with 2 or 3 drops of sulphuric acid (sp.gr. 1.84) and several cc. of hydrofluoric acid. After completely expelling the sulphuric acid, any small residue remaining is fused with a pinch of potassium pyrosulphate and dissolved in water and added to the main solution.

A few drops of strong permanganate solution (25 grams per liter) are added to the solution obtained as previously described, and it is boiled to insure complete oxidation of the phosphorus to the tribasic condition. Just sufficient sulphurous acid to decompose the excess of permanganate or separated manganese oxide is then added and the

MOLYBDENUM

boiling is continued for a few minutes longer. A slight excess of ammonia is added and the ferric hydroxide, etc., which will carry down practically all of the phosphorus, is filtered and washed thoroughly with hot water. Solution is made of the precipitate in hot dilute (sp.gr. 1.135) nitric acid and the phosphorus is precipitated with "molybdate solution" and determined as described for wulfenite.

(b) **Wulfenite.** Two (2.000) grams of the agate ground sample of wulfenite are treated with 20 cc. of nitric acid (sp.gr. 1.42) and 10 cc. of sulphuric acid (sp.gr. 1.84) in a 150 cc. beaker. The liquid is heated for a short while at a temperature just below the boiling point and then boiled until fumes of sulphuric anhydride are freely evolved. The residue is taken up with 40 cc. of water, boiled a few minutes, cooled, filtered into a 300 cc. Erlenmeyer flask, and the lead sulphate, silica, etc., are washed with cold 10% sulphuric acid and discarded.

A few drops of strong permanganate solution (25 grams per liter) are added to the filtrate from the lead sulphate and it is boiled to insure complete oxidation of the phosphorus to the tribasic condition. Just sufficient sulphurous acid is added to decompose the excess of permanganate or separated manganese oxide and the boiling is continued for a minute or two longer.

The acid solution is nearly neutralized with ammonia, partly cooled, and to it there are added one gram of tartaric acid and then a slight excess of ammonia. If less than 2 grams of sample should be employed, as in case of very high phosphorus ores, correspondingly less tartaric is to be used, since an excess tends to prevent complete precipitation of the phosphorus.

Twenty cc. of "magnesia mixture" and 4 or 5 glass beads (6 mm. diameter) are added to the warm solution and it is thoroughly chilled by immersion in a mixture of crushed ice and salt. The contents of the flask are then *vigorously* shaken (an *efficient* shaking machine can be used to advantage) for about 15 minutes, the solution being chilled several times during the shaking. The glass beads aid in starting the precipitation of the magnesium ammonium phosphate. Fifteen cc. of ammonia (sp.gr. 0.90) are next introduced and the contents of the flask are again thoroughly chilled and shaken briskly for 10–15 minutes longer. The flask is then packed in ice in a refrigerator and allowed to stand for 15 hours.

The magnesia precipitate, which may contain small amounts of basic magnesia compounds, iron, and possibly tartrates, is filtered on a 9 cm. paper and washed thoroughly with cold 10% ammonia water. The glass beads are transferred to the filter but no attempt is made to remove all of the precipitate from the flask. Under the conditions described, molybdenum, tungsten, and vanadium should all pass completely into the filtrate.

Solution is made of the magnesium ammonium phosphate in the least necessary amount of hot dilute hydrochloric acid and the paper is washed with hot water. The filtrate and washings are caught in the same 300 cc. Erlenmeyer flask and evaporated down to a volume of approximately 5 cc. After cooling, addition is made of 10 cc. of a saturated solution of sulphurous acid, the flask and its contents are permitted to stand in a warm place for 5 or 10 minutes, 20 cc. of hydrochloric acid (sp.gr. 1.20) are added and the solution is evaporated to a bulk of about 5 cc. to expel completely arsenic. Twenty cc. of nitric acid (sp.gr. 1.42) are added and the solution is again boiled down to a volume of about 10 cc. in order to eliminate practically all of the hydrochloric acid. After having been diluted with 50 cc. of water, the solution is oxidized by boiling with an excess of strong permanganate solution (25 grams per liter) and then cleared by the addition of a few cc. of sulphurous acid. The liquid is boiled for a minute or two to expel nitrous fumes, and then cooled. A piece of litmus paper is dropped into the solution and a slight excess of ammonia (sp.gr. 0.90) is added. This is followed by addition of an excess of 1 or 2 cc. of colorless nitric acid (sp.gr. 1.42). The temperature of the solution is brought to 40° C. and the phosphorus is precipitated by addition of 40 cc. of "molybdate solution" and five minutes' vigorous shaking. Determination is made of the phosphorus in the ammonium phosphomolybdate by either the Alkalimetric or the Molybdenum Reduction (Emmerton) Method.

The "**magnesia mixture**" is prepared in the following manner: Fifty-five grams of crystallized magnesium chloride, $MgCl_2 \cdot 6H_2O$, or twenty-five grams of the anhydrous salt, are dissolved in water and filtered. Fourteen grams of ammonium chloride are dissolved in water, a little bromine water and a slight excess of ammonia are added and the solution is filtered. The two solutions are mixed, sufficient ammonia is added to give a decided odor, and enough water is added to bring the volume up to one liter. The solution is allowed to stand for several days with occasional vigorous shaking and is filtered into one or more ceresin bottles. Ten cc. of the reagent will precipitate about 0.065 gram of phosphorus.

NICKEL

Ni, *at. wt.* 58.69; *sp. gr.* 8.6–8.9; *m. p.* 1452° C.; *oxides*, NiO, Ni$_2$O$_3$, Ni$_3$O$_4$.

DETECTION

After bringing the sample into solution by one of the methods described under Preparation and Solution of the Sample, silica is removed, if present, in the usual manner, by evaporating the solution to dryness in the presence of an excess of hydrochloric acid, dissolving the residue and boiling with hydrochloric acid and filtering off the silica.

Hydrogen sulphide is then passed through the solution to remove the elements precipitated by this reagent. The filtrate from this precipitation is then boiled to expel the excess of hydrogen sulphide and a little nitric acid added to oxidize any ferrous iron to the ferric state. (See page 331, Separations.) Ammonium hydroxide is then added to precipitate iron, aluminum and chromium. Cobalt, nickel, manganese and zinc are precipitated from the filtrate by adding a solution of colorless ammonium sulphide or by passing hydrogen sulphide through the ammoniacal solution. Manganese and zinc are separated from the precipitate by washing with cold hydrochloric acid of about 1.035 sp.gr. A small quantity of the precipitate is fused with borax in the loop of a clean platinum wire. A green color in the cool bead indicates nickel. Fairly small quantities of cobalt interfere with this test, so if the bead is colored blue it will be necessary to make further tests for nickel.

Dimethylglyoxime will precipitate nickel as oxime from an acetic acid solution containing sodium acetate and in this manner separate it from cobalt, manganese and zinc. After precipitating iron, aluminum and chromium and filtering them off, the solution is slightly acidified with hydrochloric acid, then is neutralized with sodium hydroxide, and acidified with acetic acid. A solution of dimethyl-glyoxime is added, when nickel, if present, will be precipitated as a flocculent red precipitate.

Nickel may be detected in the presence of cobalt by adding a solution of sodium hydroxide to the solution of cobalt and nickel until a slight precipitate is formed, then somewhat more potassium cyanide than is necessary to redissolve the precipitate and finally two volumes of bromine water. Warm gently and allow to stand for some time. If a precipitate of nickel hydroxides separates, filter, wash and test with the borax bead.

Nickel may also be detected in the presence of cobalt by precipitating the cobalt as nitrite, as described in the chapter on cobalt, and then precipitating the nickel as hydroxide with sodium hydroxide and bromine water and testing the precipitate with the borax bead.

Alpha benzildioxime added to an ammoniacal solution of nickel precipitates an intensely red salt having the composition C$_{28}$H$_{22}$N$_4$O$_4$Ni. This precipitate is very voluminous. Silver, magnesium, chromium, manganese and zinc do not interfere with this reaction.

Chapter by W. L. Savell, Paul D. Mercia and Thos. Fudge.

ESTIMATION

The determination of nickel is required, principally, in the analysis of ores, metallic nickel and its alloys, but is also required in the analysis of metallic cobalt and cobalt products as well as in a host of miscellaneous materials.

In the majority of cases the results of a nickel determination are calculated in terms of metallic nickel. Even in the determination of nickel in nickel-plating solution the results are calculated in terms of metallic nickel since this is the factor by which the solutions are controlled.

Preparation and Solution of the Sample

The materials in which nickel occurs ordinarily, may, in general, be brought into solution by treatment with acids, but in the case of some refractory ores and alloys, a fusion is required first to make the acid treatment effective. When treating ores containing sulphides or arsenides a strong oxidizing treatment is necessary to break up these compounds. Metallic nickel may be dissolved easily in nitric acid, more slowly in hydrochloric acid and still more slowly by sulphuric. Nickel alloys may be dissolved in a mixture of hydrochloric acid and nitric acid.

General Procedure for Ores. One gram of the finely powdered ore is weighed into a porcelain dish and mixed intimately with 3 grams of powdered potassium chlorate. The dish is covered with a watch-glass and 40 cc. concentrated nitric acid added slowly. The dish is allowed to stand in a cool place for a few minutes, then placed on a water bath and digested until the sample is completely decomposed, stirring the mixture frequently with a glass stirring rod, and adding a little potassium chlorate from time to time until the decomposition is complete. The watch-glass is then removed and any particles that may have spattered on it are washed back into the dish and the evaporation continued to dryness. This evaporation to dryness is repeated with the addition of 10 cc. of concentrated hydrochloric acid, and the silica dehydrated by heating for an hour or more in an air oven at 110° C. The dry residue is moistened with concentrated hydrochloric acid and the sides of the dish washed down with hot water, the mixture heated to boiling and allowed to boil for a few minutes, then withdrawn from the heat and filtered, hot, after the insoluble matter has settled.

Treat the filtrate for the removal of interfering elements as directed under Separations.

Fusion Method. The above method is used where it is desired to determine insoluble matter or "gangue." As a method of bringing the nickel in the sample into solution it is quite satisfactory and when the insoluble matter burns to a pure white ash the ignited residue may be weighed as silica, but in some cases this method does not give sufficient information regarding the composition of the gangue.

If it is necessary to make a complete analysis it is usually better to fuse the sample with the sodium and potassium carbonate mixture containing a little potassium nitrate and then treat in the usual manner to determine silica.

Potassium Bisulphate Fusion. In the treatment of nickel and cobalt oxides these are ground to a fine powder and a representative sample of 1 gram is fused with 10 grams of potassium bisulphate. This may be done in a porcelain or silica crucible or dish. The melt is extracted with water and the silica filtered off.

A small casserole has been found to be very useful for this fusion.

Solution of Metallic Nickel and Its Alloys. From 1 to 5 grams of the well-mixed drillings are treated with a minimum quantity of nitric acid and 20 cc. 1 : 1 sulphuric acid added and the solution evaporated to fumes of sulphur trioxide. Allow the fuming to continue for ten minutes. Dilute carefully with a little water and filter off the insoluble. Continue as directed in the following detailed analyses.

It may be necessary to use a mixture of nitric and hydrochloric acids to bring certain alloys into solution, after which the procedure is the same as above.

SEPARATIONS

Separation of the Ammonium Sulphide Group, Containing Nickel from the Hydrogen Sulphide Group. Mercury, Lead, Bismuth, Copper, Cadmium, Arsenic, Antimony, Tin, Gold, Molybdenum, etc.

The hydrogen sulphide group elements are precipitated from an acid solution (HCl) by H_2S, and removed by filtration, nickel, etc., passing into the filtrate.

Separation of the Ammonium Sulphide Group from the Alkaline Earths and Alkalies. Nickel is precipitated with other members of the group by passing H_2S into its ammoniacal solution, or by adding $(NH_4)_2S$ solution. The alkaline earths and alkalies are not precipitated.

Separation of Nickel from Cobalt. This procedure can be carried out in exactly the same manner as the method given for the determination of nickel by precipitation of nickel with dimethylglyoxime, since cobalt is soluble as oxime. In case more cobalt is present than nickel a larger excess of the reagent must be used. The excess of acid is best neutralized with ammonium hydroxide. If both metals are to be determined, cobalt may be determined electrolytically in the filtrate.

An alternate method is to determine the cobalt and nickel as oxides, or metal by electrolysis, together. The oxides, or plate, are dissolved in nitric acid and the nickel determined in the solution, cobalt being found by difference.

For other methods see Separation of Cobalt from Nickel, under Cobalt, page 168.

Separation of Nickel from Manganese. Nickel is precipitated by dimethylglyoxime from an acetic acid solution containing sodium acetate, manganese being determined in the filtrate.

Separation of Nickel from Zinc. Zinc does not interfere in the dimethylglyoxime precipitation of nickel when ammonium salts are present. It is advisable to precipitate the nickel in a dilute acetic acid solution, thus avoiding the addition of a large amount of ammonium salts as would be necessary if the precipitation took place in an ammoniacal solution. Zinc readily remains in solution, and may be determined in the filtrate from the nickel oxime precipitate. The following procedure is recommended:

The solution containing the two metals is neutralized with ammonium hydroxide and then made just slightly acid with acetic acid and sodium acetate added. Dimethylglyoxime solution is now added to the solution, which is nearly boiling, and the procedure given for the determination of nickel by this reagent is followed.

Separation of Nickel from Iron. Nickel cannot be separated satisfactorily from iron by precipitating the latter with ammonium hydroxide, as some of the nickel is invariably occluded by the ferric hydroxide precipitate. Two modifications of the oxime method may be used.

(1) The iron, if present as a ferric salt, is converted into a complex salt by

adding from 1 to 2 grams of tartaric acid, and the solution diluted to 200 or 300 cc., boiled and the nickel precipitated as the oxime in an ammoniacal solution by the prescribed method. Iron forms no oxime under these conditions.

The iron may be precipitated from this filtrate by colorless ammonium sulphide and the sulphide converted to ferric oxide (Fe_2O_3) by ignition.

(2) Ferric iron is reduced to the ferrous condition by warming with sulphurous acid, in a nearly neutral solution. If the original solution has an excess of acid, it is treated with a solution of sodium hydroxide until a permanent precipitate is formed. This is dissolved with a few drops of hydrochloric acid and the iron reduced by adding from 5 to 10 cc. of a saturated solution of sulphur dioxide or by passing dioxide through the solution. The solution is diluted to 200 or 300 cc. and the solution of dimethylglyoxime added in slight excess, followed by sodium acetate until a permanent precipitate of nickel oxime is formed. After adding 2 grams more of sodium acetate the solution is filtered immediately. The iron is precipitated from the filtrate by oxidizing with bromine water and adding ammonium hydroxide to precipitate the basic acetate of iron.

Procedure (1) is suitable for the determination of nickel in iron and steel.

Separation of Nickel from Aluminum. This method is the same as procedure (1) given above.

Separation of Nickel from Chromium. This separation cannot be carried out in an acetic acid solution. From 1 to 2 grams of tartaric acid are added and from 5 to 10 cc. of a 10% ammonium chloride solution, subsequently. The solution is made ammoniacal, but no precipitate should form. If the solution becomes cloudy, it is acidified with hydrochloric acid and additional ammonium chloride added and again made ammoniacal and the nickel precipitated as oxime according to directions given from this precipitation.

GRAVIMETRIC METHODS FOR THE DETERMINATION OF NICKEL
Precipitation of Nickel by Alpha Benzildioxime

The alcoholic solution of alpha benzildioxime gives an intensely red precipitate of $C_{28}H_{22}N_4O_4Ni$, when added to ammoniacal solutions containing nickel. The reaction is more characteristic for nickel than is that with dimethylglyoxime and is more delicate. In a volume of 5 cc. (according to F. H. Atack), 1 part of nickel in 2,000,000 parts of water may be detected. In the presence of 100 times as much as cobalt only a faint yellow color is produced by the cobalt. One part of nickel per million of water will cause precipitation with the compound, whereas no precipitate is formed with dimethylglyoxime under the same conditions. With glyoxime iron produces a pink color, with alpha benzildioxime ferrous salts give a faint violet color, hence do not interfere in the detection of nickel. Silver, magnesium, chromium, manganese, and zinc do not interfere. Since the nickel precipitate with this reagent is exceedingly voluminous it is advisable to have not more than 0.025 gram of nickel in the solution in which the nickel is being determined. The method is adapted to the detection and determination of minute traces of the element up to small amounts of less than 10% nickel.

Reagent, Alpha Benzildioxime. This may be prepared by boiling 10 grams of benzil (not necessarily pure) with 8 to 10 grams of hydroxylamine hydrochloride in methyl alcohol solution. After boiling for three hours the precipitate is filtered off and dried, washed with hot water and then with a small amount of 50% alcohol, and dried. This dried precipitate consists of pure benzildioxime (m.p. 237° C.). A further yield may be obtained by boiling the filtrate with hydroxylamine hydrochloride. The reagent is prepared by dissolving 0.2 gram of the salt per liter of alcohol to which is added ammonium hydroxide to make 5% solution, sp.gr. 0.96 (50 cc. per liter).

Procedure. A slight excess of the warmed solution of the above reagent is stirred into the ammoniacal solution containing nickel and the whole heated on the water bath for a few moments to coagulate the precipitate. Quantitative precipitation is complete after one minute. The liquid is filtered through a Gooch crucible, with suction, or onto a filter paper, for which a counterpoise has been selected. The counterpoise paper is treated in exactly the same manner as the one containing the precipitate. The precipitate is washed with 50% alcohol, followed by hot water, and is then dried at 110° C. In weighing the precipitate the counterpoise filter is placed in the weight pan of the balance. The precipitate contains 10.93% nickel. Weight of $C_{28}H_{22}N_4O_4Ni \times 0.1093 = Ni$.

NOTES. Acetone may be used instead of alcohol as a solvent of the reagent. The compound is more soluble in acetone than in alcohol.

The precipitate does not pass through the filter as does the compound with dimethylglyoxime.

The method is affected by the presence of nitrates, hence these must be removed by evaporation of the solution with sulphuric acid to fumes, before the addition of the reagent to the nickel solution.

In the presence of cobalt an excess of the reagent must be used, as in the case of the dimethylglyoxime precipitation.

In the presence of iron and chromium Rochelle salt, sodium citrate or tartaric acid are added to prevent precipitation of the hydroxides of these metals upon making the solution alkaline.

In the presence of manganese a fairly large excess of the reagent is required, the solution being slightly acid with acetic acid.

Zinc and magnesium are kept in solution by addition of ammonium chloride.

Large amounts of copper must be removed by precipitating with hydrogen sulphide before addition of the reagent.

The nickel salt with the reagent forms an extremely voluminous precipitate so that a concentration of 0.09 gram of nickel per 250 cc. is as high as is desirable. The process is applicable to the determination of nickel in the filtrate obtained in the separation of zinc after the removal of the hydrogen sulphide, formic acid, etc.

Method by F. W. Atack, The Analyst, **38**, 448, 318. Cockburn, Gardiner and Black, Analyst, **38**, 439, 443.

Precipitation of Nickel by Dimethyl=glyoxime

Preliminary Considerations. This method has been demonstrated by O. Brunck to be the most accurate and expeditious procedure known for nickel.[1] By this method 1 part of nickel may be detected when mixed with 5000 parts of cobalt or 1 part of nickel may be detected in 400,000 parts of water. The nickel precipitate with this reagent is almost completely insoluble in water and is only very slightly soluble in acetic acid, but is easily decomposed by strongly dissociated acids, so that the precipitation is incomplete in neutral solutions of nickel chloride, sulphate or nitrate. If, however, the free acid formed is neutral-

[1] Zeit. f. ang. Chem., **20**, 1844.

ized with sodium, potassium or ammonium hydroxides or by addition of the acetate salts of these bases, nickel will be completely precipitated, not even a trace being found in the filtrate.

" The quantitative determination of nickel in the presence of other metals is a simple operation. The nickel should be in the form of a convenient salt.

" The concentration of the solution does not matter; the precipitation can take place either in a solution of the greatest concentration, or in a very dilute solution. The reaction is not hindered by the presence of ammonium salts."

Iron, aluminum, chromium, cobalt, manganese and zinc do not interfere. Theoretically 4 parts of dimethylglyoxime, added as a 1% alcoholic solution, are necessary; a certain excess does no harm provided the alcohol volume does not exceed more than half that of the water solution containing the nickel salt, as alcohol has a solvent action on the oxime. The compound is very stable and volatilizes undecomposed at 250° C.

An excess of ammonium hydroxide is also to be avoided in the solution in which the precipitation takes place.

It has been observed that the precipitate of nickel with dimethylglyoxime may be safely ignited to the oxide NiO without loss, if the filter is first carefully charred without allowing it to take fire, then gradually heated to redness.

Procedure. Such an amount of the sample should be taken that the nickel be not over 0.1 gram, as glyoxime of nickel is very voluminous and a larger amount would be difficult to filter. If cobalt is present it should not exceed 0.1 gram in the sample taken.[1]

If hydrogen sulphide has been used to precipitate members of the second group, it is expelled by boiling the acid solution and the volume brought to 250 cc.

One or 2 grams of tartaric acid are added to prevent the precipitation of the hydroxides of iron, aluminum and chromium by ammonium hydroxide (this treatment is omitted if these are absent), and 5 to 10 cc. of a 10% solution of ammonium chloride added to keep zinc and manganese in solution, should they be present. Ammonium hydroxide is now added until the solution is slightly alkaline. If a precipitate forms, ammonium chloride is added to clear the solution, followed by ammonium hydroxide to neutralize the acid. The solution should remain clear after this treatment, otherwise the ammonium chloride is added in solution or as salt until the solution of the sample will remain clear. It is then heated to nearly boiling and the alcoholic solution of dimethylglyoxime added until the reagent is approximately seven times, by weight, the weight of nickel present. Ammonium hydroxide is now added until the solution has a distinct odor of this reagent. The precipitation of the scarlet red nickel salt is hastened by stirring. It is advisable to place the mixture on the steam bath for fifteen to twenty minutes to allow the reaction to go to completion before filtering. The precipitate is filtered off, into a platinum sponge Gooch crucible, sometimes known as a Neubauer Gooch crucible. (Other forms of Gooch crucible are used for this purpose, but the Neubauer crucible has been found to be most satisfactory.) The precipitate is dried for about two hours at 110 to 120° C. and weighed as $C_8H_{14}N_4O_4Ni$, which contains 20.32% Ni.

Weight of precipitate multiplied by 0.2032 = weight of nickel.

[1] If the sample contains more than 0.1 gram of cobalt, a large excess of ammonium hydroxide and dimethylglyoxime is necessary to prevent its precipitation, hence it is advisable to take such weights of samples that the cobalt content will be less than this weight.

In place of a Gooch crucible a tared filter paper may be used. It must be remembered, however, that a blank filter paper of the same kind as used for the precipitate must be used as a counterbalance, after treating in exactly the same manner as the one containing the precipitate. This is necessary because it has been found that filter paper loses weight during washing and drying.

Precipitation of Nickel by Electrolysis [1]

This precipitation is conducted in exactly the same manner as the one described under Cobalt for the Precipitation of Cobalt by Electrolysis, and requires that the same precautions be exercised in the practice of the method.

In the presence of cobalt the two elements may be determined together by electrolysis as described below and the deposited metal redissolved and the two elements separated by one of the methods given under Cobalt or Nickel.

Procedure. After the sample has been brought into solution by one of the methods outlined under Preparation and Solution of the Sample, the solution is evaporated with 20 cc. of 1 : 1 sulphuric acid for every gram of metal in the sample. The evaporation is continued until the solution has fumed strongly for ten minutes. Cool carefully and dilute with 20 cc. of water. Heat the solution to nearly boiling and pass hydrogen sulphide for one hour to precipitate members of the second group. This long treatment is necessary to insure complete precipitation of arsenic. Filter and boil to expel hydrogen sulphide. Add 5 cc. nitric acid to insure oxidation of iron compounds to the ferric state and add ammonium hydroxide until just slightly alkaline. Filter off the ferric hydroxide and wash with water containing a small quantity of ammonium hydroxide. To recover occluded nickel dissolve the precipitate in hydrochloric acid and reprecipitate the iron with addition of a little hydrogen peroxide. Combine the filtrates. Evaporate to about 250 cc. and add 50 cc. of strong ammonium hydroxide and electrolyze as described under Cobalt, page 170.

The increase in weight of the electrode is the weight of cobalt and nickel in the sample. The percentage of cobalt and nickel in the sample is found by multiplying the increase in weight of the electrode by 100 and dividing by the weight of the sample.

NOTE. The deposition of cobalt and nickel by the above method has been found to be the most accurate of the electrolytic methods. In the solutions containing the organic acids there is always more or less carbide deposited on the cathode with the metal. This causes high results.

Nickel in Metallic Nickel

This determination may be made in the manner described under Precipitation of Nickel by Electrolysis, separating cobalt before or after the electrolysis or by the method described under Precipitation of Nickel by Dimethylglyoxime. The latter method is recommended.

Nickel in Cobalt and Cobalt Oxide

The dimethylglyoxime precipitation is used in combination with the electrolytic precipitation. See chapter on Cobalt.

[1] W. J. Marsh, J. Phys. Chem., **18**, 705–16, 1914.

VOLUMETRIC DETERMINATION OF NICKEL
Determination of Nickel in Alloys

This method, as described by S. W. Parr and J. M. Lindgren,[1] consists of a modification of the dimethylgloxime method. The precipitation takes place in the usual manner and the precipitate is dissolved in sulphuric acid and the excess titrated with a standard solution of potassium hydroxide.

Procedure. The alloy is dissolved in nitric or hydrochloric acids and if iron, aluminum or chromium are present twice their weight of tartaric acid is added to prevent their precipitation. If chromium is present ammonium chloride is also added. If manganese or zinc is present hydrochloric acid should be used and most of the free acid evaporated. Add a few cc. of hydrogen peroxide to oxidize any ferrous iron to the ferric state. Dilute to 300 or 400 cc. and neutralize the free acid by sodium acetate. Heat the solution to nearly boiling and add five times as much dimethylglyoxime, in 1% alcoholic solution, as the nickel present. Then completely neutralize with ammonium hydroxide, using a very slight excess (or the solution may be neutralized with sodium acetate). Heat until all the nickel is precipitated. Filter and wash. Place the precipitate and filter in a beaker, add an excess of 0.05N sulphuric acid, dilute to 200 cc., heat until solution is complete and titrate back with 0.1N potassium hydroxide solution, taking the first faint yellowish tinge as the end-point. The solutions are standardized against pure nickel.

Note. Cobalt should not exceed 0.1 gram per 100 cc. and an excess should be used of the dimethylglyoxime.

Nickel in Nickel-plating Solutions

In most cases it is quite unnecessary to separate the cobalt from the nickel in making this determination and, as the principal impurity is usually iron, the best practice is to follow the method given under Precipitation of Cobalt by Electrolysis, page 170.

If chlorides or organic matter are present in the solution the preparation of the solution for electrolysis is accomplished in the following manner:

From the well-stirred solution in the plating tank, withdraw about 200 cc. and place in a small beaker. Prepare a 100-cc. burette by thoroughly cleaning it with the sulphuric acid and potassium bichromate mixture and distilled water. Wash finally with a few cc. of the nickel solution and fill the burette with the solution from the plating tank.

Run 66.7 cc. into an evaporating dish and add 2 cc. 1 : 1 sulphuric acid. Evaporate to fumes of sulphur trioxide and allow to fume strongly for ten minutes. Dissolve in a little water. Dilute to 200 cc. carefully, neutralize with a solution of ammonium hydroxide and add 50 cc. of strong ammonium hydroxide and electrolyze. (See Precipitation of Cobalt by Electrolysis.)

The increase in weight of the cathode in grams multiplied by 2 gives the weight in ounces of nickel in one United States gallon of the plating solution.

---

[1] S. W. Parr and J. M. Lindgren, Trans. Am. Brass Founders' Assoc., 5, 120–9.

Potassium Cyanide Method for Nickel[1]

The method is rapid and accurate and is especially adapted for determining nickel in steel. Iron, manganese, chromium, zinc, vanadium, molybdenum and tungsten do not interfere. Copper, however, should be removed if present. The method depends upon the selective action of potassium cyanide for nickel in preference to silver iodide, used as an indicator, the reactions taking place as indicated, the solution being slightly alkaline with ammonia—

(a) $Ni(NH_3)_6SO_4 + 4KCN = K_2Ni(CN)_4 + K_2SO_4 + 6NH_3$

(N.B. $4KCN \backsimeq Ni$)

(b) $AgI + 2KCN = KAg(CN)_2 + KI$ (N.B. $2KCN \backsimeq Ag$)

Reagents.

N/10 Silver Nitrate.—10.788 grams of pure silver are dissolved in nitric acid and made to 1000 cc. or 16.99 g. of the silver nitrate salt. (See Index for reagent.) 1 cc. $\backsimeq$ 0.01302 g. KCN.

If preferred the reagent may be made to be equivalent to about 0.001 gram Ni per cc. by dissolving 5.85 grams of $AgNO_3$ per liter.

Potassium Iodide.—25% solution.

N/10 Potassium Cyanide.—13.5 grams of pure KCN are dissolved in water, 5 grams of KOH added and the solution made to 1000 cc.

The cyanide solution is standardized against the silver nitrate solution.

If it is desired to have a solution equivalent to 0.001 g. Ni, 5 grams of KCN per liter is the approximate strength required.

Standardization of the Cyanide.—Fifty cc. of the KCN solution are diluted to about 150 cc., 5 cc. of the KI reagent added and the solution titrated with the standard $AgNO_3$ reagent until a faint permanent opalescence is obtained. A drop of the KCN solution should be sufficient to clear this. Note the number of cc. required and calculate the normality factor of the cyanide in terms of the silver nitrate reagent.

Example.—Suppose 49 cc. of the silver nitrate reagent were required for the 50 cc. of the cyanide solution, then the normality would be 49 ÷ 50 × N/10 or 0.98 N/10.

1 cc. N/10 solution is equivalent to 0.002934 gram nickel.

The reagent may be standardized against a nickel steel of the U. S. Bureau of Standards, following the procedure given below and calculating as follows:

$$\text{Nickel factor} = \frac{\text{Gram Ni in standard taken}}{(\text{cc. KCN required}) - (\text{cc. KCN equivalent to 5 cc. } AgNO_3)}$$

Citric Solution.—200 grams of $(NH_4)_2SO_4$, 150 cc. concentrated NH_4OH and 120 grams of citric acid per 1000 cc.

Procedure.[1]

One gram of the steel drillings or such an amount of material as contains not over 0.1 g. Ni, is dissolved in a beaker with 20 cc. of hydrochloric acid (1 : 1). When action ceases 10 cc. of nitric acid (1 : 1) are added and the solution boiled until the red nitrous acid fumes are driven off.

About 100 cc. of the citrate solution are added. If 2 per cent. or more of chromium is present the amount of citrate solution is doubled. The solution is now diluted to about 250 cc.

Exactly 5 cc. of the standard silver nitrate solution are now added from a pipette or burette, and then ammonium hydroxide, drop by drop, until the cloudiness caused by the silver chloride just disappears. Two cc. of the potassium iodide solution are now added.

The solution is titrated with the standard potassium cyanide solution with constant stirring until the turbidity just disappears. The end point is reached when there is no longer a distinction in clearness of the drop of the reagent and its surrounding liquid to which it is added.[1]

If the end point is passed, a measured amount of silver nitrate (5–10 cc.) is added and the cyanide titration repeated.

Calculation.—Deduct the cc. KCN equivalent of the total silver nitrate solution used from the total cc. of the KCN solution required in the titration. The remainder is the potassium cyanide required by the nickel.

The cc. KCN required by Ni multiplied by the factor for Ni = gram nickel in the sample.

1. A large excess of ammonia is to be avoided as the AgI is soluble in a large excess.

2. The presence of sulphates increases the sensibility of the end point.

3. The silver nitrate solution should not be stronger than that indicated in the method as there is danger of the iodide of silver settling out as a curdy precipitate in stronger solutions.

4. A white film is apt to form on the surface of the liquid if exposed to the air for some time. This produces no error.

Carnot's[1] Gravimetric Oxide Method for Nickel

Sodium carbonate is added to the nickel solution, in slight excess until alkaline. Nickel is now precipitated with ammonia or ammonium carbonate from a boiling solution. The precipitate is filtered, washed and ignited to oxide, NiO, and so weighed.

Determination of Nickel in Nickel Steel—Ether Extraction Cyanide Titration Method

The following procedure is recommended by the A. S. T. M.

Reagents Required

Hydrochloric Acid.—600 cc. HCl (d. 1.2) dil. with 400 cc. water.
Nitric Acid.—1000 cc. HNO₃ (d. 142) and 1200 cc. water.
Potassium Iodide.—20 grams per liter.
Silver Nitrate.—0.5 g. per liter.
Potassium Cyanide.—4.589 grams KCN per liter. The reagent is stand-

[1] A. Carnot, Jour. Chem. Soc., 1918, II, 133.

ardized against nickel steel of known nickel content as determined by the gravimetric dimethylglyoxime method.

Procedure. A sample of 1 gram of steel is dissolved by addition of 20 cc. of the dilute hydrochloric acid and then the iron oxidized by addition of 2 cc. of strong nitric acid (d. 1.42) and the solution boiled to expel the brown nitrogen fumes. After cooling the solution is transferred to an 8-oz. separatory funnel, hydrochloric acid being used to rinse out the containing beaker into the funnel. The solution is shaken with 50 cc. of ether for about 5 minutes, then allowed to settle for 1 minute and the lower clear layer drawn into a second separatory funnel (8-oz.). To the first ether solution are added 10 cc. of strong HCl (d. 1.2), cooling, and then shaking as before. After settling 1 minute the lower layer is drawn into the second separatory funnel. 50 cc. of ether are added to the second separatory funnel, the solution shaken for 5 minutes, then settled 1 minute, and the lower clear layer drawn into a 150-cc. beaker. The ether is expelled by gentle heat. 0.2 gram of potassium chlorate is added, the solution boiled to decompose the chlorate, then diluted to 100 cc. with hot water, and made faintly ammoniacal and boiled for 5 minutes. The manganese dioxide is filtered off and washed with hot water. To the filtrate are added 10 cc. of strong hydrochloric acid, the solution heated just short of boiling and copper precipitated with H_2S, and filtered off and washed with hot water. The filtrate is boiled to expel H_2S, and the boiling continued until the volume is reduced to about 100 cc. The solution is cooled, made distinctly ammoniacal, 10 cc. each of silver nitrate and potassium iodide solutions added and the solution titrated with standard potassium cyanide to a clear solution.

Determination of Bismuth in Metallic Nickel

Dissolve 20 grams of nickel drillings in 200 cc. 1 to 1 nitric acid in an 800-cc. beaker. Add 5 grams of potassium chlorate, boil for ten minutes. Cool and add 75 cc. of sulphuric acid. Evaporate until fuming strongly, cool. Add 600 cc. of water, boil gently to insure solution of all the bismuth sulphate. Cool again, filter and wash with 1 to 10 sulphuric acid. Do not allow to stand too long before filtering, some basic bismuth sulphate may separate. Pass a current of hydrogen sulphide gas through for one half hour. Bismuth, copper, arsenic, antimony, etc., are precipitated as sulphides. Filter, washing with hot water. Rinse the precipitate as completely as possible into a beaker, add 3 or 4 grams of pure potassium cyanide, warm gently for some time. Bismuth sulphide will remain undissolved. Filter through same filter as before, in order to act upon traces of sulphides that could not be washed into the beaker, and wash with hot water. Place the filter and precipitate into a 100-cc. lip beaker and heat with 10 cc. of 1 to 1 nitric acid until the separated sulphur is clean and the filter well disintegrated. Dilute a little and then filter and wash thoroughly with 1 to 1 nitric acid. Partially neutralize the filtrate with ammonia, but without producing any permanent precipitate, and then add a solution of ammonium carbonate in very slight excess. Heat nearly to boiling for some time until the bismuth carbonate has settled well, and then filter and wash with hot water. Dry the precipitate and transfer it to a small weighed porcelain crucible, removing it from the paper as completely as possible. Burn the latter carefully and add the ash to the precipitate in the crucible. Ignite the whole at a low red heat, cool and weigh as Bi_2O_3, which contains 89.68 per cent of bismuth.

Determination of Lead in Metallic Nickel

Dissolve 20 grams of nickel drillings in 200 cc. of 1 to 1 nitric acid in a 1200-cc. beaker. Evaporate to 100 cc. Add 100 cc. sulphuric acid and evaporate to heavy fumes.

Cool and add 800 cc. water. Boil gently to insure solution of sulphates, cool and add 200 cc. of 95 per cent alcohol, stir and allow to stand over night.

Filter off the lead sulphate and wash two or three times with 2 per cent sulphuric acid and once with cold water. Place filter paper, containing the lead sulphate, in a beaker and add 10 grams of slightly ammoniacal ammonium acetate and 50 cc. of water. Boil a few minutes. Filter and wash with hot water.

Filtrate, containing all the lead in solution as the acetate, is acidified slightly with acetic acid and heated to boiling. Precipitate lead by adding 20 cc. of a 5 per cent solution of potassium dichromate. Boil until precipitate changes to orange or red. Allow precipitate to settle over night.

Filter the lead chromate on a weighed Gooch crucible. Wash with water and finally with alcohol. Dry at 110° C., and weigh as $PbCrO_4$, which contains 64.10 per cent of lead.

Determination of Aluminum in Metallic Nickel

Dissolve 10 grams of nickel drillings in 100 cc. dilute nitric acid (one part water, two parts nitric acid) in a 600-cc. covered beaker, add 1 or 2 grams potassium chlorate, boil for one hour, add about 100 cc. cold water and about 5 grams ammonium chloride, make slightly alkaline with ammonia. Transfer to a 500-cc. platinum dish. Boil a few minutes, filter off iron and aluminum hydrates and wash with hot water two or three times.

The precipitate is again transferred to 500-cc. platinum dish by redissolving in 1 to 3 hydrochloric acid, add a few grams of ammonium chloride, make slightly alkaline with ammonia. Boil a few minutes, filter off iron and aluminum hydrates and wash with hot water two or three times. Repeat re-dissolving and ammonia precipitation with addition of a few grams of ammonium chloride. After final re-precipitation transfer the precipitate to 100-cc. platinum dish by washing and dissolving the precipitate in 1 to 3 hydrochloric acid. Evaporate to dryness on a steam bath. Add 2 grams pure sodium hydrate and about 2 cc. of water. Fuse at low temperature for ten minutes. Cool and add about 50 cc. of water. Place on steam bath for a few minutes. Transfer to a 500-cc. platinum dish, add 300 cc. of water. Filter into a 500-cc. casserole containing 30 cc. of hydrochloric acid. Evaporate the filtrate to dryness on the hot plate. Add 50 cc. 1 to 3 hydrochloric acid, boil and filter into a 200-cc. platinum dish. Add a few grams of ammonium chloride to the filtrate and make it slightly alkaline with ammonia. Boil a few minutes, filter on ashless paper, wash with hot water. Ignite in a small platinum crucible and weigh as Al_2O_3 which contains 53.00 per cent of aluminum.

NOTE. If the aluminum oxide should contain a small amount of chromium, it should be fused with sodium peroxide and the chromium determined by color. Deduct the weight of chromium oxide found from the weight of aluminum oxide.

NOTE. Run a blank analysis; that is, add to it all the reagents, and subject the contents to the same operations as the samples of drillings and at the same time. Deduct any alumina contained from the result of the actual assay. If this blank amounts to .004 gram, the reagents are unfit for the above analysis and the determination should be repeated with other reagents.

This method is devised for Metallic Nickel as furnished by The International Nickel Company (Huntington Works).

Method of Determining Silicon, Tungsten and Chromium in Metallic Nickel

Dissolve 1 gram of drillings in 50 cc. 1 to 1 hydrochloric acid in a 200-cc. lip beaker. Add a few drops of nitric acid at intervals until sample has decomposed. Add 5 cc. nitric acid, boil down to 10 cc. Add about 50 cc. of water and boil for five minutes and filter. Wash well with 5 per cent hot hydrochloric acid and finally with water. Evaporate filtrate to dryness, take up in 25 cc. of 1 to 1 hydrochloric acid, dilute to 50 cc., boil, filter and wash with hot water. Ignite the two precipitates in a weighed platinum crucible, being careful not to heat above a red heat. Cool and weigh.

Add 3 drops of sulphuric acid and 5 cc. hydrofluoric acid, evaporate and ignite at low red heat. Cool and weigh.

Loss equals $SiO_2 \times .4693$ equals Silicon.

The residue is impure tungstic oxide. To purify it, fuse it with 5 grams sodium carbonate, dissolve and filter off residue. Wash with hot water, ignite, weigh and subtract from weight of impure tungstic oxide.

$WO_3 \times .793$ equals Tungsten.

Add to the filtrate which contains the Chromium 15 cc. of sulphuric acid plus 15 cc. of nitric acid and evaporate to fumes.

Add 25 cc. nitric acid, dilute to 200 cc., boil until salts are all dissolved. Oxidize by adding slowly 3 grams potassium permanganate dissolved in a small amount of water until a permanent color remains, showing an excess of permanganate. Decompose the excess of permanganate by boiling twenty minutes. A precipitate of manganese dioxide should remain. If none remains, add more potassium permanganate. Keep volume of solution approximately constant. Filter off manganese dioxide, cool to room temperature, dilute to 350 cc., add an excess of standard ferrous ammonium sulphate and titrate excess with standard potassium permanganate to a pink color which remains one minute.

Having found by titration the excess of ferrous ammonium sulphate and deducted this from the total amount of the salt used, the weight of the remainder multiplied by .04427 equals Chromium, or the Iron value of the ferrous ammonium sulphate multiplied by .31 equals Chromium.

Method for Determining Arsenic, Antimony and Tin in Metallic Nickel

Dissolve 5 grams of the nickel drillings in 50 cc. of 1 to 1 nitric acid in a 145-cc. covered silica crucible. Cool and add 10 cc. sulphuric acid, evaporate until fumes of sulphuric acid are given off copiously, cool, then add 25 cc. water, evaporate until the fumes of sulphuric acid are almost all driven off, cool and add about 30 grams of potassium bisulphate (fused). Fuse and keep in a fused condition for one half hour. Cool and transfer the fusion after breaking it up in a porcelain mortar to 16-oz. distilling flask (as per drawing). Allow the outlet of the condenser to dip about one half inch into beaker half full of cold water, then add 5 grams anhydrous ferrous sulphate to the distilling flask, then add 150 cc. hydrochloric acid and distil to about 30 cc., then add 50 cc. more hydrochloric acid and distil again to about 30 cc., remove beaker containing water and distillate and place another beaker half full of cold water under the outlet of the condenser. Again add 50 cc. hydrochloric acid and distil to about 30 cc. Pass hydrogen sulphide gas through both distillates to precipitate the arsenic as arsenous sulphide. All the arsenic should be in the first distillate. The second distillate is made so as to see if all the arsenic has been distilled over. Filter the arsenous sulphide on weighed Gooch crucible, wash with hot water, then with alcohol, ether, and carbon bisulphide, dry at 100° C. and weigh as As_2S_3 which contains 60.93 per cent of arsenic.

To the solution in the distilling flask add 400 cc. hot water, then transfer to 800-cc. lip beaker, nearly neutralize with ammonia, then add 8 cc. hydrochloric acid, heat to boiling, pass stream of hydrogen sulphide gas through solution to saturation, then let stand for two or three hours, filter off antimony, tin and copper sulphides, wash precipitate quickly and thoroughly with hot water, then wash the precipitate into a 200-cc. lip beaker and wash filter with dilute sodium sulphide solution, add a few crystals of sodium sulphide to solution in the beaker, let stand in warm place for an hour, then filter off the copper sulphide, wash with dilute sodium sulphide solution, the filtrate contains all the antimony and tin. Evaporate the filtrate to about 20 cc., then pour in hot concentrated solution of oxalic acid. This will dissolve the tin sulphide and leave the antimonous sulphide undissolved. Pass hydrogen sulphide gas through the boiling solution for half an hour, filter off antimonous sulphide on weighed Gooch crucible, wash quickly and thoroughly with hot water, place Gooch crucible on steam plate and add fuming nitric acid, evaporate until dry, then heat to a dull red heat, cool and weigh as Sb_2O_4 which contains 78.97 per cent of antimony.

Add to the filtrate from the antimonous sulphide 30 cc. hydrochloric acid and 4 grams potassium chlorate, evaporate to about 25 cc., nearly neutralize with ammonia, dilute to 150 cc. with hot water, pass hydrogen sulphide gas through hot solution for half hour, filter on weighed Gooch crucible, wash quickly and thoroughly with hot water, place Gooch crucible on steam plate, add fuming nitric acid, evaporate until dry, then heat to a dull red heat, cool, and weigh as SnO_2 which contains 78.80 per cent of tin.

NOTE. Blank should be run on everything used in the above determinations and at the same time, if any arsenic, antimony or tin is found in the blank, it should be deducted.

Method for Determining Zinc in Nickel

Dissolve 5 grams of drillings in 75 cc. of 1 : 1 nitric acid in a 400-cc. beaker. Dilute to 350 cc. Cool and plate out the copper over night with .3 ampere. After the copper is all plated, pour the solution and water used in washing off cathode into a 500-cc. beaker.

Add 30 cc. of hydrochloric acid and take to dryness on a steam bath. Add 10 cc. of hydrochloric acid and again take to dryness. Take up in 100 cc. of water and 3 cc. hydrochloric acid. Boil and filter off any silica.

Add a concentrated solution of sodium carbonate to the filtrate until a distinct precipitate of nickel and zinc carbonates remain. Carefully dissolve this precipitate by adding 1 per cent sulphuric acid, stirring for at least two minutes after each addition of acid. (Great care should be taken in adding the 1 per cent sulphuric acid because if more than just enough to dissolve the carbonates is used, no zinc sulphide will precipitate.) Add 2 grams ammonium sulphate. Dilute to 400 cc. and cool.

Connect a tube to hydrogen sulphide gas and pass gas through solution using a tube which has been drawn to one millimeter diameter at the end. Pass the gas through for 40 minutes. Disconnect and allow to stand for three hours. Filter through a double ashless paper and wash two or three times with hot water, containing hydrogen sulphide.

To free the zinc sulphide of traces of nickel, dissolve in 50 cc. of 1 : 6 sulphuric acid. Dilute to 100 cc. and boil off the hydrogen sulphide gas. Cool and make slightly alkaline with ammonia.

Make acid with acetic acid. Add an equal volume of 50 per cent acetic acid.

Cool and precipitate the zinc sulphide as before. Allow to settle three hours. Filter through a double ashless paper. Wash with hot water containing hydrogen sulphide.

Place filter paper containing zinc sulphide into a small porcelain evaporating dish. Ignite and keep at a temperature of about 800° C. for one hour. Cool and weigh as zinc oxide, which contains 80.34 per cent zinc.

The procedures on pages 336c to 336f are methods of The International Nickel Company that have been contributed by Dr. Paul D. Mercia and Mr. Thos. Fudge.

NITROGEN

Element. N_2, *at.wt.* 14.01; *D.* (*air*) 0.9674; *m.p.* −210°; *b.p.* −195.5° C.; oxides, N_2O, N_2O_2, N_2O_3, N_2O_4, N_2O_5.

Ammonia. NH_3, *m.w.* 17.03; *D.* (*air*) 0.5971; *sp.gr.* liquid 0.6234; *m.p.* −77.3°; *b.p.* −38.5° C. *Crit. temp.* 130°; *liquid at 0° with 4.2 atmospheres pressure. Commercial* 28% NH_3, *sp.gr.* 0.90.

Nitric Acid. HNO_3, *m.w.*, 63.02; *sp.gr.* 1.53; *m.p.* −41.3; *b.p.* 86° C. *Boiling-point of commercial* 95% *acid is a little above* 86°, *but gradually rises to* 126° *and the strength of acid falls to* 68.9%, *sp.gr. is then* 1.42. *The acid now remains constant, the distillate being of the same strength.*

DETECTION

Element. Organic Nitrogen. Organic matter is decomposed by heating in a Kjeldahl flask with concentrated sulphuric acid as described under preparation and solution of the sample. Ammonia may now be liberated from the sulphate and so detected.

Nitrogen in Gas. Recognized by its inertness towards the reagents used in gas analysis. The element may be recognized by means of the spectroscope.

Ammonia. Free ammonia is readily recognized by its characteristic odor. A glass rod dipped in *hydrochloric acid* and held in fumes of ammonia produces a white cloud of ammonium chloride, NH_4Cl.

Moist red litmus paper is turned blue by ammonia. Upon heating the paper the red color is restored, upon volatilization of ammonia (distinction from fixed alkalies).

Nessler's Test.[1] Nessler's reagent added to a solution containing ammonia, combined or free, produces a brown precipitate, $NHg_2I \cdot H_2O$. If the ammoniacal solution is sufficiently dilute a yellow or reddish-brown color is produced, according to the amount of ammonia present. The reaction is used in determining ammonia in water.

Salts of ammonia are decomposed by heating their solutions with a strong base such as the hydroxides of the fixed alkalies or the alkaline earths. The odor of ammonia may now be detected.

Nitric Acid. *Ferrous Sulphate Test.* About 1 to 2 cc. of the concentrated solution of the substance is added to 15 to 20 cc. of strong sulphuric acid in a test-tube. After cooling the mixture, the test-tube is inclined and an equal volume of a saturated solution of ferrous sulphate is allowed to flow slowly down over the surface of the acid. The tube is now held upright and gently tapped. In the presence of nitric acid a brown ring forms at the junction of the two solutions.

[1] The reagent is made by dissolving 20 grams of potassium iodide in 50 cc. of water, adding 32 grams of mercuric iodide and diluting to 200 cc. To this is added a solution of potassium hydroxide—134 grams KOH per 260 cc. H_2O.

Chapter contributed by Wilfred W. Scott.

The test for nitrate may be made according to the quantitative procedure given for determining of nitric acid (see later). It should be remembered that ferrous sulphate should be present in excess, otherwise the brown color is destroyed by the free nitric acid. Traces of nitric acid in sulphuric produce a pink color with the sulphuric acid solution of ferrous sulphate. (See Determination of Nitric Acid—Ferrous Sulphate Method.)

Ferro- and ferricyanides, chlorates, bromides and bromates, iodides and iodates, chromates and permanganates interfere.

Diphenylamine Tests for Nitrates. $(C_6H_5)_2NH$ dissolved in sulphuric acid is added to 2 or 3 cc. of the substance in solution on a watch-glass. Upon gently warming a blue color is produced in presence of nitrates. Nitric acid in sulphuric acid is detected by placing a crystal of diphenylamine in 3 or 4 cc. of the acid and gently warming. Cl', Cl^{V}, Br^{V}, I^{V}, Mn^{VII}, Cr^{VI}, Se^{IV}, Fe''' interfere.

Copper placed in a solution containing nitric acid liberates brown fumes.

Phenolsulphonic Acid Test. See chapter on Water Analysis.

Detection of Nitrous Acid. *Acetic Acid Test.* Acetic acid added to a nitrite in a test-tube (inclined as directed in the nitric acid test with ferrous sulphate), produces a brown ring. Nitrates do not give this. If potassium iodide is present in the solution, free iodine is liberated. The free iodine is absorbed by chloroform, carbon tetrachloride or disulphide, these reagents being colored pink. Starch solution is colored blue.

Nitrous acid reduces iodic acid to iodine. The iodine is then detected with starch, or by carbon disulphide, or carbon tetrachloride.

Potassium Permanganate Test. A solution of the reagent acidified with sulphuric acid is decolorized by nitrous acid or nitrite. The test serves to detect nitrous acid in nitric acid. Other reducing substances must be absent.

ESTIMATION

Occurrence. Element. Free in air to extent of 78%+ by volume and 76% − by weight.

Air weight of 1 liter $=1.293$ grams. With oxygen as 32, air $=28.95$.

COMPOSITION OF AIR.　ON THE BASIS OF 1000 LITERS OF ATMOSPHERE

Element.	Liters per 1000 l.	Weight per 1000 l. grams	Per cent by Vol.	Per cent by Wt.
Nitrogen	780.3	975.80	78.1	75.47 −
Oxygen	209.9	299.84	21.0	23.19 −
Argon	9.4	16.76	0.9	1.296+
Carbon dioxide	0.3	0.59	0.04	0.045
Hydrogen	0.1	0.01		
Neon	0.015	0.01339		
Helium	0.0015	0.00027		
Krypton	0.00005	0.00018		
Xenon	0.000006	0.00003		

Nitrogen is found combined in nature as potassium nitrate (saltpeter), KNO_3; sodium nitrate (Chili saltpeter), $NaNO_3$, and to a less extent as calcium nitrate,

$Ca(NO_3)_2$. It occurs in plants and in animals, in the substances proteids, blood, muscle, nerve substance, in fossil plants (coal), in guano, ammonia and ammonium salts.

Free nitrogen is estimated in the complete analysis of gas mixtures. In illuminating gas the other constituents are removed by combustion and absorption and the residual gas measured as nitrogen.

Total nitrogen in organic substances is best determined by decomposition of the materials with sulphuric acid as described later, and estimating the nitrogen from the ammonia formed.

Combined nitrogen in the form of ammonia and nitric acid specially concerns the analyst. In the evaluation of fertilizers, feedstuffs, hay, fodders, grain, etc., the nitrogen is estimated after conversion to ammonia. Ammonia, nitrates and nitrites may be required in an analysis of sewages, water, and soils. Nitric acid is determined in Chili saltpeter, in the evaluation of this material for the manufacture of nitric acid or a fertilizer, the nitrate being reduced to ammonia and thus estimated.

We will take up a few of the characteristic substances in which nitrogen estimations are required, e.g., in organic substances as proteids, in soils and fertilizers; in ammonium salts, nitrates, and nitrites, free ammonia in ammoniacal liquors, nitric acid in the evaluation of the commercial acid and in mixed acids.

In general nitrogen is more accurately and easily measured as ammonia, to which form it is converted by reduction methods. Large amounts are determined by titration, whereas small amounts are estimated colorimetrically. Nitric acid and nitrates may be determined by direct titration by the Ferrous Sulphate Method outlined later. The procedure is of value in estimation of nitrates in mixed acids. The nitrometer method for determining nitrates (including nitrites), and the free acid in mixed acids, is generally used by manufacturers of explosives.

Preparation of the Sample

It will be recalled that compounds of ammonia and of nitric acid are generally soluble in water. All nitrogen compounds, however, are not included. Among those which are not readily soluble the following deserve mention: compounds of nitrogen in many organic substances; nitrogen bromophosphide, $NPBr_2$; nitrogen selenide, NSe; nitrogen sulphide, N_4S_4; nitrogen pentasulphide, N_2S_5; ammonium antimonate, $NH_4SbO_3 \cdot 2H_2O$; ammonium iodate, HN_4IO_3 (2.6 grams per 100 cc. H_2O); ammonium chlorplatinate, $(NH_4)_2PtCl_6$ (0.67 gram); ammonium chloriridate, $(NH_4)_2IrCl_6$ (0.7 gram); ammonium oxalate, $(NH_4)_2C_2O_4 \cdot H_2O$ (4.2 grams); ammonium phosphomolybdate, $(NH_4)_3PO_4 \cdot 12MoO_3$ (0.03 gram); nitron nitrate, $C_{20}H_{16}N_4 \cdot HNO_3$.

Organic Substances

By oxidation of nitrogenous organic substances with concentrated sulphuric acid, containing mercuric oxide, or potassium permanganate, the organic matter is destroyed and the nitrogen is changed to ammonia, which is held by the sulphuric acid as sulphate. Nitrates are reduced by addition of salicylic acid, zinc dust, etc., previous to the oxidation process. Practically all the procedures are based on the Kjeldahl method of acid digestion. The modification, commonly known as the Kjeldahl-Gunning-Arnold Method, is as follows:

Method in Absence of Nitrates. *Weight of Sample.* *Fertilizers* 0.7 to 3.5 grams. *Soils* 7 to 14 grams. *Meat* and *meat products* 2 grams. *Milk* 5 grams. The amount of the substance to be taken should be governed by its nitrogen content.[1]

Acid Digestion.[2] The material is placed in a Kjeldahl flask of about 550 cc. capacity. Approximately 0.7 gram of mercuric oxide or an equivalent amount of metallic mercury together with 10 grams of powdered potassium sulphate followed by 20 to 30 cc. of concentrated sulphuric acid (sp.gr. 1.84) are added. The flask is placed in an inclined position, resting in a large circular opening of an asbestos board. The flask is heated with a small flame until the frothing has ceased. (A piece of paraffin may be added to prevent extreme frothing.) The heat is then raised and the acid brought to brisk boiling, the heating being continued until the solution becomes a pale straw color, or practically water white. (In case of leather, scrap, cheese, milk products, etc., a more prolonged digestion may be required. With a good flame from one-half to one hour of acid digestion is generally sufficient to completely decompose the material.) The flask is now removed from the flame and after cooling the solution is diluted with about 200 cc. of water and a few pieces of granulated zinc added to prevent " bumping " (50 mg. or so of No. 80 granulated zinc). The solution is now alkalized strongly by addition of a mixture of sodium hydroxide and sodium sulphide solution (about 75 cc. of a mixture containing 25 grams of NaOH and 1 gram Na$_2$S). Phenolphthalein indicator added to the solution will show when the acid is neutralized. The flask is connected by means of a Hopkins distillation tube (Fig. 53) to a condenser and about 150 cc. of the solution distilled into an excess of standard sulphuric acid and the excess of the acid determined by titration with standard sodium hydroxide. (Methyl red indicator.)

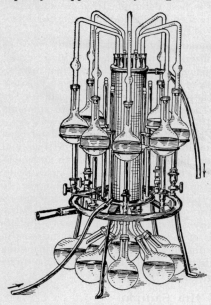

FIG. 50.

Apparatus for Determining Nitrogen.

The ammonia may be absorbed in a saturated solution of boric acid and titrated directly with standard acid. (Methyl orange indicator.)[3]

$$\text{One cc. N}/10 \text{ H}_2\text{SO}_4 = 0.001704 \text{ gram NH}_3.$$

[1] See data of approximate nitrogen content in certain nitrogenous substances, Jour. Ind. Eng. Chem., **7**, 357, 1915.

[2] Fig. 50 shows a compact apparatus with several sets of flasks and condensers, which enable half a dozen or more determinations to be made at one time.

[3] L. W. Winkler, Z. angew. Chem., **27**, 1, 630–2, 1914. E. Bernard, ibid., **27**, 1, 664, 1914.

In Presence of Nitrates. The procedure differs from the former in the preliminary treatment to reduce the nitrates. The material in the flask is treated with a mixture of 30 to 35 cc. of strong sulphuric acid containing 1 gram of salicylic acid and the mixture shaken and allowed to stand for five to ten minutes with frequent agitation. About 5 grams of sodium thiosulphate are now added and the solution heated for five minutes. After cooling, mercuric oxide or metallic mercury and potassium sulphate are added, and the solution treated as directed above.

Notes. Mercuric oxide or metallic mercury are added as a catalyzer to assist the oxidation of the organic matter. The digestion process is shortened considerably by its use. In place of mercuric oxide or the metal, copper sulphate may be used. In this case the addition of sodium sulphide is omitted. Copper sulphate acts as an indicator in the neutralization of the sample with caustic.

Potassium sulphide is added to remove the mercury from the solution and prevent the formation of mercur-ammonium compounds, which are not completely decomposed by sodium hydroxide.

Ferric chloride, $FeCl_3$, may be used in place of copper or mercury salts or oxides to assist in the oxidation of organic matter.

Soils. Available Nitrates. Five hundred to 1000 grams of the air-dried soil is extracted with 1 to 2 liters of water containing 10 to 20 grams of dextrose. Fifteen to twenty hours of leaching is sufficient. An aliquot portion is taken for analysis.

Ammonium Salts. The sample is placed in the distillation flask with splash bulb as described in the modified Kjeldahl procedure for organic substances, and the material decomposed with ammonia-free caustic solution. The ammonia is distilled into an excess of standard acid or a saturated solution of boric acid (neutral to methyl orange), and the ammonia determined as usual, either by titration of the excess of acid, or by direct titration with acid, according to the absorbent used.

Nitrates. The sample, broken down as fine as possible, is dissolved in water, decomposed with Devarda alloy and distilled as described by the modified Devarda methods given later.

Nitrites. The material, dissolved in water, is titrated with standard permanganate solution according to the procedure described later.

Mixtures of Ammonium Salts, Nitrates, and Nitrites. Ammonia is determined by distillation with caustic as usual. The nitrite is titrated with permanganate. Total nitrogen is determined by the modified Devarda methods. Nitric acid is now estimated by difference, e.g., from the total nitrogen is deducted the nitrogen due to ammonia together with the nitrogen of the nitrite and the difference calculated to the nitrate desired. The nitrate may be determined in presence of nitrite and ammonia by direct titration with ferrous sulphate. The detailed procedures may be found under the Volumetric Methods.

Nitric Acid in Mixed Acid. This is best determined by the ferrous sulphate method for nitric acid. The nitrometer method is also excellent.

SEPARATIONS

Ammonia. No special separation need be considered in the determination of ammonia. The general method has already been mentioned by which ammonia is liberated from its salts by a strong base and volatilized by heat. This effects a separation from practically all substances.

Nitric Acid. The compound may be isolated as the fairly insoluble, crystalline nitron nitrate, $C_{20}H_{16}N_4 \cdot HNO_3$ by the following procedure.

Such an amount of the substance is taken as will contain about 0.1 gram nitric acid, and dissolved in about 100 cc. of water with addition of 10 drops of dilute sulphuric acid. The solution is heated nearly to boiling and about 12 cc. of nitron acetate solution added (10 grams of nitron in 100 cc. of 5% acetic acid).[2] The solution is cooled and placed in an ice pack for about two hours, and the compound then transferred to a Gooch or Munroe crucible (weighed crucible if gravimetric method is to be followed), and after draining, it is washed with about 10 to 12 cc. of ice-water added in small portions. The nitrate may now be determined gravimetrically by drying the precipitate to constant weight at 110° C., 16.53% of the material being due to NO_3.

The base diphenyl-endo-anilo-hydro-triazole (nitron) also precipitates the following acids: nitrous, chromic, chloric, perchloric, hydrobromic, hydriodic, hydroferro- and hydroferricyanic, oxalic, picric and thiocyanic acids. Hence these must be absent from the solution if precipitation of nitric acid is desired for quantitative estimation.

Removal of Nitrous Acid. Finely powdered hydrazine sulphate is dropped into the concentrated solution. (0.2 gram substance per 5 or 6 cc.)

Chromic acid is reduced by addition of hydrazine sulphate.

Hydrobromic acid is decomposed by chlorine water added drop by drop to the neutral solution, which is then boiled until the yellow color has disappeared.

Hydriodic acid is removed by adding an excess of potassium iodate to the neutral solution and boiling until the iodine is expelled.

PROCEDURES FOR THE DETERMINATION OF COMBINED NITROGEN

Ammonia

The volumetric procedures for determination of ammonia are preferred to the gravimetric on account of their accuracy and general applicability. The following gravimetric method may occasionally be of use:

Gravimetric Determination of Ammonia by Precipitation as Ammonium Platinochloride, $(NH_4)_2PtCl_6$

Ammonia in ammonium chloride may be determined gravimetrically by precipitation with chlorplatinic acid. The method is the reciprocal of the one for determining platinum.

Procedure. The aqueous solution of the ammonium salt is treated with an excess of chlorplatinic acid and evaporated on the steam bath to dryness. The residue is taken up with absolute alcohol, filtered through a weighed Gooch crucible, and washed with alcohol. The residue may now be dried at 130° C. and weighed as $(NH_4)_2PtCl_6$, or it may be gently ignited in the covered crucible until

[1] M. Busch, Ber., **38**, 861 (1905), Treadwell and Hall, " Analytical Chemistry."
[2] Keep nitron reagent in a dark-colored bottle.

ammonium chloride has been largely expelled and then more strongly with free access of air. The residue of metallic platinum is weighed. If the ignition method is to be followed, the ammonium platinic chloride may be filtered into a small filter, the paper with the washed precipitate placed in a porcelain crucible and then gently heated until the paper is charred (crucible being covered) and then more strongly with free access of air until the carbon has been destroyed.

Factors.[1] $(NH_4)_2PtCl_6 \times 0.2400 = NH_4Cl$, or $0.08095 = NH_4$, or $\times 0.0767 = NH_3$. $Pt \times 0.5453 = NH_4Cl$, or $\times 0.1839 = NH_4$, or $\times 0.1736 = NH_3$.

VOLUMETRIC METHODS FOR DETERMINATION OF AMMONIA

Two conditions are considered:

A. Estimation of free ammonia in solution.

B. Determination of ammonia in its salts—combined ammonia.

Analysis of Aqua Ammonia

Provided no other basic constituent is present, free ammonia in solution is best determined by direct titration with an acid in presence of methyl orange or methyl red as indicator.

Procedure. About 10 grams of the solution in a weighing bottle with glass stopper is introduced into an 800-cc. Erlenmeyer flask containing about 200 cc. of water and sufficient $\frac{1}{2}$ normal sulphuric acid to combine with the ammonia and about 10 cc. in excess. The flask is stoppered and warmed gently. This forces out the stopper in the weighing bottle, the ammonia combining with the acid. Upon thorough mixing, the solution is cooled, and the excess of acid is titrated with half normal caustic.

$$\text{One cc. } \tfrac{1}{2} \text{ N. } H_2SO_4 = 0.0085 \text{ gram } NH_3$$

Factor. $H_2SO_4 \times 0.3473 = NH_3$.

NOTE. The aqua ammonia exposed to the air will lose ammonia, hence the sample should be kept stoppered. This loss of ammonia is quite appreciable in strong ammoniacal solutions.

Determination of Combined Ammonia. Ammonium Salts.

Strong bases decompose ammonium salts, liberating ammonia. This may be distilled into standard acid or into a saturated solution of boric acid (neutral to methyl orange) and titrated.

Procedure. About 1 gram of the substance is placed in a distillation flask (see Fig. 50) and excess of sodium or potassium hydroxide added and the ammonia distilled into a saturated solution of boric acid or an excess of standard sulphuric acid. Ammonia in boric acid solution may be titrated directly with standard acid (methyl orange or methyl red indicator) or in case a mineral acid

---

[1] Factors recommended by Treadwell and Hall, "Analytical Chemistry," **2**, John Wiley & Sons.

was used to absorb the ammonia, the excess of acid is titrated with standard caustic solution.

One cc. half normal sulphuric acid = 0.0085 gram NH_3.

One cc. normal acid = 0.01703 gram NH_3.

Factors. $H_2SO_4 \times 0.3473 = NH_3$ and $NH_3 \times 2.8792 = H_2SO_4$.

ANALYSIS OF AMMONIACAL LIQUOR

The crude liquor by-product from coal gas in addition to ammonia contains hydrogen sulphide, carbon dioxide, hydrochloric acid, sulphuric acid, combined with ammonia, also sulphites, thiosulphates, thiocyanates, cyanides, ferrocyanides, phenols.

Determination of Ammonia

Volatile Ammonia.   This is determined by distillation of the ammonia into an excess of standard sulphuric acid and titrating the excess of acid.   With the exception that caustic soda is omitted in this determination, the details are the same as those for total ammonia as stated in the next paragraph.

Total Ammonia.   The true value of the liquor is ascertained by its total ammonia content.   Ten to 25 cc. of the sample is diluted to about 250 cc. in a distilling flask with a potash connecting bulb, as previously described, 20 cc. of 5% sodium hydroxide are added and about 150 cc. of solution distilled into an excess of sulphuric acid.   The excess is then titrated according to the standard procedure for ammonia.

One cc. N. H_2SO_4 = 0.01703 gram NH_3.

Fixed Ammonia is the difference between the total and the volatile ammonia.

Carbon Dioxide

Ten cc. of the liquor are diluted to 400 cc. and **10 cc.** of 10% ammoniacal calcium chloride added and the mixture, placed in a flask with Bunsen valve, is digested on the water bath for two hours.   The precipitated calcium carbonate is washed, placed in a flask and an excess of N/2 HCl added and the excess acid titrated with N/2 NaOH.

N/2 HCl = 0.011 gram CO_2.

Hydrochloric Acid

Ten cc. of the liquor is diluted to 150 cc. and boiled to remove ammonia. Now hydrogen peroxide is added to oxidize organic matter, etc., the mixture being boiled to remove the excess of the peroxide.   Chlorine is titrated in presence of potassium chromate as indicator by tenth normal silver nitrate after neutralizing with dilute nitric acid.

One cc. N/10 $AgNO_3$ = 0.00364 gram HCl.

Hydrogen Sulphide

To 10 cc. of the liquor are added an excess of ammoniacal zinc chloride or acetate, the mixture diluted to about 80 cc. and warmed to 40°. After settling for half an hour the zinc sulphide is filtered off and washed with warm water (40 to 50°); the precipitate is washed from the filter into an excess of N/10 iodine solution, the sulphide clinging to the paper washed into the main solution with hydrochloric acid. The mixture is acidified and the excess iodine titrated with N/10 sodium thiosulphate.

One cc. N/10 I = 0.0017 gram H_2S or 0.0016 gram S.

Sulphuric Acid

250 cc. of the liquor is concentrated to 10 cc., 2 cc. of concentrated hydrochloric added and the mixture heated to decompose any thiosulphate, sulphide or sulphite present. The concentrate is extracted with water, filtered and made to 250 cc. The sulphuric acid is now precipitated in an aliquot portion with barium chloride.

$BaSO_4 \times 0.4202 = H_2SO_4$, or $\times 0.1374 = S$ present as H_2SO_4.

Total Sulphur. Fifty cc. of the liquor is run by means of a pipette into a deep beaker (250 cc. capacity), containing an excess of bromine covered by dilute hydrochloric acid. The mixture is evaporated to dryness on the steam bath and the residue taken up with water and diluted to 250 cc. Sulphur is now precipitated as barium sulphate as usual, preferably on an aliquot portion.

For a more complete analysis of crude liquor determining sulphite, thiosulphate, thiocyanate, hydrocyanic acid, ferrocyanic acid, and phenols the analyst is referred to Lunge, " Technical Methods of Chemical Analysis," Part II, Vol. II, D. Van Nostrand Co.

Determination of Traces of Ammonia

The determination of traces of ammonia is best accomplished by the colorimetric method with Nessler's reagent. Details of the procedure are given in the chapter on water analysis.

NITRIC ACID. NITRATES

The alkalimetric method for determining free nitric acid, and the complete analysis of the commercial product are given in the chapter on Acids. Special procedures for determining the combined acid are herein given.

Gravimetric Method for Determining Nitric Acid by Precipitation as Nitron Nitrate, $C_{20}H_{16}N_4 \cdot HNO_3$

As in case of ammonia the volumetric methods are generally preferable for determining nitric acid, combined or free. Isolation of nitric acid by precipitation as nitron nitrate may occasionally be used. The fairly insoluble, crystalline compound, $C_{20}H_{16}N_4 \cdot HNO_3$ is formed by addition of the base diphenyl-endo-anilo-hydro-triazole (nitron) to the solution containing the nitrate as directed under Separations. The precipitate washed with ice-water is dried to constant weight at 110° C. 16.53% of the compound is NO_3.

NOTE. The following acids should not be present in the solution, since their nitron salts are not readily soluble: nitrous, chromic, chloric, perchloric, hydrobromic, hydroiodic, hydroferrocyanic, hydroferricyanic, oxalic, picric and thiocyanic acids.

Solubility of less soluble nitron salts in 100 cc. of water. Nitron nitrate = 0.0099 gram, nitron bromide = 0.61 gram, iodide = 0.017 gram, nitrite = 0.19 gram, chromate = 0.06 gram, chlorate 0.12 gram, perchlorate = 0.008 gram, thiocyanate = 0.04 gram. (Treadwell and Hall, "Analytical Chemistry, Quantitative Analysis.")

VOLUMETRIC METHODS

Direct Estimation of Nitrates by Reduction to Ammonia. Modified Devarda Method [1]

An accurate procedure for the determination of nitrogen in nitrates is Allen's modification of the Devarda method. The method is based upon the quantitative reduction of nitrates to ammonia in an alkaline solution by an alloy consisting of 45 parts of aluminum, 50 parts of copper and 5 parts of zinc. The ammonia evolved is distilled into standard sulphuric acid and thus estimated. The method, originally designed for the valuation of sodium or potassium nitrates, is also of value in the determination of nitric acid, nitrites or ammonia. In the latter case the alloy is omitted.

Reagents Required. *Devarda's Alloy.* Forty-five parts aluminum, 50 parts copper and 5 parts zinc. The aluminum is heated in a Hessian crucible in a furnace until the aluminum begins to melt, copper is now added in small portions until liquefied and zinc now plunged into the molten mass. The mix is heated for a few moments, covered and then stirred with an iron rod, allowed to cool slowly with the cover on and the crystallized mass pulverized.

Standard Sulphuric Acid. This is made from the stock C.P. acid by dilution so that 1 cc. is equal to 0.0057 gram H_2SO_4, 100 cc. of acid of this strength being equivalent to approximately 1 gram of sodium nitrate. (A tenth normal acid will do, a smaller sample being taken for analysis.) Since it is necessary to standardize this acid against a standard nitrate, it is advisable to have an acid especially for this determination rather than a common reagent for general use.

Standardization of the Acid. 11.6 grams of standard potassium nitrate, equivalent to about 9.6 grams of $NaNO_3$, is dissolved and made to volume in the weighing bottle (100 cc.), and 10 cc. is placed in the Devarda flask, reduced and the ammonia distilled into 100 cc. of the acid, exactly as the following method describes. The temperature of the acid is noted and its value in terms of H_2SO_4, KNO_3 and $NaNO_3$ stated on the container. The acid expands or contracts 0.029 cc. for every degree centigrade above or below the temperature of standardization, per 100 cc.

Standard Potassium Nitrate. The purest nitrate that can be obtained is recrystallized in small crystals, by stirring, during the cooling of the supersaturated concentrated solution, and dried first at 100° C. for several hours and then

---

[1] Paper by W. S. Allen, General Chemical Company, Eighth International Congress of Applied Chemistry.

at 210° C. to constant weight. Chlorides, sulphates, carbonates, lime, magnesia and sodium are tested for and if present are determined and allowance made.

Standard Sodium Hydroxide. This should be made of such strength that 1 cc. is equal to 1 cc. of the standard acid, 2 cc. methyl red being used as indicator. Ten cc. of the acid are diluted to 500 cc. and the alkali added until the color of the indicator changes from a red to a straw color.

Methyl Red Solution. 0.25 gram of methyl red is dissolved in 2000 cc. of 95% alcohol; 2 cc. of the indicator is used for each titration. As the indicator is sensitive to CO_2, all water used must first be boiled to expel carbonic acid. (Baker & Adamson, manufacturers of methyl red.)

Sodium Hydroxide—Sp.gr. 1.3. Pure sodium hydroxide is dissolved in distilled water and boiled in an uncovered casserole with about 1 gram of Devarda's alloy to remove ammonia. This is cooled and kept in a well-stoppered bottle.

Apparatus. This is shown in the accompanying illustration, Fig. 51. It consists of the Devarda flask connected to the scrubber K, filled with glass wool. This

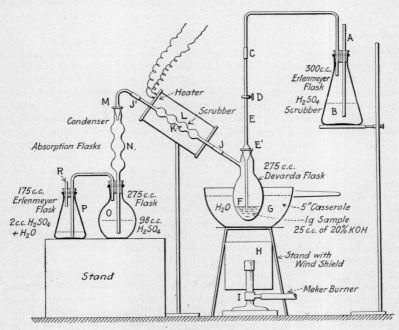

Fig. 51.—Devarda's Apparatus.

scrubber is heated by an electric coil or by steam passed into the surrounding jacket. The scrubber prevents caustic spray from being carried over into the receiving flask O. The form of the apparatus can best be ascertained from the sketch.

Weighing bottle with graduation at 100 cc. and a 10-cc. dropper with rubber bulb is used for weighing out the sample in solution. See Fig. 52.

29

Preparation of the Sample

Weight. It is advisable to take a large sample if possible, e.g., 100 grams of $NaNO_3$, 119 grams of KNO_3 or about 80 grams of strong HNO_3 (95%) or more if the acid is dilute. Solids are taken from a large sample, all lumps being broken down. After dissolving in water the sample is made up to 1 liter. (Scum is broken up by addition of a little alcohol.) One hundred cc. of this solution is placed in the weighing bottle, which has been previously weighed, being perfectly clean and dry. The difference is the weight of the 100-cc. sample.

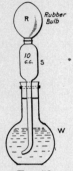

Fig. 52.

Weighing Bottle and Dropper.

Manipulation. All parts of the apparatus are washed out with CO_2-free water. All water used in this determination should be boiled to expel CO_2. Ninety-eight cc. of the standard acid is placed in flask O and washed down with 2 to 3 cc. of water. Two cc. of the standard acid is placed in flask P and washed down with 10 cc. of water and 13 to 14 drops of methyl red indicator added. Connections are made between the flasks and the scrubber. (The correction is made for the acid, the temperature being noted at the time of withdrawal.) A casserole, filled with cold water, is placed under F (see illustration). The stem E is removed from the Devarda flask and 10 cc. (or more) of the nitrate added by means of the dropper in the weighing bottle, a funnel having been inserted in the flask. The bottle reweighed gives the weight of the sample removed, by difference. The nitrate is washed down with 10 cc. of water and 25 cc. of 20% caustic added (free from NH_3), the alkali washed down with 10 cc. more of water and then 3 grams of Devarda alloy placed in the flask by means of dry funnel. The stem E is quickly replaced, the stopcock being turned to close the tube. The reaction begins very soon. If it becomes violent, the reaction may be abated by stirring the water in the casserole, thus cooling the sample. After the energetic action has abated (five minutes), the casserole with the cold water is removed and the action allowed to continue for twenty minutes, meantime heat or steam is turned on in the scrubber. E is connected at C to the flask B containing caustic to act as a scrubber. It is advisable to have a second flask containing sulphuric acid attached to the caustic to prevent ammonia from the laboratory entering the system. A casserole with hot water is placed under F and the burner lighted and turned on full. A gentle suction is now applied at R, the stop-cock D being turned to admit pure air into the evolution flask; the rate should be about 5 to 6 bubbles per second. The suction is continued for thirty minutes, hot water being replaced in the casserole as the water evaporates. The heat is now turned off and the apparatus disconnected at M and J. The contents of this elbow and the condenser are washed into the flask O. The acid in O and P poured into an 800-cc. beaker and rinsed out several times. The volume in the beaker is made up to 500 cc., 1 cc. of methyl red added, and the free acid titrated with the standard caustic. The end-point is a straw yellow.

Calculation. The cc. of the back titration with caustic being deducted, the volume of the acid remaining (e.g., combined with ammonia) is corrected to the standard condition. Expansion or contraction per 100 cc. is 0.029 cc. per each degree C. above or below the temperature at which the acid was standardized. If the acid is exactly 0.057 gram H_2SO_4 per cc., the result multiplied by 0.989 and divided by the weight of the sample taken gives per cent nitrate. (In terms of NaNO₃.)

The Weight of the Sample. Ten times the difference of the weighings of the bottle W before and after removal of the 10 cc. and the product divided by the weight of the 100 cc. of the solution equals the weight of solid taken.

Example. *Weight* of the bottle +100 cc. sample = 218 grams. Weight of the bottle = 112 grams, therefore weight of 100 cc. NaNO₃ = 106 grams.

Weight of the bottle +100 cc. sample = 218. Weight after removal of 10 cc = 207.4 grams, therefore sample taken = 10.6 grams, including the added water. Now from above the weight of the actual sample taken = $10.6 \times 10 \div 106 = 1$ gram.

Temperature Correction. Temperature of standardization = 20° C. Temperature of the sulphuric acid when taken for the analysis = 31° C. Back titration of the caustic = 2 cc. The correct volume = $(100-2) - ((31-20) \times 0.029) = 97.681$ cc. H₂SO₄ combined with ammonia from the reduced nitrate. $97.681 \times 0.989 \div 1 = 96.62\%$ NaNO₃.

Factors. H₂SO₄ × 2.06107 = KNO₃ or × 1.7334 = NaNO₃ or × 1.2850 = HNO₃.

H₂SO₄ × 0.9587 = HNO₂ or × 0.3473 = NH₃.

NH₃ × 3.6995 = HNO₃ or × 4.9906 = NaNO₃ or × 4.0513 = NaNO₂.

NaNO₃ × 1.1894 = KNO₃ and KNO₃ × 0.8408 = NaNO₃.

ANALYSIS OF NITRATE OF SODA

The following impurities may occur in nitrate of soda: KNO₃, NaCl, Na₂SO₄, Na₂CO₃, NaClO₃, NaClO₄, Fe₂O₃, Al₂O₃, CaO, MgO, SiO₂, H₂O, etc. In the analysis of sodium nitrate for determination of NaNO₃ by difference, moisture, NaCl, Na₂SO₄ and insoluble matter are determined and their sum deducted from 100, the difference being taken as NaNO₃. Such a procedure is far from accurate, the only reliable method being a direct determination of niter by the Devarda method given in detail. The following analysis may be required in the valuation of the nitrate of soda.

Determination of Moisture

Twenty grams of sample are heated in a weighed platinum dish at 205 to 210° C. for fifteen minutes in an air bath or electric oven. The loss of weight multiplied by 5 = per cent moisture. (Save sample for further tests.)

Insoluble Matter

Ten grams are treated with 50 cc. of water and filtered through a tared Gooch. The increased weight dried residue (100° C.) multiplied by 10 = per cent insoluble matter. (Save filtrate.)

Sodium Sulphate

The moisture sample is dissolved in 20 cc. hot water and transferred to a porcelain crucible. It is evaporated several times with hydrochloric acid to dryness to expel nitric acid. (Until no odor of free chlorine is noticed when thus treated.) Fifty cc. of water and 5 cc. hydrochloric acid are now added and the sample filtered. Any residue remaining is principally silica. The filtrate is heated to boiling, 10 cc. of 10% barium chloride solution added, and the precipitated sulphate filtered off, ignited and weighed.

$$BaSO_4 \times 3.0445 = \text{per cent } Na_2SO_4.$$

Iron, Alumina, Lime, and Magnesia

These impurities may be determined on a 20-gram dried sample, the material being dried and evaporated as in case of the sodium sulphate determination. The filtrate from silica is treated with ammonium hydroxide and $Fe(OH)_3$ and $Al(OH)_3$ filtered off. Lime is precipitated from the iron and alumina filtrate as oxalate and magnesia determined by precipitation as phosphate from the lime filtrate by the standard procedures.

Sodium Chloride

The filtrate from the insoluble residue is brought to boiling and magnesia, MgO (Cl free), is added until the solution is alkaline to litmus. 0.5 cc. of 1% potassium chromate (K_2CrO_4) solution is added as an indicator and then the solution is titrated with a standard solution of silver nitrate until a faint red tinge is seen, the procedure being similar to the determination of chlorides in water by silver nitrate titration. The cc. $AgNO_3 \times$ factor for this reagent $\times 10 =$ per cent NaCl.

Silver nitrate is standardized against a salt solution.

Carbonates

This determination is seldom made. CO_2 may be tested for by addition of dilute sulphuric acid to the salt. Effervescence indicates carbonates. Any evolved gas may be tested by lime water, which becomes cloudy if CO_2 is present. For details of the procedure reference is made to the chapter on Carbon.

DETERMINATION OF NITRIC NITROGEN IN SOIL EXTRACTS

Vamari-Mitscherlich-Devarda Method

Procedure. Forty cc. of water, a small pinch of magnesia and one of magnesium sulphate are added to flask D of the Mitscherlich apparatus (Fig. 53). Twenty-five cc. of standard acid and 60 cc. of neutral redistilled water are placed in flask F; 250 or 300 cc. of aqueous soil extract are placed in a 500-cc. Kjeldahl flask, 2 cc. of 50% sodium hydroxide added, the mouth of the flask closed with a small funnel to prevent spattering, and the contents of the flask boiled for thirty minutes. The water which has boiled off is replaced, and, after cooling, 1 gram of Devarda's alloy (60 mesh), and a small piece of paraffin are added and the flask connected with the apparatus; reduction and distillation are carried on for forty minutes. The receiver contents are then cooled, 4 drops of 0.02%

solution of methyl red added, the excess acid is nearly neutralized, the liquid boiled to expel CO_2, cooled to 10 to 15° and the titration completed.

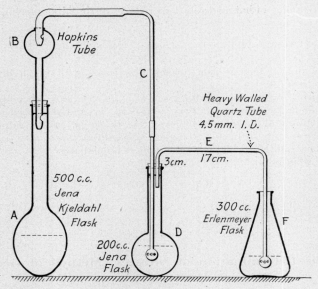

FIG. 53.—Mitscherlich's Apparatus for Nitrogen Determination.

B. S. Davisson[1] recommends an improved form of scrubber, shown in Fig. 53a to be used in place of the Hopkins bulb (Fig. 53). The bulb and adaptor are made of Pyrex glass. Steam condenses in the bulb and the condensate acts as a scrubber preventing alkali mist from being carried over with the ammonia. During the test ammonia is completely volatilized into the absorption flask. The bulb of the adaptor, shown as II, prevents back suction into the distillation flask.

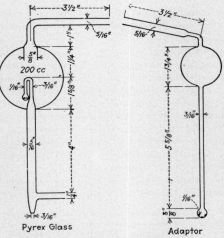

FIG. 53a.—Davisson's Scrubber.

[1] Reference B. S. Davisson, Ohio, Ag. Exp. Station, J. Ind. Eng. Chem., 11, 465 (May, 1919).

Determination of Hydroxylamine—Method of Raschig[1]

Hydroxylamine in hot acid solutions reduces ferric salts to ferrous condition quantitatively according to the reaction:

$$2NH_2OH + 2Fe_2(SO_4)_3 = 4FeSO_4 = 2H_2SO_4 + N_2O + H_2O.$$

The amount of ferrous iron formed is a measure of the hydroxylamine originally present.

Procedure. Approximately 0.1 gram of hydroxylamine salt is dissolved in a little water in an Erlenmeyer flask and 30 cc. of cold saturated solution of ferric-ammonium alum added, followed by 10 cc. of dilute sulphuric acid (1 : 4). The solution is heated to boiling and kept at this temperature for five minutes, then diluted to 300 cc. and titrated immediately with standard permanganate solution.

$$1 \text{ cc. } N/10 \text{ KMnO}_4 = 0.001652 \text{ g. } NH_2OH.$$

Iodometric Determination of Nitrates—Method of Gooch and Gruener[2]

By this method the nitrate to be estimated is treated, in an atmosphere of carbon dioxide, with a saturated solution of manganous chloride (crystallized) in concentrated hydrochloric acid, the volatile products of the reaction (nitrogen dioxide, chlorine, etc.) are now distilled and caught in a solution of potassium iodide. The iodine set free is titrated by a standard solution of thiosulphate.

Procedure. The nitrate and the manganous mixture (saturated solution of crystallized manganous chloride and strong hydrochloric acid—20 cc. per 0.2 gram sample) following it are introduced into the pipette shown in Fig. 53a (marked III) suction being applied, if necessary, at the end of the absorption train (VI). The current of CO_2 is started immediately after putting in the mixture. When the air has been replaced by CO_2, heat is applied to the retort III and the distillation continued until nearly all the liquid has passed over into the receiver IV, which is cooled by water. (See illustration.) The contents of the receivers are united and the bulbs washed out by passing the wash water directly through III and IV. Introduction of manganous chloride into the distillate does not influence the accuracy of the titration. The liberated iodine is titrated with standard sodium thiosulphate as soon as possible after

---

[1] Hydroxylamine may also be determined by reduction with an excess of titanous salt in acid solution with exclusion of air, and the excess titrated with permanganate.
Reaction: $2NH_2OH + Ti_2(SO_4)_3 = (NH_4)_2SO_4 + 4TiOSO_4 + H_2SO_4.$
For discussion of the two methods see paper by Wm. C. Bray, Miriam E. Simpson and Anna A. MacKenzie, Jour. Am. Chem. Soc., 41, 9, 1362, Sept., 1919.
[2] "Methods in Chemical Analysis," by F. A. Gooch. F. A. Gooch and H. W. Gruener, Am. Jour. Sci. (5), XLIV, 117.

admitting air to the distillate, since traces of dissolved nitric oxide reoxidized by the air would react with the iodide liberating more iodine.

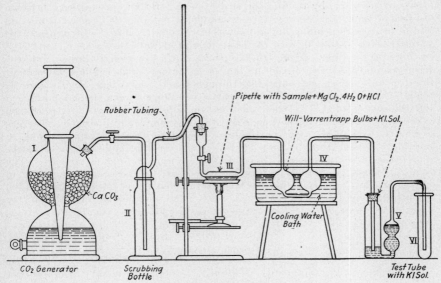

Fig. 53b.   Gooch-Gruener Apparatus.

DETERMINATION OF NITROGEN OF NITRATES (AND NITRITES) BY MEANS OF THE NITROMETER

The nitrometer is an exceedingly useful instrument employed in the accurate measurement of gases liberated in a great many reactions and has therefore a number of practical applications.   It may be used in the determination of carbon dioxide in carbonates;   the available oxygen in hydrogen dioxide;   in the valuation of nitrous ether and nitrites;   in the valuation of nitrates and nitric acid in mixed acids.

The method for the determination of nitrogen in nitrates, with which we are concerned in this chapter, depends on the reaction between sulphuric acid and nitrates in presence of mercury:

$$2KNO_3 + 4H_2SO_4 + 3Hg = K_2SO_4 + 3HgSO_4 + 4H_2O + 2NO.$$

The simplest type of apparatus is shown in the illustration, Fig. 54.   The graduated decomposition tube has a capacity of 100 cc.   It is connected at the base by means of a heavy-walled rubber tubing with an ungraduated leveling tube (b).   At the upper portion of (a) and separated from it by a glass stop-cock (s) is a bulb (c) of about 5 cc. capacity; a second stop-cock enables completely enclosing the sample, as may be necessary in volatile compounds.   The glass stop-cock (s), directly above the graduated chamber, is perforated so as to establish connection with the tube (d) when desired and the graduated cylinder (a).

Procedure. The tube (b) is filled with mercury and the air in (a) now displaced by mercury, by turning the stop-cock to form an open passage between (a) and (d) and then raising (b). A sample of not over 0.35 gram potassium nitrate or a corresponding amount of other nitrates, is introduced into (c), the material being washed in with the least amount of water necessary (1 to 2 cc.). By lowering (b) and opening the stop-cock s the solution is drawn into the decomposition chamber, taking care that no air enters. This is followed by about 15 cc. of pure, strong sulphuric acid through s_1 and s, avoiding admitting air as before. NO gas is liberated by the heat of reaction between the sulphuric acid and the water solution. When the reaction subsides, the tube (a) is shaken to mix the mercury with the liquor and the NO completely liberated. The gas is allowed to cool to room temperature and then measured, after raising or lowering (b) so that the column of mercury is the calculated excess of height above that in (a) in order to have the gas under atmospheric pressure. The excess of height is obtained by dividing the length of the acid layer in (a), in millimeters, by 7 and elevating the level of the mercury in (b) above that in (a) by this quotient; i.e., if the acid layer = 21 mm. the mercury in (b) would be 3 mm. above that in (a). The volume of gas is reduced to standard conditions by using the formula

$$V' = \frac{V(P-w)}{760(1+0.00367t)}.$$

V' = volume under standard conditions; V = observed volume; P = observed barometric pressure in mm.; w = tension of aqueous vapor at the observed temperature, expressed in millimeters; t = observed temperature.

One cc. gas = 4.62 milligrams of KNO_3, or 3.8 milligrams $NaNO_3$ or 2.816 milligrams HNO_3.

Du Pont Nitrometer Method [1]

The Du Pont nitrometer, Fig. 55, is the most accurate apparatus for the volumetric determination of nitrates. By use of this, direct readings in per cent may be obtained, without recourse to correction of the volume of gas to standard conditions and calculations such as are required with the ordinary nitrometers.

The apparatus consists of a generating bulb of 300 cc. capacity E with its reservoir F connected to it by a heavy-walled rubber tubing. E carries two glass stop-cocks as is shown in illustration. The upper is a two-way stop-cock connecting either the cup or an exit tube with the chamber. D is the chamber-reading burette, calibrated to read in percentages of nitrogen, and graduated from 10 to 14%, divided in 1/100%. Between 171.8 and 240.4 cc. of gas must be generated to obtain a reading. A is also a measuring burette, that may be used in place of D where a wider range of measurement is desired. " It is used for the measurement of small as well as large amounts of gas. It is most commonly graduated to hold 300.1 milligrams of NO at 20° C. and 760 mm. pressure and this volume is divided into 100 units (subdivided into tenths) each unit being equivalent to 3.001 milligrams of NO. When compensated, the gas from

[1] See paper by J. R. Pitman, Jour. Soc. Chem. Ind., p. 983, 1900.

ten times the molecular weight in milligrams of any nitrate of the formula RNO_3 (or five times molecular weight of $R(NO_3)_2$) should exactly fill the burette. This simplifies all calculations; for example the per cent nitric acid in a mixed acid would be

$$\frac{R63.02}{100W} = \text{per cent } HNO_3.$$

R = burette reading, W = grams acid taken."[1] C is the compensating burette very similar in form to the chamber burette D. B is the leveling bulb, by the

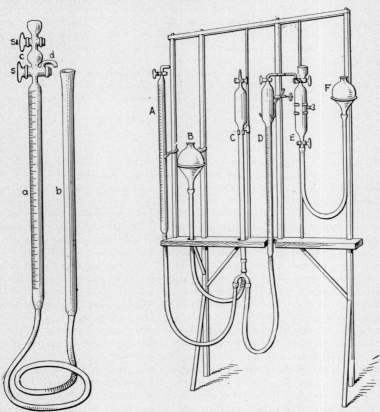

FIG. 54.—Nitrometer.   FIG. 55.—Du Pont's Nitrometer.

raising or lowering of which the standard pressure in the system may be obtained. The apparatus as shown in Fig. 55 is mounted on an iron stand. As in the more simple form of apparatus, previously described, mercury is used as the confining liquid. The parts are connected by heavy-walled rubber tubing, wired to the glass parts.

[1] A. W. Betts, Chemist, E. I. DuPont de Nemours Powder Co., in letter to author.

Standardizing the Apparatus. The apparatus having been arranged and the various parts filled with mercury, the instrument is standardized as follows: 20 to 30 cc. of sulphuric acid are drawn into the generating bulb through the cup at the top, and at the same time about 210 cc. of air; the cocks are then closed, and the bulb well shaken; this thoroughly desiccates the air, which is then run over into the compensating burette until the mercury is about on a level with the 12.30% mark on the other burette, the two being held in the same relative position, after which the compensating burette is sealed off at the top. A further quantity of air is desiccated in the same manner and run into the reading burette so as to fill up to about the same mark; the cocks are then closed, and a small piece of glass tubing bent in the form of a U, half filled with sulphuric acid (not water), is attached to the outlet of the reading burette; when the mercury columns are balanced and the enclosed air cooled down, the cock is again carefully opened, and when the sulphuric balances in the U-tube, and the mercury columns in both burettes are at the same level, then the air in each one is under the same conditions of temperature and pressure. A reading is now made from the burette, and the barometric pressure and temperature carefully noted, using the formula

$$V_t = \frac{V_o P_o (273+t)}{P_t 273},$$

the volume this enclosed air would occupy at 29.92 ins. pressure and 20° C. is found. The cock is again closed and the reservoir manipulated so as to bring the mercury in both burettes to the same level, and in the reading burette to the calculated value as well. A strip of paper is now pasted on the compensating burette at the level of the mercury, and the standardization is then complete.

Another rapid method of standardizing is to fill the compensating chamber with desiccated air as stated in the first procedure and then to introduce into the generating chamber 1 gram of pure potassium nitrate dissolved in 2 to 4 cc. of water, the cup is rinsed out with 20 cc. of 66° Béaume sulphuric acid, making three or four washings of it, each lot being drawn down separately into the bulb. The generated gas formed after vigorous shaking of the mixture, as stated under procedure, is run into the measuring burette. The columns in both burettes are balanced so that the reading burette is at 13.85 (=per cent N in KNO₃). A strip of paper is pasted on the compensating burette at the level of the mercury, and standardization is accomplished. By this method the temperature and pressure readings, and the calculations are avoided.[1]

Procedure for Making the Test. *Salts.* One gram of sodium or potassium nitrate, or such an amount of the material as will generate between 172 to 240 cc. of gas, is dissolved in a little water and placed in the cup of the generating bulb.

Liquid Acids. The acid is weighed in a Lunge pipette and the desired amount run into the funnel of the generating bulb, the amount of acid that is taken being governed by its nitrogen content.

The sample is drawn into the bulb; the funnel is then rinsed out with three or four successive washings of 95% sulphuric acid, the total quantity being 20 cc.

To generate the gas, the bulb is shaken well until apparently all the gas is

---

[1] Standardization with "C. P. KNO₃ is the better, as it is less tedious and is not subject to the correction errors that cannot be escaped when standardizing with air. The KNO₃ must be of undoubted purity."—A. W. Betts.

formed, taking care that the lower stop-cock has been left open, this cock is then closed and the shaking repeated for two minutes. The reservoir is then lowered until about 60 cc. of mercury and 20 cc. of acid are left in the generating bulb. There will remain then sufficient space for 220 cc. of gas.

NOTE. If too much mercury is left in the bulb, the mixture will be so thick that it will be found difficult to complete the reaction, a long time will be required for the residue to settle and some of the gas is liable to be held in suspension by the mercury, so that inaccurate results follow.

The generated gas is now transferred to the reading burette, and after waiting a couple of minutes to allow for cooling, both burettes are balanced, so that in the compensating tube the mercury column is on a level with the paper mark as well as with the column in the reading burette; the reading is then taken.

If exactly one gram of the substance is taken the percentage of nitrogen may be read directly, but in case of other amounts being taken, as will invariably be the case in the analysis of acids, the readings are divided by the weight of the substance and multiplied by 4.5 to obtain the per cent of nitric acid monohydrate present.

The procedure may be used for determining nitrites as well as nitrates.

Determination of HNO_3 in Oleum by Du Pont Nitrometer Method [1]

About 10 cc. oleum are weighed in a 30-cc. weighing bottle, 10 cc. 95% reagent sulphuric acid added and mixed by shaking. This mixture is transferred to the nitrometer reaction tube and the weighing bottle and nitrometer cup rinsed with three 5-cc. portions of the reagent sulphuric acid which is drawn into the reaction tube. This is vigorously shaken for three minutes and the gas then passed to the measuring tube and allowed to stand for about five minutes, after which the mercury levels are adjusted and the reading taken.

It is obvious that this determination includes any nitrous acid in the oleum.

Combined Nitric Acid

The nitric acid in nitrates may be determined by titration with ferrous sulphate. The nitrate dissolved in a little water is run into strong sulphuric acid and titrated with standard ferrous sulphate according to the procedure described for determining free nitric acid in mixed acids (Vol. II, Acids).

Determination of Free Nitric Acid

Other acids being absent, free nitric acid may be determined by titration with standard alkali. Details for the analysis of nitric acid in presence of commonly occurring impurities are given in Volume II in the chapter on Acids.

$$1 \text{ cc. } N/1 \text{ NaOH} = 0.063018 \text{ g. } HNO_3$$

[1] By courtesy of E. I. du Pont de Nemours Powder Co.

DETERMINATION OF NITRITES

Gravimetric Method of Buovold[2]

One and one fourth to 1.5 gram of $AgBrO_3$ is dissolved in 100 cc. of water and 110 cc. of 2 N. acetic acid, in an Erlnmeyer flask. 200 cc. of the nitrite solution (1 g. $NaNO_2$) are added from a burette, stirring the mixture during addition of the nitrite. A pale green precipitate is obtained. 30 cc. of H_2SO_4 (1 : 4) are added, the mixture warmed to 85°. When the yellow precipitate settles it is filtered on a Gooch and washed with hot water, then dried and weighed as $AgBr+AgCl$—chlorine is determined on a separate portion and AgCl deducted. $AgBr\times0.9070 = NaNO_2$. The method is specially applicable to nitrites high in chlorine.

Volumetric Permanganate Method

Principle. Potassium permanganate reacts with nitrous acid or a nitrite as follows:

$$5N_2O_3+4KMnO_4+6H_2SO_4=5N_2O_5+2K_2SO_4+4MnSO_4+6H_2O.$$

$$5HNO_2+2KMnO_4+3H_2SO_4=5HNO_3+K_2SO_4+2MnSO_4+3H_2O.$$

Since $2KMnO_4$ in acid solution has five available oxygens for oxidation of substances (e.g., $2KMnO_4 = K_2O.2MnO+5O$ equivalent to 10H) the molecular weights of the constituents divided by 20 in the first equation and by 10 in the second would represent the normal weights per liter, e.g., $5N_2O_3$ divided by 20 = 76 divided by 4 = 19 grams N_2O_3 per liter. $4KMnO_4$ divided by 20 or $2KMnO_4$ divided by 10 = 158.03 divided by 5 = 31.61 grams of $KMnO_4$ per liter for a normal solution. In the second equation if Na represents the univalent element we would have $5NaNO_2$ divided by 10 or 69 divided by 2 = 34.5 grams per liter. Hence 1 cc. of a normal $KMnO_4$ solution would oxidize 0.019 gram N_2O_3 or 0.0345 gram $NaNO_2$ to form N_2O_5 and $NaNO_3$ respectively.

Organic matter is also oxidized by $KMnO_4$ hence will interfere if present.

Special Reagents.

N/5 Potassium Permanganate. The solution contains 6.322 grams $KMnO_4$ per liter.

N/5 Sodium Oxalate. $Na_2C_2O_4$ reacts with $KMnO_4$ as follows:

$$5Na_2C_2O_4+2KMnO_4+8H_2SO_4 = K_2SO_4+2MnSO_4+5Na_2SO_4+10CO_2+8H_2O.$$

Hence $5Na_2C_2O_4$ divided by 10 or 134 divided by 2 = 67 grams per liter = a normal sodium oxalate solution. A N/5 solution requires 13.4 grams $Na_2C_2O_4$ per liter.

[2] Chem. Ztg. 38, 28, C. A. 8, 1250 (1914).

Preparation of the Sample.

Soluble Nitrites. Ten grams of the nitrite are dissolved in water and made to 1000 cc.; 10 cc. contain 0.1 gram of the sample.

Water-insoluble Nitrites. 0.5 to 1.0 gram of the nitrite according to the amount of nitrous acid present is taken for analysis. An excess of $KMnO_4$ solution is added, followed by dilute H_2SO_4 and the excess standard permanganate titrated with sodium oxalate according to directions given under Procedure.

Nitrous Acid in Nitric Acid and Mixed Acids. This is present generally in very small amounts so that a large sample is taken. The amount and details of the procedures are given under the special subject.

For routine work where a number of daily determinations are made, a 50-cc. burette is generally preferred.

Trial Run. If the approximate strength of the salt is not known the following test may be quickly made to ascertain whether more than 50 cc. of solution is necessary and the approximate amount of $KMnO_4$ required for oxidation.

Ten cc. of the solution together with 100 cc. of water are placed in a 4-in. casserole and about 10 cc. of dilute H_2SO_4, 1 : 1, added. Standard $KMnO_4$ from a 50-cc. burette is now run into the sample until a permanent pink color is obtained. The cc. of $KMnO_4$ multiplied by 5 = the approximate amount of permanganate solution required for oxidation of 50 cc. of sample. An excess of 5 to 10 cc. should be taken in the regular run.

Titration of Nitrite. Sufficient standard N/5 $KMnO_4$ to oxidize the sample to be titrated (as ascertained by the trial run) and 10 cc. excess are placed in a casserole. The solution is acidified with 10 cc. of dilute (1 : 4) H_2SO_4 and 50 cc. of the nitrite solution is added slowly with constant stirring. The sample is placed on a hot plate until the mixture reaches a temperature of 70° to 80° C. and 25 cc. more of the dilute H_2SO_4 added. The excess permanganate is now titrated with N/5 $Na_2C_2O_4$, the oxalate being added slowly until the permanganate color is destroyed. Five cc. excess of the oxalate are added and the exact excess determined by titrating the hot solution with N/5 $KMnO_4$ to a faint pink color. The total permanganate solution taken minus the oxalate titration = cc. $KMnO_4$ required by the nitrite.

Standard ferrous sulphate, $FeSO_4$, may be used, in place of sodium oxalate. The titration then may be conducted in the cold.

One cc. N/5 $KMnO_4$ = 0.0038 g. N_2O_3, or 0.0069 g. $NaNO_2$, or 0.0085 g. KNO_2.

Detection of a Nitrate in a Nitrite Salt

Iridium salts are colored blue by HNO_3 but no color is produced by HNO_2. Use a 0.025% solution of IrO_2 or $(NH_4)_2IrCl_6$ per 100 cc. of 98–99% H_2SO_4 and heat to boiling. The solution should be kept in a stoppered bottle. Into the hot reagent in a test tube is dropped the solid substance tested. A blue color is produced by nitrates. If the nitrite is in solution, make alkaline with KOH, evaporate to dryness and test the residue. Chlorine interferes, but not $FeCl_3$.

Determination of Pyridine in Ammonium Nitrate[1]

Dissolve 250 g. of sample in 300 cc. of distilled water, using a 1000 cc. Kjeldahl or Florence flask. Add a few drops of methyl orange and neutralize with 10% sodium hydroxide solution. Then add 15 cc. excess of 10% sodium hydroxide solution. Set up apparatus, note Fig. 55a, using 300 cc. hypobromite solution in the second flask and receiving the distillate in 25 cc. N/10

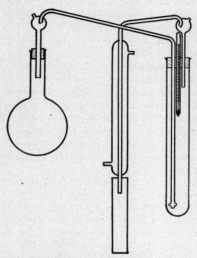

FIG. 55a.—APPARATUS FOR PYRIDINE.

The whole is set around a single ring stand. The hypobromite flask is 2 in. by 14 in. The hypobromite, 300 cc., occupies 7½ in. of the height at the start of the test.

sulfuric acid. Distil until 100 cc. of distillate have been collected. The heating should be very slow until all the ammonia, driven off, has been destroyed. This point will be indicated first by an acid reaction of the methyl orange in the first flask and second by the gradual reduction of the amount of nitrogen given off, in very small bubbles, in the hypobromite. At this point the hypobromite flask should not be warm enough to burn the hand (not above 70° to 75° C.). It is now safe to increase the heat so that boiling occurs in the hypobromite in 10 to 15 min. and 100 cc. of distillate comes over in 20 to 25 min. after active boiling starts.

Titrate the liquid in the receiver, using N/10 sodium hydroxide solution with methyl orange as the indicator.

[1] Ref. R. M. Ladd, Jour. Ind. Eng. Chem., Vol. 11, no. 6, page 552 (June, 1919). By courtesy of the author.

Record the end-point; add $\frac{1}{2}$ cc. of phenolphthalein (1 : 1000) solution and continue the titration until a red color which will persist for 30 sec. appears. Subtract the methyl orange end-point from that obtained with phenolphthalein, and multiply the difference by 0.0079. The result is the pyridine bases in grams. Methyl orange indicates pyridine plus ammonia. Phenolphthalein indicates ammonia. Difference is due to pyridine.

NOTES. Because of the fact that the methyl orange and phenolphthalein end-points are never quite the same and because an absorption of carbon dioxide by the sodium hydroxide solution may bring it about that they vary still more widely, it is necessary to standardize the solutions used to both end-points and to make a correction for their normal difference. This correction should be checked by a new standardization at least once a week. We found that with our solutions this difference was usually about 0.4 cc.

In case it is desired to use a sample of a different size, maintain the proportions indicated above, except that the total solution in the first flask should always be about 500 cc.

The hypobromite solution is made up as follows: 100 g. sodium hydroxide are dissolved in 800 cc. of water, 25 cc. of liquid bromine are added, and the mixture shaken until the bromine is entirely dissolved and made up to 1,000 cc. The solution should be made up a day in advance. It will maintain its strength for at least a week if kept in a stoppered, dark bottle. It will be brown in color. Should the brown color disappear during the distillation it would mean that an excess of ammonia is present. This should also be indicated and eliminated from the calculations by the double end-point called for, but in case this happens it is well to repeat the test, using more of the hypobromite solution.

The reactions involved and the calculations on which the proportions are based are indicated in the following equations:

$$NaOH + NH_4NO_3 \rightarrow NH_3 + H_2O + NaNO_3$$
$$40 17$$

15 cc. 10% sodium hydroxide solution contains 1.5 g. sodium hydroxide
1.5 g. sodium hydroxide will free 0.6375 g. ammonia

$$2NH_3 + 3NaBrO \rightarrow 3H_2O + N_2 + 3NaBr$$
$$34 357$$

0.638 g. ammonia is destroyed by 6.699 g. sodium hypobromite

$$2NaOH + Br_2 \rightarrow NaBrO + NaBr + H_2O$$
$$80 160 119$$

25 cc. bromine = 79.5 g. bromine
300 cc. solution contains 17.7 g. hypobromite (approx.)

DETERMINATION OF NITROGEN IN STEEL [1]

For the determination of nitrogen in steel, a modification of the method first published by A. H. Allen and modified by Prof. J. W. Langley is used.

By the following method the sample and standard distillates are prepared under similar conditions, and when treated with Nessler reagent, develop colors nearly identical in quality or tone, but proportional in intensity to the ammonia present.

If the Nessler reagent is carefully prepared and works properly, the color in sample and standard will develop almost instantly and is fully developed in less than one minute. The solutions treated with such reagent remain clear or do not cloud appreciably on standing for ten minutes; however, the comparison is best made after standing one minute, and all difficulty due to clouding avoided.

The difficulties of comparison are also reduced to a minimum by using an aliquot part of the distillate in the manner to be described instead of that corresponding to the whole sample.

Preparation of Reagents

Hydrochloric acid of 1.1 sp. gr., free from ammonia, which may be prepared by distilling pure hydrochloric acid gas into distilled water free from ammonia. To do this, take a large flask fitted with a rubber stopper carrying a separatory funnel-tube and an evolution-tube, fill it half full of strong hydrochloric acid, connect the evolution-tube with a wash-bottle connected with a bottle containing the distilled water. Admit strong sulphuric acid free from nitrous acid to the flask through the funnel-tube, apply heat as required, and distill the gas into the prepared water.

Test the acid by admitting some of it into the distilling apparatus, described farther on, and distilling it from an excess of pure caustic soda, or determine the amount of ammonia in a portion of hydrochloric acid of 1.1 sp. gr., and use the amount found as a correction.

NOTE. The ammonia-free hydrochloric acid may also be prepared as follows:
Dilute concentrated hydrochloric acid to specific gravity 1.10 and without addition of sulphuric acid distill it.
Hydrochloric acid of this strength distills without change in concentration.
The first 100-cc. distillate from one litre of acid will usually contain all the ammonia and is rejected; the portions distilled thereafter being collected for use but must, of course, be tested as usual to make sure it is free from ammonia.
Solution of caustic soda, made by dissolving 300 grams of fused caustic soda in 500 cc. of water and digesting it for 24 hours at 50° C. on a copper zinc couple, made, as described by Gladstone and Tribe, as follows:
Place from 25 to 30 grams of thin sheet zinc in a flask and cover with a moderately concentrated, slightly warm solution of copper sulphate. A thick, spongy coating of copper will be deposited on the zinc. Pour off the solution in about ten minutes and wash thoroughly with cold distilled water.
Nessler Reagent. Dissolve 35 grams of potassium iodide in a small quantity of distilled water, and add a strong solution of mercuric chloride little by little, shaking after each addition until the red precipitate formed dissolves. Finally the precipitate formed will fail to dissolve, then stop the addition of the mercury salt and filter. Add to the filtrate 120 grams of caustic soda dissolved in a small amount of water, and dilute until the entire solution measures 1 litre. Add to this 5 cc. of saturated aqueous solution of mercuric chloride, mix thoroughly, allow the precipitate formed to settle, and decant or siphon off the clear liquid into a glass-stoppered bottle.

[1] Methods of Analysis used in Laboratories of the Titanium Alloy Manufacturing Co. Contributed by L. E. Barton.

Standard Ammonia Solution. Dissolve 0.0382 gram of ammonium chloride in 1 litre of water. 1 cc. of this solution will equal 0.01 milligram of nitrogen.

Distilled Water Free From Ammonia. If the ordinary distilled water contains ammonia, redistill it, reject the first portions coming over, and use the subsequent portions, which will be found free from ammonia. Several glass cylinders of colorless glass of about 160 cc. capacity are also required.

The best form of distilling apparatus consists of an Erlenmeyer flask of about 1500 cc. capacity, with a rubber stopper, carrying a separatory funnel-tube and an evolution-tube, the latter connected with a condensing-tube around which passes a constant stream of cold water. The inside tube, where it issues from the condenser, should be sufficiently high to dip into one of the glass cylinders placed on the working table.

Method of Determination

Distillation of Sample

In a distilling flask of 1000 to 1500 cc. capacity, fitted with separatory funnel and connected with condenser, place 40 cc. prepared caustic soda solution; add 500 cc. distilled water and distill until the distillate gives no reaction with Nessler reagent.

Dissolve a 5-gram sample of the steel in 40 cc. of ammonia-free hydrochloric acid, and by means of the separatory funnel add the solution slowly to the contents of the distilling flask, washing in finally with ammonia-free water.

Distill and collect 150 cc. of distillate in a graduated flask. Cork the flask and set aside. Experience has shown that 150 cc. of distillate will contain all the nitrogen in the sample.

Preparation of Standard

After distilling the sample—the apparatus then being free from ammonia but containing the residue of sample and reagents—25 cc. of standard ammonium chloride solution and 150 cc. of ammonia-free water are added to the contents of the flask, and distillation continued until a *standard* distillate of 150 cc. is collected in a graduated flask.

As before, the single distillate will contain all the ammonia from 25 cc. of standard solution.

To the standard distillate is added 6 cc. of Nessler reagent; and since the standard ammonium chloride solution is equivalent to .00001 g. nitrogen per cc., 1 cc. prepared standard distillate is equivalent to $\dfrac{25 \times .00001}{156} = .000001$,

6 g. nitrogen per cc.= .00016% nitrogen when using one gram sample.

Comparison and Determination

To make the determination, 30 cc. of *sample distillate*, equal to *one gram* of sample, are placed in one of a pair of Nessler jars and the color developed by addition of 1 cc. Nessler reagent.

The standard and sample are allowed to stand one minute to fully develop the color.

Into the other jar the standard distillate is run from a burette until the colors in standard and sample jars are of the same intensity; the final comparison being made after bringing the contents of the jars to the same volume by addition of ammonia-free water to one or the other.

The number of cc. of standard distillate multiplied by .00016 gives the percentage of nitrogen in the steel.

30

DETERMINATION OF CONVERTER EFFICIENCY IN OXIDATION OF AMMONIA TO NITRIC ACID

In place of determining the total ammonia used and the total products of oxidation, samples may be taken, during the operation, of gases entering and leaving the converter and analyzed according to the following simple and accurate procedure suggested by Gaillard;[1] a method successfully used by the American Cyanamide Company at Warners, N. J., and by the United States nitrate plants at Sheffield and Muscle Shoals, Ala.

Principle. The gas to be analyzed is drawn into an evacuated bulb which has previously been weighed, and the increased weight due to the sample is obtained. The ammonia or nitrogen oxides in the bulb are then absorbed and titrated, and the percentage by weight of combined nitrogen in the gases is determined. The efficiency is the ratio of the combined nitrogen in the exit and inlet gases.

Sources of Error. Error may be caused by:

(a) Water condensation in the sampling tube during sampling.

(b) Air leakage into the tube during sampling.

(c) Ammonia escaping oxidation being drawn into the bulb. In presence of ammonia a cloudiness is readily observed.

(d) Changes in temperature, barometric pressure, and moisture conditions between successive weighings of the same bulb.

These errors are rendered negligible by careful manipulation. (The writer would suggest that a similar bulb tare weight be used and the procedure for weighing recommended in combustion carbon determinations be followed.)

[1] Ind. Eng. Chem., **11**, 745 (1919).

On the right hand side of the illustration below is shown the bulb in the process of evacuation by means of a vacuum pump operated by an electric motor. Attached to the system is a mercury gauge or barometer which gives the degree of evacuation of the bulb.

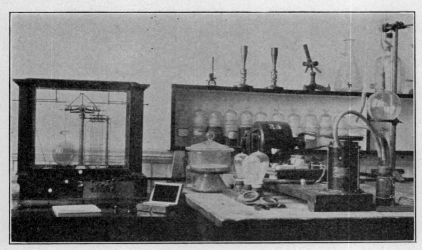

Fig. 55b.   Evacuation and Weighing of Bulbs for Converter Efficiency in Ammonia Oxidation.

On the left hand side of the illustration is shown a balance with a bulb suspended for weighing after being evacuated. It is advisable to have a bulb on the right hand arm of the balance acting as a tare weight. This counteracts the buoyancy error of the air, increasing the delicacy of weighing. If the stopcocks are not absolutely tight, the bulb will gain in weight owing to an intake of air.

PHOSPHORUS

P_4, *at.wt.* 31.02; *sp.gr.* $\begin{cases} yellow\ 1.831 \\ red\ \ \ \ \ 2.296 \end{cases}$; *m.p.* $\begin{cases} 44° \\ 725° \end{cases}$; *b.p.* $\begin{cases} 290°C \\ \ldots\ldots \end{cases}$; *oxides,*
P_2O_3, PO_2, $P_2O_.$; *acids,* H_3PO_2, H_3PO_3, H_3PO_4, HPO_3, $H_4P_2O_7$.

DETECTION

Element. Phosphorus is recognized by its glowing (phosphorescence) in the air. The element is quickly oxidized to P_2O_5; if the yellow modification is slightly warm (34° C.) the oxidation takes place with such energy that the substance bursts into flame. The red form is more stable. It ignites at 260° C.

Boiled with KOH or NaOH it forms phosphine, PH_3, which in presence of accompanying impurities is inflammable in the air.

Phosphorus oxidized to P_2O_5 may be detected with ammonium molybdate, a yellow compound, $(NH_4)_3PO_4 \cdot 12MoO_3 \cdot 3H_2O$, being formed.

Acids. *Hypophosphorous Acid, H_3PO_2,* heated with copper sulphate to 55° C. gives a reddish-black compound, Cu_2H_2, which breaks down at 100° to H and Cu. Permanganates are reduced immediately by hypophosphorous acid. No precipitates are formed with barium, strontium or calcium solutions. Zinc in presence of sulphuric acid reduces hypophosphorous acid to phosphine, PH_3.

Phosphorous Acid, H_3PO_3. Copper sulphate is reduced to metallic copper and hydrogen is evolved, no Cu_2H_2 being formed as in case of hypophosphorous acid. Permanganates are reduced slowly. Added to solutions of barium, strontium or calcium white phosphites of these elements are precipitated. Alkali phosphites are soluble in water, while hypophosphites are not readily soluble.

Orthophosphoric Acid, H_3PO_4. Ammonium · phosphomolybdate precipitates yellow ammonium phosphomolybdate from slightly nitric acid solutions. The precipitate is soluble in ammonium hydroxide.

Metaphosphoric Acid, HPO_3. Converted by nitric acid in hot solutions to the ortho form. Metaphosphoric acid is not precipitated by ammonium molybdate.

Pyrophosphoric Acid, $H_4P_2O_7$. Converted to orthophosphoric acid in hot solutions by nitric acid. No precipitate is formed with ammonium molybdate.

COMPARISON OF ORTHO, META AND PYROPHOSPHORIC ACIDS

Reagent.	Orthophosphoric acid.	Metaphosphoric acid.	Pyrophosphoric acid.
Ammonium molybdate.............	Yellow ppt.	No ppt.	No ppt.
Albumin.............................	....	Coagulated	Not coagulated
Zinc sulphate, cold, in excess.........	....	No ppt.	White ppt.
Silver nitrate in neutral solution......	Yellow ppt., Ag_3PO_4	White ppt., $AgPO_3$	White ppt., $Ag_4P_2O_7$
Magnesium salts....................	White ppt.	No ppt.	No ppt.

Chapter contributed by Wilfred W. Scott.

Phosphorous acids are distinguished from phosphoric acids by the phosphine formed with the former when acted upon with zinc.

Acid phosphates are distinguished from normal phosphates as follows: Neutral silver nitrate added to an acid phosphate liberates free nitric acid (Litmus test), the following reaction taking place:

$$3AgNO_3 + Na_2HPO_4 = Ag_3PO_4 + 2NaNO_3 + HNO_3.$$

The solution resulting when silver nitrate is added to normal phosphate solution is neutral.

$$3AgNO_3 + Na_3PO_4 = Ag_3PO_4 + 3NaNO_3.$$

ESTIMATION

The determination of the pentoxide of phosphorus is required in a large number of substances, since it is widely distributed in the form of phosphates—calcium phosphate, $Ca_3(PO_4)_2$; fluor apatite, $3Ca_3(PO_4)_2 \cdot CaF_2$; chlor apatite, $3Ca_3(PO_4)_2 \cdot CaCl_2$; vivianite, $Fe_3(PO_4)_2 \cdot 8H_2O$; wavelite, $2Al_2(PO_4)_2 \cdot Al_2(OH)_6 \cdot 9H_2O$; pyromorphite, $3Pb_3(PO_4)_2 \cdot PbCl_2$; phosphates of iron and calcium in phosphate ores, hence in slags of the blast furnace. It occurs in fertile soils, bones, plant and animal tissues.

The chemist is especially concerned in the determination of phosphoric acid (P_2O_5), in the evaluation of materials used for the manufacture of the acid—bone ash and phosphate rock (see table below). Generally, determinations of lime, iron and alumina are also desired and frequently a more complete analysis. In the analysis of phosphoric acid certain impurities occurring in the crude material used are determined, e.g., iron, lime, magnesia, sulphuric, hydrochloric and hydrofluoric acids, etc. Phosphoric acid is determined in the evaluation of phosphate fertilizers, phosphates used in medicine, phosphate baking powders, etc.

The element is determined in iron, steel, phosphor bronzes, and other alloys.

TYPICAL ANALYSES *

Substance.	Bone Ash.	Charlestown Phosphate.	Spanish Phosphorite.	Sombrero Phosphate.	Redonda Phosphates.	Canadian Phosphate.
Phosphoric oxide....	39.55	27.17	33.38	35.12	35.47	37.68
Sulphur trioxide.....	......	3.30	0.57			
Carbon dioxide......	4.43 −	4.96	4.10	7.40		
Lime................	52.46	44.03	47.16	51.33	.....	51.04
Magnesia...........	1.02	0.37	trace			
Alumina............	.....	1.44	0.89	+Fe	20.17	Fe₂O₃,
Ferric oxide........	0.17	0.43	2.59	1.02	8.85	Al₂O₃,
Fluorine, etc........	.....	2.38	4.01	....	.....	F. etc.
Alkaline salts.......	.....	0.87	.....	0.42		=6.88
Silica—sand, etc.....	0.51	5.60	3.71	2.02	9.70	4.29

* Thorpe, "Dictionary of Applied Chemistry," Longmans, Green & Co.

Preliminary Remarks. Practically all procedures for the determination of phosphorus depend upon its oxidation to ortho phosphoric acid and its precipitation by ammonium molybdate from a nitric acid solution as ammonium phospho-molybdate. It may now be determined either gravimetrically or volumetrically. Two procedures are of importance in the gravimetric determination of phosphorus; the first depends upon the direct weighing of the yellow phosphomolybdate, dried at 110° C.; the second, on the conversion of the yellow precipitate to the magnesium salt and its ignition to pyrophosphate. Two volumetric procedures, which are of special value in the determination of small amounts of phosphorus as in case of phosphorus in iron and steel, are to be recommended for their rapidity and accuracy. One of these is to dissolve the ammonium phosphomolybdate in a known amount of standard caustic, titrate the excess of alkali with standard acid, which indicates the alkali required to neutralize the molybdic acid in the yellow precipitate. From this the amount of phosphorus present may be calculated. A second procedure of equal accuracy and rapidity is to dissolve the molybdate in ammonia, add an excess of sulphuric acid, pass the warm solution through a column of zinc and titrate the reduced molybdic acid with standard potassium permanganate, the amount of permanganate required being a measure of the phosphorus present.

The impurities interfering in the procedures are silica and arsenic acid. The first may be eliminated by dehydration of the silicic acid in the solution and its removal as insoluble SiO_2 by filtration. Arsenic in small quantities does not interfere under certain conditions; in large quantities its removal is imperative.

Preparation and Solution of the Sample

Amount of the Sample Required. For accurate results it is advisable to take a fairly large sample, 5 to 10 grams, and when it has been dissolved, to dilute to a definite volume, 500 or 1000 cc. Aliquots of this solution are taken for analysis.

Iron Ores, Phosphate Rock and Minerals. Five to 10 grams of the pulverized material placed in a 3-in. porcelain dish are digested for an hour with 50 to 100 cc. of concentrated hydrochloric acid (sp.gr. 1.19), the dish being covered by a clock-glass and placed on a steam bath. The acid is now diluted with half its volume of water and the solution filtered into a porcelain dish of sufficient capacity to hold the filtrate and washings. The residue is washed with dilute hydrochloric acid (1 : 1) until free of visible iron discoloration. The filtrate and washings are evaporated rapidly on a hot plate to small volume and then to dryness over the steam bath. Meanwhile the *insoluble residue* and filter are ignited in a 20-cc. platinum crucible over a Méker burner or in a muffle furnace and the residue fused with ten times its weight of sodium carbonate. The fusion is removed by inserting a platinum wire into the molten mass, allowing to cool and then gently heating until the mass loosens from the crucible, when it may be removed on the wire. The cooled mass on the wire and that remaining in the crucible are dissolved in dilute hydrochloric acid, and the filtered solution added to the main solution. The combined solutions are evaporated to dryness, and heated gently to dehydrate the silica. The residue is taken up with a few cc. of hydrochloric acid, the solution diluted, filtered and the SiO_2 washed with dilute nitric acid solution. The combined filtrates are made up to 500 or 1000 cc. Aliquots of this solution are taken for analysis.

Iron and Steel. Five to 10 grams of the drillings or filings are dissolved in an Erlenmeyer flask with 50 to 100 cc. of dilute nitric acid, 1 : 1, more acid being added if necessary. When dissolved, a strong solution of $KMnO_4$ is added until a pink color appears; on boiling brown manganese dioxide forms in the solution if a sufficient amount of permanganate has been added. This is dissolved by adding 2% sodium sulphite solution in just sufficient quantity to dissolve the precipitate. The solution is diluted to a convenient volume for analysis. Where a number of determinations are to be made, it is advisable to weigh the amount of sample desired for the determination and to precipitate the ammonium phosphomolybdate in the flask in which the drillings have been dissolved.

Ferro-Silicon, Iron Phosphide and Acid Insoluble Alloy Steels. Decomposition is best accomplished by fusing 1–2 grams with 10–15 grams of a mixture of sodium carbonate and magnesium oxide (2 : 1). (A blank should be run on the reagents and allowance made for any phosphorus present.) The fusion is dissolved in hydrochloric acid, then taken to dryness and the SiO_2 filtered off. The SiO_2 is treated with HF and a few drops of H_2SO_4 in a platinum dish and taken to fumes. The residue is fused with Na_2CO_2, and the fusion dissolved in HCl and the solution added to the main filtrate containing the iron and phosphate, etc. This filtrate is concentrated to near dryness, 10 cc. nitric acid added and the evaporation repeated. This concentrate is diluted to about 25 cc. and phosphorus precipitated with ammonium molybdate solution as usual.

Ferro-Titanium, Metallic Titanium. The fusion, obtained as directed for ferro-silicon, is extracted with water to dissolve out the sodium phosphate. The residue is fused with sodium carbonate and again extracted with water. The water extracts of the two fusions is examined for phosphorus. The extracts are made acid with nitric acid, the solution evaporated to near dryness, nitric acid added and the concentrated solution treated as directed for steel. Iron, in this case, has been removed with titanium.

Materials Containing Tungsten. The alloy is dissolved in dil. HNO_3 and evaporated to dryness, HCl is now added and the solution again evaporated to dryness, the residue is taken up with HCl and again evaporated. This residue is extracted with dil. HCl and washed with acid ammonium nitrate. Iron and phosphorus are in solution, Si and W remain insoluble.

Ores Containing Titanium. Titanium may be recognized by the red color produced by hydrogen peroxide, H_2O_2, added to the sulphuric acid extract; also by the reduction test with zinc, which causes a play of colors, the solution becoming colorless by the reduction of iron, then, in presence of titanium, pink, purple and finally blue. (Vanadium gives similar tests.) Solutions containing titanium frequently appear milky when the solution is diluted before filtering off the insoluble residue. Since titanium forms an insoluble compound with phosphoric acid and iron oxide [1] the final residue, obtained by the method of solution for ores, phosphate rock and minerals, should be moistened with sulphuric acid and the silica expelled with hydrofluoric acid. The solution is evaporated to dryness and to SO_3 fumes, the residue fused with sodium carbonate and taken up with boiling water. TiO_2 remains insoluble, while P_2O_5 passes into the filtrate.

Determination of Phosphorus in Organic Matter. Decompose the organic matter with nitric acid in a sealed tube according to the method of Cassius described on page 145, and determine the phosphoric acid formed.

Soluble Phosphates, Phosphate Baking Powder, etc. A water extract is generally sufficient to get the material in solution. In case iron, alumina, lime and magnesia salts are present, as may occur in baking powders, an extraction with dilute 3% nitric acid is necessary. It is advisable to dissolve a 5- to 10-gram sample and take an aliquot part of the solution made up to a definite volume. Before precipitating with ammonium phosphomolybdate, 5 grams of ammonium nitrate should be added **for each** gram of the sample taken for analysis.

GRAVIMETRIC METHODS FOR DETERMINATION OF PHOSPHORUS

A. Direct Weighing of the Ammonium Phosphomolybdate Precipitation of Ammonium Phosphomolybdate

Precipitation of ammonium phosphomolybdate is common to all subsequent methods for determination of phosphorus.

Reaction.

$$H_3PO_4 + 12(NH_4)_2MoO_4 + 21HNO_3 = (NH_4)_3PO_4.12MoO_3 + 21NH_4NO_3 + 12H_2O$$

Amount of Sample Required for Analysis. In volumetric procedures the amount of sample should be such that the phosphorus content will be between 0.005 and 0.0005 gram phosphorus. In gravimetric procedure twice this amount is desirable.

Ammonium Molybdate Reagent. See pages 368, and the chapter on Reagents.

Precipitation. The free acid of the solution is nearly *neutralized* by *addition of ammonium hydroxide*. In analysis of phosphate rock or materials comparatively low in iron, it is advisable to add ammonium hydroxide in quantity sufficient to cause a slight permanent precipitate followed by just sufficient HNO_3 to dissolve the precipitate. In *iron and steel* analysis ammonium hydroxide is added until the precipitated iron hydroxide dissolves with difficulty and the solution becomes a deep amber color or cherry red. In analysis of *soluble phosphates*, litmus paper dropped into the solution indicates the neutral point. *Nitric acid* is added to the neutral or slightly acid solution, 5 cc. of acid for every 100 cc. of solution. A volume of 150 to 200 cc. of solution is the proper dilution for samples taken in amounts above recommended. To the warm solution (not over 80° C.) *ammonium molybdate is added*, 60 cc. of the reagent being required for every 0.1 gram of P_2O_5 present. The solution is stirred, or shaken, if in a flask, until a cloudy precipitate of ammonium phosphomolybdate appears. It is then allowed to settle on the steam bath at a temperature of 40 to 60° C., for one hour, then again agitated and allowed to settle in the cold for an hour longer. The filtrate should be tested with additional ammonium molybdate for phosphorus. The yellow precipitate is filtered and washed with 1% HNO_3 solution followed by a 1% solution of KNO_3, or NH_4NO_3 or $(NH_4)_2SO_4$ as the special case requires. Filtration through asbestos in a Gooch crucible is to be recommended. When a large number of determinations are to be made, as in case of iron and steel, filter paper is more convenient.

A. Direct Weighing of the Ammonium Phosphomolybdate

The sample being dissolved and the ammonium phosphomolybdate precipitated according to directions already given above, the supernatant solution is filtered through a weighed Gooch crucible and washed twice by decantation with dilute nitric acid (1%), the precipitate washed into the Gooch, followed by two washings with 1% KNO_3 or NH_4NO_3 (neutral solutions) and finally with water. The precipitate, free from contaminating impurities, is dried for two hours in an oven at 110° C., then cooled in a desiccator and weighed. Weight of precipitate $\times 0.0165 = P$, or $\times 0.03784 = P_2O_5$.

NOTE. If this procedure is to be followed it will be convenient to take 1.65 grams sample, if the phosphorus content will allow. Each 0.01 gram of precipitate will then equal 1% P.

B. Determination of Phosphorus as Magnesium Pyrophosphate

Magnesia Mixture. For precipitation of ammonium magnesium phosphate, 110 grams of magnesium chloride ($MgCl_2 \cdot 6H_2O$) are dissolved in a small amount of water. To this are added 280 grams of ammonium chloride and 700 cc. of ammonia (sp.gr. 0.90); the solution is now diluted to 2000 cc. with distilled water. The solution is allowed to stand several hours and then filtered into a large bottle with glass stopper. Ten cc. of the solution should be used for every 0.1 gram P_2O_5 present in the sample analyzed. As the reagent becomes old it will be necessary to filter off the silica that it gradually accumulates from the reagent bottle.

Procedure. The ammonium phosphomolybdate, obtained as directed (page 365), is filtered onto a $12\frac{1}{2}$ S. & S. No. 589 filter paper and washed four or five times with dilute 1% HNO_3. The precipitate is now dissolved from the filter by a fine stream of hot ammonium hydroxide, 1 : 1, catching the solution in the beaker in which the precipitation was made. The solution and washings should be not over 100 to 150 cc. Hydrochloric acid is added to the cooled solution to neutralize the excess of ammonia, the yellow precipitate, that forms during the neutralization, dissolving with difficulty, when sufficient acid has been added. To the cooled solution *cold magnesia mixture is added* drop by drop (2 drops per second) with constant stirring. Ten cc. of the reagent will precipitate 0.1 gram P_2O_5. When the solution becomes cloudy the stirring is discontinued and the precipitate allowed to settle ten minutes. *Ammonium hydroxide is added* until the solution contains about one-fourth its original volume of strong ammonia (e.g. 25 cc. NH_4OH, 90 to 100 cc. of solution). The solution is stirred during the addition and then allowed to settle for at least two hours. It is filtered preferably, through a Gooch crucible (or through an ashless filter paper), and the precipitate washed with dilute ammonium hydroxide, 1 : 4, then placed in a porcelain crucible, a few drops of saturated solution of ammonium nitrate added and the precipitate heated over a low flame till decomposed (or until the paper chars). The lumps of residue are broken up with a platinum rod and again ignited over a Scimatico or Méker burner, the heat being gradually increased. If the heating is properly conducted, the resultant ash will be white or light gray, otherwise it will be dark. The addition of solid ammonium nitrate aids the oxidation in obstinate cases, but there is danger of slight mechanical loss. The crucible is cooled in a desiccator and the residue weighed as magnesium pyrophosphate.

$$Mg_2P_2O_7 \times 0.2787 = P \quad \text{and} \quad Mg_2P_2O_7 \times 0.6379 = P_2O_5.$$

Direct Precipitation of Magnesium Ammonium Phosphate

In the absence of heavy metals whose phosphates are insoluble in an ammoniacal solution, the magnesia mixture may be added directly to the neutral solution containing the phosphate, without previous precipitation of ammonium phosphomolybdate. The magnesium ammonium phosphate is washed and ignited according to directions given above, and weighed as magnesium pyrophosphate.

The use of the Gooch crucible for the ammonium phosphomolybdate and the ammonium magnesium phosphate precipitates is recommended in preference to filter paper. See precautions on page 375.

VOLUMETRIC METHODS FOR THE DETERMINATION OF PHOSPHORUS

These volumetric procedures are especially applicable for determining small amounts of phosphorus such as are present in steel and in alloys.

C. Alkalimetric Method

The method is based on the acid character of ammonium phosphomolybdate, the following reaction taking place with an alkali hydroxide:

$$2(NH_4)_3.12MoO_3.PO_4 + 46NaOH$$
$$= 2(NH_4)_2HPO_4 + (NH_4)_2MoO_4 + 23Na_2MoO_4 + 22H_2O.$$

From the reaction 46 molecules of sodium hydroxide are equivalent to one molecule of P_2O_5, hence 1 cc. of N/10 solution of sodium hydroxide neutralizes the yellow precipitate containing an equivalent of .000309 gram of P_2O_5. (N equivalent of $P = 31 \div 23 = 1.35$.)

Solutions Required. *Nitric Acid for Dissolving:* Mix 1000 cc. of HNO_3, sp.gr. 1.42, and 1200 cc. of distilled water.

Nitric Acid for Washing: Mix 20 cc. of HNO_3, sp.gr. 1.42, and 1000 cc. of distilled water.

Ammonium Molybdate: Solution No. 1. Place in a beaker 100 g. of 85 per cent molybdic acid, mix it thoroughly with 240 cc. of distilled water, add 140 cc. of NH_4OH, sp.gr. 0.90, filter and add 60 cc. of HNO_3, sp.gr. 1.42.

Solution No. 2. Mix 400 cc. of HNO_3, sp.gr. 1.42, and 960 cc. of distilled water.

When the solutions are cold, add solution No. 1 to solution No. 2, stirring constantly; then add 0.01 gram of ammonium phosphate dissolved in 10 cc. of distilled water and let stand at least 24 hours before using.

Potassium Nitrate, 1 per cent: Dissolve 10 g. of KNO_3 in 1000 cc. of distilled water.

Phenolphthalein Indicator: Dissolve 0.2 g. of phenolphthalein in 50 cc. of 95 per cent ethyl alcohol and 50 cc. of distilled water.

Standard Sodium Hydroxide: Dissolve 6.5 g. of purified NaOH in 1000 cc. of distilled water, add a slight excess of 1 per cent solution of barium hydroxide, let stand for 24 hours, decant the liquid, and standardize it against a steel of known phosphorus content as determined by the molybdate-magnesia method, so that 1 cc. will be equivalent to 0.01 per cent of phosphorus on the basis of a 2-g. sample (see notes). A 0.1 N or N/10 solution contains 4 g. NaOH (100%) per 1000 cc.

Protect the solution from carbon dioxide with a soda-lime tube.

Ferric Chloride: Dissolve 100 g. of ferric chloride (phosphorus free) in 100 cc. of distilled water.

Standard Nitric Acid: Mix 10 cc. of HNO_3, sp.gr. 1.42, and 1000 cc. of distilled water. Titrate the solution against standardized NaOH, using phenolphthalein as indicator, and make it equivalent to the NaOH by adding distilled water. 0.1 N or N/10 solution contains 6.3 g. HNO_3 per 1000 cc.

Determination of Phosphorus in Ores and Minerals. See page 371.

Determination of Phosphorus in Iron and Steel. See page 1045, Vol. II.

Determination of Phosphorus in Copper Alloys

In a 400-cc. casserole dissolve 1 g. of copper alloy metal in 10 cc. of HNO_3, sp.gr. 1.42. Add 20 cc. of HCl, sp.gr. 1.20, and evaporate to dryness. Moisten with HCl, evaporate to dryness again, and bake to dull redness. Moisten with HCl again (add 3 cc. of ferric chloride solution unless iron is already present) and dilute to about 200 cc. with distilled water. Add NH_4OH, sp.gr. 0.90, until the basic salts of copper have dissolved and the solution has become a deep blue. Boil, allow to settle, and filter on a loosely woven filter paper. Wash with dilute ammonia and with hot water. Dissolve the precipitate on the filter with hot dilute HCl, dilute the solution to about 200 cc., add NH_4OH, sp.gr. 0.90, until the precipitate which forms at first dissolves rather slowly, and saturate with H_2S gas. Filter off and reject the precipitate. Boil the filtrate to expel H_2S, and add HNO_3, sp.gr. 1.42, until the iron is oxidized. Add NH_4OH, sp.gr. 0.90, until the solution is alkaline. Boil and filter on a loosely woven filter paper. Wash with dilute ammonia and with hot water. Dissolve the precipitate on the filter with HNO_3 (sp.gr. 1.42), receiving the solution in a 350-cc. Erlenmeyer flask. Add NH_4OH, sp.gr. 0.90, until the iron is entirely precipitated, and then add HNO_3, sp.gr. 1.42, cautiously until the solution just becomes clear. Bring the solution to a temperature of about 80° C., and add 40 cc. of ammonium molybdate at room temperature. Allow to stand for one minute, shake or agitate for 3 minutes, and filter on a 9-cm. paper. Wash the precipitate three times with the 2 per cent HNO_3 solution to free it from iron, and continue the washing with the 1 per cent KNO_3 solution until the precipitate and flask are free from acid.

Transfer the paper and precipitate to a solution flask, add 20 cc. of distilled water, 5 drops of phenolphthalein solution as indicator, and an excess of standard NaOH solution. Insert a rubber stopper and shake vigorously until solution of the precipitate is complete. Wash off the stopper with distilled water and determine the excess of NaOH solution by titrating with standard HNO_3 solution. Each cubic centimeter of standard NaOH solution represents 0.01 per cent of phosphorus.

Accuracy. Duplicate determinations should check within 0.01 per cent of phosphorus.

NOTES. The ammonium-molybdate solution should be kept in a cool place and should always be filtered before using.

All distilled water used in titrations should be freed from carbon dioxide by boiling or otherwise.

Method for Steel

Procedure. Preparation of the sample. Consult pages 364 and 366. After oxidation of the sample by adding a strong solution of $KMnO_4$ and boiling, and dissolving the precipitated MnO_2 by reduction with sodium sulphite, the greater part of the free acid is neutralized by addition of ammonia. The solution will appear a deep cherry red color. No iron precipitate should be present. Ammonium phosphomolybdate is now precipitated by addition of ammonium molybdate according to the procedure outlined on page 366. This is filtered into a Gooch crucible containing asbestos, and washed once or twice with water containing 1% nitric acid, and then several times with a 1% neutral solution of potassium nitrate until the washings are free of acid, as indicated by testing with litmus paper.   The asbestos mat containing the precipitate is transferred to a No. 4 beaker, 100 cc. of CO_2 free water added, followed by about 20 cc. of N/10 NaOH measured from a burette.   The crucible is rinsed out with 5 to 10 cc. of N/10 NaOH, the exact amount being noted and then with water, adding the rinsings to the main solution.   Phenolphthalein indicator is added, and the excess of caustic titrated with N/10 HNO_3.[1]   The total NaOH added minus the acid titration equals the cc. of the caustic required to react with the yellow precipitate.

One cc. of 0.1 N or N/10 NaOH = 0.000136 gram of P.

The exact factor should be determined as directed under Reagents.

(b) **Method for Ores.**   See pages 364–367.

D. Zinc Reduction and Titration with Potassium Permanganate

Permanganate.   Ferric-Alum Method.   This method is based on the assumption that ammonium phosphomolybdate, $(NH_4)_3.12MoO_3.PO_4$, is reduced, in acid solution, by zinc, the molybdic acid, MoO_3, forming the lower oxide Mo_2O_3, in which form it reacts with ferric iron in the receiving flask, reducing a corresponding equivalent of ferric salt to ferrous condition, being itself oxidized to MoO_3.   When the ferric solution is not placed in the receiving flask, a slight oxidation takes place, the oxide $Mo_{24}O_{37}$ apparently being formed.

Special Apparatus Required

Jones' Reductor.   Details of the reductor are given under the determination of iron by the permanganate method, also under the Volumetric Determination of Molybdenum.

Solutions Required.

Dilute Ammonia.   Mix 100 cc. of NH_4OH, sp.gr. 0.90, and 900 cc. of distilled water.

Dilute Hydrochloric Acid.   Mix 500 cc. of HCl, sp.gr. 1.20, and 500 cc. of distilled water.

Dilute Sulphuric Acid for Dissolving.   Mix 200 cc. of H_2SO_4, sp.gr. 1.84, and 800 cc. of distilled water.

Dilute Sulphuric Acid for Reductor.   Mix 500 cc. of H_2SO_4, sp.gr. 1.84, and 500 cc. of distilled water.

Ammonium Molybdate.   Solution No. 1.   Place in a beaker 100 g. of 85 per cent molybdic acid, mix it thoroughly with 240 cc. of distilled water, add 140 cc. of NH_4OH, sp.gr. 0.90, filter and add to 60 cc. of HNO_3, sp.gr. 1.42.

Solution No. 2.   Mix 400 cc. of HNO_3, sp.gr. 1.42, and 960 cc. of distilled water.

When the solutions are cold, add solution No. 1 to solution No. 2, stirring constantly, then add 0.1 g. of ammonium phosphate dissolved in 10 cc. of distilled water, and let stand at least 24 hours before using.

Acid Ammonium Sulphate.   Mix 25 cc. of H_2SO_4, sp.gr. 1.84, and 1000 cc. of distilled water, and then add 15 cc. of NH_4OH, sp.gr. 0.90.

Ferric Alum.   Dissolve 200 g. of ferric ammonium sulphate crystals in

[1] If a large quantity of yellow precipitate is present, five minutes should be allowed for the alkali to react before titrating the excess with standard acid.

1950 cc. of distilled water.   Add 50 cc. of H_2SO_4, sp.gr. 1.84, and 80 cc. of phosphoric acid, 85 per cent.

Potassium Permanganate.   Dissolve from 3.0 to 3.2 g. of $KMnO_4$ in 1000 cc. of distilled water.   Allow the solution to stand for about one week, and then filter it through an asbestos filter.   Standardize by using about 0.200 g. portions of pure sodium oxalate—29.85 cc. of 0.1 N solution.   See pages 13 and 790.

Standard for Phosphorus in Steel.   The exact value of the permanganate solution may be accurately and rapidly determined in terms of phosphorus by standardizing against a sample of standard steel containing a known amount of phosphorus, the ultimate standard being steel drillings furnished by the Bureau of Standards.   The drillings are dissolved in nitric acid, oxidized with $KMnO_4$, the excess of the reagent being destroyed by sulphite solution. Ammonia is added until the solution becomes a deep amber color.   The phosphorus is precipitated as ammonium phosphomolybdate.   The following procedure is the same as is given in the volumetric method following: The permanganate titration of the reduced molybdic acid divided into the amount of phosphorus known to be present in the solution will give the value of the permanganate in terms of phosphorus.

$$\frac{\text{Wt. of P in sample}}{\text{cc. } KMnO_4 \text{ required}} = \text{amount of P per cc. of } KMnO_4.$$

Method.   In a 400-cc. casserole dissolve 1 g. of the metal in 10 cc. of HNO_3, sp.gr. 1.42.   Add 20 cc. of HCl, sp.gr. 1.20, and evaporate to dryness. Moisten with HCl, evaporate to dryness again, and bake to dull redness. Moisten with HCl again, and dilute to about 200 cc. with distilled water and filter if cloudy.   To the solution add NH_4OH, sp.gr. 0.90, until the iron is entirely precipitated, and then add HNO_3, sp.gr. 1.42, cautiously until the solution just becomes clear, the solution having an amber color.   Bring the solution to a temperature of about 80° C., and add 40 cc. of ammonium molybdate at room temperature.   Allow to stand for one minute, shake or agitate for 3 minutes, filter on a 9-cm. paper, and wash very thoroughly (about 25 times) with acid ammonium sulphate.   Dissolve the precipitate on the paper using 50 cc. of dilute ammonia.   Add 10 cc. of H_2SO_4, sp.gr. 1.84, and immediately pass the solution through a Jones reductor, which has the reductor tube prolonged and reaching nearly to the bottom of the flask, dipping into 50 cc. of ferric-alum solution.[1]   Wash through the reductor with 150 cc. of distilled water, and follow with an additional 100 cc. of distilled water. Titrate with standard $KMnO_4$.

By this method the molybdenum in passing through the reductor is reduced entirely to the form Mo_2O_3, and is oxidized by the ferric alum to the form MoO_3, an equivalent amount of iron being reduced to the ferrous condition. As the yellow precipitate contains one atom of phosphorus to each twelve molecules of MoO_3, and as three atoms of oxygen oxidize two of molybdenum, eighteen oxygens or thirty-six irons are equivalent to one phosphorus.   Therefore, the iron value of the permanganate multiplied by the factor $P/36 \times Fe$ (or 0.01540) gives the value of the permanganate in terms of phosphorus.

Accuracy.   Duplicate determinations should check within 0.005 per cent of phosphorus.

[1] It is not required to heat the solution for reduction as is sometimes stated in text books.

NOTES. The ammonium-molybdate solution should be kept in a cool place and should always be filtered before using.

A blank determination should be made on corresponding amounts of acid and water, passing through the reductor into the usual amount of ferric-alum solution in the flask.

A small quantity of liquid should always be left in the reductor funnel, and air should never be allowed to enter the reductor.

Description of the Jones reductor is given on pages 319 and 320.

In absence of ferric iron the reduced molybdate is an amber or brown color. During titration with $KMnO_4$ the color changes to a reddish yellow and fades to a colorless solution, and then the permanganate pink color is obtained.

Calculation. CASE 1. If ferric sulphate is in the receiver ($6Mo_2O_3+18O = 12MoO_3$ in the molecule containing 1P), 18O are equivalent to 36H, hence N/10 P according to this reaction equals at.wt. P divided by $(36 \times 1000) = P$ for 1 cc. of N/10 $KMnO_4 = .0000862$ g. P.

CASE 2. No ferric salt in receiver. $Mo_{24}O_{37}+35O = 24MoO_3+2P.(35O = 70H)$. Dividing by 2 we get at.wt. P divided by $(35 \times 1000) = P$ for 1 cc. of N/10 $KMnO_4 = .0000887$.

NOTES. In case the alkalimetric method is chosen, it will be necessary to wash the precipitate free of acid by washing with neutral ammonium nitrate. (Washing with pure water is prohibited owing to the solubility of the precipitate.) A litmus paper test of the filtrate coming from the funnel is the usual practice of ascertaining whether this washing is complete. Sufficient time should be allowed for the standard alkali to react with the precipitate before addition of standard acid in the titration of the excess alkali, otherwise the results will be low; this is specially true if much "yellow precipitate" is present. If the permanganate method has been chosen, washing the precipitate with ammonium sulphate is the general practice, as the presence of nitrate salts in the precipitate would cause error in this reduction method.

As ammonium molybdate is apt to. deteriorate after standing for several weeks, it is advisable to test the reagent before use. A fresh solution should be made up every ten or fifteen days.

In precipitating phosphorus it must be remembered that overheating the solution will cause the precipitation of molybdic oxide; should this be suspected, the magnesium phosphate method will correct results.

Special Steels and Alloys. Steels containing titanium, tungsten, vanadium, etc., require a special treatment in preparing these for analysis, in the determination of phosphorus. Directions are given for these on the following page, 370a.

Ferro-Silicon, Iron Phosphide and Acid-Insoluble Alloy Steels. Decomposition is best accomplished by fusing 1–2 grams with 10–15 grams of a mixture of sodium carbonate and magnesium oxide (2 : 1). (A blank should be run on the reagents and allowance made for any phosphorus present.) The fusion is dissolved in hydrochloric acid, then taken to dryness and the SiO_2 filtered off. The SiO_2 is treated with HF and a few drops of H_2SO_4 in a platinum dish and taken to fumes. The residue is fused with Na_2CO_3, and the fusion dissolved in HCl and the solution added to the main filtrate containing the iron and phosphate, etc. This filtrate is concentrated to near dryness, 10 cc. nitric acid added, and the evaporation repeated. This concentrate is diluted to about 25 cc. and phosphorus precipitated with ammonium molybdate solution as usual.

Ferro-Titanium, Metallic Titanium. The fusion, obtained as directed for ferro-silicon, is extracted with water to dissolve out the sodium phosphate. The residue is fused with sodium carbonate and again extracted with water. The water extracts of the two fusions are examined for phosphorus. The extracts are made acid with nitric acid, the solution evaporated to near dryness, nitric acid added and the concentrated solution treated as directed for steel. Iron, in this case, has been removed with titanium.

Preparation of Cast Iron and Alloy Steels for the Determination of Phosphorus. **Cast Iron.** One gram or more of the sample is dissolved in 50 cc. of dilute nitric acid, the solution evaporated to dryness and baked at 200° C. for an hour, 15 cc. of hydrochloric acid (d. 12) are added and the solution again evaporated to dryness. 15 cc. hydrochloric acid are added to the residue and 20–30 cc. of water and the silica is filtered off and washed with water.[1] The filtrate is evaporated to pasty consistency, 15 cc. of nitric acid are added and the solution evaporated to near dryness, this treatment is repeated and the residue then taken up with 15 cc. of water. Phosphorus is now precipitated according to the directions for phosphorus in the steel solution.

Iron Containing Titanium. The material is treated as in case of cast iron. Any residue remaining from the nitric acid evaporation is treated with HF and H_2SO_4 as in case of cast iron. (Use platinum dish.) The residue remaining from the HF treatment is taken up with a little HCl and filtered. The filtrate being added to the main solution containing iron and phosphorus. The solution is heated to boiling and an ammonium acid sulphite solution is added, dropwise (2 cc. of NH_4OH saturated with SO_2 and 10 cc. NH_4OH). A precipitate will form, which dissolves. In case it does not, on stirring, the solution is cleared by adding a few drops of HCl, and the addition of sulphite continued. When all but 1–2 cc. of the reagent is added, the solution is heated. Ammonium hydroxide is now added drop by drop to the hot solution until a slight greenish precipitate is formed in the solution, which remains undissolved on stirring. Now the remaining 1–2 cc. of the sulphite is added. If a precipitate forms (titanium hydroxide) which does not redissolve on stirring, HCl is added drop by drop until the solution clears. The odor of SO_2 should be evident. If not, more sulphite should be added and the solution again cleared. 5 cc. HCl are added, CO_2 passed through the solution, which is heated to boiling to expel excess of SO_2, the iron remaining in the reduced form. Sufficient ferric chloride is now added to combine with all the phosphorus and a slight excess.

The solution is cooled under tap water and ammonium hydroxide added drop by drop until green ferrous iron precipitate redissolves, and then a white precipitate of titanium hydroxide and ferric phosphate remains and an additional drop causes a distinct reddish tint and the appearance of a green precipitate with one more. If the red color does not appear, the green precipitate is dissolved with a few drops of HCl and additional ferric chloride solution is added. The addition of ammonia is now repeated. A reddish color of excess of ferric hydroxide should be evident.

A few drops of acetic acid (d. 1.04) are added to dissolve the green precipitate, the red remaining undissolved, and 1 cc. excess. The solution is diluted to about 450 cc. with hot water, boiled 1 minute, then rapidly filtered, and washed once or twice with hot water. The filtrate passes through clear, but will become cloudy upon oxidation of iron.

The residue is dried, separated from the filter, the latter burned and its ash added to the main residue. This is mixed with 5 grams of sodium carbonate and about 0.2 g. potassium nitrate and fused in a platinum crucible for half an hour. The fusion is extracted with water and the solution filtered. The filtrate contains all the phosphorus as sodium phosphate.

[1] Traces of P in residue are recovered by treating with HF and H_2SO_4 and expelling SiO_2.

The filtrate is acidified with nitric acid and evaporated to near dryness, then taken up with a few cc. of water and to a volume of 25 cc. is added the ammonium molybdate solution according to the procedure for iron and steel. The procedure from this stage is the same as for steel.

Vanadium Steels. In presence of vanadium the ammonium phosphomolybdate will be contaminated with vanadium, so that its presence requires a special treatment. If less than 2.5% is present, the regular procedure for steel is followed with the exception that just before adding the ammonium molybdate, 5–10 cc. nitric acid are added.

If more than 2.5% of vanadium is present, the following procedure is recommended by C. M. Johnson (J. I. E. C., **11**, 113 (1919)): 1 gram of steel is treated in a covered 250-cc. porcelain dish with a mixture of 30 cc. concentrated hydrochloric acid and 30 cc. concentrated nitric acid, and the solution is heated for an hour. The cover is rinsed off into the main solution, 100 cc. of strong nitric acid is added and the solution evaporated to dryness and baked for 5 minutes at 200° C. The oxides are dissolved in 35 cc. of concentrated hydrochloric acid and the solution evaporated to about 10 cc. 10 cc. of nitric acid are added and the covered solution heated for a few minutes. The solution is filtered through asbestos, on a small wad of glass wool in a funnel. The vanadium oxide residue is washed 15 times with small portions of a solution containing 200 cc. concentrated nitric acid, 100 cc. water and 20 grams of ferric nitrate (free from phosphorus). The filtrate is concentrated to 10 cc. If a precipitate forms V_2O_4, it is filtered off and the washing repeated, the filtrate is again evaporated and filtered if necessary. If no precipitation occurs in the filtrate upon concentration, 40 cc. of nitric acid are added and the phosphorus precipitated with ammonium molybdate [1] as usual.

NOTE. Vanadium may also be precipitated by concentration to about 20 cc., neutralizing the greater part of the acid and adding ammonium chloride solid to saturation. The precipitate is washed with a saturated solution of NH_4Cl.

[1] Johnson recommends ammonium molybdate that contains 55 grams of ammonium molybdate, 50 grams of ammonium nitrate, 40 cc. ammonium hydroxide, and 700 cc. water. After heating for half an hour, the solution is diluted to 1000 cc. The solution is allowed to settle for 24 hours and is filtered. It is slightly ammoniacal.

Report of the Committee on Research and Analytical Methods— Phosphate Rock [1]

The following tentative standard methods for sampling and determination of moisture, phosphoric acid and iron and alumina in phosphate rock are recommended to the Division.

Methods of Sampling and Determination of Moisture

I. Gross Sample. *A. Car Shipments.* One hundred pounds sample per car.

1. *Sampling from the Car.* In sampling car shipments in the car at least ten scoopshovelsful, aggregating 100 lbs., shall be taken from each car at approximately equal distances from each other so as to average the car. Care shall be taken to see that each scoopful shall cover the entire face of the pile from floor to top.

2. *Sampling from the Cart or Barrow.* A small hand scoopful of 1 to 2 lbs. shall be taken from each cart or barrow either as it is being loaded or as it leaves the car.

B. Cargo Shipments. One hundred pounds minimum sample per vessel.

1. *Sampling in Hoisting Tub.* In sampling cargoes generally running from 1000 tons upward a small hand scoopful shall be taken from approximately every tenth tub before it is hoisted from the hold.

2. *Sampling from Conveyor.* If unloading is being done with automatic bucket and conveyor, periodical sections of the entire discharge of the conveyor shall be taken of such intervals and quantity as to give a sample equivalent to approximately 1 lb. per each 10 tons of cargo.

3. *Sampling from Conveying Vehicle.* Samples shall be taken with a hand scoop from various cars at such regular intervals and in such quantities as to give approximately 1 lb. for each 10 tons of cargo.

II. Laboratory Sample. The resulting gross sample obtained by any one of the methods outlined shall be crushed to pass a four-mesh screen, thoroughly mixed on a clean, hard surface and quartered down to a 10-lb. average sample.

A. Crushing. This 10-lb. sample shall all be crushed to pass an eight-mesh screen.

B. Mixing and Quartering. This eight-mesh sample shall be carefully mixed and quartered down to two 2-lb. samples.

C. Grinding. 1. *Moisture Sample.* One of these 2-lb. samples shall be held in an air-tight container. This sample is to be used for the determination of moisture.

2. *Analytical Sample.* The other 2-lb. sample shall be further mixed and quartered down to a 2- or 4-oz. sample which is then to be ground to pass a sixty-mesh screen or preferably a sixty-five mesh screen. This sample is to be used for the analytical determination.

NOTE. It is essential that the taking of the gross sample be done with small hand scoops and that the practice of taking the sample in the hand be absolutely prohibited, for it has been found that there is considerable selective action in the finer materials sifting through the fingers while a scoop retains the entire sample.

The dimensions of the screens referred to above are to be as follows:

No. of Mesh	Size of Opening Inches	Diameter of Wire Inches
4	0.185	0.065
8	0.093	0.032
65	0.0082	0.0072

III. Determination of Moisture. Moisture is to be determined on both the moisture sample and analytical sample. Of the moisture sample not less than 100 grams are to be weighed out for each determination. Of the analytical sample approximately 2 grams are to be weighed out for each determination. Both are to be dried to constant weight at a temperature of 105° C. in a well-ventilated oven, preferably with a current of dry air passing through the oven. The containers in which moisture is determined should be provided with well-fitting covers so that the samples may be cooled and weighed in the well-covered container.

31

[1] Journ. Ind. and Eng. Chem.

IV. Calculation of Results. The percentages of phosphoric acid and iron and alumina as determined on the analytical sample are to be calculated to a moisture-free basis and subsequently to the basis of the original sample as shown by the moisture content of the moisture sample.

Determination of Phosphoric Acid

Reagents. To be prepared as in Official Methods, A. O. A. C. Bureau of Chemistry, Bulletin 107 (Rev.), 1910, p. 2. Preparation of reagents (c), (d), (e) and (f), except that the ammonium nitrate solution in (d) is changed to 5% instead of 10%.

Method of Solution. To 5 grams of the sample add 30 cc. of conc. hydrochloric acid (sp.gr. 1.20) and 10 cc. of conc. nitric acid (sp.gr. 1.42) and boil down to a syrupy consistency. The residue, which should be nearly solid after cooling, is taken up with 5 cc. of conc. nitric acid and 50 cc. of water. Heat to boiling, cool, filter and make up to 500 cc. through the filter. This procedure eliminates practically all of the silica and it is necessary to filter as quickly as possible after digestion so as to avoid redissolving the silica.

Determination. Draw off an aliquot portion of 50 cc., corresponding to 0.5 gram, neutralize with ammonia, then add nitric acid until the solution is just clear. Add 15 grams of ammonium nitrate (free from phosphates), heat the solution to 50° C. and add 150 cc. of molybdate solution. Digest at 50° C. for fifteen minutes with frequent stirring. Filter off the supernatant liquid and test the filtrate with molybdate solution to see if precipitation has been complete. (If not, add more molybdate to the filtrate and digest for fifteen minutes longer.) Wash with 5 per cent ammonium nitrate solution by decantation, retaining as much of the precipitate as possible in the beaker. Dissolve the precipitate in the beaker in the least possible quantity of ammonium hydroxide (sp.gr. 0.90) and dilute this solution with several times its volume of hot water. Dissolve the remainder of the precipitate on the filter with this solution, washing beaker and filter with hot water and keeping the volume of the filtrate between 75 and 100 cc. Neutralize with hydrochloric acid, cool to room temperature and add 25 cc. of magnesia mixture from a burette, drop by drop, stirring vigorously with a rubber-tipped rod, then add 15 cc. of ammonium hydroxide (sp.gr. 0.90) and allow to stand for four hours or overnight at room temperature. The time of standing may be reduced to two hours if kept in a refrigerator or still better in an ice-water bath. Filter through a platinum or porcelain Gooch crucible, fitted with a platinum or asbestos mat carefully made and ignited to constant weight. Wash with 2.5% ammonium hydroxide until practically free from chlorides; dry, ignite, cool and weigh as magnesium pyrophosphate. If desired, filtration may be made through an ashless filter paper, igniting in the usual manner. Calculate to P_2O_5 by multiplying by 0.6378 (log 80468).

Determination of Iron and Aluminum together as Phosphates

I. Solutions Required: 1. Hydrochloric acid (1 : 1); prepared by mixing 1 part by volume of concentrated HCl (sp.gr. 1.19) with 1 part of distilled water.

2. A saturated solution of ammonium chloride, which should be filtered before use.

3. A 25% solution of ammonium acetate, faintly acid to litmus paper.

4. A solution of ammonium phosphate (10%), prepared by dissolving 20 grams of $(NH_4)_2HPO_4$ in 180 cc. of distilled water and filtering. (This should be prepared frequently in small quantity, as it attacks glass containers on standing.)

5. A standard solution of ferrous ammonium sulphate, containing iron equivalent to about 0.0100 gram of Fe_2O_3 in 10 cc. and 50 cc. conc. HCl per liter.

6. A solution of calcium and magnesium phosphates for blank determinations, prepared as follows: Dissolve 4 grams of MgO and 35 grams of $CaCO_3$ (both free of iron and aluminum) in 100 cc. concentrated HCl, add an aqueous solution of 30 grams of $(NH_4)_2HPO_4$, make up to 2 liters and filter.

7. A solution of ammonium nitrate (5%) for washing precipitates. About 400 cc. are required for each determination.

All reagents used should be as pure as practicable and all solutions should be free of suspended matter.

II. Preparation of Rock Solution. Place 2.5 grams of pulverized rock with 50 cc. of 1 : 1 HCl in a graduated 250-cc. flask, the glass of which contains less than 1%

of iron and aluminum oxides.[1] Boil gently with occasional shaking for one hour in such a manner as to avoid concentrating the solution to less than half of its original volume,[2] dilute, cool to room temperature, make up to volume and mix; filter immediately through a dry filter into a dry flask, discarding the first few cc. of the filtered solution.

Pipette a 50-cc. aliquot, representing 0.5 gram of rock, into a platinum dish and evaporate nearly to dryness.[3] Cool, take up with a few cc. of water and when the salts are loosened from the dish, add 5 cc. of 1 : 1 sulphuric acid and evaporate to fumes. Increase the temperature and evaporate nearly to dryness.[4] Cool, dilute with about 50 cc. of distilled water, add 10 cc. of conc. HCl and heat, with occasional stirring, until sulphates are dissolved. Filter into a 600-cc. Jena glass beaker through a 9-cm. paper (S. & S. No. 597), washing the paper thoroughly with dilute HCl and hot water.

III. First Precipitation with Ammonium Acetate. To the solution in the beaker, add 25 cc. of the standard iron solution when the amount of combined iron aluminum oxides in the rock does not exceed 5% and 50 cc. of the standard iron solution when the combined oxides exceed 5%.[5] Oxidize with about 3 cc. of bromine water and boil in covered beaker for about fifteen minutes to expel the excess of bromine. Rinse cover and sides of beaker with distilled water and cool to room temperature.

(Run a blank determination containing 10 cc. of 1 : 1 HCl, 25 cc. of the calcium and magnesium phosphate solution, and the same quantity of standard iron solution as is added to the rock solution.)

Add 100 cc. of saturated ammonium chloride solution,[6] 3 cc. of 10% ammonium phosphate solution, 2 drops of methyl-orange indicator and conc. ammonium hydrate (free of spangles and dissolved mineral matter) to alkaline reaction. Then add dilute HCl (about 1 : 20) drop by drop, with constant stirring, until the solution becomes faintly acid and the pink color of the methyl orange is just restored.[7] Dilute to 450 cc.[8] with distilled water, heat to boiling, and add 25 cc. of 25% ammonium acetate solution. Continue heating for about five minutes, after adding ammonium acetate, filter on a 12.5 cm. ashless filter paper (S. & S. No. 589 " White Ribbon "

[1] Experiments have shown that the solution cannot be made in flasks made of glass containing a higher percentage of alumina, because the fluorine in the rock partially dissolves the glass and adds alumina to the solution. Neither " Nonsol," " Jena " nor " Weber's " resistant glass " R " is suitable. Flasks made of glass containing little alumina, such as " Kavalier," " F Z resistant glass " or other Bohemian glass of lower alumina content have proven satisfactory. See " Chemical Glassware," P. H. Walker, J. Am. Chem. Soc., **27**, 865.

[2] This may be accomplished by heating the flask over a low Bunsen flame or on a hot plate which is just hot enough to keep the solution boiling. A glass tube about 12 ins. long by $\frac{3}{8}$ in. in diameter with a bulb in the middle makes a very satisfactory condenser when placed in the neck of the flask.

[3] It is advisable to remove as much of the HCl as possible before adding sulphuric acid so as to minimize the chances of loss by effervescence or bumping. The evaporation may be conducted in glass beakers of low alumina content. Kavalier glass has been used satisfactorily. In no case should the evaporation be conducted in porcelain.

[4] It is best to remove as much sulphuric acid as possible so that the calcium sulphate which might hold iron will dissolve readily in HCl.

[5] It has been found that when iron oxide is present in considerable excess over aluminum oxide the precipitation of the phosphates is more complete, the combined phosphates are more readily ignited to constant weight, and the precipitate does not become red on ignition.

[6] Ammonium chloride in large quantity increases the solubility of calcium and magnesium phosphates and decreases the solubility of iron and aluminum phosphates.

[7] This method of adjusting acidity was suggested by F. B. Carpenter and was found to give satisfactory results.

[8] All our work has confirmed Brown's statement (see Wiley's " Principles and Practice of Agricultural Analysis," 2d edition, 1908, Vol. II, p. 245) that the separation from calcium under the conditions of the method depends upon sufficient dilution.

is suitable), in a 3-in. rapid filtering funnel, keeping the contents of the beaker and funnel hot.[1] Wash three times with hot 5% ammonium nitrate solution, each time cutting the precipitate loose from the filter and stirring it thoroughly with the stream from the wash bottle and filling to within about $\frac{1}{4}$ in. of its upper edge. About 30 cc. are required for each washing. Return the precipitate to the precipitating beaker by washing it out of the filter with a stream of hot water. Dissolve the precipitate with dilute HCl (1 : 6), pouring about 50 cc. through the filter in successive washings and using about 25 cc. to wash down inside the beaker. Finish filter paper with distilled water.

IV. Second Precipitation with Ammonium Acetate. Cool the solution to room temperature, add 50 cc. saturated ammonium chloride solution, 4 cc. of 10% ammonium phosphate solution, 2 drops of methyl orange, and adjust acidity as before. Dilute to 300 cc. with distilled water. Heat to boiling, add 15 cc. of 25% ammonium acetate solution and continue heating for about five minutes. Filter on the same paper as used for the first filtration, scrubbing the inside of the beaker with a rubber-tipped stirring rod and rinsing with hot 5% ammonium nitrate solution. Wash the precipitate ten times with hot 5% ammonium nitrate solution, each time cutting the precipitate loose, stirring it thoroughly as before and breaking up all lumps that it may contain. About 300 cc. of wash solution are required.

As a precautionary measure, boil the filtrate and washings from both the first and second precipitates, and recover any additional precipitate.

V. Ignition of Precipitate. Transfer filter with precipitate to a weighed deep-form porcelain crucible (40 mm. in diameter is a good size) and heat gently over a low flame until the contents are dry, increase the temperature a little and continue heating until the paper is charred, increase the temperature again and continue heating until the paper is entirely burned. Ignite the *uncovered* [2] porcelain crucible for one-hour periods over blast lamp No. 4 Méker burner to constant weight, each time cooling to room temperature in desiccator before weighing. Deduct the weight of blank from each determination, and after subtracting the weight of $FePO_4$ equivalent to the amount of iron found in 0.5 gram of rock by titration, calculate the remainder to Al_2O_3. $AlPO_4 \times 0.4184 = Al_2O_3$.

Determination of Iron

I. Solutions Required. 1. *Standard Potassium Permanganate*, N/40, containing 0.79015 gram of $KMnO_4$ per liter, and having a value of 0.001996 (or practically 0.002) gram of Fe_2O_3 per cc. Standardize with pure sodium oxalate (Bureau of Standards standard sample No. 40.)

2. *Stannous Chloride.* Dissolve 50 grams of the crystallized salt in 100 cc. of hot conc. HCl and make up to 1 liter with distilled water.

3. *Mercuric Chloride.* Prepare a cold saturated solution.

4. *Manganese Solution.* (Preventive solution): (a) Dissolve 200 grams of crystallized manganese sulphate in 1000 cc. of water. (b) Pour slowly, with constant stirring, 400 cc. of conc. sulphuric acid into 600 cc. of water and add 1000 cc. of phosphoric acid of 1.3 sp.gr. Mix solutions (a) and (b).

II. Analytical Procedure. Determine iron according to Jones' and Jeffrey's modification of the Zimmermann-Reinhardt method [3] as follows: Place in a 250-cc. beaker an aliquot of the rock solution, containing not more than 5 cc. of conc. HCl, boil and reduce with the smallest possible excess of stannous chloride, added drop by drop while agitating the solution. Wash sides of beaker with distilled water and cool rapidly. Add 10 cc. of mercuric chloride solution and stir vigorously for about thirty seconds.[4] Pour the mixture into a large porcelain casserole or dish containing 20 cc. of the manganese solution in about 500 cc. of water which has just been tinted with the permanganate solution.

---

[1] The contents of the funnel will remain hot if the solution in the beaker is kept hot over a low flame and filtration is fairly rapid.

[2] Heat over Bunsen to redness before placing over blast in order to prevent loss of precipitate by blowing out of crucible.

[3] Analyst, **34** (1909), 306.

[4] Barneby has shown that only a short interval of time is necessary between the addition of mercuric chloride and manganese sulphate, if the solution is thoroughly agitated. J. Am. Chem. Soc., **36** (1914).

Titrate with N/40 permanganate solution, to original tint and correct result by the volume of $KMnO_4$ required for a blank containing the same quantity of HCl (diluted), adding 2 or 3 drops of stannous chloride to the hot solution, cooling, adding 10 cc. of mercuric chloride and titrating similarly.

When the rock solution contains carbonaceous matter it is necessary first to oxidize this with a little potassium chlorate, evaporate to dryness to eliminate chlorine, and redissolve with 5 cc. conc. HCl and about 10 cc. of water.

Calculate the Fe_2O_3 found to $FePO_4$, using the factor 1.8898, and after deducting from the weight of combined phosphates found, calculate the difference ($AlPO_4$) to Al_2O_3.

Volumetric Determination of Free Phosphoric Acid

Phosphoric acid may be titrated directly by means of standard sodium or potassium hydroxide. The choice of the indicator is important as may be seen by the following reactions:

$H_3PO_4 + NaOH = H_2O + NaH_2PO_4$ (neutral to methyl orange, acid to phenolphthalein),
$H_3PO_4 + 2NaOH = 2H_2O + Na_2HPO_4$ (neutral to phenol, and acid to methyl orange).

The slight dissociation of Na_2HPO_4 causing an alkaline reaction to phenolphthalein prevents the endpoint being sharp. This dissociation may be repressed by titrating in a cold concentrated solution containing sodium chloride. The first reaction with methyl orange indicator is more satisfactory. By the electrometric method J. S. Coye has proven that the endpoint of phenolphthalein is the full color of the alkaline salt, not a faint pink.

Reference is made to the chapter on Acidimitry and Alkalimitry, Volume 2, for titrating phosphoric acid in presence of its salts.

Sources of Error in the Determination of Phosphoric Acid

McCandless and Burton have shown [1] that discrepancies in the gravimetric determination of phosphorus in high grade materials are chiefly due to variations in the amount of HCl used in neutralizing the ammoniacal solution of the yellow precipitate. A solution made neutral, using litmus paper as indicator, has the proper neutrality. Plus errors are caused by adding the acid in excess of the neutral point, and minus errors by having ammonia in excess of neutrality.

The yellow precipitate contained in a small filter (7.5 cm.) is dissolved by adding NH_4OH drop by drop from a burette in just sufficient amount to dissolve the precipitate. The filter is washed free of the phosphate, the washing continued until the volume of the filtrate is 100–150 cc. To the cooled filtrate is added, drop by drop, dilute HCl (1 : 1), with constant stirring, until the color of the litmus paper, placed in the solution, changes to a violet, verging on blue rather than red. Magnesia mixture is now added, drop by drop, in slight excess. After 15 minutes NH_4OH (sp.gr. 0.90) is added, 12 cc. NH_4OH to 100 cc. filtrate or 18 cc. NH_4OH, if the volume is 150 cc. The precipitate is washed thoroughly with 2.5% NH_4OH water, and finally with a strong NH_4NO_3 solution in ammonia water. The Gooch and asbestos is ignited over the blast lamp before use.

[1] J. M. McCandless, J. O. Burton, Ind. Eng. Chem., *16*, 1267, Dec., 1924.

PLATINUM

Pt, *at.wt.* **195.23;** *sp.gr.* **21.48;** *m.p.* **1755° C.;** *oxides* **PtO, PtO$_2$**

DETECTION

Platinum is a gray, lustrous, soft and malleable metal. It is not altered by ignition in the air, but fuses in the oxy-hydrogen flame. It does not dissolve in any of the single acids, but a fusion with acid potassium sulphate attacks the metal slightly. The action of chlorine in general, and nitro-hydrochloric acid (aqua regia), the main solvent, converts the metal to hydrochlorplatinic acid, H_2PtCl_6, which forms many double salts, or platinichlorides. If platinic chloride is gently heated it breaks up into platinous chloride, $PtCl_2$, and chlorine.

If, however, the platinum is alloyed with silver, it dissolves in nitric acid to a yellow liquid, provided sufficient silver is present in the alloy.

The oxides can be formed by carefully igniting the corresponding hydroxides. These are very unstable, decomposing into metal and oxygen by gentle ignition.

The chlorides are the most important compounds of platinum. Two complex acids are formed with hydrochloric acid when the metal is dissolved in aqua regia.

$$PtCl_4 + 2HCl = H_2PtCl_6 \text{ (chloroplatinic acid), orange-red crystals.}$$

$$PtCl_2 + 2HCl = H_2PtCl_4 \text{ (chloroplatinous acid), only known in solution.}$$

An aqueous solution of the former is yellowish-orange, while an aqueous solution of the latter is dark brown, the former being by far the more important.

Potassium iodide precipitates platinum iodide, but it dissolves quite readily, giving a pink to a dark blood-red liquid, depending on the concentration of the solution. Nitric acid should be absent. Heat destroys this color, as well as hydrogen sulphide, sodium thiosulphate and sulphite, sulphurous acid, mercuric chloride and certain other reducing reagents.

Hydrogen sulphide precipitates black platinum disulphide, PtS_2, with the other elements of the hydrogen sulphide group. The solution should be hot, as precipitation takes place more quickly. It is difficultly soluble in ammonium sulphide. It will be found in the extract with the arsenic, antimony, tin, gold, molybdenum, etc., and is precipitated with these elements upon addition of hydrochloric acid. Platinum sulphide is soluble in aqua regia. Addition of $MgCl_2$ solution prevents formation of colloidal PtS_2.

Ammonium chloride added to a concentrated solution of platinum chloride precipitates yellow $(NH_4)_2PtCl_6$, which is slightly soluble in water, and less so in dilute ammonium chloride solution and alcohol.

Potassium chloride precipitates yellow K_2PtCl_6, which is slightly soluble in water, but insoluble in 75% alcohol.

Chapter by R. E. Hickman and Edward Wichers.

Ferrous sulphate precipitates metallic platinum on boiling from a neutral solution. Neutralize with Na_2CO_3. Free mineral acids (except dilute H_2SO_4) prevent the precipitation (difference from gold).

Stannous chloride does not reduce platinum chloride to metal, but reduces hydrochlorplatinic acid to hydrochlorplatinous acid.

$$H_2PtCl_6 + SnCl_2 = H_2PtCl_4 + SnCl_4.$$

Oxalic acid does not precipitate platinum (difference from gold).

Sodium hydroxide with glycerine reduces hydrochlorplatinic acid on warming to black metallic powder.

Formic acid precipitates from neutral boiling solutions all the platinum as a black metallic powder.

Thallium protoxide precipitates from the platinum bichloride solution a pale yellow salt, thallium platinochloride. When the salt is heated to redness it leaves an alloy of thallium and platinum.

Sodium hydroxide added to platinic chloride and then acidified with acetic acid produces a pale yellow to orange precipitate of platinic hydroxide, $Pt(OH)_4$. This dissolves in acids readily, except acetic acid.

Metallic zinc, magnesium, iron, aluminum and copper are the most important metals that precipitate metallic platinum.

$$H_2PtCl_6 + 3Zn = 3ZnCl_2 + H_2 + Pt.$$

ESTIMATION

Platinum may be present under the following conditions:

1. Native grains usually accompanied by the other so-called platinum metals, iridium, palladium, ruthenium, rhodium, osmium, and gold and silver (alloyed with one or more of the allied metals).

Ore concentrates containing the native grains as above with the base metals, iron, copper, chromium, titanium, etc. The associated minerals high in specific gravity in the gravels may be expected to appear with the platinum nuggets, such as chromite, magnetite, garnet, zircon, rutile, small diamonds, topaz, quartz, cassiterite, pyrite, epidote, and serpentine; with gold in syenite; ores of lead and silver.

2. Scrap platinum containing, oftentimes, palladium, iridium, gold, silver and iron.

3. Small amounts of platinum in the presence of large amounts of iron, silica, carbon, magnesia: platinum residues, nickel and platinum contacts, photography paper, jewelers' filings and trimmings, dental and jewelers' sweeps and asbestos, etc.

4. Platinum alloyed with silver, gold, tungsten, nickel, copper, lead, etc.

5. Platinum solutions and salts.

Preparation and Solution of the Sample

The best solvent for platinum is aqua regia. The metal is also acted upon by fusion with the fixed alkalies—sodium or potassium hydroxide and sodium peroxide or potassium or sodium nitrate; also by chlorates in the presence of

HCl. Platinum, when highly heated, alloys with other metals, as lead, tin, bismuth, antimony, silver, gold, copper, etc. The element dissolves in nitric acid when alloyed with silver. This gives a method for the determination of gold in the presence of silver and platinum alloy.

All salts of platinum are soluble in water. The less soluble salts are the chloro-platinates of potassium, ammonium, rubidium, and caesium. Heat increases the solubility while the presence of alcohol decreases the solubility.

Ores. When the free grains of platinum, gold and osmiridium are desired the following method is recommended: Five to 10 grams of the ore are taken from a well-mixed pulverized sample and placed in a large platinum dish that has been weighed. Twenty-five to 50 cc. of strong hydrofluoric acid together with 5 to 10 cc. of concentrated sulphuric acid is mixed with the ore in the dish and evaporated on the water bath, when SiF_4 and the excess of HF are expelled. The material is gently heated until SO_3 fumes are given off. This is repeated with HF if necessary. The material is washed into a casserole with about 200 cc. of hot water and digested over a water bath for fifteen or twenty minutes, and is then washed by decantation, several times pouring the supernatant liquor through a filter to save any floating material that might be washed out. The filter is cautiously burned and the residue is added to the unattacked material. This is transferred from the dish to a beaker or a porcelain dish and treated with aqua regia. The platinum and a small amount of iridium that dissolves with the platinum on account of its being alloyed can be precipitated with ammonium chloride. The remaining residue in the dish will be a small amount of sand and osmiridium. The silica is driven off with HF as described above and the bright grains weighed as osmiridium, or the sand and osmiridium are fused with silver and borax, then extracted with dilute nitric acid, leaving the osmiridium grains free from sand.

Platinum Scrap. One-half gram to a gram is dissolved in aqua regia and evaporated with HCl to get rid of the HNO_3.

If the platinum is alloyed with a large amount of copper, silver, lead and other impurities, a sample of 1 to 5 grams is dissolved in 15 to 25 cc. of HNO_3, whereby the copper, silver, lead and other impurities alloyed with the platinum as well as a large amount of platinum will dissolve. The residue after washing will be platinum and gold. These are dissolved in aqua regia as described above and the platinum precipitated with ammonium chloride. The platinum is recovered from the nitric acid solution and added to the aqua regia solution and the whole is evaporated to get rid of the HNO_3.

Small Amounts of Platinum in the Presence of Large Amounts of Iron; Iron Scale, Fe_2O_3; Sulphate of Iron, Magnesia, Sulphate of Magnesia, Silica, etc. The material is carefully weighed and the coarse scales are separated from the finer material containing the platinum by passing the fines through a 20-in. mesh or finer wire sieve. The coarse scale seldom contains platinum, but it is advisable to quarter this down to 1 kilogram or a fairly good-sized sample and test for platinum on a portion of the ground sample. This can be tested by a wet or a fire assay. The fines are quartered down to about 1 kilogram and ground to pass a 60- to 80-in. mesh sieve. One hundred to 500 grams of the material are taken for analysis. This is placed in one or more casseroles, depending on the amount taken. Each 100-gram portion is extracted by digestion on the steam bath with about 300 to 400 cc. of 10% H_2SO_4. The iron, magnesia, etc., soluble in H_2SO_4 will go into solution, leaving the platinum with the insoluble residue. Filter (a Büchner funnel may be necessary) and wash the residue with water. Test the filtrate for platinum and if any is present precipitate with zinc as described below.

After the filter is ignited in a large platinum dish, the residue is moistened with H_2SO_4, and HF is added completely covering the material. The solution is evaporated on the water bath until SO_3 fumes are given off. If necessary, repeat the treatment with H_2SO_4 and HF until all the silica is driven off as SiF_4. The residue is transferred to a casserole and digested with aqua regia according to directions given under Ores and Platinum Scrap. It is sometimes very difficult to precipitate all of the platinum in the presence of a large amount of iron, magnesia, etc., not having the solution concentrated enough for the platinum. It is advisable to reduce the platinum by iron or zinc, filter, wash with water and redissolve the black metallic platinum in aqua regia. The HNO_3 is expelled by evaporation and adding concentrated HCl from time to time and finally the platinum is precipitated with ammonium chloride.

SEPARATIONS

A careful review of the paragraph on Detection will be very helpful oftentimes in making separations from other metals and substances.

Separation of Platinum from Gold. The platinum is precipitated first with ammonium chloride, as $(NH_4)_2PtCl_6$. After the precipitate has settled it is filtered and washed free from gold with 20% ammonium chloride solution and alcohol. The gold is precipitated with a concentrated solution of ferrous sulphate or iron protochloride as metallic gold. (See also page 387a.)

Oxalic acid precipitates the gold, leaving the platinum in solution. The oxalic acid is added and the solution heated until the gold is entirely precipitated. Filter and wash the precipitate of metallic gold free from platinum. The filtrate is evaporated as far as possible without crystallizing, and the platinum is precipitated with ammonium chloride as $(NH_4)_2PtCl_6$, or it may be reduced with zinc and the black dissolved in aqua regia and treated as described above.

Separation of Platinum from Iridium. The platinum and the iridium are precipitated by iron or zinc and the black residue is washed free from impurities and the platinum is dissolved in dilute aqua regia with gentle heating, leaving the iridium as metallic iridium. The platinum solution is evaporated as described above and precipitated with NH_4Cl as $(NH_4)_2PtCl_6$.

If the platinum and iridium are precipitated together, the salt is filtered and washed with ammonium chloride solution and finally ignited. The sponge is redissolved and evaporated as above to expel the HNO_3. The platinum and the iridium are precipitated with NaOH, which brings down the platinum and iridium as $Pt(OH)_4$ and $Ir(OH)_4$. Boil this mixture with alcohol, which reduces the $Ir(OH)_4$ to $Ir(OH)_3$, but does not affect the $Pt(OH)_4$. Dissolve these hydroxides in HCl, forming $PtCl_4$ and $IrCl_3$ in solution, and the platinum is precipitated with NH_4Cl free from iridium.

See Deville-Stas-Gilchrist method on page 390.

Separation of Platinum from Palladium. Platinum is precipitated with ammonium chloride, and palladium is precipitated from the filtrate by means of dimethylglyoxime (1% alcoholic solution).

Palladium may be precipitated in presence of platinum by adding a one per cent solution of dimethylglyoxime (1% salt in 95% alcohol) to the cold, slightly acid chloride solution of the elements. If the solution is hot the palladium precipitate will be badly contaminated with platinum. (E. Wichers.)

Separation of Platinum from Ruthenium. From the chloride of platinum and ruthenium the metals are precipitated with ammonium or potassium chloride and filtered. The filter is washed with dilute ammonium chloride solution or dilute potassium chloride solution and alcohol until free from ruthenium. If a large quantity is handled it may be necessary to ignite to platinum sponge and dissolve in aqua regia, expel the HNO_3 as described above, and reprecipitate with NH_4Cl, filter and wash free from ruthenium. (See also page 387a.)

Separation of Platinum from Rhodium. The separation is accomplished by adding freshly precipitated barium carbonate to the chloride solution of platinum and rhodium, previously brought nearly to the neutral point by addition of sodium hydroxide. After boiling for two or three minutes rhodium hydroxide precipitates. The precipitate is filtered off, dissolved in HCl, the solution again nearly neutralized and the rhodium precipitation repeated.

Other platinum metals will also precipitate if present. These should be removed prior to the separation of platinum and rhodium. (E. Wichers.)

Separation of Platinum from Osmium. Both metals are reduced with zinc as a fine black powder. The metallic residue is washed and carefully ignited at a high temperature under a hood, as the fumes are poisonous and disagreeable like chlorine. The osmium will be converted into OsO_4, which is very volatile. The residue is dissolved in aqua regia and the platinum is precipitated with NH_4Cl. See Osmium.

GRAVIMETRIC METHODS FOR THE DETERMINATION OF PLATINUM

A. Weighing as Metallic Platinum

1. When the platinum contains only a small amount of impurities a sample of $\frac{1}{10}$ gram or more is taken and dissolved in aqua regia. The solution is gently heated until all is dissolved, adding another portion of aqua regia if necessary. The solution is evaporated, adding HCl from time to time in order to expel the HNO_3. Filter and evaporate again to concentrate the solution. Precipitate with ammonium chloride. After stirring, let stand until the precipitate, $(NH_4)_2PtCl_6$, settles, overnight if convenient. Filter, wash with alcohol or ammonium chloride solution and alcohol, and ignite to metal, in a reducing atmosphere, or ignite slowly in paper filter. Cool in a desiccator and weigh as metallic platinum.

$$\frac{\text{Wt. of Pt found}}{\text{Wt. of sample taken}} \times 100 = \text{per cent of Pt in the material.}$$

2. When the platinum solution contains a large amount of impurities, as iron, nickel, magnesia, etc., it is advisable to reduce the platinum to black metallic platinum with zinc, iron or magnesium as follows: The solution is made acid (2 to 5% free HCl) by adding HCl. The Zn, Fe or Mg is added in small quantities at a time until the solution becomes colorless or until the platinum is completely precipitated.[1] After action has ceased the platinum black metal is filtered onto an ashless filter paper and washed with warm dilute HCl to remove any excess Zn, Fe, or Mg that might be present. The filter and its contents are

[1] $FeCl_3$ in presence of HCl has a solvent action on platinum, hence the iron should be completely reduced.

carefully ignited and afterwards dissolved in aqua regia and treated as directed under A, 1.[1]

3. If none of the other Hydrogen Sulphide Group metals are present the platinum can be precipitated by hydrogen sulphide, filtered, washed with hot water and ignited to metal. If impurities are present in the sulphide, dissolve in aqua regia and proceed as under A, 1. The solution should be boiling and have an acidity of 3% HCl or H_2SO_4.

B. Weighing as a Salt

1. The procedure is the same as under A. The $(NH_4)_2PtCl_6$ precipitate is washed on a weighed Gooch crucible with alcohol. The crucible and contents are dried at a temperature below 100°. Cool in a desiccator and weigh as $(NH_4)_2PtCl_6$.

$$\text{Wt. of } (NH_4)_2PtCl_6 \text{ found} \times \frac{\text{Mol. wt. of Pt}}{\text{Mol. wt. of } (NH_4)_2PtCl_6} \times \frac{100}{\text{Wt. of sample}}$$
$$= \text{per cent of Pt in material.}[2]$$

2. After proceeding as described under A, the platinum is precipitated with potassium chloride as K_2PtCl_6. Transfer to a weighed Gooch crucible and wash well with alcohol. Dry below 100°, cool in a desiccator and weigh as K_2PtCl_6.

$$\text{Wt. of } K_2PtCl_6 \text{ found} \times \frac{\text{Mol. wt. of Pt}}{\text{Mol. wt. of } K_2PtCl_6} \times \frac{100}{\text{Wt. of sample}} = \% \text{ of Pt in material.}[3]$$

C. Determination of Platinum by Electro=analysis

When platinum solutions are acidulated with sulphuric acid and acted upon by a feeble current they give up the metal as a bright deposit upon the electrode. If platinum is used as the electrode, first coat it with a layer of copper and deposit the platinum upon the copper. Wash with water and alcohol and after drying weigh.

Wt. of electrode+Cu+Pt−Wt. of electrode+Cu =Wt. of Pt.

$$\frac{\text{Wt. of Pt}}{\text{Wt. of sample taken}} \times 100 = \text{per cent of Pt in material.}$$

Dr. E. F. Smith, in his work on " Electro-Analysis " recommends that the K_2PtCl_6 be dissolved in water and slightly acidulated with H_2SO_4 (2 or 3% by vol.) and after heating to about 60 to 65° and electrolyzing with N.D.$_{100}$ = .05 ampere and 1.2 volts, the platinum will be completely precipitated in from four to five hours in a perfectly adherent form. A rotating anode will precipitate the platinum much quicker.

---

[1] If iron and lead are suspected, the platinum residue is washed with 10% solution of ammonium chloride and then with 10% solution of ammonium acetate and finally with 80% alcohol.

[2] Factor $(NH_4)_2PtCl_6$ to Pt=0.4393.

[3] Factor K_2PtCl_6 to Pt=0.4013.

PALLADIUM

Element, Palladium. **Pd.** *at.wt.* **106.7;** *sp.gr.* **11.9;** *m.p.* **1549° C.;** *oxides,* PdO, PdO_2.

DETECTION

This metal is also found associated with platinum and iridium as well as ruthenium, rhodium, and osmium. It occurs in the metallic state sometimes with gold and silver. It resembles platinum as to luster and color. Palladium sponge when heated slightly gives a rainbow effect due to the formation of oxides. Hydrogen passed over the sponge restores it to the original color. It dissolves in HNO_3 and boiling H_2SO_4. HCl has little action upon it. It is readily soluble in aqua regia, forming $PdCl_2$. $PdCl_4$ is unstable.

Palladium monoxide, PdO, is formed by a long-continued heating of the spongy metal in a current of oxygen at a temperature from 700 to 840° or by heating a mixture of a palladium salt with potassium carbonate. The pure hydrated oxide is best prepared by the hydrolysis of the nitrate.

It acts as a powerful oxidizing agent to organic substances, and is reduced to metal by hydrogen peroxide.[1]

Palladium dioxide, PdO_2, is obtained in an impure hydrated form as a brown precipitate by the addition of caustic soda to potassium palladichloride. This is soluble in acids, but becomes less soluble when preserved. It can be obtained free from alkali and basic salts by the anodic oxidation of the nitrate, but it is not quite free from monoxide. The dioxide very readily decomposes into the monoxide and oxygen, and cannot be obtained in the anhydrous state. It acts as a vigorous oxidizing agent and decomposes hydrogen peroxide.

Alkalies precipitate in a concentrated solution a dark-brown precipitate soluble in an excess of the reagent. If boiled a brown palladous hydroxide is precipitated. The anhydrous oxide is black.

Ammonia added to a concentrated solution gives a flesh-red precipitate. $Pd(NH_3)_2Cl_6$, soluble in excess of ammonia. If HCl is added to this solution, the yellow compound of palladosamine chloride, $Pd(NH_3Cl)_2$, is deposited.

Sulphur dioxide precipitates the metal from the nitrate or sulphate solution but not from the chloride.

Cuprous chloride precipitates the metal from the sulphate, nitrate and chloride solution when they are not too strongly acid.

Mercuric cyanide precipitates a yellowish-white gelatinous precipitate, $Pd(CN)_2$, insoluble in dilute acids, but dissolving in ammonia and in potassium cyanide to $K_2Pd(CN)_4$.

Potassium iodide precipitates black palladous iodide, PdI_2, insoluble in water, alcohol, and ether, but soluble in an excess of reagent (Rk also ppts.).

Hydrogen sulphide precipitates black palladous sulphide, PdS, soluble in HCl and aqua regia, but insoluble in $(NH_4)_2S$.

Ferrous sulphate slowly produces a black precipitate of metallic palladium from the nitrate.

[1] "Treatise on Chemistry," Roscoe and Schorlemmer.

Ammonium chloride precipitates palladium as $(NH_4)_2PdCl_4$ from the nitrate.

Formic acid, zinc and iron reduce to metallic palladium.

Soluble carbonates precipitate brown palladous hydroxide, $Pd(OH)_2$, soluble in excess, and reprecipitated on boiling.

Phosphuretted hydrogen gas precipitates palladium phosphide. (Difference from Pt, Rh, and Ir.)

Alcohol precipitates, on boiling, metallic palladium.

Alkaline tartrates and citrates form yellow precipitates in a neutral solution from the nitrate.

Stannous chloride produces a brownish-black precipitate, soluble in hydrochloric acid to an intense green solution.

Potassium bisulphate attacks the metal readily.

An alcoholic solution of iodine dropped on the metal will turn black.

Acetylene gas passed through an acidified solution containing Pd produces a brown precipitate which, upon ignition, yields Pd. In this way Pd may be quantitatively separated from Cu.

ESTIMATION

Palladium is determined in alloys, ores, jeweler's sweeps, etc.

Preparation and Solution of the Sample

The solubility of palladium has been taken up under Detection. Palladium when alloyed with platinum, or an alloy of platinum, iridium and palladium, dissolves with the other metals in aqua regia as the chloride. When palladium is alloyed with silver the palladium and silver are dissolved in HNO_3, from which the silver can be separated.

Separations

Separation of Palladium from Platinum and Iridium. The chlorides of palladium, platinum and iridium in solution must be free from HNO_3. The platinum and the iridium are precipitated with NH_4Cl, leaving the palladium in solution. The precipitate is put on a filter and washed free from Pd with NH_4Cl solution and alcohol.

Separation of Palladium from Silver and Gold. At least three times the weight of the gold in silver should be present in the alloy in order to separate the silver and palladium from the gold. The silver and the palladium will dissolve in HNO_3, leaving the gold as the residue. This is filtered off and the silver may be precipitated with HCl. The silver chloride is filtered off and washed with hot water acidulated with HCl until free from palladium. Since AgCl tends to retain palladium it is advisable to redissolve the silver with HNO_3 after reduction of AgCl and reprecipitate the chloride to obtain a complete separation of palladium.

Separation of Palladium from Platinum. The chlorides of platinum and palladium being free from HNO_3 and having an excess of HCl is diluted with water. A 10% solution of potassium iodide is added until all of the palladium is precipitated. Avoid adding a large excess. The precipitate of PdI_2 is filtered off and washed free from platinum and alkali with water slightly acidulated with HCl. The filter is ignited to metallic sponge in a current of hydrogen. See glyoxime method on page 379.

GRAVIMETRIC METHODS FOR THE DETERMINATION OF PALLADIUM

1. The palladium is precipitated from the solution by granulated zinc, the solution having a small amount of free hydrochloric acid. The residue, after the zinc is dissolved, is put on a filter and washed free from impurities. Ignite the filter and dissolve in a small amount of aqua regia and evaporate to a syrupy consistency. Dilute with a small amount of water and add a few drops of HNO_3; precipitate the palladium with NH_4Cl crystals. Heat for a few minutes and let cool. Filter, wash with alcohol, and ignite. Reduce in hydrogen or moisten with formic acid to reduce to metal any oxide that may have formed. Dry and weigh as metallic palladium.

2. With the solution containing about one-fifth the volume of free HCl, the palladium is precipitated with 10% KI solution. Heat to nearly boiling, filter, wash free from iron, etc., with 1 : 4 HCl. Ignite, cool, reduce in hydrogen or moisten with formic acid, dry and weigh as metallic Pd.

3. The filtrate from the platinum precipitation or the nearly neutral solution containing the Pd is made to about 150 cc., and the Pd is precipitated by adding a solution of dimethylglyoxime (1% solution in alcohol). Bring to boiling and let stand overnight if convenient. Filter on a weighed Gooch crucible and wash with hot water slightly acidified with HCl, then with alcohol. Dry and weigh as $(C_8H_{14}N_4O_4)Pd$, which contains 31.67% Pd.

4. The nitric acid in the palladium solution is expelled by evaporating with HCl. Neutralize the chloride solution almost completely with sodium carbonate and mix the solution with a solution of mercuric cyanide, $Hg(CN)_2$, and heat gently for some time. Let stand until cool, overnight if convenient. A yellowish-white precipitate of $Pd(CN)_2$ is formed. Filter, wash with 1% $Hg(CN)_2$ solution, ignite and reduce in hydrogen to metal, or reduce with formic acid, dry, and weigh as metallic Pd.

5. The filtrate from the platinum precipitation is made neutral or slightly alkaline with Na_2CO_3 solution and an excess of formic acid is added. Boil until all the palladium is precipitated or the solution becomes clear. Filter, wash with hot water, ignite, reduce in hydrogen or with formic acid and weigh as metallic Pd.

DETERMINATION OF PLATINUM AND PALLADIUM[1]

Special Methods

Gold Bar. Dissolve a 100-gram sample in aqua regia, and expel the nitric acid by evaporation and the addition of small amounts of hydrochloric acid. Take up with a few cc. of dilute hydrochloric acid. If there should be present a large amount of reduced gold, add a few drops of nitric acid and heat the solution for a few minutes. Dilute to 800 cc. with water, and let it stand in a cool place until solution clears. Filter off silver chloride and wash it with cold water. Pass sulphur dioxide gas through the filtrate to reduce the gold (palladium, etc., is also reduced). Decant the clear solution on a tight filter paper, and wash several times with hot water by decantation. Then pour over the gold in the beaker, 50 cc. of nitric acid, and boil for a few minutes to dissolve the reduced palladium. Add 50 cc. of hot water and filter on the same filter paper and wash several times with hot water. Add 15 cc. of sulphuric acid to the filtrate, evaporate and heat to heavy fumes. Cool, dilute to 200 cc. with water and boil for a few minutes, and filter off any gold and lead sulphate if present. Now pass hydrogen sulphide gas through the hot solution to precipitate the sulphides of platinum and palladium, etc. Filter and wash with hot water. Place the filter paper with the precipitate in a porcelain crucible, dry, burn and ignite. Now touch the residue with the reducing flame of a Bunsen burner to reduce to metal any oxide of palladium that may have formed. Dissolve the residue with a few cc. of aqua regia, and transfer the solution to a tall 300-cc. beaker, and evaporate carefully to dryness on a steam or sand bath. Then moisten the residue with hydrochloric acid and evaporate to dryness again. Moisten the dry residue once more with hydrochloric acid and evaporate to dryness.

Now take up with 16 cc. of hydrochloric acid and 4 cc. of water, cover beaker and boil gently for a moment. Filter on a small filter paper and wash with a small stream of hot water. Discard the residue. Dilute filtrate to 60 cc., cover beaker and heat to near boiling point. Then add 16 grams of ammonium chloride and heat gently to near boiling until all ammonium chloride is in solution. Remove beaker from the hot plate, and let it stand over night in a cool place. Filter rapidly (using suction) on a tight, double filter paper, and wash with ammonium chloride solution (200 grams per liter of water). As the ammonium chloroplatinate is somewhat soluble if exposed to air, the precipitate should be covered with the wash solution all the time during filtration. Then before the ammonium chloride solution is all sucked through, wash once or twice with 95% grain alcohol. Save filtrate for palladium determination. Place the filter paper with the precipitate in a porcelain crucible, so that the precipitate does not come in contact with the sides of the crucible; if it does, a platinum mirror will form, that cannot be removed. Dry gently and smoke off the filter paper (without burning it with a flame), and finally ignite at a bright red heat. Cool and weigh metallic platinum.

Add to the filtrate from the ammonium chloroplatinate precipitate, 16 cc. of nitric acid, stir, cover beaker, placing a glass triangle under the watch glass, and let it stand over night on steam plate. When the solution is supersaturated, as indicated when half of the solution is filled with ammonia salts,

[1] Contributed by F. Jaeger, Chemist, Nichols Copper Co.

remove the beaker from the steam plate and cool the solution. Filter off the ammonium chloropalladate just like the platinum salt and wash with ammonium nitrate solution (200 grams per liter of water). Finally wash with 97% grain alcohol. The solubility of ammonium chloropalladate is greater than of the platinum salt when exposed to air, therefore great care must be taken in filtering it. Place the precipitate with the filter paper in a porcelain crucible, dry, smoke off filter paper and finally ignite at a bright red heat. When cool, reduce any oxide that may have formed with the reducing flame of a Bunsen burner. Cool and weigh metallic palladium.

To confirm that all platinum and palladium is precipitated, neutralize the filtrate from the ammonium chloropalladate with a saturated solution of sodium carbonate, then add 30 cc. of formic acid, and boil for about one hour. Any platinum or palladium if still present will be precipitated as black powder.

NOTE. All evaporations should be made on a steam or sand bath, if not incomplete precipitation of platinum will be obtained.

Refined Silver. Weigh out 1,000 gram sample and dissolve it with dilute nitric acid. Filter off the gold and any undissolved platinum, and then separate the gold from the platinum as described under "Gold Bar," and add the solution to the main filtrate. Dilute the filtrate from the gold residue, so that there will be 10 grams of silver per liter of solution, and then add a slight excess of hydrochloric acid to precipitate all the silver. Stir well and let it settle. Decant the clear solution and wash the precipitate on a Buechner funnel with cold water. Evaporate the filtrate to a small volume. Now mix the silver chloride, with about ten times its weight, with soda ash (which contains a small amount of corn starch) and dry. Place the mass in 30 grams crucibles, and fuse for about 30 minutes. Pour in molds. When cool, hammer off excess of slag, and finally boil with hydrochloric acid to clean the silver buttons. Then dissolve with dilute nitric acid and precipitate silver as silver chloride as already described. Another silver chloride precipitation will be necessary to separate all platinum and palladium from the silver.

Combine all filtrates and evaporate to dryness on steam plate. Take up with a few cc. of hydrochloric acid and water, filter off the silver chloride and wash with cold water. This small amount of silver chloride carries down considerable platinum and palladium. Therefore place the filter paper with the silver chloride in a 3-inch scorifier, dry, add 40 grams of test lead and a pinch of borax and scorify. Then cupel the lead button. Dissolve the silver buttons with dilute nitric acid and reprecipitate silver with hydrochloric acid. Finally when the pure white color of the silver chloride indicates that it is free from platinum and palladium, evaporate the filtrate to dryness on steam plate. Take up with 16 cc. of hydrochloric acid and 4 cc. of water. Boil for a minute, filter, and precipitate platinum and palladium as described under "Gold Bar."

Copper Anode Slimes. Take a 1,000-gram sample. Weigh out 3 gram portions into 3-inch scorifiers and mix each with 40 grams of test lead and a pinch of borax and litharge. Scorify and cupel. Dissolve silver buttons with dilute nitric acid, and proceed as described under "Refined Silver."

Determination of Platinum and Palladium in Refined Gold[1]

The sample may be in the shape of drillings, but from a bar it is easier to roll the gold into a thin ribbon.

Fifty grams[2] of gold sample is sufficient for gold which has been parted with sulphuric and nitric acids.

Dissolve sample in a 1,500 cc. beaker with 50 cc. of nitric acid (1.42) and 150 cc. of hydrochloric acid (1.19) using no water. Heating is not necessary. After complete solution of the sample, evaporate solution to a syrup of about 40 cc. volume, taking care not to evaporate too far, otherwise, some gold will become reduced and separate out; add 100 cc. of hydrochloric acid and re-evaporate the solution to syrup, repeating this operation four times in order to remove all nitric acid.

After the last evaporation, dilute with hot water, boil, add about 50 cc. hydrochloric acid (1.19) to clear up solution: volume of solution should be about 500 cc.

To the boiling solution[3] gradually add a mixture of 50 grams ammonium oxalate and 50 grams oxalic acid, which should precipitate all the gold, but should there be any doubt add more of the mixture of the salts. Dilute the solution to about 1,000 cc. in volume and allow to settle in a warm place over night.

Filter off the gold, washing by decantation into a 1,500 cc. beaker. For extreme accuracy, this gold may be redissolved, re-evaporated and reprecipitated, this time with sulphur dioxide gas. This is more of a precautionary measure, for as a rule, no platinum or palladium will be found with the gold.

To the solution from the gold add 5 grams of 30 mesh C.P. zinc. This precipitates any gold left in the solution along with the silver, platinum, palladium, tellurium, copper, etc.

Filter as soon as precipitation is complete and wash by decantation keeping as much of the precipitate in the beaker as possible. Ignite filter paper and transfer residue and precipitate to a 250 cc. beaker, dissolve in 10 cc. aqua regia and after complete solution, add 5 cc. sulphuric acid (1.84) evaporate to fume of SO_3 and fume well,[4] cool, dilute to 100 cc. bring solution to boiling and add 1 drop of hydrochloric acid to precipitate the silver, filter in a 400 cc. beaker, dilute filtrate to 200 cc. volume, add 5 cc. hydrochloric acid (1.19),

---

[1] Contributed by S. Skowronski,   Chemist, Raritan Copper Works, Perth Amboy, N. J.

[2] For gold which has been electrolytically refined by the Wohlwill process, 100 grams of gold should be taken as a sample, doubling the quantity of acid necessary for the solution.

[3] Sulphur dioxide gas is not recommended for the precipitation of gold as gold bullions contain a trace of tellurium, and in the presence of tellurium, palladium is precipitated as a telluride by sulphur dioxide gas. The gold precipitated with oxalic acid is free from palladium telluride and therefore may be reprecipitated with sulphur dioxide gas if a reprecipitation is thought necessary.

[4] The platinum and palladium after solution in aqua regia, and addition of sulphuric acid, should be well fumed, in order to reduce any gold remaining in the solution to the metallic condition, it is very essential that all the gold is removed at this stage, otherwise it is liable to contaminate the palladium di-methyl glyoxime.

32

and precipitate palladium with .5 grams of dimethyl glyoxime dissolved in 50 cc. of boiling water.[5]

Palladium dimethyl glyoxime, canary yellow in color, which possesses the same physical characteristics as the corresponding nickel salt, at once separates out. Allow to settle in a warm place for about five minutes. Filter in a Gooch crucible, wash with hot water, dry at 110° C. and weigh. Factor .3168.

To the filtrate from the palladium add 2 grams of 30-mesh C.P. zinc, which precipitates the platinum. Filter, ignite, precipitate and dissolve in aqua regia. Remove nitric acid by three evaporations with hydrochloric acid (1.19), taking care not to evaporate solution to dryness.[6] After the last evaporation, take up with not more than 10 cc. of water, and a few drops of hydrochloric acid. If necessary, filter, keeping volume of 10 cc. add 2 grams of ammonium chloride, stir well, add 10 cc. of alcohol, and let stand one hour with an occasional stirring.

Filter off ammonium chlor-platinate in small Gooch crucible and ignite to platinum in the usual manner.

Separation of Platinum and Gold. In place of the procedures given on page 379 it is often preferable to precipitate gold first by means of sulphur dioxide, then reprecipitating with oxalic acid from weakly acid solution to obtain gold free from platinum.

Separation of Platinum and Ruthenium. According to Deville and his co-workers the ammonium chloride separation is unsatisfactory owing to contamination of the ammonium chlor-platinate with ruthenium. The ruthenium may be separated by volatilization with chlorine passed into the alkaline solution of platinum and ruthenium. See page 387b.

[5] Palladium is best precipitated with di-methyl glyoxime in a 3-5% acid solution, gold if present will be reduced to the metallic condition and should be removed before hand. Alcohol is not recommended as the solvent for the dimethyl glyoxime, as it slows up the precipitation of the palladium. A hot water solution works quicker, and should be filtered to remove insoluble matter before addition to the palladium solution.

The precipitation should be carried out in a cold solution, since platinum will contaminate the palladium precipitate if the dimethylglyoxime reagent is added to a hot solution.

[6] Any solutions containing platinum should never be evaporated to dryness, as platinum is easily reduced in baking to the "platinous" condition which is not precipitated with ammonium chloride.

QUANTITATIVE SEPARATION OF THE ELEMENTS OF THE PLATINUM GROUP

Dissolve the material in aqua regia and filter off the insol. Ignite the residue and fuse in a nickel crucible with Na_2O_2. Cool, place in a beaker containing a little water and acidify with HCl. Combine the solutions, place in a distillation flask, make alkaline with NaOH and distil with a current of chlorine gas, catching the distillate in NaOH.

Distillate: Ru, Os. Acidify with HCl and pass in H_2S. Filter off sulphides and ignite in a combustion tube in a current of oxygen, catching the volatile OsO_4 in a solution of NaOH and alcohol. Determine Os in this solution, and weigh RuO_2 remaining in the boat.

The remaining solution is boiled to expel chlorine and then concentrated HN_4Cl solution is added and sufficient 95% alcohol to double the volume of the solution.

Precipitate: Pt, Ir, some Rh, traces of Pd. Ignite in an atmosphere of hydrogen, extract residue with dilute aqua regia (1 part acids to 4 parts water).

Residue: Ir, Rh. Fuse with $KHSO_4$, extract with water and dilute H_2SO_4. Repeat fusion and extraction. Combine filtrates, washing residue.

Residue: Ir, trace Rh.

Solution: (A) Rh. Boil with Na_2CO_3, then acidify with HCl and filter. Ignite residue and combine with Rh from (B) and (C).

Solution: Pt with traces of Pd, Ir, Rh.

Filtrate: Rh, Pd, some Pt, Ir, and any Fe, Ni, Cu that was present in sample. Nearly neutralize with NH_4OH and pass in H_2S gas.

Precipitate: sulphides of Pd, Rh, Au, Cu. Ignite and digest with HCl, filter and repeat ignition and extraction of residue. Combine extracts.

Residue (C): Rh. Combine with (A) and (B).

Filtrate: Pd, Cu. Add KCl and alcohol. If Pd is present, it will precipitate as K_2PdCl_4.

Filtrate: Ni, Fe, some Au and Rh. Evaporate to dryness with HNO_3 and ignite. Extract Fe and Ni with HNO$_3$ and ignite. Extract Fe and Ni with HCl. Ignite residue (B) and combine with (A) and (C).

Combine residues (A), (B) and (C) containing Au and Rh. Dissolve out Au by extraction with aqua regia. Rh is left as a residue.

RARER ELEMENTS OF THE ALLIED PLATINUM METALS

IRIDIUM

Element, Iridium. **Ir. at.wt. 193.1;** *sp.gr.* **22.3;** *m.p.* **23.50° C.?** *oxides,* **IrO₂, Ir₂O₃.**

DETECTION

Iridium is found associated with platinum. The element is insoluble in all acids, including aqua regia. Chlorine is the best reagent which forms the chlorides of iridium and yields compounds with other chlorides as K_2IrCl_6, which is insoluble. If the element is heated in a stream of chlorine in the presence of potassium chloride there forms a salt, K_2IrCl_6, which is sparingly soluble and is used in the separation of iridium.

The oxide, Ir_2O_3 is formed when K_2IrCl_6 is mixed with sodium carbonate and gently fused at a dull red heat.

$$2K_2IrCl_6 + 4Na_2CO_3 = Ir_2O_3 + 8NaCl + 4KCl + 4CO_2 + O.$$

The fusion is dissolved in water containing ammonium chloride; filter the residue and after ignition, to expel the ammonium chloride, is treated with dilute acid in order to remove the small quantity of alkali. A bluish-black powder is thus obtained which begins to decompose when heated above 800 degrees, and at temperatures somewhat above 1000 degrees is completely broken up into oxygen and the metal.[2]

The dioxide, IrO_2, is a black powder obtained by heating the hydroxide in a current of carbon dioxide. It is insoluble in acids.[2]

Caustic Alkalies produce in a boiling solution a dark-blue precipitate of $Ir(OH)_4$ insoluble in all acids except HCl.

Potassium chloride forms the double salt of K_2IrCl_6, which is black and is difficulty soluble in water.

Ammonium chloride precipitates black $(NH_4)_2IrCl_6$, which is difficulty soluble in water.

Hydrogen sulphide precipitates black Ir_2S_3, soluble in $(NH_4)_2S$.

Metallic zinc precipitates from an acid solution black metallic iridium.

Formic acid and sulphurous acid precipitate black metallic iridium from hot solutions.

Lead acetate gives a gray-brown precipitate.

"Treatise on Chemistry," Roscoe and Schorlemmer.

Chapter contributed by R. E. Hickman and Edward Wichers.

ESTIMATION

Substances in which iridium is determined are: platinum scrap, jewelers' sweeps, contact points, ores. Iridium is weighed as the metal.

Preparation and Solution of the Sample

Platinum scrap and contact points, etc., containing iridium dissolve with difficulty in aqua regia, depending on the amount of iridium present. The alloy is dissolved more quickly if it is rolled or hammered to a very thin sheet or ribbon. The alloy of platinum and iridium with an iridium content up to 10% dissolves in aqua regia slowly; an alloy of iridium content of 15% dissolves in aqua regia very slowly and the aqua regia will likely have to be replenished from time to time. An alloy of 25% iridium is practically insoluble in aqua regia. The filings from sweeps, etc., can be dissolved by aqua regia the same as the scrap. After expelling the HNO_3 the platinum and the iridium are precipitated together with NH_4Cl as $(NH_4)_2PtCl_6$ and $(NH_4)_2IrCl_6$. The iridium imparts a pinkish to a scarlet color to the salt.

If the iridium content is too high to be dissolved in aqua regia the metal can be mixed with NaCl, heated to a dull red heat in a porcelain or silica tube, and moist chlorine passed over the mixture. The iridium will be in the form of a chloride which dissolves in water. After filtering the solution and evaporating with HCl, the iridium as well as the platinum is precipitated with NH_4Cl or H_2S. This is a convenient way on a larger scale to dissolve osmiridium in ores. The writer has had good results with this operation.

When the iridium is contaminated with a large amount of impurities, it may be reduced from the solution with zinc, and the impurities dissolved by HNO_3 and dilute aqua regia; the residue is washed and dried as iridium.

Clean osmiridium grains are also brought into solution by a fusion of KNO_3, $NaNO_3$ or $KClO_3$ and NaOH or KOH, leaving the iridium as Ir_2O_3.

SEPARATIONS

Separation of Iridium from Platinum. See Separation of Pt from Ir.

If the platinum and iridium are alloyed with at least ten times their weight of silver or lead and the alloy dissolved in HNO_3, the silver or lead and the platinum dissolves, leaving the iridium insoluble. After washing the residue, treat with a small amount of dilute aqua regia to dissolve any platinum that may be present.

Separation from Osmium. Osmium is removed by distillation. See Os.

For other separations see under Rh and Ru. See all section 3 following.

GRAVIMETRIC METHODS FOR THE DETERMINATION OF IRIDIUM

1. By Reduction with Zinc

The solution of iridium or iridium and platinum is treated with C.P. granulated zinc and 5% free HCl. The iridium and the platinum are precipitated as fine black metal. The black metal is washed free from impurities and the platinum is dissolved in dilute aqua regia as described under the Separations. The insoluble portion is dried, ignited, reduced with hydrogen and weighed as metallic iridium.

2. By Igniting the Salt $(NH_4)_2IrCl_6$

The percentage of iridium in the salt may be judged fairly well by the color, by comparing with standard iridio-platinum salts. The salt is filtered, washed with alcohol and carefully ignited and weighed as iridio-platinum sponge metal. The percentage of iridium in the sample can be calculated from the weight of the iridium obtained. The two metals are treated as stated under 3 below.

3. By Obtaining it as a Residue

The iridium and the platinum, etc., are alloyed with at least ten times its weight of silver and the alloy dissolved in HNO_3. The residue will be a small amount of platinum, gold, if any present, and iridium. Add a small amount of dilute aqua regia, which will dissolve the gold and the rest of the platinum, leaving the iridium as a black residue. This is filtered, washed and ignited, reduced by hydrogen and weighed as metallic iridium.

One part of the iridium material is alloyed with ten parts of lead. This is packed in a graphite capsule, and the whole embedded in charcoal in an ordinary assay crucible. Heat to a high temperature in a furnace for several hours. When the crucible and contents are cold, remove the lead and clean well. Treat the lead with dilute HNO_3, thus removing the lead and leaving the iridium as the residue. Wash thoroughly and treat the residue with dilute aqua regia, which leaves the residue as pure iridium. If other metals of the platinum group are present, see separations under those metals.

4.   Determination of Iridium

Method of Deville and Stas, Modified by Gilchrist [1]

1. **Lead Fusion.** Fuse the carefully sampled platinum alloy with 10 times its weight of granular test lead for a period of one hour at a temperature of about 1000°. A covered crucible, whose outside dimensions are 4 cm. in diameter and 7 cm. in height, machined from Acheson graphite, is suitable for fusions made with 20 to 40 grams of lead. The inside of the crucible should possess a slight taper to facilitate the removal of the cooled ingot. Do not pour the fusion from the crucible, but allow it to solidify, since the iridium has largely settled to the bottom of the crucible. The crucible is best heated with an electric furnace.

[1] Raleigh Gilchrist, U. S. Bureau of Standards, Jour. Am. Chem. Soc., **45**, 2820 (1923).

2. **Disintegration with Nitric Acid.** Brush the cooled lead ingot free from carbon with a camel's hair brush and place it in a beaker. Add nitric acid of the concentration one volume of nitric acid (d. 1.42) to 4 volumes of water, using 1 cc. of acid per g. of lead. Place the beaker on the steam bath or on a hot-plate, which maintains the temperature of the solution at about 85°. Disintegration of the lead ingot is usually complete in about two hours and leaves a rather voluminous, grayish-black mass. Dilute the solution to twice its volume and decant the liquid through a double filter, consisting of a 9-cm. paper of fine texture, on which is superimposed a 7-cm. paper of looser texture. Wash the residue quite thoroughly with hot water and pass the washings through the filters. The residue is not transferred to the filters at this point. The lead nitrate solutions and washings are best caught in an Erlenmeyer flask to make easier the detection of the presence of any residue which has passed through the filters. This is done by whirling the liquid in the flask. Return the filters to the beaker without ignition.

3. **Solution of the Lead-Platinum Alloy with Aqua Regia.** Add in order —15 cc. of water, 5 cc. of hydrochloric acid (d. 1.18) and 0.8 cc. of nitric acid (d. 142) for each gram of the platinum-alloy sample taken. Heat the solution in the beaker on the steam-bath or on a hot-plate which maintains the temperature at about 85°. The lead-platinum alloy is usually completely dissolved within one and a half hours. Dilute the solution with twice its volume of water and filter through a double filter similar to the one used for the lead nitrate solution, the iridium, insoluble in aqua regia, is in the form of fine crystals, possessing a bright metallic luster and having a high density. Pass the clear solution through the filter first and then transfer the thoroughly macerated paper. It is very important to examine the beaker to see that no iridium remains. To do this the interior of the beaker is wiped with a piece of filter paper to collect any metal adhering to the sides. Then by whirling a small quantity of water in the beaker any iridium remaining gravitates towards one place whence it can be removed with a piece of paper. Wash the filters and iridium thoroughly, first with hot water, then with hot dilute hydrochloric acid (1 : 100), and lastly with hot water. The chlorplatinic acid filtrate and washings should be examined for iridium, which may have passed through the filters, in the manner described under the nitric acid treatment. The last washings should be tested for the absence of lead.

4. **Ignition and Reduction of the Iridium.** Place the washed filters and iridium in a porcelain crucible and dry, before igniting in air. After the destruction of the filter paper, ignite the iridium strongly with the full heat of a Tirrill burner. After all the carbon is burned out, cover the crucible with a Rose lid, preferably of quartz. Introduce into the crucible a stream of hydrogen, burning from the tip of a Rose delivery tube (a quartz tube preferred). After five minutes remove the burner and a few minutes later extinguish the hydrogen flame by momentarily breaking the current of hydrogen. This is best done by having a section of the rubber delivery tube replaced by a glass tube, one end of which can easily be disconnected. Allow the iridium to cool in an atmosphere of hydrogen and then weigh as metallic iridium.

NOTES. In commercial analysis no effort is made to correct the weight of iridium for small amounts of ruthenium. Correction, if desired, can be made according to the original directions of Deville and Stas. ("Procés-verbaux, Comité International des Poids et Mesures," 1877, p. 185.) The correction for iron can be made by the procedure suggested by W. H. Swanger, U. S. Bureau of Standards. The iridium is fused with zinc, the excess zinc removed with hydrochloric acid, and the zinc-iridium alloy fused with potassium pyrosulphate. The fusion is digested with dilute sulphuric acid, which leaves a residue of iridium free from iron but contaminated with silica. Silica is now removed by the usual manner and pure iridium remains. This purification is necessary if iron is present in the sample since this separates with iridium. Palladium, rhodium and gold have no effect in the determination. Ruthenium separates quantitatively with the iridium. The loss of weight of iridium during the ignition periods is insignificant.

RUTHENIUM

Element, Ruthenium. **Ru.** *at.wt.* **101.7;** *sp.gr.* **12;** *m.p.* **2450° C.?** *oxides,* Ru_2O_3, RuO_2, RuO_4.

DETECTION

This element is found in platinum ores, and as laurite, Ru_2S_3. It is barely soluble in aqua regia and insoluble in acid potassium sulphate. It dissolves when fused with KOH and KNO_3. The solution of the fusion when dissolved in water forms potassium ruthenate, K_2RuO_4, from which HNO_3 precipitates the hydroxide, which is soluble in HCl. The treatment with chlorine and KCl at a high temperature yields a salt of K_2RuCl_6. The salts that are most common are K_2RuCl_5 and K_2RuCl_6.

The oxide, Ru_2O_3, is formed when finely divided ruthenium is heated in the air, forming a blue powder which is insoluble in acids. It can also be obtained by heating the trihydroxide, $Ru(OH)_3$, in dry carbon dioxide which forms a black, scaly mass.[1]

Ruthenium dioxide, RuO_2, is obtained by roasting the disulphide or sulphate in contact with air. It is likewise formed when the metal is fused in an oxidizing atmosphere, when it burns with a sparkling smoky flame, and evolves an ozone-like smell.[1]

Ruthenium tetroxide, RuO_4, is formed in small quantities when the metal is heated at 1000° in a current of oxygen, although when heated alone it decomposes about 106°. It is prepared by passing chlorine into a solution of potassium nitrosochlororuthenate, or of potassium ruthenate or sodium ruthenate prepared by fusing the metal with sodium peroxide; the liquid becomes heated and the tetroxide distills over and is deposited in the receiver. The moist oxide quickly decomposes. In the dry state it is fairly stable, but decomposes in sunlight with the formation of lower oxides. It dissolves slowly in water, and the solution when it contains free chlorine or HCl may be kept without alteration for some days if light be excluded, but when pure slowly deposits a black precipitate.[1]

In addition to the above oxides, salts corresponding to the acidic oxides RuO_3 and Ru_2O_7 have been prepared.

Potassium hydroxide precipitates a black hydroxide easily soluble in HCl.

Hydrogen sulphide slowly produces brown Ru_2S_3.

Ammonium sulphide precipitates brownish black sulphide.

Metallic zinc precipitates metallic ruthenium, the solution first turning blue.

Potassium sulphocyanate gives on heating a dark brown solution.

Silver nitrate gives a rose red precipitate.

Mercurous nitrate produces a bright blue precipitate.

Zinc chloride produces a bright yellow precipitate which darkens on standing.

[1] "Treatise on Chemistry," Roscoe and Schorlemmer.

Potassium iodide after a time by heating precipitates the black sesqui-iodide.

ESTIMATION

Ruthenium is generally estimated in alloys and ores or residues.

Preparation and Solution of the Sample

When ruthenium is alloyed with platinum or gold, aqua regia dissolves these metals, forming the chlorides of platinum, gold and ruthenium. The ruthenium in ores is in the form of an alloy with platinum or osmiridium. This is fused with KNO_3 and KOH in a silver crucible, the osmium and the ruthenium forming salts as described above, while the iridium remains as an oxide.

Separations

Separation of Ruthenium from Platinum. The two metals are precipitated with KCl and the potassium rutheniochloride is dissolved out with cold water containing a very small amount of KCl and alcohol. The ruthenium is then precipitated from an acid solution by additions of granulated zinc.

A separation may be made by alloying with silver and dissolving the platinum and silver by HNO_3, the ruthenium remaining as the residue.

From a concentrated solution of these metals precipitate the platinum with NH_4Cl. Evaporate the filtrate with potassium nitrate to dryness and boil the residue with alcohol when the residual platinum will remain behind and the ruthenium goes into solution.

Separation of Ruthenium from Iridium. The two metals are fused with KOH and KNO_3 as described above, the ruthenium forming a salt soluble in water and the iridium remaining as an oxide.

To the solution of the two metals, sodium nitrite is added in excess, with sufficient sodium carbonate to keep the liquid neutral or alkaline. The whole is boiled until an orange color appears. The ruthenium and the iridium are converted into soluble double nitrites. Sodium sulphide is then added, small quantities at a time until the precipitated ruthenium sulphide is dissolved in the excess of alkaline sulphide. At first the addition of the sulphide gives the characteristic crimson tint due to ruthenium, but this quickly disappears and gives a bright chocolate-colored precipitate. The solution is boiled for a few minutes, and allowed to become perfectly cold and then dilute HCl cautiously added until the dissolved ruthenium sulphide is precipitated and the solution is faintly acid. The solution is filtered and the precipitate washed with hot water. The filtrate will be free from ruthenium.[1]

The fusion with KOH and KNO_3 as described above is dissolved in water in a flask or retort; chlorine is passed through this solution and thence into two or three flasks containing a solution of KOH and alcohol. The two or three flasks which form the condensing apparatus should be kept as cold as possible. The ruthenium is transformed into volatile RuO_4 which condenses in the flasks, while the iridium remains in the retort.

[1] "Select Methods in Chemical Analysis," Sir Wm. Crookes.

Separation of Ruthenium from Rhodium. The mixed solution of the two metals is treated with potassium nitrite as described above. The orange-yellow solution is evaporated to dryness upon the water bath and treated with absolute alcohol. The rhodium remains undissolved and can be filtered off and washed with alcohol. The rhodium salt can be ignited with NH_4Cl and after washing yields metallic rhodium. See Separation of Rhodium from Ruthenium.

Separation of Ruthenium from Osmium. See Separation of Osmium from Ruthenium.

GRAVIMETRIC METHODS FOR THE DETERMINATION OF RUTHENIUM

Ruthenium is weighed as the residue or metallic ruthenium after it has been separated from the other metals.

The residue containing ruthenium or osmiridium is fused in a silver crucible with five grams KOH and one gram KNO_3 at a low temperature from one-half to one hour. The mass is cooled and extracted with water. The orange-colored solution containing potassium ruthenate is gently distilled in a current of chlorine whereby the volatile ruthenium tetroxide passes over into the receivers. All connections should be ground glass so that no Ru will be reduced in the joints. The solution in the distilling flask must be kept alkaline to prevent iridium chloride from distilling over with the Ru. Add a small piece of KOH after the first distillation and distill as before. Oftentimes it will be necessary to fuse again with KOH and KNO_3 and distill as stated above. Continue to pass chlorine through the alkaline solution until all effervescence ceases. Disconnect the chlorine and draw air through the apparatus, heating the solution nearly to boiling.

1. Receivers containing KOH solution (10 to 15% KOH) and alcohol.

a. This alkaline solution containing the ruthenium tetroxide distillate is evaporated to a smaller volume and the ruthenium is precipitated by boiling with absolute alcohol. Filter, wash well with hot water, dilute HCl and again with hot water. Ignite, reduce in hydrogen and weigh as metallic Ru.

b. The alkaline solution from the receivers is made acid with HCl and the Ru is precipitated from the hot solution with hydrogen sulphide gas. Filter, wash, ignite at a high temperature, reduce in hydrogen and weigh as metallic Ru.

2. Receivers containing hydrochloric acid.

a. This acid solution containing the ruthenium tetroxide distillate is heated to nearly boiling and the ruthenium is precipitated with hydrogen sulphide gas as under 1*b*.

b. The acid solution containing the Ru is evaporated to a concentrated solution and transferred to a weighed porcelain crucible. Evaporate to dryness, bake and ignite. Reduce in hydrogen and weigh as metallic Ru.

c. Ruthenium may be estimated by precipitation with magnesium from solutions of its salts. The precipitate is washed with dilute sulphuric acid to remove excess of magnesium, dried, ignited in a current of hydrogen, cooled in carbon dioxide and weighed as metal.[1]

[1] Text Book of Inorganic Chemistry, Vol. IX, Part I. J. N. Friend.

RHODIUM

Element, *Rhodium*. Rh. *at.wt.* 102.91; *sp.gr.* 12.1; *m.p.* 1950° C.; *oxides*, RhO, Rh$_2$O$_3$, RhO$_2$.

DETECTION

Rhodium is found only in platinum ores. It is a white metal, difficultly fusible, and insoluble in acids. Rhodium, however, dissolves in aqua regia when alloyed with platinum, to a cherry red solution. It is also soluble in molten phosphoric acid and dissolves when fused with aci1 potassium sulphate with the formation of K$_3$Rh(SO$_4$)$_3$. If the metal is treated with chlorine in the presence of sodium chloride there forms a soluble salt, Na$_3$RhCl$_6$.

Rhodium monoxide, RhO, is obtained by heating the hydroxide Rh(OH)$_3$, by cupellation of an alloy of rhodium and lead, or by igniting the finely-divided metal in a current of air. It is a grey powder with a metallic appearance, and is not attacked by acids, and when heated in hydrogen is reduced with evolution of light.[1]

The oxide, Rh$_2$O$_3$, is obtained as a grey iridescent spongy mass by heating the nitrate. It is also formed as a crystalline mass when sodium rhodochloride is heated in oxygen. It is perfectly soluble in acids [1]

Rhodium dioxide, RhO$_2$, is obtained by repeated fusions of the metal with KOH and KNO$_3$. It is attacked neither by alkalies nor by acids and is reduced by hydrogen only at a high temperature.[1]

Hydrogen sulphide precipitates rhodium sulphide, when passed into a boiling hot solution containing rhodium.

Potassium hydroxide precipitates at first a yellow hydroxide, Ph(OH)$_3$ +H$_2$O soluble in an excess of the reagent. If boiled, a dark gelatinous precipitate separates. A solution of Na$_3$RhCl$_6$ does not show this reaction immediately, but the precipitate appears in the course of time. An addition of alcohol causes a black precipitate immediately.

Ammonium hydroxide produces a precipitate which dissolves in excess NH$_4$OH on heating. Addition of HCl now produces a yellow precipitate, insoluble in HCl but soluble in NH$_4$OH.

Potassium nitrite precipitates from hot solutions a bright yellow precipitate of double nitrite of potassium and rhodium.

Zinc, iron and formic acid precipitate rhodium as a black metal.

Hydrogen reduces rhodium salts.

To detect small amounts of rhodium in the presence of other metals, evaporate the solution and displace with a fresh solution of sodium hypochlorite; the yellow precipitate formed is soluble after an addition of acetic acid. After a long agitation the solution changes to an orange-yellow color and after a short time the color passes and finally a grey precipitate settles and the solution turns sky-blue.[2]

---

[1] "Treatise on Chemistry," Roscoe and Schorlemmer.
[2] Prescott and Johnson.

ESTIMATION

Rhodium is estimated mainly in ores, thermo couples and salts.

Preparation and Solution of the Sample

When rhodium is estimated in thermo couples or other alloys of platinum and rhodium the wire or sample is rolled to a thin ribbon and dissolved in aqua regia.   Both metals will go into solution, forming the chlorides of rhodium and platinum.   The aqua regia will have to be replaced from time to time, as the alloy dissolves slowly.

The rhodium from salts is precipitated with zinc and the black metallic rhodium cleaned with dilute aqua regia, filtered, washed, ignited and reduced with hydrogen.

Some alloys and ores are alloyed with silver and the silver and platinum are dissolved in HNO_3.   The residue is cleaned with aqua regia, dried, and weighed as metallic rhodium.   If the residue is ignited reduce with hydrogen.

The material or residue containing rhodium is fused with $KHSO_4$ for some time at a low red heat and the mass leached with hot water acidified with HCl. The rose-red solution contains the rhodium.   Several fusions are generally necessary.

Separations

Separation of Rhodium from Platinum.   Alloys and ores containing platinum and rhodium dissolve slowly in aqua regia as stated above.   After expelling the HNO_3 add NH_4Cl.   The precipitate is filtered and washed with dilute ammonium chloride solution, which dissolves the rhodium salt.   A very small amount of rhodium will color the filtrate pink to a rose-red color, depending on the amount of rhodium present.   A green tinge in the ammonium chloroplatinate indicates the presence of rhodium.

A solution of NaOH is added to the HCl solution of the two metals until yellow rhodium hydroxide begins to separate.   After neutralizing, the volume of the solution should be so adjusted that the estimated total content of Pt and Rh does not exceed 1 gram per 100 cc.   A mixture of equal volumes of solutions containing 90 grams of crystallized barium chloride and 36 grams of anhydrous sodium carbonate per liter, respectively, is added.   Not less than 5 cc. of each solution is taken.   After the suspension of barium carbonate is added, the solution is rapidly heated to boiling and boiled for two or three minutes.   The residue is filtered off and washed several times with a hot 2% solution of sodium chloride, after which it is returned, with the filter paper, to the original beaker and digested with 25 cc. HCl (1HCl to $4H_2O$) until solution is complete.   Dilute with water and filter off the paper pulp.   Adjust the volume to about 150 cc., heat to incipient boiling for 30 to 45 minutes while a current of hydrogen sulphide is passed in.   After the precipitation the rhodium sulphide is filtered off at once, washed with water containing a little ammonium chloride, and ignited in a weighed porcelain crucible.   The ignited sulphide is reduced and cooled in hydrogen, and weighed as metallic Rh.[1]

Separation of Rhodium from Iridium.   See Separation of Rh from Pt.

A separation can be made by adding sodium nitrite in excess to the solution of the two metals, with a sufficient quantity of sodium carbonate to make the

[1] Edward Wichers, The Jour. Am. Chem. Soc., Aug. 1924.

solution neutral or alkaline; this is boiled until the solution assumes a clear orange color. The rhodium and iridium are converted into soluble double nitrites. A solution of sodium sulphide is added in slight excess and the liquid made slightly acid. The rhodium is precipitated as dark-brown rhodium sulphide.

A solution of rhodium and iridium is evaporated with HCl and displaced with a large excess of acid sodium sulphite, $NaHSO_3$, and allowed to stand sometime when a pale yellow double salt of rhodium and sodium sulphite slowly separates out while the solution becomes nearly colorless. Wash out the precipitate, and heat with hot concentrated H_2SO_4 till the sulphurous acid is driven off. Heat the material in a crucible until rid of all free sulphuric acid. Then the iridium is dissolved out as a sulphate with a deep chrome-green color, while a double salt of sodium sulphate and rhodium oxide remains behind. This is flesh color insoluble in water and acids. Boil with aqua regia, wash, dry, heat and it decomposes into rhodium and sodium sulphate.[1]

Rhodium can also be separated from iridium, when the latter is present as an iridic salt such as $Ir(SO_4)_2$, by precipitating the mixed salts with caustic potash, dissolving the hydroxides in dilute sulphuric acid and adding caesium sulphate. The sparingly soluble rhodium caesium-alum separates in the cold, and can readily be purified by recrystallization and then by electrolysis.[2]

The residue of rhodium and iridium is melted or scorified with test lead. The lead button is cleaned and dissolved in dilute HNO_3. After filtering and washing the residue, do not ignite, but wash the contents of the filter into a beaker and fume with H_2SO_4 from one to three hours. When cool, dilute with water and let stand over night. The residue contains the iridium and a small amount of $PbSO_4$, while the solution contains the rhodium as the sulphate. To make a further separation from impurities present, the sulphate solution is made alkaline with NaOH and boiled. Let stand until cold and filter off the rhodium hydroxide. Digest with HCl until all the hydroxide has dissolved. Filter and wash with hot water. Evaporate the filtrate to dryness, dissolve in hot water and add about 15 cc. of sodium nitrite solution (40% $NaNO_2$). Heat until all action ceases, then add sodium carbonate to the hot solution until no more precipitate forms. Let cool, filter and wash with hot water. Acidify the filtrate with dilute acetic acid and add potassium chloride solution (20% KCl) until all the Rh is precipitated. Let stand over night at 50 to 60° C. When cold, filter the white precipitate, washing with 20% KCl solution containing a little $NaNO_2$. The white precipitate of potassium rhodium nitrite is digested with HCl, filtered and washed with hot water. Evaporate the HCl solution to dryness, add ammonium formate and heat until dry. Ignite, reduce in hydrogen and cool in CO_2, wash free from salts with hot water, ignite and reduce in hydrogen as before and weigh as metallic Rh.

Separation of Rhodium, Platinum and Palladium. Having the three in solution precipitate the platinum with NH_4Cl as described under Platinum. After filtering off the $(NH_4)_2PtCl_6$ precipitate, and after neutralizing the filtrate with Na_2CO_3 add mercuric cyanide to separate the palladium as $Pd(CN)_2$ as described under palladium. The filtered solution is evaporated to dryness with an excess of HCl. On treating the residue with alcohol, the double

---

[1] "Hand Book of Chemistry," Dammer.
[2] "Treatise on Chemistry," Roscoe and Schorlemmer.

chloride of rhodium and sodium is left undissolved as a red powder. By heating this in a tube through which hydrogen is passed the rhodium is reduced to the metallic state and the sodium chloride is washed out with water leaving a grey powder of metallic rhodium.

The residue containing these three metals is scorified with test lead, and the resultant lead button cupelled with silver. The silver bead is dissolved in dilute HNO_3; the solution filtered, washed with hot water, ignited, and the residue treated with dilute aqua regia to dissolve any platinum or palladium that may be present. Filter, wash with hot water, ammonia water, and again with hot water. Ignite and reduce in hydrogen as metallic Rh.

Separation of Rhodium from Ruthenium. The solution containing the two metals is treated with sodium nitrate as above and evaporated to dryness. The residue is powdered and treated in a flask with absolute alcohol. After filtering and washing with alcohol the rhodium remains undissolved.

The substance or residue containing the rhodium and ruthenium may be fused with $KHSO_4$ in a porcelain or platinum crucible causing the rhodium to go into solution as already described. The ruthenium remains insoluble.

GRAVIMETRIC METHODS FOR THE DETERMINATION OF RHODIUM

1. The solution containing rhodium is treated with zinc and the residue is washed well with hot water acidulated with HCl. The residue is then cleaned with dilute aqua regia and the black metallic rhodium is filtered off, dried, and ignited in hydrogen. Cool and weigh as metallic rhodium.

2. The metals may be melted with lead or silver and the buttons dissolved in dilute HNO_3, leaving a residue which is treated with dilute aqua regia to dissolve any platinum that might be present. The residue is treated with salt and chlorine and the melt dissolved in water as described under Iridium. The iridium is precipitated with NH_4Cl and the rhodium with zinc. The rhodium black is cleaned with dilute aqua regia, filtered, washed and ignited. Reduce in hydrogen and weigh as metallic rhodium.

3. The solution containing the rhodium is made alkaline with KOH and then acid with formic acid, boil, and the rhodium will be precipitated as finely divided metallic rhodium. After filtering proceed in the usual manner.

4. After the platinum and the palladium are eliminated, the residue of Ir, Rh and Ru is fused with $KHSO_4$ in a porcelain crucible and the melt dissolved in water. Filter, wash with hot water, and after acidulating with HCl the Rh is precipitated with C.P. powdered zinc, hydrogen sulphide gas or both. Filter, wash with hot water and ignite. Clean the residue with dilute HNO_3, then with dilute aqua regia, wash with hot water, ignite in hydrogen and weigh as metallic Rh.

5. The residue containing Ir, Rh, Ru and Os is fused with five grams KOH and one gram KNO_3, and the Ru and Os are distilled with chlorine as explained under ruthenium. The solution in the distilling flask is zinced well, filtered, washed and ignited. The residue of impure Ir and Rh is scorified with test lead, the lead button dissolved in dilute HNO_3 and the residue treated with H_2SO_4 as explained under Separation of Rh from Ir. The clear rhodium sulphate solution is treated with C.P. powdered zinc, hydrogen sulphide gas or both, and the precipitate is treated as explained above and weighed as metallic Rh.

6. The solution from the distilling flask is zinced well as explained above. The residue is filtered, washed, ignited and boiled with a few cc. of HNO_3 and boiled with H_2SO_4 for one to three hours as explained above. The rhodium sulphate solution is made alkaline with KOH and boiled with alcohol until all the rhodium is precipitated and the solution is clear. Filter, wash with hot water, dilute HNO_3 and again with hot water. Ignite, reduce in hydrogen and weigh as metallic Rh.

7. Rhodium is conveniently estimated by precipitation with magnesium from solutions of its salts. The precipitate is washed with dilute H_2SO_4 to remove excess of magnesium, dried, ignited in a current of hydrogen, cooled in carbon dioxide and weighed as metal.[1]

[1] Text Book of Inorganic Chemistry, Vol. IX, Part I. J. N. Friend.

OSMIUM

Element, Osmium. **Os.** *at.wt.* 190.9; *sp.gr.* 22.4; *m.p.* 2700° C.? *oxides,* OsO, Os_2O_3, OsO_2, OsO_4.

DETECTION

Osmium occurs with platinum ores as a natural alloy with iridium (Osmiridium) and remains undissolved in the form of hard, white metallic-looking grains when the ores are treated with aqua regia. The chlorides, $OsCl_2$ and $OsCl_4$, combine with the alkali chlorides. Through the action of HNO_3, aqua regia or heating in a stream of moist chlorine, osmic tetroxide is formed. OsO_4 is very volatile and the fumes are poisonous. It is detected readily by the odor when heated, as the fumes are highly corrosive and disagreeable like chlorine. Chlorine passed over hot osmium mixed with KCl gives K_2OsCl_6, which dissolves in cold water.

The oxy-hydrogen flame oxidizes the metal but does not melt it. When strongly heated in contact with air, the finely divided osmium burns and is converted into OsO_4, commonly called osmic acid.

Osmium monoxide, OsO, is obtained when the corresponding sulphite mixed with sodium carbonate is ignited in a current of carbon dioxide. It is a greyish-black powder insoluble in acid.[1]

The oxide, Os_2O_3, is a black insoluble powder obtained by heating its salts with sodium carbonate in a current of carbon dioxide.[1]

Osmium dioxide, OsO_2, is obtained from its salts in a similar way to the foregoing oxides. It is likewise formed when its hydroxide is heated in a current of carbon dioxide.[1]

Osmium tetroxide, OsO_4. Very finely-divided metallic osmium oxidizes slowly at the ordinary temperature, and at about 400° takes fire with formation of OsO_4. The denser the metal the higher is the temperature needed for oxidation.[1]

Hydrogen sulphide precipitates dark brown osmium sulphide, OsS_2, but only in the presence of some strong mineral acid; from an aqueous solution of osmic acid there forms a dark brownish-black sulphide, OsS_4. These are insoluble in ammonium sulphide.

Potassium hydroxide precipitates reddish-brown osmium hydroxide, $Os(OH)_4$.

Ammonium hydroxide precipitates the osmium hydroxide.

Zinc and formic acid precipitate black metallic osmium.

Hydrogen reduces osmium compounds to the metal.

Potassium nitrite added to a solution of osmic acid reduces it to osmous acid which unites with an alkali forming a beautiful red salt.

Sodium sulphite yields a deep violet coloration and a dark blue osmium sulphite separates out gradually.

Phosphorus reduces osmium from an aqueous solution.[2]

Mercury precipitates osmium from an aqueous solution of osmic acid mixed with HCl.

Stannous chloride produces a brown precipitate, soluble in HCl to a brown fluid.

[1] "Treatise on Chemistry," Roscoe and Schorlemmer.

ESTIMATION

Osmium is estimated mainly in osmiridium and platinum residues.

Preparation and Solution of the Sample

After the platinum is extracted the residue or osmiridium is mixed with two or three times its weight of common table salt. The mixture is put in a porcelain or silica tube and heated to a dull red heat; moist chlorine is then passed through the tube and thence through receivers containing KOH and C_2H_5OH to catch the Os and Ru that pass over. The mass is cooled and dissolved with water. After several treatments the entire group of platinum metals will be in solution.

The osmium material may also be fused with KOH and KNO_3 and the melt dissolved in water. The osmium will be in solution as potassium osmate, K_2OsO_4, while the iridium remains as residue.

Cold selenic acid has no appreciable action on osmium; at about 120°, however, the metal is dissolved to a colorless solution which contains selenious acid and OsO_4, but no selenate.[4]

Separations

In most cases osmium is separated from the other metals present by distillation or volatilization. See next page.

[1] "Chemical Abstracts," April 10, 1918.

GRAVIMETRIC METHODS FOR THE DETERMINATION OF OSMIUM

The osmium is very difficult to ascertain on account of the element being very volatile.

1. The potassium osmate, K_2OsO_4, solution is put in a small retort and boiled with HNO_3, the OsO_4, is conducted into receivers containing NaOH solution and C_2H_5OH. After acidifying with a little HCl the osmium can be precipitated with $Na_2S_2O_3$ as a brown precipitate of OsO_4 which is filtered, washed, dried and weighed as the oxide, or reduced in hydrogen and weighed as the metal. The osmate solution from the receivers above is heated gently and strips of aluminum are plunged in; the osmium will be deposited in metallic form, while the aluminum dissolves in the soda. Care must be taken not to add too much aluminum, as an aluminate might be precipitated which is troublesome. When the solution is discolored, the dense precipitated osmium is washed by decantation with water to remove the sodium aluminate, and then with 5% H_2SO_4 solution to remove the excess aluminum. The osmium is dried in a bell-jar filled with hydrogen, then heated to a dull redness and cooled in a current of hydrogen. The osmium is weighed as the metal. As a check the osmium may be driven off in the form of OsO_4 by heating to redness with plenty of air, or better, in a current of oxygen and weighing again.[1]

2. The osmate solution from the condensing receivers or from the fusion of KOH and KNO_3 containing the ruthenium and osmium is placed in a retort and HCl is added. A slow current of air or oxygen is passed through the retort and thence through receivers containing KOH and alcohol similar to the ones mentioned above. These receivers are kept as cold as possible. The osmium is distilled over as OsO_4 while the ruthenium remains in the retort. Combine the solutions in the receivers and proceed to determine the osmium as described above.

3. The potassium or sodium osmate solution from the receivers above or where osmium tetroxide is dissolved in potassium hydroxide solution and alcohol is heated at 40 or 50° to form potassium osmate. A slight excess of dilute sulphuric acid is added and then 10 cc. more of alcohol in order to prevent reoxidation. After ten or twelve hours, a bluish-black deposit settles, while the supernatant liquid is colorless and free from osmium. The precipitate is filtered, washed with aqueous alcohol, and converted into metallic osmium by reduction in a current of hydrogen.

4. The residue containing osmium is fused with five grams KOH and one gram KNO_3 in a silver crucible as explained under ruthenium. Add HNO_3 slowly to the distilling flask which is connected to receivers containing NaOH solution and alcohol (10% NaOH and 10% C_2H_5OH). Draw the distillate over gently with the aid of the vacuum, the same as for the chlorine distillation under ruthenium. Continue the HNO_3 until strongly acid and then boil for a short time. Transfer the alkaline solution containing the OsO_4 distillate to a beaker and pass in hydrogen sulphide gas while the solution is heating until saturation; then add HCl until the solution is distinctly acid and continue to saturate the hot acid solution with hydrogen sulphide gas. Let stand over night, filter through a weighed Gooch crucible, washing well with hot water. Ignite in hydrogen, cool in CO_2 and weigh as metallic Os.

[1] "Select Methods of Chemical Analysis," Sir Wm. Crookes.

ASSAY METHODS OF PLATINUM ORES, ETC.

Take from 10 to 30 grams of the material and place in a $2\frac{1}{2}$- or 3-in. scorifier with about 20 to 30 grams of test lead and cover with litharge   Fuse in a muffle for a half hour.   When cool clean the lead button thoroughly and dissolve the lead with dilute nitric acid (1 : 5).   When the lead is dissolved, filter, and wash the residue with hot water till free from lead.   Dry the filter and transfer the bulk of the residue to a small glass beaker.   Burn the filter, and add the ash to the main residue.   This is treated with dilute aqua regia to remove any gold, platinum, etc., that may be present.   Filter and wash thoroughly with hot water and ignite at a low temperature for a short time only, as osmium will volatilize.   Weigh as osmiridium.   (See Ir and Os.)

Take another portion of 10 to 30 grams of the material and treat with aqua regia two or three times.   This will bring the platinum and the alloys (except osmiridium) into solution.   After filtering make up the volume to 500 to 1000 cc., depending on the amount of platinum present.   Take two or three portions of 25 to 50 cc. of the solution and evaporate to nearly dryness with additions of HCl to get rid of the HNO_3.   Take up with a small amount of water and add ammonium chloride.   Digest on the water bath and let cool overnight if convenient.   Filter onto an ashless filter and wash with dilute ammonium chloride solution and alcohol.   Ignite cautiously and weigh as platinum and iridium sponge.   For the determination of iridium see under Iridium.

The filtrate from the platinum precipitation is treated with ferrous sulphate to precipitate the gold.   Digest and filter out the gold.   Ignite and alloy with silver and part for gold.

In the filtrate from the gold precipitation the palladium and rhodium are treated with zinc and HCl until the solution is colorless.   Filter, wash well with hot water and dilute HCl.

Ignite, reduce in hydrogen, cool and weigh as Pd and Rh.   Treat with HNO_3.   Filter, wash well, ignite and reduce in hydrogen, cool and weigh as metallic rhodium.   The difference is palladium, or determine this as described under palladium.

In the revision of the chapters on platinum, palladium and the other metals of the group a considerable number of suggestions by Dr. Edward Wichers of the U. S. Bureau of Standards, have been included by revising the statements of previous editions.   The chemistry of the platinum metals is a field of much variance of opinion and the methods for isolation of these elements are far from perfect.   We feel, however, that the chapters present the latest and best work on the subject.—Editor.

REACTIONS OF SALTS OF PLATINUM METALS [1]

	Ruthenium (RuCl₃).	Rhodium (RhCl₃).	Palladium (PdCl₂).	Osmium (OsCl₄).	Iridium (IrCl₄).	Platinum (PtCl₄).
Colour[2]	Dark brown	Red	Brownish yellow	Yellow	Dark brown	Yellow
Hydrogen[2] sulphide at 80° C.	Azure-blue color on prolonged treatment	Brownish black ppte., Rh_2S_3	Brownish black ppte., PdS	Brownish black ppte., OsS_4	Brownish black ppte., Ir_2S_3	Brownish black ppte., PtS_2
Ammonium sulphide	Dark brown ppte., Ru, difficultly soluble in excess	Dark brown ppte., Rh_2S_3, insoluble in excess	Black ppte., PdS, insoluble in excess	Dark ppte., insol. in excess	Brown ppte., Ir_2S_3, soluble in excess	Brown ppte., PtS_2, soluble in excess to $(NH_4)_2PtS_3$
Caustic alkalies	Black ppte. of hydrated oxide insoluble in excess	Yellow-brown ppte., $Rh(OH)_3$, soluble in excess	Yellowish brown basic salts soluble in excess	Brownish red $OsO_2.2H_2O$	Green solution Brownish black double chloride ppted.	Dark red ppte. of $PtO_2.xH_2O$
Ammonium[2] hydroxide on warming	Greenish coloring	Slow decolorization	Decolorized	Yellowish brown ppte.	Bright color	Slow decolorization
Saturated[2] NH₄Cl solution	Brown ppte.	No ppte.	No ppte.	Red ppte.	Black ppte.	Yellow ppte., $(NH_4)_2PtCl_6$
Saturated KCl solution	Violet cryst. ppte. of K_2RuCl_5	Red cryst. ppte., K_2RhCl_5	Red ppte. of K_2PdCl_4	Brown cryst. ppte., K_2OsCl_6	Brownish red ppte. of K_2IrCl_6	—
KI solution[2] (1 : 1000)	No change	No change	Dark ppte.	No change	Yellow color	Slow red-brown color
Hg(CN)₂ solution	No change	No change	White ppte., $Pd(CN)_2$	No change	No change	No change
KCNS, 1 per cent. solution	Dark violet color	Yellow color	Unchanged	Unchanged	Decolorized	Increased yellow color
Hydrazine in hydrochloric[2] acid solution	Yellow color	Yellow color	Black ppte., metallic Pd	No change	Yellow color	Black ppte., metallic Pt
Dimethyl[2] glyoxime	Yellow ppte. Ppte., Ru	No change Ppte., Rh	No change Ppte., Pd	No change Ppte., Os	No change Ppte., Ir	No change Ppte., Pt
Metallic zinc						

[1] Textbook of Inorganic Chemistry, Vol. IX, Part I. J. N. Friend.

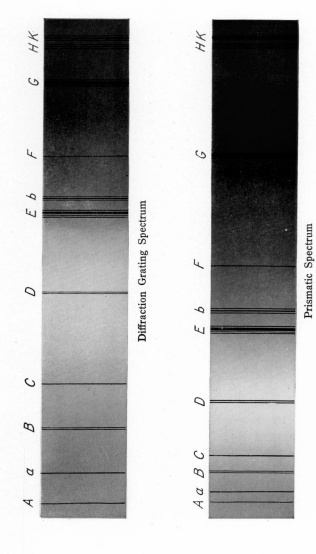

Diffraction Grating Spectrum

Prismatic Spectrum

PLATE III

POTASSIUM, SODIUM AND OTHER ALKALIES

Sodium, Na, *at.wt.* 23.00; *sp.gr.* 0.9735; *m.p.* 97.6°; *b.p.* 877.5° C.; *oxides*
Na_2O, Na_2O_2.

Potassium, K, *at.wt.* 39.10; *sp.gr.* 0.875; *m.p.* 62.5°; *b.p.* 757.5° C.; *oxides*
K_2O, K_2O_4.

Lithium, Li, *at.wt.* 6.94; *sp.gr.* 0.534; *m.p.* 186°; *b.p.* >1400° C.; *oxide* Li_2O.

Rubidium, Rb, *at.wt.* 85.44; *sp.gr.* 1.532; *m.p.* 38.5°; *b.p.* 696° C.; *oxides*
Rb_2O, Rb_2O_2, Rb_2O_3, Rb_2O_4.

Caesium, Cs, *at.wt.* 132.81; *sp.gr.* 1.87; *m.p.* 26.37°; *b.p.* 670° C.; *oxides*
Cs_2O, Cs_2O_2, Cs_2O_3, Cs_2O_4.

DETECTION

Detection of Sodium

Sodium is usually identified by the color which it imparts to the flame or by means of the spectroscope. The solution is prepared as directed under Preparation and Solution of Sample, and is freed from all constituents other than the chlorides of magnesium and the alkalies according to the methods given under Separations. With exceedingly small amounts of sodium, it may be necessary to remove the magnesium also. After acidifying with hydrochloric acid, a drop of the solution is brought into the flame by means of a loop of platinum wire. In the presence of sodium, the flame assumes an intense yellow color, which is usually sufficient to identify the element. The results may be confirmed by examining the flame in the spectroscope, when the characteristic yellow sodium line will be prominent even in the presence of traces of sodium. The spectrum of sodium is shown on Plate II. As a matter of fact, the ever-presence of the sodium line is a hindrance to the success of the method, but by observing the sudden change in the intensity of the line, little trouble will be experienced in detecting exceedingly small amounts of the metal.

Sodium may also be detected by precipitation as sodium pyroantimonate, $H_2Na_2Sb_2O_7.H_2O$, from a sufficiently concentrated neutral or weakly alkaline solution by means of a solution of acid potassium pyroantimonate. The precipitate comes down in granular or crystalline form, and its formation is hastened by rubbing the sides of the vessel with a glass rod. In making this test, magnesium must also be previously removed from the solution.

In waters and soluble salts, it is usually sufficient to test directly the concentrated solution in the flame or spectroscope.

Chapter contributed by W. B. Hicks.

Detection of Potassium

For the detection of potassium in insoluble compounds, bring the sample into solution by one of the methods given under Preparation and Solution of Sample. In other cases, prepare a strong solution of the material to be tested. Where only very small amounts of potassium are present, remove all the constituents from the solution except the chlorides of magnesium and the alkalies as directed under Separations. In the presence of considerable amounts of potassium, small quantities of other constituents will not materially interfere with the flame and spectroscopic tests. After acidifying with hydrochloric acid, bring a drop of the solution to be tested into the non-luminous flame by means of a platinum wire and observe the color produced through a Merwin color screen.[1] In the presence of potassium, a distinct reddish-violet coloration will be apparent. This must not be confused with the color caused by large amounts of sodium, which appear bluish-violet through the screen. Comparison with the coloration produced by pure salts is advisable. If necessary, confirm the results by examining the flame in the spectroscope. In the presence of a moderate amount of a volatile potassium compound, a bright red line will be readily seen in the red portion of the spectrum, and a less distinct violet line will be visible far out in the violet rays.

Potassium may be identified by precipitation as cobaltic nitrite. For this purpose place a small quantity of the solution to be examined in a test tube, acidify slightly with acetic acid, add about an equal quantity of the sodium nitrite solution, prepared by dissolving 125 grams of sodium nitrite ($NaNO_2$) in 250 cubic centimeters of distilled water, and about half as much of cobalt nitrate solution, prepared by dissolving 25 grams of cobalt nitrate ($Co(NO_3)_2 \cdot 6H_2O$) in 100 cubic centimeters of distilled water and adding 50 cubic centimeters of concentrated (glacial) acetic acid. Mix and allow the mixture to stand until effervescence ceases and the cherry-red solution is transparent. If an appreciable amount of potash is present a yellow precipitate will have settled to the bottom of the test tube. By comparing the volume of the precipitate with that produced when a known quantity of potassium chloride is used, an idea of the amount of potash present can be obtained. Ammonium salts produce a similar precipitate.

Potassium chloroplatinate, perchlorate, acid tartrate, picrate, silico-fluoride, and phospho-tungstate are all sparingly soluble in water while the corresponding sodium salts are readily soluble. Precipitation of these compounds from solution may be used in the identification of potassium.

Silicate rocks and minerals may be tested for potash by mixing the finely powdered material with an equal quantity of pure calcium carbonate, moistening with hydrochloric acid, and examining a small amount of the wet mixture on a platinum loop in the flame.

---

[1] The Merwin color screens are manufactured and sold by G. M. Flint, 84 Wendell Street, Cambridge, Mass., at 50 cents apiece, and are far superior to the ordinary cobalt glass.

Detection of Lithium

Bring the sample into solution as directed under Preparation and Solution of Sample, and separate the alkali chlorides from other constituents according to the methods under Separations. Digest the dry chlorides with amyl alcohol or with a mixture of absolute alcohol and ether, filter, and evaporate the filtrate to dryness. Moisten the residue with dilute hydrochloric acid and examine it in the spectroscope. A bright red band and a faint orange line make up the flame spectrum of lithium   Plate II.   These lie between the sodium line and the red potassium line and are easily recognized.

Lithium salts impart a carmine-red color to the flame, which is obscured by sodium, and by large amounts of potassium. But by the proper use of a color screen, the lithium flame may be recognized in the presence of large amounts of sodium.

Confirmation of the presence of lithium may be had by the formation of the sparingly soluble lithium phosphate or lithium fluoride.

Detection of Rubidium and Caesium

In the usual course of analysis, these rare elements are separated along with sodium, potassium, and lithium from all other bases. In order to detect rubidium and caesium, extract the dry chlorides of the alkali metals with a few drops of hydrochloric acid and 90% alcohol. This will dissolve most of the rare alkalies along with some sodium and potassium. Evaporate the solution to dryness, dissolve in a very small amount of water, and add chloroplatinic acid solution. Rubidium, caesium, and potassium chloroplatinates will be precipitated. Filter and wash the residue repeatedly with hot water to remove the potassium salt, which is much more soluble than rubidium and caesium chloroplatinates. During this treatment, examine the residue from time to time in the spectroscope. As the rubidium and caesium salts are concentrated through washing, their spectra will gradually become visible.

The flame spectrum of rubidium consists of two red lines at the extreme left end of the spectrum, 3 orange lines rather close together just to the left of the sodium line, and a narrow and a broad line between the violet and indigo portion of the spectrum.

Caesium has one narrow orange line, one broad yellow line just left of the sodium line, one green line, and two broad blue lines close together.

The spectra of rubidium and caesium are shown on Plate II.

ESTIMATION

The estimation of sodium and potassium is required in the analysis of rocks, clays, soils, ashes of plants, waters, brines, saline deposits, salts of the alkalies, many technical products, and in other cases. The determination of potassium is of special importance in the analysis of fertilizers. The estimation of lithium is desired in the analysis of lithium minerals, frequently in mineral waters, occasionally in rocks, and in certain other special cases. The estimation of rubidium and caesium is seldom required.

Preparation and Solution of Sample

Procedure for Rocks and Other Insoluble Mineral Products. For silicate rocks and other silicious material, bring the alkalies into solution, according to the J. Lawrence Smith or the hydrofluoric acid method, as directed on page 416. For products which are dissolved by hydrochloric acid, effect the solution by acid digestion, expel the excess of acid by evaporation, and remove other constituents as directed under Separations.

Procedure for Soils.[1] Digest 10 grams of moisture-free soil with 100 cc. of hydrochloric acid of a constant boiling-point (sp.gr. 1.115) in a 300-cc. Erlenmeyer flask fitted with a ground-glass or rubber stopper and a reflux condenser. Digest continuously for ten hours on the steam bath, shaking the flask every hour. After settling, decant the solution into a porcelain dish. Wash the insoluble residue onto a filter with hot water, and continue the washing until free from chlorides, adding the washings to the original solution for evaporation. Oxidize the organic matter present in the solution with a few drops of nitric acid and evaporate to dryness on a water bath. Moisten with hydrochloric acid and dissolve in hot water and evaporate a second time to complete dryness and until the excess of hydrochloric acid is completely removed. Moisten the cooled residue with strong hydrochloric acid and dissolve in hot water. Filter into a 250-cc. graduated flask, wash free from chlorides, and dilute to the mark. Use an aliquot of 100 cc. for the determination of the alkalies.

Procedure for Fertilizers.[2] *Potash salts.* Boil 10 grams of the sample with 300 cc. of water for thirty minutes, wash into a 500-cc. graduated flask, cool, dilute to the mark, mix and pass through a dry filter. Determine the potassium in a 25-cc. aliquot representing 0.5 gram of the original substance, according to either the modified chloroplatinate or the Lindo-Gladding method.

Mixed fertilizers. Boil 10 grams of the sample with 300 cc. of water for thirty minutes, and wash into a 500-cc. graduated flask. Add to the hot solution a slight excess of ammonia and sufficient ammonium oxalate to precipitate all the lime, cool, dilute to the mark, mix, and pass through a dry filter. Evaporate

---

[1] U. S. Dept. Agr., Bu. of Chem., Bull. **107** (revised), 14, 1907.
[2] U. S. Dept. Agri., Bu. Chem. Bull. 107 (revised), 11, 1907.

50 cc. of the filtrate to dryness and ignite gently to remove ammonium salts. Dissolve in water, filter, and determine the potassium according to the modified chlorplatinate [1] or the Lindo-Gladding method.

Organic compounds. When it is desired to determine the total potash in organic substances such as cottonseed meal, tobacco stems, etc., saturate 10 grams with strong sulphuric acid, and ignite in a muffle at low red heat to destroy organic matter. Add a little strong hydrochloric acid, warm slightly to loosen the mass from the dish, dissolve in water, filter, and determine the potassium according to the modified chlorplatinate or the Lindo-Gladding method.

If for any reason it is desired to use either the chloroplatinate or the perchlorate method in the determination of potassium, interfering substances, including sulphates, must first be removed from the solution.

Procedure for Ashes of Plants. Boil 20 grams of the sample with 300 cc. of water for thirty minutes, filter into a 500-cc. flask, and wash the residue thoroughly with hot water. Cool, dilute to the mark and mix. Take aliquots for the determination of the alkalies. The solution may also be prepared by digestion with hydrochloric acid.[2] This treatment is preferable when all the constituents of the ash are to be determined.

Procedure for Saline Residues, Soluble Salts, Brines, etc. In the case of water-soluble products, the convenience of the analyst usually determines the manner of preparing the solution. Usually it is preferable to weigh out a convenient portion, to make up the solution to definite volume, and to take an aliquot for each determination. As a general rule, a sample should be taken sufficient to give about a half gram of solids. Strong brines should be weighed and not measured.

SEPARATIONS

Separation of the Alkali Metals from other Constituents

Separation from the Hydrogen Sulphide and Ammonium Sulphide Groups of Metals

The alkali metals are usually weighed as chlorides or sulphates, and in general before undertaking their determination, all other bases and acids must first be separated from them. The hydrogen sulphide and the ammonium sulphide groups of metals are seldom to be found in solutions in which the determination of the alkalies is desired. If these are present, however, they may be readily precipitated by means of hydrogen sulphide and ammonium sulphide as detailed on pages 168 and 292.

Separation from Silica

Acidify the solution with hydrochloric acid and evaporate it in a platinum or porcelain dish on the water bath until the odor of hydrochloric acid in the dry residue can no longer be detected. Break up the dry mass with a platinum or glass rod, cool, moisten with a minimum amount of concentrated hydrochloric

---

[1] If this method is used, it will not be necessary to remove the calcium by addition of ammonia and ammonium oxalate.

[2] Lunge, "Technical Methods of Analysis," 2, 456, 1911. D. Van Nostrand Co., New York.

acid, dissolve in a small quantity of water, filter and wash the residue free from chlorides. In the presence of much silica, repeat the operation.

Separation from Iron, Aluminum, Chromium, Titanium, Uranium, Phosphoric Acid, etc.

If phosphoric acid is present in amounts insufficient to combine with all the iron, alumina, etc., or is absent altogether, heat the solution to boiling, add a few drops of nitric acid to oxidize the iron, add gradually an excess of ammonia, boil for a minute or so, allow the precipitate to settle, and filter. Wash the precipitate free from chlorides with hot water.

If phosphoric acid is present in the solution in excess of that required to combine with the iron, alumina, etc., heat the solution to boiling, oxidize with nitric acid, add a slight excess of ferric chloride solution, and precipitate with ammonia as described above.

When the precipitate is considerable, it should be dissolved in hydrochloric acid, and the precipitation repeated.

If chromates are present, these must first be reduced to the chromic salt. For this purpose, add 10 to 15 cc. of hydrochloric acid and a small amount of alcohol to the solution and heat on the water bath or hot plate for a few minutes. Heat to boiling and precipitate with ammonia as directed above. The reduction may also be done by boiling with sulphurous acid.

Separation from Sulphates

Precipitate the sulphate radical as $BaSO_4$ by the addition of a slight excess of barium chloride to the hot solution as directed on page 497 for the determination of SO_4. Remove the excess of barium chloride by addition of ammonia and ammonium carbonate.

The two operations may be combined as follows: Add a slight excess of barium chloride to the hot solution and boil for a few minutes. Then, without filtering off the $BaSO_4$, add an excess of ammonia and ammonium carbonate, allow the precipitate to settle, filter, and wash free from chlorides.[1]

Separation from Barium, Calcium and Strontium

To the not too concentrated solution, add a slight excess of ammonia and ammonium carbonate, heat to boiling, allow the precipitate to settle, filter and wash the residue a few times with hot water. Dissolve the precipitate, which is likely to contain small amounts of the alkalies, in a little dilute hydrochloric acid, and repeat the precipitation with ammonia and ammonium carbonate. Filter and wash the residue. Evaporate the combined filtrates to dryness in a platinum or porcelain dish and ignite cautiously at a very faint red heat to remove ammonium salts. Dissolve the residue in a little water, add a few drops of ammonia, ammonium carbonate, and ammonium oxalate, and allow to stand for several hours in order to precipitate the last traces of the alkaline earths. Filter and wash the residue free from chlorides.

[1] This procedure does not give faultless results as some of the potassium is carried down with the precipitate and lost.

Separation from Iron, Aluminum, Chrominum, Barium, Calcium, Strontium, Phosphates, Sulphates, etc., in One Operation

To the hot solution add a slight excess of barium chloride and boil for a few minutes. Then, without filtering off the $BaSO_4$, add an excess of ammonia and ammonium carbonate, heat to boiling, and allow the precipitate to settle. Filter and wash free from chlorides with hot water. After evaporating the filtrate to dryness, removing the ammonium salts by ignition, and dissolving the residue in a little water, precipitate the last traces of barium and calcium by addition of a few drops of ammonia, ammonium carbonate, and ammonium oxalate. By this procedure a small portion of the alkalies is retained by the precipitate and lost.

Separation from Boric Acid

Acidify the solution strongly with hydrochloric acid and evaporate to dryness. Stir up the residue with 15 to 20 cc. of pure methyl alcohol and cautiously evaporate on a water bath at not too high a temperature. Moisten the residue with a drop or two of concentrated hydrochloric acid, add 15 cc. of methyl alcohol, and again take to dryness. Repeat the evaporation with methyl alcohol a third time. This should be ample for the complete removal of half a gram of B_2O_3.

Separation from Magnesium

Mercuric Oxide Method.[1] After removing other bases and acids, evaporate the solution of the chlorides to dryness, expel ammonium chloride by gentle ignition, and dissolve the residue—except for the small amount of magnesium oxide present—by warming with a little water. Add an excess of mercuric oxide in the form of a thin paste prepared by shaking up freshly precipitated mercuric oxide in water. Evaporate the mixture to complete dryness on the water bath with frequent stirring, dry thoroughly and ignite gently at first and then more strongly until all the mercuric chloride present has been volatilized. (Be careful not to inhale the fumes.) The whole of the unchanged mercuric oxide need not be expelled by ignition. Digest the residue, composed of the excess of mercuric oxide, the precipitated magnesium oxide, and the alkali chlorides, with a small amount of hot water, filter rapidly, and wash with successive portions of hot water, first by decantation and then on the filter, but do not prolong the operation unnecessarily. If desired, determine the magnesium in the residue by expelling the mercuric oxide by ignition and weighing the magnesium oxide. Acidify the filtrate, which contains the alkalies, with hydrochloric acid, evaporate to dryness, gently ignite, cool and weigh. If the residue contains a small amount of magnesium, as it usually does, determine the magnesium in an aliquot and apply the necessary correction. The mercuric oxide should be tested for alkalies by volatilizing a portion and testing the residue.

The Barium Hydroxide Method.[1] Evaporate the solution, which may contain chlorides, sulphates or nitrates, to dryness and gently ignite to remove ammonium salts. Warm the residue with a small amount of water and treat the hot neutral solution so obtained with baryta water until no more precipitate is formed and barium hydroxide remains in slight excess. Boil, filter and wash

---

[1] Fresenius, " Quantitative Chemical Analysis," **1**, 610, 1908, John Wiley & Sons, New York.

the precipitate with hot water. If desired, determine the magnesium in the residue. Treat the filtrate, which contains the alkalies, barium and a trace of magnesium, with an excess of ammonia and ammonium carbonate to remove the barium. Acidify the filtrate with hydrochloric acid and evaporate to dryness, ignite and weigh. This residue will contain a small amount of magnesium which may be determined in an aliquot and a correction applied.

Remark. The barium hydroxide method is applicable in the presence of lithium.

The Ammonium Phosphate Method.[1] To the hot solution, add an excess of ammonia and ammonium chloride, and precipitate the magnesium by adding a slight excess of ammonium phosphate. Allow the mixture to stand an hour or so, filter and wash the residue with 2% ammonia solution. Expel most of the free ammonia from the filtrate by evaporation, acidify very slightly with hydrochloric acid, and add an excess of ferric chloride solution, which should color the solution slightly yellow. Neutralize the solution with ammonium carbonate, heat to boiling, and filter off the basic ferric phosphate, washing the residue with hot water. Evaporate the filtrate to dryness, ignite to expel ammonium salts, and determine the alkalies in the residue. Magnesium may also be separated by precipitation as magnesium ammonium arsenate [2] or magnesium ammonium carbonate.[1]

Separation of the Alkali Metals from One Another

Separation of Sodium from Potassium

After weighing the sodium and potassium together as chlorides, dissolve the residue in water and precipitate the potassium as chloroplatinate or perchlorate according to one of the methods detailed under Determination of Potassium.

Separation of Lithium from Sodium and Potassium

Extract the dry chlorides with amyl alcohol as prescribed under the Gooch method, or with alcohol saturated with hydrochloric acid gas as detailed under the Rammelsberg method.

Separation of Lithium and Sodium from Potassium, Rubidium, and Caesium

Precipitate the potassium, rubidium, and caesium as chloroplatinates as described under the chloroplatinate method for the estimation of potassium. Evaporate the filtrate to dryness and ignite gently with a little oxalic acid to reduce the platinum, or else dissolve the residue in water and pass a current of hydrogen through the hot solution to reduce the platinum. In any case, filter off the reduced platinum and determine lithium and sodium in the filtrate.

[1] Fresenius, op. cit.
[2] Browning and Drushel, Am. J. Sci. (4), **23**, 293, 1907.

METHODS FOR DETERMINATION OF SODIUM

Determination as Sodium Chloride

Sodium is commonly weighed as NaCl when it is already present as such or after conversion of other forms into the chloride. In the case of salts of volatile acids, such as nitrates for instance, the transformation is made by evaporating the solution to dryness with hydrochloric acid repeatedly or until only the chloride remains. When the sodium is present as a salt of a non-volatile acid, the latter is removed and the transformation effected according to the methods under Separations.

Usually the solution in which sodium chloride is to be determined will contain ammonium salts from some previous operation. In such cases, proceed as follows: Evaporate the sodium chloride solution, which must contain no other non-volatile substance, in a platinum dish to complete dryness on the water bath. Cover the dish with a watch-glass, and cautiously dry the residue in an air bath at 110 to 130° C. Make sure that no loss of sodium chloride is sustained by decrepitation during drying and subsequent ignition. Heat the dish and contents over a free flame held in the hand and moved back and forth under the dish in order to remove ammonium salts. But to avoid loss of sodium chloride by volatilization, take care not to heat the dish to more than a faint redness in any one spot and not to raise the temperature of the salt above incipient fusion. Cool the residue, dissolve it in a little water, and filter from the carbonaceous matter into a weighed platinum dish. Acidify the filtrate with hydrochloric acid and evaporate it to dryness on the water bath. Dry the residue at 100 to 130° C. in an air bath, ignite cautiously over a free flame, taking the precautions mentioned above to prevent loss of sodium chloride, cool in a desiccator, and weigh.

Determination as Sodium Sulphate

Sodium is often determined by weighing as Na_2SO_4 when it is present as such or after conversion of other forms into the sulphate. In the case of salts of volatile acids, the change into the sulphate is made by simply evaporating the solution with a slight excess of sulphuric acid. With salts of non-volatile acids, the transformation is effected according to the methods under Separations. When the sodium is present as an organic salt, the substance is moistened with concentrated sulphuric acid and carefully heated over a free flame until fumes cease to come off. The residue is dissolved in water and filtered from the carbonaceous matter.

As a rule the solution in which sodium sulphate is to be determined will contain an excess of sulphuric acid. In such cases, evaporate the solution to dryness in a weighed platinum dish, and cautiously ignite the dry residue until fumes cease to come off. Cool, add a lump of ammonium carbonate to the contents of the dish, and ignite a second time at dull red heat until no more fumes are given off. Cool in a desiccator and weigh as Na_2SO_4. Repeat the ignition with the addition of ammonium carbonate until a constant weight is obtained.

In case an excess of sulphuric acid is not present, evaporate the solution to dryness in a weighed platinum dish, ignite, cool in a desiccator and weigh as Na_2SO_4.

Determination by Difference

Ordinarily sodium and potassium are weighed together as chlorides or sulphates as detailed above for sodium. Potassium is then determined by one of the methods given below, and the value for sodium obtained by difference.

METHODS FOR DETERMINATION OF POTASSIUM

Determination as Potassium Chloride or Potassium Sulphate

Potassium may be weighed as chloride or sulphate. The procedure is the same as that described for sodium. Observe, however, that the potassium salts are a little more volatile than the corresponding sodium salts, so that greater care must be taken not to lose potassium by volatilization.

The Chloroplatinate Method

Application. This method is applicable in the presence of the chlorides of sodium, lithium, magnesium, calcium, and strontium.

Principle. Potassium chloroplatinate is practically insoluble in strong alcohol while the other chloroplatinates are readily soluble.

Procedure. Treat the aqueous solution of the alkali chlorides contained in a small porcelain dish with slightly more than enough chloroplatinic acid to convert all the chlorides present into the corresponding chloroplatinates. The chloroplatinic acid solution should contain the equivalent of 1 gram of platinum in each 10 cc.[1] Evaporate the solution on the steam bath to a syrupy consistency, i.e., until solidification occurs on cooling. Flood the cooled residue with a small quantity of alcohol of at least 80% strength, grind thoroughly with a pestle made by enlarging the end of a glass rod, and allow to stand one-half hour. Pour the liquid through a previously weighed Gooch crucible containing an asbestos mat, and before adding more alcohol, rub up the residue again with the glass pestle. Now continue the washing by decantation with small portions of alcohol until the wash liquid becomes colorless. Transfer the precipitate to the crucible and wash two or three times with alcohol. Dry at 130° C., cool in a desiccator, and weigh. Repeat the drying until a constant weight is obtained. Multiply the weight of K_2PtCl_6 by 0.161 to obtain the weight of K; by 0.194 to obtain K_2O; and by 0.307 to obtain KCl.

Remarks. This method is considered to be the most accurate known for the estimation of potassium. Care should be taken not to conduct the evaporation at too high a temperature nor let it go too far, as this may cause the formation of anhydrous sodium chloroplatinate, which dissolves slowly in alcohol. Too large a volume of alcohol for washing should be avoided, as K_2PtCl_6 is slightly soluble in alcohol, especially that of 80%. For this reason 95% alcohol is preferable for the washing.

Instead of using a Gooch crucible, the precipitate may be filtered on paper, dried, washed through the filter with hot water into a weighed platinum dish, evaporated to dryness, and heated at 130° C. to constant weight.

[1] For methods of preparing chloroplatinic acid from scrap platinum and from platinum residues, see Precht, Z. Anal. Chem., **18**, 509, 1879; Vogel and Haefcke, Landw. Vers. Sta., **47**, 134, 1896.

The Modified Chloroplatinate Method [1]

Application. The method is applicable in the presence of chlorides, sulphates, phosphates, nitrates, carbonates, borates and silicates, salts of sodium, barium, calcium, strontium, magnesium, iron and alumina, and is especially suited for the estimation of potassium in salines, potassium salts, and fertilizers in which only the potassium is desired.

Principle. On evaporating a solution containing potassium with a slight excess of chloroplatinic acid, the potassium is completely transformed into potassium chloroplatinate which is insoluble in strong alcohol, while any of the other chloroplatinates which may be formed are either dissolved or decomposed by alcohol, so that the excess of chloroplatinic acid may be readily removed. After dissolving the K_2PtCl_6 along with any other soluble salts contained in the residue in hot water, the platinum is precipitated from the solution by magnesium, and from the weight of platinum so obtained, the amount of potassium present is calculated.

Procedure. To the solution slightly acidified with hydrochloric acid, add chloroplatinic acid solution slightly in excess of that necessary for the complete precipitation of the potassium present and evaporate the solution on the steam bath to a syrupy consistency, i.e., until solidification occurs on cooling. Flood the cooled residue with a small quantity of alcohol of at least 80% strength, grind thoroughly with a pestle made by enlarging the end of a glass rod, and allow to stand one-half hour. The alcoholic solution should be colored if an excess of chloroplatinic acid has been used. Pour the liquid through a small filter, using suction if desired, and before adding more alcohol, rub up the residue again with the pestle. Now continue the washing by decantation with small portions of alcohol until the wash liquid becomes colorless. Three or four washings usually suffice. Transfer the precipitate to the filter and wash two or three times with alcohol.

Dissolve the precipitate of K_2PtCl_6 along with any other soluble salts present in hot water, washing it through the filter into a beaker of convenient size. To the hot solution add about 4 cc. of concentrated HCl and approximately 0.5 gram magnesium ribbon pressed into the form of a ball for every 0.2 gram potassium present, stirring the solution and holding the magnesium at the bottom of the beaker by means of a glass rod. A lump of stick magnesium weighing about 0.4 gram is preferable to the ribbon. When the action has practically ceased, add a few cc. of hydrochloric acid and allow the fluocculent platinum to settle, preferably by allowing the beaker to set for an hour on the hot plate. The supernatant liquid should be perfectly clear and limpid like water if reduction is complete. To make sure, add more magnesium, in which case the solution will darken if reduction be incomplete. To the completely reduced solution, add concentrated hydrochloric acid, and boil to dissolve any basic salts, filter on paper or a Gooch, wash thoroughly with hot water, ignite in platinum or porcelain and weigh. $Pt \times .4006 = K$ or $\times .4826 = K_2O$.

Remarks. If the solution contains very large amounts of iron, alumina, or silica, it is perferable to remove the greater part of these before proceeding to the determination of potassium. Care should be taken to insure the complete removal of the soluble chloroplatinates from the residue without the use of an

[1] Hicks, J. Ind. Eng. Chem., 5, 650, 1913.

excessive amount of alcohol, and also that the subsequent reduction of the potassium chloroplatinate with magnesium be complete.

Lindo=Gladding Method [1]

Application. This method is applicable in the presence of chlorides, sulphates, and phosphates of the alkalies and magnesium.

Principle. The potassium is precipitated as K_2PtCl_6, and the soluble chloroplatinates removed by washing with 80% alcohol. The impurities in the precipitate are then washed out by a strong solution of ammonium chloride saturated with K_2PtCl_6, and the wash solution is removed by again washing with alcohol. The purified K_2PtCl_6 is finally dried and weighed.

Procedure. To the solution, slightly acidified with hydrochloric acid, add an excess of chloroplatinic acid solution, and evaporate on the water bath to a thick paste. Treat the residue with 80% alcohol, avoiding the absorption of ammonia. Wash the precipitate thoroughly with 80% alcohol both by decantation and on the filter, continuing the washing after the filtrate is colorless. Wash finally with 10 cc. of ammonium chloride solution prepared as follows: Dissolve 100 grams of pure ammonium chloride in 500 cc. of water, add from 5 to 10 grams of potassium chloroplatinate, and shake at intervals of six to eight hours. Allow the mixture to settle over night and filter. Repeat the washing with successive portions of the ammonium chloride solution five or six times in order to remove the impurities from the precipitate. Wash again thoroughly with 80% alcohol, dry for thirty minutes at 100° C. and weigh as K_2PtCl_6. The precipitate should be perfectly soluble in water. Multiply the weight of K_2PtCl_6 by 0.161 to obtain the weight of K; by 0.194 to obtain K_2O; and by 0.307 to obtain KCl.

The Perchlorate Method [2]

Application. This method is applicable in the presence of chlorides and nitrates of barium, calcium, magnesium and the alkali metals, and also in the presence of phosphates. Sulphates should not be present.

Principle. The separation depends on the insolubility of potassium perchlorate, and the solubility of sodium and other perchlorates in 97% aclohol.

Procedure. To the neutral or slightly acidified solution, add twice as much perchloric acid as is required to convert all the bases present into perchlorates and evaporate on the water bath with stirring to a syrupy consistency. Add a little hot water and continue the evaporation with constant stirring until all the hydrochloric acid is expelled and heavy fumes of perchloric acid are given off. Avoid excessive loss of perchloric acid. Stir up the cooled mass thoroughly with 20 cc. of 97% alcohol to which 0.2% perchloric acid has been added, but avoid breaking up the potassium perchlorate crystals too finely or else they may pass through the filter. Allow the mixture to settle, and decant the alcohol off through a Gooch crucible. Repeat the washing once by decantation and then warm to remove the alcohol. Dissolve the residue in hot water, add about a half gram of perchloric acid and evaporate again until fumes of perchloric acid are given off. Wash the residue once by decantation and then several times on the filter. Remove the adhering wash-liquid by washing with pure 97% alcohol, dry at 130° C., and weigh. Multiply the weight of $KClO_4$ by 0.2825 to obtain the weight of K; by 0.3402 to obtain K_2O.

[1] U. S. Dept. Agri., Bu. Chem. Bull. 107 (revised), **11**, 1907.
[2] Wense, Zeit. Angew. Chem., 691, 1891; 233, 1892. Caspari, Zeit. Angew, Chem., 68, 1893.

Other Methods

Among the more important of other methods which have been proposed and used for the determination of potassium, may be mentioned the cobaltinitrite method,[1] which has been studied by the Association of Official Agricultural Chemists and considered to be unreliable;[2] the bitartrate method;[3] the colorimetric method;[4] and the spectroscopic method.[5]

Determination of Sodium and Potassium by Indirect Method

After removing all other constituents, weigh the sodium and potassium as chlorides. Dissolve the weighed residue in water and determine the chlorine gravimetrically by precipitation as AgCl or volumetrically by titration with standard silver nitrate (potassium chromate indicator). From the weight of the combined salts and the weight of the chlorine, calculate the amount of sodium and potassium as follows:

Let
$$x = \text{weight of NaCl} + \text{KCl}\cdot$$
$$y = \text{weight of Cl.}$$
Then
$$\text{Na} = 3.004y - 1.428x;$$
$$\text{K} = 2.428x - 4.004y.$$

The method is satisfactory when sodium and potassium are present in about equal quantities.

Determination of Magnesium, Sodium and Potassium in the Presence of One Another

In the usual course of analysis, magnesium, sodium and potassium are separated as chlorides from all other constituents. Instead of going through the tedious process of separating the magnesium from the alkalies, the magnesium, sodium, and potassium may be accurately determined in the presence of each other as follows:

Treat the solution containing these constituents with slightly more than enough sulphuric acid to convert all three bases into sulphates, evaporate it to dryness on the water bath, and ignite gently at first and then at dull red heat to break up bisulphates and expel the excess of sulphuric acid. To hasten the decomposition of the bisulphates, cool, add a lump of ammonium carbonate, and heat a second time. Cool in a desiccator and weigh. Repeat the heating with the addition of ammonium carbonate until a constant weight is obtained. Dissolve the residue in water and dilute to definite volume. Determine the potassium in one portion according to one of the methods described above, and the magnesium in a second portion as described on page 293. Deduct the weight of magnesium and potassium sulphates from the weight of the combined sulphates to obtain the amount of sodium sulphate.

[1] Addie and Wood, J. Chem. Soc., **77**, 1076, 1900; Drushel, Am. J. Sci. (4), **24**, 433, 1907; **26**, 329, 555, 1908; Bowser, J. Am. Chem. Soc., **33**, 1752, 1911.
[2] U. S. Dept. Agri. Bu., Chem., Bull. 132, 137, 152, 159.
[3] Bayer, Chem. Zeit., **17**, 686, 1893.
[4] Cameron and Failyer, J. Am. Chem. Soc., **25**, 1063, 1903; Hill, J. Am. Chem. Soc., **25**, 990, 1903.
[5] Gooch and Hart, Am. J. Sci. (3), **24**, 448, 1891.

34

METHODS FOR DETERMINATION OF LITHIUM

Determination as Lithium Chloride

Lithium may be weighed as LiCl. The procedure is practically the same as that described for sodium, but since lithium chloride is very hygroscopic, this salt must be weighed out of contact with the air. For this purpose the lithium chloride is ignited in a platinum crucible, cooled in desiccator, and the crucible and contents weighed in a glass-stoppered weighing bottle.

Determination as Lithium Sulphate

Lithium is weighed preferably as Li_2SO_4. The procedure is the same as that described for sodium, but since lithium bisulphate is easily broken up on heating, it is not necessary to ignite with ammonium carbonate.

The Gooch Method [1]

Principle. Lithium chloride is readily soluble in amyl alcohol, while sodium and potassium chlorides are not.

Procedure. Concentrate the solution as far as possible by evaporation, transfer it to a 50-cc. Erlenmeyer flask, add a small amount of amyl alcohol and heat cautiously on an asbestos plate until the water has been expelled and the boiling-point of the solution rises to about that of pure amyl alcohol (132° C.). To prevent bumping during this treatment, pass a current of dry air through the solution. When all the water has been removed, the sodium and potassium chlorides, together with some LiOH will separate from the solution. Decant the solution through a filter and wash the residue several times with hot amyl alcohol. Moisten the residue with dilute hydrochloric acid, dissolve in a little water and repeat the extraction with amyl alcohol. . If much lithium chloride is present, it will be necessary to repeat the extraction with amyl alcohol three or four times. Evaporate the combined filtrates and washings to dryness and dissolve in a little dilute sulphuric acid. Filter from the carbonaceous matter into a weighed platinum dish, evaporate to dryness, and remove the excess of sulphuric acid by gentle heating. Ignite the residue at dull redness, cool in a desiccator, and weigh as Li_2SO_4.

Remarks. For very accurate work, account must be taken of the fact that the lithium sulphate obtained according to the procedure just described always contains small amounts of potassium and sodium sulphates, if these metals were originally present. To correct for this, deduct 0.00041 gram for every 10 cc. of the filtrate exclusive of the washings in case only sodium chloride was present, or 0.00051 if only potassium chloride was present, and 0.00092 if both sodium and potassium chlorides were present.

[1] Proc. Am. Acad. Arts. Sci., **22** (N. S. 14), 177, 1886.

The Rammelsberg Method [1]

Principle. Anhydrous lithium chloride is soluble in equal parts of alcohol and ether which have been saturated with hydrochloric acid gas, while the chlorides of sodium and potassium are practically insoluble in this mixture.

Procedure. Evaporate the solution of the chlorides to dryness in a small flask provided with a two-hole stopper. During the evaporation, pass a current of dry air through the flask. Place the flask containing the dry residue in an oil or air bath and heat for half an hour at 140 to 150° C., during which time pass dry hydrochloric acid gas through the flask. Cool in a current of hydrochloric acid gas, treat the residue with a few cc. of absolute alcohol which has been saturated with hydrochloric acid gas, and add an equal volume of absolute ether. Close the flask tightly and allow it to stand with frequent shaking for twelve hours. Pour the solution through a filter, wet with the alcohol-ether mixture and wash the residue three times by decantation with the alcohol-ether mixture. Add a few more cc. of the alcohol-ether saturated with hydrochloric acid gas to the contents of the flask and allow to stand again for twelve hours. Pour the liquid through a filter, and wash the residue by decantation with the alcohol-ether mixture until the residue tested in the spectroscope shows the complete absence of lithium. Carefully evaporate the combined alcohol-ether extract to dryness on a lukewarm water bath. Dissolve the residue in sufficient dilute sulphuric acid to convert all the lithium into the sulphate, transfer the solution to a weighed platinum dish, evaporate to dryness on the water bath, and finally ignite gently. Cool the residue in a desiccator and weigh as lithium sulphate.

NOTE. Lithium may also be precipitated and weighed as Li_3PO_4,[2] or it may be precipitated as LiF [3] and then changed into the sulphate and weighed.

Spectroscopic Method [4]

Dissolve the lithium salt containing small amounts of sodium and potassium resulting from the separation by the Gooch or Rammelsberg methods in 5 or 10 cc. of water, depending on the amount of lithium present. Gradually add measured amounts of this solution to a known volume of water—testing the solution from time to time in the spectroscope—until the lithium line appears. When only traces of lithium are present, it is better to dissolve the lithium salt in a little water and dilute to the vanishing point of the lithium line. Make the spectroscopic examination as follows: Prepare a loop by winding a platinum wire four times around a No. 10 wire. Plunge the loop into the solution, and remove with the axis parallel to the surface of the water. Evaporate the drop to dryness carefully, ignite in the Bunsen flame, and observe through a good spectroscope.

Before undertaking the determination, standardize the instrument and platinum loop by carrying out the determination with known amounts of lithium.

The method gives satisfactory results when only an approximation is desired. For weighable amounts of lithium, the Gooch method is preferable.

[1] Treadwell, "Analytical Chemistry," **2**, 55, 1911. John Wiley & Sons, N. Y.
[2] Mayer, Ann. Chem. Pharm., **98**, 193, 1856. Merling, 3 Anal. Chem., **18**, 563, 1879.
[3] Carnot, 3 Anal. Chem., **29**, 332, 1890.
[4] Skinner and Collins, U. S. Dept. Agri. Bu. Chem., Bull. 153. A good bibliography is included in this bulletin.

Determination of Sodium, Potassium, and Lithium in the Pres= ence of One Another

Weigh the combined bases as sulphates, observing the precautions detailed under Determination of Sodium, dissolve in water and dilute to definite volume. In one portion determine the potassium and in a second portion determine the lithium by the Gooch or Rammelsberg method. Obtain the value for the sodium by difference.

Determination of the Alkalies in Silicates

J. Lawrence Smith method [1]

Principle. By heating the substance with 1 part ammonium chloride and 8 parts calcium carbonate, and leaching the sintered mass with water, the alkalies are obtained in solution in the form of chlorides along with some calcium, while the remaining metals are for the most part left behind as insoluble oxides, and the silica is changed to calcium silicate.

Procedure. Triturate 0.5 gram of the finely powdered mineral with an equal quantity of pure ammonium chloride in an agate mortar, add 3 grams of precipitated calcium carbonate [2] and mix intimately with the former. Transfer the mixture to a platinum crucible (preferably the J. Lawrence Smith alkali crucible), rinse the mortar with 1 gram of calcium carbonate and add to the contents of the crucible. Place the covered crucible in a slightly inclined position with the top protected from the heat of the flame This can be done by setting the crucible in a hole in a cylinder of fire clay, as shown in Fig. 56. Gradually heat the crucible over a small flame until no more ammonia is evolved, but avoid heating sufficiently to cause the evolution of ammonium chloride. This should require about fifteen minutes. Then raise the temperature until finally the lower three-fourths (and no more) of the crucible is brought to a red heat, and maintain this temperature for one hour. Allow the crucible to cool and remove the sintered cake by gently tapping the inverted crucible. Should this not be possible, digest the mass a few minutes with water to soften the cake, and then wash it into a large porcelain or platinum dish. Heat the covered dish with 50 to 75 cc. of water for half an hour, reduce the large particles to a fine powder by rubbing

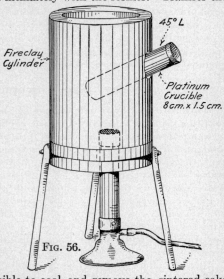

FIG. 56.

---

[1] Am. J. Sci. (3), **1**, 269, 1871; Hillebrand, U. S. Geol. Survey Bull. 422, 171, 1910.
[2] Blank determinations should be run on the calcium carbonate, and corrections made for its alkali content.

with a pestle in the dish, and decant the clear solution through a filter. Wash the residue four times by decantation, transfer it to a filter, and wash with hot water until a few cc. of the washings give only a slight turbidity with silver nitrate. To make sure the decomposition of the mineral has been complete, treat the residue with hydrochloric acid. No trace of undecomposed mineral should remain undissolved.

The aqueous extract obtained in the above operation contains the chlorides of calcium and the alkalies. To remove the calcium, treat the solution with ammonia and ammonium carbonate, heat to boiling, filter and wash the residue. As this precipitate invariably retains some alkali salts, it should be dissolved in hydrochloric acid and the precipitation repeated. Evaporate the filtrate to dryness in a platinum or porcelain dish, and expel the ammonium salts by gentle ignition over a moving flame. After cooling, dissolve the residue in a little water, and add a few drops of ammonia and ammonium oxalate to remove the last trace of calcium. After standing several hours, filter off the calcium oxalate, receive the filtrate in a weighed platinum dish, evaporate to dryness and ignite gently to remove ammonium salts. Moisten the cooled mass with hydrochloric acid to transform any carbonate into chloride, and again evaporate to dryness and ignite. Cool in a desiccator and weigh the combined chlorides. Dissolve in water, and if an insoluble residue remains, filter off, weigh and deduct from the weight of the chlorides. Determine the potassium by one of the methods already described, and obtain the value for sodium by difference.

The Hydrofluoric Acid Method

Procedure. Weigh out a half-gram sample of the finely powdered mineral, place in a small platinum dish or a large platinum crucible, moisten with a few drops of water, and add about 10 cubic centimeters of hydrofluoric acid and about 2 cubic centimeters of dilute sulphuric acid. Hold the dish with the tongs and heat it and its contents cautiously over the free flame for a few minutes. Evaporate the solution on the steam bath to complete dryness, and during this evaporation agitate the solution occasionally by giving the dish a gyratory motion with the tongs. This treatment is nearly always sufficient to decompose the mineral, but if decomposition seems incomplete add more hydrofluoric acid and evaporate a second time, agitating the mixture frequently. With refractory silicates it may be desirable to digest the mineral with the acid mixture in a covered crucible for an hour or so. Heat the dish and its contents on a radiator[1] to expel the excess of hydrofluoric and sulphuric acids. Cover the residue with dilute hydrochloric acid and digest on a steam bath until solution is complete, adding more water and acid if necessary. Flakes of organic matter which remain undissolved may be neglected. Remove the iron, alumina, calcium, etc., by the methods described under "Separations." Finally weigh the magnesium, sodium, and potassium as sulphates as described on page 413 and determine the potassium according to the modified chloro-platinate or Lindo-Gladding method. An alternative procedure is to remove sulphates and magnesium also as described under " Separations," weigh the alkalies as chlorides, and determine the potash by any of the methods already described. It should be noted however that potash is carried down with barium sulphate, and the loss thus occasioned may be appreciable.

[1] Hillebrand, W. F., U. S. Geol. Survey Bull. 422, p. 31, 1910 (reprinted, 1916). A large porcelain crucible serves well as a radiator for this purpose.

Test for Hydroxide. A small amount of the material dissolved in CO_2 free water is treated with an excess of $BaCl_2$ solution and filtered. Phenolphthalein added to the filtrate will produce the characteristic red color if an alkali hydroxide is present in the sample.

Sodium Chloride, Table Salt

U. S. Standard Specifications. Table salt must contain not over

Barium chloride...............................0.05%
Calcium and magnesium chlorides.................0.5%
Calcium sulphate..............................1.4%
Water insoluble matter........................0.1%

In addition to these impurities, salt frequently contains combined Fe_2O_3, Al_2O_3, K_2O, P_2O_5, CO_2.

Outline of Method for Examination

Moisture. A 10 gram sample is dried to constant weight at 100° C. and the loss of weight noted.

Insolubles in Water. A fifty gram sample is dissolved in 200 cc. of water and filtered on a weighed, dry filter paper. The residue is washed five times with small portions of water and then dried at 100° C. for an hour. The increased weight = water insolubles. It is advisable to run side by side a second filter, which is washed and also dried in the same way, the paper being used as a tare weight.

The filtrate from the insoluble matter is diluted to exactly 500 cc. in a volumetric flask and used for the following tests.

Sulphur Trioxide. 100 cc. of the above solution equivalent to 10 grams of sample is examined for SO_3 by addition of $BaCl_2$ sol. and the $BaSO_4$ filtered off and weighed.

Barium. If SO_3 is present above, Ba will not be found in the solution. If SO_3 is absent, test for Ba by addition of H_2SO_4 to 100 cc. of the solution, by the standard procedure. $BaSO_4 \times 0.8923 = BaCl_2$.

Lime. After removal of Fe_2O_3 and Al_2O_3 by addition of NH_4OH (100 cc. solution) CaO is determined by precipitation as oxalate by regular method.

Magnesia. MgO is determined in the filtrate from CaO by concentrating this and precipitating phosphate salt as usual.

P_2O_5. This is determined by precipitation as ammonium phosphomolybdate according to the standard procedure. $(NH_4)_2HPO_4.12MoO_3 \times .038 = P_2O_5$.

Fe_2O_3 and Al_2O_3. Determined on a 10-gram sample. See page 69. If Fe_2O_3 alone is desired determine colorimetrically. (See page 258.)

Alkalies in Alunite. The ore is mixed with pure silica and is run by the J. L. Smith method.

Alkalies Volumetric Determination. See chapter in Vol. 2.

RADIUM

Ra, at.wt. 225.95; m.p. about 700 degrees C.; half period 1580 yr.; chloride RaCl$_2$.

In 1895 Roentgen discovered X-rays. In the following year Becquerel, while studying the phosphorescent properties of some potassium uranium salt which he had prepared about fifteen years before, noticed that it was capable of producing an effect on a photographic plate without exposure to light. The plate was wrapped in several thicknesses of black paper, being perfectly opaque to ordinary light, and then placed with the uranium salt in a dark chamber. Pictures of coins and other articles were made in this way. These effects were found to be due to radiations from the disintegration products of ordinary uranium. Uranium decomposes in succession into the following products: Uranium X$_1$, Uranium X$_2$, Uranium II, Ionium, Radium, Emanation, Radium A, Radium B, Radium C, Radium D, Radium E, Radium F (polonium), Radium G (lead). There are about thirty-five members in the three radioactive families—uranium, thorium, and actinium. Uranium and thorium are the only ordinary elements that are strongly radioactive. Potassium and rubidium give off some beta radiation. The actinium series seems to be a side chain of the uranium family.

The Curies discovered radium in 1898. It is the most important of the radioactive elements.

Kinds of Radiation. Three kinds of radiation are emitted by the various radioactive substances: alpha (α) particles, beta (β) particles, and gamma (γ) rays. Rutherford showed that the alpha particles are doubly positively charged helium atoms. They have an initial velocity of from 1/20 to 1/15 that of light. The beta praticles are singly charged atoms of negative electricity, commonly known as electrons. They have an initial velocity from some substances approaching that of light. The gamma rays are very short waves in ether, and therefore have the general properties of light. They resemble X-rays but are much shorter than any X-rays that have yet been produced. They also have a much higher penetrating power. On account of the very high electrical effects of these radiations and also on account of the extremely small quantities of some of the radioactive substances it is necessary to depend for their detection and determination on the electroscope. The rays ionize the air and this discharges the electroscope.

The Lind instruments, made by the Sachs-Lawlor Company of Denver, Colorado, and distributed by the Denver Fire Clay Company of Denver, are very satisfactory for radium determinations. These instruments may be obtained from any laboratory supply house. Methods for the determination of radium are found in the second volume.

Chapter contributed by L. D. Roberts.

Detection

Radium discharges the electroscope of the alpha ray instrument. However, other radioactive elements do also. There are only three elements that give a gas called emanation as a product. These elements are actinium, radium and thorium. They can all be distinguished by collecting the gas and observing its rate of decay. The time required for one half of the gas to decay is known as the half-period. The half-period of actinium emanation is 3.9 sec., radium emanation 3.85 days, thorium emanation 54 sec. In preparing samples for the alpha ray test they should be ground to about the same mesh. If the sample causes the leaf to move faster than it moves when the chamber is empty, the specimen is radioactive. The time is taken with a stop-watch.

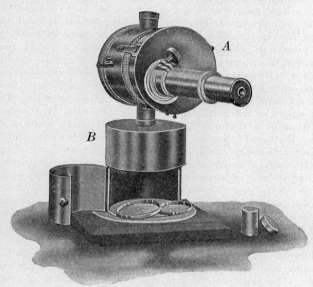

Fig. 56a.

Estimation

Radium is determined by the following methods: I, Alpha Ray; II, Emanation; III, Gamma Ray. The determination with the alpha ray instrument is only approximate and depends on having the samples in the same physical condition as much as possible. The discharge of the leaf is influenced by the fineness of the sample. Also the compound that the element is found in makes a difference. The emanating power of radium chloride is greater than that of radium sulfate, or oxalate. This method is used in the mining camps because it is simple and requires very little equipment. The emanation method is more accurate and is usually used at the plant and laboratory. The gamma ray method is used when the sample cannot be removed from its container, and also for very high grade material. Tubes of emanation prepared for therapeutic use are determined in this way.

The Alpha Ray Method

The alpha ray instrument consists of an electroscope head A (Fig. 56a) and the chamber B. Head A contains an aluminum or gold leaf attached to a metal rod and a reading telescope so placed as to read the deflection of the leaf when charged. B is the chamber to receive the sample. A spring from the metal rod in A connects this with a rod which extends into B. On the end of the latter is placed a disk. The rays discharge the leaf through this disk and rod. The rod in the chamber is insulated from the case with amberoid, made by subjecting amber powder to great pressure. A is charged by means of a battery, or an ebonite rod can be used. The battery is made by connecting seventy-five to a hundred French cells (or Eveready Tungsten No. 703) of about five volts each in series. It is best to ground the negative end and connect A by a floating wire to the positive pole of the cell which will give the desired deflection of the leaf. The case of the chamber is connected to the ground wire. A tube of distilled water is placed in the circuit to prevent the shorting of the batteries. Care should be taken not to charge the leaf enough to throw it against the case. The "natural Leak" is found by taking the time for the leaf to move over forty divisions, and calculating the divisions per second. Less than forty divisions may be taken if the time for forty divisions is quite long. If the leaf moves at the same rate over the whole scale, any part may be taken. The sample is placed in the plate for solids, and this is made exactly level full. The plate containing the sample is placed in a pan to protect the instrument from receiving active matter. The pan with its contents is now placed in the chamber B. As the charged leaf passes 8 (or some other chosen starting place) the stop watch is started. The watch is stopped at 4. The rate of discharge is calculated in divisions per second, and the natural leak is subtracted. A standard is run in exactly the same way. Since the rate of discharge is directly proportional to the amount of radium present, the amount of radium in the sample may be found by comparing its rate of discharge with that of the standard. This method gives approximate results. If the samples are of the same general character, the results may be in very good agreement, checking those obtained by the emanation method very closely.

Emanation Method

I. Carbonate Fusion. About 2 grams of a mixture of sodium and potassium carbonates are placed in a platinum boat about 2 in. long, $\frac{1}{2}$ in. wide and

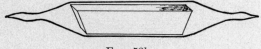

FIG. 56b.

$\frac{1}{2}$ in. deep. Larger boats may be used for low grade samples where large amounts are taken for analysis. From 0.05 to 1 gram of the sample should be weighed in the boat. The boat is now filled level full of the fusion mixture. The fusion is made over flame, or better, in an electric furnace at about 1000° C.

The boat should be put in the furnace while the furnace is cool. This will allow the fusion to dry and melt slowly, thus tending to prevent "boiling" over. The analyst will learn by practice how fast to heat up the furnace. When thoroughly fused and while still at the highest temperature, suddenly chill the fusion by dipping the boat in water, being careful not to allow water to run into the boat and wet the fusion. This chilling causes the mass to draw away from the boat and in the acid treatment slip out of the boat. The mass

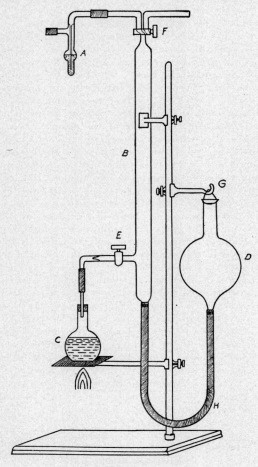

Fig. 56c.

is also made more porous so that the acid attacks it much better. If very little silica is present, the chilling is not necessary. However, it does no harm. The boat with the fusion is sealed in a glass tube as shown in Fig. 56b. The tube should be no larger than necessary. A number of tubes should be prepared in advance by having one end sealed. After standing from one to five or more days (this time depends on the amount of radium present and on the time the result is demanded, the fourth day is usually about right; this gives a little more than a 50 per cent. recovery of the emanation), the emanation surrounding the boat is drawn into an evacuated chamber. The chamber may be evacuated by means of an aspirator on the water faucet or by a pump. To draw the emanation from the tube to the evacuated chamber a rubber tube

Fig. 56d.

is placed on one end of the sealed tube and connected to the chamber with a capillary tube intervening to prevent broken glass from being drawn into the chamber, and a glass stopcock is connected by a rubber tube on the other end. The tip of the sealed tube next to the chamber is broken with a pair of pliers and the stopcock of the chamber opened for an instant. The other tip is broken and the glass stopcock opened for an instant. The stopcock of the chamber is opened again and closed after an instant. Air is let in through the glass stopcock again. After air is taken through about three times the chamber is ready to receive the emanation from the burette. It is necessary to use only a small part of the vacuum to draw the emanation from the sealed tube. The tube is broken and the boat is folded in a filter paper in such a manner that the paper will hold the boat in the neck of the flask until the flask is connected with the apparatus as shown in Fig. 56c. About 4 in. of a stick of sodium hydroxide is placed in the leveling bulb D. Boiling water is poured on this and the solution is raised in the burette about three-fourths of its height, having stopcock E closed and F open. F is now closed and the leveling bulb hung on hook G. Flask C contains 1–1 HNO_3; or if the fusion is hard to disintegrate 3 parts of acid to 2 of water is used. The boat is shaken into the nitric acid and stopcock E opened immediately. A bunsen burner flame is applied to the flask and the acid brought to boiling. The boiling is continued for 10 to 30 minutes according to the nature of the fusion. The heating must be regulated so that the solution in the burette is not driven too low. If the flask is heated too strongly the gas may be prevented from escaping through tube H by raising leveling bulb D. The burette is now connected to the chamber by means of a micro-drying bulb A containing sulfuric acid. The stopcock of the chamber is opened very slightly. Then the stopcock of the burette is opened slowly but fully. The stopcock of the chamber is now regulated till the flow of gas is such that the liquid in the burette rises steadily but not too fast. When the liquid rises to the stopcock the stopcock is turned and air let in till the level is about one-fourth down the tube. The air is drawn into the chamber till the liquid again reaches the stopcock. The air is again let in as before. The stopcock is opened into the chamber again and the liquid poured out of the leveling bulb. Air is drawn through the column till the chamber is full. The burette is disconnected, and the chamber set away to be read at the end of three hours. Just before time to make the reading the electroscope head is placed on the chamber and charged for fifteen minutes, and the chamber is opened to atmospheric pressure by opening the stopcock for only an instant, Fig. 56d. From three to ten readings with the stop watch are taken over forty small divisions, say from 8 to 4. The temperature and pressure are noted. If these vary greatly from the conditions at which the instrument was standardized, the correction of the discharge must be made. The rate of discharge will be proportional to the pressure and inversely proportional to the temperature. The discharge is calculated in divisions per second. The natural leak, or still better, the blank is subtracted from this. This discharge is compared with that of a standard.

Example of a Determination and Calculation.

0.5 g. of sample sealed Oct. 1 at 5:00 P.M.

Boiled off   Oct. 5 at 8:30 A.M.

Time of recovery 3 days, $15\frac{1}{2}$ hours

Recovery factor for 3 days, 15 hr. 0.47926   See table.
$$\frac{1}{2} \text{ hr.} \quad 0.00193$$

$$
\begin{array}{ll}
& 0.4812 \\
\text{log. of} \quad 0.4812 & -1.6822 \\
\text{colog. of } 0.4812 & .3178
\end{array}
$$

A–II Time of discharge over 40 divisions of head A on chamber II.   92.5 sec

B–II Time of discharge over 40 divisions of head B on chamber II.   100 sec.
   1.6021 log. of 40
   1.9661 log. of 92.5

-1.6360	-1.6330 log of 0.4295
.4325	-9.9450 standardization of instrument
.0030 blank	.3178
	.3010 to change $\frac{1}{2}$ g. to gram.
.4295	2.6567 to change grams to pounds
	3.3010 to change pounds to tons

        -2.1545
14.3 mg. of radium per ton.

B–II is calculated in the same way.   The standardization of this partccular instrument was -9.9825.   The last four numbers are the same in the two cases.   The number for the standardization of the instrument is the log of the number of grams of radium necessary to discharge the leaf one division per

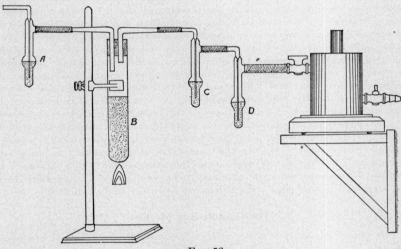

Fig. 56e.

second.   This is obtained by running through the instrument pitchblende or a radium solution the radium content of which is accurately known.   In all cases the natural leak or a blank must be subtracted.

The instrument is charged 15 minutes before the readings are made in order to allow the active deposits of Radium A, B and C to collect on the wall of the chamber.   The leaf system should be charged positively.   The readings are taken at the end of three hours after the emanation is drawn into the chamber because the activity increases to a maximum at that time.   For the first part of the period the increase is very rapid.   The maximum is actually reached at the end of four hours, but between three and five hours the change is very slow.

As soon as the readings are made the instruments must be freed from emanation by drawing dry air through them, the air being dried by sulfuric acid.

II. Bisulfate Fusion. A pyrex test tube 1.1/8 in. in diameter and 8 in. long is filled about $\frac{1}{4}$ full of fused potassium bisulfate. Sodium bisulfate or a mixture of both potassium and sodium bisulfates may be used. From 0.05 to 5 grams of the finely ground sample should be added, the amount depending on the radium content of the sample. The tube is filled about $\frac{3}{4}$ full of potassium bisulfate. In adding the last bisulfate care should be taken to carry down any of the sample adhering on the side of the tube. The mixture is now thoroughly fused, continuing till reaction is complete. The test tube is removed from the flame and held in a nearly horizontal position and slowly turned while the melt solidifies. A very small amount of barium carbonate is dropped into the tube. About $\frac{1}{4}$ of a gram is sufficient. In cases of carnotite ores, slimes, and tails the fusion will be yellow. As soon as the tube is cool enough to be handled a rubber stopper carrying two outlet tubes drawn out to rather fine tips (see figure) is inserted. If this is done as soon as possible there will be a partial vacuum in the tube when it cools and this will indicate when the emanation is taken off that there has been no leak. One of the outlet tubes should extend about one fourth way down the test tube and the other just through the stopper. After emanation has recovered at least one day the tube is connected to an evacuated chamber as shown in Fig. 56e. Micro-drying bulb A contains water or sulfuric acid and acts as an indicator to show when the current of air into the chamber is correctly regulated. B contains the fusion; C sodium hydroxide solution; D sulfuric acid. The tip near A is broken with a pair of pliers. On account of the vacuum in B air rushes through A. Immediately after breaking the first tip the second should be broken. The stopcock of the chamber is slowly opened till air bubbles rather slowly through A. The tube containing the fusion is carefully heated until all the fusion is melted. It is boiled for at least five minutes, or till the chamber is about to atmospheric pressure. During the heating, a steady current is maintained throughout the heating by regulating the stopcock of the chamber. The chamber is disconnected and set away to be read at the end of three hours. The procedure is the same from this point as in the carbonate fusion method.

The Gamma Ray Method

The sample is placed in the holder at such a distance as will cause a reasonable rate of discharge—about one division per second. A standard is then placed at the same distance and the two rates of discharge compared. The natural leak should be subtracted from each reading. The quantities of radium are proportional to the rates of discharge.

In the gamma ray instrument the lead plate and the brass case of the chamber stop all the radiation except the gamma rays.

See Fig. 56f.

Method for Solutions

A measured volume of the solution is placed in a 150 cc. pyrex flask and dilute nitric acid added, filling the flask about one-half full. A little barium nitrate should be added to prevent the precipitation of radium sulfate. The solution is boiled to expel all the emanation. In the Colorado Sc hool of Mines a special flask with a long $\frac{1}{4}$ in. neck is used. This is sealed off in the flame while the solution is nearly boiling. The ordinary flask may be used by inserting a rubber stopper carrying a tube which can be sealed. Or the neck of a pyrex flask can be drawn out and sealed. With the rubber stopper the emanation sometimes leaks. The special flasks are made by the Denver Fire Clay Company. After the emanation has recovered for one or more days the flask is connected to the burette by a light-walled rubber tube as in Fig. 56c. The tip is broken and heat is applied to the flask. On breaking the tip the rubber

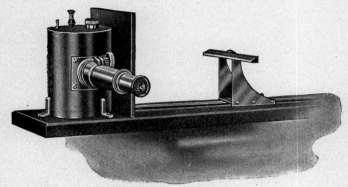

FIG. 56f.

tube collapses if the flask has not leaked. When the pressure in the tube is about atmospheric the stopcock is opened slightly from time to time until the gas goes into the burette, and then opened wide. The solution is boiled for about ten minutes. The gas is then drawn into an evacuated chamber, as described in the other methods. The calculation may be made in milligrams per liter.

Notes

The carbonate method is similar to that given in bulletin 104, U. S. Bureau of Mines.

The bisulfate method of which the method here given is a modification was first published by Howard H. Barker in the Journal of Industrial and Engineering Chemistry for July, 1918, and improved by the Radium Company of Colorado.

Radium is found principally in carnotite and pitchblende. The largest carnotite deposits in the world are in southwestern Colorado and southeastern Utah. Uranium ores contain radium; the ratio of radium to that of uranium is 3.4×10^{-7}; or, there are 3.4 parts of radium to 10,000,000 of uranium.

Radium salts correspond chemically to those of barium. In water barium sulfate is 100 times as soluble as radium sulfate.

Radium is used in the treatment of certain malignant growths and in luminous paints. Some of the radioactive elements have furnished the means of studying atomic structure.

Emanation should not be allowed to remain in the chambers any longer than is necessary. Air dried by sulfuric acid should be drawn through them.

The percent of U_3O_8 multiplied by 2.6 gives the milligrams of radium per ton in ores. This holds good unless the ratio has been disturbed by leaching or some other cause.

References.

Bulletin 70 U. S. Bureau of Mines.

Bulletin 104 U. S. Bureau of Mines.

Radioactive Substances and Their Radiations. E. Rutherford. G. P. Putnam's Sons, New York.

The Chemical Effects of Alpha Particles and Electrons. S. C. Lind. The Chemical Catalog Company, Inc., One Madison Ave., New York.

Introduction to the Rarer Elements. Philip E. Browning. John Wiley and Sons, Inc., New York.

Practical Measurements in Radioactivity. W. Makower and H. Geiger. Longmans, Green, and Co., Fourth Ave. and 30th St., New York.

The Interpretation of Radium and the Structure of the Atom. 4th Ed. Frederick Soddy. G. P. Putnam's Sons, New York.

Bulletin 16. Radium, Uranium, and Vanadium Deposits of Southwestern Colorado. Colorado Geological Survey, Boulder, Colorado.

TABLE OF RADIUM EMANATION RECOVERY (KOLOWRAT)

Days	Hours	$1 - e^{-\lambda t}$	Δ per hour 0.00	Days	Hours	$1 - e^{-\lambda t}$	Δ per hour 0.00	Days	Hours	$1 - e^{-\lambda t}$	Δ per hour 0.00
	0	0.00000	747	2	4	0.32294	504	9	8	0.81363	1377
	1	0.00747	742	2	6	0.33302	496	9	12	0.81913	1326
	2	0.01489	736	2	8	0.34295	489	9	18	0.82709	1268
	3	0.02225	730	2	10	0.35274	482	10	0	0.83470	1212
	4	0.02955	726	2	12	0.36237	475	10	6	0.84197	1159
	5	0.03681	719	2	14	0.37187	468	10	12	0.84893	1108
	6	0.04400	715	2	16	0.38122	461	10	18	0.85558	1059
	7	0.05115	709	2	18	0.39043	454	11	0	0.86193	1013
	8	0.05824	703	2	20	0.39950	447	11	6	0.86801	0968
	9	0.06527	699	2	22	0.40844	440	11	12	0.87381	0925
	10	0.07226	693	3	0	0.41725	432	11	18	0.87937	0885
	11	0.07919	688	3	3	0.43022	422	12	0	0.88467	0846
	12	0.08607	683	3	6	0.44289	413	12	6	0.88975	0809
	13	0.09290	678	3	9	0.45529	404	12	12	0.89460	0775
	14	0.09968	672	3	12	0.46741	395	12	18	0.89924	0739
	15	0.10640	668	3	15	0.47926	386	13	0	0.90367	0701
	16	0.11308	663	3	18	0.49084	378	13	8	0.90928	0660
	17	0.11971	657	3	21	0.50217	369	13	16	0.91457	0622
	18	0.12628	653	4	0	0.51325	361	14	0	0.91954	0586
	19	0.13281	648	4	3	0.52408	353	14	8	0.92423	0552
	20	0.13929	643	4	6	0.53467	345	14	16	0.92864	0519
	21	0.14572	639	4	9	0.54502	337	15	0	0.93279	0489
	22	0.15211	633	4	12	0.55514	330	15	8	0.93671	0461
	23	0.15844	629	4	15	0.56504	323	15	16	0.94039	0434
1	0	0.16473	624	4	18	0.57472	315	16	0	0.94387	0409
1	1	0.17097	620	4	21	0.58418	308	16	8	0.94713	0385
1	2	0.17717	614	5	0	0.59343	3004	16	16	0.95021	0362
1	3	0.18331	611	5	4	0.60545	2915	17	0	0.95311	0341
1	4	0.18942	605	5	8	0.61711	2829	17	8	0.95584	0321
1	5	0.19547	601	5	12	0.62842	2745	17	16	0.95841	0303
1	6	0.20148	597	5	16	0.63941	2664	18	0	0.96084	0281
1	7	0.20745	592	5	20	0.65006	2585	18	12	0.96421	0257
1	8	0.21337	588	6	0	0.66040	2509	19	0	0.96729	0235
1	9	0.21925	583	6	4	0.67044	2435	19	12	0.97010	0214
1	10	0.22508	579	6	8	0.68018	2363	20	0	0.97268	0196
1	11	0.23087	575	6	12	0.68963	2293	20	12	0.97503	0179
1	12	0.23662	570	6	16	0.69881	2225	21	0	0.97718	0164
1	13	0.24232	567	6	20	0.70771	2160	21	12	0.97914	0150
1	14	0.24799	562	7	0	0.71635	2096	22	0	0.98094	0137
1	15	0.25361	557	7	4	0.72473	2034	22	12	0.98258	0125
1	16	0.25918	554	7	8	0.73286	1974	23	0	0.98408	0114
1	17	0.26472	549	7	12	0.74076	1915	23	12	0.98545	0104
1	18	0.27021	545	7	16	0.74842	1859	24	0	0.98670	0095
1	19	0.27566	542	7	20	0.75586	1804	24	12	0.98784	0087
1	20	0.28108	537	8	0	0.76307	1751	25	0	0.98889	0080
1	21	0.28645	533	8	4	0.77007	1699	25	12	0.98985	0073
1	22	0.29178	529	8	8	0.77687	1649	26	0	0.99072	0064
1	23	0.29707	525	8	12	0.78346	1600	27	0	0.99225	0053
2	0	0.30232	522	8	16	0.78986	1553	28	0	0.99353	0044
2	1	0.30754	517	8	20	0.79607	1507	29	0	0.99459	0037
2	2	0.31271	514	9	0	0.80210	1462	30	0	0.99548	
2	3	0.31785	509	9	4	0.80795	1419			1.00000	

Figures 56b, 56c and 56e were drawn by M. Ettington.

35

SELENIUM AND TELLURIUM [1]

DETECTION

Se, *at.wt.* 79.2; *sp.gr.* $\left\{ \begin{array}{l} \textit{Cryst. } 4.82; \\ \textit{Amorp. } 4.26; \end{array} \right.$ *m.p.* 217° $\left. \right\}$ *b.p.* 690°; *oxide*, SeO_2.

Te *at.wt.* 127.5; *sp.gr.* 6.27; *m.p.* 452; *b.p.* 1390°; *oxides*, TeO_2, TeO_3.

Selenium and tellurium are commonly detected by precipitation with sulphur dioxide in hydrochloric acid solution. A selenium solution containing strong hydrochloric acid in the cold gives with either sulphur dioxide gas or the aqueous solution the amorphous red variety which, on warming, goes to the gray crystalline form. Tellurium solutions in presence of dilute hydrochloric acid with sulphur dioxide yield black elementary tellurium.

Hydrogen sulphide gives with selenious acid solutions a precipitate which is at first lemon yellow but on standing changes over to a red, due to the dissociation of yellow sulphide of selenium into sulphur and amorphous red selenium. Tellurous solutions with hydrogen sulphide give at first a red brown precipitate which rapidly darkens due to dissociation into elementary tellurium and sulphur. Both of the sulphides are soluble in alkaline sulphide solutions.

Stannous chloride, ferrous sulphate, hydroxylamine hydrochloride, hydrazine hydrochloride, phosphorous acid, or hypophosphorous acid added in the cold to selenious solutions give red elementary selenium, which goes over to the black variety on warming. Potassium iodide added in excess to a hydrochloric acid solution of either a selenite or selenate gives, in the cold, red selenium together with iodine. On warming the iodine distils and the red selenium goes over into the gray form.

A tellurium solution yields black elementary tellurium when treated with stannous chloride, hypophosphorous acid, hydrazine hydrochloride or with metals like zinc, aluminum, and magnesium.

Neutral selenious solutions with barium chloride give a precipitate of barium selenite which is soluble in hydrochloric acid. Neutral selenates with barium chloride yield insoluble barium selenate, which, like all selenates, is decomposed with the evolution of chlorine and subsequent reduction to the selenite which dissolves in the hydrochloric acid.

The few soluble alkaline tellurites give with barium chloride a white precipitate of barium tellurite which is soluble in hydrochloric acid. Barium tellurate is precipitated when a tellurate solution is treated with barium chloride. It is decomposed by hydrochloric acid, yielding chlorine and forming barium tellurite which dissolves in hydrochloric acid.

Sulphuric Acid Test. Selenium or a selenide with concentrated sulphuric, gently warmed or fuming sulphuric acid, in the cold, gives a green color, the intensity of which varies from a light green to an almost opaque greenish black, depending on the amount of selenium present. When the green solution is added to water, red elementary selenium is precipitated. This red selenium when boiled in the diluted acid changes into the gray crystalline form. The

Chapter contributed by Victor Lenher.

green color in the strong sulphuric acid is destroyed by warming the solution a few minutes. This test is not applicable to an oxidized selenium compound.

Tellurium or a telluride, but not oxidized tellurium compounds, gives in the cold with fuming sulphuric acid or with warm concentrated sulphuric acid a red color, the intensity of which depends on the amount of tellurium present. When the red solution is poured into water, black elementary tellurium is precipitated. When the red solution is warmed, sulphur dioxide is evolved, the red color disappears and if much tellurium is present, white crystals of the basic sulphate of tellurium separate.

The sulphuric acid test is frequently of no value when both of the elements are present, since the red of the tellurium may obscure the green of the selenium.

Qualitative Detection of Selenium and Tellurium in Complex Mixtures

First Method. The substance is treated with aqua regia or with a mixture of hydrochloric acid and potassium chlorate, and the free chlorine is expelled by warming at a temperature below boiling in order to avoid loss of volatile chlorides. The solution is then diluted and filtered to remove insoluble matter. Should tellurous acid precipitate on diluting with water, it can be redissolved by hydrochloric acid. The acid solution is treated with sulphur dioxide gas, or sodium acid sulphide. The formation of a precipitate indicates the possible presence of selenium, tellurium, or gold. (1) If the precipitate is allowed to settle, the liquid poured off, and the precipitate warmed with strong nitric acid, selenium and tellurium will dissolve leaving the gold insoluble. The nitric acid solution can be evaporated with hydrochloric acid to destroy the nitric acid and then treated in concentrated hydrochloric acid solution with sulphur dioxide gas. If selenium is present, it will appear as a red precipitate which on warming goes to the gray crystalline. The selenium precipitate can be filtered off through an asbestos filter and the solution when diluted with water and more sulphur dioxide added gives a black precipitate of elementary tellurium. (2) The sulphur dioxide precipitate containing possible selenium, tellurium, and gold can after washing be treated directly with hot concentrated sulphuric acid in order to get if possible the characteristic selenium or tellurium colors.

Second Method. Crude selenium or tellurium containing material from any source whether oxidized or non-oxidized can be fused with five to six times its weight of potassium cyanide. Tellurium forms potassium telluride, selenium forms selenocyanate and sulphur, which is invariably present, gives sulphocyanate. Extraction of the fused mass with water gives a purple solution if tellurium is present, the selenocyanate and sulphocyanate being colorless. The heavy metals remain insoluble, and can be removed by filtration. When a current of air is bubbled through the solution, the purple color is discharged and black elementary tellurium is precipitated. This tellurium can be filtered off and verified by the sulphuric acid test.

The selenocyanate and sulphocyanate from the air oxidation of the potassium telluride solution can be treated *under a good hood* with hydrochloric acid, when hydrocyanic acid is set free and red selenium is precipitated. The selenium can be confirmed by conversion to the black variety by heat or the sulphuric acid test can be applied.

SEPARATIONS

Separations of selenium and tellurium from other elements can be readily accomplished. Heating of the selenide and telluride combinations or mixtures in chlorine gas affords a separation from the metals whose chlorides are non-volatile.

Heating of selenites or tellurites in a current of hydrochloric acid gas forms volatile $SeO_2.2HCl$ or $TeO_2.2HCl$, while the selenates or tellurates give chlorine in addition. This treatment with hydrochloric acid gas, when applied to the oxidized selenium and tellurium compounds, is an excellent means of separation from the metals whose chlorides are non-volatile.

Both of these elements can be separated from most of the more common elements by the general principle of reducing their compounds and precipitating them in elementary form by means of sulphur dioxide, hydrazine or hydroxylamine. Gold is precipitated at the same time but can be separated by treating the well-washed precipitate with nitric acid, sp.gr. 1.25, which will dissolve the selenium and tellurium but not the gold. The nitric acid solution can then be carefully evaporated with hydrochloric acid to destroy the nitric, and convert to chloride solution.

Separations of Selenium and Tellurium

Keller's method is to separate the selenium and tellurium from each other by making use of the principle that selenium is completely precipitated by sulphur dioxide from strong hydrochloric acid solution while tellurium is not.

Procedure. The two elements are separated from the other elements by sulphur dioxide in dilute hydrochloric acid solution. The washed precipitate is dissolved in nitric acid and the solution evaporated to dryness on the water bath. The residue is dissolved in 200 cc. of hydrochloric acid, sp.gr. 1.175, and the solution warmed to expel all free nitric acid. The solution is then saturated with sulphur dioxide gas at 15–22° C. The precipitated selenium is allowed to settle, washed first by decantation with cold hydrochloric acid, sp.gr. 1.175, then with cold water to displace the acid and finally treated in the beaker with boiling water which transforms the red selenium into the black granular variety. The selenium is brought on a Gooch crucible washed with alcohol and dried at 105° C. The tellurium is precipitated in the filtrate by diluting with water, adding more sulphur dioxide and hydrazine hydrochloride as in the gravimetric method for tellurium. The tellurium is finally washed with water, then alcohol, dried at 105° C. and weighed.

Separation of Selenium and Tellurium.   Distillation Method

The following method is excellent for determining selenium and tellurium in alloys.

Procedure. The apparatus having been set up as is shown in the cut, Fig. 57, 0.5 gram of the alloy containing selenium and tellurium is placed in the distilling flask D and 30 cc. of H_2SO_4 (sp.gr. 1.84) added. All connections are made tight.

A current of dry HCl gas is allowed to flow into the distilling flask and the contents of the flask heated to 300° C. (The H_2SO_4 should not fume and the temperature should be kept below the boiling point of this acid, otherwise the acid distilling into the reservoir R would interfere with the precipitation of selenium by SO_2.) Selenium distills as selenium chloride into R, while tellurium remains in the distillation flask. During the distillation, SO_2 gas is passed into R, to reduce the selenic salt in solution and precipitate metallic selenium.

The distillation is continued for two or three hours, keeping the temperature of the distillation flask at about 300° C. (i.e., below the boiling-point of H_2SO_4). The contents of the receiver R is transferred to a 400-cc. beaker, and the distillation continued into fresh HCl to assure complete volatilization of selenium from the flask D. The contents of the receiver and any of the metal adhering to the glass wool, or the glass of the vessel, are combined. (The adhering selenium is dissolved off with a little Bromine-Potassium Bromide solution.)

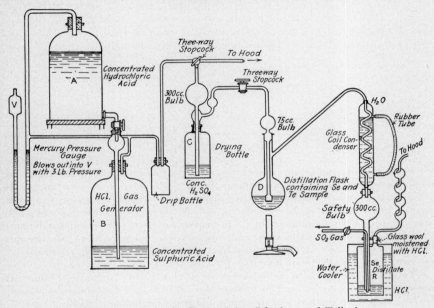

Fig. 57.—Apparatus for Determining Selenium and Tellurium.

Fig. 57 shows a convenient apparatus for routine determinations of selenium and tellurium in alloys. Hydrochloric acid gas is generated by allowing strong hydrochloric acid to flow into concentrated sulphuric acid (see A and B in drawing Fig. 57). The gas is dried by passing it through strong sulphuric acid (C in figure). A mercury pressure gauge, arranged to allow gas to blow out at a pressure of 3 or 4 pounds, prevents accident occurring due to stoppage in the system.

Determination of Selenium. The solution in the beaker is saturated with SO_2, then heated to boiling and the precipitated selenium allowed to settle several hours, or overnight. The precipitate is filtered into a weighed Gooch crucible, then washed with hot water and finally with alcohol. The residue is dried for an hour at 100° C. and weighed.

$$\frac{\text{Weight of Se} \times 100}{0.5} = \% \text{ Se.}$$

Determination of Tellurium. The residue in the distilling flask is transferred to a 600-cc. beaker containing 150 cc. of cold water. Ten cc. of 3% $Fe(NO_3)_3$ solution is added, and made ammoniacal, and then heated to boiling; the precipitate filtered off on a large filter and washed with hot water. The precipitate is dissolved in hot dilute HCl and the solution nearly neutralized with NH_4OH. The slightly acid solution is saturated with H_2S, the precipitated tellurium filtered off on an S. and S. No. 589, $12\frac{1}{2}$ cm. filter, and washed with H_2S water.

The precipitate is dissolved off the paper into a small beaker with a mixture of equal parts of HCl and bromine-potassium bromide solution.[1] The paper is washed with water keeping the volume of the solution as small as possible. The filtrate should contain 20% HCl.

Tellurium is precipitated by saturating the solution with SO_2. The precipitate, after heating to boiling, is allowed to settle for several hours and filtered onto a weighed Gooch. It is washed with hot water and then with alcohol and dried for an hour at 100° C., cooled in a desiccator and weighed.

$$\frac{\text{Weight of Te} \times 100}{0.5} = \%\text{Te}.$$

ESTIMATION

Selenium and tellurium closely resemble sulphur in chemical properties. They have crystalline and amorphous forms. The elements occur in nature frequently associated with sulphur. Selenium is frequently present in iron pyrites, hence is found in the flue dust of lead chambers of the sulphuric acid plant, and as an impurity in sulphuric acid, prepared from pyrites containing selenium.

Ores—Selenium. In copper and iron pyrites; meteoric iron. In the rare minerals clausthalite, PbSe; lehrbachite, $PbSe \cdot HgSe$; onofrith, $HgSe \cdot 4HgS$; eucairite, $CuSe \cdot Ag_2Se$; crookesite, $(CuTlAg)Se$.[1]

Tellurium. Occurs in tellurides and arsenical iron pyrites. Frequently associated with gold, silver, lead, bismuth and iron. In the minerals—altaite, PbTe; calaverite, $AuTe_2$; coloradolite, HgTe; nagyagite, $(AuPb)_2(TeSSb)_3$; petzite, Ag_3AuTe_2; sylvanite, $AuAgTe_4$; telluride, TeO_2 (tellurium ochre); tetradymite, Bi_2Te_3.[1]

[1] Thorpe, "Dictionary of Applied Chemistry."

QUANTITATIVE METHODS FOR SELENIUM

Selenium is most commonly precipitated as element by either sulphur dioxide, hydroxylamine or hydrazine. This reduction to element at the same time separates the selenium from most of the elements except tellurium and gold. When hydroxylamine hydrochloride or hydrazine hydrochloride is the precipitating agent, the material is usually most conveniently brought into hydrochloric acid solution and converted if necessary into the selenious state. From this selenious solution, which may be acid, neutral or ammoniacal, these reducing agents on boiling precipitate elementary selenium which can be brought on a Gooch crucible, washed with hot water, dried at 105°, and weighed.

Sulphur Dioxide Method

The addition of sulphur dioxide to a solution of selenious acid or a selenite which is strongly acid with hydrochloric acid is one of the oldest and best methods of precipitating elementary selenium. A selenate or selenic acid must first be reduced to a selenite or selenious acid by warming with hydrochloric acid, after which sulphur dioxide can be introduced.

It is sometimes convenient to produce the sulphur dioxide by the addition of sodium acid sulphite or of sodium sulphite. This procedure is satisfactory but should be accompanied by a blank test on the sulphite with hydrochloric acid, since the sulphites, on standing, not uncommonly give a precipitate of sulphur on acidification.

The best procedure in the precipitation of selenium by sulphur dioxide is to pass the gas slowly through the selenious containing solution which is strongly acid with hydrochloric acid. The gas should be passed into the solution at a temperature of 15–22° C. When the solution smells strongly of sulphur dioxide gas, the beaker is removed and allowed to stand for a half hour, in order that the selenium may settle. The supernatant liquor is decanted through a previously weighed Gooch crucible, and the selenium washed first by decantation with strong hydrochloric acid, after which one treatment with cold water is used to wash the precipitate in the beaker. After again decanting, hot water is poured into the beaker containing the precipitate when the flocculent red selenium turns black and granular. It is then filtered, washed with hot water, followed by alcohol and dried at 105° C. to constant weight.

If the temperature in the solution during the precipitation of the selenium rises above 22°, the selenium agglomerates and occludes impurities which cannot be washed out. If the temperature is below 15°, the precipitation is either incomplete or very much delayed.

Evaporation of selenious acid should be made on the water bath rather than at a higher temperature since there is an appreciable loss of selenium dioxide when heated above 100° C. In the reduction of selenates or selenic acid to the selenious condition by means of hydrochloric acid, the temperature must never exceed that of the steam bath or considerable selenium may be lost.

Potassium Iodide Method

Potassium iodide added to a selenious solution containing free hydrochloric acid gives a precipitate of elementary selenium, iodine being liberated simultaneously.

$$SeO_2 + 4HI = Se + 2H_2O + 2I_2.$$

With samples containing less than 0.1 gram selenium satisfactory results are obtained but with larger amounts iodine is likely to be occluded.

Procedure.   The sample containing selenious acid or a selenite in a dilution of 400 cc. is acidified with hydrochloric acid, 3–4 grams of potassium iodide are added and the iodine liberated is boiled off.   The selenium is brought on a Gooch crucible washed with hot water, dried and weighed.   Gooch and Pierce [1] suggest the use of sodium arsenite and iodine solutions in carrying out the method volumetrically.

The thiosulphate method of Norris and Fay [2] consists in treating a hydrochloric acid solution of selenious acid with a measured excess of standard sodium thiosulphate solution and then titrating the excess of thiosulphate with an iodine solution.

$$H_2SeO_3 + 4Na_2S_2O_3 + 4HCl = 4NaCl + Na_2S_4SeO_6 + Na_2S_4O_6 + 3H_2O.$$

Selenates or **selenic acid** can be analyzed by boiling with hydrochloric acid when chlorine is evolved which can be collected and titrated.

[1] Amer. Jr. Sci. (4), I, 31.
[2] Amer. Chem. Jr., 23, 119.

QUANTITATIVE METHODS FOR TELLURIUM

Tellurium can be determined gravimetrically and separated from most of the elements except selenium and gold by a number of reducing agents. The oldest method, that of Berzelius, is the use of sulphur dioxide in slightly acid solution. Black elementary tellurium is precipitated but complete precipitation is much delayed even when the solution is warm. The hydrochloric acid solution of the tellurium should always be allowed to stand twenty-four hours. The tellurium is then conveniently brought on a Gooch crucible and the filtrate further digested after the addition of more sulphur dioxide. Very frequently more tellurium settles out on standing twenty-four hours longer. After all of the tellurium is collected on a Gooch crucible, it is washed and dried at 105° C. as quickly as possible in order to avoid the slight superficial oxidation which always takes place with tellurium which has been precipitated in this manner.

Hydrazine hydrochloride used as a reducing agent for the precipitation of elementary tellurium gives fairly good results but complete precipitation as with sulphur dioxide is somewhat delayed.

The use of sulphur dioxide and hydrazine hydrochloride together is the most accurate as well as the most rapid method for the determination and is applicable to both tellurites and tellurates, as well as to the free acids.

Hydrazine Hydrochloride=Sulphur Dioxide Method

The tellurium, either as a derivative of the dioxide or as a tellurate, should be present in a solution which has an acidity of approximately ten per cent free hydrochloric acid, and it is preferable that the solution be concentrated, for otherwise the precipitate will be so finely divided that it will be difficult to wash. The solution is heated to boiling, 15 cc. of a saturated solution of sulphur dioxide is added, then 10 cc. of a fifteen per cent solution of hydrazine [1] hydrochloride followed by 25 cc. of sulphur dioxide solution. The boiling is continued until the precipitate settles in such a way that it can be easily washed, which should not take more than five minutes. The precipitated tellurium is brought on a previously weighed Gooch crucible washed with hot water until all of the chlorides are removed, after which the water is displaced by alcohol, and the crucible and contents dried at 105° C.

Tellurates or **telluric acid** can be analyzed by boiling with hydrochloric acid when chlorine is given off.

$$H_2TeO_4 + 2HCl = H_2TeO_3 + Cl_2.$$

The chlorine evolved may be passed into potassium iodide solution and the liberated iodine titrated by sodium thiosulphate or an arsenite solution.

[1] Hydrazine hydrochloride is prepared ordinarily from the sulphate according to the directions given in "Organic Chemical Reagents," Vol. III, page 42, University of Illinois Bulletin, Vol. 19, No. 6, 1921. Hydrazine sulphate is conveniently made from sodium hypochlorite and ammonia. Organic Syntheses, Vol. II, pages 37–40, Wiley & Co., 1922.

Precipitation as Tellurium Dioxide

Browning and Flint [1] utilize the insolubility of tellurium dioxide as a means of separation from the readily soluble selenious acid. Selenium and tellurium are precipitated by sulphur dioxide from a hydrochloric acid solution. The elements are filtered, washed and dissolved in hydrochloric acid containing sufficient nitric acid to effect solution and then evaporated to dryness on the water bath.

Procedure. The material is dissolved in hydrochloric acid or in a ten per cent solution of potassium hydroxide, using 2 cc. per 0.2 gram of oxide. The solution, if alkaline, is slightly acidified with hydrochloric acid and then diluted to 200 cc. with boiling water. Dilute ammonium hydroxide is now added in slight excess, and is followed by the faintest excess of acetic acid. Crystalline tellurium dioxide separates out completely on cooling and can be brought on a Gooch crucible dried at 105° C. and weighed.

NOTES

Tellurium. The element dissolves in hot concentrated hydrochloric acid. On dilution of the solution a precipitation of H_2TeO occurs. Treated with concentrated nitric acid or aqua regia H_2TeO_4 forms. With sulphuric acid the compound H_2TeO_3 forms and SO_2 is evolved. Tellurium is insoluble in carbon disulphide. The oxides TeO and TeO_2 are soluble in acids, TeO_3 being not readily soluble. All the oxides dissolve in hot potassium hydroxide solutions.

Care must be exercised to avoid overheating acid extracts in the preparation of the sample, since loss by volatilization is apt to occur; this is especially true of the halogen compounds of selenium and tellurium, the former being more volatile than the latter

Fusion Method. The finely powdered substance is intimately mixed with about five times its weight of a flux of sodium carbonate and nitrate (4 : 1) and heated gently in a nickel crucible, gradually increasing the heat, until the charge has fused. When the molten mass appears homogeneous, it is cooled and extracted with water. Sodium selenate and tellurate pass into solution and are separated from most of the heavy metals. The water extract is acidified with hydrochloric acid and boiled until no more free chlorine is evolved. (Test with starch iodide paper. Cl = blue color.) [2] Metallic selenium and tellurium may be precipitated by passing sulphur dioxide into the hydrochloric acid solution.

Keller has shown that tellurium is not precipitated by SO_2 in strong hydrochloric acid solutions (sp.gr. 1.175), whereas selenium is precipitated. Diluted with an equal volume of water (acidity 12 to 20% of above) both tellurium and selenium are precipitated by SO_2.

[1] Amer. Jr. Sci. (4), 1909, 28, 112.

METHODS AT REFINERIES
Commercial Selenium

A half-gram portion of the material ground to 100 mesh is placed in a 150-cc. beaker, and 10 cc. of water are added, followed by 15 cc. conc. nitric acid. After the sample has dissolved in the beaker, which is covered with a watch glass, it is evaporated to dryness on the water bath, and taken up in 10 cc. conc. hydrochloric acid and 20 cc. of water in the cold. The insoluble matter is filtered off and the solution received in a 400-cc. beaker. Sufficient conc. hydrochloric acid is added to make the solution 70 per cent conc. hydrochloric acid.

The selenium is precipitated at room temperature by passing a slow current of sulphur dioxide gas through the solution at the rate of two small bubbles per second, stirring frequently, to granulate the selenium. It is recommended that the temperature of the solution be maintained at 60 to 70° F. by placing the beaker in a vessel of running water.

When complete precipitation has been effected and the solution smells strongly of sulphur dioxide, the beaker is removed and allowed to settle for a half hour. The supernatant liquor is decanted through a previously weighed Gooch crucible; the precipitated selenium in the beaker is washed three times with conc. hydrochloric acid and once with cold water, decanting each time through the crucible. To the precipitate in the beaker 25 cc. of cold water are added, followed by hot water and vigorous stirring until the selenium turns black and granular. It is then filtered, washed with hot water followed by alcohol, and dried at 105° to constant weight. After weighing, the crucible can be gently heated to expel the selenium, in order to obtain a check on its purity. A residue may consist of silica or gold.

Tellurium in the filtrate is recovered by adding three gm. of powdered tartaric acid, diluting with four times its bulk of hot water, then adding 25 cc. ammonium hydroxide and saturating with sulphur dioxide gas. After the sulphur dioxide treatment, which takes but a few minutes, the solution is brought to boiling and allowed to stand for two hours on a hot plate. The granular elementary tellurium is brought on a previously weighed Gooch crucible, washed with hot water, dried at 105° and weighed.

The important laboratory details suggested by Greenwood [1] are:

1. Evaporation of the selenious acid should be made on the water bath rather than on a hot stove in order to avoid loss by volatilization. There is an appreciable loss of selenium dioxide when heated above 100° C. even in the presence of sodium and potassium chlorides. This fact has been verified in Lenher's laboratory.

2. If the temperature of the solution during the precipitation of selenium is above 70° F., the selenium agglomerates and occludes impurities, which cannot be washed out. If the temperature is below 60° F., the precipitation is either incomplete or very much delayed.

3. If the precipitated selenium is not granular, it cakes during the drying and retains moisture even at 110°.

Lead, copper and iron in selenium are determined in a sample of from 10 to 25 gm. which is dissolved in 50 to 75 cc. conc. nitric acid in a 375-cc. casserole and evaporated to dryness. The volatile selenium dioxide is expelled

---

[1] Eng. and Min. Jr. (1915), 100, 1012.

by carefully raising the temperature. The non-volatile residue is dissolved in 10 cc. conc. nitric acid and 5 cc. conc. hydrochloric acid, evaporated to 5 cc., when 5 cc. concentrated sulphuric acid are added and evaporated to fumes; it is allowed to cool, 75 cc. of water are added and allowed to stand over night, when the lead sulphate is filtered off and weighed as usual. The filtrate from the lead is treated with hydrogen sulphide, the precipitate is ignited to burn off selenium, tellurium and arsenic, the residue dissolved in nitric acid and the copper determined volumetrically. The filtrate from the precipitation of the copper is boiled to expel the hydrogen sulphide, oxidized by a few crystals of potassium chlorate, and the iron precipitated in the usual manner by ammonia.

An alternate method for the determination of **iron in selenium** is to weigh 10 gm. of the sample into a porcelain dish, ignite at a red heat until the selenium appears to be completely driven off. The residue is sometimes weighed and reported as "non-volatile matter." This residue is fused with sodium carbonate, treated with dilute sulphuric acid, the solution reduced with zinc, and the iron titrated with a weak solution of permanganate.

Insoluble in commercial selenium. 10 to 25 grams are dissolved in a 375-cc. casserole, in conc. nitric acid, and evaporated to dryness; conc. hydrochloric acid is added and the silica dehydrated. The insoluble is taken up in hydrochloric acid and water, filtered off, ignited and weighed.

Selenium and tellurium in metallic tellurium. A 0.5-gram sample of the finely powdered metal is treated with 10 cc. of conc. sulphuric acid and fumed until all the tellurium has dissolved; after which it is cooled and 20 cc. water and 50 cc. conc. hydrochloric acid are added.

Selenium is precipitated from the cold solution by passing in sulphur dioxide gas for 3 to 4 minutes, after which it is filtered through a Gooch crucible, washed three times with hydrochloric acid (2 parts conc. acid to 1 water), then with hot water and finally with alcohol, dried and weighed. To obtain a check on the purity of the selenium the Gooch crucible is ignited and reweighed.

The tellurium-containing filtrate is diluted to about 700 cc., heated to nearly boiling, a few grams of hydrazine hydrochloride added and a rapid current of sulphur dioxide gas is passed in for 15 minutes, or until the tellurium separates readily; it is then brought on to a Gooch filter, washed with hot water, finally with alcohol. The elementary tellurium is dried for one hour at 105° and weighed.

A selenium and tellurium procedure used for a number of years in one of the refineries consists in dissolving in nitric acid, adding a pinch of salt, evaporating to dryness, taking up in 25 cc. hydrochloric acid (1 : 1) and bringing to boiling. The insoluble matter is filtered off, the solution diluted to 150 cc. and sulphur dioxide passed through the boiling solution. A few grams of hydrazine hydrochloride are added and sulphur dioxide again passed in after which the solution is boiled a few minutes. The precipitated selenium and tellurium are filtered off, dissolved in nitric acid with the addition of a pinch of salt, and the solution evaporated. The residue is taken up in about 200 cc. conc. hydrochloric acid, the solution boiled about ten minutes and the selenium brought on a Gooch crucible where it is washed with hot water, then alcohol, dried and weighed.

The tellurium-containing filtrate is diluted with water to three times its volume and saturated with sulphur dioxide; a few grams of hydrazine hydrochloride are added, boiled a few minutes, the tellurium is brought on a Gooch crucible, washed with hot water, then with alcohol, dried and weighed.

Selenium and tellurium in blister or pig copper are commonly determined by the method of Keller, using samples of 50 grams or less.

Selenium and tellurium in electrolytic copper slimes are determined by treating a 0.5-gram sample of the slime in a 250-cc. beaker with 10 cc. conc. sulphuric acid, heating until the sample is decomposed and nothing remains but a white residue. After cooling, 20 cc. of water are added, followed by 2 cc. conc. hydrochloric acid, the solution is agitated to coagulate the silver chloride which is then filtered off. The acidity of the filtrate is brought up to about 32 per cent by adding conc. hydrochloric acid and the selenium and tellurium separated by sulphur dioxide, following the customary procedure.

Selenium in lead slimes. A 2-gram sample is fused with a mixture of 8 grams sodium carbonate and 2 grams nitrate in a nickel crucible. The cold fusion is extracted with water and filtered. The filtrate is acidified with hydrochloric acid and heated until chlorine is expelled. To the solution is added an equal volume of hydrochloric acid, and sulphur dioxide gas is added until the red precipitate becomes granular. The selenium can be brought on an asbestos filter, washed with hydrochloric acid, redissolved in hydrochloric acid and potassium chlorate and reprecipitated by sulphur dioxide.

Tellurium in lead slimes. 1 gram of slimes is treated with a mixture of 10 cc. conc. sulphuric acid, 10 cc. conc. nitric acid, and 20 cc. water, and evaporated to fumes. After cooling, 40 cc. water and about 2 grams tartaric acid are added. The solution is boiled and filtered. The residue is washed back into the original beaker, 5 to 10 cc. conc. sulphuric acid are added, and again evaporated to fumes; 40 cc. of water are added and the solution boiled and filtered. The two filtrates are united and treated with hydrogen sulphide gas. The sulphides are filtered on paper washed with hydrogen sulphide water, washed back into the beaker, and a little sodium bicarbonate added followed by about 4 cc. of 10 per cent sodium sulphide solution. The solution is brought to boiling, digested for 12 hours, and filtered through the filter previously used. The sulphide-containing filtrate is acidified with dilute sulphuric acid and treated with hydrogen sulphide gas to render it granular. The sulphide precipitate, after filtration, is dissolved in nitric acid with the addition of 10 cc. conc. sulphuric acid and evaporated to fumes of sulphuric acid. It is then diluted with water and boiled after adding about 2 grams of tartaric acid. The solution is cooled, diluted to about 60 cc. with water, and after adding 40 cc. conc. hydrochloric acid, is again treated with hydrogen sulphide. The precipitate is filtered off, washed with 1 : 1 hydrochloric acid, dissolved in hydrochloric acid and potassium chlorate, warmed gently to expel the excess of chlorine, tartaric acid is again added, any residue filtered off, and the filtrate made strongly acid with hydrochloric acid. The selenium is then precipitated by sulphur dioxide.

The filtrate containing the tellurium is diluted with warm water, and the tellurium precipitated and weighed as usual.

Selenium and tellurium in flue dust. 2 grams are dissolved in hydrochloric acid and potassium chlorate on a water bath, and the excess of chlorine expelled by gentle heat. The insoluble matter is filtered off and washed with concentrated hydrochloric acid, and the filtrate is treated with sulphur dioxide gas. The selenium and tellurium are brought on a Gooch crucible, washed, then dissolved in hydrochloric acid and potassium chlorate, about 2 grams of tartaric acid are added, followed by conc. hydrochloric acid, and the selenium is precipitated by sulphur dioxide, filtered off and weighed. The filtrate is diluted to three times its volume with warm water, and the tellurium is precipitated by sulphur dioxide, collected and weighed.

Flue Dust and Niter Slag

Flue dust or niter slag from Doré furnaces can be analyzed for water-soluble selenium and tellurium by boiling a 1-gram sample with water, filtering and washing with hot water, keeping the volume down to 20 cc. To this filtrate are added 200 cc. conc. hydrochloric acid; the solution is chilled with icewater; and sulphur dioxide is passed in, the selenium and tellurium being separated by the method of Keller.

The insoluble selenium and tellurium are transferred from the filter to a 50-cc. beaker, conc. nitric acid and conc. hydrochloric acid are added, and the solution evaporated at 50° C. or below. It is recommended to evaporate twice more with hydrochloric acid, keeping the temperature at 50° C. or below. The residue is taken up in hydrochloric acid (1 : 2) and filtered, after which 100 cc. conc. hydrochloric acid are added and the selenium and tellurium in the filtrate are separated by the Keller method.

Commercial Sodium Selenite

To 1 gram of the sample in a 50-cc. beaker are added 10 cc. of water and 5 drops of hydrochloric acid and shaken gently until solution is complete. After filtering out the insoluble matter, a large excess of conc. hydrochloric acid is added to the filtrate and selenium is precipitated by sulphur dioxide. In the filtrate from the selenium, the tellurium can be recovered by diluting with warm water and passing in more sulphur dioxide.

Selenic Acid

20 grams of the sample are quickly transferred to a tightly stoppered weighing bottle. After weighing, the acid is put into a liter flask, dissolved in water, and made up to the mark. A 25-cc. portion is measured out and treated with an excess of conc. hydrochloric acid and the selenium precipitated by sulphur dioxide. If tellurium is present, it can be recovered in the filtrate.

Selenium in Glass [1]

In ruby glass where selenium is present in quantities of about 0.25%, a two-gram sample is evaporated with hydrofluoric and nitric acids. This evaporation is repeated several times without ignition. The nitrates are then transferred to a small Erlenmeyer flask and the nitric acid destroyed by adding hydrochloric acid. Care must be taken not to boil the solution or selenium will volatilize. A high concentration of the hydrochloric acid must be maintained in order to hold the tellurium in solution. Sulphur dioxide is next passed into the solution when elementary selenium is precipitated, washed, dried and weighed as usual.

[1] Method furnished by the Corning Glass Works.

In glasses where selenium is used as a decolorizer and in which the selenium is present to the extent of less than .0025%, Cousen [1] recommends dissolving the glass in nitric and hydrofluoric acids and determining the selenium colorimetrically by phenyl hydrazine hydrochloride, keeping the selenium from precipitating by adding gum arabic. A yellow to a yellowish red solution is obtained which is compared with a color standard containing a known amount of sodium selenite.

Selenium and Tellurium in Refined Copper

The success and accuracy of determining small amounts of impurities, is to collect these impurities from a large sample, and to do the necessary chemical work in a small volume of solution.

Ferric hydroxide when precipitated from a copper solution has the peculiar property of forming insoluble compounds with As, Sb, Se and Te.

For complete precipitation, the iron contents must be at least thirty times the combined As, Sb, Se and Te and must be entirely precipitated from the solution.

The iron is best added by means of the ferrous salt ($FeSO_4+7H_2O$) dissolved in water and oxidized to the ferric salt with HNO_3. Roughly the iron content of $FeSO_4+7H_2O$ is 20%, or one-fifth of the weight of ferrous sulphate.

Weigh 100 grams of drillings into a 1300 cc. beaker, cover with water and add gradually 350 cc. HNO_3 (1.42). When the copper is in solution, boil out all red fumes, dilute to 700 cc. with warm water and neutralize with NH_4OH until just enough copper hydroxide has been formed to cover the bottom of the beaker; then add the required amount of oxidized ferrous sulphate in solution. Stir well, dilute to 900 cc. and boil for at least one hour, settling on the warm plate over night. Filter on a 15 cm. filter.

Treat the filtrate with ammonia until a precipitate of copper hydroxide has formed to cover the bottom of the beaker and half the quantity of oxidized $FeSO_4+7H_2O$ added previously, is again added and the solution boiled for one hour, allowing to settle over night on the warm plate. Filter on a 15 cm. filter. It is absolutely necessary that no iron salts be left in this filtrate, either in solution, or as hydrate, therefore, this filtrate had best be tested with ammonia and refiltered on a 15 cm. filter. The three precipitates on the 15 cm. papers are treated as one, dissolved in the least quantity of warm H_2SO_4 (1–1), the papers washed well with hot dil. H_2SO_4 (1–20) and filtered into a 600 cc. beaker. Make strongly ammoniacal, boil well and filter, washing the precipitate well and free from the folds into the apex of the filter. Refilter the filtrate to catch any iron that has washed through. The papers are spread open and the precipitates dissolved in 25 cc. warm HCl (1–1) in a 250 cc. beaker, using as little hot water as possible to remove the yellow stains.

Filter the solution into a 400 cc. beaker, washing with as little hot water as possible. To the filtrate add four times its volume of conc. HCl and cool.

Pass SO_2 gas through this liquid at the rate of one bubble per second, keeping the solution cool.

Filter off the precipitated Se on a tared Gooch crucible wash with cold water and alcohol, dry at 60° C. for two hours, then at 105° C. to constant weight.

Weigh as metallic Se.

Expel the Se from the crucible and reweigh as a check on the weight and the purity of the precipitate.

Tellurium

To the filtrate from the precipitated Se, add 2 grams of tartaric acid in order to keep the Sb in solution, dilute to 600 cc. with hot water, add 50 cc.[1] conc. NH_4OH, and saturate the solution with SO_2 gas. Boil for two minutes and allow to settle 4–6 hours.

Filter the Te on a tared Gooch crucible, wash with hot water and finally with alcohol, dry at 115° C. to constant weight.

Weigh as metallic Te.

VOLUMETRIC DETERMINATION OF SELENIUM AND TELLURIUM

Iodometric Determination of Selenic, or Telluric Acid—Reduction with Hydrochloric Acid and Distillation

The method depends upon the reduction of selenic or telluric acid to selenious or tellurious acid by heating with hydrochloric acid, the evolved chlorine being a measure of the acids in question. The chlorine absorbed in potassium iodide solution liberates its equivalent of iodine, which may readily be determined by titration with standard thiosulphate. The following reactions illustrate the change that takes place:

$$K_2SeO_4+4HCl = 2KCl+H_2SeO_3+H_2O+Cl_2,$$
$$K_2TeO_4+4HCl = 2KCl+H_2TeO_3+H_2O+Cl_2,$$

1 Cl. $= 1$ I $= \dfrac{Se}{2}$ or $\dfrac{Te}{2} = 63.75$ grams Te or 39.6 grams Se per liter normal solution.

According to Gooch and Evans over 30% of strong hydrochloric acid (sp.gr. 1.20) should be present. Dilute hydrochloric acid having a strength of 10% of HCl, sp.gr. 1.2, does not react with liberation of chlorine. Care must be taken not to prolong the boiling after the solution reaches a concentration of half strength, since over reduction may take place and the metals be liberated.

Procedure. The sample containing the selenate or tellurate is treated with 75 cc. of hydrochloric acid, containing 25 cc. of strong HCl, sp.gr. 1.20, per 0.2 gram of the oxides, in a distillation flask connected with a Drexel wash bottle receiver, water cooled, and charged with potassium iodide solution. A current of CO_2 is passed into the flask to sweep the liberated chlorine into the iodide solution. The sample is boiled until nearly one-third its volume has distilled into the receiver. The liberated iodine is titrated with standard thiosulphate. One cc. N/10 $Na_2S_2O_3 = 0.00396$ gram Se or 0.006375 gram Te.

[1] Te is best precipitated by SO_2 gas in a 20–30% HCl solution and the strong (80%) HCl solution from the Se precipitation is best neutralized to 25–30%, by the use of ammonia.

SILICON

Si, *at. wt.* **28.06;** *sp.gr. amor.* **2.00.;** *crys.* **2.49;** *m.p.* **1420° C.;** *oxides* **SiO, SiO₂**

DETECTION

The finely ground sample together with a small quantity of powdered calcium fluoride is placed in a small lead cup 1 cm. in diameter and depth (see Fig. 58), and a few drops of concentrated sulphuric acid added. A lead cover, with a small aperture, is placed on the cup, and the opening covered with a piece of moistened black filter paper. Upon this paper is placed a moistened pad of ordinary filter paper. The cup is now gently heated on the steam bath. At the end of about ten minutes a white deposit will be found on the under side of the black paper, at the opening in the cover, if an appreciable amount of silica is present in the material tested.

Fig. 58.

A silicate, fused with sodium carbonate or bicarbonate in a platinum dish and the carbonate decomposed by addition of hydrochloric acid with subsequent evaporation to dryness, will liberate silicon as silicic anhydride, SiO_2. The silica placed in a platinum dish is volatilized by addition of hydrofluoric acid, the gaseous silicon fluoride being formed. A drop of water placed in a platinum loop, held in the fumes of SiF_4, will become cloudy owing to the formation of gelatinous silicic acid and fluosilicic acid,

$$3SiF_4 + 3H_2O = H_2SiO_3 + 2H_2SiF_6.$$

If a silicate is fused in a platinum loop with microcosmic salt, the silica floats around in the bead, producing an opaque bead with weblike structure upon cooling.

ESTIMATION

The gravimetric procedure is the only satisfactory method for the estimation of silica. The substance in which the element is combined as an oxide or as a silicate is decomposed by acid treatment or by fusion with an alkali carbonate or bicarbonate, the material taken to dryness with addition of hydrochloric acid, whereby the compound silica is liberated. If other elements are present the silica is volatilized by addition of hydrofluoric acid and estimated by the loss of weight of the residue.

The element silicon has no important application. Its use for electrical resistance has been suggested. A rod 10 cm. long with cross section of 40 sq.mm. has a resistance of 200 ohms against a carbon rod of the same dimensions of 0.15

Chapter contributed by Wilfred W. Scott.

ohm. Impure silica finds use in fluxes in manufacture of glass; pure silica for the manufacture of silica ware. With caustic it forms an adherent sodium silicate. Silicon carbide, carborundum, is used for refractory purposes, fire brick, zinc muffles, coke ovens. Crystolon, the crystalline form, is used as an abrasive, in making grinding wheels, sharpening stones, etc.

Combined as SiO_2 and in silicates the element is very widely distributed in nature and is a required constituent in practically every complete analysis of ores, minerals, soils, etc. It is present in certain alloys, ferro-silicon, silicon carbide, etc.

The element is scarcely attacked by single acids, but is acted upon by nitric-hydrofluoric acid mixture. It dissolves in strong alkali solutions. Silica is decomposed by hydrofluoric acid and by fusion with the fixed alkali carbonates or hydroxides.

Preparation and Solution of the Sample

General Considerations. The natural and artificially prepared silicates may be grouped under two classes: 1. Those which are decomposed by acids. 2. Silicates not decomposed by acids. The minerals datolite, natrolite, olivine and many basic slags are representative of the first class, and feldspar, ortho-clase, pumice and serpentine are representative of silicates not decomposed by acids. (See more complete list under List of More Important Silicates, page 433.) The first division simply require an acid treatment to isolate the silica, the latter class require fusion with a suitable flux.

In technical analysis, in cases where great accuracy is not required, the residue remaining, after certain conventional treatments with acids, is classed as silica. This may consist of fairly pure silica or a mixture of silica, undecomposed silicates, barium sulphate and certain acid insoluble compounds. For accurate analyses this insoluble residue is not accepted as pure silica, unless impurities, which are apt to be found with the silica residue, are known to be absent from the material under examination.

Although the procedure for isolation of silica is comparatively simple, errors may arise from the following causes:

1. Imperfect decomposition of the silicate.

2. Loss of the silica by spurting when acid is added to the carbonate fusion.

3. Slight solubility of silica, even after dehydration, especially in presence of sodium chloride and magnesia.

4. Loss due to imperfect transfer of the residue to the filter paper.

5. Mechanical loss during ignition of the filter and during the blasting, due to the draft whirling out the fine, light silica powder from·the crucible.

6. Error due to additional silica from contaminated reagents or from the porcelain dishes or glassware in which the solution was evaporated. A blank of 0.01% on the sodium carbonate will make an error of 0.1% per gram sample in an ordinary fusion where 10 grams of the flux are required.

7. Error due to loss of weight of the platinum crucible during the blasting.

8. Incomplete removal of water, which is held tenaciously by the silica. Furthermore, weighing of the residue should be done quickly, as the finely divided silica tends to absorb moisture.

Two general procedures will be given for treatment of the acid decomposable and undecomposable silicates. It is frequently advisable to use these two procedures in conjunction, extracting the material first with acid, and then fusing the insoluble residue with sodium carbonate; this procedure is used when a gritty residue remains after the acid extraction. Following the general procedures for decomposition of silicates, certain special methods will be given.

List of Most Important Silicates. Silicates decomposed by acids. Allanite; allophane; analcite; botryolite; brewsterite; calamine; chabasite; croustedtitite; datolite (hydrated silicate and borate of Ca with Al and Mg); dioptase; eulytite; gadolinite; gahlenite; helvite; ilvaite (silicate ferrous and ferric iron with Al_2O_3, CaO and MgO); laumonite; melinite; natrolite (hydrated silicate of Al and Na with Fe and CaO); okenite; olivine (silicate of Fe and Mg); pectolite; prehenite (hydrated Al and Ca silicate with Fe, Mn, K, Na, etc.); teproite; wernerite; woolastonite; zaolite.

Silicates undecomposed by acids. Albite; audalusite; augite; axinite; beryl; carpholite; cyanite; diallage, epidote (silicate of Fe, Al and Ca with FeO, Mn, Mg, K, Na); euclase; feldpsar (silicate of K, Na, Al, Fe, Ca and Mg); garnet; iolite; labradorite; (micas of K and Mg); orthoclase; petalite; pinite, prochlorite; pumice; serpentine; sillimanite, talc, topaz, tourmaline (Fe_2O_3, FeO, Mn, Al, Ca, Mg, K, Na, Li, SiO_2, B_2O_3, P_2O_5, F); vesuvianite.

Preparation of the Substance for Decomposition

If the material is an ore or mineral it is placed on a steel plate within a steel ring and broken down by means of a hardened hammer to small lumps and finally to a coarse powder. A quartered portion of this is air dried and ground as fine as possible in an agate mortar and preserved in a glass-stoppered bottle for analysis.

Analyses are based on this air-dried sample. If moisture is desired it may be determined on a large sample of the original material. Hygroscopic moisture is determined on the ground, air-dried sample, by heating for an hour at 105 to 107° C.

Decomposition of the Material, General Procedures

Silicates Decomposed by Acids

Acid extraction of the silicates. 0.5 to 1 gram of the finely pulverized material placed in a beaker or casserole is treated with 10 to 15 cc. of water and stirred thoroughly to wet the powder.[1] It is now treated with 50 to 100 cc. of strong hydrochloric acid and digested on the water bath for fifteen or twenty minutes with the beaker or casserole covered by a watch-glass. If there is evidence of sulphides (pyrites), etc., 10 to 15 cc. of concentrated nitric acid are now added and the containing vessel again covered. After the reaction has subsided, the glass cover is raised by means of riders and the mixture evaporated to dryness on the water bath. (This evaporation may be hastened by using a sand bath, boiling down to small bulk at comparatively high temperature, then to dryness on the water bath. Decomposition is complete if no gritty particles remain. A flocculent residue will often separate out during the digestion, due to partially dehydrated silicic acid, hydrated silicic acid, $Si(OH)_4$ is held in solution.) The silicic acid is converted to silica, SiO_2, the residue taken up with dilute hydrochloric acid, silica filtered off, washed with water acidified with hydrochloric acid, and estimated according to the procedure given later.

---

[1] Water is added to the sample and then acid, as strong acid added directly would cause partial separation of gelatinous silicic acid, which would form a covering on the undecomposed particles, protecting them from the action of the acid.

Silicates Not Decomposed by Acids

Fusion with Sodium Carbonate or Sodium Bicarbonate. 0.5 to 1 gram of the air-dried, pulverized sample is placed in a large platinum crucible or dish in which has been placed about 5 grams of anhydrous sodium carbonate. The sample is thoroughly mixed with the carbonate by stirring with a dry glass rod, from which the adhering particles are brushed into the crucible. A little carbonate is sprinkled on the top of the mixture and the receptacle covered. It is heated to dull redness for five minutes and then gradually heated up to the full capacity of a Méker burner. When the mix has melted to a quite clear liquid, which generally is accomplished with twenty minutes of strong heating, a platinum wire with a coil on the immersed end is inserted in the molten mass, and this allowed to cool. The fusion is removed by gently heating the crucible until the outside of the mass has melted, when the charge is lifted out on the wire, and after cooling disintegrated by placing it in a beaker containing about 75 cc. dilute HCl (1 part HCl to 2 parts H_2O), covering the beaker to prevent loss by spattering. The crucible and lid are cleaned with dilute hydrochloric acid, adding this acid to the main solution. When the disintegration is complete, the solution is evaporated to dryness and silica is estimated according to directions given later.

If decomposition is incomplete, gritty material will be found in the beaker upon treatment of the fusion with dilute acid. If this is the case, it should be filtered off and fused with a second portion of sodium carbonate, and the fusion treated as directed above.

NOTES. Fusions with soluble carbonates are generally best effected with the sodium salt, except in fusions of niobates, tantalates, tungstates, where the potassium salt is preferred on account of the greater solubility of the potassium compounds. Sodium alone has an advantage over the mixed carbonates, $Na_2CO_3 + K_2CO_3$, as silica has a high melting-point and a flux, which fuses at 810° C., is more apt to cause disintegration of the silicate than the mixture, which melts at 690° C.

Prolonged blasting is undesirable, as it renders the fusion less soluble. Aluminum and iron are also rendered difficultly soluble, when their oxides are heated to a high temperature for some time.

If the melt is green, it is best to dissolve out the adhering melt from the crucible with dilute nitric acid, as a manganate (indicated by the color), if present, will evolve free chlorine by its action on HCl and this would attack the platinum.

Fluorides.[1] In presence of fluorides the melt is extracted with water (an acid extraction would volatilize some of the silica), and the extract filtered off from the insoluble carbonates. To the filtrate is added about 5 grams of solid ammonium carbonate, and the mix warmed to 40° C. and allowed to stand for several hours. The greater part of the silica is precipitated. This is filtered off and washed with water containing ammonium carbonate. Preserve this with the insoluble carbonate for later treatment. The filtrate, containing small amounts of silicic acid, is treated with 1 to 2 cc. of ammoniacal zinc oxide solution (made by dissolving C.P. moist zinc oxide in ammonia water). The mixture is boiled to expel ammonia and the precipitate of zinc silicate filtered off. The precipitate is washed into a beaker through a hole made in the filter, and the adhering material dissolved off with dilute HCl, enough being added to dissolve the remaining residue. This is evaporated to dryness and silica separated as usual. Meantime the insoluble carbonate is dissolved with HCl, evaporated to

---

[1] Sodium bicarbonate may be used in place of the carbonate with excellent results.

dryness and any silica it contains recovered. Finally all three portions of silica are combined, ignited and silica estimated as usual.

Special Procedures for Decomposing the Sample

Treatment of Iron and Steel for Silica. One gram of pig-iron castings, or 5 grams of steel are taken for analysis, both the fine and coarse drillings being taken in about equal proportion. (Fine particles contain more silicon than the coarse chips.) Twenty to 50 cc. of dilute nitric acid (sp.gr. 1.135) are added to the sample in a 250-cc. beaker or small casserole, and this covered. If the action is violent, cooling, by placing the beaker in cold water until the violent action has subsided, is advisable. Twenty cc. of 50% sulphuric acid are added and the solution evaporated on the hot plate to SO_3 fumes. After cooling, 150 cc. of water are added and 2 to 5 cc. dilute sulphuric acid. The mixture is heated until the iron completely dissolves and the silica is filtered off onto an ashless filter, washed with hot dilute hydrochloric acid (sp.gr. 1.1), and with hot water until free from iron. The residue is ignited and the silica estimated according to the procedure given later.

Pig iron and cast iron may be decomposed by digestion with a mixture of 8 parts by volume of HNO_3 (sp.gr. 1.42), 5 parts of H_2SO_4 (sp.gr. 1.84), and 17 parts of water.

Steel and wrought iron may be disintegrated by a mixture of 8 parts by volume of HNO_3 (sp.gr. 1.42), 4 parts H_2SO_4 (1.84), and 15 volumes of water.

Ferro Silicons. Dilute hydrochloric acid, 1 volume of acid (sp.gr. 1.19), with 2 volumes of water is a better solvent than the strong acid.

Steels Containing Tungsten, Chromium, Vanadium and Molybdenum. Fusion with potassium acid sulphate, $KHSO_4$, in a platinum dish, or sodium peroxide in a nickel crucible will generally decompose the material. Sodium peroxide is of special value in decomposing chromium alloys.

Silicon Carbide, Carborundum. This is best brought into solution by fusion with potassium hydroxide in a nickel crucible. Sulphuric, hydrochloric, nitric acids, or aqua regia have no effect upon this refractory material.

Sulphides, Iron Pyrites, etc. These require oxidation with strong nitric acid or a mixture of bromine and carbon tetrachloride, followed by nitric, exactly according to the procedure given for solution of pyrites in the determination of sulphur. The sample is taken to dryness and then hydrochloric acid added and the solution again evaporated. The residue is dehydrated and silica determined as usual.

Slags and Roasted Ores. Digestion with hydrochloric acid according to the first general procedure is best. The addition of nitric acid to decompose sulphides may be necessary.

Decomposition of silicates by fusion with lead oxide (method of Jannasch), and calcium carbonate and ammonium chloride (method of Hillebrand), are of value when sodium is desired on the same sample. The procedures are given under chapters on Sodium and Potassium.

NOTE. K_2CO_3 is preferred to Na_2CO_3 for fusion of tungstates, niobates and tantalates on account of the greater solubility of the potassium salts. For corundum and alumina silicates Na_2CO_3 is preferred as double salt of potassium and aluminum are less soluble than the sodium salt.[1]

Fluorides of silicon are fused with boric acid, BF_3 is volatilized, SiF_4 is not formed. P. Jannasch.[2]

[1] J. L. Smith, Am. Jour. Sci. (2), **40**, 248, 1865. C. N., **12**, 220, 1865.
[2] Ber., **28**, 2822, 1896.

PROCEDURE FOR THE DETERMINATION OF SILICON AND SILICA

As has been stated, the gravimetric method for determination of silica is the only satisfactory procedure for estimation of this substance. The oxidation of the element and its isolation have been dealt with in the section Preparation and Solution of the Sample. The following directions are for purification and final weighing of the element in the form of its oxide, silica, SiO_2.

Extraction of the Residue—First Evaporation. The residue, obtained by evaporation of the material after decomposition of the silicate, by acids or by fusion, as the case required, is treated with 15–25 cc. of hydrochloric acid (sp.gr. 1.1) covered and heated on the water bath 10 minutes. After diluting with an equal volume of water, filtration is proceeded with immediately, and the silica is washed with a hot solution consisting of 5 cc. hydrochloric acid (sp.gr. 1.2) to 95 cc. of water and finally with water. This filtration may be performed with suction. The filtrate and washings are evaporated to small volume on a sand bath and then to dryness. This contains the silica that dissolved in the first extraction.

Second Evaporation. The residue obtained from evaporation of the filtrate is dehydrated for 2 hours at 105–110° C.[1] and extracted with 10 cc. of hydrochloric acid (sp.gr. 1.1) covered and heated on the water bath for ten minutes diluted to 50 cc. with cold water and filtered immediately, without suction. The residue is washed with cold water containing 1 cc. concentrated hydrochloric acid to 99 cc. water, the washed residue containing practically all[2] the silica, that went into solution in the first extraction, is combined with the main silica residue. This is gently heated in a platinum crucible until the filters are thoroughly charred, and then ignited more strongly to destroy the filter carbon and finally blasted over a Méker burner for at least thirty minutes, or to constant weight, the crucible being covered. After cooling the silica is weighed. For many practical purposes this residue is accepted as silica, unless it is highly colored. For more accurate work, especially where contamination is suspected (silica should be white), this residue is treated further.

Estimation of True Silica. Silica may be contaminated with $BaSO_4$, TiO_2, Al_2O_3, Fe_2O_3, P_2O_5 combined (traces of certain rare elements may be present). The weighed residue is treated with 3 cc. of water, followed by several drops of concentrated sulphuric acid and 5 cc. of hydrofluoric acid, HF (hood). After evaporation to dryness, the crucible is heated to redness and again cooled and weighed. The loss of weight represents silica, SiO_2.[3]

---

[1] Dehydration of silica is aided by the presence of lime and retarded by magnesia. In presence of the latter a soluble magnesium silicate will form if the dehydration is conducted at a temperature much above 110° C., hence it is better to avoid this by taking more time and heating to 100 or 105° as recommended.

Sodium chloride has a solvent action on silica, the reaction of HCl on sodium silicate being reversible; $2HCl + Na_2SiO_3 \leftrightarrows 2NaCl + H_2SiO_3$. An evaporation of the filtrate to dryness will recover the greater part of the silica thus dissolved.

[2] Not more than 0.1% of the original SiO_2 may still be in solution.

[3] Silicic acid cannot be completely dehydrated by a single evaporation and heating, nor by several such treatments, unless an intermediate filtration of silica is made. If, however, silica is removed and the filtrate again evaporated to dryness and the residue heated, the amount of silica remaining in the acid extract is negligible. (See Article by Dr. W. F. Hillebrand, Jour. Am. Soc., **24**.)

NOTES. Lenher and Truog make the following observations for determining silica:[1]

1. In the sodium carbonate fusion method with silicates, there is always a non-volatile residue when the silica is volatilized with hydrofluoric and sulphuric acids.

2. The non-volatile residue contains the various bases, and should be fused with sodium carbonate and added to the filtrate from the silica when the bases are to be determined.

3. In the dehydration of the silica from the hydrochloric acid treatment of the fusion, the temperature should never be allowed to go above 110°.

4. Dehydrated silica is appreciably soluble in hydrochloric acid of all strengths. With the dilute acid used, this error is almost negligible.

5. Dehydrated silica is slightly soluble in solutions of the alkaline chlorides. As sodium chloride is always present from the sodium carbonate fusion, an inherent error is obviously thus introduced.

6. The dehydrated silica along with the mass of anhydrous chlorides must not be treated first with water, since hydrolysis causes the formation of insoluble basic chlorides of iron and aluminum, which do not dissolve completely in hydrochloric acid.

7. Hydrochloric acid (sp.gr. 1.1) in minimum amount should be used first to wet the dehydrated chlorides and should be followed by water to bring the volume to about 50 cc., after which the silica should be filtered off as quickly as possible.

8. Pure silica comes quickly to constant weight on ignition. Slightly impure silica frequently requires long heating with the blast flame in order to attain constant weight, and is then commonly hydroscopic.

9. Evaporations of the acidulated fusion in porcelain give practically as good results as when platinum is used.

10. Filtration of the main bulk of the silica after one evaporation is desirable, inasmuch as the silica is removed at once from the solutions which act as solvents.

11. Dehydration of the silica under reduced pressure has no advantages over the common evaporation at ordinary atmospheric pressure.

12. Excessive time of dehydration, viz., four hours, possesses no advantages.

13. Excessive amounts of sodium carbonate should be avoided, since the sodium chloride subsequently formed exerts a solvent action on the silica. The best proportions are 4–5 sodium carbonate to 1 of silicate. Less than 4 parts of sodium carbonate is frequently insufficient completely to decompose many silicates.

14. The non-volatile residue has been found to be invariably free from sodium. Pure silica, on fusion with sodium carbonate, subsequently gives no non-volatile residue.

ANALYSIS OF SILICATE OF SODA

Determination of Na₂O

Five grams of the sample are dissolved in about 150 cc. of water and heated; 1 cc. of phenolphthalein is added and then an excess of standard sulphuric acid from a burette. The excess acid is titrated with standard sodium hydroxide to a permanent pink.

$$H_2SO_4 \times 0.6321 = Na_2O.$$

Silica. Ten grams of the sample are acidified with hydrochloric acid and evaporated to dryness on the steam bath. The treatment is repeated with additional hydrochloric acid and then the residue taken up with 5 cc. of the acid and 200 cc. of water. The residue is digested to dissolve the soluble salts, filtered, washed and ignited. Silica is determined by loss of weight by volatilization of the silica with hydrofluoric and sulphuric acids. The filtrate is made to 1 liter.

Iron and Alumina. Five hundred cc. (5 grams) of the filtrate from the silica determination are oxidized with HNO_3 and the iron and alumina precipitated with ammonia, washed, ignited and weighed as Al_2O_3 and Fe_2O_3. The residue is dissolved by digestion with hydrochloric acid or by fusion with sodium acid

[1] Victor Lenher and Emil Truog, Jour. Am. Chem. Soc., **38**, 1050, May, 1916.

sulphate, and subsequent solution in hydrochloric acid. Iron is determined by titration in a hot hydrochloric acid solution with standard stannous chloride, $SnCl_2$, solution as usual. If only a small amount of precipitate of iron and alumina is present, as is generally the case, solution by hydrochloric acid is preferable to the fusion with the acid sulphate. The latter is used with larger amounts of the oxides.

Lime, CaO. This is determined in the filtrate from iron and alumina by precipitation as the oxalate and ignition to CaO.

Magnesia, MgO. This is determined in the filtrate from lime by precipitation with sodium ammonium phosphate. The precipitate is ignited and weighed as $Mg_2P_2O_7$ and calculated to MgO. Precipitate$\times 0.3621 =$ MgO.

Combined Sulphuric Acid. One hundred cc. of the filtrate from the silica determination ($=1$ gram) is treated with $BaCl_2$ solution and sulphuric acid precipitated as $BaSO_4$.

$$BaSO_4 \times 0.4202 = H_2SO_4 \quad \text{or} \quad \times 0.3430 = SO_3.$$

Sodium Chloride. Ten grams of the silicate of soda are dissolved in 100 cc. of water and made acid with HNO_3 in slight excess and then alkaline with MgO. Cl is titrated with standard $AgNO_3$ solution.

Water. This is determined either by difference or by taking 10 grams to dryness and then heating over a flame and blasting to constant weight.

NOTE. For detailed procedures for each of the above see special subject.

ANALYSIS OF SAND, COMMERCIAL VALUATION

Silica. Two grams of the finely ground material are fused in a platinum crucible with 10 grams of fusion mixture ($K_2CO_3 + Na_2CO_3$) by heating first over a low flame and gradually increasing the heat to the full blast of a Méker blast lamp. When the fusion has become clear it is cooled by pouring on a large platinum cover. The fused mass on the cover and that remaining in the platinum crucible are digested in a covered beaker with hot hydrochloric acid on the steam bath. The solution is now evaporated to dryness, taken up with a little water and 25 cc. of concentrated HCl and again taken to dryness. Silica is now determined by the procedure outlined under the general method on page 436.

Ferric Oxide and Alumina. The filtrate is oxidized with crystals of solid potassium chlorate, $KClO_3$, and iron and aluminum hydroxides precipitated with ammonia. The precipitate is filtered, washed, ignited and weighed as $Al_2O_3 + Fe_2O_3$.

Calcium Oxide. To the ammoniacal filtrate 10 cc. of ammonium oxalate solution are added, the solution heated to boiling and the precipitate allowed to settle until cold. The solution should not be over 200 cc. The calcium oxalate is filtered off, washed and ignited. The residue is weighed as CaO.

Magnesium Oxide. The filtrate from the lime is made strongly ammoniacal and 10 cc. of sodium ammonium phosphate added. The solution during the addition is allowed to stand cold for some time, three to four hours. The precipitate is filtered and washed with dilute ammonia (1 of reagent to 3 parts of water), then ignited and weighed as $Mg_2P_2O_7$. This weight multiplied by $0.3621 =$ MgO.

For more detailed directions see the individual subjects under the chapters devoted to the element.

Silicon in Cast Iron and Steel

One gram of pig iron, cost iron, and high silicon iron, or 5 grams of steel, wrought iron, and low silicon iron are taken for analysis. (By taking multiples of the factor weight 0.4693, SiO_2 to Si, the final calculation is simplified.) The sample is placed in a 250-cc. beaker and 20 to 50 cc. of dilute nitric acid added. If the action is violent, cooling the beaker in water is advisable. When the reaction subsides, 20 cc. of dilute sulphuric acid, 1 : 1, are added, the mixture placed on the hot plate and evaporated to dense white fumes. The residue is taken up with 150 cc. of water containing 2 to 5 cc. of sulphuric acid and heated until the iron completely dissolves.

The solution is filtered and the silica residue washed first with hot dilute hydrochloric acid, sp.gr. 1.1, and then with hot water added in small portions to remove the iron sulphate. The residue is now ignited and weighed as silica.

NOTE. If the ash is colored by iron oxide, silica is determined by difference, after expelling the silica by adding 4 to 5 cc. of hydrofluoric acid and a few drops of sulphuric, taking to dryness and igniting the residue.

The following acid mixtures are recommended by the U. P. Ry. For steel, wrought iron and low silicon iron, 8 parts by volume of HNO_3, sp.gr. 1.42; 4 parts of conc. H_2SO_4, sp.gr. 1.84; 6 parts HCl, sp.gr. 1.2 and 15 parts by volume of water.

Silicon in Ferro Silicon

Decompose the alloy with Eschkas mixture (Na_2CO_3–MgO) fusing in a platinum (or nickel) crucible. Dissolve the fusion in HCl and evaporate to dryness. Bake the residue at 135° and take up with HCl and water. Filter, wash and ignite to SiO_2. If titanium is present make a cold acid extraction and digest for one hour with dilute HCl. Take to fumes with 2 cc. H_2SO_4. Take up with water, filter, wash and ignite residue to SiO_2.

$$SiO_2 \times 0.4693 = Si$$

Causes of Error in Silica Determinations

F. G. Hawley (Min. Eng. Jour., March 31, 1917) states the following:

It has been shown that there are three main sources of error in the method commonly used for the determination of silica: A plus error from the small amount of SiO_2 from the flux; another plus error due to impurities retained by the SiO_2; and a minus error due to the solubility of SiO_2 in HCl.

By far the largest and most troublesome error is the one due to the solubility of the SiO_2 in the HCl. This error is much greater than most chemists realize.

The amount of freshly precipitated SiO_2 that dissolves in HCl, depends on the following conditions: First, the amount of acid present; second, the strength of the acid; third, the temperature; and fourth, the length of time the silica is in contact with the acid. There may be other conditions governing the solubility of SiO_2 in HCl, but these seem to be the principal ones. By far the most important is the amount of acid used.

Most chemists use acid of about the same strength to treat the dehydrated silica, and the assays are brought to a boil, insuring the same temperature, and are boiled for a fairly uniform length of time. Few, however, consider the importance of using a definite amount of acid. It seems to be a prevalent idea that the amount of silica dissolved by the acid solution is proportional to the amount of SiO_2 in the sample. When a small bulk of acid is used, this is certainly not the case. Experiments prove clearly that when a small amount of SiO_2 is present, the amount dissolved is proportional to the quantity of acid solution present and not to the amount of SiO_2 in the sample.

Recommended Procedure for Silica (Hawley's Method)

A modification of the peroxide method, as used for routine work, is as follows: Weigh 0.5 gram of pulp into a 30 cc. nickel crucible and add one scoopful (about 4 grams) of flux composed of equal parts of sodium peroxide and sodium hydroxide. Mix the pulp and flux, and if the sample is known to contain over 50% of SiO_2 put on a cover to prevent loss. Fuse at a low temperature, beginning much below redness and increasing very slowly until a dull red is reached.

When the fusions are made as described the nickel crucibles can be used from 20 to 40 times, but if the temperature is too high, or if there is too much sodium peroxide in the flux, they will burn out much quicker.

Remove the crucibles and partly cool; place in 4-in. casseroles, and cover the crucibles with 2-in. watch glasses so placed that a slight opening is left on one side. Through this opening squirt in 2 or 3 cc. of warm distilled water from a wash bottle. This should start a vigorous reaction between the water and the flux. As soon as the action has somewhat diminished add 3 or 4 cc. more water and continue to do this until the fused mass is disintegrated. Toward the end the water should be added with enough force to thoroughly stir the contents. If the crucible gets too cold or the water is not hot enough, the action may cease before the melt is loosened from the crucible; and if too hot the contents may boil over. With a little practice the right conditions are readily found. As soon as the action ceases and the crucible is a little over half full, rinse off the watch glass and from a large burette add about 10 cc. of 60% HCl in small portions so as to avoid too violent reaction; then add 90% HCl until it is in excess. The crucible should now be about full and everything in solution except possibly a little gelatinous silica.

With the fingers or platinum-tipped tongs, remove the crucible, rinse, and place the casserole on the hot plate to evaporate. No harm is done if it boils gently at first. When about half evaporated, place the casserole on an iron or aluminum ring so made that the bottom of the casserole is kept about one-quarter inch above the hot plate. These rings are very beneficial in preventing spitting. When the residue has become dry, cover with a watch glass and bake at about 125° C. for 30 min.; remove and when cool add 15 cc. of 60% HCl and turn about so as to moisten all parts of the residue. Allow to stand a short time and then put on the hot plate, and with cover still on, boil for 3 min. Remove and allow to cool, rinse the sides down with the minimum amount of warm distilled water, and swirl around to loosen any crust still adhering to the sides. The NaCl does not all dissolve, but will readily do so in the wash water.

It is very necessary that the casserole should not be heated after rinsing down the sides with water. After standing a few minutes, transfer the contents to a filter and wash twice with warm water; then once with hot dilute HCl and add a little to the casserole. Rub the sides and bottom of the casserole to loosen any adhering SiO_2 and rinse into the filter. This SiO_2 is ground rather fine by the rubbing and has a tendency to clog the filter, hence it is better to add this after the main portion has been partly washed. Wash the filter twice again with water, place in a crucible, partly dry, and ignite strongly for ten minutes. Weigh the SiO_2 as soon as cool, for it is hygroscopic. A correction is now made for the SiO_2 lost in solution, which under these conditions will amount to about 0.4%. A deduction is made for impurities in the SiO_2 and the SiO_2 from the flux. These gains about balance the solubility loss. Occasional tests should be made to check these losses and gains.

SILVER

Ag, *at.wt.* **107.88**; *sp. gr.* **10.50-10.57**; *m.p.* **960.5° C.**; *b.p. about* **1950° C.**; *oxides,* Ag$_2$O, Ag$_4$O, Ag$_2$O$_2$

DETECTION

A trace of silver in most substances is detected with greatest certainty by furnace assay methods.

The wet method of detection of silver most commonly practiced, depends upon observation of the properties of the precipitate formed by the addition of a not excessive amount of alkaline chloride to a cold nitric or sulphuric acid solution of the substance undergoing examination. One-tenth milligram of silver precipitated as silver chloride in a cold 200-cc. acid solution gives a very perceptible opalescence to the liquid.

Silver chloride is white when freshly precipitated, tinted pink when palladium is present; in colorless liquids on exposure to light turns brown, violet, blue or black. By agitation, heating or long standing the precipitate becomes coagulated or granular and in such a state is retained by an ordinary filter. The presence of some forms of organic matter prevents coagulation.

Silver chloride is dissolved by concentrated hydrochloric acid; raising the temperature of the acid assists the action. It is dissolved by sodium thiosulphate, alkali cyanides, mercuric nitrate, and alkaline chlorides.

From mercurous chloride, silver chloride, except when constituting a small proportion of the precipitate, is distinguished by its solubility without decomposition in ammonia. Precipitation from its ammoniacal solution is accomplished by acidifying. Lead chloride, precipitable also by hydrochloric acid, is not flocculent, does not coagulate, but dissolves quite freely by heating. Addition of hydrochloric acid to a solution of silicon, tellurium, thallium, tungsten or molybdenum may produce a precipitate, in each case, easily distinguishable from that of silver chloride, but may mask traces of the salt.

Silver, in a cold solution containing free nitric acid, only a small amount of colored salts and no mercury, may be detected through the formation of a white precipitate, similar in appearance to silver chloride, by addition of a slight excess of an alkaline thiocyanate.

When a solution of silver salt is added to a mixture of 20 cc. ammonium salicylate (20 grams. salicylic acid neutralized with ammonia, a slight excess added and the whole made up to 1000 cc.) and 20 cc. of a 5% solution of ammonium persulphate added, an intense brown color is produced, which will detect the presence of a 0.01 milligram of silver. Lead does not affect the test.

When it appears that the chloride or thiocyanate test for silver is not positive on account of the presence of other precipitable elements, the precipitate, after it

Chapter contributed by W. G. Derby.

settles, is filtered through the finest quality of paper, and the mixture of the ash of the incinerated filter with dry potassium carbonate heated on charcoal with a mouth blowpipe. If silver is present and not associated with a large amount of palladium, there will be found on the charcoal pellicles of the color characteristic of silver, which have no white or yellow sublimate when melted in the oxidizing flame of the blowpipe. The pink palladium salts of silver precipitated by a chloride or thiocyanate before the blowpipe produces metal which is dull in appearance and not readily melted.

NOTES. Silver may be recognized in a solution of concentration 1 to 240,000 by the reduction of its salts with alkaline formaldehyde.[1] Whitby's[2] method of detection and estimation of small amounts of silver depends upon the formation of a yellow color through addition of sucrose and sodium hydrate. Ammonium hydrate interferes, but bismuth, cadmium, copper, mercury of either valence, lead or zinc, in amounts equal to that of the silver, do not.[3] Maletesta and DeNola add to the solution to be tested a few drops of a solution of nitrate of chromium and then potassium hydrate to alkalinity. A brownish turbidity or black precipitate of silver oxide forms. The limit of sensitiveness is 0.5 milligram in 100 cc.

ESTIMATION

Silver is determined in copper, lead, silver, sulphur or other ores, in copper and lead furnace by-products, and in lead by furnace assay methods, in which a preliminary acid treatment of the sample is rarely employed; in native copper ore, in copper, copper alloys, gold, gold alloys and in the slime from the electrolytic refining of copper or lead by furnace methods, in which a preliminary acid treatment of the sample is employed, in silver alloys by volumetric or gravimetric methods; in mercury by a gravimetric method; in cyanide mill solution or solutions containing much organic matter by furnace process on the residue obtained by evaporation or precipitation; in silver plating electrolyte by electrolysis.

Solubility. Nitric acid, dilute or concentrated, attacks silver rapidly when hot. The presence of a soluble chloride, iodide or bromide in the solvent or substance will retard and may prevent solution. Unless oxidizing agents are present, dilute sulphuric acid has practically no action on massive silver, but hot, strong acid commences to be an active solvent at a concentration of 75% H_2SO_4. Hydrochloric acid attacks silver superficially. The action of alkaline hydrates or carbonates in solution is inappreciable; in a state of fusion, slight.

Furnace Assay Methods. These will be described in the chapter devoted to that subject.

[1] Armani and Barboni, Zeit. Chem. Ind. Kolloide, **6**, 290.
[2] Zeit. Anorg. Chem., **67**, 62; C.A., **4**, 1444.
[3] Bull. Chim. Farm., **52**, 533.

GRAVIMETRIC METHODS FOR THE DETERMINATION OF SILVER

Precipitation as Silver Chloride [1]

Introductory. Although silver might be determined as an iodide or bromide, the fact that these halides are more sensitive to light than the chloride, and decompose more readily, with liberation of the halide and the formation of sub-halides, has led to the precipitation of silver as the chloride.

Reaction. $AgNO_3 + HCl = HNO_3 + AgCl$.

Reagents. *Hydrochloric Acid.* One volume of strong HCl (sp.gr. 1.19) diluted with five volumes of water (sp.gr. of dilute HCl 1.035); 1 cc. contains 0.074 g. of HCl, equivalent to 0.219 g. of Ag.

Nitric Acid. One volume of strong HNO_3 diluted with 1.6 volumes of water (sp.gr. of acid is 1.2); 1 cc. contains 0.38 g. of HNO_3, which would dissolve 0.64 g. of Ag.

1. Preparation of the Sample. Solution. *Silver Alloys.* Place 0.5–1.0 gram of the alloy in an Erlenmeyer flask and add 5 cc. of the dilute nitric acid. Heat gently until the alloy is dissolved and the brown fumes are expelled. The solution is now diluted to about 100 cc. and the silver precipitated as stated below.

Soluble Silver Salts. The salt is weighed into a weighing bottle; 1.0–2.0 grams are sufficient for a determination. The solution is now diluted to about 100 cc. and the silver precipitated as stated below.

The Halides of Silver. These are best brought into solution by fusion with about six times the weight of the sample of sodium carbonate. This converts the silver into the carbonate and the halide combines with sodium and is dissolved out in water. The silver carbonate is washed free of the halide and then dissolved out in dilute nitric acid.

Ores of Silver. These may be brought into solution by digestion with nitric acid, the residue remaining is treated as stated above under halides of silver. Unless the ore is very high in silver, it is preferable to make the analysis by Fire Assay.

2. Precipitation of Silver Chloride. Heat the solution to boiling and add from a burette, drop by drop, 5 cc. of dilute hydrochloric acid. This is sufficient to precipitate over 1 gram of silver. The excess of acid is desired as the chloride is less soluble in free hydrochloric acid.

NOTE. The chloride is soluble in strong hydrochloric acid, hence a large excess is undesirable. Shaking or vigorously stirring the mixture will clear a cloudy solution. This is necessary to coagulate the silver chloride, as the fine suspended silver chloride will pass through the filter paper.

3. Filtration: (Procedure if Filter Paper is Used). Decant the clear solution into the filter. Test the filtrate with a drop of dilute HCl to make sure all the silver is precipitated. Now wash two or three times by decantation, using hot water containing 1 cc. of HNO_3 per 100 cc. of distilled water. Transfer the silver salt to the filter and continue washing until free from chlorides. Six to eight additional washings should be sufficient.

4. *Dry* the filter and silver salt in the oven at 100–110 degrees C.

5. *Remove* as much of the silver chloride as possible from the paper, placing the salt on a glazed sheet of paper, covering it with a watch glass.

[1] Contributed by Wilfred W. Scott.

6. *Ignite* the filter in a crucible (whose weight has been ascertained), then add to the ash a drop of nitric acid and a drop of hydrochloric acid. Heat gently to expel the acids. (Handle the crucible with tongs. Do not place on the table.)

7. *Transfer* the chloride from the glazed paper to the crucible and heat gently until the salt just begins to fuse on the sides of the crucible.

8. *Cool* in a desiccator for fifteen to twenty minutes.

9. *Weigh* as AgCl, making an allowance for the weight of the crucible. $AgCl \times 0.7526$ gives the weight of Ag in the salt.

10. Calculate the per cent silver from the weight of sample taken.

3a. **Procedure if a Gooch Crucible is Used.** Prepare a Gooch crucible with a fairly thick pad of asbestos fibre (1/8 in. thick). Wash once with alcohol and dry to constant weight at 110 degrees C. Keep a record of the weight.

4a. *Wash* the precipitate by decantation, pouring the washings through the Gooch, with application of suction. Transfer the chloride to the crucible and wash free of chlorides.

5a. *Finally* wash once with alcohol and dry at 110 degrees C. to constant weight.

6a. *Calculate* the percentage of silver as directed in the first method.

NOTES. *Solubility* of the silver halides. Milligrams of salt per 100 cc. of water. AgCl 0.00017; AgBr 0.00004; AgI 0.00001.

Interferences. Antimony, mercury, and lead interfere and should be removed if present.

Paper is separated in the first procedure as the carbon reduces the salt to metallic silver, causing low results.

Gooch. If the asbestos fibre is poor, a loss of the fibre will occur during washing of the precipitate, causing low results.

Light. Strong light will affect the salt causing the formation of the subhalide of silver and the liberation of chlorine. A drop of nitric followed by a drop of hydrochloric acid will restore the original form. This treatment is necessary only when a dark-colored salt is obtained by light action.

Large Samples. It is frequently advisable to dissolve larger samples than stated. The solution is made to 500 cc. and a portion taken for analysis.

Solubility. Nitric acid, dilute or concentrated, attacks silver rapidly when hot. The presence of a soluble chloride, iodide or bromide in the solvent or substance will retard and may prevent solution. Unless oxidizing agents are present, dilute sulphuric acid has practically no action on massive silver, but hot, strong acid commences to be an active solvent at a concentration of 75% H_2SO_4. Hydrochloric acid attacks silver superficially. The action of alkaline hydrates or carbonates in solution is inappreciable; in a state of fusion, slight.

Determination as Silver Cyanide

In the analysis of mercury, the nitric acid solution of the metal is nearly neutralized with a solution of sodium carbonate. Potassium cyanide solution is then added until the precipitate, which first forms, is dissolved. Then under a hood with strong draft, dilute nitric acid is added in slight excess of the quantity required to combine with the base in the amount of potassium cyanide present. The precipitate of silver cyanide, practically insoluble in dilute nitric or hydrocyanic acid, is coagulated by stirring or long standing and filtered from the cold solution of mercuric nitrate by use of a tared paper-bottomed Gooch crucible. The precipitate is washed with cold dilute nitric acid (1–10) until a test of the washings with hydrogen sulphide shows the absence of mercury. The crucible is dried at 100° to constant weight.

$$AgCN \times 0.8057 = Ag.$$

NOTES. Determination of silver as metal through precipitation with hypophosphorous acid [2] as silver sulphide or as silver chromate [3] are methods of doubtful technical application.

Exner,[4] using a platinum dish as the cathode and a 2-in. diameter bowl-shaped spiral anode revolving 700 R.P.M., deposited 0.4900 gram from about 125 cc. of a hot electrolyte containing 2 grams potassium cyanide in ten minutes at N.D.$_{100}$ 2 amps.

The above methods presume the absence of other metals precipitable under the conditions mentioned.

Electrolytic Method [5]

According to the strength of the silver bath 10 or 20 cc. are filtered into a tared 200-cc. platinum dish and according to the greater or smaller excess of cyanide present, $\frac{1}{2}$ to 1 gram of potassium cyanide in solution is added. The electrolyte diluted to about a half inch from the edge of the dish is kept, by a flame underneath, at a temperature of 60°–65° C. during the period of electrolysis at N.D.$_{100} = 0.08$ amp.

Complete precipitation, which requires three to three and a half hours, is recognized by test with ammonium sulphide. Without interruption of the current, by use of a siphon, displacement of the electrolyte with water is accomplished. The dish is rinsed with alcohol and ether, dried at 100°, weighed and silver obtained calculated to grams per liter or cubic foot.

NOTES. Benner and Ross[6] deposit 0.15 gram in twenty minutes with a current of 3 amperes from 50 cc. of electrolyte containing 8 grams potassium cyanide and 2 grams potassium hydrate on a 9-gram platinum gauze cathode.

[1] Z. anal. Chem., **48**, 79. L. E. Salas in communication with the author.
[2] Mawrow and Mollow, Zeit. anorg. Chem., **61**, 96.
[3] Gooch and Bosworth, Am. J. Sci., **27**, 241.
[4] J. A. C. S., Sept., 1903, 900.
[5] Langbein, "Electro-Deposition of Metals," 6th Ed.
[6] J. A. C. S., July, 1911, 1106.

VOLUMETRIC METHODS FOR DETERMINATION OF SILVER

Volhard's Thiocyanate Method [1]

This method is especially adapted to the determination of silver in cold dilute nitric acid solution. The method is based on the greater affinity of silver ions than ferric for thiocyanate ions. When the silver has been precipitated as thiocyanate, the ferric indicator reacts with the thiocyanate producing the characteristic red color.

Reactions: $AgNO_3 + KCNS = KNO_3 + AgCNS$ ppt.

$$Fe(NO_3)_3 + 3KCNS = 3KNO_3 + Fe(CNS)_3 \text{ red}$$

NOTE. Mercury and palladium, highly colored salts of cobalt and nickel, copper if over 60% in the sample, nitrous acid and chlorine interfere and should be absent.

Ferric Indicator. *Saturated Solution.* Make 100 cc. of a saturated solution of ferric ammonium sulphate or ferric sulphate. Add sufficient HNO_3 (freed from nitrous acid by heating) to clear up the solution and produce a pale yellow color, .5 cc. of this reagent is used in a test. Ferric nitrate may be used in place of sulphate.

Standard Silver Solution. A N/10 solution contains per liter 10.788 grams of silver, or 16.989 grams of $AgNO_3$. A solution containing 0.005 gram of Ag per cc. is a convenient strength.

Dissolve 1.0 gram of pure silver foil in 10 cc. of dilute HNO_3, 1 : 1.6 (sp.gr. 1.2). Boil to expel the nitrous oxides and dilute to 200 cc. 1 cc. will contain 0.005 gram of silver.

Thiocyanate Reagent. Dissolve 7.4 grams NH_4CNS or 9.2 grams of KCNS in water and dilute to 1000 cc. Standardize the solution against the standard silver solution. Half this strength is used for the weaker silver solution above.

Standardization. Measure 50 cc. of the standard silver solution into a beaker or an Erlenmeyer flask and dilute to 100 cc.

Add .5 cc. of the ferric indicator.

Titrate with the thiocyanate reagent until a permanent red tint is obtained. Each addition of the reagent will produce a temporary red color which fades immediately as long as any silver remains uncombined with the thiocyanate. A trace of excess of the reagent produces a permanent faint red color.

Note the cc. required and calculate the value of 1 cc. in terms of silver. 50 cc. of the standard silver solution contains 0.25 g. of Ag.

Some prefer to have the thiocyanate exactly equal in strength to the silver solution. Should this be desired, dilute to the necessary volume and again standardize against the silver solution.

The value of 1 cc. should be recorded on the container.

Determination of Silver in the Unknown

Weigh 0.25–0.3 gram of the alloy and dissolve in an Erlenmeyer flask by addition of 5 cc. of dilute HNO_3 (sp.gr. 1.2). Heat to expel lower oxides.

Cool, dilute to about 100 cc. and add 5 cc. of the ferric indicator.

Titrate with the standard thiocyanate reagent to a permanent faint red color.

From the cc. of the reagent used, calculate the amount of silver present in the sample taken.

Divide the result by the amount of sample taken and multiply by 100 = per cent Ag in the alloy.

[1] Contributed by Wilfred W. Scott.

GAY–LUSSAC METHOD

This very accurate method is especially adapted to the valuation of silver bullion, but may be applied in principle to the determination of silver in a nitric acid solution which contains as little as 100 milligrams of the metal, providing the volume of the solution is not so large or color so deep as to make a precipitate of silver chloride equivalent to 0.1 milligram of silver indistinguishable. Metals that interfere are mercury and tin.

The method is founded upon the almost absolute insolubility of silver chloride or bromide in cold dilute nitric acid and the property of the precipitate becoming so completely coagulated through agitation that it settles speedily, leaving a liquid sufficiently clear to permit of observance of any precipitate produced by further addition of precipitant.

Reactions: $AgNO_3 + NaCl = NaNO_3 + AgCl \downarrow$
$AgNO_3 + NaBr = NaNO_3 + AgBr \downarrow$

The use of a bromide is preferable to a chloride salt as a reagent, chiefly because on account of the greater insolubility of silver bromide, the end-point of the operation of titration is more sharply defined.

The presence of free sulphuric acid is prejudicial to a very close determination, because of the volume of liquid required to keep silver sulphate in solution, and also because the result of agitation after addition of precipitant is apt to be a fine precipitate which does not readily settle.

The factor of volume change per degree change of temperature from 15 to 21° C. is approximately 0.00012; from 20 to 26°, 0.00019; from 25 to 31° C., 0.00024.

Although the approximate precipitating value should be known by previous test, it is the better practice to determine the exact value by running two or more checks of pure silver simultaneously with each batch of assays than to apply the temperature correction factor.

Apparatus. The apparatus required consists of a pipette which will deliver approximately 100 cc. with an accuracy of not over 5 milligrams variation in weight of the standard solution at constant temperature between successive deliveries, 10 cc. burettes with glass stopcocks; and 8-oz. narrow mouth, round, flint-glass bottles with high, tightly fitting stoppers; the assay bottles should be of a quality which will endure heating in a steam bath or on a hot plate.

Fig. 59. Apparatus for Gay-Lussac Method.

Since the end-point by the Gay-Lussac method depends upon the observance of cessation of precipitation, it is evident, in order to avoid undue tediousness in its operation, that the silver content of the amount of sample taken for assay should be known within a few milligrams.

37

METHOD OF U. S. MINTS

U. S. Mint Modification of the Gay-Lussac Method for Silver [1]

This method is used in all three of the United States Mints and the U. S. Assay Office, New York City, for determining silver in ingots and fine silver has been found very satisfactory both as regards speed and accuracy.

Standard Solutions. Two standard salt solutions are regularly used in the determinations. The first is called a " normal " and the second a " decimal " solution.

The first or " normal " solution is made of such concentration that 100 cc. of it will precipitate exactly 1002 milligrams of silver. 5.43 grams of C. P. sodium chloride are dissolved in water and diluted to make one liter of solution. It is kept in a large 40-liter carboy and is siphoned off as needed.[2]

The decimal solution is made by diluting 100 cc. of the " normal " solution to a liter.

Standardization

The normal solution must be standardized at frequent intervals because of temperature changes which affect the concentration of the solution. The factor of volume change per degree change of temperature from 15 to 21 degrees C. is approximately 0.00012; from 20 to 26 degrees, 0.00019; from 25 to 31 degrees, 0.00024.

The standardization is carried out as follows:

Solution and Precipitation. A " proof " of 1004 milligrams of fine silver is carefully weighed out, placed in a glass-stoppered 8-oz. bottle and dissolved in 10 cc. of 1 : 1 nitric acid on a hot plate. Then 100 cc. of " normal " salt solution, sufficient to precipitate 1002 grams of silver, are added from an upright stationary pipette. The pipette is filled by means of a siphon controlled by a stopcock convenient to the right hand. After filling, the left forefinger is placed over the pipette, the rubber hose connection removed from the bottom, and the bottle containing the dissolved proof placed underneath, when the forefinger is removed, allowing the contents to drain into the bottle, shaking the bottle once or twice to mix the solution. Then 2 cc. of the decimal solution are added by means of a small pipette graduated in cc. and held in the hand, and the stoppered bottle is placed in the shaker.

The shaker violently agitates the solution and causes the precipitate to coagulate and settle. The bottle is removed after four minutes.

More agitation than is absolutely necessary should be avoided, due to the increasing tendency of the precipitate to become granular and settle slowly.

[1] Communicated to W. W. Scott by F. C. Bond, Humid Assayer, Denver Mint, Colorado.

[2] 40 liters are made up at one time by the Denver Mint. The strength of the solution may be regulated by the size of the pipette used. At the Denver Mint 4.82608 grams per liter are taken of the C. P. NaCl, since the pipette delivers more than 100 cc.

Titration. The bottle containing the coagulated precipitate is best placed upon a shelf in a window through which only reflected light enters, at such a height that the top of the solution is upon a level with or slightly above the eye. The shelf is backed by a blackened board which covers the window under the shelf and extends nearly to the top of the bottle.

The bottle stands a moment to allow the precipitate to settle and 1 cc. of the decimal salt solution is added from the hand pipette. The solution is shaken by moving the top of the bottle through a small arc once or twice and the reading is taken after 10 seconds. A slight white cloud forming at the top of the solution and more pronounced when viewed from below constitutes a " show " and indicates that only a small portion of the cc. added was needed to precipitate the remaining silver. This is the desired condition for a proof.

The reading is taken as a " show," " quarter," " half," " three quarters," and " one "; according to the portion of the cc. of salt solution necessary to precipitate the remaining silver. If the cloud is deep enough to indicate that all of the cc. has been used, the bottle should be placed in the shaker and the precipitate coagulated, after which another cc. is added and the reading taken as before with the addition of one cc.

The assignment of the proper value to the precipitate is difficult for the novice and experience in comparison is of much more value than any description could be. However it may be stated that a slight precipitate extending through the upper half of the solution after a slight uniform shake should be called a " quarter," a precipitate of the same appearance throughout the solution is a " half," a heavier precipitate throughout is called " three quarters," while a still denser precipitate is read " one " and should be confirmed by shaking and adding another cc., which should yield a " show," a very faint cloudiness.

The " show " of the proof influences the reading of the determinations and its appearance should be kept constantly in mind, since a " quarter " on a determination means that one quarter of a cc. more of the decimal salt solution was used in precipitating silver than was used in the proof. Thus the proof reading or " show " is taken as zero and the concentration of the " normal " solution should be adjusted so that the proof gives as light a show as possible.

Procedure. In the following determinations it is advisable to run a standard of proof silver side by side with the sample bullion for comparative purposes.

Fine Silver. For silver bullion 998 parts fine or above a sample of 1005 milligrams is weighed out, dissolved, precipitated and titrated as described under Standardization.

In case 1 cc. was added, gave a heavy precipitate, was agitated and a second cc. added which gave a " half," the reading would be $1\frac{1}{2}$ and the silver would be

$$\frac{1002+1\frac{1}{2}}{1005} = \frac{1003.5}{1005} = 998.5 \text{ fine.}$$

In case a large number of samples are to be run, tables may be prepared for each fourth of a cc. which will make the above calculation unnecessary.

Coin Ingots. In determining the silver in silver coins or in silver coin ingots as they come from the melting room, which are usually within $1\frac{1}{2}$ points of 900 fine, the sample weighed is 1115 milligrams. The color given to the solution by the copper base need not interfere with the titration.

Notes on the Method

Determinations may be made on silver bullion of almost any grade if the approximate fineness is previously determined by fire assay or the Volhard method. It is ordinary practice to weigh up the sample at the next figure even five milligrams above that calculated. Thus if it is found from preliminary assay that 1082 milligrams of bullion will contain approximately 1002 milligrams of silver, 1085 milligrams will be weighed out for a sample.

Interfering Elements. There are very few substances which will be found in bullion in sufficient quantity to interfere with the process. The presence of free sulphuric acid is detrimental to a very close determination.

The use of a bromide is considered as preferable to a chloride as a reagent but the chloride is commonly used.

An eyeshade assists in making the readings accurately.

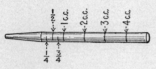

FIG. 59a.[1]

The chloride precipitate is reduced to a blue subchloride on standing in the sunlight so that the bottle should be exposed to the light as little as possible.

A set of twelve samples, with the bottles transported in a suitable wire frame, is usually run at one time.

A decimal solution of silver nitrate of equal strength with the decimal salt solution may be used for back titration, however the end-point is less distinct and it is advisable to weigh out a larger sample.

Duplicates are commonly run.

Tables giving the fineness for different classes of materials examined for each reading facilitate calculations and are recommended for use.

To determine the $\frac{1}{4}$, $\frac{1}{2}$ and $\frac{3}{4}$, a beginner should have a pipette, graduated in 1 cc., holding 5 to 7 cc. length for hand use with one cc. divided in the $\frac{1}{4}$, $\frac{1}{2}$, $\frac{1}{3}$, and he should use the same until he is familiar with the density of precipitates produced by one cc. with silver equivalent to the above fractions.

WEIGHT TAKEN 1115 MG.

	0	1	2	3	4	5	6
0	896.9	897.7	898.6	899.6	900.4	901.3	900.2
$\frac{1}{4}$	897.1	898.0	898.9	899.8	900.7	901.6	902.5
$\frac{1}{2}$	897.3	898.2	899.1	900.0	900.9	901.8	902.7
$\frac{3}{4}$	897.5	898.4	899.3	900.2	901.1	902.0	902.9

No pipette is of use in the practice of the Gay-Lussac method which shows any tendency to spatter at the beginning or ending, or yields a quickly following or clinging drop at the completion of discharge. The film of liquid adherent to the inner surface of the body of a good pipette will drain without sign of rivulet effect and be retained by the capillary of the discharge tube for at least a minute.

In the following table the left hand column represents the milligrams of bullion to be taken, the top line indicates the cc. of decimal solution required in addition to the 100 cc. of normal solution, the figures at the intersecting lines give the fineness of the bullion.

Ascertain the approximate fineness by a preliminary assay, consult the 0 column for the nearest corresponding figure slightly higher, the figure on the left of this is the weight of bullion to be taken. Now if the test required, in addition to the 100-cc. normal solution, 4 cc. decimal solution and 1115 milligrams of bullion were taken, the fineness of the bullion would be 900.4. See table under Silver Coin Bullion.

High Grade Bullion

Milli-grams of Bullion	0	1	2	3	4	5	6	7	8	9	10
1000	1000.0										
1005	995.0	996.0	997.0	998.0	999.0	1000.0					
1010	990.1	991.1	992.1	993.1	994.1	995.0	996.0	997.0	998.0	999.0	1000.0
1015	985.2	986.2	987.2	988.2	989.2	990.1	991.1	992.1	993.1	994.1	995.1
1020	980.4	981.4	982.4	983.3	984.3	985.3	986.3	987.2	988.2	989.2	990.2
1025	975.6	976.6	977.6	978.6	979.5	980.5	981.5	982.4	983.4	984.4	985.4
1030	970.9	971.8	972.8	973.8	974.8	975.7	976.7	977.7	978.6	979.6	980.6
1035	966.2	967.1	968.1	969.1	970.0	971.0	972.0	972.9	973.9	974.9	975.8
1040	961.5	962.5	963.5	964.4	965.4	966.3	967.3	968.3	969.2	970.2	971.1
1045	956.9	957.9	958.8	959.8	960.8	961.7	962.7	963.6	964.6	965.5	966.5
1050	952.4	953.3	954.3	955.2	956.2	957.1	958.1	959.0	960.0	960.9	961.9
1055	947.9	948.8	949.8	950.7	951.7	952.6	953.5	954.5	955.4	956.4	957.3

Silver Coin Bullion

	0	1	2	3	4	5	6	7	8	9	10
1095	913.2	914.2	915.1	916.0	917.0	917.8	918.7	919.8	920.5	921.5	922.4
1100	909.1	910.0	910.9	911.8	912.7	913.6	914.5	915.4	916.4	917.3	918.2
1105	905.0	905.9	906.8	907.7	908.6	909.5	910.4	911.3	912.2	913.1	914.0
1110	900.9	901.8	902.7	903.6	904.5	905.4	906.3	907.2	908.1	909.0	909.9
1115	896.9	897.8	898.6	899.5	900.4	901.3	902.2	903.1	904.0	904.9	905.8
1120	892.9	893.7	894.6	895.5	896.4	897.3	898.2	899.1	900.0	900.9	901.8
1125	888.9	889.8	890.7	891.6	892.4	893.3	894.2	895.1	896.0	896.9	897.8
1130	885.0	885.8	886.7	887.6	888.5	889.4	890.3	891.1	892.0	892.9	893.8
1135	881.1	881.9	882.8	883.7	884.6	885.5	886.3	887.2	888.1	889.0	889.9
1140	877.2	878.1	878.9	879.8	880.7	881.6	882.5	883.3	884.2	885.1	886.0
1145	873.4	874.2	875.1	876.0	876.9	877.7	878.6	879.5	880.3	881.2	882.1
1150	869.6	870.4	871.3	872.2	873.0	873.9	874.8	875.7	876.5	877.4	878.3

Recovery of Silver from Silver Residues

Convert the residues to silver chloride by treating with hydrochloric acid and filtering off the chloride and washing. Dissolve the chloride in ammonium hydroxide added in slight excess. Add sodium hyposulphite, $Na_2S_2O_4$ (not thiosulphate, $Na_2S_2O_3$). Metallic silver is formed. Thiosulphate gives silver sulphide. Photographers' residues containing "hypo" yield silver sulphide.

Combination Methods

Combination of the operations of the Gay-Lussac and Volhard methods have been devised to avoid the tediousness incident to the performance of the Gay-Lussac method by the unexperienced. By the modified methods the amount of sample to be weighed out is determined by preliminary assay, and is dissolved in the same manner as in the practice of the Gay-Lussac method, but with the added precaution to decompose nitrous acid in the silver solution by gentle boiling when completion of the titration is to be accomplished by the Volhard method.

The operation of the combination methods consists briefly of precipitation of all but a few milligrams of silver by a standard solution of alkali thiocyanate, chloride or bromide added from the Stas pipette and estimation of the excess of silver with a decimal solution of thiocyanate or by a colorimetric or nephelometric method.

The procedure favored by the writer is to use a standard solution of potassium bromide as the pipette precipitant. After the liquid is cleared by shaking, it is decanted as completely as possible into a 500-cc. Erlenmeyer flask. The precipitate is washed by five 30-cc. portions of water containing a little nitrous-free nitric acid, each portion being shaken before decanting. Using the same amount of ferric indicator as in the check assays, decimal thiocyanate solution is added until not a very deep tint remains permanent after vigorous agitation. Decinormal silver solution is then added until the tint is discharged. When the assay is sufficiently free of copper or other colored salts to permit accurate matching of tints, the decanted liquid, which may contain particles of silver bromide without interference, is titrated with decimal thiocyanate to the appearance of a tint which will match that of the check assays. Except when colored salts are present in such quantity as to make recognition of the point of bleaching of the ferric thiocyanate coloration uncertain, the extreme range of error is 0.3 part per 1000.

For colorimetric method, see Smith, I.M.M Bull. No. 28. Determination of the residual silver in the filtrate from the thoroughly washed silver bromide precipitate is practicable by use of a suitable nephelometric apparatus.[1]

Denige's Cyanide Method [2]

Silver which has been precipitated as chloride may be determined volumetrically by dissolving the precipitate with a measured quantity of a standard solution of potassium cyanide of about decinormal strength.

$$AgCl + 2KCN = KAg(CN)_2 + KCl.$$

Potassium iodide is then added and the excess of standard potassium cyanide solution determined by addition of potassium iodide and titration to the first appearance of a permanent precipitate with decimormal silver nitrate.

$$AgNO_3 + KI = AgI + KNO_3; \quad AgI + 2KCN = KAg(CN)_2 + KI.$$

Notes. If the last portion of the precipitate of silver chloride dissolves with difficulty in the potassium cyanide, the liquid may be decanted into another beaker and solution completed with ammonia. The solutions are combined.

[1] Richards and Wells, Am. Chem. J., 235, 1903; Richards, ibid., 510, 1906; Richards, Com. 8th Int. Cong. Ap. Chem., Sec. 1, 423.
[2] Clennell, " The Cyanide Handbook," 433.

Miscellaneous Volumetric Methods

Silver may be determined by addition from a burette of a portion of a known volume of its neutral or slightly acid solution to a standard solution of sodium chloride which contains a little potassium chromate or bichromate and sufficient chlorine-free magnesium oxide emulsion to neutralize free acid. The end-point is indicated by the formation of a reddish or brown precipitate.

By Pisani's Method [1] a standard solution of iodide of starch is added to a very dilute neutral solution of nitrate of silver until the fluid becomes permanently blue.

By Vogel's Modification of Pisani's Method, [2] the silver solution, which may contain free acid, is titrated with standard starch iodide solution after addition of nitric acid containing nitrous acid.

By Andrews' Modification, [3] the standard solution of starch iodide is added to a solution of silver nitrate which contains so much ferrous nitrate or sulphate that iron will be in excess of the silver present.

$$2AgNO_3 + 2Fe(NO_3)_2 + I_2 = 2AgI + 2Fe(NO_3)_3$$

By Gooch and Bosworth's Method, [4] silver is determined by precipitating with an excess of potassium chromate, dissolving the precipitate in ammonia, reprecipitating by boiling to low volume and determining iodometrically either the chromate ion combined with the silver, or that remaining after precipitating the silver with a known amount of standard potassium chromate. [5]

Nephelometric Method

This method is practicable for the determination of a small concentration of silver in a clear and colorless liquid. Less than 2 milligrams of silver can be estimated with considerable accuracy by matching the opalescence produced by a drop of hydrochloric acid with that from a known quantity in a liquid of the same volume, depth and temperature. Intensity of opalescence attains the maximum in about five minutes after precipitation. Standard silver solution is made by dissolving 500 milligrams standard silver (see Preparation at close of chapter) with several cc. of dilute nitric in a liter flask and making the solution up to the mark. For most technical determinations the apparatus may consist of clear glass cylinders (color tubes) of suitable size. More accuracy can be arrived at by use of a nephelometer of refined construction, for example [6] the combination of a projection lantern and a Duboscq colorimeter.

Preparation of Pure Silver. The volumetric methods used for the determination of high percentages of silver, employ solutions which should be standardized by metal of the highest purity. For the preparation of this metal, the electrolytic method as described below is preferred by laboratories which are suitably equipped.

[1] Robière, Bull. Soc. Chim., **17**, 306, 1915; J. S. C. I., Oct. 30, 1915, 1073.
[2] Fresenius, "Quantitative Analysis."
[3] Zeit. für Anorg. Chem., **26**, 175.
[4] Am. J. Sci., 2[7], 302.
[5] Am. Chem. Soc. Chem. Abs., Aug. 10, 1909, 1735.
[6] Wells, Am. Chem. J., **35**, 99, 508; Richards, Am. Chem. J., **35**, 510; Dienert, Compt. rend., **158**, 1117.

For the manufacture of a large quantity—several pounds—a basket-like support for the anode is made of several glass rods bent so that they will hang from the rim of a tall 1000-cc. or larger beaker or battery jar and dip into the receptacle about an inch.

Smaller anodes may be supported by the positive wire or by a cloth bag fixed in place by a string under the flare of the rim of the beaker. In any arrangement for the support of the anode, allowance of room should be made for the introduction and free movement of an L-shaped stirring rod.

The cathode may consist of sheet silver or of platinum foil, and lies flat on the bottom of the beaker. The immersed length of the silver or platinum wire leading from the cathode should be covered with rubber tubing.

Commercial silver, usually about 999 fine, may be used for the anode, but by retreatment of the deposit, very impure silver may be used, providing that the quantity of tellurium present is very low. The presence of tellurium will exhibit itself in the impossibility of obtaining the desired coarsely crystalline deposit.

Tellurium in moderate quantities may be removed by melting the silver in a crucible or scorifier, adding niter, permitting the silver to nearly freeze, raising the temperature and pouring into a hot crucible or scorifier in which the operation is repeated, preferably in a muffle furnace, until the surface of the silver is without streaks or spots when cooled to near freezing. An oxidizing atmosphere about the molten metal should be maintained. On the basis of 172 grams silver per cubic inch an anode mould for any convenient amount of silver may be shaped from 4-in. pieces of 1-in. square rod on a smooth iron plate. Just before the anode bar sets in the mould, a silver terminal strip or wire is plunged into it.

After coating the contact wire or strip and the surface of the anode about it with sealing wax, the anode is wrapped with filter paper, held firmly in place by string or rubber bands. If the anode weighs half a pound or more, the anode is also wrapped with cotton flannel which has been washed with water until free of chloride. A porous dish, cylinder or filter cone can be used instead of filter paper and cloth.

The electrolyte contains about 4% of C.P. silver nitrate and half a per cent of chlorine-free nitric acid in distilled water, and fills the beaker or jar so it wets only the lower surface of the anode.

The current, of about 0.1 ampere per square inch of cathode surface at the start, is raised after deposition has proceeded for a few minutes to the limit at which a coarsely crystalline deposit can be maintained.

Inasmuch as the electrolysis proceeds at a rate of 4 grams per ampere hour, some attention is required to break up short circuits and to pack down the rather bulky deposit. The deposit, if coarse, can be washed very easily free of electrolyte, and after heating to near redness is in the form preferred for use by many assayers.

Other methods which may be employed consist of dissolving the crude silver with nitric acid about 1.20 sp.gr. or with hot concentrated sulphuric acid, if platinum is present, separating the gold and platinum by filtration, precipitating AgCl with not too large an excess of HCl, stirring the precipitate until it coagulates, washing repeatedly with hot water until a washing is obtained which shows no precipitate with H_2S, reducing the silver chloride by contact with pure zinc, wrought iron or the silver terminal of a carbon-silver couple aluminum foil, and washing with hot dilute HCl until a test of the decanted liquid indicates absence of the precipitating element. The dried silver, mixed with about 1% of dry sodium carbonate, is packed into a clay crucible, the inside of which has been glazed with borax glass and covered with a layer of crushed charcoal.

The sodium carbonate is omitted in case it is desired to melt silver refined by electrolysis.

The silver melted in the tightly covered crucible is poured into an iron mould which has been chalked or black leaded.

By Knorr's method,[1] a solution of silver nitrate from which excess of nitric acid has been removed by evaporation is freed of metallic impurities by adding enough sodium carbonate to precipitate one-tenth of the silver, boiling and filtering. The silver in the filtrate is precipitated by sodium carbonate and the precipitate decomposed without addition of reducing reagent, by melting in a crucible. Excess sodium carbonate carried down with the precipitate of silver carbonate will cover the fusion and such as adheres tightly to the metal is readily removed by hydrochloric acid. The metal should be smelted under charcoal.

[1] Liddell, " Metallurgists and Chemists' Handbook."

If the cover of the charcoal is omitted or burned away during the fusion, the molten metal is capable of absorbing oxygen from the atmosphere to the extent of about 0 25% of its weight.   This gas is expelled during the passage of the metal into the solid state and produces a casting which cannot be rolled into smooth sheets

The most convenient size and shape of castings for rolling is but little larger than a lead pencil.   Before rolling, the casting is cleaned of particles of the mould wash   After rolling to about cardboard thickness, the sheets may be cut up into strips of convenient size and length, then digested with dilute hydrochloric acid (1 to 5 of water) washed with ammonia and finally with pure water.

The silver then should be dried and annealed by heating to redness.   It is best preserved in a glass-stoppered, salt-mouth bottle and should be exposed to laboratory atmosphere as little as possible.

The purity of each batch of silver made should be compared by use of the Gay-Lussac method with standard silver, the purity of which has been determined by analysis of a 50- or 100-gm. portion for Se and Te, As, Sb, Pb, Cu, Au, and the element employed in reducing silver chloride, if the reduction method was followed in the manufacture of the metal.

THE FIRE ASSAY FOR GOLD AND SILVER[1]

Definitions. *Fire assaying* is a branch of quantitative chemical analysis in which metals are determined in ores and metallurgical products by extracting and weighing them in the metallic state. The methods employed involve slag-melting temperatures and the use of reducing, oxidizing and fluxing reagents, and are in principle the same as those used in metallurgy.

The metals ordinarily determined by fire assaying are gold, silver and platinum. Antimony, bismuth, lead and tin can be determined in this way also, but the results are usually more or less inaccurate.

An *ore* is a mineral aggregate from which one or more metals can be extracted at a profit.

Metallurgical products include a large number of metal-bearing mixtures and compounds, ranging from high grade gold and silver bullion to very weak cyanide and sulphate solutions.

The constituents of an ore are usually divided into two general classes, the valuable minerals containing the *metals*, and the non-valuable minerals or *gangue*. A similar classification can be made in the case of many metallurgical products. In gold and silver bullion and other alloys all of the components are metallic, and the assaying problems involve simply the separation of metals.

General Outline

With ores and metallurgical products containing non-metallic elements the process consists, briefly, in the production of two liquids, liquid lead containing the valuable metals, and liquid slag containing the waste matter or gangue. The two liquids separate from each other by reason of the great difference in specific gravity. The valuable metals are separated from the lead and from each other by taking advantage of differences in chemical properties. The slag is discarded.

In the operation of the process the gold and silver, and platinum, if present, are collected from the metal-bearing portion of the ore or metallurgical product by means of molten lead reduced from litharge or lead oxide. The gangue is converted into a fusible slag by means of reagents known as fluxes.

The effectiveness of the fire assay in separating gold, silver and platinum from ores and metallurgical products depends upon two properties of these metals: first, their weak affinity for non-metallic elements, especially at high temperatures, and second, their very great affinity for molten lead. The collection of the precious metals in the lead, therefore, is the simplest part of the process. The fluxing of the gangue is much more difficult, and requires considerable knowledge and skill. If the fluxing is properly performed, the collection of the valuable metals usually takes care of itself.

Reagents. A flux is a substance which when heated in contact with some difficultly fusible compound either combines with it or takes it into solution,

---

[1] By Irving A. Palmer, Professor of Metallurgy, Colorado School of Mines.

in each case producing a compound or mixture which is easily fusible at ordinary furnace temperatures. The principal fluxes and other reagents used in fire assaying are described in the following paragraphs.

Litharge or oxide of lead, PbO, melting point 883° C., has several important uses. It furnishes the lead which collects the precious metals; it readily combines with silica, producing easily fusible silicates; and it acts as an oxidizing and desulphurizing agent. It is a very strong basic flux.

Sodium carbonate, Na_2CO_3, melting point 852° C., is a powerful basic flux. It combines with silica and alumina, producing fusible silicates and aluminates. When molten it has the property of dissolving or holding in suspension a number of refractory gangue materials. To some extent, also, it acts as an oxidizing and desulphurizing agent. Potassium carbonate, K_2CO_3, melting point 894° C., is rarely used in fire assaying because of its greater cost.

Borax glass, $Na_2B_4O_7$, melting point 742° C., is an acid flux used for combining with or dissolving the basic and some acid constituents of the gangue, producing easily fusible complex borates and mixtures of borates and other compounds. Even silica dissolves to some extent in molten borax glass.

Silica, SiO_2, melting point 1755° C., is a strong acid flux. It combines with metallic oxides and produces silicates which in many cases are considerably more fusible than silica itself.

Granulated lead or **test lead** is used in the scorification assay, which is conducted under oxidizing conditions, and in which, therefore, litharge could not be employed as a source of lead.

Lead foil or **sheet lead** is used in the assay of gold and silver bullion. It, as well as the granulated lead, should be free from silver and bismuth.

Flour is known as a reducing agent. It contains carbon, which reduces lead from litharge. Charcoal was formerly used for this purpose, but it is not so convenient.

Argol or **crude cream of tartar,** $KHC_4H_4O_6$, is both a basic flux and a reducing agent. On being heated it decomposes as follows:

$$2KHC_4H_4O_6 + heat \rightarrow K_2O + 5H_2O + 6CO + 2C.$$

It is effective in assays requiring strong reducing action and low temperatures.

Iron is sometimes used as a desulphurizing and reducing agent. It decomposes most of the heavy sulphides, yielding the metals and iron sulphide.

Potassium nitrate, KNO_3, melting point 339° C., is a powerful oxidizing agent. It is used to neutralize the effect of an excess of reducing substances in the material to be assayed. High sulphur ores, if assayed without previous roasting, require the addition of nitre to the charge. In contact with a reducing agent two molecules of potassium nitrate give up five atoms of oxygen, as shown in the following equation:

$$4KNO_3 + 5C \rightarrow 2K_2CO_3 + 3CO_2 + 2N_2.$$

The potassium oxide coming from the decomposition of the nitre acts as a basic flux.

Common salt, NaCl, melting point 819° C., is a neutral substance sometimes used as a cover for crucible fusions to exclude the air.   When molten it rests on top of the charge and does not enter into it.

FIG. 59a.   Oil Muffle Furnace.

All of the reagents used must be pure and in a finely divided condition. Sodium carbonate shows a tendency to form lumps.   These should be broken up and the entire mass put through a moderately fine screen.

Furnaces and Equipment.   The major operations in fire assaying are usually conducted in muffle furnaces.   The muffle is a box-like receptacle made of fire clay, so placed in the furnace that it is heated on top, bottom and all sides except the front.   In it are placed the refractory vessels containing the material to be assayed.   There is thus no direct contact with the fuel or products of combustion.   The fuel used may be coal, oil, gasoline or gas.

Cupels.   The separation of the precious metals from the lead alloys produced in fire assaying is effected in small shallow vessels of bone ash, known as *cupels*.   The material consists mainly of calcium phosphate, with small percentages of magnesium phosphate, calcium fluoride and calcium carbonate. It is a product of the burning of animal bones, preferably those of the sheep. It should be ground fine enough to pass a 40-mesh screen, in which case about 50 per cent of it will pass a 150-mesh screen.   The cupels are made by moistening the bone ash with a small amount of water and then compressing it in the cupel mould, which consists essentially of a ring and die.   The bone ash is forced into the shape desired at a considearble pressure, so as to insure sufficient rigidity in the cupel. The amount of water needed varies, but should be as low as possible. The cupels should be dried very slowly, so as to avoid

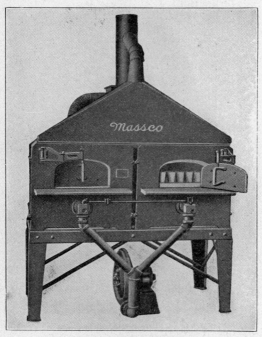

FIG. 59b. Muffle Furnace.

cracking. By using high pressures good cupels can be made from perfectly dry bone ash.

The requirements of a good cupel are that it should be infusible at ordinary furnace temperatures, that it should not be attacked by metallic oxides, that it should be porous, and that it should be sufficiently rigid to permit of considerable handling.

Cupels can be made of Portland cement, magnesia, or of mixtures of these materials with bone ash. They are considerably cheaper but not so satisfactory as those made of bone ash alone.

The Assay-Ton System. As the precious metals are bought and sold by the Troy system of weights, and ores by the Avoirdupois system, considerable time would be lost in calculating assay results, were there no way of avoiding it. To simplify the calculation Prof. C. F. Chandler, of Columbia University, invented the assay-ton system of weights. The assay-ton is equal to $29{,}166\frac{2}{3}$ milligrams. As there are $29{,}166\frac{2}{3}$ Troy ounces in an Avoirdupois ton of 2000 pounds, the number of milligrams and fractions of a milligram of precious metals found in an assay-ton of ore corresponds to the number of Troy ounces in an Avoirdupois ton.

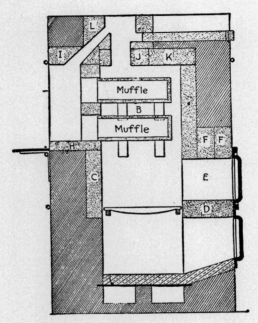

Fɪɢ. 59c.   Cross Section of Two-Muffle Assay Furnace.

Sampling.   It goes without saying that good results in assaying presuppose accurate sampling.   Silver is reported in assay certificates to the nearest tenth of an ounce;   gold usually to the nearest one hundredth of an ounce. One tenth of an ounce means one part in 291,667;   one hundredth of an ounce, one part in 2,916,667.   In the preparation of the sample, therefore, the ratio between the weight of any fractional portion and the weight of the largest particle in it must be very large, so that the accidental inclusion of a number of rich pieces in any portion shall not affect the results beyond the limits of error in assaying.   The final pulp sample should be of a fineness ranging from 80-mesh, in the case of low grade silver ores, to as fine as 200-mesh in the case of non-uniform gold ores.   Small particles of metallic gold in the material necessitate fine grinding and very thorough mixing.   It is a good rule to mix all pulp samples on a rubber cloth before weighing the portions for assay.

Balances and Weights.   The balances used in fire assaying are somewhat different from those found in chemical laboratories.   They are known as flux, pulp and assay balances.   The assay balances are for weighing the gold and silver, often exceedingly small in amount, and are the most delicate type of commercial balances made.   They should be quick in action and not liable to changes in adjustment.   The beam should be short, light and rigid.   The balance should be sensitive to .01 milligram at least. It need not have a capacity of more than .5 gram but should be accurate with that load.

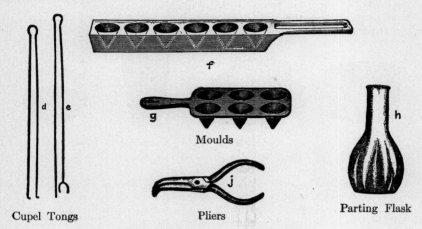

Cupel Tongs   Pliers   Parting Flask

Fig. 59d.   Apparatus for Fire Assay.

In large laboratories separate balances are furnished for weighing the gold. These balances should be adjusted before each weighing and should be handled with the greatest of care.   In the assay of gold ores, when using a half assay-ton portion, every error of .01 milligram in weighing the gold means a variation in the value of the ore of forty cents per ton.

The Crucible Assay.   This method of fire assaying is adapted to the great majority of gold and silver ores and to many metallurgical products.   The process consists in treating a weighed portion of the sample, carefully mixed with the necessary reagents, in a fire-clay crucible.   In order to do this effectively the character of the material to be assayed must be known.   Thus, ores may be oxides or sulphides.   They may be basic, acid or neutral.   They may be strongly oxidizing or strongly reducing.   Each case requires a particular method of treatment.

The amount of sample usually taken is one half assay-ton, run in duplicate. Twenty-gram fire-clay crucibles are used, that is, crucibles capable of holding twenty grams of ore and the necessary reagents.   In most cases the total charge will fill the crucible to within one inch of the top.

Lead Reduction with Oxidized Ores.   Experience has shown that the best results are obtained when the lead reduced from the charge amounts to from 25 to 30 grams.   If the ore is oxidized, a reducing agent must be added to precipitate the necessary lead.   Flour is the reagent ordinarily used, although charcoal or argol can be substituted for it.   The lead is reduced according to the following equation:

$$2PbO + C \rightarrow 2Pb + CO_2.$$

That is, 12 parts of carbon theoretically will reduce 414 parts of lead from

litharge.   Hence, the theoretical reducing power of carbon is $\frac{414}{12}$ or 34.5.
In practice, the reducing power of charcoal is found to range between 25 and
30, and that of flour from 10 to 12.   Argol has a reducing power of about 8 or 9.
In most oxidized ores, therefore, from $2\frac{1}{2}$ to 3 grams of flour will be required to
reduce from 25 to 30 grams of lead from the litharge.

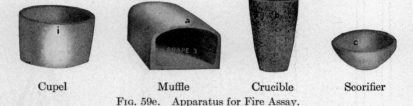

Cupel   Muffle   Crucible   Scorifier
FIG. 59e.   Apparatus for Fire Assay.

If the ore contains ferric oxide, manganese dioxide, or some other easily
reducible oxide, more flour must be added.   Some iron-manganese ores require
as much as 5 grams of flour to throw down the necessary lead.   With unknown
ores the right amount can be determined only by trial.

Lead Reduction with Sulphide Ores.   In the case of ores containing
sulphides, arsenides or other reducing substances, there will be a reduction of
lead without the addition of carbon.   In fact it is usually necessary to add an
oxidizing agent to prevent the precipitation of too much lead.   The following
reactions show the effect of a number of sulphide minerals when heated in
contact with litharge and sodium carbonate.

(1) $PbS + 3PbO + Na_2CO_3 = 4Pb + Na_2SO_4 + CO_2$,

(2) $ZnS + 4PbO + Na_2CO_3 = 4Pb + ZnO + Na_2SO_4 + CO_2$,

(3) $2FeS_2 + 15PbO + 4Na_2CO_3 = 15Pb + Fe_2O_3 + 4Na_2SO_4 + 4CO_2$.

The sodium carbonate induces the complete oxidation of the sulphur to
SO_3, with the formation of the very stable compound sodium sulphate.   In
the absence of an alkaline carbonate most of the sulphur is oxidized to SO_2
only, and the amount of lead precipitated is correspondingly decreased.

Reaction (3) shows that pyrite has a greater reducing power than flour
itself.   If, therefore, a half-assay-ton of ore consisting mainly of pyrite were
to be subjected to a crucible fusion, without the addition of some oxidizing
agent, anywhere from 100 to 150 grams of lead would be reduced.   This
would be entirely too much for the subsequent process of cupellation.   In
order to prevent the reduction of an excessive amount of lead, potassium
nitrate is added to the charge.   The following reactions show the oxidizing
power of this reagent:

(4) $2KNO_3 + 5Pb = 5PbO + K_2O + N_2$,

(5) $2FeS_2 + 6KNO_3 = Fe_2O_3 + 4SO_3 + 3K_2O + 3N_2$.

Reaction (4) shows that 202 parts of nitre will oxidize 1035 parts of lead
to litharge.   The theoretical oxidizing power of nitre, as measured against

lead, is, therefore, 5.12.   Reaction (5) when compared with reaction (3), given above, shows that 606 parts of nitre will oxidize the pyrite needed to

Fig. 59f.   Assay Balances for Small Weights.

reduce 3105 grams of lead from litharge.   Here again the oxidizing power of nitre is shown to be 5.12.   In practice, it is found to be somewhat less, more nearly 4.5.

The fire assay of sulphide ores, therefore, involves either a preliminary assay, or a calculation from the chemical analysis, in order to determine the amount of nitre to be added.   With unknown ores it is better to make a preliminary fusion, using 5 grams of the ore, 75 grams of litharge, 20 grams of sodium carbonate and 10 grams of borax glass.   The button of reduced lead is weighed and its weight divided by 5.   This gives the reducing power of the ore.   From this can be calculated the reducing power of one half assay-ton of the ore, and the amount of nitre necessary to add in order to cut down the weight of the reduced lead to about 30 grams.   An excess of silica or borax glass decreases somewhat the amount of lead by causing the formation of difficultly reducible lead silicates or borates.

Amount of Litharge.   The amount of litharge for a half assay-ton charge

38

usually ranges from 60 to 75 grams. Only about half of this is needed to produce the 25 to 30 grams of metallic lead used as the collector. The excess litharge serves to prevent the reduction of other base metals, such as antimony, bismuth, iron, copper and zinc, to help flux the silica, to act as a solvent for some of the refractory gangue materials, and to make sure that every particle of the ore in the crucible is in close proximity to one or more particles of litharge. In special cases it may be advisable to use a very large excess of litharge, as in the assay of rich gold telluride ores, zinc precipitates and saturated cupels.

Amount of Sodium Carbonate. The amount of sodium carbonate to be used depends somewhat upon the character of the ore, although the modern practice is to use about the same quantity in assaying a great variety of ores and metallurgical products. The principal function of the sodium carbonate is to flux the silica and alumina, which are nearly always present in greater or less degree. The reactions are as follows:

(6) $Na_2CO_3 + SiO_2 = Na_2SiO_3 + CO_2,$

(7) $2Na_2CO_3 + SiO_2 = Na_4SiO_4 + 2CO_2,$

(8) $Na_2CO_3 + Al_2O_3 = Na_2Al_2O_4 + CO_2.$

The two silicates and the aluminate are both quite fusible at ordinary furnace temperatures.

The silicates used in assaying and in metallurgy are usually classified according to the ratio between the oxygen in the acid radical and that in the base. Only four of these type silicates are of any practical importance. They are shown in the following table:

Sub-silicate	4RO,	SiO_2
Mono- or singulo-silicate	2RO,	SiO_2
Sesqui-silicate	4RO,	$3SiO_2$
Bi-silicate	RO,	SiO_2

In the above silicates, the ratios are $\frac{1}{2}$ to 1, 1 to 1, $1\frac{1}{2}$ to 1, and 2 to 1, respectively.

Reactions (6) and (7) show that to flux one part of silica to bi-silicate and mono-silicate requires about $1\frac{3}{4}$ parts and $3\frac{1}{2}$ parts, respectively, of sodium carbonate. If the half assay-ton of ore, therefore, consisted of almost pure quartz, it would take 25 grams of sodium carbonate to flux it to sodium bi-silicate, and 50 grams to flux it to the mono-silicate. As a matter of fact, the bi-silicate slag is satisfactory in this case. In general, the acid silicates have lower melting points but greater viscosity than the basic silicates. The excess litharge in the charge also combines with silica, and may thus produce more basic silicates. At any rate, a mixture of silicates usually has a lower melting point than that calculated from the melting points of its components.

Reaction (8) shows that 1 part of sodium carbonate is required to flux 1 part of alumina.

In practice from 30 to 35 grams of sodium carbonate are used in a half-assay-ton charge. In many cases this may seem to be a large excess. It must be remembered, however, that this reagent serves also to assist in the oxidation of the sulphides through the formation of sodium sulphate, that it has a solvent effect upon refractory oxides and other substances, and that it increases the bulk of the charge, thus protecting the ore from the action of the air and from the escape of the more volatile metals and their compounds. Being very fusible itself, an excess also serves to increase the fusibility of very refractory charges.

Amount of Borax Glass. The rational formula of borax glass, Na_2O, $2B_2O_3$, shows that it is an unsaturated compound and can take up more of the base. This base may be sodium oxide or one or more of the heavier oxides. The result of the addition is a fusible complex borate. This is shown in the following reaction:

(9) $Na_2O, 2B_2O_3 + 2CaO = Na_2O, 2CaO, 2B_2O_3$.

The compound produced is a sodium-calcium borate and shows the fluxing of .5 part of lime by one part of borax. The solvent power of borax glass for various substances has been referred to above. In practice it does not matter whether there is chemical combination or solution. What is desired is perfect liquidity at furnace temperatures.

The amount of borax glass ordinarily used in a half-assay-ton charge varies from 10 to 15 grams. If the ore is very basic and refractory, more borax should be used. As in the case of sodium carbonate, an excess of borax glass ordinarily can do no harm, as it is quite fusible.

Some assayers use silica in the assay of very basic ores. It is a good flux for iron and manganese oxides, producing fusible silicates. It is also very cheap. It cannot be used in excess, because of its very high melting point.

Assay Slags. The slags produced in crucible fusions in fire assaying are often very complex mixtures of silicates, borates, oxides and other compounds. In the molten state there can be chemical combination, solution and suspension, all at the same time. Ordinarily it is quite useless to attempt the formation of a definite silicate or borate. If a sufficient amount of the proper fluxes is used, and a high temperature at the finish, there is usually no trouble in getting a good fusion. As a general rule the greater the complexity of the slag the lower its melting point.

Weighing and Mixing the Charge

It is usually most convenient to mix the charge within the crucible. The fluxes should be put in first, the most bulky one at the bottom. They should be measured rather than weighed, in order to save time. Only the flour and nitre need to be measured accurately. The ore is carefully weighed on a pulp balance and placed on top of the fluxes. The mixing is best done by means of a steel spatula, and should be very thorough. Good mixing is shown by the uniform appearance of the charge. The fluxes should be free from lumps. The practice of using a salt or borax cover on the charge is not so common as it was. Ordinarily it is not necessary. When a salt cover is used in assaying rich ores, there is some danger of the production of volatile silver and gold chlorides. Borax glass as a cover is expensive.

Fusing the Charge. The fusion of the charge is best conducted in a muffle furnace, although it can be made in a coke furnace, or even in a blacksmith's forge. The crucibles should be placed in the muffle when the latter is at a bright red heat. The temperature is then gradually raised until at the end of 45 minutes it reaches a light yellow heat, say 1150° C. Sulphide ores should be run rather more quickly than oxide ores, so as to oxidize the sulphides before they have a chance to melt down into a matte. If the heat be raised too rapidly, there is danger of boiling over, due to the large volume of gases liberated.

The crucible fusion may be divided roughly into three stages. There is first the preliminary heating stage, accompanied by some reduction of lead from litharge, the partial fusion and decomposition of nitre if present, the partial reduction of higher oxides, and some fluxing of silica by sodium carbonate and litharge. During the second stage most of the chemical reactions take place and the entire charge seems to be in a state of violent agitation. Lead is reduced from the litharge by flour, sulphur or other reducing agent, and the multitude of small shots pick up the adjacent particles of gold and silver. Gold tellurides and silver sulphides are decomposed by litharge, setting the metals free. Sodium carbonate and borax react upon the acid and basic constituents, respectively, of the charge, and produce slags. Alumina and other oxides either combine with these reagents or dissolve in the slag mixture. There is a copious evolution of gases, such as carbon dioxide, carbon monoxide, sulphur dioxide, and nitrogen. The third stage is known as the period of quiet fusion. It is for the purpose of completing the slag-forming reactions and of rendering the slag as liquid as possible. This enables all of the small particles of lead to fall down through the slag, collecting the remaining traces of gold and silver. The latter are washed out of the slag much as a shower of rain sweeps the dust particles out of the air. The slag must be thoroughly liquid in order to insure a perfect separation from the lead.

A high temperature at the beginning of the fusion should be avoided, as it not only increases the chances of boiling over, but may cause some volatilization of compounds of the precious metals. After these metals are reduced and alloyed with lead, the temperature can be raised with less danger of loss. A row of empty crucibles or a prism of coke should be placed in front of the crucibles containing the fusions, and the muffle door should be kept closed.

The time required ranges from 40 to 55 minutes, according to conditions. A long-continued fusion at a low temperature usually means a small lead button and an imperfect collection of the gold and silver.

The period of quiet fusion should last about 10 to 15 minutes. The crucibles are then taken out of the muffle, tapped gently with a whirling motion, to collect stray shots of lead, and the contents poured into conical iron moulds. The greater part of the slag should be poured off first, so as to avoid splashing of the lead against the sides of the mould. When cold the lead buttons are taken out and hammered into rough cubes, so as to remove the adhering slag. The lead buttons are now ready for cupellation.

Crucible Charges. It is impossible to give a crucible charge that would be satisfactory in every case. Modifications in the amount and kind of reagents must be made to suit the character of the material to be assayed. However, the variations are not so great as is generally supposed, and many assayers use stock fluxes for a great variety of ores and metallurgical products. Changes are made only when the conditions seem to require them.

The following table gives the approximate amount of the different reagents used in an ordinary crucible fusion:

Ore	$\frac{1}{2}$ assay-ton
Sodium carbonate	25 to 35 grams
Borax glass	10 to 15 grams
Flour or Nitre	As required. (See Preliminary Assay)
Litharge	60 to 75 grams.

The Scorification Assay. The scorification assay is used principally in those cases in which an undue amount of interfering base metals would be reduced along with the lead if crucible fusions were made. Thus, if a crucible fusion be made upon an ore containing copper or antimony, either of these two metals will be reduced along with the lead and produce a button which is difficult to cupel. Even with sulphide ores there is a considerable reduction of the copper or antimony, as is shown in the following reactions:

(10) $Cu_2S + 3PbO + Na_2CO_3 = Cu_2 + 3Pb + Na_2SO_4 + CO_2,$

(11) $Sb_2S_3 + 9PbO + 3Na_2CO_3 = Sb_2 + 9Pb + 3Na_2SO_4 + 3CO_2.$

Nickel and cobalt are reduced in the same way.

In the scorification assay the operations are carried out under oxidizing conditions so as to prevent the reduction of the interfering metals. Metallic lead is used as the collector and is added as such. The flux is mainly litharge, coming from the oxidation of the lead, and a small amount of borax.

The operation is conducted in shallow fire-clay dishes known as scorifiers, from 2 to 3 inches in diameter. The amount of ore taken is usually $\frac{1}{10}$ assay-ton, sometimes $\frac{1}{5}$ or $\frac{1}{20}$ assay-ton. About 25 grams of granulated lead are spread over the bottom of the scorifier and the ore then added and thoroughly mixed with the lead. The mixture is then covered with about 25 grams more of granulated lead and one or two grams of borax glass. Usually from 5 to 20 portions of the ore are weighed up so as to lessen the chances of error. The scorifiers are placed in a muffle heated to redness and the door closed. As soon as the lead melts the door is opened, in order to admit air and increase the rapidity of the oxidation. The ore is seen to be floating on the lead. The latter begins to oxidize and the litharge produced in turn oxidizes the sulphides in the ore, assisted by the oxygen of the air. The temperature at this point must be low in order to prevent volatilization of gold and silver. The ore is not protected by a large bulk of fluxes as it is in the crucible assay. As the oxidation proceeds, a ring of slag, mainly litharge, begins to form around the bath of lead. The ore gradually disappears, the gold and silver going into the molten lead and the gangue combining with or dissolving in the litharge. Owing to the strong oxidizing conditions, most of the copper and practically all of the antimony present go into the slag. As the ring of slag increases, the temperature is raised. Finally, the lead becomes completely covered, and the muffle door is closed in order that the slag may become thoroughly liquid. The contents of the scorifiers is then poured into conical moulds, as in the case of the crucible fusions. The lead buttons should weigh from 15 to 20 grams. Very small buttons usually mean low results. With high copper material it is sometimes necessary to scorify two or three times and to use a large amount of lead. The buttons are cleaned and cupelled in the usual way.

The scorification assay is not adapted to ores containing volatile constituents, such as tellurides, arsenides and metallic zinc. Carbonates and highly oxidized ores are also unsuited to this method. If the ore contains much basic gangue it should not be scorified, as there is not enough acid flux to take care of it. Low grade gold ores are not usually assayed by scorification because of the small amount of ore taken. In practice the method is limited to ores and metallurgical products containing considerable quantities of antimony, copper, nickel and cobalt. It is a standard method for the assay of copper matte. Scorification is also sometimes used to reduce the size of and to purify lead buttons produced in the crucible method.

Cupellation. Cupellation is the process by which the gold and silver are separated from the lead and other base metals with which they are alloyed. The cupels have already been described. They must be thoroughly dry and somewhat heavier than the lead buttons that are to be cupelled. A bone ash cupel will absorb about its own weight of litharge, but the absorption becomes slower when the saturation point is approached.

In the cupellation process the lead is oxidized to litharge which is taken into the pores of the cupel by capillary attraction. This takes place because litharge is molten at the temperature of the operation. Most of the other base metal oxides are infusible at this temperature. When in moderate amounts, however, they dissolve in the liquid litharge and are carried into the cupel. If the lead contains much copper and antimony, the oxides of these latter metals accumulate on the cupel and may ruin the assay. Hence the need for scorification in these cases.

The cupels should be heated in the muffle for at least 20 minutes before putting in the lead buttons. Cupellations are best started at a bright red heat, say about 900° C. As soon as the buttons are put into the cupels, the muffle door should be closed. If the temperature is too low, an infusible oxide will form on the lead as soon as the latter is melted and refuse to go into the cupel. The disappearance of this film of oxide on further heating is referred to as the " *opening* " or " *uncovering* " of the lead. Sometimes it is necessary to hasten the opening by means of a burning stick of wood placed immediately over the cupel. This reduces the oxide and at the same time raises the temperature. When all of the cupellations are uncovered, the muffle door is opened and the temperature lowered rapidly to the lowest possible point at which the operation can proceed. This must be done because a temperature higher than necessary increases the loss of gold and silver. This loss occurs by absorption into the cupel and by volatilization. If the temperature falls too low, the buttons "freeze"; that is, the litharge, which melts at 883° C., solidifies on top of the liquid lead, which melts at 327° C., and the operation stops.

At first thought it would seem that a temperature slightly above 883° C. would be the proper one for cupellation. As a matter of fact the temperature of the muffle need not be above 750° C. This is due to the fact that the oxidation of the lead generates a considerable amount of heat, and the buttons are thus hotter than either the muffle or the cupels. A good indication of the right cupellation temperature is the formation of solid flakes of litharge, known as "feathers," upon the inner edge of the cupels. The volatilized litharge strikes the comparatively cool bone ash and sublimes as flake crystals.

The presence of impurities usually increases the loss of gold and silver, and adds to the difficulties of the operation. Copper or nickel in quantity may cause the buttons to freeze even at moderately high temperatures. Antimony causes the formation of a hard, infusible crust of lead antimonate which retains silver and which often splits the top of the cupel.

The surface of the lead in the cupel is convex, owing to the high surface tension of the metal. During cupellation the drops of molten litharge can be seen rolling off of the lead and disappearing into the cupel. The surface tension of the melted litharge is less than the attractive force of the bone ash. In scorification, where the vessel is not porous, the litharge forms a concave surface and climbs up the sides of the scorifier. This explains in part the high gold losses in the cupellation of lead containing gold and tellurium. Some of the gold telluride passes into the cupel just as in the case of litharge. Gold telluride is also more volatile than metallic gold.

As the operation proceeds, the lead and other base metals gradually oxidize and disappear. Copper and bismuth are less readily oxidized than lead, and hence tend to remain until most of the lead has gone. The temperature should be raised slightly at this point in order to prevent the buttons from solidifying before the base metals are completely oxidized. Small amounts of these metals usually remain, even in a well-conducted cupellation. The melting point of gold and silver being considerably higher than the temperature of the muffle, the buttons solidify soon after the base metals are gone. At the moment of solidification the buttons flash or "blick," owing to the release of the latent heat of fusion. If the buttons are large and consist mainly of silver, they may "sprout" or "spit" on being withdrawn quickly. This is due to dissolved oxygen which escapes when the button solidifies. The sprouting may be prevented by covering the silver button with a hot inverted cupel as soon as the cupellation is finished, and allowing the covered cupel to remain in the muffle for several minutes. This insures a slow cooling of the silver bead. After cooling the buttons are removed from the cupel by means of forceps and the adhering bone ash brushed off. The buttons are then weighed on an assay balance to the nearest one-tenth of a milligram. If a half-assay-ton of ore was used, the results multiplied by two equal the ounces per ton of combined gold and silver.

Cupellation involves the greatest precious metal losses of all the processes in fire assaying. For that reason it must be conducted with care and skill. The loss in silver ranges from 1 to 2 per cent, even under favorable conditions. If there are large amounts of impurities, or if a very high temperature is used, the silver loss may be several times as great. The gold loss is less than that of silver and should not be more than $\frac{1}{2}$ of 1 per cent. In all cases the percentage losses of gold and silver increase as the amounts in the ore decrease.

Parting. Parting is the separation of gold from silver in an alloy containing these metals, and is effected in fire assaying by means of nitric acid. This acid converts the silver into soluble silver nitrate, but is almost without action upon the gold. In order to part readily the alloy must contain at least three times as much silver as gold. Even at this ratio it is difficult to dissolve all of the silver. In practice, it is better to have a much larger proportion of silver, except in the assay of gold and silver bullion. If the buttons produced in the assay of an ore are known to contain enough gold to render parting difficult or impossible, they are subjected to the process known as *inquartation*. The buttons after weighing are wrapped with about 10 times their weight of pure silver foil in 3 to 5 grams of sheet lead and then cupelled. The resulting buttons are flattened and parted in the usual manner.

An important point in parting is the strength of the acid. If a strong acid is used at first, the gold in the button is liable to break up into a fine powder which is difficult to manage without loss. By using a rather weak acid, containing from 10 to 20 per cent of HNO_3, the gold has a tendency to coalesce into a coherent mass which can be washed and weighed as one piece. The treatment with weak acid is always followed by one with a stronger acid, in order to remove the last traces of silver. The second acid should be about 1.26 sp.gr., made by diluting the concentrated acid with its own volume of water. In the case of buttons containing a small proportion of gold, the first acid should be very weak, not more than 10 per cent HNO_3. With more gold a stronger acid can be used, although the weak acid is usually effective, except when the buttons are very large.

The parting may be done in porcelain capsules or in small glass flasks, known as parting flasks. Only a few cubic centimeters of acid are necessary. This should be heated to boiling and the flattened beads then dropped into it. Solution of the silver begins immediately. At the end of about 20 minutes, or when all visible action has ceased, the weak acid solution is decanted into a white casserole, carefully avoiding the loss of any gold. About 3 cubic centimeters of the stronger acid is now added to each flask or capsule and then heated almost to the boiling point. The heating is continued for at least 10 minutes, when all of the silver should be in solution. The acid is then poured off and the gold washed three times by decantation with chlorine-free water. If capsules are used, the water is drained out as completely as possible and the capsules then placed on a hot plate or in front of the muffle for drying. If parting flasks are used, a fire-clay annealing cup is inverted over the top of each completely filled flask and the flask then quickly reversed, allowing the gold to fall quietly into the annealing cup. After removing the flask by a quick side motion, the water is poured off of the gold and the cup placed on the hot plate. The final process is known as annealing. The capsules or annealing cups are placed in the muffle and heated to low redness for about 5 minutes. The heating causes the brownish-black spongy or fibrous gold to coalesce into a dense flake or bead having the characteristic yellow color of the metal. The annealing also serves to drive off any volatile impurities which may be present, and to render it easier to separate any specks of dust or dirt from the gold. After cooling the gold is weighed on a delicate balance to the nearest .01 milligram—with a little care to the nearest .005 milligram. The weight of the gold is deducted from the weight of the button before parting and the difference represents the silver in the portion taken for assay.

The Assay of Bullion

Bullion is an alloy of gold and silver with variable amounts of one or more of the base metals, and is the semi-final product of most non-ferrous metallurgical plants. In lead smelting this product is usually known as base bullion, in copper smelting as blister copper, and in amalgamation and cyanide processes as retort bullion or doré silver. The base metals may include antimony, arsenic, bismuth, cobalt, copper, lead, mercury, nickel and zinc. Small amounts of selenium and tellurium are usually present, as well as traces of the platinum group of metals. In all cases the assay of bullion resolves itself into a problem of the separation of metals from each other, there being practically no non-metallic elements present.

Bullion Sampling. The sampling of bullion involves some difficulties not encountered in the sampling of ores. Most alloys on solidifying segregate to some extent, so that the cooled metal is never uniform in composition. Whenever possible the samples should be taken from the thoroughly stirred molten alloy and then chilled quickly, either by pouring into water or by pouring into small moulds with thick metal sides and bottoms. When there is danger of oxidation, this method of course is not entirely satisfactory. In impure bullion there is often a very uneven distribution of the gold and silver, and it is necessary to drill or saw entirely through the bar in order to obtain accurate results. In copper anode plates and other forms of blister copper the plates or bars are drilled in series, so that the combined sample represents a proper percentage of drillings from all parts of each piece. Lead bullion is sometimes sampled in a similar manner, although in this case the bars are usually punched instead of being drilled. The modern tendency is to take melted samples of all metallic products, even in the case of blister copper.

The Assay of Lead Bullion. The assay of lead bullion ordinarily involves only cupellation and parting. The bullion is often impure, however, and it may then be advisable to scorify the weighed portions before cupelling. If the sample contains much copper or antimony, it should always be scorified. The precious metal loss in scorification is less than it is in cupellation, especially in the case of an impure bullion requiring a high temperature in order to cupel it.

Lead bullion is usually run in four portions of one half assay-ton each. The four silver buttons are weighed separately and if there is a satisfactory agreement in the weights, the average is taken and the result multiplied by two. The buttons are parted in pairs, thus saving time in washing and weighing. Great care should be exercised in the cupellation of lead bullion and in the subsequent parting, as the bullion is a high grade product, and ordinarily no correction is made for losses in the cupel or otherwise.

The Assay of Copper Bullion. Copper bullion may be assayed by the scorification method, but the results are satisfactory only in the case of the gold. The silver obtained is always much too low. Most of the loss can be recovered by assaying the slag and cupels, but this requires additional time and materials.

The best way to determine gold and silver in copper bullion is to use the so-called combination method, in which the copper is first removed by solution in acid. Formerly nitric acid was used for this purpose, but it was found that the results were usually low in gold. There was a tendency for some of the gold to go into solution. The nitric acid method is a convenient way of determining the silver, as the copper dissolves very rapidly in nitric acid, and the precipitation of silver as chloride is very complete.

For the assay of both gold and silver the sulphuric acid-mercuric nitrate method is recommended. One assay-ton of the finely ground, well-mixed copper borings is treated in a large beaker with 30 cc. of water and 10 cc. of a solution of mercuric nitrate containing 25 grams of mercury per liter. The beaker is well shaken so as to amalgamate the copper, and 100 cc. of concentrated sulphuric acid then added. The beaker is covered, placed on a hot plate and heated until all of the copper is dissolved. This will require from one to two hours, according to the temperature and fineness of the sample. The beaker is now removed and the solution is allowed to cool. 100 cc. of cold water are added, the mixture stirred, and then 400 cc. of boiling water added, with further stirring until all copper sulphate has dissolved. A solution of common salt is now added, just sufficient to precipitate all of the silver and mercury. Only a slight excess must be used, as silver chloride is soluble in strong sodium chloride solution. The beaker is replaced on the hot plate and the contents boiled so as to coagulate the silver chloride. The beaker is then removed, the solution diluted to 600 cc. with cold water and allowed to cool. The solution is then filtered through double filter papers, and the beaker and filter washed with hot water. The beaker should be wiped out with a filter paper and this added to the material in the filter. The filter and its contents are now transferred to a $2\frac{1}{2}$-inch glazed scorifier and the filter paper burned off at a low temperature, so as to avoid loss of silver. After the paper is burned off, 30 grams of test lead are added and the material scorified until 12 to 15 grams of lead remain. The scorifier is poured and the lead button cupelled at as low a temperature as possible. The gold and silver are parted in the usual way. The results are very accurate.

The object of the mercuric nitrate is to hasten the solution of the copper by forming a galvanic couple. It prevents also the formation of copper sulphide which is insoluble in dilute sulphuric acid.

Assays should be made in duplicate or triplicate.

The Assay of Gold and Silver Bullion

In the fire assay of gold and silver bullion a correction must always be made for the metal losses, because of the great value of the bullion and because the refining losses on a commercial scale are considerably less than the losses in assaying. The assays are, therefore, always run with a check or " proof center." The check is an artificial sample made up so as to have as nearly as possible the exact composition of the bullion to be assayed. Two checks and three portions of the bullion, all five of the same weight, are cupelled and parted under exactly the same conditions. The weights of gold and silver found in the bullion samples are then corrected by adding or subtracting the loss or gain experienced by the gold and silver in the checks. In the assay of gold bullion there is sometimes a gain in the weight of the gold, due to the imperfect elimination of the copper. This, however, should be the same in both checks and bullion samples. Results are reported in " fineness " or parts per 1000.

Silver Bullion or Doré Bullion Assay. A sample of about 500 milligrams of the bullion is accurately weighed out on the assay balance. The weight need not be an even 500 milligrams but its exact amount should be recorded. The bullion is wrapped in from 6 to 8 grams of sheet lead and cupelled with the formation of feathers of litharge. The button is covered with a hot cupel to prevent sprouting. The cupel and button are drawn out of the muffle gradually, and the button is then cleaned, flattened and parted in the usual way. To the weight of the silver a loss of from 1 to $1\frac{1}{2}$ per cent is added. The weight of the comparatively small amount of gold can be taken as it is. The weight of the original button, less the sum of the corrected silver weight and the weight of the gold, represents the amount of base metal. This is usually copper, but may be one or more of a number of other metals. The assayer can usually determine what metals are present by the appearance of the bullion and that of the cupel.

Two checks are now prepared by weighing in each case pure gold, pure silver and pure copper or other base metal in the amounts as found in the preliminary assay just described. Three 500-milligram portions of the bullion are also weighed up. The checks and bullion samples are each wrapped in an amount of sheet lead corresponding to that in the following table, taken from Bugbee's " Text-book of Fire Assaying ":

Lead Ratio in Cupellation

Fineness of Au. and Ag.	Wt. of Lead	Fineness of Au. and Ag.	Wt. of Lead
950	5 grams	750	11 grams
900	7 "	700	12 "
850	8 "	650	13 "
800	10 "	600	15 "

The table shows that as the base metal, usually copper, increases, the amount of lead must be increased.

Cupellation is performed in a single row of cupels, all of the same size. The checks are placed in the second and fourth cupels. The heat should be kept low enough to allow of the formation of feathers of litharge until at the finish, when the temperature must be raised.

The beads are cleaned, weighed, flattened and parted. The amount of gold and silver loss in the checks is determined and the proper correction applied to the weights of gold and silver found in the bullion samples.

If the bullion contains antimony, the process must include scorification, which is applied to checks and samples as well. When bismuth, selenium or tellurium is present in quantity, the silver must be separated by means of solution in nitric acid and subsequent precipitation as chloride.

Gold Bullion Assay. The assay of gold bullion is in principle the same as that of silver bullion. As the gold is usually in excess of the silver, however, the process involves inquartation, with the use of a stronger first acid than when parting ordinary silver buttons.

U. S. Mint Method. Sample portions of 500 milligrams are taken for assay. A preliminary cupellation is made as in the case of doré bullion. The amount of silver to be added for the final assay is determined by the touchstone method. The cupelled gold and silver button is rubbed on a piece of black jasper and the streak made compared with those made by alloys of known composition. This gives the fineness within 2 per cent, which is close enough. A ratio of silver to gold of 2 to 1 is used in making up the checks. If no copper is present, about 3 or 4 per cent is added as it facilitates the removal of the last traces of lead in cupellation. The cupelled buttons are flattened by hammering, annealed at a red heat, and then passed through a pair of jeweller's rolls, until they are converted into fillets about $2\frac{1}{2}$ inches long and $\frac{1}{2}$ inch wide. The fillets are again annealed and rolled up into " cornets " or spirals. Sufficient space should be left between the turns to permit of easy contact with the acid. The parting is done by boiling for 10 minutes in nitric acid of 1.28 sp.gr., and then transferring to another vessel containing acid of the same strength and boiling for 10 minutes longer. The cornets are then washed three times with distilled water, dried, annealed and weighed.

The proofs usually show a slight gain in the weight of the gold, so that the correction is made by subtracting the gain from the average weight of the gold found in the sample portions.

The gold after parting should be in one piece and have smooth edges, as otherwise there is danger of loss.

The Assay of Cyanide Solutions

A number of methods have been devised for the determination of gold and silver in cyanide solutions. Only two of these methods will be described here.

Evaporation in Lead Tray. This method is adapted to cyanide solutions containing only small amounts of base metals or other impurities. A small tray or boat is made of lead foil, capable of holding the amount of solution to be assayed. A wooden block or form is used to make the trays if many assays are required. The solution has about the same specific gravity as water, so that 29.2 cc. are assumed to be equal to an assay-ton. An amount of solution varying with its richness is put into the lead tray and slowly evaporated to dryness on the hot plate. The lead tray is then folded up and cupelled in the usual manner.

The Chiddey Method. This method, first described by Alfred Chiddey, is adapted to almost every grade and character of cyanide solutions.

From 1 to 20 assay-tons of solution are heated in a beaker or evaporating dish. To the solution is added from 10 to 20 cc. of a 10 per cent solution of lead acetate containing 40 cc. of acetic acid per liter. From $\frac{1}{2}$ to 2 grams of zinc dust or zinc shavings is then added. The gold, silver and lead immediately begin to precipitate on the zinc. The solution is heated for about 20 to 25 minutes, but not to boiling. The lead should coalesce into a spongy mass. Boiling the solution is liable to break up the sponge. The excess zinc is now dissolved by adding slowly 20 cc. of hydrochloric acid of 1.12 sp.gr. The heating is continued until effervescence ceases. It may be necessary to stir slightly in order to make sure that all zinc is dissolved. The solution is now decanted off and the lead sponge washed two or three times with water. The excess water is squeezed out of the sponge with the fingers, the sponge further dried by pressing between filter paper and then rolled into a ball with lead foil and the necessary silver for parting. A hole should be left in the lead foil for the escape of steam. The ball is then dried and cupelled.

As the lead sponge begins to break up and go into solution as soon as all of the zinc is dissolved, no time should be lost in decanting the solution after the zinc has disappeared.

Special Methods of Assay. There are many ores and metallurgical products that require special methods for the determination of the gold and silver that they contain. It is impossible to refer to these methods here. A knowledge of the composition of the sample, however, will usually enable the skilled assayer to so modify the ordinary processes as to obtain satisfactory results.

For further details in regard to standard and special methods of fire assaying the reader is referred to such works as " A Manual of Fire Assaying," by Chas. H. Fulton, published in 1911, and " A Textbook of Fire Assaying," by Edward E. Bugbee, published in 1922.

Summary. General Outline.

An assay ton (A.T.) or multiple is taken, of the 80–200-mesh ore or 20-mesh shavings of metals and is heated with a flux of lead oxide, a suitable slag forming material and a reducing or oxidizing reagent as the case requires, (see preliminary assay below), in order to obtain by reduction about 18 g. of lead from the lead oxide. The molten lead alloys with the silver and gold (and copper) and sinks to the bottom of the melt. The charge is cooled and the slag hammered off the lead and the button or cube of lead is heated on a bone ash cupel in presence of air. The bone ash absorbs the oxides of lead (and copper) leaving the precious metals in the form of a bead. As the last trace of lead is removed a play of colors is seen with a final flash of light as the alloy of silver and gold appears ("blick"). The combined gold and silver button is weighed, the silver now dissolved (parting) from the gold and the residual gold weighed. The difference between the first and second weight is due to silver.

Preliminary Assay. This is to ascertain the nature of the material as neutral, oxidizing or reducing products require different treatment. Usually one-sixth of the charge of ore is taken with the regular charge of flux, without a reducing or oxidizing agent. A typical example of a charge is one-sixth A.T. of ore, 100 g. litharge, 20 g. sodium bicarbonate and 5 g. borax. Using this charge the following cases will illustrate the method of interpretation.

A. Reducing Ores. Suppose a 2 g. Pb button is obtained. This is equivalent to $2 \times 6 = 12$ g. on a customary charge of 1 A.T. of ore. Additional reducing agent is necessary to get the 18 g. Pb desired. 1 g. charcoal reduces about 30 g. Pb, hence $(18 - 12) \div 30 = 0.2$ g. charcoal required.

On the other hand suppose a 4 g. Pb button results, equivalent to 24 g. with full charge of ore. It is necessary to add an oxidizing agent to counteract the reducing action of the ore to the extent of $24 - 16 = 6$ g. Pb excess. 1 g. $NaNO_3$ oxidizes 4 g. Pb. Therefore $6 \div 4 = 1.5$ g. $NaNO_3$ is required.

B. Neutral and Oxidizing Ores. In this case no lead is reduced. A second run must be made with say 1 gram of charcoal (30 g. Pb equivalent). If 30 g. Pb results the ore is neutral and $18 \div 30 = 0.6$ g. charcoal is required.

On the other hand suppose 28 g. Pb results, then the ore is an oxidizer and 1 A.T. of the ore will oxidize $2 \times 6 = 12$ g. Pb. To counteract this $12 \div 30 = 0.4$ g. charcoal is necessary and $18 \div 30 = 0.6$ g. charcoal to reduce 18 g. Pb, making a total of $0.4 + 0.6 = 1$ g. charcoal required.

Precaution. All reagents used should be tested by regular assay.

A 20–25 gram button of lead is generally required in the assay of ores high in silver.

Solution Assay. Place 10 assay tons of the solution in a 1000 cc. beaker. Add about 30 cc. of a 20% lead acetate solution (the amount being governed by the strength of the solution). Dilute to about twice the volume and heat. Add about 5 grams of zinc shavings, heat until solution clears. Add 30–40 cc. HCl. When clear, decant off solution, wash the lead and squeeze into a button. Assay the button.

The writer is indebted to Mr. W. G. Derby for several cuts appearing in this chapter.

DETERMINATION OF PLATINUM, PALLADIUM, GOLD AND SILVER [1]

Platinum, Palladium and Gold. Scorify the lead buttons from two or more $\frac{1}{2}$ assay ton crucible fusions together, adding at least six times as much silver as the combined weight of the Pt, Pd and Au present, and cupel *hot*. In rich materials such as slimes or concentrates, two $\frac{1}{2}$ assay ton fusions suffice, but low-grade ores may require 10 or more $\frac{1}{2}$ assay ton fusions combined for each determination.

Part the silver beads with HNO_3 (1 : 6), followed by stronger parting acid (1 : 1) and wash with water as usual. All Pd goes into solution, together with considerable Pt. The residue consists of Au plus some Pt. Dissolve residue in strong *aqua regia* and reserve the solution (solution *A*). Precipitate the silver in the nitric-acid solution—containing Ag, Pd and some Pt—with HCl. Practically all the Pt will remain in solution; but the precipitated AgCl is pink in color and contains considerable Pd. Filter off the AgCl, scorify and cupel it and part again with HNO_3 (1 : 6); all should dissolve. Re-precipitate the Ag with HCl. The liquid now contains most of the remaining Pd, but some is co-precipitated with AgCl. Filter off the AgCl and add the filtrate to the first filtrate from AgCl. Again scorify and cupel the silver chloride, dissolving the silver in nitric acid as before and re-precipitating the silver as chloride. In most cases the filtrate from this silver chloride contains all the remaining Pd. If however, the AgCl is distinctly pink, another separation must be made.

Unite all filtrates from AgCl precipitations and evaporate to small bulk, adding the *aqua-regia* solution of the Au and Pt (solution *A*). The liquid now contains all the Au, Pt and Pd present in the original ore, together with traces of Ag due to solubility of AgCl in excess of HCl, and also traces of Pb gathered from the lead retained in the silver buttons from several re-cupellations.

Evaporate the liquid to dryness on the steam bath; take up with dilute HCl (1 : 3) and evaporate again to dryness; take up with five drops of HCl and 40 cc. H_2O. Pay no attention to any insoluble residue of AgCl or $PbCl_2$.[2] Precipitate the gold by adding, say, 3 gm. of oxalic acid to the solution and boiling it. Let stand over night and filter off the Au. If Pt and Pd are high, it is necessary to re-dissolve the Au in *aqua regia*, evaporating with HCl to dryness and repeating the oxalic-acid precipitation, uniting the filtrate with that from the first gold precipitation. Burn the filter containing the gold and scorify it with six times its weight of silver and a little test lead; cupel, part and weigh the gold as usual.

To the oxalic-acid filtrates from Au add 5 cc. of HCl and make volume up to 150 cc.; heat to boiling and precipitate Pt and Pd with a rapid current of H_2S in *hot* solution, passing the current of gas for some time and keeping the solution hot during precipitation. Filter and wash the Pt and Pd sulphides

---

[1] By A. M. Smoot, Eng. and Mining Journal, April 17, 1915.

[2] In materials rich in palladium the small amount of AgCl + $PbCl_2$ may be distinctly pink in color and retain weighable quantities of Pd. If this is the case, the Pd may be recovered in the solution from the nitric acid parting of the gold. To do this, precipitate the silver in this liquid by adding HCl, filter off the silver chloride and evaporate the filtrate to dryness. Take up with a drop of HCl and a little water, let stand over night and filter through a very small filter. This liquid may be added to solution B before precipitating palladium with glyoxime.

with H_2S water containing a little HCl. Wash the precipitate from the filter with a fine water jet into the original beaker; spread the filter paper (which will contain a small amount of precipitate impossible to wash off) with the precipitate side down over the lower side of a watch glass-cover. Add *aqua regia* to the precipitate in the beaker and place the cover on the beaker; warm gently to dissolve the Pt and Pd sulphides. The fumes arising from the acid dissolve the traces of Pt and Pd adhering to the filter paper. When solution is complete and filter paper is white, remove the watch-glass cover and wash the paper with hot dilute HCl thrown against it in a fine stream.

Evaporate the *aqua-regia* solution to dryness, take up the residue with HCl and evaporate again to dryness to remove all HNO_3. Take up the residue with two or three drops of HCl and about 2 cc. of H_2O. The solution is usually perfectly clear, but it may be slightly cloudy owing to the presence of a little AgCl in it. No attention need be paid to this however. Add 5 to 10 cc. of a saturated solution of NH_4Cl, stir well and allow to stand over night. Platinum is precipitated as ammonium-platinum chloride—$(NH_4)_2PtCl_6$. Filter and wash the precipitate with 20% NH_4Cl solution. All Pd passes into the filtrate which is reserved (solution *B*). Dissolve the Pt precipitate in boiling hot 5% H_2SO_4; heat the liquid to actual boiling and precipitate with H_2S as before, filtering and washing with H_2S water. Burn the filter and precipitate at a low temperature in a scorifier; add six times as much Ag as Pt, scorifying with lead, cupel and part the silver bead containing the platinum with H_2SO_4; decant off the silver solution and wash once with strong H_2SO_4, followed by 50% H_2SO_4 until practically all the silver is washed away; finally wash with water, anneal and weigh. A minute quantity of Ag is retained with the platinum, but it can usually be neglected. In very important work where the amount of platinum is large dissolve in *aqua-regia*, evaporate the solution to dryness, take up with a drop of HCl, dilute largely with water and let the AgCl settle over night; filter on a small paper, cupel it with a little sheet lead and deduct the weight from the weight of platinum. This refinement need not be considered in materials running less than 15 or 20 oz. to the ton.

It may seem an unnecessary step to precipitate the platinum as sulphide, scorify it with silver and part it as described in the foregoing. General practice has been to ignite the ammonium-platinum-chloride precipitate and weigh the metallic residue. When this is done, however, there is danger of losing considerable platinum, which is carried away mechanically during the decomposition of the compound; furthermore, it is extremely difficult (if not impossible) to collect the finely divided residue for weighing, and the precipitate invariably contains lead and silver. Precipitation as sulphide, scorification and cupellation with excess silver and parting with sulphuric acid overcome the difficulties inherent in handling the ammonium precipitate.

The palladium is all contained in the filtrate and washings from the platinum ammonium-chloride precipitates (solution *B*). Add to this solution at least seven times as much di-methylglyoxime as there is Pd present (in any case, at least 0.1 gm. glyoxime). The precipitant should be dissolved in a mixture of two-thirds strong HCl and one-third water. Dilute the liquid to 250–300 cc., heat on a steam bath for half an hour and let stand over night. Pd is precipitated as a voluminous yellow, easily filtered glyoxime compound $C_8H_{14}N_4O_4Pd$, containing, when dried at 110° C., 31.689% of Pd. Filter the Pd precipitate on a weighed Gooch crucible and wash it, first, with dilute HCl,

39

half and half, then with warm water and finally with alcohol; dry it at 110°
to 115° C. and weigh.   The disadvantage of weighing palladium on a Gooch
crucible is overcome—at least to some extent—by the fact that the Pd compound contains a relatively small amount of Pd—less than one-third of its
weight.   This compound may also be weighed on carefully counterpoised
papers; but it is better to use Gooch crucibles, if they are available, because
of the relatively strong acid which is required for washing.   The object in
using half-and-half hydrochloric acid as a wash liquid is to dissolve out any
excess of the glyoxime precipitant.   This is easily soluble in moderately
strong HCl, but is substantially insoluble in water.

Determination of Silver in Ores and Concentrates Containing Platinum and Palladium

Make the usual crucible fusion on one-quarter, one-half or full assay ton,
according to the amount of silver present.   Instead of cupelling the lead
button, hammer it free from slag and dissolve it in dilute nitric acid.   Most
of the silver passes into solution together with palladium, and perhaps a trace
of platinum; but gold and most of the platinum remain insoluble.   The gold
and platinum retain an appreciable proportion of silver which cannot be
washed out.   Filter out the insoluble residue and wash it thoroughly with
hot dilute nitric acid, followed by hot water.   Scorify the residue once more
with a little lead and dissolve the lead button as before, filtering into the beaker
containing the first filtrate.   In this liquid precipitate the silver as AgCl by
adding standard NaCl in sufficient quantity; stir well, and if the amount of
silver is small, add about $\frac{1}{2}$ cc. of strong H_2SO_4 to form a precipitate of lead
sulphate.   Let the silver chloride, or the silver chloride plus lead sulphate,
settle over night or until the supernatant liquid is clear; filter through double
filter papers; ignite and scorify the residue of silver chloride with test lead.

If the amount of the palladium contained in the sample is small, the silver
bead obtained by cupeling the lead button obtained by scorifying the silver
chloride may be considered as sufficiently pure for ordinary purposes.   It
contains, of course, some palladium, and in accurate silver determinations the
lead button from the first silver chloride precipitation should be re-dissolved
and the silver re-precipitated, filtered and scorified as before.   The amount of
palladium retained after the second precipitation and scorification is so small
as to be negligible.

STRONTIUM

Sr″, *at.wt.* 87.62; *sp.gr.* 2.54; *m.p.* 900° C; *oxides* SrO *and* SrO_2.

DETECTION

Strontium is precipitated with barium and calcium, in the filtrate, from the ammonium sulphide group, by addition of ammonium carbonate to the ammoniacal solution. The precipitate is dissolved in acetic acid and treated with potassium dichromate, and the barium filtered off as $BaCrO_4$. Strontium and calcium in the filtrate are separated from the excess of potassium chromate by reprecipitation as carbonates by the addition of ammonium carbonate, the precipitate again dissolved in acetic acid and the excess of free acid neutralized with ammonia. Strontium may now be precipitated from the concentrated solution by boiling with an equal volume of a saturated solution of calcium sulphate.

Sodium Sulphate Test. A saturated solution of the salt added to a solution containing strontium chloride, made strongly acid with acetic acid, and the mixture boiled, will produce a distinct precipitate if strontium exceeds 0.0015 normal. Calcium does not precipitate until 1.3 normality is reached.[1]

Flame Test. Strontium, preferably in the form of the chloride in a hydrochloric acid solution, placed on a platinum loop and held in a colorless flame, colors the flame crimson. (Lithium gives a red color, calcium a yellowish-red.) The test is best confirmed by means of the spectroscope.

The Spectra of Strontium. Eight bright bands; 6 are red, 1 orange, 1 blue. Two of these, known as strontium β and γ, are red, the orange is strontium α and the blue strontium δ. The delicacy of the test is 0.6 milligram Sr per cc. The test is very much more delicate with the arc spectra, e.g., 0.03 milligram Sr per cc. See chapter on barium, Preliminary Tests under Separations.

ESTIMATION

Strontium never occurs free in nature. It is found principally in the ores celestine, $SrSO_4$, and strontianite, $SrCO_3$. It generally accompanies calcium in the various forms of calcite and aragonite. It occurs with barium in barytocelestine, and is found in barytes. It also occurs associated with barium as a silicate in brewsterite, $Al_2O_3 \cdot H_4(BaSr)O_3 \cdot (SiO_2)_6 \cdot 3H_2O$. It is found in traces in certain mineral waters and in sea-water.

The compounds of strontium are used for medicinal purposes; for red fire in pyrotechnics; for the manufacture of iridescent glass; the dioxide for bleaching purposes; the sulphide for luminous paint; the hydroxide for refining of beet-root sugar, being preferable to lime, as the saccharate of strontia is more granular.

Chapter contributed by Wilfred W. Scott.

Preparation and Solution of the Sample

The following facts regarding solubility may be of value in the determination of strontium. 100 cc. of water dissolves 1.74 grams $Sr(OH)_2 \cdot H_2O$ at 20° C. The hydroxide is less soluble than that of barium. The peroxide dissolves to the extent of only 0.008 gram per 100 cc. 20° C. One hundred cc. of water dissolves 0.0011 gram [1] $SrCO_3$ (18°); 0.0114 gram $SrSO_4$ at 18° and 0.0104 at 100°; the presence of sulphuric acid decreases this solubility, i.e., 0.00083 gram $SrSO_4$; 0.0051 gram $SrC_2O_4 \cdot H_2O$ at 18° and 5 grams at 100° C.; the presence of oxalic acid decreases this solubility. The sulphate dissolves in concentrated sulphuric acid, and is appreciably soluble in HCl, HNO_3, $HC_2H_3O_2$, NH_4Cl, NH_4NO_3, NaCl, $MgCl_2$. The carbonate and oxalate are soluble in mineral acids.

The procedure for the treatment of ores and strontium products is the same as those described for barium and calcium. We refer to the chapters on these elements for the preparation of the strontium solution.

SEPARATIONS

Separation of Strontium from Magnesium and the Alkalies. The procedure is the same as the one given in detail under barium for the separation of the alkaline earths from magnesium and the alkalies. Either the oxalic acid method or precipitation of strontium as a sulphate in presence of alcohol will accomplish this separation. If a sulphate precipitation is made it will be necessary to fuse the sulphate with sodium carbonate to get it into solution or to effect further separation from members of the ammonium carbonate group, should these be present.

Separation of Strontium from Calcium. [2] Strontium and calcium are converted into the nitrates and taken to dryness and all water expelled by heating to 140° C. for an hour or more. The nitrates are now extracted with equal parts of absolute alcohol and anhydrous ether or by boiling with amyl alcohol at 130° C. (hood). Strontium remains insoluble and calcium goes into solution as the nitrate. Strontium nitrate may require further solution in water, evaporation to dryness, heating and extraction to remove calcium completely, should this be present in large excess. The nitrate of strontium is dissolved in water and strontium determined by one of the procedures given later. See detailed procedure for separation under Barium.

Separation of Strontium from Barium. The procedure is given in detail under chapter on Barium. In brief one of the following methods may be used: Strontium and barium in a mixture of the nitrates are separated from calcium by treatment with ether-alcohol mixture, in which $Ba(NO_3)_2$ and $Sr(NO_3)_2$ are insoluble. The nitrates dissolved in water are separated by precipitating barium as $BaCrO_4$ from a faintly acetic acid solution, strontium remaining in solution.

If preferred, barium may be first removed as a chromate, strontium and calcium precipitated from an ammoniacal solution by $(NH_4)_2CO_3$ as carbonates, the carbonates converted to nitrates and $Sr(NO_3)_2$ separated from $Ca(NO_3)_2$ in an ether-alcohol solution or by amyl alcohol. Details of the separations are given under Barium.

---

[1] Treadwell claims solubility $=0.00055$, i.e., 1 part $SrCO_3$ in 18,045 parts of water.

[2] Advantage may be taken of the insolubility of strontium sulphate in ammonium sulphate in separating it from the soluble calcium salt.

GRAVIMETRIC METHODS

Strontium may be conveniently determined either as the sulphate, the carbonate or as the oxide. The first procedure is considered the best by authorities.

Determination as Strontium Sulphate, SrSO₄

Procedure. A slight excess of dilute sulphuric acid is added to the neutral solution of strontium, and then an equal volume of alcohol. The mixture is stirred well and settled for several hours, or overnight, if more convenient. The precipitate, $SrSO_4$, is filtered onto a small ashless filter and washed first with 50% alcohol containing a little sulphuric acid, then with alcohol until free of acid. The precipitate is dried and the paper and the greater part of the salt ignited separately, then combined and weighed as $SrSO_4$.

Factors. $SrSO_4 \times 0.477 = Sr$,   or $\times 0.8037 = SrCO_3$,   or $\times 0.5642 = SrO$.

Determination as Strontium Carbonate

Strontium carbonate is not readily decomposed by ignition as is calcium carbonate, so that its determination in this form may be satisfactorily made.
Procedure. The carbonate is precipitated by adding ammonium carbonate in slight excess [1] to the ammoniacal solution of strontium, heated nearly to boiling. The solution is allowed to stand for several hours and filtered cold. The washed strontium carbonate and filter are ignited gently and the cooled residue weighed as $SrCO_3$.

Factors. $SrCO_3 \times 0.5935 = Sr$,   or $\times 1.2443 = SrSO_4$,   or $\times 0.702 = SrO$.

Determination as Oxide, SrO

Strontium is precipitated as the oxalate by addition of ammonium oxalate to the slightly ammoniacal solution. The precipitate is filtered and washed with water containing ammonium oxalate. The residue is ignited and weighed as SrO.

Factors. $SrO \times 0.8456 = Sr$,   or $\times 1.7726 = SrSO_4$,   or $\times 1.4245 = SrCO_3$.

VOLUMETRIC METHODS

The volumetric methods for determining strontium presuppose its isolation from other elements.

Alkalimetric Method, Titration with Standard Acids

Either the carbonate or the oxide of strontium may be titrated with standard hydrochloric or nitric acids. The compound is treated with a known amount of standard acid added in excess, using methyl orange indicator. The solution is heated below boiling to complete the reaction and, upon cooling, the excess of acid is titrated with standard alkali.

One cc. normal acid $= 0.04381$ gram Sr, or 0.05181 gram SrO, or 0.07381 gram $SrCO_3$.

[1] N. B. Avoid a large excess of $(NH_4)_2CO_3$. NH_4Cl has a solvent action on $SrCO_3$.

Titration of the Chloride with Silver Nitrate

Strontium chloride, free from other chlorides, may be determined indirectly by titration of its combined chlorine with silver nitrate by Mohr's method, using potassium chromate indicator. One cc. N. $AgNO_3 = 0.04381$ gram Sr.

The oxide or carbonate is slightly supersaturated with hydrochloric, then taken to dryness and heated at 120° C. in the air bath to expel the excess of acid. Chlorine is determined on an aliquot portion.

SULPHUR

S, *at.wt.* **32.065;** *sp.gr.* **2.035;** *m.p.* **111°;** *b.p.* **444.53°;** *oxides* S_2O_3, SO_2, SO_3, S_2O_7; *principal acids* $H_2S_2O_4$, H_2SO_3, H_2SO_4, $H_2S_2O_3$, *and* $H_2S_2O'_8$.

DETECTION

The following tests include the detection of free sulphur and its more important combined forms.

Element. Sulphur is a polymorphous, yellow, brittle, odorless and tasteless solid; existing in the rhombic, monoclinic and triclinic crystalline forms, and also in an amorphous state. At 111° [1] it melts to a pale yellow liquid; at 180° it thickens to a dark gum-like material, containing a large percentage of amorphous sulphur; at 260° it becomes a liquid again, and at 444.53° it boils, giving off a brownish-red vapor.

The important commercial forms of elemental sulphur are: **Flowers of Sulphur,** consisting of rhombic sulphur and not less than 30% of amorphous sulphur with a small amount of occluded free acid; **Powdered Sublimed Sulphur** (often called Flour Sulphur, a confusing term that should be abolished), consisting essentially of finely ground sublimed sulphur all in the rhombic form though at times a small percentage of amorphous sulphur is present; **Refined Brimstone** and **Roll Sulphur** (in some sections termed Virgin Lump Sulphur), consisting entirely of sublimed sulphur in the rhombic form; **Powdered Brimstone** (often termed Commercial Flour, Superfine Flour and the like); and **Brimstone** or **Crude Lump Sulphur.** In these commercial sulphurs the physical form and the presence of certain small amounts of impurity are the characteristics of most importance as all the varieties named, even the Brimstones, usually contain in excess of 99.5% available sulphur.

Heated in the air sulphur burns with a blue flame, and is oxidized to SO_2, a gas with a characteristic pungent odor. This gas passed into a solution of potassium permanganate will decolorize it, if SO_2 is in excess of the amount that will react with the $KMnO_4$ in the solution.

If sulphur is dissolved in a hot alkali solution and a drop of this then placed on a silver coin, a stain of black Ag_2S will be evident, due to the action of the sulphur.

Sulphides. Hydrogen sulphide, H_2S, is liberated when a sulphide is treated with a mineral acid. This gas blackens moist lead acetate paper. H_2S has a very disagreeable odor, which is characteristic.

Sulphates. A white compound, $BaSO_4$, is precipitated in presence of free hydrochloric acid when a solution of barium chloride is added to a solution of a sulphate.

Insoluble sulphates are decomposed by boiling or fusion with alkali carbonates, forming water-soluble alkali sulphates.

Chapter contributed by Wilfred W. Scott with additions by Charles A. Newhall.

Sulphites. Sulphur dioxide, SO_2, is evolved when a sulphite is treated with hydrochloric acid. The odor of the gas is characteristic.

Sulphur dioxide decolorizes a solution of potassium permanganate. (Use very dilute solution.)

Sulphites are distinguished from sulphates by their failure to form a white precipitate, when barium chloride is added to the solution acidified with hydrochloric acid; also by the fact that H_2S is formed when zinc is added to a solution of a sulphite, acidified by hydrochloric acid.

Thiosulphates. Sulphur dioxide is evolved and free sulphur precipitated when a thiosulphate is acidified with dilute mineral acids. In presence of oxidizing agents sulphides will also liberate free sulphur.

Thiosulphates are strong reducing agents.

ESTIMATION

The determination of sulphur may be required in a great variety of substances, minerals, rocks, sulphur ores, acids, salts, water, gas, coal and other organic matter, insecticides, fungicides, stock medicants, fertilizer and other agricultural materials.

The substance occurs in nature principally in the following forms:

Element. Found free, generally mixed with earthy matter. The commercial product is exceedingly pure and may contain over 99.5% S.

Sulphur Dioxide. The gas, together with free sulphur, is found in volcanic regions.

Hydrogen Sulphide. Occurs in mineral waters and in the air, from decaying organic matter.

Sulphide Ores. Iron pyrite, FeS_2 (30 to 50% S); ferro ferric sulphide, $Fe_2O_3 \cdot 5FeS$; pyrrhotite, Fe_7S_8; copper pyrites, $CuFeS_2$; realgar, As_2S_2; orpiment, As_2S_3; galena, PbS; cinnabar, HgS; zinc blende, ZnS.

Sulphate Ores. Gypsum, $CaSO_4 \cdot 2H_2O$, very abundant; barytes, or heavy spar, $BaSO_4$; celestite, $SrSO_4$; kieserite, $MgSO_4 \cdot H_2O$; bitter spar or Epsom salts, $MgSO_4 \cdot 7H_2O$; Glauber salt, $Na_2SO_4 \cdot 10H_2O$; sulphates of alkalies in animal and plant fluids.

The gravimetric determination of sulphur, by procedures of technical importance, depends upon its precipitation as barium sulphate, $BaSO_4$, after converting it into sulphuric acid, or a soluble sulphate, if not already in this form. Oxidation of free sulphur, sulphides, sulphites, metabisulphites, thiosulphates may be accomplished by either dry or by wet methods, details of which are given under subsequent procedures.

The volumetric methods of determining sulphur depend upon titration with oxidizing agents, or by acids, or by alkalies, according to the form of the sulphur compound, or by means of a substance forming an insoluble compound with sulphuric acid. For example sulphides are treated with a strong mineral acid (HCl), the evolved H_2S absorbed in a suitable reagent, and the sulphide formed is titrated with standard iodine. Sulphites may be determined either by oxidation with iodine or by titration with an acid in presence of methyl orange. Acid sulphites or metabisulphites may be determined by the iodine titration or by titration with an alkali in presence of phenolphthalein. Thiosulphates are titrated with iodine.

Preparation and Solution of the Sample

Sulphide. Sulphides of Na, K, Cs, Rb, Ca, Sr, Ba, Mg, Mn, Fe are soluble in dilute mineral acids. The sulphides of Ag, Hg, Pb, Cu, Bi, Cd, Co, Ni require strong acids for decomposition. These are also insoluble in sodium hydroxide and potassium hydroxide solutions. As, Sb and Sn sulphides are insoluble in dilute acids, but soluble in alkalies.

Sulphate. With exception of $BaSO_4$, $CaSO_4$, $SrSO_4$ and $PbSO_4$, sulphates are soluble in water.

Thiosulphate. Nearly all are soluble in water.

Sulphite. With exception of the sulphites of the alkalies, sulphites of the metals are difficultly soluble in water, but readily decomposed by acids.

Decomposition of Sulphur Ores

The wet procedure for oxidation and decomposition of sulphur ores is given in detail under the Gravimetric Methods, page 498. This process is used for the valuation of the ore, and is applicable to a wide range of substances.

Fusion Method. One gram of the finely ground ore (80 mesh) is intimately mixed with 6 grams of zinc oxide-sodium carbonate mixture (4 parts $ZnO+1$ part Na_2CO_3), placing 2 grams more of the mixture over the charge. The material is fused and sulphur extracted according to the procedure described for coal— Eschka's method.

Sulphur in Coal, Eschka's Method

One gram of coal is intimately mixed with 3 grams of Eschka's compound, consisting of 2 parts of porous, calcined magnesia and 1 part of anhydrous sodium carbonate. The mixture, placed in a platinum crucible, is covered with about 2 grams more of Eschka's compound. The charge is placed in an open platinum crucible, which is protected from the flame by a shield, as shown in Fig. 65. If possible, a sulphur-free flame should be used to avoid contaminating the material. With proper precautions, the shield will prevent this. Heating in a crucible electric furnace completely avoids sulphur contamination. The mixture is heated very gradually, to drive off the volatile matter, the charge being stirred frequently with a platinum wire to allow free access of air. The heat is increased, after half an hour, to a dull redness. When the carbon has burned out, the gray color having changed to a yellow or light brown, the heat is removed and the crucible cooled.

The powdered fusion is digested with 100 cc. of hot water for half an hour, and the clear liquor decanted through a filter into a beaker. The residue is washed twice more with hot water, by decantation, and finally on the filter, until the volume of the total filtrate amounts to about 200 cc. About 5 cc. of bromine and a little hydrochloric acid are added, and the solution boiled. Sulphuric acid is now precipitated as $BaSO_4$ by addition of barium chloride to the hot solution, and sulphur determined by the first of the gravimetric procedures.

Sulphur in Rocks, Silicates, and Insoluble Sulphates

The material in finely powdered form is fused in a large platinum crucible with about six times its weight of sodium carbonate (sulphur free) mixed with about 0.5 gram of potassium nitrate. The charge is protected from the flame by an asbestos board or silica plate with an opening to accommodate the crucible snugly, as shown in Fig. 65. The fusion is extracted with water, the filtrate evaporated to dryness and silica dehydrated. The residue is moistened with strong hydrochloric acid, then taken up with a little water, boiled free of CO_2, and silica filtered off. The filtrate contains the sulphate, which is now precipitated as barium sulphate according to one of the standard procedures.

Barium Sulphate. This is transposed by fusion with sodium carbonate, as stated above. Barium carbonate remains in the water-insoluble residue. It is advisable to wash the residue in this case with hot sodium carbonate solution, to insure complete removal of the sodium sulphate. The filtrate is acidified with HCl, boiled free of CO_2 and $BaSO_4$, then precipitated.

Lead Sulphate. This may be transposed by digesting the compound with a strong solution of sodium carbonate saturated with CO_2, keeping the solution at boiling temperature for half an hour or more. The sulphate will be in solution and the lead is precipitated as the water-insoluble carbonate.

Strontium or calcium sulphates may be transposed by the procedure described for lead.

SEPARATIONS

Substances Containing Iron

In precipitating barium sulphate, in presence of ferric salts, from hot solutions by the gravimetric procedure commonly followed, considerable iron is carried down by the precipitate. Since $Fe_2(SO_4)_3$ loses SO_3 upon ignition, and since Fe_2O_3 weighs much less than $BaSO_4$, low results will be obtained. Hence the removal of iron is necessary, or a method should be followed in which iron does not interfere. It is found that barium sulphate precipitated from a large volume of cold solution, in which the iron has been reduced to ferrous condition, is free from iron. Details of this procedure are given in the second of the gravimetric methods, page 498.

If sulphur is to be precipitated from hot solution of comparatively small volume (200 to 400 cc.), it is necessary to remove iron. This is accomplished by precipitating this as $Fe(OH)_3$ by addition of ammonium hydroxide in decided excess (5 to 10 cc. excess of strong NH_4OH, sp.gr. 0.90). If the solution is barely neutralized with ammonia, the iron hydroxide carries down considerable of the sulphate. Even with the precaution recommended some of the combined sulphuric acid is occluded by the precipitate, so that it is necessary to recover this by dissolving the precipitate with hydrochloric acid and reprecipitating the ferric hydroxide with an excess of ammonia. The combined filtrates are now treated with barium chloride, upon acidification with hydrochloric acid, according to the procedure first given, page 497, and the sulphate determined.

Separation of Sulphur from Metals Forming an Insoluble Sulphate

This is accomplished by fusion of the compound with sodium carbonate and extraction of the mass with water. The metal remains with the residue and the sulphate of the alkali passes into solution. For details see subject under Preparation and Solution of the Sample, page 494.

Nitrates and Chlorates. These are carried down with the precipitate as barium salts if they are present in appreciable amount. They may be removed from the solution by evaporation to dryness with hydrochloric acid.

Silica. Silica will be carried down with the barium sulphate precipitate if present in appreciable amounts. It is removed by evaporation of the solution with hydrochloric acid, dehydrating the silicic acid, taking up with HCl and water and filtering.

Ammonium and Alkali Salts. These have a negligible effect on the precipitate of $BaSO_4$ if this is precipitated from a large volume, according to the second gravimetric procedure.

GRAVIMETRIC DETERMINATION OF SULPHUR

Precipitation as Barium Sulphate

Preliminary Remarks. The procedure depends upon the insolubility of barium sulphate, $BaSO_4$, in neutral or slightly acid solutions. It was formerly the general practice to precipitate the sulphur by adding a 10% barium chloride solution to the hot sulphate solution, which had been diluted from 200 to 400 cc., according to the amount of sulphur that was present (not over 0.2 gram sulphur per 100 cc.), containing 1 to 3 cc. of free concentrated hydrochloric acid per 100 cc. of solution. Special precautions were given to have the solution boiling hot, and to avoid having a volume of over 400 cc., a smaller sample being taken in high sulphur ores, rather than increase the volume. Extended experiments have shown that it is preferable to precipitate the sulphate from a large volume of cold solution. The product obtained is less apt to occlude impurities, the crystals are larger than those obtained in hot concentrated solutions, and do not pass through the filter. Precipitation may be made in presence of large amounts of iron, copper and other impurities. The procedure requires large beakers of 2- to $2\frac{1}{2}$-liter capacity, special precipitating cups, and a suction apparatus, as shown in Figs. 60, 61, 62 and 63. This apparatus may not always be available, and occasionally it is advantageous to precipitate the sulphur in a small volume, specially when the sulphur content of the material is low, hence, although the second procedure is generally recommended, the older method is also included.

I. Precipitation of Barium Sulphate from Hot Solutions

Procedure. The sulphur should be present in solution either as free sulphuric acid or as a sulphate salt. The solution is made acid by addition of hydrochloric acid (phenolphthalein indicator), and then 4 cc. added in excess (HCl, sp.gr. 1.2). After diluting to a volume of 400 cc. with hot water, the mixture is heated to boiling, and a 10% solution of barium chloride added in a fine stream,[1] through a funnel with a capillary stem, or from a burette, at the rate of 10 cc. in two to ten minutes. The reagent is added in slight excess of that required to react with the sulphuric acid or sulphate. (Ten cc. of 10% $BaCl_2.2H_2O$ solution will precipitate about 0.13 g. of sulphur.) The beaker is placed on a steam bath and the pre-

[1] E. Hintz and H. Weber recommend adding 100 cc. of N/10 $BaCl_2$ solution, boiling hot, to the hot sulphate solution all at once in place of slowly, as recommended in general practice. (See Treadwell and Hall, "Analytical Chemistry," **2**, 3d Edition, p. 469.)

cipitate allowed to settle for about two hours. The solution is filtered through a fine grade of filter paper (B. & A. grade A, or S. & S. grade No. 90), or through a tared Gooch crucible. Since the precipitate frequently passes through the filter it is advisable always to pass the solution through the same filter a second time. The precipitate is washed ten times with hot water, then dried, and ignited gently over a Bunsen burner, or in a muffle, for half an hour. (Blasting is not necessary, nor desirable.) The white $BaSO_4$ is cooled in a desiccator, and then weighed. If a filter paper has been used in place of a Gooch crucible, the ignition is best made in a porcelain crucible, with free access of air, the ignited sulphate, upon cooling, is brushed out of the crucible and so weighed.

Factors. $BaSO_4 \times 0.1373 = S$, or $\times 0.4202 = H_2SO_4$, or $\times 0.3766 = FeS$, or $\times 0.2744 = SO_2$, or $0.3430 = SO_3$, or $\times 0.4115 = SO_4$.

NOTE. If much iron or alumina is present it is advisable to precipitate the sulphate from a large volume, by the second method, rather than attempt to remove these substances. If $BaSO_4$ is present in the original material its weight should be included with that of the precipitate.

II. Precipitation of Barium Sulphate from Cold Solutions— Large Volume

Introduction. The method worked out by Allen and Bishop, General Chemical Company,[1] is especially adapted to the determination of sulphur in iron pyrites and materials high in sulphur, 30 to 50% sulphur, but by varying the amount of material used the range may be extended from smaller to greater amounts. The finely ground sample is oxidized by means of a mixture of bromine and potassium bromide, followed by nitric acid. The nitric acid is expelled by evaporation to dryness, followed by a second evaporation with hydrochloric acid, which dehydrates the silica. Iron is now reduced to the ferrous condition and the silica and residue, undissolved by addition of hot water and HCl, is filtered off. The sulphur is precipitated in a large volume of cold solution, by barium chloride solution, as $BaSO_4$ and so weighed.

Reagents. *Bromine—Potassium Bromide Solution.* 320 grams of potassium bromide are dissolved in just sufficient water to cause solution and mixed with 200 cc. of bromine, the bromine being poured into the saturated bromide solution. After mixing well the solution is diluted to 2000 cc.

Bromine—Carbon Tetrachloride Solution. Carbon tetrachloride saturated with bromine.

Barium Chloride, anhydrous, 5% solution; or crystals, 6% solution.

Procedure. *Preparation of Sample* The sample ground to pass 80-mesh sieve is carefully mixed and quartered down to 10 grams. This is dried for one hour at 100° C. and then placed in a weighing tube.

A factor weight, 1.373 grams of the sample, is placed in a deep beaker, 300 cc. capacity, $2\frac{1}{2}$ by $4\frac{1}{2}$ ins.

Oxidation of Sulphur. Ten cc. of the bromine-potassium bromide mixture for pyrrhotite ore, or bromine—carbon tetrachloride reagent for pyrites ores, are added and the beaker covered with a dry watch-glass cover. After standing

[1] Paper before Eighth International Congress of Applied Chemistry: "An Exact Method for the Determination of Sulphur in Pyrites Ores," W. S. Allen and H. B. Bishop.

fifteen minutes in the cold bath (a casserole of water will do), with occasional shaking of the beaker, 15 cc. of strong nitric acid are added and the mixture allowed to stand fifteen minutes longer, at room temperature, and then warmed on an asbestos board on the steam bath until the reaction has apparently ceased and the bromine has been volatilized. The beaker is now placed within the ring of the steam bath so that the lower portion is exposed to steam heat. The solution is evaporated to dryness, the cover of the beaker being raised above the rim by means of riders (U-shaped glass rods), Fig. 60, 1C cc. of strong hydro-

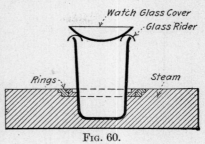

Fig. 60.

chloric acid are now added and the solution again evaporated to dryness to expel the nitric acid. The silica is dehydrated by heating in the air oven at 100° C. for one hour, or overnight if preferred.

Reduction of Iron. Four cc. of hydrochloric acid (sp.gr. 1.20), followed five minutes later by 100 cc. of hot water, are added, the sides of the beaker and the cover being rinsed into the solution. The riders being removed, the sample is gently boiled for five minutes to insure the solution of the sulphate. After cooling for about five minutes, approximately 0.2 gram powdered aluminum is stirred into the solution, keeping covered during the intervals between stirring. When the iron has been reduced, the solution becoming colorless, the sample is filtered into a 2500-cc. beaker, through a 12½ cm. filter paper (S. & S. No. 590 or B. & A. No. A). The beaker should be copped out and the residue on the filter washed nine times with hot water, filling the filter funnel and draining each time.

Precipitation of the Sulphur. The solution in the large beaker is diluted to 1600 cc. with cold water and 6 cc. HCl (sp.gr. 1.20) added, and mixed by stirring. The barium chloride solution is now added by means of a special delivering cup (Figs. 61 and 62), which should drain at the rate of 5 cc. per minute. 125 cc. of barium chloride solution are added for ores containing 30 to 50% sulphur, the factor weight being taken. The solution is not stirred while the barium chloride is being added, but when the cup has drained, the solution is mixed by stirring. The BaSO₄ is allowed to settle, two or three hours being advisable, overnight being preferred.

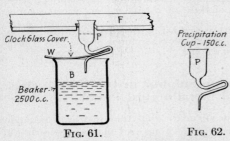

Fig. 61.   Fig. 62.

Apparatus for Precipitating Sulphur.

Filtration. The clear solution is filtered through a weighed Gooch crucible (35 cc.), using suction. This is best done by the automatic arrangement shown in Fig. 63. The beaker containing the solution is placed on a shelf; a siphon dipping to within half an inch of the precipitate at the bottom of the beaker is connected to the Gooch crucible by means of a tightly fitting stopper. The Gooch and thistle tube are best connected by heavy rubber tubing. The suction flask, or bottle, should have a capacity of about 3 liters. A Geissler stop-cock passes through the rubber stopper in the suction flask to relieve the pressure when the Gooch is to be removed. The precipitate is washed onto the asbestos mat in the crucible and washed with cold water six times, the beaker being copped out as usual.

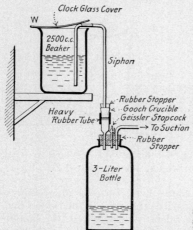

Ignition. The precipitate is dried by placing the crucible on an asbestos board over a flame for twenty-five minutes and then heated over a direct flame for thirty minutes.

Calculation. $BaSO_4 \times 10 =$ per cent S. (If factor weight is taken.)

Factor. $BaSO_4 \times 0.1373 =$ gram S.

Notes and Precautions

Although barium sulphate is only slightly soluble in water, it is appreciably soluble in the salts of the alkalies (Na, K and NH_4), and in a large excess of hydrochloric acid.

FIG. 63.—Apparatus for Filtering Barium Sulphate.

Barium sulphate occludes salts, especially nitrates and chlorides. Ferric chloride is carried down with this precipitate, though ferrous chloride is not; hence the reduction of iron is necessary. Occlusion of iron causes low results, as will be seen from the fact that with heating of $Fe_2(SO_4)_3$, SO_3 is volatilized, the salt decomposing to $Fe_2O_3 + SO_3$. With the iron reduced the precipitate burns perfectly white, whereas with ferric iron present the precipitate is invariably red or yellow. Aluminum powder used by W. H. Seamon,[1] for reduction of iron in determination of sulphur, suggested its value in the method above given.

Potassium bromide is added to the bromine mix as a diluent to prevent too vigorous a reaction. Cooling the solution is for the same purpose as a loss of sulphur will result if the reaction is violent. This is especially the case in pyrrhotite ore.

Otto Folin[2] shows that precipitation of $BaSO_4$ in a large volume of cold solution produces large crystals.

Mechanical loss and reduction of $BaSO_4$ is avoided by the Gooch crucible.

The method has been thoroughly tested in the laboratories of the Gen. Chem. Co. and has become a standard method for sulphur.

Evolution Method for Determining Sulphur in Iron, Steel, Ores, Cinders, Sulphides and Metallurgical Products

Introduction. The method depends upon the fact that hydrogen sulphide is evolved when a sulphide is acted upon by a strong acid such as hydrochloric acid. This gas, absorbed by a suitable reagent, may be determined gravimetrically[3] by weighing directly the precipitated sulphide, or by oxidation of either the hydrogen sulphide evolved or the sulphide formed in the absorbing reagent,

---

[1] Chemical Engineer, September, 1908.

[2] Journal of Biological Chem., **1**, 131–159.

[3] *Gravimetrically.* (a) Evolution of H_2S into solutions of $ZnCl_2$, KOH, $KMnO_4$, $AgNO_3$, $Hg(CN)_2$, H_2O_2, Br + HCl and subsequent oxidation to sulphate when necessary, and precipitation as $BaSO_4$. (b) Absorption of H_2S by neutral or alkaline solutions of lead, oxidation of PbS to $PbSO_4$ and weighing as such. (c) Absorption of H_2S in solutions of $AgNO_3$, $CdCl_2$, and weighing the precipitated sulphide.

and precipitating sulphur as $BaSO_4$. It may be determined volumetrically [1] by titrating the precipitated sulphide with iodine or by titrating the acid, formed by the reaction, with standard caustic. The iodine and caustic titrations may be made on the same run, or the sulphide may be weighed and the filtrate containing the free acid titrated, thus double checking results. The following reaction takes place when the gas is evolved and absorbed by neutral cadmium sulphate:

$$H_2S + CdSO_4 = CdS \text{ precipitate} + H_2SO_4 \text{ free acid.}$$

The method is especially adapted to the determination of sulphur in iron and steel or in metallurgical products containing small amounts of sulphide. It may be applied to products containing larger amounts of sulphur as sulphides or sulphates, the latter condition requiring a special preliminary treatment.

The method is not applicable for determining free sulphur or sulphur in iron pyrites.

Reagents. *Iodine Solution.* Two strengths of this reagent should be at hand for general work:

For iron and steel and low sulphur briquettes, etc. = .01 to
0.5% S . . . . N/30 I
For sulphur products containing
over 0.5% S . . . . N/10 I

Starch Solution. Made from a good grade of soluble starch, 1 gram per 200 cc. of water. Fresh solutions are desirable, as the deteriorated material produces a greenish-brown color in place of the delicate blue desired. Flocks of insoluble starch will cause the same difficulty.

Cadmium Chloride or Cadmium Sulphate Solutions. Ammoniacal Solution. Fifty-five grams of $CdCl_2 \cdot 2H_2O$ or 70 grams of the sulphate are dissolved in 500 cc. of distilled water. To this are added 1200 cc. NH_4OH (sp.gr. 0.90) and the solution diluted to 2500 cc. The solution is of such strength that 50 cc. will precipitate approximately 0.175 gram sulphur evolved as H_2S. This is equivalent to about 3.5% sulphur on a 5-gram sample.

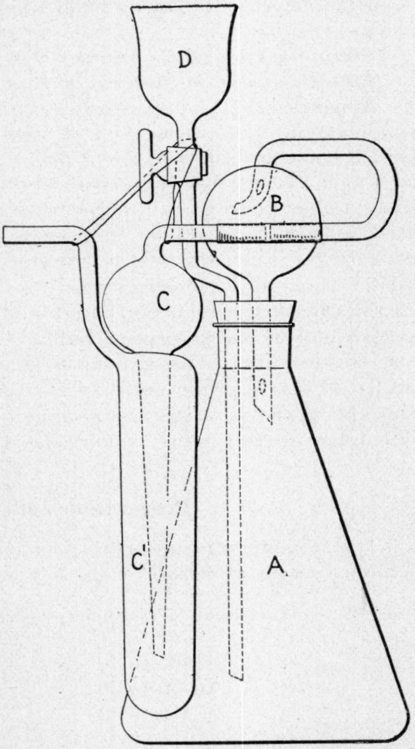

FIG. 64.—Scott's Apparatus for Determining Sulphur in Iron and Steel.[2]

[1] *Volumetrically.* (a) Absorption in a solution of KOH, $CdCl_2$ or $CdSO_4$, $ZnCl_2$ or $ZnSO_4$, Na_2HAsO_3 and titration with iodine solution. (b) Absorption in iodized KI and titration of the excess of iodine with $Na_2S_2O_3$ solution. (c) Absorption in a neutral solution of a metallic salt and titration of the liberated acid. (d) Absorption in caustic alkali and addition to an acid solution of a reducible salt, e.g., Fe_2O_3 and titration of the lower oxide, FeO.

[2] Apparatus designed by W. W. Scott.

Neutral Solution. To be used where titration with caustic is desired. Seventy grams of $CdSO_4$ are dissolved in water and made up to 2500 cc. The solution should be neutral to methyl orange, otherwise add the requisite amount of H_2SO_4 or NaOH necessary, determined by titration of an aliquot portion.

Hydrochloric Acid. One part concentrated acid to an equal volume of distilled water.

Sulphuric Acid. One volume of concentrated acid to four volumes of distilled water.

Reducing Mixture for Reduction of Sulphates. Five parts of $NaHCO_3$, 2 parts of C.P. aluminum powder and 1 part of pure carbon, best made by charring starch. A blank should be determined on this material and allowance made accordingly.

Stannous Chloride. Ten-per cent solution.

Fine Granular Aluminum or Zinc Metal. Sulphur free, 20 mesh.

Apparatus. The apparatus shown in the illustration, Fig. 64, is the author's [1] modification of the form used at Baldwin Locomotive Works. This consists of an Erlenmeyer flask *A* of about 500-cc. capacity with large base. With material in which violent foaming occurs, during the evolution of hydrogen sulphide, it is advisable to use a wash bottle with large base, in preference to an Erlenmeyer flask. Through a rubber stopper is inserted a thistle tube with glass stopcock *D*, by which the acid is introduced into the flask. The hydrogen sulphide passes through a potash connecting bulb with trap as shown. A hole blown in the side of the tube prevents liquid being swept through. Connected to the potash bulb is the absorption bulb *C*, which is suspended by a wire attached to the thistle tube. The apparatus is compact, so that on a large hot plate, 30 by 20 ins., a dozen outfits may readily be accommodated. With the use of this apparatus the writer has been able to make over seventy-five determinations of sulphur in steel in an ordinary day's run.

Preparation and Amount of Sample

The amount of material to be taken for the determination depends upon the sulphur content as shown by the following table:

Approximate % of Sulphur Present.	Amount to take for Analysis.
0.01 to 1	.5 grams
1.0 to 10	1
10.00 to 30	0.5
Above 30	0.25

The class of material will govern the method of procedure.

Iron and Steel. A 5-gram sample of drillings or finely divided material is treated directly in the evolution flask with hydrochloric acid, 1 : 1, and the hydrogen sulphide absorbed in ammoniacal cadmium chloride. The sulphide formed is titrated with iodine.

Iron Ore Briquettes and Materials Containing Sulphates. Low Sulphur. *Preliminary Reduction.* A 5-gram sample is intimately mixed with an equal weight of reducing mixture ($NaHCO_3+Al+C$) and wrapped in a 9-cm. ashless filter. The charge is placed in a 50-cc. nickel crucible with cover. The crucible

---

[1] W. W. Scott.

is inserted half way into an asbestos board or perforated silica plate (see Fig. 65) and after covering, placed over a low flame of a Méker blast burner. The flame of the blast is gradually increased during the first five minutes and the charge blasted for about twenty minutes. The crucible will appear a bright red and carbon monoxide gas escaping from under the crucible lid will burn. The loss of sulphur, however, is not appreciable. The crucible is cooled without removing the cover. When cold the fused mass is quickly pulverized and placed in the dry evolution flask containing a mat of aluminum granules or C.P. zinc dust or granulated tin. Hydrogen sulphide is best evolved with hydrochloric acid to which 4 or 5 cc. of 10% stannous chloride has been added to reduce ferric iron. The gas is absorbed in ammonical cadmium chloride and the cadmium sulphide formed titrated with iodine.

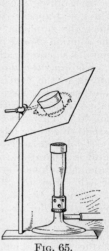

Iron Sulphide for Available H₂S. Since this product runs over 20% available hydrogen sulphide not over 0.5 gram sample should be taken. The H₂S is evolved by addition of dilute sulphuric acid, 1 : 4, in place of hydrochloric acid, and is absorbed by neutral cadmium sulphate. The acid formed by the reaction is titrated by standard N/10 NaOH.

FIG. 65.

Sodium Sulphide or Water-soluble Sulphides for Available H₂S. Ten grams dissolved in water and diluted to 1000 cc.; 50 cc. = (0.5 gram) taken for analysis.

Details of Procedure

Evolution of Hydrogen Sulphide. One-half to 1 gram of aluminum or zinc granules, 20 mesh, is placed over the bottom of the evolution flask and the sample placed above this mat of metal. The stopper with the thistle tube and condenser is inserted snugly into the neck of the flask. An absorption bulb containing about 20 cc. of distilled water is attached to the condenser. This bulb serves as a trap for the HCl that is driven out of the flask during the boiling. To this bulb is attached a second bulb containing 50 cc. of ammoniacal cadmium chloride. A third bulb may be attached if the sulphur content of the material examined is high; this, however, is seldom necessary when ammoniacal cadmium chloride is used. The rubber stopper and all rubber connections being air tight, 100 cc. of warm HCl, 1 : 1, is poured into the flask through the thistle tube, the stem of which should now dip well below the acid. The stopcock is closed during the violent action of the acid on the sample and opened when this has subsided. The acid trap prevents loss of H₂S through the thistle tube. The apparatus is now placed on the hot plate and the sample boiled vigorously for about twenty minutes. The flask is taken off the hot plate and the contents allowed to cool. At this stage it may be advisable to draw a current of air through the apparatus to sweep out any residual H₂S that may remain in the flask. Hydrogen gas is preferable to air.

Titration. (a) The contents of the bulbs are poured into a 600-cc. beaker containing about 400 cc. of distilled water. The bulbs are washed out first

40

with water and then with dilute acid.   The excess of ammonia is neutralized with concentrated HCl, 5 cc. of starch solution added and the sulphide immediately titrated with standard iodine, additional hydrochloric acid being added from time to time during the titration to insure complete decomposition of the sulphide. The liquid appears yellowish red, orange, purplish red and finally a deep blue. Since the sulphide, when present in appreciable quantity, decomposes slowly, the solution should be strongly acid at the completion of the titration, and five minutes should be allowed for a permanent end-point.

Knowing the amount of iodine necessary, a check run may be made by adding to the neutral solution an excess of iodine followed by 5 cc. of starch solution and a large excess of concentrated hydrochloric acid.   The excess of iodine is titrated with N/10 thiosulphate, $Na_2S_2O_3$, solution.   (Arsenous acid will not do.)   This procedure will prevent the loss of H_2S, which is apt to occur in samples high in sulphide.

(b) An alternate method is frequently advisable in high sulphurs.   The precipitate is separated from the solution containing ammonia by filtration.   The cadmium sulphide is now placed in the 600-cc. beaker with water and an excess of iodine run in.   Starch is added, followed by hydrochloric acid.   The excess of iodine is titrated with sodium thiosulphate, $Na_2S_2O_3$.   By this method the heat action during the neutralization of ammonia is avoided and only the precipitate is titrated.

When the iodine titration exceeds 50 cc. of N/10 iodine, a smaller amount of the sample should be taken for analysis; the iodine titration for amounts of sulphur exceeding 0.1 gram is not satisfactory, owing to a fading end-point. The method for determining available hydrogen sulphide in high sulphide products, dealing with the titration of the free acid formed during the reaction, permits of larger samples being taken.   Details of this method are given on page 509.

One cc. N/10 iodine = 0.001604 gram S.

Tenth Normal Equivalents

One cc. of N/10 iodine = 0.001704 gram H_2S
        "   "   = 0.004396 gram FeS
        "   "   = 0.003904 gram Na_2S
        "   "   = 0.003607 gram CaS
        "   "   = 0.008471 gram BaS
        "   "   = 0.00561 gram Sb_2S_3
        "   "   = 0.011959 gram PbS
        "   "   = 0.011634 gram HgS
        "   "   = 0.004782 gram CuS
        "   "   = 0.007224 gram CdS
        "   "   = 0.004872 gram ZnS
        "   "   = 0.003269 gram Zn

Combustion Method for Evaluation of Sulphide Ores.   When a sulphide ore (pyrrhotite) is heated to redness in presence of oxygen both sulphur dioxide and trioxide are evolved.   The first may be absorbed in suitable reagents and estimated volumetrically or gravimetrically.   The trioxide mist is best retained by asbestos and weighed.   The combustion furnace with silica tube used for determinations of carbon is adapted for sulphide ores.   The finely powdered dry sample, spread in a thin layer in a 3-inch porcelain boat, is placed in the red hot tube and burned in a current of oxygen, which has been purified by passing through sodium hydroxide, strong sulphuric acid and phosphorus pentoxide. The trioxide mist is removed by passing the evolved gases through an asbestos filter (P_2O_5 bulb with asbestos in one arm adjacent to the combustion tube and P_2O_5 in the other). The SO_2 is absorbed in a mixture of bromine and nitric acid, and the sulphuric acid formed is titrated after removing the reagent by evaporation; or it is absorbed in an excess of standard iodine, the excess titrated with sodium arsenite or thiosulphate, and sulphur calculated. The iodine method is preferable to the bromine, as it is more rapid and the reagent less disagreeable to handle. The gravimetric method is the most reliable. The dioxide is absorbed in chromic acid (caustic will not give correct results owing to its affinity for carbon dioxide, a product of combustion of the free and combined carbon, that are generally present in sulphide ores. Pyrrhotite frequently contains as much as 1% carbon) and weighed. The combustion method cannot be recommended for extreme accuracy. The procedure may be used for the estimation of available sulphur, but does not give the total sulphur of the ore, since .2 to .5% remains in the cinder. Error may result from the following causes: (1) Incomplete combustion of the sulphur—due to sublimation of the sulphur to cooler zones of the combustion tube, and to a fine mist of sulphur passing unburned into the asbestos, where it is retained with SO_3 and weighed as such. (2) Error due to combined water of the ore. The results are apt to be .05 to 0.5% lower than those obtained by the barium sulphate procedures, the sulphur of the cinder being included with the available sulphur.

VOLUMETRIC METHODS FOR DETERMINING SOLUBLE SULPHATES

Combined sulphuric acid in soluble sulphates is best determined gravimetrically; occasionally, however, a volumetric procedure is of technical value. A number of volumetric methods are based on the insolubility of barium sulphate. Two general procedures deserve mention: addition of barium chloride in known amount in slight excess of that required by the sulphate, and titrating the excess either with a soluble carbonate or a chromate; or addition of barium chromate and titrating the alkali chromate formed by the reaction. The sulphate is also determined by precipitation with a weak organic base benzidine, added in form of the hydrochloride salt; the benzidine sulphate, filtered off, is titrated with caustic. The typical procedures given below will meet general requirements for the volumetric determination of sulphates.

Determination of Sulphur by Titration with Barium Chloride and Potassium Chromate—Wildenstein's Method Modified [1]

Reaction.

$$Na_2SO_4 + BaCl_2 = BaSO_4 + 2NaCl \text{ and excess } BaCl_2 + K_2CrO_4 = BaCrO_4 + 2KCl.$$

Procedure. The substance containing the sulphate in solution is diluted to 50 cc. in a small flask, acidified with hydrochloric acid, if necessary, heated to

[1] See "Volumetric Analysis," Sutton, 10th Ed., p. 350.

boiling, and precipitated with a slight excess of N/4 barium chloride added from a burette (1 cc. $BaCl_2 = 0.01$ gram SO_3). The precipitate settles rapidly, so that a large excess of the reagent may readily be avoided. The mixture is cautiously neutralized with ammonia, free from carbonate (CO_2 may be precipitated with $CaCl_2$ solution), the solution heated to boiling, and N/4 potassium chromate added from a burette in .5 cc. portions, each time removing the flask from the heat, allowing the precipitate to settle and examining the clear solution. A faint yellow color will appear as soon as the excess of barium has been precipitated and a few drops of the chromate in excess are present in the solution. The value of the chromate being equivalent to the barium chloride cc. per cc., the difference between the two titrations is due to the barium chloride required by the sulphate.

One cc. N/4 $BaCl_2 = 0.01$ gram SO_3.

NOTES. Salts of the alkalies, alkaline earths (Sr and Ca) and zinc and cadmium do not interfere. Nickel, cobalt and copper, however, give colored solutions which prevent the yellow chromate being seen. Should the latter be present, the end-point may be recognized by using ammoniacal lead acetate as an outside indicator (1 vol. $NH_4OH + 4$ vols. $PbC_2H_3O_2 \cdot 3H_3O$, 5% sol.), the indicator and titrated solution being mixed drop per drop on a white tile. A yellowish red color indicates the presence of chromate.

Precipitation of the Sulphate with Barium Chromate and Titration of Equivalent, Liberated Chromate with Iodine and Thiosulphate, Hinman's Method.[1]

The sulphate, precipitated by barium chromate, liberates an equivalent amount of chromic acid, which is determined by treating with potassium iodide and titrating the liberated iodine with thiosulphate.

Reactions. $Na_2SO_4 + BaCrO_4 = BaSO_4 + Na_2CrO_4,$

$$Na_2CrO_4 + 3KI + 8HCl = 2NaCl + 3KCl + CrCl_3 + 4H_2O + 3I,$$

$$2Na_2S_2O_3 + I_2 = 2NaI + Na_2S_4O_6.$$

Procedure.[2] The solution of the sulphate, containing not over 2 per cent of SO_3, if acid, is almost neutralized with potassium hydroxide, then heated to boiling, and an excess of barium chromate solution added.[3] After boiling for one to five minutes, the hot solution is neutralized by adding calcium carbonate [4] until no further effervescence occurs. The precipitate is filtered off and washed with hot water. The combined filtrates containing the chromate liberated by the sulphate through double decomposition, is acidified with 5 cc. strong

[1] Treadwell and Hall, " Analytical Chemistry," **2**, 4th Ed., p. 716. Am. Jour. Sci. and Arts, **114**, 478.

[2] See p. 716.

[3] The barium chromate used should be free from soluble chromate, barium carbonate or soluble barium salt. The compound may be prepared by precipitating with potassium chromate added to a boiling solution of barium chloride. The precipitate is washed with boiling water containing a little acetic acid, and finally with pure water, and then dried. Four grams of the dry salt are dissolved in a liter of normal hydrochloric acid.

[4] In presence of iron, zinc and nickel, the solution is neutralized with ammonium hydroxide and an excess added; after boiling, the solution is filtered. By using calcium carbonate insoluble basic chromates of these elements would be formed, and low results for SO_3 would follow. This is avoided by the use of ammonia.

hydrochloric acid per each 100 cc. of filtrate and an excess of potassium iodide added. Iodine equivalent to the chromic acid is liberated. This is titrated with N/10 sodium thiosulphate.

One cc. of N/10 thiosulphate $=0.003269$ gram H_2SO_4.[1]

Benzidine Hydrochloride Method—Raschig

Benzidine sulphate, $C_{12}H_8(NH_2)_2 \cdot H_2SO_4$, is scarcely soluble in water containing hydrochloric acid. The weak base benzidine is neutral to phenolphthalein and the acid in its sulphate may be titrated with an alkali.[2] The method gives reliable results in the analysis of all sulphates, provided no substances are present which attack benzidine, and provided the amount of other acids and salts present is not too great.[3]

Reaction. $Na_2SO_4 + C_{12}H_8(NH_2)_2 \cdot 2HCl = 2NaCl + C_{12}H_8(NH_2)_2 \cdot H_2SO_4$ and

$$C_{12}H_8(NH_2)_2 \cdot H_2SO_4 + 2NaOH = C_{12}H_8(NH_2)_2 \cdot 2H_2O + Na_2SO_4.$$

Reagent. *Benzidine hydrochloride* is prepared by taking 6.7 grams of the free base, or the corresponding amount of the hydrochloride and mixing into a paste with 20 cc. of water in a mortar. Twenty cc. of hydrochloric acid (sp.gr. 1.12) are added and the mixture diluted to exactly 1000 cc. One cc. of this solution corresponds to 0.00357 gram H_2SO_4. The solution has a brown color. Brown flakes are likely to separate out on standing, but these do no harm.

Procedure. The sulphate solution is diluted with water so that there is at least a 50-cc. volume for each 0.1 gram sulphuric acid present. An equal volume of the reagent is vigorously stirred in, and the precipitate allowed to settle for ten minutes. The solution is filtered onto a double filter, placed on a porcelain, perforated plate in a funnel (a Büchner is O.K.), gentle suction being applied. The last portions of the precipitate are transferred to the filter by means of small portions of the clear filtrate, and the compound then washed with 20 cc. of cold water added in small portions and sucked dry with each addition. The precipitate and filter are placed in an Erlenmeyer flask, 50 cc. of water added, and the mixture shaken until homogeneous. Phenolphthalein indicator is now added, the mixture heated to about 50° C. and titrated with N/10 sodium hydroxide. When the end-point is nearly reached, the liquid is boiled for five minutes, and the titration then completed.

One cc. N/10 $=0.004904$ gram H_2SO_4.

Determination of Free Sulphuric Acid

Other free acids being absent, sulphuric acid may be accurately determined by titration with standard alkali. The method for determining sulphuric acid in presence of commonly occurring acids, and in mixed acids are given in Volume II, in the chapter on Acids, Vol. II.

1 cc. N/1 NaOH $=0.04904$ g. H_2SO_4.

[1] N/10 $Na_2S_2O_3 = \dfrac{H_2SO_4}{30} = 98.08 \div 30 = 3.269$.

[2] Method suggested by Raschig, Z. a. Chem., 617 and 818, 1903.

[3] Friedheim and Nydegger (Z. a. Chem., **9**, 1907) have found that there should not be more than 10 mol. HCl, 15 mol. HNO_3, 20 mol. $HC_2H_3O_2$, 5 mol. alkali salt, or 2 mol. ferric iron present to 1 mol. H_2SO_4. See Treadwell and Hall, "Analytical Chemistry," pp. 714–716.

DETERMINATION OF PERSULPHATES
Ferrous Sulphate Method

Ferrous salts in cold solutions are oxidized to ferric form by persulphates. Advantage is taken of this action in the quantitative determination of persulphates. **Reaction.** $2FeSO_4 + H_2S_2O_8 = Fe_2(SO_4)_3 + H_2SO_4.$

Reaction. $2FeSO_4 + H_2S_2O_8 = Fe_2(SO_4)_3 + H_2SO_4.$

Procedure.[1] About 2.5 grams of the persulphate are dissolved in water and diluted to 100 cc. Ten cc. of this solution, equivalent to one-tenth of the sample, weighed out, are placed in a flask and a considerable excess of standard ferrous sulphate solution [2] added, say 100 cc. measured out from a burette. The solution is diluted with an equal volume o ˙ hot, distilled water (70 to 80° C.), and the excess ferrous sulphate titrated with N/10 potassium permanganate. This titration is deducted from the permanganate equivalent of 100 cc. of the ferrous solution taken (if this amount was used). The difference is due to persulphate oxidation.

One cc. N/10 $KMnO_4 = 0.009708$ gram $H_2S_2O_8$; or $= 0.0114$ gram $(NH_4)_2S_2O_8$; or $= 0.01352$ gram $K_2S_2O_8.$

Oxalic Acid Method

Oxalic acid, in presence of silver sulphate, reduces persulphates in accordance with the reaction, $H_2C_2O_4 + H_2S_2O_8 = 2H_2SO_4 + 2CO_2.$

$$H_2C_2O_4 + H_2S_2O_8 = 2H_2SO_4 + 2CO_2.$$

Procedure. About 0.5 gram of the persulphate is placed in an Erlenmeyer flask, 50 cc. of N/10 oxalic acid added, together with 0.2 gram silver sulphate in 20 cc. of 10% sulphuric acid solution. The mixture is heated on the water bath for about half an hour to expel carbon dioxide. When the evolution ceases the liquid is diluted to 100 cc. with warm water and titrated warm (about 40° C.) with N/10 potassium permanganate. The excess of oxalic acid is titrated, the difference is due to oxidation by the persulphate.

For calculation see factors in previous method.

Alkali Titration of the Boiled Solution

The aqueous solutions of potassium, sodium, and barium persulphates are decomposed by boiling as follows (M = metal Na, K, or Ba):
$$2M_2S_2O_8 + 2H_2O = 2M_2SO_4 + O_2 + 2H_2SO_4.$$
Procedure. About 0.2 g. of the persulphate salt is dissolved in 200 cc. of water and the solution boiled about 15 minutes, then cooled and titrated with N/10 NaOH, using methyl orange indicator.

1 cc. N/10 NaOH = 0.02008 g. $BaS_2O_8 4H_2O$, or 0.01191 g. $Na_2S_2O_8$ or

Ammonium persulphate cannot be determined by the above method but may be determined by the ferrous sulphate method—which see above.

[1] Method suggested by Le Blanc and Eckardt, C. N., **81**, 38.
[2] About 30 grams of ferrous sulphate or ferrous ammonium sulphate crystals are dissolved in 900 cc. of water and the volume made to 1000 cc. with concentrated sulphuric acid. The reagent is standardized against N/10 potassium permanganate and the value per cc. in terms of the standard permanganate noted, the cc. permanganate solution required divided by the cc. of ferrous sulphate solution taken for titration, gives value of the reagent in terms of the permanganate.
The solutions are best verified upon a persulphate of known purity.

DETERMINATION OF SULPHUR IN COMBINATION AS SUL= PHIDES, SULPHITES, BISULPHITES, METABISULPHITES, THIOSULPHATES, SULPHATES. AND HYDROSULPHITES

Available Hydrogen Sulphide in Materials High in Sulphide Sul= phur. Iron Sulphide, Sodium Sulphide, etc.

Evolution Method. Since it is desired to obtain the H_2S that ordinarily would be obtained when the sulphide is treated with a strong acid, the mat of metallic aluminum or zinc and the addition of stannous chloride solution used in the procedure given on page 503 is omitted here.

Procedure. 0.5 to 1 gram of the sulphide is placed in the dry evolution flask. All connections are now made as directed in the general procedure. Three absorption bulbs containing neutral solution of cadmium sulphate are connected to the condenser, and supported by wires attached to the thistle tube and the arm of the condenser. All connections being tight, 100 cc. of dilute sulphuric acid, 1 : 4 are added through the thistle tube and H_2S evolved. The procedure is now the same as described on page 503.

Titration. When the evolution of the H_2S is complete, the bulbs containing the precipitate are emptied into a beaker and carefully washed out. The precipitate is now filtered and washed five or six times until free of acid. Methyl orange is added to the filtrate and the free acid titrated with $N/10$ NaOH.

The precipitate may be titrated with iodine according to (*b*) under general method of procedure, using an excess of iodine, followed by starch and acid and then titrating back with sodium thiosulphate solution. A double check may thus be obtained. See page 504.

If it is desired to weigh the CdS precipitate, it is best to evolve the H_2S into a neutral solution of cadmium salt. The precipitate formed in a neutral or slightly acid solution is crystalline and easily filtered, whereas that formed in an ammoniacal solution is gelatinous.

When a neutral $CdSO_4$ or $CdCl_2$ solution is used, H_2S should be evolved by sulphuric acid and not by hydrochloric acid, as the latter is volatile, and will pass through the condensing bulb recommended in the general procedure.

$$\text{One cc. } N/10 \text{ NaOH} = .001704 \text{ gram } H_2S$$
$$\text{`` ``} = .004396 \text{ gram FeS}$$
$$\text{`` ``} = .003904 \text{ gram } Na_2S.$$

Hydrogen Sulphide and Soluble Sulphides

Direct titration of hydrogen sulphide water, and soluble sulphides in solution may be made in absence of other substances acted upon by iodine. The solution containing the sulphide is added to an excess of $N/10$ iodine solution, made acid with hydrochloric acid, and the excess iodine titrated with $N/10$ sodium thiosulphate. The following reaction takes place:

$H_2S + I_2 = 2HI + S$. The cc. $Na_2S_2O_3$ are subtracted from cc. $I = I$ reacting with H_2S. One cc. $N/10$ iodine $= 0.00174$ gram H_2S.

NOTE. The soluble sulphide may be determined gravimetrically by oxidizing with bromine, the reagent being added until the solution is colored brownish red, the excess of the halogen removed by boiling and the sulphate precipitated as $BaSO_4$.

Determination of a Sulphide and a Sulphydrate in Presence of Each Other

When a mixture of sulphide and sulphydrate is treated with iodine the following reactions take place:

$$H_2S + I_2 = 2HI + S \quad \text{and} \quad NaHS + 2I_2 = NaI + HI + S.$$

It will be noticed that the acidity produced by the first reaction is twice that caused by the iodine action on the sulphydrate, and that the acidity in the latter titration remains unaffected. The reactions with the alkali salts is effected by addition of a standard iodine solution containing a known amount of hydrochloric acid. The reactions in this case are as follows:

$Na_2S + 2HCl = 2NaCl + H_2S$ and $NaSH + HCl = NaCl + H_2S$. The iodine reacts with the H_2S as follows: $H_2S + I_2 = 2HI + S$.

From the second set of reactions it is evident that the quantity of hydriodic acid formed by the action of iodine on the sulphide is equivalent to the hydrochloric acid required to decompose the sulphide, so that the acidity remains unchanged. On the other hand with sulphydrate, NaSH, the hydriodic acid formed by the iodine oxidation, is twice the equivalent of hydrochloric acid required to decompose the acid salt. Hence it is evident that the acidity is a measure of the quantity of sulphydrate present in the mixture. From the second set of reactions the following procedure is devised.

Procedure. To a measured amount of N/10 iodine solution containing a measured amount of N/10 hydrochloric acid (the mixture diluted to 400 cc.) is added the solution containing the sulphide and sulphydrate from a burette, until the stirred solution becomes a pale yellow color. (The cc. of solution added is noted and its equivalent of the sample calculated.) Starch is now added and the excess of the iodine titrated with N/10 sodium thiosulphate. The cc. of thiosulphate in terms of N/10 solution subtracted from the cc. N/10 iodine solution taken give cc. iodine required by the sample added. The acidity of the solution is now determined by titration with N/10 sodium hydroxide. The cc. NaOH required by the HI give total NaOH minus cc. N/10 HCl present in the iodine solution.

Calculation. *A*. Cc. N/10 iodine required by the sample minus twice the cc. of N/10 NaOH required by HI formed by the reaction multiplied by 0.003903 give weight of Na_2S, (i.e., cc. I -2 cc. NaOH) $\times 0.003904 =$ gram Na_2S.

B. Cc. N/10 NaOH required by the HI multiplied by 0.00560 gives gram weight of NaHS. Or in brief: cc. NaOH $\times 0.005608 =$ gram NaHS.

The above weights multiplied by 100 and divided by the weight of sample used in the iodine titration give per cent of constituents in the sample.

The method is of value in the analysis of alkali sulphides in absence of other compounds, which are decomposed by hydrochloric acid and which react with iodine.

Determination of Thiosulphate in Presence of Sulphide and Sulphydrate

The sulphide and sulphydrate sulphur is removed from the solution by adding an excess of freshly precipitated cadmium carbonate. The solution is filtered and diluted to a definite volume and the thiosulphate determined on an aliquot portion by running it into an excess of N/10 iodine solution and titrating the excess of iodine with N/10 thiosulphate solution.

One cc. N/10 iodine = 0.024822 gram $Na_2S_2O_3 \cdot 5H_2O$.

Determination of Sulphates and Sulphides in Presence of One Another

In one portion of the sample the sulphide is decomposed and the hydrogen sulphide expelled by boiling the solution (in presence of CO_2 replacing air in the flask) after acidifying with hydrochloric acid. The sulphate sulphur may now be precipitated as $BaSO_4$ by the usual methods.

In a second portion total sulphur is determined after oxidizing the sulphide with an excess of bromine and boiling out the excess of halogen. Total sulphur minus sulphate sulphur = sulphide sulphur.

The sulphide may be oxidized with fuming nitric acid by boiling the solution in a flask with reflux condenser. The nitric acid is expelled by evaporating the solution down to a moist residue. The sulphate is now precipitated by taking up the residue with water, adding HCl and then sufficient $BaCl_2$ to cause complete precipitation.

Determining the Sulphur in Thiocyanic (Sulphocyanic) Acid and its Salts

Oxidation of the sulphur may be accomplished as described for sulphides in the preceding method either by means of bromine or by fuming nitric acid. The sulphur is then precipitated as $BaSO_4$ as usual.

Determination of Sulphurous Acid (SO_2 in Solution) Free, or Combined in Sulphites, Acid Sulphites, Metabisulphites and Thiosulphates

Gravimetric Method, Oxidation to Sulphate and Precipitation as $BaSO_4$. Sulphur dioxide, free or combined in a soluble salt, may be oxidized to SO_3 or sulphate by means of an oxidizing agent such as chlorine, or bromine, or hydrogen peroxide (alkaline solution). The sulphuric acid or sulphate may be then precipitated and determined as $BaSO_4$ in the usual way.

Procedure. The halogen (bromine preferred) is added (in a water-saturated solution) in large excess to the sample, the free halogen then boiled out, and sulphuric acid precipitated, from a solution made slightly acid with hydrochloric acid, by addition of a solution of barium chloride, according to the standard procedure.

If hydrogen peroxide is used, the solution should be made alkaline with ammonia and the peroxide added, the excess boiled out, and the solution then made acid as directed above.

$BaSO_4 \times 0.3517 = H_2SO_3$, or $\times 0.5401 = Na_2SO_3$, or $\times 0.4458 = NaHSO_3$, or $\times 0.3387 = Na_2S_2O_3$, or $\times 0.2745 = SO_2$.

Note. If hydrogen peroxide is used, it should be tested for H_2SO_4 and allowance made accordingly.

Volumetric Methods

Titration with Iodine. Sulphurous Acid, Sulphites, Metabisulphites, Thiosulphates. Sulphurous acid, combined or free, may be titrated with iodine solution, the following reaction taking place:

$$SO_2 + 2I + 2H_2O = H_2SO_4 + 2HI.$$

The titration is accomplished by adding the solution of sulphurous acid, sulphite, or thiosulphate to the iodine, not in the reverse order, since in the latter order low results are obtained, unless the solution is very dilute (less than 0.04% SO_2).[1]

Procedure. Five grams of the sample (sulphurous acid solution titrated directly) are dissolved in a little water and transferred to a 500-cc. graduated flask, then made to volume. Each cc. of this solution contains 0.01 gram of the sample; 100 cc. of N/10 iodine, or their equivalent if the solution is stronger or weaker, are placed in a beaker together with a few drops of hydrochloric acid. A portion of the sample in a 100-cc. burette is now run into the iodine, with constant stirring, until the color of the free iodine has almost faded out; a little starch solution is now added and the titration continued to the complete fading of the blue color.

Since each cc. of the sample contains 0.01 gram of the material, it follows that the 100-cc. iodine equivalent in terms of the material titrated expressed to the fourth decimal place as a whole number, if divided by the cc. of the sample required, will give the per cent of the substance sought, provided other titratable substances are absent.

Example. Suppose *sodium sulphite* is being titrated, then since 100 cc. of N/10 iodine are equivalent to 0.6304 gram Na_2SO_3, 6304 divided by the cc. Na_2SO_3 solution required gives per cent Na_2SO_3. If 63 cc. were required the salt would be 100% pure.

NOTE. When the iodine equivalent is over unity, it is necessary to take a larger sample per 500-cc. volume to avoid having a titration of over 100 cc. For example in the *analysis of sodium thiosulphate*, a 20-gram sample is diluted to 500 cc. and a portion of this added to 100 cc. of N/10 iodine solution. In this case it must be kept in mind that each cc. of the sample contains 0.04 gram of thiosulphate and the percentage calculated accordingly upon completing the titration.

If the titration of the iodine is made in a casserole, the end-point may readily be recognized without the addition of starch.

Equivalents. 100 cc. N/10 iodine solution will oxidize:
Sodium sulphite (anhydrous), $Na_2SO_3 = 0.6304$ gram, or 0.3203 gram SO_2.
Sodium sulphite, $Na_2SO_3 \cdot 7H_2O = 1.2606$ grams.
Acid sodium sulphite, $NaHSO_3 = 0.5204$ gram.
Sodium metabisulphite, $Na_2S_2O_5$ (anhydride of $NaHSO_3$) = 0.47535 gram.
Sodium thiosulphate, $Na_2S_2O_3 \cdot 5H_2O = 2.4822$ grams.

NOTE. Hydrogen sulphide or sodium sulphide are also titrated with iodine. Equivalents for 100 cc. N/10 iodine = 0.1704 gram H_2S, or 0.3904 gram Na_2S.

[1] A secondary reaction takes place, the hydriodic acid formed reducing the SO_2 to S, e.g., $SO_2 + 4HI = 2H_2O + 2I_2 + S$. (J. Volhard, Ann. d. Chem. u. Pharm., 242, 94.) The solution, if not too dilute, will show a distinct separation of sulphur. (Treadwell and Hall, "Analytical Chemistry," 2, 3d Ed.) Raschig believes that a loss of SO_2 occurs, due to evaporation. (Z. Angew. Chem., 580, 1904.) See Sutton, "Volumetric Analysis," 10th Ed., pp. 128, 129. Gooch, "Methods in Chemical Analysis," 1st Ed., pp. 364–368.

Determination of Sodium Thiosulphate. The iodine titration is described on page 512. See also the chapter on iodine.

Acidimetric and Alkalimetric Methods

Titration of Sulphites, Acid Sulphites (Metabisulphite) or Sulphurous Acid. The choice of indicator is important as the titration with one may be different from that obtained in presence of another. For example the titration of sulphurous acid by an alkali in presence of phenolphthalein is twice the titration necessary to obtain an alkaline reaction with methyl orange. The reason for this is evident by the fact that Na_2SO_3 is neutral to phenolphthalein and alkaline to methyl orange, whereas $NaHSO_3$ is neutral to methyl orange but is acid to phenolphthalein. Advantage is taken of this in the analysis of salts containing a mixture of the normal and acid salts.

Reaction. With phenolphthalein $H_2SO_3 + 2NaOH = Na_2SO_3 + 2H_2O.$

With methyl orange $H_2SO_3 + NaOH = NaHSO_3 + H_2O.$

On the other hand if a salt is being titrated, methyl orange cannot be used for the titration of metabisulphite or acid sulphite, since these salts are neutral to this indicator, here phenolphthalein is required and an alkali titration made.

Reaction. $NaHSO_3 + NaOH = Na_2SO_3.$ $(Na_2S_2O_5 + H_2O = 2NaHSO_3.)$

Again if sodium sulphite, Na_2SO_3, is to be titrated, phenolphthalein would not do as an indicator, since Na_2SO_3 is neutral to this indicator. Here an acid titration is required with methyl orange indicator present:

$$2Na_2SO_3 + H_2SO_4 = 2NaHSO_3 + Na_2SO_4.$$

A. Sulphurous Acid

For the alkali titration of this acid it is advisable to use methyl orange as indicator, since this is not affected by carbon dioxide, which is very frequently present.

Reaction. $H_2SO_3 + NaOH = NaHSO_3.$

One cc. N/1 NaOH $= 0.06407$ gram SO_2, or $= 0.08209$ gram H_2SO_3.

B. Sodium Metabisulphite

Sodium acid sulphite does not exist in dry form, since the salt loses water and the anhydride $Na_2S_2O_5$ results. This is analogous to sulphurous acid, which exists only in water solution. It has been found that the acid sulphite solution evaporated to crystallization yields a product, which though dried with extreme care, forms the anhydride salt, $Na_2S_2O_5$. For correct report, therefore, the solid should be reported as metabisulphite, and the solution of the salt as acid sulphite.

Since metabisulphite in solution, or acid sulphite, is neutral to methyl orange, phenolphthalein indicator must be used and an alkali titration made. Carbon dioxide-free water and reagents should be used.

Reaction. $Na_2S_2O_5 + H_2O = 2NaHSO_3$ and $NaHSO_3 + NaOH = Na_2SO_3 + H_2O.$

Procedure. 9.507 grams of the finely ground powder are dissolved in about 50 cc. of cold saturated salt solution, to which has been added from a burette 50 cc. of normal sodium hydroxide. The salt solution should be made neutral to phenolphthalein. One cc. of 0.1% solution of the indicator is added and the excess acid sodium sulphite titrated with normal sodium hydroxide until a permanent faint pink color is obtained.

Since the normal equivalent of the salt has been taken for analysis the cc. alkali titration, including the 50 cc. originally present, will give the percentage directly in terms of $Na_2S_2O_5$.

NOTE. The NaCl serves to give a sharp and more permanent end-point. It may be necessary to add more of the indicator towards the end of the titration.

C. Sodium Sulphite, Na₂SO₃

Sodium sulphite, Na_2SO_3, is neutral to phenolphthalein and alkaline to methyl orange. The titration of this salt is accomplished by addition of standard acid in presence of methyl orange.

Reaction. $2Na_2SO_3 + H_2SO_4 = Na_2SO_4 + 2NaHSO_3$.

Procedure. The normal factor weight (12.6 grams) of the salt is dissolved in about 250 cc. of distilled water, 1 cc. of methyl orange added, followed by normal sulphuric acid, added from a burette until a faint orange end-point is obtained. As in the case of the metabisulphite, each cc. of normal sulphuric acid equals 1% Na_2SO_3. Hence the percentage is obtained directly from the burette reading.

NOTES. Organic coloring matter may be removed from the solution by filtering through charcoal.

If sodium carbonate is present, it will also be titrated. A correction must be applied for this. In the presence of sodium carbonate the solution will be alkaline to phenolphthalein. An approximate estimation of this may be obtained by titration with normal acid in presence of this indicator, remembering that sodium bicarbonate, $NaHCO_3$, is neutral to phenolphthalein, hence twice this titration must be deducted from the total methyl orange titration, i.e., $Na_2CO_3 + H_2SO_4$ (M.O.) = $Na_2SO_4 + H_2CO_3$ and $2Na_2SO_3 + H_2SO_4$ (P.) = $2NaHSO_3 + 2NaHCO_3$. (Alkaline hydroxides will also be titrated.) CO_2 may also be obtained by the standard procedure under carbon, the SO_2 being oxidized by addition of chromic acid. $Na_2CO_3 \times 1.5 =$ equivalent Na_2SO_3.

Sodium carbonate may be detected in a sulphite or metabisulphite by adding cold, dilute acetic acid (25%) to the dry powdered salt. An effervescence is due to the presence of carbonate, since a sulphite or metabisulphite does not effervesce under similar conditions.

Determination of Sulphites, Metabisulphites, Thiosulphates, Sulphates, Chlorides and Carbonates in Presence of One Another

1. Sodium Sulphite, Na₂SO₃

This is determined by titration with standard acid in the presence of methyl orange indicator according to the standard procedure previously described. If a carbonate is present, allowance must be made for this as stated

One cc. N/1 $H_2SO_4 = 0.126$ gram Na_2SO_3. *Calculate to per cent.*

$Na_2CO_3 \times 1.5 =$ equivalent Na_2SO_3.

2. Sodium Metabisulphite, $Na_2S_2O_5$

This is determined by titration with a standard alkali in the presence of phenolphthalein indicator according to the procedure previously described.

One cc. N/1 NaOH $=0.09507$ gram $Na_2S_2O_5$. *Calculate to per cent.*

3. Sodium Thiosulphate, $Na_2S_2O_3$

One gram of the mixed salts is placed in 100 cc. of N/10 iodine solution, and the excess of iodine titrated with N/10 sodium thiosulphate according to the standard procedure.

Calculation. $\{(cc.\ N/10\ I - cc.\ N/10\ Na_2S_2O_3) - [(\%\ Na_2S_2O_5 \times 2.104)$
$+ (\%\ Na_2SO_3 \times 1.5864)]\} \times 1.5814 = \%\ Na_2S_2O_3.$

4. Sodium Sulphate

The sample is dissolved in a little water, hydrochloric acid added, and the solution boiled to expel all of the SO_2. Barium sulphate is now precipitated and determined according to the standard procedure.

$$BaSO_4 \times 0.6086 = Na_2SO_4.$$

NOTE. The amount of the sample required is governed by the per cent Na_2SO_4 present.

5. Sodium Chloride

The sample is dissolved in water, nitric acid added and the solution boiled until all the SO_2 has either been volatilized or oxidized. The chlorine of the chloride is now precipitated with silver nitrate from a hot solution by the usual procedure.

$$AgCl \times 0.4078 = NaCl.$$

NOTE. The amount of the sample taken is governed by the per cent of NaCl present.

6. Sodium Carbonate, Na_2CO_3

Carbon dioxide is evolved from the mixture by means of chromic and sulphuric acids, the former being used to oxidize the SO_2 of the sample. The evolved gas is bubbled through a mixture of strong sulphuric and chromic acids to remove any SO_2 that may have escaped oxidation. Fig. 20. The CO_2 is absorbed either in caustic and weighed or is passed into a standard solution of barium hydroxide and titrated according to the standard procedures given under carbon.

NOTE. The amount of the sample taken is governed by the per cent of Na_2CO_3 present.

ESTIMATION OF SODIUM HYDROSULPHITE

Standard Indigo Solution. To about 150 cc. concentrated sulphuric acid in a casserole, are added 4.2 grams of indigo, slowly with stirring. The solution is kept at 80° C. for an hour in an oven, stirring once or twice during this time. After cooling the solution it is made up to four liters with distilled water. This reagent is now standardized against N/50 KMnO$_4$ solution. To do this 25 cc. of the indigo solution is diluted in a casserole with 300 cc. of water and titrated with N/50 KMnO$_4$ reagent.

1 cc. N/50 KMnO$_4$ is equivalent to 0.0015 g. indigotin.

1.505 : 1 = gram indigotine : x, where x = gram Na$_2$S$_2$O$_4$ in 25 cc. Indigo solution.

$$\frac{2 \times 10,000}{\text{cc. titration}} = \% \text{ Na}_2\text{S}_2\text{O}_3 \text{ (2.5 grams of}$$

solid) or grams per liter (25 cc. sample made up to 500 cc.).

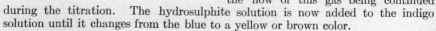

500 C.C. Flask
Na$_2$S$_2$O$_4$ Solution
Rubber Tube with Slit
Gas Line
Glass Tube
Stopper
CO$_2$
Rubber Tubing
Clamp
Glass Bead
Rubber Tube
CO$_2$
Vent
Stopper
250 C.C. Erlenmeyer
Indigo

FIG. 65a.

Procedure

Titration of Sodium Hydrosulphite Against Standard Indigo Solution. Fifty cc. of standard indigo solution are pipetted into a 300 cc. Erlenmeyer flask. The titrating apparatus as well as the 500 cc. volumetric flask are filled with CO$_2$ gas (C$_2$H$_2$ may be used in place of CO$_2$). Two and a half grams of the solid are now taken, or 25 cc. of the solution (if the material is already dissolved as a 10% solution) and placed in the 500 cc. flask and made to mark with distilled water. The flask is stoppered and connections made with the burette, etc., as shown in Fig. 65a. The burette is filled with the sample and the flask containing the indigo solution is placed under the burette as shown in the figure. The air is displaced from the apparatus by CO$_2$, the flow of this gas being continued during the titration. The hydrosulphite solution is now added to the indigo solution until it changes from the blue to a yellow or brown color.

$$\frac{\text{Factor for Indigo}}{\text{cc. titration}} = \% \text{ Na}_2\text{S}_2\text{O}_4 \text{ in solids, or grams per liter in liquids.}$$

NOTES. The hydrosulphite solution should be made alkaline with NaOH, then made up rapidly to volume and titrated in an atmosphere of CO$_2$ to prevent oxidation.

The size of the sample may be varied, but the titration should be over 10 cc.

The tip of the burette should dip below the surface of the indigo until near the end-point, then withdrawn and the titration completed with the tip above the surface.

The above method was outlined by J. H. Brackett.

DETERMINATION OF FREE SULPHUR IN A MIXTURE

Free sulphur is an essential constituent in many types of mixtures and the method of estimation will vary with the nature of the other ingredients. Hydrated lime, chalk, gypsum, dry lime sulphur, calcium arsenate, nicotene sulphate in infusorial earth carrier, sodium polysulphide, sodium chloride, epsom salts, and the usual fertilizer materials are the substances most commonly found in the mixtures now on the market. All commercial forms of sulphur are found in these mixtures and the value of the mixture usually depends largely on which form of sulphur was used. For example, an insecticide dust containing coarse crude or refined sulphur, instead of flowers or superfine, would be valueless even though the chemical analysis showed that the mixture contained the specified percentage of total sulphur. Therefore, the microscope and a little ingenuity will indicate the proper combination of methods to follow.

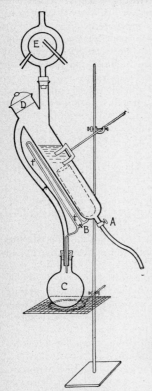

From 1 to 10 grams of the material, depending upon the amount of sulphur present, is extracted in a Soxhlet extractor (see modified form, Fig. 66) with carbon bisulphide (freshly distilled) for twelve hours. The extract is evaporated to dryness, adding 10 cc. of bromine-carbon tetrachloride mixture together with 15 cc. of nitric acid. The residue is taken up with 10 cc. of hydrochloric acid, diluted with 150 cc. of distilled water, heated to boiling and the sulphuric acid precipitated with 10% barium chloride solution, washed, dried, ignited and weighed according to the procedure for sulphur.

FIG. 66. Sanders' Extraction Apparatus.

$$\frac{BaSO_4 \times 100 \times 0.1373}{\text{Weight of sample}} = \% \text{ free rhombic sulphur.}$$

After extraction carefully dry the thimble and contents. Examine under a microscope a small portion of the dried material. Remaining sulphur if present will be in the amorphous form and have the characteristic " droplet " structure seen in flowers of sulphur. Presence of much sulphur at this stage indicates that flowers of sulphur was used in the mixture, and the proper procedure to follow will depend on the nature of the other constituents. If a soluble sulphate such as nicotene sulphate or epsom salts is indicated, then an aliquot of the residue in the extraction thimble can be leached with hot water and the sulphur determined by one of the usual methods after wet oxidization. If an insoluble material such as gypsum is indicated, then the free sulphur may be burned off in the air and the combined sulphur determined in the residue after solution by one of the standard methods; the total sulphur having first been determined in another aliquot after wet oxidization with bromine and nitric acid. Whatever procedure is followed, the content of amorphous sulphur is calculated by difference and thus the percentage of flowers established by adding the amount thus found to the CS_2 soluble sulphur.

Sanders' extraction apparatus [1] has several advantages that make this apparatus desirable for laboratory use, where a number of daily extractions are required. As may be seen from Fig. 66, by simply removing the glass stopper D the cylinder may be charged without disconnecting the apparatus, as is necessary with the Soxhlet type of apparatus. The extraction is carried on with the traps A and B closed, the siphon t–t' acting automatically as in case of the Soxhlet. With A closed and B open the apparatus may be used as a reflux condenser. The solvent liquid may be drawn off by opening A. With B closed and A open the apparatus may be used as a condenser and the ether, chloroform, carbon disulphide, etc., distilled from C. The globe-shaped Soxhlet condenser may be replaced by Allihn's or Liebig's condenser, if desired. The ball form, however, is more compact.

EVALUATION OF SPENT OXIDE FOR AVAILABLE SULPHUR

Spent oxide is the by-product of gas works, and refers to the spent Fe_2O_3 used in the scrubber for the removal of hydrogen sulphide from the gas. The FeS, as in case of pyrites, is used in the manufacture of sulphuric acid, and is evalued by its available sulphur content.

Total Sulphur. The oxide is sampled, brought into solution and the sulphur determined exactly as is given under the standard method for determination of sulphur in pyrites ore.

Residual Sulphur. Two grams of the material are ignited to expel volatile sulphur, a porcelain crucible being used. The residue is treated with strong hydrochloric acid and after digestion on the steam or water bath is diluted with water and filtered. (If SiO_2 is present evaporation to dryness is necessary.) Sulphur is determined in the filtrate as usual.

Available Sulphur. The per cent of residual sulphur is subtracted from the per cent total sulphur, the difference being available sulphur.

Iron. This may be determined on an ignited sample according to a standard procedure for iron. See chapter on Iron.

[1] J. McC. Sanders, Proc. Chem. Soc., **26**, 227–228, 1910. The Analyst, **35**, 556, 1910.

ANALYSIS OF REFINED SULPHURS AND BRIMSTONE

The impurities in commercial sulphurs and brimstones are seldom more than a few tenths per cent. Arsenic, acidity, chlorine, and the amount of amorphous sulphur present (CS_2 insoluble) are required. Also the fineness and specific volume (degrees Chancel) are often required for sulphurs used in agriculture.

Moisture. The powdered sample, weighing 50 grams, is spread out on a watch-glass and dried for an hour at 100° C., then cooled in a desiccator and weighed.

Loss of weight in grams multiplied by 2 = per cent moisture.

Arsenic. Ten grams of the material are treated with 30 cc. of carbon tetrachloride mixture (3 parts CCl_4 + 2 parts Br) and after standing for ten minutes 25 cc. strong nitric acid are added in small portions (a watch-glass covering the beaker during the intervals of addition). HNO_3 and Br are expelled by evaporation on the steam bath. Water is added and the evaporation repeated. Arsenic is now determined on the residue by the Gutzeit Method for arsenic.

NOTE. Arsenic-free reagents should be employed.

Chlorine. One hundred grams of the brimstone are extracted with hot water, the filtered extracts oxidized with 10 to 15 cc. of nitric acid and a few crystals of ammonium persulphate by boiling and treated with 5 cc. of 10% solution of silver nitrate. The solution, brought to boiling, is placed in a dark place and the silver chloride allowed to settle. This is now filtered off in a weighed Gooch crucible and chlorine calculated from the AgCl.

$$AgCl \times 0.2474 = Cl \text{ or } = 0.4078 = NaCl.$$

Mineral and Organic Impurity. Ash 100 gms. of the sample by igniting a little at a time in a tared porcelain or silica ware dish. Carry on the combustion in plenty of air and *without* the aid of any external heat except toward the last. In igniting the sample use a small pin flame gas jet such as the petroleum chemist uses in making flash and fire tests. Do not use match or taper or alcohol to ignite the sulphur as a small amount of organic matter is certain to get into the sample from these sources and cause trouble in the combustion. The sulphur once ignited will burn evenly and clean unless organic matter is present. All refined sulphurs should burn completely without the aid of any external heat. With American Gulf Coast brimstone and with oil or asphalt contaminated sulphur the organic matter present in even minute amount will cause trouble in burning unless special precautions are taken. A film of melted asphalt or oily matter forms over the surface of the molten sulphur and shuts off the air so that the flame from the burning sulphur is put out. When this film of dark oily matter is first noticed, touch it lightly with the pin flame and it will usually break or char, allowing the sulphur to burn evenly. Toward the last apply very gentle heat to the dish and thus char the organic matter but keep the temperature well below the red so that this organic material is not ignited. When all the sulphur is burnt off as indicated by no more odor of SO_2, cool the dish and weigh; this weight giving the combined organic and mineral matter. Then ignite the residue at low red heat to burn off all organic material; again cool and weigh; this weight

41

giving the mineral impurity or ash. The difference between the two weights represents the organic impurity.

Acidity. Boil 100 gms. of the pulverized sample with about 500 cc. water· The addition of a little neutral alcohol at the start will aid in wetting the sulphur which sometimes floats and causes trouble. Cool and make up to a standard volume. Pipette or filter off an aliquot and titrate with N/10 alkali, using phenolphthalein. Calculate acidity as H_2SO_4 and express as per cent acidity.

Available Sulphur. Add together moisture, organic matter, ash, arsenic, acidity, chlorine and report available sulphur as the difference.

Amorphous Sulphur (CS_2 insoluble). Weigh five to twenty grams of the finely pulverized sample into a tared Extraction Thimble and extract with carbon bisulphide. The rate of extraction is regulated so that one filling of the chamber takes about five minutes. The extraction should be complete in thirty minutes. It is important that the extraction be stopped as soon as the loss in weight of the thimble and contents becomes constant as long-continued extraction will carry some of the amorphous sulphur into solution. Note: A Soxhlet type extraction apparatus is best as other types where the thimble is not immersed in the liquid give erratic results at times on account of the tendency of the CS_2 solution of sulphur to " crawl " to the top of the thimble and there deposit out a hard scale of rhombic sulphur. When the extraction is completed, the thimble and contents are freed from CS_2 by exposure to a rapid current of dry air and then dried for thirty minutes in a water oven through which air circulates. (Finely divided sulphur, CS_2, and air is a mixture liable to spontaneous combustion, so get rid of the bulk of the CS_2 in the cold dry air current before exposing to the heat of the oven.) The weight of the contents of the thimble less ash and arsenic and organic matter is taken as the amorphous sulphur. Flowers of sulphur must contain in excess of 30% amorphous sulphur. (Flowers are often sophisticated by the addition of ground sulphur when the content of amorphous sulphur is lowered in proportion—this should not be reported as flowers but flowers with so much adulteration.) Refined lump sulphur, roll sulphur and rubber makers sulphur should contain no amorphous sulphur. Powdered sublimed sulphur usually contains some small percentage of amorphous sulphur. Note: Direct sunlight, heat and some chemical fumes cause the amorphous sulphur in flowers to revert to the soluble rhombic modification. Therefore be careful in the preparation and treatment of samples of flowers of sulphur.

Fineness and Specific Volume of Degree Chancel. Examine the sample under the microscope. Use a recessed slide and wet the specimen with alcohol or preferably with concentrated sulphuric acid. Flowers of sulphur appear as loose agglomerations of opaque yellow spherical droplets. Ground refined sulphur and ground brimstone appear as clear angular fragments almost colorless. High quality flowers consists entirely of minute droplets all of uniform size and barely touching each other. In low quality flowers the droplets are of large and irregular size and are more or less fused together. The smaller and the more uniform the size of the sulphur particles—whether the droplets of the flowers or the grains of a pulverized sulphur—the greater will be the specific volume and the Degree Chancel. This is determined by the

Chancel Sulphurimeter,[1] a tall glass tube, glass-stoppered, graduated into 100 degrees of $\frac{1}{4}$ cc. each. Five grams of the sulphur sample are accurately weighed out and dusted into the tube which is half filled with ether or alcohol. The sulphur and alcohol is strongly shaken and the tube and contents allowed to stand in a vertical position. The reading of the sulphur level is taken as soon as the subsidence ceases or at the end of an hour and this is reported as the degree Chancel. (If a Chancel Sulphurimeter is not obtainable, a tall glass-stoppered graduate of 25 cc. capacity will serve, remembering that each quarter cc. represents a degree in the Chancel scale. Of course this reading will not be as accurate as on the Sulphurimeter where the scale is larger on account of the smaller diameter of the Chancel tube.)

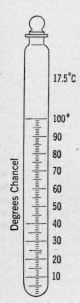

The cylindrical glass tube should have the following dimensions: The cylinder should be 23 cm. long and 15 mm. in diameter, with a scale starting from below, graduated upwards into 100 degrees, each degree being $\frac{1}{4}$ cc.; the 100 degrees (25 cc.) occupy a length of 100 mm. The cylinder is closed at the lower end, and glass stoppered, as shown in Fig. 66a. The sulphur is first passed through a sieve 1 mm. mesh, in order to break down the lumps formed with storage. As stated, a 5 gram sample is placed in the tube and this half filled with

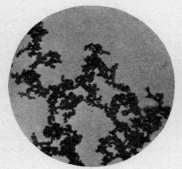 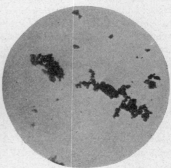

Fig. 66a.   Fig. 66b.   Flowers of Sulphur.   Fig. 66c.   Rhombic Sulphur.
Sulphurimeter.

anhydrous ether, having a temperature of 17.5° C. The sieved sample is shaken and additional ether added until the level stands 1 cc. over the 100° mark and the whole again shaken in an upright position, the sulphur allowed to settle and the degrees Chancel read off.

The procedure for analysis of refined sulphurs was contributed by Chas. A. Newhall.

[1] Lunge, Tech. Methods of Chem. Analy., Vol. I, p. 265 (1908).
Lunge, Sulphuric Acid and Alkali, Vol. I, Part 1, page 47 (1913).

QUANTITATIVE ESTIMATION OF SMALL QUANTITIES OF SULPHIDE SULPHUR; METHOD OF W. A. DRUSHEL AND C. M. ELSTON[1]

The method is a colorimetric method and consists essentially of the comparison of the depth of color of lead sulphide stains obtained from the sulphide sulphur of a given weight of a sample to be analyzed with a standard series of stains prepared from sulphide solutions of known sulphur content. A set of stains varying in depth of color from a faint yellowish brown to black representing from 0.0002% to 0.004% of sulphide sulphur may be prepared and used indefinitely for comparison. With a set of standard stains at hand the method has the advantage that within the range given the sulphide sulphur of a sample may be determined with a fair degree of accuracy in less than ten minutes.

Preparation of Standard Set of Sulphide Stains. The apparatus used for preparing standard stains and for making analyses is very simple. The inner tube of a Liebig condenser with its larger end about 18 mm. in internal diameter is cut off 15 cm. in length. The smaller end is drawn down somewhat, rounded and fitted to a sound cork stopper which in turn is fitted to a 100 cc. round-bottom flask. The condenser tube then serves as a sort of reflux condenser. To the upper and larger end of this tube a filter paper moistened with a dilute solution of lead acetate is smoothly fitted and tied, so that the steam passing up through the tube and carrying hydrogen sulphide is required to pass out through the lead acetate paper. A similar tube with the internal diameter of its larger end about 36 mm. is also prepared and used for sulphide sulphur samples containing 0.001% or more of sulphur.

A solution of sodium sulphide is made up with pure distilled water and carefully standardized. The solution is then diluted to contain exactly 0.01% of sulphide sulphur. This solution is used for making up standard solutions containing 0.0002, 0.0004, 0.0006, 0.0008, 0.001, 0.002, 0.003 and 0.004% of sulphide sulphur respectively, taking care to use distilled water free from traces of nitrites in making the dilutions. It is found that the more dilute sulphide solutions when made up with ordinary distilled water lose their sulphide content either wholly or in part on standing for several hours in stoppered bottles. This difficulty is obviated by using nitrite free distilled water in making up the solutions.

Carefully measured portions of 1 cc. to 5 cc. of the standard solutions are pipetted into the 100 cc. flask and 25 cc. of hydrochloric acid of about 0.5% strength is added. The flask is immediately attached to the condenser tube fitted with moistened lead acetate paper as previously described. The mixture is then gently boiled for a few minutes at such a rate that the steam issued not too rapidly from the upper end of the condenser tube. In this way the sulphide sulphur is quantitatively liberated as hydrogen sulphide and evenly deposited as lead sulphide on the moistened lead acetate paper. The undecomposed lead acetate is then washed out, the paper dried and labeled with the amount of sulphide sulphur present as one of the set of standard stains. In the same way complete sets in duplicate are prepared ranging in sulphide sulphur from 0.0002% to 0.004%.

[1] Am. Jour. Sci., Fourth Series, Vol. XLII, No. 248, August, 1916.

In making these estimations the larger condenser tube is used where a preliminary trial indicates that the amount of sulphide sulphur is equal to or greater than 0.001%. In all other cases the smaller tube is used. The maximum error, depending upon the amount of sulphur present, with the larger tube was 0.001% and with the smaller tube 0.0003%. These errors may be reduced by repeating the determination and taking the mean of several values found. In this way in the practical applications of the method the errors may be kept within reasonable limits.

Practical Applications. (1) In gas analysis. Twenty-five liters of air are slowly drawn through a Geissler bulb of the most modern type containing dilute potassium hydroxide solution. This solution is then washed into a measuring flash and made up to the mark with nitrite free distilled water and aliquot portions of this solution are used for determining the sulphide sulphur as previously described.

(2) In coke analysis. The simplest method of estimating sulphur in coke given by Fresenius is to boil 5 grams to 10 grams of powdered coke in dilute hydrochloric acid, and to absorb the hydrogen sulphide evolved in dilute potassium hydroxide solution. The sulphur is then oxidized to the sulphate condition by chlorine water or bromine water and weighed as barium sulphate. This method is used as a control to check up the results obtained by the colorimetric method. In this method the hydrogen sulphide is liberated and absorbed as suggested by Fresenius and aliquot portions of the sulphide containing potassium hydroxide solution are transferred to the distillation flask and the previously described procedure is followed.

(3) In paper analysis. In order that tissue paper may be used for wrapping polished metal without producing a tarnish the paper must be relatively free from sulphide sulphur. A weighed amount of paper, 1 gram to 2 grams, is cut into small pieces and transferred to the distilling flask and digested with gently boiling 0.5% hydrochloric acid, collecting the hydrogen sulphide as lead sulphide on lead acetate paper as previously described.

Sulphur Dioxide in Air

Messrs. Ries and Clark [1] have found the following method for the determination of small quantities of sulphur dioxide in the presence of excessive air to be dependable.

The sulfur dioxide was absorbed by bubbling the gas through 10 cc. of 10% solution of sodium hydroxide, 0.002 M in stannous chloride, contained in a modified 4-bulb Mitscherlich absorber. The solution was then washed into a flask, diluted to 50 cc., cooled, 50 cc. of 12 N hydrochloric acid added, the solution again cooled, and 2 cc. of carbon tetrachloride added as an indicator. (There is danger of loss of sulfur dioxide by vaporization at this point, but where the amount present is small no loss occurs.) It was then titrated with approximately 0.003 M potassium iodate, the flask being shaken very vigorously during the titration until the pink color of the carbon tetrachloride disappeared. With a little practice, the end point may be determined to a single drop. It is obviously necessary to run a blank on the caustic plus stannous chloride alone to determine the correction for the catalyst.

[1] J. Ind. and Eng. Chem., *18*, 747, 1926.

THORIUM

Th, *at.wt.* 232.15; *sp.gr.* 11.0–12.2; *m.p.* 1450°; *oxides,* ThO_2 $(ThO_3$ and Th_2O_7 known only in hydrate form)

DETECTION

In the regular analytical procedure thorium is found in the precipitate of the tri- and tetra-valent hydroxides produced by ammonium hydroxide, provided a sufficiently thorough decomposition of the sample has been obtained (see section on preparation and solution of sample). Along with rare earth elements, and a little zirconium, thorium may be obtained as oxalate by dissolving the hydroxide precipitate [2] in hydrochloric acid and adding a slight excess of oxalic acid to the hot, weakly acid (not over 0.5 normal) solution. The precipitate should be allowed to stand six or eight hours before filtration. After having been washed with water containing 2 cc. of 6 normal hydrochloric acid per 100 cc., the precipitate is rinsed into a beaker with pure water, using about 50 cc. To this mixture 5 g. of solid ammonium oxalate are added, and the mixture is heated for half an hour nearly to boiling, and well stirred. Two volumes of water are added, the mixture is allowed to stand half an hour and then filtered. To the filtrate 20 cc. of 6 normal hydrochloric acid are added. A white precipitate indicates the presence of thorium.[3] This precipitate can be put into solution by strong heating with conc. sulphuric acid, and taking up with ice-cold water. One of the following characteristic tests may be used as confirmation of thorium.

A sensitive test for thorium [4] consists in the precipitation of thorium iodate in nitric acid solution. Two reagent solutions are necessary: (I) 15 g. of potassium iodate, 50 cc. of conc. nitric acid, and 100 cc. of water. (II) 4 g. of potassium iodate, 50 cc. of conc. nitric acid, and 450 cc. of water. The solution to be tested for thorium, which must contain no hydrochloric acid, is boiled with a little sulphurous acid to reduce any cerium present. To this solution is added twice its volume of reagent (I), which causes precipitation

---

[2] It is shorter to add oxalic acid directly to the filtrate from the hydrogen sulphide group, after boiling out the excess of hydrogen sulphide. This, however, causes complications in the analysis of subsequent groups, and may also contaminate the oxalate precipitate with calcium; so it is better to first precipitate the hydroxides with (carbonate-free) ammonium hydroxide as described.

[3] Thorium oxalate is soluble in ammonium oxalate. Zirconium oxalate, and traces of the oxalates of the yttrium earths also dissolve, but the yttrium earth oxalates are reprecipitated on dilution. The addition of an excess of mineral acid to the ammonium oxalate solution of thorium and zirconium precipitates thorium oxalate, but not zirconium oxalate.

[4] Meyer and Hauser, "Die Analyze der Seltenen Erden und Erdsaüren," p. 171. Ferdinand Enke, Stuttgart (1912).

Chapter by Paul H. M.-P. Brinton.

of thorium iodate, and more or less rare earth iodates according to their concentration in the solution.   By now adding reagent (II) in volume equal to
four times the original volume, and boiling, any rare earth iodate is dissolved,
while thorium iodate (also any zirconium iodate) remains undissolved.   If
the absence of zirconium is not known with certainty, the iodate precipitate
may be boiled with 50 cc. of 10% oxalic acid solution [1] until iodine vapors are
no longer given off.   Any precipitate remaining is thorium oxalate.

Sodium pyrophosphate produces in solutions of an acid normality of
0.2–0.3 a white precipitate of thorium pyrophosphate.   To make certain that
this precipitate is entirely free from rare earths it should be redissolved, any
tetravalent cerium reduced with sulphurous acid, and then the thorium
reprecipitated as pyrophosphate as described later under Gravimetric Determination.   Zirconium and titanium would also be precipitated under these
conditions, but they would be separated from thorium on the basis of their
non-precipitation by excess of oxalic acid.

Spectrum analysis, and the determination of radioactivity [2] are also useful
in the detection of thorium.

ESTIMATION

The estimation of thorium is required chiefly in connection with the
incandescent gas mantle industry.   The main source of thorium is monazite
sand, of which the greater part comes from Brazil and India.   The South
Carolina monazite is of poor quality and of very little commercial importance.
Monazite is essentially an orthophosphate of the cerium earths, and carries
from about 4% to nearly 10% of ThO_2.   The highest grade material comes
from Travancore, India.   Thorite and thorianite, the silicate and the uranate
respectively of thorium, are now of limited importance.

Preparation and Solution of the Sample

Thorite.   The very finely pulverized sample is digested with conc. hydrochloric acid at a temperature just below boiling.   This usually effects complete
decomposition, but for safety any insoluble residue should be filtered, after
dilution, ignited, and then fused with potassium pyrosulphate.   This fusion
is taken up with 6 normal hydrochloric acid and added to the main solution.
Silica and the hydrogen sulphide group are removed in the conventional way,
leaving thorium, rare earths, etc., in the acid solution.

Monazite, Thorianite, Gas Mantle Residues, etc.   While fusion with
potassium pyrosulphate effects the decomposition, yet it is more convenient
to attack larger samples with conc. sulphuric acid.   In the case of monazite
sand, large samples are necessary to insure uniform and representative samples.
Thorianite and Carolina monazite should be finely ground, but Brazilian and
Indian monazite are fully decomposable without grinding, and filtration from
the insoluble residue is easier if the sample is taken as it comes.   A batch of
sand should be very carefully mixed before sampling, to avoid the tendency
toward segregation arising from the different sizes and densities of the constituent grains.

[1] Brinton and James, J. Am. Chem. Soc., **41**, 1084 (1919).
[2] Helmick, J. Am. Chem. Soc., **43**, 2003 (1921), gives details for the quantitative,
as well as the qualitative, determination of thorium in minerals.

The following method for the decomposition of monazite is recommended by Dr. H. S. Miner,[1] of the Welsbach Co.: 50 g. of the sand are weighed out and placed in a porcelain casserole of about 500cc. capacity. 75 cc. of conc. sulphuric acid are added, and the mixture is heated for about four hours with frequent stirring, a gentle evolution of fumes being maintained during the course of the operation. When the mass has become pasty, it is allowed to cool, and the sulphates are extracted by the addition of about 400 cc. of ice-cold water, or enough to cool the solution sufficiently so that the sulphates become soluble. This solution is decanted into a liter-graduated flask, and the remaining sulphates are extracted with small portions of cold water and decanted into the flask. A point is reached toward the end of the extraction when, due to the decreasing acidity, the small wash portions show a slight separation of rare earth phosphate. A few more extractions are made beyond this point, but these portions are not added to the graduated flask. They are temporarily preserved in a separate beaker. To the remaining sand, which has been dried, 10 cc. of conc. sulphuric acid are added, and the digestion is carried out as before, except that a somewhat higher temperature is used, enough to maintain copious evolution of white fumes, and the duration of the digestion need not exceed one and a half hours. After cooling, extraction is started with the last portions of the previous extraction liquor, i.e., those decantings which showed slight precipitates of rare earth phosphates, and which were preserved in the separate beaker. The sand is now thoroughly washed with cold water, and the washings are all decanted into the liter flask. This flask now contains all the thorium and rare earths as soluble sulphates. A little suspended silica is usually visible at this point, but this will not be mistaken for undissolved sulphates.

After cooling, the sulphate solution is made up exactly to the liter mark, thoroughly mixed, and filtered through a dry filter, discarding the first 25 or 30 cc., and receiving the remainder in a dry flask or bottle. The whole need not be filtered, and, of course, no washing is to be done. Each 100 cc. of this solution represents 5 g. of the sample.

[1] U. S. Bureau of Mines, Bulletin 212, p. 53 (1923).

SEPARATIONS

Thorium is separated from practically all elements excepting the rare earth elements and scandium by precipitation as oxalate in slightly acid solution. Zirconium may be in part precipitated along with thorium and the rare earths, especially in the absence of a sufficient excess of oxalic acid; and if considerable quantities of calcium, and to less extent strontium and barium, are present, there may be contamination unless the mineral acid concentration be kept dangerously high. In the presence of much calcium, it is better to first separate thorium from it by precipitation with freshly distilled ammonium hydroxide, and then to precipitate the thorium as oxalate. The details of the oxalate separation have been given under Detection.

The main problem, then, is the separation of thorium from the rare earth elements. The iodate method has already been given in detail. Two other much used methods will now be described.

The Pyrophosphate Method.[1] An aliquot portion of the sulphate solution, prepared as already described, and usually representing 2.5 g. of monazite sand, is diluted to about 450 cc., and 5 cc. of conc. hydrochloric acid are added. The solution is heated nearly to boiling and the iron and cerium are reduced by adding sulphurous acid solution until the yellow color is discharged. 15 cc. of sodium pyrophosphate solution (50 g. $Na_4P_2O_7.10H_2O$ in 1 liter of water) are slowly added, with constant stirring, and the mixture is then heated to gentle boiling for 5 minutes. After standing 5–10 minutes—not more—the precipitate of thorium phosphate is filtered and washed twice with water containing 1 drop of hydrochloric acid per 100 cc. A slight cloud in the first filtrate, due to atmospheric oxidation of iron (and possibly of cerium), with consequent precipitation of pyrophosphate may be neglected.[2] The filter paper and precipitate are freed from excess liquid by wrapping for a moment in cheap filter paper or blotting paper, and then dropped into a dry 250-cc. Kjeldahl flask. 15 cc. of conc. sulphuric acid and a few crystals of ammonium perchlorate are added, a small funnel is placed in the neck of the flask, and the contents are heated until the filter is disintegrated. When a clear brown solution is obtained more perchlorate [3] is added, and the heating is continued until the mixture is pure white (or very slightly yellow if much cerium was carried down in the first precipitate). The flask is placed in

---

[1] Carney and Campbell, J. Am. Chem. Soc., **36**, 1134 (1914).

[2] V. T. Jackson, U. S. Bureau of Mines, Bulletin 212, p. 65 (1923).

[3] The oxidation of the filter paper requires 10 or 15 minutes, and about 1.5 g. of perchlorate. Heating should not be continued far beyond the point at which the organic matter is fully destroyed, as the perchloric acid may break up and cause considerable foaming. Cartledge (J. Am. Chem. Soc., **41**, 49 (1919)) shows that fuming nitric acid may be satisfactorily substituted for ammonium perchlorate. In this case the precipitate and paper are shaken with 15 cc. of conc. sulphuric acid in the Kjeldahl flask for about two minutes, cooling, if necessary, until the flask is just a little too hot to be held comfortably in the hand. To the charred mixture 2–4 cc. of fuming nitric acid (sp.gr. 1.53) are added, and after a minute or two the solution is gradually raised to the boiling point. Occasionally as fumes of sulphur trioxide appear a second charring occurs, in which case 0.5 cc. more nitric acid, poured cautiously down the neck of the flask, will complete the oxidation. (If the mixture has once been white or slightly yellow, no harm is done if during cooling oxides of nitrogen redissolve in the acid, thereby restoring a yellow color.)

ice-water and then slowly, with shaking, about 100 cc. of ice-cold water are added. Complete solution may at times require several hours. Occasionally a slight cloud of suspended silica persists. This may be neglected as it will be removed in the next step.

The sulphate solution is rinsed into a solution of 30 g. of sodium hydroxide in 125 cc. of water, contained in the original beaker in which the pyrophosphate was precipitated,[1] boiled for several minutes, filtered and washed several times with hot water. The paper and precipitate are then placed in the beaker just used, 10 cc. of conc. hydrochloric acid are added, and after a few minutes of stirring, 150 cc. of water are added, and the solution is boiled. The paper shreds are filtered off and washed, and the filtrate is diluted to about 400 cc. 3 cc. of saturated sulphurous acid solution are added, the solution is heated to boiling, and the thorium is again precipitated with sodium pyrophosphate. The precipitate is washed and changed to sulphate and hydroxide in the manner just described. This second sulphate should always be perfectly white, and should dissolve entirely clear. The final hydroxide is free from rare earths, and aside from zirconium and titanium, which are precipitated as pyrophosphate, it should contain only a trace of iron as impurity. These three foreign elements will be separated from the thorium by the final precipitation of the latter as oxalate. The last hydroxide precipitate is dissolved in as little conc. hydrochloric acid as possible (never more than 10 cc.) and filtered free from paper. This chloride solution is now ready for the gravimetric determination of thorium by precipitation as oxalate.

The Thiosulphate Method.[2] 200 cc. of the sulphate solution prepared as described under Preparation of Sample, and representing 10 g. of monazite sand, are diluted to 1 liter, and poured into 150 cc. of a cold saturated solution of oxalic acid. The sulphate solution should be fed in very slowly, preferably from a separatory funnel, and vigorous stirring should be maintained to convert the gummy precipitate to the crystalline form. Only in this way is it possible to get the oxalate precipitate free from considerable amounts of phosphate. After standing not less than fifteen hours the precipitate is filtered, washed free from acid, and dried.

The filtrate and first washings are neutralized with ammonia, and hydrochloric acid is then added in an excess of from 10 to 15 cc. This precipitates the remaining rare earths (and any thorium still in solution) as oxalo-phosphates. This precipitate is also allowed to stand fifteen hours, and then it is filtered and washed with weak oxalic acid solution. This precipitate, after drying, is ignited in a porcelain dish, moistened with a little water, and dissolved by warming with conc. hydrochloric acid. The solution so obtained is filtered, largely diluted, and precipitated by the addition of a large excess of oxalic acid, warming to convert the precipitate to a crystalline form. After standing two or three hours, this oxalate precipitate is filtered, washed, and dried.

The oxalate precipitates are ignited together. If " total rare-earth oxides " are wanted, they may be weighed at this point. The oxides are transferred to a 600-cc. beaker, moistened with a little water, and dissolved by warming

---

[1] Thorium pyrophosphate adheres rather tenaciously to the glass, and this procedure eliminates the necessity for the tedious cleansing of the beaker after the first precipitation.

[2] The details here given are essentially those in use in the laboratories of the Welsbach Co. See H. S. Miner, loc. cit.

with 100 cc. of 6 normal hydrochloric acid. The solution is transferred to a 1500-cc. beaker, diluted to about 700 cc., and ammonia is added until the precipitate formed just ceases to redissolve, leaving the solution neutral or very slightly acid. Conc. hydrochloric acid is now added, a drop at a time, until the precipitate just dissolves, and then 6 to 8 drops more. 100 cc. of a boiling 30 per cent sodium thiosulphate solution are added to the boiling solution, and boiling is continued for 45 minutes. The precipitate is filtered on a $2\frac{1}{4}$-inch Büchner funnel, and washed with hot water. It is well to fold the filter paper so as to cover the whole of the inside of the funnel.

To the filtrate 10 cc. more of the thiosulphate solution are added, and it is boiled for half an hour longer. The small precipitate thus obtained is filtered, set temporarily aside and designated as " Residue No. 1." The filtrate is discarded.

The paper containing the first large precipitate is put into the original 1500-cc. beaker and boiled over a low flame with 75 cc. of 6 normal hydrochloric acid. When the thorium precipitate has dissolved and the paper is pretty well disintegrated, usually about 5 minutes, without filtering off the filter paper and sulphur, the mixture is diluted to 700 cc., and neutralized with ammonia just as before, finally having an excess of 6 to 8 drops of conc. hydrochloric acid. It is then heated to boiling and 60 cc. of boiling 30 per cent sodium thiosulphate solution are added. After boiling 30 minutes the precipitate is filtered on a Büchner funnel as before and washed. This filtrate is reserved.

The paper and precipitate are dropped into the original beaker and all the operations described in the paragraph immediately preceding are repeated, the final filtrate, however, being this time discarded.

The paper containing the thorium precipitate is boiled with 40 cc. of 6 normal hydrochloric acid. When the paper is well disintegrated, it is filtered off, well washed, and this residue is designated as " Residue No. 2." The solution is temporarily set aside.

Residues Nos. 1 and 2 are combined, ignited in a porcelain crucible, and fused with sodium pyrosulphate. The fusion is extracted with dil. hydrochloric acid and this solution is added to the reserved filtrate from the second thiosulphate precipitation. Ammonia is now added in excess, the precipitate is filtered, and dissolved in hydrochloric acid. This solution is diluted to 300 cc. nearly neutralized with ammonia, leaving only a faint acidity, heated to boiling, and precipitated by the addition of 20 cc. of boiling 30 per cent sodium thiosulphate solution. After boiling 20 minutes, the precipitate is filtered on a small Büchner funnel, washed, and dissolved in 50 cc. of 6 normal hydrochloric acid. Without filtering, the solution is diluted to 300 cc., almost neutralized with ammonia, and the thiosulphate precipitation is repeated. The last precipitate is dissolved by boiling in a mixture of 5 cc. of conc. hydrochloric acid and 20 cc. of water, the sulphur is filtered off, and the chloride solution is combined with the solution obtained at the end of the preceding paragraph.

This solution now contains all the thorium from the original sample and it is ready for the Gravimetric Determination of thorium by precipitation as oxalate.

GRAVIMETRIC DETERMINATION OF THORIUM

Thorium is nearly always precipitated as oxalate and ignited to ThO_2, in which form it is weighed. The solubility of thorium oxalate in very dilute mineral acids is slight, especially in the presence of excess oxalic acid. It is well, however, not to have over 2 cc. of conc. mineral acid per 100 cc. of solution. More strongly acid solutions may be partially neutralized with ammonium hydroxide. Thorium oxalate does not filter readily if precipitated rapidly, or from a cold solution. The solution should be boiling hot and dilute (200–500 cc.); and should be precipitated by adding very slowly a boiling-hot solution of oxalic acid, which has been saturated in the cold; or by stirring in the oxalic acid added in solid form. Both methods insure a slow rate of precipitation, but the first method seems to be the better. The precipitate is allowed to stand not less than 10 or 12 hours before filtration, and it is washed with water containing a few drops of hydrochloric acid.

The thorium oxalate is ignited with the filter paper in a porcelain or platinum crucible over the blast to constant weight. If a platinum crucible is used, the full heat of a Meeker burner is sufficient. The ThO_2 finally weighed should be pure white in color. A yellow color shows cerium earths or iron and is an indication of faulty work.

TIN

Sn, *at.wt.* **118.7;** *sp.gr.* **6.56;** *m.p.* **232°;** *b.p.* **2275°;** *oxides* **SnO₂** *and* **SnO.**

DETECTION

Tin is separated, together with arsenic, antimony, gold and platinum, from the hydrogen sulphide precipitate of the metals of the second group, by the action of yellow ammonium sulphide. (Normal ammonium sulphide does not readily dissolve the sulphides of tin.) If the ammonium sulphide solution is acidulated with hydrochloric acid and the acid solution reduced with iron, antimony, arsenic, platinum and gold are precipitated in the metallic form. The presence of tin, which is present as stannous chloride, is indicated by the reducing action of the solution on mercuric chloride, a white precipitate of HgCl or a gray precipitate of Hg being thrown down.

Reduce the hydrochloric acid solution of the sample by means of a small piece of iron wire. Treat with an excess of cold potassium hydroxide. Filter if the solution is not clear. Add an ammoniacal solution of silver nitrate. (One part AgNO₃ : 16 parts NH₄OH.) A brown precipitate of metallic silver indicates the presence of tin. Antimony, arsenic, platinum and gold are precipitated by the iron, while all of the heavy metals remaining, except lead, tin, aluminum, chromium, and zinc, are removed by the treatment with potassium hydrate.

Welch and Weber [1] recommend the following method for detection of tin: Add 10 cc. concentrated hydrochloric acid to the superficially dried precipitated sulphides from the ammonium sulphide separation. Filter off arsenic which does not decompose. Dilute filtrate to 70 cc. volume. Saturate with H₂S. Heat to expel excess H₂S. Add 5 cc. of hydrogen peroxide and heat until precipitate is redissolved. Add 5 to 10 grams of oxalic acid and pass H₂S into the hot solution. Antimony separates as a red sulphide. Filter. Filtrate contains the tin. Reduce with test lead and add mercuric chloride. White or grayish precipitate indicates presence of tin.

ESTIMATION

The estimation of tin is required in connection with the analysis of tin ores, dross, ashes, dust, tin plate, alloys such as solder, canned foods, and general analysis.

Opening Up Tin Ores

As the oxides of tin are not readily soluble in acids the tin can be most easily removed by assay. Ores, slags, dross, and ashes are first subjected to the assay process. The button obtained is then analyzed either volumetrically or gravi-

Chapter contributed by Wilfred W. Scott and B. S. Clark.

[1] Jour. Am. Chem. Soc., **38**, 5, 1011, 1916.

metrically by one of the methods given below. Having the weight of the button and the per cent of tin in it, the per cent of tin in the sample as received can be calculated.

There are two general processes of assaying, namely, the Cyanide Process and the Carbonate of Soda Process.

The Cyanide Process

The theory of this method is that the oxides are reduced to the metal by the action of potassium cyanide, the reaction being represented as follows:

$$SnO_2+2KCN =Sn+2KCNO.$$

Potassium cyanide reduces other metals also so that the button obtained is not pure.

Procedure. Take 100 grams of the sample which has been dried and finely powdered. (For complete analysis the moisture should be determined in the usual way.) Mix thoroughly with four times its weight of powdered potassium cyanide. Place about 1 in. of potassium cyanide in the bottom of a number H (height $5\frac{7}{8}$ ins., diameter $3\frac{3}{4}$ ins.) Battersea clay crucible. Place the mixture of sample and cyanide on top of the cyanide in the crucible and cover with enough more cyanide to fill the crucible to within 1 in. of the top.

Place the crucible in the assay furnace and heat slowly until it has been thoroughly warmed and the cyanide begins to melt. Then increase the heat gradually to a pure white, taking care that the cyanide does not boil over.[1] Grasp the crucible with the tongs and tap it gently on the hearth to assist in settling the metal. Continue the heating until all of the organic matter has disappeared, adding more cyanide from time to time if necessary. Near the end of the process the molten mass becomes clear and transparent and finally pasty and translucent. When this last condition appears, remove the crucible from the furnace and allow it to cool slowly at the temperature of the room.

When cool, break the crucible and slag away from the button. The appearance of the button and the slag immediately surrounding it indicates whether or not the process has been properly manipulated. The button itself should be firm and compact and the slag around it should be white or greenish in color. If the button is spongy or if the slag has a dirty black color, the assay should be discarded and a new determination made, using a fresh sample.[2]

Weight of Button =per cent Metal in Sample.

$$\frac{\text{Weight of Metal} \times \text{per cent Sn}}{100} = \text{per cent Sn in the Sample.}$$

NOTE. This process should be carried on under a hood in a segregated room, and every precaution should be taken to avoid breathing the poisonous fumes of potassium cyanide.

[1] Lunge advises that the cyanide should not be allowed to boil. He uses a small sample (10 grams). " Technical Methods of Chemical Analysis " **1**, Part 1, p. 256. It is our experience that satisfactory results are not obtained unless the extreme heat of the furnace is used.

[2] See also Mellor, " A Treatise on Chemical Analysis," p. 270, 1913.

The Sodium Carbonate Method

The sample is fused with equal parts of sodium carbonate and sulphur.[1] The fusion is then dissolved in water. The tin goes into solution as a thiostannate of sodium. Iron and copper are then separated by the addition of sodium sulphite, leaving arsenic, antimony and tin in solution.[2]

Other Methods of Opening Tin Ores

Fusion with Sodium Hydrate. The sample of ore is fused with ten times its weight of sodium hydrate. The process is carried out in an iron crucible and then transferred to nickel. The fused mass is dissolved in water and the tin determined in the usual way.[3]

Reduction by Means of Hydrogen. The ore may be reduced by strongly igniting in a porcelain tube in a current of hydrogen. The reduced metal is then dissolved in hydrochloric acid and the tin estimated by a standard method.

Fusion with Sodium Peroxide. J. Darroch and C. Meiklejohn[4] opened ores, slags, etc., by fusing with sodium peroxide in a nickel crucible. They dissolve the fused mass in hot water and acidify with hydrochloric acid. The sample is then ready for the necessary separations.

SEPARATIONS

Tin is separated from iron, aluminum, chromium, etc., by the insolubility of its sulphide in dilute hydrochloric acid. Tin, together with antimony, arsenic, platinum and gold, is separated from lead, mercury, copper, cadmium and bismuth, by the solubility of its sulphide in yellow ammonium sulphide. Antimony, arsenic, platinum and gold are precipitated as metals from a hydrochloric acid solution by the action of metallic iron, leaving tin in solution.

A few special separations are of interest.

Tin and Lead. For the analysis of an alloy of lead and tin, it is usually preferable to make the estimations on different samples. In this case, lead is estimated by Thompson's method and the tin by Baker's modification of the iodine method. Lead can also be separated from tin by the method given below for the separation of tin and copper.

Tin and Copper. This alloy can be dissolved in concentrated hydrochloric acid by the addition of potassium chlorate. A large excess of ammonium tartrate is added and the solution made alkaline with ammonia. Copper is then precipitated as sulphide by the addition of hydrogen sulphide water until no more precipitate is formed.

[1] Very finely divided carbon is sometimes preferred. Air must not be allowed to enter the crucible. Else decomposition is not complete. Mellor, "A Treatise on Chemical Analysis," 1913, p 270. If carbon is used instead of sulphur the process becomes one of reduction to the metal and is carried out in the assay furnace. The details of operation are similar to the cyanide process. The metal separates as a button in the bottom of the crucible. The button contains other metals with the tin and must be analyzed further for exact percentages.

[2] Mellor objects to the method as being tedious and dirty.

[3] Low, "Technical Methods of Ore Analysis," 3d Ed., pp. 208–213, 1908.

[4] Engineering and Mining Journal, **81**, 1177, 1906.

Tin and Antimony. Antimony is separated, in the metallic form, from the hydrochloric acid solution of the alloy, by the action of metallic iron placed in the solution. The tin may be determined by the iodine method without the removal of the antimony. If the antimony is desired, it may be filtered off and determined in the usual way.

As in the case of lead, it is usually quicker and more accurate to make these determinations on separate samples. The tin can be determined by the iodine method. The antimony can be determined volumetrically by various methods, preferably the bromate. (See chapter on Antimony.)

Tin and Phosphorus. One-half gram of the alloy is dissolved in 15 cc. of concentrated hydrochloric acid containing potassium chlorate. This is diluted to 200 cc. with water and warmed. It is then treated for a long time with hydrogen sulphide gas. The tin is all precipitated as sulphide while the phosphorus remains in solution.

Tin and Iron and Aluminum. Tin is separated from iron and aluminum by precipitation, as sulphide, from the hydrochloric acid solution.

Iron may also be separated from tin with copper, and lead by precipitation as sulphide from the alkaline ammonium tartrate solution.

Tin and Tungstic Acid. Donath and Mullner [1] separate tin oxide from tungstic acid by mixing the sample with zinc dust and strongly igniting in a covered crucible for fifteen minutes—boiling with dilute hydrochloric acid; oxidizing with potassium chlorate to change the blue tungstic oxide to tungstic acid and diluting with water. It is then allowed to stand overnight and filtered. The tin is in solution.

GRAVIMETRIC METHODS FOR THE DETERMINATION OF TIN

Determination of Tin or the Oxides of Tin by Hydrolysis

This method depends upon the precipitation of meta-stannic acid in the presence of ammonium nitrate when the stannic chloride is diluted to considerable volume and heated to boiling. It is especially applicable to the determination of tin oxide in tin paste, but may be extended to all chloride solutions of the higher oxides. The reaction involved proceeds as follows:

$$SnCl_4 + 4NH_4NO_3 + 3H_2O$$

$$= H_2SnO_3 + 4NH_4Cl + 4HNO_3. [2]$$

Stannous tin may be determined by oxidizing the chloride solution to the stannic form. The method gives concordant results and is rapid.

Procedure. For the analysis of tin paste take a catch weight of about 10 grams for a sample. Dissolve this sample by heating it in a No. 6 beaker with 300 cc. of concentrated hydrochloric acid. Transfer the acid solution to a 500-cc. volumetric flask and make up to the mark with dilute (1 : 1) hydrochloric acid.

[1] J. Chem. Soc. Absts., **54**, 531, 1888.
[2] Fresenius, "Quantitative Chemical Analysis," **1**, 406, 1903. Sodium sulphate may be used instead of ammonium nitrate. In that case the reaction is

$$SnCl_4 + 4Na_2SO_4 + 3H_2O = H_2SnO_3 + 4NaCl + 4NaHSO_4.$$

Take 50 cc. (approximately 1 gram) for a working sample. (If the determination is to be made on tin paste, the sample may be obtained directly by one of the methods described under Opening Tin Ores.) Dilute to 100 cc. with cold water. Nearly neutralize with strong ammonia and finish by adding drop by drop from a burette, dilute ammonia until a slight permanent precipitate is formed. A large amount of ammonia will tend to precipitate iron, if present, as a hydrate and to re-dissolve the meta-stannic acid.[1] Add 50 cc. of a saturated solution of ammonium nitrate. Dilute to 400 cc. with boiling water, stirring constantly. Bring the solution to an incipient boil, remove from the flame and allow the beaker to stand on the steam bath until the precipitate has settled.[2] The solution above the precipitate should be clear. Decant the supernatant liquor through a 12½ cm. S. & S. 590 filter paper and wash the precipitate by decantation[3] six times, using 200 cc. of boiling water and allowing the precipitate to settle thoroughly at each washing. Transfer the precipitate to the filter, " cop " out the beaker and wash down with hot water in the usual way. After the precipitate has been allowed to drain, transfer to a porcelain or a silica crucible and dry carefully on an asbestos board over a Bunsen flame.[4] When dry, ignite at a low temperature until the filter paper has been consumed. Increase the heat and finally blast to constant weight.

$$\frac{\text{Weight } SnO_2 \times 100 \times .7877}{\text{Weight of sample}} = \text{per cent Sn.}$$

Determination of Tin as Sulphide

The determination of tin as a sulphide involves many difficulties and should be avoided if possible. Better results can be obtained by the volumetric methods and in most cases without the necessity of preliminary separations of interfering metals. If tin must be separated as a sulphide, better results would be obtained if the precipitate were dissolved and the tin content determined by the iodine method.

Having the hydrochloric acid solution of tin after the interfering metals have been separated, to precipitate tin sulphide, neutralize with ammonia and then acidify with acetic acid. Pass hydrogen sulphide until the solution is saturated. Allow the precipitate to settle overnight. Pour the supernatant liquor off through a Gooch crucible and wash the precipitate six times by decantation, using a solution of ammonium nitrate[5] for wash water. Finally transfer to the crucible and wash free from chlorides. Dry the crucible in an oven at 100° C. Heat slowly in a Bunsen flame until[6] all the sulphur has been expelled. Care should be taken at this point to avoid forming fumes of stannic sulphide by heating too rapidly.

---

[1] Some practice is required to judge accurately the exact point when the necessary amount of ammonia has been added. The precipitate should appear white.

[2] If the boiling continues more than a few seconds the precipitate will not settle properly. Time will be saved in this case if the sample is discarded and a new determination commenced.

[3] If meta-stannic acid is washed over onto the filter at this point, clogging will result and a great deal of time will be lost.

[4] Spattering is likely to occur here, causing loss.

[5] Sulphide of tin separates as a slimy mass which tenaciously retains alkaline salts, especially in the absence of ammonium salts. Mellor, " Treatise on Chemical Analysis," p. 308, 1913.

[6] Bichloride of tin, Acker process, page 425.

42

Remove the lid of the crucible, which should be kept in place during the first part of the heating, and raise the temperature gradually, finally finishing with the blast. As sulphuric acid is usually present in some quantity, the crucible should be cooled and a small piece of ammonium carbonate should be placed in it. Repeat the ignition to drive out the acid. Cool and weigh as SnO_2.[1]

BICHLORIDE OF TIN

Bichloride of tin is of great importance in some of the industries, especially the textile. It is necessary to have exact analytical control of the processes in which this compound is used in order to insure uniform results and to certify the efficiency and economy of the process. Several methods have been developed for this purpose. The ones given below have had practical application and have proven to be satisfactory.

Stannic Acid Method.—Hot-water Precipitation. In the textile industry where bichloride of tin is used, the efficiency of the process depends directly on the neutrality of the tin liquor. If there is more than enough chlorine present in the bichloride solution to exactly oxidize all the tin to the stannic form, this excess is called " free HCl." If there is not enough chlorine present to do this, the deficiency is spoken of as "basic HCl." The difficulty of determining the " free " or " basic " HCl is apparent when it is known that $SnCl_4$ readily decomposes in water, liberating free acid. The following method has been developed especially for this purpose and has given good results.

The important point in this analysis is to determine whether the liquor has " free " HCl present or whether it is " basic " in nature. It has been found that hot water precipitates tin from the $SnCl_4$ solution as stannic hydroxide and at the same time liberates the chlorine as free HCl.

$$SnCl_4 + 4H_2O = Sn(OH)_4 + 4HCl.[2]$$

The $Sn(OH)_4$ separates in a colloidal precipitate which may be filtered off and the tin estimated as SnO_2. The liberated acid may be determined in the filtrate, and from this data the " free " or " basic " HCl can be calculated.

Procedure. For accurate work about 20 grams of the liquor should be weighed out in a tared weighing bottle, but for works control, where time is an important factor, it is sufficiently accurate to get the specific gravity of the liquor by means of a hydrometer and take a measured quantity for a sample, calculating the weight from these data.

Transfer the sample to a 100-cc. volumetric flask. Make up to volume with cold distilled water. Draw out of this solution 10 cc. (approximately 2 grams) and place in a 150-cc. tall beaker. Fill the beaker nearly full with boiling hot water, stirring continuously while the water is being poured in.[3] Place the beaker on top of the steam bath and allow the precipitate to settle. Decant the liquor

---

[1] This method is generally used only when minute traces of tin are present, and then it is considered best to dissolve the sulphide in hydrochloric acid and make the final determination by the iodine method. (See analysis of Canned Foods for " Salts of Tin," page 536.)

[2] Holleman and Cooper, " Text Book of Inorganic Chemistry," 4th Ed., 1912.

[3] If the solution is not stirred at this point, the precipitate will not settle and trouble will be experienced during the filtering process.

through an 11 cm. 590 S. & S. filter [1] and wash the precipitate six times by decantation, using hot water. Now transfer the precipitate to the filter and continue the washing until 1 drop of the filtrate gives no test for chlorine. After most of the water has drained out of the filter, place the paper and precipitate in a tared silica crucible. If there is plenty of time, dry the contents of the crucible on an asbestos board over a low Bunsen flame. In case the analysis must be made in a hurry, cover the crucible [2] and heat it very carefully over a low flame until all the water has been driven out and the paper has been charred. Then remove the cover and increase the heat to the full Bunsen flame and finally blast to constant weight. Weigh as SnO_2. Titrate the filtrate with N/1 NaOH, using methyl orange as the indicator.

Calculation:

$$SnO_2 \times .7877 = Sn$$

$$Sn \times 2.1945 = SnCl_4$$

$$SnCl_4 - Sn = Cl \text{ equiv. to } Sn$$

$$Cl \times 1.0282 = HCl \text{ equiv. to } Sn$$

$$\frac{HCl}{\text{Weight of sample}} = \text{per cent HCl equiv. to Sn}$$

$$\frac{\text{cc. N/1 NaOH} \times .03646}{\text{Weight of sample}} = \text{per cent HCl (actual).}$$

The difference between these last two figures equals " free " or " basic " HCl.

The Acker Process Method. [3] The theory of this method is practically the same as that of the hot-water method, except that in this case the liberated acid is neutralized with ammonia before the stannic hydroxide has been filtered off, the advantage being that any solution of the stannic hydroxide, by either acid or alkali, is prevented. The method is not applicable for the determination of " free " or " basic " HCl.

Procedure. Weigh out 25 cc. of the bichloride of tin solution. Transfer to a 500-cc. flask (volumetric) and make up to volume with cold water. With a standardized pipette, transfer 25 cc. of this solution to a No. 4 beaker. Dilute with hot water to precipitate most of the tin as stannic hydrate. Add 10 drops of phenolacetolin [4] (1 gram of phenolacetolin dissolved in 200 cc. of water). Titrate very carefully with dilute ammonia until the appearance of a rose-red color. Boil a few minutes on the hot plate. Allow the tin precipitate to settle. Decant through an 11-cm. filter paper (S. & S. 589, black ribbon brand). Wash rapidly with hot water without allowing the precipitate to cake down in the filter until the washings are free from chlorine. Dry the precipitate in an oven at 100° C. When dry, invert the filter into a tared porcelain crucible and heat on a gauze until the paper has disappeared. Remove the gauze and heat with the full

---

[1] Time may be saved by using a platinum cone with the filter and applying a gentle vacuum. This can be done with very little danger of breaking the paper.

[2] This precaution must be taken, else there will be a loss by decrepitation.

[3] Kindness of W. F. Dorflinger, chief chemist of Perry-Austin Manufacturing Company.

[4] Luteol may be used as indicator, giving a yellow color at the end-point. It is slightly more delicate but much more expensive.

Bunsen flame for a few minutes. Finally blast to constant weight.[1] Weigh as SnO_2.

Take the filtrate and washings and dilute them to a volume of 1000 cc. Warm 500 cc. of this solution and saturate it with hydrogen sulphide. If any tin separates, filter and ignite in a tared porcelain crucible. Moisten with a little nitric acid and heat very slowly to drive out the acid. Ignite to constant weight. Weigh as SnO_2. Add this result to the SnO_2 obtained above when calculating the final result.

Determination of Tin in Bichloride of Tin as Sulphide

This method is given as an alternative for the Acker Process Method and may be used as a check on that process. Uniform and concordant results have been obtained by the use of the two methods.[2]

Procedure. Weigh out 25 cc. of bichloride and dilute to 500-cc. volume with cold distilled water. Take 25-cc. portions of this solution for analysis. Dilute the sample to 250 cc. Saturate with hydrogen sulphide. Warm the mixture on a hot plate at a temperature of about 65° C. until the precipitate is coagulated. Test the clear supernatant liquor for unprecipitated tin by adding a little hydrogen sulphide water. Filter on an ashless filter and wash free from chlorides. Make the filtrate and washings up to 1000 cc. volume for further determinations. Dry the tin sulphide precipitate on the filter in an oven at 100° C. Remove the precipitate from the paper as completely as possible. Ignite the paper in a weighed porcelain crucible. Cool and add a few drops of nitric acid. Repeat the ignition, heating very carefully at first until the acid has nearly all been driven out. Now place the main tin precipitate in the crucible. Cover, heat gently for a few minutes, moisten with fuming nitric acid, ignite very carefully for one-half hour and then blast for fifteen minutes. Weigh as SnO_2.[3]

VOLUMETRIC DETERMINATION OF TIN

Volumetric methods for the determination of tin are based upon the reducing power of stannous compounds. They vary according to the oxidizing agent used and the details of manipulation.

Lenssen's Iodine Method as Modified by Baker.[4] This method is a modification of Lenssen's Iodine Method for the determination of tin in alkaline solutions. It is especially applicable to the determination of "salts of tin" in canned foods and to the estimation of tin coating on tin plate, but is accurate, rapid and very satisfactory for alloys and general analysis.

The method is based on the action of iodine in the presence of stannous chloride in hydrochloric acid solution. The reaction involved is:

$$SnCl_2 + I_2 + 2HCl = SnCl_4 + 2HI.$$

A. Jilek reduces tin by means of iron, filtering off precipitated Sb, Cu and excess of Fe, in an atmosphere of CO_2. The reduced tin solution is now titrated with standard iodine.

[1] If there has been any reduction, a few drops of nitric acid may be added and the ignition repeated, heating slowly at first to prevent loss by decrepitation.

Iron, lead and antimony do not interfere with the reaction. Copper in small quantities does not interfere with the determination, but if it is present in large quantities as a salt, it is likely to produce low results. Determinations made by the writer [1] show that results are accurate when less than 10% of copper, as copper chloride, is present. Larger amounts gave consistently low results. The reason for this fact centers around the difficulty of reducing all the copper to the cuprous form. If any $CuCl_2$ is left in the solution, it reacts with the potassium iodide of the iodine solution, causing the precipitation of CuI and the liberation of free iodine.

$$CuCl_2 + 2KI = CuI + 2KCl + I.$$

Copper present as the metal is not easily soluble or goes into solution in the reduced form and is not likely to disturb the determination. [2]

Solutions—Standard Tin Solution. Dissolve 5.79 grams of Kahlbaum's C. P. tin in C. P. hydrochloric acid. The solution of the tin is effected by placing about 150 cc. HCl in an Erlenmeyer flask, together with the tin, and boiling. After the tin has all been dissolved, transfer to a volumetric liter flask and make up to the mark with dilute hydrochloric acid.

$$1 \text{ cc.} = .00579 \text{ gram Sn.}$$

Standard Iodine Solution. Dissolve 12.7 grams of C. P. iodine in a water solution of 20 grams of potassium iodide. Make up to one liter and standardize against the standard tin solution. For tin plate analysis, it is convenient to adjust the iodine solution so that 1 cc. equals exactly .00579 gram of tin. Then, if a sample of the plate having a total surface of 8 sq.ins. is taken, 1 cc. of the iodine solution is the equivalent of one-tenth of a pound per base box.

Indicator. Dissolve 5 grams of pure soluble starch in 1 liter of water.

Air-free Water. Dissolve 12 grams of bicarbonate of soda in 1 liter of water. Add 20 cc. HCl and allow the resulting gas to escape. Keep in a stoppered bottle. [3]

Procedure. For practical purposes, take a sample, such that the tin content will be between .2 gram and .5 gram. A larger sample should be taken for extreme accuracy in order to decrease the possible technical error. Place the sample in flask A of the Sellars apparatus, Fig. 68, together with 100 cc. of conc. C. P. HCl. Stopper the flask and connect tubes B and D, as shown in the illustration. Boil until the metal is all dissolved. This point is indicated by the cessation of the hydrogen evolution and the appearance of large well-developed bubbles. If a sufficient amount of metallic iron is present in the sample, complete reduction is assured. If no iron was present in the sample, or if there was not enough to reduce all of the tin, make sure that the tin is all converted to the stannous form by adding aluminum foil (about 1 gram). Replace the stopper and connect as originally. Boil until normal bubbles reappear. Open cock C to allow CO_2 [4] gas to enter. Place the flask in cooling bath F without disconnecting the apparatus. After the solution has become thoroughly

[1] Mr. B. S. Clark.

[2] Sulphates must not be present. They tend to have an oxidizing effect and spoil the results.

[3] There should always be an excess of bicarbonate of soda present in order that carbon dioxide will be generated during the washing process, thus preventing air from entering the flask at any time during the analysis.

[4] Carbon dioxide generated in a Kipp apparatus is likely to contain oxygen. It is much better to use liquid CO_2 such as can be purchased in the open market.

cool, disconnect tubes B^1 and D^1 from the splash bulbs. Wash the bulbs with " air free " water, allowing the washings to drain into the bulk of the sample. Remove the stopper and wash down the sides of the flask. About 50 cc. of water should be used in the washing so that the final sample contains about 25% HCl. Add 5 cc. starch solution and titrate with the standard iodine solution.

$$\frac{\text{cc. iodine} \times .00579 \times 100}{\text{Weight of Sample}} = \text{per cent Sn,}$$

or

$$\frac{\text{cc. iodine}}{10} = \text{pounds per " base box."}[2]$$

The Sellars Apparatus. This apparatus is a device designed by Mr. W. S. Sellars of this laboratory for the purpose of facilitating the solution of tin samples out of contact with air. Added to this advantage, it is equipped with a water

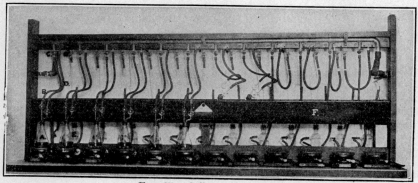

Fig. 67.—Sellars' Apparatus.

cooler. It is also constructed so that the tubes and scrubbing bottles can be cleaned by flushing with water. The use of this apparatus practically eliminates the usual sources of error in connection with the iodine method, and at the same time greatly increases the speed of the determination. Fig. 67 shows the apparatus in operation.

$A.[1]$ 300-cc. Erlenmeyer flask.

$B.[1]$ Connection with reduced pressure line from liquid carbon dioxide cylinder.

$C.[1]$ Glass manifold.

$D.[1]$ Exit connection to trap.

$E.[1]$ Water trap to prevent escape of HCl fumes and to prevent air from backing into the flask.

$F.[1]$ Cooling tank.

$G.[1]$ Low-pressure water wash-out manifold.

$H.[1]$ Perforated feed pipe to water cooler.

$K.[1]$ Outlet for cooler.

$L.[1]$ Electric hot plate.

$M.[1]$ Lead drain pipe.

[1] See Fig. 68, page 535.

[2] " Basebox "—112 sheets of tin, 14×20 ins.

Ferric Chloride Method.[1] This method depends upon the reduction of ferric chloride by stannous chloride in hot solution.

$$SnCl_2 + 2FeCl_3 = SnCl_4 + 2FeCl_2.$$

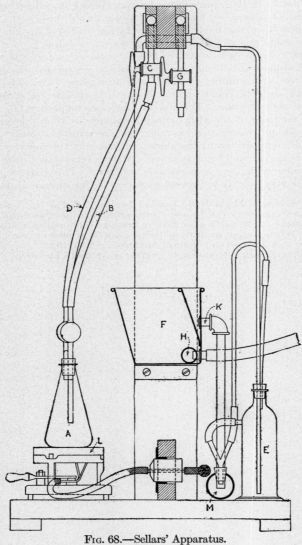

F<small>IG</small>. 68.—Sellars' Apparatus.

[1] C. Mene, Dinglers Journal, **117**, 230, 1850.   K. Pallet and A. Allart, Bul. Soc. Chim. (2) 27, 43, 438, 1877.   H. J. B. Rawlins, Chem. News, **107**, 53, 1913.   H. Nelsmann, Zeit. Anal. Chem., **16**, 50, 1877.

Antimony, copper, arsenic, bismuth, mercuric chloride, tungsten and titanium must be absent.[1]

The Standard Solution of Ferric Chloride is made by dissolving pure iron wire in hydrochloric acid. To standardize this solution, dissolve 1 gram of pure tin in 200 cc. of C. P. HCl, preventing air from coming in contact with the solution by means of a trap, or by passing carbon dioxide over it.[2] Titrate this standard sample with the ferric chloride solution. The end-point is indicated by the yellow color, due to a slight excess of the iron solution.

Procedure. Tin is first separated from the interfering metals in the usual way. If lead, copper, arsenic, antimony or bismuth are present, the sample is first reduced, in the hydrochloric solution, with iron wire. The solution is then filtered. Lead and tin remain in the filtrate. Neutralize by adding strips of zinc until the action ceases. Tin and lead are precipitated. The clear liquid should show no trace of tin with hydrogen sulphide. Allow the precipitate to settle and wash by decantation, keeping the precipitated metals in the flask. Add 150 cc. of concentrated hydrochloric acid, keeping the contents of the flask protected from the air, and bring to a boil. When everything is dissolved, titrate to a yellow color with the ferric chloride solution.[3] This part of the analysis should be done very quickly to prevent oxidation by the oxygen of the air.

[1] Lunge, " Technical Methods of Chemical Analysis," **2,** Part I, p. 267.

[2] The Sellars apparatus can be used with advantage for this purpose.

[3] The end-point can be easily identified by looking at a blue Bunsen flame through the solution. When a small quantity of ferric chloride is present, the flame appears green. Mellor, " A Treatise on Chemical Analysis," p. 310, 1913.

[4] H. A. Baker, Eighth International Congress of Applied Chemistry.

VOLUMETRIC METHOD FOR TIN IN ALLOYS

The titration of stannous solutions by iodine may be represented by the following reaction:

$$SnO + 2I + H_2O = SnO_2 + 2HI.$$

Sn is equivalent to O or to 2H, hence a normal solution contains one-half the molecular weight of Sn, or 59.35 grams per liter of solution.

Apparatus. This consists of a 300-cc. Erlenmeyer flask, with a one-hole stopper, through which passes a quarter-inch glass tube, connected with a rubber tube 12 to 15 inches in length, the other end of the rubber tubing is connected with 2–3 inches of glass tubing, which dips in the beaker containing a bicarbonate of sodium solution.

Reagents. 0.1N iodine solution. Standardize against a sodium arsenite (arsenious acid) solution. 1 cc. 0.1N I = 0.005935 g. of Sn.

Starch solution. Sulphuric and hydrochloric acids. Antimony powder.

Procedure. *Decomposition of the Sample.* Tin alloys generally decompose in hydrochloric acid, but more readily in strong, hot sulphuric acid.

A factor weight 0.5935 gram of the tin alloy is placed in a 300-cc. Erlenmeyer flask and 10 cc. of strong sulphuric acid added. The mixture is heated, preferably over a free flame, until the alloy completely disintegrates. Nearly all of the excess of free acid is expelled, keeping the flask in motion over the flame to lessen the tendency towards bumping, which is apt to occur during the concentration. The moist residue is allowed to cool.

100 cc. of (air-free) water are added followed by 50 cc. of strong hydrochloric acid and the mixture gently warmed until the solution begins to clear. The apparatus is now assembled as shown in the figure, about 15 cc. of 10% (saturated solution) sodium bicarbonate being placed in the test tube (or 50 cc. in the beaker, if this is preferred to a test tube).

About 1 gram of very finely powdered antimony metal is placed in the flask, followed by 10 cc. of saturated sodium bicarbonate solution, the stopper being removed during the addition and then immediately replaced. The air is displaced by the CO_2 generated.

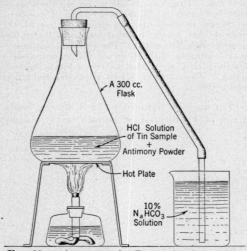

A 300 cc. Flask

HCl Solution of Tin Sample + Antimony Powder

Hot Plate

10% $N_a HCO_3$ Solution

FIG. 68a.   Apparatus for Determination of Tin

The apparatus is now placed on a hot plate, or on an inverted sand-bath dish over a flame, and the solution is gently boiled for 10 to 15 minutes. The antimony should be of such fineness as to remain suspended during the ebullition of the liquid at this stage.

The test tube is now nearly filled with saturated sodium bicarbonate and the apparatus removed to a desk for a few minutes, and then placed in a cold water-bath of running water or under tap water, until the solution cools down to near room temperature. During this cooling carbonate will be sucked back into the flask " A " to establish pressure equilibrium, CO_2 being generated in the flask. Sufficient solution will remain in the test tube to act as a seal and prevent admission of air, which would spoil results by its oxidation of the tin.

The tube B and the test tube C are disconnected and 5 cc. of starch solution added by means of a pipette passing through the hole in the stopper (which should be loosened in the throat of the flask).

Standard iodine solution is now added, the tip of the burette passing through the hole of the stopper, agitating the solution by a " swirling " motion of the flask. The end-point is a blue color, which does not fade on stirring the solution.

If a factor weight has been taken, each cc. of the iodine of 0.1N strength is equivalent to 1 per cent of tin.

NOTES. In the presence of copper a separation must be effected as over 3 per cent copper interferes in this volumetric method especially when the percentage of the tin is low. High copper alloys do not decompose readily in hydrochloric or sulphuric acids, but easily in dilute nitric acid. The solution obtained is evaporated to dryness, the residue is taken up with strong nitric acid, the oxide of tin (and antimony) remains insoluble, hot water is added and the solution filtered (hot) and the oxide washed once or twice with hot water. The tin (and antimony) is now best dissolved [1] by digesting for 3–5 minutes with 50 cc. of water saturated with SO_2 (at 60–70° C.), then 10 cc. of strong HCl are added to the solution heated to boiling and the SO_2 expelled by boiling. The solution is now ready for reduction with antimony by the method outlined above.

Tin may be reduced by iron wire or nails, the precipitated Cu, Sb and the excess of Fe filtered off, under an atmosphere of CO_2, and tin titrated with iodine.

[1] Earnest Stelling, Ind. and Eng. Chem., **16**, 346 (April, 1924).

ESTIMATION OF TIN IN CANNED FOOD PRODUCTS [1]

The tin in the canned food products is obtained as a sulphide precipitate from wet combustion, with nitric and sulphuric acids, of 100 grams food product.

The clear sulphuric acid residue is diluted, neutralized with ammonia and then rendered about 2% acid with hydrochloric acid, after which it is thoroughly saturated with hydrogen sulphide gas. This precipitate is then filtered on a Gooch crucible with a false bottom. The precipitate may contain foreign substances, such as lime, phosphorus, and silica, some lead, or even small amounts of iron, but none of these will cause any trouble subsequently in the titration, so that the labor of separating the tin completely from the precipitate is obviated.

After washing the precipitate three or four times in a Gooch crucible, it is transferred to a small porcelain dish by simply forcing out the false bottom of the Gooch crucible and its asbestos pad and rinsing off the crucible.

The precipitate, mixed with asbestos, is now transferred to a 300-cc. Erlenmeyer flask and boiled with strong hydrochloric acid, potassium chlorate being added from time to time to insure the complete breaking up and solution of the tin sulphide, as well as the elimination of the sulphur. This is accomplished in a very few minutes. A few strips of pure aluminum foil, free from tin, are then added to the flask until all of the chlorine is eliminated. The flask is then attached to the Sellars apparatus and the determination completed, according to the details given under the Iodine Method, using N/100 iodine solution.

Gravimetric Method. [1] The sample is first digested to a colorless or pale yellow solution as described under Baker's method.

Add 200 cc. water to the digested solution and pour into a 600-cc. beaker. Rinse out the Kjeldahl flask with three portions of boiling water so that the total volume of the solution is about 400 cc. Allow to cool and add 100 cc. concentrated ammonia. This amount of ammonia should render the solution nearly neutral, unless more than 50 cc. sulphuric acid have been used for digestion. The solution should be tested to see that it is still somewhat acid. In case of a large excess acid, add ammonia until just alkaline and then make about 2% acid with hydrochloric or sulphuric acid. Pass in a slow stream of hydrogen sulphide for an hour, having the covered beakers on an electric hot plate at about 95° temperature. Allow to digest on the hot plate for an hour or two.

Filter the tin sulphide on an 11-cm. filter. Wash with three portions of wash solution alternated with three portions of hot water. The wash solution is made up of 100 cc. saturated ammonium acetate, 50 cc. glacial acetic acid, and 850 cc. water. The filter papers used in this method are C. S. & S. No. 590, white ribbon.

Place the filter and precipitate in a 50-cc. beaker and digest with three successive portions of ammonium polysulphide, bringing to a boil each time and filtering through a 9-cm. filter. Wash with hot water. Acidify with acetic acid, digest on the hot plate for an hour and filter through a double 11-cm. filter. Wash with two portions of wash solution alternated with hot water and dry thoroughly in a weighed porcelain crucible. Thorough drying is essential to the success of the determination. Ignite very gently at first and later at full heat of Bunsen flame. Finally heat strongly with large burner, or Méker burner, having the crucible partly covered. Stannic sulphide must be gently roasted to the oxide, but the oxide may be heated strongly without loss, due to volatilization.

Weigh the stannic acid and convert to metallic tin by the factor .7877.

[1] H. A. Baker, Eighth International Congress of Applied Chemistry.
[2] E. L. P. Treuthardt. Association of Official Agricultural Chemists, August 15, 1915.

TITANIUM[1]

Ti, *at.wt.* 47.9; *sp.gr.* 4.5 [2]; *m.p.* 1795° C. ($\pm15°$)[3]; *oxides* TiO, Ti$_2$O$_3$, TiO$_2$, TiO$_3$.

DETECTION

The powdered ore is fused with potassium bisulphate, KHSO$_4$, until effervescence ceases. The cooled mass is dissolved in dilute sulphuric acid by boiling. **Hydrogen peroxide,** H$_2$O$_2$, added to this titanium solution, produces a yellow to orange color, according to the amount of titanium present. Hydrofluoric acid, or fluorides, destroys the color. Vanadium also produces this color with hydrogen peroxide, but the color is not destroyed by HF. The yellow color, according to Weller [4] is due to TiO$_3$ formed.

Morphine produces a crimson color with solutions of titanium in sulphuric acid.

Zinc added to hydrochloric acid solutions of titanium produces a blue color,[5] tin a fine violet solution.[6]

If **sulphur dioxide,** SO$_2$, is passed into the solution of titanium to reduce the iron, and the slightly acid solution then boiled, yellowish white metatitanic acid, TiO(OH$_{)2}$, is precipitated.

Bead Test on Charcoal. A small portion of the powdered mineral heated on charcoal with microscosmic salt and tin produces a violet-colored bead if titanium is present.

OCCURRENCE

The element is widely distributed in minerals, soils, clays and titaniferous iron ores. It is found in granite, gneiss, mica, slate, syenitic rocks, granular limestone, dolomite, quartz, feldspars and a large number of other minerals. The principal commercial minerals are:

Ilmenite, FeTiO$_3$, containing about 52.7% TiO$_2$.
Rutile, TiO$_2$, containing 90 to 100% TiO$_2$.
Titanite, CaTiSiO$_5$, containing 34 to 42% TiO$_2$.
Perovskite, CaTiO$_3$, containing about 60% TiO$_2$ and 5 to 6% Yt$_2$O$_3$.
Titaniferous ores of variable titanic oxide content.

APPLICATION

Titanic oxide is now extensively used as a white pigment for paint. For this purpose the titanic oxide is usually precipitated upon or coalesced with

---

[1] By Wilfred W. Scott and L. E. Barton.
[2] Hunter, Eighth Int. Congress Applied Chem., **2**, 125.
[3] Burgess and Waltenburg, U. S. Bureau of Standards.
[4] J. S. C. I., 1882, 506–508.
[5] Deville, C. N., **4**, 241.
[6] Cahen and Wootton, "The Mineralogy of the Rarer Metals."

precipitated barium sulphate, making a product containing about 25% titanic oxide. The white titanium pigments are characterized by their exceptional hiding power, excelling in this respect any other commercial white pigments.

Another important application of titanium is the use of ferrotitanium in the iron and steel industry. The function of the titanium is to deoxidize the steel and consequently to yield a product free from blowholes and segregation of impurities. In a steel thus purified the natural strength and resistant properties of the material are developed in the highest degree.

Titanium has also found application in the textile and leather industries. In the dyeing of leather, titanium potassium oxalate has been found particularly well adapted. The use of titanous chloride and titanous sulphate for bleaching or discharging colors is increasing. Such bleaching agents are particularly applicable for silk and wool, which are injured by the action of those bleaching agents in which chlorine is the active element.

Titanium compounds are also used for electric light filaments, arc carbons, ceramics, fine brown glazes, paint for iron and steel, etc.

ESTIMATION

Preparation and Solution of the Sample

A knowledge of the solubility of the element and its oxides is of value in the solution of the sample.

Element. This is feebly soluble in cold dilute hydrochloric or sulphuric acids; more readily so when the acids are heated. It is soluble in cold, concentrated hydrochloric acid; readily soluble in hot, concentrated hydrochloric or sulphuric acids. It is scarcely acted upon by nitric acid, but readily dissolves in hydrofluoric acid. It is soluble by fusion with the alkalies.

Oxides. Ti_2O_3, which has a black or blue color, is soluble in concentrated hydrochloric or sulphuric acids; forming, in the latter case, a violet-colored solution.[1] The oxide is insoluble in water and in ammonium hydroxide.

TiO_2 is difficultly soluble in concentrated sulphuric acid, less soluble if strongly ignited. The metatitanic acid, $TiO(OH)_2$, requires strong hydrochloric or sulphuric acid to effect solution; the orthotitanic acid, $Ti(OH)_4$, however, is readily soluble in hot or cold, dilute or concentrated acids. From titanic solutions orthotitanic acid is precipitated by ammonia, the precipitation being assisted by warming. Boiling a slightly acid solution precipitates the metatitanic acid, $TiO(OH)_2$. TiO_2 is soluble upon fusing with alkalies. TiO_2 is soluble in hydrofluoric acid, forming TiF_4, which is volatile, unless an excess of sulphuric acid is present (distinction from silica). The ignited oxide is best dissolved by fusion with $KHSO_4$ and heating the fused mass with dilute sulphuric acid solution.

Salts. Many titanic salts are decomposed in the presence of water, precipitating titanic acid, the extent of the decomposition depending on the quantity of water used. Titanic sulphate is readily soluble in water and the solution is remarkably stable unless largely diluted with water. Some of the double salts are readily soluble and their solutions stable, i.e., potassium titanium oxalate.

[1] Ebelmen, A. Ch. (3), **20**, 392, 1847.

Solution of Steel. The sample may be dissolved in hydrochloric acid (1 : 2). If a residue remains, it is treated with a mixture of equal parts of hydrofluoric and sulphuric acids and a few drops of nitric acid, in a platinum dish, and the mixture evaporated to sulphuric anhydride fumes and to complete expulsion of hydrofluoric acid. The colorimetric procedure is now used for estimating titanium. For determination of titanium in hydrochloric acid solution see page 545.

NOTE. Titanium in steel treated with ferro carbon-titanium exists in two conditions: (1) Titanium soluble in hydrochloric acid. (2) Titanium insoluble in hydrochloric acid. Of the very small amount of titanium in treated steel the greater part will usually be found in the second form. When the amount of titanium in the steel is exceedingly small, the soluble titanium frequently exceeds the insoluble, and it is then occasionally desirable to determine also that existing in the second form.

Alloys. These are dissolved in concentrated nitric acid, aqua regia or a mixture of the dilute acids. Should nitric acid be used, the excess is expelled by evaporation to dryness with hydrochloric acid. The metals of the hydrogen sulphide group are removed in an acid solution by precipitation with H_2S, and titanium determined colorimetrically in the filtrate.

Ores. One to 5 grams of the ore are treated with 10 to 50 cc. of a mixture of sulphuric and hydrofluoric acids (1 to 5), a few drops HNO_3 added, and the solution evaporated to fumes to expel HF. If a residue remains upon taking up with water containing a little sulphuric acid, it is filtered off and fused with $KHSO_4$ as directed under the fusion method.

Fusion Method for Ores. The finely powdered sample is fused with four to five times its weight of potassium bisulphate, $KHSO_4$, and the cooled fusion dissolved with dilute sulphuric or hydrochloric acid. In the presence of silica potassium fluoride is added to assist in the decomposition of the material.

(See Analysis of Titaniferous Ores, page 551.)

Titaniferous Slags. One-half gram of the finely ground sample is decomposed in a platinum dish by a mixture of 5 cc. water, 5 cc. concentrated sulphuric acid, 2 cc. nitric acid, and 10 cc. of hydrofluoric acid, the reagents being added in the order named. The solution is evaporated rapidly to SO_3 fumes to expel fluorides and the excess sulphuric acid until residue is left nearly dry. After cooling it is taken up with 40 cc. of dilute hydrochloric acid (1 : 3), which will give a clear solution containing all the constituents of the slag except silica, which has been volatilized as SiF_4. The solution is diluted to 200 cc. with cold water. Iron and titanium are precipitated by ammonia in slight excess and filtered at once without boiling. The precipitate is dissolved in cold dilute hydrochloric acid and reprecipitated with ammonia. Titanium is now separated from iron by reducing iron with SO_2 and precipitating titanium from a boiling acid solution as described on page 542.

SEPARATIONS

Details of the isolation of titanium are given in the methods for its estimation.

Separation of Titanium from the Alkaline Earths, etc. The hydroxide is precipitated when a titanium solution containing ammonium chloride is treated with ammonium sulphide, whereas barium, strontium, calcium and magnesium remain in solution. Titanium hydroxide may be precipitated by making the solution containing titanium slightly ammoniacal with HN_4OH.

Separation from Copper, Zinc, Aluminum Iron, etc. Titanium is precipitated from a slightly acid solution [1] by boiling, passing sulphur dioxide through the solution to keep the iron reduced and prevent its precipitation.

Separation from the Bivalent Metals, Manganese, Nickel, Cobalt, Zinc. Titanium is precipitated along with aluminum and iron by hydrolysis of its acetate in a hot, dilute solution, whereas manganese, nickel, cobalt and zinc remain in solution. Details of the basic acetate method are given on page 298.

Separation of Titanium from Aluminum. Small amounts of titanium from large amounts of aluminum. (One part Ti to 50 parts Al.) Cupferron, $C_6H_5(NO)N \cdot ONH_4$ added to a decidedly acid solution containing titanium and aluminum precipitates titanium, but not aluminum. The precipitate is washed by decantation and then on the filter with very dilute hydrochloric acid to remove traces of aluminum. The procedure affords a separation of titanium from chromium, nickel, cobalt, manganese, etc. Copper and iron, however, precipitate with the titanium, if present in the solution. The yellow titanium salt has the composition $(C_6H_5(NO)NO)_4Ti$. [2]

Separation of Titanium from Iron. See Gravimetric Method for Determination of Titanium, Modified Gooch Method, below.

[1] Acidity exactly 0.5% is best according to Levy, C. N., **56**, 209.
[2] Analyst, **36**, 520, 1912, method of J. Bellucci and L. Grassi.

GRAVIMETRIC METHODS

Gravimetric Determination of Titanium.   Modified Gooch Method [1]

This method is applicable to minerals and metallurgical products that are comparatively high in titanium. The method provides for the separation of titanium from iron and from aluminum and phosphoric acid with which it commonly occurs. The procedure as proposed by F. A. Gooch and modified for non-aluminous rocks by Wm. M. Thornton has been found by the author [2] to give reliable results. The details of the method with a few slight changes found to be advantageous are given below. Iron is separated from titanium by precipitation as a sulphide in presence of tartaric acid, the organic acid is destroyed by oxidation and titanium precipitated from a boiling acetic acid solution. In the presence of alumina and phosphoric acid the impure precipitate is fused with Na_2CO_3 and the impurities leached out with boiling water. In presence of zirconium, titanic acid is incompletely precipitated. Hillebrand's modification for the removal of zirconium is given in the notes.

Procedure.   Preparation of the Sample.   *Ores High in Silica.* These may be decomposed by taking to SO_3 fumes with a mixture of 10 to 15 cc. of 50% hydrofluoric acid, HF, and 3 to 4 cc. of concentrated sulphuric acid per gram of sample.

Oxides. Decomposed by fusion with sodium or potassium bisulphate. The fusion is dissolved in 10% sulphuric acid, keeping the volume as small as possible. The sample should contain not over 0.2 gram titanium.

Precipitation of Iron. To the solution containing titanium, tartaric acid, equal to three times the weight of the oxides to be held in solution, is added. This should not exceed 1 gram of the organic acid, as the subsequent removal of larger amounts would be troublesome. H_2S is passed into the solution to reduce the iron and NH_4OH added to slight alkalinity followed by a further treatment with H_2S to completely precipitate FeS. The solution should be faintly alkaline (litmus) otherwise more ammonia should be added. After filtration and washing of the ferrous sulphide with very dilute and colorless ammonium sulphide, the titanium is entirely in the iron-free filtrate.

Oxidation of Tartaric Acid. Since titanium cannot be precipitated by any reagent in the presence of tartaric acid, [3] the organic acid is oxidized by addition of 15 to 20 cc. of concentrated sulphuric acid to the sample placed in a 500-cc. Kjeldahl flask. The solution is evaporated to incipient charring of the tartaric acid. After cooling slightly, about 10 cc. of fuming nitric acid are added cautiously, a few drops at a time, and when the violent reaction has subsided the flask is heated gradually (hood), a vigorous reaction taking place accompanied by much effervescence and foaming with evolution of copious brown fumes. The organic matter gradually disappears, the effervescence becomes steady and finally ceases and white fumes of SO_3 are given off. The solution is cooled and the pale yellow syrup poured into 100 cc. of cold water, the flask washed out, adding the rinsing to the main solution. If cloudy, the solution is filtered.

---

[1] F. A. Gooch, Proc. Am. Acad. Arts and Sci., New Series, **12**, 435. Wm. M. Thornton, C. N., **107**, 2781, 123, 1913.

[2] W. W. Scott.

[3] Cupferron precipitates titanium in presence of tartaric acid.

Precipitation. Ammonia is added until the solution is nearly neutral, a point where the solution is slightly turbid, the precipitate dissolving upon vigorous stirring. If a trace of iron is suspected about 1 cc. of 10% ammonium bisulphate is added. Five cc. of glacial acetic acid followed by 15 grams of ammonium acetate or its equivalent in solution is added and the volume of the solution made up to about 350 cc. The solution is brought rapidly to boiling and maintained in ebullition for about three minutes. The titanium will precipitate in white flocculent and readily filterable condition. The precipitate is washed first with water containing acetic acid and finally with pure water. The filter and the precipitate are ignited cautiously over a low flame and finally blasted over a Méker blast for twenty minutes. The residue is weighed as TiO_2.

In the presence of large amounts of alumina and phosphoric acid, the residue above obtained is fused with sodium carbonate in a platinum dish and the fusion leached by boiling with pure water. Alumina and phosphoric acid go into solution as soluble sodium salts and titanium oxide remains insoluble in the residue.

Ignited insoluble residue $= TiO_2$.

NOTE. Titanium may be separated from aluminum by fusing the residue with potassium acid sulphate, $KHSO_4$, and precipitation of titanium in an acid solution by cupferron. Al_2O_3 is in solution.

Determination of Titanium in Ferro Carbon Titanium.
Gravimetric Method [1]

Into a 6-in. porcelain evaporating dish, weigh 0.6 gram (factor weight) of alloy.

Dissolve in a mixture of 15 cc. of dilute sulphuric acid (one acid to one water), 5 cc. of nitric acid, and 10 cc. of hydrochloric acid. Evaporate to fumes of sulphuric anhydride.

Cool and take up by boiling with 50 to 60 cc. of water and 5 to 10 cc. hydrochloric acid. Filter into a 500-cc. beaker and wash the residue with hot water and dilute hydrochloric acid.

In the filtrate precipitate iron and titanium by ammonia in slight excess. Filter without boiling and wash precipitate twice on filter with hot water.

Reject filtrate. Dissolve the precipitate in a very little dilute hydrochloric acid, washing the filter with hot water and collecting the solution and washings in the original beaker.

Nearly neutralize the solution with ammonia or ammonium carbonate; dilute to 300 cc.; saturate with sulphur dioxide gas, and boil until titanic acid is precipitated and the solution smells faintly of sulphur dioxide.

Filter and wash with hot water and dilute sulphurous acid.

Dry, ignite, and weigh as titanic oxide.

Since the factor weight of sample has been used, one milligram of titanic oxide is equal to 0.1% metallic titanium.

---

[1] Methods of analysis used in the laboratories of the Titanium Alloy Manufacturing Company.

VOLUMETRIC METHODS

The Determination of Titanium by Reduction, Addition of Ferric Salt and Titration of Reduced Iron with Potassium Per= manganate [1]

Principle. Titanic acid is reduced by means of zinc, an excess of ferric sulphate is added and the ferrous salt, formed by reduction by titanous salt, is titrated with standard permanganate. The method is more accurate than direct titration of the titanous salt with permanganate.

Reaction. $Ti_2(SO_4)_3 + Fe_2(SO_4)_3 = 2Ti(SO_4)_2 + 2FeSO_4,$

or $TiCl_3 + FeCl_3 = TiCl_4 + FeCl_2.$ [2]

Preparation of the Sample

Procedure. One to 2 grams of the ore is decomposed by hydrofluoric and sulphuric acids or by fusion with potassium bisulphate or a combination of the two according to the methods already described. Members of the H_2S group, if present, may be removed by H_2S. If iron is present it may be determined by boiling off the H_2S in the filtrate containing Fe, Ti, etc., and allowance made in the titration for titanium. If other interfering elements are present in this filtrate, titanic acid may be precipitated by boiling the slightly acid solution (sulphurous acid) according to directions given in the gravimetric method. The washed oxide is dissolved in strong H_2SO_4 and diluted as directed below.

Reduction. The solution is washed into a 100-cc. flask and diluted with water so that it will contain 10% of sulphuric acid. This acid holds titanic acid in solution and at the same time is insufficient to oxidize the reduced titanium oxide. Sufficient zinc to cause complete reduction is added and a rubber stopper carrying a Bunsen valve tube and a thistle tube with glass stop-cock is inserted in the neck of the flask. The evolved hydrogen expels the air and reduces the titanic oxide to the titanous form. Iron if present is also reduced. Gentle heat is applied until the excess of zinc dissolves. The solution is cooled and an excess of ferric sulphate added through the thistle tube, followed immediately by cold distilled water until the flask is filled to the neck. The contents of the flask is poured into a No. 6 beaker containing 150 to 200 cc. of cold distilled water and the ferrous iron, formed by the reducing action of titanous salt, is titrated with N/10 $KMnO_4$ solution.

One cc. N/10 $KMnO_4$ = 0.00481 gram Ti,   or   0.00801 gram TiO_2.

[1] H. D. Newton, A. J. Sc. (4), **25**, 130. A. F. Gooch, "Methods in Chemical Analysis."
[2] T. R. Ball and G. McP. Smith, Jour. Am. Chem. Soc., **36**, 1838, 1914.

The Determination of Titanium by Reduction with Zinc and Titration with Permanganate [1]

In a contribution from the U. S. Bureau of Standards, the authors have reported a study of this method, the results of which indicate that:

The reduction of titanium in a Jones reductor proceeds rapidly and is quantitative, provided the reduced solution is caught under a 3–5-fold excess of ferric sulphate.

The reduction is conveniently carried on in solutions containing 3–5% by volume of sulphuric acid and may be done at any temperature between 25° and 100°.

The determination is carried out as follows:

Using a Jones reductor of 19 mm. internal diameter, with zinc column 43 cm. in length, there is added in order 25–50 cc. dilute sulphuric acid, 3 to 5% by volume; 150 cc. of the titanium solution containing 3 to 5% by volume of sulphuric acid; 100 cc. more of acid and finally 100 cc. of water. The reduction is performed at a speed of about 100 cc. per minute.

The reduced solution is delivered through a tube from the reductor under and into a ferric sulphate solution containing about .02 g. of iron per cc. in 8% by volume sulphuric acid. The ferric sulphate solution should be from 3 to 5 times that theoretically required by the titanium. The reduced solution is titrated with tenth normal potassium permanganate.

The original article suggests means for separating tin, arsenic, antimony, molybdenum, iron, chromium, vanadium, tungsten and uranium, which would interfere with determination of titanium by the method described.

Volumetric Method by Reduction of Titanium and Titration with a Ferric Salt

The following volumetric method recommended by the Titanium Alloy Mfg. Co., is essentially that described by P. W. & E. B. Shimer, Proceedings of Eighth International Congress of Applied Chemistry, the method hereafter described differing principally in the form of reductor and also in a few details of operation.

Reagents.   Standard ferric ammonium sulphate solution.

Dissolve 30 grams of ferric ammonium sulphate in 300 cc. water acidified with 10 cc. of sulphuric acid; add potassium permanganate drop by drop as long as the pink color disappears, to oxidize any ferrous to ferric iron; finally dilute the solution to 1 liter.

Standardize this solution in terms of iron. The iron value multiplied by 1.4329 gives the value in titanic oxide (TiO_2); and the iron value multiplied by 0.86046 gives the value of the solution in terms of metallic titanium.

Indicator.   Saturated solution of potassium thiocyanate.

Reductor.   As a reductor a 500-cc. dispensing burette is used. The internal dimensions of the burette are $1\frac{5}{8}$ by 22 ins.

The reductor is charged with 1200 grams of 20-mesh amalgamated zinc, making a column about 12 ins. high and having an interstice volume of about 135 cc. This form of reductor is connveient, and when used as hereafter described

---

[1] G. E. F. Lundell and H. B. Knowles, Jour. Amer. Chem. Soc., Vol. 45, No. 11, page 2620 (1923).

is adapted to maintaining hot solutions, which facilitates complete reduction of the titanium.

The reductor is connected to a liter flask for receiving the reduced titanium solution, through a three-hole rubber stopper which carries also an inlet tube for carbon dioxide supply, and outlet tube for connecting with the suction pump.

Procedure. Determination of Titanium in Ferro-Carbon Titanium. One-half gram of sample is dissolved in a 6-in. porcelain evaporating dish in a mixture of 10 cc. water, 10 cc. sulphuric acid, 5 cc. of hydrochloric acid, 5 cc. of nitric acid.

The solution is evaporated to fumes of sulphuric anhydride; taken up by boiling with 50 cc. water and 10 cc. of hydrochloric acid; filtered and washed with hot water and hydrochloric acid.

The filtrate and washings should be about 100 cc. in volume.

The reductor is prepared for use by first passing through it a little hot dilute sulphuric acid followed by hot water, finally leaving sufficient hot water in the reductor to fill to the upper level of the zinc.

The hot titanic solution prepared as described above is now introduced, about 100 cc. of water being drawn from the reductor into the original beaker to bring the solution to about the upper level of the zinc. The water thus removed will not contain any titanium if the operation has been conducted as described; but it serves as a safeguard and is also *convenient* to acidify this water with 10 cc. of sulphuric and reserve it on the hot plate to be used as an acid wash after the reduction of the sample solution.

The titanium solution is allowed to remain in the reductor for ten minutes.

While the solution is being reduced the receiving flask is connected to the reductor and the air completely displaced by carbon dioxide, conveniently drawn from a cylinder of the liquefied gas.

When the reduction is complete the receiving flask is connected with the suction pump, and while still continuing the flow of carbon dioxide the reduced solution is drawn out, followed by the reserved acid wash and then three or four 100-cc. washes with hot water. The displacement of the sample solution and washing of the zinc is so regulated by means of the stop-cock that the reductor is always filled with solution or water to the upper level of the zinc.

When the washing is complete, gradually release the suction to prevent air being drawn back into the receiving flask.

Disconnect the flask, add 5 cc. of potassium thiocyanate solution as indicator and titrate immediately with standard ferric ammonium sulphate solution, adding the solution rapidly until a brownish color is produced which will remain for at least one minute.

The method is also well adapted for determining titanium in other titanium products, suitable means being employed for bringing the titanium into sulphuric acid solution.

The brown color developed at the end point indicates that an excess of ferric ammonium sulphate has been added and the depth of color is roughly proportional to such excess. Ordinarily the excess ferric ammonium sulphate amounts to about .05 cc. which is deducted from the burette reading before calculating the titanium, thus increasing the accuracy of the result. It is desirable for each operator to establish for himself, by running a few blanks, the proper amount to deduct from the burette reading as a correction.

Colorimetric Determination of Titanium with Hydrogen Peroxide

Preliminary Considerations. Hydrogen peroxide added to acid solutions of titanium produces a yellow to orange color, the depth of the color depending upon the amount of titanium present. Upon this fact the method is based. It is of especial value in determining small amounts of titanium, as it is possible to detect less than one part of the metal per hundred thousand parts of solution. Color comparisons can best be made on samples containing 0.05 to 5 milligrams of the element; larger amounts produce too deep a color for accurate comparison.

The following interferences should be made note of, e.g., molybdenum, vanadium and chromium also produce a color that would lead to error. Iron if present to the extent of 4% or over produces a color that must be allowed for; e.g., 0.1 gram Fe_2O_3 in 100 cc. of solution is equivalent to about 0.2 gm. of TiO_2 oxidized by H_2O_2 in 100 cc. of solution. Fluorides destroy the color, hence must be absent.[1] Phosphoric acid and alkali sulphates have a slight fading action,[2] hence must be allowed for by adding equivalent amounts to the standard if they are present in the sample. The addition of an excess of sulphuric acid partly counteracts the action of phosphates or alkali sulphates.[3] The color intensity is increased by increase of temperature, hence the standard and the sample

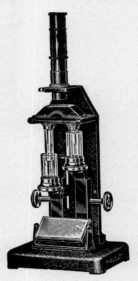

DIRECTIONS FOR THE USE OF A DUBOSCQ TYPE OF COLORIMETER

The mirror is turned so that the two halves of the field appear to be equally illuminated with the cups clean and empty. The solutions are then poured into the cups. The cup containing the standard solution is then lowered to a definite thickness of the standard solution between the bottom of the cup and the end of the plunger. With this movement the half of the field corresponding to the standard solution is seen to darken, while the other half remains luminous and colorless. If the cup containing the unknown solution is now moved in its turn, the two halves of the field are brought to the same intensity, after which the height at which the two liquid columns display this equal absorptive power is read by means of this scale. The proportion of coloring matter in two solutions is inversely proportional to the heights of the two columns necessary to obtain the same intensity of illumination, thus if the standard tube is set at 10 mm., and the solution under examination is the same intensity of color at 20 mm., the latter is just one-half the concentration of the standard. This is usually expressed by the formula:

Color of test solution ÷ Color of stand. sol. = Height of stand. sol.
                        ÷ Height of sol. to be tested.

If, therefore, the scale reading is 20 mm. for the standard, and 15 mm. for the solution to be tested, the formula reads: $\frac{20}{15} = 1.33$.

If, for example, the standard solution contains 4 ml. of coloring matter in 100 ml., the solution under test will be found to contain 4 x 1.33 = 5.32 ml. in 100 ml.

By courtesy of
Arthur H. Thomas Company,
Philadelphia, Pa.

FIG. 68b.—Colorimeter.

[1] W. F. Hillebrand, J. A. C. S., **17**, 718, 1895.   C. N., **72**, 158, 1895.
[2] P. Faber, Zeit. an. Chemie, **46**, 277, 1907.
[3] H. E. Merwin, A. J. S. (4), **28**, 119, 1909.

examined should have the same temperature.[1] Since metatitanic acid produces
no color with hydrogen peroxide, its formation must be prevented; the presence
of 5% of free H_2SO_4 accomplishes this.[2]

The procedure is very satisfactory for magnetic or other iron ores. It is
fully as accurate as the best gravimetric method and very much more rapid.

Solutions Required. *Standard Titanium Solution.* This may be prepared
by precipitations of TiO_2 from K_2TiF_6 according to the gravimetric procedure
and purification by solution and reprecipitation, the fluorine being first
removed by taking the compound to fumes with H_2SO_4 and then hydrolyzing
titanium with NH_4OH. The washed precipitate is ignited over a Méker flame
for fifteen minutes, cooled in a
desiccator and placed in tightly
stoppered bottle, since TiO_2 is
slightly hydroscopic.

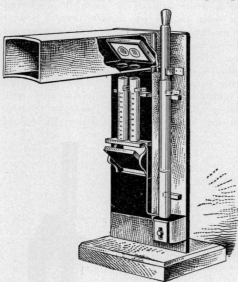

0.5 gram of TiO_2 is fused with
about twenty times its weight of
$KHSO_4$ in a platinum dish, keeping at fusion heat until the oxide
has dissolved. A high temperature is not advisable. The fusion
is dissolved in 5% sulphuric acid
by gently heating. The solution
washed into a 500-cc. graduated
flask is made up to volume with
5% H_2SO_4. One cc. contains
0.001 gram TiO_2, or 0.0006 gram
Ti.

Hydrogen Peroxide. Thirty
per cent solution. If this is not
available sodium peroxide dissolved in dilute sulphuric acid
will do.

Fig. 69.—Colorimeter.

Apparatus. Colorimeter—Fig.
69. Also see Fig. 43, page 283.

Preparation of the Sample. The solution of the sample having been
obtained by one of the procedures given under Preparation and Solution of the
Sample, the element may be determined according to the procedure given
below. If interfering substances are present, e.g., comparatively large amounts
of iron, or if tungsten, vanadium or chromium are present it will be necessary
to precipitate titanic acid by adding ammonium hydroxide to the boiling solution as directed under the gravimetric determination of the element. The
washed precipitate is dissolved in sulphuric acid.

Procedure. The sulphuric acid solution of titanium should contain 5% of
free sulphuric acid. It is poured from the beaker in which solution was effected
into a 100-cc. Nessler tube, 2 cc. of hydrogen peroxide, 30% solution are added
and the volume made up to 100 cc. with 5% sulphuric acid. The *standard* is
prepared by pouring 40 or 50 cc. of 5% sulphuric acid into a second 100-cc.
Nessler tube, adding 2 cc. of 30% hydrogen peroxide, H_2O_2, followed by sufficient

---

[1] Hillebrand.

[2] Dunnington, C. N., **64**, 302; J. A C. S., **12**, 210, 1891.

standard titanium solution to exactly match the sample and the solution made up to 100 cc. with 5% sulphuric acid. The titanium solution is added from a burette, noting the exact volume required. From this the percentage of titanium in the sample can readily be calculated. If iron is present in the sample, an equivalent amount should be added to the standard. If a colorimeter is used, a standard should be prepared which is deeper in color than the sample examined. The standard is poured into the comparison cylinder and the two tubes compared. By raising or lowering the plunger (see illustration) the standard solution is forced in or drawn out of the comparison tube. When the colors match, the cc. in the comparison tube will indicate the amount of TiO_2 present in the sample. The solution may be mixed by stirring with a platinum spiral.

Example. One-gram sample required 20 cc. of titanium standard solution, 1 cc. of which contained 0.001 gram TiO_2. Then the sample contains

$$\frac{0.001 \times 20 \times 100}{1} = 2\% \ TiO_2.$$

If the colorimeter has been used and 150 cc. of standard made by adding 30 cc. of standard titanium solution and it is found that the column of liquor in the standard comparison tube stands at 85 cc., the calculation would be as follows: 150 cc. contains 30×0.001 gram TiO_2, therefore 85 cc. are equivalent to $\frac{85 \times 0.03}{150} = 0.017$ gram TiO_2 per gram or 1.7%.

For the practical application of the colorimetric method in determining titanium in steel the following procedure is given.

NOTE. *Separation of Titanium from Iron.* J. H. Walton, Jr.[1] separates titanium from iron by fusing the finely powdered substance with three or four times as much sodium peroxide, and extracts the fusion with water. The filtrate contains the sodium pertitanate whereas the iron oxide remains on the filter paper. The filtrate is acidified with H_2SO_4 until 5% of free acid is obtained and the color of this solution compared with a standard obtained by fusing a known weight of TiO_2 with Na_2O_2 and extracting and treating with H_2SO_4 as in case of the sample.

Colorimetric Determination of Titanium in Steel Treated with Ferro=carbon Titanium [2]

The titanium in steel treated with ferro-carbon titanium exists in two conditions:

(1) Titanium *insoluble* in hydrochloric acid.
(2) Titanium *soluble* in hydrochloric acid.

Of the very small amount of titanium in treated steel the greater part will usually be found in the first form, and ordinarily the determination of titanium in this form answers every purpose of identifying and judging the quality of titanium-treated steel.

When the amount of titanium in the steel is exceedingly small, the soluble titanium frequently exceeds the insoluble and it then is sometimes desirable to determine also that existing in the second form.

[1] J. Am. Chem. Soc., **29**, 481, 1907.
[2] By L. E. Barton. Method of analysis recommended by the Titanium Alloy Manufacturing Company.

Reagents. *Peroxide Solution.* Dissolve 4 grams of sodium peroxide in 125 cc. dilute sulphuric acid (1 of acid to 3 of water), and dilute to 500 cc.

Concentrated Standard Titanium Solution. *Stock Solution.* One-fourth gram of a standard 20% carbonless ferro-titanium [1] is dissolved in 30 cc. dilute sulphuric acid (1 acid to 3 water). When solution is complete it is oxidized by the least possible quantity of concentrated nitric acid, boiled for a few minutes, cooled and diluted to such a volume that 1 cc. will contain 0.0005 gram of titanium.

When using a 5-gram sample 1 cc. is therefore equal to 0.01% titanium.

Dilute Standard Titanium Solution. This solution is made, just before making the determination, by diluting one volume of the concentrated standard titanium solution to ten volumes.

One cc. of this solution contains 0.00005 gram of titanium and is equal to 0.001% of titanium when using a 5-gram sample.

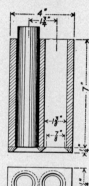

FIG. 70.

Apparatus. *Pipettes and Burettes.* The pipettes for measuring the concentrated standard solution and burette for delivering the dilute standard solution should be carefully calibrated.

Nessler Jars. These should be graduated with 50-cc. mark. It is convenient to have a set of four.

Colorimeter. The colorimeter or comparator consists of a rectangular block $2\frac{1}{4}$ by 4 by 7 ins. high—the height being about $\frac{3}{4}$ in. less than the height of Nessler jars—through which two chambers $1\frac{3}{8}$ ins. diameter and $1\frac{3}{4}$ ins. between centers are bored lengthwise—the chambers being of such diameter as to just receive the jars.

To one end of the block is fastened the base, which is $\frac{1}{4}$ in. thick and through which two $\frac{7}{8}$-in. holes are bored concentric with the chambers, thus forming a shoulder which supports the jars and also exclude light from the sides of the tubes. To prevent shadows and give better lighting the holes in the base are beveled outward at an angle of 45°. The construction will be apparent by reference to Fig. 70. The interior of the chamber is painted dead black.

(a) For Determination of Titanium Insoluble in Hydrochloric Acid

Procedure. Dissolve 5 grams of steel in 100 cc. of dilute hydrochloric acid (one of concentrated acid to two of water) by boiling gently. Wash off the cover and wash down the sides of the beaker with water and filter out the slight insoluble residue, washing with hot water and dilute hydrochloric acid until free from iron. For filtration it is advisable to use either a close-grained paper or double rapid-filtering papers such as S. & S. No. 589 white ribbon.

Ignite the residue *gently* in a platinum crucible to burn off carbonaceous matter. Treat the residue in the crucible with a mixture of 3 cc. dilute sulphuric acid (1 : 1), 2 or 3 cc. hydrofluoric acid, and a few drops of nitric acid.

---

[1] Ferro-titanium suitable for the preparation of standard titanium solutions is made and supplied by the Titanium Alloy Manufacturing Company, Niagara Falls, N. Y.

Heat and evaporate to fumes of sulphuric anhydride to complete expulsion of hydrofluoric acid.

Cool, add a few cc. of water and heat until the solution is perfectly clear. The ignited residue may also be rapidly and completely brought into solution by fusion with about 3 grams of potassium bisulphate and dissolving the fusion in water and sulphuric acid.

In either case wash the contents of the crucible into one of a pair of Nessler jars and dilute with cold water nearly to the 50-cc. mark, and in the other jar place an equal volume of distilled water.

Place the jars in the colorimeter and observe if the sample solution is colorless. If the sample solution is colored slightly yellow by iron, the water in the standard tube should be brought to the same color by addition of a few drops of a ferric solution. For this purpose a solution of ferric ammonium sulphate, 30 grams per liter, is very convenient.

If the work up to this point has been carefully performed, the addition of ferric solution will usually be unnecessary; and if more than a few drops of ferric solution are required the analysis should be rejected and a new sample started. After adjusting the color—if necessary—bring the volume of solution in both jars to the 50-cc. mark.

The volumes now being equal and the solutions practically colorless, add 2 cc. of the peroxide solution to each. If the sample contains titanium even in minute quantity it will be indicated by the immediate development of a yellow color.

Match the colors by running into the standard jar freshly prepared *dilute standard* titanium solution, keeping the volumes equal by adding an equal quantity of water to the sample, placing the jars in the colorimeter for comparison of colors.

As before stated, each cc. of the dilute standard solution is equal to 0.001% titanium when using a 5-gram sample.

The determination may be made in less than an hour and requires little attention.

(b) For Determination of Titanium Soluble in Hydrochloric Acid

For the determination of soluble titanium the filtrate from the insoluble titanium residue obtained as before described may conveniently be used.

Dilute the solution in which the iron is already in the ferrous state to 180 cc. Add 10 cc. of alum solution made by dissolving 40 grams of crystallized alum in a liter of water.

The aluminum here added is subsequently precipitated as alumina with the titanium and serves to collect quickly the exceedingly small precipitate of titanium hydroxide and facilitate its separation from the solution by filtration.

Heat the solution to about 90° C. and add ammonia or ammonium carbonate solution, stirring constantly until a slight permanent precipitate is produced. Add dilute hydrochloric acid (1 to 1) drop by drop from the wash bottle until the precipitate is just redissolved and the solution perfectly clear; then add 1 cc. more of the dilute hydrochloric acid.

Add 3 cc. of phenylhydrazine dissolved in 10 cc. hot water, which will precipitate the titanium and aluminum. Stir thoroughly and filter immediately on a 7-cm. filter paper in a Büchner funnel, using suction. Wash thoroughly with hot water.

Calcine the precipitate *gently* in a platinum crucible to destroy organic matter and dissolve the residue exactly as described under (*a*), except that 6 cc. of dilute sulphuric acid is used instead of 3 cc.

The solution, which has a very light yellow, or greenish-yellow color, is transferred to one of a pair of Nessler jars and diluted to the 50-cc. mark. About 40 cc. of water are placed in the other jar and the color of the sample solution exactly matched by addition of ferric ammonium sulphate and copper sulphate solutions, which are conveniently delivered from burettes.

For matching the original color of the solution nearly saturated solutions of ferric ammonium sulphate and copper sulphate are suitable.

Only a few drops of such solutions are required, but it is frequently necessary to use both blue and yellow to match the greenish-yellow tone of the sample solution.

The standard is finally diluted to the 50-cc. mark. The volumes now being equal and identical in color, add to each 2 cc. peroxide solution to develop the titanium color and finish the determination as before described under (*a*).

(c) For Determination of Total Titanium

The total titanium is given by the sum of the insoluble and soluble titanium determined as under (*a*) and (*b*); but if desired may be determined in one operation.

To determine total titanium, dissolve as before in hydrochloric acid and without filtering proceed as directed under (*b*) for determination of soluble titanium.

Determination of Titanium when Interfering Elements are Present

If chromium, vanadium or molybdenum is present in the steel, fuse the residue insoluble in hydrochloric acid or the calcined phenylhydrazine precipitate containing the interfering element with a mixture of sodium carbonate and a little sodium nitrate.

Dissolve the fusion in water and filter. The residue on the filter will contain the titanium, free from interfering element. Bring the residue into sulphuric acid solution by methods before described and determine the titanium as usual.

Colorimetric Determination of Titanium with Thymol [1]

Principal and Preliminary Considerations. Titanium dioxide dissolved in sulphuric acid is colored red by addition of thymol, the depth of color being directly proportional to the amount of titanium present. The intensity of the color is claimed by Lenher and Crawford to be twenty-five times that produced by hydrogen peroxide with the same amount of titanium.

As in case of hydrogen peroxide, fluorides destroy the color, hence must be absent. Dilution with water has no effect until the concentration of sulphuric acid falls below 79.4 (e.g., sp.gr. 1.725). The color then fades in direct propor-

[1] Victor Lenher and W. G. Crawford, C. N., **107**, 152, March 28th, 1913.

tion to dilution. Warm solutions are lighter in color than cold solutions with the same amount of titanium, hence the standard and the sample compared must have the same temperature. The color fades on heating but returns on cooling. The temperature should be kept below 100° C. Chlorides, phosphates and tin seem to have no effect. Tungsten, WoO_3, interferes, as it intensifies the color of the solution in direct proportion to the amount present; hence it must be removed or allowance made by adding an equivalent amount to the standard or subtracting the equivalent blank.

Special Reagents. *Thymol Solution 1%.* The thymol is dissolved in a little glacial acetic acid containing 10% ethyl alcohol, and this solution added to concentrated sulphuric acid. Addition of the thymol directly to the acid would produce a colored solution. The reagent should be kept protected from strong light, otherwise it will become colored.

Apparatus. See Colorimetric Determination of Titanium with Hydrogen Peroxide, Figs. 69, 70, also Fig. 43.

Procedure. About 0.3 gram of the material is fused with potassium acid sulphate, $KHSO_4$, and the melt dissolved in concentrated sulphuric acid. Enough thymol reagent is added so that there is present at least 0.006 gram thymol for every 0.0001 gram TiO_2. Concentrated sulphuric acid is added to bring up the volume to 50 or 100 cc. in a Nessler tube exactly as in the case of the colorimetric determination of titanium with H_2O_2. The depth of color is compared with a standard solution of titanium dissolved in a concentrated sulphuric acid added to 5 cc. of thymol solution made up to a convenient volume with concentrated sulphuric acid. The procedure is the same as described in the H_2O_2 method.

THE ANALYSIS OF TITANIFEROUS ORES [1]

Determination of Titanium

Decompose the ore by fusion with potassium bisulphate, dissolving the fusion in water, hydrochloric and sulphuric acids. If an insoluble residue remains, filter it out. Calcine the residue, add a few drops of sulphuric acid and sufficient hydrofluoric acid to dissolve silica, evaporate to fumes of sulphuric anhydride and then heat to redness.

If a residue now remains, bring it into solution directly in acids or fuse with a little potassium bisulphate, etc., finally adding the solution to the main solution obtained as before described.

If desired, the sample of ore can first be partially dissolved in hydrochloric and sulphuric acids, and the insoluble residue then fused with potassium bisulphate or treated with sulphuric and hydrofluoric acids.

Some ores may be completely decomposed by a mixture of nitric, hydrofluoric and sulphuric acids, evaporating to fumes of sulphuric anhydride in a platinum dish to free the solution from nitric and hydrofluoric acids.

The complete decomposition of the sample having been accomplished, the titanium in the solution is determined by either the gravimetric or volumetric methods for Determination of Titanium in Ferro-Carbon Titanium. Pages 542 and 547.

[1] Method of Analysis used in the laboratories of The Titanium Alloy Manufacturing Company.

Determination of Iron in Presence of Titanium

The sample is decomposed as directed under the Determination of Titanium. The sulphuric acid solution, which should have a volume of 150 to 200 cc., is saturated with hydrogen sulphide gas to reduce the iron, and filtered to separate any precipitated sulphides and free sulphur. The filtrate is collected in a flask fitted with a rubber stopper through which pass two glass tubes, one reaching nearly to the bottom for conducting gas into the solution, the other a short exit tube. Unless the solution after filtration is still highly charged with hydrogen sulphide, more gas should be passed into the solution to reduce any iron that may have been oxidized by the atmosphere during filtration. The excess hydrogen sulphide is now expelled by boiling the solution while passing a current of carbon dioxide.

When the exit gases cease to darken a piece of filter paper moistened with lead acetate solution, the flask is cooled while still passing the carbon dioxide. When the flask has partially cooled the carbon dioxide is shut off and the flask quickly cooled in running water and immediately titrated with standard permanganate solution.

Determination of Silica

This determination is conveniently combined with the determination of iron, the ore being preferably decomposed by fusion with potassium bisulphate. The fusion is dissolved and evaporated with excess sulphuric acid to fumes of sulphuric anhydride and the silica determination finished as usual—weighing, volatilizing with hydrofluoric acid, etc. If the ore contains quartz or a silicate undecomposable by treatment with potassium bisulphate and hydrofluoric acid, the residue filtered from the sulphuric acid solution should be fused with sodium carbonate and the silica then determined as usual.

Determination of Alumina

After making determination or separation of titanium by *gravimetric* method use the filtrate for determination of alumina.

Phenylhydrazine Method for Determination of Aluminum in Presence of Iron

The iron and aluminum should be in hydrochloric or sulphuric acid solution. Nearly neutralize the solution with ammonium carbonate. Pass sulphurous acid gas to complete reduction of the iron. Boil until the excess sulphurous acid is driven off and if titanic acid separates filter it out.

After filtering out titanic acid again nearly neutralize with ammonium carbonate, pass a little sulphurous acid gas and heat for a few minutes to reduce any iron that might have been oxidized during filtration. If titanium has not been detected the second treatment with sulphurous acid may be omitted. In either case the solution still containing a little free sulphurous acid is nearly neutralized with ammonium carbonate, diluted to 300 cc. and 3 cc. of phenolhydrazine added. Stir thoroughly, let settle and filter out the alumina. If the precipitate is discolored by iron, dissolve in hydrochloric acid, and repeat the reduction, neutralization and precipitation by phenylhydrazine. Ignite and weigh $Al_2O_3+P_2O_5$. Since the alumina precipitate may be contaminated by phosphoric anhydride (P_2O_5), determine it by analysis and correct the alumina determination accordingly.

Determination of Phosphorus

Phosphoric acid may be separated from titanic acid by repeatedly fusing the ore with alkali carbonate and extraction of alkali phosphate with water.

The determinations of other constituents of the ore are conducted by the usual methods of ore analysis.

ANALYSIS OF MIXED PIGMENTS CONTAINING TITANIC OXIDE

Weigh one gram sample into a 400-cc. Pyrex glass beaker, add 10 grams sodium sulphate and 40 cc. concentrated sulphuric acid (93%). Heat on hot plate for one-half hour and then increase the heat, as by placing the beaker directly over the coils of an electric hot plate and boiling for about 10 minutes. The solution should acquire a temperature of about 335° C.

Cool, dilute the solution to 300 cc., boil 20 minutes, filter while hot and wash residue and precipitate with 5% sulphuric acid. On the filter will be silica and undecomposed silicates and all the lead and barium as sulphates. This residue and precipitates can be analyzed by well-known methods if desired. The filtrate will contain the titanium, iron, aluminum, zinc and calcium.

To the filtrate while still hot add an excess of ammonia, filter and wash precipitate with hot water. Re-dissolve precipitate in hydrochloric acid (1 to 1) and again precipitate with ammonia. Filter and wash with hot water, combining filtrate with that from first separation. By this procedure the titanium, iron and aluminum will have been separated from the zinc and calcium. The use of an excess of ammonia as described would tend to carry a little aluminum into solution but in the presence of titanium and iron substantially all the aluminum will be found in the precipitate. If desired, the filtrate can be tested for presence of aluminum and then used for determination of calcium and zinc.

The precipitate of titanium, iron and aluminum hydroxide is again re-dissolved in dilute hydrochloric acid and the titanium separated and determined as before described and with all precautions given under the head of "Gravimetric Method for the Determination of Titanic Oxide."

The filtrate after separation of the titanium can be used for determination of iron and aluminum if desired.

In case it is not desired to determine iron and aluminum separately, the hydroxides obtained after second separation by ammonia can be calcined and weighed as total oxides of titanium, iron and aluminum. The titanium can be determined by the volumetric method before given and deducted from total oxides thus giving the iron and aluminum together by difference.

TUNGSTEN, TANTALUM AND COLUMBIUM [1]

TUNGSTEN

W., *at.wt.* **184.0;** *sp.gr.* **18.77;** *m.p.* **3000° C.;** *oxides,* **WO$_2$** *(brown)*; **WO$_3$** *(yellow)*; *acids,* **H$_2$WO$_4$,** *ortho tungstic;* **H$_2$W$_4$O$_{13}$,** *meta tungstic*

DETECTION

Minerals. The finely powdered material is decomposed by treating with mixed acids according to the procedure given on page 560. Tungsten is precipitated with cinchonine, the precipitate filtered off and dissolved in ammonium hydroxide, then acidified with hydrochloric acid and reprecipitated with cinchonine as described.

Tungsten oxide may be confirmed as follows:

1. The residue is suspended in dilute hydrochloric acid and a piece of zinc, aluminum, or tin placed in the solution. In the presence of tungsten a blue-colored solution or precipitate is seen, the color disappearing upon dilution with water.

2. A portion of the precipitate is warmed with ammonium hydroxide and the extracts absorbed with strips of filter paper.

(*a*) A strip of this treated paper is moistened with dilute hydrochloric acid and warmed. In the presence of tungstic acid a yellow coloration is produced.

(*b*) A second strip of paper is moistened with a solution of stannous chloride. A blue color is produced in the presence of tungsten.

(*c*) A third strip dipped into cold ammonium sulphide remains unchanged until warmed, when the paper turns green or blue if tungsten is present.

Iron, Steel and Alloys. These decomposed with strong hydrochloric acid followed by nitric acid as directed under Solution of the Sample leave a yellow residue in the presence of tungsten. If this residue is digested with warm ammonium hydroxide and the extract evaporated to dryness a yellow compound, WO$_3$, will remain if tungsten is present. This oxide may be reduced in the reducing flame to the blue-colored oxide.

---

[1] Columbium is also known as Niobium.

Chapter by Wilfred W. Scott, A. M. Smoot and J. A. Holladay.

ESTIMATION

Tungsten occurs principally as wolfram, a tungstate of iron and manganese ($FeWO_4 \cdot MnO_4$), as scheelite, a tungstate of calcium ($CaWO_4$), as ferberite, $FeO \cdot WO_3$ and hubernite, $MnO \cdot WO_3$. The best concentrate of hand-picked material contains 70 to 74% tungsten in terms of its oxide, WO_3.

The element is met with in alloys—ferro-tungsten,[1] silico-tungsten, tungsten steels containing as much as 10 to 20% of the metal, used for making high-speed, self-hardening cutting tools; tungsten powder;[1] alkali tungstates for mordanting purposes; tungstic oxide, WO_3; tungsten electric light filaments, etc.

Solution of the Sample

For solution of the sample the following facts should be kept in mind regarding solubilities.

The metal is practically insoluble in HCl and in H_2SO_4. It is slowly attacked by HNO_3, aqua regia and by alkalies. It is readily soluble in a mixture of HNO_3 and HF ($= WF_6$ or WOF_4).

Oxides. WO_2 is soluble in hot HCl and in hot H_2SO_4 ($=$ red sol.), also in KOH (red sol.). The oxide WO_3 is scarcely soluble in acids, but is readily soluble in KOH, K_2CO_3, NH_4OH, $(NH_4)_2CO_3$, $(NH_4)_2S_x$. Both the acid and the alkali solutions deposit the blue oxide on standing.

Acids. Ortho tungstates. A few are soluble in water and in acids. The alkali salts only slightly soluble. The meta tungstates are easily soluble in water. Tungstates are precipitated from alkali salts by dilute H_2SO_4, HCl, HNO_3, H_3PO_4 (aqua) as yellow $WO_3 \cdot H_2O$ or white $WO_3 \cdot 2H_2O$. Meta tungstates are not precipitated by cold acids, but are precipitated by boiling and by long standing.

Solution of Minerals. The material is best decomposed by acid treatment as described on page 560. Use of a fusion as a means of decomposition of tungsten ores preliminary to either the qualitative detection or the quantitative determination of tungsten cannot be recommended. The precipitation of tungsten by boiling with acids in presence of considerable amounts of *alkali salts* (such as result from acidification of a fusion) is absolutely worthless inasmuch as large amounts of tungsten always remain in solution. Repeated evaporations do not improve matters. In fact, when the amount of tungsten present is small, and especially if the ore contains much phosphorus, there is small likelihood that any of the tungsten will be precipitated. The use of cinchonine is necessary in order completely to precipitate tungsten under these conditions.

[1] TYPICAL ANALYSES

	TUNGSTEN POWDER	FERRO TUNGSTEN
	%	%
W,	97 to 98.7	71 to 85.5
Fe,	.5 to .6+	14 to 24.5
C,	.1 to .3+	.4 to 2.6
Si,	.3 to .7+	.1 to .4
Mn,	0 to .2	.08 to .9+
P,	— —	.008 to .02
S,	— —	.01 to .02
Al,	.2 to .5	.001 to .07
Cu,	— —	0 to .008
Mg,	0 to .3+	

SEPARATIONS

Separation of Tungsten from Silica. The oxide of tungsten, as ordinarily obtained, is frequently contaminated with silica. The removal of silica is accomplished by heating the mixture in a platinum dish with sulphuric and hydrofluoric acids and volatilizing the silica. After taking to dryness and igniting gently, the last traces of sulphuric acid are expelled by adding ammonium carbonate and again igniting.

In presence of small amounts of silica (0.1 to 0.2%) and large amounts of tungsten (75 to 85%) J. A. Holladay recommends evaporation with sulphuric and phosphoric acids, filtration to remove the bulk of the tungsten, and subsequent ignition and voltailization with sulphuric and hydrofluoric acids.

Separation from Tin. The weighed residue is mixed with six to eight times its weight of ammonium chloride (free from non-volatile residue) in a platinum crucible, placed in a larger crucible, both vessels being covered. Heat is applied until no more vapors of ammonium chloride are evolved. Additional ammonium chloride is added and the treatment is repeated three times. The fourth treatment is followed by weighing of the residue and the treatment repeated once more. If no further loss of weight takes place it is assumed that all the stannic oxide has been driven off. The inner crucible is now placed directly over the flame and heated to dull redness for a few minutes and the oxide, WO_3, weighed.

Separation of Tungsten from Tin and Antimony. Talbot's Process.[1] The mixed oxides are fused with twelve times their weight of potassium cyanide in a porcelain crucible. Tin and antimony are thrown out as metals and the soluble alkali tungstate formed. This is leached out with water and the aqueous extract boiled (hood) with an excess of nitric acid to drive off the cyanogen compounds. The tungstate is then precipitated by the usual methods. If phosphorus is present in the sample it will be found in the solution with tungsten and its removal will be necessary.

Separation of Tungsten from Arsenic and Phosphorus. Both arsenic and phosphorus may be precipitated by cold magnesia mixture in an ammoniacal solution, tungsten remaining in solution. The separation of arsenic is difficult, as it is tenaciously retained by tungsten as a complex salt. The following process is outlined by Kehrmann.[2]

One to 2 grams of the sample are fused with twice as much sodium hydroxide as is required to combine with the arsenic oxide, the resulting cake is dissolved in a little water and boiled in an Erlenmeyer flask for half an hour. After cooling, three times as much ammonium chloride as is needed to form chlorides with the alkalies present is added, and then ammonium hydroxide equal to one-fourth the volume of the solution under investigation, followed by sufficient magnesia mixture, added cold, drop by drop with constant stirring. After settling several hours, the solution is filtered and the residue washed with a weak solution of ammonia and ammonium nitrate. It is advisable to dissolve the residue in dilute acid and repeat the precipitation several times. The filtrates containing the tungsten are combined and concentrated by evaporation if necessary.

[1] J. A. Talbot, J. Sci. (2), **50**, 244, 1870.
[2] F. Kehrmann, Ber., **20**, 1813, 1887.

Separation of Tungsten from Molybdenum, Hommel's Process. The moist oxides of tungsten and molybdenum are digested with concentrated sulphuric acid and a few drops of dilute nitric acid, in a porcelain dish over a free flame for about half an hour. About three times its volume of water is added to the cooled solution, the residue, WO_3, filtered off and washed with dilute sulphuric acid (1 : 20) followed by three washings with alcohol. The residue is ignited separately from the paper and weighed with the ash of paper as WO_3.

Molybdenum is in the filtrate and may be precipitated in a pressure flask with H_2S.

Volatilization of Molybdenum with Dry Hydrochloric Acid Gas. Péchard's Process.[1] The procedure depends upon the fact that molybdenum oxide heated in a current of dry hydrochloric acid gas at 250 to 270° C. is sublimed, whereas tungsten is not affected.

The oxides of the two elements, or their sodium salts, are placed in a porcelain boat and heated in a hard glass tube, one end of which is bent vertically downward and connected with a Péligot tube containing a little water. A current of dry hydrochloric acid gas is conducted over the material, heated to 250 to 270° C. From time to time the sublimate of molybdenum ($MoO_3 \cdot 2HCl$) is driven towards the Péligot tube by careful heating with a free flame. This enables the analyst to observe whether any more sublimate is driven out of the sample and to ascertain when the tungsten is freed of molybdenum. From one and a half to two hours are generally sufficient to accomplish the separation. If sodium salt is present it is leached out of the residue, and this is then ignited to WO_3. Molybdenum may be determined in the sublimate.

Separation from Vanadium.[2] Tungstic and vanadic acids are precipitated with $HgNO_3$ and HgO, the moist precipitate dissolved in HCl and the solution largely diluted; WO_3 is precipitated free from vanadium.

Separation from Titanium.[3] The material is heated with K_2CO_3 and KNO_3, tungsten is dissolved out with water and precipitated as mercurous tungstate.

Separation of Tungsten from Iron. The procedure is given under Solution of the Sample, of Steel and Alloys. The impure oxide WO_3 is fused with Na_2CO_3 and the melt extracted with water. $Fe(OH)_3$ remains on the filter. The filtrate is evaporated to dryness with HNO_3 and the residue extracted with water. The insoluble WO_3 is washed with dilute NH_4NO_3 solution, then dissolved in NH_4OH and tungsten determined in the solution.

Separation of Tungsten from Uranium.[4] The sample is evaporated with nitric acid to near dryness, 5 cc. HNO_3 is added and the uranium is dissolved out by extraction with ether.

[1] E. Pechard, Comp. Rend., **114**, 173, 1891.
[2] Friedheim, C. N., **61**, 220.
[3] Defacqz, C. N., **74**, 293.
[4] C. A. Pierlé, Jour. Ind. Eng. Chem., **12**, 61–63, 1920.

GRAVIMETRIC PROCEDURES FOR DETERMINING TUNGSTEN

Since there is no highly commendable volumetric procedure for determining tungsten, the gravimetric methods are preferred.

The element is determined as tungstic oxide, WO_3. It may be isolated preferably by precipitation with cinchonine, or in the form of tungstic acid, ammonium tungstate, or as mercurous tungstate, in the usual course of analysis, all of which forms may be readily changed by ignition to the relatively non-volatile oxide, WO_3.

Gravimetric Determination of Tungsten in Steel and Alloys

Tungsten alloys may be decomposed by hydrochloric (or sulphuric) and nitric acids. Tungsten precipitates, carrying down chromium and a little iron. The bulk of the iron is filtered off and tungsten determined in the residue, by direct or difference methods.

Special Reagent. Cinchonine Solution. 100 grams of the alkaloid dissolved in dilute (1 : 3) HCl and made to 1000 cc. with the dilute acid. *Wash Solution.* 30 cc. of the above solution, with 30 cc. strong HCl diluted to 1000 cc.

Procedure

1. **Decomposition of the Sample.** Two grams of high tungsten alloys or 5 grams of alloys low in tungsten are dissolved in a 350-cc. beaker by addition of 20 cc. strong HCl, the beaker being placed over an asbestos mat on a hot plate or steam bath. (The temperature should be below the boiling point of HCl, since the acid should remain strong to effect decomposition.) Since the reaction is energetic the beaker is covered by a watch glass. Ten cc. of strong nitric acid are added, pouring small portions at a time through the lip of the beaker. The sample is digested until all the black particles have dissolved and only the fine greenish yellow tungstic oxide is evident. Agitation of the liquor to prevent caking assists the reaction.

2. When the sample is decomposed, the beaker is uncovered and the solution evaporated rapidly to about 20 cc. and then cautiously at low heat to about 5–10 cc. 5 cc. of HCl are added to the moist residue and the solution diluted to about 50 cc. and heated to boiling 2–3 cc. of cinchonine reagent are added. Avoid taking to dryness and baking as this will make the tungsten difficult to dissolve in NH_4OH. See step 4.

3. The mixture is filtered and the ferric chloride washed out from the tungstic residue with dilute HCl (1 : 10), *i.e.*, until the wash solution passes through the filter colorless. The filtrate is tested for tungsten by addition of more cinchonine. (See step 7.) As small amounts of tungsten come down slowly at least an hour should be given for the test.

4. A clean beaker is placed under the filter and tungsten dissolved out from the impure residue by treating this with warm ammonia water (1 : 5), the filter being half filled with each washing and thoroughly drained. It is advisable to rinse out the beaker in which the decomposition was made, with warm ammonia solution and pour this on the filter. After five washings the filter is cautiously removed and spread out on a watch glass. The residue is washed into a beaker with a stream of ammonia water and digested with about 20–30 cc. of ammonia for about 5 minutes warming to near boiling. The solution is poured through a

fresh filter into the main extract, the filter is drained and then washed twice more with warm ammonia solution (1 : 5). All the tungsten should now be in the filtrate. If much residue remains as in case of high silica samples, tungsten is apt to be present. This is recovered by acid sulphate fusion. See Notes.

5. The filtrate is boiled to expel the ammonia. When the odor has become faint, HCl is added until the solution is just acid and about 3 cc. excess is added. The total volume will be 75–100 cc., after boiling. Acidification may cause some tungsten to precipitate, but this does no harm. The ammonia is generally expelled by boiling the filtrate down to half its original volume.

6. 10 cc. of cinchonine reagent are added to the hot solution and the mixture stirred and allowed to settle until cold. If the supernatant solution is cloudy stirring up the precipitate and allowing it to settle will clear the solution. It is well to test the clear liquor with a few drops more of the reagent to ascertain whether all of the tungsten has precipitated.

7. The solution is filtered and the tungsten residue washed three or four times with dilute cinchonine reagent (washing down from the rim of the filter), and once with 5–10 cc. of water.

NOTE. The filtrate from step 3 may contain some tungsten. If the filtrate treated with cinchonine has become cloudy on standing it must be filtered, the residue washed with cinchonine reagent as in the procedure above and the filter and residue added to that containing the bulk of the tungsten.

8. The filter (or filters) is ignited in a weighed crucible to destroy the carbon, then cooled and weighed as WO_3.

$$WO_3 \times 0.7931 = W$$

NOTE. The oxide WO_3 fused with $KHSO_4$ and the cooled melt extracted with ammonium carbonate reagent should give a clear solution. If the solution is cloudy SiO_2 is indicated. This should be filtered off, ignited and its weight subtracted from the WO_3 obtained above. See optional method below.

The $KHSO_4$ fusion may be made in a porcelain or platinum crucible.

Optional Method

The impure residue obtained in step 3 of the first procedure is ignited and weighed. The residue is now fused with acid potassium sulphate, $KHSO_4$ (5–6 grams of salt). The heating is continued until effervescence ceases and the mass changes to a clear solution. The fusion is cooled by rotating the crucible so that the mass is spread in a layer over the sides. It is now dissolved by placing in about 100 cc. of ammonium carbonate solution (10% sol.) and heating to boiling. Iron, chromium and silica will remain insoluble, tungsten will be in solution. The liquor is filtered and washed 4 or 5 times with water containing a little $(NH_4)_2CO_3$. The residue is ignited, cooled and weighed. The difference of this weight and that of the impure residue is due to the WO_3.

NOTE. If the carbonate extract is boiled, acidified and cinchonine reagent added tungsten will precipitate as the cinchonine compound. See Notes.

Notes

The use of sulphuric acid in place of hydrochloric acid in the decomposition of the sample offers no advantages as to speed of decomposition. On the other hand with sulphuric acid treatment tungsten is invariably found in the filtrate (step 3) while with the HCl method it is seldom found in appreciable amounts.

Hydrofluoric acid assists in the decomposition of alloys high in silicon. A large excess must be avoided as this would attack the glass of the beaker with liberation of silicic acid, and fluosilicic acid. With large amount of HF a platinum dish should be used.

WO_3 does not dissolve readily in ammonia after it has been ignited. The moist residue of tungsten precipitated by acids is easily soluble in ammonia. If an appreciable amount of residue remains from the ammonia extraction, it is fused with $KHSO_4$, extracted with a 10% solution of $(NH_4)_2CO_3$, the CO_2 expelled by boiling and the ammoniacal solution filtered from silica, vanadium, iron, etc., and the tungsten precipitated and determined according to the procedure outlined in steps 5 to 8 inclusive.

SiO_2 is but slightly soluble in dilute ammonia solution. The silica dissolved by 100 cc., 1 : 1 NH_4OH in a 30 hour treatment on a 47% SiO_2 ore amounted to less than 0.005 g. on a gram sample. With a 30 minute treatment the amount was inappreciable. In case of doubt treat the WO_3 with HF in a platinum crucible and again ignite after expelling the acid by evaporation.

THE ANALYSIS OF TUNGSTEN ORES, CONCENTRATES AND METALLIC PRODUCTS[1]

General Considerations

Commercial tungsten minerals are marketed usually in the form of concentrates derived from treatment of lean ores by water concentration, or, sometimes, as specially selected high grade ore which has not been subjected to dressing operations other than hand sorting.

The important commercial minerals are Wolframite—iron-manganese tungstate; Hubnerite—manganese tungstate; Ferberite—iron tungstate; Scheelite—calcium tungstate. The first three grade into each other imperceptibly; commercially, they are all classed as Wolframite although the predominant mineral may, in some cases, be either manganese tungstate or iron tungstate. The determinations usually required are tungstic oxide, manganese, tin, phosphorus, sulphur, copper and arsenic sometimes molybdenum, which occurs more or less frequently in Scheelite, and occasionally bismuth and lead.

The principal minerals associated with tungsten concentrates are pyrite, arsenopyrites, cassiterite, magnetite, columbite, and always more or less quartz and silicates. Scheelite sometimes is associated with baryte.

The following analytical methods have been in practical use for a long time; they have been modified from time to time as experience showed necessary. Many suggestions from Works' chemists and others have been adopted. The descriptions are given in detail where this has been thought to be necessary because much depends upon minor points and upon the experience gained in making many hundreds of determinations.

Methods for the analysis of tungsten powders and ferro-tungsten are also given, since these are the raw materials of manufacturers of special steels.

Gravimetric Determination of Tungsten in Ores and Concentrates

Special Reagents

Cinchonine Solution. Cinchonine solution is made by dissolving 100 grams of the alkaloid in dilute HCl (1 part acid to 3 of water) and diluting to 1000 cc. with HCl of the same strength.

Cinchonine Wash Solution. 30 cc. Cinchonine solution, 30 cc. strong HCl to 1000 cc. of water.

Preparation of the Sample

Ores should be ground in agate to pass a 200-mesh screen; double screening is recommended to insure perfect fineness.

Weigh one gram into a 350 cc. beaker, add 5 cc. of water and shake to spread the ore evenly over the bottom of the beaker. Add 100 cc. of strong HCl, cover the beaker and set it to warm gently for an hour. The temperature should not exceed 60° C.—higher heating expels HCl gas. The ore is slowly decomposed and most of the tungsten is held in solution by the excess of HCl, leaving the undecomposed ore exposed to further attack by the HCl.

Stir the solution with a glass rod once or twice during this digestion to prevent

[1] By A. M. Smoot.

the formation of crusts or cakes on the bottom.   The glass rod may be left in the beaker.

After an hour increase the heat and boil until the solution is evaporated one-half or thereabouts.   The cover may be removed after boiling begins, or better, it may be supported over the beaker on three glass hooks.   After the liquid is reduced in volume to 50 or 55 cc., scrape the bottom of the beaker thoroughly with a glass rod to detach *all* caked ore and residue—this is very important, otherwise complete decomposition by the acid treatment is difficult if not impossible to accomplish. Add 40 cc. more strong HCl and 15 cc. HNO₃, replace the cover and boil until all danger of spattering (owing to the rapid expulsion of chlorine) is past, remove the cover and evaporate to a volume of 50 cc., then add 5 cc., more HNO₃, replace cover and continue boiling, finally remove cover and evaporate to a volume of 15 cc. or less.   All this boiling and evaporation requires about an hour.   Occasional stirring to break up crusts, especially when fresh additions of acid are made, is recommended.

Add 200 cc. of hot water to the concentrated solution, stir well and simmer gently just at the boiling point for half an hour.   Nearly all the tungstic acid is separated after the addition of nitric acid, and during the subsequent simmering and boiling after dilution, but a little may remain in solution.   Add 6 cc. of cinchonine solution, stir well and let stand for half an hour, or longer if convenient, thus precipitating all tungsten.

This method of attack is preferable to that formerly employed wherein less acid was used and HNO₃ was added earlier in the digestion.   Most ores are completely decomposed, so far as tungsten minerals are concerned, but it is unsafe to assume that this is the case unless the residue is pure white silica.

After the tungstic acid residue has settled well and has stood for half an hour, filter the solution through a 9 cm. paper, using a little paper pulp in the filter. Wash the residue well, first by decantation in the beaker and afterward on the paper with a dilute solution of cinchonine and HCl.

It is unnecessary to detach the tungstic acid which adheres to the sides of the beaker, but washing should be thorough to remove all iron, manganese, lime, etc., from the residue and filter paper.   Finally wash both beaker and filter *once* with cold water to displace most of the dilute cinchonine washing solution.

Procedure for Isolation of Tungsten

Wash the tungstic acid and residue back from the filter into the original beaker with a fine jet of water from a wash bottle—the residue washes out easily because of the paper pulp which prevents adherence—about 25 cc. of water should be used. Add 6 cc. of strong NH₄OH, cover the beaker and warm it gently for about ten minutes.   Tungstic acid dissolves readily, stir well and wash down the sides of the beaker with dilute ammonia, make sure that all the yellow tungstic acid has dissolved, then filter the warm solution through the same filter paper that was used at first, thus dissolving the small amount of tungstic acid that adhered thereto. Collect the filtrate in a 400 cc. beaker, wash the original beaker and the filter paper thoroughly with dilute ammonia (1 part strong NH₄OH—9 parts H₂O).   The filtrate should be clear, or at most only slightly cloudy.   The addition of about one gram of NH₄Cl to the solution before filtering and the use of a little of the same salt in the wash solution will insure a perfectly clear filtrate; but the use of this salt is not recommended unless the silicious residue tends to pass the filter in large amount.   The residue insoluble in ammonia will usually be free from tungsten; it may consist of tin oxide (cassiterite) silica and undecomposed silicates, titanium minerals or columbite; to make sure, it must be fused as described below.

Cover the 400 cc. beaker and boil it until free ammonia is expelled.   The object of boiling off the free ammonia is to minimize ammonium salts since the cinchonine tungsten compound to be precipitated comes down more quickly and completely in solutions free from ammonium salts.   Dilute the solution to 200 cc. with *hot* distilled water, acidulate with 3 cc. HCl and add 6 to 8 cc. of cinchonine solution. Stir very briskly for half a minute, which will cause the flocculent precipitate to agglomerate.   It will then settle rapidly, leaving a clear supernatant liquid.   Let it stand until cold, filter on an 11 cm. weightless ash filter (B. & A.—A grade) which has been treated to a rather copious dose of paper pulp.   If ammonium chloride is used in the prior operation, the solution should be allowed to stand for several hours, best over night, after adding cinchonine.   Wash well with dilute cinchonine solution (wash solution described above) policing the beaker with a rubber tipped rod.   Follow the cinchonine wash by one washing with cold water, transfer the filter to a small weighed platinum dish or large crucible, dry by heating on the hot plate, then burn the filter *slowly* over a bunsen burner or in a muffle followed by strong ignition until all carbon is consumed.

The use of paper pulp in the filter promotes ignition, leaving the ignited precipitate as a *porous*, friable mass, whereas if paper pulp is not used, the ignited precipitate is dense and it is sometimes difficult to burn the carbon completely.

After the carbon is practically all consumed break down the residue with a glass rod flattened at one end, wipe off the end of the rod with a small piece of moistened filter paper, adding it to the dish.   Any remaining carbon and the small piece of filter paper are quickly burned, leaving a pure yellow residue.

Moisten the residue with three drops of strong H_2SO_4, add 5 cc. of HF and evaporate slowly on the hot plate until the HF is expelled.   It is best to continue heating on the hot plate until the H_2SO_4 is expelled also; there is then no danger of spitting.   Ignite the dish cautiously at first, finally at full red heat for ten minutes, cool in a desiccator and weigh as WO_3.

The residue left after the ammonia treatment may in some cases contain WO_3, although with Scheelite and most pure ores and concentrates it does not.   It is best in all cases to examine it.   In the case of impure ores containing much insoluble residue, the residues from duplicate determinations should be examined separately. When the residue is small and light-colored, it suffices to unite the duplicates for this determination.

Ignite the filter containing the residue in a small *porcelain* crucible.   This is necessary because tin may be present, which would be reduced by the filter paper and ruin platinum.   Mix the ignited residue with five or six times its weight (in any case at least one gram) of sodium carbonate plus a very little KNO_3.   Transfer the mixture to a platinum crucible and fuse for five or ten minutes.   Leach the fusion with 50 cc. of hot water in a small beaker, filter, acidulate slightly with HCl and boil to expel CO_2.   Add 5 cc. cinchonine solution.   Let the beaker stand for several hours, best over night; long standing is essential because small amounts of tungsten are slowly precipitated by cinchonine in the presence of alkaline chlorides.

If any tungsten precipitate appears, filter it off on a 7 cm. paper and wash it with dilute cinchonine solution, followed by one wash with cold water, dissolve on the filter in warm dilute ammonia, collecting the filtrate in a small beaker, boil out the excess of ammonia, make slightly acid with HCl and reprecipitate the tungsten with cinchonine. This reprecipitation is done in a very small volume of liquid practically free from salts so the tungsten comes down quickly, let it stand for an hour, filter on a small weightless ash filter containing a little pulp, wash with dilute cinchonine solution followed by one wash with water, ignite, cool, treat with a drop of H_2SO_4 and 1 cc. of HF, evaporate, ignite and weigh. Add the percentage of WO_3 thus found to the principal amount.

Factors: $WO_3 \times 0.7931 = W$,
$$W \times 2.4739 = PbWO_4.$$

Aqua Regia Method of Watts[1]

One gram of medium grade, or 0.5 gram of high grade, *very* finely pulverized ore, is treated in a 4-oz. flask with 40–50 cc. aqua regia and kept at moderate heat, *below boiling,* on a hot plate, until the solution has evaporated to about 15 cc. The solution is shaken frequently to prevent a solid cake forming at the bottom of the flask, which would prevent complete action of the acid on the ore.

The solution, removed from the hot plate, is diluted to about 50 cc. with hot distilled water and set aside to settle for half an hour. The clear liquor is decanted through a paper filter and the residue washed in the flask twice by decantation with 25 cc. portions of water slightly acidulated with HCl.

The WO_3 in the flask is dissolved with 20 cc. of 1 : 5 NH_4OH containing 2–3 drops of HCl. The solution is decanted through the filter and any remaining residue washed by decantation twice with the ammonia reagent.

If the residue is white it is transferred to the filter. If any black particles of undecomposed ore are evident, the residue is again digested with 10–15 cc. of aqua regia and treated as in the first case, only with relatively smaller portions of solutions. The ammonia extract is added to the filter. And the filter washed down in the customary way to remove all tungsten solution, allowing it to flow into the main extract.

The filtrate containing the tungsten is evaporated to dryness in a platinum dish over a water bath. The ammonia salts expelled by heat. The residue cooled is treated with 1–2 drops of H_2SO_4 and about 2 cc. HF. The silica is expelled (in Hood), together with the acids by gently heating. The residue is now heated to dull redness for 5–10 minutes, then cooled in a desiccator and weighed as WO_3.

DETERMINATION OF COMMONLY OCCURRING SUB=
STANCES IN TUNGSTEN ORES AND CONCENTRATES

Determination of Phosphorus

Treat two grams of ore exactly as in the tungsten determination and evaporate the HCl and HNO_3 solution until only about 10 cc. of liquid remains. Add 5 cc. of strong H_2SO_4 and evaporate to fumes. Cool, take up with 25 cc. H_2O, add about two grams of tartaric acid and heat to boiling. Add 20 cc. of strong NH_4OH thus dissolving the separated WO_3. Filter into a "shaking bottle" (a glass stoppered 250 cc. bottle) and wash the residue with dilute NH_4OH. Reserve the liquid. Ignite the residue and expel SiO_2 with HF and H_2SO_4. In the case of Scheelite this residue may be rather large owing to the insolubility of $CaSO_4$. Add sufficient Na_2CO_3 to make eight or ten times the weight of the residue and fuse *thoroughly*. Leach the fusion with hot water and filter, washing the residue with dilute Na_2CO_3 solution. Acidulate the filtrate with HCl and boil to eliminate CO_2—evaporate the solution if necessary to a volume of 75 cc. and make it slightly alkaline with NH_4OH. A precipitate of $Al_2(OH)_6$ may appear at this point; if it does, make the liquid slightly acid until the precipitate dissolves, and add it to the main ammoniacal solution in the shaking bottle. The presence of tartaric acid in this solution will prevent the separation of alumina. Add 10 cc. magnesia mixture to the bottle (the whole volume of solution should not exceed 160 cc.) and cool to 5° C. or less. Add five or six glass beads to the bottle and shake in a shaking machine for five or ten minutes. The object of the glass beads and the shaking, is to start the formation of magnesium-ammonium phosphate. After shaking, add 25 cc. of strong NH_4OH and let the bottle stand in a refrigerator over night. Filter off the solution and wash the precipitate in the bottle and on the paper with dilute NH_4OH but do not attempt to remove the precipitate adhering to the sides of the bottle. The precipitate consists of magnesium ammonium phosphate and arsenate together with silica and other impurities. Tin, tungsten, molybdenum, vanadium and titanium are eliminated by precipitation with magnesia mixture in the presence of tartaric acid. Dissolve the magnesia precipitate in hot dilute HCl and evaporate to small volume, say 7 or 8 cc. Add 10 cc. of *strong* HCl and 0.5 gram NaBr and evaporate to dryness. Take up with 5 cc. HNO_3 and evaporate to dryness again. The NaBr serves to reduce As^v to As'''. The addition of strong HCl and boiling followed by evaporation to dryness eliminates As and SiO_2. Dissolve residue in 60 cc. HNO_3, S. G. 1.135, add 40 cc. of NH_4OH Sp. Gr. 0.96, cool to 35° C., add 30 cc. molybdate solution (Blair) and finish the determination by the alkalimetric method (Handy) described on page 368.

Determination of Sulphur in Ferberite, Wolframite and Scheelite

Fusion Method. Fuse two grams of pulp mixed intimately with nine grams of Na_2CO_3 and one gram KNO_3. Leach the fusion with water and wash the residue thoroughly with a dilute solution of Na_2CO_3. Acidulate the liquid with HCl and evaporate to dryness. Take up with a little HCl and 150 cc. of water, add 5 cc. of strong cinchonine solution and heat until all soluble salts are dissolved, finally boil for about five minutes. Let the solution stand over night, filter and wash with dilute cinchonine solution, using paper pulp in the filter. Neutralize the filtrate with Na_2CO_3, make slightly acid with HCl (0.5 cc. for each 100 cc. of liquid) heat to boiling and precipitate $BaSO_4$ by the slow addition of 10% $BaCl_2$

solution. Let stand over night; filter, wash and weigh. The cinchonine solutions are the same as those used in the tungsten determinations. Note that the use of cinchonine is essential in this determination because in the presence of a large excess of alkaline chlorides the complete separation of tungsten is only effected by cinchonine. If it is omitted the $BaSO_4$ precipitate will inevitable contain $BaWO_4$. Blank determinations on the reagents run parallel with the sample are essential.

Wet Method. In Ferberite and Wolframite only where sulphur is present as associated sulphides. The method is not applicable to Scheelite which may contain $BaSO_4$.

Treat two grams of the 200-mesh pulp with 20 cc. of a mixture of three parts HNO_3 and one part HCl in a 250 cc. beaker, add 0.5 cc. Br and let stand at room temperature for half an hour. Digest on a steam bath, for half an hour, remove cover and evaporate to dryness on the steam bath. Add 5 cc. strong HCl and again evaporate to dryness. Take up with 3 cc. HCl and 10 cc. H_2O, and digest until all soluble salts are dissolved. Add 250 cc. hot water and 2 cc. cinchonine solution and boil. Let the solution stand until it is cold, filter and wash the residue with water containing a few drops of HCl. Heat the liquid to boiling and precipitate S by adding 10 cc. of a 10% solution of $BaCl_2 \cdot 2H_2O$. Let stand over night, filter and wash with hot water. Ignite and weigh as $BaSO_4$. In this method, the tungsten mineral is only partly decomposed, but all sulphides of iron, copper, etc., are oxidized to sulphates. This method is useful for the analysis of impure concentrates containing much associated sulphides. It is not to be relied upon for the accurate determination of small amounts of sulphur.

Determination of Arsenic in Tungsten Ores and Concentrates

Wet Method. The method is based on the assumption that arsenic in tungsten ores and concentrates is wholly present as associated minerals which may be decomposed by acid treatment and not as complex arsenic-tungsten compounds unaffected by acids, since the treatment used does not wholly decompose the tungsten mineral or associated tin, titanium and niobium minerals.

Treat two grams of the 200-mesh pulp in a round bottom 300 cc. flask with 10 cc. strong HNO_3. Digest for half an hour over a small flame, add 7 cc. of strong H_2SO_4 and evaporate to fumes by shaking the flask over a naked flame and at the same time blowing a mild air current into it to remove HNO_3. Cool the flask, add 10 cc. of water and again evaporate to fumes of H_2SO_4; this is done to expel HNO_3 *completely* which is important. Cool the flask and add 3 cc. of water, not more, then add two grams of dry $FeSO_4$ and connect the flask to an Allihn condenser, set up vertically, by means of a rubber stopper carrying a stoppered funnel tube and a wide exit tube. The exit end of the condenser is sealed by allowing it to dip slightly below the surface of 300 cc. of *cold* water contained in a large beaker.

Add 75 cc. of *strong* HCl through the funnel tube and boil until the volume in the *flask* is reduced to 25 or 30 cc., then add 25 cc. more strong HCl and repeat the distillation. Usually two distillations serve to volatilize all arsenic which is collected in the beaker. To make sure, substitute another beaker containing water, add 25 cc. more HCl to the flask and distil again. The second distillate is tested by itself; it seldom shows any arsenic. Pass a rapid current of H_2S into the acid distillate, thus precipitating arsenic as As_2S_3. Let the precipitate settle at a temperature of 60° to 70° for an hour or two, filter through a Gooch crucible fitted with a well washed asbestos pad and wash the precipitate thoroughly with hot water, followed by alcohol, ether and finally CS_2. Dry the precipitate at 105° C.

for an hour and weigh, dissolve the As_2S_3 from the filter by passing warm NH_4OH through it, wash again with water followed by alcohol only, dry at 105° C. for an hour and weigh again. Take the weight of As_2S_3 by difference between first and second weighing. This method is preferable to weighing the Gooch crucible before filtering.

The distillation method depends on the reduction of As^v to As''' by ferrous sulphate and on volatilization as $AsCl_3$. This can only be done if the HCl employed is *strong*. Dilution of the HCl below a gravity of 1.10 will prevent the formation of $AsCl_3$, since in that case we have a solution of As_2O_3 in HCl rather than $AsCl_3$.

See also the procedure given on pages 37–39 of this volume.

Preliminary Fusion. In case the ore or concentrates contains insoluble arsenic compounds not broken up by the simple acid treatment described, it is necessary to fuse two grams of pulp with 15–20 grams of potassium pyrosulphate in a *porcelain* or *fused silica* crucible. After the melt is cold, remove it from the crucible which can usually be done easily by inverting and tapping, and break it into small pieces in a porcelain mortar. Transfer the ground melt to the 300 cc. flask, add two grams $FeSO_4$ and proceed with the distillation as above. The latter modification is unnecessary on most ores, since arsenic is rarely present except as associated arsenical minerals decomposed by acids.

Some trouble may be experienced in distilling owing to the "bumping" caused by separated WO_3 or undecomposed ore. This may be overcome by introducing a few glass beads into the flask.

Tin in Tungsten Ores and Concentrates

Fuse two grams of the finely pulverized ore in a spun iron crucible with ten grams of a mixture of equal parts of finely granulated sodium hydrate and sodium peroxide. The ore should be intimately mixed with the flux, fusion should be continued for one or two minutes after the mass is completely liquid, and a rotary motion should be given to the crucible during this period by holding it in the tongs and turning it in the flame; turning should be continued during cooling to spread the fusion.

Dissolve the fusion in 75 to 100 cc. of hot water. This is best done by placing the crucible in water in a large nickel dish since the strongly alkaline solution attacks glass or porcelain. When the mass is thoroughly disintegrated, remove the crucible and wash it; then pour the liquid into a 300 cc. beaker containing 15 grams of tartaric acid dissolved in 50 cc. of water; forthwith add 40 cc. of strong HCl. The solution will, of course, be quite hot. It should become clear on standing a few minutes except for a small amount of iron scale from the crucible, but it may require further heating to dissolve all iron hydrate. There should be no perceptible residue of undecomposed ore but it does not matter if there is a little, because it will be decomposed by the subsequent treatment. Pass a brisk current of H_2S into the warm solution until ferric iron is reduced and tin is precipitated as stannic sulphide. Filter the solution and wash the precipitate thoroughly with warm water containing H_2S. If there is a tendency for the tin sulphide to run through add a little ammonium nitrate to the wash water. It is important to remove chlorides since their presence might entail a loss of tin in the subsequent ignition.

The presence of sufficient tartaric acid prevents the precipitation of tungstic acid when the fusion is acidulated with HCl, but with large amounts of tin or with the considerable separation of sulphur caused by reduction of ferric iron in the solution some tungsten is likely to be entrained with the precipitate which no reasonable amount of washing will remove; a second fusion is therefore necessary. Ignite the sulphide precipitate in a *porcelain* crucible and transfer the residue carefully to a small spun iron crucible. The reason for this step is the presence of free sulphur and sulphides which unite with the iron of the crucible, forming iron sulphide when ignition is made in the iron crucible. The film of iron sulphide is vigorously acted upon by peroxide, causing spattering. Mix the ignited residue with about two grams of sodium peroxide and fuse as before. Dissolve the fusion, in the crucible, in 25 cc. of water, transfer to a 250 cc. beaker and boil to decompose peroxide; then add two grams of tartaric acid and 60 cc. of strong HCl. This gives a solution containing 2 vols. of strong HCl to 1 vol. of water. Iron scale from the crucible and iron hydrate dissolve immediately in acid of this strength, leaving a perfectly clear solution, but the object in adding the large excess of HCl is to obtain a solution of such acid strength that arsenic may be precipitated with H_2S while tin sulphide is not formed.

If the foregoing proportions are maintained the separation of arsenic and tin may be accomplished by passing H_2S into the solution, whereby arsenic is separated as sulphide. Pass H_2S into the hot solution for about ten minutes or until practically all ferric iron is reduced. Filter off the arsenic sulphides through a small filter and wash thoroughly with HCl (1.19 sp.gr.) diluted with one half its volume of water. Good quality filter paper will withstand HCl solutions of this strength although an asbestos filter (Gooch) may be used if preferred.

Dilute the acid filtrate with twice its volume of water, add 25 cc. of strong NH_4OH to neutralize part of the free acid, and pass a little more H_2S thus precipitating tin, antimony, copper, etc. as sulphides. Let the sulphides stand until they have settled well, then filter and wash the precipitate with H_2S water slightly acidulated with HCl. Filter the solution through a 9 cm. filter, or better through a Gooch filter, using very moderate suction. Transfer the filter to a small beaker and heat it with 10 cc. of a 10% solution of sodium monosulphide. The tin sulphide and free sulphur dissolve readily leaving, usually, a very small amount of copper sulphide. Filter the solution through a 7 cm. paper and wash the filter several times with a 1% solution of sodium sulphide. Since copper sulphide is usually exceeding small, a single treatment with sodium sulphide suffices to dissolve all the tin sulphide, but if much copper should be present a second treatment is advisable.

The alkaline filtrate and washings contained in a 450 cc. Erlinmeyer flask should measure approximately 60 cc.; add 70 cc. of strong HCl, or a little more than the volume of the liquid. The sulpho-stannate is decomposed, most of the tin sulphide will dissolve, and there will be a separation of free sulphur. If the liquid is fairly cool at the time of adding the acid, there will be no great rise in temperature; if the temperature does not exceed 50–55° C. the liberated sulphur will not agglomerate and subsequent oxidation is easily accomplished by adding about two grams of $KClO_3$ a little at a time to the cool solution, shaking after each addition. After the solution begins to clear, heat it to boiling to remove chlorine. A perfectly clear solution without agglomerations of sulphur should result. Continue boiling for about ten minutes to insure complete expulsion of chlorine. Add 70 cc. of water to the flask, making the volume approximately 200 cc. containing between one-quarter and one-third of its volume of free HCl; this proportion of free acid is important. Introduce a coil of nickel foil, $1\frac{1}{2}'' \times 7''$, rolled around a glass rod until it will just slip into the neck of the flask, heat the solution to boiling and boil gently for about 20 minutes; connect the flask by means of a double perforated rubber stopper with a current of CO_2 gas and continue boiling for a few minutes. Remove from heat and while a brisk stream of CO_2 is continued immerse the flask in cold water. When the solution is cooled to room temperature disconnect from the source of CO_2 and immediately titrate the tin with N/20 iodine solution using a few drops of starch solution as indicator. It is unnecessary to remove the nickel before titration. A "blank" should be made on the titration by boiling a solution of hydrochloric acid with a foil of nickel under the same conditions as the determination. The correction may amount to 0.2 cc. The iodine solution should be standardized against pure tin dissolved in HCl and $KClO_3$ and reduced with nickel under the same conditions as the determination.

NOTES. Antimony does not interfere; it is precipitated as metal by the nickel and remains as a black precipitate in the bottom of the flask.

Arsenic in minute proportions does not interfere. Where it is known to be present only in small amounts the steps for its elimination may be omitted. Some tungsten ores, however, contain relatively large proportions of arsenic, usually present as arsenopyrite. When a large amount of arsenic is present it should be eliminated as described, otherwise serious errors may be made.

In place of the cumbersome CO_2 apparatus mentioned for use at the end of the reduction with nickel and during cooling, the "tin reduction" apparatus devised by A. Craig and described in The Engineering and Mining Journal, Vol. 106, p. 25, may be used to great advantage.

Copper in Tungsten Ores and Concentrates

Applicable to All Ores Whether Scheelite, Wolframite or Hubnerite

Treat five grams of the fine pulp with 20 cc. of HCl in a 400 cc. beaker and digest with frequent shaking at a low temperature for an hour. Add 5 cc. of HNO_3 and boil down to a volume of, say, 10 cc., cool and add 10 cc. of H_2SO_4 and evaporate to fumes. Cool, add 150 cc. of water and boil. Filter off the separated tungstic acid. All copper is in the acid filtrate, but it is well to test the residue also to be perfectly sure.

In the case of Scheelite it is necessary to add a few cc. of HCl to the acid solution before filtering off the WO_3, since the large amount of $CaSO_4$ which separates in sulphuric acid solution alone is troublesome to wash and renders the examination of the insoluble residue difficult unless it is removed.

Neutralize the sulphuric acid solution from the WO_3 with NH_4OH, make it just acid with HCl, avoiding any greater excess than is sufficient to dissolve the iron and manganese hydrates. Add 50 cc. of a solution of sodium bisulphite (150 grams per liter) and let stand in a warm place for half an hour with frequent stirring —iron is reduced to the ferrous condition. Add 5 cc. of sodium sulphocyanate (150 grams per liter) and let stand for another hour stirring from time to time. Copper is precipitated as cuprous sulphocyanate free from molybdenum (which frequently occurs in Scheelite), arsenic or other interfering elements. Filter the cuprous sulphocyanate through a double filter, returning the first part of the filtrate to the filter since the precipitate is very fine and is apt to "run through" at first. To prevent this, add a little paper pulp to the filter.

Wash the precipitate with warm water and transfer the wet filter to a porcelain crucible, dry in an oven heated to 250° to 275°. This will char the paper gradually without danger of loss of precipitate. When the paper is thoroughly charred, heat the crucible in a muffle or over an open burner until all carbon is consumed.   Dissolve the mixture of copper sulphide and oxide remaining in the crucible in a few cc. of strong HNO_3 and finish the determination by the electrolytic or iodide methods. Very small amounts of copper are best estimated colorimetrically by comparing the blue color of the ammoniacal solution with a standard in short color tubes.   Since the exact determination of copper is sometimes very important, it is advisable always to take a large initial weight of sample (5 grams) unless copper is present in large amount—say two per cent.   This is sometimes the case in scheelites containing cupro-scheelite.   In examining the residue left by the acid treatment, dissolve the WO_3 in NH_4OH and wash the filter with dilute NH_4OH.   Ignite the filter containing the insoluble matter and undecomposed ore and fuse it with $KHSO_4$ in a porcelain crucible.   Dissolve the fusion in water and make the solution strongly alkaline with NH_4OH.   Filter off any iron hydrate and insoluble residue and observe whether or not the filtrate shows a blue color.   As little as 0.5 mg. can easily be detected in a volume of 50 cc.   If there is no blue color, further examination may be omitted.   If the qualitative test indicates copper, dissolve any iron hydrate which may have separated in dilute H_2SO_4.   Add this solution to the main liquid and slightly acidulate with HCl.   Then add sodium bisulphite and sulpho-cyanate, proceeding as in the main determination.

Manganese in Tungsten Ores and Concentrates

Treat one gram of the pulp exactly as in the tungsten determination omitting the use of cinchonine, but letting the solution stand until it is cold after decomposition has been effected.   Filter off the insoluble residue and tungstic acid and wash it *thoroughly* with dilute HCl (25 cc. per liter).   It is quite difficult to remove every trace of iron and manganese from the heavy WO_3 residue which invariably retains a little unless the precipitate and residue is washed back from the filter and agitated with the wash liquid.   The filtrate contains all Mn.

Neutralize the solution by means of Na_2CO_3 until there is a slight permanent precipitate.   Dissolve this by adding dilute H_2SO_4 (1–5) leaving the solution very slightly acid.   Add two or three grams of ZnO, ground to a cream with water, heat to boiling and titrate Mn with standard permanganate by Volhard's method. See page 304.

METALLIC TUNGSTEN AND TUNGSTEN ALLOYS

Tungsten in Tungsten Metal and Ferro=Tungsten

Treat one gram of the finely ground sample in a large (60 cc.) platinum crucible fitted with a cover with 5 cc. HF.; add HNO_3, drop by drop, until the metal dissolves. Add 3–4 cc. H_2SO_4 and evaporate on a steam bath until $HNO_3 + HF$ is expelled. Shake gently over a small Bunsen flame until H_2SO_4 fumes strongly. Cool, transfer to a 250 cc. beaker with water, finally wiping the crucible with a little filter paper. A little WO_3 sticks to the crucible; it cannot be removed by wiping. Reserve the crucible. Dilute the contents of the beaker to about 150 cc. with water, add 3 cc. HCl and boil. Remove from the stove and to the hot solution add 5 cc. cinchonine solution and let stand over night (or at least four or five hours). Filter on "ashless paper" and wash with dilute cinchonine solution. Gently ignite the precipitate in the crucible in which it was originally treated. Heat for five minutes with full Bunsen burner flame, cool and weigh. Add about 5 grams Na_2CO_3 and fuse, running the fusion around the side of the crucible to remove all WO_3. Dissolve the fusion in hot water; filter and wash five or six times with hot water. Place the filter in the crucible and ignite, add a little Na_2CO_3 and fuse again. Dissolve the fusion in water, filter and wash *very* thoroughly with hot water to remove last traces of Na_2CO_3, ignite in the same dish as at first, cool and weigh. The difference between weight of dish plus residue and weight of dish plus tungstic oxide is WO_3.

Dilute Cinchonine Solution. 30 cc. strong cinchonine solution and 30 cc. HCl to one liter.

Strong Cinchonine Solution. 100 grams cinchonine dissolved in dilute HCl (1 part acid, 3 parts water) and diluted to one liter with acid of the same strength.

Method for Phosphorus in Ferro=Tungsten and Tungsten Metal

Treat one gram of the finely powdered sample in a platinum dish, fitted with a gold cover, with 15 cc. HNO_3 (1.42 sp.gr.), add 3 cc. HF and warm gently. When action subsides, add 3 cc. more HF. After action subsides, boil, remove cover and if decomposition is not complete, add more HF and boil again. When solution is complete, wash off the cover and evaporate at a low heat to a volume of about 10 cc. then add 3–4 drops of concentrated permanganate solution and continue evaporation until crusts of WO_3 begin to form at the edges; that is, to a volume of, say, 6 cc. Add 5 cc. H_2SO_4 and evaporate on the stove at a low heat until HF and HNO_2 are expelled and H_2SO_4 fumes are given off. (Strong heat causes spattering and also cause hard, over-baked crusts to form on the bottom of the dish which resist subsequent treatment.) Cool, add 25 cc. H_2O and boil (by agitating over bunsen flame) until all soluble salts are dissolved. Destroy pink color due to excess of permanganate by adding sulphurous acid drop by drop. The pink color may not be very evident but the SO_2 is added anyway to reduce higher oxides of Mn. Boil for a minute or two after adding the SO_2. Add 1.2 grams of pure tartaric acid and when this is dissolved and the solution is cooled to a temperature of about 50° C., add 20 cc. of NH_4OH, (0.90 sp.gr.) diluted with an equal volume of water. The precipitated tungstic acid should dissolve completely, giving a clear solution. The solution is hot from the reaction between H_2SO_4 and NH_4OH. While it is still hot add 10 cc. magnesia mixture and transfer it from the Pt dish to a six-ounce glass stoppered bottle. Set the bottle in ice water and when it is *thoroughly* cooled, add four or five glass beads, say 6 mm. diameter. Stopper it tightly and shake in an efficient shaking machine for at least ten minutes. The agitation should be violent. The beads aid in starting the formation of the magnesium precipitate; after agitation add 15 cc. of strong NH_4OH and return the bottle to ice water tank or put it in a refrigerator to stand over night. Phosphorus separates as magnesium ammonium phosphate free from tungsten but containing possibly basic magnesium compounds. After standing over night, filter the solution through a 9 cm. paper containing a little paper pulp and wash the bottle and paper thoroughly by small additions of ammonia wash water (1 part NH_4OH, 3 parts H_2O). Do not attempt to remove all the precipitate from the bottle but remove the beads to the filter. Place the bottle under the filter and pour through the filter 60 cc. of HNO_3 of 1.135 sp.gr. (1 part HNO_3, 1.42; 3 parts H_2O) in five successive portions of 12 cc. each letting the acid fall on the upper edges of the filter paper. This dissolves any ammonium magnesium phosphate and washes it into the bottle where any that remained adhering to the glass is also dissolved. Add 12 cc. of NH_4OH (sp.gr. 0.90) to the liquid, cool to 35° C., add 35 cc. molybdate mixture and proceed according to either of the methods described on pages 368–370 of this volume.

NOTES. The usual proportions of HNO_3 and HF are reversed in order to provide a constant excess of HNO_3 to oxidize P. The procedure given takes a little longer than when the sample is treated with HF first and HNO_3 is added a little at a time, but solution is finally complete. It is necessary to keep the platinum dish covered after action begins, as the reaction is somewhat violent.

The platinum dishes recommended are:

    8 cm. in diameter at top.
    7.8 cm. in diameter at bottom.
    4 cm. high.

They have flat bottoms and are wire rimmed at top to give additional stiffness. They weigh 58 to 60 grams each and hold about 175 cc. The covers are made of pure gold (for economy) "dished" like a crucible cover to fit the top of the dishes closely. The dishes have small lips to aid pouring. The "tongue" of the cover overlaps the lip. Ordinary round bottomed dishes may be used but the manipulation is much more difficult; there is greater tendency to spattering and danger of local baking or overheating in evaporating to fumes of H_2SO_4. If the separated WO_3 is overheated locally, it does not dissolve readily in NH_4OH. In a flat bottomed dish, the WO_3 is spread in a thin layer and heat is applied evenly all over the bottom.

A good shaking machine should be used. The magnesia precipitate may be started by shaking the bottles by hand, but it is a tiresome job. The solution must be cold— say 8° or 10° when shaking begins.

Permanganate solution is added to insure complete oxidation of phosphorus, as in steel analysis. The color of the permanganate gradually fades in the hydrofluoric nitric solution, but after evaporating to fumes and adding water, the solution is usually slightly pink.

The amount of tartaric acid is limited to 1.20 grams since ammonium tartrate retards the formation of the magnesia precipitate. Complete precipitation can only be obtained by brisk agitation and by keeping the solution very cold followed by long standing in a cold place. By this method, determinations started at 3 p.m. may be completed by noon the next day.

Determination of Sulphur and Silicon in Ferro=Tungsten and Tungsten Metal

Fuse two grams of the finely powdered metal, intimately mixed with 8 grams of dry Na_2CO_3 and 2 grams of powdered KNO_3, in a large platinum crucible (40 cc.). The fusion is best made in an electric muffle to avoid contamination with sulphur contained in illuminating gas. Leach the fusion with 100 cc. of distilled water in a 250 cc. beaker. Filter into a 12 cm. porcelain casserole and wash the residue several times with hot distilled water. The solution contains practically all of the sulphur and most of the tungsten and silica. The residue contains some of each, but not more than traces of sulphur.

Solution. Render the liquid acid with 20 cc. HCl and evaporate to dryness on a steam bath, thus precipitating tungstic acid. After the residue is dry, heat it to $110°$–$120°$ C. in an oven to dehydrate silica, take up with 5 cc. HCl and 50 cc. water, add 3 cc. strong cinchonine solution (see tungsten determination) and boil, or heat just short of the boiling point, for half an hour. Let the solution stand until it is cold. By means of the cinchonine all WO_3 is rendered insoluble. Filter through an 11 cm. filter paper (free from ash) which has been liberally treated with an emulsion of filter paper pulp, and wash the residue with dilute cinchonine solution (see tungsten determination). Reserve the residue. The filtrate contains all the sulphur, heat it to boiling and add 10 cc. of barium chloride solution (100 grams $BaCl_2$ $2H_2O$ to 1000 cc.) boil for ten minutes, allow it to stand over night, filter on a 9 cm. filter paper (free from ash) wash repeatedly with hot water, ignite in platinum and weigh as $BaSO_4$. If the work is properly carried out every trace of WO_3 is removed from the solution and there is no danger of including $BaWO_4$ with the $BaSO_4$. If one suspects that the $BaSO_4$ is not quite pure, t should be fused with a little Na_2CO_3, leached with water and filtered, the filtrate should be acidulated with HCl, a few drops of cinchonine solution added and the solution evaporated to dryness on a steam bath. The dry residue should be dissolved in a little water, a few drops of HCl added, and any residue filtered off. In the clear filtrate $BaSO_4$ should be again precipitated as before. The weight of $BaSO_4 \times 0.1373 = S$.

A "blank" on all reagents must be made parallel with the determination; this is important since all sodium carbonate obtainable contains sulphur and there is sometimes more or less of it in the cinchonine. All evaporations should be made over steam and the operations conducted in a place free from sulphur gases.

Silicon. The residue of tungstic oxide from which the solution for the determination of sulphur was filtered contains most of the silica. The residue from leaching the original fusion contains the rest.

Wash the residue from the Na_2CO_3 fusion from the filter paper with a fine jet of water into a small casserole, add an excess of HCl and evaporate to dryness on a steam bath, heat the dry residue at $110°$–$120°$ C. to dehydrate silica and filter on a small filter paper (free from ash). Wash thoroughly with hot dilute HCl (1–10) and finally with hot water. Reserve the filter.

Wash the residue of tungstic oxide, silica and filter paper pulp from the filter into a small casserole, add 5 cc. of HCl and heat for a few minutes, then filter again through the same paper and wash it *thoroughly* with hot water. This second washing is necessary to remove sodium salts completely. (The original washing sufficed to remove all of the small amount of sulphur present.) Transfer the well-washed filter and residue to a platinum crucible, add the small filter containing

45

the little silica recovered from the Na_2CO_3 fusion residue and ignite both to constant weight in a platinum crucible. On account of the presence of filter paper pulp, the ignited residue is porous and friable; if paper pulp had not been used, the WO_3 would be dense and not easily susceptible to the subsequent treatment. Cool the ignited residue of WO_3 and SiO_2 and weigh it. Add two or three drops of H_2SO_4 and 5 or 6 cc. of pure HF. Digest at a gentle heat for some time, and then slowly evaporate off the HF. When H_2SO_4 fumes are evolved, cool and again add HF, digest as before, evaporate off the HF, ignite gently to expel H_2SO_4 and then strongly for ten or fifteen minutes. Cool and weigh. The loss in weight after expulsion is SiO_2 which multiplied by 0.4693 equals silicon.

Determination of Manganese in Tungsten and Ferro=Tungsten

The filtrate from the tungstic acid obtained by digesting the residue after evaporation to fumes of H_2SO_4 and digestion with dilute HCl (see tungsten determination) contains all the manganese. Separate iron from this solution by means of a basic acetate precipitation and wash the precipitate thoroughly with a hot dilute solution of sodium acetate, reject the precipitate, and to the filtrate, which contains all the manganese, add 5 cc. of strong NH_4OH, add about two grams of $(NH_4)_2S_2O_8$. Digest on a steam bath for 45 to 60 minutes, manganese separates as hydrated peroxide. Filter and wash with hot water. Dissolve the precipitate through the filter with ten or fifteen cubic centimeters of H_2SO_3 containing a few drops of H_2SO_4 and wash the filter paper several times with hot dilute H_2SO_4 (10%). Again filter through the same paper into a 500 cc. flask and wash with hot water. Dilute the liquid to a volume of about 200 cc. and boil it for fifteen or twenty minutes until all SO_2 is expelled. Nearly neutralize the solution with a solution of NaOH but still leave it slightly acid, complete the neutralization by adding 5 grams of zinc oxide emulsified by grinding with water, heat again to boiling and titrate manganese in the hot solution with a N/10 solution of $KMnO_4$ (Volhard's Method).

Acknowledgement is made to Mr. J. A. Holladay, Electrometallurgical Co., for cooperation in the preparation of these methods.

TANTALUM AND COLUMBIUM

Cb, *at.wt.* 93.5; *sp.gr.* 7.06; *m.p.* 1950°; *oxides* CbO, CbO_2, Cb_2O_5.
Ta, *at.wt.* 181.5; *sp.gr.* 14.49; *m.p.* 2900°; *oxides* TaO_2, Ta_2O_4, Ta_2O_5.

DETECTION

The finely powdered mineral is digested with strong hydrochloric acid, followed by concentrated nitric acid and the mixture taken to dryness. The residue is treated with hydrochloric acid, diluted with water, boiled and filtered. The residue is digested with warm ammonium hydroxide to remove tungsten and the solution filtered from the insoluble material, in which tantalum and columbium will be found, if present in the sample.

Decomposition of the material may be effected according to the procedure described for the detection of tungsten, page 554.

The residue obtained is digested, in a platinum crucible, with hydrofluoric acid and a saturated solution of potassium fluoride added. The mixture is evaporated to small volume and allowed to cool slowly. Tantalum will separate in acicular rhombic crystals (solubility—1 part of the salt in 200 parts of water) as potassium fluotantalate $2KF \cdot TaF_5$; columbium separates in plates as the double fluoride, $2KF \cdot CbF_5$, if HF is in excess, or as a double oxy-fluoride $2KF \cdot CbOF_5$, if HF is not in excess; the columbium salt being much more soluble (1 part of the salt in 12 parts of water) crystallizes after the crystals of tantalum have formed.

The crystals may be examined under a lens and then treated as follows: The needle-like crystals are heated in a shallow platinum dish or crucible cover with strong sulphuric acid to fumes, the cooled mixture is transferred to a test-tube with water and boiled to precipitate the tantalic acid. An opalescent solution is obtained when this precipitate is treated with an excess of hydrochloric acid. Metallic zinc added to this solution produces no color. A light-brown precipitate is obtained with tannic acid in the presence of tantalum. If the crystals of columbium salt are treated in the same way, metallic zinc added to the acid solution will give a blue coloration, and tannic acid an orange-red coloration. Tantalic acid fused with sodium meta-phosphate gives a colorless bead (distinction from silica). The bead moistened with $FeSO_4$ and heated in the inner flame is not colored red. Columbic acid fused in the same way gives a blue bead in the reducing flame, and a red bead by addition of $FeSO_4$, and heating in the flame.

ESTIMATION

Tantalum and columbium occur commonly with tungsten in nature. In the following minerals, however, tantalum and columbium form the more important constituents:

Columbite, $(Ta \cdot Cb)_2(Fe \cdot Mn)O_6$; pyrochlore, $RCb_2O_6R(Ti \cdot Th)_3$; hatchettolite, $2R(Cb \cdot Ta)_2O_6$ or $R_2(Cb \cdot Ta)_2O_7$; fergusonite, $R(Cb \cdot Ta)O_4$; yttrotantalite, $RR(Cb \cdot Ta)_4O_{15} \cdot 4H_2O$; samarskite, $R_3R_2(Cb \cdot Ta)_6O_{21}$.

Tantalum is used in electric light filaments; it is also used for hardening steel for drills, files, cutting edges, watch springs, and pen points. It is used in rectifiers for alternating currents.

Solution of the Sample

The statements made for solution of the sample in determinations of tungsten apply here also. It is well to keep the following facts in mind: Tantalum is insoluble in the common mineral acids—hydrochloric, nitric and sulphuric acids, but dissolves in hydrofluoric acid. Columbium is insoluble in hydrochloric, nitric and in nitro-hydrochloric acid, but dissolves in hot concentrated sulphuric acid. The oxides Ta_2O_5 and Cb_2O_5 fused with KOH form soluble salts. Cb_2O_5 (not strongly ignited) is soluble in acids, from which $(NH_4)_2S$ and NH_4OH precipitate columbic acid (containing ammonia). Freshly precipitated tantalic acid is soluble in acids, and reprecipitated by NH_4OH. The acid dissolves readily in HF.

Tantaliferous Minerals. Although decomposition may be effected by fusion with potassium acid sulphate, fusion with potassium hydroxide is recommended as being the best flux for opening the minerals. Simpson's process is as follows:[1]

Three grams of pure potassium hydroxide are fused in a nickel or silver crucible and the finely powdered mineral (0.5 gram) added, the contents mixed by gently rotating the crucible and fusion kept at a dull red heat for ten minutes longer. The crucible placed in a hole in an asbestos board, Fig. 65, is heated over a free flame for half an hour, the sample being covered. The lid is removed and allowed to cool reversed, if any material clings to this. The cooled crucible, placed in a beaker, is two-thirds filled with distilled water, and a clock-glass immediately placed over the beaker. After the violent reaction has subsided the contents of the crucible are poured into about 10 cc. of dilute hydrochloric acid (sp.gr. 1.08) in a 300-cc. beaker, and the crucible, basin and the lid washed with water, followed by about 20 cc. of the dilute acid, and again with water, adding the washings to the remaining solution. The total volume of the solution should occupy from 80 to 100 cc. A drop or two of alcohol are added to destroy any potassium manganate formed.

Separations

Isolation of Columbium and Tantalum Oxides. Separation from iron, manganese, copper, cobalt, nickel, calcium, magnesium, titanium, and tin. The solution obtained above is boiled with 5 to 10 cc. of hydrochloric acid (sp.gr. 1.16) (less acid may be used if titanium is absent). Columbium and tantalum hydroxides are precipitated. The solution is now diluted to 200 cc. and boiled for fifteen minutes longer to make sure that the precipitation is complete. After settling, the clear solution is decanted through a close-grained filter and the residue, having been transferred to the filter, is washed with dilute hydrochloric acid (sp.gr. 1.08) until the washings give no indication of iron. The residue may contain tantalum, columbium, tungsten, silica, antimony and tin. The greater part of the tin, titanium, and all of the iron, manganese, cobalt, nickel, copper, calcium and magnesium are removed in the filtrate.

NOTES. If the filtrate becomes turbid, it is advisable to dilute the solution and repeat the boiling to recover the columbium and tantalum that may still be in solution.

In the presence of appreciable amounts of titanium a soluble double chloride of columbium and titanium is formed, so that the precipitation of columbium

[1] E. S. Simpson, Chem. News, **99**, 243, 1909.

is not complete. (See L. Weiss and Landecker, Chem. News, 101, 2, 13, 26, 1910.) The formation of this compound is hindered by the addition of an oxidizing agent—sodium nitrate—to the alkali.

Removal of Tin, Antimony, Tungsten and Silica. *Tungsten* is removed by digesting the moist precipitate with ammonium hydroxide or sulphide, tungsten being soluble in these reagents. Antimony and Tin are also removed.

Silica is volatilized by heating the residue with sulphuric and hydrofluoric acids according to the standard procedure.

Tin. The oxide may be reduced with hydrogen passed over the heated residue within a boat placed in a combustion tube. The tin may now be dissolved out with hydrochloric acid.

Determination of Columbium and Tantalum

The insoluble residue obtained, freed from other elements by the procedures outlined, is ignited at a red heat for fifteen or twenty minutes and the residue weighed as $Cb_2O_5 + Ta_2O_5$.

Separation of Columbium and Tantalum, Selenium Oxychloride Method.[1]

The method depends on the solubility of columbium oxide and the comparative insolubility of tantalum oxide in a mixture of sulphuric acid and selenium oxychloride as found by Victor Lenher.[2]

The method worked out by H. B. Merrill is as follows:—

The oxides are separated together with titanium dioxide from the other elements by the usual methods and the total percentage of titanium and columbium pentoxides (with titanium) is determined.

A weighed sample (0.2—0.3 g.) of the ignited oxides is boiled with 50 cc. of a 1 : 1 mixture of selenium oxychloride and strong sulphuric acid (sp. gr. 1.84) in an Erlenmeyer flask, on a sand bath for half an hour, avoiding heating to voluminous fumes. After cooling the solution is carefully decanted from the residue, with suction, through asbestos in a Gooch crucible that has been weighed. The filtrate is poured into a large volume of water and the solution boiled. A white precipitate indicates the presence of columbium. The residue in the flask is again extracted with 20 cc. of the reagent for fifteen minutes, the extract being poured through the asbestos filter and the filtrate tested as before for columbium. The process is repeated until the filtrate upon hydrolysis gives only a faint precipitate, due to traces of dissolved tantalum pentoxide. Three or four extractions are usually sufficient.

The undissolved tantalum pentoxide is transferred to the crucible with a jet of water from a wash bottle, washing unnecessary. The crucible is ignited and weighed, the gain in weight representing tantalum pentoxide.

Columbium (with titanium) is determined by difference. If titanium is present it is determined colorimetrically in a separate sample and its amount deducted from columbium.

[1] Henry Baldwin Merrill, Jour. Am. Chem. Soc. **43**, 2378 (Nov. 1921).
[2] Victor Lenher, Jour. Am. Chem. Soc. **43**, 21 (1921).

URANIUM

U, *at.wt.* **238.17;** *sp.gr.* **18.7;** *m.p.* **<1850° C.;**[1] *oxides* UO_2, UO_3,
(oxide U_3O_8, *formed by ignition* $= UO_2 + 2UO_3$*)*

DETECTION

The mineral is warmed with a slight excess of nitric acid (1 : 1) until decomposition is complete. The solution is diluted with water and then an excess of sodium carbonate added and the mixture boiled and filtered. Sufficient nitric acid is added to neutralize the carbonate, and after expelling the CO_2 by boiling, sodium hydroxide is added to the filtrate. A yellow precipitate is formed in presence of uranium. The precipitate is insoluble in an excess of the reagent, but dissolves in ammonium carbonate.

Uranous salts are green or blue and form green or bluish-green solutions, from which alkalies precipitate uranous hydroxide, reddish brown, insoluble in excess, but readily dissolved by ammonium carbonate. Uranous salts are strong reducing agents.

Uranyl salts ($UO_2 \cdot R_2$) are yellow. Alkali carbonates give a yellow precipitate, soluble in excess. UO_2 is regarded as a basic radical, known as "uranyl." The radical migrates to the cathode, upon electrolysis of a uranyl solution. Uranyl salts are more stable than uranous and are better known.

Potassium ferrocyanide, $K_4Fe(CN)_6$, added to uranous or uranyl solutions gives a reddish brown precipitate (or a red color in dilute solutions). The precipitate dissolves in a large excess of HCl. If sufficient ferrocyanide is present the color changes to green on boiling. Addition of sodium hydroxide to the ferrocyanide precipitate of uranium changes the color to yellow. (Distinction from cupric ferrocyanide. Ferrocyanide gives a green precipitate with vanadium, the color deepens on addition of nitric acid. A blue color is produced with ferric iron. No color change with chromates. Distinction from vanadium, chromium and iron.)

Barium carbonate precipitates the uranic ion completely (distinction from the ions of nickel, cobalt, manganese, zinc).

Disodium hydrogen phosphate added to uranyl solutions, in presence of alkali acetates or free acetic acid gives a yellowish white precipitate, $UO_2HPO_4 \cdot x\ H_2O$, soluble in mineral acids. Warming promotes precipitation.

Tartaric acid, certain organic compounds, hydroxylamine hydrochloride, ammonium carbonate, prevent precipitation of uranium by alkalies and ammonia.

Oxides UO_2, brown or black; UO_3, brick red; $UO_2(OH)_2$, yellow. All oxides are converted to U_3O_8 on ignition with free access of air.

Chapter by Wilfred W. Scott and Albert H. Low.

ESTIMATION

The element occurs in the following minerals:[2]

Pitchblende, or uraninite, containing 40 to 90% U_3O_8.
Autunite, $Ca(UO_2)_2P_2O_8 \cdot 8H_2O$, contains 55 to 62% UO_3.
Torbernite, $Cu(UO_2)_2 \cdot P_2O_5 \cdot 8H_2O$, contains 57 to 62% UO_3.
Carnotite, a vanadate of potassium and uranium, $V_2O_5 \cdot U_2O_3 \cdot K_2O \cdot 3H_2O$.
Samarskite, a urano-tantalate of iron and yttrium, etc., 10 to 13% UO_3.
Fergusonite, a columbate of cerium, uranium, yttrium, calcium and iron.

Nearly all the silicates, phosphates and zirconates of the rare earths contain uranium.

The element is used in the ceramic industry for producing yellow, brown, gray, and velvety-black tints. It produces canary-yellow glass. It is used as a mordant in dyeing of silk and wool. It also finds use in photography. The metal is used in cigarette-lighters and self-lighting burners.

Preparation and Solution of the Sample

The element dissolves in hydrochloric and in sulphuric acids; less readily in nitric acid. It is insoluble in alkaline solutions.

The oxide, UO_2, dissolves in nitric acid and in concentrated sulphuric acid.

U_3O_8 is readily soluble in nitric acid, but dissolves with difficulty in hydrochloric acid. V_2O_5 dissolves with difficulty in nitric acid but easily in hydrochloric (red colored solution); U_3O_8 is readily soluble in a mixture of glacial acetic-nitric acids (100 : 5), V_2O_5 and Fe_2O_3 (ignited) are practically insoluble in this reagent.

The salts, UF_4 and $UO_2(HPO_4)_2 \cdot 4H_2O$, are insoluble in water, but dissolve in strong mineral acids.

Solution of Ores. One gram or more of the ore is dissolved with 15 to 20 cc. of aqua regia, by placing the mixture first on the steam bath for ten to fifteen minutes and then gently boiling over a low flame or on the hot plate. The solution is taken to dryness, silica dehydrated as usual, the residue treated with 10 cc. of hot dilute hydrochloric acid and diluted to about 50 cc. with hot water and the silica filtered off. Uranium passes into the filtrate. The solution is now treated as directed under Separations. If much silica or acid-insoluble matter is present, this should be treated in a platinum dish with strong hydrofluoric acid, and evaporated twice on the steam bath with hydrochloric acid to expel HF. The residue, dissolved with hydrochloric acid and water, is added to the first portion of solution obtained.

Carnotite. Solution of the ore is readily effected by boiling with nitric acid to which a little hydrofluoric acid is added. One gram of ore with 20 cc. nitric acid and 5 cc. hydrofluoric acid at boiling temperature will be completely decomposed in five minutes. Some authorities recommend addition of sulphuric acid with ores containing barium to break up the combination of barium and uranium. If the lead acetate separation of vanadium is used, the sulphuric acid should be expelled previously to this separation. Consult the gravimetric procedures following the section on "Separations."

SEPARATIONS

Separation of Uranium from Copper, Lead, Bismuth, Arsenic, Antimony and the Other Members of the Hydrogen Sulphide Group. The solution containing uranium, etc., having an acidity of about 5 cc. strong HCl per 100 cc. of solution, is saturated with hydrogen sulphide and allowed to settle and again saturated with H_2S. The sulphides are filtered off and washed. The filtrate and washings contain the uranium that was present in the sample.

Separation of Uranium from Iron and from Elements Having Water-insoluble Carbonates. The filtrate from the hydrogen sulphide group is concentrated to about 150 cc., and 15 cc. of hydrogen peroxide added. The solution is now neutralized with sodium carbonate and about 3 grams added in excess. After boiling for about twenty minutes, renewing the water evaporated, the hydroxide of iron, insoluble carbonates, etc., are filtered off, washed with hot water and the filtrate set aside for the determination of uranium. To recover any occluded uranium the precipitate is dissolved in just sufficient nitric acid to effect solution, and iron again precipitated by addition of hydrogen peroxide and sodium carbonate and boiling as directed above. The combined filtrates from this precipitate are concentrated to about 250 cc.

Separation of Uranium from Vanadium. Procedure 1. *To be Used in the Gravimetric Determination of Uranium.* The solution obtained as directed, under the previous separation, is acidified with nitric acid, adding a slight excess, and CO_2 expelled by boiling. The acid is now neutralized with ammonia in slight excess, then re-acidified with nitric acid in slight excess, finally adding about 4 cc. of the strong acid additional. Vanadium is now precipitated as lead vanadate by adding 10 cc. of a 10% solution of lead acetate, followed by sufficient strong ammonium acetate solution (1 vol. strong NH_4OH + 1 vol. H_2O + sufficient glacial acetic acid to neutralize the NH_4OH) to neutralize the free nitric acid (about 20 cc.). The precipitated lead vanadate is allowed to settle for a couple of hours on the steam bath, and is then filtered off (returning the first portions if not perfectly clear) and washed well with hot water. The uranium passes into the filtrate.

The excess of lead remaining in the filtrate is next removed. A marked excess of ammonia is added to the hot filtrate, which is then boiled for about a minute and filtered. No washing required. The precipitate contains all the uranium as ammonium uranate, perhaps some ferric and aluminum hydroxides and a portion of the lead. Most of the lead passes into the filtrate, which is discarded. The precipitate is now treated on the filter with a strong hot solution of ammonium carbonate containing a little free ammonia until all the uranium is dissolved and then washed with the same solution diluted. Most of the admixed lead and other impurities remain on the filter, the lead as carbonate. Sufficient strong hydrogen sulphide water is now added to the filtrate, or the gas is passed to precipitate all the lead remaining. The mixture is heated to boiling and then allowed to stand until clear. Any iron present is precipitated with the lead. Finally, the uranium solution is filtered off and the precipitate washed with hydrogen sulphide water containing a little ammonium carbonate. The filtrate is boiled to expel the hydrogen sulphide and ammonium carbonate and concentrated to 200–250 cc. Uranium is now precipitated and determined by the gravimetric procedure given on page 581a.

Procedure 2. *To be Used in the Volumetric Determination of Uranium.* The separation of vanadium from uranium may be effected by precipitation of the latter as a phosphate according to the following procedure. The solution is heated and allowed to run in a small stream through a funnel with constricted stem, into a boiling solution of 15 grams of ammonium acetate, 5 grams of micro-cosmic salt dissolved in 100 cc. of water containing about 5 cc. of glacial acetic acid. A rod, with a cup-shaped tip, placed in the solution prevents bumping. The mixture is allowed to boil for a few minutes, the beaker is then removed from the heat and the precipitate allowed to settle. This is now transferred to a filter after first decanting off the clear solution. It is washed once with hot water, then washed back into the beaker and dissolved in a small amount of hot dilute nitric acid, the precipitate clinging to the filter being dissolved off by the acid, which is allowed to run through the filter into the beaker. This nitric acid solution containing the vanadium is diluted to about 75 cc. and the uranium (together with aluminum if present) again precipitated as the phosphate according to the procedure described. The precipitate is again transferred to the filter previously used, and washed off with hot water four or five times. Vanadium passes into the filtrate. The phosphate is now dissolved off the filter with 15 cc. of hot dilute sulphuric acid (1 : 3), and uranium determined by titration with permanganate according to the directions given under the volumetric method described later.

Glacial Acetic Method for Separating Uranium from Vanadium. * Uranium nitrate or oxide dissolves readily in a mixture of glacial acetic acid and nitric acid, 20 parts of the former to 1 part of the latter. The nitrate and oxide of vanadium are insoluble in the reagent. Addition of water causes vanadium to dissolve. See method for determining uranium in Carnotite under the "Gravi-metric" methods.

Separation of Uranium from Molybdenum, Tungsten and Vanadium. * The residue obtained by evaporating a mixture of uranyl nitrate and nitric acid with ammonium molybdate, or sodium tungstate, or sodium vanadate to dryness, is slightly moistened with nitric acid (5 cc. $HNO_3 - 1.42$) and extracted with ether; uranium dissolves completely while molybdenum, tungsten and vanadium remain insoluble. The evaporation is conveniently conducted in a glass boat of a size that may be placed in the paper thimble of a Soxhlet extraction apparatus, commonly used with volatile solvents. The extraction is generally complete after the ether has siphoned over five or six times.

* Peligot, Ann. Chem. Phys., (3) 5 (1842). C. A. Pierlé, Jour. Ind. Eng. Chem. 21, 60 (1920).

GRAVIMETRIC DETERMINATION OF URANIUM AS THE OXIDE, U_3O_8

Procedure. The filtrate containing the uranium, as obtained according to the method given, under "Separations," is made slightly acid with nitric acid and boiled for a short time to ensure the absence of CO_2, then ammonia is added in marked excess, the mixture boiled for about a minute more and filtered. No washing necessary. Either paper or a weighed Gooch crucible may be used for the filtration. The precipitate is dried and ignited to the oxide U_3O_8, in which form it is weighed.

$$U_3O_8 \times 0.8482 = U.$$

NOTES. The purity of the oxide may be ascertained by dissolving in HNO_3 and testing for vanadium with H_2O_2 and for Al_2O_3 by adding $(NH_4)_2CO_3$.

Treadwell recommends that the oxide be reduced by hydrogen passed over the red-hot residue, the brown UO_2 being formed. The oxide is cooled in a current of hydrogen.

GRAVIMETRIC METHOD FOR URANIUM IN ORES *

Take 0.5 gram of the finely ground ore, or more, according to richness. Treat by heating gently in an 8-oz. "copper flask" with nitric or hydrochloric acid, or both, together with about 1–2 cc. of hydrofluoric acid, to effect complete solution of the uranium. Sometimes galena is present, in which case it is best to start with 10 cc. or more of hydrochloric acid and heat until the galena is decomposed. Whenever hydrochloric acid is used, boil almost to dryness to expel most of it before continuing. To this residue, or to the original ore, if hydrochloric acid appeared unnecessary, add 10 cc. of nitric acid and 1–2 cc. of hydrofluoric acid. Boil very gently to effect complete decomposition, and finally to approximate dryness. Allow to cool, add 3 cc. of nitric acid and 50 cc. of hot water, and see that everything soluble is dissolved.

Now make slightly alkaline with ammonia, then just acid with nitric acid, and again alkaline with a little solid ammonium carbonate, followed by about 5 cc. of strong ammonia and 3–4 grams more of ammonium carbonate.

Boil for about a minute and then filter, having a wetted wad of absorbent cotton in the apex of the filter. Wash twice with hot water. Boil and concentrate the filtrate in a covered beaker during the next step.

Dissolve the precipitate on the filter with a little hot dilute nitric acid, receiving the filtrate in the original flask. Again neutralize and precipitate as before, washing this second precipitate well with hot water. Add the filtrate to the first one and continue the concentration to 150–200 cc. Now acidify with nitric acid, and then, in case of doubt, add about 1 cc. of hydrogen peroxide. A reddish brown color indicates vanadium.

A. *Vanadium Present.*—Boil to expel any remaining CO_2, make just alkaline with ammonia, then just acid with nitric acid, finally adding about 4 cc. of the latter in excess. The appearance of the liquid is usually a sufficient indication of the neutralization points. Now add 1 gram of lead acetate crystals and then sufficient ammonium acetate solution (about 20 cc.) to neutralize the nitric acid and precipitate the lead vanadate. Boil for about 10 minutes and then filter through a double filter, returning the first portions if not perfectly

* By Albert H. Low.

clear. Wash with hot water. Receive the filtrate in a large beaker. If bulky, boil down to perhaps 200–250 cc. Now add ammonia in marked excess and boil for a minute to expel any CO_2. Filter hot,[1] paying no attention to a turbid filtrate unless it is yellowish (in which case wash the precipitate once with hot water, re-acidify the filtrate with nitric acid, heat to boiling and again precipitate with ammonia, filtering through the previous precipitate). No washing required. Place the last beaker under the funnel and fill the latter with a strong hot solution of ammonium carbonate, to which some free ammonia has been added. Usually one filling is sufficient to dissolve all the uranium and leave a white residue of lead carbonate, perhaps slightly discolored by a trace of iron. Wash with hot water, using a little more of the ammonium carbonate solution, if apparently necessary. Add to the filtrate sufficient strong hydrogen sulphide water to precipitate all the remaining lead (ordinarily 25 cc. of strong hydrogen sulphide water), or pass the gas for a short time. This also removes traces of iron. Heat to boiling, then allow to stand and settle. Filter, washing with hydrogen sulphide water containing some ammonium carbonate. Boil to expel the sulphide, then acidify with nitric acid and boil off all CO_2. Continue according to C.

B. *Vanadium Absent.*—Boil the nitric acid solution sufficiently to expel all CO_2, then add ammonia in marked excess and boil a little longer to expel any CO_2 in the ammonia. Filter the hot mixture, returning the first portions if not perfectly clear. No washing required. Dissolve the uranium on the filter with hot ammonium carbonate solution, as described in the last paragraph, and continue from this point as in the same situation above. Do not omit the hydrogen sulphide treatment, for, even in the absence of lead, there will usually be traces of iron to be removed. Continue according to C.

C. Add ammonia in marked excess, boil well for one minute and then filter through an ashless filter, returning the first portions if not clear. No washing required. Ignite filter and precipitate thoroughly in a porcelain crucible and weigh, after cooling, as U_3O_8. Impurities are usually present.

Dissolve the residue in the crucible by warming with a little nitric acid. Dilute and test for vanadium with hydrogen peroxide. A faint brownish tinge may be neglected. Rinse the solution into a small beaker, add solid ammonium carbonate in excess, boil a minute or two and then filter through a small filter, washing with hot water. The residue on the filter may consist of alumina and other insoluble matter. Ignite filter and residue in the original crucible, weigh and deduct the weight from that of the impure U_3O_8 previously found.

Ammonium Acetate Solution.—Eighty cc. of strong ammonia, 100 cc. of water and 70 cc. of 90 per cent. glacial acetic acid.

NOTE.—A yellow filtrate from the ammonium uranate indicates incomplete precipitation. This may be due to a deficiency in ammonium nitrate, as ammonium uranate is perceptibly soluble in pure water. Add a gram or so of ammonium nitrate to the filtrate, boil and refilter. Or, better, dissolve the precipitate on the filter with dilute nitric acid, so that the mixed filtrates will be markedly acid, and repeat the precipitation with ammonia. The filtrate should be colorless.

GLACIAL ACETIC ACID METHOD FOR DETERMINING URANIUM IN CARNOTITE*

The following method depends upon the fact that uranium nitrate or oxide is soluble in a mixture of glacial acetic and nitric acids in the proportion of 20 parts by volume of the former to 1 part of the latter, while vanadic nitrate and oxide (V_2O_5) are not.[1]

Procedure

Half a gram or more of carnotite ore according to its richness (ground to pass 100-mesh sieve) is taken for analysis and digested at boiling temperature with 25 cc. dilute HNO_3 (1 : 1) and 1–2 cc. HF. (An amount that will fill a small crucible lid.) The solution is rapidly evaporated to dryness and baked gently to expel water, but not ignited.

Fifteen to 20 cc. of glacial acetic-nitric reagent (20 : 1) are added, rinsing down the sides of the beaker to remove any adhering material, using a policeman if necessary. (The reagent may be conveniently handled in a small wash bottle, the transferring of precipitates and washing with the reagent being necessary, no water being used at this stage.) The residue transferred to a filter is washed with the reagent five or six times, using small portions of the mixture.

The filtrate and glacial washings are rapidly evaporated to dryness and the residue again extracted with glacial-nitric acids.[2] This extract, free from vanadium,[3] is evaporated to dryness, and gently heated over a free flame until the residue turns dark.[4] Ten cc. of nitric acid and 40 cc. of water are added and the mixture heated to dissolve the uranium.

The greater part of the free nitric acid is neutralized by addition of ammonia (no permanent precipitate should form). Solid ammonium carbonate is added (covering the beaker during the intervals between additions of the carbonate as loss will occur through effervescence in an uncovered beaker) until a precipitate forms that remains undissolved on stirring. 2–3 grams additional ammonium carbonate and 5 cc. of ammonium hydroxide are now added and the solution warmed to coagulate the hydroxides of iron and aluminum. Uranium passes into solution.

The precipitate is filtered off and washed with hot water. The filtrate and washings (concentrated by boiling to about 150 cc.–200 cc. if the volume is large) is acidified with nitric acid (uranium precipitates andre dissolves).[5] Carbon dioxide is expelled by boiling and a decided excess of ammonium hydroxide added. The boiling is now continued until uranium precipitates completely. If the supernatant solution is yellow, it is again acidified with nitric acid, followed by an excess of ammonium hydroxide and the boiling repeated. This generally effects complete precipitation.[6]

The precipitated uranium is filtered off, washing being unnecessary. The filter and precipitate are placed in a crucible, the greater part of the water expelled by drying and the material then ignited. The greenish black residue is weighed as U_3O_8.

This residue should be soluble when boiled with nitric acid.[7] If it is not, contamination by iron and aluminum is indicated. Any residue remaining should be filtered off, then washed free of uranium with hot water, and ignited. Its weight is subtracted from the uranium oxide to obtain the true value of U_3O_8.

Notes. 1. C. A. Pierlé, Jour. Ind. Eng. Chem. 12, 61, 1920.

* Method by Wilfred W. Scott.

2. Uranium dissolves completely in the glacial-acetic mixture. A small amount of the vanadium may dissolve, hence the extract is evaporated and the residue again extracted. A smaller quantity of the reagent may be used in this second extraction. 25 cc. of the reagent will dissolve, at boiling temperature, about 4.5 grams of U_3O_8 in five minutes and about 0.003 gram of V_2O_5. The proportion of glacial acetic acid to nitric acid should not fall below 10 : 1, otherwise vanadium will dissolve in appreciable amount. The author uses a round bottom flask filled with cold water, placing this over the beaker to act as a condenser to prevent loss of acetic acid.

3. The acetic acid extract filters rapidly. The red colored residue contains the vanadium and practically all of the silica, iron and alumina.

4. The residue is ignited to destroy organic matter which would prevent precipitation of uranium by ammonia.

5. Uranium carbonate precipitates and then dissolves in the acid when present in excess. CO_2 must be expelled as this prevents precipitation of uranium by ammonium hydroxide.

6. A colored solution indicates the presence of uranium. The nitrate formed by acidification with HNO_3 and making alkaline with ammonia with additional boiling insures complete precipitation of uranium.

7. Nitric acid dissolves uranium oxide very easily. The oxide of iron is practically insoluble. Vanadic oxide difficultly soluble.

8. Hydrogen peroxide added to the nitric extract will produce a reddish brown color if vanadium is present.

Wash water used in transferring the uranium precipitate should contain ammonium nitrate to prevent solution of uranium.

DETERMINATION OF URANIUM IN CARNOTITE

The following method, worked out by C. E. Scholl,[1] is adapted for eliminating the difficulties in an accurate determination of uranium, caused by the presence of iron, alumina and vanadium, which commonly occur in carnotite.

Procedure. To the sample of material containing about 0.2 gram uranium oxide, are added 25 to 50 cc. 1 : 1 nitric acid and heat applied until all the uranium is in solution. If necessary, allow to stand on the water bath to keep warm overnight. After diluting with warm water to 250 cc. the solution is filtered. Ferric chloride, equivalent to about three times the weight of vanadium present, is added. To the cold or slightly warm solution solid sodium carbonate is now added in small portions until all the acid is neutralized and then an excess of 1 gram added, the beaker being kept covered during the intervals between the additions. The sample is placed on a hot plate and heated to about 90° C., but not to boiling. After keeping hot for 15 minutes, the solution is filtered and washed. The residue contains all the iron and vanadium and the greater part of the aluminum. The filtrate is neutralized cautiously with nitric acid until uranium begins to precipitate. The greater part of the CO_2 is removed by boiling. Sodium hydroxide is now added in excess and after boiling 15 minutes the solution is filtered. The filtrate contains the remainder of the aluminum and the vanadium not previously precipitated. The precipitate containing the uranium is dissolved in dilute nitric acid and heated to 90° C. and an excess of NH_4OH added and the solution boiled. The precipitate is filtered off, ignited and weighed as uranium oxide, U_3O_8.

The oxide is tested for its purity with nitric acid as in the previous method.

[1]Jour. Ind. Eng. Chem., Vol. 11, No. 9, p. 842, Sept., 1919.

VOLUMETRIC DETERMINATION OF URANIUM BY REDUCTION AND OXIDATION

Introduction. The determination of uranium by oxidation of the lower oxide UO_2 to UO_3 may be accomplished by means of potassium permanganate in precisely the same manner as in the determination of iron, the Jones reductor being used for the reduction of the uranic salt to the uranous form. The metal must be in solution either as a sulphate, a chloride or an acetate, but not as a nitrate. If present as a chloride the usual preventative solution of phosphoric acid and manganous sulphate solution must be present as in case of the titration of a chloride of iron, hence a sulphate solution is to be preferred. Although the degree of reduction varies with conditions, it is found that with brief contact with the oxygen of the air the oxide UO_2 is formed.[1]

Procedure. Solution. The method for preparation of the sample, isolation of the uranium, has been given under Preparation and Solution of the Sample and Separations. The solution from the ammonium carbonate precipitate is acidified with sulphuric acid and boiled to expel the CO_2.

Reduction. The uranium sulphate solution, diluted to a volume of 100 to 150 cc., containing one-sixth of its volume of sulphuric acid, is heated nearly to boiling and the organic matter that may be present oxidized by addition of just sufficient potassium permanganate solution to produce a faint pink color. Fifteen to 20 cc. of dilute sulphuric acid are passed through the 18-in. column of zinc in the Jones reductor, followed by the hot uranium sulphate solution, flowing very slowly, fifteen to twenty-five minutes being required for 0.2 gram uranic oxide, thirty to forty minutes for 0.3 gram of the oxide, care being taken that the liquid in the reductor always covers the zinc.[2] The uranic solution is followed by 10 to 15 cc. of dilute 1 : 6 solution of sulphuric acid.

Titration. The olive-green solution is poured into a beaker or casserole. The lower oxides are immediately oxidized to UO_2 by the air, as seen by the slight change of color to sea green. The hot solution is now titrated with tenth normal permanganate. The solution during titration gradually becomes more and more yellowish green, as the highest oxidation is approached, until a faint pink color is obtained. With large amounts of uranium the color appears a yellowish pink.

One cc. $N/10$ $KMnO_4 = 0.01193$ gram U.

NOTE. 55.85 grams Fe is equivalent to 119.25 grams U.—Sutton.

C. A. Pierlé[2] claims that high results are obtained by this method.

[1] Oxidation of lower oxides by air to UO_2''. O. S. Pulman, Jr. Am. Jour. Sc. (4), **16**, 229.

[2] Hydrogen dioxide formed by nascent hydrogen in contact with air would vitiate results.—Gooch.

Determination of Uranium in Alloy Steels and Ferro=Uranium [1]

The following method provides for the analysis of steels containing Cr, Mo, V, W, Co, Ni, C, Mn, Si, Al and Ti.

A 2 g. sample is dissolved in 75 cc. of 1 : 1 hydrochloric acid. After solution is complete the solution is oxidized by the dropwise addition of nitric acid. In the case of samples where tungsten is present an easily filterable product is obtained by diluting to 300 cc. and boiling for 15 min. The tungstic oxide is then filtered out and washed, the filtrate and wash waters being returned to the original beaker for evaporation to dryness, followed by baking at a moderate temperature. On dissolving the residue with 50 cc. of 1 : 1 hydrochloric acid and diluting with hot water, a solution is obtained from which the balance of the silica and the last traces of tungsten can be separated by filtering. The two precipitates after washing are available for the determination of tungsten and silicon by the usual methods. Filtrates and wash waters from these precipitates are combined and evaporated to a syrupy consistency in preparation for the extraction of most of the iron with ether. In the absence of tungsten the original solution is evaporated to dryness and baked with the object of removing silica. After the extraction of the iron, the aqueous layer is evaporated to a small volume to free it from the excess of acid. It is then diluted to a volume of 150 cc. with hot water, and an excess of sodium carbonate in the form of a saturated solution is added. This solution is boiled and, after settling, filtered, the precipitate being washed with hot water. The precipitate consists of the hydroxides of chromium, iron, manganese, cobalt, nickel, copper, and aluminum, if all of these elements are present, together with traces of silica, titanic oxide, phosphorus, and vanadium compounds. The filtrate contains uranium, molybdenum, vanadium, and traces of the elements which occur chiefly in the precipitates.

Bulky precipitates should be dissolved in hydrochloric acid and reprecipitated one or more times with sodium carbonate solution to insure a complete separation of the uranium.

All filtrates from the precipitate are cautiously acidified with sulphuric acid and boiled long enough to insure the complete removal of all carbon dioxide. Ammonia free from carbonate is then added in slight excess. Boiling precipitates the uranium, much of the vanadium, and traces of impurities. The molybdenum is left in the filtrate. Steels contain only small amounts of phosphorus and the contamination of the uranium from this source is usually negligible. If the amount of phosphorus is large, it may be necessary to dissolve the precipitate in nitric acid, and after suitable oxidation, precipitate the phosphoric acid with ammonium molybdate. The phosphorus can then be removed as ammonium phosphomolybdate. The uranium and vanadium may be reprecipitated from this filtrate along with the manganese, if permanganate is used to oxidize the phosphorus, by adding a few drops of sulphuric acid, a small amount of ammonium persulphate, and enough carbonate-free ammonium hydroxide to give an excess. The precipitate obtained by boiling the solution is in the condition corresponding to the first uranium precipitate mentioned above.

The impure uranium precipitate containing phosphorus in negligible

[1] By G. L. Kelley, F. B. Myers and C. B. Illingworth, Jour. Ind. & Eng. Chem., Apr., 1919.

amounts, or free from it, is transferred to a beaker with a little water and solid ammonium carbonate added. On heating this solution under conditions and for a time calculated to result in only a partial decomposition of the ammonium carbonate, the uranium and vanadium go into solution leaving the manganese, iron, and other impurities undissolved. The filtrate is acidified with sulphuric acid and boiled until free from CO_2, when a slight excess of carbonate-free ammonium hydroxide is added. This precipitates only the uranium and vanadium.

The combined precipitates of uranium and vanadium are ignited at dull redness in a platinum crucible, allowing free access of air to reoxidize any reduced material. The ignited residue is weighed as $U_3O_8 + V_3O_5$. In general, only a small part of the vanadium is present in this precipitate, thus making it unavailable for the vanadium determination. It is necessary, however, to determine the vanadium to correct the weight of uranium oxide. This may be done by almost any of the several known methods for determining vanadium. To this end we determine the vanadium after reduction with hydrochloric acid by permanganate titration, and by oxidation with ammonium persulphate and silver nitrate, followed by electrometric titration. The latter method is the more certain and convenient, but the former gives entirely satisfactory results. For the purpose of either method the precipitate is dissolved in 50 cc. of concentrated hydrochloric acid and evaporated with 30 cc. of sulphuric acid (sp. gr. 1.58) until fumes appear. When the titration is to be completed with permanganate, the sulphuric acid solution is diluted to 250 cc. with hot water and titrated at 80° C. to the first pink color. At the same time like quantities of sulphuric and hydrochloric acids are evaporated, diluted, and titrated in similar fashion to obtain a blank correction for the vanadium. When the titration is to be made electrometrically, the sulphuric acid solution is diluted to 250 cc. with hot water, oxidized with silver nitrate and ammonium persulphate and titrated with ferrous sulphate. The weight of vanadium so found is multiplied by 1.784 to convert it into the corresponding weight of the oxide V_2O_5. This weight is subtracted from the weight of the residue $U_3O_8 + V_2O_5$. The corrected weight of the oxide U_3O_8 is converted into the corresponding weight of uranium by multiplying by 0.8483 from which the percentage of uranium can be calculated.

VANADIUM

V, *at.wt.* 50.96; *sp.gr.* 6.025; *m.p.* 1720° C.; *oxides* V_2O, V_2O_2, V_2O_3, V_2O_4, V_2O_5; *vanadates—meta* $NaVO_3$, *ortho* Na_3VO_4, *pyro* $Na_4V_2O_7$, *tetra* $Na_3HV_6O_{17}$, *hexa* $Na_2H_2V_6O_{17}$.

DETECTION

Ammonium Sulphide or Hydrogen Sulphide passed into an ammoniacal solution of vanadium precipitates brown V_2S_5, soluble in an excess of alkali sulphide and in alkalies, forming the brownish-red thio- solution, from which the sulphide may be reprecipitated by acids.

Reducing Agents. Metallic zinc, sulphites (SO_2), oxalic acid, tartaric acid, sugar, alcohol, hydrogen sulphide, hydrochloric acid, hydrobromic and hydriodic acids (KI) reduce the acid solutions of vanadates with formation of a *blue-colored* liquid. (See Volumetric Methods.) Reduction is hastened by heating.

Hydrogen Peroxide added to a cold acid solution of vanadium produces a *brown color*, changing to blue upon application of heat.

Solid Ammonium Chloride added to a neutral or slightly alkaline solution of a vanadate precipitates the colorless, crystalline salt, NH_4VO_3, insoluble in ammonium chloride. The ammonium metavanadate ignited is decomposed, ammonia volatilizing and the red pentoxide of vanadium remaining as a residue.

The colorless ammonium vanadate solution becomes yellow when slightly acidified. Acids produce a red color when added to the solid salt.

The oxide, V_2O_5, is distinguished from Fe_2O_3 by the fact that it fuses very readily with the heat of Bunsen burner, whereas the oxide of iron, Fe_2O_3, is infusible in the heat of a blast lamp. M.p. $V_2O_5 = 658°$ C.; m.p. $Fe_2O_3 = 1548°$ C.

Comparison of Vanadium and Chromium Salts. Vanadium, like Chromium, forms a soluble salt upon fusion with sodium carbonate and potassium nitrate or with sodium peroxide. The solution of vanadates and of chromates are yellow or orange; the color of the chromate becomes more intense when strongly acidified, whereas that of the vanadate is reduced. The yellow color of the vanadate solution is destroyed by boiling with an excess of alkali, but may be restored by neutralizing the alkali with acid. The chromate color is not destroyed. (Yellow with alkalies, orange in acid solution.) Silver nitrate produces a dark-maroon precipitate with a soluble chromate and an orange-colored precipitate with a vanadate; mercurous nitrate produces a red-colored precipitate with chromates and a yellow with vanadates. *Vanadates are also distinguished from chromates by the reduction test;* reducing agents such as a soluble sulphite, or sulphurous acid added to acid solutions, form *a blue-colored liquid with vanadates and a green color with chromates. Ammonium hydroxide* added in excess to the cold reduced solutions

---

[1] Reduction with zinc is rapid with vanadates, much less vigorous with chromates. V_2O_5 reduced to V_2O_2, color changes to blue, green, lavender and finally violet. SO_2 or H_2S reduces V_2O_5 to V_2O_4. V_2O_2 forms vanadyl salts.

Chapter by Wilfred W. Scott.

gives a brown color, or a brown to dirty green precipitate with *vanadium*, and violet or lavender color or a light green-colored precipitate with *chromium*, depending upon the concentration of the solutions. Hydrogen peroxide added to the reduced cold acid solutions changes the vanadium blue to reddish brown; the chromium green remains unchanged.

Detection of Vanadium in Steel. Five grams of the sample are dissolved in dilute nitric acid, the nitrous fumes boiled off, the solution cooled, and an excess of sodium bismuthate added. After filtering through an asbestos filter an excess of concentrated ferrous sulphate solution is added, and the solution divided into two equal parts in test-tubes. To one portion 10 cc. of hydrogen peroxide are added and to the other 10 cc. of water. If vanadium is present the peroxide solution will show a deeper color than the untreated solution. A deep red color is produced with high vanadium steels and a brownish-red with low. Since titanium also causes this color, it would interfere, if it were not for the fact that the color produced with titanium is destroyed by hydrofluoric acid and fluorides, whereas that of vanadium is not. In presence of titanium, 5 cc. of hydrofluoric acid are added to the treated sample.

The brown color produced by hydrogen peroxide, with vanadium solutions, will remain in the water portion when shaken with ether. The ether layer is colored a transient blue in presence of chromium.

ESTIMATION

The materials in which the estimation of vanadium is desired may be surmised from the following facts: Industrial application. Vanadium is used in special iron and steel alloys. It increases the strength of steel as well as the compression power, without loss of hardness, and increases the resistance to abrasion; hence vanadium steels are used in locomotive and automobile cylinders, pistons, bushings and in all parts of machines subject to jar. It is used in high-speed tools, vanadium bronzes for gears, trolley wheels, etc. It is used in indelible inks, and in the form of alkali vanadates and hypovanadates it serves as a mordant for aniline black on silk, for calico printing and like uses. Vanadium salts are used in ceramics where a golden glaze is desired.

The element occurs widely distributed in minute quantities. It is found in iron ores, hence occurs in blast-furnace slags as the oxide, V_2O_5. The principal ores are:

Patronite, a sulphide of vanadium containing 28 to 34% V_2O_5, associated with pyrites and carbonaceous matter; the principal source of vanadium.

Vanadinite, $(PbCl)Pb_4(VO_4)_3$, containing 8 to 21% V_2O_5.

Carnotite, $K_2O \cdot 2UO_2 \cdot V_2O_5 \cdot 3H_2O$, contains 19 to 20% V_2O_5.

Descloizite, $(PbZn)_2NVO_5$, contains 20 to 22% V_2O_5.

Roscoelite, a vanadium mica with variable composition.

Eusynchite, contains 17 to 24% V_2O_5.

Cuprodescloizite, $(PbZnCu)_2(OH)VO_4$, contains 17 to 22% V_2O_5.

Calciovolborthite, $(CuCa)_2(OH)VO_4$, contains 37 to 39% V_2O_4.

Vanadium occurs in ores of copper and lead, it is present in certain clays and basalts, in soda ash, phosphate soda, and in some hard coals.

Preparation and Solution of the Sample

In decompositon of the material for analysis the following facts regarding the solubility of the metal, its oxides and principal salts, will be helpful:

Element. The metal is not attacked by aqueous alkalies, but is soluble by fusion with potassium or sodium hydroxide, and sodium carbonate containing potassium nitrate. It is insoluble in dilute hydrochloric and sulphuric acids. It dissolves in concentrated sulphuric acid and in dilute and concentrated nitric acid forming blue solutions.

Oxides. V_2O_2 is easily soluble in dilute acids, giving a lavender-colored solution.

V_2O_3 is insoluble in hydrochloric and sulphuric acids, and in alkali solutions. It dissolves in hydrofluoric acid, and in nitric acid.

V_2O_4 is easily soluble in acids, forming blue-colored solutions. It dissolves in alkali solutions.

V_2O_5 is soluble in acids, alkali hydroxide and carbonate solutions. Insoluble in alcohol and acetic acid.

Salts. Ammonium meta vanadate, NH_4VO_3, is slightly soluble in cold water, readily soluble in hot water. The presence of ammonium chloride renders the salt less soluble. The vanadates of lead, mercury and silver are difficultly soluble in water. These are dissolved, or are transposed by mineral acids, the vanadium going into solution; i.e., lead vanadate treated with sulphuric acid precipitates lead sulphate and vanadic acid passes into solution.

General Procedure for Decomposition of Ores. One gram (or more) of the finely divided material is placed in a large platinum crucible together with five times its weight of a mixture of sodium carbonate and potassium nitrate ($Na_2CO_3 = 10$, $KNO_3 = 1$). The product is heated to fusion over a blast lamp and, when molten, about 0.5 to 1 gram more of the nitrate added in small portions. (Caution—platinum is attacked by KNO_3. A large excess of Na_2CO_3 tends to prevent this.) The material should be kept in quiet fusion for ten to fifteen minutes, when most of the ores will be completely decomposed. The cooled fusion is extracted with boiling water, whereby the vanadium goes into solution. Arsenic, antimony, phosphorus, molybdenum, tungsten and chromium pass into solution with the vanadium. These must be removed in the gravimetric determination of this element. (Iron remains insoluble in the water extract.)

Should there be any undecomposed ore, the residue from the water extract will be gritty. If this is the case, a second fusion with the above fusion mixture should be made.

Small amounts of occluded vanadium may be recovered from the water-insoluble residue by dissolving this in nitric acid and pouring the solution into a boiling solution of sodium hydroxide. Vanadium remains in solution.

Vanadium may be determined volumetrically after removal of the hydrogen sulphide group, by titration with potassium permanganate according to the procedure given later. The isolation and determination of vanadium by the gravimetric procedures are given in detail later.

Ores and Material High in Silica. The sample is treated in a platinum dish with about ten times its weight of hydrofluoric acid (10 to 50 cc.) and 2 to 5 cc. of strong sulphuric acid. The silica is expelled as SiF_4 and the hydrofluoric acid driven off by taking the solution to SO_3 fumes. The residue is extracted with hot water containing a little sulphuric acid. Any undissolved residue may be brought into solution by fusion with potassium acid sulphate, $KHSO_4$, and extraction with hot water containing a little sulphuric acid. By this treatment the iron passes into solution with vanadium.

Products Low in Silica. Decomposition may be effected by fusion in a nickel crucible with sodium peroxide and extraction with water. The water should be added cautiously, as the reaction is vigorous. One gram of the finely divided ore is intimately mixed with 3 to 4 grams of Na_2O_2 and 1 gram of the peroxide placed on the charge. The material is then fused as stated.[1]

Iron and Steel. The solution of the sample, isolation of vanadium and its volumetric determination are given at the close of the chapter.

Alloys. These may be decomposed with nitric acid, or aqua regia. The isolation of vanadium with mercurous nitrate or lead acetate are given under the gravimetric methods.

SEPARATIONS

Fusion with sodium carbonate and potassium nitrate and extraction of the melt with water effect a separation of vanadium from most of the metals, which remain insoluble as carbonates or oxides. Arsenic, molybdenum, tungsten, chromium and phosphorus, however, pass into the filtrate with vanadium.

Removal of Arsenic. This element generally occurs in vanadium ores. It may be removed when desired, by acidifying the water extract of the fusion with sulphuric acid, and after reducing arsenic with SO_2, precipitating the sulphide, As_2S_3 with H_2S gas. Vanadium passes into the filtrate.

Removal of Molybdenum. The procedure is similar to that used for arsenic, with the exception that the sulphide of molybdenum is best precipitated under pressure. The solution in a pressure flask is treated with H_2S. The flask is stoppered and heated in the steam bath. It is advisable to resaturate the solution with H_2S before filtering off the sulphide.

Separation from Phosphoric Acid. In the gravimetric procedure phosphorus and vanadium are precipitated together as mercuric vanadate and phosphate. The mercury is expelled by heat and the oxides V_2O_5 and P_2O_5 weighed. (V_2O_5 in presence of P_2O_5 does not melt as it does in pure form, but only sinters.) The oxides are fused with an equal weight of sodium carbonate, the melt dissolved in water, then acidified with sulphuric acid and vanadium reduced to the vanadyl condition by SO_2 gas. The excess of SO_2 is expelled by boiling and passing in CO_2. Phosphoric acid is now precipitated with ammonium molybdate (50 cc. of a solution containing 75 grams ammonium molybdate dissolved in 500 cc. of water and poured into 500 cc. nitric acid—sp. gr. 1.2) in presence of a large amount of ammonium nitrate and a little free nitric acid. It is advisable to dissolve the precipitate in ammonia and reprecipitate in presence of additional ammonium molybdate and nitrate by acidifying with nitric acid. The equivalent P_2O_5 is deducted from the weight of the combined oxides, the difference being due to V_2O_5.

NOTE. Vanadium must be completely reduced to the vanadyl form, as vanadic acid will precipitate with phosphoric acid.

[1] Direct reduction and titration of vanadium in presence of a large accumulation of salts leads to erroneous results. The vanadium should be separated by precipitation with lead acetate.

Separation of Vanadium and Chromium. A volumetric procedure for determining vanadium and chromium in the presence of one another is given. If a separation is desired the following procedures may be used:

A. The solution is acidified with nitric acid. If hydrochloric acid is present it is expelled by taking to near dryness twice with nitric acid, the residue is taken up with water and SO_2 gas passed in to completely reduce the vanadium. This solution is poured into a boiling solution of 10% sodium hydroxide. After boiling a few minutes, the solution is filtered and the residue washed. The filtrate contains vanadium, the residue chromium. It is advisable to pour the filtrate into additional caustic to remove the small amount of chromium that passes into the solution.

B. One hundred cc. of the neutral solution is made acid with about 15 cc. of glacial acetic acid and hydrogen peroxide added. The solution is boiled for a few minutes. Chromium is thereby reduced to Cr_2O_3, whereas vanadium appears as V_2O_5. Lead acetate will now precipitate lead vanadate, the reduced chromium remaining in solution. The lead vanadate now treated with strong sulphuric acid is decomposed upon heating. Addition of water precipitates $PbSO_4$, the vanadium remaining in solution.

GRAVIMETRIC METHODS

The following procedures presuppose that vanadium is present in the solution as an alkali vanadate, the form in which it occurs in the water extract from a fusion with sodium carbonate and potassium nitrate, as is described in the method of solution of ores containing vanadium. Chromium, arsenic, phosphorus, molybdenum and tungsten, if present in the ore will be found in this solution.

Mercurous Nitrate Method for Determination of Vanadium— Gravimetric [1]

Principle. A nearly neutral solution of mercurous nitrate precipitates vanadium completely from its solution. The dried precipitate ignited forms the oxide, V_2O_5, mercury being volatilized.

Procedure. To the alkaline solution or an aliquot portion of the water extract from the sodium carbonate potassium nitrate fusion nearly neutralized with nitric acid [2] (the solution should remain slightly alkaline) is added drop by drop, a nearly neutral solution of mercurous nitrate in slight excess of that necessary to precipitate completely the vanadium present, as may be determined by allowing the precipitate to settle and adding a few drops more of the reagent. The mixture is heated to boiling and then placed on the water bath or steam plate and the gray-colored precipitate allowed to settle. The precipitate is washed several times

---

[1] Method of Rose. J. W. Mellor, " A Treatise on Quantitative Inorganic Analysis."
[2] Should the alkaline solution of the vanadate be made acid, nitrous acid, from the nitrate fusion, will be liberated and cause reduction of the vanadate to the vanadyl salt, in which form it is not precipitated by mercurous nitrate; hence great care should be used in neutralizing the alkaline solution to avoid making it acid. It is a good practice to measure the acid added, having determined on an aliquot portion the amount necessary to add to neutralize the solution. This is readily accomplished when a comparatively large sample has been prepared for analysis and an aliquot portion taken for analysis, several determinations being made on the same fusion.

with water containing a few drops of mercurous nitrate, washing once or twice by decantation and finally on the filter paper. The precipitate is dried, then ignited in a porcelain crucible in a hood over a Bunsen burner to a red heat. The fused red residue is V_2O_5.

$$V_2O_5 \times 0.5604 = V.$$

Gravimetric Method of Determining Vanadium by Precipitation with Lead Acetate [1]

Principle. From a weakly acetic acid solution, vanadium is quantitatively precipitated by lead acetate. The precipitate is dissolved in nitric acid, lead removed as a sulphate, and vanadium determined in the filtrate by taking to dryness and igniting to the oxide, V_2O_5.

Procedure. To the alkaline solution or an aliquot portion obtained by extraction of the carbonate fusion of the ore with water, just sufficient amount of nitric acid is added to nearly neutralize the alkali present, as in the case of the method described for precipitation of vanadium by mercurous nitrate, and then a 10% solution of lead acetate is added in slight excess with continuous stirring. The precipitate is allowed to settle on the steam bath. The vanadate, first appearing orange colored, will fade to white upon standing. The lead vanadate is filtered and washed free of the excess of lead acetate with water containing acetic acid. The precipitate is washed into a porcelain dish with a little dilute nitric acid, and brought into solution by warming the lead salt with nitric acid. To this, the ash of the incinerated filter is added. Sufficient sulphuric acid is added to precipitate completely the lead, and the solution taken to small volume on the water bath and then to SO_3 fumes, but not to dryness. About 100 cc. of water are added and the mixture filtered; lead sulphate will remain upon the filter and the vanadium will be in solution. The lead sulphate is washed free of vanadium (i.e., until the washings no longer give a brown color with hydrogen peroxide).

The filtrate containing all the vanadium is evaporated to small volume in the porcelain dish, then transferred to a weighed platinum crucible and evaporated to dryness on the water bath and finally the residue (V_2O_5) heated to a dull redness over a Bunsen flame.

$$V_2O_5 \times 0.5604 = V.$$

NOTES. Lead may be separated from the vanadium by passing H_2S through the nitric acid solution, the excess of H_2S volatilized by boiling and the liberated sulphur filtered off. The filtrate is evaporated to dryness and the vanadium ignited with a few drops of nitric acid to the oxide V_2O_5.

Lead may also be separated as lead chloride in the presence of alcohol, the solution taken to dryness and vanadium oxidized by addition of nitric acid and ignited to V_2O_5.

[1] Method by Roscoe, Ann. Chem. Pharm., Supplement **8**, 102, 1872. Treadwell and Hall, " Analytical Chemistry," p. 305.

VOLUMETRIC PROCEDURES FOR THE DETERMINATION OF VANADIUM

Reduction of the Vanadate, V_2O_5, to Vanadyl Condition, V_2O_4, and Reoxidation with Potassium Permanganate

Principle. Vanadium in solution as a vanadate is reduced to the vanadyl salt by H_2S or SO_2, the excess of the reducing agent expelled and the solution titrated with standard $KMnO_4$, vanadium being oxidized to its highest form, V_2O_5.

Reactions. *a.* $V_2O_5 + SO_2 = V_2O_4 + SO_3$. *b.* $V_2O_4 + O = V_2O_5$. Hence

$$N/10 \text{ sol.} = \frac{\text{At. wt. V}}{10} \text{ grams to the liter.}$$

Procedure. An aliquot portion of the solution containing vanadium, as obtained by one of the procedures given for the solution of the sample, is taken for analysis; dilute sulphuric acid (1 : 1) is added to acid reaction and 5 cc. of acid per 100 cc. of solution added in excess. The vanadium content should be not over 0.5 gram V when a tenth normal permanganate is used for the titration. *If arsenic or molybdenum is present* these may be removed from the solution by passing in H_2S. The insoluble sulphides are filtered off and washed with H_2S water. The filtrate is boiled down to two-thirds its volume and the sulphur filtered off. In the absence of members of the H_2S group, this portion of the procedure is omitted.

Oxidation with $KMnO_4$. The solution containing the vanadium is oxidized by adding, from a burette, tenth normal potassium permanganate to a faint permanent pink. If the solution has been treated with H_2S, the vanadium is in the vanadyl condition, and the amount of permanganate required to oxidize the solution completely will give a close approximate value for the vanadium present, each cc. of N/10 $KMnO_4$ being equivalent to 0.0051 gram vanadium.

Reduction. The vanadite is now reduced to vanadyl salt by passing through the acid solution, containing approximately 5% free sulphuric acid, a steady stream of SO_2 gas. Reduction may also be accomplished by adding sodium metabisulphite, or sodium sulphite, to the acid solution. The excess SO_2 is now removed by boiling (a current of CO_2 passed into the hot solution will assist in the complete expulsion of the SO_2).

NOTE. $KMnO_4$ is reduced by SO_2.

Test for Iron. A drop test with potassium ferricyanide, $K_3Fe(CN)_6$, on a white tile will give a blue color in the presence of ferrous iron. Since ferrous iron will titrate with potassium permanganate, its oxidation is necessary. This is accomplished by adding tenth normal potassium dichromate solution cautiously to the cold liquid until no blue color is produced by the spot test with $K_3Fe(CN)_6$ outside indicator. If the sample is sufficiently dilute, the blue color of the vanadyl solution will not interfere in getting the point where the iron is completely oxidized. Care must be taken not to pass this end-point, otherwise Va_2O_4 will also be oxidized and the results will be low.

NOTE. The action of the dichromate is selective to the extent that iron is first oxidized and then V_2O_4. If the amount of iron present is large a separation must be effected. In case a sodium carbonate potassium nitrate fusion has been made and vanadium has been extracted by water, iron will not be present. A special procedure for determination of vanadium in steel is given.

Potassium Permanganate Titration. N/10 $KMnO_4$ is now cautiously added until a pink color, persisting for one minute, is obtained. During the titration the solution changes from a blue color to a green, then a yellow and finally a faint pink. The reaction towards the end is apt to be slow if made in a cold solution.

NOTES. In absence of chromium, it is better to make the titration in a hot solution, 60 to 80° C., the end-point being improved by heat. In case an excess of permanganate has been added, the excess may be determined by a back titration with tenth normal thiosulphate. The solution may be rerun, if desired, by repeating the reduction with SO_2 and the titration with $K_2Cr_2O_7$ and $KMnO_4$.

One cc. N/10 $KMnO_4 = 0.0051$ gram V, or $= 0.00912$ gram V_2O_5.

For solutions containing less than 0.5% vanadium a weaker permanganate reagent should be used. A fiftieth normal permanganate solution will be found to be useful for materials low in vanadium.

The author obtained excellent results by the above procedure on materials containing small amounts of iron and chromium; with amounts equal to that of vanadium present in the solution no interference was experienced. The titration with potassium permanganate is made in cold solutions if chromium is present, as the permanganate will oxidize chromium in hot solutions. Potassium permanganate added to samples containing chromic salts, and the mixture boiled, will oxidize these quantitatively to chromates. This reaction does not take place in cold solutions to any appreciable extent during a titration and only slowly in warm solutions.

Volumetric Determination of Vanadium by Reduction with Zinc to V_2O_2

The procedure proposed by Gooch and Edgar is to reduce vanadic acid, in presence of sulphuric acid, by zinc to the oxide, V_2O_2; oxidation of the unstable V_2O_2 by the air is anticipated by means of ferric chloride or sulphate, in the receiver of the Jones reductor, the highest degree of reduction being registered by the ferrous salt formed by the reaction of the reduced vanadate on the ferric salt, i.e., $V_2O_2 + 3Fe_2O_3 = 6FeO + V_2O_5$. Compounds reduced by zinc and oxidized by $KMnO_4$ must be absent or allowed for.

Procedure. The Jones reductor is set up as directed in the procedure for the determination of iron by zinc reduction. The receiver attached to the tube containing the column of zinc is charged with a solution of ferric alum in considerable excess of that required for the oxidation of the reduced vanadium. (The amalgamated zinc is cleaned by passing through the column, a dilute solution of warm sulphuric acid. The final acid washings should show no further reducing action on permanganate when the reductor is clean.)[1] Gentle suction is applied, and through the column of clean amalgamated zinc are passed in succession—100 cc. of hot water, 100 cc. of 2.5% sulphuric acid, and then the solution of vanadic acid diluted to 25 cc. in a 2.5% sulphuric acid solution, and finally 100 cc. of hot

[1] Corrections should be made for the action of zinc upon the reagents without the vanadic acids, as it is almost impossible to get a condition where no blank is obtained with permanganate. The reductor is cleaned first by passing about 500 cc. of dilute 2.5% sulphuric acid through the column of zinc. A blank is now obtained with the same quantity of reagents as is used in the regular determination, only omitting the vanadium, and this is deducted from the titration obtained for each sample reduced.

water. To the receiver is added a volume of 4 cc. of syrupy phosphoric acid to decolorize the solution. The reduced iron salt is now titrated with $N/10$ $KMnO_4$.

One cc. $N/10$ $KMnO_4 = 0.0017$ gram V, or $= 0.00304$ gram V_2O_5.

Determination of Vanadium in Steel

The following method is used in analyzing the Bureau's vanadium and chrome-vanadium steels. The procedure was worked out by J. R. Cain and L. F. Witmer of the U. S. Bureau of Standards.

Five to 10 grams of drillings are dissolved in hydrochloric acid (1 : 1), a few drops of hydrofluoric are added, and the solution is boiled for a few minutes. The insoluble matter is filtered off, ignited, fused with a little sodium carbonate, the fusion dissolved in water and added to the main filtrate. This is then oxidized with the minimum amount of nitric acid needed, and boiled till free from fumes. The iron is extracted with ether and the excess of ether removed from the aqueous layer by evaporation on the steam bath. After concentration on the bath, strong nitric acid is added to the solution and it is evaporated to dryness. The residue is dissolved in strong nitric acid, the solution is diluted with water and nearly neutralized with strong sodium hydroxide solution. It is then poured slowly into 150 to 200 cc. of a 10% sodium hydroxide solution, stirring vigorously. The solution is filtered, and the series of operations are repeated with the precipitate until it is free from vanadium, as shown by dissolving it in nitric acid and testing with hydrogen peroxide. In the latter treatments the amount of sodium hydroxide solution used may be smaller. From the combined filtrates the vanadium is precipitated with mercurous nitrate solution, after making nearly but not quite neutral with dilute nitric acid. After settling, the precipitate is collected on paper and washed with dilute mercurous nitrate solution. The filter is burned off in a platinum crucible and the precipitate ignited till all the mercury is expelled. The impure vanadium pentoxide left is fused with a little sodium carbonate, the fusion is dissolved in water and filtered (on asbestos) from insoluble matter. A second precipitation with mercurous nitrate is then made. Sometimes a further fusion and precipitation may be necessary in order to get a product sufficiently pure for the next step, which is a final fusion with sodium carbonate. The fusion is dissolved in dilute sulphuric acid and the vanadium is reduced by sulphur dioxide gas and titrated against $N/50$ permanganate after complete expulsion of the excess of reducing agent.

Volumetric Determination of Molybdenum and Vanadium in Presence of One Another

Sulphur dioxide reduces V_2O_5 to V_2O_4, but does not reduce molybdic acid provided the sample contains 1 cc. of free sulphuric acid per 50 cc. of solution and not more than 0.2 gram of molybdic acid. By means of amalgamated zinc V_2O_5 is reduced to V_2O_2 and MoO_3 to Mo_2O_3. Upon these two reactions the determination is based according to the procedure worked out by Edgar.[1] Details of the method are given in the chapter on Molybdenum, page 321.

[1] Graham, Edgar, Am. Jour. Sci. (4), 25, 332. Gooch, "Methods in Chemical Analysis," John Wiley & Sons.

Volumetric Determination of Vanadium, Arsenic or Antimony in Presence of One Another. Edgar's Method [1]

Tartaric or oxalic acid reduces V_2O_5 to V_2O_4, but does not act upon arsenic or antimony. On the other hand SO_2 causes the reduction of all three. Therefore if aliquot portions of the solution are taken, one portion being treated with tartaric acid and vanadium determined by titration with iodine, and another portion reduced with SO_2 and again titrated with iodine, the difference between the two titrations is due to the cc. of reagent required for the oxidation of the reduced arsenic or antimony. [2]

Reactions. $V_2O_4 + I_2 + H_2O = V_2O_5 + 2HI.$
$As_2O_3 + V_2O_4 + 3I_2 + 3H_2O = As_2O_5 + V_2O_5 + 6HI.$
$Sb_2O_3 + V_2O_4 + 3I_2 + 3H_2O = Sb_2O_5 + V_2O_5 + 6HI.$

Vanadium. One portion is boiled with about 2 grams of tartaric or oxalic acid, until the solution turns the characteristic blue of vanadium tetroxide. After cooling, the solution is nearly neutralized with potassium bicarbonate, and an excess of standard iodine solution added. Neutralization is now completed, an excess of bicarbonate added, and after fifteen to thirty minutes the excess iodine titrated with standard arsenious acid, starch being used as an indicator. This titration measures the vanadium present.

Arsenic or Antimony. A second portion of the solution is placed in a pressure flask and acidified with sulphuric acid. A strong solution of sulphurous acid is added, the flask closed and heated for an hour on the steam bath. After cooling, the flask is opened and the solution transferred to an Erlenmeyer flask and the excess of SO_2 removed by boiling, a current of CO_2 being passed through the liquid. The cooled solution is treated with bicarbonate, iodine added and the titration conducted exactly as described for determination of vanadium in the first portion. The difference between the first titration and the second is a measure of the cc. required for oxidation of arsenic or antimony.

Determination of Vanadium and Iron in Presence of Each Other

The solution slightly acidified with sulphuric acid is treated with sulphurous acid, the excess expelled and the reduced vanadium and iron titrated with standard potassium permanganate. [4]

$$10FeO + 5V_2O_4 + 6H_2SO_4 + 4KMnO_4 = 5Fe_2O_3 + 5V_2O_5 + 2K_2SO_4 + 4MnSO_4 + 6H_2O.$$

The solution is now reduced with zinc in the Jones reductor and again titrated with permanganate. [2] V_2O_5 is reduced by zinc to V_2O_2, the sample being caught

---

[1] G. Edgar, Am. Jour. Sci. (4), **27**, 299.
[2] Gooch, "Methods of Chemical Analysis."
[3] Graham, Edgar, Am. Jour. Sci., (4), **26**, 79.
See Am. Jour. Sci., (4), **27**, 174, also Gooch, "Methods in Chemical Analysis," p. 510, for procedure determining iron, chromium and vanadium, in presence of one another.
[4] When the color has changed from a bluish-green to greenish-yellow the solution is heated to 70 to 80° C., and the permanganate titration completed in a hot solution.

in ferric alum solution (details for determining of vanadium by reduction with zinc are given under the volumetric methods for this element).

$$10FeO + 5V_2O_2 + 12H_2SO_4 + 8KMnO_4 = 5Fe_2O_3 + 5V_2O_5 + 4K_2SO_4 + 8MnSO_4 + 12H_2O.$$

The difference between the two titrations multiplied by 0.00456 = vanadic acid (V_2O_5) originally present.

Iodometric Method for Estimation of Chromic and Vanadic Acids in Presence of One Another

The following procedure developed by Edgar,[1] is given by Gooch ("Methods of Chemical Analysis").

In carrying out the operation, the alkali salts of the chromic and vanadic acid are put into the Voit flask of the distillation apparatus shown in the cut, Fig. 71.

One or 2 grams of potassium bromide are added, the flask is connected with the absorption apparatus containing a solution of potassium iodide made alkaline with sodium carbonate or sodium hydroxide, and the whole apparatus is filled with hydrogen gas. Fifteen to 20 cc. of concentrated hydrochloric acid are added through the separatory funnel and the solution is boiled for ten minutes, an interval of time found to be enough for the completion of the reduction. A slow current of hydrogen is maintained to avoid back suction of the liquid from the Drexel bottle. The apparatus is disconnected, the Voit flask placed in a beaker containing cold water, and the alkaline solution in the absorption

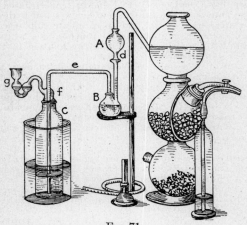

FIG. 71.

apparatus cooled by running water. The contents of the trap are washed into the Drexel bottle and the solution therein is made slightly acid with hydrochloric acid. The liberated iodine is titrated with approximately N/10 sodium thiosulphate and the color is brought back by a drop or two of N/10 iodine solution, after the addition of starch.

Alkaline potassium iodide is again placed in the absorption apparatus and the latter connected with the Voit flask. The current of hydrogen is turned on and, after the air has been expelled, the apparatus is disconnected momentarily, 1 or 2 grams of potassium iodide are added to the solution in the Voit flask, and connections made again. Through the separatory funnel 10 cc. to 15 cc. of concentrated hydrochloric acid and 3 cc. of syrupy phosphoric acid are added and the solution in the reduction flask is boiled to a volume of 10 cc. to 12 cc. The absorption apparatus is removed and cooled, hydrochloric acid is added and the liberated iodine titrated with approximately N/10 sodium thiosulphate.

[1] Graham Edgar, Am. Jour. Sci. (4), **26**, 333.

The iodine determined in the first titration corresponds to a reduction of the chromic and vanadic acids according to the equation

$$V_2O_5 + 2CrO_3 + 8HBr = V_2O_4 + Cr_2O_3 + 4Br_2 + 4H_2O,$$

while in the second case the iodine corresponds to a reduction of the vanadium tetroxide to trioxide as indicated in the equation

$$V_2O_4 + 2HI = V_2O_3 + I_2 + H_2O.$$

The second titration, therefore, determines the vanadic acid present, and the difference between the first and second furnishes the necessary data for the calculation of the chromium.

DETERMINATION OF VANADIUM IN FERRO=VANADIUM

Standard Methods of the American Vanadium Company [1]

Dissolve 0.510 gram of the alloy (100 mesh) in a 250-cc. beaker with 50 cc. dilute sulphuric acid (1 : 2) and 10 cc. (concentrated) nitric acid. If the alloy does not decompose, when heated, add a few cc. of hydrofluoric acid. Take down to copious white fumes. Cool, add 50 cc. dilute sulphuric acid (1 : 2) and water. Heat until all salts are in solution and transfer to a white casserole containing 100 cc. dilute sulphuric acid (1 : 2). Dilute the solution to 400 cc. with H_2O and heat to 60° C. The solution is ready to titrate.

Add potassium permanganate until a deep red is obtained. Just discharge the red color with ferrous ammonium sulphate.

Get the neutral point by alternating the permanganate and ferrous ammonium sulphate until one drop of the ferrous sulphate just discharges the pink color.

Now add N/10 ferrous ammonium sulphate from a burette until the vanadium is reduced and then 3 cc. in excess.

Titrate the excess of ferrous ammonium sulphate with N/10 potassium bichromate, using potassium ferricyanide as an indicator.

From the cc. of ferrous ammonium sulphate used, subtract the cc. of bichromate used. The number of cc. used gives the per cent of vanadium in the alloy.

The relation between ferrous ammonium sulphate and bichromate is established by adding 150 cc. sulphuric acid (1 : 2) to a casserole, diluting to 400 cc.

Find the neutral point and then add 25 cc. ferrous ammonium sulphate and titrate with bichromate until the blue spot is just discharged.

EXAMPLE

Blank.

Ferrous ammonium sulphate used.......................... 25   cc.
Potassium bichromate used............................... 24.6 "
                        25.00
                        24.60
                        ———
                        .40 ÷ 25 = −0.016 factor.

Alloy.

Ferrous ammonium sulphate used.......................... 40.00
Potassium bichromate used............................... 2.40
40 cc. × −0.016... =   .64 cc.
40.00 − 0.64... = 39.36 "
Correction on ferrous ammonium sulphate:
      39.36 − 2.4 .... = 36.96% V.
Solutions used:
      N/10 potassium bichromate.
      N/10 ferrous ammonium sulphate.
Potassium ferricyanide, a crystal the size of a pea in 50 cc. of water.
Potassium permanganate, 5 grams per liter.

[1] Methods developed in the Bridgeville Laboratory. By courtesy of the American Vanadium Company.

Determination of Vanadium in Vanadium Ores

Weigh 0.51 gram of the finely powdered ore in a $1\frac{1}{2}$-in. diameter iron crucible filled three-fourths full of sodium peroxide. Fuse. Dissolve the fusion in water and add 100 cc. H_2SO_4 (1 : 2) in excess and evaporate until white fumes come off. Cool and dilute and filter. Gas the filtrate, which should be about 400 cc., until all H_2S metals are precipitated. Boil and filter. Boil the filtrate until all H_2S is off. Transfer to a 500-cc. casserole and add 50 cc. H_2SO_4 (1 : 2) and heat to above 60° C. Titrate as in the determination of vanadium in ferro-vanadium.

Determination of Vanadium by Precipitation with Phosphomolybdate

The alloy or steel is dissolved in nitric acid and treated, dropwise, with strong potassium permanganate until a precipitate of MnO_2 is obtained on boiling. Ammonium bisulphite solution is now added drop by drop until the manganese dissolves. The solution is boiled until the nitrogen oxides are expelled. 10 cc. of a 5 per cent ammonium phosphate solution and 10 grams of ammonium nitrate are added, the solution heated to boiling, and 50 cc. of ammonium molybdate solution added. The solution is shaken for 3–5 minutes to precipitate all of the vanadium and phosphorus. The precipitate is separated by decanting onto an asbestos filter. The precipitate is dissolved in conc. H_2SO_4, the iron oxidized by HNO_3, the nitric expelled by fuming off and the vanadium now reduced by adding hydrogen peroxide and heating. The addition of peroxide and heating is repeated. The green or blue solution is heated to 80° and titrated with 0.1N $KMnO_4$ as usual.

Vanadium in Steel—A. S. T. M. Method

SOLUTIONS REQUIRED

"*Dilute Hydrochloric Acid (2 : 1)*.—Mix two parts of HCl (sp.gr. 1.19) with one part of water.

"*Dilute Nitric Acid (sp.gr. 1.20)*.—Mix 380 cc. of HNO_3 (sp.gr. 1.42) and 620 cc. of distilled water.

"*Dilute Hydrochloric Acid (1 : 9)*.—Mix 100 cc. of HCl (sp.gr. 1.19) and 900 cc. of distilled water.

METHOD

" (a) (When Cobalt does not exceed 1 per cent.)

"Treat 2 g. of the steel in a porcelain dish with 70 cc. of dilute HCl (2 : 1) and heat until all action ceases. Cautiously add 10 cc. of dilute HNO_3 (sp.gr. 1.20) and evaporate the solution to dryness or to low bulk. Treat the residue with 45 cc. of dilute HCl (2 : 1) and heat until the soluble portion dissolves. Add an equal volume of water, boil for three minutes, filter and wash the residue thoroughly with dilute HCl (1 : 9) until free from iron. Evaporate the solution to small volume and destroy hydrochloric acid by small additions of nitric acid (sp.gr. 1.42) followed by boiling. If a residue appears, filter the solution. From this point proceed for chromium as in the Determination of Chromium in Chrome-Nickel Steel by the Persulphate Oxidation Method and for Vanadium as in the Determination of Vanadium in Chrome-Vanadium Steel by Reduction with Ferrous Sulphate and Titration with Permanganate.

" (b) (When Cobalt exceeds 1 per cent.)

"See the Determination of Vanadium in Chrome-Vanadium Steel by the Electrolytic Separation Method.

"Proceed as described in this method until the bicarbonate precipitate has been washed with hot water. Return the paper and precipitate to the flask and add 60 cc. of dilute HCl (2 : 1). Heat the solution, add 10 cc. of HNO_3 (sp.gr. 1.20) and evaporate the solution to dryness or to a small volume. Treat the residue with 45 cc. of dilute HCl (2 : 1) and proceed as described above.

NOTES

"Tungsten interferes with the determination of vanadium by the above method and must be removed. While chromium can be determined in the presence of tungsten, it is conveniently determined in the same sample as is vanadium and the end point is easier to get when tungsten has been removed.

"In case hydrochloric acid is present when silver nitrate is added, enough of the latter must be added to take care of this before the regular addition is made.

"The pink color of cobalt interferes with the end point readings. If under 1 per cent, the color does not seriously affect the results, if over this, it should be removed."

Determination of Vanadium in Steel (When Chromium is present)

Dissolve 5.1 grams of steel in a covered 400-cc. beaker with 60 cc. of HCl (concentrated). After total solution, add concentrated HNO_3 sufficient for complete oxidation. Evaporate to a syrupy consistency, add 40 cc. HCl (concentrated) and evaporate to about 20 cc. Cool and transfer contents to a separatory funnel, washing with dilute HCl (2 HCl : 1 H_2O). Add 100 cc. ether, cork and shake for some time, cooling funnel under tap water while shaking. Remove cork, place funnel in stand and allow it to stand for at least five minutes. Run out the lower layer of the separation into the original 400-cc. beaker.

Evaporate the ether off. Add 5 cc. HNO_3 (concentrated) and just bring to a boil, Stir out all nitrous fumes, make alkaline with NaOH (saturated solution). Make just acid with HNO_3 (concentrated) cool solution.

Add above solution to a solution containing 300 cc. cold water and 5 cc. of NaOH (saturated solution). Boil and filter, washing with hot water thoroughly. Make filtrate just acid with HNO_3 (concentrated). Add 40 cc. of a saturated solution of lead acetate. (If lead precipitate forms just clear solution by adding HNO_3 drop by drop and bring to a boil.) Add 60 cc. of ammonium acetate. Boil for twenty minutes. The vanadium is precipitated as lead vanadate.

Filter the lead vanadate onto a Munktell paper, washing with hot water. Put filter containing lead vanadate in a small porcelain dish and burn off paper at a low heat. Add a little HNO_3 and evaporate on the hot plate, then put the dish in the cold end of a muffle to drive off the remaining HNO_3. Avoid baking. Dissolve in HCl (concentrated) and transfer the solution to a 400-cc. beaker. Add 60 cc. dilute H_2SO_4 (1 : 2). Oxidize thoroughly with KMnO4 (5 grams to a liter.) Add 40 cc. HCl (concentrated) and evaporate to dense white fumes. Cool, add 40 cc. of water and again take to white fumes. Cool, add 150 cc. of water, cool, and titrate with N/50 KMnO . Each cc. of permanganate used is equal to 0.00102 gram of vanadium, or in this case, having used a ten-factor weight, each cc. represents 0.02% vanadium.

NOTES.

Vanadium [1] is used chiefly in steel for purposes requiring great toughness and torsional strength, common applications being in automobile parts, gears, piston rods, tubes, boiler plates, transmission shafts, gun barrels, and forgings that have to withstand great stresses and wear. The vanadium content in steel usually ranges from 0.1 to 0.4%. Vanadium is also used occasionally in certain tungsten alloy steels for making high-speed steel. Introduced in small proportions, it reduces considerably the amount of tungsten required in steel for a given hardness and toughness. Vanadium differs from tungsten in that it has a good effect not only on tool steel, but on structural steel as well.

Chrome-vanadium steels and chrome-vanadium-molybdenum steels are the latest developments in structural alloy steels that have gained an extensive market. Almost all these steels are made in the open-hearth furnace, chromium and vanadium alloys being added shortly before casting. In their physical properties these steels are much like chrome-nickel steels, but they have a greater contraction of area. Most of the chromium-vanadium steels made go into automobiles. Some manufacturers prefer them because of their greater freedom from the surface imperfections—notably seams —which the steels containing nickel are likely to have.

[1] Machinery, June, 1924.

Determination of Vanadium in Ores, Mine and Crude Mill Samples[1]

Reagents Indicator. $K_3Fe(CN)_6$. Make up just before using in a drop bottle which has been washed out with NaOH and then with water. Place a small crystal of the salt in the bottle and wash several times to remove the oxidized coat, pouring out the washings, then add about 30 cc. of distilled water and dissolve for use.

Ferrous ammonium sulphate, $FeSO_4(NH_4)_2SO_.6H_2O$. 135 grams of the salt are dissolved in a mixture of 333 sulphuric acid and 3000 cc. of water.

Standardize the solution as follows: Take 25 cc. of the solution and add 25 cc. H_2SO_4. Dilute to 400 cc. and titrate with standard $K_2Cr_2O_7$ reagent using $K_3Fe(CN)_6$ indicator according to the procedure used in the determination of iron.

Potassium permanganate,—$KMnO_4$. 3.162 grams per liter.

Potassium dichromate, $K_2Cr_2O_7$. 4.903 grams per liter.

Standardize with 1/2 gram of iron wire. (1 cc. = .056 g. Fe.)

PROCEDURE. Solution of the Sample.

Take 2.04 grams of the ore, ground to pass through a 60-mesh screen. Fuse with (about 20 times its volume), sodium peroxide in a 25 cc. iron crucible. (If the mixture can be thoroughly cintered, it is better than a complete fusion.) Take up the mass with water in a 600 cc. beaker. Add a solution of 50 cc. strong sulphuric acid diluted with 100 cc. of water. Then dilute to 400 cc. with water.

Titration. Heat to about 80° F. and add just sufficient of the permanganate reagent to make the solution pink. Titrate back with a few drops of ferrous ammonium sulphate reagent until the color of the permanganate just disappears The solution will now appear a green or blue color. Now add the "ferrous" solution until the ferric cyanide spot test shows a faint blue color. Then 2 cc. excess of the "ferrous" solution. Titrate back the excess with the standard potassium dichromate reagent until a drop of the indicator shows no color on the spot plate, with a drop of the solution.

Calculate the ratio of the ferrous solution and the dichromate to a common basis. Deduct the cc. dichromate from the cc. "ferrous" added to the bluish green solution above, and divide the result by 4. With an exactly N/10 solution the result will be the percent vanadium in the sample.

Lead Acetate Method. The solution obtained from the fusion of the ore (see Solution of the sample) is acidified with acetic acid. Then sufficient lead acetate is added to completely precipitate the vanadium (usually 2 or 3 grams, according to the amount of vanadium in the sample), the solution is stirred, allowed to settle and the lead vanadate filtered off and treated as follows.

Take the residue on the filter (this at first appears yellow) and dissolve in a very dilute solution of nitric acid. *The acid should be hot.* Transfer to an 800 cc. beaker.

Add about 10 cc. of concentrated sulphuric acid and evaporate to fumes of SO_3. All traces of nitrous oxide must be expelled. Dilute to 600 cc. with distilled water, add 25 cc. of concentrated sulphuric acid, hat and titrate with standard permanganate according to the first of the volumetric procedures.

Determination of Vanadium in Cupro-vanadium, Brasses and Bronzes

Dissolve 1.020 grams of cupro-vanadium in aqua regia. Evaporate to small bulk and add excess of peroxide of hydrogen. Dilute to 600 cc. and add ammonia until all copper goes into solution. Heat to boiling and add sufficient barium chloride solution to precipitate all the vanadium. Boil and filter. Wash all copper out of filter with hot ammonia water. Transfer the filter to a beaker, add 100 cc. 1 : 2 sulphuric acid, boil and filter on close filter paper. Titrate the filtrate with N/10 ferrous ammonium sulphate and N/10 potassium bichromate the same as in the case of the ferro alloy, except that this being a two-factor weight, the result must be divided by 2.

Vanadium copper, brasses and bronzes are treated in the same manner except that a ten-factor weight is used and the titration carried out with N/50 solution instead of N/10.

[1] Communicated to the author by Theodore Marvin, Dupont Powder Company.

VOLUMETRIC–PHOSPHOMOLYBDATE METHOD FOR DETERMINATION OF VANADIUM

Reagents

Ammonium Molybdate. See page 1363 for phosphorus in steel.

Ammonium Phosphate. 50 g. salt per liter of water.

Acid Ammonium Sulphate. 50 cc. strong H_2SO_4, 950 cc. water and 15 cc. strong ammonium hydroxide. Use hot, 80° C.

Nitric Acid. 100 cc. strong HNO_3 and 1200 cc. water.

Nitric Acid for Washing. 20 cc. strong HNO_3 per liter.

Potassium Permanganate, Standard. 0.35 g. salt per liter of solution. Standardized against sodium oxalate. Adjust so that 1 cc. will equal 0.0005 g. vanadium, or 0.02 per cent on a 2.5-g. sample. 1 g. $Na_2C_2O_4 = 0.7612$ g. V.

Potassium Permanganate, for oxidation. 25 g. salt per liter of solution.

Sodium Bisulphite, for reduction. 30 g. salt per liter of solution.

Procedure for Steel

A sample of 2.5 g. of steel in a 300-cc. beaker or Erlenmeyer flask are dissolved in 50 cc. of the nitric acid, and to the boiling solution are added 6 cc. of the permanganate oxidation solution, the boiling being continued until MnO_2 precipitates. The precipitate is now dissolved by cautious additions of sodium bisulphite solution and the boiling continued until no brown fumes are evident. Now 5 cc. of ammonium phosphate solution are added and 10 g. ammonium nitrate. The solution is removed from the heat and 50 cc. of ammonium molybdate reagent immediately added. After standing for 1 minute, the solution is agitated for 3 minutes, then allowed to settle and the clear solution decanted through an asbestos filter, the residue is washed three times with hot acid ammonium sulphate reagent, decanting each time through the filter. The flask containing the bulk of the residue is placed under the filter. (The washings are best conducted with suction, using a bell jar filter.) The precipitate on the filter is dissolved by successive portions of hot strong H_2SO_4, catching the solution in the vessel containing the bulk of the precipitate. The precipitate is now dissolved by heating and to the solution a few drops of the nitric acid are added and the heating continued to strong fumes.

The solution is cooled and hydrogen peroxide added in small quantities with vigorous shaking after each addition, until the solution takes on a deep brown color. The solution is again heated for 4 or 5 minutes, then cooled and 100 cc. of water added, the solution again heated to about 80° C. and titrated to a permanent pink color with standard potassium permanganate.

NOTE. If the peroxide treatment followed by heating does not result in a clear green or blue color, the solution should be evaporated to strong sulphuric acid fumes and the peroxide treatment repeated. The presence of nitric acid interferes with the reduction of vanadium.

VANADIUM IN STEEL—ETHER EXTRACTION—HYDRO-CHLORIC ACID REDUCTION METHOD

Reagents

Hydrochloric Acid.   600 cc. strong HCl and 400 cc. water.
Sulphuric Acid.   Equal volumes of strong H_2SO_4 and water.
Other Reagents.   See Phosphomolybdate Precipitation Method for Vanadium.

Procedure

A sample of 2.5 g. of steel in a 250-cc. beaker is dissolved in 50 cc. of the HCl, and then small portions of HNO_3 added to oxidize the iron. After expelling the brown fumes by heating, the solution is cooled and transferred to an 8-oz. separatory funnel, together with the rinsings (small portions of HCl) of the beaker. Now 50 cc. of ether are added and the mixture shaken for 5 minutes. After settling for 1 minute the clear lower layer is drawn into another separatory funnel. The first funnel is treated with 10 cc. of strong HCl, again shaken vigorously and the settling repeated, the lower layer being added to the solution in the second separatory funnel. The combined solutions in the second separatory funnel are treated with 50 cc. of ether and shaken for 5 minutes, allowed to settle 1 minute and the clear lower layer drawn into a 150-cc. beaker. This aqueous solution is warmed gently to expel the ether, 25 cc. of the H_2SO_4 (1 : 1) added and the mixture concentrated to strong fumes. After cooling, 25 cc. of water are added followed by a slight excess of potassium permanganate solution and the sample heated to boiling. 15 cc. of strong HCl are added and heat applied until the solution again fumes. The heating is continued for 10 minutes. After cooling, 100 cc. of water are added, the solution heated to 80° C. and titrated with standard potassium permanganate reagent to a permanent pink color.

NOTES. In heating the solution to expel the brown fumes of oxides of nitrogen, the solution should not be boiled.

In presence of chromium, the pink color will fade on standing owing to the oxidation of chromium. The oxidation of chromium is reduced by titrating the solution cold, but only ten seconds are allowed for the pink color to remain. A blank must be run with the same amount of chromium and allowance made for its oxidation. The blank is conveniently made by putting a suitable amount of chrome steel or chrome-nickel steel through the recommended procedure. By varying the amounts of steel and hence the amount of chromium in solution, data for a charted curve may be obtained that will be convenient for a blank deduction.

47

Preparation of a Gooch Crucible

Asbestos Fibre.—The asbestos for use in Gooch crucibles should be carefully selected. The fibres should be moderately stiff, not the "cottony" type. Cut the fibre into pieces about $\frac{1}{4}$ inch long. Ignite the asbestos in a platinum dish at low red heat. Cool and transfer to a clean porcelain mortar and mascerate to a pulp with strong hydrochloric acid. Dilute with water and transfer to a large beaker containing 600–800 cc. of water. Stir thoroughly, allow to settle and pour off the milky water. Repeat the washing with water until the milkiness, due to powdered fibres, is scarcely evident. Now filter off the asbestos onto a Buechner funnel. Again wash with water until free of acid. Transfer to a wide mouth bottle, add water in sufficient amount to form with the stirred up fibre a thin suspension of asbestos. This is now ready for use. If preferred the asbestos may be dried and kept in this form until desired.

Preparation of the Filter. The crucible, either of platinum or porcelain, having a perforated bottom, is placed in a funnel tube and the apparatus set up as is shown in Fig. 71a. The suction bottle holding the Gooch is attached to a second bottle, if a water filter pump is used to obtain the vacuum, as there is danger of water being sucked into the apparatus from the tap. Suction is now applied and a small amount of the finely suspended asbestos is poured into the crucible, in amount sufficient to form a thin pad of the material about $\frac{1}{16}$ inch thick over the bottom of the Gooch. The felt is washed with distilled water, the asbestos drawn down hard. It is possible to see light through the bottom of a properly made filter. The crucible is placed in an

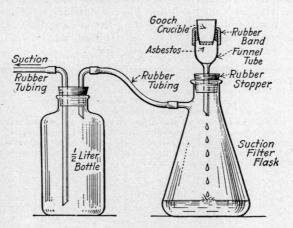

FIG. 71a.   Filtering Through a Gooch Crucible.

oven and the filter dried to constant weight at 110° C. For $BaSO_4$ and $AgCl$ determinations it is advisable to make the filter about twice the above thickness to prevent the precipitate from passing through. Whenever the Gooch crucible is used, suction should be applied before pouring material into the crucible and the suction continued during the washing of the precipitate.

ZINC

Zn, at.wt. 65.38; *sp.gr.* 6.48 *to* 7.19; *m.p.* 419°; *b.p.* 920°; *ZnO oxide.*

DETECTION

The finely powdered material, when heated on charcoal in the reducing flame of a blowpipe, gives an incrustation, yellow when hot—white when cold. On moistening with cobalt nitrate solution and re-igniting, the mass is greenish-yellow. Materials containing above 5% Zn will give positive tests. With experience, less can be detected, but for smaller amounts the regular procedure as given under Titration in Acid Solution, Separating Zn as ZnS, should be followed, using samples as follows: For material containing 0.01–0.05%, 10 to 20 grams; 0.05–0.10%, 5 to 10 grams; 0.10–0.5%, 5 grams; 0.5% on up, 2 grams to 0.5 gram, depending on per cent of zinc present.

In case the material is of interest, only if it carries higher than several per cent of zinc, a shorter and easier wet test is to bring the material into solution by means of hydrochloric or nitric acid, add bromine water and then precipitate iron, aluminum and manganese with ammonia, as given under heading of Determination of Zinc in Acid Solution, Separating Iron, Aluminum and Manganese with Ammonia and Bromine, filter, wash and make the filtrate acid with hydrochloric acid, 10 cc. excess added for each 100 cc. of solution, and potassium ferrocyanide added. Zinc, if present, gives the characteristic precipitate. Copper interferes and if present must be separated with hydrogen sulphide, as given under heading Procedure for Copper-bearing Ores.

In case manganese and copper are known to be absent, a still shorter test may be used: To the solution of the zinciferous material add 2 or 3 grams of citric acid per 200 cc. solution, then make ammoniacal, add ferrocyanide—a white precipitate indicates zinc.

ESTIMATION

The determination of zinc is called for in the buying and selling of ores for smelters, refuse material, e.g., from galvanizing plants, foundries, brass mills, and blast furnaces, in manufacture of brass, white metals, and alloys in general, paints and pigments, zinc chloride for preservation purposes, and in the control work in smelting of zinc and lead ores.

Preliminary. The method to be followed in the estimation of zinc will depend largely on the nature of the material in which it occurs, the quantity present, and the experience of the analyst. Each of the methods outlined will give correct results only on the materials for which they are indicated, there being but one method recommended which is applicable to all zinciferous mate-

Chapter by F. G. Breyer, L. S. Holstein and L. A. Wilson.

rials. It cannot be emphasized too strongly that each step has a definite purpose (which may not be at once apparent to the analyst making only an occasional zinc determination), and no part of the procedure should be varied or omitted, excepting after abundant experience.

Preparation of Sample

The representative sample should be ground to pass a 100-mesh screen or finer. If the material contains shot metal, it should be screened out and the percentage present calculated. It is then treated as given under heading Material Containing Metallics, page 602.

Moisture Determination in the Pulp

One of the commonest causes of differences in zinc ore analysis is the failure to take moisture determinations on the pulp sample.

In order that analyses made on the same pulp at different times and in different laboratories may be compared it is absolutely necessary that all determinations be corrected to a dry basis. It is not sufficient that the sample be dried before or after having been pulped, but a sample for moisture must be weighed out at the same time as the sample for analysis, and the analytical result corrected for the per cent of moisture found at the time of weighing. This is especially true on roasted zinc ores which contain sulphates of zinc, iron and lime and which take up moisture quite rapidly under ordinary atmospheric conditions.

The usual temperature for drying should be 110° C., but on special ores, e.g., those containing sulphates, it is necessary to dry at 250° C. unless it is first shown that there is no loss of water above 110° C.

The determination is best made by weighing approximately two grams in a small glass-stoppered weighing tube and drying to constant weight, the weighing tube being closed with the glass stopper as soon as the tube is taken from the drying oven.

SEPARATIONS

Silica. Evaporate with hydrochloric acid or take to fumes of sulphuric acid. The dehydration with sulphuric acid is complete and gives silica that is easily filtered and washed.

Cadmium, Lead, Arsenic, Antimony, Bismuth and Copper. Aluminum may be used to separate all the metals, except cadmium, the latter being only partially separated. The procedure is as given in the standard method.

The separation may also be made as follows: Evaporate the solution of the zinciferous material to fumes with 7 cc. of 1 : 2 sulphuric acid. Cool, take up in about 50 cc. of water and warm, add 10 cc. of 10% sodium thiosulphate, boil until evolution of sulphur dioxide ceases, then filter. Cadmium if present is not precipitated. It should be separated by the procedure given under Titration in Acid Solution Separating Zinc as Sulphide.

Iron, Aluminum and Manganese. This separation may be effected by precipitation with ammonia and bromine, providing the quantities present are small. When large amounts are present the basic acetate procedure is followed, or, better, the zinc separated as sulphide in dilute sulphuric acid solution, page 605.

Nickel and Cobalt. When nickel or cobalt are present, the only safe procedure is to separate the zinc as zinc sulphide in dilute sulphuric acid solution, as described under the standard method. Weiss has shown conclusively that zinc can be precipitated free from either cobalt or nickel under the conditions there outlined.

METHODS OF ANALYSIS

I. Gravimetric methods.
II. Electrolytic methods.
III. Titration with standard solution of $K_4Fe(CN)_6$.
    (a) In acid solution.
    (b) In alkaline solution.
    (c) In acid solution, separating Zn as ZnS. (Standard method.)

GRAVIMETRIC METHODS

Weighing as Zinc Oxide

In this case the procedure is the same as in the volumetric method, in which zinc is separated as zinc sulphide up to point where the zinc sulphide is filtered off and washed. It is now ignited in a weighed crucible and heated to 800 to 900° C. in a muffle for one hour and weighed as ZnO. Factor $ZnO \times 0.8034 = Zn$.

The precipitate of zinc sulphide may also be filtered on a Gooch crucible, and ignited as above.

Weighing as Sulphate

The zinc sulphide is dissolved in hydrochloric acid. Sulphuric acid is added and the solution evaporated in a weighed crucible, all excess acid fumed off and the resulting zinc sulphate finally ignited at a dull red heat and weighed. $ZnSO_4 \times .405 = Zn$.

Electrolytic Methods

The determination is best made from an alkaline electrolyte or one slightly acid with acetic acid and containing a considerable amount of sodium acetate. The alkaline electrolyte tends to give high results, due to the presence of zinc oxide or hydroxide in the deposit. The best results are obtained with a solution weakly acid with one of the weaker organic acids. The procedure for the acetate electrolyte is as follows:

The zinc is separated from other elements by precipitating with hydrogen sulphide in dilute sulphuric acid solution, as given under the standard method. The precipitate is filtered and washed, dissolved in hot hydrochloric acid— 5 cc. 1 : 1 sulphuric acid added and the whole evaporated to fumes to expel hydrochloric acid. Cool and dilute, neutralize with sodium hydrate solution, make slightly alkaline, then acidify with acetic acid, and add about 5 grams of sodium acetate. The volume of solution should now be about 100 to 125 cc. Electrolyze with a platinum gauze electrode with 0.5 ampere at 5 volts.

The electrolytic methods, on account of the special apparatus needed, the experience and care necessary to get reliable results, and the unavoidable errors involved in their use, are less desirable than the gravimetric oxide method and *still less* desirable than the ferrocyanide method.

VOLUMETRIC METHODS

Titration in Acid Solution, Separating Iron, Aluminum, and Manganese with Ammonia and Bromine

General. This method is especially adapted to material low in silica, alumina, iron, and manganese. When the operator gains experience in manipulation, it is, possible to obtain good results on samples higher in these elements, but its haphazard use with materials high in these impurities is one of the chief causes of the common inaccuracy of zinc work. If copper or cadmium are present in quantities, the titration in acid solution, separating Zn as ZnS, is to be preferred for accurate work.

Procedure for Ores. One-half or 1 gram (depending on the per cent of zinc present) is weighed in a 250-cc. beaker. Fifteen cc. of hydrochloric [1] acid (sp.gr. 1.2) are added, a cover-glass put on, and the ore agitated to prevent caking. Boil down to a volume of about 5 cc.[2] cool, wash down cover-glass and sides of beaker with a jet of water. Add 10 cc. of saturated bromine water, 5 grams of ammonium chloride and 15 cc. of ammonia water (sp.gr. 0.90) and boil vigorously for a minute or two. Filter off the precipitated hydroxides, and wash four times with hot water, containing 50 grams ammonium chloride and 25 cc. ammonia per liter. The precipitate is now washed from the filter into the beaker in which the original precipitation was made, and the precipitate dissolved in strong hydrochloric acid. Ten cc. of ammonia (sp.gr. 0.9) are added,[*] the solution boiled, filtered and washed as before, the filtrate being combined with the first filtrate.[3] The solution is then diluted to 250 cc., heated to boiling, and 4 drops of ammonium sulphide solution added to destroy oxidizing agents [4] and precipitate small amounts of copper and cadmium. The solution is neutralized with hydrochloric acid, the resolution of the precipitated zinc sulphide serving in lieu of an indicator. Ten cc. excess of concentrated hydrochloric acid are added [5] and the solution titrated, not below 75° C., with standard ferrocyanide, using uranium nitrate (10% solution)[6] as an external indicator.

Standardization of the Ferrocyanide Solution

The potassium ferrocyanide is standardized by weighing out portions of C.P. zinc that will give a titration of approximately the same number of cc. as the sample. Dissolve in 15 to 20 cc. of hydrochloric acid and dilute to about 225 cc.

---

[1] Nitric acid should be added in case of sulphide ores.

[2] In case of siliceous ore, it is advisable to evaporate to dryness, and on unknown material to evaporate slowly, in order to make sure of complete solution of the zinc. Certain siliceous and oxide ores are difficultly soluble in hydrochloric acid, and frequently cause low results, where rapid decomposition is the routine.

[3] In case of high silica, alumina, iron, and manganese materials, three precipitations are necessary.

[4] It is necessary to destroy all oxidizing agents, as they will react with the ferrocyanide.

[5] The excess of hydrochloric acid should be carefully measured. A burette is very useful in neutralizing the solution.

[6] The strength of the uranium nitrate is a matter of personal preference, some using a saturated solution. On the other hand some prefer ammonium molybdate. The strength of solution given above, however, is recommended as the first choice of a large number of experienced zinc chemists.

[*] In presence of manganese 10 cc. of saturated bromine water are added.

Then add 37 cc. strong ammonia,[1] taking care to avoid spattering, heat to boiling, add 4 drops of ammonium sulphide, neutralize and add 10 cc. excess hydrochloric acid and titrate.

General Notes

The precipitate with ammonia carries down zinc. This is especially true with siliceous material or material high in iron and alumina. By working with hot ammoniacal ammonium chloride solution and making two or three precipitations, the amount held can usually be made negligible.

The precautions in regard to adding ferrocyanide and keeping conditions of standardization and titration the same, hold here as in all ferrocyanide titrations.

Titration in Alkaline Solution

General. This procedure is designed for rapid routine work on roasted or oxidized ores, especially those high in silica, alumina, iron, and manganese. It should only be used on unroasted sulphides, copper, or high cadmium-bearing ores, when the operator has had *long* experience. It is designed to give the zinc content of materials soluble in hydrochloric or nitric acid. For materials containing insoluble zinc, the titration in acid solution, in which zinc is separated as sulphide, is preferred.

Procedure for Common Ores. The following method is recommended: The weight of ore to be taken will depend on the approximate amount of zinc present. For material above 50%, take 1 gram; from 10 to 50%, 2 grams; 5 to 10%, 4 grams; and below 5%, 5 to 10 grams. Weigh the sample into a tall 400-cc. beaker, cover with water and add 25 cc. concentrated hydrochloric acid, rotating the beaker to prevent caking. In case sulphides are present, nitric acid also should be added. Place on a hot plate or steam bath and evaporate to dryness.[2] Now add 50 cc. concentrated nitric acid, cover with a watch crystal and boil off all nitrous fumes. When these have disappeared, add about 3 to 4 grams $KClO_3$ and boil until chlorine fumes do not show.[3] Cool, wash off the watch crystal and sides of the beaker, and dilute to about 100 cc. Wash into 500 cc. graduated flask, make up to the mark and shake well. Filter through a close 24-cm. qualitative paper and without waiting for the whole to run through, measure out 250 cc. of the clear filtrate[4] into a 600-cc. beaker. Add ferric nitrate solution, if necessary, so as to bring the iron content up to about 300 to 400 milligrams, i.e., if only a small amount is present, add 10 cc.; if 10 to 15% is present, add 5 cc., and proceed exactly as under Standardization.

Procedure for Copper-Bearing Ores. Either method is recommended: *Separation of Copper by Aluminum.* The sample is treated as usual up to

[1] The amounts of acid and ammonia used throughout should be carefully measured, so as to keep the amount of ammonium salts approximately the same. This is very important in order to avoid errors, due to varying blanks.

[2] The temperature of the hot plate should not be over 120° C., as $ZnCl_2$ is appreciably volatile at higher temperatures.

[3] Any oxidizing agent such as chlorine or chlorine oxides acts on the ferrocyanide.

[4] The graduated flasks should be standardized against one another, i.e., the 500 cc. should be twice the volume of the 250 cc.

the point where manganese has been separated and 250 cc. of the clear filtrate measured out. Add 25 cc. 1 : 1 sulphuric acid and evaporate to strong fumes, cool, dilute to 100 cc., add a gram or two of 20-mesh zinc-free aluminum. Heat until all the copper separates, filter, wash and proceed with the filtrate as in the regular method, after oxidizing iron with a few drops of nitric acid.

Separation of Copper by Hydrogen Sulphide. After separation of the manganese with chlorate, sulphuric acid is added and the solution taken to fumes, as in above. Cool, dilute to 100 cc., and add sulphuric acid so that 12% is present. Warm slightly and pass hydrogen sulphide through the solution. Filter off the copper sulphide, wash, boil H_2S out of the filtrate, and titrate as usual, after adding ferric nitrate and citric acid.

Material Containing Cadmium. If the material contains cadmium in quantities sufficient to warrant separation (0.15% or more), it is best to use the titration in acid solution, separating zinc as sulphide.

Material Containing Carbonaceous Matter. If the material under examination contains carbonaceous matter, coal, etc., it must be separated by taking to dryness with hydrochloric acid. Take up in acid and water, filter and wash, and evaporate the filtrate to dryness. Take up in nitric acid and proceed as in the regular method.

If the carbonaceous material is not removed, the manganese does not separate cleanly, due to the reducing action of carbonaceous compounds.

Procedure for Material Containing Metallics. On account of the lack of uniformity in the case of metallic zinciferous material containing lead and iron, it is well to work on large samples. Five or 10 grams of the metallics reduced to as fine a size as possible are weighed out and dissolved in nitric acid. The nitrous fumes are boiled off and the whole made up to 500 cc. or 1000 cc. Fifty or 100 cc. are now pipetted off into a 600-cc. beaker and the zinc titrated as usual. In case the metallic portion contains manganese, which is unusual, it can be separated by the regular procedure. Copper is separated as given under Copper-bearing Ores. Material containing cadmium should be analyzed by other methods, as given under Standard Procedure.

Solutions

Potassium Ferrocyanide. 34.8 grams pure salt in 1000 cc. water. One cc. = approximately .010 g. Zn. This solution should be allowed to stand about four weeks before using.

Ferric Nitrate. One part salt in 6 parts water, It is well to add a little nitric acid to prevent hydrolysis.

Citric Acid. One part acid in 3 parts water. One hundred cc. of nitric acid should be added to each liter to prevent mould growth.

Standardization. The factor for the standard solution varies slightly, as would be expected, with the amount of ferrocyanide used, so that it is best to have at least three sets of factors, one at 40 cc., one at 20 cc., and one at 10 cc.

Weigh into 600-cc. beakers at least three portions of C.P. zinc. Dissolve the metal in about 10 cc. nitric acid, first covering with water. Boil off the nitrous fumes and dilute to 250 cc. with distilled water.   Add 10 cc. of ferric nitrate solution, and 15 cc. citric acid solution, make faintly ammoniacal, using a piece of litmus paper as indicator.   Then add a measured excess of ammonia, as follows: 40 cc. factor, 20 cc. excess; 20 cc. factor, 10 to 12 cc. excess; and for low titrations make only faintly ammoniacal.   Heat the solution to a full boil, and titrate immediately with the standard ferrocyanide, stirring the solution *thoroughly* and adding ferrocyanide not too rapidly.   The titration is completed when a drop of solution gives a bluish-green coloration with a drop of 50% acetic acid on a spot plate.   To prevent passing the end-point, or until the operator is experienced, a portion (50 cc.) of the solution may be held back in a small beaker, the end-point passed, and the titration completed after adding the part in the small beaker.

General Notes

A standard zinc solution may be used in case the end-point is passed.   However, this is not to be recommended as a usual practice.   In any case it should be very dilute, so that 1 cc. = 0.001 gram zinc.

The ferrocyanide should be added gradually and the solution stirred constantly, to prevent occlusion of ferrocyanide or zinc solution by the heavy precipitate.

A moisture sample should be weighed at the same time as the sample for analysis.

The variation of factor with amount of zinc titrated is more marked in this method than in the titration in acid solution.   Hence, it is necessary that standards be run covering the whole range of zincs to be titrated.   It will be found that the factors from 30 to 50 cc. are almost the same and from 15 to 30 cc. slightly lower, from 5 to 15 cc. still lower.

The zinc used as a standard should be carefully examined for foreign particles and oxidized zinc.   In case stick zinc is used, the surface should be scraped clean before cutting.   Merck's and Kahlbaum's stick zinc, as well as Baker & Adamson's, Eimer & Amend's, or J. T. Baker's powdered zinc answer the purpose as regards metallic impurities.   It is desirable to check the factor by means of a standard ore.

The standard of the ferrocyanide solution should be frequently checked, at least once every ten days.   A solution of such a strength that 1 cc. equals 10 milligrams of zinc has in glass a temperature coefficient sufficient to decrease the factor 0.2 % per 5° C. rise in temperature, so care should be taken that no sharp change of temperature occurs between standardization and titration.

The factors in alkaline and acid solution are not identical.   In alkaline solution the precipitate closely approaches the normal ferrocyanide, while in acid solution there is formed a double ferrocyanide of zinc and potassium.   If blank determinations are made, and the titrations corrected for the blank, a very nearly constant factor will be obtained which can be used over the entire range.

Standard Method

Titration in Acid Solution—Separating of Zinc as Sulphide

General.   The method of separating zinc as sulphide in a solution slightly acid with sulphuric acid is of almost universal application, and can be used on any class of zinciferous material that has come under the author's observation.   The steps fit together, so that copper and cadmium are easily separated and any zinc in the insoluble state, e.g., spinels, etc., can readily be looked for.   The method of decomposing (taking to fumes of sulphuric acid) tends to take into solution material that would be overlooked in the rapid decompositions effected in the preceding methods.   Moreover, the use of the internal indicator gives a very sharp end-point, so that this method is fully as accurate as any gravimetric method.   The method is more time consuming than the ones already given, but it is not designed for rapid routine work, but rather as a standard procedure that will give absolutely reliable results on all classes of material. This method is also recommended for routine work in case the analyst is called on to make only occasional zinc analyses.

Standardization of the Ferrocyanide Solution

NOTE. The standardization of the solution is given first, on account of the method of titration.

Weigh into tall 400-cc. beakers several portions of C.P. zinc, using about 0.35 gram. Cover with water and dissolve in 10 cc. hydrochloric acid (sp.gr. 1.2). Now add 13 cc. ammonia (sp.gr. 0.9), make acid with hydrochloric acid, and add 3 cc. excess. Add 0.03 or 0.04 milligram of ferrous iron in the form of a ferrous sulphate solution and dilute to about 200 cc. with distilled water. Heat to boiling and titrate as follows: About one-quarter of the solution is reserved in a small beaker and the ferrocyanide added to the main solution with vigorous stirring. The solution takes on a blue color, which changes to a creamy white when an excess of ferrocyanide is added. Now add a few cc. more and pour in the reserved portion of zinc solution, excepting about 5 or 10 cc. Add ferrocyanide until the end-point is reached and add about $\frac{1}{2}$ cc. more. The last of the reserved zinc solution is then poured into the main beaker, washing out the small beaker with a portion of the main solution, and the ferrocyanide added drop by drop until the blue color fades sharply to a pea green with one drop of ferrocyanide.[1] This is the end-point. Repeat until satisfactory standards are obtained.

Procedure. Weigh into a tall 150-cc. beaker an amount of sample so that it gives a titration of about 40 cc., i.e., 5 grams for a 10% ore to $\frac{1}{2}$ gram for 60% ore and over. Moisten with water and add 10 cc. of hydrochloric acid (sp.gr. 1.20), cover with a watch-glass. In case of sulphides it is necessary to add nitric acid. Boil moderately on a hot plate for half hour or so. Remove and wash down cover-glass and sides of beaker, add 10 cc. of 1 : 1 H_2SO_4 and evaporate to strong fumes of sulphuric acid. In case of very siliceous material, it is well to break up the silica with a glass rod before adding the sulphuric acid. After fuming, the solution is cooled and diluted to 40 to 50 cc. and about a gram of 20-mesh aluminum added. Cover with a watch crystal and boil until water white (about ten to fifteen minutes). This will reduce the iron and precipitate all the hydrogen sulphide metals, except cadmium.[2] The silica and precipitated metals are filtered off and washed with hot water.

Add 5 cc. of 1 : 1 sulphuric to the filtrate and dilute to 100 cc. Pass a rapid stream of hydrogen sulphide through the solution for fifteen minutes. Add dilute ammonia, a drop at a time until yellow cadmium sulphide precipitates. Then heat the solution to 70 to 90° C. and continue to pass hydrogen sulphide for a few minutes. Filter at once through a close paper previously packed by washing with a polysulphide, an acid and water.[3] The precipitate is washed with cold 8 to 10% sulphuric acid and finally with hot water. The filtrate is boiled to remove hydrogen sulphide, cooled, neutralized with potassium hydroxide

---

[1] It is only by adding an *excess* of ferrocyanide that one is assured of a precipitate of normal composition.

[2] Cadmium is partially precipitated, but goes back in solution.

[3] All the cadmium is separated, except about 0.05%, which does not interfere with the titration at the given acidity.

solutions, or sodium hydroxide solution, to within an acidity of a couple of drops of 20% sulphuric acid. Methyl orange is used as an indicator. Add from 2 to 4 cc. of 5%[1] sulphuric acid per 100 cc. of solution according to the amount of zinc present. Cool thoroughly.[2] A rapid stream of hydrogen sulphide is now passed through the solution for forty minutes.[3] Allow the precipitate to settle ten or fifteen minutes, filter and wash with cold water. A hole is punched in the filter paper and the sulphide washed back into the beaker in which it was precipitated. The filter paper and glass tube are then washed with 10 cc. of hydrochloric acid in hot water, catching the washings in the same beaker. Boil off the hydrogen sulphide, add 13 cc. of ammonia (sp.gr. 0.9), neutralize with hydrochloric acid, add 3 cc. excess and dilute to 200 cc. Heat to boiling and titrate as under Standardization. When cadmium is absent or present in quantities less than 0.05, the procedure is of course shortened considerably.

To Separate Cadmium Electrolytically. After filtering off the silica and precipitated hydrogen sulphide metals, add 1 cc. of 1 : 1 sulphuric acid, dilute to 125 cc. and electrolyze with 0.8 to 1.0 ampere per 100 sq.cm. of electrode surface for $1\frac{1}{2}$ hours at 2.95 to 3.05 volts. Proceed with the residual solution as above. As in all electrolytic separations the current must be carefully watched.

Procedure with Material Containing Insoluble Zinc

Proceed as usual up to point where the solution is to be reduced. Filter off the silica and insoluble material, wash with hot water and proceed with the filtrate as usual. Burn the insoluble residue in a platinum crucible, taking the usual precautions in case lead is present. Fume off the silica with hydrofluoric and sulphuric acids and fuse with acid potassium sulphate. Dissolve in water and sulphuric acid and proceed as in the regular method. The solution may be added to the main portion or titrated separately.

Discussion on Separating Zinc as Zinc Sulphide and Titrating in Acid Solution

Precipitation. The method of precipitating zinc as sulphide in sulpnuric acid solution was investigated by G. Weiss (Inaugural Dissertation, München, 1906), and the work confirmed by F. G. Breyer. The main points of Weiss' paper are as follows:

1. "Sulphate solutions are preferable to chlorides." A N/10 chloride solution is not completely precipitated by H_2S. Furthermore, the precipitate of sulphide from HCl solution when quantitative is not crystalline and easy to filter like that obtained from sulphate solution.

2. "The concentration of a sulphate solution is without influence on the completeness of precipitation from N/10 down. That is for solutions containing at most 400 milligrams ZnO per 100 cc."

3. "Sulphate solutions of 400 milligrams ZnO per 100 cc. may be N/100 acid with H_2SO_4 before beginning the precipitation." Even at acidity N/20 before

---

[1] Bear in mind at this point the acid liberated by the action of H_2S in the zinc sulphate. See Discussion below.

[2] In cold solution the precipitate is more granular and easier to filter.

[3] The hydrogen sulphide should pass through at a rate of at least eight bubbles per second.

precipitation less than a milligram of zinc remains unprecipitated. According to Weiss, if the solution were diluted to 300 cc. 1.3 grams of H_2SO_4 could be added or $6\frac{1}{2}$ cc. of 20% H_2SO_4, and still have the precipitation complete. Even if as much as 10 cc. of 20% acid were added the loss would still be only a little more than 1 milligram. Precipitating 300 milligrams from 100 cc., however, only 100 milligrams or $\frac{1}{2}$ cc. of 20% acid could be added. This means that when the solution becomes more acid than 550 milligrams of H_2SO_4 per 100 cc. the precipitation of ZnS ceases. Knowing approximately the zinc content of a solution one can easily calculate the H_2SO_4, freed when the $ZnSO_4$ is converted into ZnS, and the difference between 550 milligrams and this calculated H_2SO_4 is the amount of acid that may be added when precipitating from 100 cc. of solution. For two hundred cc. of course more acid can be added, being the difference between 1.100 grams and the calculated H_2SO_4 freed from the $ZnSO_4$. One and one-half times the amount of Zn judged to be present is close enough for the H_2SO_4 freed.

4. " The precipitation, under the above given conditions, is incomplete when a slow current of hydrogen sulphide is used (about four bubbles per second). One must work with as fast a stream as possible without causing mechanical losses (at least eight bubbles per second)." Weiss is the first one to discuss this all-important question in the precipitation of ZnS. His explanation of the efficacy of the rapid stream of H_2S is as follows:

The precipitation takes place according to the following equation:

$$ZnSO_4 + H_2S \leftrightarrows ZnS + H_2SO_4.$$

Equilibrium is reached, i.e., the velocity becomes equal in both directions, and precipitation ceases when the amount of H_2SO_4 per 100 cc. reaches a certain point, under a given set of conditions. Let these conditions remain exactly the same with the exception of the H_2S and have the active mass of that increased. The equilibrium will be displaced from left to right and as a consequence ZnS will come down in the presence of more acid than before. H_2S is not very soluble in water at room temperature, but if one increases the surface of contact between the two the H_2S is dissolved much more rapidly and consequently the mass of H_2S active at any time greatly increased. This is exactly what is accomplished when the zinc solution is constantly kept full of bubbles of H_2S. One can easily see how greatly increased the mass of H_2S would be in the extreme case, when the solution is all foam.

5. " A strong current of gas, like that called for above, will precipitate the usual amounts of zinc used in analytical operations in forty minutes."

6. " At temperatures above 50° the precipitation is incomplete; furthermore, at room temperature the ZnS comes down in a form suitable for filtration."

7. " Water only is required for washing the precipitates."

End-point. The change of color from blue to pea green is very sharp. It should be observed by looking down through the solution and not from the side. The change in color may be explained as follows: The ferrocyanide, having stood for three or four weeks, has oxidized slightly to ferricyanide, due to dissolved oxygen in the water. The few tenths of a milligram of ferrous iron added acts with this ferricyanide giving the ferro-ferricyanide blue as long as the ferrocyanide is not in excess. When it is in excess the blue is decomposed and gives the colorless ferro-ferrocyanide.

Rapid Ferrocyanide Method for Determining Zinc in Ores

Outside Indicator

Removal of Interfering Elements

1. 0.5 gram of ore. 8-oz. Erlenmeyer flask. 5 cc. of HCl, 10 cc. HNO₃. Boil gently almost to dryness. Remove from heat. Add 12 cc. HNO₃ and 5 grams (measured) of KClO₃. Boil gently just to dryness, finishing by manipulating flask (in holder) over free flame.

2. Add 35 cc. of Extraction Solution and heat to boiling, boiling very gently until disintegration is complete. Now add 10–25 cc. of saturated bromine water, according as manganese contents appear low or high, as indicated by brown color of residue.

3. Boil a minute or two longer and then filter through an 11 cm. filter containing a small moistened wad of absorbent cotton in the apex. Receive filtrate in a 400 cc. beaker. Wash out the flask with hot water. Remove adhering residue with rubber-tipped glass rod, or dissolve it with a few drops of HCl, then add 5 cc., or an excess, of NH₄OH and rinse into filter, finally washing out flask several times with hot water. Now wash filter and residue 10 times with hot Wash Solution.

Precipitation of H₂S Group

4. Add a little litmus solution to filtrate as indicator, stir and cautiously add HCl just to acidity, then 3 cc. in excess. Dilute, if necessary, to 200–250 cc. with hot water and heat nearly to boiling. Now add 50 cc. of saturated H₂S water and then the hot liquid is ready for titration.

Titration of Zinc

5. Pour off about half the liquid as a reserve and titrate the balance until the end-point is passed. Use a spot-plate in which about 2 drops of a 15% solution of uranium nitrate have previously been placed in each depression. Transfer the zinc solution to the spot-plate with a glass tube instead of a rod, taking only a drop or two for each test, except for the final tests of the titration, when about $\frac{1}{4}$ of a cc. should be taken. After the first end-point is passed add a portion of the reserve and again pass the end-point. Repeat this, each time with more caution, until the reserve is reduced to about 5 cc. Now titrate, 6 drops at a time, until the end is again observed, then pour the entire liquid, or most of it, over the 5 cc. of the remaining reserve and then back into the same beaker again and finish the titration 2 drops at a time until the end-point, or brown tinge, is plainly apparent. Read the burette. Allow a couple of minutes for the tests to fully develop and then deduct from the burette reading for as many tests as show and for 1 drop additional. Multiply the number of cc. of ferrocyanide solution used by the percentage value of 1 cc.

Note.—The above method, suggested by A. H. Low, is commonly used in zinc ore assays in the middle west.—Editor.

Reagents

6. Extraction Solution. 200 grams of commercial ammonium chloride dissolved in a mixture of 500 cc. of strong ammonia and 750 cc. of water.

Wash Solution. 100 grams of commercial ammonium chloride, 50 cc. of strong ammonia. Dissolve and dilute to 1 liter.

7. Standard Ferrocyanide. $K_4Fe(CN)_6.3H_2O$ 21.6 grams to the liter. 1 cc. = about 0.005 gram Zn, or 1%. Standardize on about 0.2 gram of pure zinc. Dissolve in 10 cc. of HCl. Dilute somewhat, neutralize as above, complete the dilution, heat and titrate precisely as described above. No filtration or H_2S water necessary.

NOTES

Potassium chlorate is added to precipitate the manganese and to dilute the mass with salts facilitating the subsequent zinc extraction. The presence of HCl is undesirable, hence an excess of HNO_3 should be present to expel HCl during evaporation. Bromine water is added to insure complete precipitation of manganese.

The degree of acidity is important as it has a direct influence on the end point. Two drops (*i.e.* one drop excess) are necessary to produce a color at the end point, hence the deduction of one drop is made.

The zinc compound formed has approximately the following composition— $K_2ZnFe(CN)_6.Zn_2Fe(CN)_6$, the precipitating reagent is $K_4Fe(CN)_6.3H_2O$, two molecules of this reagent (422.37 g. ×2) precipitate three atoms of zinc (65.38×3) hence a reagent containing 21.54 grams is equivalent to 5 grams of zinc or to 0.005 g. Zn per cc.

In presence of Fe and M.O. indicator the color is lavender changing to pink at the end point. If phenolphthalein is the indicator, the end point is a pea green (from robin's egg blue) becoming white or gray with an excess of ferrocyanide reagent (H_2S present in large amount interferes).

Determination of Zinc in Alloys. See chapter in volume 2.

False Endpoint. During the titration with potassium ferrocyanide with too rapid addition of the reagent a false endpoint is obtained. This is recognized by the fact that an additional amount of reagent causes no deepening of the brown color with the uranium indicator. With additional reagent, heating and stirring no further brown color is produced. When the true endpoint is reached an addition of more ferrocyanide results in a deeper color of the spot test. The solution stirred and boiled still gives the test.

If considerable copper and lead are present in the ore, it is advisable to remove these from the solution. The addition of paper pulp assists in the settling of the sulphides, enabling a test to be made with the clear supernatant solution. With the dark suspended sulphides a sharp endpoint can not be obtained, so that it is advisable to allow the solution to settle and test using the clear liquor.

DETERMINATION OF SMALL AMOUNTS OF ZINC

The following method is applicable to samples containing 0.05% Zn or less.

Procedure. A large sample, 10 or 20 grams, is brought into solution by the standard procedure, taken to fumes of sulphuric acid and the zinc precipitated as sulphide after separating groups 5 and 6 by the procedures given under Standard Method, filtered and dissolved in hydrochloric acid. The sample is now washed into a 100-cc. Nessler tube, 5 cc. of ferrocyanide added and the whole made up to the mark, mixed by pouring into a beaker and then back into the tube. A standard containing the same amount of acid is made up and a standard zinc chloride solution added until the turbidity of standard and unknown are the same. From the amount of zinc added to the standard the percentage can be calculated. The standard zinc solution is made up by dissolving C.P. zinc in hydrochloric acid and diluting so that 1 cc. is equal to 1 milligram of zinc.

SPECIAL METHODS
Determination of Metallic Zinc in Zinc Dust

Discussion. There have been various methods proposed for determining the metallic zinc content of zinc dust. Most of these are based upon its reducing power. The latter may be determined by any one of many ways, although the results from different methods will not be concordant, due to the inaccuracies inherent with most of the methods. Potassium bichromate, iodate, ferric sulphate, and iodine have been used for measuring the reducing power of zinc dust. Fresenius also proposed dissolving the zinc dust in dilute sulphuric acid and after drying passing the hydrogen over heated copper oxide in a combustion tube, absorbing the water formed in a calcium chloride tube and weighing.

There have also been methods devised based on the volume of hydrogen evolved when a sample of zinc dust is dissolved in dilute acid. Several investigators have concluded from comparative investigations that the gasometric determination of the hydrogen evolved gives the most consistently accurate results. The best arrangement of apparatus for carrying out this hydrogen evolution method is shown in Fig. 72. The time required for a determination is about $1\frac{1}{2}$ hours.

Procedure. One gram of zinc dust is weighed and transferred as rapidly as possible to a small Erlenmeyer flask A, of 100 or 200 cc. capacity, in which is placed a piece of sheet platinum about 1.5 cm. square. About 5 g. of clean unoxidized ferrous sulphate crystals are added on top of the zinc dust and the flask nearly filled with distilled water saturated at room temperature with hydrogen gas.

The object of adding the sheet platinum and ferrous sulphate is to increase the rate of hydrogen evolution by catalytic action. A further reason for adding the ferrous sulphate on top of the zinc dust sample is to coagulate the latter as much as possible when it becomes wetted, and thus prevent the floating of more than an unappreciable amount of the sample.

The rubber stopper containing separatory funnel B and connecting tube C is tightly inserted into the neck of the flask. A little distilled water is poured into B and the three-way stop-cock in C turned to connect the flask with the downward outlet. Enough water is now run in from the separatory funnel to displace all the air in the flask and the connecting tube through the bore in its stopcock. The stopcock in C is now turned so that the downward outlet is in connection with the measuring tube D. By raising the leveling bottle E, containing 10 per cent. sulphuric acid also saturated with hydrogen at room temperature, all the gas in D is displaced. The stopcock in C is now turned through 90 deg. so as to connect the decomposing flask A with the measuring tube D. The system is hence completely filled with liquid and ready for the generation of hydrogen. The measuring tube D has a total capacity of 400 cc. and is graduated from 250 to 400 cc. by 0.25 cc.

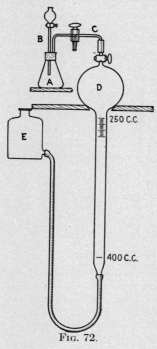

FIG. 72.

Thirty cubic centimeters of 1 : 1 sulphuric acid are now poured into the separatory funnel. A small portion of this acid is allowed to run into the decomposing flask until a brisk but not too rapid evolution of hydrogen takes place. The acid, being much heavier than water, settles to the bottom of the flask and the action commences immediately. The gas evolved, together with some solution and a very small amount of zinc dust passes over into the measuring tube, displacing the acid there. When the action in the decomposing flask has slowed down, more strong acid is introduced until all has been added. During this time the acid in the measuring tube and flask is shaken so as to wash down the particles of zinc dust from the upper parts of the flask and tube now filled with gas. The particles in the measuring tube on coming in contact with the 10 per cent. sulphuric acid are readily dissolved and generate their portion of hydrogen.

When all the zinc dust has been dissolved, water is run in from the separatory funnel to force the hydrogen over into the measuring tube and to fill the flask and connecting tube with water through to the stopcock which is then closed. After leveling with the leveling bottle the volume of hydrogen generated from the 1-g. sample at the prevailing atmospheric conditions is read from the measuring tube. The percentage of metallic zinc in the sample is then calculated from the following expression:

$$\text{Per cent. of Metallic Zinc} = \frac{V \times (P-p) \times 0.29196}{(1+0.00367t)760},$$

in which V = volume of gas in measuring tube at atmospheric conditions, P = barometric pressure, p = vapor tension of water above 10 per cent. sulphuric acid at room temperature, and t = room temperature.

Necessary Precautions. To obtain results of the highest accuracy, it is necessary when weighing out samples of zinc dust which are very finely divided, to keep the time of exposure as small as possible in order to minimize the oxidation that takes place with the oxygen of the air. It is also highly important when samples are to be held, that they be kept in ground glass stoppered bottles, completely filled, and sealed with paraffin or wax.

The two variables most likely to affect the results are temperature and barometric pressure. A change in the barometric pressure is practically always extended over a reasonable length of time. A careful reading of the barometer when the volume of gas in the measuring tube is read will eliminate any error from this source. A temperature change, on the other hand, affects not only the volume of gas, according to Charles' law, but also affects the vapor tension of water and hence the actual pressure of the hydrogen when measured.

The rubber connection between the connecting and measuring tubes must be of heavy rubber and should be shellacked.

The vapor tension of water is slightly lower above 10 per cent. sulphuric acid than above pure water, as shown in Fig. 73 and for accurate work should be used in place of the ordinary vapor tension tables.

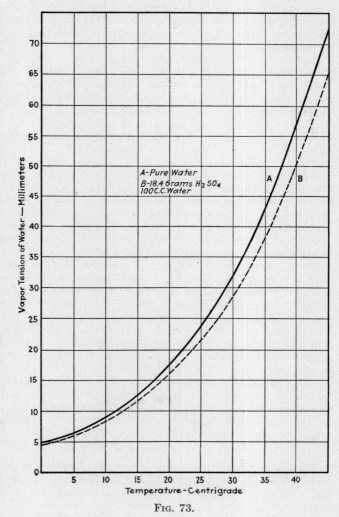

A-Pure Water
B-18.4 Grams $H_2 SO_4$
100 C.C. Water

Vapor Tension of Water — Millimeters

Temperature-Centrigrade

FIG. 73.

The result obtained should be corrected for any metallic impurities, as Fe, Al, etc., which evolve hydrogen when dissolved in sulphuric acid.

48

DETERMINATION OF IMPURITIES IN SLAB ZINC

LEAD

Electrolytic Method

The sample is thoroughly mixed on a sheet of paper and a magnet passed over to remove iron particles. Any pieces showing discolorations are discarded. The pieces of spelter for the weighed sample are removed from the paper by lifting and not pouring. The amount of sample taken is determined by the amount of lead in the spelter.

For No. 1 or High Grade Spelter a sample of 17.286 grams is weighed into a 600 cc. beaker, 200 cc. of distilled water added, and 60 cc. of concentrated nitric acid added gradually until solution is complete. The solution is boiled down to 200 cc., expelling all nitrous fumes, and then transferred to a 250 cc. beaker, washing out original beaker without unnecessarily increasing volume. Four or five drops of 5% silver nitrate solution are added, to precipitate any traces of chlorine present, and the solution electrolyzed for one and one-half to two hours, with initial temperature of 60–70° C. and current of 1.8 amperes.

For No. 2 or Intermediate grade, a sample of 8.643 grams is weighed into a 250 cc. beaker, 100 cc. of distilled water added, and 40 cc. of concentrated nitric acid added. The solution is boiled until all nitrous fumes have been expelled and diluted to 200 cc. with distilled water, 4 or 5 drops of 5% silver nitrate added and the solution electrolyzed as given under High Grade Spelter.

For the three lower grades of spelter, Brass Special, Selected and Prime western, a sample of 17.286 grams is weighed into a 600 cc. beaker, covered with 200 cc. of water and 60 cc. of concentrated nitric acid added. This solution is boiled down to 200 cc. transferred to a 500 cc. graduated flask and the flask filled to the graduation. After thorough mixing, an aliquot portion of 125 cc. (equivalent to 4.3215 grams) is poured into a 250 cc. beaker, 10 cc. of concentrated nitric acid added, boiled for ten minutes, diluted to 200 cc., with distilled water, silver nitrate added and electrolyzed as given for high grade spelter.

The anode is made of sheet platinum with wire stem and is sand blasted. No gold solder should be used on these cylinders. The surface area is 135 sq. cm. A spiral of platinum wire is used as a cathode. A spiral of aluminum wire may also be used, although its life is much shorter.

When electrolysis is complete, the anode is removed from the solution as rapidly as possible, washed three times with distilled water, once with alcohol, dried in an air bath at 210° C. for thirty minutes, cooled and weighed.

The weight of PbO_2 can be calculated directly to percentage of lead as factor weights of sample were taken.

Lead Acid Method

Weigh 10 grams of the sample into a 400 cc. beaker and add 120 cc. of "lead acid." When all but about 10% of the zinc is dissolved, filter and wash with lead acid. Retain the filtrate. Wash the metallics back into the beaker and dissolve in nitric acid. Add 40 cc. of "lead acid" and evaporate to strong fumes. Cool and add 35 cc. of water, which is the amount evaporated from the "lead acid," and heat to boiling. Add the filtrate containing most of the zinc and a little lead sulphate, stir and allow to settle over night. Filter on a Gooch crucible, wash with lead acid, a mixture of alcohol and water (1 : 1), finally with alcohol and ignite inside a porcelain crucible and weigh as lead sulphate.

Lead Acid. Add 1 gram of lead acetate in 300 cc. of water to dilute sulphuric acid (300 cc. acid to 1800 cc. of water). Shake well, allow to cool and settle. Filter off the precipitated lead sulphate. By the use of this sulphuric acid saturated with lead, the solubility of lead sulphate need not be considered, the solution being brought back to the same concentration each time.

NOTE. The rejection limits for spelter of the American Society for Testing Materials consider five grades as follows:

Grade	Pb Not Over	Fe Not Over	Cd Not Over	Sum Pb + Fe + Cd Not Over	Al
No. 1. High Grade	.07%	.03%	.07%	.10%	None
No. 2. Intermediate	.20	.03	.50	.50	"
No. 3. Brass Special	.60	.03	.50	1.0	"
No. 4. Selected	.80	.04	.75	1.25	"
No. 5. Prime Western	1.60	.08			

IRON

Hydrogen Sulphide Method

Ten grams or more of the sample of spelter, which has been passed under a magnet to remove any metallic iron particles, is weighed out into a 250 cc. beaker and dissolved with 50 cc. of concentrated hydrochloric acid. After standing several minutes until violent action has ceased, about 1 cc. of potassium chlorate solution (50 grams per liter) is added and boiled until all chlorine fumes have been driven off. It is then cooled, 50 cc. of water added, neutralized with ammonia, adding a large excess of the latter, boiled for two or three minutes and then allowed to settle. The solution is filtered, the precipitate on the filter washed with dilute ammonia water and finally with hot water. The precipitate is dissolved into a 500 cc. Erlenmeyer flask, using 10 cc. 1 : 4 sulphuric acid. The paper is washed thoroughly with hot water. After diluting to a volume of about 300 cc. hydrogen sulphide is passed through for 5 minutes to reduce the iron. The excess of hydrogen sulphide is boiled off, being careful to exclude the entrance of any air. When no trace of hydrogen sulphide is present, as indicated by no test with lead acetate paper, the solution is cooled rapidly, and titrated with a standard solution of potassium permanganate, 1 cc. of which is equivalent to about 0.00034 grams of iron. A blank determination is also run in order to determine the amount of potassium permanganate required to show the pink color on titration. The permanganate solution is standardized against Bureau of Standards Sodium Oxalate.

Colorimetric Method

Solutions. *Ammonium Sulphocyanate.* One part salt to 2 parts of water.
Potassium Chlorate. One part salt to 20 parts of water.
Standard Iron Solution I. 500 grams of low iron granulated zinc is dissolved in 800 to 1000 cc. of concentrated hydrochloric acid, sufficient amount of potassium chlorate solution added to oxidize the iron present, boiled several minutes to expel chlorine, cooled and made up to 2500 cc. This solution is standardized by measuring off 50 cc. portions, corresponding to 10 grams of zinc, and determining the iron content by the Hydrogen Sulphide Method.

Standard Iron Solution II. 0.7 grams of ferrous ammonium sulphate is dissolved in water, 10 cc. dilute sulphuric acid added, and the iron oxidized with permanganate. The solution is diluted to 1000 cc. each cc. being equivalent to .0001 gram iron.

Procedure. Ten grams of sample are dissolved and oxidized as given under the Hydrogen Sulphide Method. The solution is cooled and diluted to about 25 cc. and transferred to a comparison tube,[1] 2 cc. of the sulphocyanate solution are added, and the volume brought up to 100 cc. and mixed by pouring into the beaker and back into the tube. The red color is compared with the color produced on adding 2 cc. of sulphocyanate to 50 cc. of standard iron solution I,[2] and diluting to 100 cc. Standard Iron Solution II is added until the colors are the same, and the iron present in the unknown calculated. In case the iron is over 0.030, it should be determined by the Hydrogen Sulphide Method.

CADMIUM
Sulphide Method

The determination of cadmium in spelter is best carried out by either of the two following methods, according as the amount to be determined is low or high.

Cadmium in Spelter (Cd —.001% to .02%)

A sample of 500 grams is taken and placed in a 2000 cc. flask with 500 cc. of water and 100 cc. of 1 : 1 sulphuric acid. Sulphuric acid (1 : 1) is added in small quantities from time to time, to maintain solution of the zinc (but never any violent action), until about 90% of the sample has been dissolved. This requires about 750 cc. of 1 : 1 sulphuric acid. When the evolution of hydrogen has become slow, after the last addition of acid, the solution is filtered off, retaining as much of the undissolved metallics as possible in the flask. The metallics are washed twice with water, decanting and pouring the small amount of metallics on the filter. The metallics on the filter are then washed from the paper into the original flask and all dissolved in nitric acid. The solution is washed into a 600 cc. beaker and sufficient sulphuric acid (1 : 1) added to convert all the metals present to sulphates, leaving an excess of free acid of 10–15 cc. This requires about 95 to 100 cc. of acid. The solution is evaporated until all nitric acid has been expelled.

After cooling, water is carefully added and the beaker heated until all the soluble salts have been dissolved. The lead sulphate is allowed to settle, filtered off and washed. Enough water must be added to keep all zinc sulphate in solution after cooling. Hydrogen sulphide is passed through the filtrate for 15–20 minutes. No precipitate will appear at first, so that a drop or two of ammonia is added and repeated at intervals until a considerable amount of zinc sulphide has been precipitated. The sulphides are then filtered off and washed with cold water. The film of sulphides on the side of the original beaker and the sulphides on the paper are dissolved in 1 : 2 hydrochloric acid, washing with water, and catching the solution in a tall 400 cc. beaker. 15 cc. of 1 : 1 sulphuric acid are added and evaporation carried down to fumes. After cooling, 200 cc. of water is added, hydrogen sulphide passed through and ammonia added as above to produce a pre-

---

[1] Clear glass test-tubes $\frac{3}{4}$ in. in diameter, and holding 110 cc. make good comparison tubes.

[2] The zinc content of the standard and unknown must be approximately the same. (See references, Bureau of Standards Bulletin, No. 1, Vol. 3.)

cipitate of cadmium sulphide. The precipitate is treated as before and a third precipitation made. In the case of very low cadmium, a fourth precipitation is advisable. The last precipitate is filtered at once on a weighed Gooch crucible, washed with cold water, dried at 110° C. and weighed as CdS or the cadmium may be weighed as metallic cadmium, by dissolving the last precipitate and electrolyzing as given by the following method:

The last precipitation of cadmium as CdS is filtered off, washed with cold water, and dissolved in dilute hydrochloric acid as in previous precipitations, catching solution and washings in original beaker. Twenty-five cc. of 1 : 1 sulphuric acid are added and evaporation carried to dense fumes. After cooling, a small amount of water is added, the solution neutralized with ammonia (using methyl orange as indicator), then made just acid with sulphuric acid (sp.gr. 1.09) and 3 cc. added in excess. The solution is then transferred to a 250 cc. beaker, (washing out the original beaker), diluted to cover the cylinder (beaker $\frac{2}{3}$ full), and electrolyzed over night using a current of one ampere. The cylinder with cadmium deposit is removed, washed with water and alcohol, dried at 110° C., removed from oven as soon as dry, cooled and weighed.

Cadmium in Spelter (Cd −0.2% to 1.0%)

A sample of 10 (or 25) grams is taken and entirely dissolved in a tall 400 cc. beaker with nitric acid, the solution then treated with 25 (or 50) cc. of 1 : 1 sulphuric acid and evaporated to expel all nitric acid. After cooling, water is carefully added and the beaker heated until all the soluble salts have been dissolved. The lead sulphate is allowed to settle, filtered off and washed. The filtrate is diluted to 200 cc. and hydrogen sulphide gas passed through for 15–20 minutes. No precipitate will appear at first, so that a drop or two of ammonia is added and repeated at intervals until a considerable amount of zinc sulphide has been precipitated. The sulphides are filtered off and washed with cold water. They are then dissolved with 1 : 2 hydrochloric acid, catching in the original beaker, 15 cc. of 1 : 1 sulphuric acid added and taken to fumes. This is again diluted to 200 cc. hydrogen sulphide gas passed through, and ammonia added as before to produce a precipitate of cadmium sulphide. This is treated as above and a third precipitation made. The cadmium sulphide from this is filtered at once on a weighed gooch, washed with cold water, dried at 110° C., and weighed as CdS, or the cadmium may be weighed as metallic cadmium, by dissolving the last precipitate and electrolyzing as given under Method for Low Cadmium.

General Remarks

A very retentive filter paper must be used for the cadmium sulphide precipitates.

In dissolving the cadmium sulphide precipitates from the filter paper, the funnel should be covered to guard against loss by spraying. The number of treatments with hydrochloric acid depends upon the amount of sulphides on the paper. The paper may be finally washed with water after all sulphides have dissolved, if desired.

It is not absolutely necessary to wash the final precipitate of cadmium sulphide on the gooch, as this has been accomplished in the transfer of the precipitate from the beaker. This has been included in the methods to safeguard against free acid remaining on the Gooch.

ANALYSIS OF ZINC CHLORIDE SOLUTION

The methods to be used for the analysis of zinc chloride solution or fused zinc chloride are essentially those given under the various chapters for the various elements. It seems advisable, however, to include a set of methods of analysis suitable for the commercial evaluation of zinc chloride.

I. Specific Gravity at 15° C.

The specific gravity is determined by means of a picnometer. The volume is brought to the graduation after the solution in the picnometer has been brought to 15° C., using a water bath. The weight of this volume of boiled distilled water is determined at 15° C., and the specific gravity of the zinc chloride solution calculated, compared to water at 15° C.

II. Zinc (Manganese and Copper Absent)

About 25 grams of the well shaken solution is weighed out in a weighing bottle and transferred to a 500 cc. graduated flask. Sufficient nitric acid is added to clarify the solution upon dilution. The flask is filled to the mark with distilled water and thoroughly mixed.

A portion, approximately equivalent to one gram of $ZnCl_2$ is accurately measured from a pipette or a burette and the zinc determined by titration with a standard potassium ferrocyanide solution as given under procedure for zinc in ores.

Zinc (Manganese or Copper Present)

If manganese or copper is present an aliquot portion is measured out and the manganese or copper separated according to the methods given under procedures for ores, before titration with potassium ferrocyanide.

III. Chlorine

Another portion of this solution, approximately equivalent to .5 grams $ZnCl_2$ is measured off into a 500 cc. Erlenmeyer flask, 15 cc. of distilled water, 100 cc. of standard N/10 silver nitrate solution and 40 cc. of nitric acid are added to the flask and boiled until all nitrous fumes are driven off. After cooling, the excess silver nitrate is titrated with standard N/10 ammonium thiocyanate solution using 5 cc. (1–6) ferric nitrate solution as an indicator. A blank is run at the same time and the amount of chlorine determined from the difference in volumes of ammonium thiocyanate required. The factor for the standard ammonium thiocyanate solution is best determined with C.P. sodium chloride.

IV. Sulphuric Anhydride (SO_3)

Twenty-five cc. of the original well shaken solution of zinc chloride are measured off with a pipette into a 400 cc. beaker, diluted to 300 cc. with hot water and a few drops of hydrochloric acid added. Any insoluble matter is filtered off, five cc. of bromine water are added to the filtrate and the solution boiled until excess bromine is all driven off. The SO_3 in the filtrate is precipitated with 25 cc. hot 10% barium chloride solution. After standing on the steam plate for 3 hours, the barium sulphate is filtered off, ignited and weighed. The weight of the sample is determined from the specific gravity.

V. Iron (Fe)

A portion of the well shaken solution, equivalent to 10 grams of zinc is evaporated to a syrupy consistency and the iron determined by the colorimetric method as under spelter.

In case the iron is too high to estimate colorimetrically, it is separated with ammonia, filtered off, washed with hot water and dissolved in hot dilute sulphuric acid. This solution is cooled, run through a Jones reductor and titrated with standard potassium permanganate solution, or the iron may be determined by the hydrogen sulphide method as given under analysis of spelter.

VI. Iron and Aluminum ($Fe_2O_3 + Al_2O_3$)

Either 10 or 20 cc. of the original well shaken solution are transferred to a 400 cc. beaker, diluted with 150 cc. of water and hydrochloric acid added to a very faint excess. (2 drops concentrated acid.) A rapid stream of hydrogen sulphide is passed through the solution for 30 to 40 minutes. The precipitate of zinc sulphide is filtered off and washed thoroughly. The filtrate is boiled for about 15 minutes to remove hydrogen sulphide, cooled, sufficient bromine water added to more than oxidize all the iron, and then boiled to remove the excess bromine. Ammonium hydroxide is now added in slight excess, the precipitate of iron and alumina filtered off and washed with hot water. The precipitate is dissolved from the paper with hot hydrochloric acid (1 : 4) re-precipitated, filtered off, washed free from chlorides, ignited and weighed.

VII. Manganese (Mn)

Either 10 or 20 cc. of the original well shaken solution are transferred to a 400 cc. beaker, 25 cc. of sulphuric acid (1 : 1) added and evaporated to practically complete expulsion of all excess sulphuric acid. Nitric acid (1 : 3) is now added and the manganese determined according to the bismuthate method.

VIII. Lime (CaO)

Twenty-five cc. of the well shaken solution are measured off with a pipette, a few drops of hydrochloric acid added, and the solution diluted to 150 cc. Twenty grams of ammonium chloride and a few drops of bromine water are added. The iron and manganese are precipitated by ammonia and filtered off after bringing to boiling. The filtrate is evaporated to 150 cc., the lime precipitated with 25 cc. ammonium oxalate solution (saturated solution) and allowed to stand for 3 hours. The calcium oxalate is filtered off and washed four times with hot water. A hole is then punched in the filter paper, the precipitate washed into a 400 cc. beaker, with boiling water, 10 cc. of sulphuric acid (1 : 1) poured over the paper and the paper washed with boiling water. The solution is diluted to 150 cc. with hot water and titrated with standard potassium permanganate solution.

IX. Magnesia (MgO)

The filtrate and washings from the lime determination is made slightly acid with hydrochloric acid and 15 cc. of a saturated solution of microcosmic salts added. The solution is cooled and the magnesia precipitated by the slow addition of ammonia. Sufficient ammonia must be added to hold all zinc in solution, an excess of about 50 cc. being required. The precipitate after standing 12 hours is filtered off, and redissolved in hot 1–4 hydrochloric acid. Twenty grams of ammonium chloride are added to this solution, then a few drops of microcosmic solution, and precipitation affected with ammonia as before, the excess of ammonia being only 10 cc. in this case. After standing 12 hours, the precipitate is filtered off, washed six times with 1 : 10 ammonia water, or until free from chlorides, ignited and weighed as $Mg_2P_2O_7$.

X. Alkalies ($NaCl+KCl$)

A sample of about 10 grams is taken and made up to a volume of 500 cc. From this a portion equivalent to approximately 2 grams is measured off, slightly acidified with hydrochloric acid and the zinc precipitated with hydrogen sulphide. After filtering, the filtrate is tested with ammonium sulphide, and any zinc, iron and manganese precipitated is filtered off. The filtrate is boiled for about thirty minutes to drive off the hydrogen sulphide, and cooled. After cooling, the solution is made slightly acid with hydrochloric acid, a few drops of bromine water added and the excess of bromine boiled off. The solution is diluted to about 200 cc. with hot water, and 10 cc. of hot 10% $BaCl_2$ solution added, to precipitate SO_3. Without filtering off the barium sulphate, the solution is made ammoniacal, one gram of ammonium carbonate and 5 cc. of ammonium oxalate solution added and the precipitates allowed to settle in a warm place. The precipitate is filtered off, washed with hot water, and the combined filtrate and washings evaporated to dryness in a porcelain dish. The ammonium salts are completely driven off by ignition over a low flame. The residue remaining is dissolved in a small amount of water with two drops of hydrochloric acid, transferred to a weighed platinum dish, the porcelain dish being washed with a minimum amount of water, evaporated to dryness on a hot plate, and after cooling weighed.

Any magnesia which may be present with the alkali chlorides is determined, calculated to $MgCl_2$ and deducted from the weight of salts in the dish. This difference is expressed as combined alkali chlorides $KCl+NaCl$.

XI. Copper

Twenty cc. of solution are taken, 5 grams of ammonium chloride and 20 cc. of ammonium hydroxide added, diluted to 100 cc. in a color comparison tube and compared with a zinc chloride solution of equal strength, to which a measured amount of standard copper solution is added to give the same depth of color. If iron interferes with the color comparison, it should be filtered off before diluting to volume. When the amount of copper present is over 0.05%, the determination should be carried out by some other method. See chapter on copper.

XII. Barium (Ba)

If sulphuric anhydride (SO_3) is found to be present, it is not worth while to make a determination for barium. If no sulphuric anhydride is present, barium should be looked for and determined by diluting 10–20 cc. to 300 cc. with water, adding slight excess of hydrochloric acid heating to boiling and precipitating the barium with ammonium sulphate solution (10 per cent.).

ANALYSIS OF FUSED ZINC CHLORIDE

The analysis of fused zinc chloride for zinc, etc., is carried out by the same methods as given under Zinc Chloride Solution, working on a solution of the fused salt in water. This solution is made up by rapidly transfering approximately 20 grams of fused salt to a weighing bottle, weighing, dissolving in water in a 2000 cc. graduated flask to which a few drops of nitric acid has been added to prevent precipitation of basic zinc chloride, and filling to the mark. Aliquot portions are taken from this solution for the various determinations.

Special determinations are sometimes called for with fused zinc chloride which is to be used for some special purpose. It is very essential that such analyses be carried out carefully according to the method prescribed in order that the results, which are largely empirical, may be comparable. One analysis of this sort which is commonly called for is Basicity expressed in some empirical way to give a measure of the relative quantities of the basic zinc chloride, which will settle out upon dissolving fused zinc chloride in water. It may be expressed as the volume of standard hydrochloric acid (usually N/2) required to neutralize 10 cc. of a 40° Baume solution of the fused zinc chloride, diluted with 300 cc. of water, when using Methyl orange as indicator, or the weight of basic chloride obtained by treating 10 grams of the sample, weighed in a weighing bottle, with 400 cc. of water, agitating to complete all possible solution, filtering off insoluble on a gooch crucible and washing with water until combined filtrate and washings total just 1000 cc.

Ammonia (NH₃)

It is often of value to know if ammonium chloride is present. A sample of 5–10 grams is weighed and transferred to a 500 cc. distilling flask, 100 cc. of water, 50 cc. of sodium hydroxide (20 per cent.) and a small quantity of granulated zinc[1] added. The ammonia and water are distilled over into an absorption bottle containing a measured quantity of standard acid. The excess of acid is titrated with standard alkali solution, using methyl orange.

See chapter on nitrogen, page 343, also analysis of paint pigments, volume 2.

[1] The addition of granulated zinc aids in the expulsion of the ammonia by hydrogen formed with the sodium hydroxide.

Determination of Zinc in Pig Lead [1]

Solutions Required

Nitric Acid (1 : 4).—Mix 200 cc. of HNO_3 (sp. gr. 1.42) with 800 cc. of distilled water.

Sulphuric Acid (1 : 1).—Carefully pour, with stirring, 500 cc. of H_2SO_4 (sp. gr. 1.84) into 500 cc. of distilled water.

Acidulated Hydrogen Sulphide Water.—Add 20 cc. of HCl (sp. gr. 1.19) to 1000 cc. of distilled water and saturate with hydrogen sulphide.

Ammonium Thiocyanate Solution (2 per cent).—Dissolve 20 g. of NH_4CNS in 1000 cc. of distilled water.

Hydrochloric Acid (1 : 3).—Mix 100 cc. of HCl (sp. gr. 1.19) and 300 cc. of distilled water.

Standard Zinc Solution (0.1 mg. of zinc per cc.).—Dissolve exactly 0.1 g. of U. S. Bureau of Standards pure zinc in 5 cc. of HCl (sp. gr. 1.19) and dilute to exactly 1000 cc. with distilled water.

Potassium Ferrocyanide Solution.—Dissolve 34.8 g. of $K_4Fe(Cn)_6.3H_2O$ in 1000 cc. of distilled water.

Method

Dissolve 222.23 g. of the sample in 1100 cc. of HNO_3 (1 : 4), using a 1300-cc. beaker. When the lead is dissolved, transfer the solution to a 2000-cc. graduated flask and add slowly 150 cc. of H_2SO_4 (1 : 1). Cool, fill the flask to the mark and then pour the solution into a clean 3000-cc. flask provided with a rubber stopper. Rinse the measuring flask with exactly 50 cc. of water, which is equivalent to the volume of lead sulphate which is present. Mix the solution thoroughly by shaking, allow the precipitate to settle and filter through a dry filter until 1800 cc. of filtrate has been obtained.

Place exactly 1800 cc. of filtrate (equivalent to a 200-g. charge) in a No. 9 porcelain evaporating dish and evaporate the solution to approximately 100 cc. Transfer the solution to a 600-cc. beaker, neutralize with ammonia, and then add 5 cc. of HCl (sp. gr. 1.19) for every 100 cc. of solution. Warm the solution and pass in a rapid current of hydrogen sulphide until it is saturated. Digest for 30 minutes on the steam bath, add an equal volume of water and again saturate with hydrogen sulphide. Filter and wash with acidulated H_2S water.

Discard the precipitate and evaporate the filtrate in glassware containing no zinc (such as Pyrex) until the volume of the solution is approximately 100 cc.

Neutralize the solution with ammonium hydroxide, add 5 g. of citric acid, and warm until the acid is dissolved. Add small portions of calcium carbonate to the hot citric acid solution until about 1 g. of calcium citrate has separated and then pass in a rapid current of H_2S as the solution is allowed to cool. Allow the solution to stand for from 2 to 4 hours, part of the time on a water bath, until the supernatant liquid is clear.

Collect the precipitate on a filter, wash with a 2 per cent solution of ammonium thiocyanate and then dissolve the precipitate in hot dilute hydrochloric acid (1 : 3). If the solution has a reddish color (due to iron), the zinc must

[1] Standard Method of the A. S. T. M.

be reprecipitated as above.   If the solution is clear, evaporate it to dryness on the steam bath, take up the residue in 3 cc. of HCl (sp. gr. 1.19), add 20 cc. of water and filter if not perfectly clear.

Transfer the solution (Note 2) to a 50-cc. Nessler jar and dilute to 45 cc. Prepare other Nessler jars containing 3 cc. of HCl (sp. gr. 1.19), definite volumes of standard zinc solution, and diluted to 45 cc.   Add 5 cc. of potassium ferro-cyanide solution to each jar, mix quickly, and compare the turbidities by viewing longitudinally as the jars are held over a sheet of fine print.   Add more of the standard zinc solution from a burette to the jar which approximates the turbidity of the unknown most closely, until the turbidities match each other, and calculate the percentage of zinc on the basis of a 200-g. sample or the aliquot portion taken.

NOTES.   1. All glassware that contains zinc must be avoided and in umpire work a blank test should be carried along with the test.

2. The whole solution can be used if the lead contains no more than 0.002 per cent of zinc.   If more zinc is present, it is best to take such an aliquot portion of the solution as will give approximately 4 mg. of zinc and then to add enough HCl to provide 3 cc.

3. For further details concerning the turbidmetric test, consult the "Determination of Small Quantities of Zinc" by Mr. Bodansky, in the *Journal of Industrial and Engineering Chemistry*, Vol. 13, pp. 696–697 (1921).

4. The addition of calcium carbonate with the formation of a precipitate of calcium citrate serves the purpose of giving a clear filtrate, and prevents the loss of colloidal sulphide.

ZIRCONIUM [1]

Zr, at.wt. 91.2; **sp.gr.** 4.1 (amorphous) to 6.2 (coherent); **m.p.** about 1700; [2] **oxides,** ZrO_2 (Zr_2O_5 and ZrO_3 in hydrated form).

DETECTION

The material to be tested is brought into solution by one of the methods given in the section on the Preparation of the Sample. In the regular course of qualitative analysis the zirconium will be found in the main precipitate formed either by ammonium hydroxide alone or by the combined action of ammonium hydroxide and ammonium sulphide. This precipitate can be dissolved in sulphuric acid, the acidity then adjusted so that 10% of the solution, by volume, is conc. sulphuric acid, 10 cc. of 3% hydrogen peroxide and 1 cc. more of conc. sulphuric acid added, and then a large excess of Na_2HPO_4 or $(NH_4)_2HPO_4$ solution, to which has also been added 1/10 its volume of conc. sulphuric acid. A white precipitate proves the presence of zirconium, since no other elements (titanium having been peroxidized) have phosphates which are insoluble in such an acid concentration. [3] This is the simplest and surest method of detecting zirconium. In the regular course of analysis thorium and the rare earths would probably have been precipitated from the acid solution of the ammonium hydroxide precipitate by an excess of oxalic acid (see " Detection " in chapter on thorium). The filtrate would contain almost all of the zirconium, [4] along with titanium, beryllium and the more common elements of the group. The oxalic acid in this filtrate can be destroyed by evaporation with sulphuric acid, and then zirconium, titanium and aluminum precipitated by boiling the weakly acid solution (containing 8 or 10 drops more sulphuric acid than is needed to redissolve the first faint precipitate formed by gradually adding dilute ammonia) with sodium thiosulphate. This precipitate may be brought into solution by fusing with potassium pyrosulphate, and taking up with sulphuric acid. The phosphate test as just outlined may then be applied.

Turmeric paper on drying after having been moistened with a slightly acid (HCl or H_2SO_4) solution containing zirconium shows a reddish-brown color, similar to that due to boric acid. Titanium gives a like reaction in the tetravalent condition, but if the acidified solution is reduced by adding a piece of zinc, and then promptly tested (before titanium is reoxidized), the test is specific for zirconium.

[1] Paul H. M.-P. Brinton, Professor of Analytical Chemistry, University of Minnesota.

[2] Considerable doubt exists as to the reliability of the physical constants of zirconium. If the contentions of Coster and Hevesy that practically all zirconium contains varying amounts of hafnium (more justly called celtium) is true, it is not surprising that such discordant results have been recorded by different investigators.

[3] Biltz and Mechlenburg, Z. angew. Chem., **25**, 2110 (1912); Lundell and Knowles, J. Am. Chem. Soc., **41**, 1801 (1919).

[4] Oxalic acid produces a precipitation of zirconium oxalate in dilute mineral acid solution, but the precipitate is soluble in excess of oxalic acid. (It is also soluble in ammonium oxalate; and is not reprecipitated by addition of hydrochloric acid, thus differing from thorium.)

From aluminum zirconium may be differentiated by the solubility of zirconium basic carbonate in excess of a strong, cold solution of ammonium carbonate (on boiling the hydroxide separates out again); and by the insolubility of zirconium oxyiodate, produced by an alkali iodate in a slightly acid solution. (Thorium and tetravalent cerium also give the iodate test.)

ESTIMATION

The main ores of zirconium are baddeleyite, essentially zirconium dioxide; and zircon, essentially zirconium orthosilicate. While zircon is the most widely distributed ore of zirconium, the most important commercially is now the Brazilian baddeleyite, which, mixed with a little zircon, comes under the trade name of zircite, and contains around 75% of ZrO_2. The determination of zirconium in furnace-lining material, high temperature laboratory ware, enamels, and steel is at times called for.

Preparation and Solution of the Sample

Acid treatment will not decompose ordinary zirconium ores. The following fluxes are most commonly used.

Sodium peroxide is used in a nickel crucible which has been lined by fusing some sodium carbonate,[1] and causing it to cool in an even layer on the sides and bottom. This prevents in large measure the attack of the crucible by the peroxide, and avoids the subsequent introduction of large amounts of nickel into the solution. Several grams of sodium peroxide are melted in the crucible, after lining with sodium carbonate, and allowed to solidify on the bottom. This prevents particles of the ore from being caught in the carbonate lining and remaining unfused. The sample of finely ground ore (0.5 g.) is mixed with 8–10 g. of sodium peroxide in the crucible thus prepared, and fused at low red heat over a small flame, by gently swirling the crucible, held in tongs. 5–10 minutes should suffice. When cool the crucible is placed in a large platinum dish, or porcelain casserole, and covered with warm water, keeping the vessel covered with a watch glass until danger of loss from effervescence is over, the solution is boiled until the carbonate lining has dissolved, and the crucible is then removed. The solution is next made decidedly acid with hydrochloric acid, and boiled until carbon dioxide is expelled. This should give a clear solution containing the zirconium and all the other constituents of the ore. If a very small amount of undecomposed ore is found here, it should be filtered off, ignited, fused with a little potassium pyrosulphate, dissolved in 5% sulphuric acid, filtered if necessary, and added to the main solution.

Potassium hydrogen fluoride is applied by Smith and James [2] in the following manner: The finely ground ore is fused with 12 to 15 times its weight of potassium hydrogen fluoride. The latter may be prepared from potassium carbonate or fluoride by treating with a slight excess of hydrofluoric acid, and evaporating over a small flame until a clear fused mass is obtained. When cool the melt may be broken up and preserved for use. The mixture

[1] Private communication from J. A. Holladay, Electro Metallurgical Co., Niagara Falls, N. Y.

[2] Smith and James, J. Am. Chem. Soc., **42**, 1764 (1920).

of ore and fluoride, in a platinum disn, is carefully heated over a small flame. When the mixture has softened it is stirred with a platinum rod, and the size of the flame is gradually increased, finally heating over a Meeker burner until the mass just fuses to a clear liquid.[1] The cooled melt is treated with 50 cc. of 1 : 1 sulphuric acid, gently heated until nearly all water is expelled, and then more strongly until abundant fumes are evolved. The cold residue is boiled with water. This solution contains all the zirconium. If it is to be used for the selenite method of Smith and James, the zirconium is precipitated (along with many other bases) by ammonium hydroxide, filtered, washed free from sulphates—since sulphates are undesirable for the selenious acid precipitation—and dissolved in hydrochloric acid.

Borax is strongly recommended by Lundell and Knowles [2] for the decomposition of all ores of zirconium. This flux is particularly suited to the decomposition of samples in which zirconium only is to be determined, and by precipitation with cupferron, since boric acid does not interfere with this reagent. If other elements of the sample are to be determined, the boric acid must be removed by volatilization as methyl borate, so in such cases it is better to employ the sodium peroxide fusion. In using the borax method Lundell and Knowles recommend the following procedure: 4 g. of the flux are melted in a platinum crucible and allowed to cool. About 0.3 g. of the finely ground ore is placed on top of the fused borax, the crucible is covered, and heated over a Meeker burner until thorough fusion has taken place, which does not ordinarily require more than half an hour. During the operation it is well to stir the melt occasionally with a short platinum rod or stiff wire, which is allowed to remain in the crucible, and which may be conveniently handled with the crucible tongs. When the decomposition of the ore is complete, the platinum rod is removed, and it is put into the beaker to be used for the solution of the melt. As the melt cools, the crucible is gently twirled in order to run the fusion up on the sides in a thin layer. The cooled melt is dissolved in 150 cc. of 1 : 5 hydrochloric acid in a 250-cc. beaker, by filling the crucible with acid, inverting in such a manner that one edge of the crucible rests on the crucible cover, which has been placed on the bottom of the beaker—thus allowing free circulation of the solvent—and then gently warming on the steam bath. The solution is transferred to a platinum dish or porcelain casserole, 20 cc. of 1 : 1 sulphuric acid are added, and the mixture is evaporated until heavy fumes escape.[3] The cooled solution is next diluted to about 100 cc., the impure silica is filtered off and washed with hot water. This solution is allowed to stand warm over night. If the ore contained interfering amounts of phosphorus, it will be thrown down as zirconium phosphate by this digestion. If a precipitate appears, it is filtered and washed with 5% ammonium nitrate solution. This precipitate will contain zirconium,

---

[1] Experiments in the writer's (Brinton's) laboratory by Mr. Tohru Kameda indicate that there is no loss of zirconium by volatilization if the process is carried out as here described. Continuing the heating for 5 minutes after clear fusion has been attained may cause a loss of as much as 2 mg. of ZrO_2.

[2] Lundell and Knowles, J. Am. Chem. Soc., **42**, 1439 (1920).

[3] During the evaporation boric acid will separate and form a crystal skin over the surface, thus retarding evaporation. Occasional stirring during this period will hasten the evaporation. As the sulphuric acid becomes more concentrated by evaporation, the boric acid eventually dissolves again.

and there is still apt to be a little zirconium in the impure silica first filtered off. Also there may be some phosphoric acid in the filtrate; so to this filtrate 5 g. of ammonium chloride, and then ammonium hydroxide in slight excess are added, and after boiling for several minutes, the precipitated hydroxides are filtered and washed with 2% ammonium nitrate solution. This filtrate is discarded. The precipitate and paper are digested in 100 cc. of hot 5% sulphuric acid solution, and the filter shreds and any insoluble residue are filtered and well washed with hot water. This solution, containing most of the zirconium of the sample, is temporarily set aside. The last residue filtered, the impure silica, and any zirconium phosphate obtained by the over-night digestion are all ignited in the original platinum crucible. The ignition residue is moistened with water, 1 cc. of 1 : 1 sulphuric acid, and 5 cc. of hydrofluoric acid are added, and the mixture is heated until all sulphuric acid has been expelled. The residue in the crucible is fused with a little sodium carbonate, digested in water, and filtered, washing with hot water. The filtrate is discarded. The insoluble residue is again ignited in the platinum crucible, fused with potassium pyrosulphate; and, after cooling, the melt is dissolved in hot 5% sulphuric acid. This solution is added to the main solution, thus giving one solution in which is contained all the zirconium originally present in the sample.

SEPARATIONS

From Members of the Copper and Tin Groups. The members of these groups are precipitated free from zirconium by hydrogen sulphide in slightly acid solution.

From Iron, Nickel, Cobalt, Manganese, and Zinc. The iron, in acid solution, is reduced by hydrogen sulphide gas, tartaric acid equal to 5 times the approximate weight of the oxides present is added, the solution is made ammoniacal, and then hydrogen sulphide is passed into the solution until the sulphide precipitate has coagulated. The precipitate is filtered on close-grained paper, care being taken to keep the funnel well filled, so that oxidation of the sulphides and consequent " running through " is avoided. The sulphides are promptly washed, using the same precaution, with water containing a little ammonium chloride and ammonium sulphide. If the cupferron precipitation of zirconium is to be used, the tartrate need not be destroyed, and it is sufficient to acidify to a total acidity of 10% of conc. sulphuric acid (by volume), and to boil out the hydrogen sulphide. If for further work it should be necessary to remove the tartrate, the solution is acidified with 10 cc. of conc. sulphuric acid, evaporated to a volume of about 50 cc., 10 cc. of conc. nitric acid added, and the solution evaporated to fumes of sulphur trioxide. After cooling, 10 cc. of conc. nitric acid are again added, and the mixture once more evaporated to fumes. At times a third addition of nitric acid and evaporation may be necessary to completely destroy the tartaric acid.

From Titanium. Zirconium may be separated from titanium by precipitating the zirconium with Na_2HPO_4 or $(NH_4)_2HPO_4$ in a 10% (by volume) sulphuric acid solution, in presence of hydrogen peroxide. See Gravimetric Determination as phosphate.

From Thorium and Rare Earth Elements. These elements may be precipitated with an excess of oxalic acid, leaving the zirconium in the filtrate. (See " Detection " in chapter on thorium.) For complete separation the rare earth oxalates should be boiled with conc. sulphuric acid until decomposed, diluted, nearly neutralized with ammonia, and again precipitated with an excess of oxalic acid. The combined filtrates contain all the zirconium. (A method adapted to the determination of small amounts of rare earth oxides in the cupferron precipitate is given later.)

From Aluminum, Chromium, and Uranium. Zirconium is quantitatively separated from these three elements by precipitation with cupferron in 10% (by volume) sulphuric acid solution. The uranium must be in the hexavalent condition or it will contaminate the zirconium precipitate. (Nitric acid should not be used to oxidize uranium, as it decomposes cupferron. Uranium will practically always be in the hexavalent condition without special oxidation. If in doubt, the 10% sulphuric acid solution may be boiled with a little hydrogen peroxide.)

From Molybdenum. By precipitation of molybdenum sulphide in acid solution. (See chapter on molybdenum.)

From Tungsten. By precipitation of tungsten in acid solution by cinchonine hydrochloride. (See chapter on tungsten.)

From Vanadium. By fusion of the mixed oxides with sodium peroxide or sodium carbonate, and leaching out the soluble sodium vanadate with water.

GRAVIMETRIC DETERMINATION OF ZIRCONIUM

By Precipitation with Cupferron. The solution, which has a volume of 300–400 cc., and contains 10% (by volume) of conc. sulphuric acid (any tartaric acid present does not interfere), is cooled to about 10° C., and then an excess of cupferron solution (6 g. dissolved in 100 cc. of cold water and filtered) is added. The formation of a fine white precipitate (nitrosophenyl-hydroxylamine) which redissolves shows that an excess of the reagent has been added. The zirconium precipitate is white and curdy, but any titanium present is also quantitatively precipitated, and it will impart a yellow color to the precipitate. A brownish color indicates that the previous separation from iron has been faulty. The precipitate is filtered and washed with cold 1 : 10 hydrochloric acid solution. The precipitate and paper are carefully ignited in a weighed platinum crucible, very slowly until the rush of gases from decomposition of the organic matter has ceased; and then to constant weight over a blast or over a large Meeker burner. The weight thus found represents all the zirconium and titanium dioxides, and some rare earth oxides, if these elements have not previously been completely removed.

To correct for titanium and rare earths the ignited oxides are fused with potassium pyrosulphate, dissolved in 1 : 10 sulphuric acid, and diluted to exactly 100 cc. in a graduated flask. By means of a *dry* 50-cc. pipette exactly one-half of the solution is taken for the determination of titanium, and the pipette and the 100-cc. flask are completely rinsed into another vessel, thus giving the other half for the determination of rare earths.

Small amounts of titanium are determined by the colorimetric method; while for larger amounts the zinc reductor—permanganate titration method —is used. For details of these see chapter on titanium.

Rare earths may be determined according to Hillebrand's method [1] as follows: " Precipitate the hydroxides with an excess of potassium hydroxide, decant the liquid, wash by decantation with water once or twice and then slightly on the filter. Wash the precipitate from the paper into a small platinum dish, treat with hydrofluoric acid, and evaporate nearly to dryness. Take up in 5 cc. of 5% (by volume) hydrofluoric acid. If no precipitate is visible, rare earths are absent. If a precipitate is present, collect it on a small filter held by a perforated platinum or rubber cone and wash it with from 5 to 10 cc. of the same acid. Wash the crude rare earth fluorides into a small platinum dish, burn the paper in platinum, add the ash to the fluorides and evaporate to dryness with a little sulphuric acid. Dissolve the sulphates in dil. hydrochloric acid, precipitate the rare earths by ammonia, filter, re-dissolve in hydrochloric acid, evaporate the solution to dryness, and treat the residue with 5 cc. of boiling hot 5% oxalic acid. Filter after fifteen minutes, collect the oxalates on a small filter, wash with not more than 20 cc. of cold 5% oxalic acid, ignite and weigh as rare earth oxides which are to be deducted from the weight of the cupferron precipitate."

While the recovery of the rare earths by this method is not absolutely complete, the error is negligible for the amounts ordinarily found.

The sum of the weights of rare earth oxides and titanium dioxide found are subtracted from the weight of the ignited cupferron precipitate to obtain the true weight of ZrO_2.

By Precipitation with Selenious Acid.[2] The zirconium and other bases should be in about 5% (by volume) hydrochloric acid solution, free from sulphuric and nitric acids, and should be in a volume of about 750 cc. The solution is heated to boiling, and an excess of $12\frac{1}{2}\%$ selenious acid solution (usually 20–40 cc.) is added. After the precipitate of zirconium selenite has settled it is filtered, washed with 3% hydrochloric acid solution (Smith and James directions; the writer prefers a wash liquid containing 40 cc. of $12\frac{1}{2}\%$ selenious acid reagent in 750 cc. of 3% hydrochloric acid), ignited to constant weight over a strong Meeker burner or blast, and weighed as ZrO_2. For a discussion of the possible interference of large amounts of iron, of titanium, and of phosphoric acid, the original paper of Smith and James should be consulted.

By Precipitation with Ammonium Phosphate. This method is suitable for separations, and for the determination of small amounts of zirconium such as are encountered in rock analysis. In view of the probability that almost all minerals of zirconium contain appreciable amounts of celtium (" hafnium "), which gives practically all the reactions of zirconium, but which has a very much higher atomic weight, an uncertainty as to the factor ZrO_2/ZrP_2O_7 exists, so the pyrophosphate can hardly be recommended as a weighing form for large amounts of zirconium. The following details are recommended for observance in using this method.[3]

A. Volume of Solution. From 25 cc. for small amounts (i.e., 0.0005 g. ZrO_2) to 200 cc. for amounts ranging around 0.1 g. ZrO_2.

[1] W. F. Hillebrand, The Analysis of Silicate and Carbonate Rocks, U. S. Geol. Survey Bulletin 700, p. 176; Lundell and Knowles, loc. cit., p. 1446.
[2] Smith and James, loc. cit.
[3] Lundell and Knowles, J. Am. Chem. Soc., **41**, 1806 (1919); Hillebrand, op. cit., p. 173; Nicolardot and Reglade, Compt. rend., **168**, 349 (1919).

B. Acidity.   Twenty per cent sulphuric acid by weight.

C. Hydrogen Peroxide.   Sufficient to keep titanium peroxidized; 10 cc. will do no harm.

D. Precipitant.   Secondary ammonium phosphate in excess.   From 10 to 100 times the theoretical requirement, as expressed by the ratio $Zr : P_2O_5$, should be used here.   The large excess is desirable when small amounts of zirconium are determined.

E. Precipitation Conditions.   (1) Temperature: Cold or tepid, preferably 40–50° C.   (2) Time: Two hours for amounts of ZrO_2 in excess of 0.005 g. Six hours, or longer, for smaller amounts.

F. Filtration.   Warm solution, decant as far as possible to avoid clogging the filter.

G. Washing.   This should be done with cold 5% ammonium nitrate, since the phosphate is appreciably soluble in pure water.

H. Ignition.   Ignite very carefully in a partially covered platinum crucible over a low flame until carbon is destroyed, then blast or heat over a Meeker burner for 15 minutes.

RECOMMENDED METHOD FOR THE DETERMINATION OF ZIRCONIUM IN COMMERCIAL ZIRCONIUM ORES

The following combination gives a method essentially as recommended by the U. S. Bureau of Standards.   Details will be found in the respective sections of this chapter.

Decomposition by fusion of 0.3 g. sample with borax, following the procedure under " Preparation and Solution of the Sample " until the sulphuric acid solution containing all the zirconium originally present in the sample is obtained.

Separation from members of the copper and tin groups by precipitation with hydrogen sulphide in an approximately 1% sulphuric acid solution; and then, with intervening filtration, separation from iron, manganese, etc., by treatment with tartaric acid, ammonia, and hydrogen sulphide, as detailed under "Separations," finally obtaining the zirconium in 10% (by volume) sulphuric acid solution.   (Removal of tartaric acid is not necessary.)

Precipitation with cupferron, with subsequent corrections for any titanium and rare earths present, just as outlined in the " Gravimetric Determination."

Determination of Zirconium in Steel.   See chapter on iron.

Determination as Zirconium Oxide

With pure salts the zirconium may be precipitated completely as the hydroxide by the addition of ammonia, settling and finally igniting and weighing as the oxide, ZrO_2.

VOL. I

PART II

QUALITATIVE TESTS OF SUBSTANCES

QUALITATIVE TESTS OF SUBSTANCES

BLOWPIPE AND FLAME TESTS OF SOLIDS

Blowpipe Tests on Charcoal

Heat a small portion of the material on charcoal in the reducing flame, using a blowpipe. Scoop out a round hole in the charcoal, place a little of the substance in the cavity, and direct the inner flame of the blowpipe against it at an angle of thirty degrees.

RESULT OF TEST.	INFERENCE.
Melts and runs into the charcoal	Alkalies, K, Na, etc.
An alkaline residue on charcoal	Ca, Sr, Ba, Mg.
A residue which, when moistened with a drop of $Co(NO_3)_3$ and heated in O. F., produces a color which is blue	Aluminum, silicon.
Produces a color which is green	Zinc, tin, antimony.
Produces a color which is red	Barium.
Produces a color which is pink or rose-red	Manganese.
Deflagrates	Nitrates, chlorates.
Leaves an incrustation which is white near flame	Antimony.
White, garlic odor	Arsenic.
Dark red	Silver.
Red to orange	Cadmium.
Lemon yellow (hot), light yellow (cold)	Lead.
Orange yellow (hot), light yellow (cold)	Bismuth.
Yellow (hot), white (cold)	Zinc or tin, latter nonvolatile.

Blowpipe Tests. — Substance fused with Na_2CO_3 on Charcoal. Place a small amount of the substance on charcoal with a little sodium carbonate, and fuse, using reducing flame.

RESULT OF TEST.	INFERENCE.
Metallic globules, without incrustation	
Yellow flakes	Gold.
Red flakes	Copper.
White globule, moderately soft	Silver.
Metallic globules, with incrustation	
White, moderately soft beads	Lead or tin (volatilized lead leaves yellow coat).
White, brittle beads	Bismuth or antimony (yellowish).
Yellow in O. F.	Chromium.
Green in O. F.	Manganese.
A substance (in R. F.) which, when moistened and placed on a silver coin, leaves a brown or black stain	Sulphur compounds.

TEST.	INFERENCE.
Dark gray magnetic powder which, when moistened on a filter paper with a drop of dil. HCl and HNO_3, and gently dried over a flame, leaves a stain which is faint pink, turning blue	Cobalt.
Green stain, turning yellow	Nickel.
A stain turned blue by $K_4Fe(CN)_6$	Iron.

In place of using charcoal the above tests may be made with a splinter of wood covered with a coating of fused Na_2CO_3. The test is made by dipping the heated splinter into a mixture of the powdered substance with fused sodium carbonate and plunging for a moment in the reducing flame. Examine the material on the splinter, scrape off on a piece of glazed paper and examine.

Blowpipe Test. Substance moistened with cobalt nitrate solution and ignited.

COLOR OF RESIDUE OR INCRUSTATION.	INFERENCE.
Brick red	BaO
Pink	MgO
Gray	SrO, CaO.
Yellowish green	ZnO
Dark muddy green	Sb_2O_5
Bluish green	SnO
Blue	Al_2O_3, SiO_2

Flame Test

Flame Test. Moisten a platinum wire in conc. HCl, dip into the powdered substance and insert into a Bunsen flame. If sodium is prominent, examine through a blue glass. (Test the cobalt glass to see if it is effective in cutting out the yellow sodium light by examining a sodium flame through it.)

FLAME COLOR.	COLOR THROUGH BLUE GLASS.	ELEMENT.
Carmine red	Purple	Lithium
Dull red	Olive green	Calcium
Crimson	Purple	Strontium
Golden yellow	Absorbed	Sodium
Greenish yellow	Bluish green	Barium, molybdenum
Green		Cu, $-PO_4$, $-B_2O_3$,
Blue		Cu, Bi, Pb, Cd, Zn, Sb, As
Violet	Violet red	Potassium

The platinum wire should be cleaned before making the test. This can be accomplished by dipping it into conc. HCl and holding it in the Bunsen, or, better, a flame of a blast lamp, until the flame is no longer colored. Repeatedly dipping into the HCl may be necessary.

Examine the flame through a spectroscope, if available, and compare the spectra with a spectra chart. Mere traces of the alkali and alkaline earth metals can be detected in this way by their characteristic spectral lines.

Behavior of Substances fused with Microcosmic Salt and Borax Beads

A clear bead is formed by fusing the flux on a loop of platinum wire. Dip the bead in the finely powdered substance to be examined, and heat again—first in the oxidizing flame; second in the reducing or inner flame. Metallic salts are mostly changed to oxides. In the Table—h. signifies hot; c., cold; sups., supersaturated with oxide; s. s., strongly saturated; h. c., hot and cold.

Color of the Bead.	With Microcosmic Salt, Sodium Ammonium Hydrogen Phosphate.		With Sodium Tetraborate (Borax).	
	In outer or oxidizing Flame.	In inner or reducing Flame.	In outer or oxidizing Flame.	In inner or reducing Flame.
Colorless.	Si (swims undissolved). Al, Mg, Ca, Sr, Ba, Sn (s. s., opaque). Ti, Zn, Cd, Pb, Bi, Sb (not sat.).	Si (swims undissolved). Al, Mg, Ca, Sr, Ba (sups. not clear). Ce, Mn, Sn.	h. c.: Si, Al, Sn (sups. opaque). Al, Mg, Sr, Ca, Ba, Ag (not sat.). Zn, Cd, Pb, Bi, Sb, Ti, Mo.	Si, Al, Sn (s. s. opaque). Alkaline earths and earths. h. c.: Mn, Ce. h.: Cu.
Yellow or Brownish.	h. (s. s.): Fe, U, Ce. c.: Ni.	h.: Fe, Ti. c.: Ni.	h., not sat.: Fe, U. h., sups.: Pb, Bi, Sb.	h.: Ti, Mo.
Red.	h. (s. s.): Fe, Ni, Cr, Ce.	c.: Cu. h.: Ni, Ti with Fe.	h.: Fe, Ce. c.: Ni.	c.: Cu (sups. opaque).
Violet or Amethyst.	h. c.: Mn.	c.: Ti.	h. c.: Mn. h.: Ni with Co.	c.: Ti.
Blue.	h. c.: Co. c.: Cu.	h. c.: Co. c.: W.	h. c.: Co. c.: Cu.	h. c.: Co.
Green.	h.: Cu, Mo; Fe with Co or Cu. c.: Cr.	c.: Cr. h.: U, Mo.	c.: Cr. h.: Cu, Fe with Co.	Cr. sups.: Fe.
Gray and Opaque.	Ag, Pb, Sb, Cd, Bi, Zn, Ni.			The same as with microcosmic salt.

Solvents in order of preference—H_2O, dil. HCl, conc. HCl, dil. HNO_3, conc. HNO_3, aq. reg. Refractory Substances—H_2SO_4+HF, fusions with Na_2CO_3, $KHSO_4$, NaOH or KOH. Analysis—— Neutral or slightly acid solutions of the salts of

Pb		
Hg'		
Ag		
As'''		
As^V		
Sb^V		
Sb'''		
Sn^{IV}		
Sn''		
Au'''		
Pt^{IV}		
Mo^{VI}		
Hg''		
Pb		
Bi		
Cu'		
Cu''		
Cd		
Al		
Cr'''		
Cr^{VI}		
Fe''		
Fe'''		
Co		
Ni		
Mn''		
Mn^{VII}		
Zn		
Ba		
Sr		
Ca		
Mg		
K		
Na		
NH_4		

Add hydrochloric acid.

Precipitate. (1) Silver Group.

$PbCl_2$ (w)	
HgCl (w)	
AgCl (w)	

Add hot water. Res. Sol.

$PbCl_2$	Tests (a) +H_2SO_4=$PbSO_4$ white. (b) +H_2S=PbS	
HgCl	Add NH_4OH	Residue—NH_2HgCl+Hg black.
AgCl		Solution—$(NH_3)_2(AgCl)_2$ acidify wi

Solution.

Pass hydrosulphuric acid gas into the warm solution.

(2) Tin and Copper Group. Precipitate.

As_2S_3 (y)	
As_2S_5 (y)	
Sb_2S_5 (o)	
Sb_2S_3 (o)	
SnS_2 (y)	
SnS (br)	
Au_2S_3 (br)	
PtS_2 (bk)	
MoS_3 (br)	

(2) Copper Group.

HgS (bk)	
PbS (bk)	
Bi_2S_3 (br)	
Cu_2S (bk)	
CuS (bk)	
CdS (y)	

Add yellow ammonium sulphide.

Solution.

$(NH_4)_4As_2S_5$
$(NH_4)_3AsS_4$
$(NH_4)_3SbS_4$
......
$(NH_4)_2SnS_3$
Solution.
Solution.
$(NH_4)_2MoS_4$

Residue.

HgS
PbS
Bi_2S_3
Cu_2S
CuS
CdS

Add dilute hydrochloric acid. Precipitation of Tin Group.

As_2S_3
As_2S_5
Sb_2S_5
......
SnS_2
......
Au_2S_3
PtS_2
MoS_2

Dissolve in HCl with the aid of $KClO_3$.

H_3AsO_4
$SbCl_3$
......
$SnCl_4$
$AuCl_3$
$PtCl_4$
$MoCl_6$

Digest in hot dilute HNO_3.

HgS Dissolve in nitrohydroc
$Pb(NO_3)_2$
$Bi(NO_3)_3$
$Cu(NO_3)_2$
......
$Cd(NO_3)_2$

Add 2 or 3 drops of dilute H_2SO_4. Ppt. Sol.

$PbSO_4$
$Bi(NO_3$
$Cu(NO_3$
$Cd(NO_3$

(3) Iron Group. Precipitate.

$Al(OH)_3$ (w)
$Cr(OH)_3$ (g)
$Fe(OH)_3$ (r)

Boil with NaOH. Res. Sol.

$NaAlO_2$	Add NH_4Cl or render
$Cr(OH)_3$	Fuse on platinu
$Fe(OH)_3$	KNO_3 and Na_2CO
	Extract with wa

Cr''' becomes Cr'', Fe''' becomes Fe'' and Mn^{VII} becomes Mn''.

Boil to expel H_2S, add HNO_3 and boil to oxidize ferrosum; and add ammonium chloride and ammonium hydroxide.

(4) Zinc Group. Precipitate.

CoS (bk)
NiS (bk)
MnS (pk)
ZnS (w)

Digest with cold dilute HCl. Res. Sol.

CoS
NiS
$MnCl_2$
$ZnCl_2$

Solution.

$MnCl_2.2NH_4Cl$, $(NH_4)_2ZnO_2$, and $MgCl_4.NH_4Cl$ are formed; also soluble double compounds of Co and Ni with ammonium hydroxide.

Add hydrosulphuric acid to the strongly ammoniacal solutions.

(5) Calcium Group. Precipitate.

$BaCO_3$
$SrCO_3$
$CaCO_3$

Add NH_4OH and $(NH_4)_2CO_3$.

Solution.

Add Na_2HPO_4 to a portion, and if a precipitate is obtained, use $(NH_4)_2HPO_4$ with the other portion and fil-

Abbreviations: (w)=white; (y)=yellow; (o)=orange; (br)=brown; (bk)=

THE METALS. ANALYSIS OF THE SOLUTION.

$+K_2CrO_4 = PbCrO_4$ yellow. (d) $+KI = PbI_2$ yellow.

gCl white.

AsH₃ gas ...... SbH₃ gas+Sb	Conduct the gas into AgNO₃ sol. Also obtain mirror and spots.	H_3AsO_3 } Remove $AgNO_3$ with $CaCl_{21}$ and add H_2S{ As_2S_3 Lemon yellow. Gützeit Test—AsH_3 colors $HgCl_2$ paper a deep maroon. See method on page 40. $SbAg_3$ } Dissolve in hot HCl, dilute, filter and add H_2S{ Sb_2S_3 Orange.

Sn .. Au Pt	Digest with warm HCl.	$SnCl_2$ } Test with $HgCl_2$. { $HgCl$, White; or Hg Gray. Sb Au Pt	Dissolve in nitro-hydro-chloric acid.	$SbCl_5$ } $SbCl_5$ reject or test in Marsʰ apparatus. $AuCl_3$ $PtCl_4$

Boil with solution of FeSO₄. { Au / Dissolve in nitrohydrochloric acid, evaporate to dryness with excess of NH_4Cl and digest with alcohol. } ignite to Au°, Yellow. { $AuCl_3$. NH_4Cl Evaporate and ignite to Au°, Yellow. / Pt ... $(NH_4)_2PtCl_6$ Ignite to Pt°, Gray.

Blue to green-brown \ Evaporate to dryness with excess of HNO_3. Dissolve res. in NH_4OH and add to an
or black solution. / excess of HCl. Test this sol. with Na_2PHO_4 } Ammonium phosphomolybdate, Yellow.

nd test with (a) $SnCl_2$ = White HgCl or Gray to Black Hg. (b) Au wire = Hg on wire.

formation of PbI_2 or $PbCrO_4$. See Pb above.

Bi(OH)₃	Add hot K_2SnO_2 pouring over ppt. on filter.} Bi Black.
Cu(OH)₂.2NH₄OH.2NH₄NO₃	Deep blue solution evidence of copper. For traces add $HC_2H_3O_2$ and test with $K_4Fe(CN)_6${ $Cu_2Fe(CN)_6$ Red-brown.
Cd(OH)₂2NH₄OH.2NH₄NO₃	Add KCN till blue color disappears, then H_2S{ CdS Lemon-yellow.

HCl and precipitate with $(NH_4)_2CO_3${ $Al(OH)_3$ White, gelatinous.

Sol. { K₂CrO₄ and / Na₂CrO₄ } Acidify with $HC_2H_3O_2$ and add $Pb(C_2H_3O_2)_2${ $PbCrO_4$ Lemon-yellow.
Res. Fe(OH)₃ } Dissolve in HCl and add KCNS { $Fe(CNS)_3$ Blood red. Test original solution (acid)
with KCNS for Feʹʹʹ and with $K_3Fe(CN)_6$ for Feʹʹ{ $Fe_3[Fe(CN)_6]_2$ Blue.

CoCl₂	a. Test with borax bead. Blue bead. c. Or evaporate +H_2SO_4 and add nitroso-β-naphthol. b. Add $NaHCO_3$ and H_2O_2, Green solution. Co—Red precipitate. Test with borax bead.
NiCl₂	a. Test with borax bead. Brown bead. c. Or make sol. ammoniacal and add 1% b. Heat with Br and NaOH { $Ni(OH)_3$ } add KI. Free I in CS_2. alcoholic sol. nitrosobetanaphthol = $[(CH_3)_2C_2N_2O_2H]_2Ni$ Red.
Mn(OH)₂	Boil with PbO_2 and HNO_3 } $HMnO_4$, Purple.
Na₂ZnO₂	Add H_2S } ZnS White. Ppt. is insoluble in dilute acetic acid.

BaCrO₄	Add NH₄OH & (NH₄)₂CO₃.	Dissolve in HCl and add H_2SO_4{ $BaSO_4$ White. SrCO₃ CaCO₃	Dissolve in HC₂H₃O₂.	$Sr(C_2H_3O_2)_2$ $Ca(C_2H_3O_2)_2$
Sr(C₂H₃O₂)₂				
Ca(C₂H₃O₂)₂				

Divide into 2 portions: 1. Add $CaSO_4$ set aside 10 minutes { $SrSO_4$ White.
Moisten $SrSO_4$ with HCl and apply flame test.
2. Add K_2SO_4, boil, set aside ten minutes.
Filter and add { CaC_2O_4. White, soluble in HCl.
$(NH_4)_2C_2O_4$

Precipitate $MgNH_4PO_4$ White.

K—Apply flame test using cobalt glass. Violet.

Na—After removal of Mg apply flame test, yellow.

NH₄—To the original solution add KOH in strong excess, warm (note odor) and test with moist litmus
paper; pass gas into Nessler's reagent K_2HgI_4 sol.{ NHg_2I, Brown.

=red; (g) =green; (pk) =pink. Ppt. =precipitate. Res. =residue. Sol. =solution.

GENERAL SUMMARY OF TESTS FOR ACIDS

Acids.	Detecting Reagents.	Reactions resulting from Test.
Acetates	H_2SO_4 (conc.)	Odor of vinegar
Arsenates	(a) $(NH_4)_2MoO_4 + HNO_3$	Yellow precipitate
	(b) Magnesia mixture	White granular precipitate
	(c) Reduced on $C + Na_2CO_3$	Garlic odor, arsenic mirror
Arsenites	(a) Magnesia mixture	No reaction
	(b) $H_2S + HCl$	Yellow precipitate
Bromides	(a) H_2SO_4 (conc.)	Red Br vapor
	(b) Chlorine water $+ CS_2$	Reddish color, due to Br
Borates	H_2SO_4 (conc.) + alcohol	Green flame
Carbonates	Dilute acids	CO_2 evolved. Limewater test
Chlorates	(a) H_2SO_4 (conc.)	Explosive liberation of $Cl + ClO_4$
	(b) Heated alone	O given off
Chlorides	$AgNO_3 + HNO_3$	White precipitate, sol. in NH_4OH
Chromates	(a) H_2SO_4 (conc.)	O liberated (sol. yellow to green)
	(b) HCl	Chlorine of HCl liberated
	(a) Alcohol + NaOH	Reduced and $Cr(OH)_3$ precipitated
Cyanides	H_2SO_4 (conc.)	HCN (POISON). Odor, bitter almonds
Ferricyanides	$FeSO_4 + HCl$	Turnbull's blue precipitate
Ferrocyanides	$FeCl_3 + HCl$	Prussian blue precipitate
Fluorides	H_2SO_4 (conc.)	HF gas liberates silicic acid from glass rod with drop of H_2O
Hypochlorites	Dilute acids	Cl liberated, yellow gas
Iodides	(a) H_2SO_4 (conc.)	Violet vapor of iodine
	(b) Chlorine water $+ CS_2$	Violet color to CS_2
Nitrates	$FeSO_4 + H_2SO_4$ (conc.)	Brown ring
Nitrites [1]	Dilute acids	N_2O_3 brown evolved
Oxalates	H_2SO_4 (conc.)	$CO + CO_2$ evolved
Permanganates	Reducing agents	Decolorized
Phosphates	$HNO_3 + (NH_4)_2MoO_4$ at 40°	Yellow precipitate
Silicates	(a) Fused with Na_2CO_3 and HCl added	Silicic acid precipitated
	(b) HF	SiF_4 gas liberated
Sulphates	$HCl + BaCl_2$	White precipitate of $BaSO_4$
Sulphides	Dil. acids	H_2S gas blackens $Pb(C_2H_3O_2)_2$
Sulphites	Dilute acids	SO_2 gas
Sulphocyanides	$FeCl_3$	Deep red color
Thiosulphates	Dilute acids	SO_2 gas + free S
Tartrates	Ignited	Char. Odor of burnt sugar
Organic acids	Heated	Generally char.

[1] Nitrites $+ KI + CS_2 =$ violet color in CS_2 due to free I.

TABLES OF REACTIONS
BASES AND ACIDS

TABLES OF REACTIONS OF THE BASES.

HYDROGEN CHLORIDE GROUP.

REAGENT.	LEAD, $Pb(NO_3)_2$.	MERCURY, $HgNO_3$.	SILVER, $AgNO_3$.
Hydrochloric acid, HCl.	Lead chloride, $PbCl_2$, white ppt. Slightly sol. in cold water. Solubility in 100 c.c. H_2O, $0° = 673$ mg. $100° = 3340$ mg. Converted into the insol. basic salt by NH_4OH.	Mercurous chloride, $HgCl$, white ppt. Sol. in hot HNO_3 and in aqua regia. NH_4OH converts it to $HgCl \cdot NH_2 + Hg$, black. 100 c.c. H_2O dissolves 0.31 mg. (cold), 10 mg. (hot).	Silver chloride, $AgCl$, white ppt. Insol. in acids. Sol. in NH_4OH, KCN, and $Na_2S_2O_3$. AgCl darkens in the light. Solubility in 100 c.c. H_2O, 0.15_2^{200} mg., 2.2^{1000} mg.
Ammonium hydroxide, NH_4OH.	$(PbO)_2Pb(NO_3)_2$. Basic salt, white ppt. Insol. in excess. Only slightly sol. in water.	Mercuric ammonium salt and mercury, $HgNH_2NO_3 + Hg$, black ppt. Insol. in excess.	Silver oxide, Ag_2O, brown ppt. Sol. in excess Sol in KCN. 4.3^{200} mg. in 100 c.c. H_2O. Sol. in KCN.
Hydrogen sulphide, H_2S.	Lead sulphide, PbS, black ppt. Insol in $(NH_4)_2S_x$. Sol. in HNO_3. Cold water dissolves 0.1 mg. Hot water, insol.	Mercuric sulphide, $HgS + Hg$, black ppt. Slightly sol. in HNO_3. Sol. in aqua regia. Insol. in water.	Silver sulphide, Ag_2S, black ppt. Insol in $(NH_4)_2S_x$. Sol. in hot HNO_3, 0.02 mg. in cold water. Sol. in conc. H_2SO_4.
Potassium chromate, K_2CrO_4.	Lead chromate, $PbCrO_4$, yellow ppt. Slightly sol. in HNO_3. Sol. in NH_4OH. Solubility 0.02^{180}. Insol. in hot water and in acetic acid.	Mercurous chromate, Hg_2CrO_4, brick red. Slightly sol. in HNO_3. Sol. in aqua regia. Sol. in hot water and in KCN.	Silver chromate. Ag_2CrO_4, dark red ppt. Sol. in HNO_3 and in NH_4OH. 2.8^{180} mgs. in 100 c.c. H_2O. Sol. in KCN.

Potassium ferrocyanide, $K_4Fe(CN)_6$.	Lead ferrocyanide, $Pb_2Fe(CN)_6$, white ppt. Insol. in cold water.	Mercurous ferrocyanide, $Hg_4Fe(CN)_6$, white ppt.	Silver ferrocyanide, $Ag_4Fe(CN)_6$, yellowish white ppt. Insol. in acids and in NH_4OH. Sol. in KCN.
Sodium carbonate, Na_2CO_3.	Basic lead carbonate, $2PbCO_3 \cdot Pb(OH)_2$, white ppt., "white lead." Insol. in hot and cold water.	Basic salt, yellow ppt, becoming black.	Silver carbonate, Ag_2CO_3, white ppt. Sol. in NH_4OH, 3.1^{100}, 50 mg. at 100° in H_2O. Sol. in $Na_2S_2O_3$.
Sodium hydroxide, NaOH.	Lead hydroxide, $Pb(OH)_2$, white ppt. Sol. in excess. Sol. in HNO_3. Insol. in NH_4OH. Sol. in KOH. Slightly sol. in water.	Mercurous oxide, Hg_2O, black ppt. Sol. in HNO_3. Insol. in NH_4OH and alkalies. Sol. in glacial acetic acid.	Silver oxide, Ag_2O, brown ppt. Sol. in NH_4OH and in HNO_3. 4.3 mg. at 20° in H_2O. Sol. in KCN.
Stannous chloride, $SnCl_2$.	Lead chloride, $PbCl_2$. Sol. in hot water. (See above.)	Mercury, Hg, dark gray ppt. Sol. in HNO_3, conc. H_2SO_4. Insol. in HCl.	Silver chloride, AgCl, white ppt. (See above.)
Sulphuric acid, H_2SO_4.	Lead sulphate, $PbSO_4$, white ppt. Insol. in excess. Slightly sol. in HNO_3. Sol. in NaOH. $NH_4C_2H_3O_2$. 4.2 at 20°.	Mercurous sulphate, Hg_2SO_4, white ppt. Slightly sol. in water, 200 mg. Sol. in H_2SO_4, HNO_3.	Silver sulphate, Ag_2SO_4, white ppt. Formed only in conc. solutions. Sol in H_2SO_4 and in HNO_3. 580 mg. in H_2O. Insol. in alkalies.
Miscellany.	Zn ppts. Pb in crystalline form. Pb sol. in HNO_3. Hot conc. H_2SO_4.	Pptd. in acid solutions by Cu, Zn, from its salts as metallic Hg. SO_2 reduces mercurous salts to Hg, which collects as globules on boiling solution. Hg sol. in HNO_3. Insol. in HCl.	Insoluble salts, AgBr, AgI. Ag is displaced from its salts in crystalline form by Zn, Cu, Hg, and in gray form by SO_2, $SnCl_2$, $FeSO_4$, etc. Ag sol. in HNO_3. Insol. in alkalies.

The numerals indicate milligrams of the substance that will dissolve in 100 c.c. of water at stated temperature. Reference to Van Nostrand's Chemical Annual, edited by Olsen.

THE HYDROGEN SULPHIDE GROUP.

Insoluble Subgroup.

Sol. = Soluble.　Insol. = Insoluble.

REAGENT.	BISMUTH, $BiCl_3$.	CADMIUM, $CdSO_4$.	COPPER, $CuSO_4$.	MERCURY (= ic), $HgCl_2$.
Hydrogen sulphide, H_2S.	Bismuth sulphide, Bi_2S_3, brown ppt. Sol. in HNO_3. Insol. in $(NH_4)_2S_x$ and in KCN. Cold water 100 c.c. dissolves 0.018 mg.	Cadmium sulphide, CdS, yellow ppt. Sol. in HNO_3, H_2SO_4 (hot dil.). Insol. in $(NH_4)_2S_x$, KCN. Cold H_2O, 0.13 mg. Hot H_2O forms coloidal solution.	Copper sulphide, CuS, black ppt. Sol. in HNO_3, KCN. Slightly sol. in $(NH_4)_2S_x$. Insol. in H_2SO_4 (hot dil.). Cold H_2O, 0.033 mg.	Mercuric sulphide, HgS, white → yellow → red → brown → black. Insol. in HNO_3, $(NH_4)_2S_x$. Sol. in $Br + KClO_3$ or aqua regia, Na_2S. Cold H_2O, 2.5 mg.
Ammonium hydroxide, NH_4OH.	Bismuth hydroxide, $Bi(OH)_3$, white ppt. Insol. in excess. Changed by boiling to Bi_2O_3. Insol. in H_2O. Insol. in alkalies.	Cadmium hydroxide, $Cd(OH)_2$ white ppt. Sol. in excess. Insol. in water, e.g., 0.26^{250} mg. Sol. in NH_4 salts.	Basic copper, Ammonium sulphate. Sol. in excess = a deep blue. $Cu(NH_3)_4''SO_4''·H_2O$, characteristic test.	Amido mercuric chloride, $HgNH_2Cl$, white ppt.
Potassium chromate, K_2CrO_4.	Bismuth chromate, $(BiO)_2CrO_4$, yellow ppt. Sol. in HNO_3. Insol. in KOH. (See Pb.)	Basic chromate, $Cd_2(OH)_2CrO_4$, yellow. Insol. in NaOH.	Copper chromate, $CuCrO_4$, reddish brown ppt. Sol. in NH_4OH, forming a green solution.	Mercuric chromate, $HgCrO_4$, reddish yellow ppt. Sol. in HNO_3. Slightly sol. in water.
Potassium cyanide, KCN.		Cadmium cyanide, $Cd(CN)_2$, white ppt. Sol. in excess, = $Cd(CN)_2(KCN)_2$. From this H_2S ppts. CdS, yellow.	Copper cyanide, $Cu(CN)_2$, greenish yellow ppt. Sol. in excess = $Cu(CN)_2(HCN)_2$. No pptn. by H_2S.	$Hg(CN)_2$. Sol. in water. 12500 mg.

	Bismuth	Cadmium	Copper	Mercury
Potassium ferrocyanide, $K_4Fe(CN)_6$.	Bismuth ferrocyanide, $Bi_4(Fe(CN)_6)_3$, white ppt. Insol. in HCl.	Cadmium ferrocyanide, $Cd_2Fe(CN)_6$, yellowish white ppt. Sol. in HCl. Insol. in H_2O.	Copper ferrocyanide, $Cu_2Fe(CN)_6$, reddish brown ppt. Slightly sol. in NH_4OH. Insol. in acids.	
Sodium carbonate, Na_2CO_3.	Basic bismuth carbonate, $(BiO)_2CO_3$, white ppt. Insol. in water. Sol. in acids. Insol. in Na_2CO_3.	Cadmium carbonate, $CdCO_3$, white ppt. Insol. in excess. Sol. in NH_4OH. Sol. in acids.	Basic copper carbonate, $Cu_2(OH)_2CO_3$, blue. Changed to black CuO on boiling.	Mercuric basic carbonate, $HgCO_3(HgO)_3$, reddish brown ppt. Changed to yellow HgO on boiling. Insol. in H_2O.
Sodium hydroxide, NaOH.	Bismuth hydroxide, $Bi(OH)_3$, white ppt. See above.	Cadmium hydroxide, $Cd(OH)_2$, white ppt. Insol. in excess. (See above.)	Copper hydroxide, $Cu(OH)_2$, light blue. Insol. in excess. Changed to black CuO on boiling.	Mercuric hydroxide, $Hg(OH)_2$. Easily changed to HgO, yellow. Insol. in excess. In presence of $NH_4Cl = HgNH_2Cl$, white ppt.
Sulphuric acid, H_2SO_4.	*No precipitate.	No precipitate.	No precipitate.	No precipitate.
Stannous chloride, $SnCl_2$.	Darkens. Precipitate of BiOOH changes to Bi_2O_3.		Cuprous chloride, CuCl, white ppt. Sol. in HCl, NH_3 aq., NH_4Cl. Insol. in H_2O.	Murcurous chloride, HgCl, white ppt. In excess = gray, Hg.
Miscellany.	In water Bi salt precipitates as BiOCl, white. Reduced by Na_2SnO_2 to metallic Bi, black. Bi sol. in HNO_3, conc. H_2SO_4.	Na_2HPO_4 ppts. $Cd_3(PO_4)_2$, white. Sol. in NH_4OH and in dilute acids. Cd sol. in acids. Sol. in NH_4NO_3.	Na_2HPO_4 precipitates $Cu_3(PO_4)_2$ greenish blue. Sol. in NH_4OH. KI ppts. Cu_2I_2, white $+ I =$ brown. Cu sol. in HNO_3, hot conc. H_2SO_4.	Precipitated by Cu. KI ppts. HgI_2, red. Sol. in excess. Hg sol. in HNO_3, conc. H_2SO_4. Insol. in HCl.

* A precipitate forms on long standing.

	ARSENIC, As$\cdots$, As$\cdots\cdots$.		ANTIMONY, Sb$\cdots$, Sb$\cdots\cdots$.	
	(ous) K_3AsO_3.	(ic) KH_2AsO_4.	(ous) $SbCl_3$.	(ic) $KSbO_3$.
Hydrogen sulphide, H_2S.	Arsenic trisulphide, As_2S_3, yellow ppt. Sol. in alkalies, $(NH_4)_2S_x$, $(NH_4)_2S$. Insol. in conc. HCl.	Arsenic trisulphide + S. $As_2S_3 + S_2$, yellow. The ppt. forms slowly by heat.	Antimony trisulphide, Sb_2S_3. orange ppt. Sol. in alkalies, $(NH_4)_2S_x$, $(NH_4)_2S$, HCl (conc.). 0.17 mg.	Antimony pentasulphide, Sb_2S_5, orange ppt. Sol. in alkalies, $(NH_4)_2S_x$, $(NH_4)_2S$, HCl (conc.).
Ammonium hydroxide, NH_4OH.			Antimonious hydroxide, $Sb(OH)_3$, white ppt. Sol. in excess.	Ammonium metantimonate, NH_4SbO_3. Very slightly sol. in excess.
Copper sulphate, $CuSO_4$.	Copper arsenite, $CuHAsO_3$, yellowish green ppt. Sol. in NH_4OH, NaOH, HNO_3.	Copper arsenate, $Cu_3(AsO_4)_2$, greenish blue ppt. Sol. in NH_4OH and in HNO_3.	Antimony oxychloride, white, SbOCl, caused by dilution. Insol. alk. Sol. HCl, CS_2.	Copper antimonate, brown ppt.
Mercuric chloride, $HgCl_2$.	Mercuric arsenite, $Hg_3(AsO_3)_2$, white ppt. Sol. in acids.		Antimony oxychloride, caused by dilution. Sol. in conc. HCl.	
Silver nitrate, $AgNO_3$.	Silver arsenite, Ag_3AsO_3, yellow ppt. Sol. in HNO_3, NH_4OH, $HC_2H_3O_2$.	Silver arsenate, Ag_3AsO_4, reddish brown ppt. Sol. in HNO_3 and NH_4OH.	Silver chloride and antimony trioxide, $AgCl + Sb_2O_3$, white ppts.	Silver antimonate, Ag_2SbO_3, white ppt. Sol. in NH_4OH.
Miscellany.	Magnesia mixture. No ppt. Arsenic sol. in HNO_3, Cl_2, H_2O, aq. reg., hot alkalies. Marsh test (Zn + HCl, etc.)	Magnesia mixture ppts. $MgNH_4AsO_4$, white crys. ppt. Sol. in acetic acid. AsH_3 flame deposits arsenic. Sol. in NaOCl. Sol. in $(NH_4)_2S$. Residue insol. in HCl (conc.).	KOH ppts. $Sb(OH)_3$. Na_2CO_3 ppts. $Sb(OH)_3$ Marsh test (Zn + HCl).	Sb. sol. in hot conc. H_2SO_4 and in aq. reg. SbH_3 in flame deposits antimony. Insol. in NaOCl.

* See Van Nostrand's Chemical Annual for solubility of salts.

Subgroup.

Tin, Sn··, Sn····.		Platinum, Pt····.	Gold, Au···.
(ous) SnCl₂.	(ic) SnCl₄.	PtCl₄.	AuCl₃.
Stannous sulphide, SnS, dark brown. Sol. in alkalies. Difficultly sol. in (NH₄)₂Sₓ. Sol. in HCl (conc.). 100 c.c. H₂O diss. 0.002 mg.	Stannic sulphide, SnS₂, yellow ppt. Sol. in alkalies, (NH₄)₂Sₓ, (NH₄)₂S and alkali carbonates. HCl (conc.). H₂O = 0.02 mg.	Platinic sulphide, PtS₂, dark brown ppt. Difficultly sol. in aqua regia. Insol. in HCl (conc.).	Gold sulphide, Au₂S₃, black ppt. Sol. in alkali sulphides, aqua regia. Insol. in HCl (conc.).
Stannous hydroxide, SnO(OH)₂ Insol. in excess. Darkens on cooling. Insol. in H₂O. Sol. in dilute acids, alk.	Stannic hydroxide, Sn(OH)₄. Slightly sol. in excess.	Ammonium chloroplatinate. (NH₄)₂PtCl₆, yellow ppt. Sol. in large excess. 679²⁰° mg.	Fulminating gold, Au₂O₃. 2 NH₃, yellow ppt., Insol. in excess.
Cuprous chloride, 2 CuCl, white ppt. Sol. in acids. Reduction by SnCl₂.			
Mercurous chloride, HgCl, white ppt. Insol. in cold HCl (conc.). Reduction by SnCl₂.			
Silver chloride and silver, AgCl + Ag. Reduction by SnCl₂.	Silver chloride, AgCl.	Silver chloride and platinum oxide, AgCl + PtO, brown ppt.	Silver chloride and gold oxide. AgCl + Au₂O₃, brown ppt.
KOH ppts. Sn(OH)₂, Na₂CO₃ ppts. Sn(OH)₂. Insol. in excess.	KOH ppts. Sn(OH)₂. NaCO₃ ppts. Sn(OH)₂. Insol. in excess.	KOH ppts. K₂PtCl₆. Na₂CO₃ gives no ppt. Pt sol. in aq. r., fused alk.	SnCl₂ solution ppts. "Purple of Cassius," red ppt. Au sol. in KCN, aq. reg.
Metallic Sn deposited by Zn in Marsh test.	Stannic salts reduced by H, generated by Sn.	Zn ppts. Pt, black, from its salts. Also see Electromotive Series, p. 10.	Zn ppts. Au from its salts.

50

The Ammonium Sulphide Group.

Numbers refer to mgs soluble in 100 c.c. cold water.

REAGENT.	ALUMINUM, Al··· $Al_2K_2(SO_4)_4$.	CHROMIUM, Cr··· $Cr_2K_2(SO_4)_4$.	IRON, Fe·· FERROUS, $FeSO_4$.	Fe··· FERRIC, $FeCl_3$.
Ammonium sulphide, $(NH_4)_2S$.	Aluminum hydroxide, $Al(OH)_3$, white flocculent ppt. Sol. in acids. H_2S gas liberated.	Chromium hydroxide, $Cr(OH)_3$, grayish green, ppt. H_2S liberated. Sol. in acids and alkalies.	Iron sulphide, FeS, black ppt. sol. in acids. Oxidizes in the air to $FeSO_4$ and finally to a brown basic ferric sulphate.　0.89 mg.	Iron sulphide, FeS(+S), black ppt. Sol. in acids. S remains undissolved.
Ammonium hydroxide, NH_4OH.	Aluminum hydroxide, $Al(OH)_3$, white. Very slightly sol. in excess of reagent.	Chromium hydroxide, $Cr(OH)_3$, grayish green. Slightly sol. in excess, forming a reddish solution when cold and concentrated.	Ferrous hydroxide, white ppt. becoming green, then reddish brown, in the air and in presence of NH_4Cl. $Fe(OH)_3$ slowly forms. Sol. in NH_4Cl.　6.7 mgs.	Ferric hydroxide, $Fe(OH)_3$, reddish brown ppt. Insol. in excess of reagent.
Ammonium carbonate, $(NH_4)_2CO_3$.	Aluminum hydroxide, $Al(OH)_3$, white ppt. CO_2 gas.	Chromium hydroxide, $Cr(OH)_3$, grayish green. Sol. in excess of reagent CO_2 liberated.	Ferrous carbonate, $FeCO_3$, white ppt. sol. in excess. Slowly changed to hydroxide.	Basic salt changing to $Fe(OH)_3$, reddish brown ppt.
Barium carbonate, $BaCO_3$.	Aluminum hydroxide, $Al(OH)_3$, white.	Basic salt, CO_2, liberated, and $Cr(OH)_3$ formed.	Iron not pptd. in ferrous form by $BaCO_3$.	Same as above.
Borax bead, $Na_2B_4O_7 \cdot 10 H_2O$.		OF, yellowish green when hot, changing to emerald-green, cold.	OF, yellow. RF, green.	OF, yellow. RF, green.
Hydrogen sulphide, H_2S.			In presence of sodium acetate, FeS pptd.	Fe··· reduced to Fe·· with liberation of free S. (Also reduced by $SnCl_2$.)

Reagent				
Potassium ferricyanide, $K_3Fe(CN)_6$.		(All Cr compounds oxidized to compounds of chromic acid; e.g. $2 Na_2CO_3 + 3 KNO_3 + 2 Cr(OH)_3 = 2 Na_2CrO_4 +$ etc.)	Ferrous ferricyanide (Turnbull's blue), $Fe_3(FeC_6N_6)_2$, dark blue ppt. Insol. in HCl. Decomposed by NaOH to $Fe(OH)_2$.	Reddish brown color produced.
Potassium ferrocyanide, $K_4Fe(CN)_6$.	White ppt. forms slowly.		Potassium ferrous ferrocyanide, $K_2Fe_3(FeC_6N_6)_2$, bluish white, oxidized in air to blue.	Ferric ferrocyanide, $Fe_4(FeC_6N_6)_3$, (Prussian blue), dark blue. Insol. in mineral acid. NaOH forms $Fe(OH)_3+$.
Sodium hydroxide, $NaOH$.	* Aluminum hydroxide, $Al(OH)_3$, white. Sol. in excess, forming $Na'_3 - AlO_3'''$. $(NaAlO_2)$. Repptd. by NH_4Cl.	* Chromium hydroxide, $Cr(OH)_3$, greenish ppt. Sol. in excess, forming green solution $NaCrO_2$. Repptd. by boiling or by addition of NH_4Cl.	* Ferrous hydroxide, $Fe(OH)_2$, white becoming $Fe(OH)_3$. Insol. in excess. Non-volatile organic substance prevents pption.	* Ferric hydroxide, $Fe(OH)_3$, reddish brown ppt. Insol. in excess. Sol. in mineral acids.
Sodium phosphate, Na_2HPO_4.	$AlPO_4$, white. Sol. in NaOH. Insol. in $HC_2H_3O_2$.	$CrPO_4$, green ppt. Sol. in mineral acids and in NaOH.	$Fe_3(PO_4)_2$, white ppt. becoming blue in the air.	$FePO_4$, yellowish white, scl. in excess. Insol. in $HC_2H_3O_2$.
Miscellany.	Sodium acetate in excess boiled with Al salt ppts. basic $Al(OH)_2(C_2H_3O_2)$ in neutral solutions.	$NaC_2H_3O_2$ forms no ppt. unless Fe and Al are present, in which case Cr partially pptd. by boiling.	KCNS, no color. KCN ppts. $Fe(CN)_2$, brown. Sol. in excess.	$KCNS =$ red $Fe(SCN)_3$. Boiling with $NaC_2H_3O_2$ in neutral solutions a red brown ppt. formed of $Fe(OH)_2(C_2H_3O_2)$.

* Presence of non-volatile organic substances, tartrates, citrates, and sugar prevents precipitation.

The Ammonium Sulphide Group — *Continued*

REAGENT.	COBALT, CoCl₂.	NICKEL, NiCl₂.	MANGANESE, MnSO₄.	ZINC, ZnSO₄.
$(NH_4)_2S$.	Cobalt sulphide, CoS, black ppt. Insol. in $HC_2H_3O_2$. Very slightly sol. in HCl. Sol. in aqua regia and warm HNO_3. 0.38 mg.	Nickel sulphide, NiS, black ppt. Slightly sol. in excess, forming a brown solution. Very slightly sol. in HCl. Sol. in aqua regia. 0.36 mg.	Manganese sulphide, MnS, buff-colored ppt. Sol. in HCl, $HC_2H_3O_2$. Oxidizes in the air. 0.6 mg.	Zinc sulphide, ZnS, white ppt. Insol. in $HC_2H_3O_2$. Sol. in HCl. Presence of NH_4Cl aids pption.
NH_4OH.	Blue basic salt, sol. in excess, forming red solution. No ppt. formed in presence of NH_4Cl. Solution becomes red.	Nickel hydroxide, $Ni(OH)_2$, green ppt. Sol. in excess, forming a blue solution. No ppt. formed in presence of NH_4Cl.	Manganese hydroxide, $Mn(OH)_2$, white ppt. In presence of NH_4Cl a dark brown ppt. slowly forms by oxidation, $Mn(OH)_3$.	Zinc hydroxide, $Zn(OH)_2$, white ppt. Sol. in excess, $ZnSO_4(NH_3)_4$ formed. In presence of NH_4Cl no ppt. forms.
$(NH_4)_2CO_3$.	Basic carbonate, red, lilac, or pink ppt. Sol. in excess, forming red solution, becoming brown by air oxidation.	Basic carbonate of variable compositions, green ppt. Sol. in excess.	Manganese carbonate, $MnCO_3$, white ppt. Sol. in excess. Boiling aids pption.	Basic zinc carbonate, usually $Zn_2(OH)_2CO_3$, white ppt. Sol. in excess. (Acid carbonate ppts. $ZnCO_3$ white.)
$BaCO_3$.	No ppt. in cold.	No ppt. in cold.	No ppt. in cold.	No pption. in cold. (Na_2CO_3 ppts. above aided by boiling.)
Borax bead.	Blue.	Violet, hot. Yellowish brown, cold.	OF, violet-red, hot. Amethyst-red, cold. RF, colorless.	
H_2S.	From neutral or alkaline solutions, CoS, black. No ppt. in acid solutions.	From neutral or alkaline solutions, NiS. No ppt. in acid solutions.	MnS forms slowly from alkaline solutions. No ppt. in acid solutions.	From neutral or alkaline solutions, or in solutions acidified with $HC_2H_3O_2$, ZnS is pptd.

	Cobalt	Nickel	Manganese	Zinc
$K_3Fe(CN)_6$.	Cobalt ferricyanide, $Co_3(FeC_6N_6)_2$, dark brown ppt. Insol. in HCl.	Nickel ferricyanide, $Ni_3(FeC_6N_6)_2$, yellowish green ppt. Insol. in HCl.	Manganese ferricyanide, $Mn_3(FeC_6N_6)_2$, brown ppt. Insol. in HCl.	Zinc ferricyanide, $Zn_3(FeC_6N_6)_2$, yellowish brown ppt. Sol. in HCl and in NH_4OH.
$K_4Fe(CN)_6$.	Cobalt ferrocyanide, $Co_2Fe(CN)_6$, green ppt. becoming greenish blue. Insol. in HCl. Sol. in KCN.	Nickel ferrocyanide, $Ni_2Fe(CN)_6$, light green ppt. Insol. in HCl.	Manganese ferrocyanide, $Mn_2Fe(CN)_6$, faint red ppt. Diff. sol. in HCl. Easily sol. in H_2SO_4, HNO_3.	Zinc ferrocyanide, $Zn_2Fe(CN)_6$, white ppt. Insol. in dilute acids and in NH_4OH.
NaOH.	Basic salt, blue. Boiled with excess = red. Insol. in excess. Sol. in $HC_2H_3O_2$, HCl, NH_4OH.	Nickel hydroxide, $Ni(OH)_2$, apple green ppt. Insol. in excess. Sol. in $HC_2H_3O_2$, HCl, NH_4OH. Oxidized by Br to Ni $(OH)_3$.	Manganese hydroxide, $Mn(OH)_2$, white ppt. turning brown. Insol. in excess. Sol. in NH_4Cl.	Zinc hydroxide, $Zn(OH)_2$, white ppt. Sol. in excess, forming Na_2ZnO_2. Repptd. by boiling.
Na_2HPO_4.	$Co_3(PO_4)_2$, blue ppt. Sol. in NH_4OH. Sol. in dil. acids. Sol. in H_3PO_4.	$Ni_3(PO_4)_2$, green ppt. Sol. in NH_4OH. Sol. in dil. acids.	$Mn_3(PO_4)_2$, white ppt. Sol. in NH_4OH, mineral acids, and $HC_2H_3O_2$. Boiled with $NH_4OH + NH_4Cl = MnNH_4PO_4$, rose colored.	$Zn_3(PO_4)_2$, white ppt. Sol. in excess and in dil. acids and in NH_4OH.
Miscellany.	HCN ppts. reddish brown, $Co(CN)_2$. Sol. in excess. Addition of Br+NaOH = $K_3Co(CN)_6$. Co not pptd. (Ni is pptd.)	HCN ppts. greenish yellow, $Ni(CN)_2$. Sol. in excess. Repptd. by HCl. Pptd. by Br + NaOH as $Ni(OH)_3$; black + CNBr (poison gas).	$PbO_2 + HNO_3 + Mn$ salt warmed = red $HMnO_4$. $Na_2CO_3 + KNO_3$ fused on Pt = green, Na_2MnO_4.	KCN ppts. $Zn(CN)_2$, white. Sol. in excess. From this solution $(NH_4)_2S$ ppts. ZnS, white.

For solubility of salts, see tables in D. Van Nostrand's *Chemical Annual*, edited by Olsen.

The Ammonium Carbonate Group

Solubility in milligrams per 100 c.c. of water cold, "c", and hot, "h".

Reagent.	Barium, BaCl₂.	Calcium, CaCl₂.	Strontium, SrCl₂.	Soluble in Presence of NH₄ Salts. Magnesium, MgSO₄.
Ammonium carbonate, $(NH_4)_2CO_3$.	Barium carbonate, $BaCO_3$, white ppt. Sol. in acids. Slightly sol. in NH_4Cl. Precipitation aided by excess of NH_4OH and by boiling. "c" 2.2 mg.; "h" 6.5 mg.	Calcium carbonate, $CaCO_3$, white ppt. Sol. in acids. Slightly sol. in NH_4Cl. Rendered less sol. by boiling with NH_4OH. "c" 1.3 mg.; "h" 88 mg.	Strontium carbonate, $SrCO_3$, white ppt. Sol. in acids. Slightly sol. in NH_4Cl. Rendered less sol. by boiling with excess of NH_4OH. "c" 1.1 mg.	Basic magnesium carbonate $MgCO_3 + Mg(OH)_2$, white ppt. on warming. No ppt. formed if NH_4 salts are present. But if absent, solubility only 10.6 mgs.
Ammonium hydroxide, NH_4OH.				Magnesium hydroxide, $Mg(OH)_2$. Sol. in NH_4Cl.
Ammonium oxalate, $(NH_4)_2C_2O_4$.	Barium oxalate, BaC_2O_4, white ppt. Sol. in HCl. Slightly sol. in $HC_2H_3O_2$ and water. "c" 9.3; "h" 22.8.	Calcium oxalate, CaC_2O_4, white ppt. Sol. in HCl. Almost insol. in $HC_2H_3O_2$ or in $H_2C_2O_4$. "c" 0.68 mg.; "h" 1.4 mg.	Strontium oxalate, SrC_2O_4, white ppt. Sol. in HCl. Slightly sol. in $HC_2H_3O_2$ and water. "c" 5.1 mg. Sol. in hot solutions.	
Ammonium sulphate,$(NH_4)_2SO_4$, or sulphuric acid, H_2SO_4.	Barium sulphate, $BaSO_4$, white ppt. Insol. in H_2O and in acids. "c" 0.17 mg.; "h" 0.3 mg.	Calcium sulphate, white ppt. Somewhat sol. in H_2O. Sol. in $(NH_4)_2SO_4$. Insol. in alcohol. "c" 179 mg.; "h" 178 mg.	Strontium sulphate, $SrSO_4$, white ppt. Slightly sol. in H_2O and in $(NH_4)_2SO_4$. "c" 11.4 mg.; "h" 10.4 mg.	
Potassium chromate, K_2CrO_4.	Barium chromate, $BaCrO_4$, yellow ppt. Insol. in $HC_2H_3O_2$ in presence of K_2CrO_4. Sol. in HCl, HNO_3. "c" 0.38 mg.; "h" 43 mg.			

	Magnesium hydroxide, Mg(OH)$_2$, white ppt. Sol. in NH$_4$Cl.	Strontium hydroxide, Sr(OH)$_2$, white ppt. Slightly sol. in water; "c" 410 mg.; "h" very soluble.	Calcium hydroxide, Ca(OH)$_2$, white ppt. Difficultly sol. in water. Sol. in NH$_4$Cl. "c" 170 mg.; "h" 80 mg.	Barium hydroxide, Ba(OH)$_2$, white ppt. formed only in conc. solutions. "c" 5560 mg.; "h" very soluble.
Sodium hydroxide, NaOH.				
Sodium phosphate, Na$_2$HPO$_4$.	Magnesium hydrogen phosphate, MgHPO$_4$, white ppt. Boiled = Mg$_3$(PO$_4$)$_2$. In presence of NH$_4$Cl and NH$_4$OH a white crystalline ppt. of MgNH$_4$PO$_4$ slowly forms. Sol. in acetic acid.	Strontium hydrogen phosphate, SrHPO$_4$, white ppt. Sol. in acids. Insol. in H$_2$O.	Calcium hydrogen phosphate, CaHPO$_4$, white ppt. "c" = 28 mg. Sol. in acids. In presence of NH$_4$OH, Ca$_3$(PO$_4$)$_2$ pptd. "c" 3 mg. to 8 mg.	Barium hydrogen phosphate, BaHPO$_4$, white ppt. "c" 10–20 mg. Sol. in acids. If NH$_4$OH present, BaNH$_4$PO$_4$ formed.
Flame.	White.	Crimson.	Yellowish red.	Yellowish green.
Spectra.		Several orange and red lines. Brilliant blue line.	Sharp orange line α and bluish line β.	Green α and β bands. Fainter yellow and red bands.
Miscellany.	No ppt. by H$_2$SiF$_6$. Boiled with Na$_2$CO$_3$ = white Mg$_3$(OH)$_2$(CO$_2$)$_2$. No ppt. if NH$_4$Cl present.	No ppt. by H$_2$SiF$_6$.	No ppt. by H$_2$SiF$_6$.	H$_2$SiF$_6$ ppts. BaSiF$_6$. Insol. in alcohol and dilute acids. "c" 26 mg.; "h" 90 mg.

The Soluble Metal Group.

REAGENT.	AMMONIUM, NH$_4$Cl.	LITHIUM, LiCl.	POTASSIUM, KCl.	SODIUM, NaCl.
Hydrofluosilicic acid, H$_2$SiF$_6$.			Potassium fluosilicate, K$_2$SiF$_6$. Transparent ppt. Slightly sol. in water. "c" = 120 mg.; "h" 955 mg.	Sodium fluosilicate, Na$_2$SiF$_6$, white ppt. Somewhat sol. in water. "c" = 650 mg.; "h" 2460 mg.
Nessler's reagents, HgI$_2$(KI)$_2$. KOH.	Reddish brown or yellow, according to amount of ammonia. Test very delicate.			
Platinic chloride, PtCl$_4$, H$_2$PtCl$_6$.	Ammonium chloroplatinate, (NH$_4$)$_2$PtCl$_6$, yellow ppt. Slightly sol. in water. Insol. in alcohol. "c" 670.0 mg. Very sol. in hot water.		Potassium chloroplatinate, K$_2$PtCl$_6$, yellow ppt. Slightly sol. in water. Insol. in alcohol or ether. "c" 480 mg.; "h" 5180 mg.	
Potassium pyroantimonate, H$_2$K$_2$Sb$_2$O$_7$.				Sodium pyroantimonate, Na$_2$H$_2$Sb$_2$O$_7$, white cryst. ppt. Best formed in slightly alkaline solutions. NaSbO$_3$ + aqua. "c" = 31 mg.
Sodium carbonate, Na$_2$CO$_3$.	Ammonia gas evolved on boiling.	Lithium carbonate, Li$_2$CO$_3$, white ppt. Slightly sol. in water. Less sol. in hot than in cold. "c" = 1539 mg.; "h" 728 mg.		

	Lithium	Ammonium	Potassium	Sodium
Sodium cobaltic nitrite, $Co(NO_2)_3 \cdot 3NaNO_2$.		In acetic acid solutions (see Na), $Co(NO_2)_3 \cdot 3NH_4NO_2$, yellow ppt. Sol. in inorganic acids.	Potassium cobalt nitrite, $K_3Co(NO_2)_6$, yellow, or $K_2NaCo(NO_2)_5$ in presence of an excess of sodium. Insol. in acetic acid. Sol. in inorganic acids. Hastened by warming. Solution should have acetic acid present. 70 mg. 25°.	No ppt. in acetic acid solutions of sodium salts.
Sodium or potassium hydroxide, NaOH or KOH.		Ammonia gas evolved when salt is warmed with NaOH or KOH.		
Sodium phosphate, Na_2HPO_4.	Lithium phosphate, Li_3PO_4, white ppt. Slightly sol. in water. Sol. in HCl. "c" = 40 mg.			
Tartaric acid, $H_2C_4H_4O_6$.		Monoammonium tartrate, $NH_4HC_4H_4O_6$, white cryst. ppt. Hastened by shaking. Slightly sol. in H_2O.	Monopotassium tartrate, $KHC_4H_4O_6$, white cryst. ppt. Hastened by stirring. Somewhat sol. in H_2O. "c" 370 mg.	
Flame and spectrum.	Red flame. Bright crimson line with feeble orange line.		Violet flame. A red and blue line.	Yellow flame. Single yellow line.

REACTIONS OF THE ACIDS.

Inorganic Acids.

Acids.	Silver Nitrate, $AgNO_3$.	Barium Chloride, $BaCl_2$.	Calcium Chloride, $CaCl_2$.	Lead Acetate, $Pb(C_2H_3O_2)_2$.	Characteristic Reactions.
Arsenic, H_3AsO_4, arsenates.	Reddish brown ppt., Ag_3AsO_4. Sol. in NH_4OH, HNO_3.	White ppt., $Ba_3(AsO_4)_2$. Sol. in acids.	White ppt. $Ca_3(AsO_4)_2$. Sol. in acetic acid.	Lead acetate ppts. white salt. Insol. in acetic acid. Sol. in HNO_3.	$(NH_4)_2MoO_4$ produces a yellow ppt. $MgSO_4 + NH_4Cl + NH_4OH =$ white ppt.
Arsenious, H_3AsO_3, arsenites.	Yellow ppt., Ag_3AsO_3. Sol. in NH_4OH, HNO_3.	White ppt., $Ba_3(AsO_3)_2$. Sol. in acids.	White ppt. Sol. in $C_2H_4O_2$.	Lead acetate ppts. white arsenious salt. Sol. in acetic acid, HNO_3.	Marsh's test with both forms—-ous and -ic—gives arsenic mirror. Sol. in NaOCl. H_2S produces yellow As_2S_3 and As_2S_5. Gutzeit test.
Boric, H_3BO_3, $(H_2B_4O_7)$, borates.	White ppt. Sol. in NH_4OH, HNO_3.	White ppt. Not readily sol. in water. Sol. in acids.	White ppt. Sol. in acids.	White ppt. Caused by lead acetate. Sol. in excess.	$H_2SO_4 + C_2H_5OH$ colors flame green. Turmeric paper dipped in boric acid salt acidified with HCl, dried, turns red.
Carbonic, H_2CO_3, carbonates.	Grayish white ppt., Ag_2CO_3. Sol. in HNO_3. $3.1^{15°}$ mg.	White ppt., $BaCO_3$. Sol. in acids. 2 2 mg.	White ppt., $CaCO_3$. Sol. in acids. 2.2 mg.	Lead acetate ppts. white lead. Sol. in HNO_3.	Effervesces with dilute inorganic acids, HCl, H_2SO_4, HNO_3, etc., CO_2 gas being evolved. Limewater clouded by CO_2, $CaCO_3$ being formed.

Chloric, HClO$_3$, chlorates.			H$_2$SO$_4$, conc. warmed with salt causes explosion.	Heated on charcoal deflagrates. H$_2$SO$_4$ evolves yellow gas, ClO$_2$.
Chromic, H$_2$CrO$_4$, chromates.	Dark red ppt., Ag$_2$CrO$_4$. Sol. in HNO$_3$. 2.8$^{18°}$ mg.	Yellow ppt. BaCrO$_4$. Sol. in HCl, HNO$_3$. Insol. in acetic acid. 0.38 mg.	Pb(C$_2$H$_3$O$_2$)$_2$ ppts. yellow PbCrO$_4$. Sol. in NaOH. 0.0288$^{°}$ mg.	Reduced to green CrCl$_3$ by warming with alcohol and HCl.
Hydriodic, HI, iodides.	Yellow ppt., AgI$_2$. Very difficultly sol. in NH$_4$OH. 0.035^{215} mg.		Soluble lead salt ppts. PbI$_2$, yellow. Sol. in hot water. 39,000 mg.	Liberated from its salts by HNO$_2$, or Cl water. Imparts in free form violet color to CS$_2$, blue to starch.
Hydrobromic, HBr, bromides.	Light yellow ppt., AgBr. Difficultly sol. in NH$_4$OH.		Pb(C$_2$H$_3$O$_2$)$_2$ppts. PbBr$_2$, white. Sol. in hot water. 455 mg.	Liberated from its salts by Cl, colors CS$_2$ reddish yellow.

Inorganic Acids.

Acids.	$AgNO_3$.	$BaCl_2$.	$CaCl_2$.	$Pb(C_2H_3O_2)_2$.	Characteristic Reactions.
Hydrochloric, HCl, chlorides.	White ppt., AgCl. Insol. in HNO_3. Sol. in NH_4OH, KCN. 0.152 mg.			$PbCl_2$, white, is pptd. Sol. in hot water. Insol. in alcohol. 673 mg.	$K_2Cr_2O_7 + H_2SO_4$ (conc.), gives CrO_2Cl_2. $MnO_2 + H_2SO_4$ gives Cl gas. $KMnO_4 + HCl$ evolves Cl.
Hydrocyanic, HCN, cyanides.	White ppt., AgCN. Sol. in NH_4OH, KCN. 0.021 mg.			White ppt. with sol. Pb salt, $Pb(CN)_2$. Sol. in HNO_3.	Warmed with $NaOH + FeSO_4 + FeCl_3$ and acidified with HCl, "Prussian Blue" formed. Warmed with $(NH_4)_2S_x = NH_4CNS$, which produces a blood-red color with $FeCl_3$.
Hydroferricyanic, $H_3Fe(CN)_6$, ferricyanides.	Orange ppt., Ag_3FeCy_6. Sol. in NH_4OH, KCN. Insol. in HNO_3.			On warming, PbO_2 is pptd.	Dark blue, Turnbull's blue, ppt. with $FeSO_4(Fe··)$.
Hydroferrocyanic, $H_4Fe(CN)_6$, ferrocyanides.	White ppt. Sol. in KCN. Insol. in HNO_3.			White ppt. Sol. in HNO_3.	Prussian Blue with $FeCl_3$ ($Fe···$). Red ppt. with copper salts ($Cu··$).
Hydrofluoric, HF, fluorides.		White ppt., BaF_2. Sol. in HCl. 163^{18} mg.	White ppt., CaF_2. Sol. in HCl. Insol. in acetic acid. 1.6 mg.	$Pb(C_2H_3O_2)_2$ ppts. white salt. Sol. in HNO_3.	H_2SO_4 evolves HF, which etches glass.

Hydrofluosilicic, H_2SiF_6, fluosilicates.		White ppt., $BaSiF_6$. Insol. in HCl. $26^{17°}$ mg.			Decomposed by H_2SO_4 into HF and SiF_4. Ppts. K in concentrated solution, as K_2SiF_6, white.
Hydrosulphuric, H_2S, sulphides.	Black ppt., Ag_2S. Sol. in hot HNO_3. 0.02 mg.			Black ppt., PbS. Sol. in hot HNO_3. 0.1 mg.	Most sulphides decomposed by strong inorganic acids, with odor of rotten eggs (see $AgNO_3$ and $Pb(NO_3)_2$ tests). Colors sodium nitro prussiate. violet.
Hypochlorous, HClO, hypochlorites.	White ppt., AgClO. Sol. in HNO_3.			White ppt., becomes brown on boiling.	Cl evolved when salt is treated with HCl and many other acids. Ppts. MnO_2 black from solution of $MnSO_4$.
Iodic, HIO_3, iodides.	White ppt., $AgIO_3$. Sol. in NH_4OH. Reduced by SO_2 to AgI. 4.4 mg.	White ppt., $Ba(IO_3)_2$. Sol. in HCl. 8 mgs.		White ppt. Sol. in HNO_3.	With acetic acid and KI, free iodine formed. CS_2 colored violet. Starch colored blue.
Metaphosphoric, HPO_3, metaphosphates.	White ppt. Sol. in NH_4OH, HNO_3.	White ppt., Sol. in excess of metaphosphate.	White ppt. Sol. in dil. acids.	White ppt. Sol. in HNO_3.	Coagulates albumen. Boiled with $HNO_3 = H_3PO_4$.

Inorganic Acids.

Acids.	AgNO₃.	BaCl₂.	CaCl₂.	Pb(C₂H₃O₂)₂.	Characteristic Reactions.
Nitric, HNO_3, nitrates.					H_2SO_4 (conc.) poured into a mixture of a nitrate salt with $FeSO_4$ produces a brown ring at upper surface of heavier H_2SO_4.
Nitrous, HNO_2, nitrites.	White ppt., $AgNO_2$. Sol. in hot water. 330 mg.				With an iodide, when acidified with HCl, liberates iodine, which will color CS_2 violet.
Phosphoric, H_3PO_4, phosphates.	Yellow ppt., Ag_3PO_4. Sol. in NH_4OH and HNO_3. 1.93 mg.	White ppt., $Ba_3(PO_4)_2$. Sol. in acids.	White ppt. Sol. in inorganic and in acetic acids.	White ppt. with Pb salt. Sol. in NaOH. Insol. in NH_4OH. Insol. in $HC_2H_3O_2$. 0.014 mg.	$(NH_4)_2MoO_4 + HNO_3$ ppts. at 40°, yellow phospho ammonium molybdate. Magnesia mixture produces a white ppt., $MgNH_4PO_4$.
Phosphorus, H_3PO_3, phosphites.	White ppt. On warming, causes reduction of silver salt to Ag.	White ppt. Sol. in acids, acetic acid.	White ppt. Sol. in NH_4Cl.	White ppt. Insol. in $HC_2H_3O_2$. Sol. in HNO_3.	Reducing action. Very concentrated solutions heated evolve PH_3.
Pyrophosphoric, $H_4P_2O_7$, pyrophosphates.	White ppt. Sol. in NH_4OH, HNO_3.	White ppt. Sol. in HCl. 10 mg.	White ppt. Sol. in excess of the pyrophosphate.	Pb salt same as Ca salt.	Does not coagulate albumen as does the metaphosphoric acid. Boiled with HNO_3 changes to H_3PO_4.

Acid					
Silicic, H_2SiO_3, silicates.	Yellow ppt. Sol. in HNO_3.	White ppt.	White ppt.	White ppt. Insol. in HNO_3.	Heated with Na_2CO_3, evolves CO_2. SiO_2 skeleton with $NaPO_3$ bead. Decomposes when evaporated to dryness, SiO_2 separating.
Sulphocyanic, HCNS, thiocyanates or sulphocyanates.	White ppt., AgCNS. Difficultly sol. in NH_4OH. 0.021 mg.			White ppt.	$FeCl_3$ produces a blood-red color, which is destroyed by $HgCl_2$.
Sulphuric, H_2SO_4, sulphates.	Conc. $AgNO_3$ produces white ppt., Ag_2SO_4. Sol. in water. 580 mg.	White ppt., $BaSO_4$. Insol. in acids. 0.172 mg.	White ppt. Sol. in water and acids.	White ppt. Sol. in NaOH, $NH_4C_2H_3O_2$, $(NH_4)_2C_4H_4O_6$.	Sulphates heated with Na_2CO_3 on charcoal in reducing flame from Na_2S, which blacken a silver coin when it is moistened.
Sulphurous, H_2SO_3, sulphites.	White ppt., Ag_2SO_3. Decomposed by boiling into $Ag + Ag_2SO_4 + SO_2$.	White ppt. Sol. in HCl. 19.7 mg.	White ppt. Sol. in HCl.	White ppt.	Reducing agent. Reduces KI (starch sol. test). Decolorizes $KMnO_4$. SO_2 evolved when salt is acidified with HCl.
Thiosulphuric, $H_2S_2O_3$, thiosulphates.	$Ag_2S_2O_3$, white ppt. Sol. in $Na_2S_2O_3$. Decomposed by boiling, forming $H_2SO_4 + Ag_2S$.	Conc. solution produces white ppt.		White ppt. On boiling becomes gray. $PbSO_4 + PbS$.	Decomposed by HCl to $SO_2 + S$. $Na_2S_2O_3$ dissolves AgCl.

Organic Acids.

Acids.	AgNO₃.	CaCl₂.	FeCl₃.	H₂SO₄ conc.	Special Tests.
Acetic, $H \cdot C_2H_3O_2$.	White ppt. Sol. in hot water.		Reddish brown solution. Ppts. on boiling of ferric acetate $= Fe_2(OH)_2A_4$. Sol. in HCl.	Heated, gives odor of vinegar.	H_2SO_4 + alcohol = ethyl acetate, recognized by characteristic odor.
Benzoic, C_6H_5COOH.	White ppt. C_6H_5COOAg. Sol. in hot water.		Buff-colored ppt. $(CH_5COO)FeOH$. Sol. in HCl.	Dissolves without charring or evolution of gas.	$Pb(C_2H_3O_2)_2$ ppts. white compound. m.p. 121°, sublimes when heated. $H_2SO_4 + C_2H_5OH$ heated = ethyl benzoate.
Carbolic, C_6H_5OH.			Deep violet. Color destroyed by acetic acid (not destroyed in case of salicylic acid).		Bromine water, even in very dilute solutions, gives a white ppt. sol. in NaOH, KOH. C_6H_5OH + 1 c.c. $H_2SO_4 + NaNO_2$ warmed = deep green or blue color.
Citric, $C_3H_4OH(COOH)_3$.	White ppt. Sol. in NH₄OH. No reduction on heating.	White ppt. Less sol. in hot than cold water. Sol. in NH₄Cl. Insol. in NaOH. Crystalline form insol. in NH₄Cl.			White ppt. with lead acetate. Sol. in ammonium citrate. Prevents pption. of $Fe(OH)_3$ by alkalies. $CdCl_2$ ppts. $Cd(C_6H_5O_7)_2$. Insol. in hot water. Sol. in acetic acid. (Cd salts, no ppt. with tartrates.)
Formic, $HCOOH$.	White ppt. in conc. solutions. Becomes dark from reduced silver salt.		Red ppt. Color destroyed by HCl.	With reducing agent heated, gives CO, which burns with blue flame.	H_2SO_4 (conc.) + ethyl alcohol = ethyl formate, pleasant characteristic odor.

Acid					
Gallic, $C_6H_2(OH)_3COOH$.	Metallic Ag from reduction.		Blue-black ppt. Sol. in excess = green.		Melts at 200°. The alkaline solutions absorbs O. Limewater or $Ba(OH)_2$ produces a blue ppt.
Lactic, $C_2H_4OHCOOH$.	Reduction results. Ag formed (no action on Fehling's sol.).			Charring on heating with evolution of CO.	Decolorizes $KMnO_4$ and efferverscence takes place, with odor of acetaldehyde.
Malic, $C_2H_3OH(COOH)_2$.	White ppt.	White ppt. only in presence of strong alcohol (distinction from citric).			Lead acetate ppts. white salt. Sol. in hot water. Prevents pption. of $Fe(OH)_3$ by alkalies.
Oxalic, $H_2 \cdot C_2O_4$.	White ppt. Sol. in HNO_3, NH_4OH.	White ppt. Insol. in acetic acid. Sol. in HCl, HNO_3.		Heated, CO_2 and CO evolved.	H_2SO_4 (dilute) + MnO_2 gives CO_2. Destroys color of $KMnO_4$ when heated with that reagent in presence of dilute H_2SO_4.
Salicylic, $C_6H_4OHCOOH$.	White ppt. Sol. in hot water, $C_6H_4OHCOOAg$.		Deep violet color. Destroyed by mineral acids.	Dissolves. Prolonged heating darkens solution and gas is evolved.	Lead acetate ppts. a white salt. Acid m.p. 156°. HNO_3 heated with salt produces yellow picric acid. Color is intensified by caustic soda.
Tannic, $C_{14}H_{10}O_9$.	White ppt.		Blue-black color (ink).		Lead acetate gives a yellow ppt., with astringent taste. The acid ppts. a solution of glue. Limewater produces a gray ppt.
Tartaric, $H_2 \cdot C_4H_4O_6$.	White ppt. Sol. in excess of tartrate, HCl, NH_4OH. Reduction on heating.	White crystalline ppt. Action similar to magnesium precipitation.			Chars when heated. Odor of burnt sugar. Prevents pption. of $Fe(OH)_3$ by alkalies. Silver Nitrate test, page 104.

51

SOLUBILITY TABLE

Since no salt is absolutely insoluble, the term "insoluble" is only relative. For solubility of the salts formed, see Van Nostrand's *Chemical Annual*, edited by Professor John C. Olsen.

Cation \ Anion	F'	Cl'	Br'	I'	CN'	NO₃'	ClO₃'	C₂H₃O₂'	S''	CO₃''	SiO₃''	SO₄''	CrO₄''	BO₃'''/BO₂'	PO₄'''	AsO₄'''	AsO₃'''	Fe(CN)₆''''	Fe(CN)...'''	C₂O₄''
K·	W	W	W	W	W	W	W	W	W	W	W	W	W	W	W	W	W	W	W	W
Na·	W	W	W	W	W	W	W	W	W	W	W	W	W	W	W	W	W	W	W	W
Li·	W	W	W	W	W	W	W	W	W	W	W	W	W	W	W	W	W	W	W	W
Ba··	wa	W	W	W	wA	W	W	W	W	A	A	I	A	A	A	A	A	—	wA	—
Sr··	wa	W	W	W	W	W	W	W	W	A	A	I	wA	A	A	A	A	—	W	—
Ca··	wa	W	W	W	W	W	W	W	W	A	A	wa	wA	A	A	A	A	W	W	—
Mg··	wa	W	W	W	W	W	W	wA	W	A	A	W	W	wA	A	A	A	W	W	—
Al···	W	W	W	W	—	W	W	W	—	—	wa	W	—	A	A	A	—	—	—	
Mn··	A	W	W	W	A	W	W	wA	A	A	A	A	W	W	A	A	A	A	I	A
Zn··	wA	W	W	W	A	W	W	W	A	A	A	A	W	W	A	A	A	—	A	wa
Cr···	W	W	W	W	A	W	W	wA	—	—	A	W	A	A	A	A	—	—	—	
Cd··	wA	W	W	W	A	W	W	W	A	A	A	W	W	wA	A	—	—	—	—	
Fe··	wA	W	W	W	wa	W	W	W	A	A	A	W	—	A	A	A	A	I	I	
Fe···	W	W	W	—	—	W	W	A	—	—	A	W	W	A	A	A	A	W	I	
Co··	wA	W	W	W	wa	W	W	W	A	A	A	W	A	A	A	A	A	I	I	—
Ni··	wA	W	W	W	wa	W	W	W	A	A	A	W	A	A	A	A	A	I	I	—
Sn··	W	W	—	W	—	W	W	A	A	—	—	W	A	A	A	—	—	I	I	—
Sn····	W	W	—	W	—	—	—	—	A	—	—	—	—	—	A	A	A	—	I	—
Pb··	A	W	W	W	A	W	W	W	A	A	A	I	I	A	A	A	—	wA	A	—
Cu··	A	W	W	I	I	W	W	W	A	A	—	W	W	A	A	A	—	—	I	—
Sb···	W	A	A	wA	—	—	—	A	A	—	—	A	A	—	A	A	—	—	—	—
Bi···	W	A	A	A	—	A	W	A	A	A	A	A	A	A	A	A	—	—	—	—
Hg·	—	I	I	I	—	W	W	A	—	A	—	wA	A	—	A	A	A	—	—	—
Hg··	wA	W	W	A	W	W	W	A	A	A	—	W	wA	—	A	A	A	—	—	—
Ag·	W	I	I	I	I	W	W	A	A	A	A	wA	A	A	A	A	A	A	I	
Pt····	—	W	W	I	W	W	—	—	A	—	—	W	—	—	—	—	A	—	—	—
Au··	—	W	W	A	W	—	—	—	A	—	—	—	—	—	—	—	—	—	—	—

ABBREVIATIONS. — W = soluble in water; A = soluble in acids; wA = slightly soluble in water, readily soluble in acids; wa = difficultly soluble in water and in acids; I = insoluble in water and acids.

The metals are arranged in order of their electromotive series.

VOL. I

PART III
FACTORS, ACID AND ALKALI TABLES AND OTHER USEFUL DATA

TABLES AND USEFUL DATA

I.—MELTING-POINTS OF THE CHEMICAL ELEMENTS [1]

Reproduced from Circular No. 35 (2d edition) of U. S. Bureau of Standards.

Element.	C.	F.	Element.	C.	F.
Helium	< -271	< -456	Neodymium	840?	1544
Hydrogen	-259	-434	Arsenic	850?	1562
Neon	$-253?$	-423	Barium	850	1562
Fluorine	-223	-369	Praseodymium	940?	1724
Oxygen	-218	-360	Germanium	958	1756
Nitrogen	-210	-346	SILVER	960.5	1761
Argon	-188	-306	GOLD	1063.0	1945.5
Krypton	-169	-272	COPPER	1083.0	1981.5
Xenon	-140	-220	Manganese	1260	2300
Chlorine	-101.5	-150.7	Samarium	1300–1400	2370–2550
Mercury	-38.9	-38.0	Beryllium		
Bromine	-7.3	$+18.9$	(glucinum)	1350?	2462
Caesium	$+26$	79	Scandium	?	
Gallium	30	86	Silicon	1420	2588
Rubidium	38	100	NICKEL	1452	2646
Phosphorus	44	111.2	Cobalt	1480	2696
Potassium	62.3	144	Yttrium	1490	2714
Sodium	97.5	207.5	Chromium	1520	2768
Iodine	113.5	236.3	IRON	1530	2786
	S_I 112.8	235.0	PALLADIUM	1549	2820
Sulphur	S_{II} 119.2	246.6	Zirconium	1700?	3090
	S_{III} 106.8	224.2	Columbium		
Indium	155	311	(Niobium)	1700?	3090
Lithium	186	367	Thorium	$\begin{cases} >1700 \\ <Pt \end{cases}$	$\begin{cases} >3090 \\ <Pt \end{cases}$
Selenium	217–220	422–428			
TIN	231.9	449.4	Vanadium	1720	3128
Bismuth	271	520	PLATINUM	1755	3191
Thallium	302	576	Ytterbium	?	
CADMIUM	320.9	609.6	Titanium	1800	3272
LEAD	327.4	621.3	Uranium	<1850	<3362
ZINC	419.4	786.9	Rhodium	1950	3542
Tellurium	452	846	Boron	2200–2500?	4000–4500
ANTIMONY	630.0	1166	Iridium	2350?	4262
Cerium	640	1184	Ruthenium	2450?	4442
Magnesium	651	1204	Molybdenum	2500?	4500
ALUMINUM	658.7	1217.7	Osmium	2700?	4900
Radium	700	1292	Tantalum	2850	5160
Calcium	810	1490	TUNGSTEN	3000	5430
Lanthanum	810?	1490	Carbon	$\begin{cases} >3600 \\ \text{for p.}=1 \text{ at.} \end{cases}$	$\begin{cases} >6500 \\ \text{for p.}=1 \text{ at.} \end{cases}$
Strontium	$>Ca <Ba?$				

II. OTHER TEMPERATURE STANDARDS

Temperatures of Flames.[2]

	Cent.	Deg. of Accuracy.
Bunsen, open..............................	1100	Within 100° C.[3]
Meker......................................	1500[1]	
Petrol blow lamp	1600[1]	
Oxyhydrogen with H_2+O.. .:................	2420	Within 100° C.
Oxyacetylene..............................	2400[1]	
Thermat...................................	2500	
Electric arc................................	3500	Within 150° C.
Sun.......................................	6000	Within 500° C.

[1] Metallurgical and Chemical Engineering, Vol. XIII, No. 5, May, 1915.
[2] Measurement of High Temperatures, G. K. Burgess and H. Le Chatelier.
[3] Pyrometer tests of the open Bunsen flame gave temperatures ranging from 1000° to 1100° C.

III. OTHER TEMPERATURE STANDARDS

Approximate Temperatures by Colors.

	Cent.	Fahr.
First visible red...................................	525	977
Dull red...	700	1292
Cherry red.......................................	900	1652
Dull orange......................................	1100	2012
White...	1300	2372
Dazzling white...................................	1500	2732

Substance.	Phenomenon.	C.	F.	Variation with pressure (pressure in mm. of Hg.)
Oxygen.........	Boiling.......	−183.0	−297.4	C. $°-183.0+0.01258$ (p. 760) -0.0000079 (p. 760)
Carbon dioxide..	Sublimation in inert liquid..	− 78.5	−109.3	C. $°=-78.5+0.017$ (p. 760)
Sodium sulphate $Na_2SO_4+10H_2O$	Transformation into anhydrous salt.	32.384	90.291	
Water..........	Boiling.......	100	212	C. $°-100+0.03670$ (p. 760) -0.00002046 (p. 760)
Naphthalene....	Boiling.......	217.96	423.73	C. $°=217.96+0.058$ (p. 760)
Benzophenone ..	Boiling.......	305.9	582.6	C. $°-305.9+0.063$ (p. 760)
Sulphur........	Boiling.......	444.6	832.3	C. $°=444.6+0.0908$ (p. 760) -0.000047 (p. 760)
Ag_3Cu_2	Eutectic Freezing.....	779	1434	
Sodium chloride.	Freezing......	801	1472	

IV.—ELECTROMOTIVE ARRANGEMENT OF THE ELEMENTS

Each element is positive to the element placed after it and negative to the element placed above.

1. Cs	14. Al	27. Ge	40. Os	53. Mo
2. Rb	15. Zr	28. In	41. Au	54. V
3. K	16. Th	29. Ga	42. H	55. Cr
4. Na	17. Ce	30. Bi	43. Sn	56. As
5. Li	18. Di	31. U	44. Si	57. P
6. Ba	19. La	32. Cu	45. Ti	58. Se
7. Sr	20. Mn	33. Ag	46. Cb	59. I
8. Ca	21. Zn	34. Hg	47. Ta	60. Br
9. Mg	22. Fe	35. Pd	48. Te	61. Cl
10. Be	23. Ni	36. Ru	49. Sb	62. F
11. Yt	24. Co	37. Rh	50. C	63. N
12. Er	25. Cd	38. Pt	51. Bo	64. S
13. Sc	26. Pb	39. Ir	52. W	65. O

ACID AND ALKALI TABLES

V.—HYDROCHLORIC ACID

By W. C. FERGUSON

Degrees Baumé.	Sp. Gr.	Degrees Twaddell.	Per Cent HCl.	Degrees Baumé.	Sp. Gr.	Degrees Twaddell.	Per Cent HCl.
1.00	1.0069	1.38	1.40	14.25	1.1090	21.80	21.68
2.00	1.0140	2.80	2.82	14.50	1.1111	22.22	22.09
3.00	1.0211	4.22	4.25	14.75	1.1132	22.64	22.50
4.00	1.0284	5.68	5.69	15.00	1.1154	23.08	22.92
5.00	1.0357	7.14	7.15	15.25	1.1176	23.52	23.33
5.25	1.0375	7.50	7.52	15.50	1.1197	23.94	23.75
5.50	1.0394	7.88	7.89	15.75	1.1219	24.38	24.16
5.75	1.0413	8.26	8.26	16.0	1.1240	24.80	24.57
6.00	1.0432	8.64	8.64	16.1	1.1248	24.96	24.73
6.25	1.0450	9.00	9.02	16.2	1.1256	25.12	24.90
6.50	1.0469	9.38	9.40	16.3	1.1265	25.30	25.06
6.75	1.0488	9.76	9.78	16.4	1.1274	25.48	25.23
7.00	1.0507	10.14	10.17	16.5	1.1283	25.66	25.39
7.25	1.0526	10.52	10.55	16.6	1.1292	25.84	25.56
7.50	1.0545	10.90	10.94	16.7	1.1301	26.02	25.72
7.75	1.0564	11.28	11.32	16.8	1.1310	26.20	25.89
8.00	1.0584	11.68	11.71	16.9	1.1319	26.38	26.05
8.25	1.0603	12.06	12.09	17.0	1.1328	26.56	26.22
8.50	1.0623	12.46	12.48	17.1	1.1336	26.72	26.39
8.75	1.0642	12.84	12.87	17.2	1.1345	26.90	26.56
9.00	1.0662	13.24	13.26	17.3	1.1354	27.08	26.73
9.25	1.0681	13.62	13.65	17.4	1.1363	27.26	26.90
9.50	1.0701	14.02	14.04	17.5	1.1372	27.44	27.07
9.75	1.0721	14.42	14.43	17.6	1.1381	27.62	27.24
10.00	1.0741	14.82	14.83	17.7	1.1390	27.80	27.41
10.25	1.0761	15.22	15.22	17.8	1.1399	27.98	27.58
10.50	1.0781	15.62	15.62	17.9	1.1408	28.16	27.75
10.75	1.0801	16.02	16.01	18.0	1.1417	28.34	27.92
11.00	1.0821	16.42	16.41	18.1	1.1426	28.52	28.09
11.25	1.0841	16.82	16.81	18.2	1.1435	28.70	28.26
11.50	1.0861	17.22	17.21	18.3	1.1444	28.88	28.44
11.75	1.0881	17.62	17.61	18.4	1.1453	29.06	28.61
12.00	1.0902	18.04	18.01	18.5	1.1462	29.24	28.78
12.25	1.0922	18.44	18.41	18.6	1.1471	29.42	28.95
12.50	1.0943	18.86	18.82	18.7	1.1480	29.60	29.13
12.75	1.0964	19.28	19.22	18.8	1.1489	29.78	29.30
13.00	1.0985	19.70	19.63	18.9	1.1498	29.96	29.48
13.25	1.1006	20.12	20.04	19.0	1.1508	30.16	29.65
13.50	1.1027	20.54	20.45	19.1	1.1517	30.34	29.83
13.75	1.1048	20.96	20.86	19.2	1.1526	30.52	30.00
14.00	1.1069	21.38	21.27	19.3	1.1535	30.70	30.18

V.—HYDROCHLORIC ACID (Continued)

Degrees Baumé.	Sp. Gr.	Degrees Twaddell.	Per Cent HCl.	Degrees Baumé.	Sp. Gr.	Degrees Twaddell.	Per Cent HCl.
19.4	1.1544	30.88	30.35	22.5	1.1836	36.72	36.16
19.5	1.1554	31.08	30.53	22.6	1.1846	36.92	36.35
19.6	1.1563	31.26	30.71	22.7	1.1856	37.12	36.54
19.7	1.1572	31.44	30.90	22.8	1.1866	37.32	36.73
19.8	1.1581	31.62	31.08	22.9	1.1875	37.50	36.93
19.9	1.1590	31.80	31.27	23.0	1.1885	37.70	37.14
20.0	1.1600	32.00	31.45	23.1	1.1895	37.90	37.36
20.1	1.1609	32.18	31.64	23.2	1.1904	38.08	37.58
20.2	1.1619	32.38	31.82	23.3	1.1914	38.28	37.80
20.3	1.1628	32.56	32.01	23.4	1.1924	38.48	38.03
20.4	1.1637	32.74	32.19	23.5	1.1934	38.68	38.26
20.5	1.1647	32.94	32.38	23.6	1.1944	38.88	38.49
20.6	1.1656	33.12	32.56	23.7	1.1953	39.06	38.72
20.7	1.1666	33.32	32.75	23.8	1.1963	39.26	38.95
20.8	1.1675	33.50	32.93	23.9	1.1973	39.46	39.18
20.9	1.1684	33.68	33.12	24.0	1.1983	39.66	39.41
21.0	1.1694	33.88	33.31	24.1	1.1993	39.86	39.64
21.1	1.1703	34.06	33.50	24.2	1.2003	40.06	39.86
21.2	1.1713	34.26	33.69	24.3	1.2013	40.26	40.09
21.3	1.1722	34.44	33.88	24.4	1.2023	40.46	40.32
21.4	1.1732	34.64	34.07	24.5	1.2033	40.66	40.55
21.5	1.1741	34.82	34.26	24.6	1.2043	40.86	40.78
21.6	1.1751	35.02	34.45	24.7	1.2053	41.06	41.01
21.7	1.1760	35.20	34.64	24.8	1.2063	41.26	41.24
21.8	1.1770	35.40	34.83	24.9	1.2073	41.46	41.48
21.9	1.1779	35.58	35.02	25.0	1.2083	41.66	41.72
22.0	1.1789	35.78	35.21	25.1	1.2093	41.86	41.99
22.1	1.1798	35.96	35.40	25.2	1.2103	42.06	42.30
22.2	1.1808	36.16	35.59	25.3	1.2114	42.28	42.64
22.3	1.1817	36.34	35.78	25.4	1.2124	42.48	43.01
22.4	1.1827	36.54	35.97	25.5	1.2134	42.68	43.40

Sp. Gr. determinations were made at 60° F., compared with water at 60° F. From the Specific Gravities, the corresponding degrees Baumé were calculated by the following formula: Baumé = 145 − 145/Sp. Gr.

Atomic weights from F. W. Clarke's table of 1901. O = 16.

ALLOWANCE FOR TEMPERATURE:

10−15° Bé. − 1/40° Bé. or .0002 Sp. Gr. for 1° F.
15−22° Bé. − 1/30° Bé. or .0003 "   " " 1° F.
22−25° Bé. − 1/28° Bé. or .00035 "   " " 1° F.

AUTHORITY — W. C. FERGUSON.

This table has been approved and adopted as a Standard by the Manufacturing Chemists' Association of the United States.

W. H. BOWER,   JAS. L. MORGAN,
HENRY HOWARD,   ARTHUR WYMAN,
A. G. ROSENGARTEN,

New York, May 14, 1903.   *Executive Committee.*

VI.—HYDROCHLORIC ACID

LUNGE AND MARCHLEWSKI

Specific Gravity. $\frac{15°}{4°}$ in Vacuo.	Per Cent HCl by Weight.	1 Liter contains Grams HCl.	Specific Gravity $\frac{15°}{4°}$ in Vacuo.	Per Cent HCl by Weight.	1 Liter contains Grams HCl.	Specific Gravity $\frac{15°}{4°}$ in Vacuo.	Per Cent HCl by Weight.	1 Liter contains Grams HCl.
1.000	0.16	1.6	1.075	15.16	163	1.145	28.61	328
1.005	1.15	12	1.080	16.15	174	1.150	29.57	340
1.010	2.14	22	1.085	17.13	186	1.152	29.95	345
1.015	3.12	32	1.090	18.11	197	1.155	30.55	353
1.020	4.13	42	1.095	19.06	209	1.160	31.52	366
1.025	5.15	53	1.100	20.01	220	1.163	32.10	373
1.030	6.15	64	1.105	20.97	232	1.165	32.49	379
1.035	7.15	74	1.110	21.92	243	1.170	33.46	392
1.040	8.16	85	1.115	22.86	255	1.171	33.65	394
1.045	9.16	96	1.120	23.82	267	1.175	34.42	404
1.050	10.17	107	1.125	24.78	278	1.180	35.39	418
1.055	11.18	118	1.130	25.75	291	1.185	36.31	430
1.060	12.19	129	1.135	26.70	303	1.190	37.23	443
1.065	13.19	141	1.140	27.66	315	1.195	38.16	456
1.070	14.17	152	1.1425	28.14	322	1.200	39.11	469

COMPOSITION OF CONSTANT BOILING HYDROCHLORIC ACID*

Pressure mm. of Mercury.	Per Cent of HCl.	Grams constant boiling distillate for 1 mol. HCl.
770	20.218	180.390
760	20.242	180.170
750	20.266	179.960
740	20.290	179.745
730	20.314	179.530

Temperature of constant boiling hydrochloric acid is 108.54° at 763 mm. Specific gravity 1.09620^{25}.

* Hulett and Bonner, Jour. Am. Chem. Soc. xxxi, 390.

VII.—NITRIC ACID

By W. C. Ferguson

Degrees Baumé.	Sp. Gr. 60° F. / 60°	Degrees Twaddell.	Per Cent HNO$_3$.	Degrees Baumé.	Sp. Gr. 60° F. / 60°	Degrees Twaddell.	Per Cent HNO$_3$.
10.00	1.0741	14.82	12.86	21.25	1.1718	34.36	28.02
10.25	1.0761	15.22	13.18	21.50	1.1741	34.82	28.36
10.50	1.0781	15.62	13.49	21.75	1.1765	35.30	28.72
10.75	1.0801	16.02	13.81	22.00	1.1789	35.78	29.07
11.00	1.0821	16.42	14.13	22.25	1.1813	36.26	29.43
11.25	1.0841	16.82	14.44	22.50	1.1837	36.74	29.78
11.50	1.0861	17.22	14.76	22.75	1.1861	37.22	30.14
11.75	1.0881	17.62	15.07	23.00	1.1885	37.70	30.49
12.00	1.0902	18.04	15.41	23.25	1.1910	38.20	30.86
12.25	1.0922	18.44	15.72	23.50	1.1934	38.68	31.21
12.50	1.0943	18.86	16.05	23.75	1.1959	39.18	31.58
12.75	1.0964	19.28	16.39	24.00	1.1983	39.66	31.94
13.00	1.0985	19.70	16.72	24.25	1.2008	40.16	32.31
13.25	1.1006	20.12	17.05	24.50	1.2033	40.66	32.68
13.50	1.1027	20.54	17.38	24.75	1.2058	41.16	33.05
13.75	1.1048	20.96	17.71	25.00	1.2083	41.66	33.42
14.00	1.1069	21.38	18.04	25.25	1.2109	42.18	33.80
14.25	1.1090	21.80	18.37	25.50	1.2134	42.68	34.17
14.50	1.1111	22.22	18.70	25.75	1.2160	43.20	34.56
14.75	1.1132	22.64	19.02	26.00	1.2185	43.70	34.94
15.00	1.1154	23.08	19.36	26.25	1.2211	44.22	35.33
15.25	1.1176	23.52	19.70	26.50	1.2236	44.72	35.70
15.50	1.1197	23.94	20.02	26.75	1.2262	45.24	36.09
15.75	1.1219	24.38	20.36	27.00	1.2288	45.76	36.48
16.00	1.1240	24.80	20.69	27.25	1.2314	46.28	36.87
16.25	1.1262	25.24	21.03	27.50	1.2340	46.80	37.26
16.50	1.1284	25.68	21.36	27.75	1.2367	47.34	37.67
16.75	1.1306	26.12	21.70	28.00	1.2393	47.86	38.06
17.00	1.1328	26.56	22.04	28.25	1.2420	48.40	38.46
17.25	1.1350	27.00	22.38	28.50	1.2446	48.92	38.85
17.50	1.1373	27.46	22.74	28.75	1.2473	49.46	39.25
17.75	1.1395	27.90	23.08	29.00	1.2500	50.00	39.66
18.00	1.1417	28.34	23.42	29.25	1.2527	50.54	40.06
18.25	1.1440	28.80	23.77	29.50	1.2554	51.08	40.47
18.50	1.1462	29.24	24.11	29.75	1.2582	51.64	40.89
18.75	1.1485	29.70	24.47	30.00	1.2609	52.18	41.30
19.00	1.1508	30.16	24.82	30.25	1.2637	52.74	41.72
19.25	1.1531	30.62	25.18	30.50	1.2664	53.28	42.14
19.50	1.1554	31.08	25.53	30.75	1.2692	53.84	42.58
19.75	1.1577	31.54	25.88	31.00	1.2719	54.38	43.00
20.00	1.1600	32.00	26.24	31.25	1.2747	54.94	43.44
20.25	1.1624	32.48	26.61	31.50	1.2775	55.50	43.89
20.50	1.1647	32.94	26.96	31.75	1.2804	56.08	44.34
20.75	1.1671	33.42	27.33	32.00	1.2832	56.64	44.78
21.00	1.1694	33.88	27.67	32.25	1.2861	57.22	45.24

VII.—NITRIC ACID (Continued)

Degrees Baumé.	Sp. Gr. $\frac{60°}{60°}$ F.	Degrees Twaddell.	Per Cent HNO_3.	Degrees Baumé.	Sp. Gr. $\frac{60°}{60°}$ F.	Degrees Twaddell.	Per Cent HNO_3.
32.50	1.2889	57.78	45.68	40.75	1.3909	78.18	63.48
32.75	1.2918	58.36	46.14	41.00	1.3942	78.84	64.20
33.00	1.2946	58.92	46.58	41.25	1.3976	79.52	64.93
33.25	1.2975	59.50	47.04	41.50	1.4010	80.20	65.67
33.50	1.3004	60.08	47.49	41.75	1.4044	80.88	66.42
33.75	1.3034	60.68	47.95	42.00	1.4078	81.56	67.18
34.00	1.3063	61.26	48.42	42.25	1.4112	82.24	67.95
34.25	1.3093	61.86	48.90	42.50	1.4146	82.92	68.73
34.50	1.3122	62.44	49.35	42.75	1.4181	83.62	69.52
34.75	1.3152	63.04	49.83	43.00	1.4216	84.32	70.33
35.00	1.3182	63.64	50.32	43.25	1.4251	85.02	71.15
35.25	1.3212	64.24	50.81	43.50	1.4286	85.72	71.98
35.50	1.3242	64.84	51.30	43.75	1.4321	86.42	72.82
35.75	1.3273	65.46	51.80	44.00	1.4356	87.12	73.67
36.00	1.3303	66.06	52.30	44.25	1.4392	87.84	74.53
36.25	1.3334	66.68	52.81	44.50	1.4428	88.56	75.40
36.50	1.3364	67.28	53.32	44.75	1.4464	89.28	76.28
36.75	1.3395	67.90	53.84	45.00	1.4500	90.00	77.17
37.00	1.3426	68.52	54.36	45.25	1.4536	90.72	78.07
37.25	1.3457	69.14	54.89	45.50	1.4573	91.46	79.03
37.50	1.3488	69.76	55.43	45.75	1.4610	92.20	80.04
37.75	1.3520	70.40	55.97	46.00	1.4646	92.92	81.08
38.00	1.3551	71.02	56.52	46.25	1.4684	93.68	82.18
38.25	1.3583	71.66	57.08	46.50	1.4721	94.42	83.33
38.50	1.3615	72.30	57.65	46.75	1.4758	95.16	84.48
38.75	1.3647	72.94	58.23	47.00	1.4796	95.92	85.70
39.00	1.3679	73.58	58.82	47.25	1.4834	96.68	86.98
39.25	1.3712	74.24	59.43	47.50	1.4872	97.44	88.32
39.50	1.3744	74.88	60.06	47.75	1.4910	98.20	89.76
39.75	1.3777	75.54	60.71	48.00	1.4948	98.96	91.35
40.00	1.3810	76.20	61.38	48.25	1.4987	99.74	93.13
40.25	1.3843	76.86	62.07	48.50	1.5026	100.52	95.11
40.50	1.3876	77.52	62.77				

Specific Gravity determinations were made at 60° F., compared with water at 60° F.

From the Specific Gravities, the corresponding degrees Baumé were calculated by the following formula:

$$\text{Baumé} = 145 - \frac{145}{\text{Sp. Gr.}}.$$

Baumé Hydrometers for use with this table must be graduated by the above formula, which formula should **always** be printed on the scale.

Atomic weights from F. W. Clarke's table of 1901. $O = 16$.

ALLOWANCE FOR TEMPERATURE:

At 10° — 20° Bé. — 1/30° Bé. or .00029 Sp. Gr. = 1° F.
20° — 30° Bé. — 1/23° Bé. or .00044 " " = 1° F.
30° — 40° Bé. — 1/20° Bé. or .00060 " " = 1° F.
40° — 48.5° Bé. — 1/17° Bé. or .00084 " " = 1° F.

AUTHORITY—W. C. FERGUSON.

This table has been approved and adopted as a Standard by the Manufacturing Chemists' Association of the United States.

W. H. BOWER,   JAS. L. MORGAN,
HENRY HOWARD,   ARTHUR WYMAN,
New York, May 14, 1903.   A. G. ROSENGARTEN,   *Executive Committee*

VIII.—NITRIC ACID

LUNGE AND REY

Specific Gravity 15° 4° in vacuo	100 parts by weight contain		1 liter contains grams		Specific Gravity 15° 4° in vacuo	100 parts by weight contain		1 liter contains grams	
	% N_2O_5	% HNO_3	N_2O_5	HNO_3		% N_2O_5	% HNO_3	N_2O_5	HNO_3
1.000	0.08	0.10	1	1	1.195	27.10	31.62	324	378
1.005	0.85	1.00	8	10	1.200	27.74	32.36	333	388
1.010	1.62	1.90	16	19	1.205	28.36	33.09	342	399
1.015	2.39	2.80	24	28	1.210	28.99	33.82	351	409
1.020	3.17	3.70	33	38	1.215	29.61	34.55	360	420
1.025	3.94	4.60	40	47	1.220	30.24	35.28	369	430
1.030	4.71	5.50	49	57	1.225	30.88	36.03	378	441
1.035	5.47	6.38	57	66	1.230	31.53	36.78	387	452
1.040	6.22	7.26	64	75	1.235	32.17	37.53	397	463
1.045	6.97	8.13	73	85	1.240	32.82	38.29	407	475
1.050	7.71	8.99	81	94	1.245	33.47	39.05	417	486
1.055	8.43	9.84	89	104	1.250	34.13	39.82	427	498
1.060	9.15	10.68	97	113	1.255	34.78	40.58	437	509
1.065	9.87	11.51	105	123	1.260	35.44	41.34	447	521
1.070	10.57	12.33	113	132	1.265	36.09	42.10	457	533
1.075	11.27	13.15	121	141	1.270	36.75	42.87	467	544
1.080	11.96	13.95	129	151	1.275	37.41	43.64	477	556
1.085	12.64	14.74	137	160	1.280	38.07	44.41	487	568
1.090	13.31	15.53	145	169	1.285	38.73	45.18	498	581
1.095	13.99	16.32	153	179	1.290	39.39	45.95	508	593
1.100	14.67	17.11	161	188	1.295	40.05	46.72	519	605
1.105	15.34	17.89	170	198	1.300	40.71	47.49	529	617
1.110	16.00	18.67	177	207	1.305	41.37	48.26	540	630
1.115	16.67	19.45	186	217	1.310	42.06	49.07	551	643
1.120	17.34	20.23	195	227	1.315	42.76	49.89	562	656
1.125	18.00	21.00	202	236	1.320	43.47	50.71	573	669
1.130	18.66	21.77	211	246	1.325	44.17	51.53	585	683
1.135	19.32	22.54	219	256	1.330	44.89	52.37	597	697
1.140	19.98	23.31	228	266	1.3325	45.26	52.80	603	704
1.145	20.64	24.08	237	276	1.335	45.62	53.22	609	710
1.150	21.29	24.84	245	286	1.340	46.35	54.07	621	725
1.155	21.94	25.60	254	296	1.345	47.08	54.93	633	739
1.160	22.60	26.36	262	306	1.350	47.82	55.79	645	753
1.165	23.25	27.12	271	316	1.355	48.57	56.66	658	768
1.170	23.90	27.88	279	326	1.360	49.35	57.57	671	783
1.175	24.54	28.63	288	336	1.365	50.13	58.48	684	798
1.180	25.18	29.38	297	347	1.370	50.91	59.39	698	814
1.185	25.83	30.13	306	357	1.375	51.69	60.30	711	829
1.190	26.47	30.88	315	367	1.380	52.52	61.27	725	846

VIII.—NITRIC ACID (Continued)

Specific Gravity 15°/4° in vacuo	100 parts by weight contain		1 liter contains grams		Specific Gravity 15°/4° in vacuo	100 parts by weight contain		1 liter contains grams	
	% N_2O_5	% HNO_3	N_2O_5	HNO_3		% N_2O_5	% HNO_3	N_2O_5	HNO_3
1.3833	53.08	61.92	735	857	1.495	78.52	91.60	1174	1369
1.385	53.35	62.24	739	862	1.500	80.65	94.09	1210	1411
1.390	54.20	63.23	753	879	1.501	81.09	94.60	1217	1420
1.395	55.07	64.25	768	896	1.502	81.50	95.08	1224	1428
1.400	55.97	65.30	783	914	1.503	81.91	95.55	1231	1436
1.405	56.92	66.40	800	933	1.504	82.29	96.00	1238	1444
1.410	57.86	67.50	816	952	1.505	82.63	96.39	1244	1451
1.415	58.83	68.63	832	971	1.506	82.94	96.76	1249	1457
1.420	59.83	69.80	849	991	1.507	83.26	97.13	1255	1464
1.425	60.84	70.98	867	1011	1.508	83.58	97.50	1260	1470
1.430	61.86	72.17	885	1032	1.509	83.87	97.84	1265	1476
1.435	62.91	73.39	903	1053	1.510	84.09	98.10	1270	1481
1.440	64.01	74.68	921	1075	1.511	84.28	98.32	1274	1486
1.445	65.13	75.98	941	1098	1.512	84.46	98.53	1277	1490
1.450	66.24	77.28	961	1121	1.513	84.63	98.73	1280	1494
1.455	67.38	78.60	981	1144	1.514	84.78	98.90	1283	1497
1.460	68.56	79.98	1001	1168	1.515	84.92	99.07	1287	1501
1.465	69.79	81.42	1023	1193	1.516	85.04	99.21	1289	1504
1.470	71.06	82.90	1045	1219	1.517	85.15	99.34	1292	1507
1.475	72.39	84.45	1068	1246	1.518	85.26	99.46	1294	1510
1.480	73.76	86.05	1092	1274	1.519	85.35	99.57	1296	1512
1.485	75.18	87.70	1116	1302	1.520	85.44	99.67	1299	1515
1.490	76.80	89.60	1144	1335					

IX.—PHOSPHORIC ACID AT 17.5°

Specific Gravity.	Per Cent. P_2O_5.	Per Cent. H_3PO_4.	Specific Gravity.	Per Cent. P_2O_5.	Per Cent. H_3PO_4.	Specific Gravity.	Per Cent. P_2O_5.	Per Cent. H_3PO_4.
1.809	68.0	93.67	1.462	46.0	63.37	1.208	24.0	33.06
1.800	67.5	92.99	1.455	45.5	62.68	1.203	23.5	32.37
1.792	67.0	92.30	1.448	45.0	61.99	1.198	23.0	31.68
1.783	66.5	91.61	1.441	44.5	61.30	1.193	22.5	30.99
1.775	66.0	90.92	1.435	44.0	60.61	1.188	22.0	30.31
1.766	65.5	90.23	1.428	43.5	59.92	1.183	21.5	29.62
1.758	65.0	89.54	1.422	43.0	59.23	1.178	21.0	28.93
1.750	64.5	88.85	1.415	42.5	58.55	1.174	20.5	28.24
1.741	64.0	88.16	1.409	42.0	57.86	1.169	20.0	27.55
1.733	63.5	87.48	1.402	41.5	57.17	1.164	19.5	26.86
1.725	63.0	86.79	1.396	41.0	56.48	1.159	19.0	26.17
1.717	62.5	86.10	1.389	40.5	55.79	1.155	18.5	25.48
1.709	62.0	85.41	1.383	40.0	55.10	1.150	18.0	24.80
1.701	61.5	84.72	1.377	39.5	54.41	1.145	17.5	24.11
1.693	61.0	84.03	1.371	39.0	53.72	1.140	17.0	23.42
1.685	60.5	83.34	1.365	38.5	53.04	1.135	16.5	22.73
1.677	60.0	82.65	1.359	38.0	52.35	1.130	16.0	22.04
1.669	59.5	81.97	1.354	37.5	51.66	1.126	15.5	21.35
1.661	59.0	81.28	1.348	37.0	50.97	1.122	15.0	20.66
1.653	58.5	80.59	1.342	36.5	50.28	1.118	14.5	19.97
1.645	58.0	79.90	1.336	36.0	49.59	1.113	14.0	19.28
1.637	57.5	79.21	1.330	35.5	48.90	1.109	13.5	18.60
1.629	57.0	78.52	1.325	35.0	48.21	1.104	13.0	17.91
1.621	56.5	77.83	1.319	34.5	47.52	1.100	12.5	17.22
1.613	56.0	77.14	1.314	34.0	46.84	1.096	12.0	16.53
1.605	55.5	76.45	1.308	33.5	46.15	1.091	11.5	15.84
1.597	55.0	75.77	1.303	33.0	45.46	1.087	11.0	15.15
1.589	54.5	75.08	1.298	32.5	44.77	1.083	10.5	14.46
1.581	54.0	74.39	1.292	32.0	44.08	1.079	10.0	13.77
1.574	53.5	73.70	1.287	31.5	43.39	1.074	9.5	13.09
1.566	53.0	73.01	1.281	31.0	42.70	1.070	9.0	12.40
1.559	52.5	72.32	1.276	30.5	42.01	1.066	8.5	11.71
1.551	52.0	71.63	1.271	30.0	41.33	1.062	8.0	11.02
1.543	51.5	70.94	1.265	29.5	40.64	1.058	7.5	10.33
1.536	51.0	70.26	1.260	29.0	39.95	1.053	7.0	9.64
1.528	50.5	69.57	1.255	28.5	39.26	1.049	6.5	8.95
1.521	50.0	68.88	1.249	28.0	38.57	1.045	6.0	8.26
1.513	49.5	68.19	1.244	27.5	37.88	1.041	5.5	7.57
1.505	49.0	67.50	1.239	27.0	37.19	1.037	5.0	6.89
1.498	48.5	66.81	1.233	26.5	36.50	1.033	4.5	6.20
1.491	48.0	66.12	1.228	26.0	35.82	1.029	4.0	5.51
1.484	47.5	65.43	1.223	25.5	35.13	1.025	3.5	4.82
1.476	47.0	64.75	1.218	25.0	34.44	1.021	3.0	4.13
1.469	46.5	64.06	1.213	24.5	33.75	1.017	2.5	3.44

X.—SULPHURIC ACID

By W. C. Ferguson and H. P. Talbot

Degrees Baumé.	Specific Gravity $\frac{60°}{60°}$ F.	Degrees Twaddell.	Per Cent H_2SO_4.	Weight of 1 Cu. Ft. in Lbs. Av.	Per Cent O. V.*	Pounds O. V. in 1 Cubic Foot.
0	1.0000	0.0	0.00	62.37	0.00	0.00
1	1.0069	1.4	1.02	62.80	1.09	0.68
2	1.0140	2.8	2.08	63.24	2.23	1.41
3	1.0211	4.2	3.13	63.69	3.36	2.14
4	1.0284	5.7	4.21	64.14	4.52	2.90
5	1.0357	7.1	5.28	64.60	5.67	3.66
6	1.0432	8.6	6.37	65.06	6.84	4.45
7	1.0507	10.1	7.45	65.53	7.99	5.24
8	1.0584	11.7	8.55	66.01	9.17	6.06
9	1.0662	13.2	9.66	66.50	10.37	6.89
10	1.0741	14.8	10.77	66.99	11.56	7.74
11	1.0821	16.4	11.89	67.49	12.76	8.61
12	1.0902	18.0	13.01	68.00	13.96	9.49
13	1.0985	19.7	14.13	68.51	15.16	10.39
14	1.1069	21.4	15.25	69.04	16.36	11.30
15	1.1154	23.1	16.38	69.57	17.58	12.23
16	1.1240	24.8	17.53	70.10	18.81	13.19
17	1.1328	26.6	18.71	70.65	20.08	14.18
18	1.1417	28.3	19.89	71.21	21.34	15.20
19	1.1508	30.2	21.07	71.78	22.61	16.23
20	1.1600	32.0	22.25	72.35	23.87	17.27
21	1.1694	33.9	23.43	72.94	25.14	18.34
22	1.1789	35.8	24.61	73.53	26.41	19.42
23	1.1885	37.7	25.81	74.13	27.69	20.53
24	1.1983	39.7	27.03	74.74	29.00	21.68

Sp. Gr. determinations were made at 60° F., compared with water at 60° F.

From the Sp. Grs., the corresponding degrees Baumé were calculated by the following formula: Baumé = 145 − 145/Sp. Gr.

Baumé Hydrometers for use with this table must be graduated by the above formula, which formula should always be printed on the scale.

* 66° Baumé = Sp. Gr. 1.8354 = Oil of Vitriol (O. V.).

1 cu. ft. water at 60° F. weighs 62.37 lbs. av.

Atomic weights from F. W. Clarke's table of 1901. O = 16.

H_2SO_4 = 100 per cent.

	% H_2SO_4		% O. V.		%60°
O. V.	= 93.19	=	100.00	=	119.98
60°	= 77.67	=	83.35	=	100.00
50°	= 62.18	=	66.72	=	80.06

X.—SULPHURIC ACID (Continued)

APPROXIMATE BOILING POINTS

Degrees Baumé.	* Freezing (Melting) Point. F.
0	32.0
1	31.2
2	30.5
3	29.8
4	28.9
5	28.1
6	27.2
7	26.3
8	25.1
9	24.0
10	22.8
11	21.5
12	20.0
13	18.3
14	16.6
15	14.7
16	12.6
17	10.2
18	7.7
19	4.8
20	+ 1.6
21	− 1.8
22	− 6.0
23	−11
24	−16

APPROXIMATE BOILING POINTS

50° B,	295° F.
60° "	386° "
61° "	400° "
62° "	415° "
63° "	432° "
64° "	451° "
65° "	485° "
66° "	538° "

FIXED POINTS

Specific Gravity.	Per Cent H_2SO_4.	Specific Gravity.	Per Cent H_2SO_4.
1.0000	.00	1.5281	62.34
1.0048	.71	1.5440	63.79
1.0347	5.14	1.5748	66.51
1.0649	9.48	1.6272	71.00
1.0992	14.22	1.6679	74.46
1.1353	19.04	1.7044	77.54
1.1736	23.94	1.7258	79.40
1.2105	28.55	1.7472	81.32
1.2513	33.49	1.7700	83.47
1.2951	38.64	1.7959	86.36
1.3441	44.15	1.8117	88.53
1.3947	49.52	1.8194	89.75
1.4307	53.17	1.8275	91.32
1.4667	56.68	1.8354	93.19
1.4822	58.14		

Acids stronger than 66° Bé. should have their percentage compositions determined by chemical analysis.

* Calculated from Pickering's results, Jour. of Lon. Ch. Soc., vol. 57, p. 363.

AUTHORITIES — W. C. FERGUSON; H. P. TALBOT.

This table has been approved and adopted as a standard by the Manufacturing Chemists' Association of the United States.

W. H. BOWER,
HENRY HOWARD,
JAS. L. MORGAN,
ARTHUR WYMAN,
A. G. ROSENGARTEN,
Executive Committee

New York, June 23, 1904.

52

X.—SULPHURIC ACID (Continued)

Degrees Baumé.	Specific Gravity $\frac{60°}{60°}$ F.	Degrees Twaddell.	Per Cent H_2SO_4.	Weight of 1 Cu. Ft. in Lbs. Av.	Per Cent O. V.	Pounds O. V. in 1 Cubic Foot.
25	1.2083	41.7	28.28	75.33	30.34	22.87
26	1.2185	43.7	29.53	76.00	31.69	24.08
27	1.2288	45.8	30.79	76.64	33.04	25.32
28	1.2393	47.9	32.05	77.30	34.39	26.58
29	1.2500	50.0	33.33	77.96	35.76	27.88
30	1.2609	52.2	34.63	78.64	37.16	29.22
31	1.2719	54.4	35.93	79.33	38.55	30.58
32	1.2832	56.6	37.26	80.03	39.98	32.00
33	1.2946	58.9	38.58	80.74	41.40	33.42
34	1.3063	61.3	39.92	81.47	42.83	34.90
35	1.3182	63.6	41.27	82.22	44.28	36.41
36	1.3303	66.1	42.63	82.97	45.74	37.95
37	1.3426	68.5	43.99	83.74	47.20	39.53
38	1.3551	71.0	45.35	84.52	48.66	41.13
39	1.3679	73.6	46.72	85.32	50.13	42.77
40	1.3810	76.2	48.10	86.13	51.61	44.45
41	1.3942	78.8	49.47	86.96	53.08	46.16
42	1.4078	81.6	50.87	87.80	54.58	47.92
43	1.4216	84.3	52.26	88.67	56.07	49.72
44	1.4356	87.1	53.66	89.54	57.58	51.56
45	1.4500	90.0	55.07	90.44	59.09	53.44
46	1.4646	92.9	56.48	91.35	60.60	55.36
47	1.4796	95.9	57.90	92.28	62.13	57.33
48	1.4948	99.0	59.32	93.23	63.65	59.34
49	1.5104	102.1	60.75	94.20	65.18	61.40
50	1.5263	105.3	62.18	95.20	66.72	63.52
51	1.5426	108.5	63.66	96.21	68.31	65.72
52	1.5591	111.8	65.13	97.24	69.89	67.96
53	1.5761	115.2	66.63	98.30	71.50	70.28
54	1.5934	118.7	68.13	99.38	73.11	72.66
55	1.6111	122.2	69.65	100.48	74.74	75.10
56	1.6292	125.8	71.17	101.61	76.37	77.60
57	1.6477	129.5	72.75	102.77	78.07	80.23
58	1.6667	133.3	74.36	103.95	79.79	82.95
59	1.6860	137.2	75.99	105.16	81.54	85.75

X.—SULPHURIC ACID (Continued)

Degrees Baumé.	* Freezing (Melting) Point. °F.
25	− 23
26	− 30
27	− 39
28	− 49
29	− 61
30	− 74
31	− 82
32	− 96
33	− 97
34	− 91
35	− 81
36	− 70
37	− 60
38	− 53
39	− 47

ALLOWANCE FOR TEMPERATURE

At 10° Bé. .029° Bé. or .00023 Sp. Gr. = 1° F.
" 20° " .036° " .00034 " = 1° "
" 30° " .035° " .00039 " = 1° "
" 40° " .031° " .00041 " = 1° "
" 50° " .028° " .00045 " = 1° "
" 60° " .026° " .00053 " = 1° "
" 63° " .026° " .00057 " = 1° "
" 66° " .0235° " .00054 " = 1° "

Degrees Baumé.	Freezing Point °F.	Per Cent 60° Baumé.	Pounds 60° Baumé in 1 Cubic Foot.	Per Cent 50° Baumé.	Pounds 50° Baumé in 1 Cubic Foot.
40	− 41	61.93	53.34	77.36	66.63
41	− 35	63.69	55.39	79.56	69.19
42	− 31	65.50	57.50	81.81	71.83
43	− 27	67.28	59.66	84.05	74.53
44	− 23	69.09	61.86	86.30	77.27
45	− 20	70.90	64.12	88.56	80.10
46	− 14	72.72	66.43	90.83	82.98
47	− 15	74.55	68.79	93.12	85.93
48	− 18	76.37	71.20	95.40	88.94
49	− 22	78.22	73.68	97.70	92.03
50	− 27	80.06	76.21	100.00	95.20
51	− 33	81.96	78.85	102.38	98.50
52	− 39	83.86	81.54	104.74	101.85
53	− 49	85.79	84.33	107.15	105.33
54	− 59	87.72	87.17	109.57	108.89
55	..	89.67	90.10	112.01	112.55
56	..	91.63	93.11	114.46	116.30
57	.. Below 40	93.67	96.26	117.00	120.24
58	..	95.74	99.52	119.59	124.31
59	− 7	97.84	102.89	122.21	128.52

X.—SULPHURIC ACID (Continued)

Degrees Baumé.	Specific Gravity $\frac{60° F.}{60°}$	Degrees Twaddell.	Per Cent H_2SO_4.	Weight of 1 Cu. Ft. in Lbs. Av.	Per Cent O. V.	Pounds O. V. in 1 Cubic Foot.
60	1.7059	141.2	77.67	106.40	83.35	88.68
61	1.7262	145.2	79.43	107.66	85.23	91.76
62	1.7470	149.4	81.30	108.96	87.24	95.06
63	1.7683	.153.7	83.34	110.29	89.43	98.63
64	1.7901	158.0	85.66	111.65	91.92	102.63
64¼	1.7957	159.1	86.33	112.00	92.64	103.75
64½	1.8012	160.2	87.04	112.34	93.40	104.93
64¾	1.8068	161.4	87.81	112.69	94.23	106.19
65	1.8125	162.5	88.65	113.05	95.13	107.54
65¼	1.8182	163.6	89.55	113.40	96.10	108.97
65½	1.8239	164.8	90.60	113.76	97.22	110.60
65¾	1.8297	165.9	91.80	114.12	98.51	112.42
66	1.8354	167.1	93.19	114.47	100.00	114.47

Degrees Baumé.	Freezing (Melting) Point.	Per Cent 60° Baumé.	Pounds 60° Baumé in Cubic Foot.	Per Cent 50° Baumé.	Pounds 50° Baumé in Cubic Foot.	
60	+12.6	100.00	106.40	124.91	132.91	
61	27.3	102.27	110.10	127.74	137.52	
62	39.1	104.67	114.05	130.75	142.47	
63	46.1	107.30	118.34	134.03	147.82	
64	46.4	110.29	123.14	137.76	153.81	
64¼	43.6	111.15	124.49	138.84	155.50	
64½	41.1	112.06	125.89	139.98	157.25	
64¾	37.9	113.05	127.40	141.22	159.14	
65	33.1	114.14	129.03	142.57	161.17	
65¼	24.6	115.30	130.75	144.02	163.32	
65½	13.4	116.65	132.70	145.71	165.76	
65¾	− 1	118.19	134.88	147.63	168.48	
66	−29	119.98	137.34	149.87	171.56	

XI.—SULPHURIC ACID TABLE
94–100% H_2SO_4
By H. B. BISHOP

Bé.	Sp. Gr. at 60° F.	Per Cent. H_2SO_4	Wt. 1 Cu. Ft.	Allowance for Temperature.
66	1.8354	93.19	114.47	At 94% .00054 sp.gr. =1° F.
66.12	1.8381	94.00	114.64	" 96 .0053 " =1° F.
66.23	1.8407	95.00	114.80	" 97.5 .00052 " =1° F.
66.31	1.8427	96.00	114.93	" 100 .00052 " =1° F
66.36	1.8437	97.00	114.99	
66.36	1.8439	97.50	114.99	
66.36	1.8437	98.00	114.99	
66.30	1.8424	99.00	114.91	
66.16	1.8391	100.00	114.70	

FUMING SULPHURIC ACID EQUIVALENTS

Total SO₃	Equivalent H₂SO₄	Per Cent H₂SO₄	Per Cent Free SO₃	Total SO₃	Equivalent H₂SO₄	Per Cent H₂SO₄	Per Cent Free SO₃
81.63	100.00	100	0	90.82	111.25	50	50
81.82	100.23	99	1	91.00	111.48	49	51
82.00	100.45	98	2	91.18	111.70	48	52
82.18	100.67	97	3	91.37	111.93	47	53
82.37	100.90	96	4	91.55	112.15	46	54
82.55	101.13	95	5	91.73	112.37	45	55
82.73	101.35	94	6	91.92	112.60	44	56
82.92	101.58	93	7	92.10	112.82	43	57
83.10	101.80	92	8	92.29	113.05	42	58
83.29	102.03	91	9	92.47	113.28	41	59
83.47	102.25	90	.1	92.65	113.50	40	60
83.65	102.47	89	11	92.84	113.73	39	61
83.84	102.70	88	12	93.02	113.95	38	62
84.02	102.92	87	13	93.20	114.17	37	63
84.20	103.15	86	14	93.39	114.40	36	64
84.39	103.38	85	15	93.57	114.62	35	65
84.57	103.60	84	16	93.76	114.85	34	66
84.75	103.82	83	17	93.94	115.08	33	67
84.94	104.05	82	18	94.12	115.30	32	68
85.12	104.27	81	19	94.31	115.53	31	69
85.31	104.50	80	20	94.49	115.75	30	70
85.49	104.73	79	21	94.67	115.97	29	71
85.67	104.95	78	22	94.86	116.20	28	72
85.86	105.18	77	23	95.04	116.42	27	73
86.04	105.40	76	24	95.22	116.65	26	74
86.22	105.62	75	25	95.41	116.88	25	75
86.41	105.85	74	26	95.59	117.10	24	76
86.59	106.07	73	27	95.78	117.33	23	77
86.78	106.30	72	28	95.96	117.55	22	78
86.96	106.53	71	29	96.14	117.77	21	79
87.14	106.75	70	30	96.33	118.00	20	80
87.33	106.98	69	31	96.51	118.22	19	81
87.51	107.20	68	32	96.69	118.45	18	82
87.69	107.42	67	33	96.88	118.68	17	83
87.88	107.65	66	34	97.06	118.90	16	84
88.06	107.87	65	35	97.25	119.13	15	85
88.24	108.10	64	36	97.43	119.35	14	86
88.43	108.33	63	37	97.61	119.57	13	87
88.61	108.55	62	38	97.80	119.80	12	88
88.80	108.78	61	39	97.98	120.03	11	89
88.98	109.00	60	40	98.16	120.25	10	90
89.16	109.22	59	41	98.35	120.48	9	91
89.35	109.45	58	42	98.53	120.70	8	92
89.53	109.67	57	43	98.71	120.92	7	93
89.71	109.90	56	44	98.90	121.15	6	94
89.90	110.13	55	45	99.08	121.37	5	95
90.08	110.35	54	46	99.27	121.60	4	96
90.27	110.58	53	47	99.45	121.83	3	97
90.45	110.80	52	48	99.63	122.05	2	98
90.63	111.02	51	49	99.82	122.28	1	99
				100.00	122.50	0	100

Compiled from the table by H. B. Bishop, Van Nostrand's Chemical Annual, 1913.

XII.—ACETIC ACID AT 15°

OUDEMANS

Specific Gravity.	Per Cent $H.C_2H_3O_2$.	Specific Gravity.	Per Cent $H.C_2H_3O_2$.	Specific Gravity.	Per Cent $H.C_2H_3O_2$.	Specific Gravity.	Per Cent $H.C_2H_3O_2$.
0.9992	0	1.0363	26	1.0623	51	1.0747	76
1.0007	1	1.0375	27	1.0631	52	1.0748	77
1.0022	2	1.0388	28	1.0638	53	1.0748	78
1.0037	3	1.0400	29	1.0646	54	1.0748	79
1.0052	4	1.0412	30	1.0653	55	1.0748	80
1.0067	5	1.0424	31	1.0660	56	1.0747	81
1.0083	6	1.0436	32	1.0666	57	1.0746	82
1.0098	7	1.0447	33	1.0673	58	1.0744	83
1.0113	8	1.0459	34	1.0679	59	1.0742	84
1.0127	9	1.0470	35	1.0685	60	1.0739	85
1.0142	10	1.0481	36	1.0691	61	1.0736	86
1.0157	11	1.0492	37	1.0697	62	1.0731	87
1.0171	12	1.0502	38	1.0702	63	1.0726	88
1.0185	13	1.0513	39	1.0707	64	1.0720	89
1.0200	14	1.0523	40	1.0712	65	1.0713	90
1.0214	15	1.0533	41	1.0717	66	1.0705	91
1.0228	16	1.0543	42	1.0721	67	1.0696	92
1.0242	17	1.0552	43	1.0725	68	1.0686	93
1.0256	18	1.0562	44	1.0729	69	1.0674	94
1.0270	19	1.0571	45	1.0733	70	1.0660	95
1.0284	20	1.0580	46	1.0737	71	1.0644	96
1.0298	21	1.0589	47	1.0740	72	1.0625	97
1.0311	22	1.0598	48	1.0742	73	1.0604	98
1.0324	23	1.0607	49	1.0744	74	1.0580	99
1.0337	24	1.0615	50	1.0746	75	1.0553	100
1.0350	25						

XIII.—MELTING POINTS OF ACETIC ACID
RUDORFF, Ber. **3**, 390.

100 gr. $H.C_2H_3O_2$ mixed with gr. water.	100 parts by weight contain parts water.	Melting (solidifying) point °C.	100 gr. $H.C_2H_3O_2$ mixed with gr. water.	100 parts by weight contain parts water.	Melting (solidifying) point °C.
0.0	0.0	16.7°	8.0	7.407	6.25°
0.5	0.497	15.65	9.0	8.257	5.3
1.0	0.990	14.8	10.0	9.090	4.3
1.5	1.477	14.0	11.0	9.910	3.6
2.0	1.961	13.25	12.0	10.774	2.7
3.0	2.912	11.95	15.0	13.043	−0.2
4.0	3.846	10.5	18.0	15.324	−2.6
5.0	4.761	9.4	21.0	17.355	−5.1
6.0	5.660	8.2	24.0	19.354	−7.4
7.0	6.542	7.1			

Boiling point 100% acid 117 8°.

XIV.—AQUA AMMONIA
According to W. C. Ferguson

Degrees Baumé.	Sp. Gr. $\frac{60°}{60°}$ F.	Per Cent NH_3.	Degrees Baumé.	Sp. Gr. $\frac{60°}{60°}$ F.	Per Cent NH_3.	Degrees Baumé.	Sp. Gr. $\frac{60°}{60°}$ F.	Per Cent NH_3.
10.00	1.0000	.00	16.50	.9556	11.18	23.00	.9150	23.52
10.25	.9982	.40	16.75	.9540	11.64	23.25	.9135	24.01
10.50	.9964	.80	17.00	.9524	12.10	23.50	.9121	24.50
10.75	.9947	1.21	17.25	.9508	12.56	23.75	.9106	24.99
11.00	.9929	1.62	17.50	.9492	13.02	24.00	.9091	25.48
11.25	.9912	2.04	17.75	.9475	13.49	24.25	.9076	25.97
11.50	.9894	2.46	18.00	.9459	13.96	24.50	.9061	26.46
11.75	.9876	2.88	18.25	.9444	14.43	24.75	.9047	26.95
12.00	.9859	3.30	18.50	.9428	14.90	25.00	.9032	27.44
12.25	.9842	3.73	18.75	.9412	15.37	25.25	.9018	27.93
12.50	.9825	4.16	19.00	.9396	15.84	25.50	.9003	28.42
12.75	.9807	4.59	19.25	.9380	16.32	25.75	.8989	28.91
13.00	.9790	5.02	19.50	.9365	16.80	26.00	.8974	29.40
13.25	.9773	5.45	19.75	.9349	17.28	26.25	.8960	29.89
13.50	.9756	5.88	20.00	.9333	17.76	26.50	.8946	30.38
13.75	.9739	6.31	20.25	.9318	18.24	26.75	.8931	30.87
14.00	.9722	6.74	20.50	.9302	18.72	27.00	.8917	31.36
14.25	.9705	7.17	20.75	.9287	19.20	27.25	.8903	31.85
14.50	.9689	7.61	21.00	.9272	19.68	27.50	.8889	32.34
14.75	.9672	8.05	21.25	.9256	20.16	27.75	.8875	32.83
15.00	.9655	8.49	21.50	.9241	20.64	28.00	.8861	33.32
15.25	.9639	8.93	21.75	.9226	21.12	28.25	.8847	33.81
15.50	.9622	9.38	22.00	.9211	21.60	28.50	.8833	34.30
15.75	.9605	9.83	22.25	.9195	22.08	28.75	.8819	34.79
16.00	.9589	10.28	22.50	.9180	22.56	29.00	.8805	35.28
16.25	.9573	10.73	22.75	.9165	23.04			

ALLOWANCE FOR TEMPERATURE

The coefficient of expansion for ammonia solutions, varying with the temperature, correction must be applied according to the following table:

Degrees Baumé.	Corrections to be Added for Each Degree Below 60° F.		Corrections to be Subtracted for Each Degree Above 60° F.			
	40° F.	50° F.	70° F.	80° F.	90° F.	100° F.
14° Bé	.015° Bé	.017° Bé	.020° Bé	.022° Bé	.024° Bé	.026° Bé
16°	.021 "	.023 "	.026 "	.028 "	.030 "	.032 "
18°	.027 "	.029 "	.031 "	.033 "	.035 "	.037 "
20°	.033 "	.036 "	.037 "	.038 "	.040 "	.042 "
22°	.039 "	.042 "	.043 "	.045 "	.047 "	
26°	.053 "	.057 "	.057 "	.059 "		

XV.—SODIUM HYDROXIDE SOLUTION AT 15°

LUNGE

Specific Gravity.	Degrees Baume.	Degrees Twaddell.	Per Cent Na₂O.	Per Cent NaOH.	1 Liter contains Grams	
					Na₂O.	NaOH
1.007	1.0	1.4	0.47	0.61	4	6
1.014	2.0	2.8	0.93	1.20	9	12
1.022	3.1	4.4	1.55	2.00	16	21
1.029	4.1	5.8	2.10	2.70	22	28
1.036	5.1	7.2	2.60	3.35	27	35
1.045	6.2	9.0	3.10	4.00	32	42
1.052	7.2	10.4	3.60	4.64	38	49
1.060	8.2	12.0	4.10	5.29	43	56
1.067	9.1	13.4	4.55	5.87	49	63
1.075	10.1	15.0	5.08	6.55	55	70
1.083	11.1	16.6	5.67	7.31	61	79
1.091	12.1	18.2	6.20	8.00	68	87
1.100	13.2	20.0	6.73	8.68	74	95
1.108	14.1	21.6	7.30	9.42	81	104
1.116	15.1	23.2	7.80	10.06	87	112
1.125	16.1	25.0	8.50	10.97	96	123
1.134	17.1	26.8	9.18	11.84	104	134
1.142	18.0	28.4	9.80	12.64	112	144
1.152	19.1	30.4	10.50	13.55	121	156
1.162	20.2	32.4	11.14	14.37	129	167
1.171	21.2	34.2	11.73	15.13	137	177
1.180	22.1	36.0	12.33	15.91	146	183
1.190	23.1	38.0	13.00	16.77	155	200
1.200	24.2	40.0	13.70	17.67	164	212
1.210	25.2	42.0	14.40	18.58	174	225
1.220	26.1	44.0	15.18	19.58	185	239
1.231	27.2	46.2	15.96	20.59	196	253
1.241	28.2	48.2	16.76	21.42	208	266
1.252	29.2	50.4	17.55	22.64	220	283
1.263	30.2	52.6	18.35	23.67	232	299
1.274	31.2	54.8	19.23	24.81	245	316
1.285	32.2	57.0	20.00	25.80	257	332
1.297	33.2	59.4	20.80	26.83	270	348
1.308	34.1	61.6	21.55	27.80	282	364
1.320	35.2	64.0	22.35	28.83	295	381
1.332	36.1	66.4	23.20	29.93	309	399
1.345	37.2	69.0	24.20	31.22	326	420

XV.—SODIUM HYDROXIDE SOLUTION AT 15° (Continued)

Specific Gravity.	Degrees Baumé.	Degrees Twaddell.	Per Cent Na₂O.	Per Cent NaOH.	1 Liter contains Grams	
					Na₂O.	NaOH
1.357	38.1	71.4	25.17	32.47	342	441
1.370	39.2	74.0	26.12	33.69	359	462
1.383	40.2	76.6	27.10	34.96	375	483
1.397	41.2	79.4	28.10	36.25	392	506
1.410	42.2	82.0	29.05	37.47	410	528
1.424	43.2	84.8	30.08	38.80	428	553
1.438	44.2	87.6	31.00	39.99	446	575
1.453	45.2	90.6	32.10	41.41	466	602
1.468	46.2	93.6	33.20	42.83	487	629
1.483	47.2	96.6	34.40	44.38	510	658
1.498	48.2	99.6	35.70	46.15	535	691
1.514	49.2	102.8	36.90	47.60	559	721
1.530	50.2	106.0	38.00	49.02	581	750

XVI.—VAPOR TENSION OF WATER IN MILLIMETERS OF MERCURY −2° TO +36° C.

ACCORDING TO REGNAULT, BROCH, AND WEIBE

°C.	0	.1	.2	.3	.4	.5	.6	.7	.8	.9
	mm.	mm.	mm.	mm.	mm.	mm.	mm.	mm.	mm.	mm.
−2	3.958	3.929	3.900	3.872	3.844	3.815	3.787	3.760	3.732	3.705
−1	4.258	4.227	4.197	4.166	4.136	4.106	4.076	4.046	4.016	3.987
−0	4.579	4.546	4.513	4.481	4.448	4.416	4.384	4.352	4.321	4.289
0	4.579	4.612	4.646	4.679	4.713	4.747	4.782	4.816	4.851	4.886
1	4.921	4.957	4.992	5.028	5.064	5.101	5.137	5.174	5.211	5.248
2	5.286	5.324	5.362	5.400	5.438	5.477	5.516	5.555	5.595	5.635
3	5.675	5.715	5.755	5.796	5.837	5.878	5.920	5.961	6.003	6.046
4	6.088	6.131	6.174	6.217	6.261	6.305	6.349	6.393	6.438	6.483
5	6.528	6.574	6.620	6.666	6.712	6.759	6.806	6.853	6.901	6.949
6	6.997	7.045	7.094	7.143	7.192	7.242	7.292	7.342	7.392	7.443
7	7.494	7.546	7.598	7.650	7.702	7.755	7.808	7.861	7.914	7.968
8	8.023	8.077	8.132	8.187	8.243	8.299	8.355	8.412	8.469	8.526
9	8.584	8.642	8.700	8.759	8.818	8.877	8.937	8.997	9.057	9.118
10	9.179	9.240	9.302	9.364	9.427	9.490	9.553	9.616	9.680	9.745
11	9.810	9.875	9.940	10.006	10.072	10.139	10.206	10.274	10.342	10.410
12	10.479	10.548	10.617	10.687	10.757	10.828	10.899	10.970	11.042	11.114
13	11.187	11.260	11.333	11.407	11.481	11.556	11.631	11.706	11.782	11.859
14	11.936	12.013	12.091	12.169	12.247	12.326	12.406	12.486	12.566	12.647
15	12.728	12.810	12.892	12.974	13.057	13.141	13.225	13.309	13.394	13.480
16	13.565	13.651	13.738	13.825	13.913	14.001	14.090	14.179	14.269	14.359
17	14.450	14.541	14.632	14.724	14.817	14.910	15.003	15.097	15.192	15.287
18	15.383	15.479	15.575	15.672	15.770	15.868	15.967	16.066	16.166	16.266
19	16.367	16.469	16.571	16.673	16.776	16.880	16.984	17.088	17.193	17.299
20	17.406	17.513	17.620	17.728	17.837	17.947	18.057	18.167	18.278	18.390
21	18.503	18.616	18.729	18.844	18.959	19.074	19.190	19.307	19.424	19.542
22	19.661	19.780	19.900	20.021	20.142	20.264	20.386	20.510	20.634	20.758
23	20.883	21.010	21.137	21.264	21.393	21.522	21.652	21.782	21.913	22.045
24	22.178	22.311	22.446	22.581	22.716	22.853	22.990	23.128	23.266	23.406
25	23.546	23.686	23.828	23.970	24.113	24.257	24.401	24.547	24.693	24.839
26	24.987	25.135	25.284	25.434	25.584	25.736	25.888	26.041	26.195	26.349
27	26.505	26.661	26.818	26.976	27.134	27.294	27.454	27.615	27.777	27.939
28	28.103	28.267	28.432	28.599	28.766	28.933	29.102	29.271	29.442	29.613
29	29.785	29.958	30.132	30.307	30.482	30.659	30.836	31.015	31.194	31.374
30	31.555	31.737	31.919	32.103	32.288	32.473	32.660	32.847	33.036	33.225
31	33.416	33.607	33.799	33.992	34.187	34.382	34.578	34.775	34.973	35.172
32	35.372	35.573	35.775	35.978	36.182	36.387	36.593	36.800	37.008	37.217
33	37.427	37.638	37.851	38.064	38.278	38.493	38.710	38.927	39.146	39.365
34	39.586	39.807	40.030	40.254	40.479	40.705	40.933	41.161	41.390	41.621
35	41.583	42.085	42.319	42.554	42.791	43.028	43.266	43.506	43.747	43.989

XVII.—USEFUL DATA OF THE MORE IMPORTANT INORGANIC COMPOUNDS *

Substance.	Formula.	Molecular or Atomic Weight.	Normal Coefficient 1 c.c. = gm.	Solubility in 100 Gms. Water.	Indicator.
Acetic acid.........	$HC_2H_3O_2$	60.04	.06004		
anhydride........	$(CH_3CO)_2O$	102.07	.051035		
Aluminium.........	Al	27.10	.009033		
chloride..........	Al_2Cl_6	266.96	.04449	69.87[15°]	P.
chloride..........	$Al_2Cl_6 \cdot 12H_2O$	483.15	.08053	40	P.
oxide.............	Al_2O_3	102.20	.01703	insol.	
phosphate........	$AlPO_4$	122.14	.04071		
sulphate..........	$Al_2(SO_4)_3$	342.38	.05706	36.1[20°]	P.
sulphate..........	$Al_2(SO_4)_318H_2O$	666.67	.11111	87	P.
Ammonia..........	NH_3	17.03	.01703	....	M.
Ammonium.........	NH_4	18.04	.01804	....	M.
chloride..........	NH_4Cl	53.50	.05350	29.4[0°]	M.
hydroxide........	NH_4OH	35.05	.03505	....	M.
nitrate...........	NH_4NO_3	80.05	.08005	118[0°]	M.
oxalate...........	$(NH_4)_2C_2O_4 \cdot H_2O$	142.11	.07105[13]		
sulphate..........	$(NH_4)_2SO_4$	132.14	.06607	71[0°]	M.
Antimony..........	Sb	120.20	.06010		
chloride (tri)......	$SbCl_3$	226.58	.11329[3]		
chloride (penta)...	$SbCl_5$	297.50	.14875[3]		
oxide (tri)........	Sb_2O_3	288.40	.07210[3]		
oxide (penta).....	Sb_2O_5	320.40	.08010[3]		
Arsenic............	As	74.96	.03748		
oxide............	As_2O_5	229.92	.03832 [1]	150	
oxide............	As_2O_5	229.92	.05748 [3]		
Arsenious oxide.....	As_2O_3	197.92	.03299	1.7[16°]	
Arsenious oxide.....	As_2O_3	197.92	.04948 [3]	1.7[16°]	
Barium............	Ba	137.37	.068685		
carbonate........	$BaCO_3$	197.37	.098685	.0022[20°]	M.
chloride..........	$BaCl_2$	208.29	.104145	30.9[0°]	
chloride..........	$BaCl_22H_2O$	244.32	.12216	36.2[0°]	
hydroxide........	$Ba(OH)_2$	171.38	.08569		
hydroxide........	$Ba(OH)_28H_2O$	315.51	.15775	5.15[15°]	
oxalate...........	$BaC_2O_4 \cdot H_2O$	243.40	.12170		
oxide............	BaO	153.37	.076685	1.5[0°]	
sulphate..........	$BaSO_4$	233.43	.11672	.000172[0°]	
sulphite..........	$BaSO_3$	217.43	.10872		
peroxide..........	BaO_2	169.37	.08469 [3]	insol.	
Boric acid..........	H_3BO_3	61.92	.06192		
Bromine...........	Br	79.92	.07992	4.17[0°]	
Cadmium carbonate.	$CdCO_3$	172.405	.08620	insol.	
chloride..........	$CdCl_2$	183.32	.09166	140[20°]	
chloride..........	$CdCl_22H_2O$	219.35	.109675	168[20°]	
sulphide..........	CdS	144.46	.07223	insol.	
Calcium...........	Ca	40.07	.020035		
acetate...........	$Ca(C_2H_3O_2)_2$	158.14	.07907		
bicarbonate.......	$Ca(HCO_3)_2$	162.10	.08105		
carbonate........	$CaCO_3$	100.075	.050037	.0013	M.
chloride..........	$CaCl_2$	110.99	.055495	59.5[0°]	
chloride..........	$CaCl_26H_2O$	219.09	.109543	117.4[0°]	
fluoride..........	CaF_2	78.07	.039035		
hydroxide........	$Ca(OH)_2$	74.09	.037045	.17[0°]	
oxalate...........	$CaC_2O_4 \cdot H_2O$	146.10	.07305		

[1] Precipitation reagents. [2] Acids and bases. [3] Oxidizing and reducing agents.

M. Methyl orange. P. Phenolphthalein. Temp. C.

* Compiled and arranged by R. M. Meiklejohn.

XVII.—USEFUL DATA OF THE MORE IMPORTANT INORGANIC COMPOUNDS (Continued)

Substance.	Formula.	Molecular or Atomic Weight.	Normal Coefficient 1 c.c. = gm.	Solubility in 100 Gms. Water.	Indicator.
Calcium oxide.......	CaO	56.07	.028035	.13⁰°	
sulphate.........	CaSO₄	136.13	.06807	.179⁰°	
sulphide.........	CaS	72.13	.03607	.15¹⁰°	
Carbon.............	C	12.005	.003001	insol.	
dioxide...........	CO₂	44.005	.022003¹	179.67 cc.⁰°	
dioxide...........	CO₂	44.005	.044005²	"	P.
Chlorine...........	Cl	35.46	.03546	150 cc.⁰°	
Chromic anhydride..	CrO₃	100.00	.033333³	163.40⁰°	
oxide............	Cr₂O₃	152.00	.025333 ³	insol.	
Citric acid.........	H₃C₆H₅O₇	192.09	.06403	133.	
Cobalt.............	Co	58.97	.029485		
Copper.............	Cu	63.57	.031785		
oxide............	CuO	79.57	.07957		
sulphate.........	CuSO₄	159.63	.15963	20⁰°	
sulphate.........	CuSO₄5H₂O	249.71	.24971	31.61⁰°	
sulphide.........	CuS	95.63	.047815	.000033	
Cyanogen..........	CN	26.013	.026013		
Ferric oxide........	Fe₂O₃	159.68	.07984 ³		
phosphate.......	FePO₄·4H₂O	222.94	.07431		
Ferrous oxide.......	FeO	71.84	.07184 ³	insol.	
sulphate.........	FeSO₄	151.90	.15190 ³		
sulphate.........	FeSO₄7H₂O	278.01	.27801 ³	32.8⁰°	
ammon'm sulphate.	FeSO₄(NH₄)₂SO₄6H₂O	392.14	.39214	18⁰°	
Formic acid........	HCOOH	46.02	.04602		
Hydrobromic acid...	HBr	80.93	.08093	221.2⁰°	
Hydrochloric acid...	HCl	36.47	.03647	82.5¹⁰°	
Hydrocyanic acid....	HCN	27.02	.02702		
Hydrofluoric acid....	HF	20.01	.02001	264	
Hydrofluosilicic acid.	H₂SiF₆	144.32	.02405 ²		
Hydroiodic acid.....	HI	127.93	.12793		
Hydrogen peroxide...	H₂O₂	34.016	.017008		
Hydrogen sulphide...	H₂S	34.076	.017038	437 cc.⁰°	
Iodine.............	I	126.92	.12692	.0182¹¹°	
Iron...............	Fe	55.84	.05584		
Lead...............	Pb	207.20	.10360		
carbonate........	PbCO₃	267.205	.13360	.00198	M.
chromate........	PbCrO₄	323.20	.16160	.00002¹⁸°	
oxide...........	PbO	223.20	.11160		
peroxide.........	PbO₂	239.20	.11960		
sulphate.........	PbSO₄	303.26	.15163		
sulphide.........	PbS	239.26	.11963	.0001	
sulphite.........	PbSO₃	287.26	.14363 ¹		
Magnesium.........	Mg	24.32	.01216		
carbonate........	MgCO₃	84.32	.04216	.0106	
chloride.........	MgCl₂	95.24	.04762	52.2⁰°	M.
chloride.........	MgCl₂6H₂O	203.34	.10167	167	M.
oxide............	MgO	40.32	.02016	.00062	M.
sulphate.........	MgSO₄	120.38	.06019	26.9⁰°	M.
sulphate.........	MgSO₄7H₂O	246.49	.123245	76.9⁰°	M.
Malic acid.........	H₂C₄H₄O₅	134.07	.06703		
Manganese.........	Mn	54.93	.027465		
chloride.........	MnCl₂	125.85	.062925	62.16¹⁰°	
peroxide.........	MnO₂	86.93	.043465	insol.	

¹ Precipitation reagents.   ² Acids and bases.   ³ Oxidizing and reducing agents.
M. Methyl orange.   P. Phenolphthalein.   Temp. C.

XVII.—USEFUL DATA OF THE MORE IMPORTANT INORGANIC COMPOUNDS (Continued)

Substance.	Formula.	Molecular or Atomic Weight.	Normal Coefficient 1 c.c. = gm.	Solubility in 100 Gms. Water.	Indicator.
Manganese sulphate.	$MnSO_4$	150.99	.075495	53.2[00]	
Mercuric chloride....	$HgCl_2$	271.52	.13576	5.73[00]	
Mercurous chloride..	$HgCl$	236.06	.23606 [3]		
Mercury..........	Hg	200.6	.2006		
Nickel............	Ni	58 68	.02934		
Nitric acid........	HNO_3	63.02	.06302 [2]		
Nitric acid........	HNO_3	63.02	.021006 [3]		
Nitrogen trioxide....	N_2O_3	76.02	.019005 [3]		
pentoxide........	N_2O_5	108.02	.05401 [2]		
pentoxide........	N_2O_5	108.02	.018033 [3]		
Nitrous acid........	HNO_2	47.02	.04702		
Ntirogen..........	N	14.008	.01401		
Oxalic acid........	$H_2C_2O_4$	90.03	.04501		
Oxalic acid........	$H_2C_2O_42H_2O$	126.06	.06303	4.9[00]	
Phosphoric acid.....	H_3PO_4	98.06	.09806 [2]	v. sol.	M.
Phosphoric acid.....	H_3PO_4	98.06	.04903 [2]	v. sol. 0	P.
Phosphoric acid.....	H_3PO_4	98.06	.03268	v. sol.	
Phosphorous acid (ortho)...........	H_3PO_3	82.06	.02735		
pentoxide........	P_2O_5	142.08	.02368		
Potassium.........	K	39.10	.03910		
bicarbonate.......	$KHCO_3$	100.11	.10011	22.4[00]	M.
bitartrate........	$KHC_4H_4O_6$	188.16	.18816	.37[00]	P.
bromate..........	$KBrO_3$	167.02	.16702		
biodate..........	$KH(IO_3)_2$	389.95	.38995 [2]		
biodate..........	$KH(IO_3)_2$	389.95	.03249 [3]		
bromide..........	KBr	119.02	.11902	53.48[00]	
carbonate........	K_2CO_3	138.20	.06910	89.4[00]	M.
carbonate........	K_2CO_3	138.20	.06910 [2]		
chlorate..........	$KClO_3$	122.56	.020427 [3]	3.3[00]	
Potassium chloride...	KCl	74.56	.07456	28.5[00]	
chromate.........	K_2CrO_4	194.20	.06473 [3]	61.5[00]	
cyanide..........	KCN	65.11	.06511 [1]	v. sol.	
dichromate.......	$K_2Cr_2O_7$	294.20	.07355 [1]		
dichromate.......	$K_2Cr_2O_7$	294.20	.14710 [2]	4.9[00]	P.
dichromate.......	$K_2Cr_2O_7$	294.20	.04903 [3]	4.9[00]	
ferrocyanide......	$K_4Fe(CN)_6$	338.32	.36832 [3]		
ferrocyanide......	$K_4Fe(CN)_63H_2O$	422.37	.422.37[3]	27.8[12o]	
fluoride..........	KF	58.10	.05810		
hydroxide........	KOH	56.11	.05611	107[15o]	
iodate...........	KIO_3	214.02	.03567	4.74[00]	
iodide...........	KI	166.02	.16602	126.1[00]	
manganate.......	K_2MnO_4	197.13	.049283[3]		
nitrate..........	KNO_3	101.11	.033703	13.3[00]	
nitrite...........	KNO_2	85.11	.08511	300[15.5o]	
oxalate..........	$K_2C_2O_4 \cdot H_2O$	184.23	.09211		
oxide............	K_2O	94.20	.04710	v. sol.	
perchlorate.......	$KClO_4$	138.56	.01732 [3]		
permanganate.....	$KMnO_4$	158 03	.031606	2.83[00]	
sulphide..........	K_2S	110.26	.05513	sol.	
sulphocyanate.....	$KCNS$	97.17	.09717	177.2[00]	
tartrate..........	$K_2C_4H_4O_6$	226.25	.11312	sol.	
Silver............	Ag	107.88	.10788		

[1] Precipitation reagents.   [2] Acids and bases.   [3] Oxidizing and reducing agents.

M. Methyl orange.   P. Phenolphthalein.   Temp. C.

XVII.—USEFUL DATA OF THE MORE IMPORTANT INORGANIC COMPOUNDS (Continued)

Substance.	Formula.	Molecular or Atomic Weight.	Normal Coefficient 1 c.c. = gm.	Solubility in 100 Gms. Water.	Indicator.
Silver bromide......	AgBr	187.80	.18780		
chloride.........	AgCl	143.34	.14334		
nitrate..........	AgNO$_3$	169.89	.16989	122$^{0°}$	
Sodium...........	Na	23.00	.02300		
bromide.......	NaBr	102.92	.10292	79.5$^{0°}$	
bicarbonate.......	NaHCO$_3$	84.01	.08401	6.90$^{0°}$	M.
carbonate........	Na$_2$CO$_3$	106.005	.053002	7.1$^{0°}$	M.
chlorate..........	NaClO$_3$	106.46	.01774^3		
chloride..........	NaCl	58.46	.05846	35.7$^{0°}$	
cyanide..........	NaCN	49.01	.04901	sol.	
hydroxide........	NaOH	40.01	.04001	133.3$^{18°}$	
iodide...........	NaI	149.92	.14992 ^{1}	158.7$^{0°}$	
nitrate...........	NaNO$_3$	85.01	.02834	72.9$^{0°}$	
nitrite............	NaNO$_2$	69.01	.06901	83.3$^{20°}$	
oxalate...........	Na$_2$C$_2$O$_4$	134.01	.06700	3.22$^{15.5°}$	
oxide............	Na$_2$O	62.00	.03100	decomp.	
phosphate (mono).	NaH$_2$PO$_4$	120.06 ^{2}	.12006	v. sol.	M.
phosphate (disod)..	Na$_2$HPO$_4$	142.05 ^{2}	.14205	....	P.orM.
phosphate (disod)..	Na$_2$HPO$_4$12H$_2$O	358.24 ^{2}	.35824	6.3$^{0°}$	P.orM.
phosphate (trisod).	Na$_3$PO$_4$	164.04 ^{2}	.16404	....	P
phosphate........	Na$_3$PO$_4$·12H$_2$O	380.24	.38024^3		P
sulphide..........	Na$_2$S	78.06	.03903	15.4$^{10°}$	
sulphite..........	Na$_2$SO$_3$	126.06	.06303		
thiosulphate......	Na$_2$S$_2$O$_3$5H$_2$O	248.20	.24820	74.7$^{0°}$	
thiosulphate......	Na$_2$S$_2$O$_3$·5H$_2$O	248.20	.24820^3		
Stannic oxide.......	SnO$_2$	150.70	.15070		
Stannous chloride....	SnCl$_2$	189.62	.09481	83.9$^{0°}$	
chloride.........	SnCl$_2$2H$_2$O	225.65	.112825	118.7$^{0°}$	
oxide............	SnO	134.70	.06735	insol.	
Sulphur dioxide.....	SO$_2$	64.06	.03203	7979 cc.$^{0°}$	
trioxide..........	SO$_3$	80.06	.04003		
Sulphuric acid......	H$_2$SO$_4$	98.076	.049038		
Sulphurous acid.....	H$_2$SO$_3$	82.076	.04104 ^{2}		
Tartaric acid........	H$_2$C$_4$H$_4$O$_6$	150.07	.075035	115$^{0°}$	
Tin...............	Sn	118.70	.05935		
Titanium chloride...	TiCl$_3$	154.48	.15448 ^{3}		
Zinc..............	Zn	65.37	.037685		
carbonate........	ZnCO$_3$	125.37	.062685	.001$^{15°}$	
chloride..........	ZnCl$_2$	136.29	.068145	209$^{0°}$	
oxide............	ZnO	81.37	.040685	.001	
sulphate..........	ZnSO$_4$	161.43	.080715	43.02$^{0°}$	
sulphate..........	ZnSO$_4$7H$_2$O	287.54	.14377	115.2$^{0°}$	
sulphide..........	ZnS	97.43	.048715	.00069	

[1] Precipitation reagents.   [2] Acids and bases.   [3] Oxidizing and reducing agents.
M. Methyl orange.   P. Phenolphthalein.   Temp. C.

XVIII.—CONVERSION FACTORS *

A'	A	B	B'	A'	A	B	B'
1.7408	Ag	AgBr	0.5744	0.3055	AgNO2	HNO2	3.2729
2.1857	Ag	AgBrO3	0.4575	0.2470	AgNO2	N2O3	4.0487
1.3287	Ag	AgCl	0.7526	0.3709	AgNO3	HNO3	2.6959
1.2411	Ag	AgCN	0.8057	0.3441	AgNO3	NaCl	2.9061
2.1765	Ag	AgI	0.4595	0.7970	AlCl3	Cl	1.2547
1.4265	Ag	AgNO2	0.7010	1.1023	AlCl3	H2SO4	0.9072
1.5748	Ag	AgNO3	0.6350	1.3925	AlF3	CaF2	0.7182
1.0742	Ag	Ag2O	0.9310	0.5303	Al2O3	Al	1.8856
1.2935	Ag	Ag3PO4	0.7731	2.6121	Al2O3	AlCl3	0.3828
1.4034	Ag	Ag4P2O7	0.7126	2.2336	Al2O3	Al2P2O7	0.4477
0.2316	Ag	As	4.3175	2.3902	Al2O3	Al2P2O8	0.4184
0.7408	Ag	Br	1.3498	3.3501	Al2O3	Al2(SO4)3	0.2985
0.7502	Ag	HBr	1.3330	6.5232	Al2O3	Al2(SO4)3 ·18H2O	0.1533
0.3287	Ag	Cl	3.0423				
0.3380	Ag	HCl	2.9582	2.1088	Al2O3	FeO	0.4742
1.1859	Ag	HI	0.8433	1.5625	Al2O3	Fe2O3	0.6400
1.1765	Ag	I	0.8500	2.8792	Al2O3	H2SO4	0.3473
1.1033	Ag	KBr	0.9064		Al2O3	K2Al2(SO4)4	
0.6911	Ag	KCl	1.4469	9.2859		·24H2O	0.1077
1.1361	Ag	KClO3	0.8802	1.6067	Al2O3	Na2Al2O4	0.6224
1.2844	Ag	KClO4	0.7786	8.8738	Al2O3	(NH4)2Al2	
0.6036	Ag	KCN	1.6568			(SO4)4·24H2O	0.1127
1.5390	Ag	KI	0.6498	1.8804	Al2O3	(SO3)3	0.5318
0.9540	Ag	NaBr	1.0482	2.3501	Al2O3	(SO3)3	0.4255
0.5419	Ag	NaCl	1.8453	0.2219	AlPO4	Al	4.5070
1.3897	Ag	NaI	0.7196	0.4185	AlPO4	Al2O3	2.3902
0.4256	AgBr	Br	2.3498	0.5816	AlPO4	P2O5	1.7193
0.6811	AgBr	BrO3	1.4681	0.6224	Al2P2O7	P2O5	1.6067
0.4311	AgBr	HBr	2.3206	0.5816	Al2P2O8	P2O5	1.7193
0.6337	AgBr	KBr	1.5779	2.0453	Al2(SO4)3	BaSO4	0.4889
0.8893	AgBr	KBrO3	1.1244	0.8594	Al2(SO4)3	H2SO4	1.1636
1.2556	AgBr	AgBrO3	0.7964	1.4723	Al2(SO4)3	NaHCO3	0.6792
0.5480	AgBr	NaBr	1.8247	0.7015	Al2(SO4)3	SO3	1.4255
0.7526	AgCl	Ag	1.3287	0.7506	Al4C3	Al	1.3322
1.1852	AgCl	AgNO3	0.8437	1.4154	Al4C3	Al2O3	0.7065
0.2474	AgCl	Cl	4.0423				
0.2545	AgCl	HCl	3.9305	4.3175	As	Ag	0.2316
0.5202	AgCl	KCl	1.9225	1.6403	As	AsO3	0.6096
0.8550	AgCl	KClO3	1.1695	1.8538	As	AsO4	0.5394
0.4078	AgCl	NaCl	2.4519	1.3201	As	As2O3	0.7575
0.3732	AgCl	NH4Cl	2.6793	1.5336	As	As2O5	0.6521
0.4754	AgCl	ZnCl2	2.1034	1.6417	As	As2S3	0.6091
0.4863	AgCN	KCN	2.0564	2.0692	As	As2S5	0.4833
0.2020	AgCN	HCN	4.9551	3.3862	As	2I	0.2953
0.6174	AgI	NH4I	1.6197	2.5391	As	MgNH4AsO4 ·½H2O	0.3938
0.5448	AgI	HI	1.8354				
0.5406	AgI	I	1.8500	2.0715	As	Mg2As2O7	0.4827
0.7450	AgI	IO3	1.3423	2.1875	As	NaH2AsO4	0.4571
0.6428	AgI	I2O3	1.5558	2.4809	As	Na2HAsO4	0.4031
0.7790	AgI	I2O7	1.2836	1.9233	As2O3	MgNH4AsO4 ·½H2O	0.5199
0.7071	AgI	KI	1.4143				
0.9115	AgI	KIO3	1.0971	1.5691	As2O3	Mg2As2O7	0.6373
0.6385	AgI	NaI	1.5661	1.1253	As2O3	Mg2P2O7	0.8886

$$A \times A' = B \qquad B \times B' = A$$

* Compiled by Wilfred W. Scott and R. M. Meiklejohn. See page 698.

XVIII.—CONVERSION FACTORS (Continued)

A'	A	B	B'	A'	A	B	B'
1.1616	As_2O_3	As_2O_5	0.8609	0.8111	BaO_2	Ba	1.2329
1.2435	As_2O_3	As_2S_3	0.8042	0.4912	$BaSiF_6$	Ba	2.0359
1.5674	As_2O_3	As_2S_5	0.6380	0.7448	$BaSiF_6$	$BaCl_2$	1.3427
2.5652	As_2O_3	$4I$	0.3898	0.6271	$BaSiF_6$	BaF_2	1.5948
1.3493	As_2O_5	As_2S_5	0.7411	0.5484	$BaSiF_6$	BaO	1.8235
1.3507	As_2O_5	$Mg_2As_2O_7$	0.7403	0.4076	$BaSiF_6$	F	2.4533
1.6556	As_2O_5	$MgNH_4AsO_4$ $\cdot\frac{1}{2}H_2O$	0.6040	0.4292	$BaSiF_6$	HF	2.3297
1.4264	As_2O_5	NaH_2AsO_4	0.7011	0.5160	$BaSiF_6$	H_2SiF_6	1.9379
1.6177	As_2O_5	Na_2HAsO_4	0.6182	0.3729	$BaSiF_6$	SiF_4	2.6814
2.2081	As_2O_5	$4I$	0.4529	0.2156	$BaSiF_6$	SiO_2	4.6380
				0.8108	BaS	Ba	1.2334
				1.2833	BaS	$BaSO_3$	0.7792
1.5394	Au	$AuCl_3$	0.6496	1.3777	BaS	$BaSO_4$	0.7258
2.0898	Au	$HAuCl_4$ $\cdot 4H_2O$	0.4785	1.4725	BaS	BaS_2O_3	0.6791
1.8172	Au	$KAu(CN)_4$ $\cdot H_2O$	0.5503	0.6318	$BaSO_3$	Ba	1.5828
				1.0736	$BaSO_3$	$BaSO_4$	0.9315
				1.1475	$BaSO_3$	BaS_2O_3	0.8715
3.2018	B	B_2O_3	0.3123	0.4889	$BaSO_4$	$Al_2(SO_4)_3$	2.0453
11.5596	B	KBF_4	0.0865	0.5885	$BaSO_4$	Ba	1.6993
1.2292	B_2O_3	BO_2	0.8135	0.8923	$BaSO_4$	$BaCl_2$	1.1207
1.6877	B_2O_3	BO_3	0.5925	1.0466	$BaSO_4$	$BaCl_2\cdot2H_2O$	0.9554
1.1146	B_2O_3	B_4O_7	0.8972	0.8455	$BaSO_4$	$BaCO_3$	1.1827
1.7743	B_2O_3	H_3BO_3	0.5636	1.0854	$BaSO_4$	$BaCrO_4$	0.9213
3.6103	B_2O_3	KBF_4	0.2770	1.1197	$BaSO_4$	$Ba(NO_3)_2$	0.8931
2.7347	B_2O_3	$Na_2B_4O_7$ $\cdot 10H_2O$	0.3657	0.6571	$BaSO_4$	BaO	1.5219
0.4915	KBF_4	H_3BO_3	2.0347	0.7256	$BaSO_4$	BaO_2	1.3782
0.7575	KBF_4	$Na_2B_4O_7$ $\cdot 10H_2O$	1.3202	0.8599	$BaSO_4$	$Ba_3(PO_4)_2$	1.1629
1.8594	Ba	$Ba(C_2H_3O_2)_2$	0.5378	0.7258	$BaSO_4$	BaS	1.3778
1.8444	Ba	$BaCrO_4$	0.5420	1.0688	$BaSO_4$	BaS_2O_3	0.9356
1.6993	Ba	$BaSO_4$	0.5885	0.5147	$BaSO_4$	$CaSO_3$	1.9429
1.4368	Ba	$BaCO_3$	0.6960	0.5832	$BaSO_4$	$CaSO_4$	1.7147
1.8162	Ba	BaS_2O_3	0.5506	0.3766	$BaSO_4$	FeS	2.6554
0.7002	Ba	S_3	1.4283	0.1457	$BaSO_4$	H_2S	6.8503
0.9335	Ba	S_4	1.0712	0.3517	$BaSO_4$	H_2SO_3	2.8441
1.1207	$BaCl_2$	$BaSO_4$	0.8923	0.4202	$BaSO_4$	H_2SO_4	2.3801
0.4709	$BaCl_2$	H_2SO_4	2.1238	0.7465	$BaSO_4$	K_2SO_4	1.3395
1.2837	$BaCO_3$	$BaCrO_4$	0.7790	1.0164	$BaSO_4$	$K_2Al_2(SO_4)_4$ $\cdot24H_2O$	0.9839
0.7770	$BaCO_3$	BaO	1.2869	0.5834	$BaSO_4$	$KHSO_4$	1.7141
0.0608	$BaCO_3$	C	16.4411	0.1727	$BaSO_4$	MgO	5.7894
0.2229	$BaCO_3$	CO_2	4.4853	1.0560	$BaSO_4$	$MgSO_4\cdot7H_2O$	0.9469
8.1696	$C(in\ BaCO_3)$	H_2SO_4	0.1223	0.6469	$BaSO_4$	$MnSO_4$	1.5458
0.5420	$BaCrO_4$	Ba	1.8444	0.4541	$BaSO_4$	Na_2CO_3	2.2022
0.6053	$BaCrO_4$	BaO	1.6520	0.4458	$BaSO_4$	$NaHSO_3$	2.2432
0.9213	$BaCrO_4$	$BaSO_4$	1.0854	0.2656	$BaSO_4$	Na_2O	3.7651
0.2052	$BaCrO_4$	Cr	4.8725	0.3344	$BaSO_4$	Na_2S	2.9904
0.3947	$BaCrO_4$	CrO_3	2.5337	0.5401	$BaSO_4$	Na_2SO_3	1.8515
0.3000	$BaCrO_4$	Cr_2O_3	3.3338	0.3387	$BaSO_4$	$Na_2S_2O_3$	2.9524
1.4138	$BaCrO_4$	$Cr_2(SO_4)_3$ $\cdot18H_2O$	0.7073	0.6086	$BaSO_4$	Na_2SO_4	1.6431
				1.3804	$BaSO_4$	Na_2SO_4 $\cdot10H_2O$	0.7244
0.7666	$BaCrO_4$	K_2CrO_4	1.3046	0.5661	$BaSO_4$	$(NH_4)_2SO_4$	1.7665
0.5807	$BaCrO_4$	$K_2Cr_2O_7$	1.7224	1.2991	$BaSO_4$	$PbSO_4$	0.7697
0.5723	$Ba(OH)_2$	H_2SO_4	1.7473	0.1373	$BaSO_4$	S	7.2812
				0.2744	$BaSO_4$	SO_2	3.6439

$$A \times A' = B \qquad B \times B' = A$$

XVIII.—CONVERSION FACTORS (Continued)

A'	A	B	B'	A'	A	B	B'
0.3430	BaSO₄	SO₃	2.9155	0.5133	CaF₂	Ca	1.9483
0.4115	BaSO₄	SO₄	2.4301	0.7182	CaF₂	CaO	1.3924
0.4174	BaSO₄	ZnS	2.3957	1.7438	CaF₂	CaSO₄	0.5735
1.2318	BaSO₄	ZnSO₄·7H₂O	0.8118	0.4867	CaF₂	F	2.0545
				0.5126	CaF₂	HF	1.9508
See Gl.	Be			1.8485	CaF₂	H₂SiF₆	0.5410
				1.0760	CaF₂	NaF	0.9294
1.6681	Bi	BiAsO₄	0.5995	0.3459	Ca(HCO₃)₂	CaO	2.8908
1.1154	Bi	Bi₂O₃	0.8965	0.5429	Ca(HCO₃)₂	CO₂	1.8419
0.5994	BiAsO₄	Bi	1.6683	0.4119	CaHPO₄	CaO	2.4276
0.8017	BiOCl	Bi	1.2474	0.9338	CaHPO₄	Ca₂P₂O₇	1.0709
1.8658	BiOCl	Bi(NO₃)₃		0.5219	CaHPO₄	P₂O₅	1.9161
		·5H₂O	0.5359	0.7175	Ca(H₂PO₄)₂	NaHCO₃	1.3938
0.8942	BiOCl	Bi₂O₃	1.1184	0.6067	Ca(H₂PO₄)₂	P₂O₅	1.6483
1.1024	BiOCl	BiONO₃	0.9072	1.0833	CaH₂P₂O₇	Ca(H₂PO₄)₂	0.9231
0.8965	Bi₂O₃	Bi	1.1154	0.2594	CaH₂P₂O₇	CaO	3.8553
1.4955	Bi₂O₃	BiAsO₄	0.6687	0.7773	CaH₂P₂O₇*	NaHCO₃	1.2866
1.2328	Bi₂O₃	BiONO₃	0.8112	0.6573	CaH₂P₂O₇	P₂O₅	1.5214
2.0867	Bi₂O₃	Bi(NO₃)₃		0.2773	Ca(HSO₃)₂	CaO	3.6063
		·5H₂O	0.4792	0.6336	Ca(HSO₃)₂	SO₂	1.5783
1.0691	Bi₂O₃	Bi₂SO₃	0.9354	0.6582	Ca(NO₃)₂	N₂O₅	1.5192
0.8122	Bi₂S₃	Bi	1.2313	0.7146	CaO	Ca	1.3993
0.9059	Bi₂S₃	Bi₂O₃	1.1039	2.8204	CaO	Ca(CH₃CO₂)₂	0.3546
				1.9795	CaO	CaCl₂	0.5052
1.3498	Br	Ag	0.7408	1.7847	CaO	CaCO₃	0.5603
1.7935	Br	AgCl	0.5576	1.3924	CaO	CaF₂	0.7182
2.3498	Br	AgBr	0.4256	2.8908	CaO	Ca(HCO₃)₂	0.3459
1.0126	Br	HBr	0.9875	4.1766	CaO	Ca(H₂PO₄)₂	0.2394
0.1001	Br	O	9.9900	3.8553	CaO	CaH₂P₂O₇	0.2594
0.8433	BrO₃	Ag	1.1858	3.6063	CaO	Ca(HSO₃)₂	0.2773
1.4681	BrO₃	AgBr	0.6811	1.8446	CaO	Ca₃P₂O₈	0.5421
				2.1427	CaO	CaSO₃	0.4667
2.7699	Ca	CaCl₂	0.3610	2.4280	CaO	CaSO₄	0.4119
1.3993	Ca	CaO	0.7146	3.0707	CaO	CaSO₄·2H₂O	0.3257
1.7690	Ca	Cl₂	0.5653	2.1417	CaO	CH₃CO₂H	0.4670
3.3973	Ca	CaSO₄	0.2944	0.7848	CaO	CO₂	1.2743
0.7800	Ca₃(AsO₄)₂	Mg₂As₂O₇	1.2820	1.7494	CaO	H₂SO₄	0.5716
0.3546	Ca(CH₃CO₂)₂	CaO	2.8203	2.0852	CaO	NaCl	0.4796
0.7594	Ca(CH₃CO₂)₂	CH₃CO₂H	1.3169	1.8905	CaO	Na₂CO₃	0.5229
0.6202	Ca(CH₃CO₂)₂	H₂SO₄	1.6124	2.5336	CaO	Na₂SO₄	0.3947
0.9016	CaCl₂	CaCO₃	1.1091	0.5718	CaO	S	1.7489
0.5052	CaCl₂	CaO	1.9795	1.4280	CaO	SO₃	0.7003
1.2265	CaCl₂	CaSO₄	0.8153	0.5421	Ca₃P₂O₈	CaO	1.8446
0.6390	CaCl₂	Cl₂	1.5650	0.7178	Ca₃P₂O₈	Mg₂P₂O₇	1.3932
0.4004	CaCO₃	Ca	2.4975	12.0994	Ca₃P₂O₈	(NH₄)₃PO₄· 12MoO₃	0.0826
1.1091	CaCO₃	CaCl₂	0.9016				
1.6197	CaCO₃	Ca(HCO₃)₂	0.6174	0.4579	Ca₃P₂O₈	P₂O₅	2.1839
0.5603	CaCO₃	CaO	1.7847	3.2362	CaS	BaSO₄	0.3091
0.4397	CaCO₃	CO₂	2.2743	1.9429	CaSO₃	BaSO₄	0.5147
1.3604	CaCO₃	CaSO₄	0.7351	0.4667	CaSO₃	CaO	2.1427
1.7204	CaCO₃	CaSO₄·2H₂O	0.5813	1.7147	CaSO₄	BaSO₄	0.5832
1.3810	CaCO₃	K₂CO₃	0.7241				
1.0593	CaCO₃	Na₂CO₃	0.9441	0.8153	CaSO₄	CaCl₂	1.2265
0.7288	CaCO₃	HCl	1.3720	0.5735	CaSO₄	CaF₂	1.7438
0.7182	CaF₂	AlF₃	1.3925	0.4119	CaSO₄	CaO	2.4280

$$A \times A' = B \qquad B \times B' = A$$

53

* Phenolphthalein indicator.

XVIII.—CONVERSION FACTORS (Continued)

A'	A	B	B'	A'	A	B	B'
0.2791	CaSO$_4$	F	3.5824	0.2607	C$_4$H$_4$O$_6$HK	H$_2$SO$_4$	3.8370
0.2939	CaSO$_4$	HF	3.4021				
0.7205	CaSO$_4$	H$_2$SO$_4$	1.3880	3.0423	Cl	Ag	0.3287
0.5881	CaSO$_4$	SO$_3$	1.7003	4.0423	Cl	AgCl	0.2474
0.3257	CaSO$_4$·2H$_2$O	CaO	3.0707	4.7910	Cl	AgNO$_3$	0.2088
0.8088	CaWO$_4$	WO$_3$	1.2364	1.2547	Cl	AlCl$_3$	0.7970
				3.5726	Cl	BaCrO$_4$	0.2799
1.4296	Cb	Cb$_2$O$_5$	0.6995	2.6645	Cl	CH$_2$ClCO$_2$H	0.3753
				0.5653	Cl	Ca	1.7690
1.6310	Cd	CdCl$_2$	0.6131	1.5650	Cl$_2$	CaCl$_2$	0.6390
2.1033	Cd	Cd(NO$_3$)$_2$	0.4754	4.7454	Cl	CsCl	0.2107
1.1423	Cd	CdO	0.8754	0.5358	Cl	F	1.8663
1.2853	Cd	CdS	0.7780	1.0284	Cl	HCl	0.9724
1.8547	Cd	CdSO$_4$	0.5392	1.3831	Cl	H$_2$SO$_4$	0.7230
0.2853	Cd	S	3.5059	3.5792	Cl	I	0.2794
1.4277	CdO	CdCl$_2$	0.7004	1.1027	Cl	K	0.9069
1.1251	CdO	CdS	0.8888	2.1027	Cl	KCl	0.4756
1.6235	CdO	CdSO$_4$	0.6159	3.4563	Cl	KClO$_3$	0.2893
1.2690	CdS	CdCl$_2$	0.7880	3.9075	Cl	KClO$_4$	0.2559
1.6365	CdS	Cd(NO$_3$)$_2$	0.6110	0.1957	Cl	Li	5.1095
1.4430	CdS	CdSO$_4$	0.6930	0.3429	Cl	Mg	2.9162
0.2359	CdS	H$_2$S	4.2393	1.3429	Cl	MgCl$_2$	0.7447
0.2220	CdS	S	4.5059	2.8672	Cl	MgCl$_2$·6H$_2$O	0.3488
				0.5685	Cl	MgO in MgCl$_2$	1.7590
2.1340	Ce	Ce$_2$(C$_2$O$_4$)$_3$ 3H$_2$O	0.4686	1.2257	Cl	MnO$_2$	0.8158
2.7685	Ce	Ce(NO$_3$)$_4$	0.3612	0.6486	Cl	Na	1.5417
4.0385	Ce	Ce(NO$_3$)$_4$ (NH$_4$NO$_3$)$_2$ ·H$_2$O	0.2476	1.6486	Cl	NaCl	0.6066
				3.0023	Cl	NaClO$_3$	0.3331
1.1711	Ce	Ce$_2$O$_3$	0.8539	3.4535	Cl	NaClO$_4$	0.2896
1.2282	Ce	CeO$_2$	0.8142	0.5088	Cl	NH$_4$	1.9656
3.0548	Ce	Ce(SO$_4$)$_3$	0.3274	1.5088	Cl	NH$_4$Cl	0.6628
2.2542	CeO$_2$	Ce(NO$_3$)$_4$	0.4436	0.2256	Cl	O	4.4325
1.0487	Ce$_2$O$_3$	CeO$_2$	0.9536	2.4098	Cl	Rb	0.4150
2.3640	Ce$_2$O$_3$	Ce(NO$_3$)$_4$	0.4230	3.4098	Cl	RbCl	0.2933
				0.8368	Cl	Sn in SnCl$_4$	1.1949
0.3753	CH$_2$ClCO$_2$H	Cl	2.6645	1.8369	Cl	SnCl$_4$	0.5444
0.5446	CH$_3$CHOH CO$_2$H	H$_2$SO$_4$	1.8366	3.9216	Cl	PbCl$_2$	0.2550
				4.5572	Cl	PbCrO$_4$	0.2194
1.3169	CH$_3$CO$_2$H	Ca(CH$_3$CO$_2$)$_2$	0.7594	2.6737	Cl	SnCl$_2$	0.3740
1.3663	CH$_3$CO$_2$H	CH$_3$CO$_2$Na	0.7319	1.9217	Cl	ZnCl$_2$	0.5204
0.8499	CH$_3$CO$_2$H	(CH$_3$CO)$_2$O	1.1766	0.8934	ClO$_3$	KCl	1.1194
0.8167	CH$_3$CO$_2$H	H$_2$SO$_4$	1.2244	1.7175	ClO$_3$	AgCl	0.5823
0.8828	CH$_3$CO$_2$H	Na$_2$CO$_3$	1.1327	0.7005	ClO$_3$	NaCl	1.4276
0.5164	CH$_3$CO$_2$H	Na$_2$O	1.9368	1.4412	ClO$_4$	AgCl	0.6939
1.7255	CH$_3$CO$_2$H	Pb	0.5796	0.7496	ClO$_4$	KCl	1.3339
0.5979	CH$_3$CO$_2$Na	H$_2$SO$_4$	1.6729	0.5878	ClO$_4$	NaCl	1.7013
0.3779	CH$_3$CO$_2$Na	Na$_2$O	2.6463				
0.8660	CH$_3$CO$_2$Na	Na$_2$SO$_4$	1.1547	4.1472	CN	Ag	0.2411
0.9609	(CH$_3$CO)$_2$O	H$_2$SO$_4$	1.0407	5.1472	CN	AgCN	0.1943
1.0897	C$_2$H$_2$O$_4$	H$_2$SO$_4$	0.9177	2.8577	CNS	AgCNS	0.3499
0.7782	C$_2$H$_2$O$_4$·2H$_2$O	H$_2$SO$_4$	1.2851	0.7240	CNS	(CN)$_2$S	1.3813
0.6537	C$_4$H$_6$O$_6$	H$_2$SO$_4$	1.5298	4.9361	Co	Co(NO$_3$)$_2$· 6H$_2$O	0.2026
1.1197	C$_4$H$_6$O$_6$	NaHCO$_3$	0.8931	7.6706	Co	Co(NO$_2$)$_3$ (KNO$_2$)$_3$	0.1304

$$A \times A' = B \qquad B \times B' = A$$

XVIII.—CONVERSION FACTORS (Continued)

A'	A	B	B'
1.2714	Co	CoO	0.7866
1.3618	Co	Co₃O₄	0.7343
2.6290	Co	CoSO₄	0.3804
4.7677	Co	CoSO₄·7H₂O	0.2097
1.0711	CoO	Co₃O₄	0.9336
6.0332	CoO	Co(NO₂)₃	
		(KNO₂)₃	0.1657
2.0679	CoO	CoSO₄	0.4836
4.4853	CO₂	BaCO₃	0.2229
3.4853	CO₂	BaO	0.2869
2.9473	CO₂	Ba(HCO₃)₂	0.3393
0.2727	CO₂	C	3.6656
2.2743	CO₂	CaCO₃	0.4397
1.8419	CO₂	Ca(HCO₃)₂	0.5429
1.2743	CO₂	CaO	0.7847
0.6364	CO₂	CO	1.5713
3.9178	CO₂	CdCO₃	0.2552
1.3636	CO₂	CO₃	0.7333
7.3997	CO₂	Cs₂CO₃	0.1351
2.6327	CO₂	FeCO₃	0.3798
2.0220	CO₂	Fe(HCO₃)₂	0.4951
2.2287	CO₂	H₂SO₄	0.4487
3.1409	CO₂	K₂CO₃	0.3184
2.1409	CO₂	K₂O	0.4671
1.6790	CO₂	Li₂CO₃	0.5956
1.5442	CO₂	LiHCO₃	0.6476
0.6790	CO₂	Li₂O	1.4727
1.9164	CO₂	MgCO₃	0.5218
1.6629	CO₂	Mg(HCO₃)₂	0.6013
0.9164	CO₂	MgO	1.0913
2.6121	CO₂	MnCO₃	0.3828
2.0108	CO₂	Mn(HCO₃)₂	0.4973
1.6119	CO₂	MnO	0.6203
2.4089	CO₂	Na₂CO₃	0.4151
1.9093	CO₂	NaHCO₃	0.5238
3.2283	CO₂	Na₂SO₄	0.3098
1.4091	CO₂	Na₂O	0.7097
2.1836	CO₂	(NH₄)₂CO₃	0.4579
6.0720	CO₂	PbCO₃	0.1647
5.2472	CO₂	Rb₂CO₃	0.1906
3.3283	CO₂	RbHCO₃	0.3006
4.2472	CO₂	Rb₂O	0.2354
3.3551	CO₂	SrCO₃	0.2981
2.3823	CO₂	Sr(HCO₃)₂	0.4198
2.3550	CO₂	SrO	0.4246
2.8493	CO₂	ZnCO₃	0.3510
4.8725	Cr	BaCrO₄	0.2052
1.4615	Cr	Cr₂O₃	0.6842
6.2154	Cr	PbCrO₄	0.1609
3.7346	Cr	K₂CrO₄	0.2678
2.8288	Cr	K₂Cr₂O₇	0.3535
3.0163	CrO₂	BaCrO₄	0.3315
1.1905	CrO₂	CrO₃	0.8400

A'	A	B	B'
3.8476	CrO₂	PbCrO₄	0.2599
2.5337	CrO₃	BaCrO₄	0.3947
1.9420	CrO₃	K₂CrO₄	0.5149
1.4710	CrO₃	K₂Cr₂O₇	0.6798
3.2320	CrO₃	PbCrO₄	0.3094
4.2526	Cr₂O₃	PbCrO₄	0.2351
1.3158	Cr₂O₃	CrO₃	0.7600
0.2670	Cs	Cl	3.7453
1.2670	Cs	CsCl	0.7893
1.0602	Cs	Cs₂O	0.9432
1.2258	Cs	Cs₂CO₃	0.8157
1.3617	Cs	Cs₂SO₄	0.7344
0.3945	Cs₂PtCl₆	Cs	2.5359
0.4182	Cs₂PtCl₆	Cs₂O	2.3918
1.1950	Cs₂O	CsCl	0.8368
1.2843	Cs₂O	Cs₂SO₄	0.7786
0.4996	Cs₂PtCl₆	CsCl	2.0015
0.4834	Cs₂PtCl₆	Cs₂CO₃	2.0686
0.9305	Cs₂SO₄	CsCl	1.0747
0.9003	Cs₂SO₄	Cs₂CO₃	1.1107
1.2517	Cu	CuO	0.7989
1.1258	Cu	Cu₂O	0.8882
1.2522	Cu	Cu₂S	0.7986
2.5112	Cu	CuSO₄	0.3982
3.9283	Cu	CuSO₄·5H₂O	0.2546
0.5226	CuCNS	Cu	1.9137
0.6541	CuCNS	CuO	1.5288
2.0062	CuO	CuSO₄	0.4985
1.2327	CuO	H₂SO₄	0.8112
3.1383	CuO	CuSO₄·5H₂O	0.3186
0.3353	CuS	S	2.9828
0.9996	Cu₂S	CuO	1.0004
0.8991	Cu₂S	Cu₂O	1.1122
3.1371	Cu₂S	CuSO₄·5H₂O.	0.3188
0.8746	Er₂O₃	Er	1.1433
2.4533	F	BaSiF₆	0.4076
2.0545	F	CaF₂	0.4868
3.5827	F	CaSO₄	0.2791
1.0531	F	HF	0.9496
1.2660	F	H₂SiF₆	0.7899
1.9342	F	K₂SiF₆	0.5170
2.2105	F	NaF	0.4524
1.6518	F	Na₂SiF₆	0.6054
2.2701	Fe	FeCl₂	0.4405
2.9051	Fe	FeCl₃	0.3442
4.8410	Fe	FeCl₃·6H₂O.	0.2066
3.1851	Fe	Fe(HCO₃)₂.	0.3140
1.2865	Fe	FeO	0.7773
1.4298	Fe	Fe₂O₃	0.6994
1.3820	Fe	Fe₃O₄	0.7236

$$A \times A' = B \qquad A \times A' = A$$

XVIII.—CONVERSION FACTORS (Continued)

A'	A	B	B'	A'	A	B	B'
2.7020	Fe	FePO₄	0.3701	0.6938	GeO₂	Ge	1.4414
1.5743	Fe	FeS	0.6352	0.2739	K₂GeF₆	Ge	3.6510
2.1483	Fe	FeS₂	0.4655				
2.7205	Fe	FeSO₄	0.3676	0.3626	GlO	Gl	2.7582
4.9789	Fe	FeSO₄·7H₂O	0.2008	3.1881	GlO	GlCl₂	0.3137
3.5807	Fe	Fe₂(SO₄)₃	0.2793	7.0607	GlO	GlSO₄·4H₂O	0.1416
7.0225	Fe	FeSO₄(NH₄)₂SO₄6H₂O	0.1424	8.9363	H	H₂O	0.1119
0.5643	Fe(titr. equiv.)	HNO₃	1.7721	0.5636	H₃BO₃	B₂O₃	1.7743
				1.3330	HBr	Ag	0.7502
0.7820	Fe	Na₂Cr₂O₇	1 2788	2.3206	HBr	AgBr	0.4309
0.3798	FeCO₃	CO₂	2.6327	3.9305	HCl	AgCl	0.2544
0.4948	Fe(HCO₃)₂	CO₂	2.0220	1.3720	HCl	CaCO₃	0.7288
0.4742	FeO	Al₂O₃	2.1088	0.9724	HCl	Cl	1.0284
1.6125	FeO	FeCO₃	0.6202	1.2893	HCl	HNO₂	0.7756
2.4757	FeO	Fe(HCO₃)₂	0.4039	1.7280	HCl	HNO₃	0.5787
1.1114	FeO	Fe₂O₃	0.8998	1.1255	HCl	H₂SO₃	0.8885
2.1002	FeO	FePO₄	0.4761	1.3448	HCl	H₂SO₄	0.7436
1.2237	FeO	FeS	0.8172	2.0445	HCl	KCl	0.4891
1.6698	FeO	FeS₂	0.5989	1.2915	HCl	K₂O	0.7743
2.1146	FeO	FeSO₄	0.4729	1.6030	HCl	NaCl	0.6239
3.8700	FeO	FeSO₄·7H₂O	0.2584	1.4533	HCl	Na₂CO₃	0.6881
1.3653	FeO	H₂SO₄	0.7324	1.4669	HCl	NH₄Cl	0.6817
1.1146	FeO	SO₃	0.8972	1.7861	HCl	SnCl₄	0.5599
0.6400	Fe₂O₃	Al₂O₃	1.5625	2.8089	HCNS	AgCNS	0.3560
2.0318	Fe₂O₃	FeCl₃	0.4922	2.0589	HCNS	CuCNS	0.4857
1.4509	Fe₂O₃	FeCO₃	0.6892	3.9510	HCNS	BaSO₄	0.2531
2.2277	Fe₂O₃	Fe(HCO₃)₂	0.4489				
0.9666	Fe₂O₃	Fe₃O₄	1.0346	1.0658	HCO₂H	H₂SO₄	0.9383
1.8898	Fe₂O₃	FePO₄	0.5292				
1.1011	Fe₂O₃	FeS	0.9082	2.3297	HF	BaSiF₆	0.4292
1.5028	Fe₂O₃	FeS₂	0.6655	1.9508	HF	CaF₂	0.5126
1.9027	Fe₂O₃	FeSO₄	0.5256	3.4021	HF	CaSO₄	0.2939
3.4822	Fe₂O₃	FeSO₄·7H₂O	0.2872	0.9496	HF	F	1.0531
4.9118	Fe₂O₃	FeSO₄(NH₄)₂SO₄·6H₂O	0.2036	2.4510	HF	H₂SO₄	0.4080
				1.8368	HF	K₂SiF₆	0.5444
2.5041	Fe₂O₃	Fe₂(SO₄)₃	0.3993	3.6065	HF(2HF)	H₂SiF₆	0.2773
1.8428	Fe₂O₃	H₂SO₄	0.5427	1.2022	HF(6HF)	H₂SiF₆	0.8318
1.5041	Fe₂O₃	SO₃	0.6648				
0.4708	FePO₄	P₂O₅	2.1239	1.1768	Hg	HgCl	0.8498
2.6554	FeS	BaSO₄	0.3766	1.3535	Hg	HgCl₂	0.7388
0.3877	FeS	H₂S	2.5795	1.0798	Hg	HgO	0.9261
1.1155	FeS	H₂SO₄	0.8963	1.1599	Hg	HgS	0.8622
0.3648	FeS	S	2.7417	1.1125	HgCl	HgNO₃	0.8989
0.6654	FeS₂	Fe₂O₃	1.5025	0.9856	HgCl	HgS	1.0146
0.5346	FeS₂	S	1.8709	0.4017	HgCl	SnCl₂	2.4898
0.6457	FeSO₄	H₂SO₄	1.5487	0.5519	HgCl	SnCl₄	1.8121
0.5271	FeSO₄	SO₃	1.8973	0.8694	HgCl₂	HgCl	1.1502
0.3528	FeSO₄·7H₂O	H₂SO₄	2.8345	0.8569	HgCl₂	HgS	1.1670
0.7358	Fe₂(SO₄)₃	H₂SO₄	1.3590	1.0898	HgO	HgCl	0.9176
0.1114	FeSO₄(NH₄)₂SO₄·6H₂O	Na₂Cr₂O₇	8.9802	1.1316	Hg₂O	HgCl	0.8837
				1.0860	HgS	Hg(CN)₂	0.9208
				1.1287	HgS	HgNO₃	0.8860
0.7450	Ga₂O₃	Ga	1.3424	1.3952	HgS	Hg(NO₃)₂	0.7167
0.5931	Ga₂S₃	Ga	1.6860	0.8966	HgS	Hg₂O	1.1153

$$A \times A' = B \qquad B \times B' = A$$

XVIII.—CONVERSION FACTORS (Continued)

A'	A	B	B'	A'	A	B	B'
0.9310	HgS	HgO	1.0741	2.0391	H_2SiF_6	H_2SO_4	0.4904
1.2751	HgS	$HgSO_4$	0.7843	1.5279	H_2SiF_6	K_2SiF_6	0.6545
				0.7227	H_2SiF_6	SiF_4	1.3837
0.8433	HI	Ag	1.1859	0.9860	H_2SiF_6	SiF_6	1.0141
1.8354	HI	AgI	0.5448	0.7699	H_2SiO_3	SiO_2	1.2988
0.4170	HI	Pd	2.3979				
1.4092	HI	PdI_2	0.7097	2.8441	H_2SO_3	$BaSO_4$	0.3517
2.5868	HI	TlI	0.3866	0.8885	H_2SO_3	HCl	1.1255
				1.1949	H_2SO_3	H_2SO_4	0.8369
3.2729	HNO_2	$AgNO_2$	0.3055	0.7805	H_2SO_3	SO_2	1.2812
0.7756	HNO_2	HCl	1.2893				
1.0431	HNO_2	H_2SO_4	0.9587	0.9072	H_2SO_4	$AlCl_3$	1.1023
0.6382	HNO_2	NO	1.5667	0.3473	H_2SO_4	Al_2O_3	2.8789
0.3623	HNO_2	NH_3	2.7605	1.1636	H_2SO_4	$Al_2(SO_4)_3$	0.8594
0.5627	HNO_3	Cl	1.7772	1.7473	H_2SO_4	$Ba(OH)_2$	0.5723
0.5787	HNO_3	HCl	1.7280	2.3800	H_2SO_4	$BaSO_4$	0.4202
0.7782	HNO_3	H_2SO_4	1.2850	0.1223	H_2SO_4	$C(in BaCO_3)$	8.1696
1.6045	HNO_3	KNO_3	0.6233	1.6124	H_2SO_4	$Ca(CH_3CO_2)_2$	0.6203
0.2223	HNO_3	N	4.4986	0.5716	H_2SO_4	CaO	1.7494
1.3490	HNO_3	$NaNO_3$	0.7413	1.3878	H_2SO_4	$CaSO_4$	0.7205
0.2702	HNO_3	NH_3	3.7000	1.8366	H_2SO_4	CH_3CHOH CO_2H	0.5446
0.8489	HNO_3	NH_4Cl	1.1780				
3.5232	HNO_3	$(NH_4)_2PtCl_6$	0.2838	1.2244	H_2SO_4	CH_3CO_2H	0.8169
0.4762	HNO_3	NO	2.0999	1.6729	H_2SO_4	CH_3CO_2Na	0.5979
0.6032	HNO_3	N_2O_3	1.6579	1.0407	H_2SO_4	$(CH_3CO_2)_2O$	0.9609
0.7301	HNO_3	N_2O_4	1.3697	0.9177	H_2SO_4	$C_2H_2O_4$	1.0894
0.8571	HNO_3	N_2O_5	1.1668	1.2851	H_2SO_4	$C_2H_2O_4 \cdot 2H_2O$	0.7782
1.5488	HNO_3	Pt	0.6457	1.5298	H_2SO_4	$C_4H_6O_6$	0.6537
				0.7230	H_2SO_4	Cl_2	1.3831
0.8163	H_3PO_4	HPO_3	1.2251	0.4487	H_2SO_4	CO_2	2.2287
0.9081	H_3PO_4	$H_4P_2O_7$	1.1012	1.6275	H_2SO_4	$CuSO_4$	0.6144
1.0002	H_3PO_4	H_2SO_4	0.9998	0.5694	H_2SO_4	Fe	1.7564
1.1356	H_3PO_4	$Mg_2P_2O_7$	0.8806	0.7324	H_2SO_4	FeO	1.3653
0.3165	H_3PO_4	P	3.1593	0.5427	H_2SO_4	Fe_2O_3	1.8428
0.7244	H_3PO_4	P_2O_5	1.3804	0.8963	H_2SO_4	FeS	1.1155
0.3767	H_2PtCl_6 $\cdot 6H_2O$	Pt	2.6541	1.5487	H_2SO_4	$FeSO_4$	0.6457
				2.8345	H_2SO_4	$FeSO_4 \cdot 7H_2O$	0.3528
2.4074	H_2S	As_2S_3	0.4154	1.3590	H_2SO_4	$Fe_2(SO_4)_3$	0.7358
1.3495	H_2S	As_2O_5	0.7410	0.9651	H_2SO_4	H_3AsO_4	1.0362
6.8506	H_2S	$BaSO_4$	0.1460	0.7436	H_2SO_4	HCl	1.3448
4.2393	H_2S	CdS	0.2360	0.9177	H_2SO_4	$H_2C_2O_4$	1.0894
2.5795	H_2S	FeS	0.3877	0.9383	H_2SO_4	HCO_2H	1.0658
2.8782	H_2S*	H_2SO_4	0.3474	0.4080	H_2SO_4	HF	2.4510
0.9408	H_2S	S	1.0629	0.9587	H_2SO_4	HNO_2	1.0436
1.8799	H_2S	SO_2	0.5319	1.2850	H_2SO_4	HNO_3	0.7782
2.3495	H_2S	SO_3	0.4258	0.9998	H_2SO_4*	H_3PO_4	1.0002
				0.3474	H_2SO_4*	H_2S	2.8782
0.6129	H_2SeO_3	Se	1.6315	0.8369	H_2SO_4*	H_2SO_3	1.1949
				0.4904	H_2SO_4	H_2SiF_6	2.0388
1.9379	H_2SiF_6	$BaSiF_6$	0.5160	2.4293	H_2SO_4	$K_2Al_2(SO_4)_4$ $\cdot 24H_2O$	0.4116
0.5410	H_2SiF_6	CaF_2	1.8485	3.8370	H_2SO_4	$KHC_4H_4O_6$	0.2607
0.7899	H_2SiF_6	F	1.2660	2.0617	H_2SO_4	KNO_3	0.4850
0.2773	H_2SiF_6	2HF	3.6065	0.9604	H_2SO_4	K_2O	1.0413
0.8318	H_2SiF_6	6HF	1.2022	1.1442	H_2SO_4	KOH	0.8740

$$A \times A' = B \qquad B \times B' = A$$

* Phenolphthalein indicator.

XVIII.—CONVERSION FACTORS (Continued)

A'	A	B	B'	A'	A	B	B'
1.1241	H_2SO_4......	K_2SiF_6.....	0.8896	2.6074	I..........	Tl........	0.6222
1.7767	H_2SO_4......	K_2SO_4......	0.5628				
0.4111	H_2SO_4......	MgO.......	2.4325	1.2090	In.........	In_2O_3......	0.8271
1.2274	H_2SO_4......	$MgSO_4$.....	0.8147	1.4187	In.........	In_2S_3......	0.7047
1.1920	H_2SO_4......	NaCl......	0.8389				
1.0808	H_2SO_4†.....	Na_2CO_3.....	0.9252	1.1243	Ir.........	Ir_2O_3......	0.8895
1.3661	H_2SO_4......	$Na_2C_2O_4$....	0.7320				
2.6711	H_2SO_4......	$Na_2Cr_2O_7$...	0.3744	3.5437	K.	$KClO_4$....	0.2822
1.7132	H_2SO_4†.....	$NaHCO_3$....	0.5837	1.2046	K.	K_2O....	0.8302
2.4583	H_2SO_4*.....	NaH_2PO_4....	0.4085	6.2169	K.	K_2PtCl_6....	0.1609
2.8964	H_2SO_4†.....	Na_2HPO_4....	0.3453	2.2285	K.	K_2SO_4....	0.4487
2.1222	H_2SO_4......	$NaHSO_3$.....	0.4712	0.0571	$K_2Al_2(SO_4)_4$ $\cdot 24H_2O$	Al..........	17.51
1.7334	H_2SO_4......	$NaNO_3$......	0.5769				
0.6321	H_2SO_4......	Na_2O.......	1.5820	0.1077	ditto	Al_2O_3......	9.2859
0.8158	H_2SO_4......	NaOH......	1.2258	0.4556	ditto	H_2O......	2.1949
1.1151	H_2SO_4......	Na_3PO_4.....	0.8968	0.4116	ditto *	H_2SO_4.....	2.4293
0.7959	H_2SO_4......	Na_2S.......	1.2564	0.4134	ditto (total)	H_2SO_4(+Al +K)	2.4190
0.9598	H_2SO_4......	Na_2SiF_6.....	1.0418	0.1352	ditto	S..........	7.4004
2.5706	H_2SO_4......	$Na_2S_2O_3$....	0.3890	0.3374	ditto	SO_3..........	2.9635
1.4485	H_2SO_4......	Na_2SO_4.....	0.6904	0.0824	ditto	K..........	12.136
0.3473	H_2SO_4......	NH_3.......	2.8792	0.0993	ditto	K_2O..........	10.075
1.0909	H_2SO_4......	NH_4Cl.....	0.9167	0.9756	$K_2Al_2(SO_4)_4$..	$NaHCO_3$....	1.0251
0.5310	H_2SO_4......	$(NH_4)_2O$.....	1.8850	0.7575	KBF_4.......	$Na_2B_4O_7$ $10H_2O$	1.3202
0.6948	H_2SO_4......	$(NH_4)_2S$.....	1.4392				
1.3473	H_2SO_4......	$(NH_4)_2SO_4$...	0.7422	0.9064	KBr	Ag......	1.1033
2.3268	H_2SO_4......	$(NH_4)_2S_2O_8$..	0.4298	1.5779	KBr	AgBr.....	0.6338
0.7750	H_2SO_4......	N_2O_3.......	1.2903	0.6715	KBr	Br......	1.4892
0.9382	H_2SO_4......	N_2O_4.......	1.0659	0.3285	KBr	K......	3.0440
1.1013	H_2SO_4......	N_2O_5.......	0.9080	1.4469	KCl......	Ag......	0.6911
0.0275	H_2SO_4‡ (Y ppt.)	P..........	36.336	1.9225	KCl......	AgCl.....	0.5202
0.4829	H_2SO_4......	P..........	1.0709	0.4756	KCl......	Cl......	2.1027
0.0629	H_2SO_4‡ (Y ppt.)	P_2O_5..........	15.878	0.4891	KCl......	HCl.....	2.0445
0.3267	H_2SO_4......	S..........	3.0590	1.6438	KCl......	$KClO_3$....	0.6084
0.6532	H_2SO_4 †....	SO_2.......	1.5309	1.8584	KCl......	$KClO_4$....	0.5381
1.3064	H_2SO_4*.....	SO_2.......	0.7655	2.5233	KCl......	$KHC_4H_4O_6$...	0.3963
0.8163	H_2SO_4......	SO_3.......	1.2250	3.2602	KCl......	K_2PtCl_6.....	0.3069
1.0912	H_2SO_4......	$S_2O_5Cl_2$.....	0.9164	1.1686	KCl......	K_2SO_4.....	0.8557
1.1881	H_2SO_4......	SO_3HCl.....	0.8417	1.3090	KCl......	Pt........	0.7639
0.4084	H_2SO_4......	TiO_2........	2.4488	1.1696	$KClO_3$...	AgCl......	0.8550
0.6665	H_2SO_4......	Zn........	1.5004	0.2893	$KClO_3$...	Cl......	3.4563
0.8296	H_2SO_4......	ZnO......	1.2054	0.3843	$KClO_3$...	K_2O......	2.6021
1.6459	H_2SO_4......	$ZnSO_4$......	0.6076	0.3399	$KClO_4$...	K_2O......	2.9418
				1.6568	KCN	Ag......	0.6036
				2.0564	KCN	AgCN.....	0.4863
0.8500	I.........	Ag.........	1.1765	2.2014	KCN	AgCl.....	0.4543
1.8500	I.........	AgI.......	0.5405	0.3184	K_2CO_3...	CO_2......	3.1409
0.2794	I.........	Cl........	3.5792	0.8120	K_2CO_3...	KOH......	1.2315
1.0079	I.........	HI........	0.9921	1.2610	K_2CO_3...	K_2SO_4.....	0.7930
1.3081	I.........	KI........	0.7645	1.3046	K_2CrO_4...	$BaCrO_4$....	0.7666
1.6863	I.........	KIO_3......	0.5930	0.5149	K_2CrO_4...	CrO_3.....	1.9420
1.5594	I.........	$NaIO_3$......	0.6413	1.7224	$K_2Cr_2O_7$...	$BaCrO_4$....	0.5807
0.4967	I.........	Na_2SO_3....	2.0135	0.6800	$K_2Cr_2O_7$...	CrO_3.....	1.4706
0.4203	I.........	Pd........	2.3790	1.1388	$K_2Cr_2O_7$...	Fe........	0.8782
1.4204	I.........	PdI_2......	0.7040	0.3200	$K_2Cr_2O_7$...	K_2O........	3.1231

$$A \times A' = B \qquad B \times B' = A$$

* Phenolphthalein   † Methyl orange   ‡ Titration of yellow ppt.

XVIII.—CONVERSION FACTORS (Continued)

A'	A	B	B'	A'	A	B'	B'
0.4395	$KHCO_3$	CO_2	2.2752	0.8896	K_2SiF_6	H_2SO_4	1.1241
0.8703	$KHCO_3$	K_2SO_4	1.1490	0.5270	K_2SiF_6	KF	1.8976
1.7141	$KHSO_4$	$BaSO_4$	0.5834	0.4730	K_2SiF_6	SiF_4	2.1141
0.6399	$KHSO_4$	K_2SO_4	1.5628	0.2735	K_2SiF_6	SiO_2	3.6567
0.7645	KI	I	1.3081	0.3903	K_2SiO_3	SiO_2	2.5622
0.2355	KI	K	4.2460	0.5628	K_2SO_4	H_2SO_4	1.7767
0.4995	$KMnO_4$	Mn_2O_3	2.0022	0.4487	K_2SO_4	K	2.2285
0.4826	$KMnO_4$	Mn_3O_4	2.0721	0.8557	K_2SO_4	KCl	1.1686
0.8981	$KMnO_4$	$NaMnO_4$	1.1134	0.7930	K_2SO_4	K_2CO_3	1.2610
0.4850	KNO_3	H_2SO_4	2.0617	0.5405	K_2SO_4	K_2O	1.8500
2.4041	KNO_3	K_2PtCl_6	0.4159	2.7897	K_2SO_4	K_2PtCl_6	0.3585
0.8408	KNO_3	$NaNO_3$	1.1894	0.6327	K_2SO_4	K_2S	1.5804
0.2968	KNO_3	NO	3.3694	0.4595	K_2SO_4	SO_3	2.1765
0.5342	KNO_3	N_2O_5	1.8721	0.6059	K_3AsO_4	$Mg_2As_2O_7$	1.6503
0.8741	KOH	H_2SO_4	1.1440				
1.2315	KOH	K_2CO_3	0.8120	1.1727	La	La_2O_3	0.8528
0.8394	KOH	K_2O	1.1913				
0.7131	KOH	$NaOH$	1.4024	6.1096	Li	$LiCl$	0.1637
0.4004	K_2MnO_4	Mn_2O_3	2.4975	5.3231	Li	Li_2CO_3	0.1879
0.7117	K_2HAsO_4	$Mg_2As_2O_7$	1.4050	2.1527	Li	Li_2O	0.4645
0.7529	K_2O	Cl	1.3282	5.5648	Li	Li_3PO_4	0.1798
0.4671	K_2O	CO_2	2.1409	1.2965	$LiCl$	Li_2SO_4	0.7713
0.7743	K_2O	HCl	1.2915	0.3523	$LiCl$	Li_2O	2.8381
1.0413	K_2O	H_2SO_4	0.9604	0.5956	Li_2CO_3	CO_2	1.6790
0.8302	K_2O	K	1.2046	2.4727	Li_2O	Li_2CO_3	0.4044
2.5270	K_2O	KBr	0.3957	3.6794	Li_2O	Li_2SO_4	0.2718
1.5830	K_2O	KCl	0.6317	2.5850	Li_2O	Li_3PO_4	0.3870
2.1255	K_2O	$KHCO_3$	0.4705	0.9569	Li_3PO_4	Li_2CO_3	1.0451
1.4671	K_2O	K_2CO_3	0.6816	1.7595	Li_3PO_4	$LiHCO_3$	0.5682
3.9949	K_2O	$KHC_4H_4O_6$	0.2504	1.4234	Li_3PO_4	Li_2SO_4	0.7023
3.1231	K_2O	$K_2Cr_2O_7$	0.3200	0.1262	Li_2SO_4	Li	7.9207
3.5248	K_2O	KI	0.2837	0.7712	Li_2SO_4	$LiCl$	1.2966
1.1913	K_2O	KOH	0.8394	0.7282	Li_2SO_4	SO_3	1.3732
2.1466	K_2O	KNO_3	0.4658				
2.0616	K_2O	K_2CrO_4	0.4851	2.9162	Mg	Cl	0.3429
5.1609	K_2O	K_2PtCl_6	0.1938	1.6579	Mg	MgO	0.6032
1.8500	K_2O	K_2SO_4	0.5406	4.5789	Mg	$Mg_2P_2O_7$	0.2184
0.6582	K_2O	Na_2O	1.5194	4.9498	Mg	$MgSO_4$	0.2020
0.8500	K_2O	SO_3	1.1765	0.7447	$MgCl_2$	Cl	1.3429
0.1609	K_2PtCl_6	K	6.2169	0.3488	$MgCl_2 \cdot 6H_2O$	Cl	2.8672
0.3068	K_2PtCl_6	KCl	3.2602	0.5218	$MgCO_3$	CO_2	1.9164
0.2844	K_2PtCl_6	K_2CO_3	3.5177	0.6013	$Mg(HCO_3)_2$	CO_2	1.6629
0.4159	K_2PtCl_6	KNO_3	2.4041	5.7894	MgO	$BaSO_4$	0.1727
0.1938	K_2PtCl_6	K_2O	5.1609	1.7590	MgO	Cl	0.5685
0.3585	K_2PtCl_6	K_2SO_4	2.7897	1.0913	MgO	CO_2	0.9164
1.9521	K_2PtCl_6	$K_2Al_2(SO_4)_4 \cdot 24H_2O$	0.5120	2.4325	MgO	H_2SO_4	0.4111
				2.3621	MgO	$MgCl_2$	0.4234
2.0545	K_2PtCl_6	$K_2Cr_2(SO_4)_4 \cdot 24H_2O$	0.4964	2.0914	MgO	$MgCO_3$	0.4782
				3.6294	MgO	$Mg(HCO_3)_2$	0.2755
0.4013	K_2PtCl_6	Pt	2.4906	2.7619	MgO	$Mg_2P_2O_7$	0.3621
0.6931	K_2PtCl_6	$PtCl_4$	1.4427	2.9859	MgO	$MgSO_4$	0.3349
0.8785	K_2PtCl_6	$PtCl_4 \cdot 5H_2O$	1.1383	3.5236	MgO	Na_2SO_4	0.2838
0.5170	K_2SiF_6	F	1.9342	0.7951	MgO	S	1.2576
0.5444	K_2SiF_6	HF	1.8368	1.9859	MgO	SO_3	0.5036
0.6545	K_2SiF_6	H_2SiF_6	1.5279	1.3932	$Mg_2P_2O_7$	$Ca_3(PO_4)_2$	0.7178

$$A \times A' = B \qquad B \times B' = A$$

XVIII.—CONVERSION FACTORS (Continued)

A'	A	B	B'	A'	A	B	B'
0.8806	$Mg_2P_2O_7$	H_3PO_4	1.1356	0.5303	$MnSO_4$	SO_3	1.8858
0.2184	$Mg_2P_2O_7$	Mg	4.5789				
1.2784	$Mg_2P_2O_7$	$Mg(CH_3CO_2)_2$	0.7822	1.5000	Mo	MoO_3	0.6667
0.8552	$Mg_2P_2O_7$	$MgCl_2$	1.1692	2.0022	Mo	MoS_3	0.4995
1.8260	$Mg_2P_2O_7$	$MgCl_2\cdot6H_2O$	0.5477	3.8250	Mo	$PbMoO_4$	0.2615
0.7572	$Mg_2P_2O_7$	$MgCO_3$	1.3206	1.3348	MoO_3	MoS_3	0.7492
1.3141	$Mg_2P_2O_7$	$Mg(HCO_3)_2$	0.7610	1.3617	MoO_3	$(NH_4)_2MoO_4$	0.7344
0.3621	$Mg_2P_2O_7$	MgO	2.7619	1.0863	MoO_3	$(NH_4)_3PO_4$	
1.0811	$Mg_2P_2O_7$	$MgSO_4$	0.9250			$(MoO_3)_{12}$	0.9205
2.2135	$Mg_2P_2O_7$	$MgSO_4\cdot7H_2O$	0.4516	2.5500	MoO_3	$PbMoO_4$	0.3923
1.0781	$Mg_2P_2O_7$	NaH_2PO_4	0.9276	1.8727	$MoO_4\cdot(NH_4)_2$	$PbMoO_4$	0.5342
1.4731	$Mg_2P_2O_7$	Na_3PO_4	0.6789				
3.2170	$Mg_2P_2O_7$	Na_2HPO_4		3.3564	N	HNO_2	0.2979
		$\cdot12H_2O$	0.3109	4.4986	N	HNO_3	0.2223
3.4144	$Mg_2P_2O_7$	Na_3PO_4		6.0757	N	KNO_2	0.1646
		$\cdot12H_2O$	0.2929	7.2179	N	KNO_3	0.1385
1.2755	$Mg_2P_2O_7$	Na_2HPO_4	0.7840	4.9263	N	$NaNO_2$	0.2030
1.2758	$Mg_2P_2O_7$	Na_2SO_4	0.7838	3.2841	N	NO_2	0.3045
1.0336	$Mg_2P_2O_7$	$(NH_4)H_2PO_4$	0.9675	2.7131	N	N_2O_3	0.3686
1.1865	$Mg_2P_2O_7$	$(NH_4)_2HPO_4$	0.8428	3.2841	N	N_2O_4	0.3045
0.2787	$Mg_2P_2O_7$	P	3.5877	3.8551	N	N_2O_5	0.2594
0.6379	$Mg_2P_2O_7$	P_2O_5	1.5676	6.0685	N	$NaNO_3$	0.1648
1.9389	$MgSO_4$	$BaSO_4$	0.5158	1.2159	N	NH_3	0.8227
0.8147	$MgSO_4$	H_2SO_4	1.2274	6.9674	N	Pt	0.1435
2.0476	$MgSO_4$	$MgSO_4\cdot7H_2O$	0.4884	2.8576	N	SO_3	0.3499
0.6651	$MgSO_4$	SO_3	1.5036				
0.3248	$MgSO_4\cdot7H_2O$	SO_3	3.0786	3.4748	Na	Br	0.2878
0.9469	$MgSO_4\cdot7H_2O$	$BaSO_4$	1.0560	1.5417	Na	Cl	0.6486
				5.5182	Na	I	0.1812
2.0923	Mn	$MnCO_3$	0.4779	4.4747	Na	$NaBr$	0.2235
1.2913	Mn	MnO	0.7744	2.5417	Na	$NaCl$	0.3934
1.5826	Mn	MnO_2	0.6319	2.3044	Na	Na_2CO_3	0.4340
1.4369	Mn	Mn_2O_3	0.6959	1.8261	Na	NaF	0.5476
1.3884	Mn	Mn_3O_4	0.7203	3.6525	Na	$NaHCO_3$	0.2740
2.5846	Mn	$Mn_2P_2O_7$	0.3869	6.5183	Na	NaI	0.1534
1.5838	Mn	MnS	0.6314	1.3478	Na	Na_2O	0.7419
0.3828	$MnCO_3$	CO_2	2.6121	1.7395	Na	$NaOH$	0.5750
1.2352	$MnCO_3$	$Mn_2P_2O_7$	0.8096	2.7404	Na	Na_2SO_3	0.3649
0.4973	$Mn(HCO_3)_2$	CO_2	2.0108	3.0885	Na	Na_2SO_4	0.3238
0.6204	MnO	CO_2	1.6119	0.6224	$Na_2Al_2O_4$	Al_2O_3	1.6067
1.6203	MnO	$MnCO_3$	0.6172	0.1399	$Na_2Al_2(SO_4)_4$		
1.2256	MnO	MnO_2	0.8159		$\cdot24H_2O$	S	7.1493
2.1287	MnO	$MnSO_4$	0.4698	0.6925	$Na_2B_4O_7$	B_2O_3	1.4441
2.0016	MnO	$Mn_2P_2O_7$	0.4996	1.2287	$Na_2B_4O_7$	H_3BO_3	0.8139
1.1128	MnO	Mn_2O_3	0.8987	0.3657	$Na_2B_4O_7$		
1.2266	MnO	MnS	0.8153		$\cdot10H_2O$	B_2O_3	2.7347
1.1289	MnO	SO_3	0.8859	0.6488	ditto	H_3BO_3	1.5412
2.0022	Mn_2O_3	$KMnO_4$	0.4995	1.3202	ditto	KBF_4	0.7575
0.9662	Mn_2O_3	Mn_3O_4	1.0349	1.0482	$NaBr$	Ag	0.9540
1.9130	Mn_2O_3	$MnSO_4$	0.5227	1.8247	$NaBr$	$AgBr$	0.5480
0.4062	Mn_2O_3	S	2.4619	0.7765	$NaBr$	Br	1.2878
1.1014	Mn_3O_4	MnO_2	0.9080	1.8454	$NaCl$	Ag	0.5419
2.5848	Mn_3O_4	K_2MnO_4	0.3869	2.4519	$NaCl$	$AgCl$	0.4078
1.7357	MnS	$MnSO_4$	0.5761	2.9061	$NaCl$	$AgNO_3$	0.3441
1.5458	$MnSO_4$	$BaSO_4$	0.6469	0.4796	$NaCl$	CaO	2.0852

$$A\times A'=B \qquad B\times B'=A$$

XVIII.—CONVERSION FACTORS (Continued)

A'	A	B	B'	A'	A	B	B'
0.6066	NaCl	Cl	1.6486	1.6003	NaH₂PO₄*	NaH₂PO₄·4H₂O	0.6249
0.6239	NaCl	HCl	1.6030	0.2582	NaH₂PO₄*	Na₂O	3.8728
0.8389	NaCl	H₂SO₄	1.1920	0.2585	NaH₂PO₄*	P	3.8678
1.8211	NaCl	NaClO₃	0.5491	0.5917	NaH₂PO₄*	P₂O₅	1.6900
2.1632	NaCl	NaClO₄	0.4623	0.3453	NaH₂PO₄†	H₂SO₄	2.8964
0.9066	NaCl	Na₂CO₃	1.1030	2.5220	Na₂HPO₄	Na₂HPO₄·12H₂O	0.3965
1.4370	NaCl	NaHCO₃	0.6959	0.4365	Na₂HPO₄	Na₂O	2.2911
1.2149	NaCl	Na₂HPO₄	0.8231	0.9366	Na₂HPO₄	Na₄P₂O₇	1.0677
1.7803	NaCl	NaHSO₃	0.5617	0.5001	Na₂HPO₄	P₂O₅	0.9995
0.5303	NaCl	Na₂O	1.8858	0.7565	Na₂H₂P₂O₇	NaHCO₃	1.3219
1.2151	NaCl	Na₂SO₄	0.8230	0.2792	Na₂H₂P₂O₇	Na₂O	3.5822
1.1656	NaCl	ZnCl₂	0.8579	0.6397	Na₂H₂P₂O₇	P₂O₅	1.5632
0.3331	NaClO₃	Cl	3.0023	0.3395	Na(NH₄)HPO₄·4H₂O	P₂O₅	2.9441
0.2896	NaClO₄	Cl	3.4535	0.8968	Na₃PO₄	H₂SO₄	1.1151
2.2022	Na₂CO₃	BaSO₄	0.4541	0.6789	Na₃PO₄	Mg₂P₂O₇	1.4731
0.9441	Na₂CO₃	CaCO₃	1.0593	0.5669	Na₃PO₄	Na₂O	1.7639
0.5290	Na₂CO₃	CaO	1.8905	1.2991	Na₃PO₄	Na₂SO₄	0.7698
1.1327	Na₂CO₃	CH₃CO₂H	0.8828	2.3179	Na₃PO₄	Na₃PO₄·12H₂O	0.4314
0.4151	Na₂CO₃	CO₂	2.4091	0.4331	Na₃PO₄	P₂O₅	2.3091
0.6881	Na₂CO₃	HCl	1.4533	2.2432	NaHSO₃	BaSO₄	0.4458
0.9253	Na₂CO₃	H₂SO₄	1.0807	0.4712	NaHSO₃	H₂SO₄	2.1222
1.3038	Na₂CO₃	K₂CO₃	0.7670	0.5617	NaHSO₃	NaCl	1.7803
0.4340	Na₂CO₃	Na	2.3044	0.2979	NaHSO₃	Na₂O	3.3570
1.1030	Na₂CO₃	NaCl	0.9066	0.9134	NaHSO₃	Na₂S₂O₅	1.0948
1.5850	Na₂CO₃	NaHCO₃	0.6309	0.3081	NaHSO₃	S	3.2460
1.9637	Na₂CO₃	NaHSO₃	0.5092	0.6156	NaHSO₃	SO₂	1.6244
0.7549	Na₂CO₃	NaOH	1.3247	1.5662	NaI	AgI	0.6385
0.5849	Na₂CO₃	Na₂O	1.7097	0.8466	NaI	I	1.1812
0.7320	Na₂C₂O₄	H₂SO₄	1.3661	0.1534	NaI	Na	6.5183
1.9951	Na₂CrO₄	PbCrO₄	0.5014	0.2068	NaI	Na₂O	4.8361
1.2788	Na₂Cr₂O₇	Fe	0.7820	0.6413	NaIO₃	I	1.5594
0.3744	Na₂Cr₂O₇	H₂SO₄	2.6711	0.7413	NaNO₃	HNO₃	1.3490
1.2367	Na₂Cr₂O₇	Na₂CrO₄	0.8086	0.5769	NaNO₃	H₂SO₄	1.7334
0.9294	NaF	CaF₂	1.0760	1.1894	NaNO₃	KNO₃	0.8408
0.9135	Na₂HAsO₃	Mg₂As₂O₇	1.0947	0.8117	NaNO₃	NaNO₂	1.2319
0.4031	Na₂HAsO₄	As	2.4809	0.3647	NaNO₃	Na₂O	2.7423
0.6182	Na₂HAsO₄	As₂O₅	1.6177	0.2004	NaNO₃	NH₃	4.9911
0.8349	Na₂HAsO₄	Mg₂As₂O₇	1.1978	0.3530	NaNO₃	NO	2.8327
0.6793	NaHCO₃	Al₂(SO₄)₃	1.4721	0.6353	NaNO₃	N₂O₅	1.5740
1.3938	NaHCO₃*	Ca(H₂PO₄)₂	0.7175	1.2258	NaOH	H₂SO₄	0.8158
1.2866	NaHCO₃*	CaH₂P₂O₇	0.7773	0.7748	NaOH	Na₂O	1.2906
0.8931	NaHCO₃	C₄H₆O₆	1.1197	1.1766	NaOH	Na₂SiF₆	0.8499
0.5238	NaHCO₃	CO₂	1.9092	2.0400	NaOH	(NH₄)₃PO₄(MoO₃)₁₂	0.4902
0.5838	NaHCO₃	H₂SO₄	1.7129	3.7651	Na₂O	BaSO₄	0.2656
1.0251	NaHCO₃	K₂Al₂(SO₄)₄	0.9756	1.9365	Na₂O	CH₃CO₂H	0.5164
2.2395	NaHCO₃	KHC₄H₄O₆	0.4465	2.6463	Na₂O	CH₃CO₂Na	0.3779
0.9284	NaHCO₃	KMnO₄	1.0760	1.5820	Na₂O	H₂SO₄	0.6321
0.2740	NaHCO₃	Na	3.6525	0.7419	Na₂O	Na	1.3478
0.6309	NaHCO₃	Na₂CO₃	1.5851	1.8858	Na₂O	NaCl	0.5303
1.4291	NaHCO₃*	NaH₂PO₄	0.6997	1.7097	Na₂O	Na₂CO₃	0.5849
1.3219	NaHCO₃*	Na₂H₂P₂O₇	0.7565				
0.3690	NaHCO₃	Na₂O	2.7099				
0.4085	NaH₂PO₄*	H₂SO₄	2.4482				
0.6997	NaH₂PO₄*	NaHCO₃	1.4291				

$$A \times A' = B \qquad B \times B' = A$$

* Phenolphthalein   † Methyl orange.

XVIII.—CONVERSION FACTORS (Continued)

A'	A	B	B'	A'	A	B	B'
1.3548	Na_2O	NaF	0.7381	1.1663	Nd	Nd_2O_3	0.8574
2.7099	Na_2O	$NaHCO_3$	0.3690				
2.2911	Na_2O	Na_2HPO_4	0.4365	3.6999	NH_3	HNO_3	0.2704
3.8728	Na_2O	NaH_2PO_4	0.2582	2.8792	NH_3	H_2SO_4	0.3473
3.5822	Na_2O	$Na_2H_2P_2O_7$	0.2792	4.9969	NH_3	KNO_2	0.2001
3.3570	Na_2O	$NaHSO_3$	0.2979	5.9364	NH_3	KNO_3	0.1685
2.7423	Na_2O	$NaNO_3$	0.3647	0.8227	NH_3	N	1.2159
1.2906	Na_2O	$NaOH$	0.7748	4.0517	NH_3	$NaNO_2$	0.2468
3.2916	Na_2O	$NaPO_3$	0.3038	4.9911	NH_3	$NaNO_3$	0.2005
1.7639	Na_2O	Na_3PO_4	0.5669	2.8207	NH_3	$(NH_4)_2CO_3$	0.3545
1.2592	Na_2O	Na_2S	0.7942	3.1409	NH_3	NH_4Cl	0.3184
2.0334	Na_2O	Na_2SO_3	0.4918	3.8785	NH_3	$(NH_4)_2HPO_4$	0.2578
2.2915	Na_2O	Na_2SO_4	0.4364	6.7570	NH_3	$NH_4H_2PO_4$	0.1480
2.5503	Na_2O	$Na_2S_2O_3$	0.3921	2.0577	NH_3	NH_4OH	0.4860
3.0665	Na_2O	$Na_2S_2O_5$	0.3261	13.035	NH_3	$(NH_4)_2PtCl_6$	0.0767
2.1458	Na_2O	$Na_4P_2O_7$	0.4660	3.8790	NH_3	$(NH_4)_2SO_4$	0.2578
1.1458	Na_2O*	P_2O_5	0.8727	3.1714	NH_3	N_2O_5	0.3153
1.2915	Na_2O	SO_3	0.7743	0.1127	$(NH_4)_2Al_2$		
2.9904	Na_2S	$BaSO_4$	0.3344		$(SO_4)_4 \cdot 24H_2O$	Al_2O_3	8.8738
1.8506	Na_2S	CdS	0.5404	0.0376	ditto	NH_3	26.6235
0.4365	Na_2S	H_2S	2.2908	0.1414	ditto	S	7.0719
1.2564	Na_2S	H_2SO_4	0.7959	0.3531	ditto	SO_3	2.8320
0.7942	Na_2S	Na_2O	1.2592	2.6793	NH_4Cl	$AgCl$	0.3732
0.4108	Na_2S	S	2.4348	0.6628	NH_4Cl	Cl	1.5088
1.8515	Na_2SO_3	$BaSO_4$	0.5401	0.6817	NH_4Cl	HCl	1.4669
0.6511	Na_2SO_3	H_2SO_3	1.5359	0.9167	NH_4Cl	H_2SO_4	1.0909
0.3890	Na_2SO_3	H_2SO_4	2.5706	0.9675	$(NH_4)H_2PO_4$	$Mg_2P_2O_7$	1.0336
2.0135	Na_2SO_3	I	0.4966	0.1480	$(NH_4)H_2PO_4$	NH_3	6.7570
0.4918	Na_2SO_3	Na_2O	2.0334	0.8428	$(NH_4)_2HPO_4$	$Mg_2P_2O_7$	1.1865
0.5082	Na_2SO_3	SO_2	1.9677	0.2578	$(NH_4)_2HPO_4$	NH_3	3.8785
0.2541	$Na_2SO_4 \cdot 7H_2O$	SO_2	3.9356	0.4896	$(NH_4)_2MoO_4$	Mo	2.0425
2.9524	$Na_2S_2O_3$	$BaSO_4$	0.3387	0.0165	$(NH_4)_3PO_4$		
1.5696	$Na_2S_2O_3$	$Na_2S_2O_3$			$(MoO_3)_{12}$	P	60.476
		$\cdot 5H_2O$	0.6371	0.0378	ditto	P_2O_5	26.424
1.6431	Na_2SO_4	$BaSO_4$	0.6086	0.2838	$(NH_4)_2PtCl_6$	HNO_3	3.5232
0.3947	Na_2SO_4	CaO	2.5338	0.4393	$(NH_4)_2PtCl_6$	Pt	2.2748
1.1547	Na_2SO_4	CH_3CO_2Na	0.8660	0.9188	$(NH_4)_2PtCl_6$	$PtCl_6$	1.0884
0.6904	Na_2SO_4	H_2SO_4	1.4484	1.4392	$(NH_4)_2S$	H_2SO_4	0.6948
1.2267	Na_2SO_4	K_2SO_4	0.8152	1.7665	$(NH_4)_2SO_4$	$BaSO_4$	0.5661
0.2838	Na_2SO_4	MgO	3.5236	0.7422	$(NH_4)_2SO_4$	H_2SO_4	1.3473
0.7838	Na_2SO_4	$Mg_2P_2O_7$	1.2758	0.2120	$(NH_4)_2SO_4$	N	4.7166
0.3238	Na_2SO_4	Na	3.0883	0.2578	$(NH_4)_2SO_4$	NH_3	3.8790
0.8230	Na_2SO_4	$NaCl$	1.2151	0.5753	$(NH_4)_2SO_4$	N_2O_3	1.7383
0.7461	Na_2SO_4	Na_2CO_3	1.3403	0.4298	$(NH_4)_2S_2O_8$	H_2SO_4	2.3268
0.5913	Na_2SO_4	NaF	1.6912				
0.6739	$Na_2S_2O_5$	SO_2	1.4839	4.9225	Ni	$NiC_8H_{14}N_4O_4$	0.2032
1.6904	Na_2SO_4	$NaHSO_4$	0.5916	4.9556	Ni	$Ni(NO_3)_2$	
0.4364	Na_2SO_4	Na_2O	2.2915			$\cdot 6H_2O$	0.2018
0.5495	Na_2SO_4	Na_2S	1.8198	1.2727	Ni	NiO	0.7858
0.8874	Na_2SO_4	Na_2SO_3	1.1269	2.6371	Ni	$NiSO_4$	0.3792
1.8877	Na_2SO_4	$Na_2SO_47H_2O$	0.5297	4.7863	Ni	$NiSO_4 \cdot 7H_2O$	0.2089
2.2681	Na_2SO_4	Na_2SO_4		3.0643	NiO	$NiC_8H_{14}N_4O_4$	0.3263
		$\cdot 10H_2O$	0.4409	2.0720	NiO	$NiSO_4$	0.4826
1.3383	Na_2SO_4	$Na_2S_2O_5$	0.7472	1.8150	$NiSO_4$	$NiSO_4 \cdot 7H_2O$	0.5510
0.5636	Na_2SO_4	SO_3	1.7743				

$$A \times A' = B \qquad B \times B' = A$$

* Phenolphthalein.

XVIII.—CONVERSION FACTORS (Continued)

A'	A	B	B'	A'	A	B	B'
1.5667	NO	HNO_2	0.6382	1.6900	P_2O_5	NaH_2PO_4	0.5917
2.0999	NO	HNO_3	0.4762	1.5632	P_2O_5	$Na_2H_2P_2O_7$	0.6397
2.8362	NO	KNO_2	0.3526	2.9441	P_2O_5	$Na(NH_4)$	
3.3694	NO	KNO_3	0.2968			$HPO_4 \cdot 4H_2O$	0.3395
2.2997	NO	$NaNO_2$	0.4348	0.8729	P_2O_5	Na_2O	1.1458
2.8327	NO	$NaNO_3$	0.3530	2.3091	P_2O_5	Na_3PO_4	0.4331
				26.424	P_2O_5	$(NH_4)_3PO_4$	
4.0487	N_2O_3	$AgNO_3$	0.2470			$(MoO_3)_{12}$	0.0378
0.9594	N_2O_3	HCl	1.0423	1.8727	P_2O_5	$Na_4P_2O_7$	0.5340
1.6579	N_2O_3	HNO_3	0.6032	5.0278	P_2O_5	$U_2P_2O_{11}$	0.1989
1.0798	N_2O_3*	H_2SO_3	0.9261				
1.2903	N_2O_3	H_2SO_4	0.7750	0.5797	Pb	CH_3CO_2H	1.7255
2.2392	N_2O_3	KNO_2	0.4466	1.5700	Pb	$Pb(C_2H_3O_2)_2$	0.6369
0.3686	N_2O_3	N	2.7131	1.3424	Pb	$PbCl_2$	0.7449
1.8156	N_2O_3	$NaNO_2$	0.5508	1.2897	Pb	$PbCO_3$	0.7754
0.7895	N_2O_3	NO	1.2666	1.2479	Pb	$(PbCO_3)_2Pb$	
1.2104	N_2O_3	N_2O_4	0.8262			$(OH)_2$	0.8014
0.7928	N_2O_4	HCl	1.2617	1.5601	Pb	$PbCrO_4$	0.6410
1.3697	N_2O_4	HNO_3	0.7301	1.0773	Pb	PbO	0.9283
0.8920	N_2O_4*	H_2SO_3	1.1210	1.1545	Pb	PbO_2	0.8662
1.0659	N_2O_4	H_2SO_4	0.9382	1.1643	Pb	$Pb(OH)_2$	0.8589
0.3045	N_2O_4	N	3.2841	1.1549	Pb	PbS	0.8659
0.6523	N_2O_4	NO	1.5332	1.4639	Pb	$PbSO_4$	0.6831
0.6752	N_2O_5	HCl	1.4810	0.1548	Pb	S	6.4629
1.1668	N_2O_5	HNO_3	0.8571	0.2551	$PbCl_2$	Cl_2	3.9216
0.7599	N_2O_5*	H_2SO_3	1.3159	0.1647	$PbCO_3$	CO_2	6.0722
0.9080	N_2O_5	H_2SO_4	1.1013	0.1612	$PbCrO_4$	Cr	6.2154
1.8721	N_2O_5	KNO_3	0.5342	0.2352	$PbCrO_4$	Cr_2O_3	4.2526
1.5740	N_2O_5	$NaNO_3$	0.6353	0.4551	$PbCrO_4$	$K_2Cr_2O_7$	2.1971
0.3153	N_2O_5	NH_3	3 1710	0.5014	$PbCrO_4$	Na_2CrO_4	1.9951
0.5556	N_2O_5	NO	1.7997	0.6410	$PbCrO_4$	Pb	1.5600
0.7038	N_2O_5	N_2O_3	1.4209	1.1733	$PbCrO_4$	$Pb(C_2H_3O_2)_2$	
						$\cdot 3H_2O$	0.8523
1.3353	Os	OsO_4	0.7489	0.7999	$PbCrO_4$	$(PbCO_3)_2Pb$	
						$(OH)_2$	1.2501
3.1593	P	H_3PO_4	0.3165	0.6905	$PbCrO_4$	PbO	1.4482
3.5877	P	$Mg_2P_2O_7$	0.2787	0.7401	$PbCrO_4$	Pb_2O_4	1.3512
3.8678	P	NaH_2PO_4	0.2585	0.9383	$PbCrO_4$	$PbSO_4$	1.0657
60.476	P	$(NH_4)_3PO_4$		0.5642	$PbMoO_4$	Pb	1.7722
		$(MoO_3)_{12}$	0.0165	0.2615	$PbMoO_4$	Mo	3.8250
2.2887	P	P_2O_5	0.4369	0.3923	$PbMoO_4$	MoO_3	2.5500
5.8936	P_2O_5	Ag_3PO_4	0.1697	0.9283	PbO	Pb	1.0772
1.7193	P_2O_5	$AlPO_4$	0.5815	1.2461	PbO	$PbCl_2$	0.8025
1.7193	P_2O_5	$Al_2P_2O_3$	0.5816	1.1972	PbO	$PbCO_3$	0.8353
1.9161	P_2O_5	$CaHPO_4$	0.5219	1.4842	PbO	$Pb(NO_3)_2$	0.6738
1.6483	P_2O_5	$Ca(H_2PO_4)_2$	0.6067	1.3589	PbO	$PbSO_4$	0.7359
1.5214	P_2O_5	$CaH_2P_2O_7$	0.6573	0.8659	PbS	Pb	1.1549
1.7893	P_2O_5	$Ca_2P_2O_7$	0.5589	0.9328	PbS	PbO	1.0720
2.1839	P_2O_5	$Ca_3P_2O_8$	0.4579	0.1340	PbS	S	7.4629
1.2139	P_2O_5	$FePO_4$	0.4708	0.7697	$PbSO_4$	$BaSO_4$	1.2991
1.1552	P_2O_5	H_3PO_3	0.8657	1.2507	$PbSO_4$	$Pb(C_2H_3O_2)_2$	
1.3804	P_2O_5	H_3PO_4	0.7244			$\cdot 3H_2O$	0.7995
1.0709	P_2O_5	H_2SO_4	0.4829	0.8525	$PbSO_4$	$(PbCO_3)_2Pb$	
1.5676	P_2O_5	$Mg_2P_2O_7$	0.6379			$(OH)_2$	1.1731
1.4364	P_2O_5	$NaPO_3$	0.6962	0.7887	$PbSO_4$	PbO_2	1.2679
1.9995	P_2O_5	Na_2HPO_4	0.5001	0.7535	$PbSO_4$	Pb_3O_4	1.3272

$$A \times A' = B \qquad B \times B' = A$$

* Phenolphthalein

XVIII.—CONVERSION FACTORS (Continued)

A'	A	B	B'
0.7889	$PbSO_4$	PbS	1.2676
0.2640	$PbSO_4$	SO_3	3.7879
3.7270	Pd	K_2PdCl_6	0.2683
2.0024	Pd	$PdCl_2 \cdot 2H_2O$	0.4994
3.3791	Pd	PdI_2	0.2959
2.1623	Pd	$Pd(NO_3)_2$	0.4625
2.3979	Pd	HI	0.4170
2.3790	Pd	I	0.4203
0.7096	PdI_2	HI	1.4092
0.7041	PdI_2	I	1.4204
0.9703	PdI_2	IO_3	1.0306
2.4903	Pt	K_2PtCl_6	0.4015
2.6541	Pt	$H_2PtCl_6 \cdot 6H_2O$	0.3768
2.2748	Pt	$(NH_4)_2PtCl_6$	0.4396
1.7266	Pt	$PtCl_4$	0.5792
2.1881	Pt	$PtCl_4 \cdot 5H_2O$	0.4570
1.4423	$PtCl_4$	K_2PtCl_6	0.6934
1.3177	$PtCl_4$	$(NH_4)_2PtCl_6$	0.7589
1.1383	$PtCl_4 \cdot 5H_2O$	K_2PtCl_6	0.8785
1.1703	Pr	Pr_2O_3	0.8545
3.7382	Rh	Na_3RhCl_6	0.2675
2.0338	Rh	$RhCl_3$	0.4917
1.6775	Rb	$AgCl$	0.5961
0.4150	Rb	Cl	2.4098
1.4150	Rb	$RbCl$	0.7067
1.3510	Rb	$RbCO_3$	0.7402
1.0937	Rb	Rb_2O	0.9144
1.5622	Rb	Rb_2SO_4	0.6402
3.3871	Rb	Rb_2PtCl_6	0.2952
1.1855	$RbCl$	$AgCl$	0.8435
0.2933	$RbCl$	Cl	3.4098
1.2686	Rb_2CO_3	$RbHCO_3$	0.7883
1.2939	Rb_2O	$RbCl$	0.7729
1.4284	Rb_2O	Rb_2SO_4	0.7001
0.4178	Rb_2PtCl_6	$RbCl$	2.3938
0.3987	Rb_2PtCl_6	Rb_2CO_3	2.5071
0.5060	Rb_2PtCl_6	$RbHCO_3$	1.9762
0.3229	Rb_2PtCl_6	Rb_2O	3.0972
5.2848	S	BaS	0.1892
6.7820	S	$BaSO_3$	0.1475
7.2810	S	$BaSO_4$	0.1373
6.0334	S	BaS_2O_3	0.1657
6.3143	S	$BaS_2O_3 \cdot H_2O$	0.2397
2.2117	S	$BaS_4 \cdot H_2O$	0.4521
1.0629	S	H_2S	0.9408
3.0591	S	H_2SO_4	0.3270
2.7417	S	FeS	0.3648
1.8709	S	FeS_2	0.5346
3.4392	S	K_2S	0.2908
2.4348	S	Na_2S	0.4108

A'	A	B	B'
6.4630	S	Pb	0.1547
1.9981	S	SO_2	0.5005
2.4972	S	SO_3	0.4005
1.7616	S_3	BaS	0.5677
1.3212	S_4	BaS	0.7569
3.6439	SO_2	$BaSO_4$	0.2745
1.5783	SO_2	$Ca(HSO_3)_2$	0.6336
1.2812	SO_2	H_2SO_3	0.7805
1.5309	SO_2†	H_2SO_4	0.6532
0.7655	SO_2*	H_2SO_4	1.3064
1.9677	SO_2	$NaHSO_3$	0.6156
	SO_2	Na_2SO_3	0.5082
3.9356	SO_2	$Na_2SO_3 \cdot 7H_2O$	0.2541
1.2498	SO_2	SO_3	0.8001
0.4255	SO_3	Al_2O_3	2.3501
1.4255	SO_3	$Al_2(SO_4)_3$	0.7015
2.9155	SO_3	$BaSO_4$	0.3430
0.7003	SO_3	CaO	1.4280
1.7003	SO_3	$CaSO_4$	0.5881
1.2250	SO_3	H_2SO_4	0.8163
2.1765	SO_3	K_2SO_4	0.4595
0.5036	SO_3	MgO (in $MgSO_4$)	1.9859
1.5033	SO_3	$MgSO_4$	0.6651
1.8858	SO_3	$MnSO_4$	0.5303
0.7743	SO_3	Na_2O	1.2915
1.7743	SO_3	Na_2SO_4	0.5636
1.1416	SO_3	66° O. V	0.8760
2.0162	SO_3	$ZnSO_4$	0.4960
2.4298	SO_4	$BaSO_4$	0.4116
0.3130	$SO_3 \cdot HCl$	HCl	3.1953
0.8417	$SO_3 \cdot HCl$	H_2SO_4	1.1881
0.4561	$S_2O_5Cl_2$	H_2SO_4	2.1926
2.7649	Sb	$KSbOC_4H_4O_6 \cdot \frac{1}{2}H_2O$	0.3617
1.8850	Sb	$SbCl_3$	0.5305
1.1997	Sb	Sb_2O_3	0.8336
1.3328	Sb	Sb_2O_5	0.7503
1.4002	Sb	Sb_2S_3	0.7142
1.2662	Sb	Sb_2O_4	0.7897
1.1109	Sb_2O_3	Sb_2O_5	0.9001
1.3462	Sb_2S_3	$SbCl_3$	0.7428
1.6315	Se	H_2SeO_3	0.6129
1.8336	Se	H_2SeO_4	0.5454
1.4040	Se	SeO_2	0.7123
1.6060	Se	SeO_3	0.6227
2.1307	Si	SiO_2	0.4693
2.6961	Si	SiO_3	0.3709
3.2615	Si	SiO_4	0.3066
2.6814	SiF_4	$BaSiF_6$	0.3729
1.3837	SiF_4	H_2SiF_6	0.7227
2.1141	SiF_4	K_2SiF_6	0.4730
1.0141	SiF_6	H_2SiF_6	0.9860

$$A \times A' = B \qquad A \times A' = A$$

* Phenolphthalein.   † Methyl orange.

XVIII.—CONVERSION FACTORS (Continued)

A'	A	B	B'	A'	A	B	B'
4.6380	SiO₂	BaSiF₆	0.2156	0.5000	Tl₂PtCl₆	Tl	1.9999
1.2988	SiO₂	H₂SiO₃	0.7699	0.5869	Tl₂PtCl₆	TlCl	1.7038
3.6567	SiO₂	K₂SiF₆	0.2735	0.5738	Tl₂PtCl₆	Tl₂CO₃	1.7435
2.5622	SiO₂	K₂SiO₃	0.3903	0.8111	Tl₂PtCl₆	TlI	1.2329
2.0282	SiO₂	Na₂SiO₃	0.4930	0.6520	Tl₂PtCl₆	TlNO₃	1.5337
1.7296	SiO₂	SiF₄	0.5782	0.5196	Tl₂PtCl₆	Tl₂O	1.9244
2.6554	SiO₂	SO₃	0.3766	0.6178	Tl₂PtCl₆	Tl₂SO₄	1.6187
1.5975	SiO₂	Si(OH)₄	0.6259	0.8094	Tl₂SO₄	Tl	1.2354
2.3495	SiO₂	ZnSiO₃	0.4256				
				0.8817	UO₂	U	1.1342
0.5975	Sn	Cl₂	1.6737	0.8482	U₃O₈	U	1.1789
1.5974	Sn	SnCl₂	0.6261	0.9621	U₃O₈	UO₂	1.0394
1.9010	Sn	SnCl₂·2H₂O	0.5260	1.7884	U₃O₈	UO₂(NO₃)₂	
2.1949	Sn	SnCl₄	0.4555			·6H₂O	0.5591
1.1348	Sn	SnO	0.8812	0.6668	U₂P₂O₁₁	U	1.4990
1.2696	Sn	SnO₂	0.7877	0.7566	U₂P₂O₁₁	UO₂	1.3221
0.3740	SnCl₂	Cl	2.6738				
0.3847	SnCl₂	HCl	2.6000	1.3137	V	V₂O₂	0.7612
0.8421	SnCl₂	Fe₂O₃	1.1875	1.4706	V	V₂O₃	0.6800
1.1899	SnCl₂	SnCl₂·2H₂O	0.8404	1.6275	V	V₂O₄	0.6145
0.5444	SnCl₄	Cl	1.8369	1.7843	V	V₂O₅	0.5604
0.5599	SnCl₄	HCl	1.7861	0.5604	V₂O₅	V	1.7843
1.4974	SnO₂	SnCl₂·2H₂O	0.6678	1.2638	V₂O₅	VO₄	0.7913
1.7288	SnO₂	SnCl₄	0.5784				
2.4389	SnO₂	SnCl₄		0.8785	Yt₂O₃	Yt	1.1383
		(NH₄Cl)₂	0.4100	0.7882	Y₂O₃	Y	1.2687
0.8938	SnO₂	SnO	1.1188				
				2.4739	W	PbWO₄	0.4042
1.6848	Sr	SrCO₃	0.5935	0.8519	WO₂	W	1.1739
1.1826	Sr	SrO	0.8456	0.7931	WO₃	W	1.2609
2.0963	Sr	SrSO₄	0.4770	1.9621	WO₃	PbWO₄	0.5099
1.5300	SrO	SrCl₂	0.6536				
0.4357	SrSO₄	SO₃	2.2943	1.5004	Zn	H₂SO₄	0.6665
				1.2247	Zn	SO₃	0.8165
1.9767	Ta	TaCl₅	0.5059	2.0849	Zn	ZnCl₂	0.4796
1.2204	Ta	Ta₂O₅	0.8194	7.6222	Zn	ZnHg(CNS)₄	0.1312
				1.2448	Zn	ZnO	0.8034
1.5177	Te	H₂TeO₄	0.6587	2.3315	Zn	Zn₂P₂O₇	0.4289
1.8003	Te	H₂TeO₄2H₂O	0.5554	1.4906	Zn	ZnS	0.6709
1.2509	Te	TeO₂	0.7994	2.1035	ZnCl₂	AgCl	0.4754
1.3765	Te	TeO₃	0.7265	0.5204	ZnCl₂	Cl	1.9217
				0.8579	ZnCl₂	NaCl	1.1656
0.8790	ThO₂	Th	1.1379	1.2054	ZnO	H₂SO₄	0.8296
1.4155	ThO₂	ThCl₄	0.7065	0.9840	ZnO	SO₃	1.0162
2.2271	ThO₂	Th(NO₃)₄		1.6749	ZnO	ZnCl₂	0.5970
		·6H₂O	0.4492	1.1541	ZnO	ZnCO₃	0.6490
				2.9427	ZnO	Zn₂P₂O₇	0.3398
1.6652	Ti	TiO₂	0.6005	1.1975	ZnO	ZnS	0.8351
				1.9840	ZnO	ZnSO₄	0.5040
1.1738	Tl	TlCl	0.8519	3.5340	ZnO	ZnSO₄·7H₂O	0.2830
1.1470	Tl	Tl₂CO₃	0.8718	2.3957	ZnS	BaSO₄	0.4174
1.6220	Tl	TlI	0.6165	2.9510	ZnS	ZnSO₄·7H₂O	0.3389
1.3040	Tl	TlNO₃	0.7669	0.4256	ZnSiO₃	SiO₂	2.3495
1.0392	Tl	Tl₂O	0.9623	0.4960	ZnSO₄	SO₃	2.0162
0.7786	Tl₂CrO₄	Tl	1.2843				
0.6775	TlHSO₄	Tl	1.4759	0.7390	ZrO₂	Zr	1.3532
0.6165	TlI₂	Tl	1.6222				

$$A \times A' = B \qquad B \times B' = A$$

NOTE.—The editor will welcome additional factors not appearing in these tables. Although a large portion of these factors have appeared in previous editions the entire list has received careful revision, using the atomic weights of 1920. Credit for the arrangement of this section is due to R. M. Meiklejohn. The factors have been checked by three men working independently.

Example of Method for Using Factors. Suppose the product weighed is 0.8535 gram AgCl and the equivalent weight of Cl is desired; hunt up the factors AgCl–Cl. This may be found on the first page of the conversion factors, a little below the middle of the page. Using the formula $A \times A' = B$, and substituting the values for A (weight of AgCl) and A' (factor) we have $0.8535 \times 0.2474 = 0.21124$ gram Cl. If, on the other hand, the weight of Cl were known to be, say, 0.2501 gram and the weight of the equivalent AgCl were desired, we would use the formula $B \times B' = A$ and, substituting the values for B and B', we would have $0.2501 \times 4.0423 = 1.01098$ gram AgCl.

VOLUMETRIC FACTORS

$$1 \text{ c.c. N/2 \ HCl} = .018235 \text{ gram HCl}$$
$$1 \text{ c.c. N/10 HCl} = .003647 \text{ gram HCl}$$
$$1 \text{ c.c. N/2 \ KOH} = .028055 \text{ gram KOH}$$
$$1 \text{ c.c. N/6 \ KOH} = .04706 \text{ gram oleic acid} = .008173 \text{ gram } H_2SO_4$$
$$1 \text{ c.c. N/10 KOH} = .00561 \text{ gram KOH}$$
$$1 \text{ c.c. } K_2Cr_2O_7 = 3.8633 \text{ gram per liter} = .010 \text{ gram I}$$
$$1 \text{ c.c. N/10 } Na_2S_2O_35H_2O = .0248 \text{ gram } Na_2S_2O_35H_2O = .012692 \text{ gram I}$$

$$
\begin{aligned}
1 \text{ c.c. NO} &= 0.0021 && \text{gram } HNO_2 \\
&= 0.0028143 && \text{gram } HNO_3 \\
&= 0.0045154 && \text{gram } KNO_3 \\
&= 0.003796 && \text{gram } NaNO_3 \\
&= 0.00134 && \text{gram NO} \\
&= 0.001696 && \text{gram } N_2O_3 \\
&= 0.0024119 && \text{gram } N_2O_5
\end{aligned}
$$

A'	A	B	B'	A'	A	B	B'

XIX.—COMPARISON OF CENTIGRADE AND FAHRENHEIT SCALE

° C.	−100	−0	+0	+100	+200	+300	+400	+500	+600	+700	+800	+900	° C.
	°F.	F.	F.	F.	F.	F.	F.	F.	F.	F.	F.	F.	
0	−148	+ 32	32	+212	392	572	752	932	1112	1292	1472	1652	0
5	−157	+ 23	41	221	401	581	761	941	1121	1301	1481	1661	5
10	−166	+ 14	50	230	410	590	770	950	1130	1310	1490	1670	10
15	−175	+ 5	59	239	419	599	779	959	1139	1319	1499	1679	15
20	−184	− 4	68	248	428	608	788	968	1148	1328	1508	1688	20
25	−193	− 13	77	257	437	617	797	977	1157	1337	1517	1697	25
30	−202	− 22	86	266	446	626	806	986	1166	1346	1526	1706	30
35	−211	− 31	95	275	455	635	815	995	1175	1355	1535	1715	35
40	−220	− 40	104	284	464	644	824	1004	1184	1364	1544	1724	40
45	−229	− 49	113	293	473	653	833	1013	1193	1373	1553	1733	45
50	−238	− 58	122	302	482	662	842	1022	1202	1382	1562	1742	50
55	−247	− 67	131	311	491	671	851	1031	1211	1391	1571	1751	55
60	−256	− 76	140	320	500	680	860	1040	1220	1400	1580	1760	60
65	−265	− 85	149	329	509	689	869	1049	1229	1409	1589	1769	65
70	−274	− 94	158	338	518	698	878	1058	1238	1418	1598	1778	70
75	−283	−103	167	347	527	707	887	1067	1247	1427	1607	1787	75
80	−292	−112	176	356	536	716	896	1076	1256	1436	1616	1796	80
85	−301	−121	185	365	545	725	905	1085	1265	1445	1625	1805	85
90	−310	−130	194	374	554	734	914	1094	1274	1454	1634	1814	90
95	−319	−139	203	383	563	743	923	1103	1283	1463	1643	1823	95
100	−328	−148	+212	392	572	752	932	1112	1292	1472	1652	1832	100

° C.	−200	−100	+100	+200	+300	+400	+500	+600	+700	+800	+900	+1000	° C.

C°	1100	1200	1300	1400	1500	1600	1700	1800	1900	2000
F°	2012	2192	2372	2552	2732	2912	3092	3272	3452	3632

Degrees C.×1.8+32 = Degrees F.　　　　Degrees F.−32÷1.8 = Degrees C.

Absolute zero, −273° C. = −459° F.

COMPARISON OF CENTIGRADE AND FAHRENHEIT SCALE FOR EVERY 1° C. FROM 0° TO 100° C.

C.	0	10	20	30	40	50	60	70	80	90	C.
	F.	F.	F.	F.	F.	F.	F.	F.	F.	F.	
0	32	50	68	86	104	122	140	158	176	194	0
1	33.8	51.8	69.8	87.8	105.8	123.8	141.8	159.8	177.8	195.8	1
2	35.6	53.6	71.6	89.6	107.6	125.6	143.6	161.6	179.6	197.6	2
3	37.4	55.4	73.4	91.4	109.4	127.4	145.4	163.4	181.4	199.4	3
4	39.2	57.2	75.2	93.2	111.2	129.2	147.2	165.2	183.2	201.2	4
5	41.0	59	77	95	113	131	149	167	185	203	5
6	42.8	60.8	78.8	96.8	114.8	132.8	150.8	168.8	186.8	204.8	6
7	44.6	62.6	80.6	98.6	116.6	134.6	152.6	170.6	188.6	206.6	7
8	46.4	64.4	82.4	100.4	118.4	136.4	154.4	172.4	190.4	208.4	8
9	48.2	66.2	84.2	102.2	120.2	138.2	156.2	174.2	192.2	210.2	9
C.	9	19	29	39	49	59	69	79	89	99	C.

100° C. = 212° F.

XX.—RELATION OF BAUMÉ DEGREES TO SPECIFIC GRAVITY AND THE WEIGHT OF ONE UNITED STATES GALLON AT 60° F.—LIQUIDS LIGHTER THAN WATER

Baumé.	Specific Gravity.	Pounds in Gallon.	Baumé.	Specific Gravity.	Pounds in Gallon.	Baumé.	Specific Gravity.	Pounds in Gallon.	Baumé.	Specific Gravity.	Pounds in Gallon.
10	1.0000	8.33	31	0.8695	7.24	52	0.7692	6.41	73	0.6896	5.75
11	0.9929	8.27	32	0.8641	7.20	53	0.7650	6.37	74	0.6863	5.52
12	0.9859	8.21	33	0.8588	7.15	54	0.7608	6.34	75	0.6829	5.69
13	0.9790	8.16	34	0.8536	7.11	55	0.7567	6.30	76	0.6796	5.66
14	0.9722	8.10	35	0.8484	7.07	56	0.7526	6.27	77	0.6763	5.63
15	0.9655	8.04	36	0.8433	7.03	57	0.7486	6.24	78	0.6730	5.60
16	0.9589	7.99	37	0.8383	6.98	58	0.7446	6.20	79	0.6698	5.58
17	0.9523	7.93	38	0.8333	6.94	59	0.7407	6.17	80	0.6666	5.55
18	0.9459	7.88	39	0.8284	6.90	60	0.7368	6.14	81	0.6635	5.52
19	0.9395	7.83	40	0.8235	6.86	61	0.7329	6.11	82	0.6604	5.50
20	0.9333	7.78	41	0.8187	6.82	62	0.7290	6.07	83	0.6573	5.48
21	0.9271	7.72	42	0.8139	6.78	63	0.7253	6.04	84	0.6542	5.45
22	0.9210	7.67	43	0.8092	6.74	64	0.7216	6.01	85	0.6511	5.42
23	0.9150	7.62	44	0.8045	6.70	65	0.7179	5.98	86	0.6481	5.40
24	0.9090	7.57	45	0.8000	6.66	66	0.7142	5.95	87	0.6451	5.38
25	0.9032	7.53	46	0.7954	6.63	67	0.7106	5.92	88	0.6422	5.36
26	0.8974	7.48	47	0.7909	6.59	68	0.7070	5.89	89	0.6392	5.33
27	0.8917	7.43	48	0.7865	6.55	69	0.7035	5.86	90	0.6363	5.30
28	0.8860	7.38	49	0.7821	6.52	70	0.7000	5.83	95	0.6222	5.18
29	0.8805	7.34	50	0.7777	6.48	71	0.6965	5.80	..	......	...
30	0.8750	7.29	51	0.7734	6.44	72	0.6930	5.78	..	......	...

XX.—(a) RELATION OF BAUMÉ DEGREES TO SPECIFIC GRAVITY— LIQUIDS HEAVIER THAN WATER

Baumé Degrees.	Specific Gravity.	Baumé Degrees.	Specific Gravity.	Baumé Degrees.	Specific Gravity.	Baumé Degrees.	Specific Gravity.
0.0	1.0000	6.0	1.0432	24.0	1.1983	42.0	1.4078
0.1	1.0007	7.0	1.0507	25.0	1.2083	43.0	1.4216
0.2	1.0014	8.0	1.0584	26.0	1.2185	44.0	1.4356
0.3	1.0021	9.0	1.0662	27.0	1.2288	45.0	1.4500
0.4	1.0028	10.0	1.0741	28.0	1.2393	46.0	1.4646
0.5	1.0035	11.0	1.0821	29.0	1.2500	47.0	1.4796
0.6	1.0042	12.0	1.0902	30.0	1.2609	48.0	1.4948
0.7	1.0049	13.0	1.0985	31.0	1.2719	49.0	1.5104
0.8	1.0055	14.0	1.1069	32.0	1.2832	50.0	1.5263
0.9	1.0062	15.0	1.1154	33.0	1.2946	51.0	1.5426
1.0	1.0069	16.0	1.1240	34.0	1.3063	52.0	1.5591
1.5	1.0105	17.0	1.1328	35.0	1.3282	53.0	1.5761
2.0	1.0140	18.0	1.1417	36.0	1.3303	54.0	1.5934
2.5	1.0175	19.0	1.1508	37.0	1.3426	55.0	1.6111
3.0	1.0211	20.0	1.1600	38.0	1.3551	56.0	1.6292
3.5	1.0247	21.0	1.1694	39.0	1.3679	57.0	1.6477
4.0	1.0284	22.0	1.1789	40.0	1.3810	58.0	1.6667
5.0	1.0357	23.0	1.1885	41.0	1.3942	59.0	1.6860
						60.0	1.7059

54

XXI.—COMPARISON OF METRIC AND CUSTOMARY UNITS (U. S.).

Length

1 millimeter, mm.	=0.03937 inch.	1 inch	=25.4001 millimeters.
1 centimeter, cm.	=0.39371 inch.	1 inch	= 2.54001 centimeters.
1 meter, m.	=3.28083 feet.	1 foot	= 0.304801 meter.
1 meter	=1.09361 yards.	1 yard	= 0.914402 meter.
1 kilometer	=0.62137 (U. S.) mile.	1 mile	= 1.60935 kilometers.

Areas

1 square millimeter, sq.mm.	= 0.00155 sq.in.	1 sq.in.	=645.16 sq.mm.
1 square centimeter, sq.cm.	= 0.1550 sq.in.	1 sq.in.	= 6.452 sq.cm.
1 square meter, sq.m.	=10.764 sq.ft.	1 sq.ft.	= 0.0929 sq.m.
1 square meter	= 1.196 sq.yd.	1 sq.yd.	= 0.8361 sq.m.
1 square kilometers	= 0.3861 sq.mi.	1 sq.mi.	= 2.5900 sq.km.
1 hectare	= 2.471 acres.	1 acre	= 0.4047 hectare

Volumes

1 cubic millimeter, cu.mm.	= 0.000061 in.	1 cu.in.	=16,387.2 cu.mm.
1 cubic centimeter, cc.	= 0.06103 cu.in.	1 cu.in.	= 16.3872 cc.
1 cubic meter	=35.314 cu.ft.	1 cu.ft.	= 0.02832 cu.m.
	=61,028 cu.ins.		=28.32 liters.
1 cubic meter	= 1.3079 cu.yd.	1 cu.yd.	= 0.7645 cu.m

Capacities

1 cubic centimeter, cc.	=0.03381 (U.S.) liquid ounce.	1 ounce	=29.574 cc.
1 cubic centimeter	=0.2705 (U. S.) apothecaries' dram.	1 dram	=3.6967 cc.
1 cubic centimeter	=0.8115 (U. S.) apothecaries' scruple.	1 scruple	=1.2322 cc.
1 liter	=1.05668 (U. S.) liquid quarts.	1 quart	=0.94636 liter.
1 liter	=0.26417 (U. S.) gallon.	1 gallon	=3.78543 liters.
1 liter	=0.11351 (U. S.) peck.	1 peck	=8.80982 liters.
1 hectoliter	=2.83774 (U. S.) bushels.	1 bushel	=0.35239 hectoliter.

Masses

1 gram	=15.4324 grains.	1 grain	= 0.06480 gram.
1 gram	= 0.03527 avoirdupois ounce.	1 ounce (av.)	=28.3495 grams.
1 gram	= 0.03215 troy ounce.	1 ounce (troy)	=31.10348 grams.
1 kilogram	= 2.20462 pounds (av.)	1 pound (av.)	= 0.45359 kilogram.
1 kilogram	= 2.67923 pounds (troy).	1 pound (troy)	= 0.37324 kilogram.

Table of Equivalents, U. S. Bureau of Standards. For British Imperial Weights and Measures see Van Nostrand's Chemical Annual.

Avoirdupois Weight

The system of weights in ordinary use by which common or heavy articles are weighed.

16 drams	=1 ounce	= 28.35 grams.
16 ounces	=1 pound	=453.59 grams.
25 pounds	=1 quarter	= 11.34 kilograms.
4 quarters	=1 hundred weight	= 45.359 kilograms.

1 avoirdupois pound contains 7000 grains.
1 avoirdupois ounce contains 437.5 grains.

Apothecaries' Weight

The system of weights employed in weighing medicines.

$$
\begin{array}{lll}
1 \text{ grain} & = & 0.0648 \text{ gram.} \\
20 \text{ grains} = 1 \text{ scruple} & = & 1.296 \text{ grams.} \\
3 \text{ scruples} = 1 \text{ drachm} & = & 3.888 \text{ grams.} \\
8 \text{ drachms} = 1 \text{ ounce} & = & 31.103 \text{ grams.} \\
12 \text{ ounces} = 1 \text{ pound} & = & 373.236 \text{ grams.}
\end{array}
$$

1 apothecaries' (or troy) pound contains 5760 grains.
1 apothecaries' (or troy) ounce contains 480 grains.

Fluid Measure

$$
\begin{array}{llll}
\mathbf{1} \text{ minim} & = & .06161 & \text{cubic centimeter.} \\
60 \text{ minims} & = 1 \text{ fluid drachm} = & 3.696 & \text{cubic centimeters.} \\
8 \text{ fluid drachms} = 1 \text{ fluid ounce} & = & 29.573 & \text{cubic centimeters.} \\
16 \text{ fluid ounces} & = 1 \text{ pint} & = 473.179 & \text{cubic centimeters.} \\
8 \text{ pints} & = 1 \text{ gallon} & = & 3.785 & \text{liters.}
\end{array}
$$

1 gallon contains 231 cubic inches.

The minim, fluid drachm, fluid ounce and pint are the fluid measures employed by apothecaries.

Other Data

1 cubic foot of water weighs 62.37 pounds.

1 gallon (U. S.) of water weighs 8.33 pounds (the British gallon is 20% more than the U. S.).

1 liter of water weighs approximately 2.2 pounds.

Areas of Plane Figures

Area of any triangle $= \frac{1}{2}$ base multiplied by the altitude.

Rectangle. Area = base multiplied by altitude, i.e., multiply the length of one side by the length of a perpendicular side.

Parallelogram. Area = multiply the length of one side by the vertical distance to the parallel side.

Trapezoid. Area = multiply half the sum of the parallel sides by the perpendicular distance between the two.

Circle. Area $= 0.7854d^2$, or πr^2 or $\frac{1}{2}Cr$, or $\frac{1}{4}Cd$.

$\pi = 3.1416$, r = radius, C = circumference, d = diameter.

Volumes of Solids

Regular Prism. Area $= \frac{1}{2}nrah$, or Bh.

n = number of sides, r = perpendicular from center of base to sides of the base, h = height of prism, B = area of base.

Lateral area $= 2\pi rh$.

Right Circular Cylinder. Volume $= \pi r^2h$ also Bh. Lateral area $= 2\pi rh$.

Regular Pyramid. Volume $= \frac{1}{3}$ altitude multiplied by area of base.

Right Circular Cone. Volume $= \frac{1}{3}\pi rh$. Lateral area $= \pi rs$.

r = radius of base, s = slant height or $\sqrt{r^2 + h^2}$.

Sphere. Volume $= 4/3\pi r^3$, or $4.189r^3$. Area $= 4r^2$ or $3.1416d^2$.

Barrel or Cask. Approximate gallons $= .0034n^2h$ where n = mean diameter of barrel. h = height, both measurements in inches.

XXII.—A TABLE OF CONSTANTS

THE UNITED GAS

All Volumes of Gases and Vapors are given at 60° F. and 30" pressure.

		I.	II.	III.	IV.	V.	VI.	VII.	VIII.	IX.	X.	XI.	XII.

NAME OF GAS OR VAPOR.	SYMBOL OR FORMULA.	Molecular Weight.	Sp. Gravity Gas or Vapor at 60° F. Air=1.0.	Boiling-point °Fahr.	Sp. Gravity Liquid at 60° F. Water=1.0.	Sp. Heat Eq. Wts. at Const. Pr. Water=1.0.	Cubic Feet per Pound.	Weight 1 Cubic Foot in Pounds.	Calories per Molecular Wt. in Grams.	British Thermal Units Per Cu. Ft.	British Thermal Units Per Pound.
Carbon to CO	C	12	0.8292				15.749	.06350	29,000	276.2	4,350
Carbon to CO₂	C	12	0.8292				15.749	.06350	96,960	923.5	14,544
Carbonic Ox..	CO	28	0.9671			0.2450	13.503	.07407	67,960	323.5	4,368
Hydrogen....	H	1	0.0692			3.0490	188.620	.00530	68,360	326.2	61,523
Methane.....	CH_4	16	0.5529			0.5929	23.626	.04234	211,930	1009.0	23,838
Ethane......	C_2H_6	30	1.0368				12.594	.07940	370,440	1764.4	22,226
Propane.....	C_3H_8	44	1.5206	− 13°			8.587	.11645	529,210	2521.0	21,651
Butane......	C_4H_{10}	58	2.0045	+ 33°			6.514	.15350	687,190	3274.0	21,326
Pentane.....	C_5H_{12}	72	2.4883	+100°	0.6273		5.248	.19055	847,110	4035.6	21,177
Hexane......	C_6H_{14}	86	2.9721	+156°	0.6640		4.393	.22760	999,200	4759.8	20,914
Ethylene....	C_2H_4	28	0.9676			0.4040	13.495	.07410	333,350	1588.0	21,430
Propylene....	C_3H_6	42	1.4514				8.997	.11115	492,740	2347.2	21,120
Butylene.....	C_4H_8	56	1.9353	+ 23°			6.747	.14820	650,620	3099.2	20,913
Amylene.....	C_5H_{10}	70	2.4191	+102°	0.6511		5.398	.18525	807,630	3847.2	20,767
Acetylene....	C_2H_2	26	0.8984				14.534	.06880	310,050	1476.7	21,465
Allylene.....	C_3H_4	40	1.3823				9.447	.10585	467,550	2227.1	21,046
Crotonylene..	C_4H_6	54	1.8661	+64°			6.998	.14290			
Benzene......	C_6H_6	78	2.6953	+177°	0.8846	0.3754	4.845	.20640	799,350	3807.5	18,447
Toluene......	C_7H_8	92	3.1792	+230°	0.8720		4.107	.24345	955,680	4552.0	18,699
Xylene......	C_8H_{10}	106	3.6630	+287°	0.8692		3.565	.28050			
Mesitylene....	C_9H_{12}	120	4.1468	+326°			3.149	.31755	1,282,310	6108.0	19,235
Naphthalene..	$C_{10}H_8$	128	4.4230	+424.4°	1.1517		2.952	.33870			
Hydrogen Sul.	H_2S	34	1.1769			0.2423	11.096	.09012	140,900	672.2	7,549
Ammonia....	NH_3	17	0.5888			0.5083	22.178	.04509	90,560	432.8	9,598
Hydrocy. acid.	HCN	27	0.9348				13.968	.07159	158,620	757.0	10,575
Cyanogen....	C_2N_2	52	1.8000				7.258	.13779	259,620	1238.2	8,986
Carbon Bi-Sul.	CS_2	76	2.6298	+114.8°			4.965	.20139	265,130	1264.6	6,279
Methyl Alc...	CH_4O	32	1.1121	+131.2°	0.8027		11.742	.08516	182,230	872.9	10,250
Ethyl Alcohol.	C_2H_6O	46	1.5894	+172.9°	0.7946	1.4534	8.216	.12172	340,530	1622.0	13,325
Carbonic Acid.	CO_2	44	1.5195			0.2163	8.593	.11637			
Water.......	H_2O	18	0.6217	+ 212°	1.0000	0.4805	21.004	.04761			
Sulphur Diox.	SO_2	64	2.2128			0.1553	5.901	.16945			
Oxygen......	O	16	1.1052			0.2174	11.816	.08463			
Nitrogen.....	N	14	0.9701			0.2438	13.460	.07429			
Air.........		..	1.0000			0.2374	13.059	.07658			

Series labels in column II: Paraffin Series C_nH_{2n+2}; Olefin Series C_nH_{2n}; Acetyl. Series C_nH_{2n-2}; Aromat. Series C_nH_{2n-6}.

Heat of Combustion (Columns X–XII).

AUTHORITIES AND METH.

In Column IX. the figures given in Hempel's "Gas Analysis," p. 375, were selected for the fundamental weight of Oxygen, Nitrogen, Hydrogen, Carbonic Oxide and Air.

The formula used for the conversion to English units, is—grams per liter at 0° C. and 760 mm. × .05922 = pounds per cu.ft. at 60° F. and 30" pressure. The derivation of the factor employed is

$$.05922 = \frac{28.316 \times .0022046 \times 30.00 \times 492}{29.92 \times 520}$$

The weights of the compound gases are calculated from these data by Avogadro's law.

Column IV. is calculated by the formula: $\text{sp.gr.} = \dfrac{\text{wt. 1 cu. ft. gas}}{\text{wt. 1 cu.ft. air}}$, and the figures thus obtained agree with

the theoretical formula: $\text{sp.gr.} = \dfrac{\text{mol. wt.}}{28.94}$.

FOR CERTAIN GASES AND VAPORS

IMPROVEMENT COMPANY

The Temperature of Products of Combustion is reduced to 18° C. = 64.4° F.

XIII.	XIV.	XV.	XVI.	XVII.	XVIII	XIX.	XX.	XXI.	XXII.	XXIII.	XXIV.	XXV.	XXVI.
Cu. Ft. per Cu. Ft. of Combustible.					Pounds per Pound of Combustible.					Heat of Formation at Const. Pres.			
Req. for Combustion.		Products of Combustion.			Req. for Combustion.		Products of Combustion.			Calories per Molecular Wt. in Grams.	B.t.u.		NAME OF GAS OR VAPOR.
Air.	Oxygen.	CO_2	H_2O.		Air	Oxygen.	CO_2	H_2O			Per Cu. Ft.	Per Pound.	
4.785	1.0	..	...	CO—2.0	5 771	1.333+	.....	...	CO—2.333+	.....	.....	......	Carbon to CO
9.570	2.0	2.0	..	......	11.541	2.666+	3.666+	...		.....	.....	......	Carbon to CO_2
2.393	0.5	1.0	..	......	2.471	.571	1.571	...		.....	+138.4	+1869.2	Carbonic Ox.
2.393	0.5	..	1.0	......	34.624	8.000	.....	9.000		.....	.....	......	Hydrogen
9.570	2.0	1.0	2.0	......	17.312	4.000	2.750	2.250		+21,750	+103.1	+2435.6	Methane
16.748	3.5	2.0	3.0	......	16.156	3.733	2.933	1.800		+28,560	+136.0	+1713.6	Ethane
23.925	5.0	3.0	4.0	......	15.737	3.636	3.000	1.636		+35,110	+167.2	+1436.3	Propane
31.103	6.5	4.0	5.0	......	15.520	3.586	3.034	1.552		+42,450	+202.2	+1317.3	Butane
38.280	8.0	5.0	6.0	......	15.386	3.555	3.055	1.500		+47,850	+227.9	+1196.2	Pentane
45.458	9.5	6.0	7.0	......	15.295	3.534	3.069	1.465		+61,080	+290.9	+1278.4	Hexane
14.355	3.0	2.0	2.0	......	14.836	3.428	3.142	1.286		−2,710	−12.9	−174.2	Ethylene
21.533	4.5	3.0	3.0	......	14.836	3.428	3.142	1.286		+3,220	+15.3	+138.1	Propylene
28.710	6.0	4.0	4.0	......	14.836	3.428	3.142	1.286		+10,660	+50.7	+342.6	Butylene
35.888	7.5	5.0	5.0	......	14.836	3.428	3.142	1.286		+18,970	+113.7	+614.1	Amylene
11.963	2.5	2.0	1.0	......	13.313	3.076	3.384	0.692		−47,770	−227.5	−3300.7	Acetylene
19.140	4.0	3.0	2.0	......	13.850	3.200	3.300	0.900		−39,650	−188.8	−1784.2	Allylene
26.318	5.5	4.0	3.0	......	14.105	3.259	3.259	1.300		....	.....	......	Crotonylene
35.888	7.5	6.0	3.0	......	13.313	3.076	3.384	0.692		−12,510	−47.3	−229.3	Benzene
43.065	9.0	7.0	4 0	......	13.547	3.130	3.348	0.782		−3,520	+16.7	−68.8	Toluene
50.243	10.5	8.0	5.0	......	13.720	3.170	3.311	0.849		....	.....	......	Xylene
57.420	12.0	9.0	6.0	......	13.850	3.200	3.300	0.900		+490	+2.3	+7.3	Mesitylene
57.420	12.0	10.0	4.0	......	12.984	3.000	3.437	0.563		....	.....	......	Naphthalene
7.178	1.5	..	1.0	SO_2—1.0	6.111	1.412	.....	0.529	SO_2—1.883	+4,740	+22.6	+250.9	Hydrogen Sul.
3.589	0.75	..	1.5	N—0.5	6.111	1.412	.....	1.588	N—0.823	+11,890	+56.7	+1259.0	Ammonia
5.981	1.25	1.0	0.5	N—0.5	6.410	1.481	1.630	0.333	N—0.518	−27,480	−131.1	−1832.0	Hydrocy. acid
9.570	2.0	2.0	..	N—1.0	5.323	1.230	1.692	...	N—0.538	−65,700	−313.2	−2273.9	Cyanogen
14.355	3.0	1.0	..	SO_2—2.0	5.466	1.263	0.579	...	SO_2—1.684	−26,010	−124.0	−616.0	Carbon Bi-Sul.
7.178	1.5	1.0	2.0	......	6.492	1.500	1.375	1.125		+51,450	+246.4	+2894.0	Methyl Alc.
14.355	3.0	2.0	3.0	......	9.033	2.087	1.913	1.174		+58,470	+278.5	+2288.0	Ethyl Alcohol.
......	......	...	...	......	......	......	...	...		+463.1	+3979.1		Carbonic Acid
......	......	...	...	......	......	......	...	...		+327.1	+6870.4		Water
......	......	...	...	......	......	......	...	...		+337.3	+1999.1		Sulphur Diox.
......	......	...	...	......	......	......	...	...		......	....		Oxygen
......	......	...	...	......	......	......	...	...		......	....		Nitrogen
......	......	...	...	......	......	......	...	...		......	....		Air

ODS OF CALCULATION.

Columns V. and VI. are taken chiefly from Lunge's " Coal Tar and Ammonia."

Column VII. is from Ganot's " Physics," edition 1896, page 445.

Column X. and XXIII. are from Julius Thomsen's " Thermochemical Investigations," and his results are translated into English units in columns XI.–XII. and XXIV.–XXV.

Columns XIII. and XVIII. are calculated on the assumption that

air = 20.9% oxygen + 79.1% nitrogen by *Volume.*

air = 23.13% oxygen + 76.87% nitrogen *by Weight.*

XXIII.—SPECIFIC GRAVITY OF GASES *

Name.	Formula.	Molecular Weight.	Specific Gravity, Air = 1.		Weight in Grams of 1 Liter at 0°, 760 mm. at Sea Level, Lat. 45°.[2]
			Calculated.[1]	Observed.	
Acetylene............	C_2H_2......	26.026	0.8993	0.9056	1.1624
Air.................			1.0000		1.2926
Ammonia.............	NH_3.......	17.034	0.5856	0.5971	0.7608
Argon................	A.........	39.88	1.3780	1.379	1.7812
Arsine...............	AsH_3.....	77.984	2.6933	2.695	3.4830
Bromine.............	Br_2........	159.84	5.5230	5.524(227.9°)	7.1390
Butane..............	C_4H_{10}......	58.10	2.0075	2.01	2.5949
Carbon dioxide.......	CO_2.......	44.005	1.5205	1.52908	1.9654
Carbon monoxide.....	CO.......	28.005	0.9677	0.96716	1.2508
Carbon oxysulphide....	COS....	60.065	2.0754	2.1046	2.6827
Chlorine.............	Cl_2........	70.92	2.4505	2.4901	3.1675
Cyanogen............	C_2N_2......	52.03	1.7977	1.8064	2.3338
Ethane..............	C_2H_6......	30.058	1.0386	1.0494	1.3425
Ethylene............	C_2H_4......	28.042	0.9689	0.9852	1.2524
Fluorine.............	F_2.........	38.0	1.3130	1.26	1.6972
Helium..............	He........	4.00	0.1382	0.131	0.1786
Hydrobromic acid....	HBr.......	80.928	2.7963	2.71	3.6145
Hydrochloric acid.....	HCl.......	36.468	1.2601	1.2681	1.6288
Hydrofluoric acid.....	HF........	20.008	0.6913	0.7126	0.8936
Hydroiodic acid......	HI........	127.928	4.4203	4.3757	5.7137
Hydrogen............	H_2........	2.016	0.06965	0.06952	0.09004
Hydrogen selenide....	H_2Se......	81.216	2.8063	2.795	3.6274
Hydrogen sulphide....	H_2S.......	34.076	1.1774	1.1906	1.5220
Hydrogen telluride....	H_2Te......	129.516	4.4752	4.489	5.7846
Krypton.............	Kr........	82.92	2.8252	2.868	3.7035
Methane.............	CH_4......	16.037	0.5541	0.5545	0.7163
Neon................	Ne........	20.2	0.6980	0.6963	0.9022
Nitric oxide.........	NO........	30.01	1.0370	1.0367	1.3404
Nitrous oxide........	N_2O.......	44.02	1.5210	1.5298	1.9661
Nitrogen............	N_2.......	28.02	0.9682	0.96737	1.2515
atmospheric........	N_2+A etc..			0.97209	1.25718
Nitrogen dioxide......	NO_2......	46.01	1.5898	1.60(135°)	2.0550
" "	N_2O_4......	92.02	3.1788	2.65(26.7°)	4.1099
Nitrosyl chloride......	NOCl.....	65.47	2.2622	2.31	2.9241
Oxygen..............	O_2........	32.00	1.1057	1.1053	1.4292
Phosphine...........	PH_3.......	34.064	1.1770	1.1829	1.5214
Propylene...........	C_3H_6......	42.063	1.4534	1.498	1.8787
Silicon fluoride.......	SiF_4......	104.3	3.6039	3.60	4.6584
Sulphur dioxide.......	SO_2.......	64.06	2.2135	2.2638	2.8611
Xenon..............	X.........	130.2	4.4988	4.526	5.8152

* D. Van Nostrand Chemical Annual.—Olsen.

[1] Obtained by dividing the weight of one liter of the gas by 1.2926 (*i.e.*, by the weight of one liter of air).

[2] Obtained by multiplying $\frac{1}{2}$ the molecular weight of the gas by $\frac{1}{16} \times 1.429234$ (*i.e.*, by $\frac{1}{16} \times$ weight of one liter of oxygen).

USEFUL MEMORANDA

Gas Calculations

Conversion Formulae for Gas Volumes. Conversion from existing to standard conditions,

$$V = \frac{V' \times 273 \times (P - w)}{273 + t \times 760}.$$

Gas at standard volume to new conditions of temperature and pressure

$$V' = \frac{V \times 273 + t \times 760}{273 \times (P - w)}.$$

V = standard volume, V' = volume under other conditions of temperature and pressure (other than 0° C. and 760 mm. Hg pressure).

t = temperature of the gas, P = barometric pressure in mm. Hg + static pressure of the gas. w = water vapor pressure at temperature and pressure that the gas is measured or to which it is to be converted.

Petot Formulae for Measuring the Velocity of a Gas in a Flue.

A. Simple formula for fair approximations $V = 42\sqrt{h}$.

B. More exact formula $V = 1290 \dfrac{\sqrt{\frac{1}{2}}h(1 + .00217t)}{BM}$.

V = velocity in feet per second. h = gauge reading in terms of water, i.e., the total differential in terms of vertical reading. In a gauge inclined 1 : 10, with ether in place of water, 10″ differential = 10 × .72 ÷ 10 = .72. In the formula B ½ the reading is taken as the deflection due to velocity pressure is half the total deflection (proven by repeated tests, suction is producted by the gas flow on the straight petot arm). Allowance is made for this in formula A. t = temperature of the gas measured. B = barometric pressure + plus static pressure of the gas in terms of inches of mercury (air standard 29.92″). M = specific gravity of the gas compared with the hydrogen molecule H_2 = 2.016.

To obtain the specific gravity of a gas, multiply the molecular weight of each constituent by its per cent ratio in the mixture, divide by 100 and finally divide the sum of these results by 2.016.

Examples. The sp.gr. of a gas having the composition 87% nitrogen, 5% oxygen, and 8% sulphur dioxide we would find

28.02 (N_2) × 87 ÷ 100 = 24.38
32.00 (O_2) × 5 " = 1.6

64.06 (SO_2) × 8 " = 5.12 total = 31.1 sp.gr. = $\dfrac{31.1}{2.016}$ = 15.42.

The sp.gr. of air containing 78% nitrogen, 21% oxygen, 1% argon would be

28.02 × .78 = 21.86,
32.00 × .21 = 6.72,
39.88 × .1 = 0.4 total = 28.98 sp.gr. = 14.38.

Consult table XXIII.

The gram molecular weight of any gas occupies 22.4 liters.

Definitions

Heat

Calorie is 1/100 of the heat required to raise the temperature of one gram of water from 0° C. to 100° C.

One kilogram calorie = 1000 gram calories or 3.968 B.t.u.

Latent heat of evaporation is the quantity of heat required to convert 1 gram of a liquid into vapor without change of temperature.

Latent heat of fusion is the quantity of heat required to change a gram of the substance from a solid to a liquid state without changing temperature.

Specific heat. The British thermal unit (B.t.u.) = 1/180 part of the heat required to raise the temperature of one pound of water from 32° F. to 212° F.

Specific heat of a body is the quantity of heat required to raise the temperature of a unit weight of that body one degree, C.

Photometry

Lux. This is the practical unit of illumination and is the light received from a source of 1 meter distance.

Foot-candle. This is the illumination received by a standard candle at one foot distance. The candle loses in burning 120 grams per hour.

1 foot-candle = 10.76 lux.

Candle-power. The comparison of the illumination of a light with one standard candle. A simple method for determination is to place the light and the burning candle 60 inches apart (2 meters and 100 inches distance also taken). A screen with a spot of oil is so placed between the lights as to receive a uniform illumination on each side, which is readily seen by the spot. The square of the distance from the screen to the light tested divided by the square of the distance from the screen to the standard candle = the candle-power of the light. When gas is tested the flame is so regulated that it burns 5 cubic feet of gas an hour. The standard candle burns 120 grams per hour as stated above.

Units in Electricity

Ampere is the unit of current flow. An ampere will deposit from a silver nitrate solution 1.118 milligrams of silver (.001118 g.) per second.

Coulomb is the unit of quantity. One coulomb will deposit .001118 g. Ag (time factor not considered as in case of ampere). 96,494 coulombs will deposit the gram equivalent of an element from a solution of its salt, i.e., 107.88 grams of silver, $\dfrac{63.57}{2}$ grams of cupric copper Cu^{++} $\dfrac{55.84}{3}$ of ferric iron Fe^{+++}, 63.57 of cuprous copper Cu^{+}, $\dfrac{55.84}{2}$ gram of ferrous iron Fe^{++}, etc. One gram equivalent of an element carries 96,500 (round numbers) coulombs.

Electromotive force—E.M.F.—is the force causing electricity to move.

Erg. This is the unit of power and is the force which applied to the standard gram body would give that body an acceleration of 1 cm. per second.

Dyne = the force which acting on one gram for one second produces a velocity of one centimeter per second.

Joule.   This is the practical unit of electrical energy and is produced when a steady current of one coulomb per second, i.e., one ampere, passes through a resistance of one ohm for one second.

Ohm.   This is the resistance that is offered at 0° C. by a column of mercury 106.3 centimeters long, being 1 square millimeter in cross section and weighing 14.4521 grams.

Volt.   This is the electromotive force which produces one ampere of current through a conductor having a resistance of one ohm.

    If any two of the factors are known the third can be found by the formula $I = E/R$; I = ampere current, E = electromotive force (volts) and R = ohms resistance.

Watt.   One watt is produced when one ampere of current flows under a pressure of one volt (E.M.F.).

    One watt = .00134 horse power, = 44.25 foot pounds, per minute, = 14.33 gram calories per minute, = .057 B.t.u. per minute.

Watt-hour and kilowatt-hour are the units of energy used in commercial electrical work.

    Watt-hour = 3600 joules, = 2655.4 ft. lbs., = 859975 gram calories, = 3.412 B.t.u. = .001341 h.p. hr.

    One kilowatt = 1000 watts.

Other Terms

Foot-pound is the unit of work done in raising a weight of one pound through a distance of one foot.   Work = weight multiplied by height, or force multiplied by distance.

Horse power = 550 foot pounds.

Specific gravity of a substance is the ratio of the weight of a given volume of that substance compared with the weight of an equal volume of water.

    a. Solid heavier than water.   Weight of solid in the air divided by the difference between this weight and its weight in water = sp.gr. of solid.

    b. Solid lighter than water.   The weight of the solid in the air divided by this weight + the loss of weight of a sinker due to the buoyancy of the solid = sp.gr.

    c. Solid soluble in water.   Determine gravity in a liquid in which it is insoluble whose specific gravity is known, multiply the gravity in terms of the liquid by the gravity of the liquid = sp.gr. in terms of water.

    d. Liquids.   Compare with water by weighing in a specific gravity bottle; the bottle being filled with water is weighed, the bottle, drained and dried, is filled with the liquid and again weighed ; the weights are compared on subtracting the weight of the empty bottle.

    e. The loss of weight of an insoluble substance immersed in the liquid is divided by the loss of weight of the substance immersed in water = sp.gr.

    f. The specific gravity is determined by the hydrometer.   See chapter on Acids, Volume 2.

VOLUME AND WEIGHT CONVERSION TABLE

Multiply by

To Convert From	To Cu. In.	To Cu. Ft.	To Cu. Yd.	To Fl. Oz.	To Pint.	To Quart.	To Gallon.	To Grain.	To Oz. Troy.	To Oz. Av.	To Lb. Troy.	To Lb. Av.	To CC. or G.	To Ltr. or Kg.	To Cu. M.
Cu. In.	1.00000	$.0_{3}5787$	$.0_{4}2143$	.554112	.034632	.017316	.004329	252.891	.526857	.578037	.043905	.036127	16.3871	.016387	$.0_{4}1639$
Cu. Ft.	1728.00	1.00000	.037037	957.505	59.8442	29.9221	7.48052	436996.	910.408	998.848	75.8674	62.4280	28316.9	28.3169	.028317
Cu. Yd.	46656.0	27.0000	1.00000	25852.6	1615.79	807.896	201.974	117990_{3}	24581.0	26968.9	2048.42	1685.56	764556.	764.556	.764556
Fl. Oz.	1.80469	.001044	$.0_{4}3868$	1.00000	.062500	.031250	.007813	456.390	.950813	1.04318	.079234	.065199	29.5736	.029573	$.0_{4}2957$
Pint.	28.8750	.016710	$.0_{3}6189$	16.0000	1.00000	.500000	.125000	7302.23	15.2130	16.6908	1.26775	1.04318	473.177	.473177	$.0_{3}4732$
Quart.	57.7500	.033420	.001238	32.0000	2.00000	1.00000	.250000	14604.5	30.4260	33.3816	2.53550	2.08635	946.354	.946354	$.0_{3}9463$
Gallon.	231.000	.133681	.004951	128.000	8.00000	4.00000	1.00000	58417.9	121.704	133.527	10.1420	8.34541	3785.42	3.78542	.003785
Grain.	$.0_{2}3954$	$.0_{5}2288$	$.0_{6}8475$	.002191	$.0_{3}1369$	$.0_{4}6850$	$.0_{4}1712$	1.00000	.002083	.002286	$.0_{3}1736$	$.0_{3}1428$	.064799	$.0_{4}6479$	$.0_{7}6479$
Oz. Troy	1.89805	.001098	$.0_{4}4068$	1.05173	.065733	.032867	.008217	480.000	1.00000	1.09714	.083333	.068571	31.1035	.031104	$.0_{4}3110$
Oz. Av.	1.72999	.001001	$.0_{4}3708$	.958608	.059913	.029957	.007489	437.500	.911457	1.00000	.075955	.062500	28.3495	.028350	$.0_{4}2835$
Lb. Troy	22.7766	.013181	$.0_{3}4882$	12.6208	.788800	.394400	.098600	5760.00	12.0000	13.1657	1.00000	.822857	373.242	.373242	$.0_{3}3732$
Lb. Av.	27.6799	.016018	$.0_{3}5933$	15.3378	.958611	.479306	.119826	7000.00	14.5833	16.0000	1.21528	1.00000	453.593	.453593	$.0_{3}4536$
CC or Gram	.061024	$.0_{4}3531$	$.0_{5}1308$	.033814	.002113	.001057	$.0_{3}2642$	15.4323	.032151	.035274	.002679	.002205	1.00000	.001000	.000001
Liter or Kg.	61.0237	.035315	.001308	33.8140	2.11337	1.05669	.264172	15432.3	32.1507	35.2739	2.67923	2.20462	1000.00	1.00000	.001000
Cu. M.	61023.7	35.3146	1.30795	33814.0	2113.37	1056.69	264.172	154320_{3}	32150.7	35273.9	2679.23	2204.62	1000000.	1000.00	1.00000

Note. The small subnumeral following a zero indicates that the zero is to be taken that number of times; thus, $.0_{3}1428$ is equivalent to .0001428.

Values used in constructing table:

1 inch = 2.540001 cm. 　　　　　　　1 lb. av. = 453.5926 g. 　　　　　　　1 lb. av. = 7000 grains.

∴ 1 cu. in. = 16.387083 cc. = 16.387083 g H_2O at 4°C = 39°F. 　　∴ 1 gal. = 8.34541 lb. 　　　∴ 1 gallon = 58417.87 grains.

231 cu. in. = 1 gallon = 3785.4162 g. 　　∴ 1 lb. av. = 27.679886 cu. in. H_2O at 4°C.

ENERGY CONVERSION FACTORS

Multiply by

To Convert From	B. T. U.	P. C. U.	Cal.	Ft. Lbs.	Ft. Tons	Kg. M.	HP Hrs.	KW Hrs.	Joules	Lbs. C	Lbs. H₂O
B. T. U.	1.00000	.555556	.251996	778.000	.389001	107.563	$.0_33929$	$.0_32931$	1055.20	$.0_36876$	.001031
P. C. U.	1.80000	1.00000	.453593	1400.40	.700202	193.613	$.0_37072$	$.0_35276$	1899.36	$.0_31238$	.001855
Calories	3.96832	2.20462	1.00000	3091.36	1.54368	426.844	.001559	.001163	4187.37	$.0_32729$	.004089
Ft. Lbs.	.001285	$.0_37141$	$.0_33239$	1.00000	.000500	.138255	$.0_55050$	$.0_53767$	1.35625	$.0_38840$	0.1325
Fr. Tons	2.57069	1.42816	.647804	2000.00	1.00000	276.511	.001010	$.0_37535$	2712.59	$.0_31768$	.002649
Kg. M.	.009297	.005165	.002343	7.23301	$.0_33617$	1.00000	$.0_53653$	$.0_32725$	9.81009	$.0_36394$	$.0_59580$
HP Hrs.	2544.99	1413.88	641.327	1980000	990.004	273747	1.00000	.746000	2685473	.175044	2.62261
KW Hrs.	3411.57	1895.32	859.702	2654200	1327.10	360959	1.34041	1.00000	3599889	.234648	3.51562
Joules	$.0_39477$	$.0_35265$	$.0_32388$	.737311	$.0_33687$	.101937	$.0_33724$	$.0_32778$	1.00000	$.0_46518$	$.0_49766$
Lbs. C	14544.0	8080.00	3665.03	1131503_3	5657.63	1564396	5.71434	4.26285	153470_3	1.00000	14.9876
Lbs. H₂O	970.400	539.111	244.537	754971	377.487	104379	.381270	.284424	1023966	.066744	1.00000

"P. C. U." refers to the "pound-centigrade unit." The ton used is 2000 pounds. "Lbs. C" refers to pounds of carbon oxidized, 100 % efficiency, equivalent to the corresponding number of heat units. "Lbs. H₂O" refers to pounds of water evaporated at 100° C. = 212° F. at 100% efficiency. The sub-numeral following a zero indicates that the zero should be taken the number of times indicated. Thus: 0_33653 is equivalent to .000003653.

INTER-CONVERSION TABLE

Multiply by

To Convert from	Grains per Cu. Ft.	Ounces per Cu. Ft.	Pounds per Cu. Ft.	Grams per Cu. Ft.	Grains per Gallon	Ounces per Gallon	Pounds per Gallon	Grains per Cu. In.	Ounces per Cu. In.	Grains per Liter
Grains per Cu. Ft.	1.00000	.002286	$.0_31428$	.064799	.133680	$.0_33056$	$.0_41910$	$.0_35787$	$.0_41323$	.002288
Oz. per Cu. Ft.	437.500	1.00000	.062497	28.4954	58.4848	.133678	$.0_38355$	.253180	$.0_35786$	1.00103
Lbs. per Cu. Ft.	7000.00	16.0000	1.00000	453.593	935.757	2.13885	.133680	4.05090	$.0_39258$	16.0163
G. per Cu. Ft.	15.4324	.035274	.002205	1.00000	2.06302	.004715	$.0_32947$	.008931	$.0_42041$	.035310
Grains per Gallon	7.48050	.017098	.001069	.484727	1.00000	.002286	$.0_31428$	.004329	$.0_49894$	.017116
Oz. per Gallon	3272.72	7.48050	.467510	21.2068	437.500	1.00000	.062499	1.89392	.003429	7.48798
Lbs. per Gallon	52363.6	119.690	7.48014	3393.09	7000.00	16.0000	1.00000	30.3028	.069261	119.808
Grains per Cu. In.	1728.00	3.94960	.246840	111.972	231.000	.527990	.032990	1.00000	.002285	3.95380
Oz. per Cu. In.	756000	1728.00	108.000	48987.8	101061	231.000	14.4370	437.500	1.00000	1729.73
G. per Liter	437.050	.998950	.062432	28.3203	58.4220	.133540	$.0_38346$	.252910	$.0_35780$	1.00000

INTER-CONVERSION TABLE

To Convert From	Multiply by				
	Lbs. per Sq. In.	Tons per Sq. In.	Kg. per Sq. Cm.	M. Tons per Sq. Cm.	Atmospheres
Lbs. per Sq. In....	1.00000	.000446	.070307	$.0_47031$	.068041
Tons per Sq. In....	2240.00	1.00000	157.488	.157488	152.412
Kg. per Sq. Cm....	14.2233	.006350	1.00000	.001000	.967768
Metric Tons per Sq. Cm.........	14223.0	6.34969	1000.00	1.00000	967.768
Atmospheres......	14.6970	.006558	1.03329	.001033	1.00000

1 gram = 15.43 grams

1 gram = .0468 grams

1 m. = 3.28 ft.

The above tables were obtained by courtesy of du Pont Ne Mours Company through the kindness of Dr. E. C. Lathrop.

													PROPORTIONAL PARTS.								

log 12 = 1.0792   log 144 = 2.1584
colog 12 = 8.9208-10   colog 144 = 7.8416-10

Nos.	0	1	2	3	4	5	6	7	8	9	1	2	3	4	5	6	7	8	9
10	0000	0043	0086	0128	0170	0212	0253	0294	0334	0374	4	8	12	17	21	25	29	33	37
11	0414	0453	0492	0531	0569	0607	0645	0682	0719	0755	4	8	11	15	19	23	26	30	34
12	0792	0828	0864	0899	0934	0969	1004	1038	1072	1106	3	7	10	14	17	21	24	28	31
13	1139	1173	1206	1239	1271	1303	1335	1367	1399	1430	3	6	10	13	16	19	23	26	29
14	1461	1492	1523	1553	1584	1614	1644	1673	1703	1732	3	6	9	12	15	18	21	24	27
15	1761	1790	1818	1847	1875	1903	1931	1959	1987	2014	3	6	8	11	14	17	20	22	25
16	2041	2068	2095	2122	2148	2175	2201	2227	2253	2279	3	5	8	11	13	16	18	21	24
17	2304	2330	2355	2380	2405	2430	2455	2480	2504	2529	2	5	7	10	12	15	17	20	22
18	2553	2577	2601	2625	2648	2672	2695	2718	2742	2765	2	5	7	9	12	14	16	19	21
19	2788	2810	2833	2856	2878	2900	2923	2945	2967	2989	2	4	7	9	11	13	16	18	20
20	3010	3032	3054	3075	3096	3118	3139	3160	3181	3201	2	4	6	8	11	13	15	17	19
21	3222	3243	3263	3284	3304	3324	3345	3365	3385	3404	2	4	6	8	10	12	14	16	18
22	3424	3444	3464	3483	3502	3522	3541	3560	3579	3598	2	4	6	8	10	12	14	15	17
23	3617	3636	3655	3674	3692	3711	3729	3747	3766	3784	2	4	6	7	9	11	13	15	17
24	3802	3820	3838	3856	3874	3892	3909	3927	3945	3962	2	4	5	7	9	11	12	14	16
25	3979	3997	4014	4031	4048	4065	4082	4099	4116	4133	2	3	5	7	9	10	12	14	15
26	4150	4166	4183	4200	4216	4232	4249	4265	4281	4298	2	3	5	7	8	10	11	13	15
27	4314	4330	4346	4362	4378	4393	4409	4425	4440	4456	2	3	5	6	8	9	11	13	14
28	4472	4487	4502	4518	4533	4548	4564	4579	4594	4609	2	3	5	6	8	9	11	12	14
29	4624	4639	4654	4669	4683	4698	4713	4728	4742	4757	1	3	4	6	7	9	10	12	13
30	4771	4786	4800	4814	4829	4843	4857	4871	4886	4900	1	3	4	6	7	9	10	11	13
31	4914	4928	4942	4955	4969	4983	4997	5011	5024	5038	1	3	4	6	7	8	10	11	12
32	5051	5065	5079	5092	5105	5119	5132	5145	5159	5172	1	3	4	5	7	8	9	11	12
33	5185	5198	5211	5224	5237	5250	5263	5276	5289	5302	1	3	4	5	6	8	9	10	12
34	5315	5328	5340	5353	5366	5378	5391	5403	5416	5428	1	3	4	5	6	8	9	10	11
35	5441	5453	5465	5478	5490	5502	5514	5527	5539	5551	1	2	4	5	6	7	9	10	11
36	5563	5575	5587	5599	5611	5623	5635	5647	5658	5670	1	2	4	5	6	7	8	10	11
37	5682	5694	5705	5717	5729	5740	5752	5763	5775	5786	1	2	3	5	6	7	8	9	10
38	5798	5809	5821	5832	5843	5855	5866	5877	5888	5899	1	2	3	5	6	7	8	9	10
39	5911	5922	5933	5944	5955	5966	5977	5988	5999	6010	1	2	3	4	6	7	8	9	10
40	6021	6031	6042	6053	6064	6075	6085	6096	6107	6117	1	2	3	4	5	6	8	9	10
41	6128	6138	6149	6160	6170	6180	6191	6201	6212	6222	1	2	3	4	5	6	7	8	9
42	6232	6243	6253	6263	6274	6284	6294	6304	6314	6325	1	2	3	4	5	6	7	8	9
43	6335	6345	6355	6365	6375	6385	6395	6405	6415	6425	1	2	3	4	5	6	7	8	9
44	6435	6444	6454	6464	6474	6484	6493	6503	6513	6522	1	2	3	4	5	6	7	8	9
45	6532	6542	6551	6561	6571	6580	6590	6599	6609	6618	1	2	3	4	5	6	7	8	9
46	6628	6637	6646	6656	6665	6675	6684	6693	6702	6712	1	2	3	4	5	6	7	7	8
47	6721	6730	6739	6749	6758	6767	6776	6785	6794	6803	1	2	3	4	5	5	6	7	8
48	6812	6821	6830	6839	6848	6857	6866	6875	6884	6893	1	2	3	4	4	5	6	7	8
49	6902	6911	6920	6928	6937	6946	6955	6964	6972	6981	1	2	3	4	4	5	6	7	8
50	6990	6998	7007	7016	7024	7033	7042	7050	7059	7067	1	2	3	3	4	5	6	7	8
51	7076	7084	7093	7101	7110	7118	7126	7135	7143	7152	1	2	3	3	4	5	6	7	8
52	7160	7168	7177	7185	7193	7202	7210	7218	7226	7235	1	2	2	3	4	5	6	7	7
53	7243	7251	7259	7267	7275	7284	7292	7300	7308	7316	1	2	2	3	4	5	6	6	7
54	7324	7332	7340	7348	7356	7364	7372	7380	7388	7396	1	2	2	3	4	5	6	6	7

π = 3.1416   log π = 0.4971   colog π = 9.5029-10

Computed by H. G. Shaw, Ph.D.

A convenient chart mounted on cardboard may be obtained of the above logarithms and the following antilogarithms from L. E. Knott Apparatus Co., 79–83 Amherst Street, Cambridge, Mass.

| | log 60 = 1.7782 | | | | | | | log .55 = 9.7404-10 | | |
| | colog 60 = 8.2218-10 | | | | | | | colog .55 = 0.2596 | | |

Nos.	0	1	2	3	4	5	6	7	8	9	PROPORTIONAL PARTS.								
											1	2	3	4	5	6	7	8	9
55	7404	7412	7419	7427	7435	7443	7451	7459	7466	7474	1	2	2	3	4	5	5	6	7
56	7482	7490	7497	7505	7513	7520	7528	7536	7543	7551	1	2	2	3	4	5	5	6	7
57	7559	7566	7574	7582	7589	7597	7604	7612	7619	7627	1	2	2	3	4	5	5	6	7
58	7634	7642	7649	7657	7664	7672	7679	7686	7694	7701	1	1	2	3	4	4	5	6	7
59	7709	7716	7723	7731	7738	7745	7752	7760	7767	7774	1	1	2	3	4	4	5	6	7
60	7782	7789	7796	7803	7810	7818	7825	7832	7839	7846	1	1	2	3	4	4	5	6	6
61	7853	7860	7868	7875	7882	7889	7896	7903	7910	7917	1	1	2	3	4	4	5	6	6
62	7924	7931	7938	7945	7952	7959	7966	7973	7980	7987	1	1	2	3	3	4	5	6	6
63	7993	8000	8007	8014	8021	8028	8035	8041	8048	8055	1	1	2	3	3	4	5	5	6
64	8062	8069	8075	8082	8089	8096	8102	8109	8116	8122	1	1	2	3	3	4	5	5	6
65	8129	8136	8142	8149	8156	8162	8169	8176	8182	8189	1	1	2	3	3	4	5	5	6
66	8195	8202	8209	8215	8222	8228	8235	8241	8248	8254	1	1	2	3	3	4	5	5	6
67	8261	8267	8274	8280	8287	8293	8299	8306	8312	8319	1	1	2	3	3	4	5	5	6
68	8325	8331	8338	8344	8351	8357	8363	8370	8376	8382	1	1	2	3	3	4	4	5	6
69	8388	8395	8401	8407	8414	8420	8426	8432	8439	8445	1	1	2	2	3	4	4	5	6
70	8451	8457	8463	8470	8476	8482	8488	8494	8500	8506	1	1	2	2	3	4	4	5	6
71	8513	8519	8525	8531	8537	8543	8549	8555	8561	8567	1	1	2	2	3	4	4	5	5
72	8573	8579	8585	8591	8597	8603	8609	8615	8621	8627	1	1	2	2	3	4	4	5	5
73	8633	8639	8645	8651	8657	8663	8669	8675	8681	8686	1	1	2	2	3	4	4	5	5
74	8692	8698	8704	8710	8716	8722	8727	8733	8739	8745	1	1	2	2	3	4	4	5	5
75	8751	8756	8762	8768	8774	8779	8785	8791	8797	8802	1	1	2	2	3	3	4	5	5
76	8808	8814	8820	8825	8831	8837	8842	8848	8854	8859	1	1	2	2	3	3	4	5	5
77	8865	8871	8876	8882	8887	8893	8899	8904	8910	8915	1	1	2	2	3	3	4	4	5
78	8921	8927	8932	8938	8943	8949	8954	8960	8965	8971	1	1	2	2	3	3	4	4	5
79	8976	8982	8987	8993	8998	9004	9009	9015	9020	9025	1	1	2	2	3	3	4	4	5
80	9031	9036	9042	9047	9053	9058	9063	9069	9074	9079	1	1	2	2	3	3	4	4	5
81	9085	9090	9096	9101	9106	9112	9117	9122	9128	9133	1	1	2	2	3	3	4	4	5
82	9138	9143	9149	9154	9159	9165	9170	9175	9180	9186	1	1	2	2	3	3	4	4	5
83	9191	9196	9201	9206	9212	9217	9222	9227	9232	9238	1	1	2	2	3	3	4	4	5
84	9243	9248	9253	9258	9263	9269	9274	9279	9284	9289	1	1	2	2	3	3	4	4	5
85	9294	9299	9304	9309	9315	9320	9325	9330	9335	9340	1	1	2	2	3	3	4	4	5
86	9345	9350	9355	9360	9365	9370	9375	9380	9385	9390	1	1	2	2	3	3	4	4	5
87	9395	9400	9405	9410	9415	9420	9425	9430	9435	9440	0	1	1	2	2	3	3	4	4
88	9445	9450	9455	9460	9465	9469	9474	9479	9484	9489	0	1	1	2	2	3	3	4	4
89	9494	9499	9504	9509	9513	9518	9523	9528	9533	9538	0	1	1	2	2	3	3	4	4
90	9542	9547	9552	9557	9562	9566	9571	9576	9581	9586	0	1	1	2	2	3	3	4	4
91	9590	9595	9600	9605	9609	9614	9619	9624	9628	9633	0	1	1	2	2	3	3	4	4
92	9638	9643	9647	9652	9657	9661	9666	9671	9675	9680	0	1	1	2	2	3	3	4	4
93	9685	9689	9694	9699	9703	9708	9713	9717	9722	9727	0	1	1	2	2	3	3	4	4
94	9731	9736	9741	9745	9750	9754	9759	9763	9768	9773	0	1	1	2	2	3	3	4	4
95	9777	9782	9786	9791	9795	9800	9805	9809	9814	9818	0	1	1	2	2	3	3	4	4
96	9823	9827	9832	9836	9841	9845	9850	9854	9859	9863	0	1	1	2	2	3	3	4	4
97	9868	9872	9877	9881	9886	9890	9894	9899	9903	9908	0	1	1	2	2	3	3	4	4
98	9912	9917	9921	9926	9930	9934	9939	9943	9948	9952	0	1	1	2	2	3	3	4	4
99	9956	9961	9965	9969	9974	9978	9983	9987	9991	9996	0	1	1	2	2	3	3	3	4

| g = 32.16 | log g = 1.5073 |
| | colog g = 8.4927-10 |

Mants.	0	1	2	3	4	5	6	7	8	9	1	2	3	4	5	6	7	8	9
											PROPORTIONAL PARTS.								
.00	1000	1002	1005	1007	1009	1012	1014	1016	1019	1021	0	0	1	1	1	1	2	2	2
.01	1023	1026	1028	1030	1033	1035	1038	1040	1042	1045	0	0	1	1	1	1	2	2	2
.02	1047	1050	1052	1054	1057	1059	1062	1064	1067	1069	0	0	1	1	1	1	2	2	2
.03	1072	1074	1076	1079	1081	1084	1086	1089	1091	1094	0	0	1	1	1	1	2	2	2
.04	1096	1099	1102	1104	1107	1109	1112	1114	1117	1119	0	1	1	1	1	2	2	2	2
.05	1122	1125	1127	1130	1132	1135	1138	1140	1143	1146	0	1	1	1	1	2	2	2	2
.06	1148	1151	1153	1156	1159	1161	1164	1167	1169	1172	0	1	1	1	1	2	2	2	2
.07	1175	1178	1180	1183	1186	1189	1191	1194	1197	1199	0	1	1	1	1	2	2	2	2
.08	1202	1205	1208	1211	1213	1216	1219	1222	1225	1227	0	1	1	1	1	2	2	2	3
.09	1230	1233	1236	1239	1242	1245	1247	1250	1253	1256	0	1	1	1	1	2	2	2	3
.10	1259	1262	1265	1268	1271	1274	1276	1279	1282	1285	0	1	1	1	1	2	2	2	3
.11	1288	1291	1294	1297	1300	1303	1306	1309	1312	1315	0	1	1	1	2	2	2	2	3
.12	1318	1321	1324	1327	1330	1334	1337	1340	1343	1346	0	1	1	1	2	2	2	3	3
.13	1349	1352	1355	1358	1361	1365	1368	1371	1374	1377	0	1	1	1	2	2	2	3	3
.14	1380	1384	1387	1390	1393	1396	1400	1403	1406	1409	0	1	1	1	2	2	2	3	3
.15	1413	1416	1419	1422	1426	1429	1432	1435	1439	1442	0	1	1	1	2	2	2	3	3
.16	1445	1449	1452	1455	1459	1462	1466	1469	1472	1476	0	1	1	1	2	2	2	3	3
.17	1479	1483	1486	1489	1493	1496	1500	1503	1507	1510	0	1	1	1	2	2	2	3	3
.18	1514	1517	1521	1524	1528	1531	1535	1538	1542	1545	0	1	1	1	2	2	2	3	3
.19	1549	1552	1556	1560	1563	1567	1570	1574	1578	1581	0	1	1	1	2	2	3	3	3
.20	1585	1589	1592	1596	1600	1603	1607	1611	1614	1618	0	1	1	2	2	2	3	3	3
.21	1622	1626	1629	1633	1637	1641	1644	1648	1652	1656	0	1	1	2	2	2	3	3	3
.22	1660	1663	1667	1671	1675	1679	1683	1687	1690	1694	0	1	1	2	2	2	3	3	3
.23	1698	1702	1706	1710	1714	1718	1722	1726	1730	1734	0	1	1	2	2	2	3	3	4
.24	1738	1742	1746	1750	1754	1758	1762	1766	1770	1774	0	1	1	2	2	2	3	3	4
.25	1778	1782	1786	1791	1795	1799	1803	1807	1811	1816	0	1	1	2	2	2	3	3	4
.26	1820	1824	1828	1832	1837	1841	1845	1849	1854	1858	0	1	1	2	2	3	3	3	4
.27	1862	1866	1871	1875	1879	1884	1888	1892	1897	1901	0	1	1	2	2	3	3	3	4
.28	1905	1910	1914	1919	1923	1928	1932	1936	1941	1945	0	1	1	2	2	3	3	4	4
.29	1950	1954	1959	1963	1968	1972	1977	1982	1986	1991	0	1	1	2	2	3	3	4	4
.30	1995	2000	2004	2009	2014	2018	2023	2028	2032	2037	0	1	1	2	2	3	3	4	4
.31	2042	2046	2051	2056	2061	2065	2070	2075	2080	2084	0	1	1	2	2	3	3	4	4
.32	2089	2094	2099	2104	2109	2113	2118	2123	2128	2133	0	1	1	2	2	3	3	4	4
.33	2138	2143	2148	2153	2158	2163	2168	2173	2178	2183	0	1	1	2	2	3	3	4	4
.34	2188	2193	2198	2203	2208	2213	2218	2223	2228	2234	1	1	2	2	3	3	4	4	5
.35	2239	2244	2249	2254	2259	2265	2270	2275	2280	2286	1	1	2	2	3	3	4	4	5
.36	2291	2296	2301	2307	2312	2317	2323	2328	2333	2339	1	1	2	2	3	3	4	4	5
.37	2344	2350	2355	2360	2366	2371	2377	2382	2388	2393	1	1	2	2	3	3	4	4	5
.38	2399	2404	2410	2415	2421	2427	2432	2438	2443	2449	1	1	2	2	3	3	4	4	5
.39	2455	2460	2466	2472	2477	2483	2489	2495	2500	2506	1	1	2	2	3	3	4	5	5
.40	2512	2518	2523	2529	2535	2541	2547	2553	2559	2564	1	1	2	2	3	4	4	5	5
.41	2570	2576	2582	2588	2594	2600	2606	2612	2618	2624	1	1	2	2	3	4	4	5	5
.42	2630	2636	2642	2649	2655	2661	2667	2673	2679	2685	1	1	2	2	3	4	4	5	6
.43	2692	2698	2704	2710	2716	2723	2729	2735	2742	2748	1	1	2	3	3	4	4	5	6
.44	2754	2761	2767	2773	2780	2786	2793	2799	2805	2812	1	1	2	3	3	4	4	5	6
.45	2818	2825	2831	2838	2844	2851	2858	2864	2871	2877	1	1	2	3	3	4	5	5	6
.46	2884	2891	2897	2904	2911	2917	2924	2931	2938	2944	1	1	2	3	3	4	5	5	6
.47	2951	2958	2965	2972	2979	2985	2992	2999	3006	3013	1	1	2	3	3	4	5	5	6
.48	3020	3027	3034	3041	3048	3055	3062	3069	3076	3083	1	1	2	3	4	4	5	6	6
.49	3090	3097	3105	3112	3119	3126	3133	3141	3148	3155	1	1	2	3	4	4	5	6	6

Mants.	0	1	2	3	4	5	6	7	8	9	1	2	3	4	5	6	7	8	9
											PROPORTIONAL PARTS.								
.50	3162	3170	3177	3184	3192	3199	3206	3214	3221	3228	1	1	2	3	4	4	5	6	7
.51	3236	3243	3251	3258	3266	3273	3281	3289	3296	3304	1	2	2	3	4	5	5	6	7
.52	3311	3319	3327	3334	3342	3350	3357	3365	3373	3381	1	2	2	3	4	5	5	6	7
.53	3388	3396	3404	3412	3420	3428	3436	3443	3451	3459	1	2	2	3	4	5	6	6	7
.54	3467	3475	3483	3491	3499	3508	3516	3524	3532	3540	1	2	2	3	4	5	6	6	7
.55	3548	3556	3565	3573	3581	3589	3597	3606	3614	3622	1	2	2	3	4	5	6	7	7
.56	3631	3639	3648	3656	3664	3673	3681	3690	3698	3707	1	2	3	3	4	5	6	7	8
.57	3715	3724	3733	3741	3750	3758	3767	3776	3784	3793	1	2	3	3	4	5	6	7	8
.58	3802	3811	3819	3828	3837	3846	3855	3864	3873	3882	1	2	3	4	4	5	6	7	8
.59	3890	3899	3908	3917	3926	3936	3945	3954	3963	3972	1	2	3	4	5	5	6	7	8
.60	3981	3990	3999	4009	4018	4027	4036	4046	4055	4064	1	2	3	4	5	6	6	7	8
.61	4074	4083	4093	4102	4111	4121	4130	4140	4150	4159	1	2	3	4	5	6	7	8	9
.62	4169	4178	4188	4198	4207	4217	4227	4236	4246	4256	1	2	3	4	5	6	7	8	9
.63	4266	4276	4285	4295	4305	4315	4325	4335	4345	4355	1	2	3	4	5	6	7	8	9
.64	4365	4375	4385	4395	4406	4416	4426	4436	4446	4457	1	2	3	4	5	6	7	8	9
.65	4467	4477	4487	4498	4508	4519	4529	4539	4550	4560	1	2	3	4	5	6	7	8	9
.66	4571	4581	4592	4603	4613	4624	4634	4645	4656	4667	1	2	3	4	5	6	7	9	10
.67	4677	4688	4699	4710	4721	4732	4742	4753	4764	4775	1	2	3	4	5	7	8	9	10
.68	4786	4797	4808	4819	4831	4842	4853	4864	4875	4887	1	2	3	4	6	7	8	9	10
.69	4898	4909	4920	4932	4943	4955	4966	4977	4989	5000	1	2	3	5	6	7	8	9	10
.70	5012	5023	5035	5047	5058	5070	5082	5093	5105	5117	1	2	4	5	6	7	8	9	11
.71	5129	5140	5152	5164	5176	5188	5200	5212	5224	5236	1	2	4	5	6	7	8	10	11
.72	5248	5260	5272	5284	5297	5309	5321	5333	5346	5358	1	2	4	5	6	7	9	10	11
.73	5370	5383	5395	5408	5420	5433	5445	5458	5470	5483	1	3	4	5	6	8	9	10	11
.74	5495	5508	5521	5534	5546	5559	5572	5585	5598	5610	1	3	4	5	6	8	9	10	12
.75	5623	5636	5649	5662	5675	5689	5702	5715	5728	5741	1	3	4	5	7	8	9	10	12
.76	5754	5768	5781	5794	5808	5821	5834	5848	5861	5875	1	3	4	5	7	8	9	11	12
.77	5888	5902	5916	5929	5943	5957	5970	5984	5998	6012	1	3	4	5	7	8	10	11	12
.78	6026	6039	6053	6067	6081	6095	6109	6124	6138	6152	1	3	4	6	7	8	10	11	13
.79	6166	6180	6194	6209	6223	6237	6252	6266	6281	6295	1	3	4	6	7	9	10	11	13
.80	6310	6324	6339	6353	6368	6383	6397	6412	6427	6442	1	3	4	6	7	9	10	12	13
.81	6457	6471	6486	6501	6516	6531	6546	6561	6577	6592	2	3	5	6	8	9	11	12	14
.82	6607	6622	6637	6653	6668	6683	6699	6714	6730	6745	2	3	5	6	8	9	11	12	14
.83	6761	6776	6792	6808	6823	6839	6855	6871	6887	6902	2	3	5	6	8	9	11	13	14
.84	6918	6934	6950	6966	6982	6998	7015	7031	7047	7063	2	3	5	6	8	10	11	13	15
.85	7079	7096	7112	7129	7145	7161	7178	7194	7211	7228	2	3	5	7	8	10	12	13	15
.86	7244	7261	7278	7295	7311	7328	7345	7362	7379	7396	2	3	5	7	8	10	12	13	15
.87	7413	7430	7447	7464	7482	7499	7516	7534	7551	7568	2	3	5	7	9	10	12	14	16
.88	7586	7603	7621	7638	7656	7674	7691	7709	7727	7745	2	4	5	7	9	11	12	14	16
.89	7762	7780	7798	7816	7834	7852	7870	7889	7907	7925	2	4	5	7	9	11	13	14	16
.90	7943	7962	7980	7998	8017	8035	8054	8072	8091	8110	2	4	6	7	9	11	13	15	17
.91	8128	8147	8166	8185	8204	8222	8241	8260	8279	8299	2	4	6	8	9	11	13	15	17
.92	8318	8337	8356	8375	8395	8414	8433	8453	8472	8492	2	4	6	8	10	12	14	15	17
.93	8511	8531	8551	8570	8590	8610	8630	8650	8670	8690	2	4	6	8	10	12	14	16	18
.94	8710	8730	8750	8770	8790	8810	8831	8851	8872	8892	2	4	6	8	10	12	14	16	18
.95	8913	8933	8954	8974	8995	9016	9036	9057	9078	9099	2	4	6	8	10	12	15	17	19
.96	9120	9141	9162	9183	9204	9226	9247	9268	9290	9311	2	4	6	8	11	13	15	17	19
.97	9333	9354	9376	9397	9419	9441	9462	9484	9506	9528	2	4	7	9	11	13	15	17	20
.98	9550	9572	9594	9616	9638	9661	9683	9705	9727	9750	2	4	7	9	11	13	16	18	20
.99	9772	9795	9817	9840	9863	9886	9908	9931	9954	9977	2	5	7	9	11	14	16	18	20

COMMON MINERALS

			Specific Gravity
Aluminum	Al		2.60
Andalusite	Al_2SiO_5	Silicate of aluminum	3.16– 3.20
Anglesite	$PbSO_4$	Lead sulphate	6.12– 6.39
Anthracite		Hard Coal	1.32– 1.70
Antimony	Sb		6.71
Apatite	$3Ca_3P_2O_8$, CaF_2	Phosphate of lime	3.17– 3.23
Aragonite	$CaCO_3$	Carbonate of lime	2.94
Argentite	Ag_2S	Silver sulphide	7.20– 7.36
Arsenic	As		5.73
Arsenolite	As_2O_3	White arsenic	3.70– 3.72
Asphaltum			1.0 – 1.80
Atacamite	$CuCl_23Cu(OH)_2$	Chloride of copper	3.75
Azurite	$Cu_3(OH)_2(CO_3)_2$	Blue carbonate of copper	3.77– 3.83
Barite	$BaSO_4$	Barium sulphate	4.3 – 4.6
Bauxite	$Al_2O_32H_2O$	Hydrate oxide of aluminum	2.55
Beryl	$Be_3Al_2Si_6O_{18}$	Silicate of beryllium	2.63– 2.80
Biotite		Magnesia-iron mica	2.70– 3.10
Bismuth	Bi		9.80
Bismuthinite	Bi_2S_3	Sulphide of bismuth	6.4 – 6.50
Bituminous Coal		Soft Coal	1.14– 1.40
Bornite	Cu_3FeS_3	Sulphide of copper and iron	4.90– 5.40
Cadmium	Cd		8.60
Calamine	$H_2Zn_2SiO_5$	Silicate of zinc	3.40– 3.50
Calcite	$CaCO_3$	Carbonate of lime	2.7
Cassiterite	SnO_2	Dioxide of tin	6.8 – 7.10
Cerargyrite	AgCl	Horn silver	5.55
Cerussite	$PbCO_3$	Carbonate of lead	6.46– 6.57
Chalcocite	Cu_2S	Copper glance	5.5 – 5.8
Chalcopyrite	$CuFeS_2$	Copper pyrite	4.1 – 4.3
Chromite	$FeCr_2O_4$	Chromic iron	4.32– 4.57
Chromium	Cr		6.50
Chrysolite	$(MgFe)_2SiO_4$	Silicate of magnesia and iron	3.27– 3.37
Cinnabar	HgS	Sulphide of mercury	8.0 – 8.2
Cobalt			8.6
Cobaltite	CoAsS	Sulph-arsenide of cobalt	6.0 – 6.30
Copper	Cu		8.8 – 8.90
Corundum	Al_2O_3	Oxide of aluminum	3.95– 4.10
Cryolite	Na_3AlF_6	Fluoride of aluminum and sodium	3.00
Cuprite	Cu_2O	Red copper ore	5.85– 6.15
Cyanite	Al_2SiO_5	Aluminum silicate	3.56– 3.67
Diamond	C		3.50
Dolomite	$(CaMg)CO_3$	Carbonate of lime and magnesia	2.80– 2.90
Enargite	$CuAsS_4$		4.45
Epidote	$HCa_2(AlFe)_3Si_3O_{13}$	Silicate of iron alumina and lime	3.25– 3.5
Fluorite	CaF_2	Fluor spar	3.2
Franklinite		Oxide of zinc, manganese and iron	5.07– 5.22
Galena	PbS	Sulphide of lead	7.43
Garnet			3.15– 4.3
Gold	Au		15.6 –19.3
Graphite	C		2.09– 2.23
Gypsum	$CaSo_4+2H_2O$	Sulphate of lime	2.3
Hematite	Fe_2O_3	Red oxide of iron	4.9 – 5.3
Ice	H_2O		0.916
Iodyrite	AgI	Iodide of silver	5.6 – 5.7
Iridium	Ir		22.42
Iron	Fe		7.86

55

COMMON MINERALS—*Continued*

			Specific Gravity
Kaolinite	$2H_2OAl_2O_32SiO_2$	Silicate of alumina	2.6
Lead	Pb		11.37
Limonite	$2Fe_2O_33H_2O$	Brown oxide of iron	3.6 – 4.0
Magnesite	$MgCo_3$	Carbonate of magnesia	3.0 – 3.12
Magnetite	$FeO,\ Fe_2O_3$	Magnetic oxide of iron	5.16– 5.18
Malachite	$Cu_2(OH)_2CO_3$	Green carbonate of copper	3.9 – 4.0
Manganese	Mn		7.39
Manganite	$Mn_2O_3H_2O$	Hydrated manganese oxide	4.2 – 4.4
Monazite			4.8 – 5.1
Marcasite	FeS_2	White iron pyrite	4.85– 4.90
Mercury	Hg		13.6
Millerite	NiS	Nickel sulphide	5.3 – 5.6
Mimetite	$3PB_3As_2O_8PbCl_2$	Lead arsenate	7.0 – 7.25
Muscovite	$H_2KAl_3(SiO_4)_3$	Potash mica	2.76– 3.0
Naphtha			0.60– 0.756
Niccolite	$NiAs$	Nickel arsenide	7.33– 7.67
Nickel	Ni		8.9
Opal	SiO_2nH_2O		1.9 – 2.3
Orpiment	As_2S_3	Yellow sulphide of arsenic	3.4 – 3.5
Orthoclase	$KAlSi_3O_8$	Potash feldspar	2.46– 2.6
Ozocerite		Mineral wax	0.85– 0.90
Palladium	Pd		11.3 –11.8
Platinum	Pt		14.0 –19.0
Proustite	Ag_3AsS_3	Light red silver ore	5.57– 5.64
Pyrargyrite	Ag_3SbS_3	Dark red silver ore	5.77– 5.86
Pyrite	FeS_2	Iron sulphide	4.95– 5.10
Pyrolusite	MnO_2	Dioxide of manganese	4.82
Pyromorphite	$3Pb_3P_2O_8PbCl_2$	Lead phosphate	6.5 – 7.1
Pyrrhotite	$Fe_{11}S_{12}$	Magnetic pyrite	4.58– 4.64
Quartz	SiO_2		2.65– 2.66
Realgar	AsS	Red sulphide of arsenic	3.55
Rhodochrosite	$MnCo_3$	Carbonate of manganese	3.45– 3.6
Rhodonite	$MnSiO_3$	Silicate of manganese	3.40– 3.68
Rutile	TiO_3	Dioxide of titanium	4.2
Serpentine	$H_4Mg_3Si_2O_9$	Silicate of magnesia	2.50– 2.65
Siderite	$FeCO_3$	Carbonate of iron	3.8 – 3.9
Silver	Ag		10.1 –11.1
Smaltite	$CoAs_2$	Arsenide of cobalt	6.4 – 6.6
Smithsonite	$ZnCo_3$	Carbonate of zinc	4.30– 4.45
Sphalerite	ZnS	Sulphide of zinc	3.9 – 4.
Spinel	$MgAl_2O_4$	Aluminate of magnesia	3.5 – 4.1
Stephanite	Ag_8SbS_4	Brittle silver	6.2 – 6.3
Stibnite	Sb_2S_3	Sulphide of antimony	4.5 – 4.6
Sulphur	S		2.08
Sylvanite	$(Au,\ Ag)Te$	Telluride of gold and silver	7.9 – 8.3
Talc	$H_2Mg_3Si_4O_{12}$	Silicate of magnesia	2.7 – 2.8
Tephroite	Mn_2SiO_4	Silicate of manganese	4.0 – 4.1
Tetrahedrite	$4Cu_2S,\ Sb_2S_3$	Gray copper	4.4 – 5.1
Tin	Sn		7.29
Topaz		Fluo-silicate of alumina	3.4 – 3.6
Tourmaline		Silicate of alumina, iron and magnesia	2.98– 3.20
Willemite	Zn_2SiO_4	Silicate of zinc	3.9 – 4.18
Wolframite	$(Fe,\ Mn)WO_4$	Tungstate of iron and manganese	7.2 – 7.5
Wulfenite	$PbMoO_4$	Molybdate of lead	6.7 – 7.0
Zinc	Zn		7.15
Zincite	ZnO	Zinc oxide	5.43– 5.7
Zircon	$ZrSiO_4$	Silicate of zerconium	4.70

HEAT OF COMBUSTION

Compiled by N. F. WILSON, JR.

Lefax

Heat of Combustion of Various Materials

Material	Burned To	Calories per gm.	B.T.U. per lb.	Authority
Alcohol, Ethyl	CO_2+H_2O liquid	7,183	12,931	Favre & Silberman
Alcohol, Ethyl	CO_2+H_2O liquid	6,850	12,530	Andrews
Alcohol, Methyl		5,322	9,579	Anonymous
Alcohol, Amyl	CO_2+H_2O liquid	8,933	16,079	Anonymous
Antimony	SbO_4	961	1,730	Dulong
Asphalt		9,532	17,159	Slossen & Colburn
Benzol, C_6H_6 Gas	CO_2+H_2O liquid	10,070	18,126	Berthelot
Benzol, C_6H_6 Gas	CO_2+H_2O liquid	9,650	17,370	Anonymous
Benzol, C_6H_6 Liquid	CO_2+H_2O liquid	10,030	18,054	Stohman
Cane Sugar		3,961	7,130	Berthelot
Carbon, Bisulphide	CO_2	3,401	6,122	Favre & Silberman
Carbon, Crystallized	CO	2,405	4,329	Berthelot
Carbon, Crystallized	CO_2	7,859	14,146	Berthelot
Carbon, Amorphous	CO	2,489	4,480	Berthelot
Carbon, Amorphous	CO_2	8,137	14,647	Berthelot
Carbon, Amorphous	CO_2	8,080	14,544	Berthelot
Carbon, Vapor	CO_2	11,328	20,390	Berthelot
Carbon, Vapor Diamond	CO_2	11,134	20,041	Berthelot
Carbonic, Oxide CO	CO_2	5,640	10,152	Thomsen
Cellulose	CO_2+H_2O liquid	4,208	7,574	Berthelot
Charcoal	CO	2,473	4,451	Favre & Silberman
Charcoal	CO	2,442	4,396	Berthelot
Charcoal	CO_2	8,080	14,544	Favre & Silberman
Charcoal	CO_2	8,137	14,647	Berthelot
Charcoal, Beech	CO_2	7,140	12,852	Schwackhöfer
Charcoal, Soft	CO_2	7,071	12,723	Schwackhöfer
Charcoal, Sugar	CO_2	8,040	14,472	Favre & Silberman
Coal, Anthracite		7,800	14,040	Various
Coal, Bituminous		8,500	15,300	Average of Various
Coal, Coke		7,000	12,600	Average of Various
Coal, Lignite		6,900	12,420	Average of Various
Cyanogen	CuO	5,195	9,351	Dulong
Coke, Gas	CO_2	8,047	14,485	Favre & Silberman
Coke, Petroleum	CO_2	8,017	14,430	Mohler
Copper	CuO	590	1,062	Thomsen
Dynamite, 75%		1,290	2,322	Roux & Sarran
Gas, Acetylene C_2H_2	CO_2+H_2O liquid	12,142	21,855	Berthelot
Gas, Acetylene C_2H_2	CO_2+H_2O liquid	11,527	20,749	Thomsen
Gas, Coal		4,440	7,990	Anonymous
Gas, Coal		7,370	12,266	Anonymous
Gas, Ethylene C_2H_4	CO_2+H_2O liquid	11,858	21,344	Favre & Silberman
Gas, Ethylene C_2H_4	CO_2+H_2O liquid	12,072	21,730	Berthelot
Gas, Ethylene C_2H_4	CO_2+H_2O gas	11,293	20,327	Berthelot
Gas, Methane CH_4	CO_2+H_2O liquid	13,063	23,513	Favre & Silberman
Gas, Methane CH_4	CO_2+H_2O liquid	13,344	24,019	Berthelot
Gas, Methane CH_4	CO_2+H_2O gas	12,066	21,719	Berthelot
Gas, Petroleum		10,800	19,440	Anonymous
Gas, Producer		773	1,391	Anonymous
Gas, Producer		1,370	2,466	Anonymous

Heat of Combustion of Various Materials

Material	Burned To	Heat Produced		Authority
		Calories per gm.	B.T.U. per lb.	
Gas, Naphthalene	CO_2+H_2O liquid	9,793	17,637	Anonymous
Gas, Naphthalene	CO_2+H_2O liquid	9,690	17,442	Berthelot
Gas, Naphthalene	CO_2+H_2O gas	9,354	16,837	Berthelot
Gas, Water		2,350	4,230	Anonymous
Gas, Water		3,032	5,458	Anonymous
Glycerin	CO_2+H_2O liquid	4,316	7,769	Stohman
Graphite	CO_2	7,901	14,222	Berthelot
Gunpowder		750	1,350	Various
Hydrogen	H_2O liquid	34,462	62,032	Favre & Silberman
Hydrogen	H_2O liquid	34,180	61,524	Thomsen
Hydrogen	H_2O liquid	34,500	62,100	Berthelot
Hydrogen	H_2O gas	28,800	51,840	Thomsen
Hydrogen	H_2O gas	29,150	52,470	Berthelot
Iron	Fe_3O_4	1,702	3,064	Dulong
Iron	Fe_2O_3	1,582	2,848	Anonymous
Magnesium	MgO	6,077	10,939	Anonymous
Nickel	NiO	1,006	1,811	Dulong
Oil, Cotton Seed		9,500	17,100	Anonymous
Oil, Castor		8,848	15,926	Anonymous
Oil, Coal, Heavy		8,900	16,020	St. C. Deville
Oil, Olive		9,862	17,751	Dulong
Oil, Linseed		9,430	16,974	Anonymous
Oil, Rape		9,489	17,080	Stohman
Oil, Schist		9,000	1,620	Anonymous
Oil, Sperm		10,000	18,000	Gibson
Paraffin	CO_2+H_2O liquid	11,140	20,050	Stohman
Paraffin	CO_2+H_2O gas	10,340	18,612	Stohman
Peat		5,940	10,692	Bainbridge
Petroleum, Crude		11,094	19,969	Mohler
Petroleum, Refined		11,045	19,881	Mohler
Phosphorus	PO_5	4,509	8,116	Andrews
Phosphorus	PO_5	4,394	7,909	Abria
Pitch		8,400	15,120	Anonymous
Silicon	SiO_2	7,407	13,333	Berthelot
Stearic Acid	CO_2+H_2O liquid	9,374	16,873	Stohman
Starch	CO_2+H_2O liquid	4,228	7,610	Berthelot
Sulphur, Rhombic	SO_2	2,221	3,998	Favre & Silberman
Sulphur, Rhombic	SO_2	2,166	3,899	Berthelot
Sulphur, Monoclinic	SO_2	2,241	4,034	Thomsen
Tallow		9,500	17,100	Stohman
Tin	SnO_2	1,233	2,219	Dulong
Tin	SnO_2	1,144	2,059	Andrews
Turpentine		10,852	19,533	Favre & Silberman
Wood, Beech, 12.9% H_2O		4,168	7,502	Gottlieb
Wood, Birch, 11.8% H_2O		4,207	7,572	Gottlieb
Wood, Oak, 13.3% H_2O		3,990	7,182	Gottlieb
Wood, Pine, 12.2% H_2O		4,422	7,960	Gottlieb
Zinc	ZnO	1,301	2,342	Andrews
Zinc	ZnO	1,298	2,336	Dulong

The Calculation of the Heat of Combustion of Mixed Fuels

The heating value of a mixed fuel is calculated from its constituents as found from ultimate analysis and any one of several formulas may be used for the determination. Dulong's formula is probably the most used and is given below along with Mahler's which has the same basic form. Practically every formula used is similar to the two given below.

$$\text{B.T.U. per lb. of Fuel} = 14650\,C + 62100\left(H - \tfrac{1}{8}O\right) \qquad \textit{Dulong}$$

$$\text{B.T.U. per lb. of Fuel} = 14650\,C + 62100\,H - 5400(O + N) \qquad \textit{Mahler}$$

where C = Parts Carbon in Fuel
  O = Parts Oxygen in Fuel
  H = Parts Hydrogen in Fuel
  N = Parts Nitrogen in Fuel

Example: It is desired to find how many British Thermal Units will be developed by a pound of coal the ultimate analysis of which is:

Carbon = .7216
Oxygen = .0785
Hydrogen = .0496
Nitrogen = .0166

$$\text{B.T.U.} = (14650 \times .7216) + 62100\left(.0496 - \frac{.0785}{8}\right) = 13{,}043.02 \qquad \textit{Dulong}$$

$$\text{B.T.U.} = (14650 \times .7216) + (62100 \times .0496) - 5400(.0785 + .0166) = 13{,}138.06$$
$$\textit{Mahler}$$

The calorific value of the above coal as determined by tests of U. S. Geological Survey was 12,958 B.T.U.

Heats of Fusion of Chemical Elements and Inorganic Compounds

Physikalisch-Chemische Tabellen, 1912

Landolt-Boernstein

Substance	Melting Point Deg. C.	Heat of Fusion in Kg. Calories		Authority and Date
		1-kilogram	1-gram-atom	
Bismuth	266.8	12.64	2.63	Person, 1849
"		12.4	2.58	Mazzotto, 1891
Bromine	−7.32	16.18	1.293	Regnault, 1849
Cadmium	320.7	13.7	1.54	Person, 1848
Chlorine	−103.5	22.96	0.814	Estreicher & Staniewski, 1910
Copper		43.0	2.74	J. W. Richards, 1903
Gallium	13	19.1	1.33	Berthelot, 1878
Iron, cast—white		32–34		Gruner, 1874
Iron, cast—gray		23		Gruner, 1874
Lead	325	5.86	1.21	Rudberg, 1830
"	326.2	5.37	1.11	Person, 1849
"		5.37	1.11	Mazzotto, 1891
"	322.4	5.32	1.11	Spring, 1886
"		6.45	1.34	Robertson, 1903
Mercury		2.84	0.57	Person, 1848
"	−38.7	2.75	0.55	Pollitzer, 1911
"	−38.7	2.85	0.57	Koref, 1911
Palladium	1500	36.3	3.86	Violle, 1878
Phosphorus	27.35	4.74	0.147	Pettersson, 1881
"	29.73	4.74	0.147	Pettersson, 1881
"	40.05	4.97	0.154	Pettersson, 1881
"	44.2	5.034	0.156	Person, 1848
Platinum	1779	27.2	5.3	Violle, 1877
Potassium	58	15.7	0.61	Joannis, 1887
"		13.61	0.532	Bernini, 1906
Silver	999	21.1	2.28	Person, 1848
Sodium	96.5	31.7	0.73	Joannis, 1887
"		17.75	0.408	Bernini, 1906
Sulphur	115	9.37	0.300	Person, 1848
" mono.	119	10.4	0.33	Wiegand, 1908
Thallium	290	7.2	1.47	Robertson, 1903
Tin, ordinary, white	228	13.3	1.6	Rudberg, 1830
" " "	232.7	14.25	1.70	Person, 1849
" " "	227.3	14.65	1.74	Spring, 1886
" " "		13.06	1.62	Mazzotto, 1891
" " "		14.05	1.67	Robertson, 1903
Zinc	415.3	28.1*	1.84	Person, 1849
"		28.0	1.8	Mazzotto, 1891

* Not entirely reliable.

Inorganic Compounds

Substance	Melting Point Deg. C.	Heat of Fusion in Kg. Calories		Authority and Date
		1-kilo-gram	1-gram-mol.	
Aluminum bromide, $AlBr_3$		10.47	2.79	Kablukow, 1908
Ammonia, NH_3	−75	108.1	1.84	Massol, 1902
Antimony tribromide, $SbBr_3$	94.6	9.76	3.51	Tolloczko, 1901
Antimony trichloride, $SbCl_3$	73.2	13.29	3.01	Tolloczko, 1901
Arsenic tribromide, $AsBr_3$	31.0	8.93	2.81	Tolloczko, 1901
Barium chloride, $BaCl_2$	958.9	27.8	5.8	Plato, 1907
Caesium hydroxide, $CsOH$	272.3	10.7	1.61	v. Hevesy, 1910
Calcium chloride, $CaCl_2$	773.9	54.6	6.06	Plato, 1907
Calcium chloride, $CaCl_2.6H_2O$	28.5	40.7	8.9	Person, 1849
Calcium nitrate, $Ca(NO_3)_2.4H_2O$	42.4	33.49	7.94	Pickering, 1891
Carbon dioxide (5.10 atm.)	−56.29	43.8	1.93	Kuenen & Robson, 1902
Hydrogen peroxide, H_2O_2		2.70	9.18	de Forcrand, 1900

NOTES. For aluminum and iodine the only value available is the heat necessary to bring 1 kg. of the substance from 0° C. to the molten condition. It is, for aluminum, 239.4 kg. calories per kg. and for iodine 11.7 kg. calories per kg.

The value given by Smith is 334.21±0.08 joules. He assumes the mean calorie to equal 4.1832 joules.

Heat, Fusion, Chemical Elements, Inorganic Compounds

Substance	Heat of Fusion		Authority and Date
	1-kilo-gram	1-gram-mol.	
Ice	75.99	1.369	Pettersson, 1881
"	76.03	1.375	Pettersson, 1881
"	75.94	1.368	Pettersson, 1881
"	76.60	1.380	Pettersson, 1881
"	77.71	1.400	Pettersson, 1881
"	78.26	1.411	Zakrzewski, 1892
"	79.25	1.428	Person, 1848
"	79.06	1.424	Regnault, 1844
"	79.25	1.428	Regnault, 1844
"	80.025	1.442	Bunsen, 1870
"	79.24	1.428	Desains, 1843
" (See Note 2)	79.896 ± 0.02	1.440	A. W. Smith, 1903
"	79.61	1.435	Bogojawlenski, 1905
"	79.2	1.427	Leduc, 1906
"	79.67	1.436	W. A. Roth, 1908

Heat, Fusion, Chemical Elements, Inorganic Compounds (*Continued*)

Substance	Heat of Fusion		Authority and Date
	1-kilo-gram	1-gram-mol.	
Iodine chloride, ICl	14.15	2.30	Berthelot, 1880
" " ICl α	16.42	2.66	Stortenbeker, 1892
" " ICl β	14.0	2.27	Stortenbeker, 1892
Lead bromide, PbBr$_2$	12.34	4.53	Ehrhardt, 1885
" "	9.9	3.65	Goodwin & Kalmus, 1909
Lead chloride, PbCl$_2$	20.90	5.81	Ehrhardt, 1885
" "	18.5	5.15	Goodwin &Kalmus, 1909
Lead iodide, PbI$_2$	11.50	5.30	Ehrhardt, 1885
Lithium nitrate, LiNO$_3$	88.5	6.10	Goodwin & Kalmus, 1909
Mercuric iodide, HgI$_2$	9.79	4.44	Guinchant, 1907
Nitric acid, HNO$_3$	9.54	0.601	Berthelot, 1877
Nitric oxide, N$_2$O$_5$	76.67	8.28	Berthelot, 1874
" " N$_2$O$_4$	32.2–37.2	2.96–3.42	Ramsay, 1890
Phosphoric acid, hypo. H$_4$P$_2$O$_6$	51.23	8.30	Joly, 1886
" " ortho. H$_3$PO$_4$	25.71	2.521	Thomsen, 1905
Phosphorous acid, hypo. H$_3$PO$_2$	35.00	2.31	Thomsen, 1905
" " ortho. H$_3$PO$_3$	37.44	3.072	Thomsen, 1905
Potassium chloride, KCl	86.0	6.41	Plato, 1906
dichromate, K$_2$Cr$_2$O$_7$	29.8	8.77	Goodwin & Kalmus, 1909
fluoride, KF	108.0	6.27	Plato, 1907
hydroxide, KOH	28.6	1.61	v. Hevesy, 1910
nitrate, KNO$_3$	47.37	4.79	Person, 1848
nitrate, KNO$_3$	25.5	2.57	Goodwin & Kalmus, 1909
Rubidium hydroxide, RbOH	15.8	1.62	v. Hevesy, 1910
Silver bromide, AgBr	12.6	2.37	Goodwin & Kalmus, 1909
chloride, AgCl	21.3	3.05	Goodwin & Kalmus, 1909
chloride, AgCl	30.7	4.40	Robertson, 1903
nitrate, AgNO$_3$	17.6	2.99	Guinchant, 1907
nitrate, AgNO$_3$	15.2	2.58	Goodwin & Kalmus, 1909
Sodium chlorate, NaClO$_3$	49.6	5.25	Goodwin & Kalmus, 1909
chlorate, NaClO$_3$	48.4	5.15	Foote & Levy, 1907
chloride, NaCl	123.5	7.22	Plato, 1906
chromate, Na$_2$CrO$_4$-10H$_2$O	36.0	12.3	Berthelot, 1878
chromate, Na$_2$CrO$_4$-10H$_2$O	39.2	13.4	Berthelot, 1878
fluoride, NaF	186.1	7.82	Plato, 1907
hydroxide, NaOH	40.0	1.60	v. Hevesy, 1910
nitrate, NaNO$_3$	62.97	5.355	Person, 1848
nitrate, NaNO$_3$	45.3	3.69	Goodwin & Kalmus, 1909
phosphate, Na$_2$HPO$_4$. 12H$_2$O	66.8	23.9	Person, 1849
sulphate, Na$_2$SO$_4$. 10H$_2$O	51.2	16.5	Cohen, 1894
thiosulphate, Na$_2$S$_2$O$_3$. 5H$_2$O	37.6	9.3	v. Trentinaglia, 1876
Strontium chloride, SrCl$_2$	25.6	4.06	Plato, 1906
Sulphuric acid, H$_2$SO$_4$	8.77	0.860	Berthelot, 1874
· H$_2$SO$_4$	24.031	2.358	Pickering, 1891
H$_2$SO$_4$	22.82	2.239	Knietsch, 1909
H$_2$SO$_4$	25.98	2.559	Bronsted, 1910
H$_2$SO$_4$. H$_2$O	31.72	3.68	Berthelot, 1874
H$_2$SO$_4$. H$_2$O	38.97	4.52	Luginin & Dupont, 1911
H$_2$SO$_4$. H$_2$O	39.92	4.63	Pickering, 1891
H$_2$SO$_4$. H$_2$O	34.91	4.05	Hammerl
H$_2$SO$_4$. H$_2$O	36.08	4.18	Hammerl
H$_2$SO$_4$. H$_2$O	38.38	4.45	Bronsted, 1910
Thallium bromide, TlBr	12.7	3.61	Goodwin & Kalmus, 1909
chloride, TlCl	16.6	3.98	Goodwin & Kalmus, 1909

Brief Summary of Some Important Chemical Laws and Hypotheses

§ 19. **1. Avogadro's Hypothesis.** Equal volumes of all gases at the same pressure and temperature contain an equal number of molecules.

2. Boyle's (or Mariotte's) Law. The volume of all gases at a constant temperature is inversely proportional to the pressure.

3. Common Ion Effect—Repression of Ionization. Ionization is repressed by adding to the solution a salt which has a common ion with that of the solute.

4. Complex Ions. These consist of a group of elements, possessing characteristics distinct from the elements of which they are composed. Example $K_4Fe(CN)_6$ ionized $= K_4$ and $Fe(CN)_6$. $KClO_3$ ionized $= K$ and ClO_3.

5. Conservation of Mass, Law of. The total weight of matter resulting from a combination or decomposition is always equal to the sum of the weights of the substances taking part in the reaction. In all chemical transformations mass remains constant.

6. Constant Proportion, Law of. The elements combine with one another in absolutely fixed relative proportions by weight.

7. Dalton's Atomic Theory. All matter consists of an aggregate of minute particles, or atoms, which are chemically indivisible.

8. Dalton-Henry's Law. The pressure exerted by a mixture of gases occupying a given volume is equal to the sum of the separate pressures which the different gases would exert if they alone occupied the given volume. Every gas behaves with respect to its own particular properties just as if it alone was present.

9. Dulong-Petit's Law. All elements in the solid state have the same atomic heat. That is, elements taken in proportion to their atomic weights require equal quantities of heat in order to be raised to the same temperature.

10. Electrolytic Dissociation Theory of Arrhenius. All substances which form solutions capable of conducting an electric current, the electrolytes, exist in solution, in part at least, as dissociated ions, atoms or atomic groups, carrying a definite charge of electricity. Each positive ion (anion) involves the presence of a negative ion (cation) carrying an equivalent amount of electricity.

When two oppositely charged poles are placed in such a solution, the positively charged anode attracts the negative particles in the solution and repels the positive, while the negatively charged cathode attracts the positively charged particles and repels those negatively charged, a flow of electricity thus being produced.

11. Electromotive or Potential Series. Metals placed in solution tend to pass from a free element to the ionic condition, the more positive elements will displace the less positive from their ionic condition. See list of elements given in order of activity, in the table in the latter part of this volume.

12. Faraday's Law. In equal periods of time a current of definite strength separates the ions from the solutions of electrolytes in quantities by weight, which stand in the same ratio to one another as their equivalent weights, *i.e.*, their atomic weight divided by their valence.

The strength of an electric current can be measured by determining the weight of silver or copper deposited at the cathode in a given time from a solution of silver or copper, or by measuring the volume of hydrogen or oxygen produced from water by the action of the current.

13. **Gay-Lussac's Law (Charle's Law).** At constant pressure, volumes of all gases increase on warming in the same proportion for every one degree. The coefficient of expansion is 1/273 (0.003665). That is to say, for 1° C. increase in temperature the gas expands 1/273 of its volume. A volume of 273 cc. of gas at 0° C. would become 274 cc. at 1° C.

14. **Hesse's Law Thermo Law of Conservation.** The evolution of heat which accompanies a chemical process is always the same whether the process takes place in one step or whether it passes through a number of intermediate processes.

15. **Lavoisier-Leplace's Law.** Every compound has a certain heat of formation which is equal to its heat of decomposition.

16. **Mass Action, Law of (Guldberg-Waage's Law).** The speed of reaction between two substances in solution is directly proportional, at any moment, to the molar concentrations of these reacting substances in solution, and to a constant, which is characteristic of the chemical nature of the reacting components, and of the temperature.

17. **Multiple Proportion, Law of.** If two elements combine in more than one proportion, the masses of the one which combine with a given mass of the other bear a simple rational relation to one another, *i.e.*, are always a whole multiple of the lowest.

Example: N_2O, N_2O_2, N_2O_3, N_2O_4, N_2O_5 : the five oxides of nitrogen.

18. **Neuman-Kopp's Law.** Molecular heat corresponds to the sum of atomic heats of the elements which constitute the molecule.

19. **Osmotic Pressure.** A substance in solution produces the osmotic pressure, at a given temperature, which it would exert, if it were contained as a gas, at the same temperature, in the volume occupied by the pure solvent of the solution.

20. **Periodic Law of Mendeleeff.** The properties of the elements are periodic functions of their atomic weights.

21. **Reversible Reaction.** Compounds in solution resulting from a chemical reaction in turn react forming the original compounds present. This reversibility is prevented by removal from solution of one of the resulting compounds, by formation of an insoluble compound, which precipitates from solution, or by the formation of a gas which escapes. (See " Common Ion Effect.")

22. **Law of Electrostatic Force. Coulomb's Law.**

$$F = \frac{1 q_1 q_2}{C d_2}.$$

F = force acting between two charged bodies.
q_1 and q_2 = quantities of electricity.
C = specific inductive capacity or dielectric constant of the medium.

23. **Law of Jule.**

$$h = rc^2.$$

h = amount of heat evolved in a given time.
r = resistance to the passage of the current.
c = strength of the current.

24. **Ohm's Law.**

$$C = \frac{E}{R}.$$

C = strength of current.
E = electromotive force.
R = resistance.

PREPARATION OF REAGENTS

This chapter includes a list of reagents generally used in the analytical laboratory. Certain special reagents having a limited use are given in the body of the text under the subject with which they are concerned. The alphabetical arrangement will be found convenient for reference. The term per cent is used for convenience and does not represent true percent but rather the grams of reagent per 100 cc. of solution. For example, a ten per cent solution of sodium chloride means 10 grams of NaCl dissolved in water and made to 100 cc. A ten per cent of alcohol solution refers to volume and means 10 cc. of alcohol diluted with water to 100 cc.

COMMON LABORATORY DESK OR SHELF REAGENTS

Reagent Name	Per Cent	Vol. R.	Vol. H_2O	Res. sp.gr.	Approx. N.	Gm. per cc.
Acetic a. glacial.	100	. .	. .	1.06	17 N	1.06 g.
Acetic a.	80	. .	. .	1.07	15 N	0.84 g.
Acetic a. dil.	28	1	2	1.04	5 N	0.28 g.
Hydrochloric a. conc.	40	. .	. .	1.20	13 N	0.48 g.
Hydrochloric a. dil.	16	1	1.5	1.08	5 N	0.19 g.
Nitric a. conc.	96	. .	. .	1.50	23 N	1.43 g.
Nitric a. commer.	70	. .	. .	1.42	16 N	1.00 g.
Nitric a. dil.	23	1	2	1.14	5 N	0.31 g.
Sulphuric a. conc.	95	. .	. .	1.84	36 N	1.77 g.
Sulphuric a. dil.	14	1	6	1.09	5 N	0.25 g.
Ammonium hyd. conc.	28	. .	. .	0.90	15 N	0.26 g.
Ammonium hyd. dil.	9	1	2	0.96	5 N	0.08 g.
Sodium hyd.	40	. .	. .	1.44	14 N	0.58 g.
Sodium hyd. dil.	14	1	2	1.07	5 N	0.20 g.

R = strong reagent. Res. = resulting. N = Normal. Gm. per cc. = grams of 10% reagent.

Reagents Used for Precipitations and Volumetric Determinations

Volumetric reagents should be standardized at 20° C. if possible. The temperature at which the standardization is made should be noted on the container as the strength of the reagent per cc. will vary with a change of temperature, the concentration increasing with contraction at a lower temperature, and expansion causing decrease with a rise in temperature. It is not advisable to attempt to make the reagent exactly a normal equivalent as the strength may change slightly from this factor. It is necessary to establish the exact factor of the solution, however, by careful standardization. The "factor" is the ratio between a given number of cc. of the solution in question and the number of cc. of a theoretically correct solution. For example, if 50 cc. of the reagent were found to be equivalent to 55 cc. of a theoretically normal solution, the normality factor of the reagent in question would be $55 \div 50 = 1.1$ N and all titrations would have to be multiplied by this factor to obtain the cc. in terms of a normal solution.

Normalities are expressed either by a figure before N or after it, thus

0.2 N or N/5=fifth normal. 0.1 N or N/10=tenth normal. 0.5 N or N/2 = half normal. .02 N or N/50=fiftieth normal, etc.

Acetic Acid, Glacial. C.P., 99.5% pure. The determination of its strength should be made by titration and not by specific gravity, as the 98% and 80% acid have the same specific gravity, 1.067. The determination of the melting-point gives results equally good with those obtained by titration and requires less time. It is made after the manner of the "titer test" for acids, the tube being half filled, chilled to 10° to 11° C., and further chilled by placing the outside bottle in ice-water; the temperature of the super-cooled acid rises to its melting-point, where it remains stationary for some time. The melting-points of acids of various strengths are as follows:

100%, 16.75° C.; 99.5%, 15.65°; 99%, 14.8°.

Alcohol. Commercial "Cologne Spirits." For the preparation of alcohol free from aldehyde for alcoholic potash, cologne spirits are treated with silver oxide as follows: $1\frac{1}{2}$ grams of silver nitrate are dissolved in 3 cc. of water, added to 1 liter of alcohol and thoroughly shaken; 3 grams of potassium hydrate are dissolved in 15 cc. of warm alcohol and, after cooling, added to the alcoholic silver nitrate and thoroughly shaken again, best in a tall bottle or cylinder. The silver oxide is allowed to settle, the clear liquid siphoned off and distilled, a few bits of pumice, prepared by igniting and immediately quenching under water, being added to prevent bumping. Alcohol for use in the free acid determination is prepared by placing 10 to 15 grams of dry sodium carbonate in the reagent bottle, taking care to filter it before use.

Alcohol, Amyl. C.P.

Alcohol Potash Solution. See Potassium Hydroxide.

Alizarin S Test. The reagent used is a 0.1% filtered solution of commercial alizarin S, the sodium salt of alizarin monosulphonic acid (yellow with acids, purple with alkalies).

Alkaline Tartrate Solution. See Tartrate Solution.

Ammonium Acetate. 70% solution. Use the salt or cautiously add 1000 cc. of ammonium hydroxide to 1200 cc. of glacial acetic acid.

Ammonium Carbonate. 250 grams of the salt per liter. Add 100 cc. of strong ammonium hydroxide. The solution contains approximately 22% of $(NH_4)_2CO_3.NH_4CO_2NH_2$.

Ammonium Chloride. 10% solution, 100 grams of NH_4Cl per liter.

Ammonium Chloride *Solution Standard.* Dissolve 3.82 grams of ammonium chloride in 1 liter of distilled water. Dilute 10 cc. of this to 1 liter with ammonia-free water. 1 cc.=0.00001 gram of nitrogen. Used in water analysis.

Ammonium Molybdate *Reagent Standard.* 4.75 grams of the salt are dissolved in water and made up to 1 liter. One cc. with a half-gram sample is equal approximately to 1% of Pb. Standard in lead determination.

Ammonium Molybdate. One hundred grams of pure molybdic acid are thoroughly mixed with 400 cc. of cold distilled water and 80 cc. of strong ammonia (sp.gr. 0.90) added. When the solution is complete, it is poured slowly and with constant stirring into a mixture of 400 cc. of strong nitric acid (sp.gr. 1.42) and 600 cc. of distilled water. This order of procedure should be followed, as the nitric acid poured into the ammonium molybdate solution will cause the precipitation of a difficultly soluble oxide of molybdenum and necessitate filtration. Fifty milligrams (.05 gram) of microcosmic salt, dissolved in a little water, are added to clarify the reagent, the precipitate agitated, then allowed to settle for twenty-four hours and the clear solution decanted through a filter into a large reagent bottle. Sixty cc. of the reagent should be used for every 0.1 gram of P_2O_5 present in the solution analyzed.

Ammonium Nitrate. 20% solution. 200 grams of NH_4NO_3 per liter.

Ammonium Oxalate. 4% solution. 40 grams of $(NH_4)_2C_2O_4.2H_2O$ per liter.

Ammonium Phosphate. 10% solution. 100 grams of $(NH_4)_2HPO_4$ per liter. Sodium ammonium phosphate Microcosmic salt may be used in place of ammonium phosphate.

Ammonium Sulphate. 25% solution. 250 grams of $(NH_4)_2SO_4$ per liter.

Ammonium Sulphide (Colorless). Saturate 750 cc. of strong ammonia with hydrogen sulphide gas and add 500 cc. of strong ammonia and 1000 cc. of water.

Ammonium Polysulphide (Yellow). To a solution of ammonium sulphide made according to the directions above add about 75 grams of flowers of sulphur. Shake well.

Ammonium Thiocyanate (" sulphocyanate "). See Thiocyanate.

Arsenic Standard Solution. One gram of resublimed arsenous acid, As_2O_3, is dissolved in 25 cc. of 20% sodium hydroxide solution (arsenic free) and neutralized with dilute sulphuric acid. This is diluted with fresh distilled water, to which 10 cc. of 95% H_2SO_4 has been added, to a volume of 1000 cc. Ten cc. of this solution is again diluted to a liter with distilled water containing acid. Finally 100 cc. of the latter solution is diluted to a liter with distilled water containing acid. One cc. of the final solution contains 0.001 milligram of As_2O_3. Used in arsenic determinations by Gutzeit and Marsh methods.

Arsenite. *Tenth Normal Arsenious Acid.* As_2O_3 is equivalent to $2I_2$, i.e., to 4H, hence $\frac{1}{4}$ the gram molecular weight of arsenious oxide per liter will give a normal solution: $198 \div 4 = 49.5$.

4.95 grams of pure arsenious oxide is dissolved in a little 20% sodium hydroxide solution and the excess of the alkali is neutralized with dilute sulphuric acid, using phenolphthalein indicator, the solution being just decolorized. Five hundred cc. of distilled water containing about 25 grams of sodium bicarbonate are added. If a pink color develops, this is destroyed with a few drops of weak sulphuric acid. The solution is now made to volume, 1000 cc. The reagent is standardized against a measured amount of pure iodine. The oxide may be dissolved directly in sodium bicarbonate solution. For iodine titrations. A standard reducing agent.

NOTE. Commercial arsenious oxide is purified by dissolving in hot hydrochloric acid, filtering the hot saturated solution, cooling, decanting off the mother liquor, washing the deposited oxide with water, drying and finally subliming.

Standard Antimony Solution. A stock solution is made up by weighing out 0.553 gm. of $KSbOC_4H_4O_6$ which is dissolved in distilled water and made up to 2000 cc. which represents 1 cc. = .0001 gm. of Sb.

From the above stock solution take 100 cc. and make up to 1000 cc.; this solution now equals − 1 cc. = .00001 gm. of Sb, which is used for making the standard stains and introducing into checks. For standard antimony stains.

Barium Chloride. A 10% solution. 100 grams of $BaCl_2.2H_2O$ per liter.

Barium Chloride, anhydrous, 5% solution; or crystals, 6% solution. Reagent used for sulphur determinations.

Barium Hydroxide. 5% solution. 50 grams of $Ba(OH)_2.8H_2O$ per liter.

Alpha Benzildioxime. This may be prepared by boiling 10 grams of benzil (not necessarily pure) with 8 to 10 grams of hydroxylamine hydrochloride in methyl alcohol solution. After boiling for three hours the precipitate is filtered off and dried, washed with hot water and then with a small amount of 50% alcohol, and dried. This dried precipitate consists of pure benzildioxime (m.p. 237° C.). A further yield may be obtained by boiling the filtrate with hydroxylamine hydrochloride. The reagent is prepared by dissolving 0.2 gram of the salt per liter of alcohol to which is added ammonium hydroxide to make 5% solution, sp.gr. 0.96 (50 cc. per liter). Reagent used in the determination of nickel.

Benzidine Hydrochloride is prepared by taking 6.7 grams of the free base, or the corresponding amount of the hydrochloride, and mixing into a paste with 20 cc. of water in a mortar. Twenty cc. of hydrochloric acid (sp.gr. 1.12) are added and the mixture diluted to exactly 1000 cc. One cc. of this solution corresponds to 0.00357 gram of H_2SO_4. The solution has a brown color. Brown flakes are likely to separate out on standing, but these do no harm.

Reagent used in the determination of sulphur.

Bromine. The commercial article; also a N/3 solution, made by dissolving 26.6 grams of bromine in 1 liter of carbon tetrachloride.

Bromine—Potassium Bromide Solution. 320 grams of potassium bromide are dissolved in just sufficient water to cause solution and mixed with 200 cc. of bromine, the bromine being poured into the saturated bromide solution. After mixing well the solution is diluted to 2000 cc. Used for sulphur determinations.

Bromine—Carbon Tetrachloride Solution. Carbon tetrachloride saturated with bromine. Used for sulphur determination.

Bismuth Standard Solution. One gram of metallic bismuth is dissolved in the least amount of dilute nitric acid (1 : 1) that is necessary to keep it in solution and diluted to 1000 cc. in a graduated flask. One hundred cc. of this solution is diluted to 1000 cc. One cc. of this diluted solution contains 0.0001 gram of bismuth.

Calcium Chloride. 10% solution. 100 grams of $CaCl_2.6H_2O$ per liter.

Acid Calcium Chloride Solution. Saturate with calcium chloride a mixture of 90 parts of water and 10 parts of concentrated hydrochloric acid (specific gravity 1.2).

Calcium Hydroxide (Lime Water). A saturated water solution of $Ca(OH)_2$. Keep tightly stoppered. Decant or filter before use.

Cinchonine Solution. Cinchonine solution is made by dissolving 100 grams of the alkaloid in dilute HCl (1 part acid to 3 of water) and diluting to 1000 cc. with HCl of the same strength.

Cinchonine Wash Solution. 30 cc. of Cinchonine solution, 30 cc. of strong HCl diluted to 1000 cc. Used in tungsten determinations.

Cinchonine Potassium Iodide Solution. Ten grams of cinchonine are dissolved by treating with the least amount of nitric acid that is necessary to form a viscous mass and taking up with about 100 cc. of water. The acid is added a drop at a time, as an excess must be avoided. Twenty grams of potassium iodide are dissolved separately and cinchonine solution added.

The resulting mixture is diluted with water to 1000 cc. After allowing the reagent to stand forty-eight hours, any precipitate formed is filtered off and the clear product is ready for use. The reagent preserved in a glass-stoppered bottle keeps indefinitely. It should be filtered free of suspended matter before use.

Copper Sulphate—Alkaline Tartrate Solution. See Fehling's solution.

Citric Acid. One part of acid of 3 parts in water. One hundred cc. of nitric acid should be added to each liter to prevent mould growth.

Citrate of Ammonia. 25 grams of the salt per 50 cc. of water.

Copper Solution Standard. One gram of purest electrolytic copper is dissolved in 20 cc. of dilute nitric acid, sp.gr. 1.2, and the solution diluted to 1000 cc. For standardizing the thiosulphate to be used with high-grade copper ores, crude copper, blister copper, etc., a copper solution containing ten times the above amount of metallic copper is prepared. Standard in the iodide determination of copper.

Caustic. Standard solution. See Sodium Hydroxide.

Cuprous Chloride, Acid, for Gas Analysis. The directions given in the various text-books being troublesome to execute, the following method, which is simpler, has been found to give equally good results: Cover the bottom of a two-liter bottle with a layer of copper oxide or "scale" $\frac{3}{8}$ in. deep, place in the bottle a number of pieces of rather stout copper wire reaching *from top to bottom*, sufficient to make a bundle an inch in diameter, and fill the bottle with common hydrochloric acid of 1.10 sp.gr. The bottle is occasionally shaken, and when the solution is colorless, or nearly so, it is poured into the half-liter reagent bottles, containing copper wire, ready for use. The space left in the stock bottle should be immediately filled with hydrochloric acid (1.10 sp.gr.).

By thus adding acid or copper wire and copper oxide when either is exhausted, a constant supply of this reagent may be kept on hand.

The absorption capacity of the reagent per cc. is, according to Winkler, 15 cc. of CO; according to Hempel 4 cc.

Care should be taken that the copper wire does not become entirely dissolved and that it extend from the top to the bottom of the bottle; furthermore the stopper should be kept thoroughly greased the more effectually to keep out the air, which turns the solution brown and weakens it.

Cuprous Chloride, Ammoniacal, for Gas Analysis. The acid cuprous chloride is treated with ammonia until a faint odor of ammonia is perceptible; copper wire should be kept in it similarly to the acid solution. This alkaline solution has the advantage that it can be used when traces of hydrochloric acid vapors might be harmful to the subsequent determinations, as, for example, in the determination of hydrogen by absorption with palladium. It has the further advantage of not soiling mercury as does the acid reagent.

Absorption Capacity, 1 cc. absorbs 1 cc. of CO.

Cuprous chloride is at best a poor reagent for the absorption of carbonic oxide; to obtain the greatest accuracy where the reagent has been much used, the gas should be passed into a fresh pipette for final absorption, and the operation continued until two consecutive readings agree exactly. The compound formed by the absorption—possibly Cu_2COCl_2—is very unstable, as carbonic oxide may be freed from the solution by boiling or placing it *in vacuo;* even if it be shaken up with air, the gas is given off, as shown by the increase in volume and subsequent diminution when shaken with fresh cuprous chloride.

Devarda's Alloy. Forty-five parts of aluminum, 50 parts of copper and 5 parts of zinc. The aluminum is heated in a Hessian crucible in a furnace until the aluminum begins to melt, copper is now added in small portions until liquefied and zinc now plunged into the molten mass. The mix is heated for a few moments, covered and then stirred with an iron rod, allowed to cool slowly with the cover on and the crystallized mass pulverized.

Dimethylglyoxime. 1% solution. 1 gram of the salt per 100 cc. of ethyl alcohol. Used in nickel determinations.

Diphenyl Carbazide Reagent. One-tenth of a gram of the compound is dissolved in 10 cc. of glacial acetic acid and diluted to 100 cc. with ethyl alcohol.

Diphenyl carbazide may be made by heating a mixture of 15 grams of urea with 50 grams of phenyl hydrazine four hours, finishing at 155° C. The solid product is crystallized three times with alcohol. A light straw-colored product is obtained. A white product is obtained if the urea is cut down to 5 grams, the yield, however, is only 25 per cent of that obtained by the first method and the compound possesses no advantages. Used in chromium determinations.

Fehling's *Solution for Determination of Sugar.* (*a*) *Copper sulphate solution.* 34.639 g. of crystallized copper sulphate are dissolved in water and made up to 500 cc. (*b*) *Alkaline tartrate solution.* Dissolve 173 g. of Rochelle salts and 125 g. of potassium hydroxide in water and dilute to 500 cc.

Ferric Chloride. 10% solution. 100 grams of $FeCl_3.6H_2O$ per liter.

Ferric Chloride Solution Standard. (10 grams of iron dissolved in 200 cc. of HCl, oxidized with HNO_3 and made to one liter.) 1 cc.=0.01 g. of Fe.

Ferric Indicator. Saturated solution of ferric ammonium alum. Should this not be available, $FeSO_4$ may be oxidized with nitric acid, and the solution evaporated with an excess of H_2SO_4 to expel the nitrous fumes. A 10% solution is desired. Five cc. of either of these reagents are taken for each titration. Used in titrating silver by Volhard's method.

Standard Iron Solution. A ferric solution, the iron content of which has been determined, is diluted and divided so as to obtain 0.0004 gram of Fe. This is made up to 2 liters with water containing 200 cc. of iron-free, C.P. H_2SO_4. One hundred cc. of this solution, together with 10 cc. of normal ammonium sulphocyanate solution, is used as a standard. One hundred cc. contains 0.00002 gram of Fe.

Normal sulphocyanate contains 76.1 grams of NH_4CNS per liter.

Used in standardization of the reagent for the determination of iron.

Standard Iron Solution. 8.6322 grams of ferric ammonia alum is dissolved in dilute hydrochloric acid and made up to one liter. The iron is determined in 100 cc. portions by the dichromate method. One cc. will contain about 0.001 gram of Fe.

Ferric Nitrate. One part of salt in 6 parts of water. It is well to add a little nitric acid to prevent hydrolysis.

Ferrous Sulphate Standard. Reagent for Nitric Acid Determinations.
A. Reagent to be Used in Titration of Nitric Acid in Sulphuric Acid, Oleum, etc. 176.5 grams of $FeSO_4.7H_2O$ are dissolved in about 400 cc. of water, and 500 cc. of about 60% H_2SO_4 (1 vol. of 66° Bé. acid per 1 vol. of H_2O) are added with constant stirring, and the solution (cooled if necessary) made up to 1000 cc. 1 cc. will be equivalent to $0.02 \pm$ gram of HNO_3, the exact value being determined by standardization.

B. Reagent for Titration of Nitric in Phosphoric or Arsenic Acid. Ferrous sulphate to be used should be made up as follows: 264.7 grams of $FeSO_4.7H_2O$ is dissolved in 500 cc. of water, 50 cc. of 66° Bé. H_2SO_4 (93.2%) added and the solution made up to 1000 cc. 1 cc. will be equal to approximately 0.02 gram of HNO_3. The exact strength is ascertained by titrating a known amount of nitric acid in phosphoric or arsenic acid upon warming to 40° or 50° C.

Ferrous Ammonium Sulphate. 0.1 N solution contains 39.214 grams per liter. Dissolve 39.5 grams in water and dilute to 1000 cc. Clear with a few drops of sulphuric acid. Standardize against potassium permanganate. Pipette out exactly 50 cc. of the ferrous solution, add 5 cc. strong sulphuric acid and titrate to a pink color with 0.1 N potassium permanganate. The cc. permanganate required multiplied by the factor of normality of the permanganate solution divided by 50 will give the normality of the ferrous solution. The reagent keeps well if stoppered, but soon loses strength if the bottle is opened frequently.

Hydrochloric Acid, Arsenic-free. The commercial acid is treated with potassium chlorate to oxidize the arsenic to its higher form and the acid distilled. The distilling apparatus may be arranged so that constant distillation takes place, acid from a large container dropping slowly into a retort containing potassium chlorate, fresh hydrochloric acid being supplied as rapidly as the acid distills. See Fig. 5, page 48.

Hydrochloric Acid. Desk reagent. See beginning of chapter.

Tenth-normal Hydrochloric Acid. This may be standardized by any of the accepted methods, or as follows: Twenty cubic centimeters of the approximately N/10 acid is measured out with a pipette, and the silver chloride precipitated by an excess of silver nitrate solution in a volume of 50 to 60 cc. After digesting at 70 to 80° C., until the supernatant liquid is clear, the chloride is filtered off on a tared Gooch filter and washed with water containing 2 cc. of nitric acid per 100 cc. of water until freed from silver nitrate. After drying to constant weight at 130° C., the increase of weight over the original tare is noted and from this weight, corresponding to the silver chloride, the strength of the hydrochloric acid is calculated, after which it is adjusted to the strength prescribed. The standardization should be based upon several concordant determinations using varying amounts of acid.

Hydrogen Peroxide. Thirty per cent solution. If this is not available, sodium peroxide dissolved in dilute sulphuric acid will do.

Hydrogen Sulphide. See page 732.

Indicators. See pages 744–745.

Iodine Solution. Standard 0.1 N. Dissolve 20–25 grams of potassium iodide in as little water as possible. To this add 12.7 grams of resublimed iodine (theoretical amount 12.692 grams), and when dissolved make up to a liter, preferably in a dark colored bottle as the light affects the solution.

Standardization. For general use standardize the iodine against 0.1 N sodium thiosulphate, or 0.1 N arsenous acid solution, taking 50 cc. of the reagent, carefully measured from a burette or pipette, adding a solution of starch and then the iodine solution until a blue color is obtained. Establish the factor for the iodine in terms of a true 0.1 solution. Note the temperature of the solution and make record on the bottle.

For tin plate analysis standardize the iodine solution against standard tin

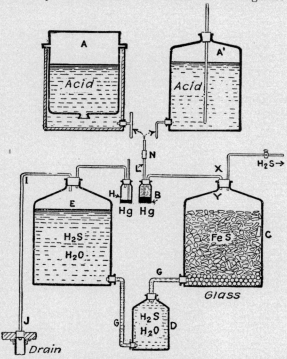

FIG. 74. HYDROGEN SULPHIDE GAS GENERATING APPARATUS

The apparatus is assembled with A empty at any convenient height greater than the desired gas pressure, above E and C. The minimum pressure of gas, in inches of water, desired at F is then decided upon. Six inches is satisfactory for most work. Water is added at E and air is allowed to escape through F until water just reaches the iron sulphide in C. F is then closed and E is filled with water. F is carefully opened and water is allowed to flow from E to C until the difference between the levels in the two bottles equals the desired minimum lead in inches of water. F is then closed; A is filled with dilute acid and the mercury layer in B is adjusted until the pressure of the acid column is just sufficient to overcome the back pressure of the mercury column in B and the air pressure in C. This is best accomplished by adding slightly more mercury than necessary and then carefully raising L until the dilute acid begins to drop on the sulphide. A short rubber tubing insert at N facilitates this operation. After the mercury layer is correctly adjusted, F is opened and the apparatus is ready to function. No further adjustments are necessary and operation is entirely automatic. To obtain hydrogen sulphide open cock F.

solution. It is convenient to adjust the iodine solution so that 1 cc. equals exactly .00579 gram of tin. Then, if a sample of the plate having a total surface of 8 sq. in. is taken, 1 cc. of the iodine solution is the equivalent of one-tenth of a pound per base box. Use starch as indicator.

Lead Acetate. 10% solution. 100 grams of $Pb(C_2H_2O_1)_2.3H_2O$ per liter. If cloudy, clear with a few cc. of acetic acid.

Standard Lead Solution. A convenient solution may be made by dissolving 0.1831 gram of lead acetate, $Pb(C_2H_3O_2)_2.3H_2O$, in 100 cc. of water, clearing any cloudiness with a few drops of acetic acid and diluting to 1000 cc. If 10 cc. of this solution are diluted to 1000 cc., each cc. will contain an equivalent of 0.000001 gram Pb.

Harcourt suggests a permanent standard made by mixing ferric, copper and cobalt salts. For example, 12 grams of $FeCl_3$ together with 8 grams of $CuCl_2$ and 4 grams of $Co(NO_3)_2$ are dissolved in water, 400 cc. of hydrochloric acid added and the solution diluted to 4000 cc. 150 cc. of this solution together with 115 cc. of hydrochloric acid (1 : 2) diluted to 2000 cc. will give a shade comparable to that produced by the standard lead solution above, when treated with the sulphide reagent. The exact value per cc. may be obtained by comparison with the lead standard.

Lead Acetate Test Paper for Removal of H_2S. Large sheets of qualitative filter paper are soaked in a dilute solution of lead acetate and dried. The paper is cut into strips $7+5$ cm.

Lead Acetate Cotton. Preparation of. A roll of absorbent cotton is opened and saturated with a 10% solution of lead acetate and surplus drained off, then hung on a line to dry in a warm place away from the influence of hydrogen sulphide. Do not dry in an oven. When dry, place in a stoppered bottle until used.

Magnesium Ammonium Chloride. Magnesia Mixture. For precipitation of ammonium magnesium phosphate, 110 grams of magnesium chloride $(MgCl_2.6H_2O)$ are dissolved in a small amount of water. To this are added 280 grams of ammonium chloride and 700 cc. of ammonia (sp.gr. 0.90); the solution is now diluted to 2000 cc. with distilled water. The solution is allowed to stand several hours and then filtered into a large bottle with glass stopper. Ten cc. of the solution should be used for every 0.1 gram P_2O_5 present in the sample analyzed. As the reagent becomes old it will be necessary to filter off the silica that it gradually accumulates from the reagent bottle.

Magnesia Wash Solution. For washing magnesium ammonium phosphate precipitate. 100 grams of NH_4NO_3 are dissolved in water, 335 cc. of strong NH_4OH added and the solution diluted to 1000 cc.

Manganous Sulphate Solution. 48 grams of manganous sulphate are dissolved in 100 cc. of distilled water.

Mercuric Chloride. Saturated solution of $HgCl_2$ (60 to 100 grams per liter).

Mercuric Chloride. 5% solution. 50 g. of $HgCl_2$ per liter.

Sensitized Mercuric Chloride (or Bromide Paper). 20×20 in. Swedish Filter Paper No. 0 is cut into four equal squares. For use in the large Gutzeit apparatus the paper is dipped into a 3.25% solution of mercuric chloride (mercuric bromide may be used in place of the chloride) or if it is to be used in the small Gutzeit apparatus, it is dipped into a 0.35% mercuric chloride solution. (The weaker the solution, the longer and less intense will be the stain.) The paper should be of uniform thickness, otherwise there will be an irregularity in length of stain for the same amounts of arsenic. See chapter on Arsenic.

Methyl Red Solution. 0.25 gram of methyl red is dissolved in 2000 cc. of 95% alcohol; 2 cc. of the indicator are used for each titration. As the indicator is sensitive to CO_2, all water used must first be boiled to expel carbonic acid.

Nessler's Solution. Dissolve 50 grams of potassium iodide in the smallest possible quantity of cold water. Add a saturated solution of mercuric chloride until a faint show of excess is indicated. Add 400 cc. of 50% solution of potassium hydrate. After same has clarified by sedimentation, make up to 1 liter with water, allow to settle and decant. For determining NH_3 in water.

Nitric Acid. Desk reagents. See beginning of the chapter.

Nitric Acid. Tenth Normal Solution. The acid is standardized against the caustic solution and should be of such strength that 1 cc. of HNO_3 is equal to 1 cc. of NaOH. Phenolphthalein indicator is used. Approximately 6.7 cc. of 95% HNO_3 diluted to 1000 cc. $= N/10$ HNO_3 solution.

Nitric Acid Standard—Ferrous Sulphate Method for Nitrates. *Standardization of the Acid.* 11.6 grams of standard potassium nitrate, equivalent to about 9.6 grams of $NaNO_3$, are dissolved and made to volume in the weighing bottle (100 cc.), and 10 cc. are placed in the Devarda flask, reduced and the ammonia distilled into 100 cc. of the acid. The temperature of the acid is noted and its value in terms of H_2SO_4, KNO_3 and $NaNO_3$ stated on the container. The acid expands or contracts 0.029 cc. for every degree centigrade above or below the temperature of standardization.

Oxalic Acid. 0.1 N solution. 6.303 grams of $H_2C_2O_4.8H_2O$ per liter.

Standardization. Titrate 50 cc., measured by a burette, of the solution with 0.1 N NaOH in presence of phenolphthalein indicator. Multiply cc. of caustic used by its factor (titration in presence of phenolphthalein) and divide by 50 to get the factor for the oxalic acid.

Titrate the oxalic acid, hot, with 0.1 N $KMnO_4$ to a faint pink color. Convert cc. of $KMnO_4$ to exact normality and divide by 50 to obtain the factor for the oxalic acid.

Palladous Chloride. Five grams of palladium wire are dissolved in mixture of 30 cc. of hydrochloric and 2 cc. of nitric acid, this evaporated just to dryness on a water bath, redissolved in 5 cc. of hydrochloric acid and 25 cc. of water, and warmed until solution is complete. It is diluted to 750 cc. and contains about 1% of palladous chloride. It will absorb about two-thirds of its volume of hydrogen.

Paraffin. Dissolve in carbon tetrachloride and use solution for reagent labels.

Peroxide Solution. Dissolve 4 grams of sodium peroxide in 125 cc. of dilute sulphuric acid (1 of acid to 3 of water) and dilute to 500 cc. Used in titanium colorimetric determinations.

Standard Potassium Dichromate. The reaction of potassium dichromate with iron may be represented as follows:

$$K_2Cr_2O_7 + 6FeCl_2 + 14HCl = 6FeCl_2 + 2CrCl_3 + 2KCl + 7H_2O.$$

When oxygen reacts with ferrous salts, the following reaction takes place:

$$6FeCl_2 + 6HCl + 3O = 6FeCl_3 + 3H_2O.$$

Comparing this reaction with that of dichromate, it is evident that a normal solution of dichromate contains one-sixth of the molecular weight of $K_2Cr_2O_7$ per liter, namely, 49.033 grams. For general use it is convenient to have two strengths of this solution, N/5 for ores high in iron and N/10 for products containing smaller amounts. See Introduction, § 25.

Standardization. For N/5 solution 9.807 grams of the recrystallized dehydrated salt are dissolved and made up to one liter; N/10 potassium dichromate contains 4.9033 grams of the pure salt per liter. It is advisable to allow the solution to stand a few hours before standardization. The Sibley iron ore furnished by the U. S. Bureau of Standards, Washington, D. C., is recommended as the ultimate standard. For accurate work it is desirable to use a chamber burette with graduations from 75 to 90 cc. in tenths and from 90 to 100 in twentieths of a cc. A titration of 90 to 100 cc. of the dichromate would require 0.9 to 1.1 grams of iron for a fifth normal solution and half this amount for a tenth normal solution of dichromate. If the ore contains 69 per cent. of iron, 1.4 grams should be taken for a N/5 solution. The ore is best dissolved in strong HCl, adding a few drops of stannous chloride solution and heating just below boiling. In case of an ore or iron ore briquette, containing silica in an appreciable amount, a carbonate fusion of the residue may be necessary. Reduction and titration of the ore is done exactly as prescribed in chapter on Iron.

The equivalent iron in the ore divided by the cc. titration required for complete oxidation gives the value in terms of grams per cc., e.g., 1.4 grams of ore containing 69.2% of Fe required a titration of 95 cc. of $K_2Cr_2O_7$ solution, then

$$1 \text{ cc.} = \frac{(69.2 \times 1.4)}{100} \div 95 = 0.0102 \text{ gram of Fe.}$$

Optional Methods for Standardization of Potassium Dichromate

It is not always possible to use the method of standardization with a standard ore, and chamber burettes may not be available. The following methods are reliable.

First. Optional Method. Standardization against Pure Iron. About 0.25 gram of pure iron wire are dissolved in an Erlenmeyer flask with Bunsen valve. Before adding the iron 20 cc. of the dilute sulphuric acid (5 N desk reagent) are added and about 2 grams of sodium bicarbonate. This displaces the air with CO_2. The wire is dropped in and the rubber stopper carrying the valve inserted. When the reaction has ceased and the iron is in complete solution, the dichromate solution is added until the ferrous iron is just oxidized to ferric state, the end-point being recognized by an outside indicator (1% solution of potassium ferricyanide being placed in drops on a white tile with depressions). Ferrous iron produces a blue color, ferric iron only a yellowish color with the indicator.

Second. Optional Method. 50 cc. of the potassium dichromate solution are pipetted or measured out by a burette into a 300-cc. glass-stoppered bottle. 150 cc. of water are added and 5 cc. of conc. sulphuric acid. After cooling under a tap, 10 cc. of 25% potassium iodide are added, the liberated iodine is now titrated with 0.1 N thiosu.pnate soiution, using starch indicator. The potassium dichromate factor is obtained by multiplying the cc. of thiosulphate by its normality factor and dividing by 50. See Introduction, pages 11, 14.

Potassium Dichromate. Four per cent solution. 40 grams of $K_2Cr_2O_7$ per liter.

Potassium Chromate. Ten per cent solution of neutral potassium chromate.

Potassium Ferricyanide, $K_3Fe(CN)_6$. The salt should be free of ferrocyanide, as this produces a blue color with ferric salts, which would destroy the end-point. It is advisable to wash off the salt before using. A crystal the size of a pinhead dissolved in 50 cc. of water is sufficient for a series of determinations. The solution should be made up fresh for each set of determinations.

Potassium Ferrocyanide, *Standard Solution for Zinc Determination.* 21.55 grams of the crystallized salt per liter. (1 cc. of 0.5 gram sample is equal to about 1%.)

Standardization. Low's Method. Weigh carefully 0.2 gram of pure zinc, place in an 8-oz. flask and add 10 cc. of conc. hydrochloric acid (sp.gr. 1.2). When the zinc has dissolved, dilute with 25 cc. of water, add a few drops of litmus solution, and make slightly alkaline with ammonia. Again acidify with hydrochloric acid, adding 3 cc. in excess. Dilute to about 250 cc. with hot water, heat to nearly boiling and titrate with the ferrocyanide solution, using a 15% uranium nitrate solution as an outside indicator, making the tests as in case of the dichromate titration of iron. A brown tinge obtained by adding a drop of the titrated solution to a drop of the uranium nitrate on a white tile is the end-point desired. It is advisable to divide the solution titrated, adding the reserve in portions until the entire amount has been titrated. This avoids overrunning the end-point.

Potassium Ferrocyanide *for Zinc Determinations.* 43.1 grams of pure salt in 1000 cc. of water. One cc. = approximately 0.010 Zn. This solution should be allowed to stand about four weeks before using.

Standardization. New Jersey Zinc Co. Method. Weigh into tall 400-cc. beakers several portions of C.P. zinc, using about 0.35 gram. Cover with water and dissolve in 10 cc. of hydrochloric acid (sp.gr. 1.2). Now add 13 cc. of ammonia (sp.gr. 0.9), make acid with hydrochloric acid, and add 3 cc. excess. Add 0.03 or 0.04 milligram of ferrous iron in the form of a ferrous sulphate solution and dilute to about 200 cc. with distilled water. Heat to boiling and titrate as follows: About one-quarter of the solution is reserved in a small beaker and the ferrocyanide added to the main solution with vigorous stirring. The solution takes on a blue color, which changes to a creamy white when an excess of ferrocyanide is added. Now add a few cc. more and pour in the reserved portion of zinc solution, excepting about 5 or 10 cc. Add ferrocyanide until the end-point is reached and add about $\frac{1}{2}$ cc. more. The last of the reserved zinc solution is then poured into the main beaker, washing out the small beaker with a portion of the main solution, and the ferrocyanide added drop by drop until the blue color fades sharply to a pea-green with one drop of ferrocyanide. This is the end-point. Repeat until satisfactory standards are obtained.

Potassium fluoride solution; made by dissolving 100 grams of potassium fluoride in about 1200 cc. of hot, CO_2-free water, then neutralizing the solution with hydrofluoric acid or potassium hydroxide as the reagent may require, using 5 cc. of phenolphthalein as indicator. Dilute sulphuric acid may be used in place of hydrofluoric acid in the final acid adjustment to get a neutral product. One cc. of the solution in 10 cc. of CO_2-free water should appear a faint pink. The concentrated mix is filtered if necessary and then diluted to 2000 cc. with CO_2-free water. The gravity will now be approximately 1.32 or about 35° Bé. One cc. contains 0.5 g. of potassium fluoride.

Potassium Hydroxide. (a) For carbon dioxide determination, 500 grams of the commercial hydrate are dissolved in 1 liter of water.

Absorption Capacity. One cc. absorbs 40 cc. of CO_2.

(b) For the preparation of potassium pyrogallate for special work, 120 grams of the commercial hydrate are dissolved in 100 cc. of water.

Potassium Iodide Solution. Dissolve 250 grams of potassium iodide free from iodate in distilled water and dilute to 1000 cc.

Potassium Nitrate Standard. The purest nitrate that can be obtained is recrystallized in small crystals, by stirring, during the cooling of the supersaturated concentrated solution, and dried first at 100° C. for several hours and then at 210° C. to constant weight. Chlorides, sulphates, carbonates, lime, magnesium and sodium are tested for and if present are determined and allowance made.

Used as a standard in the Devarda method for nitrate determination.

Potassium Permanganate. For oxidation purposes. Two per cent solution filtered free of dioxide through asbestos is required.

Potassium. Permanganate, Standard Solutions. As in case of potassium dichromate, it is convenient to have two standard solutions, N/5 and N/10. (0.2 N and 0.1 N.)

Since commercial potassium permanganate is seldom pure, it is necessary to determine its exact value by standardization. This is commonly accomplished by any of the following methods:

(a) By a standard electrolytic iron solution.

$$2KMnO_4 + 10Fe + 9H_2SO_4 = 2MnSO_4 + K_2SO_4 + 5Fe_2(SO_4)_3 + 8H_2O + H_2.$$

(b) By ferrous salt solution, e.g., $(NH_4)_2SO_4.FeSO_4.6H_2O$.

(c) By oxalic acid or an oxalate.

Reaction. $2KMnO_4 + 5Na_2C_2O_4 + 8H_2SO_4$
$$= K_2SO_4 + 2MnSO_4 + 5Na_2SO_4 + 10CO_2 + 8H_2O.$$

From the reactions above it is evident that $2KMnO_4$ will oxidize 10Fe or $5Na_2C_2O_4$. $2KMnO_4 = K_2O + 2MnO + 5O$, hence a normal solution would contain one-fifth of the molecular weight of $KMnO_4 = 31.6$ grams of the pure salt. Hence a N/5 solution would contain 6.32 grams per liter and a N/10 solution 3.16 grams.

It is advisable to dissolve the potassium permanganate in about 500 cc. of hot water and filter the solution through asbestos to remove any dioxide of manganese that may be present, as MnO_2 aids in the decomposition of $KMnO_4$ solution. The reagent should be kept tightly sealed in a dark bottle well protected from the light. This solution should stand two or three days before standardization. The solution is diluted to 1000 cc. for a 0.2 N solution or to 2000 cc. for a 0.1 N solution, if 6.32 grams have been taken.

Standardization. (a) *Iron Wire in Sulphate Solution.* Standard iron wire, as furnished by chemical supply houses, with the percentage of iron given (usually about 99.8% of Fe), gives results, when properly treated, which are practically identical with those obtained by the use of sodium oxalate.

Samples of 0.2000 gm. are weighed out and the weights multiplied by the factor for iron, giving the true weight of iron. 50 cc. of 20% sulphuric acid are placed in a 250-cc. Erlenmeyer flask, which is fitted with a rubber stopper carrying a Bunsen valve. Into the acid is dropped from one to two grams of sodium bicarbonate in order that the air may be replaced by carbon dioxide and then the sample of wire is introduced and the flask stoppered by the Bunsen valve. The whole is then boiled gently until the iron is completely dissolved, and the boiling continued for a minute or two more. The solution is now cooled, and, after dilution to 150 cc., titrated to the first permanent pink.

$$\frac{\text{Net weight of iron}}{\text{cc. used}} = \text{iron standard.}$$

The atmosphere of CO_2 and the Bunsen valve are employed so that oxygen may be kept out of contact with the hot solution. Under these conditions the iron goes into solution completely as $FeSO_4$. The boiling after complete solution is necessary in order that all traces of hydrocarbons from the carbon present in the iron may be expelled. Hydrocarbons reduce permanganate.

The complete reaction is:

$$2KMnO_4 + 10FeSO_4 + 8H_2SO_4 \rightarrow 5Fe_2(SO_4)_3 + K_2SO_4 + 2MnSO_4 + 8H_2O.$$

(b) *Standardization of $KMnO_4$ against sodium oxalate* is recommended as the most accurate procedure. The salt has no water of crystallization and is not hygroscopic. It can be obtained from the Bureau of Standards with a guarantee of purity. Traces of moisture can be expelled by heating the salt to 120° C. for two hours, then cooling in a desiccator.

The chemistry of the process is as follows:

$$2KMnO_4 \rightarrow K_2O + 2MnO + 5O,$$
$$Na_2C_2O_4 + H_2SO_4 \rightarrow H_2C_2O_4 + Na_2SO_4,$$
$$H_2C_2O_4 + O \rightarrow H_2O + 2CO_2,$$
and so $5O + 5H_2C_2O_4 \rightarrow 5H_2O + 10CO_2.$

From these expressions it is seen that

$2KMnO_4 \backsim 5O \backsim 5H_2C_2O_4 \backsim 5Na_2C_2O_4$, and the complete reaction is

$2KMnO_4 + 5Na_2C_2O_4 + 8H_2SO_4 \rightarrow K_2SO_4 + 2MnSO_4 + 8H_2O + 10CO_2.$ Since $2KMnO_4 \backsim 10Fe$, from the reaction $2FeO + O \rightarrow Fe_2O_3$,

$$2Na_2C_2O_4 \backsim 10Fe.$$

Weigh 0.2 to 0.3 gram of pure sodium oxalate, place in a beaker and add 200 cc. of water and about 5 cc. strong sulphuric acid. Heat to boiling and titrate with the potassium permanganate solution. The reaction starts slowly, but with the progress of the titration the action becomes vigorous. Towards the end of the reaction the pink color fades less rapidly and finally a permanent pink color is obtained with one drop of the reagent.

Calculation of Normality. Since 67 grams of sodium oxalate per 1000 cc. of solution is a normal solution, the cc. equivalent of the amount taken is obtained by dividing by 0.067. This value divided by the titration of the oxalate with the reagent being standardized will give the normality of the reagent.

Example. Suppose 0.268 gram of sodium oxalate required a titration of 50 cc. of the permanganate solution. Since 67 g. is equivalent to 1000 cc. N solution of sodium oxalate, then 0.268 is equivalent to

$$0.268 \times 1000 \text{ divided by } 67 = 4 \text{ cc. N sol.}$$
$$4 \text{ divided by } 50 = 0.08 \text{ N.}$$

Potassium Permanganate.[1] **Preparation of N/100 Solution.** Dissolve 0.40 g. of pure potassium permanganate crystals in one liter of redistilled water in a thoroughly clean Florence flask which has been rinsed with the same water. Digest at or near the boiling point for 36 hours. A funnel covered with a watch-glass may be used as a reflux condenser. Cool and allow to stand over night. Without disturbing the sediment of manganese oxides, filter with gentle suction through a 3-in. Büchner funnel lined with ignited asbestos. Both funnel and filter flask should be rinsed with redistilled water. Transfer the permanganate solution to a glass-stoppered bottle free from traces of organic matter. The solution should be kept in the dark when not in use. If the asbestos becomes clogged with oxides, these may be dissolved out with hot concentrated hydrochloric acid, followed by washing with redistilled water without disturbance of the pad.

After standing two or three days, this permanganate solution may be conveniently standardized against N/50 oxalic acid (0.1261 g. of pure crystals to 100 cc.) or sodium oxalate of similar strength. To 10 cc. of the oxalic acid solution add 10 cc. of 10 per cent sulphuric acid which has been treated with just sufficient permanganate solution to give it a faint pink color. Place in a water-bath at 65° C. for a few minutes. Then titrate at once to a definite pink color which persists for at least a minute. Correct for the blank obtained by titrating 10 cc. of the sulphuric acid and the same volume of water to the same end-point. On account of the sensitivity of this reagent it is desirable to check it up rather frequently.

Potassium or Ammonium Thiocyanate (*Sulphocyanate*) *0.1 N Solution.* About 8 grams of ammonium or 10 grams of potassium salt are dissolved in water and diluted to one liter. The solution is adjusted by titration against the N/10 silver nitrate solution. It is advisable to have 1 cc. of the thiocyanate equivalent to 1 cc. of the silver nitrate solution. Owing to the deliquescence of the thiocyanates the exact amount for a N/10 solution cannot be weighed.

Potassium Pyrogallate. Except for use with the Orsat or Hempel apparatus, this solution should be prepared only when wanted. The most convenient method is to weigh out 5 grams of the solid acid upon a paper, pour it into a funnel inserted in the reagent bottle, and pour upon it 100 cc. of potassium hydrate containing 120 grams KOH. The acid dissolves at once, and the solution is ready for use.

Attention is called to the fact that the use of potassium hydroxide purified by alcohol has given rise to erroneous results.

Absorption capacity. One cc. absorbs 2 cc. of O.

Salt Solution Standard. 16.48 grams of fused C.P. sodium chloride are dissolved in 1 liter of distilled water. 100 cc. of this solution diluted to 1 liter gives a standard solution, each cc. of which contains .001 gram of chlorine.

Schiff's Fuchsin Bisulphite Reagent. See Index.

[1] *Jour. of Ind. and Eng. Ch.*, Feb. 1918.

Silver Nitrate Standard Solution. 4.8 grams of dried silver nitrate crystals are dissolved in one liter of distilled water. Each cc. of this solution is equivalent to approximately .001 gram of chlorine, standardized against the Standard Salt Solution.

NOTE. N/50 solutions of both sodium chloride and silver nitrate can be used where it is inconvenient to make too many standard solutions, using the proper factors.

Silver Nitrate *N/10.* This solution contains 10.788 grams of Ag or 16.989 grams of AgNO₃ per liter. The silver nitrate salt, dried at 120° C., or pure metallic silver may be taken, the required weight of the latter being dissolved in nitric acid and made to volume, or 17.1 grams of the salt dissolved in distilled water and made to 1000 cc. The solution is adjusted to exact decinormal strength by standardizing against a N/10 sodium chloride solution, containing 5.846 grams of pure NaCl per liter.

Silver, Pure, Preparation of. The volumetric methods used for the determination of high percentages of silver employ solutions which should be standardized by metal of the highest purity. For the preparation of this metal, the electrolytic method as described below is preferred by laboratories which are suitably equipped.

By Knorr's method, a solution of silver nitrate from which excess of nitric acid has been removed by evaporation is freed of metallic impurities by adding enough sodium carbonate to precipitate one-tenth of the silver, boiling and filtering. The silver in the filtrate is precipitated by sodium carbonate and the precipitate decomposed without addition of reducing agent, by melting in a crucible. Excess sodium carbonate carried down with the precipitate of silver carbonate will cover the fusion and such as adheres tightly to the metal is readily removed by hydrochloric acid. The metal should be smelted under charcoal.

For standardizing the decimal solution, 1 gram of standard silver is dissolved in a funnel-closed liter flask by 20 cc. of equal volumes of nitric acid and water. After boiling to decompose nitrous acid the solution is made up to mark with water at room temperature.

When using the standard thiocyanate solution with the object of obtaining very accurate results, it is good practice, instead of using a standard solution of silver, to obtain the standard by titrating, simultaneously with the assays, solutions of known quantities of standard silver approximately equal in amount and dissolved in the same manner as the assays. Volhard's Method for Silver.

Sodium Bismuthate. Prepare as follows: Heat 20 parts of caustic soda nearly to redness in an iron or nickel crucible, and add, in small quantities at a time, 10 parts of basic bismuth nitrate, previously dried in a water oven. Then add 2 parts of sodium peroxide and pour the brownish-yellow fused mass on an iron plate to cool; when cold, break it up in a mortar, extract with water, and collect on an asbestos filter. The residue, after being washed four or five times by decantation, is dried in the water oven, then broken up and passed through a fine sieve.

Sodium Carbonate. 10% solution. 100 grams of Na₂CO₂ per liter.

Sodium Hydroxide. Desk reagent. See the beginning of this chapter.

Sodium Hydroxide. Normal solution. 1 cc. of the reagent is equivalent to 0.04904 g. of H₂SO₄. Use rubber stoppers for bottles with sodium compounds.

Preparation of Standard Solution. Make up a stock concentrated solution of sodium hydroxide by dissolving sodium hydroxide in water in the proportion of 200 grams of NaOH to 200 cc. of water. Allow this solution to cool and settle in a stoppered bottle for several days. Decant the clear liquid from the precipitate of sodium carbonate into another clean bottle. Add clear barium hydroxide solution until no further precipitate forms. Again allow to settle until clear. Draw off about 175 cc. and dilute to 10 liters with freshly boiled distilled water. Preserve in a stock bottle with a large guard tube filled with soda lime. Determine the exact strength by titrating against pure benzoic acid (C_6H_5COOH) or against standard sulphuric acid, using phenolphthalein as indicator. This solution will be approximately one-fourth normal, but do not attempt to adjust it to any exact value. Determine its exact strength and make proper corrections in using it. See chapter on Alkalimetry and Acidimetry.

This should be made of such strength that 1 cc. is equal to 1 cc. of the standard acid, 2 cc. of methyl red being used as indicator. Ten cc. of the acid are diluted to 500 cc. and the alkali added until the color of the indicator changes from a red to a straw color.

Sodium Hydroxide. Tenth Normal Solution. For determination of phosphorus by the alkali volumetric method, consult chapter on Phosphorus.

Sodium Hydroxide Solution, Alcoholic. Dissolve pure sodium hydroxide in 95 per cent alcohol in the proportion of about 22 grams per 1000 cc. Let stand in a stoppered bottle. Decant the clear liquid into another bottle, and keep well stoppered. This solution should be colorless or only slightly yellow when used; it will keep colorless longer if the alcohol is previously treated with NaOH (about 80 g. to 1000 cc.), kept at about 50° C. for 15 days, and then distilled.

N/5 Sodium Oxalate. $Na_2C_2O_4$ reacts with $KMnO_4$ as follows:

$$5Na_2C_2O_4 + 2KMnO_4 + 8H_2SO_4 = K_2SO_4 + 2MnSO_4 + 5Na_2SO_4 + 10CO_2 + 8H_2O.$$

Hence $5Na_2C_2O_4$ divided by 10 or 134 divided by 2 = 67 grams per liter = a normal sodium oxalate solution. A N/5 solution requires 13.4 grams $Na_2C_2O_4$ per liter.

Sodium Sulphide. Ten per cent solution, made from colorless crystals. Sodium sulphide may be made by saturating a strong solution of sodium hydroxide with hydrogen sulphide gas, and then adding an equal volume of the sodium hydroxide. The solution is diluted to required volume, allowed to stand several days, and filtered.

Sodium metabisulphite. Solid salt of $Na_2S_2O_5$.

Sodium Thiosulphate. Used in copper determinations. 7.5 grams of the salt, $Na_2S_2O_3.5H_2O$, are dissolved and made to 2 liters with water. The solution is standardized against a copper solution containing 1 gram of pure copper per liter. 1 cc. = 0.001 gram of Cu. Approximately the same amount of copper in the same volume of solution is taken as is to be determined by the assay. For high-grade copper ores and crude copper, etc., it is advisable to prepare a standard thiosulphate solution ten times the above strength. The copper solution is made slightly ammoniacal and then acid with acetic acid. Potassium or sodium iodide crystals, free from iodate, are added and the liberated iodine titrated with the standard thiosulphate. (See Procedure under Copper.)

Sodium Thiosulphate, Standard 0.1 N Solution for General Analysis

Sodium thiosulphate reacts with iodine as follows:

$$I_2+2Na_2S_2O_3=2NaI+Na_2S_4O_6.$$

1 gram molecule of thiosulphate is equivalent to 1 atom of iodine$=1$ atom of hydrogen, hence a tenth normal solution is equal to one-tenth the molecular weight of the salt per liter, e.g., 24.822 grams of $Na_2S_2O_3.5H_2O$; generally a slight excess is taken—25 grams of the crystallized salt. It is advisable to make up 5 to 10 liters of the solution, taking 125 to 250 grams of sodium thiosulphate crystals and making up to volume with distilled water, boiled free of carbon dioxide. The solution is allowed to stand a week to ten days, and then standardized against pure, resublimed iodine.

Standardization. About 0.5 gram of the purified iodine is placed in a weighing bottle containing a known amount of saturated potassium iodide solution (2 to 3 grams of KI free from KIO_3 dissolved in about $\frac{1}{2}$ cc. of H_2O), the increased weight of the bottle, due to the iodine, being noted. The bottle and iodine are placed in a beaker containing about 200 cc. of 1% potassium iodide solution (1 gram KI per 200 cc.), the stopper removed with a glass fork and the iodide titrated with the thiosulphate to be standardized.

Calculation. The weight of the iodine taken, divided by the cc. of thiosulphate required, gives the value of 1 cc. of the reagent; this result divided by 0.012692 gives the normality factor.

NOTE. The thiosulphate solution may be standardized against iodine, which has been liberated from potassium iodide in presence of hydrochloric acid by a known amount of standard potassium bi-iodate, a salt which may be obtained exceedingly pure.

$$KIO_3.HIO_3+10KI+11HCl=11KCl+6H_2O+6I_2.$$

A tenth normal solution contains 3.2496 grams of the pure salt per liter. (One cc. of this will liberate 0.012692 gram of iodine from potassium iodide.) The purity of the salt should be established by standardizing against thiosulphate, which has been freshly tested against pure resublimed iodine.

About 5 grams of potassium iodide (free from iodate) are dissolved in the least amount of water that is necessary to effect solution, and 10 cc. of dilute hydrochloric acid (1 : 2) are added, and then 50 cc. of the standard bi-iodate solution. The solution is diluted to about 250 cc. and the liberated iodine titrated with the thiosulphate reagent; 50 cc. will be required if the reagents are exactly tenth normal.

Optional Methods of Standardization. The thiosulphate may be standardized against standard 0.1 N potassium dichromate (4.9033 grams of pure $K_2Cr_2O_7$ per liter of solution) as follows. Place in a beaker 100 cc. of distilled water and 10 cc. of a 25% solution of potassium iodide. Now add from a burette 50 cc. of the standard N/10 potassium dichromate and 10 cc. of conc. HCl, and titrate the liberated iodine with standard N/10 thiosulphate reagent. When the yellow color has almost disappeared, add about 2 cc. of starch solution and continue the titration until the blue changes to a sea-green. The cc. of dichromate taken divided by the cc. of thiosulphate required will give the factor in terms of tenth normal solution.

Tenth normal potassium permanganate may be used to liberate the iodine in place of the dichromate, if desired, and the thiosulphate titration made as usual.

It is best to leave the thiosulphate solution as it is after determining its exact iodine value, rather than to attempt to adjust it to exactly decinormal strength. Preserve in a stock bottle provided with a guard tube filled with soda lime.

Sodium Thiosulphate Solution. N/100 solution is made as needed from the N/10 stock solution.

Sulphuric Acid. Desk reagents. See first part of this chapter.

Sulphuric Acid. Standard. See chapter on Acidimetry and Alkalimetry.

Sulphuric Acid Solution, Half Normal. Add about 15 cc. of sulphuric acid (1.84 specific gravity) to distilled water, cool and dilute to 1000 cc. Determine the exact strength by titrating against freshly standardized sodium hydroxide or by any other accurate method. Either adjust to exactly half normal strength or leave as originally made, applying appropriate correction.

Stannous Chloride Solution. The reagent is prepared by dissolving 2 grams of stannous chloride crystals in hot concentrated hydrochloric acid and making up to 1 liter. The solution should be kept in a dark bottle to which the titrating burette is attached in such a way that the liquid may be siphoned out into this, as shown in the illustration, Fig. 83. The air entering the bottle passes through phosphorous or pyrogallic acid to remove the oxygen. In this way, protected from the air, the reagent will keep nearly constant for several weeks. It is advisable, however, to restandardize the solution about every ten to fifteen days. One cc. will be equivalent to about 0.001 gram of Fe.

Stannous Chloride, Strong. Sixty grams of the crystallized salt are dissolved in 600 cc. of strong HCl and made up to one liter. The solution should be kept well stoppered.

Starch Solution. Five grams of potato or arrowroot starch are rubbed to a paste in cold water, 200 cc. of hot water stirred in, the starch dissolved by boiling, and then poured into about two liters of hot water and boiled for a few minutes.

Preservatives. Various preservatives are used for preserving the starch solution. A few cc. of 5% NaOH solution added to the boiling solution above will preserve the starch.

Salicylic acid, 10 cc. of 1% solution per liter is a good preservative.

Chloroform, a few drops per liter of starch solution.

Zinc chloride solution, a few cc. of 10% solution per liter.

Sodium chloride and acetic acid. The starch solution is prepared by adding to 500 cc. of a saturated sodium chloride solution (filtered) 100 cc. of 80% acetic acid and 3 grams of starch, mixing cold. The solution is now brought to boiling and boiled about two minutes. The solution thus prepared keeps indefinitely.

Tartrate Solution, Alkaline. Twenty-five grams of C.P. sodium potassium tartrate, $NaKC_4H_4O_6.4H_2O$, is dissolved in 50 cc. of water. A little ammonia is added and then sodium sulphide solution. After settling some time the reagent is filtered. The filtrate is acidified with hydrochloric acid, boiled free of H_2S and again made ammoniacal and diluted to 100 cc.

Thymol Solution 1%. The thymol is dissolved in a little glacial acetic acid containing 10% of ethyl alcohol, and this solution added to concentrated sulphuric acid. Addition of the thymol directly to the acid would produce a colored solution. The reagent should be kept protected from strong light, otherwise it will become colored.

Zinc, Amalgamated. This is best prepared by dissolving 5 grams of mercury in 25 cc. of concentrated nitric acid with an equal volume of water, 250 cc. of water are added and the solution poured into 500 grams of shot zinc, 20-mesh. When thoroughly amalgamated, the solution is poured off and the zinc dried.

57

PREPARATION AND USES OF CHEMICAL INDICATORS

Cochineal. Yellowish red with acids, violet with alkalies. Cannot be used in presence of iron, alumina, or acetates. Cochineal may be used for the titration of ammonia and for alkali carbonates and bicarbonates in cold solution. Carminic acid is the essential compound in this indicator and is preferable to the indicator made from cochineal.

Preparation. One part of the crushed insect is extracted with 10 parts of 25% alcohol. The extract when filtered is ready for use.

Lacmoid. Red with acids, blue with alkalies.

Preparation. It is advisable to use the C.P. reagent. 0.2 gram of the purified lacmoid is dissolved in 100 cc. of alcohol. The commercial product may be purified by dissolving in alcohol (96%) and filtering. The filtrate is evaporated over strong sulphuric acid in an evacuated desiccator and the residue used.

The reagent is used in the titration of strong acids and bases and ammonia. It is not suited for titration of weak acids or nitrous acid.

Litmus. Red with acids, blue with alkalies.

Preparation. A solution obtained by extracting the crushed cubes with hot water may be used after allowing the insoluble matter to settle out. With such a solution the color change is not sharp. The best method of preparation is as follows: The unbroken cubes are digested for some time with several portions of warm (85%) alcohol. The alcoholic solution contains undesirable compounds and is thrown away. The residue is boiled with water, allowed to stand several days and the solution siphoned off. This blue solution is made slightly acid with acetic acid, evaporated to a syrup, and then mixed with a large amount of 95% alcohol. By this means the desired coloring matter (azolitmin) is precipitated. It is filtered off, washed with hot alcohol and dissolved in sufficient hot water to give a solution 3 or 4 drops of which will impart a distinct blue color to 50 cc. of water. Tincture of litmus should be preserved in a wide-mouthed bottle loosely stoppered with a plug of cotton so as to exclude dust while permitting the access of air. Litmus gives good results with mineral acids, oxalic acid and caustic alkalies; if used in titrating carbonates, the solution must be boiled to expel CO_2 as it affects the indicator.

Methyl Orange. Pink with acids, yellow with alkalies.

Preparation. 0.2 g. is dissolved in 100 cc. of hot water. If not clear after cooling, it should be filtered. Methyl orange is not very sensitive to acids. For this reason it should not be used with organic acids. Because of its low sensitiveness to acids it may be used for the titration of carbonates in cold solution. It gives good results with ammonia.

Phenacetolin. Color with mineral acids is golden yellow. Color with alkali carbonates and ammonia is pink.

Preparation. Dissolve 0.2 gram in 100 cc. of 50% alcohol.

Phenolphthalein. Colorless with acids, red with alkalies.

Preparation. One-tenth of a gram is dissolved in 100 cc. of 85% alcohol. Phenolphthalein is very sensitive to acids, less sensitive to alkalies. Can be used with mineral acids and with most organic acids. Carbonates give indistinct end-points unless solution is boiled to expel CO_2. Useless with ammonia.

Rosolic Acid. Yellow with acids, violet red with alkalies.

Preparation. Dissolve 0.2 grams in 100 cc. of 85% alcohol. Rosolic acid can be used with mineral acids including sulphurous acid and with oxalic acid, but not with other organic acids. In neutralizing sulphurous acid with ammonia this indicator changes when the normal sulphite is formed. It is somewhat affected by carbonic acid.

Tumeric Paper. Yellow with acids, brown with alkalies.

Preparation. Pieces of the root are first digested with several portions of water to remove undesired matter, and then with alcohol. Paper is moistened with the alcoholic solution and dried in the dark. It is used in titrating highly colored solutions where an inside indicator cannot be used, and for the detection of borates. When moistened with a solution containing free boric acid and then dried, a brown color results which is not destroyed by dilute sulphuric or hydrochloric acids and is changed by alkali hydroxide to a bluish-black color.

Preparation of Compounds

Cupferron. A liter of water, 30 grams of ammonium chloride and 60 grams of nitrobenzine are stirred by means of a mechanical stirrer to an emulsion. While stirring 80 grams of zinc dust are added in small portions, keeping the temperature of the mixture between 15° and 18° C. by addition of shaved ice. When the odor of nitrobenzine has disappeared and the precipitate of zinc hydroxide appears gray sufficient zinc dust has been added, about half an hour reduction usually being required. The cooled solution is filtered and the precipitate washed with ice water. From the filtrate and washings phenylhydroxylamine crystals are now salted out by means of NaCl added to saturation of the solution at 0° C.

The white needle-like crystals are filtered off with suction and dried between filter paper. (Caution: Solutions of phenylhydroxylamine are active skin poisons. Wash off with water and alcohol.)

The crystals are dissolved in about 500 cc. of ether, the solution filtered through a dry filter, then cooled to 0° C. and saturated with dry ammonia gas. A little more than the theoretical amount (one mole) of fresh amyl alcohol is added as quickly as possible. Heat is generated and snow white crystals of cupferron precipitate out. The mother liquor is filtered off and the crystals washed with ether and dried between filter papers. The cupferron should be kept in a closely stoppered bottle containing a small lump of ammonium carbonate.

Cupferron, $C_6H_5N_2O_2NH_4$ will precipitate copper, iron and titanium from strongly acid solutions. Copper salt is gray, ferric salt red and the titanium salt is yellow. The separation of iron and titanium in acid solutions from aluminum, chromium, manganese, cobalt, nickel, zinc and alkali earth metals makes it a valuable laboratory reagent. Zirconium will precipitate with iron and titanium if present. The reagent is not very stable in acid solution. Precipitations are made from cold solutions.

Copper precipitated with iron by means of cupferron may be dissolved out with ammonium hydroxide and reprecipitated by acidifying the filtrate, containing the copper, with acetic acid. The reagent is removed from the copper by washing with a 1% solution of sodium carbonate.

General References

Allen, A. H., Organic Analysis, v. 1–8, 4th Ed., 1909–1913. P. Blakiston's Son & Co.

Arnold and Mandel, Compendium of Chemistry. John Wiley & Sons, New York. 1904.

Authenrieth, W., and Warren, W. H., Laboratory Manual for the Detection of Poisons. 1921. P. Blakiston's Son & Co., Philadelphia, Pa.

Baskerville, C., and Curtman, L. J., Quantitative Analysis, 1910. The Macmillan Co.

Blair, A. A., The Chemical Analysis of Iron, 7th Ed., 1918. Lippincott & Co., Philadelphia, Pa.

Brearly, H., and Ibbotson, F., The Analysis of Steel Works Materials, 1902. Longmans, Green & Co., London and New York.

Browning, P. E., Introduction to the Rarer Elements, 3d Ed., 1919. John Wiley & Sons, New York.

Cahen, E. and Wootton, W. O., The Mineralogy of Rarer Metals, 1920. J. B. Lippincott Co.

Carins, F. A., Quantitative Chemical Analysis, 3d Ed., 1896. H. Holt & Co., New York.

Clowes and Coleman, Quantitative Analysis, 5th Ed. P. Blakiston's Son & Co.

Classen, A., Quantitative Analysis. Translated by N. H. Harriman, 1902. Geo. Wahr, Pub., Ann Arbor, Mich.

Comey, A. M., Dictionary of Chemical Solubilities. 1921. Macmillan & Co.

Crookes, W., Select Methods in Chemical Analysis, 4th Ed., 1905. Longmans, Green & Co., New York, London. 1905.

Dennis, L. M., Gas Analysis. Macmillan & Co., New York. 1913.

Fresenius, K. M., Quantitative Chemical Analysis, 6th Ed. Translated by A. I. Cohn, 1911. John Wiley & Sons, New York. 1912.

Ibid., Qualitative Chemical Analysis. Trans. 17 Ed., 1921, by C. A. Mitchell. J Wiley and Sons.

Fulton, C. M., Fire Assaying, 2d Ed., 1911. McGraw-Hill Co., New York. 1911.

Furman, H. Van F., Practical Assaying, 5th Ed., 1901. John Wiley & Sons, New York.

Gardner, H. A., Paint Technology and Tests, 1911. McGraw-Hill Book Co., New York.

Gardner, H. A., and Schaeffer, J. A., Analysis of Paints and Painting Materials, 1911. McGraw-Hill Book Co., New York.

Griffin, Technical Methods of Analysis, 1921. McGraw-Hill.

Gill, A. H., Gas and Fuel Analysis for Engineers, 3d Ed., 1903. J. Wiley & Sons, New York.

Gill, A. H., A Short Hand-book of Oil Analysis, 6th Ed., 1911. J. B. Lippincott Co., 1919, Philadelphia and London.

Gooch, F. A., Methods in Chemical Analysis, 1912. John Wiley & Sons, New York.

Hempel, W., Gas Analysis. Macmillan & Co., New York.

Hillebrand, W. F., The Analysis of Silicate and Carbonate Rocks, Bull. 422, U. S. Geological Survey, 1910. Washington Gov. Print. Office.

Hopkins, B. Smith, Chemistry of the Rarer Elements, D. C. Heath and Co., New York.

Johnson, C. M., Rapid Methods for the Chemical Analysis of Special Steels, 2d Ed., 1920, 1914. John Wiley & Sons, New York.

Julian, F. A., Text Book of Quantitative Analysis, 1902. Ramsey Publishing Co., St. Paul, Minn.

Langbein, Electro Deposition of Metals, 1920. H. C. Baird & Co.

Leach, A. E., Food Inspection and Analysis, 4th Ed., 1920. John Wiley & Sons, New York.

Levy, S. I., The Rare Earths, 1915. Edward Arnold Pub.

Lewkowitsch, J., Chemical Analysis of Oils, Fats, Waxes, 1913–1916. The Macmillan Co.

Liddell, Metallurgists and Chemists' Handbook. 1918.

Low, A. H., Technical Methods of Ore Analysis, 7th Ed., 1919. John Wiley & Sons, New York.

Lunge, G., The Manufacture of Sulphuric Acid and Alkali, 4th Ed., 1913. Van Nostrand Co., New York.

Lunge, G., Technical Methods of Chemical Analysis. Trans. Edited by Chas. A. Keane. Gurney & Jackson, London. D. Van Nostrand Co., New York.

Mahin, E. G., Quantitative Analysis, 1919. McGraw-Hill Book Co.

Marks, L. S., Mechanical Engineers' Handbook, 1916, McGraw-Hill Book Co.

Mead, R. K., Portland Cement, 2d Ed., 1911. The Chemical Pub. Co., Easton, Pa. Ibid., Chemists' Pocket Manual, The Chemical Pub. Co., Easton, Pa.

Mellor, J. W., Treatise on Quantitat?e Inorganic Analysis, 1913. Chas. Griffin & Co., Ltd., London.

Olsen, J. C., Quantitative Chemical Analysis, 5th Ed., 1921. D. Van Nostrand Company, New York.

Perkin, F. M., Electro Chemistry. Longmans, Green & Co., London and New York. 1905.

Prescott, A. B., and Johnson, O. C., Qualitative Chemical Analysis, 7th Ed., 1920. Revised by J. C. Olsen. D. Van Nostrand Co., New York.

Popoff, Stephen, Quantitative Analysis, 1924. P. Blakiston's Son and Co., Philadelphia, Pa.

Rhead, E. L., and Sexton, A. H., Assaying and Metallurgical Analysis, 2d Ed., 1911. Longmans, Green & Co., London and New York.

Rogers, A., Manual of Industrial Chemistry, 3d Ed., 1921. D. Van Nostrand Co.

Roscoe, H. E., and Schorlemmer, C., Treatise on Chemistry, 1911. Macmillan & Co.

Schoeller, W. R., and Powell, A. R., The Analysis of Minerals and Ores of the Rarer Elements. Chas. Griffin and Co., London.

Scott, W. W., Qualitative Analysis, 4th Ed., 1921. D. Van Nostrand Co., New York. Ibid., Technical Methods of Metallurgical Analysis, D. Van Nostrand Co., New York.

Segerblom, W., Tables of Properties. Exeter Book Pub. Co., Exeter, N. H.

Seidell, A., Solubilities of Inorganic and Organic Substances, Second Ed., 1919. D. Van Nostrand Co., New York.

Smith, E. F., Electro-analysis. P. Blakiston's Son & Co. 1918.

Stillman, T. B., Engineering Chemistry, 5th Ed., 1916. The Chemical Publishing Co., Easton, Pa.

Sutton, F. A., Systematic Handbook of Volumetric Analysis, 10th Ed., 1911. P. Blakiston's Son & Co., Philadelphia.

Talbot, H. P., An Introductory Course in Quantitative Chemical Analysis, 1899. Macmillan & Co., New York.

Thorpe, E., A Dictionary of Applied Chemistry, Five Volumes, 1913. Longmans, Green & Co., London and New York.

Treadwell, E. P., Analytical Chemistry. Translated by W. T. Hall, 6th Ed., 1924. J. Wiley & Sons, New York.

Van Nostrand's Chemical Annual. Edited by J. C. Olsen, 4th Issue, 1918. D. Van Nostrand Company, New York.

White, C. H., Methods in Metallurgical Analysis, 1915. D. Van Nostrand Co., New York.

Wiley, H. W., Principles and Practice of Agricultural Analysis, 1914. The Chemical Pub. Co., Easton, Pa.

NOTE

The lapse of paging between the end of Volume
I and the beginning of Volume II, is to allow for
the expansion of Volume I, without changing the
folios of Volume II.

INDEX

SUBJECT INDEX

Abbé refractometer, 1125

Abraham, H., bituminous substances, 1289–1350

Abraham's ductility test, 1304

Absorption bulbs:
carbon dioxide, Fleming, Geissler, Gerhardt, Liebig, Vanier, 111, 114

gas analysis, Friedrich, Hanjkus, Nowicki-Heinz, Varrentrapp, Winkler, Wolff, 1237

Absorption spectrum, carbon monoxide in air, 1274

Accuracy in methods of gas analysis, 1251

Accuracy, limit of, in alloy analysis, 1041

Acetanilid, 1732

Acetate extraction of lead, 273, 279

Acetates, 1547

Acetic acid, complete analysis of acetone, formic acid, furfurol, hydrochloric acid, metals, sulphuric acid, sulphurous acid, 1545–1547

method for nitrite, 338

specific gravity table, 673, 1548

Acetic anhydride, analysis of, 1544

Acetin method for det. glycerol, 1755

Acetone, analysis of, 1759

extraction of rubber, 1576, 1585, 1587

in acetic acid, 1546

Acetyl value for oils, 1135

Acetylene flame, temperature of, 658

properties of, 1227g

Acid number in soap analysis, 1604

Acid, free in soap, 1600

Acidimetric and alkalimetric methods for metabisulphites, sulphites, sulphurous acid, 513

phosphorus in steel, 1364

Acidimetry and alkalimetry, 1491–1560

Acid number, Chinese wood oil (Tung oil), 1169

linseed oil, 1165

varnish, 1174

Acid number, linseed oil, 1165

Acidity in explosives, 1394

in rubber, 1570

in water, 1429

Acids, chapter on, 1491

analysis of acetic. See subject above.

carbolic, 1549

carbonic, 1549

Acid, citric, 1549

chlorsulphonic, 1540

fluosilicic, 1507

formic, 1542

hydrochloric, determination of total acidity, arsenic, barium, chloride, chlorine, nitric acid, sulphuric acid, silica, total solids, 1501

hydrofluoric, determination of acidity, hydrofluosilicic acid, sulphuric acid, sulphurous acid, 1506, 1509

nitric, determination of acidity, free chlorine, hydrochloric acid, iodine, nitric and nitrous acids, non-volatile solids, sulphuric acid, 1509

procedure for determining, in arsenic acid, ferrous sulphate method, 1516

in oleum and mixed acids, 1515

in phosphoric acid, 1516

in sulphuric acid, 1514

nitrous, permanganate titration of, 1510, 1530

oleum and mixed acids, complete analysis, determination of total acidity, lower oxides, nitric acid, sulphuric acid and free SO_3, calculations, table, 1530–1540

organic, 1542–1551

phosphoric acid, analysis of, 1521

sulphuric acid, analysis of, 1524

det. lead, iron, arsenic, zinc, selenium, hydrochloric and nitric acids, 1524–1528

tartaric acid, 1550

Acids—arsenic in acids, 49

corrections for in calorimetry, 1653

formulæ for diluting or strengthening of, 1533

free acids in aluminum salts, estimation of, 12

in aluminum salts, test for, 13

in presence of iron salts, estimation of, 1551

indicators for determination of, 1491

number in oil analysis. See Oils.

reactions—tables of, 630, 646–653

standards, preparation of benzoic, hydrochloric, sulphuric, 1493–1495

1